THE BIRDS OF THE IBERIAN PENINSULA

Eduardo de Juana and Ernest Garcia

CHRISTOPHER HELM
LONDON

Published in 2015 by Christopher Helm, an imprint of
Bloomsbury Publishing Plc, 50 Bedford Square, London WC1B 3DP

www.bloomsbury.com

Bloomsbury is a trademark of Bloomsbury Publishing Plc

Bloomsbury Publishing: London, New Delhi, New York and Sydney

A CIP catalogue record for this book is available from the British Library

ISBN (print) 978-1-4081-2480-2
ISBN (epub) 978-1-4729-0590-1
ISBN (epdf) 978-1-4729-0591-8

10 9 8 7 6 5 4 3 2 1

Commissioning Editor: Jim Martin

Design: Rod Teasdale

Printed and Bound in China

Cover art by Juan M. Varela Simó. Front: Spanish Imperial Eagle *Aquila adalberti*; Spine: Black-eared Wheatear *Oenanthe hispanica*; Back: Great Bustard *Otis tarda*.

MIX
Paper from responsible sources
FSC® C104723
FSC
www.fsc.org

CONTENTS

1. PREFACE

Our knowledge of the birds of Iberia has increased dramatically since the middle of the twentieth century and the scope of the regional ornithological literature has expanded at the same time. Whereas the historical literature tended to be anecdotal, often subjective and seldom quantitative, the modern equivalent manifests a scientific approach to all aspects of the subject. This book is our attempt to consolidate the wealth of material that is now available. Its principal aim is to provide a review of the Iberian avifauna during the early years of the 21st century that will be a useful baseline for future studies.

The unwritten job description for a book of this kind is that it should be an improvement on what is already available. As it happens, the last work to consider the Iberian avifauna as a whole was *'Aves Ibéricas'* by Mario Díaz, Benigno Asensio & José Luis Tellería published in Spanish in two volumes: the non passerines in 1996 and the passerines in 1999. That book is a good summary of the information that was then available but it antedates the most recent national atlas surveys of breeding and wintering birds, many of the regional atlases and all the recent censuses of bird populations, as well as a large diversity of informative publications that have become available post-2000. Portugal has a more recent avifaunal study: *'Aves de Portugal. Ornitologia do território continental'* by Paulo Catry, Helder Costa, Gonçalo Elias and Rafael Matias, published in 2010 in Portuguese. The scope of this latter book is similar to that of the present work but of course it only deals with Portuguese territory.

'The Birds of the Iberian Peninsula' is the first regional avifaunal work to be published in English, a circumstance which we are confident will make it more easily accessible to a wider readership. It covers all four of the political entities that comprise modern Iberia: Spain, Portugal, Andorra and Gibraltar. It also includes the Balearic Islands and other continental archipelagos of the Iberian Peninsula. We have not simply extended the material covered in *'Aves Ibéricas'* but rather have reviewed all the key elements of the literature, both recent and historical, including that published in the many regional bird reports. The available information is extensive and we have had to be selective to a considerable extent. We have concerned ourselves in particular with drawing together recent data on Iberian bird distribution and populations, giving particular attention to past changes and current trends and their causes. Breeding and wintering ranges and habitat selection are considered in detail as are the patterns of occurrence of migrants. Phenological data is presented for species whose occurrence in the region is seasonal. Conservation issues, in particular those that affect the key elements of the Iberian avifauna, are reviewed briefly. In general, we have not considered matters of taxonomy, identification or breeding biology, except in a few specific cases. Behavioural studies are quoted very selectively, with an emphasis on those that shed light on the principal topics covered.

2. ACKNOWLEDGEMENTS

We acknowledge most gratefully the enormous and continuing contribution to our knowledge of the Iberian avifauna that has been made by the authors and fieldworkers who undertook the numerous studies on which this book is based. We trust that we have done their work justice and that our efforts to avoid errors have been successful: the responsibility for any remaining mistakes remains our own. Authors of books and papers are cited in the usual way but severe space limitations have made it impossible to name authors of sections of collaborative works or observers of specific records.

Jim Martin and Nigel Redman, of Bloomsbury Publishing, have been instrumental in bringing the manuscript to publication and Chris Harbard commented on an earlier version of the text. Jim originally suggested the idea of this book to EG, who secured the willing assistance of EdJ in a collaborative venture. A number of other individuals have helped us with specific information, often in response to our enquiries, and we are most grateful to them all for their help. They are: Pep Arcos, Keith Bensusan, Luis Carrera, Francisco Chiclana, José Luis Copete, Fernando de Juana, Norman Elkins, José Rafael Garrido, Jorge Garzón, Ricard Gutiérrez, Manuel Máñez, Antoni Margalida, Albert Martínez Vilalta, Rubén Moreno-Opo, the late Mario Mosquera, Alejandro Onrubia, Andrew Paterson, Charles Perez, Javier Prieta, Manuel Rendón, Beatriz Sánchez and Antonio Sandoval. Our good friend Juan Varela painted the cover illustrations and we gratefully acknowledge the contribution of the following photographers, who provided their work without charge: Paul Acolina, Stephen Daly, Karin Faber, Andrew Fortuna, Charles Perez, Gabriel Sierra (Tatavasco Imágenes) and Harry Van Gils. We are particularly grateful for the assistance provided by SEO/Birdlife, especially by Juan Carlos del Moral, Blas Molina and the personnel of the Investigations and Bird Monitoring Unit. The interest and support of the recent and current directors of SEO/Birdlife, Alejandro Sánchez and Asunción Ruiz is also gratefully acknowledged.

3. A BRIEF HISTORY OF IBERIAN AVIFAUNAL STUDIES

As so often seen elsewhere in Europe, the earliest accounts of interest relating to the Iberian avifauna are treatises on falconry and works dedicated to hunting. The 14th century account by the chancellor Pero López de Ayala (1386), *'El libro de la Caza de las Aves'*, partly based on the famous *'De Arti Venandi cum Avibus'* of the Holy Roman Emperor Frederick II, stands out among these in Spain. For Portugal there is *'A Arte da Caça de Altaneria'* by Diogo Fernandes Ferreira (1616), published during the reign of Felipe II. Other noteworthy 17th century works include *'Décadas de la Historia de la insigne y coronada Ciudad y Reino de Valencia'*, by Gaspar de Escolano (1611): which has interesting information on the birds of the Albufera de Valencia; *'Primera parte de la historia natural y moral de las aves'* by Francisco Marcuello (1617), and *'Arte de ballestería y montería'* by Alonso Martínez de Espinar (1644).

Avifaunal studies as such did not emerge until the mid 19th century, when a series of investigators began to publish accounts of the birds found in natural history collections as well as, in some cases, of those that they themselves observed or collected in the field (Table 3.1). These accounts, especially the earlier ones, contain numerous errors and imprecisions and they are often extremely short on detail, all of which greatly diminishes their usefulness. A notable exception, in our view, is the *'Estudio ornitológico del Real Sitio de San Ildefonso'* by José María de Castellarnau (1877), which is based on field observations and offers very interesting information on the birds of the Central Range.

Table 3.1. The principal avifaunal accounts published in Spain and Portugal between the mid 19th century and the early 20th century.

Author	Year	Region	Title
Francisco de los Ríos Naceyro	1850	Galicia	*Catálogo de las aves observadas en las cercanías de Santiago y otros puntos de Galicia*
Ignacio Vidal	1851	Valencia	*Catálogo de las aves de la Albufera*
Mariano de la Paz Graells	1853	Madrid	*Catálogo metódico de las aves observadas hasta el día en el Área de la Fauna Matritense*
Antonio Machado Núñez	1854	Andalucía	*Catálogo de las aves observadas en algunas provincias de Andalucía*
Ángel Guirao	1859	Murcia	*Catálogo metódico de las aves observadas en una gran parte de la provincia de Murcia*
Víctor López-Seoane	1861	Granada	*Catálogo de las aves observadas en Andalucía*
José Vicente Barbosa du Bocage	1862	Portugal	*Instrucções practicas sobre o modo de colligir, preparar e remeter productos zoologicos para o Museu de Lisboa*
Francisco Barceló	1866	Balearic Islands	*Catálogo metódico de las aves observadas en las islas Baleares*
Vicente Mompó	1876	Valencia	*Catálogo de las aves de la provincia de Valencia*
Albino Giraldes	1879	Portugal	*Catálogo das aves de Portugal existentes actualmente no Museu de Coimbra*
Estanislao Vayreda	1883	Northern Catalonia	*Fauna ornitológica de la provincia de Gerona*
Salvador Calderón	1896	Andalucía	*Aves de Andalucía en el Museo de Sevilla*
Carlos de Bragança, King of Portugal	1903	Portugal	*Catalogo illustrado das aves de Portugal*
Mauricio Hernández Ponsetí	1911	Menorca Island	*Catálogo de las aves observadas en la isla de Menorca*
Celestino Graíño	1913	Asturias	*Fauna ornitológica de la provincia de Asturias*
Julián Aldaz	1918	Basque Country	*Catálogo de las aves observadas en Guipúzcoa y Vizcaya*
Luis Iglesias	1927	Galicia	*As aves de Galicia*
João Alves dos Reis Júnior	1931	Portugal	*Catálogo sistemático e analítico das aves de Portugal*

Despite their various shortcomings, these and other similar works permitted the emergence of the first general accounts of Iberian birds by the late 19th century, specifically the *'Catálogo de las aves de España, Portugal e islas Baleares'* by Ventura Reyes Prosper (1886) and the *'Aves da Península Ibérica e especialmente de Portugal'* by Manuel Paulino d'Oliveira (1896). However, we consider *'Aves de España'* by José Arévalo y Baca, published in Madrid in 1887, to be much more useful. This last work incorporated the field observations of some foreign ornithologists, notably Lord Lilford and Howard Irby, with whom Arévalo kept contact.

The contributions of ornithologists from elsewhere in Europe were certainly fundamental during the early stages of the chronicling of the Iberian avifauna. The earliest of these writers was the German Alfred Edmund Brehm, the son of Christian Ludwig Brehm, one of the most important European ornithologists of his day. Alfred Brehm, himself a famous naturalist and author, undertook a long journey through Spain between 1856 and 1857, in the company of his brother Reinhold. They travelled in Barcelona, Valencia, Murcia, Málaga, Granada, Madrid and Toledo, making contact with various local naturalists (Reig 2004a). The outcome of their journey was the publication in 1857 of the *'Vorläufige Zusammenstellung der Vögel Spaniens, mit kritischer Benutzung der bisher von spanischen Ornithologen herausgegebenen Verzeichnise'* (A provisional list of the birds of Spain, with a critical use of the catalogues previously published by Spanish ornithologists) and the description of a new species, the Thekla Lark, *Galerida theklae*, named after Christian's daughter. During their journey the Brehm brothers also saw an unfamiliar eagle at the El Pardo estate, Madrid, which their father was to describe as a new species, the Spanish Imperial Eagle, *Aquila adalberti*, in 1861. Alfred Brehm revisited Spain in 1879, in the company of the Archduke Rudolf of Austria. His work has gone relatively unnoticed, however, when compared to that of the British ornithologists who succeeded him.

One of the earliest and most influential of the British ornithologists who took an interest in Iberian birds was Lord Lilford, who visited Spain on a number of occasions and published his observations in *'Ibis'*, mainly in 1885 and 1886: *Notes on the ornithology of Spain'*. Major contributions were also made around this period by Howard Saunders in *'A list of the birds of Southern Spain'* (1871) and Lieutenant Colonel L. Howard Irby, with his *'Ornithology of the Straits of Gibraltar'* (1875 and 1895).

There is no doubt that the existence of the British enclave of Gibraltar, as well as the strong commercial links with the United Kingdom associated with the sherry trade, proved to be highly conducive to the ornithological exploration of the extreme south of the Peninsula. Particularly notable examples of the work of British naturalists of the Victorian period are books by Abel Chapman and Walter J. Buck: *'Wild Spain'* (1893) and *'Unexplored Spain'* (1910), and by Willoughby Verner: *'My life among the wild birds in Spain'* (1909), as well as the accounts published in *'Ibis'* by Heatley Noble (1902), Hubert Lynes (1912) and J. H. Stenhouse (1921). Links associated with the wine trade, this time of port wine, were also conducive to expanding knowledge of the Portuguese avifauna, particularly through the contributions of William C. Tait: *'A list of the birds of Portugal'* (1887) and *'The Birds of Portugal'* (1924), and of Henry W. Coverley, who published various articles in *'Ibis'* between 1931 and 1939. Other works by British ornithologists that merit particular mention are the account of the birds of Andorra and the eastern Pyrenees by William Eagle Clarke (1889); *'On the birds of Central Spain, with some notes on those of South-East Spain'* by Harry F. Witherby (1928) and those by Claud B. Ticehurst and Hugh Whistler (1925–1935) on a series of visits to Navarra, Cataluña, Galicia, Valencia, the Balearic islands of Ibiza and Formentera and a good part of Portugal. The sojourns of the German Leo Boxberger also provided interesting information, published in *Ornithologische Gelegenheitsbeobachtungen aus den östlichen Spanien* (1921), *Winterbeobachtungen aus Malaga* (1931) and *Beiträge zur Brutvogelfauna der Provinz Malaga* (1934).

Our knowledge of the Balearic avifauna owes much to the contributions of another Briton, Captain Philip Winchester Munn, who lived at Alcudia, Mallorca, from 1919 to 1949 and who published *'The Birds of the Balearic Islands'* (1931) and numerous papers, mainly in *'Ibis'*. Visits to the islands by the German Adolf von Jordans in 1914, 1921 and 1927 also proved fruitful and other interesting contributions to Balearic ornithology were made by Alexander von Homeyer (1862–1863), Lilford (1865), Saunders (1871), Gosse (1919–1920) and Witherby (1919).

Indigenous Iberian field ornithologists did not emerge much before the mid 20th century. A long hiatus in activity following the Spanish civil war (1936–39) was briefly interrupted by the publication in 1945 of the *'Sinopsis de las aves de España y Portugal'* by Augusto Gil Lletget, of the Natural Sciences Museum in Madrid, but this is little more than a summary of published material. By then, however, Francisco Bernis had commenced publishing notes and short articles based on his personal ornithological observations.

Francisco Bernis Madrazo (1916–2003) is justifiably regarded as the father of Spanish ornithology (Fernández 2004). For many years he was the general secretary of the Sociedad Española de Ornitología (SEO) as well as being a zoology professor of the Universidad Complutense de Madrid. Bernis was well aware of the backward state of ornithology as a scientific discipline in Spain and from the first he worked to establish links with ornithologists in other countries and to set up both the material and the human infrastructure that would enable its development. He was the driving force behind the creation of the SEO in 1954 and of the establishment of its journal, *Ardeola*. Early milestones in the history of the society were the publication of a first checklist of Spanish birds, the *'Prontuario de la avifauna española'* (Bernis 1954a); the creation of the Centro de Migración de Aves in 1956, to administer bird ringing; the institution of biennial national ornithological conferences from 1968 onwards; participation in international winter censuses of aquatic birds, also from 1968 onwards, and the taking of initial steps in the area of bird conservation, with the formation of the National Section within the International Council for Bird Preservation (ICBP), in 1963. In 1973 Bernis also set in train the first breeding bird atlas project for Spain. His numerous publications notably include *Migración en aves, tratado teórico práctico* (1966b); *Aves migradoras ibéricas* (1966–71); *La migración de las aves planeadoras en el Estrecho de Gibraltar* (1980) and *Aves de los medios urbano y agrícola en las mesetas españolas* (1988b).

A second major founding figure in the history of Spanish ornithology was José Antonio Valverde Gómez (1926–2003), author of *Aves del Sahara Español* (1957) and *Vertebrados de las Marismas del Guadalquivir* (1960). Valverde is especially known for his lifetime work in ensuring the conservation of Doñana, which was acquired in 1963 with the financial support of the World Wildlife Fund. He established the biological research institution there, the Estación Biológica de Doñana, and was its first Director.

The first breeding bird atlas in Spain was eventually published in 1997 under the coordination of Francisco J. Purroy. Earlier it encouraged the successive publication of a series of regional or provincial atlases, beginning with La Rioja (de Juana 1980), Cataluña (Muntaner *et al.* 1983) and Navarra (Elósegui 1985), which gradually began to shed light on the details of the distribution of the Spanish avifauna. There was a relatively rapid increase in the numbers of Spanish adherents to ornithology and birding in the 1970s, many of them doubtless encouraged by the spectacular television programmes of the late Félix Rodríguez de la Fuente, so that before too long it was possible to embark upon some relatively ambitious national censuses: of the wintering populations of Common Cranes (Fernández-Cruz *et al.* 1981), gulls and terns (Bermejo *et al.* 1986) and waders (Alberto & Purroy 1981), and of the breeding population of the Griffon Vulture (SEO 1981). The Spanish rarities committee was established in 1984, and operated during 1988–94 as part of a pan-Iberian body that also included Portugal. Conservation work also accelerated with the founding in 1978 of the Coordinadora para la Defensa de las Aves (CODA) and, in 1986, of the Comité de Protección de las Aves within SEO, the latter coinciding with the accession of Spain and Portugal to the European Community. The first inventory of Important Bird Areas for Spain (de Juana 1990) was drawn up in support of the 1979 Birds Directive and was a major step towards protecting the best such locations in the country.

A considerable amount of ornithological research in Spain today is carried out by national or regional state institutions as well as by academics and non-governmental organisations but SEO/Birdlife remains the prime mover in avifaunal studies. Recent achievements in this area have been the establishment of a common birds census programme, the 'Programa de Seguimiento de Aves Comunes Reproductoras' (SACRE), in 1995; the publication of the second breeding bird atlas (Martí & del Moral 2003); the compilation of 38 species monographs during 2004–12; the compilation and analysis of the waterbird censuses of 1980–2009 (González & Pérez-Aranda 2011) and the publication of a first atlas of wintering birds (del Moral *et al.* 2012a). A Migration Atlas, based on ringing recoveries, is in preparation: its preliminary findings are available online (www.migraciondeaves.org). Many Spanish regions have regional societies that publish periodical bird reports and other publications. Particularly noteworthy among these is the work of the Institut Català d'Ornitologia (www.ornitologia.org, which has recently published both a second breeding atlas of Catalan birds (Estrada *et al.* 2004) and a regional wintering bird atlas (Herrando *et al.* 2011).

The corresponding situation in Portugal evolved more slowly (Catry *et al.* 2010). The founding father of Portuguese ornithology was Joaquim R. dos Santos Júnior (1901–90), the director of the Instituto de Zoologia Dr. Augusto Nobre at the Universidade do Porto. Santos Júnior founded the Sociedade Portuguesa de Ornitologia (SPO) in 1964 and this society published the journal '*Cyanopica*' from 1968 to 1989, began scientific bird ringing in the country and established an ornithological reserve at Mindelo, Vila do Conde. However, the society was gradually taken over by '*passarinheiros*', people whose principal interest was in cagebirds, who took over the committee in 1989. Ornithological activity then came under the ambit of a state institution, the 'Centro do Estudos de Migrações e Protecção de Aves' (CEMPA). Under the direction over many years of António M. Teixeira, CEMPA conducted the winter censuses of aquatic birds, took charge of bird ringing, began the first campaigns to study raptor migration at Sagres, initiated an atlas of Portuguese breeding birds (Rufino 1989a) and, from 1990 onwards, published the journal '*Airo*'.

This refreshed Portuguese ornithological environment coincided with a wider awareness of environmental concerns that led to the emergence of several non-governmental organisations, including the 'Liga para a Protecção da Natureza' (LPN) and the 'Sociedade Portuguesa para o Estudo das Aves' (SPEA). SPEA, the current Portuguese ornithological society, was created in 1993 under the presidency of João Carlos Farinha. It is now the Portuguese partner of BirdLife International and conducts or coordinates all essential ornithological work, including the holding of biennial conferences and the publication of the journals '*Pardela*', since 1995, and '*Anuário Ornitológico*', since 2003. The major landmarks of this recent stage in the history of Portuguese ornithology include the launching of a separate national rarities committee, the Comité Português de Raridades, in 1995 and the publishing of the *Atlas das aves invernantes do Baixo Alentejo* (Elias *et al.* 1998), the inventory of Important Bird Areas '*Zonas importantes para as aves em Portugal*' (L. T. Costa *et al.* 2003) and the second national atlas of breeding birds (Equipa Atlas 2008). The publication of '*Aves de Portugal, ornitologia do territorio continental*' (Catry *et al.* 2010) was a key advance in bringing current knowledge of the Portuguese avifauna up to date.

Ornithology in Gibraltar has a long history, dating back to the observations there of visible migration of raptors, hirundines and other birds across the Strait by a British clergyman, the Reverend John White, in the early 1770s (Holt-White 1901). Expatriate British residents at Gibraltar, most notably Howard Irby in the late 19th century (as noted above), Helen Rait-Kerr in the 1930s and 1940s and Sir Gerald Lathbury in the 1960s, all made a significant contribution to our knowledge of the birds of Gibraltar itself and of its hinterland. Thereafter local ornithologists estabished the Gibraltar Ornithological Society in 1976. This society, now the Gibraltar Ornithological and Natural History Society (GONHS), became the local partner of Birdlife International in 2001. It publishes the journal '*Alectoris*' and, since 2001, the annual *Gibraltar Bird Report*. The GONHS maintains a bird observatory and conducts bird ringing in Gibraltar, using rings supplied by the British Trust for Ornithology. Publications by Gibraltarian ornithologists include '*The birds of Gibraltar*' (Cortés *et al.* 1980) and '*Birds of the Strait of Gibraltar*' (Finlayson 1992).

4. IBERIA: GEOGRAPHY AND CLIMATE

THE IBERIAN PENINSULA AND ITS ISLANDS

The Iberian peninsula is the massive pentagonal promontory that comprises the southwestern extremity of the Eurasian landmass. It closely approaches North Africa at the Strait of Gibraltar, only 14km wide at its narrowest point, which links the Mediterranean Sea and the Atlantic Ocean. It is a region of great physical, climatic and biological diversity (Lautensach 1967, Vilá Valentí 1968). It has also been subject to human, or rather hominin, influences for hundreds of thousands of years. Modern humans have wrought profound changes on its topography and, most of all, on the distribution of vegetation and natural habitats. Political influences have also been long-lasting. Iberia occupies a strategic position on the western fringe of Europe, bordering Africa and dominating the Strait of Gibraltar, until recently the sole access route by sea to the Mediterranean. Its peoples have come in turn under the sway of foreign visitors, settlers and invaders, notably including the Phoenicians, the Greeks, the Carthaginians, the Romans and the Moors.

Political boundaries have come and gone but they endure today in the form of the four separate national entities that make up modern Iberia. By far the largest is Spain (504,880 km^2, about 85% of the Peninsula), followed by Portugal (91,630 km^2, 15%). The Principality of Andorra (465 km^2) and the British territory of Gibraltar (7 km^2) are tiny adjuncts by comparison. The Balearic Islands (4,992 km^2) are an overseas region of Spain. Iberia thus extends over 596,982 km^2 in total (The Times atlas), larger than France (543,965 km^2) and more than twice the size of the United Kingdom (244,755 km^2). The Pyrenean watershed marks the traditional northeastern limit of the Iberian Peninsula and by this definition, the French region of Cerdagne (540 km^2, part of the Pyrénées-Orientales department) also falls within Iberia, being south of that line. A further complication is the existence of Llívia, a tiny (13km^2) enclave of Spanish territory, technically a part of Girona province that lies within Cerdagne and is separated from the rest of Spain by 2km of French territory. The geographical coverage of this book (Fig. 1) includes all the territory of mainland Iberia and also that of the Balearic Islands and the smaller archipelagos off the Iberian coast: including the Columbretes, Berlengas and Cíes islands and a number of smaller islets, which together form a convenient biological grouping. The Chafarinas Islands off the coast of Morocco are a Spanish territory but not part of Iberia. The distant territories of Macaronesia that are a political part of Spain (the Canary islands) or Portugal (the Azores and Madeira) are not included, since the avifaunas of these oceanic islands have been shaped by influences other than those that have affected the Iberian mainland. Cerdagne (and Llívia) are technically included but they do not figure in censuses or atlas work in Spain. However, the Eyne plateau in Cerdagne is a major census point for migrants flying south into Catalonia in autumn. The Arán valley, Lleida, is also included since it is part of Spain, despite lying north of the Pyrenean watershed and thus, strictly speaking, outside the Iberian Peninsula.

Figure 4.1. Geographical coverage of this book. The Iberian Peninsula, Balearic Islands and other associated continental archipelagos.

The geography of mainland iberia

The Iberian peninsula is distinguished by its high average elevation. Indeed, the mean altitude of Spain (650m) is the second-highest of any European country, after lofty Switzerland (mean altitude 1,300m). Very little of the land lies below 100m: just the coastal fringes and the lowest reaches of the river valleys. Instead the Peninsula is traversed by a series of mountain chains (*cordilleras* in Spanish, *cordilheiras* in Portuguese) separated by more or less elevated plateaus (Fig. 4.2) (Muñoz Jiménez & Sanz Herráiz 1995).

The Pyrenees provide a clear boundary, and a significant biological barrier, between the Peninsula and the rest of Europe. They run for over 450km southeastwards from the Basque coastlands at the southeastern corner of the Bay of Biscay across Navarra and Aragón to approach the Mediterranean in Catalonia. The Spanish/French border follows the watershed. The range rises comparatively abruptly on the French side but lower mountains, the pre-Pyrenean sierras, run parallel to the southern flank from eastern Navarra to western Catalonia and extend the montane area very considerably on the Spanish side. The western Pyrenees reach widely above 1,000m in Navarra, rising above 2,000m in the centre and east of the range, where the highest peaks exceed 3,000m. The highest summits of all are in eastern Aragón, where Aneto (3,404m) is the tallest of the entire range. A chain of low mountains, the Catalan coastal range, also flanks the Mediterranean coast in Catalonia, extending from the Pyrenees to the Ebro.

The second major mountain system of northern Iberia is the Cantabrian Range (Cordillera Cantábrica), which runs roughly parallel to the Biscay coast. It is made up of lower coastal ranges and a second much higher chain further inland. The whole Cordillera extends some 300km from the Basque Mountains (Montes Vascos), which link the range to the Pyrenees, across Cantabria, Asturias and the northernmost reaches of Castilla y León to Galicia. The lower Galician mountains provide further elevated terrain in the west, together with the Montes de León and the uplands of Trás-os-Montes and Alto Douro in northeastern Portugal. The crowning glory of the Cantabrian Range is the Picos de Europa, a region of high tops straddling eastern Asturias, western Cantabria and northern León province. The limestone that makes up the central massif is heavily eroded into jagged peaks and vertiginous gorges. The major summits of the Picos rise well above 2,000m, the highest being Torrecerredo (2,648m).

The Iberian Range (Sistema Ibérico) runs roughly northwest/southeast for some 500km from eastern Castilla y León and La Rioja (Sistema Ibérico Norte) across eastern Castilla-La Mancha and southern Aragon (Sistema Ibérico Sur) as far as the borders of the Comunidad Valenciana. This range provides a southern boundary to the Ebro basin, which is bordered in the north by the Pyrenees and pre-Pyrenees and in the east by the Catalan coastal range. The watershed divides the Ebro valley from the basins of the other major central Iberian rivers, which drain into the Atlantic. The Iberian Range is a complex series of mountains of medium elevation, generally below 1,500m. Few peaks exceed 2,000m, the highest being the summit of the isolated Moncayo massif in the north (2,316m).

A higher range, the Central Range (Sistema Central) runs east/west from the Iberian Range across the south of Castilla y León and northern Madrid towards the Portuguese border in Salamanca and Cáceres. It reaches as far west as central Portugal, where it includes the Serra da Estrela, whose summit (1,993m) is the highest point in the country. It includes such well known landmarks as the Sierra de Guadarrama (Madrid/Segovia) and the Sierra de Gredos (Ávila/Cáceres). The core of the range rises widely above 2,000m, to 2,429m on Peñalara in the Sierra de Guadarrama and to 2,592m at Pico Almanzor in the Sierra de Gredos. The Central Range is flanked along much of its southern side by a parallel range of lower mountains, comprising the Sierra de Las Villuercas in eastern Cáceres and the Montes de Toledo, south of Madrid.

The Central Range divides the central Spanish plateau, the Meseta, into its northern and southern sectors, the Meseta Norte and Meseta Sur. The Meseta or tableland, so named after the Spanish for table (*mesa*), is itself at an elevation of 600–750m, highest in the north and sloping gently towards the west so that its rivers drain into the Atlantic. The Meseta Norte makes up most of Castilla y León and is a vast area of seemingly endless plains, much of it farmland devoted to cereal crops especially. The Meseta Sur includes the province of Madrid and the plains of La Mancha. A long, low range of gentle, wooded hills, the Sierra Morena, runs across the south of the region, from Alentejo and northern Algarve in southwestern Portugal, eastwards as far as Albacete province, Castilla-La Mancha. The Sierra Morena straddles southern Extremadura and southern Castilla-La Mancha and northern Andalucía: where it crosses the north of Huelva, Sevilla, Córdoba and Jaén provinces. Its highest elevation is the Sierra Madrona (1,332m), Córdoba/Ciudad Real.

The Betic Range (Betic Cordillera) is the southernmost of the principal Iberian ranges. It extends from the the Strait of Gibraltar in Cádiz province through Málaga, Granada and Jaén, and into Murcia. The western sector of the range includes the Serranía de Ronda (1,919m). Further east the range rises to its highest elevations in the Sierra Nevada, Granada/Almería. Although the western Betic mountains rise widely above 1,000m, with peaks above 1,500m, the central part of the system is much higher. The Sierra Nevada is highest of all and its two crowning peaks, Mulhacén (3,481m) and Pico Veleta (3,392m) are the greatest elevations in the entire Iberian peninsula, higher than the loftiest summits of the Pyrenees. For all that, the Sierra Nevada is a gentle, rounded massif and a road runs all the way to the top.

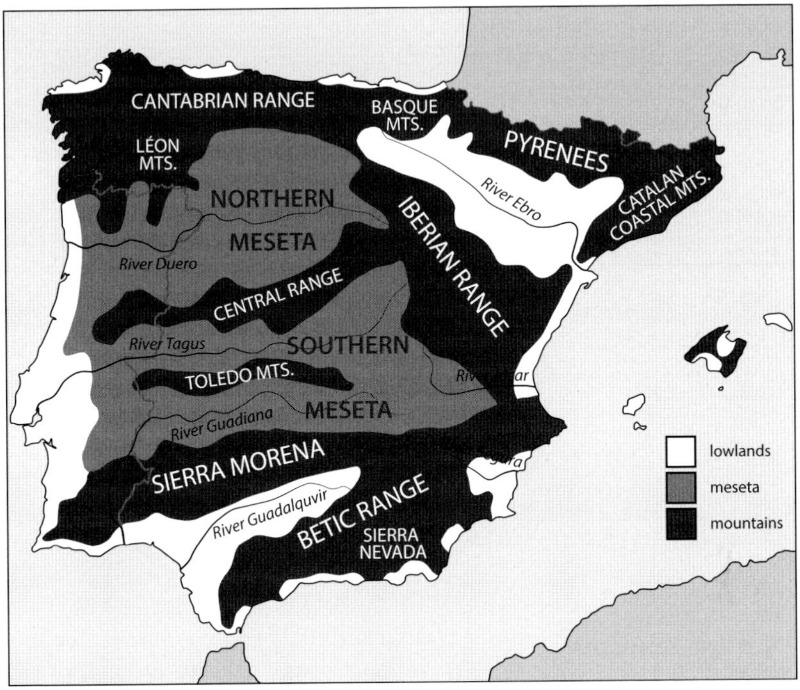

Figure 4.2. Principal geographical features of Iberia

The principal rivers of Iberia are the Tagus (1,007km), Ebro (910km), Duero (895km), Guadiana (778km) and Guadalquivir (657km). Of these only the Ebro drains into the Mediterranean, where its delta is a vast triangular outpushing of silt jutting from the coast of Tarragona. Numerous smaller rivers, notably the Júcar, Turia and Segura, drain the coastal sierras of the eastern and southern Mediterranean hinterland but none of them compares with the Ebro, whose massive basin makes up a major depression that nestles between the Pyrenees and the Iberian Range. The Ebro rises in the foothills of the Cantabrian Range but it is fed by many rivers, large and small, running down from the high mountains and often swollen by meltwater from the summits.

The other major rivers reach the Atlantic in the west or south-west. The northernmost is the Duero (Douro in Portuguese), which drains the Northern Meseta in Castilla y León and flows across northern Portugal to its estuary at Porto. The Tagus (Tajo in Spanish, Tejo in Portuguese) runs east across Castilla-La Mancha and northern Extremadura, between the Central Range and the Montes de Toledo, to cross central Portugal before turning south-west to reach the estuary, one of the most important Iberian wetlands, at Lisbon. The Guadiana runs roughly parallel and to the south of the Tagus across southern Castilla-La Mancha and southern Extremadura. The river then delimits much of the international border between southern Portugal and Andalucía, as it runs southwards to its estuary on the south-west coast. The Guadalquivir is sandwiched between the western Sierra Morena and the Betic mountains, reaching the Atlantic at Doñana. Its wedge-shaped valley, the Plain of Andalucía is fertile farmland. The Guadalquivir delta, the Marismas del Guadalquivir, is the foremost of the Iberian wetlands and is of the greatest conservation importance, both regionally and globally.

The Atlantic and Biscay coastlands, and also the Mediterranean coasts to a lesser extent, are also crossed by numerous lesser rivers, whose estuaries (the Rías) are often of ornithological significance. The principal ones are noted under 'Habitats'. The coastal plains themselves vary in extent but are mainly narrow. Those of the Biscay coastlands are particularly narrow, and non-existent in much of northern Galicia, since the northern reaches of the Cantabrian Range approach the sea in places. The Mediterranean coastal strip is also relatively narrow, especially in the central Costa del Sol where the Sierra Nevada foothills reach the sea, and in the north-east where some of the Catalan coastal mountains also reach the Mediterranean. The Cádiz lowlands are broader and can be seen as a southerly extension of the Plain of Andalucía. The Huelva coastal lowlands are narrower. The Portuguese coastline is elevated north of Porto but there is a coastal plain of varying width around the remainder of the country. This broadens in Ribatejo and the west of Baixo Alentejo into the Tagus depression, which encompasses the Tagus and Sado estuaries and their environs.

The Balearic Islands

The Balearic Islands (Illes Balears) are geologically an eastwards extension of the Betic Cordillera of southern continental Iberia, being the summits of high mountains that were submerged when the Mediterranean basin flooded at the end of the Miocene, some five million years ago. The geology is calcareous and sea cliffs and gorges are a typical feature. The islands have a combined land area of 4,992km^2 and are home to over 1.1 million people, not including the masses that flock there on holiday. The four largest islands are Mallorca, Menorca, Ibiza (Eivissa) and Formentera (Fig. 4.3). Nearly 80% of the human population is on Mallorca and most of the remainder are divided between Menorca and Ibiza.

Figure 4.3 The Balearic archipelago

The archipelago also comprises a number of lesser islands but most of these are very small uninhabited islets and rocks. Many are nonetheless important as the nesting sites of Balearic Shearwaters, other seabirds and Eleonora's Falcons. The larger islets include Cabrera and its satellite islands, south-east of Mallorca; Dragonera, south-west of Mallorca; Conillera, west of Ibiza, and Illa de l'Aire, south-east of Menorca. The Cabrera archipelago is protected as a National Park that includes 1,318ha of land and 8,703ha of marine reserve. The four principal Balearic Islands have been profoundly modified by touristic developments ever since they became byword for mass tourism in the 1960s. None of these changes have affected Cabrera, a military zone, and the island is celebrated as a staging point for migrant birds, which have included many rarities: it is to Iberia what Fair Isle is to the British Isles.

The principal geographical features of the islands that are of ornithological importance include a number of important wetlands, sea cliffs and mountains, these last often harbouring forest and scrub formations. Specific sites are mentioned within the relevant species accounts but the major ones are summarised below. In addition, the easternmost location of the islands within our region means that certain species that are at least uncommon on passage on the Iberian mainland occur in the Balearics regularly on passage: the Red-footed Falcon and Icterine Warbler are examples.

Mallorca is distinguished by the Sierra de Tramuntana, which occupies about a fifth of the island, including all the north coast, where sheer cliffs front the Mediterranean. The Tramuntana extends from the western extremity of Mallorca north-eastwards to Cape Formentor. There are a number of summits over 1,000m, the highest being the Puig Major (1,445m) in the central sector of the range. The habitats include extensive Iberian Holm Oak and pine forests. Black Vultures nest in the north-east, between Sóller and Cape Formentor, and both Ospreys and Eleonora's Falcons nest along the coast. The major wetland of the archipelago is the Albufera de Mallorca, a freshwater marsh with lagoons and extensive reedbeds in the north-east, just inland from Alcudia Bay. The Albufera is a Natural Park (1,700ha) and a Ramsar site. The saltpans of the Salobrar de Campos in the south are another important site on Mallorca. Dragonera island off the north-west of the island has a large Audouin's Gull colony.

Menorca is less developed than Mallorca and the whole island is a World Biosphere Reserve. The landscape includes pastures and woodlands of Aleppo Pine, olive and Iberian Holm Oak. The north coast, with its steep sea cliffs, is of special interest ornithologically, with large colonies of Cory's Shearwaters, as well as Ospreys and Audouin's Gulls. The Albufera d'Es Grau in the west is a significant wetland.

Ibiza has wooded mountains in the north and west especially and imposing sea cliffs in the north-east. Conillera and its surrounding islets off western Ibiza has a large Eleonora's Falcon colony and nesting Balearic Shearwaters. The saltpans in the south of the island are important for waders.

Formentera supports most of the Balearic Shearwaters of the archipelago, and hence the world, particularly at Cape de La Mola in the south-east. The saltpans in the north, and the adjacent brackish lagoons: the Estanys Pudent and des Peix, are important for waders and other waterbirds.

Other peninsular archipelagos

A number of other small islands or groups of islands occur in the continental waters of the Iberian peninsula. They do not compare with the Balearic Islands in terms of size and none of them is permanently inhabited. All, however, are of ornithological importance, chiefly as the locations of seabird colonies. The principal ones are the Cíes Islands off western Galicia; the Berlengas off Cape Raso, Portugal; Alborán Island in the central Alborán basin, which is the westernmost sector of the Mediterranean south of eastern Andalucía; and the Columbretes Islands, off the Levant coast of Castellón.

CLIMATE

The Bioclimatic Zones: Temperature and Precipitation

The Iberian peninsula has a varied climate, reflecting the existence and interaction of the very different marine influences of the Atlantic and Mediterranean, its southerly location, its close proximity to the arid vastnesses of the Sahara, its great diversity of relief and its large size (Font Tullot 1983). The last of these allows the Iberian interior to experience a continental climate that is relatively hot in summer and relatively cold yet dry in winter.

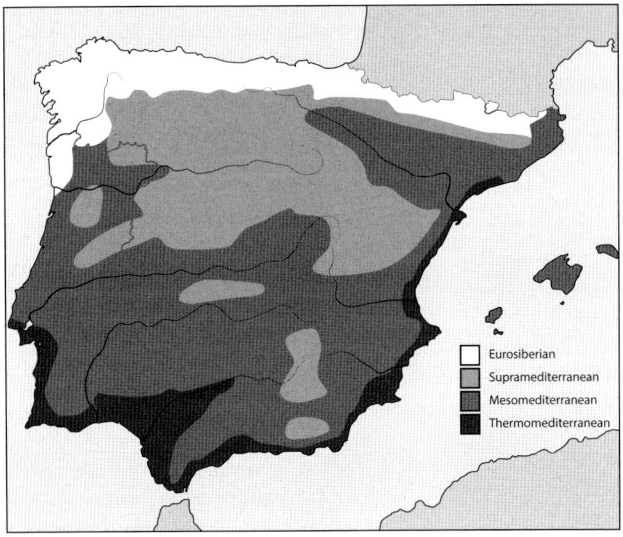

Figure 4.4. Bioclimatic zones of the Iberian Peninsula.

The bioclimatic zones that may be identified in Iberia (Fig. 4.4) are of considerable significance to the distribution of a majority of bird species there and therefore frequent reference to them is made in the species accounts. The principal division is between the cooler, more humid, Eurosiberian zone in the north and the Mediterranean zone in the remainder of the Peninsula. Both zones are well defined by climatic characteristics and by the distribution and species composition of their vegetation.

The Eurosiberian Zone takes in what is sometimes known as 'Green Spain' and it also includes the far north of Portugal. It otherwise comprises most of Galicia, the Biscay coastlands, the Cantabrian Range and the Pyrenees. It can be subdivided into its lowland, montane and subalpine/alpine components, but this classification is of much lesser significance to bird distribution than to that of plant species. This is the bioclimatic zone that corresponds with the deciduous and coniferous forests of the wider Palearctic, including the whole of temperate Europe. Rainfall is generally high and there is no summer drought period.

Table 4.1. Characteristics of the principal bioclimatic zones and sub-zones of Iberia (After Peinado & Rivas-Martínez 1987).

Bioclimatic zone	Sub-zone	Mean annual temperature (ºC)	Characteristic natural vegetation
EUROSIBERIAN	Alpine	<3	Grassland/dwarf shrubs
	Sub-alpine	3–6	Grassland/heathland with conifers
	Montane	6–10	Beech, Scots Pine, Silver Fir
	Lowland	>10	Pedunculate/Sessile Oak
MEDITERRANEAN	Crioromediterranean	<4	Grassland/dwarf shrubs
	Oromediterranean	4–8	Scots Pine, broom scrub, grassland
	Supramediterranean	8–13	Pyrenean/Lusitanian/Iberian Holm Oak, Junipers, Laurel-leaved Cistus scrub, heathland, dwarf shrub scrub
	Mesomediterranean	13–17	Iberian Holm/Cork Oak, Maritime/Stone Pine, Gum Cistus, Lentisc/Olive scrub
	Thermomediterranean	17–19	Iberian Holm/Cork Oak, Stone/Aleppo Pine, Lentisc/Olive scrub, grassland, pseudosteppe

The Mediterranean Zone is characterised by its hot, dry summers and cool (or cold) wet winters. It includes the remainder of Iberia and also the Balearic Islands. Its interface in the north with the Eurosiberian zone corresponds with the rain shadow of the Cantabrian Range, where the cool, humid influences of the Atlantic no longer act. It thus extends north to the southern foothills of the range and also of the Pyrenees. The temperature and rainfall regimes within the Iberian Mediterranean Zone still differ markedly between different regions in relation to latitude, elevation and humidity. It is possible to identify three principal subzones that are of ornithological significance: the supramediterranean, the mesomediterranean and the thermomediterranean – and these terms are widely used in the text. Two other subzones, the oromediterranean and the crioromediterranean, corresponding respectively with the subalpine and alpine subzones of the Eurosiberian, are of very limited extent, at the highest elevations of the highest Mediterranean mountains, and again are most notably of botanical importance. The precise definitions, and particularly the exact distributions, of the subzones differ between authorities but their essential characteristics are summarised in Table 4.1.

The Thermomediterranean is the most restricted of the three principal Mediterranean subzones. It includes the coastal strip of Portugal south of the Tagus, the Guadalquivir depression and the whole of coastal Andalucía, and the Levant coast as far north as the Ebro delta. This part of Iberia enjoys the mildest winters and the hottest summers. Frost or snow are very rare but summer maximum temperatures frequently exceed 30°C and are regularly above 40°C in the Guadalquivir valley especially. Annual rainfall varies considerably, being relatively high in the west, where the winter Atlantic depressions normally bring heavy rain, and much lower in the east. The driest part of Iberia is in the south-east, in the rain shadow of the Sierra Nevada; in Almería and southern Murcia, where semi-desert conditions occur and the landscape is broadly reminiscent of nearby north-west Africa.

The Mesomediterranean extends over most of central and eastern Portugal, the southern Meseta, the Sierra Morena and Betic Range, the coastal mountains of the Levant and the Ebro depression. Summers are hot and dry here, especially inland, but winters are cool, with heavy rains in the west that become progressively sparser eastwards. The Ebro depression is particularly dry, being in the rain shadow of the northern and central mountain ranges.

The Supramediterranean extends across the northern meseta and includes all but the highest levels of the Central and Iberian Ranges, as well as the pre-Pyrenean mountains. There are also outposts of the Supramediterranean further south at the higher elevations of the Montes de Toledo and the Betic Range. Summers are hot and dry but winters are cold, with periods of snowfall and sub-zero temperatures even on the meseta.

Altitude influences ambient temperatures at any season. Low-lying areas anywhere in Iberia enjoy mild to hot weather in summer. Conversely, the high tops of all the main mountain systems, with the exception of the Betic Mountains, are snow-capped in winter and late into the spring, with snowfields remaining on sheltered, mainly north-facing, slopes throughout the year. The Sierra Nevada in the Betic Range also lives up to its name during most winters but hardly any snow lingers here in summer. Lower levels of the Betic Range, such as the Serranía de Ronda, are only snow-covered in some years. In general, however, the extent and duration of very cold or snowy weather shows considerable between-year variation in all regions.

Rainfall also varies markedly between seasons, between different regions and between years, as described for the period 1971–2000 by the Iberian Climate Atlas (Agencia Estatal de Meteorología 2011), whose figures we quote here. The highest annual precipation, exceeding 2,200mm, occurs in the north-west, in the Serra do Gerês in northern Portugal and in southwestern Galicia, and also in northeastern Navarra. The opposite extreme is seen in the arid south-east: in Almería, Murcia and southern Alicante, where annual precipitation is less than 300mm, and even lower, below 200mm, at Cabo de Gata, Almería.

Seasonal variation in rainfall is more marked in southern Iberia, where there is typically little or no precipitation in summer, than in the north. The wettest month everywhere is December and the driest is July. The west-to-east passage of weather fronts and Atlantic depressions, chiefly in winter, results in a corresponding precipitation gradient across the Peninsula (Fig. 4.5). The north-west receives most precipitation at all times but especially in winter, when the western half of the Peninsula is also relatively wet, with heavy precipitation affecting much of Portugal,

Figure 4.5. Distribution of mean annual precipitation in Iberia during 1971–2000. Source: Iberian Climate Atlas 2011.

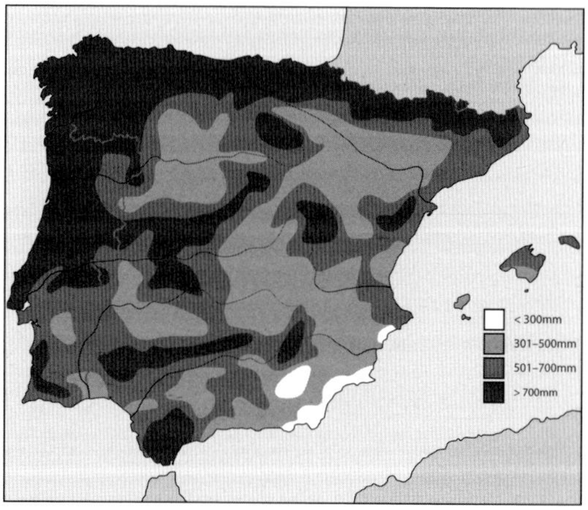

the Cantabrian mountains, the western Central Range and Montes de Toledo, the western Sierra Morena and the western Betic Mountains. Rain shadow effects result in comparatively sparse winter rainfall in the south-east, the Ebro depression, the central Meseta Norte and the eastern Meseta Sur. Summer rainfall patterns follow a north/south gradient, with significant summer rainfall only in the Eurosiberian Zone and the Iberian Range, and extremely sparse or nil rainfall in the southern half of the Peninsula, where a summer drought is experienced. Local and sometimes intense thunderstorms may nevertheless produce brief outbreaks of very heavy rain in summer anywhere inland, sometimes causing flash flooding in the dried-up watercourses.

Climate exerts a profound influence on plant and animal life everywhere and this is clearly evident in the Iberian peninsula. Climatic factors are central to determining the distribution and nature of plant cover, both natural and artificial, across the region and this has clear effects on the distributions of animal species, birds included. The seasonality of rainfall affects waterbirds and those dependent on waterside habitats markedly in Iberia. The drying-up of natural wetlands in the south in summer means that many locations that are alive with waterfowl in winter are deserted in summer. However, the proliferation of artifical, permanent waterbodies in the form of reservoirs has allowed a number of species to breed where formerly they were absent.

Winds

The prevailing wind patterns are the major determinant of the Iberian climate and weather. These winds are principally influenced by the North Atlantic Oscillation, the changes in the relative strengths and positions of the low-pressure system over Iceland (the Icelandic Low) and the high pressure system over the Azores (the Azores high). These control the direction and strength of westerly winds into Europe, including Iberia. In winter the Azores high is usually centred at lower latitudes resulting in moist westerly winds over Iberia bringing winter rainfall throughout the region. The Azores high expands northeastwards in summer and blocks westerlies except in northern Iberia, resulting in the drought period that typifies summer in the Mediterranean region.

The underlying pattern is disrupted at times by polar air masses affecting the north and north-west in winter and by tropical or subtropical air masses from the south, west and south-west in summer. The track of the westerly storms that especially affect seabird activity in Iberian Atlantic waters in autumn and winter is strongly influenced by the position and strength of the North Atlantic jet stream. A more southerly position of this high-altitude westerly wind zone brings gales and rain as it steers Atlantic depressions over southern Iberia and into the Mediterranean. A particular Iberian wind pattern that has important implications especially for bird migration is the east wind or levanter. These easterlies are typically associated with periods of high pressure over western Europe and low pressure to the south or southwest. Episodes of levanter may also arise from high pressure over the Balearic Islands or during the approach of a cold front from the west towards the Strait of Gibraltar. They generally blow from the north-east along the Levant coast and from the east through the Alborán basin, the Strait of Gibraltar and the coastal south-west. In the Balearic Islands similar conditions produce the prevalent northerly winds. The levanter may occur at any time of year but is most persistent between July and October and in March, when it may continue for periods of up to two weeks. However, it blows strongest between October and May, when easterly gales may affect the Strait in particular. The levanter is associated with hazy weather, and coastal fogs in summer. At the Strait it may produce long periods of drizzling rain, as well as the infamous levanter cloud that adorns the Rock of Gibraltar.

Climate Change

Climate change is as old as climate itself but recent concerns that anthropogenic influences may be causing global warming, or at least contributing to it, have become highly topical. Much of what has been written on the subject is speculative to a greater or lesser degree and this of course applies to the possible changes that might result within Iberia should significant planetary warming occur. The most general prognostication is that the warmest parts of the region that are furthest from the North Atlantic: the arid south-east of the peninsula, will become hotter and drier still, leading to an expansion of the semi-desert regions. In general, regional climate model projections for the end of the 21st century (Pérez & Boscolo 2010) indicate pronounced increases in mean seasonal temperature, by up to a possible maximum of 6°C in summer and 2–3°C in winter (2-3°C), as well as a decrease in precipitation throughout the year, resulting in increasingly arid conditions in most of the Iberian Peninsula. It has been specifically suggested that a forecast increase in winter storm activity over the northeastern Atlantic will affect the Iberian Peninsula to a lesser extent than the countries bordering the North Sea. However, higher storm activity over the adjacent Atlantic is likely to lead to a 2–4% increase in maximum wind speeds in northwestern Spain by the end of the 21st century, and the number of days with high winds in Galicia could increase by up to 10% (WWF 2006). There has been additional speculation that Hurricane Vince, said to be Europe's first-ever tropical cyclone, which developed to the south-west of the Iberian Peninsula in early October 2005 and made landfall as a tropical depression on the Huelva Coast on 11 October, may have been a herald of unusually stormy weather to come. However, its claimed uniqueness may not have been justified since there is evidence that a similar storm struck southwestern Iberia in October 1842 (Vaquero *et al.* 2008).

5. HABITATS

The Iberian Peninsula offers a great diversity of habitats, all of which have been modified to a greater or lesser extent by human activity. Some of them, such as urban settlements, plantations of exotic trees, croplands and ricefields, are entirely artificial. The region was never covered by a permanent ice-cap during the last glaciation and instead acted as a glacial refuge for many plant and animal species, resulting in the very high floristic diversity of the region, with over 5,000 species of vascular plants, including many endemics. The natural habitats of the Peninsula were formerly far more extensive, and uniform, than those encountered today. In particular, vast monotonous tracts of Mediterranean forest, dominated by the Iberian Holm Oak, would have extended across most of the Mediterranean bioclimatic region, including the now largely treeless Meseta. The Mediterranean forests have been fragmented and largely replaced by open country since prehistoric times, when human populations expanded around the Mediterranean basin and drastically modified the region. Portugal has about 25% forest cover and Spain about 19%, the latter representing an improved situation following widespread afforestation during the 20th century (MARM 2008). These changes will have had profound effects on the native fauna and flora but they clearly benefited open-country species, which must once have been far more locally distributed if they were present at all.

A diversity of classification systems have been used to describe the vegetation of Iberia. That used here is based on Rivas Martínez (1987) and Sainz Ollero *et al.* (2010). We also then describe all other habitats, whether natural, seminatural or artificial, that are important to birds. Detailed descriptions of the sites mentioned here and within the species accounts are outside the scope of this book but summaries of all the mainland Spanish sites and their avifauna, with maps of the majority, may be found in de Juana (1994), Garcia & Paterson (2008) and Rebane & Garcia (2008). The scientific names of all plant and animal species, other than birds, mentioned in the text are given in Appendix 3.

MEDITERRANEAN EVERGREEN PLANT COMMUNITIES

1. Lowland and hill forests.

About two-thirds of the Iberian Peninsula were once covered in Holm Oak forest, much of which has long been felled. Most of the remaining tree cover takes the form of semi-natural woodlands, many of them with a characteristic understorey of native shrubs and herbaceous plants. There are also extensive plantations of exotic species, notably pines and eucalyptus.

Large areas, in the south and west especially, are covered by parkland-like woodlands (*dehesas* in Spanish, *montados* in Portuguese) in which more or less widely spaced Holm Oaks are interspersed with cereal crops or are used for grazing. Holm Oak dehesas are especially associated with pig husbandry, the pigs browsing on branch trimmings and consuming the acorns in autumn – before conversion into the celebrated Ibérico hams and other porky products. The Cork Oak has a more restricted natural distribution, occurring in warm, humid regions, chiefly in the south-west and in the Ampurdán (Empordà) region in the northern Catalan hills. Most Cork Oak woodlands are managed for the cork, which is harvested every eight years or so, and many take the form of dehesas, which are characteristic of Alentejo, western Andalucía and southern Extremadura. Both Holm and Cork Oak dehesas may have a limited shrub layer: of such species as junipers, brooms, lavenders, Cistus and Mediterranean Mezereon. However, woodlands in hilly terrain especially may have a much denser undergrowth of similar species.

Natural or semi-natural woodlands of several pine species are an important feature in many regions. The Stone Pine is particularly characteristic of Doñana and the southwestern coastlands in general. The Aleppo Pine replaces it on the Mediterranean coastal hills and has also been planted in southern Portugal. The Maritime Pine is also widespread on coastal hills as well as on the Meseta, and has been extensively planted for forestry, notably in Portugal and Galicia. Pine woodlands have a typical understorey of broom, Cistus and heathers.

All the Mediterranean woodlands are important habitats for a wide range of bird species, at all times of year. All have a diverse breeding passerine and near-passerine community, comprising both resident and migrant species. The cereal dehesas are an important nesting habitat for the Black-shouldered Kite and the Azure-winged Magpie is typical of Holm Oak dehesas especially. The latter are also important to wintering Cranes. The Red-necked Nightjar is typical of lowland pinewoods in the south.

2. Scrub

Scrub takes a wide diversity of forms that together cover vast areas of uncultivated, once forested, land as well as comprising the shrub and herb layers of the more open woodlands. The taller formations may include small tree species such as the Olive and Strawberry Tree, as well as tree heaths, Lentisc, Cistus and brooms. Lower formations, no more than a metre or so tall, may comprise great expanses of Cistus, brooms, heaths, thymes and/or lavenders, among others. A characteristic feature of the Mediterranean coastal lowlands is the palmetto scrub, largely comprised by tracts of the Dwarf Fan Palm. Soil characteristics as well as local climatic conditions largely determine the type of scrub that develops. Acid soils favour heaths and cistus, notably the Gum Cistus, which covers huge areas in southern Portugal, western Andalucía and Extremadura. Very heavily overgrazed areas develop a poor cover of unpalatable species such as Asphodels or *Lygos* brooms. The latter forms a type of scrub (*retamar*) that is widespread in the centre and south-west but which is poorly attractive to bird species. This is in contrast to other forms of scrub, which support a diversity of passerines and others, notably including a diversity of *Sylvia* warblers. Many of its components, but especially the wild olive (*acebuche*) and the lentisc, fruit in winter and attract great numbers of wintering passerines, including Blackcaps, Robins and Song Thrushes.

3. Sub-montane and montane forests

The cooler wetter environments of the higher reaches of the Mediterranean mountains favour the Lusitanian Oak in the Pyrenees, the Iberian Range, the higher levels of the Meseta, and the west and south-west, and the Pyrenean Oak in the Cantabrian, Central and Iberian Ranges, as well as the Scots Pine also in the Central and Iberian Ranges and the Corsican Pine in the hills of central and eastern Spain. Both the Scots and Black Pines are typical of higher elevations, except in the Betic Range, where the native high altitude forests are now only represented by relict stands of the Spanish Fir (*pinsapo*) although there are pine plantations, such as those that now support breeding Citril Finches in the Sierra Nevada.

The Spanish Juniper forms woods in the uplands of central Spain that are especially extensive on the páramos of the Iberian Range and its neighbourhood in Teruel and Soria. Juniper woodlands also occur in the Ebro valley in Zaragoza, on the Meseta Norte in Burgos and Segovia and on the Meseta Sur in Cuenca and Ciudad Real. There are also juniper copses in Madrid and on the southern versant of the Cantabrian Range in León. Juniper berries are important to wintering thrushes especially.

The Sweet Chestnut is not native to Iberia but occurs as extensive plantations in humid, often montane areas in central Iberia: for example in the Sierras de las Batuecas and Gredos in the Central Range; in the south in the western Sierra Morena, the Serranía de Ronda and the Sierra Nevada, and in Girona and Barcelona in Catalonia. It is also abundant in much of Portugal and in northern Iberia. All these forests or woodlands have shrub and herb layers of varying complexity and all represent important bird habitats.

4. Subalpine and alpine habitats.

The high tops of the Central, Iberian and Betic Ranges, at 1,600–2,000m, are characterised by the 'hedgehog zone', where the largely bare terrain is interspersed by dome-shaped, often spiny shrubs, many of them leguminous species. In the Mediterranean region, only the Sierra Nevada has an alpine zone, where some 40 endemic plants species occur. The Bluethroat is a notable breeding species in montane scrub in central and northern Iberia.

EUROSIBERIAN PLANT COMMUNITIES

The high rainfall here, generally over 1,000mm annually, makes for heavy leaching and acid soils.

1. Lowland forests

The Pedunculate and Sessile Oaks are characteristic of low and medium levels in the north of the Peninsula, where the Atlantic climate holds sway. Forests of this type are typical of the lower northern slopes of the Cantabrian mountains. There are much more recent and very extensive plantations of eucalyptus species, particularly in Galicia and western Asturias, and of Monterey Pines and other exotic pines throughout the Cantabrian range and the Basque mountains. These exotic plantations have a dense undergrowth of native shrubs and hence they are not devoid of ornithological interest. The Downy Oak also forms important forests in the lower reaches of the Pyrenees: the Pyrenean Oak extends only to Basque Mountains but not to the Pyrenees, despite its name. The Middle Spotted Woodpecker is a distinctive member of the breeding community of the Eurosiberian oakwoods.

2. Montane forests

The Beech is the dominant forest component above about 600m and extends to altitudes of about 1,900m. It is the principal forest tree of the northern slopes of the Cantabrian Range and also occurs in the Pyrenees and the northern Iberian Range, growing as pure stands or as mixed beech-pine forest, often intermingled with ash, elm, field maple or yew. The Scots and Black Pines are characteristic of high elevations of the Pyrenees, with the Mountain Pine at the highest levels there. The Silver Fir can also be found in the Pyrenees, typically on wet, north-facing slopes, often mixed with Beech. Natural pine forests are almost entirely lacking in the Cantabrian Range, where the beeches and oaks give way to birches higher up. The White-backed Woodpecker is confined to old-growth beech forests in the western Pyrenees and the Boreal Owl inhabits the high Pyrenean coniferous forests.

3. Heathland and grassland

Forest clearance has resulted in the formation of wide expanses of heathland in the northern coastlands and hills. This is *Calluna* and *Erica* heath, with gorses and brooms, quite similar to the heathlands of northwestern Europe but more floristically diverse. Grazed areas remain as grassy pastures. The distinctive breeding birds of the northern heathlands include the Hen Harrier, Cirl Bunting and Dartford Warbler. Some of the coastal heaths, for example on the Cabo Peñas peninsula, Asturias, are important for passerine migrants, in autumn particularly.

4. Subalpine and alpine habitats

The subalpine regions of the northern mountains, above the treeline, sustain dwarf juniper scrub but are now mostly grassy pastureland subject to heavy grazing by sheep, cattle, horses and wild ungulates. Such grasslands cover large areas of the higher uplands of the Cantabrian and Pyrenean mountains. A 'hedgehog zone', as found on mountaintops further south, is also encountered on the drier, south-facing Pyrenean peaks. True alpine plant communities, of widely scattered dwarf shrubs and herbs, including many endemic species, occur only in the Pyrenees, above 2,600m. Characteristic birds of the high pastures include both the Red-billed and Yellow-billed Choughs and Water Pipit, with the Grey Partridge and Ortolan Bunting in flanking scrub areas.

OTHER NATURAL HABITATS

The sea

The sea is one of the principal shapers of the Iberian avifauna, not only because it provides the habitats of seabirds and shorebirds but also because of its indirect effects on the climate of the entire region. The Iberian coastline is over 3,300km long, about half of it in the Mediterranean. Most of the peninsula is surrounded by a continental shelf some 10–40km wide, where water depths range down to 200m (Fig. 5.1). The shelf is narrower along the southern Mediterranean coast and in the Strait of Gibraltar and wider: extending to 80km or more, on the eastern Mediterranean coastline of Valencia and Catalonia. The Balearic island groups of Mallorca/Menorca and Ibiza/Formentera are each bordered by similar shelves separated from each other and the Iberian mainland by deeper water. The continental shelves generate upwellings, especially in spring and summer, resulting in plankton booms that sustain large shoals of anchovies and other bait fish, which in turn provide important food supplies for seabirds and cetaceans. More local upwellings, often associated with undersea reefs and sea mounts, also attract seabirds to particular areas, for example the Strait of Gibraltar; the Gorringe Bank: 130 miles off southwestern Portugal and the Galicia Bank: 200 miles west of Galicia. Pelagic seabirds congregate around the margins of the continental shelf in the Atlantic with some regularity, and they may include such regionally scarce species as Fea's, Zino's and Bulwer's Petrels that are very seldom seen from land.

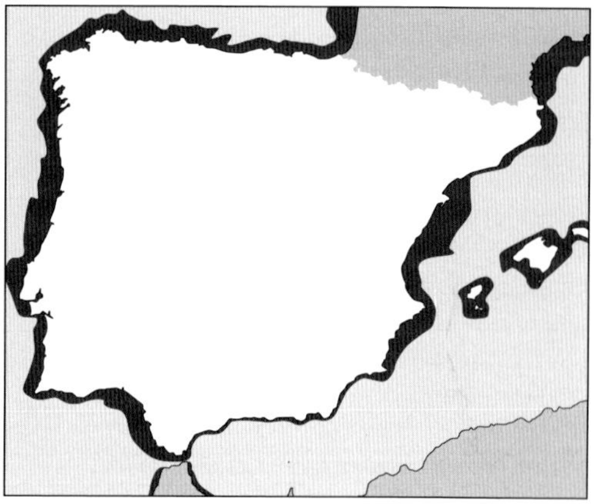

Figure 5.1. The Iberian continental shelf. 200m contour.

Sea cliffs and rocky coasts

A rocky coastline predominates on the Iberian north coast, and in the north-west in Galicia and northernmost Portugal, south to Porto, although this is interrupted by some sandy beaches and many small coves. The northern coastline is punctuated at intervals by rocky headlands, some of which provide excellent vantage points for seawatching. The most renowned of these are Cape Bares, A Coruña; Punta La Vaca on the Cape Peñas peninsula, Asturias, and Cape Higuer, Gipuzkoa. There are numerous rocky islets immediately offshore all along the north coast, some large enough to support some littoral vegetation. The Sisargas Islands and those of the Galician Islands National Park: Cíes and Ons, lie off the north-west coast. Sheer cliffs are frequent in the north-west and they include two of the highest in Europe, Vixía Herbeira (620m) near San Andrés de Teixido, A Coruña, and Jaizkibel (547m), between Pasaia and Cape Higuer, Gipuzkoa. The Shag has its principal Iberian breeding populations on this coast and there are also large colonies of Yellow-legged Gulls. European Storm-petrels breed on some of the islets. The north-west has residual breeding colonies of Kittiwakes and Common Guillemots, the latter on the verge of extinction after a long period of decline.

A rocky coastline is also characteristic of parts of southwestern and southern Portugal: locally in Estremadura and more widely in Baixo Alentejo and western and central Algarve. The rocky Berlengas Islands are the only continental Portuguese islands that have important seabird colonies, including all the Cory's Shearwaters that breed in Portugal and the sole known breeding colony of Madeiran Storm-petrels in Europe. Rocky islets and stacks off the southwestern and southern coasts of Algarve are of little importance to seabirds, other than Yellow-legged Gulls.

The Atlantic coast of Andalucía, from the Portuguese border to Cape Trafalgar is predominantly sandy and low-lying but the north coast of the Strait has long rocky stretches and promontories. The sandy cliffs between Cape Trafalgar and Barbate once supported breeding Ospreys and a cliff-nesting colony of Cattle Egrets, and are still attractive to Yellow-legged Gulls. The Rock of Gibraltar itself has a mainly rocky shoreline and has sheer sea cliffs on the eastern and southern coasts, rising to the summit at 425m.

The Mediterranean coastline of Iberia is extensively rocky in the north of the Costa Brava, from the French border south to Roses: in particular around the Cape Creus peninsula, and then again from l'Estartit south to Blanes. A rocky coastline predominates further south around Cape La Nao, Alicante, and at Cape Palos, Murcia. The most prominent sea cliffs in this sector are those of the Peñón de Ifach, at Calpe, Alicante. The Costa del Sol, from Cape Gata, Almería to Gibraltar has rocky stretches in Granada and eastern Málaga provinces and some rocky headlands in the west.

The offshore islands of the Mediterranean coast are much more important ornithologically than the mainland rocky coasts. The principal ones are the Medes Islands, off l'Estartit, Girona; the Columbretes islands, Castellón; Benidorm island, Alicante; Isla Grosa, Murcia, and Alborán island, Almería, off the south coast. All are inhabited by the ubiquitous Yellow-legged Gulls but Shags nest on the Medes and Columbretes Islands, Audouin's Gulls have important colonies on the Columbretes, Isla Grosa and Alborán, Eleonora's Falcons have one of their principal Iberian haunts on the Columbretes and European Storm-petrels nest on Benidorm Island.

Sandy beaches, dunes,

Sandy beaches may be found all around the Peninsula but the finest examples are the broad, long strands of the Atlantic coast, in central Portugal from Porto to Peniche, in the Portuguese south-west from Setúbal to Sines and in the eastern half of Algarve. Similar beaches make up most of coastal southwestern Andalucía, from the Guadiana estuary to Cape Trafalgar, and north-west of Tarifa. Long sandy beaches occur along much of the Mediterranean coast, except in the rocky stretches mentioned above. These beaches are narrower than the Atlantic equivalents, very largely because they experience a much more limited tidal range. Dune systems are most highly developed on Atlantic coasts, with notable examples at Liencres, Cantabria; Corrubedo, A Coruña; in Portugal on the coasts of Beja Litoral, Alentejo and eastern Algarve, and at Doñana.

Sandy beaches, and their dune systems, are particularly important to Kentish Plovers and, more locally, Little Terns, as well as to non-breeding flocks of Sanderlings. Snow Buntings may be encountered in dune areas in the north-west in winter. Sandy beaches are of course highly attractive to the tourist industry and enormous stretches have been transformed irrevocably by hotel developments and the building of holiday villas and retirement homes, to the detriment of Kentish Plovers in particular. Such developments have had their greatest impact in the Mediterranean but the effects of tourism have also been felt on the Atlantic coasts, notably in eastern Algarve.

Coastal lagoons, estuaries and coastal marshes

Coastal wetlands provide some of the most significant wildlife habitats in Iberia, some of them being of international importance for their breeding bird populations and/or for the numbers of certain species that rely on them on passage or in winter. The foremost of these is the Guadalquivir delta, where the Marismas – the seasonally inundated salt flats – form part of an extraordinary complex of natural and man-modified habitats that include freshwater and brackish lagoons, freshwater and salt marshes, creeks, canals, saltpans, ricefields, a 35km sandy beach, the River Guadalquivir itself, and a range of dry land habitats including Stone Pine woodlands and the largest sand dune system in Iberia, all spread over an area of over 1,300km^2. Much of this area is fortunately protected as the Doñana National Park or as part of the peripheral Natural Park. The Guadalquivir delta is of critical importance to waterfowl, waders and a great diversity of landbirds, as is made clear in the species accounts.

The Ebro delta is much smaller (320km^2) but is still the second largest wetland in Spain. It enjoys some protection as a Natural Park and, like Doñana, is a key site for breeding waterfowl, herons and waders. It is also important for its breeding gulls and terns, in particular for Audouin's Gull, which has its largest global colony there. The Tagus estuary, with the Sado estuary nearby, is the most important coastal wetland in Portugal, and is a key staging and wintering area for waders especially.

Many smaller rivers drain the Peninsula and their estuaries are often of ornithological importance, particular for wildfowl, herons, waders, gulls and terns. Several of those in the north and north-west especially have drowned valleys, the Rías, which run some way inland. The more important of these other estuaries on the Biscay and Atlantic coasts include the following, which are listed from east to west across the north coast and then from north to south: the Ría de Bidasoa, Gipuzkoa; Ría de Gernika, Bizkaia, Marismas de Santoña, Cantabria; Ría de San Vicente, Cantabria; Ría de Villaviciosa, Asturias; Ría de Aviles, Asturias; Ría del Eo, Asturias/Lugo; Ría de Ortigueira and other Rías Altas, A Coruña; Ría de Arousa, Ría de Vigo and other Rías Bajas, Pontevedra; Miño/Minho estuary, Pontevedra/Viana do Castelo; Ria de Aveiro, Aveiro; Mondego estuary, Coimbra; Alvor estuary, Faro; Ria de Faro, Faro; Piedras estuary, Huelva; Odiel estuary, Huelva, and Barbate estuary, Cádiz.

Estuaries within the Mediterranean are generally much less important for birds than those on the Atlantic, largely because they lack salt marshes or mudflats on account of the short tidal range. The Ebro delta is an outstanding exception but the Llobregat delta, Barcelona, is also an important site for breeding, passage and wintering aquatic birds. Further south, the Guadalhorce estuary, Málaga, and the Guadiaro and Palmones estuaries, Cádiz, also attract waterbirds at all seasons, although in small numbers.

Coastal lagoons provide important waterbird habitats in many regions and are especially important along the Mediterranean east coast, where there are some major wetlands. Such lagoons may be associated with freshwater marshes and the important ones invariably offer reedbeds and other emergent vegetation. Some have been lost to development but most of the remaining ones are protected and are managed as reserves. Principal examples from the Levant coast include the Aiguamolls de l'Empordà, Girona; the Prat de Cabanes–Torreblanca, Castellón; the Marjal del Moro, Valencia; the Albufera, Valencia; the Marjal de Pego-Oliva, Valencia/Alicante; the Clot de Galvany, Alicante; El Hondo, Alicante; the Lagunas de La Mata and Torrevieja, Alicante; and, the largest of all, the Mar Menor, Murcia. The Mediterranean coastline of Andalucía is very poor in wetlands, the few examples of significance being the Cañada de Las Norias, Almería: which sometimes attracts very large numbers of White-headed Ducks; the Albufera de Adra, Almería, and the pools at the Guadalhorce estuary, Málaga.

Similar habitats along the Atlantic coast include the Laguna de El Portil, Huelva, and a series of Portuguese coastal lagoons of great importance for their attractiveness to nesting waders and terns and to both passage and wintering waterfowl and waders. These include the Lagoas de Salgados and de Santo André, Algarve; de Melides, Baixo Alentejo, de Albufeira and de Óbidos, Estremadura and the Barrinha de Esmoriz, Beira Litoral. Some are permanently or seasonally linked by channels to the sea. Coastal lagoons in Galicia include those at the Dunas de Corrubedo, the Laguna de Traba and the Laguna de A Frouxeira, all in A Coruña. Natural lagoons of any size are lacking from the Biscay coastlands, partly due to the narrowness of the coastal strip.

Inland cliffs, gorges and rocky terrain

Rugged terrain may be encountered in nearly all parts of Iberia in the form of sheer mountain cliff faces and river gorges. These are especially abundant in the eastern half of the Peninsula and in Mallorca, where limestone rock predominates, but some impressive river gorges are also found in quartzite areas of Extremadura and eastern Portugal, for example along the international border stretches of the Duero/Douro and Tagus rivers. Many of the characteristic and important bird species of the region depend on rocky habitats, often for nesting. Cliffs and gorges, or their associated scree slopes, are very often inhabited by a guild of 'rock-loving' or rupestral species of which the more widespread typical examples include the Griffon and Egyptian Vultures, Golden and Bonelli's Eagles, Peregrine Falcon, Black Stork, Rock Dove, Eagle Owl, Alpine Swift, Crag Martin, Blue Rock Thrush, Red-billed Chough and Raven. Such notable species as the Lammergeier, the Wallcreeper and the Yellow-billed Chough are also associated with similar habitats in the north and the Black Wheatear is characteristic of rocky terrain in the south and east.

Lakes and marshes

There are numerous inland lakes and ponds of diverse sizes throughout the Peninsula, although very many were drained and converted to agricultural use during the middle years of the 20th century especially. The lost wetlands included three of the largest and most important, the Laguna de Antela in Ourense (some 3,600ha), the Laguna de La Nava or 'Mar de Campos' in Palencia (up to 5,000ha in wet years) and the Laguna de La Janda in southern Cádiz (4,000ha). The present Laguna de La Nava is a restored fragment of the former lake that is now itself one of the most important wetlands of the northern Meseta. The original site of the Laguna de La Janda is now ricefields and both the former basin and the surrounding plain retain very high ornithological interest.

The Eurosiberian zone has few natural lakes, despite the high rainfall there, probably because of the steepness of much of the terrain. There are a number of small glacial lakes in the upper reaches of the Cantabrian Range and the Pyrenees but these seldom attract many birds. However, Lago Ercina at Covadonga in the Picos de Europa is one of the very few known regular Iberian nesting sites of the Common Teal. The more interesting of the natural lowland wetlands in the region include the Salburúa marshes and the Laguardia lagoons, both in Álava.

The natural wetlands of the Iberian interior vary considerably in extent in summer depending on the amount of rainfall during the previous winter and spring. Many in the centre and south dry up completely, some of them every year and others only during drought years. Their attractiveness to birds fluctuates accordingly but, when water levels permit, these wetlands host numerous breeding waterbirds, and large numbers use them on passage and in winter. Many of these lakes, but particularly those that are prone to drying up, are brackish to a greater or lesser extent. The principal natural/semi-natural lakes and marshes of the Mediterranean bioclimatic zone in Spain include the following, most of which figure in some of the species accounts.

In the Ebro basin: Las Cañas reservoir, originally a natural lake, Navarra; the Laguna de Pitillas, Navarra – the foremost breeding site of the Great Bittern in Iberia; the Laguna de Sariñena, Huesca; the Saladas de Chiprana, Zaragoza, the Laguna de Gallocanta, Zaragoza/Teruel – best known for its importance as a staging area for migrating Common Cranes, and two recently restored lagoons that already support good numbers of birds: the Estany d'Ivars, Lleida, and Laguna del Cañizar at Villarquemado, Teruel.

On the northern Meseta: the Lagunas de Villafáfila, Zamora; the Laguna de La Nava and Laguna de Boada, Palencia, and the Laguna de El Oso, Ávila.

On the southern Meseta: the lagoons at Villafranca, El Taray, Miguel Esteban, Pedro Muñoz and Manjavacas which are among the 100+ lakes, many of them ephemeral, of La Mancha Húmeda Natural Park and Biosphere Reserve, in the provinces of Ciudad Real, Albacete, Toledo and Cuenca, which together make up one of the main sites for wildfowl and other waterbirds in the Iberian Peninsula; and the Tablas de Daimiel National Park, Ciudad Real – also of great importance for aquatic species despite having been seriously reduced in extent by over-abstraction of water from its aquifer.

In Andalucía: the Cádiz lagoons – including the Lagunas del Puerto de Santa María, Puerto Real, Medina and Espera – especially important in some cases for the presence of breeding Red-knobbed Coots, Marbled Ducks and White-headed Ducks as well as supporting numerous other aquatic birds all year round; the Córdoba lagoons – including the Lagunas de Zóñar, del Rincón, Amarga, Dulce, del Conde and de Tíscar – variously important for waterbirds, including the White-headed Duck; and the Laguna de Fuentedepiedra, Málaga – location of the largest Greater Flamingo colony in Iberia; the Lagunas de Campillos, Málaga.

There are few remaining natural wetlands inland in Portugal. They include the seasonally inundated marshlands (*pauis*) that are riparian in nature and described below.

Rivers, streams and riparian vegetation

Watercourses are naturally of great importance to such species as the Common Sandpiper, Dipper and Grey Wagtail that are especially linked with the faster-flowing, perennial streams and rivers. Little Ringed Plovers and Little Terns may breed on shingle islands and Bee-eaters and Sand Martins may tunnel in sandy banks. However, riparian vegetation is also of great importance. In some regions this may take the form of riverine woodland, made up of alders, willows, elms, poplars and ashes, as well as a dense undergrowth of shrubs and herbaceous plants. Particularly fine examples are found along the Ebro in Navarra and Aragón, where the ox-bow lakes (*galachos*) and riparian woods (*sotos*) are especially associated with Penduline Tits, Golden Orioles and numerous other passerines, both resident and migrant. The Lesser Spotted Woodpecker is also associated with riparian woods locally.

Riparian vegetation may provide almost the only natural habitat across the agricultural expanses of the meseta. In its sparser forms, particularly in the warmer regions, it may take the form of clumps of tamarisks or oleanders, often growing on islands in the stream. Marginal reedbeds and stands of reedmace also occur where the flow is not normally rapid. The Nightingale is everywhere especially associated with densely vegetated watercourses of all sizes, as are the Cetti's and Western Olivaceous Warblers. Riparian vegetation also often accommodates heronries.

In Portugal extensive freshwater marshes, often permanent but with fluctuating water levels, were an important feature of the river valleys. These *pauis* (singular *paul*) have mainly been lost to agriculture and those that remain are subject to management of water levels. Some of these latter are still among the most important wetlands for birds and other wildlife in Portugal. The most celebrated include the Paul do Boquilobo, Santarém, the Paul de Tornada, Leiria; and the Pauis de Arzila, Madriz and Taipal, Coimbra, in the Mondego valley.

Pseudosteppes, salt flats and arid terrain

True steppes are characterised by a continental climate and low precipitation and are essentially grasslands. By this definition, the 'steppes' of Alentejo, Extremadura and the Spanish meseta are properly termed pseudosteppes, since they consist of pastures or expanses of dwarf shrubs mixed with cereal crops (rather than grasslands proper) and they experience significant winter rainfall in most years. Such habitats are nonetheless extremely valuable ornithologically for their community of open-country species, notably bustards, sandgrouse and larks that are comparatively rare elsewhere in Europe.

Salt flats or salt steppes occur in the Ebro depression, eastern Castilla-La Mancha, and parts of the arid south-east in Murcia and Almería. They are very dry regions with a sparse covering of dwarf shrubs and grasses. The Bardenas Reales, Navarra; the Belchite steppes, Zaragoza, and Los Monegros, Huesca/Zaragoza, retain some of the best examples of this habitat, although much land has been lost to irrigation. Characteristic species here include Dupont's and Lesser Short-toed Larks, both sandgrouse and Stone Curlews.

The driest Iberian region, as mentioned above, is in eastern lowland Almería and southern Murcia. The mean annual precipitation of less than 200mm is low enough to produce true desert conditions, which are most evident in the badlands of the Desierto de Tabernas, Almería. Parts of the arid south-east are dissected by heavily eroded gullies, the *ramblas*, which are usually dry but are prone to occasional flash floods. The Trumpeter Finch is a keynote species in this region.

MAN-MADE HABITATS

Virtually all habitats in Iberia, as generally in Europe, have been modified by human activities, most particularly by the replacement of climax forests by open country devoted to agriculture. All of the artificial and semi-artifical habitats of Iberia have impacts on the avifauna, both positive and negative. Very often it is the case that one community's loss is another community's gain, with lowland forest and woodland birds being the biggest losers and open country species the biggest gainers.

Forestry plantations

Forest crops in Iberia principally take the form of pine, eucalyptus and poplar plantations and also the cultivation of the Iberian Holm Oak and the Cork Oak. The oak woodlands have a semi-natural character and support a diverse avifauna, as described above.

Eucalyptus plantations, chiefly of *E. globulus*, have their greatest extent in the west: especially in Portugal, Galicia, western Andalucía and Extremadura, and across the Biscay coastlands, where they are cropped for wood pulp for the paper industry. These plantations have displaced natural woodlands and heathland over vast areas. They have an extensive understorey of native shrubs, however, and thus support a diversity of passerine species, as well as providing nest sites for corvids and forest raptors. Much of Doñana was threatened with afforestation with eucalyptus before it was designated a reserve. Large areas of what is now Monfragüe National Park were planted with eucalyptus that has since been removed. The southern eucalyptus woodlands have very little undergrowth and are a poor habitat for birds, although Spanish Sparrows use them for their colonies in the south-west and they again provide nest sites for species as diverse as White Storks, Black Kites and egrets.

Poplars of a number of hybrid clones are planted widely for timber and wood chips in Mediterranean Iberia. These plantations often take the form of geometrical stands of uniform age, generally sited in lowlands near watercourses. They may be the only significant tree presence in agricultural regions such as the northern meseta and, as such, they may permit such species as the Hobby and Scops Owl to breed there. More widely, they are used for nesting by a diversity of other raptors, and such passerines as the Golden Oriole.

Pine plantations, especially of exotic Monterey Pines but also of Black Pines, which are native, are widespread in northeastern Spain and both the Black Pine and the Scots Pine have been planted widely in the Mediterranean bioclimatic region both in Spain and Portugal. These plantations attract the local avifauna of coniferous woodlands, and nestboxes are often placed to attract hole-nesters to the plantations, where natural holes are largely lacking. In general, however, the uniform age and species-composition of plantations makes for lower species diversity and density than would be found in natural woodlands.

Orchard crops

Enormous areas of the Iberian peninsula are devoted to the cultivation of orchard crops. The most economically important are olives and grapevines, with oranges and other citrus crops in the east and south. Almonds and cherries are also important locally, and apples are the typical crop of the Cantabrian lowlands, where they sustain a large cider industry. Many other fruits, some of them exotic, are grown in the warmer regions especially.

The small black fruits of the wild Olive are an important resource for many passerines in winter. The much larger cultivated fruits, grown especially for the oil, are planted widely at low elevations in the Mediterranean bioclimatic zone, in southern and south-central Spain, the Ebro depression and in much of Portugal other than in the northern coastlands. Olive groves range from the intensively managed, which may be cleared of undergrowth and are subjected to agrochemical treatments, to neglected or abandoned ones, that have a much more natural aspect, with abundant undergrowth and, invariably, a much denser avifauna. The Rufous-tailed Scrub-robin and Western Olivaceous Warbler are both particularly associated with olive and citrus groves, in the Guadalquivir depression especially. Other characteristic breeding birds include the Turtle Dove and Woodchat Shrike. Fruiting cultivated olives are also attractive to wintering passerines and large flocks of starlings and thrushes may congregate there, sometimes attaining pest status.

Traditional vineyards may attract such passerines as larks, starlings and finches, and sometimes such open-country birds as sandgrouse and the Stone-curlew. Most modern vineyards are intensively managed, however, and these have little to offer birds.

Non-irrigated farmland

Non-irrigated or 'traditional' agriculture is the primary landscape feature of the Spanish meseta and it extends over large areas in all suitable regions of Iberia, where soils, water availabilty, elevation and gradient allow. Non-intensive farming, characterised by the presence of small fields – often separated by hedgerows in the north especially – and areas of both short-term and long-term fallows, supports the best populations of farmland birds, including great numbers of larks, finches and buntings all year round and also such charismatic birds as storks, cranes, bustards, sandgrouse, harriers and stone-curlews. Cereal crops are especially important to nesting Great Bustards and Montagu's Harriers. Farmland birds invariably decline or are displaced by intensive agriculture and this has become an increasing problem in Iberia, particularly since Spain and Portugal acceded to the European Union.

Irrigated farmland

Crop irrigation has become possible in large tracts of the drier sectors of the Peninsula since the proliferation of reservoirs (see below). Some of the crops, notably cotton, are of no value to birds but others, such as alfalfa, attract some species. In winter, bird densities are much higher in irrigated farmland than elsewhere. The principal consequence for birds of the spread of crop irrigation, however, is the loss of such habitats as cereal fields, pastures and steppe habitats, leading to the disappearance of many species from the areas affected. Such important birds as the Little Bustard and Pin-tailed Sandgrouse have been displaced in this way.

Pastures

Montane grasslands have been described above but there are large areas of permanent and semi-permanent rough pasture in many lower-lying areas of the Peninsula, particularly in the south-west. The best of these, from the ornithological point of view, have a diverse flora with many weed species that sustain larks, pipits, finches and buntings in abundance, especially in winter. Often this demands that grazing intensity should not be too severe. Such areas, including the grassy pseudosteppes of Extremadura, are also a principal habitat for bustards, sandgrouse and stone-curlews among others. The pastures were traditionally exploited seasonally for grazing sheep between autumn and spring, with the herds travelling for long distances along drove roads to spend the summer in the mountains of the north and centre. Such a practice has almost disappeared and sheep are now present on the pastures year-round, which of course raises problems of overgrazing.

Wet pastures are much rarer and most have been lost to agriculture. They originate from seasonally inundated grasslands on flood plains. Parts of the Marismas del Guadalquivir have something of this character but a better surviving example is provided by the water meadows (*lezirias*) of the lower Tagus in Portugal. These are grazing lands criss-crossed by reedy ditches and dykes, where horses and cattle are raised. They attract open country and wetland birds at all seasons.

Ricefields

Rice has been cultivated in Spain since at least the 10th century but ricefields have become much more widespread since the mid 20th century, both in Spain and Portugal. Rice is grown in regions that combine warm temperatures and readily available water. Some 60% of the Ebro delta is ricefields and rice is grown extensively in the Guadalquivir delta; in the lower valleys of the Mondego, Tagus and Sado rivers in Portugal; in the Guadiana valley in Extremadura; at the Albufera in Valencia; in the Ebro valley in Aragón and southern Navarra, and at La Janda, Cádiz.

Ricefields have many of the attributes of freshwater marshland, with particular features that make them especially attractive to birds. The fields proper are shallow lagoons with short emergent vegetation – the rice. They are supplied by ditches, which often support reedbeds. The fields themselves are separated by dykes, many of which become covered with thin vegetation and provide nest sites for such species as Black-winged Stilts and Moorhens. Ricefields provide food for birds in the form of the plants themselves, which are locally attractive to Purple Swamphens, and in the form of spilt grain, which becomes available in autumn and winter after the harvest, in late summer. The fields also provide food indirectly in the form of invertebrates, including numerous aquatic insects, worms and, especially, crayfish. These last are the alien Red Swamp Crayfish, which thrive in the ricefields, where they burrow into the dyke walls. The crayfish are the principal attraction of the ricefields to storks, herons, Glossy Ibises and Lesser Black-backed Gulls, among others, which flock there in very large numbers. Passage and wintering waders are also often very numerous, attracted both by the rice and the invertebrates. In general, ricefields can be seen as an asset to birds, especially where agrochemicals are used sparingly, or not at all. In the latter case, the SEO/Birdlife sponsored organic rice project at Riet Vell, in the Ebro delta, is a model example: its paddies are thronged with birds.

Saltpans

The harvesting of sea salt from the evaporation of brackish or sea water in shallow lagoons under the glare of the hot Mediterranean sun has been practised in Iberia since at least the time of the Phoenicians but it probably peaked in the 19th century, when high demand for salt for food preservation preceded the introduction of refrigeration. Saltpans went into decline thereafter and many of the surviving examples exist only because they are protected as wildlife reserves, although there is still an active salt industry in a few areas. Others have been converted into fish farms. Saltpans are typical of estuarine regions in the thermomediterranean, where the hot dry summers are ideal for evaporating the water. Inland examples are not unknown, however: the Laguna de Fuentedepiedra once had an active salt extraction industry.

The principal aggregations of saltpans are, in Portugal: at the Ria de Aveiro; the Mondego, Tagus, Sado and Guadiana estuaries, the Ria Formosa and Castro Marim, and, in Spain: at Bonanza and the Marismas de Trebujena in the Guadalquivir estuary; Cádiz Bay; Cabo de Gata and Roquetas de Mar, Almería; San Pedro del Pinatar, Murcia; Torrevieja, La Mata and Santa Pola, Alicante, and the Ebro delta, Tarragona.

Reservoirs and artificial waterbodies

Lakes and other wetlands of any size are naturally scarce in much of Iberia away from the coasts. Some regions, notably Extremadura, have an almost total lack of such habitats except on a very small scale. The widespread and continuing construction of artificial waterbodies, particularly the numerous reservoirs that appeared during the second half of the 20th century, has transformed the landscape in many areas and at the same time has had a diversity of impacts on birds and other wildlife.

Reservoirs form behind dams constructed across rivers, often where these pass through gorges. The resulting lakes are thus often very long and relatively narrow, although others are able to expand laterally. In general, reservoirs that are hemmed in by rocky terrain tend to be deep everywhere and these are the least attractive to birds, although they may support a few Great Crested Grebes and perhaps Cormorants in winter. The construction of such reservoirs has also meant the loss of rocky stream and gorge habitats in the affected areas, to the detriment of other species. In contrast, other reservoirs are much more like natural lakes and these offer a range of water depths and may have areas of reedbeds and other emergent vegetation. The best examples have islands, which may house colonies of such diverse species as Cattle Egrets, Gull-billed Terns and Collared Pratincoles, as well as nesting waterfowl. Such bird-friendly reservoirs are widespread. They include the Ebro reservoir, Cantabria/Burgos; Ullibarri-Gamboa, Álava; Las Cañas, Navarra; San José, Valladolid; Azud de Riolobos, Salamanca; Santillana, Madrid; Los Canchales, Badajoz; Arrocampo-Almaraz, Cáceres; Sierra Brava, Cáceres; Azután, Toledo; Castrejón, Toledo, and Alqueva, Alentejo/Badajoz, among many others.

Many reservoirs are important for waterfowl and other aquatic or waterside species, from Ospreys and Purple Swamphens to Reed Warblers. The shallows may host diverse waders on passage and in winter. Very large concentrations of wildfowl winter at some locations, notably at the Sierra Brava reservoir, which regularly sees one of the largest gatherings of wintering ducks in Iberia. Very large numbers of gulls, geese and Cranes roost at reservoirs near their winter feeding areas.

Such other artificial wetlands as gravelpits are often also attractive to waterbirds but even irrigation ditches, dewponds and watering holes for livestock are important for birds generally, especially in summer when water may be scarce over large areas.

Cities, towns and villages

Urban habitats of various types are freely available across Iberia. They range from modern, bustling cities with high-rise buildings to long-established towns and villages with plenty of old, sometimes medieval, buildings as well as more recent edifices. Such habitats also include the many historic buildings: castles, cathedrals, churches, aqueducts and others, both those in use and those that only survive as ruins. In general, older buildings are more bird-friendly than newer ones; indeed the latter are often built so as to deliberately exclude birds from entering the eaves.

Birds are a very obvious feature in many Iberian towns and villages. Typically these human settlements also accommodate large numbers of such breeding species as White Storks, feral Rock Doves, Collared Doves, Common and Pallid Swifts, Barn Swallows, Blackbirds, Spotless Starlings and House Sparrows, with Lesser Kestrels and Jackdaws also a feature in some areas. Yellow-legged Gulls have become familiar if not entirely welcome nesters on rooftops in many coastal towns. Less obvious bird inhabitants may often include Barn, Tawny, Scops and Little Owls, and both Eagle Owls and Peregrine Falcons are resident in some towns and cities, including Madrid. Town parks and gardens accommodate many other familiar passerine birds, at all times of year. Rubbish dumps also attract often very large numbers of gulls, White Storks, Cattle Egrets and other scavengers, in winter especially.

6. THE IBERIAN AVIFAUNA

There is a steadily expanding interest in birds, other wildlife and environmental issues in general within Iberia, to an extent that is certainly not universal within the Mediterranean region. The evidence includes the increasing membership of the national and regional ornithological societies and other conservation bodies, and the establishment and maintenance of an expanding range of national parks, nature reserves, important bird areas and other protected zones, together with the implementation of national and European environmental protection legislation. Iberia has also long been a destination of choice for birders from other parts of Europe and from further afield, attracted by the opportunity to see a wide range of species often in beautiful surroundings.

The avifauna of the Iberian Peninsula and the Balearic Islands reflects the strategic location of the region at the southwestern extremity of Eurasia, its proximity to the African continent, its variety of terrestrial and aquatic habitats, its long coastline, its considerable altitudinal range, its broad climatic range and the good state of conservation of much of the natural environment (Bernis 1972, de Juana 2002). Nevertheless, in terms of species richness, the avifauna is comparable with that of other sizable areas of Europe. Spain, for example, has some 254 regular breeding species whereas France has 259 and Greece 237. The diversity of the Iberian avifauna may be tempered by its location at a geographical extreme of a continental land mass. The existence of a peninsular effect, in which species richness declines with distance from a continental core, has been suggested, not without controversy (Busack & Hedges 1984), to explain why such patterns occur (Simpson 1964, de Juana 2002). Any peninsular effect is frequently obscured by environmental variables and when its possible application in Iberia was examined in the context of Iberian forest species (Ramírez & Tellería 2003) it became apparent that whereas distance to the continent and not environmental variables are the main correlates to species richness in Atlantic Iberia, it is the environmental variables, especially mean annual rainfall but also floristic diversity (Tellería & Santos 1994), that best account for species richness in Mediterranean Iberia.

BREEDING SPECIES

Only a small proportion of the avifauna is unique to the region and hence of particularly high biological significance and conservation value. There are only four endemic species and all four are the outcome of 20th century reclassifications of subspecies of more widely distributed taxa. They are the Balearic Shearwater *Puffinus mauretanicus*, the Spanish Imperial Eagle *Aquila adalberti*, the Iberian Azure-winged Magpie *Cyanopica cooki* and the Balearic Warbler *Sylvia balearica*; formerly treated respectively as races of the Manx Shearwater *Puffinus puffinus*, the Imperial Eagle *Aquila heliaca*, the (Oriental) Azure-winged Magpie *Cyanopica cyanus* and the Marmora's Warbler *Sylvia sarda*. All four are geographically separated from the former parent taxa, by the span of nearly two continents in the case of the magpie, and their genetic distinctiveness is not in question. There is also the Iberian Chiffchaff *Phylloscopus ibericus*, formerly treated as a subspecies of the Common Chiffchaff *P. collybita* (which also breeds in Iberia), whose breeding range outside Iberia extends marginally into southwestern France and the far north-west of Morocco. In addition, and more controversially in certain cases, some 21 endemic subspecies have been identified within the Peninsula and/or the Balearics, together with a further 19 near-endemic subspecies whose geographical range extends a short way into adjacent parts of France or Morocco (Table 6.1) or the western Mediterranean islands. As ever with subspecies, different authorities disagree regarding the validity of certain forms, dismissing some as clinal variants rather than discrete entities: the Iberian forms of the Red-legged Partridge and the Sky Lark are examples. Other subspecies are much more strongly differentiated from related forms and are widely accepted as valid: some, such as the Iberian Green Woodpecker, may even be candidates for separate species status themselves. A number of historical subspecies that no longer attract contemporary support are not included in the Table. We have chosen not to concentrate unduly on taxonomic issues within this work but some of the more important matters relating to species identity and subspecies validity are raised within the appropriate species accounts.

The uniqueness of endemic species and subspecies makes them of the highest conservation importance at a global level. However, about 100 other Iberian breeding species are of special interest for a variety of reasons. Most of them are included within the Spanish and/or Portuguese Red Books (Appendix 2). Ten species have their sole European breeding populations in Iberia: the Madeiran Storm-petrel, Red-knobbed Coot, Black-bellied Sandgrouse, Red-necked Nightjar, White-rumped Swift, Little Swift, Dupont's Lark, Black Wheatear, Western Olivaceous Warbler and Trumpeter Finch. The Cream-coloured Courser, which has bred successfully in Spain on at least two occasions in recent years, may need to be added to this list in due course. Three other species have their entire western European populations in Iberia: the Marbled Duck, White-headed Duck and the Rufous-tailed Scrub Robin, and the Iberian populations of the Thekla Lark represent nearly the entire European population.

The Iberian peninsula is a sizable landmass and so it is only to be expected that the regional populations of many widespread species rank among the largest in Europe. However, the Iberian populations – particularly those of Spain – of

Table 6.1. Endemic* and near-endemic subspecies of Iberia.

Species	Subspecies	Range
Rock Ptarmigan	*L. m. pyrenaica*	Pyrenees. Extends into French sector
Western Capercaillie	*T. u. cantabricus**	Cantabrian mountains
	T. u. aquitanicus	Pyrenees. Extends into French sector
Red-legged Partridge	*A. r. hispanica**	N. & W. Iberia
	*A. r. intercedens**	S. & E. Iberia, Balearics
Grey Partridge	*P. p. hispaniensis*	N. Spain, Pyrenees. Extends into French sector
Small Buttonquail	*T. s. sylvaticus*	SW Iberia & coastal N. Morocco.
Pin-tailed Sandgrouse	*P. a. alchata*	Mainland Iberia and SE France
Eurasian Eagle Owl	*B. b. hispanus**	Mainland Iberia
Green Woodpecker	*P. v. sharpei*	Spain, Portugal, Andorra. Extends into French Pyrenees
Great Spotted Woodpecker	*D. m. hispanus**	Mainland Iberia
Lesser Short-toed Lark	*C. r. apetzii**	E. and SW Spain and SE Portugal
Crested Lark	*G. c. pallida**	Mainland Spain and Portugal
Thekla Lark	*G. t. theklae*	Mainland Iberia and extreme S. France
Eurasian Skylark	*A. a. guillelmi**	N. Portugal and NW Spain
	*A. a. sierrae**	S. Portugal and C. and S. Spain
	*A. a. cantarella**	NE Spain
Yellow Wagtail	*M. f. iberiae*	Mainland Iberia and Balearics. Extends into S. France
Bluethroat	*L. s. azuricollis**	Mainland Spain and Portugal
Black Redstart	*P. o. aterrimus**	S. and W. Iberia
Black Wheatear	*O. l. leucura*	Iberia and extreme S. France
Spotted Flycatcher	*M. s. balearica**	Balearic Islands
Pied Flycatcher	*F. h. iberiae**	Spain
Long-tailed Tit	*A. c. taiti*	N. Iberia and SW France
	A. c. irbii	S. Iberia and Corsica
Crested Tit	*L. c. weigoldi**	W. & S. mainland Iberia
Coal Tit	*P. a. vieirae**	Mainland Iberia
Blue Tit	*C. c. ogliastrae*	S. Iberia, Corsica and Sardinia
	*C. c. balearicus**	Mallorca
Great Tit	*P. m. mallorcae**	Balearic Islands
	P. m. corsus	S. Iberia, Corsica and Sardinia
Eurasian Nuthatch	*S. e. hispaniensis*	Mainland Iberia and N. Morocco
Southern Grey Shrike	*L. m. meridionalis*	Mainland Iberia and S. France
Eurasian Jay	*G. g. fasciatus**	Mainland Iberia
Eurasian Magpie	*P. p. melanotos*	Mainland Iberia and French Pyrenees
European Greenfinch	*C. c. vanmarli*	W. Iberia and NW Africa
Common Crossbill	*L. c. balearica**	Balearic Islands
Common Bullfinch	*P. p. iberiae*	N. Iberia and SW France
Common Reed Bunting	*E. s. lusitanica**	Atlantic Iberia
	E. s. witherbyi	Mediterranean Iberia, Balearic Islands, S. France and Atlantic Morocco

some 39 species are much more noteworthy because they are by far the largest in Europe west of Russia (BIE). These are the Cattle Egret, Black-shouldered Kite, Egyptian Vulture, Griffon Vulture, Black Vulture, Booted Eagle, Bonelli's Eagle, Lesser Kestrel, Purple Swamphen, Little Bustard, Great Bustard, Eurasian Stone-curlew, Collared Pratincole, Kentish Plover, Audouin's Gull, Gull-billed Tern, Black-bellied Sandgrouse, Pin-tailed Sandgrouse, Great Spotted Cuckoo, Eurasian Hoopoe, Calandra Lark, Greater Short-toed Lark, Thekla Lark, Red-rumped Swallow, Tawny Pipit, Black-eared Wheatear, Melodious Warbler, Dartford Warbler, Spectacled Warbler, Subalpine Warbler, Western Orphean Warbler, Western Bonelli's Warbler, Iberian Chiffchaff, Short-toed Treecreeper, Southern Grey Shrike, Woodchat Shrike, Spotless Starling, Rock Sparrow, European Serin, Citril Finch and Rock Bunting. In addition, the Iberian populations of the Black Stork, White Stork, Glossy Ibis, Whiskered Tern, European Roller, Spanish Sparrow and Corn Bunting are the largest in western Europe. Among these, the Iberian populations of the two bustards, Audouin's Gull, the Iberian Chiffchaff, the two shrikes, the Spotless Starling and the Citril Finch are also of global significance.

A number of additional species whose European distribution is more or less restricted have important populations in Iberia. They are the Greater Flamingo, Red Kite, Lammergeier, Short-toed Eagle, Montagu's Harrier, Golden Eagle, Peregrine Falcon, Pied Avocet, Slender-billed Gull, Barn Owl, Eurasian Scops Owl, Eagle Owl, Little Owl, European Nightjar, Pallid Swift, European Bee-eater, Eurasian Wryneck, Crested Lark, Crag Martin, Alpine Accentor, European Stonechat, Sardinian Warbler, Crested Tit, Short-toed Treecreeper, Northern Raven and Ortolan Bunting.

Certain species are of regional interest in that they have outpost populations in Iberia. Among these, the Capercaillie, Ptarmigan and Grey Partridge are represented by endemic subspecies. However, 12 other species are also represented by populations that are very locally distributed in Iberia: the Levantine Shearwater, Eleonora's Falcon, Black-legged Kittiwake, Common Guillemot, Short-eared Owl, Boreal Owl, Black Woodpecker, Middle-spotted Woodpecker, Ring Ouzel, Marsh Tit, Lesser Grey Shrike and Rook.

Lists such as these illustrate the importance of the Iberian avifauna on a European level as well as on a global scale in some cases. However, the circumstances of each species are different and are described in the respective species accounts.

BREEDING BIRDS OF THE IBERIAN CLIMATIC REGIONS

The Mediterranean climatic region extends over the greater part of the Iberian landmass and so the character of much of the Iberian avifauna is Mediterranean. Certain breeding species are especially typical of the warmest sectors. For example, the Rufous-tailed Scrub-robin and the Western Olivaceous Warbler are especially typical of the thermomediterranean region and many others, including the European Bee-eater, Red-necked Nightjar, Pallid Swift, Thekla and Lesser Short-toed Larks and Zitting Cisticola, are largely confined to the thermo- and mesomediterranean. Others, including the Stock Dove, Calandra Lark, Tawny Pipit and Subalpine Warbler, have a Mediterranean distribution largely in the mesomediterranean and the supramediterranean. The Eurosiberian climatic region in the north also has its characteristic avifaunal elements (Table 6.2), comprising species that within Iberia are entirely restricted to the region or that have an exclusively montane distribution if they extend south of it.

Most of the species in Table 6.2 enjoy wide distributions across the central and northern latitudes of the Palearctic region. The populations of a few: the Ptarmigan, Capercaillie, Boreal Owl and White-backed Woodpecker may be relicts from the period of the last glaciation when they and others ranged more widely in southern Europe.

RECENT CHANGES IN THE IBERIAN BREEDING AVIFAUNA

The composition of the Iberian breeding avifauna underwent some important changes during the 20th century. Human influences, both positive and negative in nature, have been implicated to a greater or lesser extent in practically all of them. They involve losses and gains of some species and marked increases or declines in the abundance of others.

Very few breeding species have disappeared completely from Iberia although at least three now seem on the verge of regional extinction. The Small Buttonquail, Common Guillemot and the Lesser Grey Shrike declined inexorably during the 20th century and all three now have tiny populations that seem likely to disappear soon – and may already have done so in the case of the buttonquail. In addition, the Black-legged Kittiwake, a recent colonist, has also declined markedly and is vulnerable to disappearance. The Hazel Grouse had a marginal presence in the Spanish Pyrenees until at least the late 19th century but it is now probably extinct there even on the French side (ODF). The Common Crane last nested in Spain around 1954 and had ceased to do so in Portugal during the 19th century. The White-tailed Eagle, the Lanner Falcon and the Demoiselle Crane were recorded breeding in the 19th century, and the Ruddy Shelduck may also have done so then, but none of these species may have been established elements of the avifauna. However, one formerly extinct species has since become re-established of its own accord: the Glossy Ibis ceased to breed during the mid 20th century but nesting birds returned to Spain in 1993 and to Portugal in 2005 and the regional population is now both sizable and increasing.

Table 6.2. The distinctive bird species of the Eurosiberian climatic region in Iberia.

Species	Distribution		
	Pyrenees	Cantabrian Range	Northern Iberian Range
Rock Ptarmigan	X		
Western Capercaillie	X	X	
Grey Partridge	X	X	X
Eurasian Woodcock	X	X	X
Tengmalm's Owl	X		
Black Woodpecker	X	X	
Middle Spotted Woodpecker	X	X	
White-backed Woodpecker	X		
Tree Pipit	X	X	X
Whinchat	X	X	X
Ring Ouzel	X	X	X
Marsh Tit	X	X	X
Wallcreeper	X	X	
Eurasian Treecreeper	X	X	
Red-backed Shrike	X	X	X
Yellow-billed Chough	X	X	
White-winged Snowfinch	X	X	
Common Bullfinch	X	X	X
Yellowhammer	X	X	X

Table 6.3. Recent established natural additions to the Iberian avifauna.

Species	Year of first confirmed breeding	Species	Year of first confirmed breeding
Greylag Goose	Post-2000	Great Black-backed Gull	2005
Common Shelduck	1972	Black-legged Kittiwake	1975
Red-crested Pochard	19th century	Sandwich Tern	1971
Tufted Duck	1988	Common Tern	1956
Madeiran Storm-petrel	1981	Stock Dove	1927
Great Cormorant	1973	Collared Dove	1960
Great White Egret	1997	White-rumped Swift	1966
Black-shouldered Kite	1865	Little Swift	2000
Black-headed Gull	1960	Red-rumped Swallow	1921
Mediterranean Gull	1988	Common Starling	1950s
Lesser Black-backed Gull	1973	Trumpeter Finch	1971

At least 22 species have established breeding populations in Iberia between the late 19th century and the present day: most of them arrived during the second half of the 20th century or more recently (Table 6.3). Some may actually be recolonisers since there is reason to believe, although there is no definite evidence, that the Greylag Goose, the Great Cormorant and the Great White Egret, for example, may have bred in Iberia in the Middle Ages, when such species were probably more widely distributed in western Europe than has been the case more recently. In addition, Dupont's Lark was first proved to breed in Iberia as recently as 1967 but it may have been overlooked earlier.

At least a further ten species have self-sustaining Iberian breeding populations that have been established by individuals that have escaped from captivity or have been deliberately released into the wild (Table 6.4). The Barbary Partridge population of Gibraltar most probably also belongs in this category. This list is likely to increase further since a number of other species have feral populations that are close to becoming fully established in parts of the region (see Appendix 1).

Table 6.4. Established alien breeding species in Iberia (Category C1).

Species	Spain	Portugal
Common Pheasant	X	
Rose-ringed Parakeet	X	X
Monk Parakeet	X	
Red-billed Leiothrix	X	
Crested Myna		X
Black-headed Weaver		X
Yellow-crowned Bishop		X
Common Waxbill	X	X
Red Avadavat	X	X
Chestnut Munia		X

Gains and losses apart, a diversity of breeding species have seen large increases in their Iberian breeding populations since the mid 20th century and a similar number have shown corresponding large declines (Tables 6.5, 6.6). The species included in both tables are those for which there is long-term data of population change. Recent data from both the Spanish and Portuguese common birds censuses has indicated significant changes in the abundance of a number of other species, principally passerines, and these are described in the species accounts. Some of the species that have increased; for example the White-headed Duck, many herons, the White Stork, most of the raptors and the Purple Swamphen, have recovered and sometimes exceeded their earlier numbers following a period of serious decline (de Juana 2004). Among the species that have declined since 1950, the Lesser Kestrel and perhaps the Red Kite have seen recent promising increases in their breeding populations. However, the Portuguese populations of the Capercaillie, Grey Partridge, Red-knobbed Coot and Common Guillemot declined to extinction during this period.

As noted above, the historical changes in the composition of the Iberian avifauna and in the population sizes of its constituent species have undoubtedly been due in large measure to human activity. There is some risk in attempting to evaluate the past circumstances of particular species, given the lack of precise information, and especially of quantitative data, from any periods earlier than the 1950s. Nevertheless, some clear tendencies can now be discerned in the regional evolution of certain ecological groupings (Mayol *et al.* 2003, de Juana 2004, Catry & Pacheco 2008).

Waterbirds stand out among groups that have shown the greatest abundance changes over recent decades. In the main, they have shown clear positive trends in recovering from what was a parlous state of conservation, the outcome of large-scale wetland drainage and also of direct persecution: by hunting and for the commercial plumage trade. In the case of Spain, it is calculated that some 60% of the total wetland surface was drained between 1948 and 1990. The great majority of wetlands are now protected areas: the 68 Spanish sites listed under the Ramsar convention up to February 2012 extend over 285,000ha in total. The loss of lakes and lagoons has been compensated for to some extent by the proliferation of reservoirs and by the expansion of rice cultivation. The result has been that numerous aquatic species that were formerly regarded as gravely threatened have now shown large increases in their breeding populations, so that they are not now regarded as under threat or are just 'Vulnerable'. They include the White-headed Duck, Great Crested Grebe, Great White Egret, Grey Heron, Glossy

Table 6.5. Species whose Iberian breeding populations have shown a large overall increase since 1950.

Gadwall	Common Buzzard
White-headed Duck	Spanish Imperial Eagle
Great Crested Grebe	Booted Eagle
Night Heron	Eleonora's Falcon
Cattle Egret	Purple Swamphen
Little Egret	Pied Avocet
Grey Heron	Slender-billed Gull
Black Stork	Audouin's Gull
White Stork	Yellow-legged Gull
Spoonbill	Gull-billed Tern
Eurasian Griffon Vulture	Common Woodpigeon
Black Kite	Collared Dove
Lammergeier	Zitting Cisticola
Eurasian Black Vulture	Sardinian Warbler
Short-toed Eagle	Penduline Tit
Western Marsh Harrier	Spanish Sparrow

Table 6.6. Species whose Iberian breeding populations have shown a large overall decline since 1950.

Marbled Duck	Common Guillemot
Ferruginous Duck	Black-bellied Sandgrouse
Rock Ptarmigan	Pin-tailed Sandgrouse
Western Capercaillie	European Turtle Dove
Grey Partridge	Little Owl
Common Quail	Greater Short-toed Lark
Great Bittern	Rufous-tailed Scrub Robin
Red Kite	Black-eared Wheatear
Lesser Kestrel	Black Wheatear
Baillon's Crake	Common Whitethroat
Red-knobbed Coot	Moustached Warbler
Little Bustard	Lesser Grey Shrike
Kentish Plover	Southern Grey Shrike
Common Snipe	Woodchat Shrike
Whiskered Tern	Red-billed Chough
Black Tern	Common Reed Bunting

Ibis, Eurasian Spoonbill, Greater Flamingo, Purple Gallinule, Pied Avocet, Slender-billed Gull, Sandwich Tern and Common Tern. Nevertheless, a few waterbird species; notably the Marbled Duck, Ferruginous Duck, Great Bittern and Red-knobbed Coot, still have dangerously small breeding populations. Moreover, many wetlands, even such vital ones as the Marismas del Guadalquivir, the Ebro delta and the Albufera de Valencia, are still under serious threat at times. They are affected by such problems as contamination by pesticides, eutrophication resulting from nutrient run-off from farmland, excessive water abstraction for crop irrigation, siltation of lagoons and the uncontrolled burning of reedbeds. The requirements of the EU Water Framework Directive (2000) offer an opportunity to remedy such problems.

A very much improved situation has also been evident among nearly all raptor species. Raptors suffered intense persecution until the mid 1960s and they were also seriously affected by exposure to DDT and other organochlorine pesticides and by the indiscriminate use of poison baits, especially strychnine (whose use remained legal in Spain until 1983), to control vermin. Direct persecution is no longer a serious problem for most species but other threats to raptors remain. These include high mortality from electrocution on electricity pylons, a problem that is being reduced by design changes to offer safe perches. Mortality from collision with wind turbines is also a serious problem locally for some species, notably Griffon Vultures. Poison baits remain a threat, particularly those phytosanitary products such as aldicarb and carbofuran that are used illegally to combat vole plagues. Such baits seem to have had a particularly serious impact on the populations of the Egyptian Vulture and the Red Kite, both of which commonly consume small dead animals. In contrast, the Griffon and Black Vultures and the Lammergeier, which chiefly take large carrion, have not been seriously affected and all three have shown very significant population increases. The situation of the Spanish Imperial Eagle has also improved markedly, with breeding numbers increasing from some 50 pairs in the 1960s and 1970s to at least 365 pairs in 2012. Notable increases in numbers have also been seen in the populations of the Black-shouldered Kite, Marsh Harrier, Lesser Kestrel and Eagle Owl. Only one Iberian raptor, the Osprey, is classed as 'Critically Endangered' as a breeding species but this species too is increasing, assisted by reintroduction projects on the Spanish mainland.

The common breeding bird monitoring programmes (SACRE and NOCTUA) have revealed positive overall tendencies among Iberian forest birds. These reflect the recent improvement in woodland and forest cover in the Peninsula, where historical deforestation was very extensive, particularly at the lower elevations in Mediterranean Spain. Recent decades have seen massive reforestation and afforestation, as well as much natural regeneration of forest cover following the abandonment of marginal farmland, leading to a much improved level of tree cover in the region. The major beneficiaries have included some of the scarcer woodpeckers, notably the Black, Middle Spotted and Lesser Spotted Woodpeckers, which were formerly very rare. However, one forest species, the Capercaillie, continues to decline to an alarming extent: its demise may be partly due to climatic change.

In contrast, the same monitoring programmes that have revealed an upturn in the fortunes of forest birds have also shown significant decline in many open-country species, particularly those dependent on areas given over to agriculture or livestock. The population trends shown by those species that are especially associated with pseudosteppe habitats, for which the Iberian Peninsula offers an important refuge on a pan-European scale, are a particular source of concern. Such species as the Little Bustard, the Black-bellied and Pin-tailed Sandgrouse and the Greater Short-toed Lark, all of which were common or abundant until recently, are now classed as 'Vulnerable' in Spain. In many cases, the negative trends of such species are explicable in terms of the same types of threats that have affected farmland birds throughout Europe, arising from agricultural intensification. The spread of crop irrigation, often at the expense of cereal farmland and other non-irrigated crops, has also been very harmful to the Iberian 'steppe' species. Remedying these problems will probably require substantial changes to the agricultural policies of the European Union. Some such agri-environmental measures have been introduced and have already had positive outcomes in Spain, in particular those designed to maintain or increase the extent of fallows, pastures and leguminous cover within the great expanses of non-irrigated crops. The Great Bustard population has been increasing significantly in recent years, apparently as a result of these improvements. Nevertheless, the population declines of many other farmland and open country species continue to give cause for concern, in particular those of the European Turtle Dove, Dupont's Lark and the Lesser Grey Shrike.

A further group of species whose conservation status is relatively unfavourable is comprised by the breeding seabirds. Some of these once suffered intensive exploitation in the form of human consumption of eggs and young, for example in the Balearic Islands. This is no longer the case but some Iberian seabirds, for example Cory's Shearwater, have since been suffering from mortality at sea associated with becoming the bycatch of long-line fisheries. Some seabird colonies also suffer predation by rats and feral cats. The Common Guillemot is now 'Critically Endangered' as a breeding species in Iberia, as is the Balearic Shearwater, an Iberian endemic species. In contrast, the Iberian populations of Audouin's Gull, once a globally threatened species, now number many thousands of pairs following spectacular increases since the mid 20th century.

WINTERING SPECIES

Winters are relatively mild in the coastal, low-lying areas of Iberia, in most of the Guadalquivir valley and in the western regions that are most subject to the influence of the Atlantic Ocean. The region is therefore attractive as the winter quarters of many birds that migrate south-west from central and northern Europe, where freezing temperatures and long-lasting snow cover are regular in winter (Tellería 1988). Iberia also often receives mid-winter influxes of birds displaced by cold snaps from their wintering grounds in France, the Low Countries and the British Isles.

The principal non-passerine species that regularly winter in very large numbers in Iberia include a range of waterfowl, other aquatic species, several waders and a diversity of gulls, among others (Table 6.7). It is very much apparent that the numbers of some species that appear in any given winter are highly variable. In some cases, notably the Northern Lapwing and Golden Plover, the numbers that arrive reflect the incidence and severity of cold weather elsewhere in Europe. Wintering waterfowl numbers, on the other hand, are often determined by the wetness of the Iberian winter. In particular, a dry year in southern Spain often sees a great reduction in the numbers of waterfowl that winter at the Marismas del Guadalquivir.

Table 6.7. Non-passerine species that regularly winter in Iberia in very large numbers.
Counts or estimates of the post-1990 Spanish wintering population are given as an indicator of the numbers involved. Most of these species also have a large winter presence in Portugal. Resident species whose populations are not increased by winter immigration are excluded.

Species	Post-1990 wintering population in Spain	Species	Post-1990 wintering population in Spain
Greylag Goose	55,000-145,000	Black-winged Stilt	up to 15,000
Eurasian Wigeon	50,000-100,000+	Pied Avocet	5,000-29,000
Eurasian Teal	30,000-155,000	European Golden Plover	100,000
Mallard	140,000-315,000	Northern Lapwing	100,000-500,000+
Northern Pintail	6,000-48,000	Little Stint	up to 21,000
Northern Shoveler	55,000-163,000	Dunlin	up to 125,000
Red-crested Pochard	6,000-24,000	Common Snipe	100,000+
Common Pochard	14,000-61,000	Eurasian Woodcock	500,000+
Great Cormorant	70,000	Black-tailed Godwit	10,000-70,000
Greater Flamingo	60,000+	Black-headed Gull	300,000-500,000
Red Kite	35,000	Mediterranean Gull	40,000+
Common Coot	76,000-140,000	Lesser Black-backed Gull	300,000+
Common Crane	150,000+	Common Woodpigeon	several million

The abundance of the most numerous wintering passerine species (Table 6.8) can only be crudely estimated but some species appear annually in very large numbers and most of these are widely distributed across the region during the season. Most of the species in the table are frugivores or seed-eaters, both of which find abundant seasonal resources in Iberia in winter. The remainder are insectivores that are able to overwinter in the milder regions of the Peninsula.

Smaller numbers of a considerable diversity of species also arrive to winter in the region, again in numbers that often fluctuate considerably between years (Table 6.9).

A number of Iberian wintering species stand out on account of their regional abundance in winter having declined steadily since the middle of the 20th century, to the extent that two of them, the Bean Goose and the Rook, can now best be regarded as former winterers. Such species are the 'short-stoppers', birds which now regularly winter in numbers further north in Europe than used to be the case, for a diversity of reasons that probably includes climatic amelioration. Other species that illustrate the phenomenon include the Common Scoter, Tufted Duck, Red Kite, Common Crane, Lesser Black-backed Gull and Stock Dove, although short-stopping by Cranes (in France) and by Lesser Black-backed Gulls (in northwestern Europe) has been offset by large population increases in both species that have seen their Iberian wintering populations also increase. The White Stork,

1. Dehesa of Iberian Holm Oaks, a frequent landform of the south-west. Here the ground between the widely spaced trees is being used for grazing by fighting bulls. Iberian Azure-winged Magpies *Cyanopica cooki* are among the many typical species of holm oak dehesas. Crops, especially cereals, are grown in some dehesas. The Black-shouldered Kite *Elanus caeruleus* favours crop dehesas (Ernest Garcia) .

2. Holm Oak dehesas and the Sierra de Guadarrama from Colmenar Viejo, Madrid. The cooler and more humid climate of the Iberian mountain ranges creates suitable conditions for many species that avoid the hot dry summers of the meseta. In the Guadarrama they include the Bluethroat *Luscinia svecica azuricollis* and Red-backed Shrike *Lanius collurio* (Eduardo de Juana).

3. Montane heathland in the Cantabrian Range at Leitariegos, León. A typical habitat of the Iberian subspecies of the Grey Partridge *Perdix perdix hispaniensis* (Eduardo de Juana).

4. The Algarve coast, looking northwards from Cape St Vincent, the southwesternmost extremity of Iberia. Shags *Phalacrocorax aristotelis*, Peregrine Falcons *Falco peregrinus* and Red-billed Choughs *Pyrrhocorax pyrrhocorax* share the cliffs with the ubiquitous Yellow-legged Gulls *Larus michahellis* (Ernest Garcia).

5. The Strait of Gibraltar viewed from the Spanish shore east of Tarifa. The mountain dominating the Moroccan side is Jebel Musa. The Strait is only 14km across at its narrowest point and is an important flyway for birds migrating between the European and African continents, in particular storks, raptors and other soaring migrants that avoid long sea crossings. For seabirds it provides the only link between the Mediterranean and the Atlantic (Stephen Daly).

6. Europa Point, the southern tip of Gibraltar, during an easterly gale. The Rock of Gibraltar offers an excellent watchpoint for observing seabird, raptor and passerine migration (Ernest Garcia).

7. The seabird observatory at Cape Estaca de Bares, Galicia. This promontory on the southern shore of the Bay of Biscay is the northernmost point of the Iberian Peninsula and a celebrated watchpoint for seabird migration, particularly in autumn (Ernest Garcia).

8. The Sado estuary, Setúbal, looking eastwards from the Tróia peninsula. One of the principal inlets of the Atlantic coast, attracting numbers of waterfowl and waders both in winter and during passage periods (Ernest Garcia).

9. Eastward view of the central massif of the Picos de Europa in the Cantabrian Range, viewed in summer from Ponga, Asturias. The forests flanking the mountains provide a refuge for Capercaillies *Tetrao urogallus* and Brown Bears *Ursus arctos* whereas the rocky heights are the nesting habitat of such species as the Wallcreeper *Tichodroma muraria*, Yellow-billed Chough *Pyrrhocorax graculus* and Snowfinch *Montifringilla nivalis* (Ernest Garcia).

10. The rockface of Peña Falcón on the Tagus reservoir in Monfragüe National Park, Cáceres, Extremadura. The cliffs accommodate a wide range of rupicolous species, including a colony of Griffon Vultures *Gyps fulvus*, Black Storks *Ciconia nigra*, Eagle Owls *Bubo bubo* and Black Wheatears *Oenanthe leucura* (Ernest Garcia).

11. A section of the restored Laguna de La Nava, Palencia, one of the important wetlands of the northwestern meseta. It attracts large numbers of waterfowl, especially in winter (Eduardo de Juana).

12. The eroded badlands and salt steppe of Las Bardenas Reales, Navarra, in the Ebro valley. This is an important location for steppe species including Dupont's Lark *Chersophilus duponti* and both sandgrouse (Eduardo de Juana).

13. Cereal steppe in Extremadura, at Valencia de las Torres, Badajoz. An important habitat for both the Great Bustard *Otis tarda* and the Little Bustard *Tetrax tetrax* (Eduardo de Juana).

14. The pastures of La Serena, Badajoz, Extremadura, grazed by merino sheep, are among the most important locations for steppe species in Iberia, including bustards, sandgrouse and larks (Eduardo de Juana).

15. A ricefield in December in the Vegas Altas of the Guadiana river, Badajoz, Extremadura, following stubble burying. Large numbers of gulls and herons are among the birds attracted to seek crayfish and other invertebrates in the recently disturbed mud (Eduardo de Juana).

16. Traditional saltpans on the western side of the Guadiana estuary at Castro Marim, Algarve. Saltpans have provided a useful substitute for long-lost coastal lagoons and similarly attract such breeding birds as the Black-winged Stilt *Himantopus himantopus*, Collared Pratincole *Glareola pratincola* and Little Tern *Sternula albifrons* (Ernest Garcia).

17. Marbled Duck *Marmaronetta angustirostris*. Once numerous but now a scarce and declining species that is almost entirely confined to southern and eastern Spain. Its continued presence may depend on immigration from Morocco (Karin Faber).

18. An adult male White-headed Duck *Oxyura leucocephala*. This species reached the brink of extinction in Iberia but has recovered its numbers and is increasing, although it remains largely confined to southern and eastern Spain (Eduardo de Juana).

19. Balearic Shearwater *Puffinus mauretanicus*. An endemic Iberian species. The Balearic Islands are the sole known breeding location but it occurs seasonally on all Iberian coasts. Much of the population leaves the Mediterranean after the breeding season and most spend the summer and early autumn off Atlantic coasts and in the Bay of Biscay (Stephen Daly).

20. Male Little Bittern *Ixobrychus minutus*. This is one of a sizable guild of heron species that may be encountered at Iberian wetlands, many of which have recently seen substantial increases in their breeding populations (Stephen Daly).

21. Glossy Ibises *Plegadis falcinellus*, and a Grey Heron *Ardea cinerea*, at the La Janda ricefields, Cádiz. The Ibises are now a familiar sight at southern and eastern wetlands and have a large and increasing breeding population, at Doñana especially, following recolonisation of the Peninsula during the late 20th century (Stephen Daly).

22. Greater Flamingoes *Phoenicopterus roseus*. Flamingoes breed very locally in southern and eastern Spain, their principal colony by far being at the Laguna de Fuentedepiedra, Málaga, where thousands of pairs breed in a good year. Non-breeders are frequent at wetlands of southern Portugal and southwestern Spain (Eduardo de Juana).

23. Honey Buzzard *Pernis apivorus*. The Iberian Peninsula is a major flyway for migrating flocks of this species, which form spectacular concentrations at the Strait of Gibraltar especially. It is also a frequent migrant in the Balearic Islands. There is a sizable breeding population, chiefly in the northern half of the Peninsula (Paul Acolina).

24. Black-shouldered Kite *Elanus caeruleus*. Probably a relatively recent colonist of the Peninsula but apparently confined to Portugal until the mid 1970s, since when it has become increasingly widespread across Spain and has reached southwestern France (Tatavasco Imágenes).

25. Lammergeier *Gypaetus barbatus*. The Pyrenees are the Iberian and European stronghold of this spectacular species. It was formerly much more widespread in the region and there are reintroduction projects in Andalucía and the Cantabrian Range (Stephen Daly).

26. Griffon Vulture *Gyps fulvus*. Widespread, locally numerous and increasing across the Peninsula and a recent colonist of the Balearic Islands (Harry Van Gils).

27. Rüppell's Vulture *Gyps rueppellii*. One of several tropical African species that now occurs regularly in Iberia (Charles Perez).

28. Black Vulture *Aegypius monachus*. The increasing breeding population of southwestern Spain and the adjacent fringe of Portugal is by far the largest in Europe. There is a small population in Mallorca (Tatavasco Imágenes).

29. Short-toed Eagle *Circaetus gallicus*. Widespread and locally common across the Peninsula and a numerous migrant at the Strait of Gibraltar (Stephen Daly).

30. Spanish Imperial Eagle *Aquila adalberti*. An endemic Iberian species that was in danger of extinction by the mid 20th century. It is now increasing steadily as a result of conservation measures (Stewart Finlayson).

31. Bonelli's Eagle *Aquila fasciata*. A characteristic inhabitant of rocky terrain across the warmer regions of Spain and Portugal (Stephen Daly).

32. Lesser Kestrel *Falco naumanni*. The Iberian population, concentrated in the southwest and the Ebro valley, is recovering after a period of marked decline. Most nest in holes in buildings (Paul Acolina).

33. Purple Swamphen *Porphyrio porphyrio*. Now resident and locally common in wetlands of the warmer parts of the Peninsula and in the Balearic Islands. It underwent marked range expansion during the second half of the 20th century, assisted by local introductions (Paul Acolina).

34. Red-knobbed Coot *Fulica cristata*. An African species with a marginal and precarious presence in southern and eastern Spain, despite attempts to reinforce the population with introduced birds (Andrew Fortuna).

35. Little Bustards *Tetrax tetrax*. Still widespread and locally common in mainland Iberia but declining rapidly in the face of agricultural intensification (Stephen Daly).

36. Great Bustard. About half of the global population inhabits the Meseta, Extremadura and southern Portugal. Increasing despite some range contraction (Tatavasco Imágenes).

37. Black-winged Stilt *Himantopus himantopus*. A noisy and unmissable presence at wetlands almost throughout the region. Large numbers breed and many remain in winter (Andrew Fortuna).

38. Collared Pratincole *Glareola pratincola*. A locally common summer visitor to southwestern Iberia that also nests locally in Castilla-La Mancha and on the Levant coast. (Stephen Daly).

39. Slender-billed Gulls *Larus genei*. Breeds locally at a few sites in southern and eastern Spain and locally common at Mediterranean and southwestern wetlands (Stephen Daly).

40. Audouin's Gulls *Larus audouinii*. Iberia supports the bulk of the global breeding population, following spectacular range expansion and population increase in the region during the second half of the 20th century. The Ebro delta houses the largest colonies (Stephen Daly).

41. Black-bellied Sandgrouse *Pterocles orientalis*. Widespread but local across the interior of Spain and adjacent parts of Portugal (Tatavasco Imágenes).

42. Pin-tailed Sandgrouse *Pterocles alchata*. Widespread but local in semi-arid country across south-central Spain (Tatavasco Imágenes).

43. Scops Owl *Otus scops*. Widespread throughout the region. Mainly a summer visitor but some remain in winter (Tatavasco Imágenes).

44. Eagle Owl *Bubo bubo*. Increasingly widespread across mainland Iberia. It is typical of rocky hillsides but also nests in wooded areas and locally even in cities (Tatavasco Imágenes).

45. Pallid Swift *Apus pallidus*. A common and locally numerous summer visitor to southern Iberia as well as the Levant coasts and the Balearic Islands, and locally elsewhere (Charles Perez).

46. European Bee-eaters *Merops apiaster*. A widespread and locally common summer visitor to most of the region (Andrew Fortuna).

47. Roller *Coracias garrulus*. A summer visitor to western, central and southeastern Spain and eastern Portugal, and locally elsewhere. Many use nestboxes erected for them in open country (Tatavasco Imágenes).

48. Hoopoe *Upupa epops*. Widespread and common throughout most of the region. Many winter in the southwest and the Mediterranean coastlands. (Tatavasco Imágenes).

49. Dupont's Lark *Chersophilus duponti*. The scarce and local population, largely found in eastern Spain, is the only one in Europe (Tatavasco Imágenes).

50. Lesser Short-toed Lark *Calandrella rufescens*. The local resident population in eastern and southwestern Spain and southeasternmost Portugal is the only one in Europe (Andrew Fortuna).

51. Red-rumped Swallow *Cecropis daurica*. A 20th century colonist of Iberia that is now widespread and common (Eduardo de Juana).

52. Alpine Accentor *Prunella collaris*. Breeds commonly on the highest mountains in Spain and Andorra but is much more widespread throughout the region at lower elevations in winter, then occurring even on rocky coasts (Tatavasco Imágenes).

53. Rufous-tailed Scrub-robin *Cercotrichas galactotes*. A scarce, local and declining summer visitor to the southern third of Spain and adjacent areas of Portugal. It inhabits vineyards, olive and citrus groves and expanses of scrub (Ernest Garcia).

54. Blue Rock Thrush *Monticola solitarius*. A common resident of rocky areas in the warmer parts of the Peninsula and Balearic Islands. It often nests on large stone buildings such as castles (Harry Van Gils).

55. Zitting Cisticola *Cisticola juncidis*. A common resident of open country throughout the region but most abundant where winters are mild (Andrew Fortuna).

56. Southern Grey Shrike *Lanius meridionalis*. Widespread and locally common across the warmer sectors of mainland Iberia, although declining in some regions (Tatavasco Imágenes).

57. Woodchat Shrike *Lanius senator*. A common summer visitor to the Mediterranean climatic region (Harry Van Gils).

58. Iberian Azure-winged Magpie *Cyanopica cooki*. Fossil and genetic evidence has established that this is a native Iberian species. It inhabits Portugal and west-central Spain (Tatavasco Imágenes).

59. Yellow-billed Chough *Pyrrhocorax graculus*. A locally common resident of the Pyrenees and the Cantabrian Range (Tatavasco Imágenes).

60. Spanish Sparrow *Passer hispaniolensis*. A common breeder in southern and western Spain and adjacent eastern Portugal but it occurs only locally or sporadically elsewhere (Eduardo de Juana).

61. Rock Sparrow *Petronia petronia*. Widespread in mainland Iberia and in the western Balearic Islands (Tatavasco Imágenes).

62. Rock Bunting *Emberiza cia*. A common resident of broken terrain across most of the Iberian mainland (Tatavasco Imágenes).

63. Ortolan Bunting *Emberiza hortulana*. A summer visitor to northern mainland Iberia, inhabiting open bushy terrain on sunny hillsides in the uplands (Eduardo de Juana).

64. Corn Bunting *Emberiza calandra*. One of the most abundant and typical inhabitants of open country, especially farmland, throughout the region (Andrew Fortuna).

Table 6.8. Passerine species that regularly winter in Iberia in very large numbers.
Resident species whose populations are not increased by winter immigration are excluded.

Species	Wintering population	Species	Wintering population
Eurasian Skylark	several million	Blackcap	several million
Eurasian Crag Martin	?	Common Chiffchaff	several million
Meadow Pipit	several million	Common Starling	several million
White Wagtail	several million	Common Chaffinch	variable
European Robin	several million	Brambling	variable
Black Redstart	?	European Serin	?
Eurasian Stonechat	?	European Greenfinch	?
Common Blackbird	variable	European Goldfinch	several million
Fieldfare	variable	Eurasian Siskin	Irruptive in some years
Song Thrush	several million	Common Linnet	?
Redwing	variable	Common Reed Bunting	?

Table 6.9. Other principal Iberian wintering species.
Resident species whose populations are not increased by winter immigration are excluded.

Common Scoter	Hen Harrier	Common Greenshank	Little Gull
Red-throated Diver	Common Buzzard	Green Sandpiper	Caspian Tern
Black-throated Diver	Osprey	Wood Sandpiper	Sandwich Tern
Great Northern Diver	Merlin	Common Sandpiper	Razorbill
Great Crested Grebe	Sanderling	Ruddy Turnstone	Common Guillemot
Slavonian Grebe	Temminck's Stint	Grey Phalarope	Atlantic Puffin
Black-necked Grebe	Purple Sandpiper	Arctic Skua	Stock Dove
Northern Gannet	Ruff	Great Skua	Ring Ouzel
Little Egret	Jack Snipe	Common Gull	Snow Bunting
Grey Heron	Bar-tailed Godwit	Great Black-backed Gull	
White Stork	Eurasian Whimbrel	Herring Gull	
Western Marsh Harrier	Spotted Redshank	Black-legged Kittiwake	

the Black Stork and the Common Buzzard are among other species for which short-stopping has meant that some populations travel no further than Iberia instead of continuing to sub-Saharan Africa. Many White Storks, including some of northern European origin, travel no further than southwestern Spain. Some Black Storks also remain to winter in Iberia. The passage of Common Buzzards across the Strait of Gibraltar has declined enormously since the 1980s, which means that migrant northern European populations of this abundant species now travel no further than the Peninsula itself.

The mildness of winters in southern Iberia, particularly within the predominantly coastal thermomediterranean region, permits a further regional phenomenon, the early return of birds that have previously travelled as far as sub-Saharan Africa. The principal species involved are the White Stork: which begins to return from Africa in November and December, the Black-tailed Godwit: which returns in large numbers from January onwards, and the Great Spotted Cuckoo, which arrives from November and December onwards. It has also become apparent that some hirundines, in particular the Barn Swallow and the House Martin, are present in southernmost Iberia in some numbers from January onwards, at least in some seasons. This phenomenon is not fully understood and it is not entirely clear whether such hirundines are early returners or whether some at least have overwintered in Iberia.

A further category of Iberian wintering species is comprised by those that are almost entirely summer visitors to the Palearctic region from winter quarters in sub-Saharan Africa but which nevertheless still have small or very small wintering populations in Iberia (Table 6.10). The Hoopoe also belongs in this category but its wintering population is substantial. Such overwintering is a phenomenon of long-term standing in certain cases, for example in the Scops Owl and Wryneck, but it appears to be a recent and increasingly frequent trend in others, notably the terns, the raptors and the European Nightjar. Each case is discussed in the relevant species account.

Table 6.10. Trans-Saharan migrant species that have small wintering populations in Iberia.

Garganey	Osprey	Common Tern
Common Quail	Lesser Kestrel	Little Tern
Little Bittern	Little Crake	Whiskered Tern
Squacco Heron	Baillon's Crake	Black Tern
Purple Heron	Little Ringed Plover	European Scops Owl
Black Kite	Curlew Sandpiper	European Nightjar
Egyptian Vulture	Marsh Sandpiper	Pallid Swift
Short-toed Eagle	Wood Sandpiper	Eurasian Wryneck
Booted Eagle	Gull-billed Tern	Yellow Wagtail

MIGRANTS AND MIGRATION

Migrant species

The great majority of those species that are represented by populations that travel to winter in Iberia also include through-migrants, which cross the Peninsula to and from northwestern Africa in particular. In many cases: such as the White Stork, the Marsh Harrier and the Common Chiffchaff, the wintering range extends from the western Palearctic south to sub-Saharan Africa. In others, the Hen Harrier, Common Crane and the Common Starling for example, northwestern Africa marks the usual southern limit of the winter quarters and only a small proportion of those that enter Iberia in autumn continue that far south.

Migrant species that normally winter entirely in sub-Saharan Africa comprise a distinct category. Many are summer visitors to Iberia and most of these have other western European populations that overfly the Peninsula twice-yearly on passage (Table 6.11). A few occur in Iberia solely on passage, although they may then linger in the region for some time (Table 6.12)

Migration

A great deal is known about bird migration in Iberia. Information gleaned from the bird ringing programmes both within the region and in other countries has provided often ample evidence of the origins and destinations of many of the migrants. More recently, remote tracking devices have begun to provide fascinating and sometimes surprising insights into the behaviour of the migrants and the routes that they follow. Records of visible migration, both of birds actually on the move and those that appear seasonally at staging locations, have all helped to develop a fuller picture of the phenomenon. The species accounts summarise the often complex movements that affect the Peninsula and they highlight the principal gaps in current knowledge. They also confirm that migration is a highly flexible phenomenon, with the routes followed, the numbers of individuals involved and the extent of the behaviour itself all liable to alter even in the short to medium term, as species adjust to environmental and climatic changes. Such long-lasting alterations in migratory behaviour are of course additional to the seasonal variation that is imposed by the effects of wind and weather on the migrants.

A great deal of migration, in Iberia as elsewhere, is invisible to human observers because it takes place at night or at high altitudes, or both. Radar studies have shown the massive scale of such movements and satellite-tracking studies are now revealing that nocturnal journeys are performed by at least some individuals of species that have hitherto been regarded as diurnal migrants, even among some raptors such as the Osprey and the Hobby. Visible movements are nevertheless compelling sights in parts of the Peninsula: most obviously in coastal regions, at the Pyrenean passes and at the Strait of Gibraltar. Some of the general aspects of such movements are summarised here but more specific details are given in the species accounts.

Table 6.11. Trans-Saharan migrant species that have breeding populations in Iberia.

Garganey	Common Tern	Common Redstart
Common Quail	Little Tern	Whinchat
European Storm-petrel*	Whiskered Tern	Northern Wheatear
Little Bittern	Turtle Dove	Black-eared Wheatear
Squacco Heron	Great Spotted Cuckoo	Rufous-tailed Rock Thrush
Purple Heron	Common Cuckoo	Grasshopper Warbler
Black Stork	Scops Owl	Savi's Warbler
White Stork	European Nightjar	Reed Warbler
Honey Buzzard	Red-necked Nightjar	Great Reed Warbler
Black Kite	Alpine Swift	Western Olivaceous Warbler
Egyptian Vulture	Common Swift	Melodious Warbler
Short-toed Eagle	Pallid Swift	Spectacled Warbler
Montagu's Harrier	White-rumped Swift	Subalpine Warbler
Booted Eagle	European Bee-eater	Western Orphean Warbler
Osprey	European Roller	Common Whitethroat
Lesser Kestrel	Wryneck	Garden Warbler
Eurasian Hobby	Greater Short-toed Lark	Bonelli's Warbler
Eleonora's Falcon	Sand Martin	Iberian Chiffchaff
Little Crake	Barn Swallow	Spotted Flycatcher
Baillon's Crake	Red-rumped Swallow	Pied Flycatcher
Collared Pratincole	House Martin	Golden Oriole
Little Ringed Plover	Tawny Pipit	Red-backed Shrike
Common Sandpiper	Tree Pipit	Woodchat Shrike
Slender-billed Gull	Yellow Wagtail	Ortolan Bunting
Gull-billed Tern	Rufous-tailed Scrub Robin	

* winters in the southern Atlantic

Table 6.12. Migrant species that occur regularly in Iberia only on passage. Species marked * breed in the region at least occasionally but only in very small numbers.

Great Shearwater	Red-necked Phalarope	Red-throated Pipit
Sooty Shearwater	Long-tailed Skua	Aquatic Warbler
Manx Shearwater	Sabine's Gull	Sedge Warbler
Wilson's Storm-petrel	Lesser Crested Tern*	Icterine Warbler
Red-footed Falcon	Roseate Tern	Lesser Whitethroat
Corn Crake	Arctic Tern	Willow Warbler*
Grey Phalarope	Black Tern*	Wood Warbler*

Seabird movements

Much seabird passage invariably takes place far offshore but coastal movements also occur and these may take on impressive proportions under particular circumstances of wind and weather. The largest and most species-diverse coastal movements occur on the Atlantic coasts, including those of the Bay of Biscay (Table 6.13). Northward passage along Portuguese and Galician coasts is evident in spring, when some movements also occur eastwards off the Biscay coasts. However, it is the post-breeding southbound movements that regularly result in the most spectacular passages along both northern and western Iberian coasts. These latter movements occur, for different species, at various times between late July and early December. Strong onshore winds, from the west and especially the north-west, often accompany the eastward passage of North Atlantic weather systems, including occasionally the remnants of hurricanes during late summer and autumn. Such conditions may drive large numbers of southbound seabirds: including shearwaters, Gannets and skuas, into the eastern Bay of Biscay, from where they often coast westwards close inshore past such headlands as Cape Higuer in the Basque country, Punta La Vaca on the Cape Peñas promontory in Asturias and, most notably, Cape Estaca de Bares in northeastern Galicia. Much of what is known of the scale and timing of seabird movements in northern Iberia is due to the regular seawatching programmes that have been established at the last two of these sites especially in recent years. The autumn movements also involve the abundant shearwaters that gather in late summer and autumn in the Bay of Biscay to exploit the annual shoaling of anchovies and other small fish there. Such shearwaters include Great and Sooty Shearwaters that enter the Bay during their southward passage in the north-east Atlantic, Cory's and Manx Shearwaters from eastern Atlantic colonies, and Balearic Shearwaters that travel to the Bay from their Mediterranean breeding grounds immediately after nesting.

Seabird movements may also be seen along Mediterranean coasts although the species-composition differs somewhat from that in Atlantic waters (Table 6.13) and the volume of passage is notably much inferior in most species. Such regular Atlantic migrants as the Great, Sooty and Manx Shearwaters, the Long-tailed Skua, the Arctic Tern and the Common Guillemot are all rare or absent in the Mediterranean. Conversely, Audouin's Gull in particular is often abundant along Mediterranean coasts but becomes progressively rarer west of the Strait on Iberian Atlantic coasts.

The Strait of Gibraltar sees a diversity of movements of seabirds entering and leaving the Mediterranean. Several species that breed in the Mediterranean but winter in the eastern or south Atlantic travel via the Strait: Scopoli's Shearwater, the Lesser Crested Tern and probably most of the Mediterranean nesting populations of the European Storm-petrel are the principal examples. The Balearic Shearwater also travels to and from its post-breeding feeding and moulting grounds in the northeastern Atlantic via the Strait. Similarly, the Strait is the conduit for the migratory flows of Gannets, Great Skuas, Razorbills and Puffins that have wintering populations in the western Mediterranean. Movements through the Strait are not always visible, whether wholy or in part, from either the Iberian or African coasts but they are often apparent when birds are flying into headwinds, especially those with an onshore component. Thus Scopoli's Shearwaters leaving the Mediterranean in October and November are most likely to be seen from the Iberian shore when strong southwesterlies are blowing but they are most often seen there on their return journey in March when they are confronted by strong easterly or southeasterly winds.

Seabird movements are not only associated with migration. Seabirds are highly susceptible to weather conditions at sea and influxes into coastal waters sometimes occur during stormy weather, which is most likely to occur between mid September and early April and particularly in winter. At such times pelagic species that are present or wintering offshore may become much more visible to land-based observers. Particularly severe conditions are liable to produce 'wrecks' of Leach's Petrels and Black-legged Kittiwakes especially as well as scarcer birds such as Northern Fulmars and Little Auks at times. Such events occur most often on Atlantic coasts but storm-driven birds may appear in the Mediterranean and even inland across the Peninsula. A series of monthly marine transects to points some 22 miles off Punta Galea, Vizcaya, in the eastern Bay of Biscay, between October 2001 and October 2003 (Ocio & Astigarraga 2008) has identified that skuas, shearwaters other than the Balearic Shearwater; Mediterranean, Sabine's and Little Gulls, Kittiwakes, and auks other than Razorbills tend to be mainly or entirely pelagic in that region, and hence unlikely to appear onshore in normal circumstances.

Since 2005 the Iberian Seawatching Network, known regionally as the RAM (Red de observación de Aves y Mamíferos marinos / Rede de observação de Aves e Mamíferos marinhos) has established a synchronised protocol for monitoring seabirds (and cetaceans) from selected key points around the entire Iberian coastline. Watches are conducted simultaneously for a three-hour period on one day per month. There are over 50 watchpoints that have participated in the RAM but not all of them are employed regularly. The combined data comprises another valuable daabase on seabird activity around Iberia, to which we have referrred. The RAM data is available from the project website (www.telefonica.net/web2/redavesmarinas/).

Table 6.13. The principal seabird species recorded on passage at principal watchpoints on the Iberian coasts.

Species	Cape Estaca de Bares (and other points on the Biscay coasts)	Cape Carvoeiro (and other headlands on the Portuguese west coast)*	Europa Point, Gibraltar (and the Strait of Gibraltar)	Cape Creus, Girona (and other Mediterranean watchpoints)
Common Scoter	X	X		
Cory's Shearwater	X	X	X	X
Great Shearwater	X			
Sooty Shearwater	X	X		
Manx Shearwater	X	X		
Balearic Shearwater	X	X	X	X
Levantine Shearwater				X
Northern Gannet	X	X	X	X
Great Cormorant	X	X	X	X
European Shag	X	X	X	X
Pomarine Skua	X	X	X	X
Arctic Skua	X	X	X	X
Long-tailed Skua	X			
Great Skua	X	X	X	X
Black-headed Gull	X	X	X	X
Slender-billed Gull			X	X
Mediterranean Gull	X	X	X	X
Audouin's Gull			X	X
Common Gull	X			
Lesser Black-backed Gull	X	X		X
Herring Gull	X			
Yellow-legged Gull	X	X	X	X
Great Black-backed Gull	X			
Black-legged Kittiwake	X	X		
Little Gull	X		X	X
Gull-billed Tern			X	X
Lesser Crested Tern			X	
Sandwich Tern	X	X	X	X
Common Tern	X	X	X	X
Arctic Tern	X	X		
Little Tern	X			
Black Tern	X		X	X
Common Guillemot	X			
Razorbill	X	X	X	X
Atlantic Puffin	X		X	X

*After Sengo *et al.* (2012).

Trans-Pyrenean movements

The massive influxes of Wood Pigeons into Iberia from France across the Pyrenean passes in autumn have long been the target of hunters. However, the wider importance of the Pyrenean valleys for bird migration was not fully appreciated until a pioneering study by Lack & Lack (1953) who found that, contrary to their expectations, the bird migration into Iberia in October was not largely concentrated at the Biscay coastal lowlands, north of the Pyrenees, but also occurred on a large scale across the mountains themselves. The visible movements reported by the Lacks involved pigeons, raptors, Sky Larks, hirundines and pipits in particular. Trans-Pyrenean passage is now monitored regularly, particularly by the French organisation '*Organbidexka Col Libre*', who have studied the autumn movements in particular since the early 1980s. After due reconnaisance, their efforts have been concentrated at three locations in the western Pyrenees, in the French Basque country: Organbidexka in the French Pyrenees north of Navarra and Lindux and Lizarietta on the Navarran/French border, as well as from the Eyne plateau at the head of the Tet and Aude valleys in the eastern French Pyrenees, bordering Catalonia. These four locations were together considered to see an autumn movement of 60,000–70,000 raptors of 15 species, with a further nine raptor species occurring irregularly. They also saw an annual passage of about one million pigeons, some 25–30% of the total pigeon passage (Urcun & Bried 1998a). Concerted campaigns have found larger numbers of raptors using the passes in recent years together with a wide range of other species, although the numbers seen at particular locations are considerably affected by wind and weather conditions. Recent observations and annual reports from the Pyrenean passes are reported online on the migration website of the French Birdlife Partner, the LPO, the Ligue pour la Protection des Oiseaux (www.migraction.net). Reference to the principal movements is made in the relevant species accounts. The movements have principally been monitored during the post-breeding season, with daily watches kept from mid July to mid November in some years. The spring movements have not been studied intensively in the same way.

Landbird migration at the Strait of Gibraltar

The Strait of Gibraltar affords the possibility of a short sea-crossing between Iberia and North Africa to those species that tend to avoid long journeys over water. These especially include soaring migrants that employ thermals and relief-induced upcurrents to facilitate their journeys and that are not well suited to sustained flapping flight, which is necessary for lengthy sea crossings. In addition to the White and Black Storks, and very variable numbers of Common Cranes, the principal soaring bird species that use the Strait of Gibraltar flyway are the raptors listed in Table 6.14. Further information on the migration of each of these is given in the species accounts. Bernis (1973, 1980) and Finlayson (1992) in particular offer further detail on such aspects of the movements as local routes in relation to wind direction and the daily pattern of migration.

Table 6.14. Raptor species recorded annually on passage across the Strait of Gibraltar.
Those marked * only occur in small numbers. Irregular migrants and vagrants are excluded here.

European Honey Buzzard	Western Marsh Harrier	Osprey
Black Kite	Hen Harrier*	Lesser Kestrel
Red Kite*	Montagu's Harrier	Common Kestrel
Egyptian Vulture	Sparrowhawk	Merlin*
Eurasian Griffon Vulture	Common Buzzard	Eurasian Hobby
Rüppell's Vulture*	Long-legged Buzzard*	Eleonora's Falcon*
Black Vulture*	Lesser Spotted Eagle*	Lanner Falcon*
Short-toed Eagle	Booted Eagle	Peregrine Falcon*

The migrations of storks and raptors across the Strait are a spectacular phenomenon and have attracted a great deal of attention. An early consideration was that it might be possible to monitor the size of the source populations, and any fluctuations in their numbers, by counting migrants at the Strait. The only comprehensive attempt to do this was the campaign by the Spanish raptor migration group, the '*Grupo Español de Migración de Rapaces*' (GEMRA) that tried to count the southward passage of raptors and storks across the entire northern coastline of the Strait in 1972, 1974, 1976 and 1977 (Bernis 1973, 1980). That project involved a considerable number of observers, over 50 during the best-covered season (1976), and of course it succeeded in counting very large numbers of birds, serving at least to suggest an order of magnitude for the population sizes of some of the species involved, specifically those that are entirely or principally summer visitors to Europe. The campaign also highlighted the very great logistical problems involved in attempting a complete count of the movements.

The principal difficulty arises from the broad front across which some species cross the Strait. This problem is most acute in spring, when birds may arrive from Cape Trafalgar in the west to Gibraltar in the east, a distance of over 60km.The front is even broader in the case of the Honey Buzzard, which may arrive in spring as far west as Doñana and may cross the Sea of Alborán in the east, missing the Strait altogether. The movements are always concentrated at the downwind end of the Strait except in calm conditions. Monitoring the entire passage is very labour-intensive, requiring a large group of observers spaced out at strategic points across the many kilometres of potential migratory front. Moreover, the passage seasons are extremely long: from February to June in 'spring' and July to November in 'autumn' for raptor migration, which again makes very heavy demands on manpower.

A second major difficulty with respect to counting the southbound passage as the birds depart from the northern Strait coastline is that very large numbers do not cross as soon as they reach the area. If strong winds are blowing the birds will linger in the hinterland of the Strait waiting for a break in the weather and it becomes impossible to avoid recounting the same individuals as the numbers build up. Counting from the northern shoreline is more easily achieved in spring, when birds often arrive lower down and generaly do not linger after making landfall. However, the problems of manpower availability and the broad arrival front: even wider in spring than in autumn, remain. (The opposite applies to the African shore, where it might be best to count the autumn arrivals rather than the spring departures).

In our opinion, it probably is not possible to use counts of migrants at the Strait to assess the absolute size of the source populations of most of the species that use the flyway. The White Stork and the Griffon Vulture in autumn would be the most likely exceptions since both of these tend to depart across a narrow sector of the north shore: however, major fractions of the populations of both species are non-migratory. What is possible, however, is to obtain an indication of population trends by monitoring rates of passage at particular locations over periods of many years. This has been done at Gibraltar itself, where the rates of spring passage during 1967–77 have been compared with those during 1990–2004, showing significant increases between the two periods in the abundance of Black Kites, Eurasian Sparrowhawks and Booted Eagles, for example (Bensusan *et al.* 2007). Observations at the Strait also have the potential to detect phenological changes in the passage periods of all visible migrants, not just soaring birds.

The Strait also sees very large movements of swifts and hirundines, aerial insectivores that may perhaps choose to travel overland as much as possible in order to feed on the move, even though such birds can and do travel great distances over water when the situation demands. Certain other migrants, especially finches, are also numerous at the Strait on passage. A large diversity of other species can be seen on passage across the region but they are broad-front migrants, many of them nocturnal travellers, which show a limited tendency to concentrate at the Strait.

7. INTRODUCTION TO THE SPECIES ACCOUNTS

1. Species included

Species accounts are provided for every species that has been known to occur or to have occurred in a natural state within the Iberian peninsula during historical time, as well as for those alien species that are believed to have established self-sustaining populations. In short, full treatment is given to species that are included in the national lists of all or any of the political territories that make up Iberia within Categories A, B or C of the Association of European Records and Rarities Committees (AERC, http://www.aerc.eu/) list. Category D species, those whose wild origin has not yet been established, are considered together after the principal list. Exotic species (Category E) are summarised in Appendix 1.

2. Nomenclature

English, Scientific, Spanish and Portuguese names are given for all species. English and scientific names, and the species sequence, are those favoured by the AERC (Crochet & Joynt 2012), although British usage has been adopted in the few English names (e.g. 'diver' instead of 'loon') where the AERC list deviates from that of the British Ornithologists' Union (2006). Spanish names are those of the *Lista de las aves de España* (Gutiérrez *et al.* 2012). Portuguese names follow *Nomes portugueses das aves do Paleárctico Ocidental* (Costa *et al.* 2000).

3. Principal sources

The Iberian ornithological literature is extensive and, of course, steadily expanding. It includes two national breeding bird atlases for both Spain and Portugal and one for Andorra, a Spanish winter atlas, regional atlases, existing national and regional avifaunas including the recent *Aves de Portugal*, group and single-species monographs for a wide range of species but especially raptors, seabirds and waterfowl, regional bird reports for many (but not all) the Spanish autonomous regions and several Spanish provinces (see Appendix 5), the Portuguese and Gibraltar bird reports, the SEO/Birdlife journal *Ardeola* and its magazine *Aves y Naturaleza* (formerly *La Garcilla*), the SPEA magazine *Pardela* and its antecedents *Airo* and *Cyanopica*, papers in the Iberian and international ornithological literature, the bulletins of working groups such as the Iberian Seabird Group and the Exotic Species Group, and a wide diversity of local publications. In addition there are reports on bird ringing, including the Spanish *Revista de Anillamiento*, and we have been able to consult the ringing and recovery archives of SEO and GONHS. In general, we have not cited records that have only been published on the internet, such as on the SEO/Birdlife newsgroup '*Avesforum*', since these lack any corroboration or vetting: the more interesting of these observations usually appear later in Regional Bird Reports or other publications. An exception is the very recent rarity records that have appeared on the 'Rare birds in Spain' or 'Rare birds in Portugal' websites that are pending assessment by the respective rarities committees: we draw attention to the most important of these. In general, we have covered published material available up to December 2013.

Citation of sources

We have cited specific sources in the usual style but some publications have been consulted so frequently that we have abbreviated the references, as listed below, in order to improve the flow of the text. The full citations are listed in the References list.

Abbreviation	Data source
AAE	*Las aves acuáticas en España, 1980–2009.* González & Pérez-Aranda (2011).
ACP	*O estado das aves comuns em Portugal 2010.* Meirinho *et al.* (2011).
ACRE	*Las aves comunes reproductoras en España.* Carrascal & Palomino (2008).
ANA	*Atles dels ocells nidificants d'Andorra/Atlas of the breeding birds of Andorra.* ADN. (2002).
AP	*Aves de Portugal.* Catry *et al.* (2010).
AI1	*Aves Ibéricas I. No-paseriformes.* Díaz *et al.* (1996).
AI2	*Aves Ibéricas II. Paseriformes.* Tellería *et al.* (1999).
AMI	*Aves migradoras Ibéricas.* Bernis (1966–1970).
BG	*The Birds of Gibraltar.* Cortes *et al.* (1980).
BE	*Birds in Europe.* Birdlife International (2004).
BM	*The Birds of Morocco.* Thévenot *et al.* (2003).
BSG	*Birds of the Strait of Gibraltar.* Finlayson (1992).
BWP	*Handbook of the Birds of the Western Palearctic.* Cramp *et al.* (1979–1994).
CA1	First Catalan breeding bird atlas. Muntaner *et al.* (1983).
CA2	Second Catalan breeding bird atlas 1999–2002. Estrada *et al.* (2004).

CMA	Centro de Migración de Aves (Spain)
CWA	Catalan Winter Atlas 2006–09. Herrando *et al.* (2011).
GAE	Exotic species group (Grupo de Aves Exóticas, SEO/Birdlife).
GONHS	Gibraltar Ornithological and Natural History Society records database.
HBW	*Handbook of the Birds of the World*. Del Hoyo *et al.* (1992–2011).
LR	(*Libro Rojo*) Spanish Red Data Book. Madroño *et al.* (2004).
LV	(*Livro Vermelho*) Portuguese Red Data Book. Cabral *et al.* (2005).
OF	*Nouvel inventaire des oiseaux de France*. Dubois *et al.* (2008)
PA1	First Portuguese breeding bird atlas. Rufino (1989a).
PA2	Second Portuguese breeding bird atlas. Equipa Atlas (2008).
RBS	Rare birds in Spain website (www.rarebirdspain.net)
RBP	Rare birds in Portugal website (www.avesdeportugal.info/raridades).
SA1	First Spanish breeding bird atlas. Purroy (1997).
SA2	Second Spanish breeding bird atlas. Martí & del Moral (2003).
SMA	Preliminary Spanish Migration Atlas. SEO/Birdlife (2012).
SWA	Spanish Winter Atlas 2007–10. Del Moral *et al.* (2012).

References to specific records that are published in the Bird News section (*Noticiario Ornitológico*) of the Spanish journal *Ardeola*, in Spanish regional bird reports, in the Portuguese and Gibraltarian equivalents, and elsewhere are cited directly in the text and are not listed in the Bibliography. The names of the publications cited for such records are given in full with the exception of the following, which are abbreviated as shown:

Anuário Ornitológico (Portugal)	*Anu. Ornitol.*
Gibraltar Bird Report	*GBR*
Boletín GIAM (Bulletin of the *Grupo Ibérico de Aves Marinas*).	*GIAM*

References to the various regional, usually annual, Spanish bird reports are cited as *A. O.* (for Anuario Ornitológico) followed by the region name, the year covered and the page numbers. Thus *A. O. Balears* 2007: 15 is a reference to page 15 in the 2007 bird report from the Balearic Islands.

The principal data sources for the Figures are the record summaries in *Ardeola* (Noticiarios Ornitológicos) and in the *Anuário Ornitológico*, including the annual rarity reports, as well as the various regional bird reports. Other sources are given in the Figure captions.

Ringing recoveries

The sources of information on ringing recoveries are not always cited in the text since the same few sources apply in most cases. These are, for Spain and Portugal combined, those given and analysed by Francisco Bernis in his multivolume work *Aves Migradoras Ibéricas* (AMI: 1966–71) and the additional records of recoveries of birds ringed as pulli between the late 1970s and early 1990s summarised in *Aves Ibéricas*, Volumes 1 and 2 (AI1: Díaz *et al.* 1976, AI2: Tellería *et al.* 1999). In addition, later ringing recoveries have largely been obtained from the Spanish annual ringing reports of the *Oficina de Anillamiento* published in the journal *Ecología* (Dirección General de Conservación de la Naturaleza), the ringing archives of SEO/Birdlife: including the *Preliminary Spanish Migration Atlas* (PM: SEO/Birdlife 2012), the summaries of Portuguese ringing records in *Aves de Portugal* (ADP: Catry *et al.* 2010) and the Gibraltar ringing archives of GONHS. Papers analysing ringing recoveries for particular species are cited in the usual way. We have not examined every single ringing recovery involving Iberia and so all recovery totals need to be regarded as minima.

Historical records

Historical sources in the literature have been consulted and cited where this is informative. They include a considerable number of papers and some books written especially during the second half of the 19th century and early in the 20th century. Such sources sometimes allow the past distribution and abundance of particular species to be compared with the present and the more recent past. They have their limitations, however. Early accounts are essentially anecdotal and often take the form of annotated lists of birds seen by the authors or their contacts in particular regions. There are, unfortunately, no historical equivalents of the scientifically-designed, quantitative surveys that are conducted today. The 19th century authors used such terms as 'abundant' and 'common' liberally without definition and some of their statements need to be regarded with caution.

It is also the case that field identification skills were often limited, and that optical aids were poor and frequently not used at all. Often the emphasis was on collecting specimens of birds and their eggs and some of the accounts make sobering reading as bird after bird is systematically slaughtered. Some references to collecting appear hypocritical to our modern eyes. For example, Chapman (1884) says of the Spanish Imperial Eagle at Doñana '*There are still a few pairs of this superb Eagle in the district, though their numbers are sadly thinned by the greed of collectors since I first met with them there in 1872. I obtained a pair of magnificent adults from their nest...*'

Mere sight records were often not regarded as valid. Such sight records nonetheless often figure in the published material and it is not always easy to decide, from our contemporary viewpoint, which are acceptable and which are probably erroneous. What, for example, should we make of Howard Saunders' statements that the Short-toed Eagle was common in winter in the marshes of southern Spain, that the Grey-headed Woodpecker *Picus canus* occurred in the hills, that he had seen the Rock Nuthatch *Sitta neumayer* and its 'unmistakable' nest, that the Willow and Wood Warblers were present in Spain in winter and that he had no doubt that the Olive-tree Warbler *Hippolais olivetorum* bred abundantly near Málaga (Saunders 1871)? Saunders is given as an example but similar highly implausible statements can be found in many other sources. Lilford (1866) identified a pair of Grey-headed Woodpeckers, as well as Green Woodpeckers, in the grounds of the Casa de Campo, Madrid. Such observations, which were almost certainly mistaken, probably result from identification errors and, sometimes, from undue reliance on the reliability of informants. In the case of the Grey-headed Woodpecker a lack of familiarity with the Iberian form of the Green Woodpecker, which was only recognised as distinct and described by Saunders himself in 1872, clearly caused confusion, as pointed out by Witherby (1928). Some identification errors are obvious. For example, Munn (1925) considered Eleonora's Falcon to be very rare in Mallorca but he saw flocks of up to 22 'Peregrines' hawking for insects around the Albufera. He also thought that Willow and Wood Warblers, as well as Chiffchaffs, wintered in Mallorca and Menorca (Munn 1921, 1924). The European ranges of many Mediterranean species were not properly known in the 19th century and we should not be too critical of the apparent shortcomings of some of our pioneering predecessors, who worked under difficult conditions and without the ample resources that are now available. On the contrary, we owe a great deal to them. However, we have decided in most cases not to cite historical records that we consider very likely to have been erroneous and instead only to include those that were probably correct, partly for reasons of space.

Rarity records

We include and analyse all rarity records that have been reviewed and considered acceptable (homologated) by the rarities committees of Spain, Portugal and Gibraltar for birds observed in their respective territories up to the end of 2010. Accepted 2011 records from Spain and Gibraltar are also considered. Subsequent records up to the end of 2012, and a few of the most interesting records from 2013, are summarised but all of these were pending assessment by the rarities committees when this book was published. The accounts often also mention accepted records for the same species from the countries north and south of Iberia: Britain and Ireland, France and Morocco; as well as the Azores and Italy in some cases, in order to place the Iberian records in context. Otherwise, records that have not been evaluated by the relevant rarities committee are not included, to preserve the consistency in acceptance standards that is a principal reason for the existence of such review bodies and to help sustain the credibility that such committees deserve and require. An exception is made in a few cases where non-homologated records have been published elsewhere and this circumstance is indicated together with our own evaluation of those records.

Citations are not included for accepted records except where these are the subject of specific publications. The records themselves are summarised in the rarities reports published annually in *Ardeola* and *Noticiário Ornitológico*, and periodically in the Gibraltar Bird Report. The rarities reports are listed in Appendix 5, which also gives the contact addresses of the rarities committees, which may be used to submit records.

5. Censuses and estimates of population sizes and population trends

A diversity of censuses and estimates of national populations are available, especially for Spain. There the populations of a number of individual species (e.g. the White Stork, Black Vulture and Marbled Duck) are censused regularly as are those of certain guilds (e.g. wintering aquatic birds). Estimates of national breeding populations in Spain were attempted for all species as part of both breeding atlas projects, which took place during 1975–95 (Purroy 1997) and 1998–2002 (Martí & del Moral 2003). Thereafter, an intensive survey of the breeding season populations of 95 common, mainly passerine, species was undertaken in mainland Spain during 2004–06, as part of the SACRE programme, the Spanish Common Birds Census (Carrascal & Palomino 2008). Similar nationwide censuses have not yet been attempted in Portugal but the Portuguese Common Birds Census has monitored breeding population samples since 2004 to investigate numerical trends (Meirinho *et al.* 2011).

The earliest of all Iberian censuses was the breeding White Stork survey organised by Francisco Bernis in 1948. Earlier workers, especially those of the 19th century gave their impressions of abundance, sometimes using such terms as 'vast numbers' without giving any numerical information. Some later estimates, from the second half of the 20th century, have clearly been little more than informed guesses. Recent studies have been far more thorough, using tried and tested methodology and statistically-evaluated results, and no doubt this scientific approach will continue from now on. At the time of writing, however, quantitative information on long-term numerical trends is not yet available for Iberian species, although there are strong indications for many of them.

In essence, the first Spanish breeding atlas survey (SA1: Purroy 1997) used line transects to assess bird densities and the results were then extrapolated across the known extent of the appropriate habitats in Spain. These results were updated using additional information by Birdlife International and published in *'Birds in Europe'* (Birdlife International 2004). The second Spanish breeding atlas survey (SA2: Martí & del Moral 2003) was based on observers' estimates for the 10km-squares that they surveyed. The SACRE-based population estimates (ACR: Carrascal & Palomino 2008) were by far the most systematic of the national surveys, employing point counts and correcting the results for differences in the ease of detectability of different species. Full details of the complex methodology and of the statistical analyis of this last survey are provided by Carrascal & Palomino (2008) alongside the results for each species.

Census results, population estimates and reported population trends are all included within the individual species accounts. However, it is very much the case that the results of all these surveys must be contemplated with a considerable degree of caution. Indeed, many recent results have been the subject of controversy and strong criticism (Blanco *et al.* 2012, 2014). A comprehensive evaluation of the main Spanish breeding bird censuses has been made by Murgui (2011a). He notes that the most recent national estimates are on average nearly five times as high as earlier ones, and more than ten times greater for 19% of species and identifies the possible shortcomings in the methods employed by all the different surveys.

Bird population estimation is a developing science and the available results from Spain are best seen as a work in progress that will be refined in future. Carrascal (2011) has offered a thoughtful defence of the methodology and results of the SACRE survey. We have not taken it upon ourselves to adjudicate between conflicting datasets but instead have quoted the available figures, with our own comments where we have had reason to make them. It is likely that some of the recent, very large, estimates for the breeding populations of some species in Spain will be confirmed in due course, whereas some others may be revised downwards or upwards – irrespective of any actual changes in abundance. Such improved estimates will result from refinements to the methodology, perhaps especially from ensuring that the sampling procedure is truly comprehensive and representative. For the present it remains important to remember that none of the extrapolated estimates is set in stone. However, it is worth bearing in mind that Spain alone covers over half a million square kilometres, so that a species that manages even a low mean density of just a single pair per square kilometre will still be represented there by over one million individuals.

Trends in abundance and distribution of both breeding and wintering birds are monitored by a number of programmes. In Spain, the principal and most long-standing of these is the common birds census, SACRE, mentioned above. SACRE (Programa de Seguimiento de Aves Comunes Reproductoras en España) commenced in 1996 and trends for a diversity of species are available for the period 1998–2012 (Escandell 2013). These, and their Portuguese equivalents (ACP), are included in the species accounts for widespread species whose observed trends are significant statistically but less marked trends and apparent indications of population stability are not normally highlighted except in the cases of common species, where sample sizes are large and well-dispersed. The SACRE results have been presented by Escandell (2013), showing the overall national trend for Spain as a whole and the separate trends for Andalucía, central Spain (Castilla y León, Castilla-La Mancha, Madrid and Extremadura), the east (the Levant regions including the eastern Pyrenees) and the north (Galicia, the Biscay regions, the western Pyrenees and the Cantabrian Range). The SACRE programme will undoubtedly produce very useful long-term results but the significance of the present data must be viewed against a background of a relatively short span of survey years for most SACRE site surveys. Although some 500 SACRE sites have been monitored for five or more years during 1998–2012 their distribution across Spain is very uneven, with concentrations around Madrid, Barcelona and the Basque Country and very poor coverage of the northwestern meseta, much of the southern meseta and the south-east in general (Escandell 2011). Nevertheless, the SACRE population trends for 1998–2012 show a similar pattern to those reported in some other European countries: there have been marked declines in species associated with agricultural landscapes but, conversely, some woodland species have shown large increases.

A similar programme that monitors wintering birds in Spain (SACRE – invierno) began in winter 2008/09 and will eventually allow trends in these populations to be detected. A separate programme that monitors nocturnal species in Spain (programa Noctua) was initiated in 1998. Other current monitoring schemes in Spain include a constant-effort ringing programme (programa Passer), a phenological programme (Aves y Clima – Red Fenológica) and the annual national census of wintering aquatic birds, conducted in January (Censo nacional de aves acuáticas invernantes): which has run since the 1960s, although intermittently at first. The principal findings of all these surveys are also included in the relevant species accounts, together with the results of national surveys of particular species or species-groups, which have been published as monographs. The most important of these projects have their equivalents in Portugal, notably in the form of the common birds census (CAC: Censo de Aves Comuns), Noctua-Portugal, a migrant phenology project (Projeto Chegadas) and a winter coastal bird census (Projeto Arenaria).

6a. Format of the species accounts – non-vagrants

All the principal species accounts follow a similar format but we have retained the freedom to depart from this where the need arose. We decided to dispense with subheadings since all too often they compelled us to rearrange material to fit the format in ways that were less than ideal from a narrative point of view. Each account includes the following, where relevant:

1. English, Scientific, Spanish and Portuguese names, in that order.
3. Subspecies. Those that occur or have been recorded in Iberia, with a brief summary of their regional status. Species for which no subspecies are listed are monotypic.
4. Breeding population estimates. These are summarised for passerine and near-passerine species only. The sources of the Spanish estimates are given in each case. Those from Portugal and Andorra are from Birdlife International (2004) unless otherwise stated. Gibraltar estimates are from GONHS. Non-passerines have a diversity of population estimates, which are discussed in the species accounts.
5. Breeding distribution. Distribution within Iberia, including a map of the breeding range.
6. Breeding habitat.
7. Breeding population. The most recent available estimates of the size of the breeding population, together with previous estimates (if any), population trends and density estimates in the principal habitats. Mainland Spain, the Balearic Islands and Portugal are treated separately in turn, followed by Andorra and Gibraltar.
8. Wintering. Wintering distribution: including a distribution map, habitats and wintering populations.
9. Migration. Phenology and routes. Origins of migrant and/or wintering populations as revealed by ringing recoveries or observations of marked individuals. The temporal distribution of records is shown graphically in many cases, using the same format as for winter records where necessary.
10. Conservation issues. Species-specific information.

Examples of maps of breeding and wintering distributions

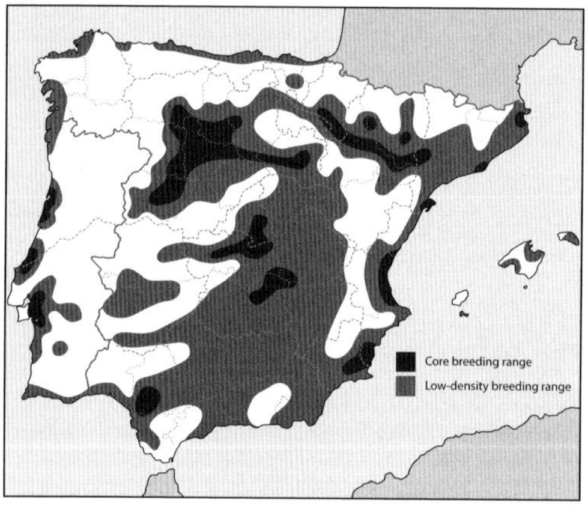

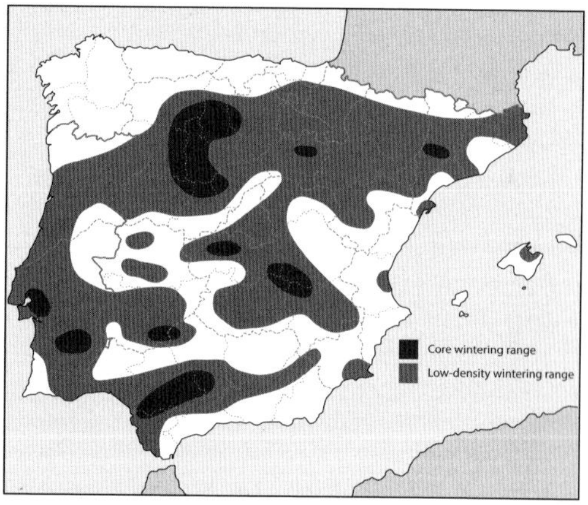

Figures showing the temporal distribution of records, of wintering species in particular, often have a particular format in which initial observations are shown together with subsequent records of the same birds where these remained at particular sites for long periods. Such figures generally show the numbers of records per month. Hence 'First' records show the first month in which particular records occurred and 'All' records show the additional months in which those birds that were long-stayers were also recorded. The need for this is shown in the following example (Fig. 7.1).

The actual number of records per month (irrespective of the number of birds involved on each occasion), 293 records in total in this case, is represented by the 'First' columns. Some birds remained for two or three months and these additional months are added to the 'First' records to give 'All' the records for each month. Hence, for example, a bird or flock that arrives in November and remains until January is included under November, December and January under 'All' records but it only figures under 'November' as a 'First' record. This is necessary to give a full picture of winter phenology especially. In the example, the 'First' columns alone suggest a peak presence in November followed by a decline in abundance through the winter. In fact, what actually occurs is that new arrivals peak in November but further arrivals occur later in the winter, while the earlier birds remain in Iberia. Peak winter abundance is thus in January in this case.

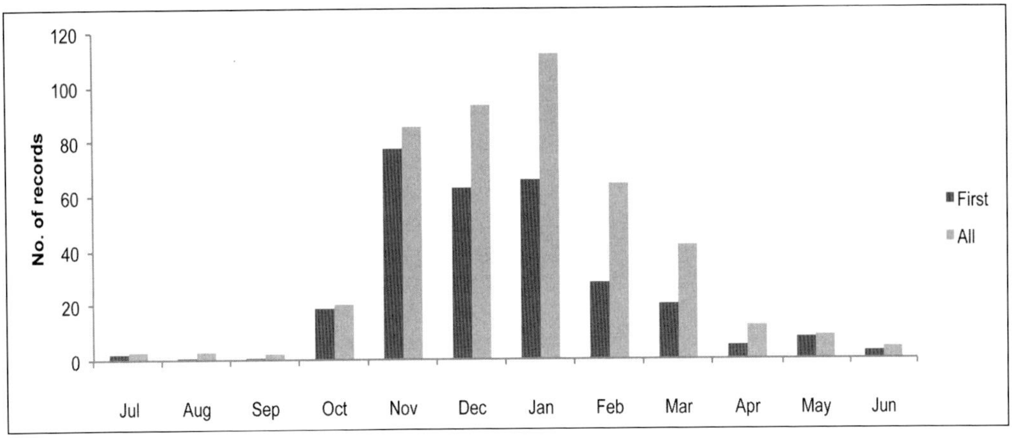

Figure 7.1. Monthly distribution of records of Greater Scaups in Spain 1983–2010. N = 293 records.

6b. Format of the species accounts – vagrants

The species accounts for vagrant species have a somewhat different format, as follows:

1. English, Scientific, Spanish and Portuguese names in that order.
2. Status in the western Palearctic, with particular emphasis on the countries north and south of Iberia: Britain and Ireland, France and Morocco; as well as Italy in some cases.
3. Overview of historical and recent records in Iberia. Trends.
4. Numbers of birds involved and ages/sexes.
5. Temporal distribution of records.
6. Regional distribution of records. May include a map, giving record totals for each region.

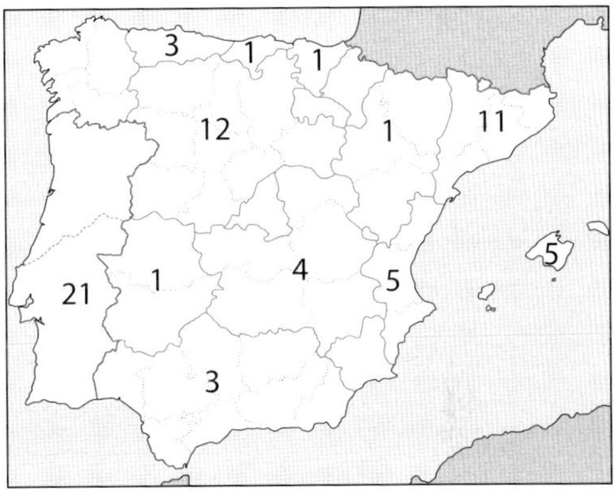

7. Locations

A degree of familiarity with the political geography of the Iberian Peninsula is necessary in order to understand our accounts of bird distributions and movements especially. The following notes are mainly provided to help those who are not already familiar with the region but we also point out a number of conventions that we have decided to apply.

7.1 The political and administrative regions of Spain and Portugal

Spain

Mainland Spain is divided into 15 politically autonomous regions, the Balearic Islands comprising a 16th region (Fig. 7.2). The autonomous regions differ greatly in size. Some of them also, notably the Basque Country (País Vasco) and Catalonia, are ancestral nations in their own right, whose ancient territories extended into what is now France. Both these regions, and also – to differing extents – Galicia, Asturias and the Balearic Islands, nurture a fiercely independent nature in some quarters. Such political considerations are remote from the scope of this book. However, they mean in practice that although the Spanish language (Castilian) is universally used in Spain, the major regional languages – Català (Catalan) and its variants; Euskara (Basque) and Galego (Galician) – are regionally prominent. Certain regional names have become increasingly pre-eminent in official usage for these latter regions especially and so we have used such names in preference to their Castilian equivalents, but we are not making any political comment by so doing. In practice, there are not too many such names involved. The principal ones are Gipuzkoa and Bizkaia (for the Basque provinces of Guipúzcoa and Vizcaya), A Coruña and Ourense (for the Galician provinces and cities of La Coruña and Orense), and Girona and Lleida (for the Catalan provinces and cities of Gerona and Lérida). For the autonomous regions of Catalonia, the Basque Country and the Balearic Islands we have used these English names instead of the Castilian or regional ones.

The autonomous regions are divided in most cases into provinces (Figure 7.3), most of which bear the name of the provincial capital. The regions of Cantabria, Navarra, La Rioja, Murcia and Madrid each comprise a single province. There are 48 provinces in total. It is important to realise that in Spain the provincial name is often used to refer to the entire province, and not just the provincial capital. Thus 'Sevilla' is used to mean Sevilla province and not just the city of Sevilla, which is properly referred to as 'Sevilla capital'. We have retained this convention in the text, mainly because it avoids frequently repeating the word 'province'..

Figure 7.2. Autonomous regions of Spain

Figure 7.3. Provinces of Spain.

Portugal

The 11 traditional administrative provinces (Figure 4a) have been replaced for administrative purposes by 18 smaller regions (Figure 4b). Nevertheless, the traditional names remain widely used and, for our purposes, they are helpful when describing the broader patterns of bird distribution. Conversely, the new names are useful for locating the positions of specific sites more precisely. We have thus used both sets of names according to need. Many of our species accounts, and especially our maps showing the geographical distribution of rarities, divide Portugal into just two sectors: northern and southern. The dividing line is that which separates the provinces of Beira Alta and Beira Litoral in the north from Beira Baixa, Ribatejo and Estremadura in the south.

Figure 7.4. Administrative divisions of Portugal

7.4a. Traditional provinces

7.4b. Current administrative districts.

7.2 Descriptions of species distributions and the locations of specific records

References in the species accounts to particular locations, whether as part of descriptions of species' distributions or when indicating where particular observations were made, give the site name (e.g. Azud de Riolobos) where relevant, followed by the Province/District in which it is located (Salamanca in this case). Those with a good working knowledge of Iberian geography will need no more information but, otherwise, Tables 7.2 and 7.3 list the Spanish provinces and Portuguese districts respectively against the wider regions to which they belong.

Table 7.2. Alphabetical list of Spanish provinces, showing the autonomous regions to which they belong.

Province	Region	Province	Region
A Coruña	Galicia	Illes Balears	Balearic Islands
Álava (Araba)	Basque Country	Jaén	Andalucía
Albacete	Castilla-La Mancha	La Rioja	La Rioja
Alicante	Comunidad Valenciana	León	Castilla y León
Almería	Andalucía	Lleida	Catalonia
Asturias	Asturias	Lugo	Galicia
Ávila	Castilla y León	Madrid	Madrid
Badajoz	Extremadura	Málaga	Andalucía
Barcelona	Catalonia	Murcia	Murcia
Bizkaia	Basque Country	Navarra	Navarra
Burgos	Castilla y León	Ourense	Galicia
Cáceres	Extremadura	Palencia	Castilla y León
Cádiz	Andalucía	Pontevedra	Galicia
Cantabria	Cantabria	Salamanca	Castilla y León
Castellón	Comunidad Valenciana	Segovia	Castilla y León
Ciudad Real	Castilla-La Mancha	Sevilla	Andalucía
Córdoba	Andalucía	Soria	Castilla y León
Cuenca	Castilla-La Mancha	Tarragona	Catalonia
Girona	Catalonia	Teruel	Aragón
Gipuzkoa	Basque Country	Toledo	Castilla-La Mancha
Granada	Andalucía	Valencia	Comunidad Valenciana
Guadalajara	Castilla-La Mancha	Valladolid	Castilla y León
Huelva	Andalucía	Zamora	Castilla y León
Huesca	Aragón	Zaragoza	Aragón

Table 7.3. Alphabetical list of Portuguese districts, showing their situation in the former provinces.
* Former names do not correspond exactly with current districts in all cases.

District	Former provinces *	District	Former provinces *
Aveiro	Beira Litoral	Leiria	Estremadura
Beja	Baixo Alentejo	Lisboa	Estremadura
Braga	Minho	Portalegre	Alto Alentejo
Bragança	Trás-os-Montes e Alto Douro	Porto	Douro Litoral
Castelo Branco	Beira Baixa	Santarém	Ribatejo
Coimbra	Beira Litoral	Setúbal	Baixo Alentejo
Évora	Alto Alentejo	Viana do Castelo	Minho
Faro	Algarve	Vila Real	Trás-os-Montes e Alto Douro
Guarda	Beira Alta	Viseu	Beira Alta

7.3 Locations of sites that are frequently mentioned in the text.

A significant number of locations, many of them wetland sites, are mentioned very frequently and in a large number of species accounts. We have avoided repeated mention of the geographical location of these sites by listing that information in Table 7.4. All other sites are identified where mentioned in the text by province (Spain) or district (Portugal). The location of the numbered sites is shown in Figure 7.5. The terms 'Marismas del Guadalquivir' and 'Doñana' are often used interchangeably in everyday language and we have used this broader interpretation of 'Doñana' in the texts in order to save space. Strictly speaking 'Doñana' is the Coto de Doñana National Park plus the protected peripheral natural park, which together make up a large proportion but not all of the Marismas. We have referred to the National Park specifically where this is important.

Table 7.4. Locations of frequently mentioned sites. NP = National Park.

Number	Site	Province	Region
1	Aiguamolls de l'Empordà	Girona	Catalonia
2	Albufera de Mallorca	Mallorca	Balearic Islands
3	Albufera de Valencia	Valencia	Comunidad Valenciana
4	Berlengas Islands	Leiria	Estremadura
5	Brazo del Este	Sevilla	Andalucía
6	Cádiz bay	Cádiz	Andalucía
7	Cape Bares	A Coruña	Galicia
8	Cape St Vincent	Faro	Algarve
9	Doñana NP	Huelva/Sevilla	Andalucía
10	Ebro delta	Tarragona	Catalonia
11	El Hondo reservoir	Alicante	Comunidad Valenciana
12	Gijón bay	Asturias	Asturias
13	Guadalhorce estuary	Málaga	Andalucía
14	La Mancha wetlands	Ciudad Real/Toledo/Cuenca	Castilla-La Mancha
15	Gallocanta (Laguna de Gallocanta)	Zaragoza/Teruel	Aragón
16	La Nava (Laguna de La Nava)	Palencia	Castilla y León
17	Villafáfilla (Lagunas de Villafáfilla)	Zamora	Castilla y León
18	Llobregat delta	Barcelona	Catalonia
19	Mar Menor	Murcia	Murcia
20	Santoña (Marismas de Santoña)	Cantabria	Cantabria
21	Marismas del Guadalquivir	Huelva/Sevilla/Cádiz	Andalucía
22	Monfragüe NP	Cáceres	Extremadura
23	Odiel estuary	Huelva	Andalucía
24	Picos de Europa NP	Cantabria/Asturias /León	Cantabria/Asturias/Castilla y León
25	Punta La Vaca headland	Asturias	Asturias
26	Ria de Aveiro	Aveiro	Beira Litoral
27	Ria Formosa	Faro	Algarve
28	Sado estuary	Setúbal	Estremadura/Baixo Alentejo
29	Santander bay	Cantabria	Cantabria
30	Sierra Brava reservoir	Cáceres	Extremadura
31	Sierra del Guadarrama NP	Madrid/Segovia	Madrid/Castilla y León
32	Sierra Nevada	Granada/Almería	Andalucía
33	Daimiel (Tablas de Daimiel NP)	Ciudad Real	Castilla-La Mancha
34	Tagus estuary	Lisboa/Setúbal/Santarém	Estremadura/Ribatejo

Figure 7.5. Locations of frequently mentioned sites.

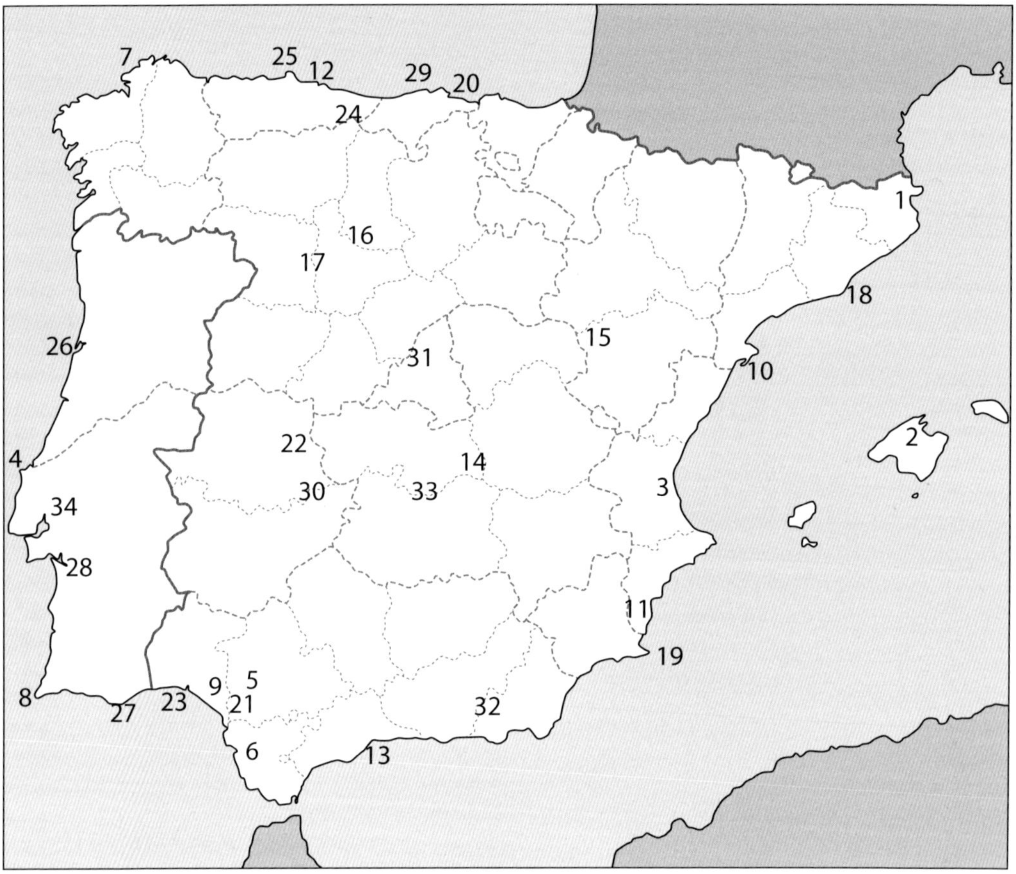

SPECIES OF THE IBERIAN PENINSULA

THE SYSTEMATIC LIST

BLACK SWAN *Cygnus atratus*

(Cisne negro, Cisne-preto)

Recorded in Spain (Category C) and Portugal (Category E).

Excluding obviously released birds, we are aware of 51 records in Spain and five in Portugal during 1986–2010, involving at least 88 individuals. The frequency of records has increased rapidly, from 19 during 1986–99 to 37 during 2000–10, and they have been annual since 1995. The Portuguese records were all in 2009–10. Some 70% of records have involved single birds, the largest groups being five at Santoña in January 1995 and five at Figueira da Foz, Coimbra, in September 2010 (*Ardeola* 45: 119, *Anu. Ornitol.* 8: 95). There have also been several breeding attempts: in the Basque Country in 2001, Cantabria in 2002 and 2004, and Catalonia in 2004 (*Ardeola* 49: 178, SA2, GAE).

The observations of this Australasian species in Spain have been attributed to escapees (AI1), some possibly having originated from elsewhere in Europe (Jutglar & Masó 1999, Clavell 2006). Breeding records apart, observations are chiefly in winter and in the north, supporting the hypothesis that the birds arrive from other European countries. The increase in record frequency also accords well with increases in naturalised populations elsewhere in Europe, including France (Dubois 2007).

MUTE SWAN *Cygnus olor*

(Cisne Vulgar, Cisne-mudo)

Largely absent from Iberia until quite recently, other than in collections of ornamental waterfowl. Observations in the wild seem very often to involve birds originating from other European countries. Breeding has become increasingly frequent.

The Mute Swan was once regarded as an extremely rare winter visitor to Spain (Bernis 1954a). There are references to its occurrence there as long ago as the 17th and 18th centuries (Sargatal & del Hoyo 1989, Dies *et al.* 1999) and to the arrival of flocks in Catalonia during the 1829–30 and 1890–91 winters (Maluquer 1964a). There are also historical references to its presence in Portugal in the 15th century and to the occurrence of 'swans' there during the late 19th and early 20th centuries (AP). There were numerous observations during the first half of the 20th century in Spain, notably those associated with an irruption during the harsh winter of 1962–63. Some dozens appeared during January 1963 in Catalonia and the Biscay coastlands, where most were shot (*Ardeola* 8: 269–270, 11: 145; Maluquer 1964a).

The Iberian rarities committees considered the Mute Swan during 1984–2002, when there were 27 records in total. It was evident by then, however, that many records were not being submitted since observations were attributed to introduced or escaped birds (de Juana 2006). Further observations have been published in *Ardeola* and the GAE bulletins, and the Portuguese records committee accepted 17 records during 2000–09. Setting aside obviously introduced birds, there is a published sample of 71 records (133 birds) during 1982–2007.

The frequency of such records has increased rapidly. Mute Swans have been widely reported but tend to appear along coasts, especially in the Atlantic sector. Occurrences peak in winter, suggesting arrivals from elsewhere in Europe. Although the French population is relatively small, it is apparently increasing and now extends along the Atlantic coast almost to the Spanish border. Moreover, one of a flock of eight seen in Menorca during winter 1989–90 had been ringed in Hungary (*Ardeola* 39: 74). Although 70% of records were of single birds, they included ten flocks of 4–8 individuals. Adults and subadults made up 64% of those whose ages were recorded.

Introduced birds nested at the Aiguamolls de l'Empordà, Girona, in 1989 (Sargatal & del Hoyo 1989) and at the Roquetas saltpans, Almería, in 1993 (*Ardeola* 42: 101). The most recent Catalan atlas found breeding Mute Swans at four sites, while the regional wintering population there grew from six birds in 2002 to 49 in 2009 (CA2, CWA). In Cantabria, Mute Swans have nested since 2001 at the wetlands around Santander Bay and the Dunas de Oyambre (SA2) and a small population has built up there so that there were 68 birds in Cantabria in January 2013 (*Ardeola* 60: 507). Nesting in Galicia was recorded for the first time at the Ría de Pontevedra in 2006 and there were 20 birds there in 2007 (GAE 2006, 2007). It seems likely that the Mute Swan will become progressively established as a breeding species across the Iberian peninsula.

TUNDRA SWAN *Cygnus columbianus*

(Cisne Chico, Cisne-pequeno)

C. c. bewickii **(Bewick's Swan). Occurs sporadically in Spain in winter, most often in the extreme north-east. Vagrant to Portugal.**

Most of the European wintering population is in the north-west but there is a regular winter presence in France, mainly in Camargue. The earliest known occurrence in Spain was in the cold winter of 1890–91, when seven were shot at the Llobregat delta (Maluquer 1964a). The harsh winter of 1962–63 saw a small influx into France and several reached Catalonia, where at least four were killed on the Girona coast (Maluquer 1964a). By the early 1980s there had been 18 records in Catalonia and two in Menorca, Balearic Islands (Ferrer *et al.* 1986). The Spanish rarities committee accepted six records during 1984–2007 as well as a 1976 record from Galicia (*Ardeola* 37: 111). A record of two individuals at the Miño/Minho estuary, Galicia/Portugal (Fernández de la Cigoña 1986) may not be entirely reliable but is the only one reported from Portugal (AP). One of two birds seen at the Aiguamolls de l'Empordà and elsewhere in Girona in January–February 2012 had been ringed on Kashin Island, Russia, in August 2011 (RBS).

Extreme dates for 19 reliable Iberian records (at least 56 birds) are 21 October to late February, with a clear peak in December–January (Fig. 1). Some 84% have been in Catalonia, notably at the Ebro delta and the Aiguamolls de l'Empordà. There have also been isolated occurrences in Aragón (*Ardeola* 37:111), and in the Balearic Islands and Galicia, as noted above. Half the records are of single birds but there have been seven groups of 2–3 birds and four larger flocks: the largest of 13 birds.

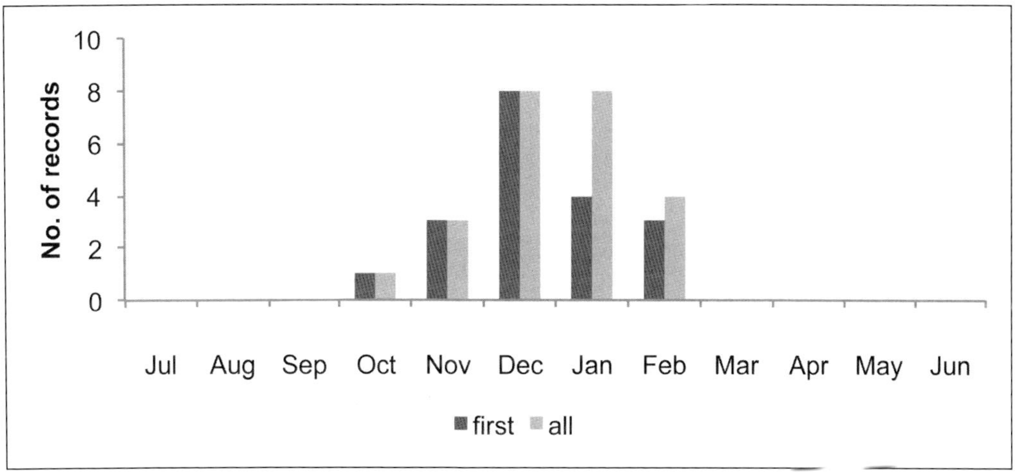

Figure 1. Records by month, 1939–2007 (n = 19)

WHOOPER SWAN *Cygnus cygnus*

(Cisne Cantor, Cisne-bravo)

A rare and irregular winter visitor to Spain, most often appearing in the coastal north-west. Vagrant to Portugal.

Its western European winter quarters are chiefly in the British Isles, where mainly Icelandic birds occur, and in north-central Europe, where birds from Scandinavia, Finland and Russia are found. It is rare in France, where it is only regular in the north-east. There is only one recent record from North Africa.

One was found in Mallorca prior to 1877 (Ferrer *et al.* 1986); another was reported near Sevilla (Irby 1895); three were at Ivars d'Urgell, Catalonia, during winter 1890–91 (Maluquer Sostres 1952), and one was at the Albufera de Valencia in 1907 (Bernis 1956). There are also several records of swans from the Portuguese Atlantic coast between the late 19th century and 1925 but the species involved was not always clear (AP). We are aware of 21 Spanish records prior to 1980: five each in Galicia and Catalonia, four each in the Comunidad Valenciana and the Balearic Islands, and one each in Andalucía, Navarra and the Basque Country. The Spanish rarities committee accepted 36 more records during 1984–2009. Subsequent records include an influx of at least 21 individuals into Galicia and the Biscay coastlands in early November 2012, some of which remained into 2013

(RBS). There are two accepted Portuguese records, both at border localities: at the Miño estuary in 1990, and at Castro Marim in 2009–2010 and two others are more recent (RBP).

The monthly distribution of 49 accepted records shows a marked concentration in winter, with some 65% of first observation dates in October–December (Fig. 2). The birds were predominantly in the northern coastlands (Fig. 3). Extreme dates are 13 August and 12 May, with the exception of an adult that first appeared at a reservoir in Ciudad Real on 31 May. The longest-stayers were two adults that remained on a reservoir in Asturias from October 1991 to April 1992 (*Ardeola* 43: 106).

Historical observations were frequently in the peninsular north-east, where Catalonia, the Comunidad Valenciana and the Balearic islands together account for 62% of pre-1981 records, but there have been only three records there since. Most recent observations have been in the coastal north-west. An adult at the Ría del Eo in 1990 wore an Icelandic ring and we may speculate that the birds seen in north-west Spain indeed originate from Iceland, whereas those that used to occur in the north-east originated from Scandinavia/Siberia: as do most of those that reach France.

Iberian occurrences have been very irregular. There were six in winter 1990–91, probably involving at least 22 birds, and five in winter 2004–05. There were none during 1997–99 but there have been annual records, apparently increasingly frequent, since 2000. The 49 records considered involved a total of 116 individuals. About half were of single birds but there are 16 records of three or more birds, including two of seven and one of eight, which seem likely to have been family parties. The larger groups were all in the north-west.

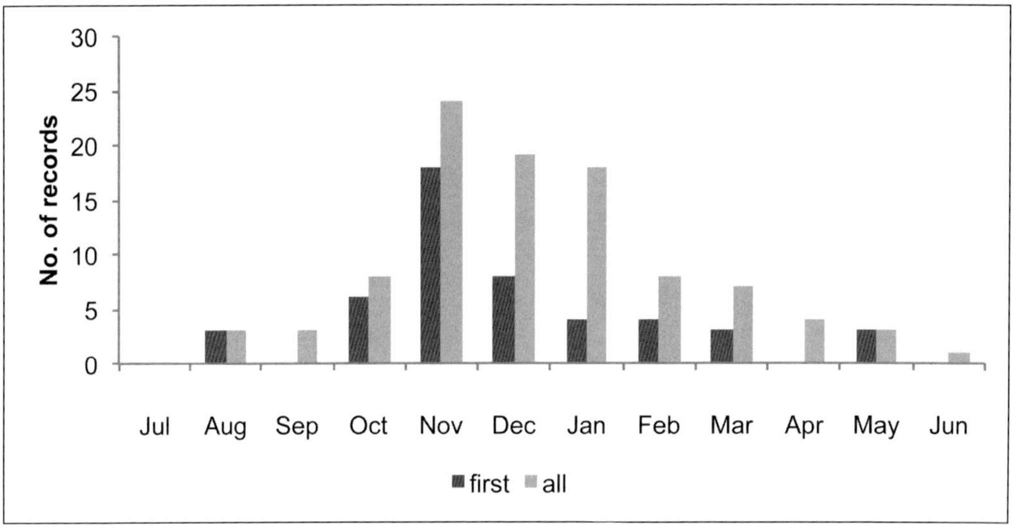

Figure 2. Records by month, 1891–2010 (n = 49)

Figure 3. Records by region, 1891–2010 (n = 49)

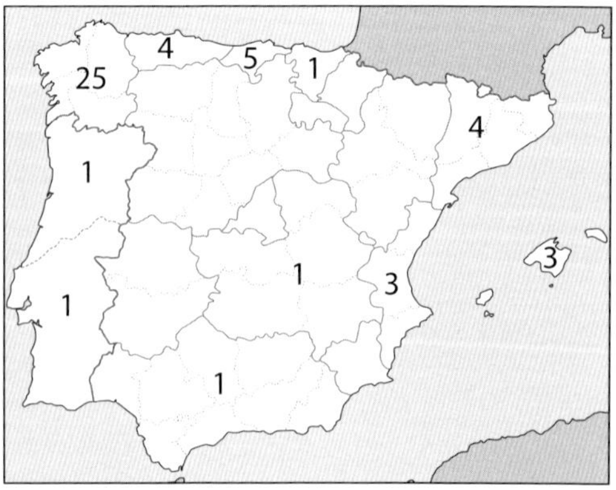

BEAN GOOSE *Anser fabalis*

(Ánsar Campestre, Ganso-campestre)

A. f. rossicus. **(Tundra Bean Goose). A rare and local winter visitor, chiefly to northern Spain. Vagrant to Portugal and the Balearic Islands.**

A. f. fabalis. **(Taiga Bean Goose). Occasional.**

The species was formerly common in winter on the northern Meseta. However, its winter quarters are now chiefly in central Europe, including northeastern France. Hence, it has been included in the Spanish rarities list since 2006.

The central Duero valley in Castilla y León was the principal wintering area of the Tundra Bean Goose and flocks totalling several thousand occurred there in the mid-1960s (Bernis 1964a). This wintering population in western Europe had gone largely unnoticed, although Castellarnau (1877) had commented that it was very common in wet years on the plains of Segovia from November to March, causing so much damage to crops that guards, '*ganseros*', were posted to keep the birds away. Bernis (1964a) found that this goose was well known to local people, and they commented that it had declined locally. He estimated the 1963–64 wintering population at 2,000–4,000 birds, mainly at the Lagunas de Villafáfila. At that time there were also small winter flocks at a single locality in Toledo and in Aragón, and occasional individuals or small groups appeared at the Ebro delta and the Albufera de Valencia (Arévalo 1887, Bernis 1964a, Ferrer *et al.* 1986). Individuals strayed to the Balearic Islands on very rare occasions (Munn 1924, 1930).

Wintering individuals were also recorded at various sites in southern Spain during the 19th and 20th centuries, notably at Doñana, where a few occurred almost annually until at least the 1980s (Llandres & Urdiales 1990). The Bean Goose was often recorded in Portugal during the late 19th and early 20th centuries (AP) in variable numbers but it was reported to outnumber the Greylag Goose in some seasons. It is now only a vagrant in Portugal and there have been just six records (seven birds) since 1993 (AP). It was apparently common in winter in both Morocco and Algeria during the late 19th century (Heim de Balsac & Mayaud 1962), but there are only five 20th century records from Morocco, the last in 1994 (BM).

The draining of the Laguna de la Nava in the mid 20th century left Villafáfila and the nearby Ricobayo reservoir as the only regular wintering area. The Bean Goose flocks wintered there in the company of Greylag Geese, feeding during the day in the fields around the Villafáfila wetlands and roosting on the reservoir (Rodríguez & Palacios 1991). A few are still recorded there in most winters. The goose roost was censused annually from 1969 and the counts proved to be well-timed to document the decline of the wintering population almost to extinction. Their numbers ranged from 2,800 to 4,500 birds during 1972–82, but there were only 1,400 in 1983 and 100 in 1993, and counts have been in single figures since 1998 (LR, AAE).

A few were reported on rare occasions during the 1990s from locations as far apart as Galicia, the Basque Country, Navarra, La Rioja, Aragón, Mallorca, the Comunidad Valenciana, Salamanca and Extremadura. The only notable count outside Zamora during this period was of two flocks totalling 39 birds at the Balsa de Loza, Navarra, on 10 April 1994: presumably migrants on their way back north (*A. O. Navarra 1993–94*: 107).

The Bean Goose population was reduced to just a few individuals by the start of the 21st century. Occasional individuals still appear at diverse locations, often with Greylag Geese. Published records from Spain, including those accepted during 2006–09 total just 26 (53 birds) post 2000, 17 of them of single birds. The largest groups were of six at Castropepe, Zamora, on 8 December 2003 and five at the Sierra Brava reservoir, Cáceres, on 13 January 2002 (*Ardeola* 49: 178, 51: 546). Extreme dates for all these records were 25 October and 2 March, but 19 were in December and January. Most were within the range of the movements of migratory Greylag Geese: in the Basque Country, across the northern meseta and south to Doñana. There is only one accepted record in Portugal post-2000: at Ponta da Erva in the Tagus estuary on 17 November 2002. There were at least a further 19 Spanish records and a Portuguese one up to the end of 2012 (RBS, RP).

Individual Taiga Bean Geese were reported occasionally and still occur. Recent records include single individuals at the Albufera de Mallorca in January 1997 (Clavell 2002); at the Balsa de Loza, Navarra, in February 1999 (*A. O. Navarra 1999*: 78); at Villafáfila in January 2001 and December 2001 (*Ardeola* 49: 178) and at Baldaio, A Coruña, in January 2001 (*A. O. Galicia 2001*: 52). Single birds at the Tagus estuary during winter 1998–99 and on November 2002, and two there on 4 December 1999, were also *A. f. fabalis* (AP).

PINK-FOOTED GOOSE *Anser brachyrhynchus*

(Ánsar piquicorto, Ganso-de-bico-curto)

A rare winter visitor to Spain. Vagrant to Portugal.

Pink-footed Geese from Iceland and Greenland winter in the British Isles and those from Spitsbergen on eastern North Sea coasts. They only occasionally appear in France, chiefly on Channel coasts during cold spells: 21 French ringing recoveries are all of Spitsbergen birds (OF).

There are at least 59 Iberian records (121 birds) up to 2010. The earliest was an Icelandic-ringed individual recovered in Asturias in November 1953 (AMI), an autumn in which Icelandic birds were also recovered in the Azores and Canary Islands (Sáez-Royuela 1957, Rydzewski 1958), followed soon afterwards by one taken at Doñana in February 1954 (*Ardeola* 1: 121). Two were at Pendueles, Asturias, in January 1955 (Bernis 1964b), and there are three published records of single birds for the 1960s and 1970s (*Ardeola* 12: 229, 15: 122, 22: 111). The Spanish rarities committee accepted 58 records (99 birds) during 1984–2011. There are only two accepted Portuguese records: one at the Tagus estuary on February 2000 and one near Peniche, Leiria, in October–November 2010, and at least two post-2010 observations, both in November 2012 (RBP).

Records have become increasingly frequent with annual sightings since 1996, probably reflecting the massive increase of the western population since the 1950s (BirdLife International 2012a). Most are accompanying Greylags. Extreme dates of accepted records are 23 September and 3 April, most (86%) first appearing during November–February (Fig. 4). Some 40% of records are from Castilla y León, in particular at the Lagunas de Villafáfila, La Nava and Boada, where large numbers of Greylag Geese winter. The Biscay coastlands and Galicia also account for 40% of records (Fig. 5). Most records (80%) are of single birds but there are 11 of 2–4 birds and a remarkable flock of 16 was at Santoña from September 2004 to January 2005 (*Ardeola* 53: 166). Up to 13 adults and seven first-winter birds have been identified.

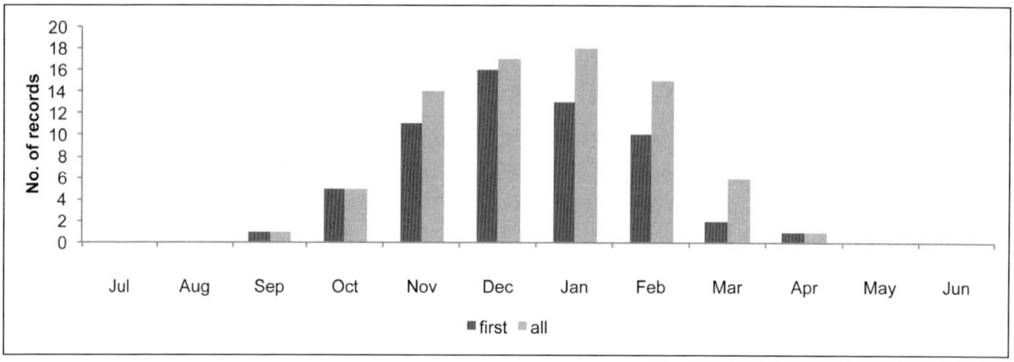

Figure 4. Records by month, 1953–2010 (n = 59)

Figure 5. Records by region, 1953–2010 (n = 59)

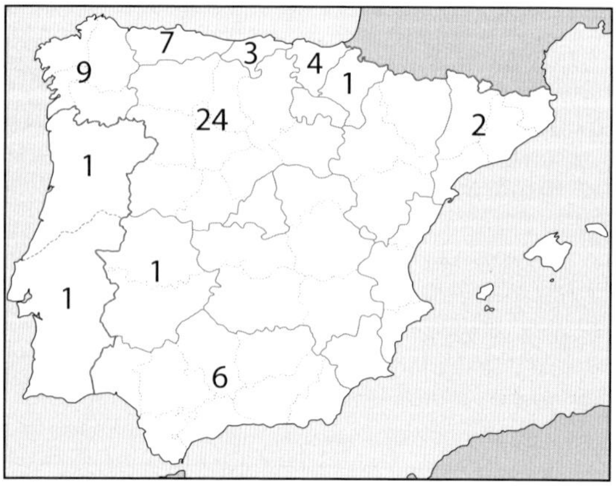

GREATER WHITE-FRONTED GOOSE *Anser albifrons*

(Ánsar Careto, Ganso-grande-de-testa-branca)

***A. a. albifrons*. A rare winter visitor to Spain. Vagrant to Portugal.**
***A. a. flavirostris* (Greenland White-fronted Goose). Vagrant to Spain.**

Over a million birds winter in The Netherlands, Germany and some nearby countries. Very few reach Iberia, however, and they are uncommon in France except during invasions in cold spells. It is a vagrant in North Africa, where there are five records from Morocco.

This goose was reported in the 19th century from Doñana and the Laguna de La Janda, Cádiz: where one was shot on 8 January 1876 (Irby 1895). The only known records during the first half of the 20th century were at the Albufera de Alcudia, Mallorca, in September 1921 (Munn 1925) and in northern Portugal near Porto, in February 1926 (Reis Júnior 1931). There were five further observations at Doñana during the 1960s and 1970s (Hiraldo 1971a, Ree 1973a), together with three in Asturias and one in Cantabria (*Ardeola* 22: 111; Álvarez-Balbuena *et al.* 2000).

The Spanish rarities committee accepted 26 records during 1984–93 (de Juana 2006), since when the nominate subspecies has been removed from the Spanish rarities list. They included the first two known occurrences in the region of the subspecies *flavirostris*: single birds at Villafáfila in March 1990 and December 1991. A third record of *flavirostris* has since been accepted: an adult at the Ría de Villaviciosa from November 2009 to February 2010. During the same period (1984–93) a number of authors described the species as scarce but regular at Doñana (García *et al.* 1989, Llandres & Urdiales 1990, Persson 1995) although most of these records were not submitted. There are five accepted recent records for Portugal, in addition to the historical record from Porto: at the Tagus estuary in November 1994 and December 1998 (2 birds), at the Sado estuary in January 1998 (2), at Figueira da Foz, Coimbra, in March 2010, and at Esposende, Braga, in December 2010.

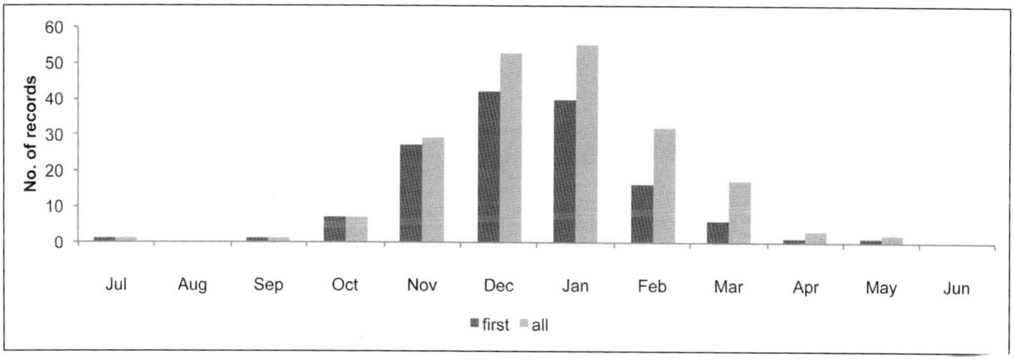

Figure 6. Records by month, 1921–2011 (n = 142)

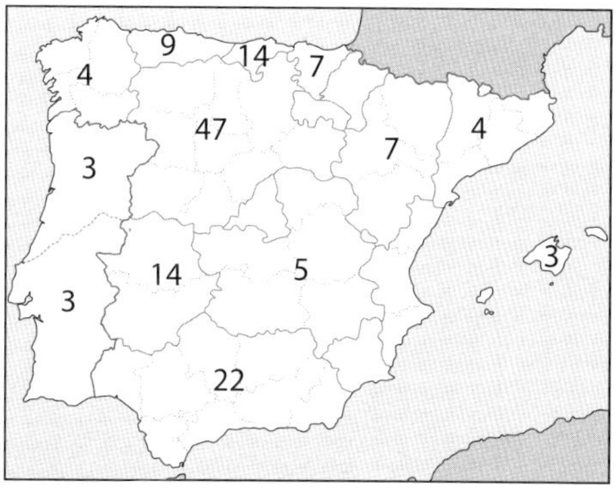

Figure 7. Records by region, 1921–2011 (n = 142)

We have analysed a sample of 142 separate records, involving at least 466 individuals, up to winter 2010–11. There were only ten records during the 1980s but there have been annual occurrences since 1989 and records have become increasingly frequent: there were 16 records (88 birds) during winter 2010–11: when a cold spell produced a large influx into northern France (Paepegaey 2012). The 2012–13 winter saw the largest recorded influx of this species into the northern meseta and western Spain, when numbers at La Nava peaked at 94 in January (*Ardeola* 60: 151). Although 44% of records involve single birds, and 28% are of pairs or trios, quite a few recent observations have been of small flocks. The age of 146 individuals was reported and 62% were adults and 38% first-year birds. Most records have been in winter and 77% of initial observation dates fall in November–January (Fig. 6). With the exception of bizarre and perhaps suspect observations in Catalonia in July and Mallorca in September, the extreme dates are 16 October and 7 May. In common with records of other rare geese, the observations are concentrated in the northern Meseta, Andalucía and Extremadura – where large numbers of Greylag Geese occur– as well as in the northern coastlands (Fig. 7).

LESSER WHITE-FRONTED GOOSE *Anser erythropus*

(Ánsar Chico, Ganso-pequeno)

Vagrant to Spain.

This rare species has always been a vagrant to western Europe, where occasional individuals have appeared in winter. Irby (1895) mentions one taken near Sevilla in February 1878 and an observation of two at Doñana in March 1882. Much later, a confiding adult was captured there in March 1971 (*Ardeola* 16: 254–255). There are five recent accepted Spanish records of unringed birds: in Menorca in March 1997; at the Laguna de La Nava in February 1998, at the Laguna de Boada in December 1998 and at the Lagunas de Villafáfila in January 2002 and again in January–February 2003 (de Juana 2006). In addition, a ringed bird at Villafáfila in November 2001 was of captive Swedish origin (*Ardeola* 50: 154). There are also records of single birds in the La Nava area during winter 2011–12 and at the Ebro reservoir during late October to December 2012. Occurrences are thus in winter and mainly on the northern Meseta.

GREYLAG GOOSE *Anser anser*

(Ánsar Común, Ganso-bravo)

A. a. anser. **Widespread and locally abundant in winter in Spain and Portugal. Some breed in Spain. A scarce winter visitor to Mallorca.**
A. a. rubrirostris. **Vagrant.**

The Iberian peninsula is a major wintering ground for the Greylag Goose. The great majority winter at Doñana, their traditional site (Chapman & Buck 1893, Bernis 1964a), but there are also significant winter concentrations at the Lagunas de Villafáfila, La Nava and Boada in the Duero valley; at the Tagus estuary in Portugal and, increasingly, on the Guadiana floodplain in Extremadura. Much smaller numbers winter, often irregularly, at other wetlands throughout the Peninsula.

Small numbers remain in summer. Some may be injured birds that have been unable to migrate but there are very recent indications that a wild breeding population is becoming established. The 2007 Spanish census of breeding waterfowl (Palomino & Molina 2009) found ten pairs and 53 apparently wild adult individuals, at four sites in western and southwestern Spain: Villafáfila (one pair), the Laguna de La Carrizosa, Ciudad Real (three pairs/12 adults present), the Huéznar reservoir in the Sierra Norte, Sevilla (three pairs/33 adults) and Doñana (three pairs). Wild non-breeders have over-summered at Villafáfila annually since the 1970s but they first bred there in 2006, when two pairs fledged goslings. The 12 apparently wild adults at La Carrizosa in 2007 were accompanied by three broods and a colony of some 30 birds was reported at the same site in 2010 (*Ardeola* 57: 517). Initially unsuccessful breeding by up to three pairs of unknown but probably wild origin has been recorded since 2001 on the Huéznar reservoir, Sevilla (*Ardeola* 54: 396). The three pairs at Doñana may have included some individuals of captive origin (Palomino & Molina 2009). There have since been reports of successful breeding by birds of unknown origin at Daimiel in 2009; on the Llobregat river, Barcelona, in 2010; on the Alqueva reservoir, Badajoz, in 2011, and at the Salburua wetlands, Álava, in spring 2012 (*Ardeola* 57: 517, 58: 481, 59: 413). There are no comparable records of nesting in Portugal (AP), although the greater part of the Alqueva reservoir is in eastern Alentejo.

The importance of the Peninsula as a Greylag winter refuge grew during the late 20th century. The Doñana wintering population was 5,000–10,000 birds during the mid 20th century (AMI). Mörzer Bruyns (1975) mentions counts there of up to 25,000 birds. However, a census in 1989 found 82,000 birds wintering in Spain, including 71,191 birds at Doñana (Troya & Bernués 1990), and a further 867 in Portugal: all but seven at the Tagus estuary (Rufino 1989b). The dominance of Doñana as a wintering site was almost absolute formerly and Bernis (1964a) considered that only a few hundred birds wintered elsewhere. Since then the sites in the central Duero basin and the Tagus estuary have gained importance. At the former, Villafáfila began to attract large numbers in 1979 (Rodriguez & Palacios 1991): the 1989 census found 10,219 individuals there, 12% of the Iberian total.

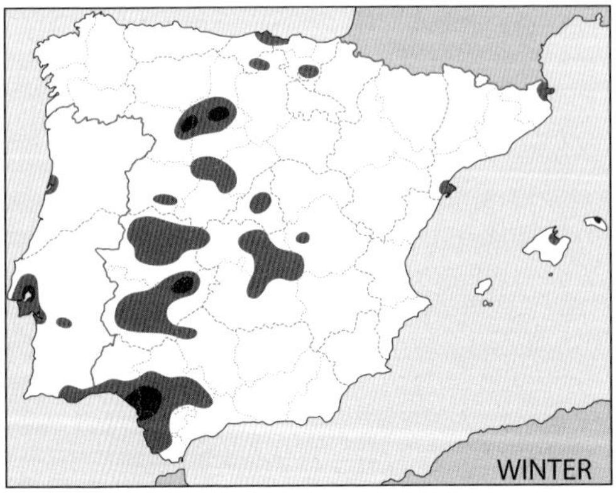

The January wildfowl counts of 1990–2009 confirm that

the wintering population is highly concentrated at just a handful of sites, with Doñana and the lakes of the Duero basin together holding nearly 90% of the birds. The total Spanish wintering population during this period averaged 91,666 birds (range 55,154 birds–145,458; AAE). Wintering numbers seem to have been fairly stable since the 1990s although some decline has been reported at Doñana and, latterly, at Villafáfila, as birds have moved to Extremadura and other sites on the meseta: mainly the recently restored lagoons of La Nava and Boada (Jubete & Martín 2009). The importance of these latter areas is highlighted by recent observations: 22,533 were counted at La Nava on 18 December 2009 (*Ardeola* 57: 216).

Very few Greylags wintered in Extremadura until the 1980s but their numbers have increased dramatically since the 1990s, especially in the Guadiana valley, where the flocks feed over the stubble on harvested ricefields and roost on the reservoirs, notably that of Sierra Brava. Counts at the Vegas Altas ricefields, increased from 1,200 in winter 1996–97, to at least 3,000 in winter 1998–99 and 15,000 in the same region in January 2010 (*A. O. Extremadura 1998*: 139–140, *Ardeola* 57: 216).

Small flocks may also appear at wetlands as far apart as Galicia, the Levant coast and Mallorca in winter but they only amount to a small fraction of the wintering population. Large numbers wintered in southernmost Spain at the Laguna de la Janda, north-west of Tarifa, in the late 19th century and early 20th century (Irby 1895, Verner 1909) until the lake was drained in the mid 1960s. Variable numbers cross the Strait and winter in Morocco, but only a few hundred birds are involved in good years (BM).

The wintering population at the Tagus estuary, chiefly at Punta da Erva, varies from year to year, ranging from a few hundred to several thousand birds. It comprises the great majority of the Portuguese wintering population but there are occasional records elsewhere, for example at the Ria de Aveiro, the Ria Formosa and Castro Marim (AP). These usually involve small groups but gatherings of several hundred have been reported recently from several wetlands in Baixo Alentejo (*Anu. Ornitol.* 6: 47, 7: 73). A count of 53 at Castro Marim on 26 January 2008 (*Anu. Ornitol.* 8: 73) was a high number for Algarve.

The Greylag flocks enter Iberia chiefly at the western end of the Pyrenees, via Navarra and the Basque Country, and cross the Peninsula south-south-westwards. The Duero basin is an important staging area for the majority that winter in Doñana. The Spanish wintering population is fluid and considerable interchange occurs between Doñana and the Duero lakes during the course of the winter, sometimes in response to poor weather (SWA). Greylags arrive in numbers in October and remain until February, some lingering into March (Sáez-Royuela & Santos 1985). Counts of thousands at Doñana are only known between November–February (Rendón *et al.* 2008). The main influx at the Tagus estuary is later, in December (AP). The departure route from Iberia lies somewhat to the east of that used in autumn (Bernis 1964a).

Ringing recoveries, and reports of birds with field-readable marks, show that the birds originate from northern and central Europe: chiefly Denmark, Fennoscandia, Germany and Poland (AMI). A total of 421 Greylag Geese marked in Spain have been reported elsewhere, all of them again from northern and central Europe and most (70%) from The Netherlands and Germany, where many are likely to have been in transit to or from breeding areas further north. Six recoveries from Britain are perhaps exceptional and include neck-collared birds marked in Doñana that were found in May 1995 and during June–August 2004 (SWA).

The wintering flocks forage on agricultural land, including cereal stubble and ricefields, roosting on nearby lakes and reservoirs (Rodriguez & Palacios 1991). Those at Doñana consume large quantities of *Scirpus* sedge rhizomes in the marismas (Amat *et al.* 1991); they are well known for their habit of flying at dawn to the massive sand dunes within the reserve, where they ingest quantities of the sand to serve as grit (Bernis 1964a).

Eastern Greylags *A. a. rubrirostris* have been reported on the Levant coast. One wintered at the Llobregat delta in 1993–94 and six were at the Marjal de Sueca, Valencia, from 28 October to 19 November 1994 (Clavell 2002). Dies *et al.* (1999) considered that many of the Greylags seen at the Albufera de Valencia are of this subspecies but this remains unconfirmed. One of this race was at the Ría de Villaviciosa on 8 November 2005 (*Ardeola* 53: 192). A recent Portuguese record: one at Castro Marim on 19 January 2008 (RBP) has also been assigned to *rubrirostris*.

BAR-HEADED GOOSE *Anser indicus*

(Ánsar Indio, Ganso-de-cabeça-listada)

Recorded in Spain (Category C) and once in Portugal (Category D).

Occurrences in Europe of this Asian species are attributed to escapes and to the increasing feral populations that exist in several countries. Individuals were killed by hunters at Doñana in 1955, and also in 1965 and 1969 (*Ardeola* 2: 186, 11: 145; Castroviejo 1971), and at least six were present there during winter 1985–86 (García *et al.* 1989). Spanish records were considered by the rarities committee during 1994–2005. There is one record from Portugal: two birds at the Tavira saltpans, Faro, in February 2000 (*Anu. Ornitol.* 3: 4). We are aware of 40 records up to 2010, most of them during November–February. Bar-headed Geese are generally observed with Greylag Geese and most occur within the passage and wintering regions of that species. Thus all the observations in Castilla y León (18) have been at the Lagunas de Villafáfila, La Nava and Boada, and all of those in Andalucía (9) are from Doñana. Solitary birds account for over 80% of records but there have been two family groups.

GREATER CANADA GOOSE *Branta canadensis*

(Barnacla Canadiense, Ganso-do-canadá)

The few recorded in Spain (Category C) and Portugal (Category D), generally in winter, are doubtless from the large feral European populations, although escapes from collections may also occur.

There is an early reference to one at Sevilla in 1945 (AMI). In 1970 a Swedish-ringed individual was captured at El Puerto de Santa María, Cádiz; three were seen at the Valdecañas reservoir, Cáceres, and another was at Doñana (*Ardeola* 15: 125, 17–18: 52; Ree 1973a). The Spanish rarities committee considered the species during 1984–2005, when 18 records were accepted. There are also five accepted records for Portugal up to 2010. In all there are at least 31 Iberian records, involving 40 individuals up to 2010. A pair with five juveniles at Sotos de Alfaro, La Rioja, in September 2013 may have nested there (*Ardeola* 60: 507).

Most appear in winter (Fig. 8) which suggests arrivals from elsewhere in Europe. Many observations have been within the regular passage and wintering areas of the Greylag Goose (Fig. 9): in particular, there are five records from the Lagunas de Villafáfila and La Nava and three from Doñana. The longest stay on record was by a bird at Lagoa dos Salgados, Faro, from September 2002 until March 2003 (*Anu. Ornitol.* 3: 5). Most observations have been of single birds but there are six records of pairs or trios.

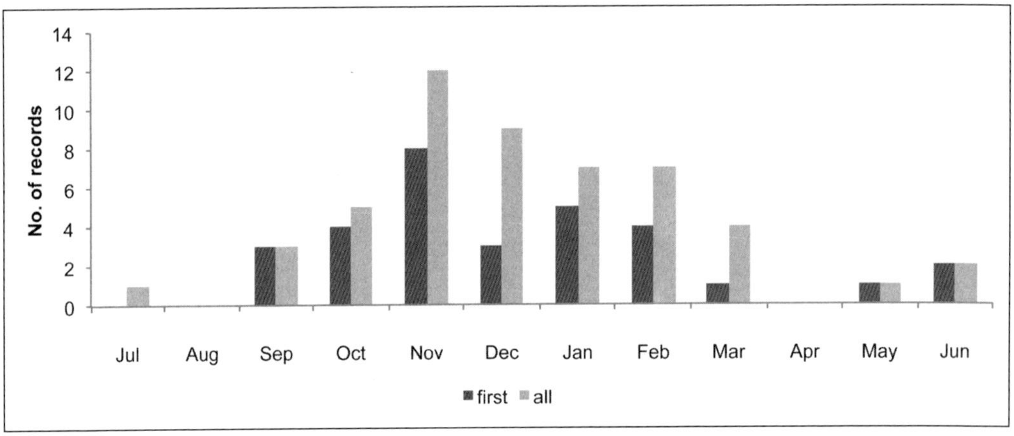

Figure 8. Records by month, 1970–2011 (n = 31)

Figure 9. Records by region, 1970–2011 (n = 31)

BARNACLE GOOSE *Branta leucopsis*

(Barnacla Cariblanca, Ganso-marisco)

A scarce winter visitor to Spain, occasional in Portugal.

Both wild and feral populations winter in northern latitudes. Usually hardly any winter in France: an average of only 37 birds during 1997–2006, although severe cold spells sometimes bring several thousand into the country (OF). It is a rare vagrant to Morocco and to the Azores.

The Barnacle Goose was formerly regarded as a vagrant to Iberia, most likely to appear in severe winters (Bernis 1966; Sáez-Royuela & Santos 1985). A late 19th century record of one obtained near Sevilla was even then regarded as a possible escape (Irby 1895). It was also cited as occurring in Galicia (Iglesias 1927). Further observations or shot individuals were reported from Spain during the mid 20th century: at Valdoviño, A Coruña, in February 1948 (Trigo de Yarto 1960); the Albufera de Valencia in January 1951 (Bernis 1956a); Pineda de la Sierra, Burgos, in December 1952 (*Ardeola* 1: 122); and Doñana in November 1954 (*Ardeola* 2: 181). One shot in Valencia in January 1957 had been ringed as a juvenile in Greenland in August 1955 (Ferrer *et al.* 1986).

Iberian records have become increasingly frequent and there have been annual observations in Spain since 1990–91, but the species has never been considered by the Spanish rarities committee. Only ten records have been accepted by the Portuguese rarities committee, the earliest in 1997. There were a further three Portuguese records during 2011–12. We can consider a sample of 132 records up to winter 2010–11 that involved 371 individuals. Records have become increasingly frequent, reaching 135 individuals (36% of the total) in winter 2010–11, when a cold spell in November also brought exceptional numbers into France (Paepegaey 2012). The increase may be due to the recent establishment of naturalised populations in central Europe, at least in part, but the wild populations are also known to have increased greatly in recent decades (BE 2004).

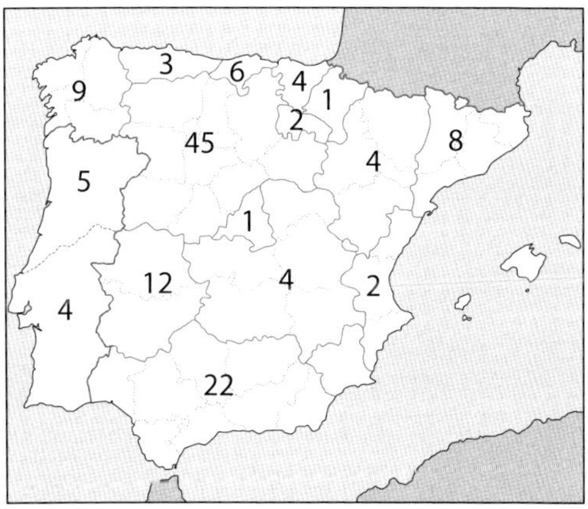

Figure 10. Records by region, 1948–2011 (n = 132)

The geographical spread of the records (Fig. 10) is not restricted to coastal regions, as might perhaps be expected (AMI). Almost half have been inland, 34% of them in Castilla y León, where Villafáfila is the principal favoured site. In this respect, as well as in the relative frequency of reports from Doñana, the pattern of occurrence resembles that of Greylags rather than of Brent Geese. Barnacle Geese occur very largely in winter and 88% of first observation dates lie in November–February, with 58% in December–January (Fig. 11). Occurrences were always in winter in the past but spring and summer records have occurred since 2005; such unseasonal records are also becoming increasingly frequent in France (OF) and seem most likely to involve birds from the naturalised populations.

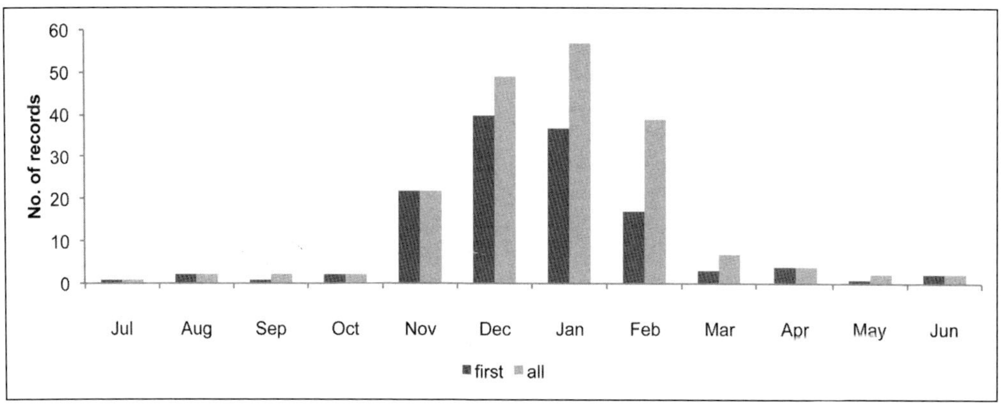

Figure 11. Records by month, 1948–2011 (n = 132)

BRENT GOOSE *Branta bernicla*

(Barnacla Carinegra, Ganso-de-faces-pretas)

***B. b. bernicla*. (Dark-bellied Brent Goose). Scarce and irregular in winter.**

***B. b. hrota*. (Pale-bellied Brent Goose). Very scarce and irregular in winter on Atlantic coasts.**

***B. b. nigricans*. (Black Brant). Vagrant to Spain.**

The European winter quarters are all in the north-west. The nominate race is common on French Atlantic coasts south to the Gironde, but only some 500 *B. b. hrota* occur in France in winter on average (OF), largely in Normandy and Brittany. A few Black Brants reach Europe in winter and 47 individuals had been found in France by 2005 (Frémont *et al.* 2007). Further south the Brent Goose has occurred as a vagrant to Morocco, where 40 individuals have been found including at least one *hrota* (Bergier *et al.* 2013), as well as one each in Mauritania and Senegal (Trotignon *et al.* 1980, Isenmann *et al.* 2010). There are also records from the Canary Islands and Azores. Subspecies other than the nominate are assessed by the Spanish rarities committee. The species is a rarity in Portugal.

The earliest reports of Brent Geese in Iberia all came from northern Portugal during the late 19th and early 20th centuries: at Figueira da Foz, in December 1879; at the Douro estuary in December 1884; at the Ría de Aveiro in February 1896 and at Vila do Conde in January 1918 (Tait 1887, Reis Júnior 1931, Themido 1933). These suggested that a few individuals might occur in some winters along the Biscay and Atlantic coasts of Iberia (AMI), which was amply confirmed in winter 1968–69 when 30 appeared at the Ría del Eo, Asturias; six at the Ría de Ortigueira, A Coruña, and 20 at the Tagus estuary (Fournier & Fournier 1972, Pagezy & Trotignon 1972). That winter was then exceptional since there were very few additional records until the 1990s, despite increasing numbers of observers. In addition to some from the north and west coasts, these pre-1990 records included single observations at Doñana in January 1968 (Hiraldo 1971a), at Gallocanta in October 1976 (Lucientes 1977) and at San Pedro del Pinatar, Murcia, in January 1979 (López-Jurado & Ibáñez 1981).

The 1991–92 winter was again exceptional, with 310 birds at 26 localities: including 36 in Santander bay; 45 at Gijón; 44 at O Grove, Pontevedra, and 73 in the Ría de Aveiro (*Ardeola* 40: 89–90 & 41: 93; AP), including a flock of 11 *hrota* in Cádiz bay (*Ardeola* 41: 105). Brent Geese have appeared annually in Iberia since the 1990–91 winter, although the numbers involved have varied considerably between years. Noteworthy influxes also occurred in 1999–2000 (43 birds, including at least 39 *hrota*, at seven sites); 2005–06 (63 birds at eleven sites); 2009–10 (71 birds at 12 sites, including a flock of 12 juvenile *bernicla* at the Lagoa de Óbidos, Leiria (*Anu. Ornitol.* 8: 9); and 2013–14 (at least 94 at Santoña alone).

Allowing for obvious duplications, we can consider 129 Iberian records up to 2011, involving 677 birds in total. Of these, 29 records (86 birds) were accepted as *hrota*. Most of the remainder, and the great majority, were probably of the nominate race. Both subspecies have been seen together on several occasions, for example at Foz, Lugo, in November 1994 (*Ardeola* 43: 107) and during three consecutive winters at the Lagoa de Óbidos: in November–December 2004, December 2005, and February–April 2006. Recorded age-classes comprised 33 adults and 33 first-winters of *hrota* and 15 adults and 23 first-winters of *bernicla*. Only one was identified as *B. b. nigricans*: a bird at Villafáfila on 4 January 2002 (*Ardeola* 49: 179).

Brent Geese are mainly found at estuaries and coastal lagoons. There are records year-round, but 85% of first sightings are in October–February (Fig. 12). The Biscay and Atlantic coasts, from the Basque Country to Andalucía, together account for 88% of the records and inland sightings are uncommon (Fig. 13). All observations of *B. b. hrota* have been at coastal sites, almost 90% of them in Galicia or Portugal, which probably reflects their westerly origins.

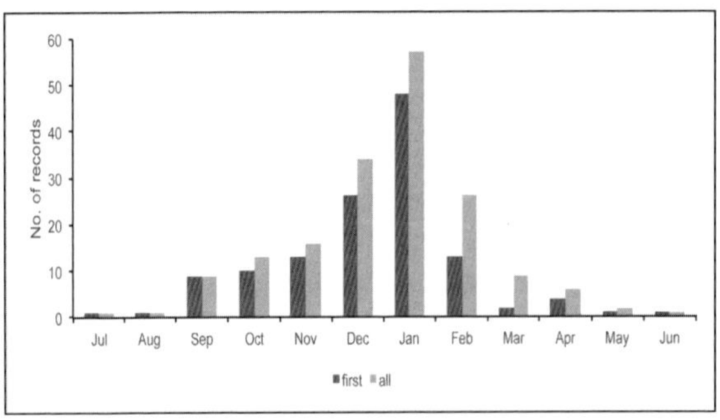

Figure 12. Records by month, 1879–2010 (n = 129)

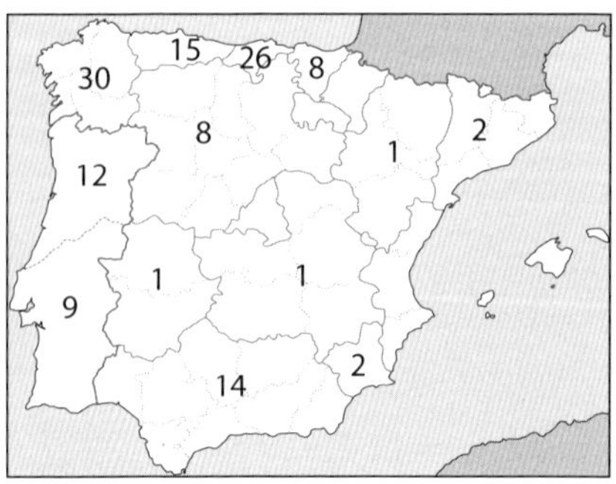

Figure 13. Records by region, 1879–2010 (n = 129)

RED-BREASTED GOOSE *Branta ruficollis*

(Barnacla Cuellirroja, Ganso-de-peito-ruivo)

Vagrant to Spain.

The Red-breasted Goose occurs with some frequency in Europe, notably in Holland (Van den Berg & Bosman 1999). There were 49 records in France up to 2010 and some 77 individuals have been recorded in Britain.

There is an early record from Doñana, where one was shot in October 1969 (Castroviejo 1971). There are also unconfirmed observations from there in January 1984 (Llandres & Urdiales 1990) and in 1984–85 (García *et al.* 1989). Eight records were accepted in Spain during 1988–2007 but there are no known occurrences in Portugal. The nine Spanish records comprised eight single birds and one pair, these often accompanying Greylags. One at the Laguna de Traba, A Coruña, in January 1995 is the only coastal record. The other records comprised five in Castilla y León: four at Villafáfila and one at La Nava; one in Aragón at Gallocanta, and one in Extremadura at the Sierra Brava reservoir. Most have been in winter, as elsewhere in Europe, the extreme dates being 14 November and 3 February, excluding the October bird at the Marismas.

EGYPTIAN GOOSE *Alopochen aegyptiaca*

(Ganso del Nilo, Ganso do Egypto)

Occurs in Spain (Category C) and Portugal (Category E). May be colonising the region.

The Egyptian Goose, now well-established in northwestern Europe, nests in southern Mauritania but there are no accepted records from Morocco and there is no evidence that those seen in Iberia may come from Africa.

Single birds hunted in April 1968 at the Albufera de Valencia (Pechuán 1969) and in winter 1979 at Navahermosa, Toledo (*Ardeola* 13: 244) are the earliest Iberian records. The species was on the Spanish rarities list until 2005 by when there were 29 records. Observations have since become increasingly frequent: there were 16 up to 2000, 21 during 2001–04 and 55 during 2005–08. In addition, excluding family parties, the number of birds involved per observation increased from 1.6 pre-2000 to 2.5 in 2005–08. Records have also become increasingly frequent in Portugal, with over 20 during 2002–10 (Matias 2010, 2011).

There are records year-round (Fig. 14), although winter observations predominated in the early years, suggesting arrivals from elsewhere in Europe rather than recent escapes (de Juana 2006). The observations have a broad geographical spread (Fig. 15) but are relatively numerous in the east and south. Breeding has occurred in Catalonia since 2004, when a pair with goslings was found at Osona (GAE 2003/2005). In 2006 nesting occurred again at Osona and also at Vallès and La Selva (GAE 2006). At least seven pairs nested with some success in Extremadura in 2012 and 2013, most at La Serena reservoir (*Ardeola* 59: 413, 60: 508). In Portugal, a pair with goslings was seen at Évora in April 2007 (*Anu. Ornitol.* 7: 97). Family parties have included 5–9 goslings.

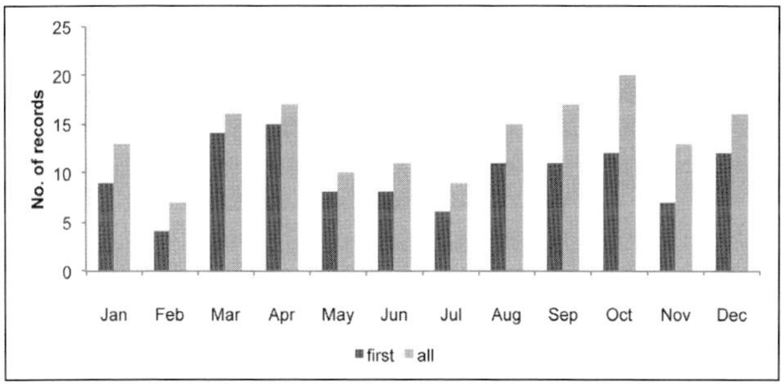

Figure 14. Records by month, 1993–2010 (n = 117).

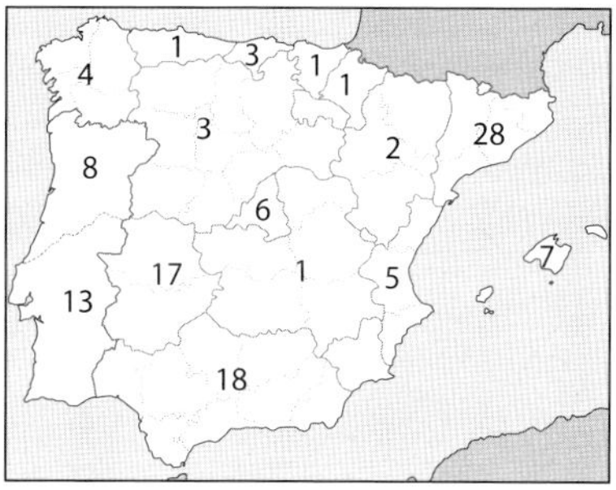

Figure 15. Records by region, 1993–2010 (n = 117).

RUDDY SHELDUCK *Tadorna ferruginea*

(Tarro Canelo, Pato-casarca)

Formerly a regular visitor to Spain from Morocco and also recorded in Portugal. More recent occurrences may mainly involve feral individuals from elsewhere in Europe. Breeding has become increasingly frequent.

The Ruddy Shelduck is an irregular visitor to western Europe, arriving especially in late summer. There are also small feral European populations, probably of captive origin. There is a significant breeding population in the Moroccan uplands, where the winter population is estimated at 2,000 individuals (BM).

It may have nested in southern Spain in the 19th century, at Doñana and Fuentedepiedra (Arévalo 1887, Irby 1895) but there is no certain evidence of this (Valverde 1960). However, it certainly was a regular post-breeding visitor to the Marismas, arriving in August and remaining until February or March. This wintering population almost certainly originated from Morocco but it seldom reached 200 individuals (Valverde 1960), although 500 were counted in 1962 (Hidalgo 1989). It then became progressively rarer and there were only about ten birds each year during 1968–72 (Ree 1973a) and only one or two appeared by the early 1980s (Sáez-Royuela & Santos 1985). The only other known records from Spain until the 1980s were of single birds at wetlands in Valencia: a male taken during September 1954 or 1955 (Bernis 1956a) and one seen in May 1963 (Ferrer *et al.* 1986). In Portugal there are four old specimens in Lisbon museum, three of them obtained in May 1888 (Tait 1924).

The Spanish rarities committee considered the species in 1984–2002, accepting 75 records during this period. Records became annual and increasingly frequent post-1987 (Fig. 16). The Portuguese rarities committee accepted 34 records up to 2010, the earliest in 1990, but retains the species in Category D (Matias *et al.* 2007). We have based our review on 199 records from 1954–2010, excluding birds of obvious captive origin (often colour-ringed).

Occurrences are well-distributed throughout the year but they are relatively few in May–June and tend to increase during the late summer, peaking in September, with a second peak around March. However: July-October observations made up 54% of records pre-2000 but only 28% thereafter. A summer influx has also been noted in France (OF) and in Europe generally (Vinicombe & Harrop 1999) and has been taken to indicate a possible influx of wild birds that are avoiding the summer drought or performing a moult migration. The recent tendency for Iberian observations to occur more uniformly throughout the year may be due to the establishment of a very small feral breeding population in the region.

Ruddy Shelduck observations are also widely spread geographically (Fig. 17) although with a recent trend towards the southwestern quadrant of the peninsula. Thus whereas Andalucía, southern Portugal, Extremadura, Madrid and Castilla-La Mancha together accounted for 38% of records up to 2000, these same regions saw 66% of records thereafter.

There are a number of breeding records from the Madrid region, perhaps initially involving the free-flying birds from the collection at the Casa de Campo zoo. There a pair bred on the Aulencia river in 2004 (*Ardeola* 51: 547) and several pairs have nested since 2002 at a golf course at Villanueva de la Cañada (GAE 2007). Young birds have also been reported at the Ensenada da Ínsua, A Coruña, in August in 1995 and 1996 (*Ardeola* 44: 124 & 46: 133); the Tagus estuary in August 2000 (*Anu. Ornitol.* 1: 11) and Etxabarri-Viña, Álava, in June 2013 (*Ardeola* 60: 508). Single birds or pairs birds comprised 78% of the sampled records and there were some 20 records of flocks of 5–10 individuals that may often have been family groups. Exceptional gatherings have occurred at the Santillana reservoir, Madrid: 41 birds on 8 December 2010 and 40 on 17 December 2011 (*Ardeola* 59: 168).

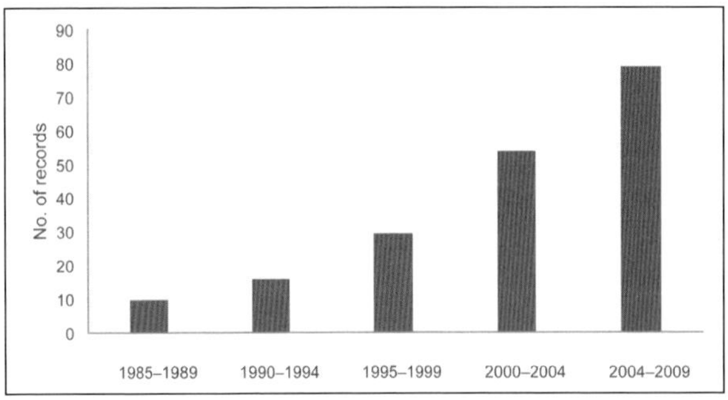

Figure 16. The increasing frequency of records during 1985–2009 (n = 188)

Figure 17. Records by region, 1954–2010 (n = 199).

COMMON SHELDUCK *Tadorna tadorna*

(Tarro Blanco, Tadorna)

A scarce and local breeding resident in Spain and the Balearic Islands. Has recently bred in Portugal. More widespread and numerous in winter.

The Common Shelduck is best known as a winter visitor to Iberia but it has become established as a breeding species relatively recently. The bulk of the breeding population is at east coast wetlands but some have nested inland since at least the 1980s and others nest in the south, with a few pairs also in the Balearic Islands (Table 1). It does not breed in Galicia or the Biscay coastlands, except for a minimal presence at Santoña. Breeding pairs typically inhabit saline lagoons, saltpans, coastal marshes and estuaries but some also occur at reservoirs and freshwater lakes. Nests are sited in rabbit burrows and other cavities in sand dunes, the banks of watercourses and rocky areas, or beneath dense low vegetation.

Perhaps the earliest observation of breeding birds was of a pair with five full-grown young at Doñana in July 1962 (Bernis 1963a) but they could have nested elsewhere. The first unequivocal record was the finding of a dead duckling at the Ebro delta in 1972 (*Ardeola* 20: 336) and breeding has continued there each year since, the delta population rising to some 60 pairs by 2004 (LR), the largest in Iberia. The colonisation follows increased wintering on the Levant coast by juveniles originating from the Camargue (Robledano & Calvo 1989): where the breeding population increased from some 50 pairs in the early 1960s to at least 500 pairs post-2000 (OF).

A second breeding site was discovered at the Mar Menor in 1976 and the species colonised several wetlands of coastal Alicante: the Santa Pola saltpans, Laguna de la Mata and El Hondo reservoir, during 1976–85. Breeding was proved at Punta Entinas, Almería, in 1987. Several pairs have nested at the Albufera de Valencia since 2002 (*A. O. C. Valenciana 2009*: 8–9). The first breeding at a non-coastal site was in 1984 at the Laguna Salada de Chiprana, Aragón. There has since been a breeding presence at Doñana and in the southern Meseta, as well as occasional, sometimes sporadic, nesting elsewhere (Table 1).

Table 1. Establishment and distribution of breeding Common Shelducks. Sources: LR, PA2, ([1] *Ardeola* 51: 547. [2] *Ardeola* 50: 155. [3]PA2. [4]*Ardeola* 58: 482).

Region	Main Sites	Colonised	Population *c.* 2004
Catalonia	Ebro delta	1972	60 pairs
Murcia	Mar Menor	1976	Up to 20 pairs
Comunidad Valenciana	Calpe and Santa Pola saltpans; Lagunas de Torrevieja and La Mata	1976–85	Up to 30 pairs
Balearic Islands	Ibiza saltpans; Salobrar de Campos, Mallorca; Estany Pudent, Formentera[2]	Mid 1980s	2–4 pairs
Aragón	Laguna Salada de Chiprana; Gallocanta	1984	1–8 pairs
Andalucía	Doñana, & Odiel; Fuentedepiedra	1987 in Almería. Doñana from 1998.	Up to 6 pairs in total.
Castilla-La Mancha	Lagunas de la Mancha in Albacete, Toledo, Ciudad Real and Cuenca	1989	17–25 pairs in 2001
Algarve[3]	Castro Marim	2000	2–3 pairs
Cantabria	Santoña	2001	1 pair
Castilla y León	Villafáfila[1]	2004	2 pairs
Extremadura	Valdecañas reservoir[4]	2011	18 pairs. Also bred 2012, 2013

The total population grew from 1–2 pairs in 1972–76 to 65–70 pairs in 1995 and to at least 125–150 pairs by 2004. Further increases are probable. A pair bred successfully for a second time on the north coast at Santoña in 2007 (*Ardeola* 54: 389). Nesting was reported from Menorca in 2007 and two pairs raised 20 young there in 2010 (*Ardeola* 55: 289, 57: 518).

Breeding in Portugal was first proved in 2000, when pairs with young were seen in eastern Algarve: at Castro Marim and the Esteiro de Leziria (PA2). Two to three pairs nested in this region annually during 2000–07 and there have since been indications

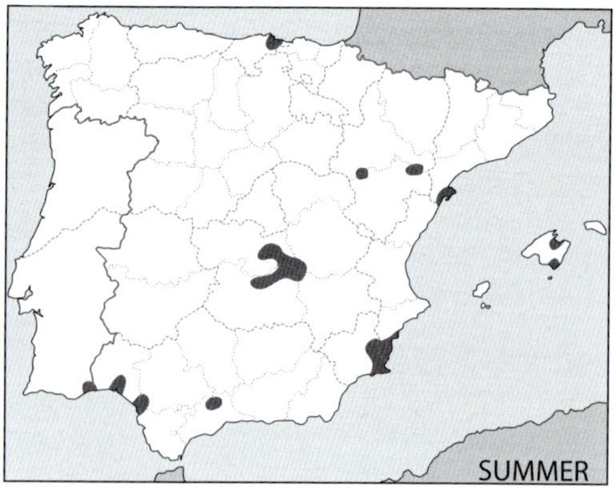

of breeding further west in the region, in the coastal saltmarshes of Tavira, Olhão and Faro. In particular, two females with 23 ducklings were seen at Salgados do Fialho on 23 May 2008 (AP) and 20 adults and 30 juveniles were at Castro Marim on 28 June 2008 (AP). Breeding at the Tagus estuary was confirmed in 2009 at Ponta da Erva, where a pair with ten ducklings was seen (*Anu. Ornitol.* 8: 54).

The breeding populations of the eastern and southern coasts are probably resident but variable numbers of immigrants enter Spain in winter, perhaps chiefly from southern France. The wintering population was estimated at 1,000–4,000 birds, exceptionally up to 6,000, in the 1970s (SA1). The January waterfowl counts during 1990–2009 found 3,905 Shelducks on average (range 1,334– 9,516, AAE).

The principal wintering site is Doñana, where 5,296 were censused in 1989 (AI1), 88% of the Iberian wintering population that year. Wintering birds may be seen there from October to April although large numbers are only present during December–February in some years, with a very marked peak in January (Rendón *et al.* 2008). Here as elsewhere the between-year variation in wintering numbers is very considerable, partly reflecting variation in habitat suitability: the January counts ranged from just two birds to 5,328 during 1990–2009. However, the relative importance of Doñana to wintering Shelducks has declined since the 1970s, as that of the Ebro delta has grown: on average just under 50% of Spanish winterers were at Doñana during 1990–2009 (AAE). Other Andalucian coastal wetlands that attract significant numbers in some winters are the Cádiz bay wetlands (up to 670 birds), the Cerrillos saltpans, Almería (up to 306) and Odiel (up to 175) (AAE).

The only other coastal wintering sites in Spain that regularly hold more than a few birds are the Ebro delta, the Alicante coastal wetlands, the Albufera de Valencia and the Mar Menor, but even here only a few hundred birds in total occurred in most years until the 1990s. Only 120 Shelducks were counted at the Ebro delta in 1980 but the January wintering population there during 1990–2009 has averaged 789 birds, 18.4% of the Spanish wintering population, with a maximum count of 2,142 in 2009 (AAE). Counts of 204 at the Mar Menor saltpans on 17 January 1988 and 389 at El Hondo, Alicante, on 14 December 1996 were then notable but 1,200 were at La Mata saltpans on 7 December 2009 (*Ardeola* 36: 237, 44: 247, 57: 217) and numbers have peaked at 1,728 at El Hondo in January (AAE).

Inland wintering now occurs at some of the lakes of La Mancha and also at Fuentedepiedra, although fewer than 100 birds are present at such sites on average, with 300–400 on occasion (AAE). Smaller numbers have appeared in winter at Gallocanta; Villafáfila; Navalvillar de Pela, Badajoz; and the Sierra Brava reservoir, Cáceres, among other inland locations. Only a few individuals usually winter on northwestern or northern coasts. However, more may appear when there are cold snaps further north in Europe, as noted in Asturias in January 1987, when some numbers appeared at a several sites, with a maximum of 98 at the Ría de Villaviciosa (*Ardeola* 35: 300). There was a peak winter count of 32 at Santoña in 2009 (SWA).

Mainly small numbers occur in the Balearics in winter, fewer than 100 birds on average in January, notably at the Salobrar de Campos, Mallorca (up to 292), and at saltpans in Ibiza and Formentera (up to 110) (AAE).

In Portugal, the Tagus estuary attracts from a few dozen to several hundred birds in winter: a count of 659 here at Benavente, Ribatejo, on 10 February 2007 (*Anu. Ornitol.* 6: 47) is exceptional and a count of 335 at Alcochete, Setúbal, on 25 January 2008 (*Anu. Ornitol.* 7: 73) is also notable. The Castro Marim reserve also holds a regular wintering population of several dozen birds (AP), occasionally more: there were 112 there on 9 January 2007 (*Anu. Ornitol.* 6: 47). The wintering population is generally present from November to February, with stragglers remaining into March. A few individuals may appear at other coastal sites in some years, from the Cávado estuary in the north-west south through Estremadura, Ribatejo and Alentejo to Algarve. Inland records are very rare in Portugal (AP).

In general, wintering numbers have increased very considerably since the mid 20th century, as also noted in France (OF). Small flocks may now appear almost anywhere in winter, including on the meseta and in Extremadura, although the Shelduck remains rare at any season in the north-west. There are 12 Spanish recoveries of birds ringed in France, chiefly in Camargue, and three others from Belgium. Spanish-ringed birds have been recovered in France (2) and The Netherlands. A few thousand winter in Morocco (BM) and many of these presumably overfly Iberia on passage.

MANDARIN DUCK *Aix galericulata*

(Pato Mandarín, Pato-mandarim)

Recorded in Spain (Category C) and Portugal (Category E).

The Mandarin was introduced to England from the 18th century onwards and the increasing population there has been estimated at 7,000 birds in winter (Musgrove *et al.* 2011). Smaller feral populations elsewhere in Europe are also increasing. There were some 30 pairs and 115–160 individuals in France in 2006 (Dubois 2007).

The earliest Spanish records were of individuals hunted at the Albufera de Valencia and at Almenara, Castellón, in 1917 and 1942 respectively (Bernis 1956a). During the 1960s one was taken in Cantabria in November and a hunter killed three males out of a flock of 'ducks' at the Ebro delta in February (*Ardeola* 9: 138, 15: 126–127). Other early records noted in *Ardeola* were in November 1975, January 1981 and April 1982. The Rarities Committee accepted 24 records during 1984–2003. The Mandarin has not been considered by CR(SEO) since 2005 but there are numerous subsequent observations of apparently wild birds. There are three recent records in Portugal: in April 2008, November 2008 and October 2010 (Matias 2010, 2011). Records have become increasingly frequent: there were 44 up to 2003 but 62+ during 2004–08. There have also been several instances of breeding since 2006, from the Río Turia, Valencia; by the Tagus in Toledo, and at Calzadilla, Cáceres (GAE 2006, 2008; *La Garcilla* 127: 28, *Ardeola* 57: 217).

Records are predominantly in winter, 67% of first observations falling between November and January and none during May–July. Two-thirds of observations are also in the north, notably in Catalonia, Galicia and Asturias (Fig. 18). Clavell (2002) also noted that records from Catalonia were especially frequent in the north, in the Pyrenees. It thus seems likely that most of the birds seen in Spain originate from feral populations elsewhere in Europe.

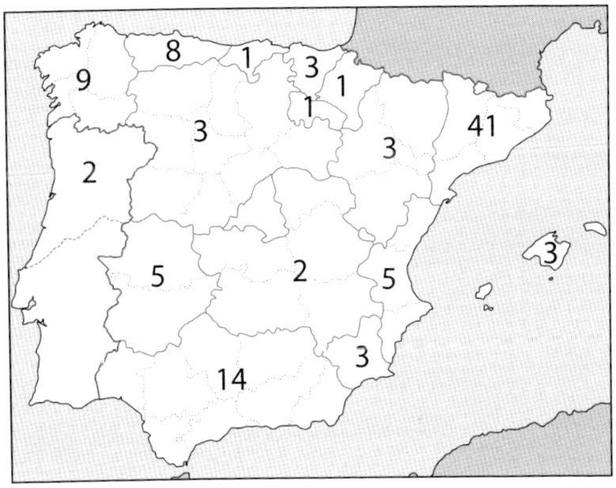

Figure 18. Records by region in Spain, 1942–2008 (n = 104)

Most Spanish records (63%) involve single birds but there are observations of pairs (20%), trios and small groups, as well as single records of a flock of 11 and another of ten birds on reservoirs in Asturias. The occurrence of such flocks hints at a feral origin rather than recent escapes. Records of the showy males are twice as frequent as those of females, no doubt because the latter are much more easily overlooked: the imbalance is much lower where flocks of four or more have been seen: 29 males v. 22 females.

EURASIAN WIGEON *Anas penelope*

(Silbón Europeo, Piadeira-comum)

Common throughout on passage and in winter, locally abundant.

Over 100,000 Wigeon winter in the Iberian peninsula in some years, making this the third most abundant of the wintering duck species. Winterers are mainly present from October to March, but large numbers only occur during November–February, although the earliest arrivals are in September or even August and some linger into the spring (AI1, AP). A few individuals are reported from time to time in summer and some of these are likely to have been birds that were injured by hunters.

Wintering Wigeon are attracted to estuaries and coastal marshes that offer opportunities for grazing nearby as well as for consuming floating vegetation. Inland lakes and reservoirs hold smaller numbers in similar habitats, the birds grazing on the margins and roosting out on open water. Flocks may appear at almost any suitable wetland but most of the population

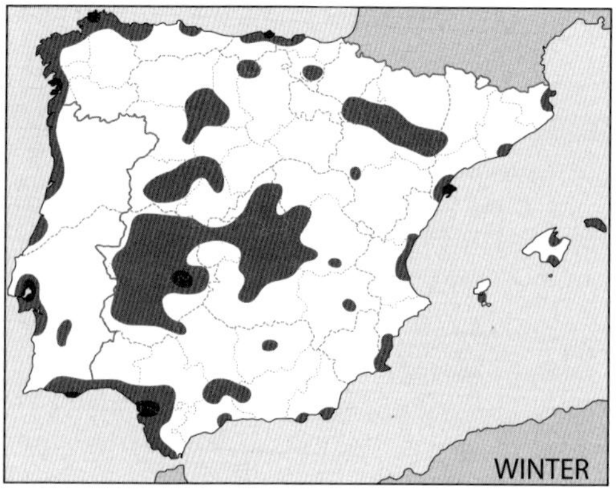

is concentrated at just a few sites. The 1989 census (Rufino 1989b, Troya & Bernues 1990) found 130,000 birds in Iberia but some 100,000 were at Doñana, the next largest concentrations being far smaller: 6,363 birds at the Ebro delta and 3,500 birds at the Marismas de Santoña. The number present in any one year is greatly influenced by cold weather further north, which drives birds south into Iberia, and locally also by the extent of inundation of the Iberian wetlands in a given season. Thus the total Spanish wintering population during 1990–2009 averaged 46,934 birds but ranged from 21,753 to 123,695 individuals (AAE). A few hundred winter in the Balearics, chiefly in Mallorca.

The numbers wintering in the south, at Doñana, have shown a marked recent decline, even though western Andalucía remains the principal Iberian wintering site. Large numbers occur within the Marismas only during November–February, peaking in January-February (Rendón *et al.* 2008). The mean January population at Doñana was around 40,000 birds until 2000 but averaged fewer than 10,000 during 2000–09 (Rendón *et al.* 2008, AAE). Cádiz bay still attracts large numbers: for example 7,600 were counted in February 2005 (*Ardeola* 52: 210) and the maximum January count there during 1990–2009 was 10,545 birds (AAE).The Ebro wintering population has also declined in recent decades: it averaged 3,555 during 1990–2009 (AAE), but the wintering populations elsewhere in eastern Spain and in the north have remained generally stable or have shown some increase. For example, the numbers wintering at Santoña, in the estuarine marshes of the Asón, have tended to increase since counts there began, following the protection of the site. They peaked at 3,500 in 1989, 6,210 in January 2002 , 4,667 in December 2006 and 6,258 in December 2007 (*Ardeola* 53: 193, 55: 132, 56: 153).

The numbers of Wigeon that winter inland have increased markedly since the 1980s, although here too there are marked between-year fluctuations. In particular, the expansion of rice cultivation in the Guadiana valley in Extremadura, and the availability of suitable reservoirs there for roosting, have seen large numbers remaining to winter there. Over 8,000 have been censused in January at the Sierra Brava reservoir and counts of 1,000+ have been made at several other reservoirs in the region recently.

The Portuguese wintering population is concentrated at the Tagus estuary and the Ria Formosa (AP). The mean wintering population in the country was 4,414 birds during 1987–89 (AI1), 11,257 birds during 1992–95 and 10,001 birds during 2005–07 (AP).

There are 33 Iberian recoveries in Iberia of foreign-ringed Wigeon. They comprise birds ringed in The Netherlands (21), England (5), Russia (2: from the Caspian Sea), Kazakhstan (1), Denmark (2), Finland (1) and Iceland (1). Many were ringed as adults on passage but in general these recoveries show that individuals from across a broad sector of the breeding range may appear in the Peninsula in winter. Birds ringed in Spain have been recovered in Russia (7), France (5), Kazakhstan (1), Iceland (1), Lithuania (1), Italy (1) and Turkey (1). A few have travelled prodigious distances either to or from Iberia: one ringed in Kazakhstan flew at least 5,600km to Jaén. Another, ringed in Doñana in November 1970, was recovered 3,200km to the east at Lake Burdur in Turkish Anatolia in January 1972. Thousands cross the Peninsula to winter in Morocco (BM) and small numbers continue as far south as Senegal (Borrow & Demey 2001).

AMERICAN WIGEON *Anas americana*

(Silbón Americano, Piadeira-americana)

Very rare. Recorded in Spain and Portugal, mostly in winter.

There are at least five European recoveries of birds ringed in Canada, and there were 757 European records by the start of the 21st century, this despite the females being easily overlooked (Votier *et al.* 2003). They include at least 57 records from France and eight from Morocco, as well as 18 from the Azores, one from Madeira and 11 from the Canaries.

The earliest records are from Doñana in October 1971 (*Ardeola* 19: 15–16, Ree 1973a) and at the Laguna de Manjavacas, Cuenca, in April 1988 (*Ardeola* 37: 113). By 2010 there were 32 accepted records for Spain and four for Portugal: the earliest there in 1996. They involved 45 individuals, including four records of pairs and three of trios. Males make up a clear majority (76%) of Iberian records, as elsewhere in Europe. At least four more Spanish records and one Portuguese record occurred up to 2012.

The American Wigeon arrives in the British Isles chiefly in autumn but arrivals in Iberia are predominantly in winter (Fig. 19), extreme dates being 26 September and 28 May. The longest documented stay was by three at São Jacinto, Aveiro, from November 1996 to March 1997 (*Pardela* 8: 9, 11: 8). Observations in successive winters at the Ria de Aveiro, and at Ribadeo and the Ría de Foz, Lugo, and the Ensenada da Ínsua, A Coruña, suggest returning individuals. Most are found with Eurasian Wigeons at Atlantic coastal estuaries, particularly in the peninsular north-west (Fig 20). A few additional inland observations include one at the Ebro reservoir, Cantabria, and three at the Lagunas de La Nava and Boada, Palencia.

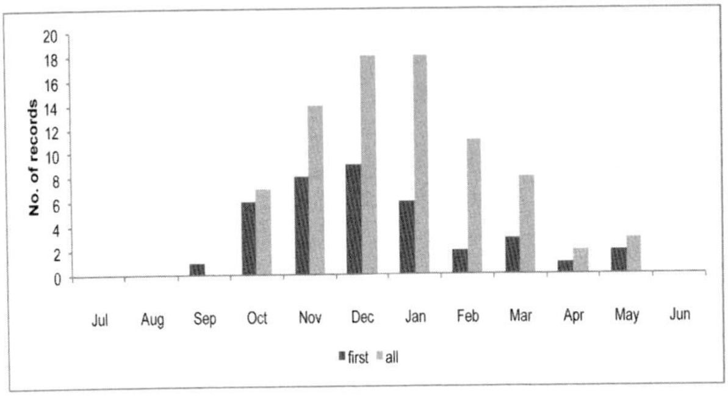

Figure 19. Records by month, 1971–2010 (n = 38)

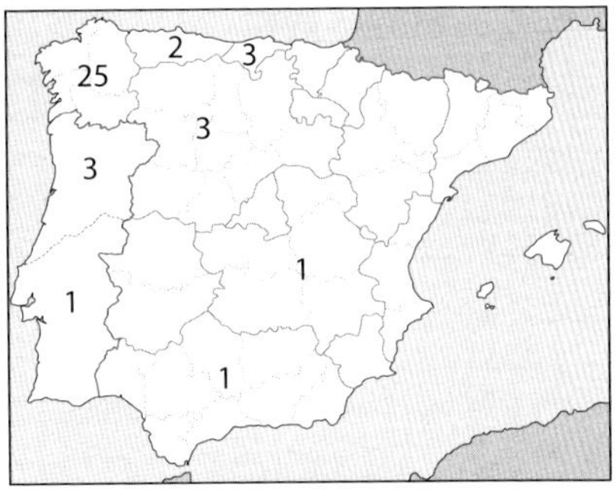

Figure 20. Records by region, 1971–2010 (n = 38)

FALCATED DUCK *Anas falcata*

(Cerceta de Alfanjes, Pato-falcado)

Appearances in Europe are generally attributed to escapes from captivity but arrivals of wild birds cannot be discounted entirely. Some winter as far west as northeastern India and what are likely to have been genuine vagrants have been recorded in Turkey, Jordan and Iraq. European records are predominantly in winter, when birds are often found accompanying Eurasian Wigeons (Vinicombe & Cottridge 1996).

There are three accepted records for Spain, all of single males: at the Ría de Camariñas/Laguna de Traba, A Coruña, from 12 December 1992 to 8 May 1993; at the Ría del Eo, Lugo/Asturias in December 1999, and at the Llobregat delta, Barcelona, from 17 January to 30 March 2004. An eclipse male that was hunted at the Paul do Boquilobo, Santarém, on 25 December 1995 is the sole accepted Portuguese record, assigned to Category D, but there has since been an observation of two males at the Paul do Taipal, Coimbra, on 9 January 2011 (RP). These occurrences, and the six known for France (OF), are clearly in winter, nine of the ten initial observations falling between November and February.

GADWALL *Anas strepera*

(Ánade Friso, Frisada)

A widespread resident in Spain and Portugal. Some breed in the Balearic Islands. Common on passage and most numerous in winter.

The Gadwall is increasingly widespread and common in mainland Iberia but most abundant in the southwestern quadrant. It nests on shallow, clean, freshwater lakes and in marismas, with notable concentrations at Doñana, the wetlands of La Mancha, the Ebro delta and Gallocanta. Many scattered pairs occur widely elsewhere on small pools and other wetlands.

It has bred sporadically at the Albufera in Mallorca since 1993 (Clavell 2002) but now seems well established there: 50–60 pairs nested in 2003 (*Ardeola* 52: 210). It was first found nesting in Menorca in 2009, when two or three broods were recorded at the Albufera des Grau and at least two pairs bred on the island in 2011 (*Ardeola* 57: 518, 59: 414).

Bernis (AMI) considered the Spanish breeding population to be 'minuscule' and earlier references mention only wintering birds or rare instances of nesting, the latter generally at Doñana (Saunders 1871, Arévalo 1887, Chapman & Buck 1893, Irby 1895). Valverde (1960) estimated that 25–30 pairs bred in the Marismas in favourable years. The Gadwall has since increased conspicuously in numbers and breeding distribution in both Spain and Portugal, as indeed elsewhere in western Europe (BWP, Hagemeijer & Blair 1997, OF). The Spanish population was estimated at 650–1,120 pairs during 1975–95 (SA1). In 1998–2002

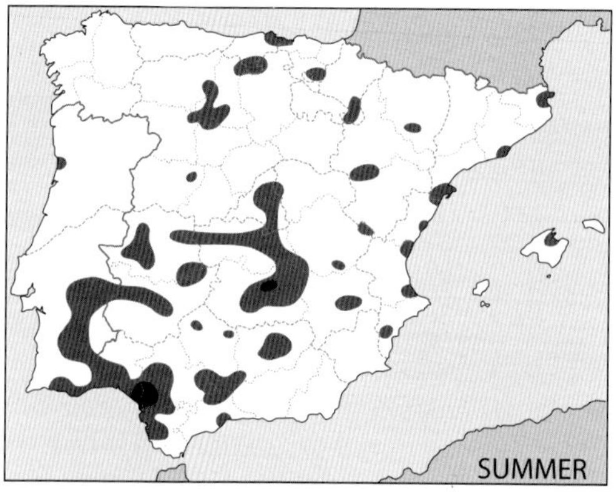

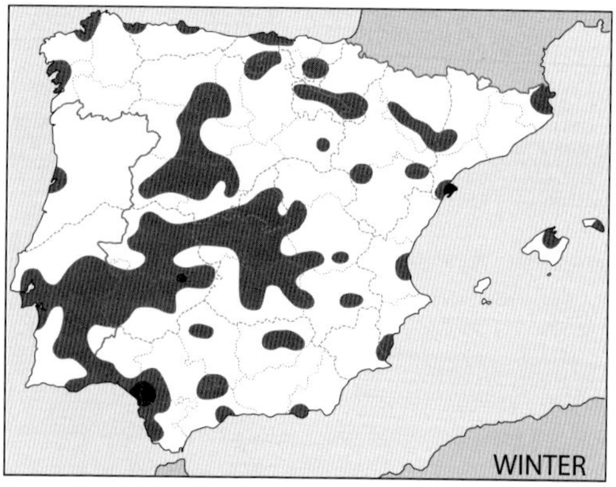

it was assessed at 2,511–3,872 pairs (SA2), although some of the apparent increase may have resulted from more intensive surveying. Breeding numbers may have since stabilised but are prone to fluctuate, especially in the south where water levels in wetlands are periodically affected by droughts. The survey of breeding waterfowl in Spain in 2007 found 3,819 individual Gadwalls and estimated a minimum population of 4,700 birds (Palomino & Molina 2009).

Until the 1980s the Portuguese population was restricted to a few scattered locations in Alentejo. Since then the breeding range has expanded widely throughout central Alentejo and extended to the coastal Algarve wetlands during the mid 1990s: where the Gadwall was first found nesting at Ria Formosa in 1994 (AP). Breeding has also occurred in recent years at Aveiro, in the coastal north-west: at the Barrinha de Esmoriz and in the Reserva Natural das Dunas de São Jacinto (AP). The total population has been estimated at 150–250 pairs (BE 2004) and is continuing to increase.

The breeding population is probably largely resident but some Gadwalls arrive from further north to winter in the region and a minority continue south to North Africa. The wintering birds begin to arrive in September and are chiefly present in October–March (AMI), peaking in the south in December–February (Rendón *et al.* 2008). The same sites that hold the bulk of the nesting population also attract wintering birds but they then frequent a wider range of habitats including estuaries, reservoirs and ricefields. The Iberian wintering population was of the order of 8,000–10,000 birds during the 1980s, reaching 14,000 birds in 1989 (Rufino 1989b, Troya & Bernués 1990). It was probably much smaller previously since the Gadwall ranks very low in the waterfowl hunting statistics compiled by Bernis (1964b). It has shown a further moderate increase since the 1980s. In Spain the mean January population during 1990–2009 was 13,575 birds (range 7,203–24,448 individuals; AAE). In Portugal wintering numbers increased from an average of 304 birds during 1987–89 (AI1) to 696 birds during 1992–95 and 2,498 birds during 2005–07 (AP).

The regional distribution of the wintering population has shown some significant changes since the 1980s. In particular, there has been a significant shift in Spain away from Andalucía and into the Guadiana ricefields, presumably due to the recent expansion of the latter and the availability, since the 1990s, of the adjacent Sierra Brava reservoir, Cáceres, which is used for feeding and roosting by many waterfowl species. Up to 6,000 Gadwalls have been counted there in January and the mean winter count during 1990–2009 of 2,077 birds is exceeded only by those at the Ebro delta (2,709) and Doñana (2,589) (AAE). Some other Extremaduran reservoirs also attract significant numbers of Gadwalls in winter: the Gargáligas, Orellana and Valdecañas reservoirs have all seen recent maximum counts of over 1,000 birds (AAE). Wintering numbers at Mediterranean coastal wetlands have remained fairly stable but clear increases have occurred in Castilla y León, Galicia and the Biscay coastlands. The construction of the Alqueva reservoir is also implicated in the recent increase in the Portuguese wintering population (AP). A gathering of 1,298 individuals at Faro on 19 October 2008 (*Anu. Ornitol.* 7: 73) was exceptionally large for Portugal: at least some seem likely to have been migrants.

At least 46 foreign-ringed Gadwalls have been recovered in Iberia: including birds ringed in Britain, Germany, the Low Countries and the Baltic States, with a few from France, the Czech Republic, Poland and Sweden. There are 61 foreign recoveries of birds ringed in Iberia, 79% from France and most of the remainder from the British Isles and Northern Europe. Two Spanish-ringed birds have been recovered in Morocco. Single juveniles ringed at Doñana were recovered in Croatia and Finland.

BAIKAL TEAL

(Cerceta del Baikal, Marrequinha-formosa)

Anas formosa

Occasional appearances in the wild in Europe have hitherto been attributed to escapes from captivity. However, birds of proved wild origin have been found in Denmark and in England (Fox *et al.* 2007, Votier *et al.* 2009). European records are predominantly in winter, which supports the possibility of wild origins. Two males have been found in Spain at the Albufera de Valencia: on 5 January 1974 and 19 January 1983.

EURASIAN TEAL *Anas crecca*

(Cerceta Común, Marrequinha-comum)

***A. c. crecca*. A very scarce breeding species in Spain. Has bred Portugal. Common and locally abundant throughout the Peninsula on passage and in winter. Some winter in the Balearic Islands.**

The Eurasian Teal has only a tiny breeding population in Iberia, most often on small lakes with dense marginal vegetation, sometimes at up to 1,800m in the mountains (SA2). Regular breeding is only known from three areas in Spain: the Río Louro valley in Galicia; Lago Ercina in the Picos de Europa, and the glacial lakes of the Sierras de Cebollera and Urbión, La Rioja/Soria (SA2). Breeding elsewhere is sporadic and in the south, as at Doñana, it is dependent on years of exceptional rainfall (Valverde 1960). Nonetheless, a considerable range of locations have seen occasional nesting. They include the Lagunas de Louro and de A Frouxeira, A Coruña; the San Andrés and La Furta reservoirs, and the Ría de Villaviciosa, Asturias; the Ebro reservoir, Cantabria/Burgos; Villafáfila; La Nava; the Castrejón reservoir, Toledo; the Ebro delta; the Marjal del Moro and Albufera de Valencia; the lakes and reservoirs of the Sierra Segundera, Zamora; and Gallocanta.

The Spanish breeding population was estimated at 7–68 pairs in 1985–97 and 7–42 pairs in 1998–2001 (SA2). The Río Louro population averaged 19 pairs in the 1980s but declined to 13 pairs in the 1990s and to seven pairs in the early 21st century. The Lago Ercina population was estimated at a dozen pairs but is also thought to have declined. The Sierras de Cebollera and Urbión had ten pairs in 2001. Doñana may support a maximum of 14–20 pairs in wet years. The 2007 census of breeding waterfowl in Spain found 14 pairs; including five at Lago Ercina and four in the glacial lakes of the Iberian Range, but breeding was not proved at any of them (Palomino & Molina 2009). It has very rarely been found nesting in Portugal and the only recent reported instance involves a female with six large young at Herdade da Louçana, Alto Alentejo, on 30 June 2001 (AP).

Iberia is a major wintering area for Eurasian Teal originating from northern and central Europe, with particular influxes occurring in hard winters (AMI). Most ringing recoveries are in January–February (AMI, Sáez-Royuela & Santos 1985). The majority (254) of 502 recoveries of foreign-ringed Teal recovered in Iberia analysed by Bernis (AMI) were ringed as wintering birds in the Camargue in southern France, and many of the 165 recoveries of birds ringed in the Low Countries, or the 49 ringed in England, would similarly have been birds that were hatched elsewhere. However, Iberian recoveries up to 2010 include a few individuals from as far as Iceland (3), Lapland (4) and Russia (4): these last including two from the Astrakhan reserve at the Volga delta on the Caspian.

Table 2. The wintering population of Eurasian Teal in different regions of Spain during 1978–89 and 1990–2009. Mean January counts (AI1, AAE).

Region	Wintering population 1978–89	Wintering population 1990–2009	% change	Key sites
Andalucía	69,232	30,126	-56	Doñana, Cádiz bay
Catalonia	8,020	15,629	+95	Ebro delta, Aiguamolls de l'Empordà
Ebro Basin	4,437	8,087	+82	Lagunas de Gallocanta, Sariñena & Pitillas
Castilla-La Mancha & Madrid	3,547	7,029	+98	La Mancha wetlands
C.Valenciana & Murcia	2,014*	2,481	+23	Mar Menor, Levant coast wetlands
Galicia	1,140	4,076	+258	Rías Altas and Bajas
Castilla y León	1,091	3,508	+222	Villafáfila, La Nava
Extremadura	542	14,020	+2,487	Sierra Brava and other reservoirs
Biscay coastlands	525	3,161	+502	Santoña, Ebro reservoir
Balearic Islands	304	1,343	+342	Albufera de Mallorca

* Includes Almería

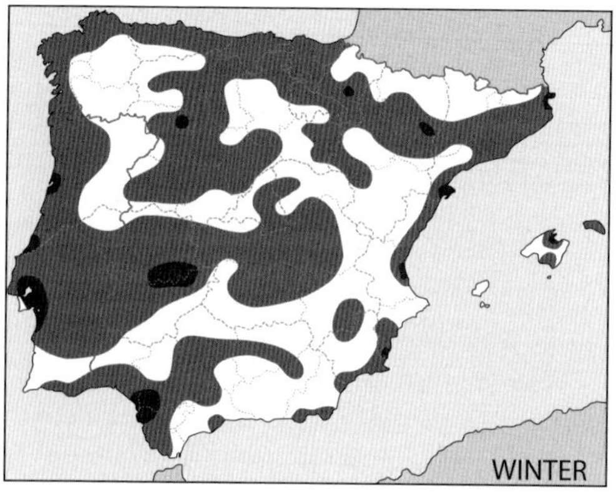

Birds ringed in Iberia to 2010 (n=114) have principally been recovered in France (74%) with the remainder mainly across north-central Europe east to Russia. One ringed in Spain has been found in Tunisia. A recovery of a juvenile male that was ringed at Daimiel in October 2003 and found 5,276km away in May 2005 on the river Obi in Siberia at 66°41'N, 67°59'E, is especially noteworthy.

The numbers that arrive each year vary but have exceeded 200,000 birds: the Iberian wintering population in 1989 was *c.* 210,000 birds (Rufino 1989b, Troya & Bernués 1990). In 1989, and as was then usual, Doñana held most birds, 159,110 individuals. There were also 12,541 at Gallocanta, 8,139 at the Ebro delta and 5,052 at the Tagus estuary. The wintering numbers have been lower more recently. In Spain during 1990–2009 the January counts found 85,548 birds on average (range 31,848–155,704; AAE). The distribution of the Spanish wintering birds has shifted markedly away from Andalucía since the 1980s (Table 2), all other areas showing moderate to large increases in contrast. Such increases were particularly evident in the Guadiana valley in Extremadura during the 1990s, corresponding with the expansion of rice cultivation there, but wintering numbers in this region have since tended to decline: at their peak up to 30,940 birds were counted roosting at the Sierra Brava reservoir. Teal are harder to census than some other ducks due to their tendency to frequent small water bodies, even streams, and to conceal themselves in aquatic vegetation.

The Portuguese wintering population averaged only 1,690 birds during 1976–81, mainly divided between the Paul do Boquilobo and the Tagus estuary (AP). By 1987–89 it averaged 9,790 birds (AI1). It then increased markedly to average 21,413 birds during 1992–95, concentrated at the Tagus and Sado estuaries. Habitat changes associated with fish farming at the latter location have since diminished its importance for this and other wintering waterfowl and the mean Portuguese wintering population during 2005–07 was 13,156 birds (AP).

The migrants arrive in September–November and depart in February–April (Rendón *et al.* 2008, AP). Two flyways have been identified. One enters Iberia along the Atlantic coast and these birds winter predominantly in the north, from Galicia to Navarra, and in the southwestern quadrant of the Peninsula. The other migratory stream is from central Europe and enters to the east of the Pyrenees, these birds wintering chiefly on eastern coasts (Asensio & Carrascal 1992). On average, males arrive earlier than females. Many continue south to North Africa, where thousands winter in Morocco (BM).

GREEN-WINGED TEAL *Anas carolinensis*

(Cerceta Americana, Marrequinha-americana)

Very rare. Recorded in Spain and Portugal, mainly in winter.

Only adult males in breeding plumage are readily identifiable but this is still the most frequent of the Nearctic waterfowl in the Western Palearctic. There were 605 records up to 2003 for Britain and 161 up to 2010 for Ireland. There are at least five accepted records from Morocco, ten from the Canaries, four from the Azores and two from Madeira. The pattern of occurrence suggests regular, if small-scale, migration to the region rather than vagrancy (de Juana & Garcia 2010).

The first Iberian observations were relatively recent: in Spain at Ponteceso, A Coruña, from 22 January to 5 February 1989, and in Portugal at the Sado estuary on 23 December 1990 (*Ardeola* 38: 154, 39: 76). Since then the number of records has increased rapidly, as elsewhere in Europe, so that by 2010 there were 46 accepted records for Spain and 11 for Portugal, excluding two or three hybrids with the Eurasian Teal found in Navarra and Catalonia. There were 17 records during the 1990s and 40 from 2000. Nine separate individuals were identified in winter 2006–07 alone. Several individuals apparently returned to some sites in successive winters, as has also been noted quite often in Britain and in Europe generally. In particular, one bird returned to Cospeito, Lugo for seven winters between 1991 and 1998, and one was at the Aiguamolls de l'Empordà, Girona, for four winters between 2002 and 2006. Most records have been of single birds but there are three of two males together and one of a group three males: at the Miño estuary (Galicia/Portugal) during October–December 2002.

Midwinter records predominate (Fig. 21) and 84% of first observations have been during November–March, with extreme dates of 11 October and 10 May. Green-winged Teals reach Britain and Ireland in autumn and many remain there to winter, whereas spring records predominate further east and north in Europe, suggesting that birds return to North America on a more easterly and northerly route than they follow in autumn. The longest stayers have been single birds at the Laguna de La Furta, Asturias: from October 2002 until April 2003; at the Ensenada da Ínsua, A Coruña, between November 2007 and May 2008, and at the Ría de Villaviciosa from October 2007 to March 2008. As many as 49% of records have been in Galicia and most fall within the northern half of Iberia (Fig. 22). There are no records for Andalucía, where the Green-winged Teal might be expected to occur among the great numbers of Eurasian Teals that winter at Doñana. Very few have been inland.

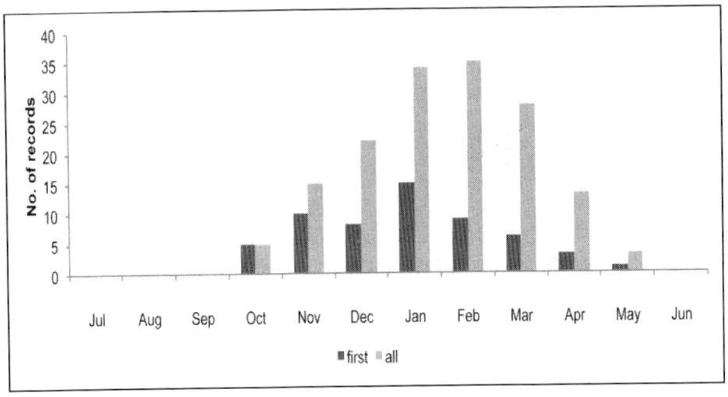

Figure 21. Records by month, 1989–2010 (n = 57)

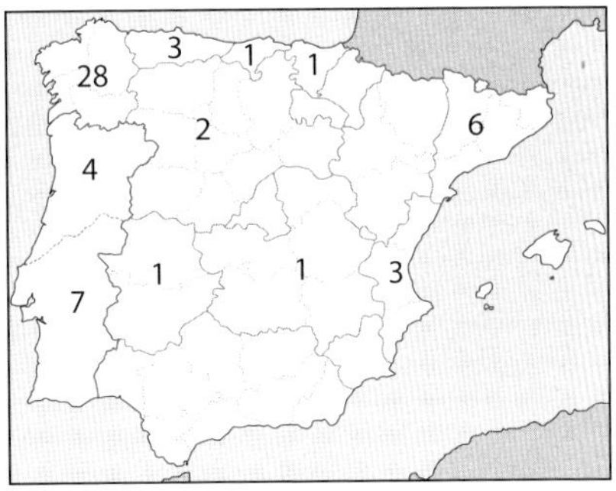

Figure 22. Records by region, 1989–2010 (n = 57)

MALLARD

Anas platyrhynchos

(Ánade Azulón, Pato-real)

***A. p. platyrhynchos*. Widespread and common as a breeding species, on passage and in winter throughout the region.**

The Mallard breeds in nearly all parts of the Peninsula and is only absent from arid sectors of southeastern and eastern Spain. It is also thinly distributed in mountainous regions, notably in northern and central Portugal (PA2) and the higher elevations of the principal mountain systems (SA2), partly because it avoids fast-flowing streams and rivers. It nests in the principal Balearic Islands: Mallorca, Menorca and Ibiza, and was first proved to breed on Formentera in 2005 (*Ardeola* 53: 378).

The Spanish population was estimated at 65,450–100,000 pairs in 1975–95 (SA1) and may be increasing (SACRE 2011). The largest breeding numbers are in Andalucía and in the Duero and Ebro basins (Palomino & Molina 2009) but the population fluctuates substantially from year to year according to rainfall. Populations at the key sites in good years include *c.*15,000 pairs at Doñana, 16,000–20,000 pairs at the Ebro delta, 2,500 pairs in the Comunidad Valenciana and 8,000 pairs in the wetlands of Castilla y León (SA2). The 2007 survey of breeding waterfowl in Spain estimated the Mallard population to number at least 170,000 individuals (Palomino & Molina 2009). Post-breeding concentrations at particular sites, often related to moulting, may be considerable. A count of 6,939 at the Azud de Riolobos, Salamanca, on 18 August 2002 (*Ardeola* 50: 343) is noteworthy for this inland site. The Portuguese population has been estimated at 3,000–10,000 pairs (BE 2004). It showed a moderate increase during 2004–09 (ACP).

The breeding population is augmented in winter by variable numbers of migrants from northern and central Europe, some of which continue south to winter in North Africa. The numbers that arrive are probably relatively low since there are comparatively few ringing recoveries in Iberia, bearing in mind that large numbers are ringed elsewhere in Europe and that

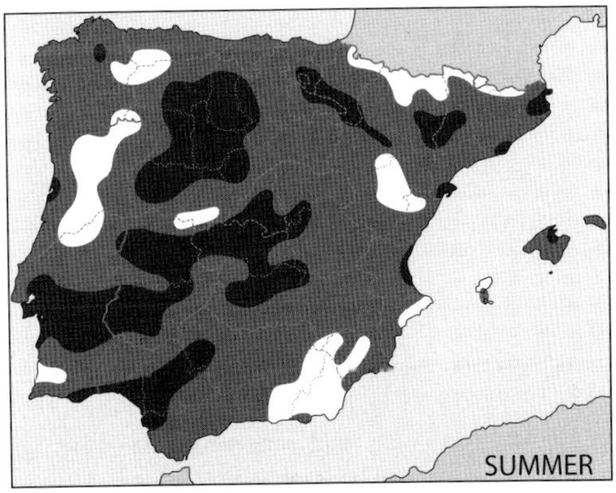

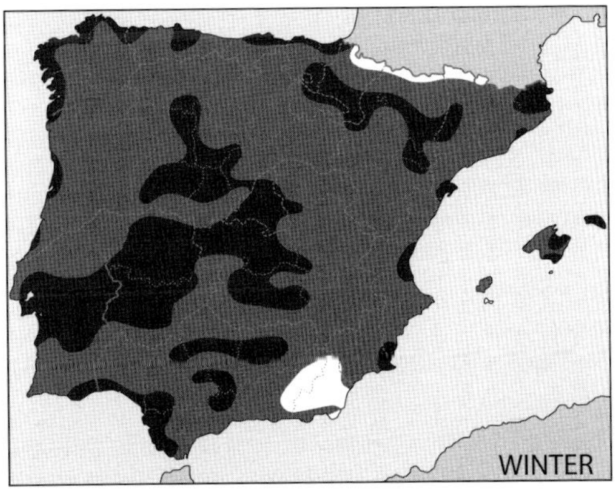

the Mallard is a quarry species. Recoveries include birds ringed in France, the Low Countries, Sweden, Denmark, England and the former Czechoslovakia.

Many of the breeding birds are fairly sedentary. Recoveries of 231 ducklings ringed in Doñana gave a mean dispersal distance of 142km (Sáez-Royuela & Santos 1985). However, 56 Doñana-ringed ducklings have been recovered abroad: most in France (28) and Morocco (10) but some as far afield as the British Isles and Russia, showing that some individuals are highly dispersive. Three female ducklings ringed at Doñana have been recovered several autumns later in Belarus (3,300km); just east of the Urals at Sverdlovsk, Russia (5,100km) and in Kazakhstan (6,100km). An adult male ringed at Doñana in July 1986 was recovered in Ukraine (2,966km) in October 1989. The majority of 126 Spanish-ringed Mallards recovered outside Iberia have been in France/Low Countries (68%) and in North Africa (21%; one in Algeria and 26 in Morocco). The longer movements of some Iberian-bred individuals could involve abmigration resulting from pairing with immigrant individuals in winter. Similarly, 2,792 Mallards marked in Portugal during 1993–99, chiefly at Aveiro, gave mainly fairly local recoveries but 15 were found abroad, ten of them in France and the remainder in Ireland, England, Spain (2) and Russia (AP). Migrants are present between September and April, but mostly during December–February (Sáez-Royuela & Santos 1985).

This is the most abundant and the most widespread of Iberian wintering waterfowl, although Iberian-bred birds are likely to comprise the majority of the wintering population, as noted above. The total Iberian wintering population numbered *c.* 210,000 birds in 1989 (Rufino 1989b, Troya & Bernués 1990), a year that followed a decade of steady increase in wintering numbers. Subsequent further increases have been seen in all areas. Hence during 1990–2009 the mean January population in Spain alone averaged 205,503, but annual counts ranged from 139,247 to 313,823 individuals (AAE). Wintering Mallard are very widely dispersed throughout the region, at wetlands large and small, and hence are difficult to census comprehensively. The principal concentrations in 1989 were of 42,822 birds at the Ebro delta; 15,923 at the Orellana reservoir, Badajoz, and 11,400 at the Tablas de Daimiel (AI1). During 1990–2009 the Ebro delta continued to support the largest wintering population, averaging 42,419 birds, but the next largest concentrations were at the Sierra Brava reservoir, Cáceres, averaging 10,510 birds (AAE). Other principal Spanish wetlands: including Doñana, the Albufera de Valencia and the Aiguamolls de l'Emporda, have seen mean January concentrations of 1,000–9,000 birds but all the above locations together accounted for less than 40% of the wintering population, the remaining birds being scattered between numerous small sites. Reservoirs attract considerable numbers of birds that often use them as roosts. For example, the reservoirs in the Guadiana valley: now especially the Sierra Brava reservoir, serve as daytime roost sites for flocks that feed at night on the ricefields (Pérez Chiscano 1975, AAE). A count of 20,000+ at the Sierra Brava reservoir on 18 December 2005 (*Ardeola* 54: 170) is noteworthy. Mean numbers wintering in the Balearic Islands, chiefly at the Albufera de Mallorca, increased fivefold since 1978–89, to average 3,686 during 1990–2009 (AAE).

The Portuguese wintering population is concentrated at the Tagus and Sado estuaries; the Paul do Taipal, Baixo Mondego; the Reserva Natural das Dunas de São Jacinto, Aveiro; and, latterly, at the Albufeira de Alqueva (AP). The wintering numbers have shown a progressive increase since the 1970s: from a mean of 5,859 birds during 1976–81, to 9,395 during 1992–95 and 17,639 during 2005–07 (AP). This increase, which has been paralleled by other wintering waterfowl in Portugal, has been attributed to reduced hunting pressure as well as more thorough recent censusing.

AMERICAN BLACK DUCK *Anas rubripes*

(Ánade Sombrío, Pato-escuro-americano)

Vagrant to Spain.

A vagrant to Europe, mostly to Iceland and the British Isles. There are records of 38 individuals in Britain and 17 in Ireland. A Canadian bird has been recovered in France and there are another four records there (OF). There are also several records from the Azores and one from the Canary Islands.

There are 12 accepted records from Iberia up to 2011, all of them in Spain, although many may have involved the same individuals returning in successive winters. The first was a male that was at the Ría de Villaviciosa, Asturias, between 11 January 1993 and 28 November 1995. The second, and the only one away from the northwestern coastlands of Galicia and Asturias, was at the Llobregat delta, Barcelona, on 13 November 1996. Subsequently, what seems to have been the same individual was found in northern Galicia for four successive winters between 1996–97 and 1999–2000, initially at the Lagunas de Cospeito, Lugo, but at the Ría de Foz and Ría de Ortigueira during the latter two seasons (de Juana 2006). This same bird may also have been the one found at the Ría de Foz during the 2005–06 winter and in winters thereafter until February 2010. A male was at the Ría de Villaviciosa in September 2006. These records may thus have involved just four individuals. The period of stay is unclear for the earliest record but the remainder have clearly been predominantly in winter, nine of the 14 first dates of observation falling between 16 October and 23 January. There was again a male at Cospeito, from December 2011 to February 2012.

NORTHERN PINTAIL *Anas acuta*

(Ánade Rabudo, Arrábio-comum)

***A. a. acuta*. A very scarce and irregular breeding species in Spain. Has bred Portugal. Common and locally abundant throughout the Peninsula on passage and in winter. A few winter in the Balearic Islands.**

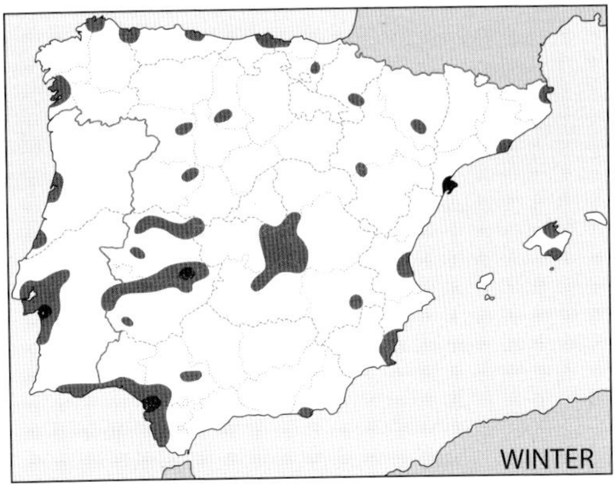

The Pintail is very scarce and irregular as a nesting species in Iberia, where it only nests with any frequency at Doñana, although even there it may not breed at all during drought years (Molina *et al.* 2005). The first historical evidence of breeding in the Marismas is due to Noble (1902). Valverde (1960) considered that it could be quite common there in some years and up to 45 pairs have been claimed to breed during years of good or above-average rainfall (García *et al.* 2000). Elsewhere, wet seasons in Castilla-La Mancha saw 18–26 pairs in 1997 and 9–16 pairs in 1998 but there were none at all in 2001 (SA2) or subsequently. The 2007 survey of breeding waterfowl found only 5–8 pairs in Spain in what was a dry year, with proved nesting only by two pairs at Doñana and three pairs on the Río Moros in Segovia (Palomino & Molina 2009).

Opportunistic breeding has been reported occasionally at a diversity of other sites scattered throughout Spain, including Villafáfila; the Laguna de la Zaida, near Gallocanta, Zaragoza; the Cañada de las Norias, Almería; the marjal de Pego-Oliva, Valencia; the Ullibarri reservoir, Álava; the Charcones de Miguel Esteban and Lagunas de Pedro Muñoz, Toledo; and the Lagunas de Manjavacas and de El Hito, Cuenca (SA2). The only record of breeding in Portugal was the finding of a nest with eggs at the Sado estuary in 1943 (AP). In general, most breeding birds are found at freshwater lakes but the Doñana population nests on the '*vetas*': low-lying islands in the marismas.

The Pintail is far more evident on passage and in winter. Migrants are apparent in September and October, returning in March and April. Large numbers are only present at Doñana during November–February (Rendón *et al.* 2008). Individuals ringed as ducklings in Sweden (5), Finland, Russia and Iceland have been recovered in Iberia (AMI, AI1) and most other recoveries are of birds ringed in northern Europe, chiefly in the Low Countries (62) and Denmark (14), but also in England (10), France (5), Sweden (5), Italy (3), Finland (1), Iceland (1) and Russia (15). The last of these include 11 recoveries of birds ringed as adults at the Volga delta on the Caspian Sea (AMI). Single Pintails ringed in Senegal and Mali have been recovered in Spain. There are 17 extra-regional recoveries of birds ringed in Spain: from France (15), Finland and Estonia. Many of those that winter in northern and sub-Saharan Africa probably cross the Peninsula during their journeys.

Doñana is the chief Iberian wintering site. Chapman (1888) considered Pintails the most abundant of the very numerous wintering dabbling ducks there. In 1989 some 25,000 wintered in Iberia and 17,600 were at Doñana (Rufino 1989b, Troya & Bernués 1990). Other major concentrations that year were 4,220 at the Albufera de Valencia and 996 at the Ría del Eo (Lugo/Asturias). The January populations in Spain during 1990–2009 averaged 25,678 birds (range 6,460–48,441; AAE). However, Pintails are relatively scarce in comparison with some other dabbling duck species and the majority of the wintering population congregates at just a few sites, with small numbers scattered elsewhere. Just four locations together held over 75% of the Spanish wintering birds during 1990–2009: Doñana (60%); the Ebro delta; and the Sierra Brava and Gargáligas reservoirs, Extremadura. However, Pintails may congregate at new locations in some years and may abandon others where they are sometimes common. For example, 7,000 were at the Charca de La Copa, Logrosán, Cáceres, on 15 January 2008 (*Ardeola* 56: 346). Overall, the Spanish wintering population has shown a moderate increase since the 1970s although with marked interannual fluctuations at particular sites (AAE). The largest increases have been in Extremadura, where the expansion of rice cultivation and reservoirs has proved attractive to Pintails: the Sierra Brava reservoir held a remarkable 13,250 birds on 14 January 2012 (*Ardeola* 59: 169).

The Portuguese wintering population increased from a mean of 859 birds during 1976–81 to 2,678 during 1992–95 and 8,070 during 2005–07 (AP). The main wintering site in the 1970s and 1980s was the Paul do Boquilobo but since then most have wintered at the Sado estuary (AP).

GARGANEY

(Cerceta Carretona, Marreco)

Anas querquedula

A very scarce and often sporadic breeding species in Spain and Portugal. Frequent on passage and some remain in winter.

The Garganey is a very scarce breeding species in Iberia and only nests at all regularly at two locations: Doñana and the Aiguamolls de l'Empordà, and then only in years when rainfall has been favourable. Up to 30 pairs may breed at Doñana in good years and the Aiguamolls population ranged from zero to 12 pairs during 1984–2002 (LR). The 2007 census of breeding waterfowl in Spain coincided with dry conditions in the south. Only 5–8 pairs of Garganey were found nesting in Spain in 2007: one at Las Tablas de Daimiel, two at the Ebro delta, one at the Laguna de La Nava and up to three pairs at the Albufera de Mallorca (Palomino & Molina 2009). The Mallorca record was the first recorded instance of nesting in the Balearic Islands; another pair nested there in 2008 (*Ardeola* 55: 289, 56: 346).

Single pairs, or groups of 2–3 pairs, may nest opportunistically at wetlands scattered throughout the Peninsula. Typically they nest for one or two years and then disappear. In Spain such localities have included Gallocanta; the La Mancha wetlands, including Daimiel; Villafáfila; the Lagunas de Villadangos and Chozas, León; the Laguna de La Nava; the Ebro and Llobregat deltas; the central levant coastal wetlands, especially El Marjal del Moro, Valencia; the Lagunas de Campillos, Málaga, and the Charcas de Salburua, Álava. Nesting attempts in Portugal are rare and irregular and most records date from the mid 20th century at locations in Ribatejo (AP). A recent instance of confirmed breeding in Portugal is provided by female with nine ducklings seen at Elvas, Alto Alentejo, in late May 2007 (*Anu. Ornitol.* 7: 125).

Overall the Garganey may have declined considerably as a breeding species in Iberia but there is no evidence that it has ever been numerous there. A claim that it was an abundant breeder at Daimiel in the late 19th century (LR) is unsubstantiated. Valverde (1960) considered that no more than 100 pairs ever nested at Doñana even in years of exceptional water levels and only rare references to a breeding season presence anywhere in Spain are made by 19th century authors: in particular Irby (1875) was told by Lilford that he saw a pair in Doñana in May 1872. Nesting Garganey favour shallow freshwater wetlands with plenty of floating and submerged vegetation and expanses of open water. The few wintering birds also occur in similar habitats, as well as in coastal lagoons and saltpans.

The Garganey is largely a summer visitor to Europe, wintering in great numbers in sub-Saharan Africa, notably in the Senegal and Niger deltas (Borrow & Demey 2002, Zwarts *et al.* 2009). The first migrants appear in southern Spain in mid February but most pass in March–April (Fig. 23). Tait (1877) saw them on sale at Porto in February and March. Migrants then often appear at wetlands throughout the Peninsula in small groups and flocks of up to 30 but passage is concentrated along Mediterranean coasts. Some substantial gatherings have been recorded at this season, especially in the south: for example, in 2008 1,000 were seen from Tarifa beach on 14 March and 450 flew north-west there on 15 March, with 900 seen in flight over the Strait of Gibraltar also on 15 March (*Ardeola* 55: 289). Several thousand were reported on passage at Doñana on 9 March 2012 (*Ardeola* 59: 169). The return passage is relatively inconspicuous everywhere (Ferrer *et al.* 1986). It extends from late July to early October.

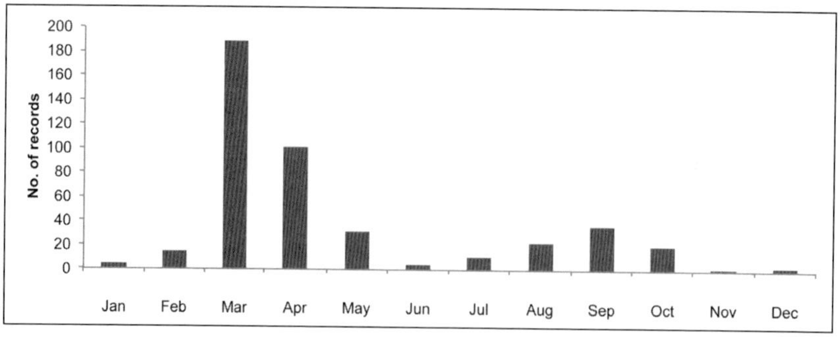

Figure 23. Monthly distribution of Garganeys observed in Spain in the north-west (Galicia & Asturias, 1992–2004, n = 216 records) and centre-west (Madrid, Toledo, Ciudad Real & Extremadura, 1991–2008, n = 226 records).

Garganey observations in northwestern and west-central Spain (Fig. 23) show a marked peak in observations during the spring passage period, a much less prominent peak in autumn, and very few records at other times. A similar distribution applies elsewhere in Iberia. All of the few winter occurrences in this sample were in the north-west, perhaps because the Atlantic coastlands are milder than the meseta at this time, but there are too few observations to allow firm conclusions to be drawn. Elsewhere the wintering population is small and largely comprises scattered individuals at diverse locations, chiefly in the south. The numbers involved vary from year to year: in 1989 only five individuals were found in winter in the whole of Spain, all of them at a reservoir in Murcia (AI1). However, up to 100 winter at Doñana in some years (García *et al.* 2000): recent January counts there included 54 in 2002 and 34 in 2007 (SWA). A count of 30 at a pool at the Badina de Escudera, Navarra, in January 2010 is noteworthy (SWA). Some January records possibly involve early returning migrants rather than over-winterers.

There are some 23 recoveries of Garganeys ringed outside Iberia: four from southern France, 15 from the Low Countries and single birds from England, the Czech Republic, Belarus and Russia. There are also at least three recoveries of birds ringed in Spain: from France, Italy and Dagestan, Russia.

BLUE-WINGED TEAL *Anas discors*

(Cerceta Aliazul, Marreca-d'asa-azul)

Very rare. Recorded in Spain and Portugal.

The numerous records from the Western Palearctic, including at least 16 ringing recoveries, suggest regular, if small-scale, migration to the region rather than vagrancy (de Juana & Garcia 2010). There are observations of some 240 individuals in Great Britain, 94 in Ireland, 66 in France and 33 in Morocco, as well as a considerable number from the Azores, the Canaries and Senegal. The temporal pattern of occurrence suggests that autumn arrivals are mainly in the British Isles, with the spring return further east on average, through such countries as The Netherlands, Denmark and Sweden, where most appear in April.

The earliest Iberian records were in Spain where birds were taken at the Laguna de Medina, Cádiz, in December 1964, and at Cospeito, Lugo, in July 1969 (*Ardeola* 22: 108). Three more were captured during the 1970s; at the Ebro Delta in February 1971 (Maluquer 1973) and at Doñana in February and December 1972 (*Ardeola* 50: 132, Amat 1979a). The same period also saw five recoveries of birds ringed in Canada; four in Spain: in 'Alicante' in September 1969, at Santa Pola in October 1971, at the Ebro delta in January 1974 and at La Jonquera, Girona in October 1977, and one in Portugal: at Braganza in January 1974 (Dennis 1981). Subsequently, up to 2010, there have been 33 accepted records for mainland Spain and 11 for mainland Portugal. In total there are 52 Iberian records, involving 49 single birds and three pairs. The known-sex individuals comprised 37 males and 12 females, the latter perhaps being overlooked sometimes since they are comparatively inconspicuous. Record frequency has declined recently, particularly in view of the increase in observer effort: there were 15 during 1998–2007 but 21 during the previous decade. This decline has also been noted in France and in Europe generally (de Juana 2006).

The pattern of Iberian occurrences (Fig. 24) differs from those of the Green-winged Teal, American Wigeon and Ring-necked Duck, for which winter records predominate. There are records in every month and although quite a few are in winter there are even more in spring (Fig. 24). Thus there are 18 records in which the first dates were in March–April, but 12 in which they were first noted in December–January. There may also be a lesser peak in occurrences in autumn, in September–October. These observations, and the fact that most Blue-winged Teals have been short-stayers, suggests that many of those seen in Iberia are on passage to or from winter quarters further south, perhaps in the Sahel (de Juana & Garcia 2010). The geographical distribution of the records (Fig. 25) also differs from that of the Nearctic species mentioned. There are relatively few in the north: only 20% have been in Galicia and the Biscay coastlands. Instead there are more in the Levant and Balearic Islands, and even more in the south, in Andalucía and southern Portugal.

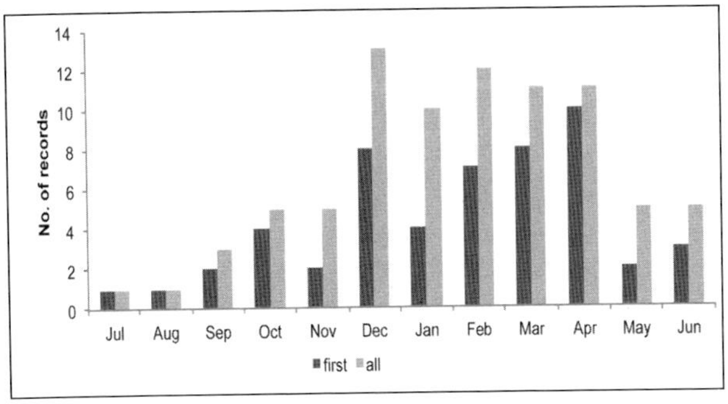

Figure 24. Records by month, 1964–2010 (n = 52)

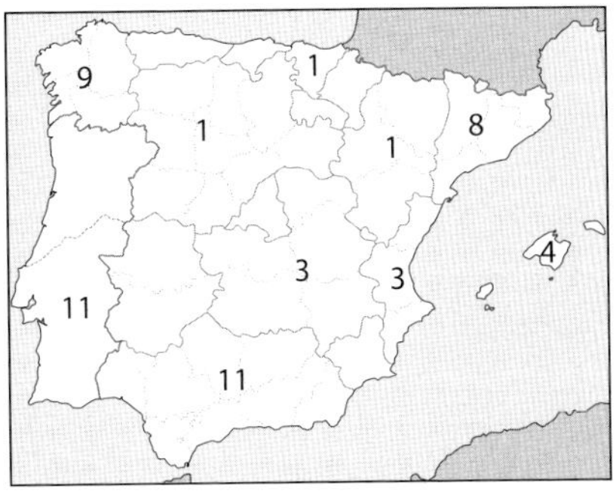

Figure 25. Records by region, 1964–2010 (n = 52)

NORTHERN SHOVELER

Anas clypeata

(Cuchara Común, Pato-colhereiro-comum)

A very scarce, local but gradually increasing breeding species in Spain, Portugal and the Balearic Islands. Common and widespread on passage and in winter, when locally abundant.

The small Iberian breeding population fluctuates considerably from year to year according to the degree of inundation of wetlands. It has probably increased. Breeding Shovelers frequent shallow waters with plenty of emergent and submerged vegetation, including ponds, lakes, gravel pits, marshes and riparian backwaters. Some nest on islands in reservoirs. They range more widely in winter when they also appear in saltpans and coastal lagoons and many congregate to feed in ricefields. There are no breeding records in the historical literature and Bernis (AMI) referred to just sporadic cases of breeding in scattered localities. Valverde (1960) considered that no more than 6–10 pairs bred at Doñana in the most favourable years. The Spanish population was estimated at 30–50 pairs in 1997 (SA1) and at a minimum of 215 pairs in 1998–2003 (SA2), when an increase in the number and geographical spread of occupied sites was also noted.

The 2007 Spanish breeding waterbirds census (Palomino & Molina 2009) found 729 birds at the sampled wetlands and estimated a breeding season population of up to 3,600 individuals, most of them probably non-breeders. Most were thinly and widely dispersed within the principal river basins. The remainder were scattered throughout mainland Spain, where the Shoveler breeds or has bred recently at many major and some minor wetlands including the lagoons of Cádiz and Jaén; La Albuera pools, and the Valuengo and Orellana reservoirs, Badajoz; the Valdecañas and Tozo reservoirs, Cáceres; Daimiel and other wetlands of La Mancha; the Ebro and Llobregat deltas; the Aigüamolls de l'Empordà; the Albufera de Valencia; El Hondo and the Santa Pola saltpans, Alicante; the Cañada de Las Norias, Albufera de Adra, charcones de Punta Entinas and the Saladar de los Canos, Almería; the gravel-pits and watercourses of Madrid; the Ullibarri reservoir and Salburua ponds, Álava; the Balsa de Villamañán, León; the Ría de Villaviciosa, Asturias; the Pitillas lagoon, Navarra; Villafáfila; La Nava; and Gallocanta (SA2, LR). Water levels were low in the south in 2007 but there have been up to 50 pairs at Doñana in good years (García *et al.* 2000).

A few breed at the Albufera de Mallorca in the Balearics, where nesting was first recorded in 2001 and also occurred in 2002, 2003, and 2008–10 (*Ardeola* 50: 155, 52: 210, 58: 483). Very few do so in Portugal, where occasional pairs are reported in some years, chiefly in Alentejo and Algarve (AP): notably at the Lagoa dos Salgados, where annual nesting by up to four pairs was reported during 2001–07 (SA2, PA2): a female with ducklings was at this site on 3 June 2009 (*Anu. Ornitol.* 8: 54).

Iberian breeders may be resident but migrants begin to appear in August, with peak arrivals in October–November. They depart in March–April, with stragglers in May. The species is ubiquitous on passage and in winter, when most suitable wetlands attract at least a few. The Iberian wintering population is very large: the Shoveler is then second in abundance only to the Mallard. In 1989 it comprised *c.*145,000 birds: including 81,510 at Doñana, 20,296 in the Albufera de Valencia and 13,660 in the Ebro delta (Troya & Bernués 1990). During 1990–2009 the Spanish wintering population averaged 109,484 birds (range 55,377–162,994; AAE). The wintering population may have been much smaller during the first half of the 20th century since the Shoveler occupied a relatively low position in the ranked waterfowl hunting statistics of the period (Bernis 1964b).

Wintering Shovelers are very widely dispersed but the largest concentrations prior to the 1990s were in the south and east. Since 1990 Extremadura has also become extremely important as a wintering region, and the numbers that congregate there at the Sierra Brava reservoir are second only to those seen at Doñana, a far more extensive location (Table 3). The Levant coastal wetlands retain their importance for wintering Shovelers but gatherings of up to several thousand may be observed at many locations in the southern half of Spain especially. Considerable numbers may gather on occasion at quite small lakes: for

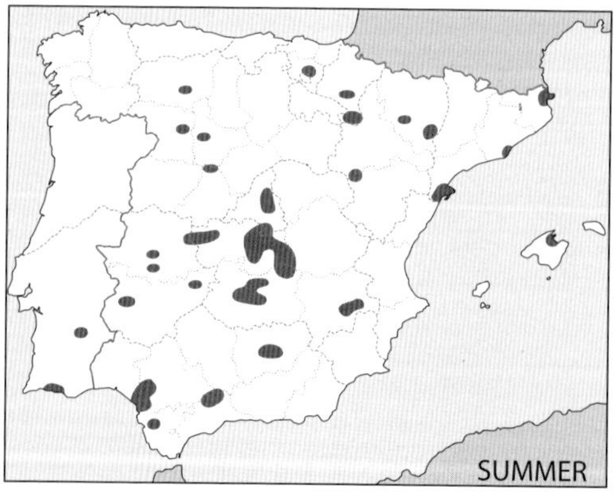

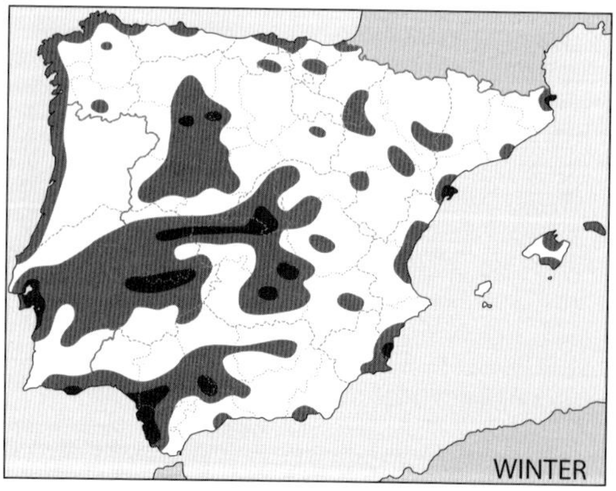

example there were 1,192 at the Azud de Riolobos, Salamanca on 19 March 2002 and 4,000 on the lake at the Dehesa de Abajo, Sevilla, on 27 December 2007 (*Ardeola* 50: 344, 55: 133). The wintering numbers everywhere fluctuate considerably between years but the long-term trend shows an increase in the Spanish wintering population.

Table 3. January counts of Shovelers at their five principal wintering sites in Spain during 1990–2009 (AAE).

Site	January wintering numbers			% of national mean
	Minimum	Maximum	Mean	
Doñana	4,685	105,550	44,713	31.6
Sierra Brava reservoir, Cáceres	145	45,750	15,551	11.0
Ebro delta	2,103	22,135	12,122	8.6
Albufera de Valencia	380	28,732	7,914	5.6
El Hondo, Alicante	1,000	13,914	6,592	4.7

The numbers wintering in the Balearic Islands were formerly small but have increased markedly in recent years, especially at the Albufera de Mallorca, where January counts during 1990–2009 have ranged from 300 to 2,675 individuals.

The numbers wintering in Portugal have remained broadly stable, with interannual fluctuations. The main wintering concentrations are at the Tagus/Sado estuaries, the Ria Formosa, Faro, and the Paul do Taipal, Coimbra, at each of which 1,000–4,000 may occur. The mean Portuguese wintering population was 4,436 birds during 1987–89 (AI1), 12,096 birds during 1992–95: with a maximum of 15,737 in January 1995, and 8,109 birds during 2005–07 (AP). The numbers that winter in Morocco have increased in recent decades (BM) and many thousands of Shovelers cross Iberia to winter there.

Some 224 recoveries of birds of extra-regional origin are principally of birds ringed in northern and central Europe, in the Low Countries (58%); Scandinavia, Finland and the Baltic States (24%), Britain (11%) and France/Germany (9%). However, 81 extra-regional recoveries of Iberian-ringed birds have been mainly from France (48%) and Russia (28%) and they include three from Italy and one from Morocco. At least half of the 23 Russian recoveries have been well to the east: they include that of an adult female ringed at Doñana in January 1971 that was recovered 5,500km away in asiatic Russia in August 1972 and another adult female also from Doñana that was ringed in January 1965 and recovered in the same region the following August.

MARBLED DUCK — *Marmaronetta angustirostris*

(Cerceta Pardilla, Pardilheira)

A very scarce and local breeding species in southern and eastern Spain. Partial migrant. Has bred in Mallorca. Very rare in Portugal.

The once numerous Marbled Duck population in Spain is at a low ebb and the species is classed as 'Critically endangered' there (LR, Ballesteros *et al.* 2008). The two principal breeding areas are Doñana and the Levant coast. The numbers present in each of these regions fluctuate considerably from year to year but are small. The total breeding population is 30–200 pairs.

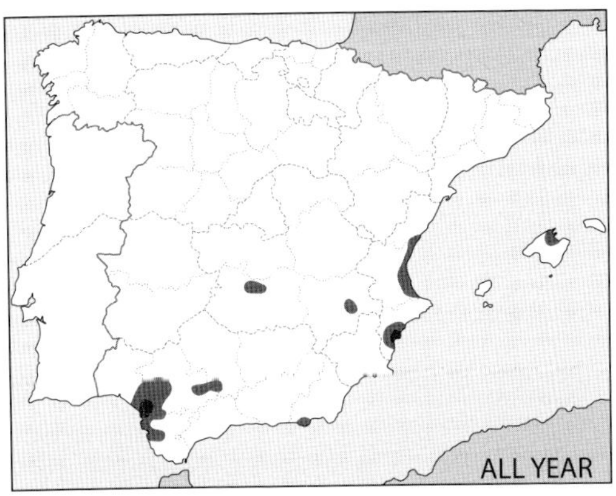

The chief breeding sites within Doñana are currently at Veta La Palma, Las Nuevas, El Codo de la Esparaguerra and the Brazo del Este. The Levant coast population is mainly at El Hondo and the Santa Pola saltpans but nesting, sometimes by several pairs, also occurs in some years at El Hondo de Amorós, the Clot de Galvany and the Marjal de Pego-Oliva, Alicante; the Cañada de Las Norias and Río Antas estuary, Almería; the Albufera, Marjal del Moro and the Marjal de Xeresa-Xeraco, Valencia; and the Marjal de Almenara, Castellón. Birds are sometimes present at the Mar Menor in the breeding season, but nesting in Murcia was not proved until 2007. The Marbled Duck is rarely reported from

Catalonia but a pair with four ducklings was at the Ebro delta in June 2009 (*A. O. Catalunya 2009*: 31).

Away from these regions there have been rare instances of nesting at lagoons in Castilla-La Mancha: at El Taray (1970), Daimiel (1991) and Pétrola (1998). There are also historical records of nesting at lakes in Andalucía outside the Marismas: at the Lagunas de la Cigarrera (1978) and de Las Turquillas (1998), Sevilla; the Lagunas del Taraje (1910, 1964) and de Los Tollos (1991), Cádiz, and at Fuentedepiedra (1967). Sporadic nesting by up to two pairs has also occurred in Mallorca, at Es Salobrar in 1969 and at the Albufera (Green *et al.* 2003, AOB 2005). Four pairs nested at the Albufera in 2007 (Table MD1).

The Marbled Duck is attracted to shallow waters with abundant submerged and emergent vegetation, and in particular to brackish temporary or seasonal lagoons (LR). It typically forages for seeds and aquatic invertebrates in water no more than 20cm deep, often at night. The clutch size is unusually large, 11.8 eggs on average, greater than in any other European duck, and close parental care ensures that the ducklings have a very high probability of surviving to fly (Green 1998). These positive attributes may be offset by the fact that the species is apparently short-lived.

The current situation makes it hard to imagine that the Marbled Duck was the most numerous breeding duck at Doñana in the late 19th century (Valverde 1964a). Saunders (1871) described it as abundant there all year round but Chapman & Buck (1893) said that it was absent between November and March. No fewer than 500 were shot in the Marismas in a single day in August 1926 and flocks of several thousand could be observed in the early decades of the 20th century (Valverde 1964a). Nevertheless, its numbers there varied markedly between years: hunting statistics for a group of lagoons show that 357 birds were bagged in 1926–27 and 245 in 1928–29, but none in 1925–26, one in 1927–28 and three in 1929–30 (Valverde 1964a). A decline soon set-in and Valverde (1960) considered that the breeding population at Doñana in an average year in the 1950s was no more than 100–200 pairs. However, Hidalgo (1991) suggested that 300–500 pairs were present during the 1950s and 1960s. The population in 1984–88 was put at 150–250 pairs but this may have been an overestimate since the number of broods actually seen was much lower (LR). Since 1991 the nesting population there has ranged from zero pairs in 1995, a severe drought year, to 30–50 pairs in 1997.

The Levant population may have been around 200 pairs during the 1960s (Navarro & Robledano 1995) but there is little historical data for this region. There were 10–30 pairs during 1985–94, increasing to 31–48 pairs in 1995 and 84–133 pairs in 1997 but declining to 28–33 pairs in 2001 (LR). Other sites in eastern Spain have held only a few pairs, the largest counts being seven pairs at the Cañada de Las Norias in 1997 and six pairs at the Albufera de Valencia in 1997 and 1998.

The population is the subject of up to five censuses annually. The 2007 breeding season census, conducted in a year when water levels at the Marismas were unsatisfactory, found a total of 97 pairs, including 37% in Andalucía and 58% on the Levant coast (Table 4). The latter included the first proved instance of nesting in Murcia, by one pair at Mazarrón.

Table 4. Breeding distribution and productivity of the Marbled Duck in Spain in 2007. After Ballesteros *et al.* 2008 (with corrections)

Region	Province	No. of pairs	No. of broods	No. of ducklings
Andalucía	Almería	3	3	26
	Cádiz	5	3	29
	Huelva	2	0	0
	Sevilla	26	10	74
Com. Valenciana	Alicante	47	42	184
	Valencia	9	8	61
Murcia	Murcia	1	1	3
Balearic Islands	Mallorca	4	3	16
TOTAL		97	70	393

The Marbled Duck is now little more than a vagrant to Portugal, with just 12 accepted records, of single birds or groups of up to three, in the southern half of the country during 1988–2009 (AP, CPR 2008–09). There are a further four southern records (nine birds) from 2010–12. There are no proved instances of nesting in Portugal although it may have done so in the distant past: it was reported in the country with some frequency during the first half of the 20th century but was considered rare thereafter (AP).

The principal causes of the decline in this formerly locally abundant species include loss of habitat and habitat degradation (LR, Green 2010). The latter includes infiltration by agrochemicals and poor water quality: which has resulted in episodes of mass mortality due to salmonellosis and botulism. To this may be added the effects of illegal hunting and lead poisoning from ingested shotgun pellets. A perhaps surprising problem has resulted from the massive infiltration of the marismas by the introduced American crayfish; this now supports a thriving trapping industry but the traps are known to kill numbers of ducklings especially. Another introduction, of the Purple Swamphen to the Levant wetlands, has had the unexpected effect of reducing the availability of *Scirpus* rush seeds, an important source of food to the ducks, since the gallinules eat the plants. Loss of submerged vegetation as a result of the activities of carp and flamingoes has also been a problem for Marbled Ducks locally.

The Iberian Marbled Ducks are part of a metapopulation that includes birds nesting in the countries of the Maghreb. Those breeding at Doñana tend to disperse elsewhere in Andalucía after nesting, for example to the Laguna de Medina and other lagoons in Cádiz. Similarly, outside the breeding season those in the Levant may occur in small numbers on wetlands throughout Almería, Alicante and Valencia, but they tend to be most numerous at sites where they also nest. Significant gatherings are sometimes reported in autumn at or near the principal breeding sites. There was a spate of such records during 1996–98, when the population seemed to enjoy a temporary recovery after a prolonged period of drought, due perhaps in part to recruitment from North Africa. Thus there were over 120 at El Hondo from September 1996 to February 1997, peaking at 286 on 19 October 1996 (*Ardeola* 44: 247), 535 at El Hondo on 8 September 1997 and 594 at Veta la Palma, Doñana, on 27 October 1998 (Green 2010). Such large numbers are still sometimes recorded: there were 544 at El Hondo in September 2007.

Arévalo (1887) noted that large numbers appeared at the Albufera de Valencia in July and August in some years but that they did not nest there and instead immediately began their post-breeding moult. Gatherings of any size at the Ebro delta, for example the 80–100 there on 12 September 1968 (Maluquer 1971), no longer occur. The Catalan bird reports give 26 records during 1996–2009, most of them from the Ebro delta but some also at the Llobregat delta and, more rarely, the Aiguamolls de l'Empordà: most are of single birds but up to seven were present at the Ebro delta in September 1999 and May 2005 and there were five at the Llobregat delta in August 2001. A few appear from time to time at other sites that are not too distant from the breeding range, including lagoons in Madrid and southern Extremadura and also at the Guadalhorce estuary, Málaga. There were 22 at Daimiel on 9 January 2011 and nine were there on 6 April 2011 (*Ardeola* 58: 483). The 2007 censuses (Fig. 26) show how the numbers present in Spain can vary considerably between different times of the year.

The January censuses of wintering waterfowl in Spain during 1990–2009 confirm the scarcity of the Marbled Duck in mid-winter (AAE). The mean count during this period was just 75 birds (range 2–245). The vast majority of observations were from Doñana and nearby lagoons in western Andalucía (59%) and from El Hondo and other wetlands of the Comunidad Valenciana (39%), in short, from sites within the breeding range (AAE).

It rarely appears further north: such exceptional records include one at the Ría de Villaviciosa from 20 January until 18 July 1978; two at Las Cañas reservoir, Navarra on 31 October 1981 and one at the same site on 15 and 23 August 1983; two in the Isabel La Católica park in Gijón, Asturias on 24 December 1988 and one there on 13 January 1991; two at a pool between Aldeanueva and Pascualete, Cáceres on 6 June 2003; two at La Almendra reservoir, Salamanca, on 9 January 2011 and single birds at Arnedo, La Rioja, on 28 November 2009; at Ribadesella, Asturias, on 29 April 2010 and at Cecebre, A Coruña, on 9 September 2010 (Álvarez-Balbuena *et al.* 2000; *Ardeola* 29: 178, 37: 331, 38: 331, 50: 344, 57: 218, 57: 519).

Ringing recoveries show that some Marbled Ducks travel to winter in northwestern Africa: there are five recoveries of Spanish-ringed birds from Morocco and three from Algeria, all during November–March. In addition, an adult female ringed

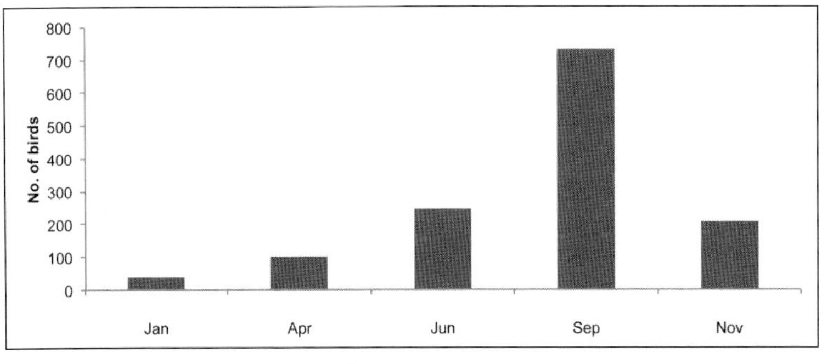

Figure 26. Results of five monthly censuses of the Marbled Duck in Spain in 2007 (Ballesteros *et al.* 2008)

at the Guadalquivir estuary in October 1990 was recovered at Ain, France (1,343km), in September 1992. There are several recoveries and observations of marked birds from Doñana on the Levant coast but a large-scale marking programme at El Hondo, Alicante, did not result in any Doñana sightings, suggesting that interchange between the western and eastern Iberian populations is limited. Some movement from Doñana to the Levant is nevertheless thought to occur in drought years (Green & Navarro 1997). It is believed that seasonal influxes at Doñana in some autumns are of birds of Moroccan origin, which probably accounts for the large interannual fluctuations in numbers noted there both in the past (Noble 1902, Valverde 1964a) and more recently (Ballesteros *et al.* 2008). In general, interchange between the Maghreb and Iberian populations may occur at any time of year as the birds respond to changes in habitat availability (LR). At present Iberia may represent a 'sink' in the population dynamics of this Marbled Teal metapopulation, and this has worrying implications should Moroccan wetlands continue to deteriorate.

RED-CRESTED POCHARD *Netta rufina*

(Pato Colorado, Pato-de-bico-vermelho)

A scarce and local resident in Spain and Portugal, and in Mallorca in the Balearic Islands.

This species is a relatively recent arrival not just in Iberia but in the whole of Europe. The core populations are in the south of the Asian Palearctic, from Turkey east to China. It seems to have begun its westward spread around 1830 and this process accelerated during the mid 20th century (BE). In the late 19th century some bred at the Albufera de Valencia and larger numbers occurred there in winter (Saunders 1871, Arévalo 1887) but it appears to have been entirely absent from Andalucía at that time (Irby 1875, Chapman & Buck 1893). It seems to have begun to colonise the Ebro delta progressively during 1930–50 (Ferrer *et al.* 1986) and Doñana during 1935–40 (Valverde 1960). The La Mancha wetlands were also occupied at around this period: it bred commonly there at Damiel by the 1940s (Gil Lletget 1945) and Bernis (1952) found 60 pairs at the Laguna del Taray in spring 1951.

The Red-crested Pochard is largely restricted in the breeding season to southern and eastern Iberia. The great majority are on the Mediterranean coast: at the Ebro delta and at most wetlands south to Almería, in the many wetlands of La Mancha and at Doñana and other wetlands in southern and central Andalucía. There is also a small population in Aragón and at one site in Soria, as well as occasional nesting at scattered locations elsewhere, including southern Portugal. The breeding sites are wetlands that offer large expanses of open water but also abundant fringing and submerged vegetation. Both freshwater and brackish water sites are used but rivers are seldom visited. Some of the favoured lagoons dry up in drought years so that the breeding population fluctuates considerably locally (Amat 1984).

The small population at the Albufera de Mallorca is entirely or chiefly the result of the release there in 1991 of 60 juveniles originating from Doñana and the Ebro delta (SA2, LR). However, the Red-crested Pochard was thought to be a rare breeder in Mallorca in the past (Bernis *et al.* 1958). Nesting in Menorca was first reported in 2009, when one pair bred at Es Mercadal, and a female with five ducklings was at the Albufera des Grau on 12 July 2011 (*Ardeola* 57: 519, 59: 415). It remains a rare visitor to the other Balearic islands.

The total Spanish population was put at 3,000–5,000 birds during the 1970s (Noval 1975, Szijj 1975) and at 5,400–8,600 pairs during the 1990s (SA1). It was considered to oscillate around 4,000 pairs at the start of the 21st century (Table 5). More recently, the Spanish breeding population was censused in spring 2007 (Palomino & Molina 2009), when 8,172 birds were counted and a mean population estimate of 14,400 individuals was calculated.

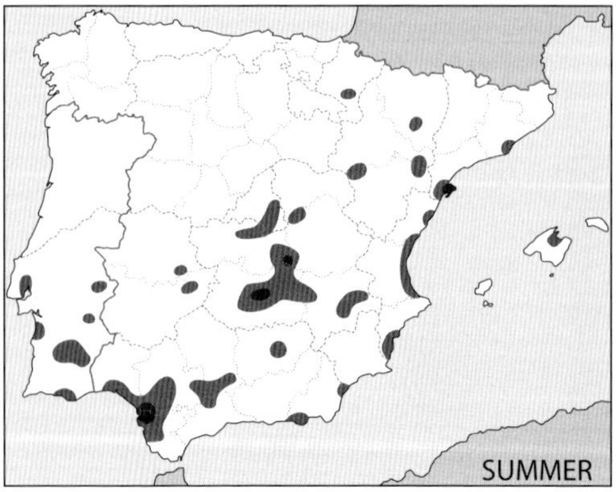

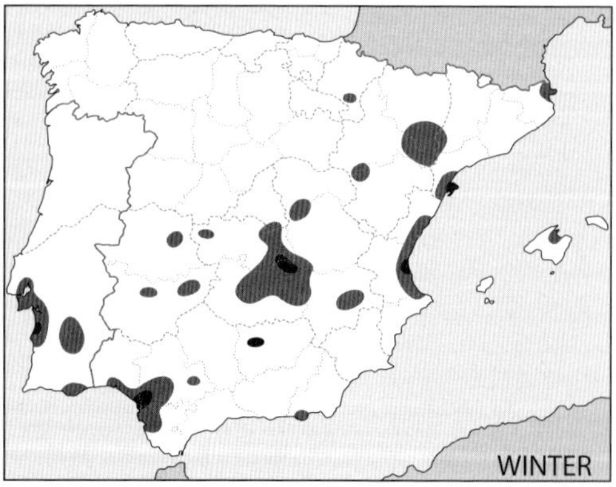

Table 5. The breeding population of the Red-crested Pochard in the Spanish regions in the early 21st century. Chiefly after SA2 and LR.

Region	Population c. 2002 (pairs)	Main sites	Notes
Andalucía	400–800	Doñana	Fluctuating
Aragón	<30	Estanca de Chiprana, Zaragoza	200–400 pairs at Gallocanta in the 1970s.
Castilla-La Mancha	1,500 in wet years	El Retamar, Ciudad Real; Manjavacas, Cuenca; El Taray, Toledo	Daimiel was the chief site up to the 1970s prior to its deterioration.
Castilla y León	A few	Monteagudo reservoir, Soria	
Catalonia	Up to 2,000	Ebro delta	Increasing.
Madrid	A few	Scattered pairs	
Comunidad Valenciana	c. 300. Range 160–500 during 1988–2000.	El Hondo & Santa Pola saltpans, Alicante; Albufera de Valencia	Declining. The Albufera alone held c. 1,000 pairs in 1964.
Extremadura	A few	Orellana reservoir	Since 1988
Galicia	0	Castrelo de Miño reservoir	Bred once; 1992
Cantabria	1 (2009)[1]	Marismas Blancas	First breeding record.
La Rioja	0	Ebro valley wetlands	Last bred in early 1990s.
Murcia	1–3	Alfonso XIII reservoir	Last bred 1996
Navarra	4–5 pairs	Las Cañas reservoir	Declining.
Balearic Islands	80–90 (2003)[2]	Albufera de Mallorca	60 juveniles introduced in 1991

[1] *Ardeola* 56: 153. [2] *A.O. Balears 2005.*

The Portuguese population has been estimated at 40–100 pairs (BE) but is increasing and expanding its range, which remains almost entirely south of the Tagus (AP). The Red-crested Pochard formerly nested largely at the Tagus estuary (PA1) but in recent decades it has become relatively widespread at wetlands in Alentejo and at some sites in Algarve, including ponds on golf courses there (PA2). It was virtually unknown in Portugal until quite recently: the earliest documented record was in 1967, at the Lagoa de Santo André, in coastal Baixo Alentejo, which remains a favoured breeding site. The first breeding record came from the Tagus estuary, however, in 1980 (AP).

The Iberian birds are essentially resident but there is an exceptional ringing recovery in Cork, Ireland (1,650km) in January 1977 of one ringed as a duckling at Doñana in July 1975. The Iberian birds typically have moult migrations to favoured sites where considerable numbers gather after the breeding season. The entire population at breeding sites in Andalucía, which tend to dry up in summer, participates, the males leaving before the females. However, where water levels are not so variable, as in the Ebro delta, only the males depart and the females moult in the breeding areas (Amat *et al.* 1987). The Ebro reservoir, Cantabria, and the Laguna de Gallocanta were important moult locations during the 1970s and 1980s, attracting thousands in late summer. For example, 5,000 were at the Ebro reservoir in July 1994 (*Ardeola* 45: 120) and 11,000 were counted at Gallocanta in August 1997 (LR). Only a few hundred birds have appeared at either site more recently so a count of 900 at the Ebro reservoir on 9 October 2011 (*Ardeola* 59: 169) is notable. Moulting birds have tended to be more dispersed in recent years but the Orellana reservoir, Badajoz, attracts 400–1,500 birds, which begin to arrive there in June (*Ardeola* 36: 239). A gathering of 2,400 at the Ullibarri reservoir, Álava, on 5 September 2008 (*Ardeola* 55: 290) is also noteworthy.

The Iberian birds are joined by variable numbers of migrants from central Europe in winter. There are small scattered breeding populations, chiefly in France, Germany, Austria, the Netherlands and the Czech Republic, that are partly migratory and some include birds that winter in Iberia. There are at least 15 recoveries of birds ringed in Germany, four from Denmark, nine from France and one from The Netherlands. All these recoveries have been in Spain and Spanish-ringed birds have given nine recoveries from France as well the exceptional recovery from Ireland.

The total Iberian wintering population was put at 20,000 birds in 1989 (Troya & Bernués 1990, SA1). January counts in Spain during 1990–2009 averaged 14,694 birds (range 6,420– 24,216; AAE) during the late 1970s and in the 1980s. As always with waterfowl, there are marked year-to-year fluctuations in the numbers wintering at particular locations but the distribution of the Spanish wintering population has changed markedly since the 1980s (Table 6). In addition, the wintering population has shown a gradual decline during the same period.

Table 6. The mean wintering population of Red-crested Pochards in different regions of Spain during 1978–89 and 1990–2009. Figures are means of January counts (AI1, AAE)

Region	Wintering population 1978–89	Wintering population 1990–2009	% change	Key sites
Ebro Basin	8,013	165	-98	Gallocanta
C. Valenciana & Murcia	7,363	5,071	-31	Levant coast wetlands
Castilla-La Mancha & Madrid	2,833	2,477	-13	La Mancha wetlands
Andalucía	2,095	2,505	+20	Doñana
Catalonia	1,831	4,236	+131	Ebro delta
Biscay coastlands	65	8	-88	Ebro reservoir
Extremadura	30	132	+340	Orellana reservoir
Castilla y León	4	2	-50	Villafáfila, La Nava
Galicia	1	1	0	
Balearic Islands	0	143	-	

Gallocanta has ceased to be a major wintering site. Counts of over 35,000 birds were recorded there at their peak in 1978. They chiefly arrived in November and remained until February. Most recent years have seen only a few hundred birds, and often many fewer, but over 1,000 were present in 1989–90. The population has always been concentrated at just a few wintering locations and since 1990 some two-thirds have been shared between three extensive sites: the Ebro delta, the Albufera de Valencia and Doñana. The Ebro delta retains significant numbers during the moult period and the population increases further in winter. The winter population of the delta during 1996–98 was estimated at around 1,200 birds but there were 8,375 birds in 1999 and 5,059 in 2001 and this has recently been the principal wintering location in Spain (LR). The other important wintering sites include the wetlands of the Comunidad Valenciana, whose recent population has ranged from 3,000 to 10,000 birds, chiefly at the Albufera de Valencia but also at El Hondo and the Santa Pola saltpans, the Marjal de Almenara, and the Marjal del Moro (LR). The mean count for Doñana is around 2,000 birds but the numbers present there have ranged from zero to over 15,000 in recent winters. Similar variation is evident at the lagoons of Castilla-La Mancha, with a mean wintering population in the 1990s of 2,086 birds. In this region, the Buendía reservoir saw an average January population of 572 individuals during 1990–2009, but the annual counts ranged from just one bird to 4,112 (AAE). Elsewhere, Red-crested Pochards may be encountered in very small numbers almost anywhere in winter, with records from Galicia, Asturias, Cantabria, the Basque Country, Castilla y León, Extremadura, La Rioja and Navarra. A few also occur in Murcia.

The Portuguese wintering population numbers a few hundred individuals at locations south of the Tagus. The Lagoa de Santo Andre, Setúbal, is the principal wintering site and has attracted as many as 517 birds, in January 1993 (AP). The mean wintering population in Portugal during was 169 birds during 1987–89 (AI1), 509 birds during 1992–95 and 306 birds during 2003–05 (AP). A count of 120 at Évora on 14 March 2009 is noteworthy (*Anu. Ornitol.* 8: 55).

COMMON POCHARD *Aythya ferina*

(Porrón Europeo, Zarro-comum)

A widespread but scarce breeding species in Spain and Portugal, and in the Balearic Islands. More numerous and locally abundant on passage and in winter.

The Common Pochard has a modest-sized but widespread and possibly increasing population. It is most abundant in the south, centre and east of the Peninsula and much more local in the north (SA2). Doñana is by far the principal breeding location. The other wetlands of Huelva, Cádiz and the Guadalquivir valley; the Ebro delta and the coastal wetlands of the Levant and the south-east; and the wetlands of La Mancha and Madrid, support very small breeding numbers and nesting there is not always regular. Known breeding sites elsewhere include Gallocanta; the Laguna de Pitillas, Navarra; Villafáfila and the Ebro reservoir. Isolated pairs and small groups may be found nesting, often sporadically, almost anywhere in the Peninsula. It also breeds in the Balearic Islands. Historical accounts seldom mention more than occasional nesting, notably at the Albufera de Valencia (Saunders 1871, Arévalo 1887), and even in the 1960s Bernis (AMI) noted that in Spain there was just a small breeding population at Doñana with sporadic nesting elsewhere. Nevertheless, Valverde (1960) considered it a very common breeding species in years with good water levels. The limited evidence thus indicates that it is a more abundant breeding species than used to be the case.

The Spanish population was estimated at 3,000–3,500 pairs by Purroy *et al.* (1997) but the 2003 atlas found over 5,000 pairs. Breeding numbers fluctuate from year to year, especially in the drier south. For example, the population of Doñana has been estimated to vary from 3,000 to 5,000 pairs and it is the most abundant of the breeding waterfowl there in some years (García *et al.* 2000). However, the atlases provide evidence of an ongoing overall increase in numbers and in the geographical spread of the population. The spring 2007 census of waterfowl (Palomino & Molina 2009) found 3,285 birds and estimated the Spanish population at 16,600 individuals, with the largest concentrations in Andalucía and in the Tagus, Guadiana and Ebro basins, but the great majority of those found in some regions were clearly migrants that had yet to depart or were summering non-breeders. Thus, although 1,900 birds were estimated to be present in the Ebro basin and Catalonia during the census period, the regional bird reports for Catalonia, Aragón, La Rioja and Navarra offer solid evidence of a combined breeding population there of at most 40–50 pairs.

The breeding population at the Albufera de Mallorca numbered 2–5 pairs during 2000–03 but increased to 30 pairs by 2008. Several bred for the first recorded time in Menorca in 2008 (*Ardeola* 49: 180, 50: 155, 52: 210, 56: 346).

The small Portuguese breeding population has been estimated at 20–50 pairs (BE 2004) and it too has become more widespread in recent decades. In the 1980s a few pairs bred regularly in the Tagus valley and occasionally elsewhere in the south. Since the 1990s breeding has become fairly regular at the Paul do Boquilobo and the Lagoa dos Salgados: with 8–12 pairs at the latter in 2002 and six females with ducklings there in July 2009, with at least occasional pairs nesting at irrigation dams in Alentejo and at coastal lagoons and ornamental ponds in Algarve (PA2, AP, *Anu. Ornitol.* 8: 55).

The Iberian birds are dispersive. The mean distance travelled by 27 birds ringed as ducklings in Doñana and recovered elsewhere was 212km and they included individuals that moved as far as Morocco, Italy, France, Germany and Poland (Sáez-Royuela & Santos 1985). There are also recoveries of Spanish-ringed birds from Switzerland (3), the Czech Republic, Finland, Ukraine (2) and Russia (4), these last including an adult male ringed at Doñana in July 1975 and recovered in the Urals (5,109km) in May 1979. Very large numbers of Pochards from northern and central Europe winter in Iberia, especially in cold years when northern lakes freeze over. There are at least 95 Iberian recoveries of birds ringed outside the region. Many (31%) were marked in France, mainly at the Camargue during passage periods, but others came from countries in central and northern Europe: including Britain, Denmark, the Baltic states (24%), the Czech Republic, Switzerland, Germany, Italy, Hungary and Russia. Some of the thousands of Pochards that winter in Morocco probably cross Iberia on passage. There are 16 recoveries of Spanish-ringed birds from Morocco and also one from Algeria.

The Iberian wintering population in 1989 was *c.*60,000 birds (Rufino 1989b, Troya & Bernués 1990), with a notable concentration of 24,015 birds at Gallocanta that year. Eight other Spanish sites held more than 2,000 birds during 1972–89: the Ullibarri-Gamboa reservoir, Álava; the Ebro delta; the Albufera de Valencia; El Hondo, Alicante; Daimiel and the Peñarroya reservoir, Ciudad Real; Fuentedepiedra; Odiel and Doñana (SA1). The wintering population in Spain had declined considerably

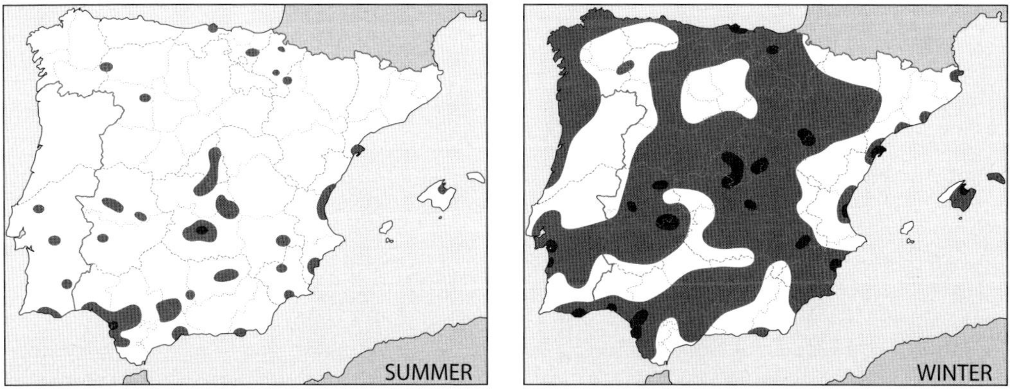

by 1990 and a continued moderate decline has been detected since then in some regions. In particular, Gallocanta has ceased to attract any numbers since the early 1990s and the populations of the Comunidad Valenciana have also declined (AAE). The Spanish January population during 1990–2009 averaged 29,220 birds (range 14,421–61,152; AAE). Pochards are well dispersed in winter across much of Iberia and no Spanish location held more than 7% of the population during this period, with the initial exception of Gallocanta (AAE). The wintering birds begin to arrive in September–October but peak numbers are present during December–February. Departures are chiefly in March–April.

The Portuguese wintering population has also declined progressively. The species was abundant in winter at the Ria de Aveiro in the early 20th century but it is now scarce there. The formerly substantial wintering concentrations at the Sado estuary disappeared after 1993, when habitat changes associated with fish farming were implemented. Wintering birds are now widely dispersed, with some concentration at the Paul do Boquilobo, the Lagoa de Santo André and the Alqueva reservoir (AP). An influx in winter 1978–79 resulted in a national count of 8,220 Pochards, including 5,800 at Boquilobo. However, the mean wintering numbers in Portugal have declined from 2,555 birds during 1976–81 to 1,260 during 1992–95 and 1,013 during 2005–07 (AP). A count of 580 at the Lagoa de Santo André on 10 February 2007 (*Anu. Ornitol.* 6: 48) is noteworthy in relation to recent national totals.

The numbers wintering in the Balearic Islands are generally modest but were higher than usual in 2003, when there were maxima of 402 at the Albufera de Mallorca and 352 at the Albufera des Grau, Menorca (*Ardeola* 52: 210).

RING-NECKED DUCK *Aythya collaris*

(Porrón Acollarado, Caturro)

Very rare. Recorded in Spain and Portugal, mainly in winter.

This species has appeared frequently in Europe, especially since the late 1970s. By 2003 there were records of at least 495 individuals in Britain and 137 in France. There are at least 46 accepted records for the Canary Islands, 27 for the Azores up to 2009 and two for Madeira, as well as 24 records for Morocco. Many birds have been long-stayers and others have apparently returned to the same sites in successive winters. There is a recovery in Wales of a bird ringed in Canada and one ringed in England was found in Greenland. Together with the tendency for birds to appear in winter, it may be that their occurrences represent small-scale regular migration than vagrancy.

The first Iberian record was a male that remained in a city park at Gijón, Asturias, in January–March 1978, and which returned there the following winter (*Ardeola* 24: 254–255, Noval 1981). There have been annual observations since 1994 and by 2010 there were 73 accepted records for Spain and 16 for Portugal. At least ten further Spanish records and two Portuguese records were reported up to 2012. Allowing for some probable duplication there are some 87 accepted Iberian records to 2010. They involved some 145 individuals, among which 70 were identified as males and 68 as females. Although 70% of records involved single birds, there are seven of groups of four or more. The largest flocks were of seven at the Laguna de Traba, Galicia, in December 2002, and eight on a reservoir at Zuares del Páramo, León, in November 2001. Record frequency increased considerably until the mid-2000s, peaking at 11 records in winter 2003–04.

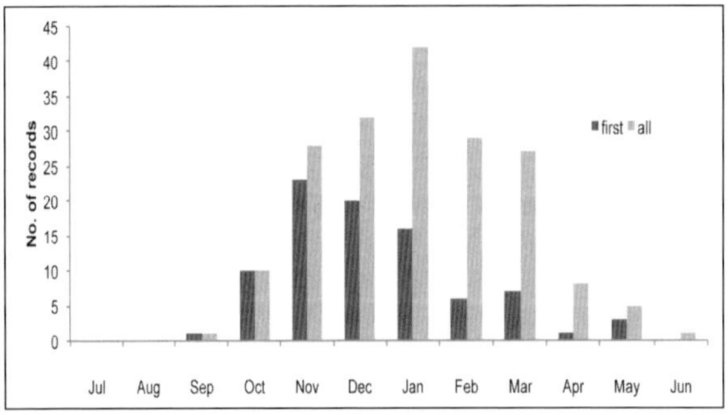

Figure 27. Frequency of records, 1991–2010 (n = 83)

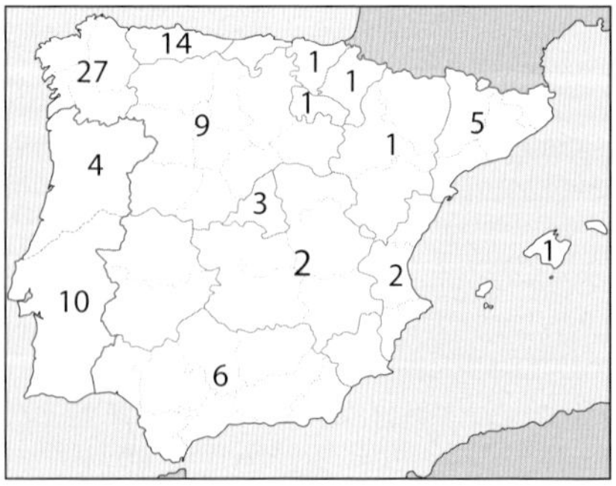

Figure 28. Records by region, 1986–2010 (n = 87)

Winter records predominate and peak in January (Fig. 27). Some 80% of observations fall during October–January, with extreme dates of 30 September and 16 June. Three long-stayers remained on reservoirs in Asturias respectively from October 1994 to April 1995, from November 1996 to June 1997 and from November 1999 to May 2000. Some individuals return to the same localities in successive winters: for example, the same male probably accounted for the records at Alcarràs, Lleida, during December 2009–February 2010, November 2011–January 2012 and November 2011–March 2012. Records are concentrated in the northwestern quadrant of Iberia (Fig. 28), 62% in total being within Galicia, Asturias, northern Portugal and Castilla y León. There are very few from the Mediterranean coastlands and only one in the Balearic Islands: at the Albufera de Mallorca in December 2007 (*Ardeola* 56: 315).

FERRUGINOUS DUCK *Aythya nyroca*

(Porrón Pardo, Pêrra)

Extremely rare and in danger of extinction as a breeding species in Iberia. Very scarce on passage and in winter throughout the region, including the Balearic Islands.

The Ferruginous Duck now has an extremely precarious position among the breeding birds of Iberia, its minimal population being in danger of extinction. It is listed as 'Critically endangered' in Spain (LR). It declined inexorably during the 20th century, very probably because of habitat deterioration. The Iberian birds may form part of a metapopulation totalling a few thousand individuals that nests in Tunisia, Algeria and Morocco (LR). The principal populations range from eastern Europe across the southern Palearctic to northwestern China.

The Ferruginous Duck has exacting habitat requirements that cannot easily be met in modern Iberia. It only nests at freshwater sites that offer very high water quality and it is intolerant both of salinity and eutrophication (Amat & Soriguer 1982, LR). It requires an abundance of submerged macrophytes and abundant emergent vegetation, notably stands of reedmace, and it typically forages close to such vegetation rather than in open water. The few remaining breeding birds are restricted to western Andalucía: in particular Doñana, and, now occasionally, to the coastal wetlands of Valencia and the lakes of La Mancha.

The Spanish population was estimated at 1–10 pairs by SA1 and SA2 found only six pairs. In recent years breeding has only occurred occasionally at a few locations. The 2007 Spanish breeding waterfowl survey found only two pairs, both of them in Huelva: at the Lagunas de El Portil and the Laguna de Palos (Ballesteros *et al.* 2008). The present situation reflects the final stages of a long-term decline. Doñana alone may have held some 500 pairs at the beginning of the 20th century but these had declined to 'a dozen' by the 1950s (Valverde 1960). The early 20th century also produced records of nesting at the Laguna de Medina and the (since drained) Laguna del Torero in Cádiz, and the Laguna de Palos, Huelva. Breeding was only confirmed in the marismas on three occasions during 1970–92. In total 124 captive-bred birds were released at Doñana and the nearby Laguna de El Portil, during 1992–96 and several pairs bred in the area during 1993–2001, but without any resulting long-term recovery in the population (Callaghan 1999, LR): records from these sites are omitted from Table 7.

In the Comunidad Valenciana the Ferruginous Duck bred at the Albufera de Valencia during the first half of the 20th century and perhaps up to the 1970s. Since then it has nested there in 1988 and 1996. It bred at the Clot de Galvany in 1990, and at the Marjal del Moro in 2001 at least, having been present at the latter site since 1992. Daimiel in Castilla-La Mancha had a nesting population in the early 20th century that dwindled to two pairs during 1980–84 with none thereafter until a single pair nested in 1991. Away from these 'traditional' areas, a pair bred at the Ebro delta in 1962 and three females hatched young at Es Mercadal, Menorca, in 2012 (*Ardeola* 60: 509).

The Ferruginous Duck was apparently not infrequently recorded in Portugal during the late 19th and early 20th centuries and it was said to have nested in the lower Guadiana valley, but there are no confirmed breeding records from that period (AP). Much more recently, a pair bred in coastal Algarve at the Lagoa dos Salgados in 2001 and a pair was reported annually there during at least 2002–07, with breeding proved in 2004 and 2007 (AP, PA2).

Spanish-bred birds may be resident or partly migratory: two ringed in Doñana have been recovered in eastern Spain and one travelled from there to Larache, northern Morocco. Some Ferruginous Ducks that are probably of central European origin occur in winter in the Peninsula and Balearic Islands, and others may pass through on migration to and from sub-Saharan winter quarters, but the sole Iberian recovery of an extra-regional bird was of one ringed at the Camargue, France (Sáez-Royuela & Santos 1985).

The Spanish wintering population has been estimated at 5–40 birds, with marked interannual variation (SA1, Callaghan 2001), many fewer than in France where the wintering population averaged 156 birds during 2003–08 (Legendre *et al.* 2012). The annual January counts during 1990–2009 found just 22 on average in Spain (range 7–65; AAE). Most winter records involve single birds, or groups of two or three. However, there were 15 at San Andrés reservoir, Asturias, on 27 December 1976 (Álvarez-Balbuena *et al.* 2000), 22 at the Laguna de San Juan, Madrid, on 13 December 1987 (Juan 2002); 17 at El Porcal gravel pits, Madrid, on 4 November 1989 (Juan 2002) and 48 at the Lagunas de Espera, Sevilla, on 3 December 1989 (Raya 1993). Similar gatherings have been lacking more recently so 18 at the Albufera des Grau, Menorca, on 28 October 2009 is a noteworthy count (*A. O. Balears 2009*: 114). Larger numbers may formerly have been more frequent in the Balearics: Munn (1932) noted that they were plentiful at the Albufera de Mallorca during winter 1930–31, when many were shot.

Winter sightings, during October–March, clearly predominate (Fig. 29). June–August records are chiefly of non-breeders observed in Andalucía, the Levant coastlands and the Balearic Islands. In the sample examined there were 628 records (1,126 birds) during 1974–2010, although movements may have given rise to some duplication. Within this sample, 544 individuals were sexed and of these 71% were males and 29% were recorded as females. Only 64 of the females were seen in the absence of accompanying males so very likely many females (and immature birds) are overlooked.

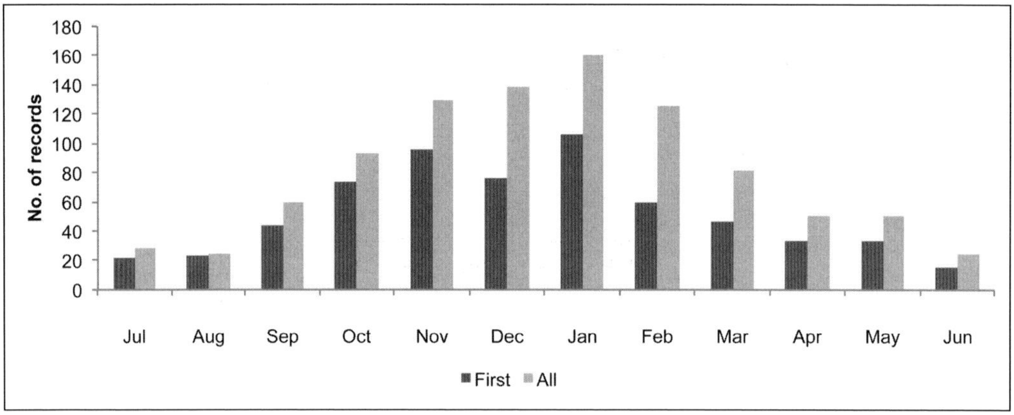

Figure 29. Records by month of Ferruginous Ducks in Spain, 1974–2010 (n = 628).

Overall, there is a marked predominance of records from the Levant coast in Catalonia, the Comunidad Valenciana and the Balearic Islands, which together account for some 45% of the records and 48% of the individuals in the available sample (Table 7). Andalucía, where a few pairs breed, and Castilla-La Mancha, where some may also breed, together produced only 13% of sightings. Elsewhere, the gravel pits of Madrid accounted for 19% of all the individuals reported and, more surprisingly, several sites in Asturias: notably La Furta and La Granda reservoirs, attracted considerable numbers. The regional spread of observations must be regarded with some caution, however, since there are marked inter-regional differences in both coverage and documentation of observations.

Table 7. Records by region of Ferruginous Ducks in Spain, 1994–2010. N = 628 records, 1,126 individuals.

Region	No of records	No of birds	Region	No of records	No of birds
Andalucía	57	77	Castilla y León	18	23
Aragón	54	82	Extremadura	19	40
Asturias	45	79	Galicia	16	20
Balearic Islands	77	173	La Rioja	2	2
Cantabria	5	6	Madrid	55	121
Catalonia	89	158	Murcia	20	49
Castilla-La Mancha	27	43	Navarra	12	20
Comunidad Valenciana	116	214	Basque Country	16	19

The Ferruginous Duck is rare in Portugal, with just some 75 documented records to 2010 (excluding the nesting location of the Lagoa dos Salgados), but it is now recorded annually there. Records are largely from coastal lagoons, especially south of the Tagus (AP, *Anu. Ornitol.* 7 & 8). The occurrence pattern is identical to that seen in Spain, most occurring in October–March (AP). A count of 15 at Loulé, Algarve, on 1 December 2006 (*Anu. Ornitol.* 6: 48) was exceptional: most Portuguese records, as in Spain, involve single birds, pairs or very small groups (AP).

Ferruginous Duck x Common Pochard hybrids are occasionally reported and provide a pitfall for the unwary. Reports of such hybrid pairs within Spain include a female paired with a male Common Pochard that raised three young at the Albufera de Valencia in 2001 (*A. O. C. Valenciana 2000–2001–2002*) and a similar pair at Daimiel in 2007 that produced two hybrid offspring (*A. O. Ciudad Real 2006–2007*: 156).

TUFTED DUCK *Aythya fuligula*

(Porrón Moñudo, Negrinha)

An extremely scarce and local breeding species in northern Spain. Has bred Portugal. Widespread and locally common in winter.

The Tufted Duck has a tiny breeding population in mainland Iberia that seems to be of very recent origin. The first proved instance of breeding was in 1988, at the Rosarito reservoir, Toledo (Muñoz-Pulido *et al.* 1990) and this was followed by nesting at Castro de Rei, Lugo in 1995 (*Ardeola* 42: 216). Most records of definite or suspected nesting come from the north of the Peninsula: from Galicia and the Biscay lowlands. The 1998–2002 atlas (SA2) found breeding birds in only seven squares: in the provinces of León, Zamora, Lugo, Orense, Álava (2 squares) and – a southern outpost – in southeastern Cáceres. The total number of pairs involved was just 12, including five in the Salburua wetlands, Álava. Breeding has also been proved in Asturias where single pairs bred at reservoir sites on seven occasions during 1977–2000 (Álvarez-Balbuena *et al.* 2000, *A. O. Asturias*).

The most regular reports of breeding have been at Villafáfila, where Tufted Ducks nested annually during 1997–2007 at the artificial lagoons at the information centre there: as many as ten pairs were present in 1998 and four in 2004, but only one pair bred in 2007 (*Ardeola* 51: 547, 55: 134; Palomino & Molina 2009). The Spanish waterfowl census of 2007 found only one other pair: at the Sierra Brava reservoir, Cáceres (Palomino & Molina 2009). Pairs have occasionally been reported during the breeding season in Portugal, north of the Tagus, but there is only one confirmed nesting record for the country: a pair at Herdade de Maria Ribeira, Elvas, Alto Alentejo, in 1998 (AP). Nesting Tufted Ducks occur in similar habitat to Common Pochards, at lakes and reservoirs offering water of some depth and with abundant marginal and submerged vegetation.

Several thousand Tufted Ducks may winter in the Peninsula, the actual numbers involved varying considerably from year to year. Small numbers also occur then in the Balearic Islands, where Munn (1921, 1924) noted them at the Albufera de Mallorca and in Menorca. Up to 2,000 winter in Morocco in some years (BM) and many of these may cross the Peninsula on passage. There are at least 43 Iberian recoveries of foreign-ringed birds: from Central Europe (17), the British Isles (9), Iceland (4), France (7) and Northern Europe (6). Ten extra-regional recoveries of Iberian-ringed birds have been in France (4), Britain, the Netherlands, Finland, Poland, and Russia (2), these last including one from as far east as the Urals.

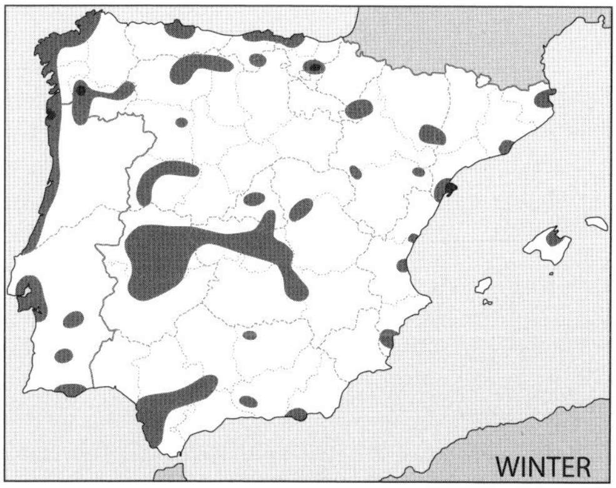

As many as 5,000–6,000 birds were present in winter in 1989, when the largest concentrations were in Galicia and Portugal, with 800 at the Rio Miño estuary and 847 at the Castrelo de Miño reservoir (Rufino 1989b, Troya & Bernués 1990). In 1982 the principal location was the Ullibarri reservoir, Álava. The January counts of wintering waterfowl in Spain during 1990–2009 found 3,129 birds on average (range 1,388–6,464; AAE): the largest concentrations again at the Castrelo reservoir and Miño estuary, 364 and 190 birds on average respectively, with maximum counts of 1,699 and 660 birds respectively (AAE). In general, the largest wintering concentrations are reported in the northern and northwestern coastlands but the population is well scattered across much of the northern two-thirds of the Peninsula (Table 8).

The Tufted Duck is generally scarce further south but small concentrations occur in some winters. For example, there were 100 at the Laguna del Comisario, Cádiz, in January 1978 (Tellería 1981) and 200 on the Orellana reservoir, Badajoz, in winter 1992–93 (Prieta *et al.* 2000), with up to 399 counted in January at the latter site subsequently (AAE). Wintering birds may have been more numerous in the south in the 19th century since they were then plentiful at wetlands in the environs of the Strait of Gibraltar in some winters, and occasionally seen in the Bay of Gibraltar (Irby 1875), and Arévalo (1887) noted them in spring and autumn off the Málaga beaches. These latter observations of Tufted Ducks on the open sea are surprising and perhaps the identifications cannot be regarded as entirely trustworthy.

Table 8. The wintering population of Tufted Ducks in different regions of Spain during 1978–89 and 1990–2009. Mean January counts (AI1, AAE)

Region	Wintering population 1978–89	Wintering population 1990–2009	% change	Key sites
Galicia	993	999	+1	Miño estuary, Castrelo de Miño reservoir
Castilla-La Mancha & Madrid	434	576	+33	La Mancha wetlands; El Porcal gravelpits, Madrid
Biscay coastlands	415	508	+22	Gijón; Ullibarri and Ebro reservoirs
Catalonia	220	208	-5	Ebro delta
Andalucía	184	125	-32	Doñana, Lagoons of western Andalucía
Castilla y León	165	206	+25	Villafáfila, La Nava
Extremadura	146	409	+180	Orellana reservoir
C. Valenciana & Murcia	98*	62	-37	Mar Menor, Levant coast wetlands
Ebro Basin	63	99	+57	Gallocanta
Balearic Islands	15	86	+473	Albufera de Mallorca

*includes Almería

A gradual decline in wintering numbers has been apparent in Spain post-2000, especially at those localities that usually attract most birds. For example, fewer than 100 individuals have wintered at the Castrelo de Miño reservoir in recent years (AAE). The Portuguese wintering population has declined consistently since the early 20th century. Tait (1887, 1924) regarded it as the commonest duck in the country and it was abundant in winter in the northern coastlands during the first half of the 20th century, notably at the Ria de Aveiro (AP). Even in December 1963 there were 1,500–2,000 at the Ria de Aveiro and a further 500 at the Lagoa de Santo André. Other important wintering sites included the Minho estuary; the Paul do Boquilobo and the Ria Formosa. However, the mean number wintering in Portugal during 1992–95 was only 626 birds and it declined further to 126 during 2005–07 (AP). In contrast, at Herdade do Espargueiro, Mourão, 272 were counted on 15 November 2008 and 462 on 24 January 2009 (*Anu. Ornitol.* 7: 74, 8: 55).

Wintering birds are chiefly present in Iberia from October to March. Influxes coinciding with cold spells further north seem likely but remain unconfirmed. A few summering non-breeders are recorded very occasionally, even at southern locations.

GREATER SCAUP *Aythya marila*

(Porrón Bastardo, Negrelho-comum)

***A. m. marila*. Very scarce but annual in winter in northern Iberia, rare elsewhere.**

Breeds in Iceland and across the northern Holarctic. Many winter around the North Sea but very few reach Iberia, where the total wintering population seems unlikely to exceed 50 birds.

Irby (1873, 1875) described it as a rare visitor to the Strait of Gibraltar in winter and saw one in the Bay of Gibraltar in December 1871. However, the species is not otherwise noted in the historical literature. Bernis (1956) refers to single specimens obtained at the Ebro delta and the Albufera de Valencia in the early 1950s and Maluquer (1960) saw a male at the Ebro delta in March 1956. In 1961 one was seen on the Ebro reservoir, Cantabria, in January (Bernis & Valverde 1972) and one was killed on the Laguna de Sobrado dos Monxes, A Coruña (*Ardeola* 8: 273) in November. In terms of actual records, the Greater Scaup was effectively no more than a vagrant until the latter part of the 20th century but whether it has since become commoner or was overlooked in the past cannot be determined. There were 46 records in Spain between 1956 and 1982, fewer than two per year (Muntaner & Ferrer 1983) but there were at least 293 records (581 birds) during 1983–2010, about 11 records per year.

In Spain the great majority are present only in mid-winter, November–February (Fig. 30). Portuguese records show a similar pattern of occurrence (AP). The earliest arrivals are in October and very few are reported after March. However, some occasionally linger into the spring and there have been very rare instances of summering birds: including a female that remained at San Andrés reservoir, Asturias, from 15 June to 15 August 1979 (Álvarez-Balbuena *et al.* 2000) and a male that was far inland at the Azúd de Riolobos, Salamanca, from at least 14 May 2002 to 25 April 2003 (*Ardeola* 50: 344).

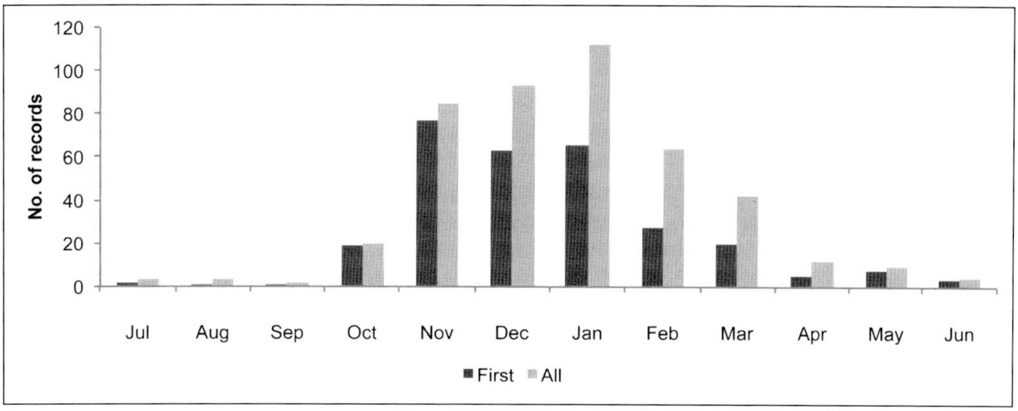

Figure 30. Records by month in Spain, 1983–2010 (n = 293)

The great majority of observations are from the north-west, in particular from the rías and reservoirs of Galicia and Asturias: including the Miño estuary; the Rías de Ortigueira, del Eo and Villaviciosa; and the Gijón area reservoirs. Galicia and Asturias together accounted for over 60% of all records and individuals reported in Spain during 1983–2010. Some are recorded far inland, often on reservoirs, particularly in the west of Castilla y León: from where there are 16 records. However, inland reports are less than annual: they include at least four from Navarra, five from Aragón, four each from Madrid and Castilla-La Mancha and 11 from Extremadura. A few are found annually on the east coast, most regularly in Catalonia, only occasional individuals occurring in the Comunidad Valenciana. There are only five records of single birds in Andalucía since 1994. There are also very few records south of Iberia: only 18 from Morocco up to 2010, largely of single birds (Bergier *et al.* 2011).

Flock sizes are small. The mean of published observations during 1994–2010 was just 1.6 birds, with maxima of seven at the Selga de Ordás reservoir, León, in December 2000; seven at Umia-O Grove, Pontevedra, during October–November 2003, and seven at the Ría del Eo in January 2006. In addition, the Spanish winter atlas survey produced an exceptional record of up to 17 in Santander bay in January 2009 and a count of ten in the Ría de Vigo, Pontevedra, in January 2011 (SWA). There are also a few older records of larger than average numbers. Álvarez-Balbuena *et al.* (2000) include a count of 40 at the Ría del Eo on January 1987. The 1983–2010 records included 450 individuals whose sex was recorded and 250 (56%) were males. Only 110 individuals were identified as first-winters and 45 of these (41%) were recorded as males. It is very likely that some first-winter females are not always distinguished from adults. There is a single ringing recovery: a bird ringed in Austria, presumably during a previous winter, was shot at Malpartida de Cáceres during the 1990–91 winter (*Ardeola* 42: 216).

The Greater Scaup is rarely recorded in Portugal and there are just 15 published records (31 birds) during 1996–2008 (AP, *Anu. Ornitol.* 7), most of them from coastal wetlands, notably the Lagoa de Santo André, Baixo Alentejo. There was an exceptional gathering of up to six at the Vilamoura wetland, Faro during January 2009, two of which remained there until at least 7 March (*Anu. Ornitol.* 8: 55–56).

It is only a vagrant to the Balearics. Munn (1926) saw an immature bird at Alcudia in November 1924. Clavell (2002) gives four records: three birds at the Albufera des Grau, Menorca, in December 1985, one there in December 1988 and two at the Albufera de Mallorca in November 1991. The fourth record: of 40 birds at Binigaus beach, Menorca, on 6 January 1988 (*A. O. Balears 1988*: 23) is highly exceptional if the identification was correct. Since then there has been one bird, again at Es Grau, in December 2002 (*A. O. Balears 2002*: 103)

LESSER SCAUP *Aythya affinis*

(Porrón Bola, Negrelho-americano)

Very rare. Recorded in Spain and Portugal, mainly in winter.

It was first found in Europe in the late 1980s but by 2008 there had been 157 individuals reported in Britain, 22 in Ireland and 16 in France. There are also nine accepted records from the Azores, 12 from the Canary Islands and four from Morocco.

The earliest Iberian observation was at Laxe, A Coruña, in January 1991. By 2010 there were 16 accepted records in mainland Spain and four in Portugal. They involved 24 individuals: 18 males and seven females. These comprise 18 records of single birds, a pair and a group of four: this last at the Selga de Ordás reservoir, León, in December 2006. Extreme dates are 17 October and 22 April but 70% of first observations were in November–January (Fig. 31). Records are concentrated in the north-west, more than half of them in Galicia (Fig. 32). There were at least three 2011–12 records from Spain and one from Portugal.

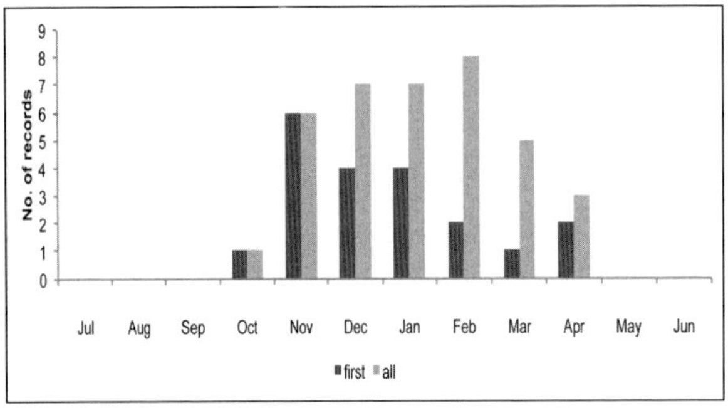

Figure 31. Records by month, 1991–2010 (n = 20)

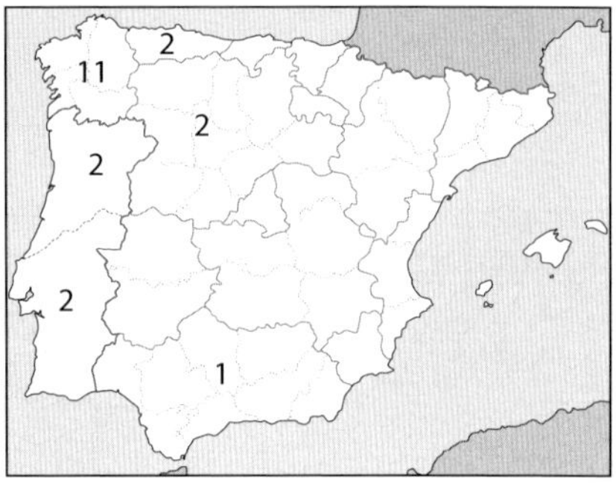

Figure 32. Records by region, 1991–2010 (n = 20)

COMMON EIDER *Somateria mollissima*

(Eider Común, Êider-grande)

***S. m. mollissima*. A very scarce winter visitor in variable numbers to Biscay and Catalan coasts. Very rare elsewhere.**

The Common Eider was regarded as no more than a winter vagrant to Iberia by Bernis (AMI) and there are hardly any historical records. The earliest documented record was in Portugal where a small flock appeared at Cascais in autumn 1891, from which a female was shot by the King, Dom Carlos (Gil Lletget 1945, AP). In Spain there is an early vague reference to occurrence in Catalonia by Fuset (1913) but the first specimen records, and the only records for their respective decades, were captures of single males at Valdoviño, A Coruña, in early March 1951 and in Santander bay on 26 March 1968 (*Ardeola* 6: 368, 13: 246–247).

The number of records in Spain began to increase from the late 1970s, for example in Catalonia and in Asturias (*Ardeola* 22: 111–112, 28: 152) but the first recorded occurrence of some numbers was during the 1988–89 winter, when there was an influx into France (OF) and 212 were counted in total in Spain and Portugal in January 1989 (Rufino 1989b, Troya & Bernués 1990). Thereafter there have been frequent and locally regular sightings in recent years, usually of very small numbers, and the species has never been included in the Spanish rarities list. The great majority are found in the eastern Biscay coastlands: from the Basque Country and Santoña, west to central Asturias. There are notably fewer further west; in western Asturias and around Galicia. The northeastern Mediterranean coastlands also attract some birds in winter, perhaps originating from the Camargue where 20–80 winter regularly (OF). There are annual records from Catalonia, mainly from the Ebro delta northwards, as well as less regular observations from the Comunidad Valenciana.

Records elsewhere are very rare. The Common Eider is a vagrant to Portugal and there have only been nine published records during 1980–2007, as well as a report of one at the Minho estuary on 22 January 2011 (RP). Thus far, only four records have been submitted and accepted by the Portuguese rarities committee (AP, CPR 2008–09). Five of the recent records have been in coastal Algarve and there are also three from Cape Raso, Estremadura: only two records, from Beira Litoral and the recent Minho bird, have been in the north of the country, a circumstance that may reflect the distribution of observers more than of the birds themselves.

Observations in southern Spain are also very rare. There are seven published records in total from Murcia, Andalucía and Gibraltar, with extreme dates of 22 October and 22 January. Inland sightings are exceptionally rare. There was a male at the

Almoguera reservoir, Guadalajara, on 3 February 1985 during a widespread cold snap (*Ardeola* 33: 205); a female was at the Balsa del Pulguer, Navarra, on 6 October 1988 (*Ardeola* 36: 240) and a male was at the Ullibarri-Gamboa reservoir, Álava, on 31 January and 1 February 2008 (*Ardeola* 55: 134).

The Common Eider is accidental in the Balearic Islands, with only two known records: a male at Palma de Mallorca from January to 15 April 1987 (*A. O. Balears 1987*: 15) and one bird at the Port of Maó, Menorca, in December 1988 (*A. O. Balears 1988*: 23).

The majority of observations fall between September and March (Fig. 33). Some birds remain for extended periods at particular sites. Regular wintering locations include Gijón Bay, Asturias and Santoña on the Biscay coast, and the Ebro and Llobregat deltas in the Mediterranean, but individuals or small groups may settle to winter at any sheltered coastal sites in the north or north-east. Occasional individuals have been reported summering in the same regions. A more remarkable instance of over-summering occurred at Santoña where a flock of 16 was present throughout 1991 (*Ardeola* 40: 91).

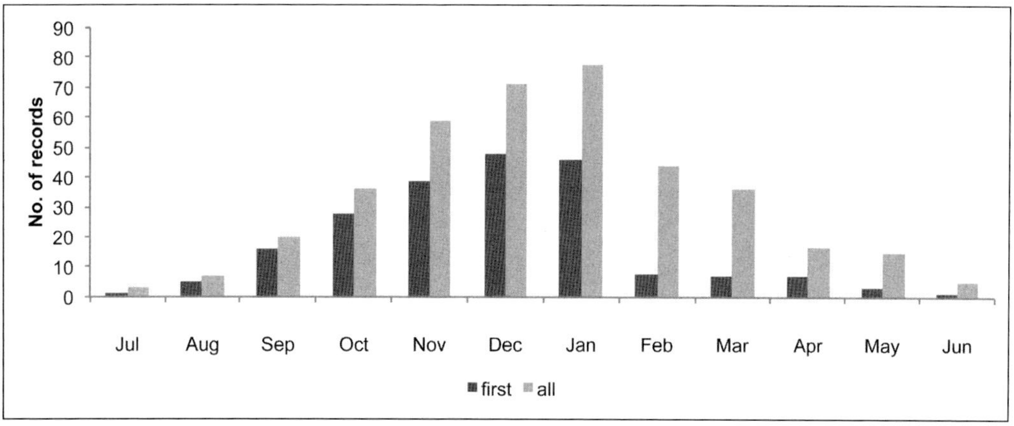

Figure 33. Records by month, 1990–2010 (n = 209).

The post-1989 records show that the numbers reaching Iberia vary markedly between winters. Most years see very few birds indeed but there were influxes in 1989–90, 1990–91, 1995–96 and 2002–03 when some sizable gatherings were reported. By far the largest on record involved 300 birds at Zarautz, Gipuzkoa in November 1990 (*A. O. Gipuzkoa 1990*: 12). An influx in autumn and winter 1988–89 affected the Mediterranean coasts but neither the Atlantic ones (*Ardeola* 38: 332) nor the Balearic Islands: there were 212 individuals present during January, including 93 at the Ebro delta, 54 at the Aigüamolls de l'Empordà and another 54 at the Albufera de Valencia (AI1).

The records from the north and northeastern coasts include 276 sexed individuals, of which 109 were males. Only nine of 68 males whose age was noted were reported as adults. The inference is that immature individuals, of both sexes, predominate among birds that reach Iberia.

KING EIDER *Somateria spectabilis*

(Eider Real, Èider-real)

Vagrant to Spain.

A very rare winter visitor to Britain, with 226 individuals recorded mainly in Scotland, and 18 have been found in Ireland. There are only ten records from France and none from North Africa.

There are five accepted records for Spain, four of them in the north: three in Galicia and one in Catalonia, and one in the southeast, in Murcia. The first three were first-winter males found at the Ría de Corme y Laxe from 9–16 March 1980, at the same site from 10 December 2000 to 1 April 2001, and at the Ría de Arousa on 1 February 2004. The Catalan record was of an adult male seen at the Ebro delta on the remarkable date of 21 June 1987 (Martínez-Vilalta 1989). The Murcia record was also surprisingly untimely: it involved a first summer female that was at the Mar Menor lagoon from 14 May–6 June 2010.

LONG-TAILED DUCK *Clangula hyemalis*

(Pato Havelda, Pato-rabilongo)

Annual in Iberia in winter, although always in very small numbers and principally on northern coasts.

Many winter in northwestern Europe, particularly in the Baltic, but they become progressively scarcer further south. The annual wintering population in France during 1988–2006 ranged from just 11 to 100 birds, these mostly on Atlantic coasts south to the Gironde estuary (OF, Zucca et al. 2010).

The earliest reports were in December 1921, when one was shot at the Albufera de Valencia (Pardo 1922) and another was seen at Aveiro, in northern Portugal (Tait 1924). One was taken at the Ebro delta in 1923 and two more were obtained there around 1925 (Bernis 1956a). Single males were killed at the Albufera de Valencia in December 1955 (Bernis 1956a) and November 1957 (Sala 1958), and the first record from Andalucía was in January 1958, when a female was seen at the Roquetas de Mar saltpans, Almería (Valverde 1959). The earliest observations on the Biscay coasts were at Gijón in December 1978 and at Santoña in February 1980 (*Ardeola* 28: 152). The Spanish rarities committee considered the species during 1984–93 (de Juana 2006) and accepted 47 records. Many others have occurred subsequently. There are also 11 accepted records from Portugal during 1991–2009, and two more from 2011–12.

There are annual observations for all winters since 1982–83, totalling 132 records (177 birds). Peak occurrences were in 1988–89 (12 records, 15 birds) and 1991–92 (12 records, 27 birds), when more than usual were also found in France (OF). Although 83% of records involved single individuals, there have been groups of up to six, these last appearing during winter 1991–92, at Sanxenxo, Pontevedra, and at Santoña. The sample included 79 females and 47 males: among these 14 of the females were adults and nine were first-winters, and the males included 14 adults/second-winters and 21 first-winters.

It is principally found in winter, in bays, estuaries and coastal inlets as well as on coastal lagoons, reservoirs and saltpans. Some 86% of first observations have been during October–January, with extreme dates of 3 October and 13 May (Fig. 34): with the surprising exceptions of a probable female at the Lagunas de Espera, Cádiz, during July–August 1990 (*Ardeola* 39: 76), an adult male at Vinaròs, Castellón, in July 1991 (*Ardeola* 40: 184) and a first summer male at a couple of reservoirs in Corvera, Asturias, during June–September 1999 (*A. O. Asturias* 6: 34). At least half a dozen long-stayers have remained at particular locations for 4–5 months. Records show a concentration on northern coasts (Fig. 35) and 63% in total have been in Galicia, Asturias, Cantabria or the Basque Country. There are nonetheless quite a number from Mediterranean coasts and Andalucía. A decayed corpse was found in the Balearic Islands in May 1986 and two females were seen in December 1994 at the Salobrar de Campos, Mallorca, where one remained into January 1995 (*A. O. Balears 1994*: 109, *A. O. Balears 1995*: 107). Only two have been far inland: at Gallocanta in March 1992 (*A. O. Aragón 1991–1992*: 32) and at Villafáfila, from November 2006 to March 2007 (*Ardeola* 54: 171).

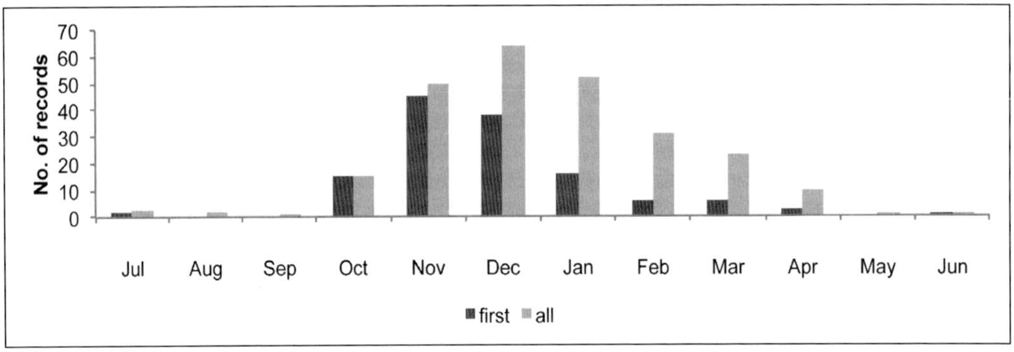

Figure 34. Records by month, 1921–2010 (n = 132)

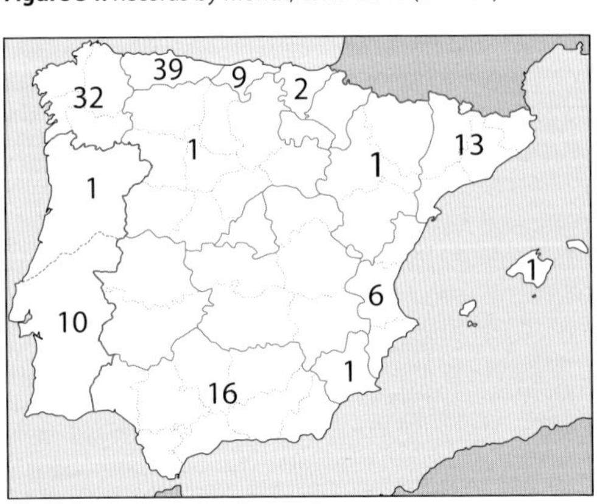

Figure 35. Records by region, 1921–2010 (n = 132)

COMMON SCOTER *Melanitta nigra*

(Negrón Común, Negrola-comum)

Common on passage and in winter on Atlantic coasts, much scarcer in the Mediterranean. Rare inland.

The Common Scoter nests across the northern Palearctic. Most of the population remains in northern waters but large numbers follow Atlantic coasts to winter locally off Iberia and also off Morocco (BM), with some reaching northern Mauritania (Isenmann *et al.* 2010). Flocks are often seen on passage from headlands and other coastal sites across the Biscay coasts, and along Atlantic coasts south to the Strait of Gibraltar. Very few pass through the Strait into the Mediterranean, however.

The southward autumn passage occurs from late July to early December, but chiefly in October–November. The largest numbers are reported from well-watched north-coastal vantage points, such as Cape Estaca de Bares and Punta La Vaca. A huge passage was seen by Snow *et al.* (1955) at Cape Bares on 7 October 1954: flock after flock passed for many hours and a sample count recorded 5,650 in two hours in the middle of the day, declining to about 800 per hour in the late afternoon. A total of 22,679 were counted in 200 hours of watch at the same location during October 1969, a mean rate of 113 birds per/hour (Huyskens & Maes 1971). More recent autumn counts from Cape Bares have tended to be lower and the Common Scoter is said to have declined sharply as a migrant there since the 1960s, when a peak passage rate of 1,361 birds/hour was counted on 29 September 1967 (Sandoval *et al.* 2009). In contrast, the passage rate of 272 birds/hour on 11 October 2009 was the highest recorded in the previous ten years (Sandoval *et al.* 2010). On the basis of sample counts, the mean annual post-breeding daytime passage at Cape Bares during 2004–08 was of the order of 40,000 birds (Sandoval *et al.* 2010). Elsewhere, a total of 13,674 were counted on westward passage in autumn 1991 at Punta La Vaca, with peak rates of over 1,000 birds/hour (García Sánchez 1993). It is very likely that many more birds pass at night. The return movements are less evident, or less well observed, but occur in March–April. There is a count of 1,534 birds northbound in two hours at Cape Silleiro, Pontevedra, on 1 April 1989 (*Anuario Histórico das Aves de Galicia*: 28).

Wintering birds are very unevenly distributed and the bulk of the population is concentrated off the Portuguese coast north of Cape Mondego: especially off the Ria de Aveiro in north-central Portugal, and in the Gulf of Huelva, off Doñana. Several thousand birds winter at each of these sites, which offer relatively shallow waters and a sandy seabed with abundant bivalve molluscs, the principal diet item. Aerial surveys off the Ria de Aveiro in 1992 found some 5,000 birds, many of them well offshore (Rufino & Neves 1990) and the total numbers wintering in Portuguese waters were estimated at 6,000 that year. Aerial counts in other years have found larger numbers. A count of 18,643 Common Scoters was made in winter 1968–69, the great majority in northern and central coastal waters. The Portuguese wintering population in the 1990s was estimated to vary from 6,000 to 27,000 birds according to year (AP). Relatively few winter elsewhere off Portugal but a few hundred occur on the Algarve coast, in the vicinity of the Guadiana estuary (AP). There is evidence that the recent wintering population in Portugal, though substantial, is a fraction of the numbers that occurred during the late 19th and early 20th centuries, when contemporary authors referred to the Common Scoter being 'extremely abundant' and occurring in 'immense flocks' (Tait 1887, AP). An adult ringed at the nest in Iceland in 1928 and recovered in Portugal several years later (*Ardeola* 2: 79–86) is the only known Iberian recovery.

The numbers that winter off Spanish coasts clearly vary considerably from year to year and it is difficult to detect any recent trends, although marked declines have certainly occurred locally. The largest count in the Gulf of Huelva was 3,235 birds in January 1983 (Ena & Purroy 1984) although up to 5,000 have been estimated to winter there in some years (García *et al.* 1989). Smaller numbers occur all around Atlantic coasts, but there are local concentrations of up to a few hundred birds, and sometimes more, off the Rias Baixas, Pontevedra; in Gijón bay; Santander bay and at Santoña (AI1, Paterson 1997). The numbers wintering off the Rias Baixas have sometimes been much larger: 6,000 birds were counted in the O Grove inlet in January 1976 (Carbonell & Muñoz-Cobo 1980).

The Common Scoter is present but very thinly spread off Mediterranean coasts. Small flocks of up to ten birds, and occasionally up to 40, are widespread. Much larger concentrations have occasionally been seen off Catalonia, particularly at the Ebro delta: including 316 in January 1976 (Ena & Purroy 1982) and 388 in November 1998 (*Ardeola* 49: 181–182). Up to 140 have been reported wintering in Málaga Bay in some years (Paterson 1997). There are only 13 records for the Balearic Islands, all of single birds or groups of up to four individuals (Clavell 2002; *A. O. Balears 2003*: *96, 2007*: 118, *2009*: 84). All have been between late October and January with the exception of a moribund female that appeared at Ibiza on 24 May 2003.

There are occasional sightings of individuals in summer, even in the south, as well as very rare records inland. The latter usually involve single individuals but there is an extraordinary report of a flock of 50 that appeared at a pool near Badajoz city, Extremadura, during winter 1978–79: most left in late January but one male remained until April (*Ardeola* 28: 152).

BLACK SCOTER *Melanitta americana*

(Negrón Americano, Negrola-americana)

Vagrant to Spain.

Seldom found in Europe: there are eight records for Britain and ten for France, most of them in winter. All were males, females being indistinguishable from those of the Common Scoter in the field.

The three Spanish records are all from northern coasts: an adult male at the Ría de Navia, Asturias, on 1 November 1986, a male at the Ría de Villaviciosa, Asturias, from 29 October to 5 November 1994 and a first-winter male at the Ría de Corme y Laxe, Galicia, from about 10 February to 24 March 2002.

SURF SCOTER *Melanitta perspicillata*

(Negrón Careto, Negrola-de-lunetas)

Very rare. Recorded in Spain and Portugal, mainly in winter.

A regular winter visitor to northwestern Europe, off Scotland in particular. There are records of 399 individuals in Britain in 1958–2003 and of 168 in Ireland up to 2010. There are also 62 accepted records for France up to 2009 and six from the Azores, but none from North Africa.

The first known Iberian occurrence was at the Ría de Corme y Laxe, A Coruña, during January–March 1983. There are 19 accepted Spanish records up to 2009, involving 20 birds, and at least five further observations up to 2012. There are also two accepted Portuguese records: at the Sado estuary, Setúbal, during January–February 2000, and at Cape Raso, Lisboa, in March 2001. Records have averaged one per two winters since 1982–83 but there were birds at three sites in winter 1999–2000. Some 90% were first observed between October and January, extreme dates being 1 October and 30 April. All have been found on Atlantic coasts, especially off Galicia. The most southerly observations were in Huelva province: two birds off Doñana beach on February 1991, and a first-winter male at Odiel between January and April 2008.

VELVET SCOTER *Melanitta fusca*

(Negrón Especulado, Negrola-d'asa-branca)

Regular but very scarce in winter, chiefly on northern and northeastern coasts. Vagrant to the Balearic Islands.

This species winters around the Baltic and North Seas, the northern French coasts and south to the Black and Caspian Seas. Small numbers reach the Camargue in southern France, where up to 20 winter annually – occasionally over 100 – and flocks of several hundred occur there on passage in spring (OF).

Irby (1883) obtained a female in Santander Bay on 21 November 1876 but later references to this species in Iberia are otherwise lacking until quite recently. Bernis (1956a) examined one collected at Sueca, Valencia, in December; four were seen and one collected in Roses Bay, Girona, in December 1958 (Balcells 1961) and one was taken from a group of three in the Ría de Vigo, Pontevedra, in December 1961 (*Ardeola* 8: 273–274). Very few were reported thereafter until the mid-1980s, so that the Velvet Scoter was regarded as extremely rare in Spain by Santos and Sáez-Royuela (1985). It is on the Portuguese rarities list and the earliest record there seems to have been one at São Jacinto, Aveiro, on 18 February 1978 (*Airo* 2). There are only ten published Portuguese records up to 2011, totalling 25 individuals: comprising seven records north of the Mondego estuary and single birds seen in the Tagus estuary, the Sado estuary and off eastern Algarve (AP, RBP). Four records have been accepted by the Portuguese rarities committee.

Virtually all Velvet Scoter records fall within September–March and the great majority are reported during November–January (Fig. 36). The large number of new sightings in January suggests that many arrive late in Iberia after beginning the winter further north. There is an exceptional late spring record of seven at the Ebro delta on 26 May 1989 and an unusual report of a flock of eight flying north-west at Punta La Vaca on 10 July 1991 (*Ardeola* 36: 240, 42: 216). A sighting of three at Cape Matxitxako, Bizkaia, on 29 July 2002, is also unseasonal (*A. O. Bizkaia 2002–2003*: 85).

The birds on the Biscay coastlands are reported chiefly from Asturias and Cantabria (Fig. 37). In Asturias there are sometimes some in winter in Gijón Bay and occasionally a small flock has been present: there were 38 there in winter 1991–92 (*Ardeola* 40: 91) but only two were found in the January census in Asturias in 2008 and none at all in 2006–07 (*A. O. Asturias*). The other principal wintering site is Santoña, where the numbers reported range from just a few individuals to ten or more: 16 were present during January–April 2008 (*A. O. Cantabria*). A few are also reported in winter in Santander Bay in some years. There are annual reports from Galicia in winter but only very small numbers are involved: 45 published records for 1993–2004 involved a total of just 82 individuals. Most observations in Galicia are of single birds or twos and threes, exceptionally as many as seven. The few reports from the Basque Country include a flock of 30 off San Sebastián on the early date of 17 September 1994 (*A. O. Gipuzkoa 1994*: 25).

The great majority of the Mediterranean records are from coastal Catalonia; from Roses Bay south to the Ebro delta. Most sightings are of a few birds or flocks of up to ten, often in the company of Common Scoters, but Roses Bay has held a larger wintering flock in some years, with 29 in January 1989 and 45 in January 1991 (Paterson 1997; *Ardeola* 40: 91), although counts there have not exceeded ten birds post-2000 (*A. O. Catalunya*). Most of the birds that reach Catalonia and the southern French coast probably do so overland across central Europe since there is no known passage through the Strait of Gibraltar or across mainland Iberia: although most of the few inland records are from the Ebro valley. Just a few individuals are reported further south: in the Comunidad Valenciana, Murcia or Andalucía, in some winters. There are exceptional counts of seven at Cullera Bay on 21 January 1989; up to ten off Gandía during January–February 2009, and six at the Mar Menor during January–February 2013 (*A. O. C. Valenciana 2009*: 17; *Ardeola* 36: 240, 60: 160). The few Andalucían records include two from the Strait: three birds on 27 March 1993 and one on 11 November 2004 (*Ardeola* 43: 245, 52: 426). There is just one record from the Balearics: a female in Maó harbour, Menorca, from 8 January to 17 March 1985 (*Ardeola* 34: 280; *A. O. Balears 1985–1986*: 12). Inland records are very rare but there was one at the Laguna de Lor, Navarra, in December 1988; one at Saucedilla, Cáceres, in January 2002, four birds at the Saladas de Alcañiz, Teruel, from December 2003 to January 2004; one at the Lagunazo de Moncayuelo, Zaragoza, in March 2012 and one at El Perdiguero reservoir, La Rioja, in March 2012 (*Ardeola* 36: 241, 54: 171, 59: 171; *A. O. Extremadura 2001–2003*: 126).

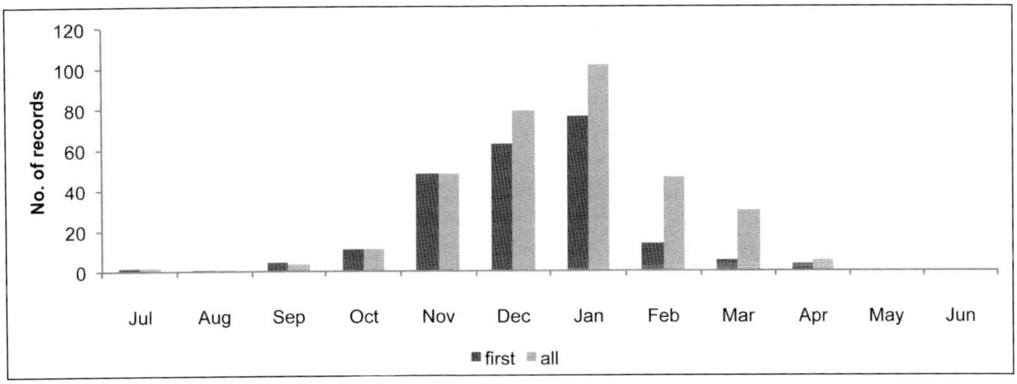

Figure 36. Records by month, 1990–2010 (n = 226).

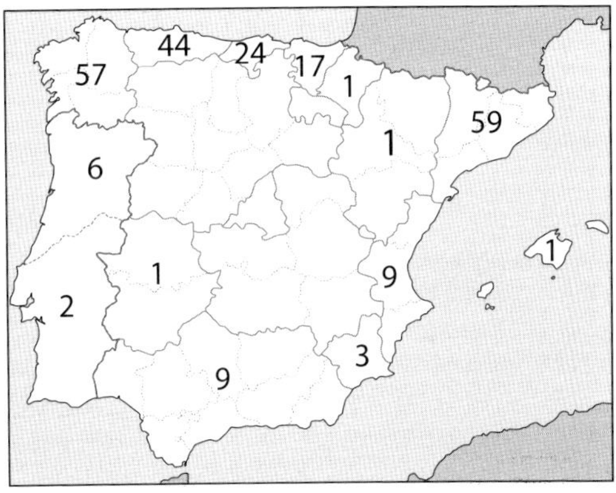

Figure 37. Records by region, 1990–2009 (n = 234).

The total Iberian wintering population may have comprised 100–200 birds during some years in the 20th century: for example, there were site counts of 50 in 1982 and 125 in 1989 (SWA), but it has probably not exceeded 50 birds post-2000 and there were no more than 20 during 2007–10, with none at all seen in some years in such traditional areas as Asturias and northern Catalonia (COA, CWA, SWA). Clearly this formerly rare or at least very uncommon wintering species attained peak abundance in Iberia during the 1980s but has since become scarce once again. A similar trend is shown by the French wintering population, which increased to a historical maximum of nearly 9,000 birds in 1986 but declined thereafter to average 700 in January during 1997–2006, of which only a few dozen were in the south (OF). This progressive scarcity may be due to a decline in the source populations but also to a tendency for wintering further north, enabled by recent mild winters (Jensen & Lutz 2007).

WHITE-WINGED SCOTER *Melanitta deglandi*

(Negrón Aliblanco, Negrola-d'asa-branca-americana)

Vagrant to Spain.

An adult male of the Asian subspecies *stejnegeri* was seen in December 2011 in Galicia, first at the Ensenada de San Simón, Vigo, and later on at A Lanzada beach, O Grove. There were eight Western Palearctic records of this species by 2008: in Iceland (5), Finland, France and Poland (Haas 2012).

BUFFLEHEAD *Bucephala albeola*

(Porrón Albeola, Olho-dourado-de-touca)

Vagrant to Spain and Portugal.

The relatively few observations in Europe have often been attributed to escapes but their occurrence on the Atlantic seaboard in winter, including multiple records in some winters, points to a natural origin for many. There are 14 accepted records for Britain, two for Ireland, five for France and two for the Azores.

There are four accepted records for Iberia. The first was a female at the Laguna de Traba, A Coruña, on 25 December 1992, which remained there until at least 9 January 1993. What may have been the same individual was subsequently at the Tagus estuary from 24 January to 7 February 1993. Another female was at El Astillero, Santander, from 25 November 1997 to 10 January 1998. The latter locality also attracted a male that remained there from 7 January to at least 21 March 2010. Subsequently an adult male was photographed on the Sopeira reservoir, Huesca, in December 2012 (RBS). All the dates and most of the locations suggest birds of natural origin.

BARROW'S GOLDENEYE *Bucephala islandica*

(Porrón Islándico, Olho-dourado da Islândia)

Vagrant to Spain.

A first-winter female was at the Laguna de Traba, A Coruña, from 26 January to 22 March 1997 (Rabuñal 2000). In addition, a young male is said to have been captured at the Albufera de Valencia on 22 December 1871 (Saunders 1871). Occurrences in mainland Europe have often been attributed to escapes from captivity. However, three observations from Britain are accepted as being of natural origin. There are also two 19th century records from France and three more recent ones there: in January 1972, February 1983 and December 1989 (OF).

COMMON GOLDENEYE *Bucephala clangula*

(Porrón Osculado, Olho-dourado-comum)

***B. c. clangula*. Very scarce and irregular in Spain in winter. Vagrant to Portugal and the Balearic Islands.**

Most Common Goldeneyes winter in northern Europe and very few venture as far south as Iberia. They most frequently appear in the Biscay coastlands, from Galicia east to the Basque Country (Fig. 38). There are also fairly frequent sightings in the Mediterranean coastal regions, chiefly in Catalonia, but there are at least eight records from the Comunidad Valenciana, one from Murcia, and seven from Andalucía. There is a scattering of inland records, often from reservoirs, with rare observations from Navarra, La Rioja, Aragón, Madrid, Castilla y León and Castilla-La Mancha as well as three records of single birds from the Cádiz lagoons, Andalucía.

There are only two records of single birds in the Balearic Islands, in Maó harbour, Menorca, on 18–26 January 1987 and at the Albufera de Mallorca during 2–18 December 1997 (*Ardeola* 34; 280; *A. O. Balears 1997*: 204). The Goldeneye is a rare vagrant to Portugal known only from three pre-1921 specimens and three recent records of single birds: in the Tagus estuary on 15 December 1991; at Cabo Raso, Cascais, on 24 November 2006, and at Peniche, Leiria, during 10–14 January 2009 (AP, CPR 2008–09).

The Common Goldeneye was regarded as a rare, or at best scarce, winter visitor by Bernis (AMI) and there are few pre-1950 records, possibly at least in part because of under-recording. Saunders (1871) saw one in Málaga market in January 1868 and Irby (1883) saw an adult male at Santander on 3 December 1876. During the early 20th century, in addition to the three Portuguese specimens, Fuset (1913) refers to one at Barcelona University and Bernis (1956) found two in Valencia, one of them obtained at the Albufera around January 1947. There were two further records in 1956: at Santillana reservoir, Madrid, on 4 March, and at the Ebro delta: where two females were killed in winter (*Ardeola* 3: 189–190, 9: 139). We are not aware of any records from the 1960s and just six birds were reported during the 1970s, all of them killed in winter: three at the Ebro delta; one at El Taray lagoon, Toledo and two at El Vellón reservoir, Madrid (*Ardeola* 15: 118, 20: 331, 22: 110, 24: 255). Records in Spain became much more frequent during the 1980s and 1990s, although some recent decline has been noted subsequently and only nine individuals were reported during 2007–10 (SWA).

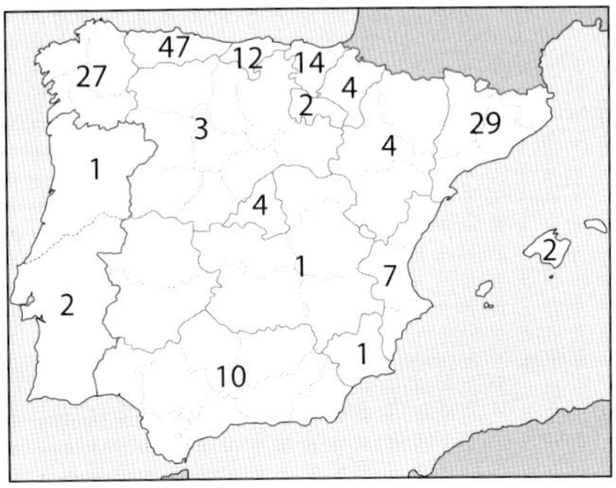

Figure 38. Records by region, 1975–2010 (n = 170)

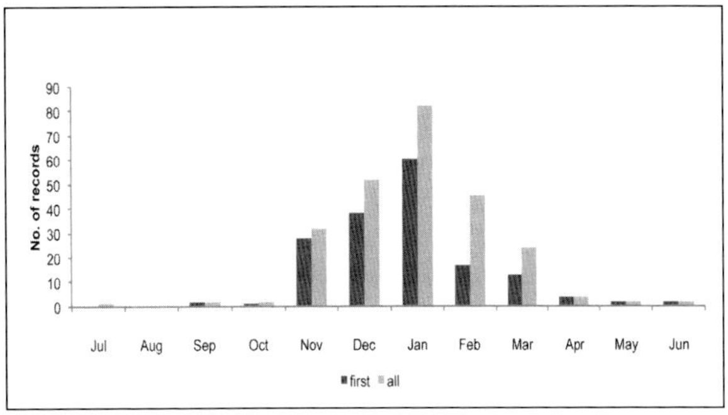

Figure 39. Records by month in Spain, 1975–2010 (n = 167).

The wintering birds are chiefly present during November– February (Fig. 39). The increase in new records in December and, especially, in January suggests that many of the individuals involved spend the earlier part of the winter further north. The most unseasonal records include a male at the Rambla de Morales, Almería, on 9–10 June 2004 and a female at La Granda reservoir, Asturias, from 28 June to 7 July 1989 (*Ardeola* 51: 547; Álvarez-Balbuena *et al.* 2000).

Wintering Goldeneyes may be found on sheltered coastal inlets and lagoons, such as the Ría de Villaviciosa, Santoña, the Aiguamolls de l'Empordà and the Llobregat Delta, each of which has produced seven or more records. The deep waters of reservoirs, such as the Ullibarri reservoir, Álava, and La Granda reservoir, Asturias, have also provided multiple sightings. Some 70% of observations are of single birds and two birds were involved in a further 15% of records. There are eight of four or more birds, all from the northern coastal regions of Asturias, Cantabria and the Basque Country. A group of seven at Luanco, Asturias, on 5 November 1991 (*Ardeola* 38: 332) is highly noteworthy but by far the largest group on record is the bizarre flock of 14 males seen (and photographed) a short distance inland from the Mediterranean on the Pedret reservoir, Girona, on 13 March 2006, some of which remained there until 20 March (*A. O. Catalunya 2008*: 5). Adult males, which are unmistakable, are usually very rarely reported. Not including the 14 at Pedret, they account for only 18 individuals among the 149 birds whose sex was reported. No fewer than 131 birds (88%) were reported as females. The 149 included six identified as first-winter males, six first-winter females, 11 identified as females/first-winters and 13 first-winters.

SMEW *Mergellus albellus*

(Serreta Chica, Merganso-pequeño)

Very rare. Recorded in Spain and Portugal, mainly in winter.

Wintering birds occur in northern Germany, Denmark and the Netherlands, with small numbers in England. Some 200–300 have wintered in France in recent years, mainly in the north-east (OF). Unknown in Morocco but there are two records for Algeria.

Irby (1895) considered it to be an occasional winter visitor to the Strait and there are specific early records from the Balearic Islands: at the Albufera de Mallorca, in December 1929 (Munn 1932) and the Albufera de Valencia (Arévalo 1887, Dies *et al.* 1999). One is said to have been taken in Portugal near Alcochete in December 1921 and the King, Dom Carlos, described it as a rare visitor in December–January, but there are no recent or confirmed records for the country (AP). Two were taken at the

Aiguamolls de l'Empordà in February 1954 (Palaus 1960) and a female was obtained at the Ebro delta in February 1956 (Bernis 1964b). There are also records from the Ebro delta in April 1969 (Maluquer 1971), the Grado reservoir, Huesca, in November 1973 (Lucientes 1977) and in Galicia in January 1983 (*Ardeola* 30: 102). There were 25 accepted records from Spain during 1984–2010. Only five winters have produced more than a single record: 1984–85, 1986–87, 1996–97, 2005–06 and 2009–10. There are also two reports from February 2012. We are also aware of a number of unsubmitted records from the north-east (Clavell 2002).

The 31 records available up to 2010 involved 44 individuals. Single birds made up some 71% of records and there are eight records of two together but only one of a flock: of up to seven at the Ría de Villaviciosa in January–February 1985. Sexed individuals comprised nine males and 22 females (some of which may have been first-winter birds of either sex). Almost 80% of first occurrences have been during November–February (Fig. 40) and the extreme dates are 26 September and 22 May. This excludes an odd record of one that was at the Azud de Riolobos, Salamanca, in August–September 2002. Most records were single-day observations and only three stays of over one month are known. This suggests that birds enter Iberia only briefly, perhaps in response to cold spells further north, and they occur almost entirely in the north (Fig. 41). One at the Guajaraz reservoir, Toledo, in December 1990 (*Ardeola* 40: 184) is the only record in the southern half of Spain.

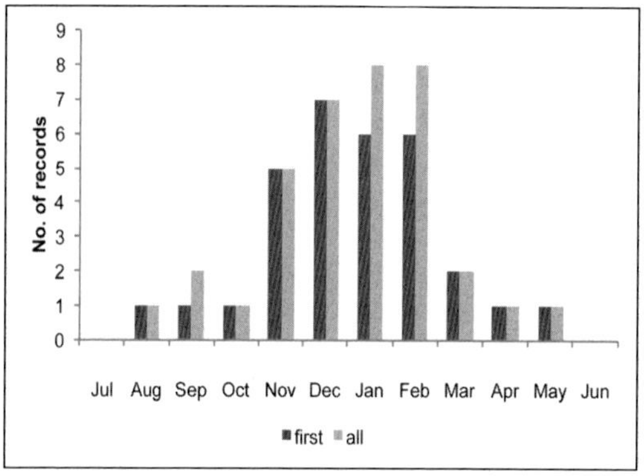

Figure 40. Records by month in Spain, 1954–2010 (n = 31)

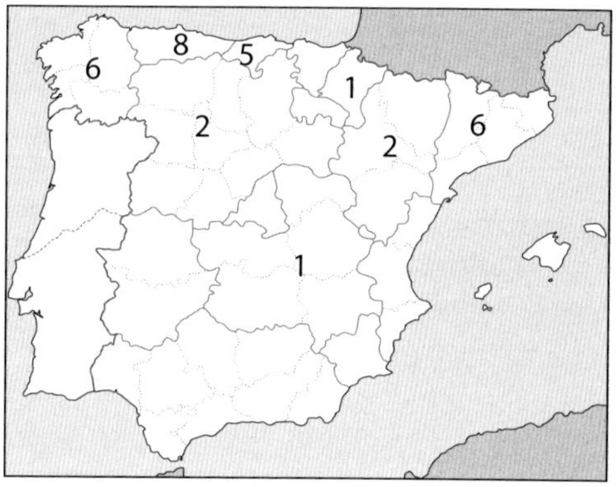

Figure 41. Records by region in Spain, 1954–2010 (n = 31)

HOODED MERGANSER *Mergus cucullatus*

(Serreta capuchona, Merganso-capuchinho)

A male appeared after a cold spell on a pool at Santa Cristina del Páramo, León, during 8–24 December 2009. The few European records of this Nearctic species have tended to be attributed to escapes, despite their concentration on Atlantic coasts and the existence of four records from the Azores and one in the Canary Islands. There were 23 records from France by 2010.

RED-BREASTED MERGANSER *Mergus serrator*

(Serreta Mediana, Merganso-de-poupa)

Scarce but locally common on coasts in winter. Occasional inland.

The Red-breasted Merganser breeds across the northern Holarctic. The winter quarters in western Europe are chiefly in the north-west and only small numbers occur on Iberian coasts. Nevertheless, it is regular at a few locations, including Santoña; the Rías Bajas of Galicia: in particular that of Arousa; the Sado estuary; Odiel; the Mar Menor, and the Ebro delta. There are no ringing recoveries to indicate the possible origin of the wintering birds.

A census in 1989 found some 400 individuals in Iberia, including 122 at the Sado estuary, 91 in the Mar Menor and 60 at Odiel (Rufino 1989b, Troya & Bernués 1990, AP). The Spanish regional bird reports for 1995–99 give a total of 179 records: 93 (485 birds) from Atlantic and Biscayan waters and 89 (206 birds) from the Mediterranean, and show a trend of diminishing abundance from west to east across the north coast and from north to south in the Mediterranean. A count in 2002 found 388 birds in Spain: including 127 in Catalonia, 190 in Galicia and 48 in Andalucía, and the Spanish wintering population during 2007–10 was estimated at 300 birds (SWA). The numbers found at the principal wintering sites vary considerably from season to season.

December/January counts at Arousa have ranged from 135 in December 2003 to over 300 in January 1998 (*A. O. Galicia*). The Ebro delta wintering population was only 37 in January 1999 but there were 126 there in January 2002 (*A. O. Catalunya*). The Mar Menor was formerly the principal wintering location in Spain and January counts there during 1972–90 averaged 174 birds, but ranged from nil to 353 (SWA). Recent counts at the Mar Menor have been smaller, following a decline in fish prey there, and the Spanish winter atlas counts of 2007–10 found an average of 48 Red-breasted Mergansers in Murcia, with a range of 31 to 65 birds (SWA). Overall, the wintering population in Spain has declined by over 40% since the early 1990s (SWA).

The Portuguese wintering population averaged 184 individuals during 1993–96, with a peak count of 358 in January 1993, but it had declined to an average of 87 birds during 2005–07 (AP). Although the majority favour the Sado estuary, some are seen with some regularity elsewhere, notably in the Minho and Tagus estuaries, the Albufeira lagoon and at the Ria Formosa, Algarve (AP).

The Red-breasted Merganser is recorded annually in the Balearic archipelago but only in tiny numbers. The regional bird reports give 91 records during 1985–2010, 47 of which were of single birds, the remainder mainly involving 2–3 individuals. The largest recorded count was just six, in Alcudia Bay, Mallorca, during winter 1999–2000 (*A. O. Balears 1999*: 60). It may have been more numerous in winter in the past since Munn (1921) saw several small groups wintering in Alcudia Bay in 1920–21.

All wintering sites are in estuaries or other sheltered coastal waters. Winterers are largely reported in November–January, departing during February–April. However, there are extreme records on 5 May and 26 August. Small numbers of migrants may sometimes be observed on passage from coastal watchpoints. For example, 104 flew west at Punta La Vaca during watches in autumn 1997, with peak passage from mid-November to early December, athough the earliest observation was on 9 September (*A.O. Asturias 2003*: 77). There is no significant passage through the Strait: there are only seven recent records for Gibraltar, involving 11 birds in total (GONHS), but Irby (1895) reported 'considerable numbers' in the Bay of Gibraltar in December and January in some winters.

Red-breasted Mergansers are seldom reported well inland but there are at least 15 records from scattered sites in Madrid, Cáceres, Ciudad Real, Álava, Cantabria, Zaragoza, León, Navarra, Valladolíd and Palencia. Most involved single birds but there were five at the Torre de Abraham reservoir, Ciudad Real, on 25 March 1989 and three on the Río Órbigo, León, on 6 April 2012 (*Ardeola* 36: 241, 59: 171). None of the birds were reported on more than two consecutive days so they were clearly in transit across the Peninsula.

GOOSANDER *Mergus merganser*

(Serreta Grande, Merganso-grande)

Occasional in winter in Spain and Portugal.

In France 220–250 pairs nest in the Alps and small numbers arrive to winter, largely in the north-east. There are only five records for Morocco, the most recent in 1988.

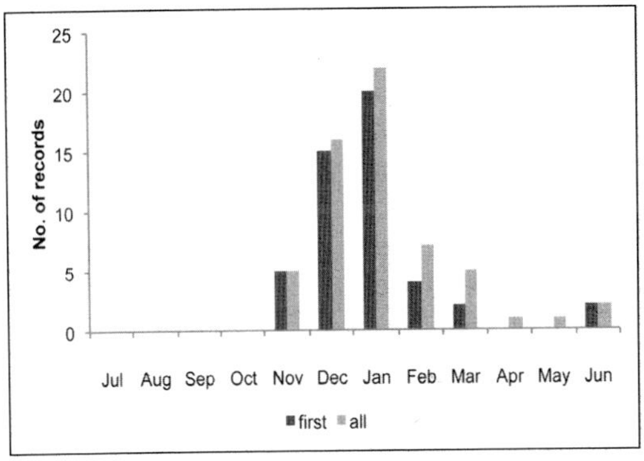

Figure 42. Records by month, 1959–2010 (n = 48)

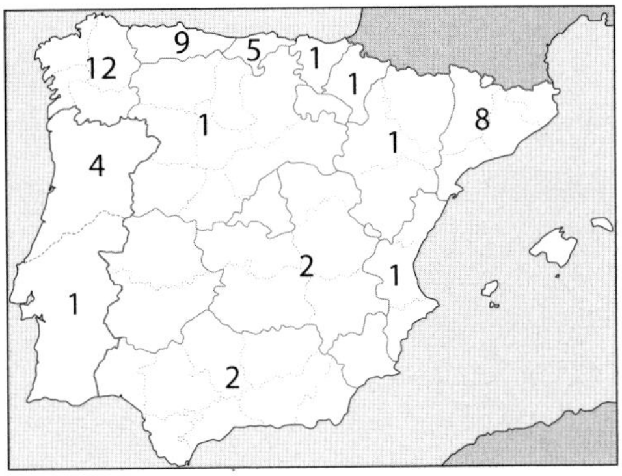

Figure 43. Records by region, 1959–2010 (n = 48)

There are a few records for Spain from the 19th century and the first half of the 20th century: from Galicia (Ríos 1850), Málaga (Arévalo 1877), Santander (Irby 1883), the Albufera de Valencia (Bernis 1956a), the Balearic Islands (Munn 1921) and Barcelona (Fuset 1913). Thereafter there are several documented records up to the mid 1980s, chiefly from the Biscay coasts and coastal Catalonia (Sáez-Royuela & Santos 1985, Ferrer *et al.* 1986). These include a notable arrival in January 1979, which coincided with an influx into France, when Goosanders appeared at Santoña; the Ebro reservoir, Cantabria (11 birds); the Ría de Villaviciosa (14 birds), and the Tordera estuary, Girona (*Ardeola* 28: 152, Ferrer *et al.* 1986, Álvarez-Balbuena *et al.* 2000). The rarities committees accepted 33 records for Spain during 1984–2010, five of which were during winter 1996–97, when there were also more Goosanders in France than usual (OF). Another exceptional influx into Iberia during occurred during winter 2010–11, when at least 30 were reported at widely spread locations between December and March. Most were on the Biscay coasts or in Catalonia/Valencia, but there were six in Portugal and some also appeared inland: in Aragón, Extremadura and Madrid (*Ardeola* 59: 363, 60: 447–449). Five Spanish records (11 birds) and two Portuguese records (3 birds) from this influx had been accepted by 2012. These recent records apart, there are only three earlier accepted records for Portugal, of single birds at the Minho estuary in February 1997, the Cávado estuary in January 2000 and Figueira de Castelo Rodrigo, Guarda, in December 2008. We are also aware of a number of unsubmitted records, chiefly from Catalonia (Clavell 2002). There were records in five winters during 1990–99 and in seven winters during 2000–10.

The 48 records to 2010 involved 113 birds. Single birds occurred in 69% of cases and there were seven records of pairs or trios and eight records of over three birds. The largest flock was of 16 at the Ría de Villaviciosa in January 1987. Some 73% of first observations have been in December–January, but there are observations from November to June (Fig. 42). Most have been in northern coastal regions (Fig. 43), and just two were in Andalucia: three birds at Odiel in December 2001, and a female at a reservoir in Alhama de Granada in June 2005.

RUDDY DUCK *Oxyura jamaicensis*

(Malvasía Canela, Pato-rabo-alçado-americano)

Recorded in Spain (Category C) and Portugal (Category C).

The nominate race has been established in the wild in Britain since the 1960s. This feral population had reached some 6,000 pairs by 2000 (Hughes et al. 1999, Henderson 2009) and gave rise to new populations on the nearby continent (Muñoz-Fuentes et al. 2006), where Ruddy Ducks were first seen in 1965 and have bred since 1977. There were some 35 breeding pairs in France by 2006 (Perennou 2003, Dubois 2007). The first observation in Spain was in 1983, but hybrids with White-headed Ducks were reported from 1991. Such hybrids are fertile and thus threaten the genetic integrity of the native White-headed Ducks, so a programme to eliminate Ruddy Ducks and hybrids was put in place in Spain (Torres-Esquivias & Moreno-Arroyo 2000). The current international plan for conservation of the White-headed Duck now includes the eradication of this alien species from Europe as one of its aims (Hughes *et al.* 2006). Culling began in 2005 in Britain and there were only an estimated 90 birds left in the country by April 2011 (Musgrove *et al.* 2011). Culling in France reduced the population there to 13–16 pairs by 2010 (de Seynes et al. 2011).

The first Ruddy Duck recorded in Spain was a female shot at the Ebro delta during winter 1983–84 (Ferrer *et al.* 1986). A male then appeared on the Laguna de Ablitas, Navarra, in November 1985, and a female was shot at the Albufera de Valencia in December 1985 (*Ardeola* 34: 128, 48:125). Thereafter there have been annual records since 1988. The rarities committee considered the species until 2006, accepting 40 records involving 85 birds. There are also nine accepted records from Portugal up to 2010, the earliest in 1995. We consider here a total of 52 records (98 birds) that we believe to be representative. However, at least 175 were culled in Spain during 1984–2008, in addition to 66 hybrid birds (Henderson 2009). Only one of the records accepted by the Spanish rarities committee undoubtedly involved hybrids: two at El Hondo in 1977 (*Ardeola* 46: 135). The majority of the records in our sample (67%) involved solitary individuals and they included 55 males and 28 reported as females. The relative scarcity of the latter seems likely to be due to confusion with the comparatively similar females or immatures of the White-headed Duck. Observations of Ruddy Ducks became increasingly frequent until the late 1990s, peaking at 11 records in 1997. More recently they have declined sharply, very possibly due to the effectiveness of the British cull.

Iberian records have occurred in every month but peak in winter, with 50% of first observations during November–January (Fig. 44). Some long-stayers remained into spring but the temporal distribution pattern is certainly influenced by culling. In general, the Iberian occurrence pattern recalls that of British birds in France, where they arrive (or used to arrive) from July onwards but principally in October–January (OF).

Records are widely spread throughout the region (Fig. 45) but although the northern half of Iberia accounted for only 21 observations, these involved some 60% of the individuals, whereas all the records from Andalucía or southern Portugal were of single birds. In the north, the Ullibarri-Gamboa reservoir, Álava, produced 12 records comprising a total of 46 birds: including 11 in January–February 1997, eight in winter 1997–98 and ten in December 1998. The same site also saw the first known Iberian breeding record: a female appeared with a duckling in July 1998 (*Ardeola* 46: 135, 49: 155). El Hondo reservoir, Alicante, has also attracted Ruddy Ducks with some regularity: 11 have occurred there, seven of them in early 1997 (*Ardeola* 46: 135), and breeding was reported in 1992 (Torres-Esquivias 2008).

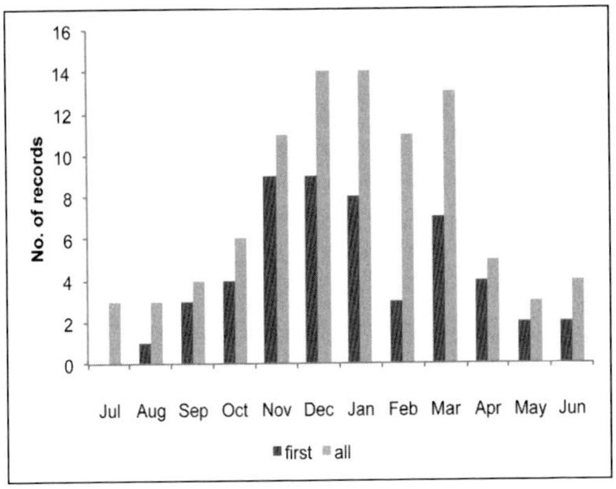

Figure 44. Records by month, 1985–2010 (n = 52)

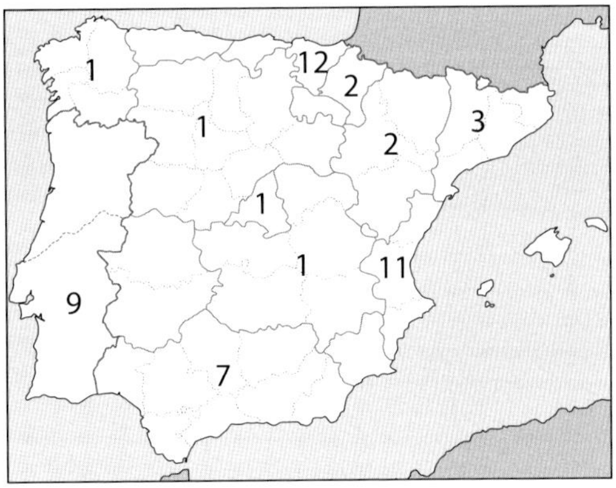

Figure 45. Records by region, 1985–2010 (n = 52)

WHITE-HEADED DUCK *Oxyura leucocephala*

(Malvasía Cabeciblanca, Pato-rabo-alçado-europeu)

A scarce resident or partial migrant in southern and eastern Spain. Introduced (unsuccessfully) to Mallorca. Very rare in Portugal.

The fluctuating fortunes of the Iberian population of the White-headed Duck have attracted interest from well outside the boundaries of the Peninsula. In particular, the saga of its interaction with the alien Ruddy Duck (*q.v.*) has raised ongoing concerns and some controversy, especially regarding the management of established introduced species.

The Iberian White-headed Ducks form part of an outpost western population of a species whose chief distribution is in south-central Asia, from Turkey east to northwestern China. The Spanish birds are the largest contingent in the western population, which also includes those nesting in Morocco, Algeria and Tunisia. The Spanish sub-population of about 2,500 birds is increasing and the North African subpopulation of 400–600 birds appears to be stable but the main central Asian population of 5,000–10,000 birds is declining (Hughes *et al.* 2006).

The Spanish birds are fairly sedentary but some post-breeding dispersal occurs. Extralimital appearances include at least two records of single birds from Aragón and a record of two males at the Laguna de Lor, Navarra. There was a surprising gathering of up to 23 birds at the Laguna de La Nava, Palencia, during August–November 2000, the first record for the province (*A. O. Palencia 1998–2001*: 65). The early 2000s also saw postnuptial gatherings at the Azud de Riolobos, Salamanca, where numbers in September–October peaked at 43 birds in 2000, 77 in 2001, 109 in 2002 and 116 in 2003, before local habitat deterioration deterred the birds so that there were only 14 in 2004, seven in 2005 and none thereafter (Rouco 2007). Occurrences in Catalonia are less than annual: the regional bird reports record six records during 1996–2008, nearly all at the Ebro delta and involving just a few individuals, with the exception of a flock of 16 at the delta during August–September 2000.

The White-headed Duck is very rarely found in Portugal. One or two 19th century records have been followed by nine records (11 individuals) at sites south of the Tagus during 1994–2009 (AP, CPR 2008–09) and a further observation in 2010 (RBP). Regular breeding in Morocco ceased in the 1950s but records there have increased since the 1990s, in parallel with increases in the Spanish population, which suggests that some have crossed there from Spain (BM). The rare occurrences in France have also become more frequent, although at least some of the birds found there may be of eastern origin (OF).

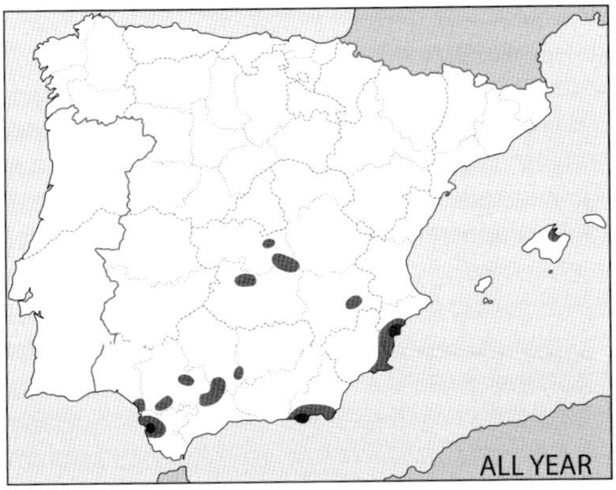

Breeding birds seek out freshwater or brackish lagoons, sometimes quite small ones, offering abundant peripheral vegetation such as reeds or reedmace, in which the nests are sited. After nesting the birds flock together to moult and to winter at the same sites but also at larger waters, including lakes and reservoirs that need not necessarily offer much marginal cover. The essential requirement at all times is a good population of aquatic plants and invertebrates, which the ducks obtain by diving. Much feeding is nocturnal. Autumn and winter flocks in recent years have numbered 100 birds or more. A favoured site is the Laguna de Medina, Cádiz, where breeding was recorded in the 1880s (Chapman 1884) and where the autumn flock size has recently reached 600 birds (*Ardeola* 50: 345). The largest known concentrations on record are very recent: about 1,000 were present at the Cañada de Las Norias, Almería, during winter 2010–11 (*Ardeola* 58: 198).

The former size of the population is uncertain but such historical references as there are give the impression of scarcity. Saunders (1871) found it 'resident and common' and nesting in the Guadalquivir delta near Sanlúcar but gave no indication of actual numbers. Chapman (1884) encountered six on the Laguna de Santa Olalla, Doñana, in May 1872 but Noble (1902) saw none in over a month spent in the Marismas. Valverde (1960) considered that the combined population of Doñana and the Cádiz lagoons was just 25 pairs. Away from western Andalucía, Bernis (AMI) only noted that the odd pair might nest in La Mancha and in the Levant coastal wetlands. Torres-Esquivias (2008) put the Spanish population in 1950 at around 500 birds, although he does not give the basis of this estimate. Irrespective of any former numbers, it is likely that the Spanish population was less than 100 or so individuals by the start of the 1970s and there is a much-quoted count of just 22 birds in 1977 (Torres-Esquivias 1982), probably the nadir of the species in Spain. The Laguna de Zóñar, Córdoba, was the sole known breeding site in 1977 and the species was then clearly on the brink of extinction in Spain. That same year saw an attempted reintroduction at Doñana, where Tom Gullick released 16 ducklings obtained from Pakistan (Rey Benayas *et al.* 2012). The fate of these is unclear.

A hunting ban from 1979 at Zóñar, after the lake was acquired by a group of conservationists, seems to have given the birds enough leeway to begin a slow build-up in numbers during the 1980s. The designation of many Andalucian lagoons as protected areas must also have been extremely helpful. White-headed Ducks gradually colonised lagoons in Cádiz and Sevilla, and later in Huelva and Jaén, as well as in Córdoba during the 1980s, although not all sites were occupied regularly. In 1988 they also bred in Almería and there were sightings in Castilla-La Mancha. The breeding range extended further to include Alicante, Ciudad Real, Toledo and Málaga in 1990. Birds were introduced to Mallorca, at the Albufera, in 1996 but this introduction has been unsuccessful and only one bird was found in the islands during the breeding season in 2007 (Ballesteros *et al.* 2008). By 1998 the Spanish population had risen to 1,339 individuals and the breeding range included wetlands in Cuenca and Albacete. In 2007 the breeding distribution included 74 wetlands dispersed between Andalucía, the Comunidad Valenciana and some wetlands in Castilla-La Mancha.

The numerical evolution of the Spanish population (Figure 46) has been closely followed since 1977. Quite large interannual fluctuations have been reported and these are thought to be associated with the effects on the breeding lagoons of year-to-year variation in rainfall. Thus there was a particularly successful breeding season in 2000 after which 4,489 birds were censused (Torres-Esquivias & Moreno-Arroyo 2000). However, no more than 2,700 birds could be found by the start of the 2001 breeding season and these declined further to 2,300 by the start of 2002 (Torres-Esquivias 2002). Thereafter peak annual numbers have remained within the range of 2,000–2,700 individuals, although there was an unexplained temporary decline to 1,601 birds in 2005. The January 2011 census found 2,546 birds (SWA).

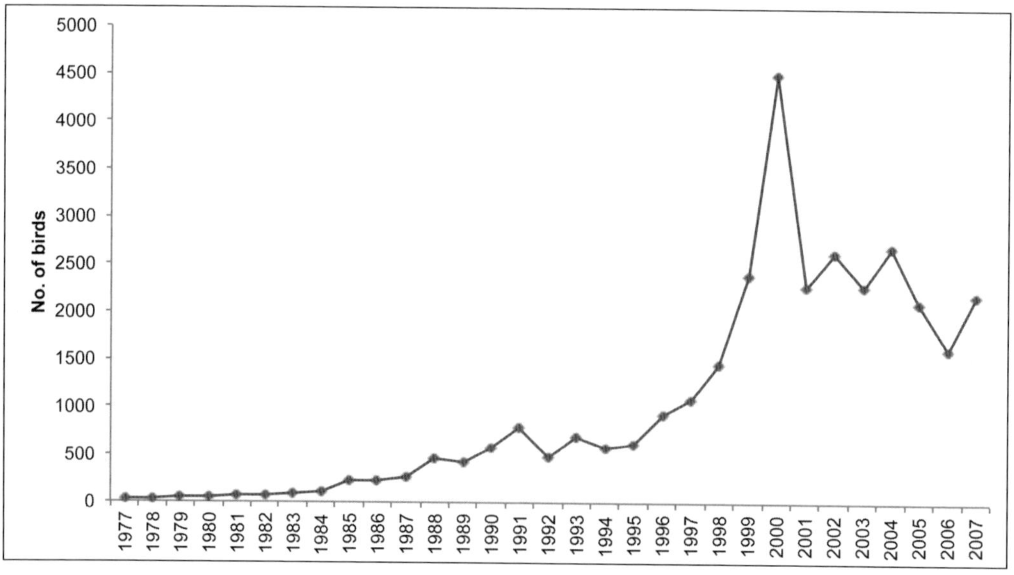

Figure 46. Fluctuations in the White-headed Duck population, 1977–2007 (after Torres-Esquivias 2008).

The entire population has been censused five times a year since 1990: in January, April, June, September and November

(Ballesteros *et al.* 2008). These counts have revealed considerable variation in numbers during the course of the year. For example, in 2007 the April and June counts were almost one third lower than in January, and about half of the September count. This pattern may simply reflect an apparent decline during the breeding season, when the females are often hidden in marginal vegetation, followed by an increase resulting from the re-emergence of the adults and recruitment of ducklings to the population, followed by a decline occasioned by natural mortality and/or wider dispersion. These numerical changes are, however, accompanied by local redistributions in the population so that, for example, in 2007 there were 198 in Huelva in January but none at all there in June, and in the whole of Castilla-La Mancha in the same year there were only 33 White-headed Ducks in January but 357 in April and 346 in June.

The January counts of wintering waterfowl in Spain during 1990–2009 found 1,276 birds on average (range 417–2,469; AAE). Some 80% were divided between 17 sites, almost all of them in western Andalucia, the lower Guadalquivir valley, coastal Almería and Murcia, but with the largest contingent then wintering at El Hondo (Table 9). A few were at lagoons in Castilla-La Mancha. Winter site occupancy is very variable, especially at the smaller lagoons that may be dry in some winters.

Table 9. January counts of White-headed Ducks at their ten principal wintering sites in Spain during 1990–2009 (AAE).

Site	January wintering numbers			% of national mean
	Minimum	Maximum	Mean	
El Hondo, Alicante	0	1,506	331	19.0
Laguna de Medina, Cádiz	0	559	168	9.7
Albufera de Adra, Almería	52	248	140	8.0
Cerrillos saltpans, Almería	0	318	135	7.7
Campotejar waterworks, Murcia	42	169	113	6.5
Cañada de Las Norias, Almería	0	227	104	6.0
Doñana NP	0	232	83	4.8
Doñana natural park south	0	195	64	3.7
Lebrija-Las Cabezas lagoons, Sevilla	40	40	40	2.3
Cartagena waterworks, Murcia	0	155	37	2.1

The Iberian White-headed Ducks face a number of threats. Hunting was probably the principal cause of their near disappearance from Spain, and was responsible for their extinction in other Mediterranean countries (Hughes *et al.* 2006). The Iberian population has also suffered from habitat loss, poisoning from intake of lead shot, outbreaks of botulism and genetic contamination from interbreeding with Ruddy Ducks. These problems have been addressed in recent years to varying extents. Most of the key sites are now protected, which has reduced losses to hunting. However, some of these, notably El Hondo, remain contaminated by lead shot from former hunting seasons. Water quality is often a problem, especially in drought years when some natural lagoons dry up completely and others become very shallow, favouring botulism. Introductions of alien fish species have had serious effects on White-headed Ducks and other aquatic species that require benthic vegetation and its associated invertebrates. Carp and American Perch have had dramatic effects at some sites. The perch eat ducklings and the carp destroyed the aquatic vegetation at the Laguna de Zóñar in the 1970s and, more recently, at the Laguna de Medina. Both these important sites were temporarily abandoned by the ducks until the fish problem was controlled (Fernández-Delgado *et al.* 2008). The threat from genetic introgression with the Ruddy Duck should recede now that the principal source populations of the latter have been exterminated.

HAZEL GROUSE *Tetrastes bonasia*

(Grévol Común, Galinha-montês)

Formerly resident in the Pyrenees.

The fossil record shows that the Hazel Grouse was a member of the late Pleistocene avifauna of Cantabria and there are Neolithic remains from the Madrid region in the Central Range (Arribas 2004). There are no known fossils from the Iberian Pyrenees but a remnant population survived there until at least the end of the 19th century, possibly later. One was seen in the French Pyrenees in Haut-Garonne as recently as 1991 and a feather was found there in 1992 (Catusse *et al.* 1992).

Saunders (1884) saw one in the woods of Roncesvalles, Navarra, and there seem to have been quite a few observations in the French Pyrenees at around that time (Purroy 1976, Ferrer *et al.* 1986), when the Hazel Grouse still seems also to have been common in the Arán valley, Lleida. However, a comment by Vayreda (1883) regarding its presence elsewhere in Catalonia does not appear to be reliable (Maluquer 1973). None of four reports from the Spanish Pyrenees during the 20th century: in Aragón at Oza-Canfranc on 29 June 1959 (Aragüés & Lucientes 1980) and at Ordesa around 1960 (Sampietro *et al.* 1998) and on 1 November 1990 (Villacampa 1993), and in the Irati forest, Navarra, on 24 May 1972 (Purroy 1976) seems to be entirely trustworthy. A reintroduction project began in the Arán valley in September 2011 (*Quercus* 308: 15) with the release of several pioneering individuals for a telemetric study. A summary of this project is available on the Spanish Ministry of Agriculture, Food and Environment website.

ROCK PTARMIGAN *Lagopus muta*

(Lagópodo Alpino, Lagopo-cinzento)

***L. m. pyrenaica*. A very local, scarce and declining resident.**

This endemic subspecies is confined to the highest levels of the Pyrenees. Its Spanish range extends over about 1,470km^2 between the Larra massif in the extreme east of Navarra and that of Ripollés in Girona. It is also present in Andorra and has a wider distribution within the French Pyrenees, where its range covers some 1,860km^2 (SA2, LR). There is one major discontinuity within the Spanish range, a gap of at least nine kilometres at the Fosa de Cerdanya that appears to isolate the easternmost population: east of Puigmal, Girona, from the core population in Lleida and Huesca (LR).

Ptarmigan occur year-round on alpine pastures and in alpine scrub, scree and snowfields, above 1,800m during the breeding season on the north-facing French slopes but usually above 2,000m on the south-facing Iberian ones (Afonso & Canut 2006). Radio-tracking has revealed that any seasonal movements are very limited and that the birds remain in winter even on windy ridges and summits (Martínez-Vidal *et al.* 2011). Densities in spring average 0.3–0.5 pairs/100ha but reach 3–4 males/100ha in the most favourable areas (LR).

The Pyrenean population is the most southerly in Europe and its situation is precarious. The Spanish population post-2000 was put at 442–738 pairs (LR), with a further 30–60 pairs in Andorra (BE 2004). However, a study in the Aigüestortes i Estany de Sant Maurici National Park (Afonso & Canut 2006) has found interannual fluctuations. The majority in Spain are in Huesca and Lleida (SA2), with a handful of pairs in easternmost Navarra. There are estimated to be up to 2,000 pairs in the contiguous population in the French Pyrenees (OF). The population has shown a historical decline with the disappearance of the species from outlying massifs in both the Spanish and French Pyrenees, although the core Spanish populations are considered to be stable (LR). The Ptarmigan occurred in the Cantabrian Range until the first quarter of the 20th century (García-Dory 1982). It was distributed much more widely in Iberia during cold climatic periods in prehistory: fossil remains are known from Atapuerca, Burgos, and from a couple of sites in the Central Range near Madrid (Arribas 2004).

WESTERN CAPERCAILLIE *Tetrao urogallus*

(Urogallo Común, Tetraz-real)

***T. u. cantabricus*. An endangered and declining resident of the Cantabrian Range. Extinct in Portugal.**

***T. u. aquitanicus*. A very scarce and declining resident of the Pyrenees.**

Both Iberian subspecies are smaller and darker than the northern, nominate subspecies but each is distinctive. Cantabrian males have pale grey necks and show white blotching on the breast and belly whereas Pyrenean males are darker, with dark necks and no white underpart markings; Cantabrian females are greyish but those in the Pyrenees have brown plumage tones (Castroviejo 1975, SA1).

The Capercaillie stands out among the most threatened species in Iberia with a history of decline since the 19th century. The Cantabrian race is now confined to the Cantabrian Range in Asturias and León, with a residual presence in Cantabria. The Pyrenean race is more abundant on the northern, French, side of the range but there is a very small population in Navarra and western Huesca and a much larger separate one that extends from the east of Huesca to Girona, with its core in Lleida (LR).

The Cantabrian race occurs at 1,200–1,600m, chiefly in beechwoods but also in oakwoods, birchwoods and in conifer plantations (Quevedo *et al.* 2006, González *et al.* 2010). The Pyrenean form occurs at 1,700–2,000m and the preferred habitat on north-facing slopes is also beechwoods and mixed beech-spruce forests: on south-facing slopes it occurs in Black Pine and Scots Pine woods (Canut *et al.* 2004). It requires woods and forests with a well-developed shrub layer, including blaeberry, an important food plant. The summer and winter distributions are similar and there are no significant seasonal shifts in altitude (SWA).

The Cantabrian race was restricted to the western part of the range by 2005–06, when it included about 250 males: 150 in Asturias, 80–90 in León and just four in Cantabria (SA2; Robles *et al.* 2006). At that time only 220 leks were known to be in use, 31% of the historical total. In contrast, a census in 1982 found 334 occupied leks and 582 males: 291 in Asturias, 243 in León, 23 in Cantabria, 22 in Lugo and 3 in Palencia (Del Campo & García-Gaona 1983). This race formerly enjoyed a much wider range that extended into Portugal and the northern Iberian Range. It was present in the north-west of Portugal in the Serra da Gerês and the Serra Amarela in the 18th and 19th centuries but was already rare by the 19th century and became extinct in the country early in the 20th century (Catry 1999, AP). In the Iberian Range it was present in La Demanda (Burgos/La Rioja) and Cebollera (Soria/La Rioja) sierras, probably until the mid 19th century (Castroviejo *et al.* 1974). More recent contractions in range led to its disappearance from Palencia in the 1980s. There were 14 occupied lek sites in Galicia in 1982, in the Sierra de Ancares in the east, but it ceased to breed in the region by the late 1990s (LR). Similarly, it was formerly far more widespread and abundant in Cantabria than at present: the number of occupied leks there fell from 55 in the 1950s to 14 in 1982 and to three in 1999. In all, this subspecies suffered a decline of at least 45% during 1982–2002 (LR).

The Pyrenean form is more numerous, largely thanks to the relatively large population on the French side of the range, but it too is declining (LR). The breeding range is fairly continuous in the northern, French, Pyrenees but it is fragmented on the southern, Spanish, side, comprising two subpopulations. The great majority of the Spanish birds are in an area that extends from the Alto Ter in Girona in the south east across Lleida (and Andorra) to the Cinca valley in Huesca. The remainder comprise a tiny subpopulation to the north-west, in the extreme east of Navarra and the adjacent part of Huesca. The degree of connectivity, if any, between the Spanish and the French/Andorran populations is unknown.

The total Pyrenean population in 2005 was believed to comprise some 2,600 males of which 562–573 were in Spain and 209 in Andorra, with *c.* 1,800 in France. The entire French Pyrenean population during 2005–07 has been estimated at 2,300–5,000 adults (OF). This population too has shown a progressive decline in recent decades, although not to the same extent as in the Cantabrian Range. Thus 91.7% of the known Pyrenean leks were still occupied in 2005. However, the outpost population in Navarra had declined to just two males found at one lek and is clearly in danger of disappearance.

The decline of the Iberian Capercaillies has attracted considerable attention and it is evident that a number of factors are involved. It seems clear that the extensive range contraction that took place during the 20th century was brought about by large-scale habitat loss, particularly in the Cantabrian Range (Obeso & Bañuelos 2003). Asturian forests in the early 21st century cover less than 25% of their original extent and much of what remains is fragmented (Garcia *et al.* 2005). Asturian beechwoods declined by 59% during 1952–62 and oakwoods were reduced by 46% during 1957–62 (Castroviejo *et al.* 1974). Hunting during the 20th century also had a serious impact on populations that were already too small to withstand such persecution (Castroviejo *et al.* 1974): it was banned in 1979. Human activity, in particular the construction of such infrastructures as roads, ski runs, wind farms and electricity pylons, is also implicated in lek displacement and some of these obstructions may cause deaths from collisions (Rodriguez-Muñoz 2011).

An important negative factor in more recent years, again at least with respect to the Cantabrian Range, is habitat modification through overgrazing by domestic livestock and also by deer and wild boar. Together these herbivores destroy the shrub layer, depriving the birds of food sources and secure nest sites, and the boar are an additional threat to nests: the proximate cause of decline seems to be very poor breeding success. A negative correlation between increase in livestock and Capercaillie presence has been demonstrated in the Cantabrian Range (Blanco-Fontao *et al.* 2011). The suggested link between such overgrazing and Capercaillie decline has not been investigated fully, however, and remains conjectural, as does any supposed influence of climate change (Rodriguez-Muñoz 2011). A study of a core population at Muniellos, Asturias (Suárez-Seoane & García-Rovés 2004) found that, locally, proximity to human activity was a more significant predictor of lek abandonment than habitat features. Capercaillie conservation in Spain, attempting to halt and reverse the species' decline, centres on habitat management and restoration but measures have been frustrated by ongoing habitat destruction and increased disturbance in many areas (Rodríguez-Muñoz 2011).

RED-LEGGED PARTRIDGE *Alectoris rufa*

(Perdiz Roja, Perdiz-comum)

***A. r. hispanica*. A common resident in northern and western Iberia.**
***A. r. intercedens*. A common resident in southern and eastern Iberia and in the Balearic Islands.**

The Red-legged Partridge is an extremely important quarry species for many Iberian predators, ranging from the Iberian Lynx to raptors such as Bonelli's Eagle. Healthy partridge numbers are thus an important requirement for the conservation of some of the most charismatic and scarce species of the region. However, the partridge itself merits concern and recognition since it is one of the few endemic species of Europe. The two Iberian subspecies differ chiefly in plumage tone and the variation is likely to be clinal, as well as confused by the large-scale release of captive-bred birds for hunting.

The Red-legged Partridge is common and locally abundant throughout most of the region. It is only scarce or absent in much of the cooler, wetter eurosiberian region and is thus relatively uncommon in Galicia, the Biscay coastlands and the coastal north-west of Portugal (SA2; PA2). It mainly occurs at altitudes below 1,500m although some are found locally to 2,500m, descending lower in winter (AMI). More unusually, nesting birds occur to nearly 3,000m in the Sierra Nevada (Garzón & Henares 2012). It is characteristic of warmer, drier regions where it favours open habitats including non-irrigated farmland, areas of low scrub interspersed with open ground, shrubby pastures and open areas with trees: such as olive groves, vineyards and Pyrenean Oak woodlands. (Lucio & Purroy 1992, Borralho *et al.* 1999, Fortuna 2002). High densities have also been reported from the marginal scrub of wetlands (ACRE), perhaps because such sites often comprise oases of dense low cover in more open country. Habitat-selection during winter is broader and densities may then exceed 50 birds/km^2 in some farmland areas (Tellería *et al.* 1998, AI1). The Spanish population has been estimated at nearly ten million birds (ACRE) but the size of the large Portuguese population has not been assessed closely. Only a few pairs occur in Andorra.

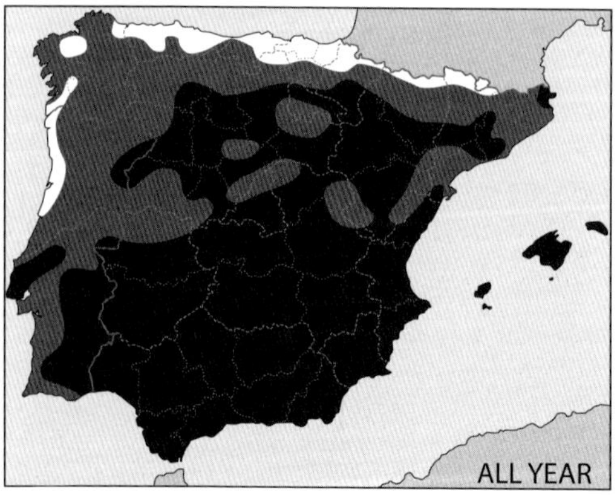

The status of the species in Iberia is somewhat confused since there have been local indications of decline, sufficient indeed for the species to have been included in the Spanish red data book, albeit under the heading of Data Deficient (LR). The Red-legged Partridge is a prime quarry species of the 900,000 licensed hunters in Spain (and the many in Portugal), being classified along with rabbits and hares as small game or *'caza menor'* (Ballesteros 1998). Accordingly its numbers are boosted by the release of captive-bred birds, estimated at three million annually (Martínez *et al.* 2002), which complicates efforts to monitor any changes in abundance. Such releases may take considerable pressure off wild birds but they are highly unlikely to respect the subspecific differentiation of the wild population and some have involved hybrids with Rock Partridges *A. graeca* and Chukar Partridges *A. chukar* (Negro *et al.* 2001, AP). However, despite these releases, the shooting bags per hunter have declined in 65% of the 39 Spanish provinces in which the species is hunted (SA2).

Overall, it seems that the species is secure but agricultural intensification, and scrub encroachment where farmland has been abandoned, threaten its abundance locally. Partridges have also suffered indirectly from the effects of viral haemorrhagic disease on rabbits since the resulting drastic decline in rabbits has put them under greater pressure from hunters: this may prove to be a temporary problem as rabbit populations recover.

BARBARY PARTRIDGE *Alectoris barbara*

(Perdiz Moruna, Perdiz-mourisca)

***A. b. barbara*. A scarce resident in Gibraltar (Category C).**

Gibraltar is the only Iberian location where it occurs. Its origins there are unknown but it seems likely that birds were introduced from Morocco. There are no records that this actually occurred, however, although the importation of 'game from Barbary' for shooting purposes was authorised by General Bland, then Governor of Gibraltar, in 1749 (BG). The Barbary Partridge was not mentioned specifically but no other species is likely to have been involved. *Alectoris* partridges figure in the late Pleistocene fossil record from Gibraltar (Garrod *et al.* 1928, Eastham 1968) but the species involved cannot be identified and could have been the Red-legged Partridge, which has never been recorded in Gibraltar.

The first ornithologist to note the Barbary Partridge in Gibraltar was the Reverend John White in 1771 (in Holt-White 1901). It was seen there by Reid (1871–1890), Irby (1875) and all subsequent visitors. The population has managed to persist despite extensive and progressive loss of habitat to urbanisation and to dense scrub encroachment on the Upper Rock. The birds have benefited recently from the return of the massive Great Sand Slopes on the eastern side of the Rock to their natural state, after having been covered with corrugated iron sheeting during most of the 20th century to collect rainwater. The area is now a steep open expanse with a thin cover of native vegetation and it is frequented by partridges despite also proving attractive to large numbers of Yellow-legged Gulls that nest there. The other habitats comprise scree slopes, firebreaks and areas of low scrub with some open ground. The birds are seldom seen outside the Upper Rock nature reserve and its immediate margins, the eastern slopes and Windmill Hill in the south.

A number of published records of Barbary Partridges from nearby in Spain have prompted suggestions that some birds may stray from Gibraltar occasionally. These records are all from around the same period (*Ardeola* 4: 193, 6: 367–369, 6: 382; Brosse & Jacquemard-Brosse 1962) so perhaps are more likely to be the result of an introduction to a shooting estate in the hinterland. It is improbable that any partridges will leave Gibraltar now since the isthmus and adjacent town of La Linea comprise several kilometres of largely urbanised ground.

The Gibraltar population fluctuates but probably seldom reaches 100 individuals. Coveys are small and rarely exceed ten birds. The species is clearly vulnerable and possibly declining. There is no significant hunting but nests and chicks are subject to predation by rats, feral cats and Yellow-legged Gulls. The decline in habitat quality has restricted its range and reversing this should be the aim of future conservation measures.

GREY PARTRIDGE *Perdix perdix*

(Perdiz Pardilla, Charrela-comum)

***P. p. hispaniensis*. Darker than the nominate race and with a larger 'horseshoe' mark on the belly. A scarce resident in the mountains of northern Spain and in Andorra. Extinct in Portugal.**

The Grey Partridge is a montane species in Iberia with a fragmented range that includes the Cantabrian Range, the northern Iberian Range and the Pyrenees. Here it is always found above 1,000m and more usually above 1,500–1,800m (Lucio *et al.* 1992, LR) although heavy snowfall drives it to lower levels. The Cantabrian breeding range is continuous from the Sierra de Híjar and Puerto de Pozazal in central Cantabria eastwards across the whole of Asturias to the Sierra de Caurel in the far east of Galicia. There is also a separate outpost population in southeastern Galicia and the Montes de León: in the Sierras de Sanabria-Segundera, Cabrera, Teleno and the Montes Aquilianos, as well as a residual presence in southern Cantabria/northern Burgos, in the Montes de Samo and Valnera. The core population of the Iberian Range is in La Demanda, Urbión, Castrejón and Cebollera sierras, with an isolated presence further south in the Sierra del Moncayo. The Pyrenean range extends

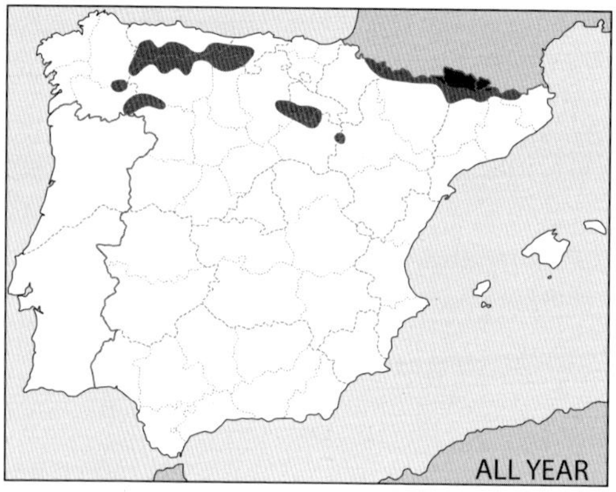

continuously from the upper reaches of the Roncal and Salazar valleys in easternmost Navarra across Huesca and Lleida to the headwaters of the Muga river in Girona (SA2, LR). The Grey Partridge is also present in the French Pyrenees: an estimated minimum of 3,000 pairs (Novoa 2008).

The typical habitat of the Grey Partridge is a mosaic of montane pastures, stony ground and well developed patches of scrub. The latter may include such species as heaths, junipers, rhododendrons and blaeberries but *Genista* and *Cystisus* broom scrub seems to be especially favoured (LR). Dense scrub is important for nesting and more open formations for foraging.

The Spanish population has been estimated at 2,000–6,000 pairs (SA1), with some 35–70 pairs in Andorra. It is thinly spread, with seldom more than 1–3 pairs/100ha but 5–8 pairs/100ha occur locally in particularly favoured locations (SA2). The species is in long-term decline and the periphery of the range has contracted in many areas, especially at lower levels (Lucio *et al.* 1992). Its range is thought to have declined by over 40% in La Rioja between 1951 and 2000, by 50% in Galicia during 1991–97 and by 60% in Catalonia in 1990–2000 (LR).

The Grey Partridge used to breed in the mountains of northern Portugal but it has been extinct there since the mid 20th century (Pimenta & Santarém 1996, Catry 1999, AP). It was abundant in the 19th century in the upper reaches of Minho and Trás-os-Montes and the range extended just south of the Douro in the vicinity of Tarouca and Mêda district. It rapidly declined during the first half of the 20th century and the last known records from the Peneda-Gerês National Park and the Serra da Nogueira, at the core of its former distribution, were in the 1950s. It may have survived a little longer in the Montesinho Natural Park (AP). The possibility remains that occasional birds may cross into northeasternmost Portugal from the Sanabria region in the Montes de León and one is claimed to have been shot there near Bragança in late 2003 (AP).

The decline seems to be due to diminishing habitat quality resulting from overgrazing, scrub invasion, afforestation and cessation of hill farming. Disturbance, hunting – some of it still legal – and increases in opportunistic predators such as foxes and wild boar are also implicated. The species may have suffered genetic contamination from released birds that are liable to reduce the fitness of the indigenous race and may disrupt its genetic identity. At least 3,600 captive-bred Grey Partridges were released into the breeding range in Spain between 1986 and 1991 (Heredia 1992). Climate change may be also affecting a species that avoids areas of mediterranean climate (LR).

COMMON QUAIL *Coturnix coturnix*

(Codorniz Común, Codorniz)

***C. c. coturnix*. A common summer visitor or partial migrant throughout most of the Peninsula and in the Balearic Islands.**

The Quail is common and locally abundant across the greater part of the Iberian Peninsula but especially in cereal-growing areas. It ranges to altitudes of about 1,000m and occasionally higher, often breeding on alpine pastures: at over 2,000m in the Sierra Nevada (Garzón & Henares 2012). In Spain it is only absent as a breeding species from the arid south-east but it is relatively scarce in the Biscay coastlands (SA1, SA2). In Portugal it is also very scarce in the north-west, the region of maximum maritime influence, and is most abundant in the farming regions of Alentejo (PA2). It nests on all the principal Balearic Islands and in Andorra.

It is chiefly a summer visitor but a few thousand birds may remain in winter, chiefly in the irrigated areas of the Guadiana valley in Extremadura and in the Guadalquivir valley, with occasional winter records elsewhere, chiefly in the major river basins and in coastal areas (SWA). Winter-sown cereal crops are now the preferred habitat but the birds also occur in other open areas, such as meadows, that have sufficient low herbaceous cover to provide some shelter from their many predators. In summer some move into taller crops, such as maize, sunflowers and sorghum. Local breeding numbers are highly sensitive to rainfall. Interannual variation in breeding abundance may thus be huge in such regions as the dry croplands and pasturelands of Extremadura, where the Quail is abundant during wet years and virtually absent in very dry ones (de Juana & García 2005). Breeding success is strongly linked to spring rainfall, which is important for the growth of cereal crops, being high in wet years and reduced during drought periods. There is also a direct positive relationship between the average stay duration of the Quail at a site and the accumulated rainfall during the breeding period, which in turn affects the timing of the cereal harvest: earlier

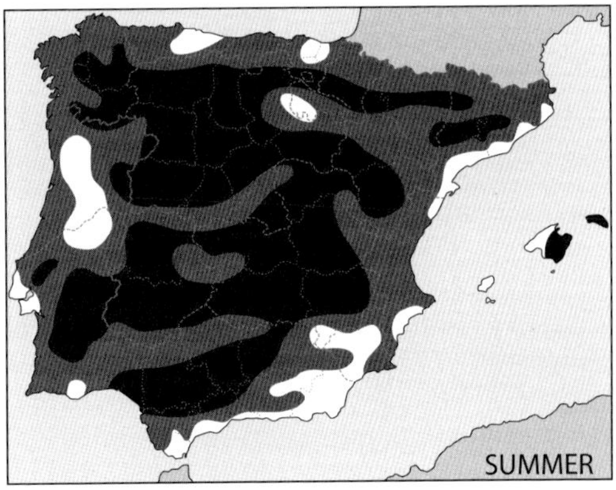

in dry years and later in wet years (Puigcerver *et al.* 1999).

The 'wet-my-lips' song of the Quail remains a characteristic sound of the Iberian farmlands but there can be little doubt that it was far more numerous in the past. For example, Smith (1868) described it as 'excessively abundant' in Portugal, where the markets were glutted with them; Arévalo (1887) considered it extremely abundant in Spain, and Irby (1895) refers to incredible numbers on passage in southern Andalucía. The species has declined throughout Europe (BE) so migrants were also much more abundant in the past. There are more recent indications of decline over some years in some regions (Bernis AMI): including Andalucía, Madrid, Castilla-La Mancha and Catalonia, as well as suggestions of increases elsewhere, such as in Aragón, the Comunidad Valenciana and Extremadura (LR). The overall population trend is difficult to establish on account of the marked interannual fluctuations but the Quail showed a significant 38% decline on SACRE plots during 1998–2010 (Escandell 2011).

The Iberian population is of outstanding importance on a European scale: 33–57% of the breeding population of the European Union is thought to be in Spain (Guyomarc'h & Perennou 2009), where there are an estimated 320,000–435,000 pairs (SA1). The Portuguese population has been estimated to be with 5,000–50,000 pairs and a few breed in Andorra (BE).

A large proportion of the European breeding birds, including most of the Iberian populations, winter in the African Sahel region. The remainder winter in North Africa or in southern Iberia. Migrants return to Iberia during March–April. Some, however, breed in North Africa during early spring and they or their offspring may then migrate to Europe, where they may breed (again in the case of the adults) later in the season. Some Iberian Quail are thought also to migrate further north in Europe to nest late in the summer. Quail 'invasions' of northern Europe are linked with good early breeding seasons in Iberia and North Africa resulting from wet springs there. Iberian Quail are thus at the core of the Atlantic metapopulation of the species, that comprises birds that breed at various times from the western Sahara, Morocco and northern Algeria north through France and the Low Countries to Britain and Ireland (Guyomarc'h *et al.* 1998).

Migrants cross Iberia during both seasons: during March–May and August–October, when birds may appear on coasts and in other habitats in which they do not nest. Passage in the Balearics extends from early March to early June, peaking from mid April to mid May; the autumn movements there are in September–December, peaking in October (Martínez & Suárez 2008). Birds ringed in France, Belgium, The Netherlands, Switzerland, Italy and Morocco have been recovered in Spain and Portugal (AMI, AP). There are also 23 foreign recoveries of birds ringed in Spain: 15 in France, three in Morocco, two in The Netherlands and one each in Belgium, Italy and Croatia.

Large-scale recent ringing of Quails in Spain has been sponsored by hunting bodies (Real Federación Española de Caza and FEDENCA) so that 21,246 birds had been marked with ICONA rings by 2007. Initial results indicate that most birds that have bred in the south or in coastal Mediterranean areas move later to northern and/or higher regions, where cereal harvesting is later and where they may still find food and shelter in field margins, pastures and along streams. Birds are thus concentrated in northern Spain when hunting begins in late summer and just six provinces there account for over 75% of August recoveries: Burgos (18), Palencia (11) and Soria (5) on the Meseta, and Teruel (14), Zaragoza (11) and Huesca (5) in the Ebro valley. Rodríguez-Teijeiro *et al.* (2009) show that hunting intensity is maximal in the provinces of Burgos: over 250,000 Quails hunted per year, Palencia and Teruel, and that in Spain later mowing dates mean larger hunting bags.

Quail populations have been adversely affected in some areas by agricultural intensification: the loss of cereal crops to irrigated farmland, olive groves, vineyards and other unsuitable forms of land use, as well as by the effects of pesticides and excessive hunting. The annual bag in Spain is around 1.4 million birds, although this includes an unknown number of released Japanese Quails *C. c. japonica* or hybrids (*Anuarios de Estadística Agraria*, Ministerio de Agricultura). These last comprise an additional threat, since released birds hybridise with the wild ones and may reduce their fitness (Rodríguez-Tejeiro *et al.* 2003; LR). In particular, Japanese Quail are sedentary and this trait may be helped to spread in the wild populations through hybridisation, since hybrids are fertile, with unforeseeable ecological consequences (Guyomarc'h & Perennou 2009). The tendency for Iberian Quails to become sedentary seems to be increasing (Guyomarc'h 1992).

COMMON PHEASANT *Phasianus colchicus*

(Faisán Vulgar, Faisão)

Present locally in Spain (Category C) and Portugal (Category E), and in Majorca and Ibiza in the Balearic Islands, following mainly recent and recurring introductions. There are very few, if any, naturalised birds.

The Common Pheasant can be encountered in scattered locations in Spain, Portugal and in the Balearics but its presence is due to repeated releases of captive birds on shooting estates. Annual releases in Spain are of the order of 100,000 birds: there were 103,000 in 2007 and 115,000 in 2008 (*Anuarios de Estadística Forestal*). Seasonal releases in Catalonia alone averaged 28,000 birds during 2004–08 (Sazatornil & Sardà-Palomera 2011). Such releases often occur only a few days, or even hours, before most of the birds are shot, so they have little opportunity to become naturalised. Apparently wild pheasants are most likely to be encountered in northern Catalonia: in Barcelona and Girona: where there may be 400–800 breeding pairs (CA1) and 2,100–2,400 individuals in winter (Sazatornil & Sardà-Palomera 2011); in Madrid, the Basque Country and Galicia. It is not known whether these populations are self-sustaining or whether they depend on regular influxes of released birds (SA2). Released birds have occasionally nested in Portugal but there do not appear to be any naturalised birds there (AP).

Pheasants are known to have been present on some royal estates in Spain during the mid and late 19th century. Arévalo (1887) noted that they had been present years earlier in the Casa de Campo royal park in Madrid and that around 1887 King Alfonso XII had ordered releases at La Granja de San Ildefonso, Segovia. Witherby (1928) heard them there and also saw 'a fair number' in the Jardines de Aranjuez, where they are still common. Gil Lletget (1945) regarded all introductions in Spain as relatively recent. However, Alonso Martínez de Espinar, in his *Arte de Ballestería y Montería* 1644, speaks of pheasant hunting in Spain and he describes the species well and states that it nested in woodlands, riparian woods and in humid regions. The species has nonetheless been present in other western European countires for much longer (BWP).

RED-THROATED DIVER *Gavia stellata*

(Colimbo Chico, Mobelha-pequena)

A very scarce winter visitor to Spain and Portugal. Vagrant to the Balearic Islands.

The Red-throated Diver winters in large numbers around Britain and Ireland and in the North Sea but becomes progressively scarcer further south. It is the rarest of the divers that winter annually in Iberia, and the population may not reach 100 individuals in some years. It may have been commoner in the past since Saunders (1871) considered it regular in autumn and winter and 'saw many between Gibraltar and Cádiz'. Irby (1875) collected it in the Strait of Gibraltar and noted that 'some species of diver' was sometimes very common in winter at the (western) end of the Strait. Similarly there is at least anecdotal evidence that it was relatively common off the Portuguese west coast in the late 19th and early 20th centuries (Smith 1868, AP). However, Arévalo (1887) considered it rare off Valencia, where it only appeared during very severe winters.

There are very few recent records from the Strait and no confirmed recent records from Morocco (BM). Red-throated Divers are now most regularly recorded from Atlantic coasts of northern Spain and in the Mediterranean off Catalonia. The regional bird reports show an annual presence off Galicia, Asturias, Cantabria and the Basque Country. The number of birds involved is difficult to ascertain but the reports note 88 records in total for Galicia during the 14 winters from February 1992 to December 2004 (6.3 records/winter) and 36 records for Asturias during the ten winters from January 1992 to December 2000 (3.6 records/winter). Most of these have been of single birds but some small groups occurred in the shelter of rías and bays. In Asturias these included up to 14 in Gijón Bay during winter 1993–94. In Galicia, there have been up to six in Cariño harbour during January–March 1999, and up to five in both the Ría de Ortigueira and at Nigrán, Pontevedra, during winter 2002–03. Santoña has recently attracted the largest overwintering concentrations, peaking at ten birds in winter 2009–10 and 27 in winter 2010–11 (*A. O. Cantabria*). Most records from the Basque Country are of single birds, an exception being ten at El Abra–Getxo, Vizcaya, on 7 December 1998 (*A. O. Bizkaia 1998–1999*: 60).

There are relatively few recent observations from Portugal, with only some 20 published records in the 30 years up to 2007 (AP). These generally involved single birds seen at various points of the entire coastline from Minho to Algarve, but there were three together in the Minho estuary in early January 2007 (CPR 2008–09) and there is a record of a gathering of 20+ off the Ria de Aveiro on 23–26 February 1999 (*Anu. Ornitol.* 1: 7). Up to five were in the Sado estuary in January 2012 (RBP). However, both Llandres & Urdiales (1990) and Garrido (1996) considered it the most frequent of the divers off the coast of Huelva, occurring in pairs or groups of up to four: so it would appear that a wintering presence off the south-west Iberian coast is not unusual.

Most Mediterranean records are from the Catalan coast. The Catalan bird reports give 61 records between February 1996 and November 2009, about four records per winter. Roses Bay and the Ebro and Llobregat deltas are favoured locations and have attracted groups of up to three birds in some years. There were six together at the Ter estuary in winter 1999–2000. The Red-throated Diver is reported less than annually from the Comunidad Valenciana, all records being of single individuals. It is also noted less than annually in the south-east and south: published records comprise 11 from Málaga, one from Almería and two from Gibraltar, all of single birds. All but one of the Andalucían and Gibraltar records were during 1991–98. There are five records from the Balearic Islands (Ferrer *et al.* 1986, Clavell 2002), the most recent being two in Alcudia bay in December 1999.

Red-throated Divers are rare inland where they may appear on reservoirs, usually singly. Records up to 2007 include at least two each from Asturias and Álava and single observations in Teruel, Cuenca, Navarra, Valladolid and Badajoz.

Very few arrive before November and some do not seem to appear until January, both on northern coasts and in the Levant (Fig. 47). There is an exceptionally early record of one at Cape Prieto, Asturias, on 27 September 1965 (Huyskens & Maas 1966). Most depart by the end of February but there are records of stragglers into May, these including three at Cudillero, Asturias, on 7 May 1989 (*GIAM* 6: 6) and one at Txingudi, Gipuzkoa, on 11 May 1999 (*A. O. Gipuzkoa 1999–2000*: 5).

Figure 47. Records by month in northern Spain (Galicia, Asturias, Cantabria and the Basque Country: 1992–2004, n = 239 records) and on the Levant coast (Catalonia and the Comunidad Valenciana: 1987–2002, n = 132 records).

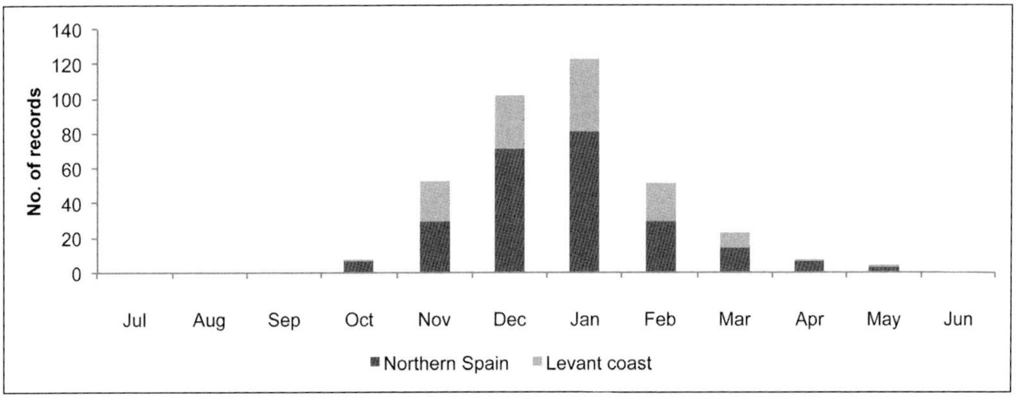

BLACK-THROATED DIVER *Gavia arctica*

(Colimbo Ártico, Mobelha-de-garganta-preta)

***G. a. arctica*. A very scarce winter visitor to Spain and Portugal. Vagrant to the Balearic Islands.**

This is the second most numerous of the regular diver species but it is certainly not the most widespread since the great majority of records come from northern Atlantic coasts in Galicia and, in the Mediterranean, from northern Catalonia. It is considered the rarest of the divers in Portugal and is normally very rare further south. The pattern of records suggests that birds reach the coasts of northern and western Iberia via the Atlantic but that those that occur in northern Mediterranean waters arrive there overland, probably via France. The rarity of the Black-throated Diver in southern waters, including the well-watched Strait of Gibraltar, strongly suggests that Mediterranean wintering birds do not enter or leave via the Strait.

The available records suggest a total wintering Iberian population of 100–200 individuals. The Biscay coast population and wintering numbers in Galicia during 1995–99 have been estimated at 100 and 50 birds respectively and up to 60 wintered in Catalonia during 2006–09 (CWA, SWA). However, more thorough prospecting may yet find higher numbers of this and other divers.

It is annual in the north-west and along the Biscay coast but the great majority of records, and all the largest concentrations from northern waters, are reported from Galicia. The Galicia bird reports give 101 records in total during the 14 winters from February 1992 to December 2004 (7.2 records/winter). Of these, 58 involved just one or two birds but some reports are of gatherings of long-stayers that remained for several months in favoured locations in A Coruña: notably on the Baldaio coast, off Carnota, and off Basoñas beach on the south shore of the Ría de Muros. Groups of up to ten individuals are probably not unusual there and larger flocks have included up to 17 off Basoñas beach during November 2001–March 2002 and 15 off Carnota in February 1992 (*A. O. Galicia*). Very few are reported further south than the O Grove peninsula, where A Lanzada beach is also a regular haunt. Most reports from the Biscay coasts are of single birds, or groups of 2–3, found overwintering at well-watched sites such as Gijón Bay, Santoña and Txingudi, Gipuzkoa. Up to 17 were in present in Gijón during winter 1992–93 but only a few have been found there in more recent seasons (*A. O. Asturias*).

It may not be as rare in Portugal as the extreme dearth of published records seems to suggest: the only recent records are two seen off the Ria de Aveiro on 27 December 1997 and one at the Lagoa de Santo André, Setúbal, on 30 December 2000; one at the Lagoa de Albufeira, Setúbal, in January–February 2010 and one at the Douro estuary on 21 October and 6 November 2010 (AP; *Anu. Ornitol.* 8: 10). Further south, the few published records from coastal Andalucía comprise two first winter birds off Doñana in January 1987; single birds off the Málaga coast in November 1982, February 1992 and February 1997 and an exceptional flock of up to 20 in Almería Bay in January–February 1960 (Paterson 1997, *A. O. Málaga 1995–1997*: 24; *Ardeola* 6: 373–374). Irby (1895) considered it occasional in the Strait of Gibraltar in winter and Arévalo (1887) noted that a specimen obtained in the Strait was in the Sevilla university collection, but we are not aware of any later records from the Strait.

The largest known concentrations of wintering Black-throated Divers in Iberia are in the Mediterranean off the northern Catalan coast. The winter censuses there during 2006–10 found 26–41 individuals, mostly in Roses bay and the nearby Ter estuary (CWA). About 20 are reported there in many winters but numbers peaked at 30 in January 2001 and 29 in January 2004. There is also a record of at least 40 off Pals beach near the Ter estuary on 1 March 1980 (*Ardeola* 28: 149). A few occur regularly further south in Catalonia, where they are most often reported around the Ebro delta. The regional bird reports note 91 records from Catalonia between January 1996 and December 2009, seven records per winter, of which 53 involved 1–2 birds. The Black-throated Diver is found less than annually off the Comunidad Valenciana: the regional bird reports give only 18 records between January 1986 and February 2010, two involving two birds together and the remainder of single individuals.

The Black-throated Diver is very rare in the Balearic Islands, perhaps surprisingly so in view of its relative frequency in Catalonia and off the southern French coast. There are only four published records (Ferrer *et al.* 1986, Clavell 2002), most recently one at La Savina, Formentera in January 1985 (*A. O. Balears 1985–1986*: 6). It is also very rare inland in Iberia, other than on coastal lagoons: but there are records of single birds at reservoirs in Navarra, Madrid, Cáceres (2) and Málaga.

The wintering birds arrive late in the season and most initial observations are in December both in the north and in Catalonia (Fig. 48). The great majority of records from the Comunidad Valenciana have been single-day observations in December. The earliest arrivals are not normally before November and some do not appear to arrive before January. Wintering birds depart from February onwards. It is likely that the numbers of this and other divers reaching Iberia vary from year to year according to the severity of conditions further north. In addition to an inland July observation in Navarra (*Ardeola* 41: 193), there are also unseasonal reports of single birds at Gozón, Asturias, on 18 August 1983; at Llanes, Asturias, on 9 August 1986, and at the Ter estuary, Girona, on 10 September 2000 (Paterson 1997, *Ardeola* 36: 231, *A. O. Catalunya 2000*: 46).

Figure 48. Records by month in northern Spain (Galicia, Asturias, Cantabria and the Basque Country: 1992–2009, n = 246 records) and in Catalonia (1996–2009, n = 171 records).

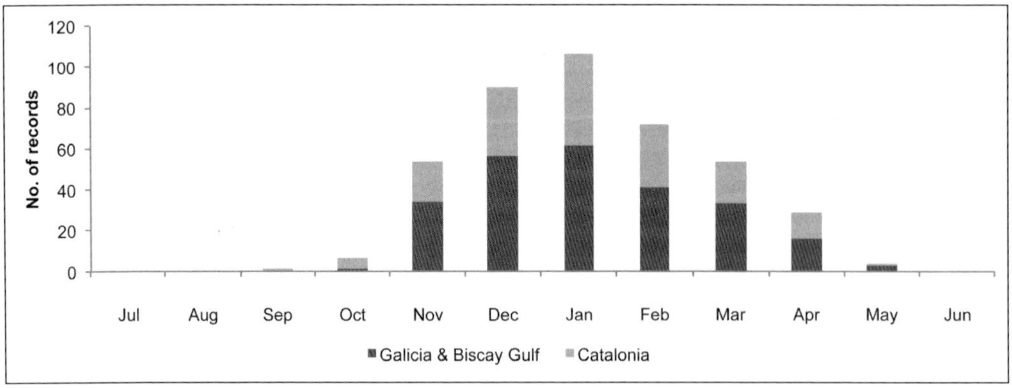

PACIFIC DIVER *Gavia pacifica*

(Colimbo del Pacífico, Mobelha do Pacífico)

Vagrant to Spain.

A first-year bird was photographed at the Marismas de Santoña on 7 December 2009 (López Velasco 2010). The first European record was in Britain as recently as January 2007 (Mather 2010). It was followed by four more British records up to 2009 and one in Ireland in 2010.

GREAT NORTHERN DIVER *Gavia immer*

(Colimbo Grande, Mobelha-grande)

Scarce but regular in winter in Spain and Portugal. Vagrant to the Balearic Islands.

Several thousand Great Northern Divers winter off the coasts of northwestern Europe, with a small minority ranging further south. This is nevertheless the most numerous of the three diver species that may be encountered regularly in Iberia. The total winter population in the late 1990s was put at 270–340 birds (LR) but there have been fewer recently: an estimate of 140–240 individuals in Spain during 2008–10 (SWA). It is most numerous on northern Atlantic and Biscay coasts, especially off Galicia and Asturias, with small numbers appearing off the northern Mediterranean coasts and only sporadic records in Portugal, southern Spain or inland. Most records involve single birds or just a few: double-figure gatherings seem to be confined to the Biscay coasts.

It is unclear whether the Mediterranean Iberian records involve birds that enter via the Strait or whether some or most of them reach the east coast overland through France or Spain. However, the latter seems probable since there is no discernible passage through the Strait and the Great Northern Diver is seen much more often on the Spanish east coast than in the south.

The mean winter population off Galicia during 1996–2001 was 123 birds but it declined by half following the *'Prestige'* oil spill in 2002–03, although numbers recovered post-2005 (de Souza *et al.* 2010). The spill is known to have killed at least 65 birds, most of them in Galicia, and its lasting effect indicates that some birds show marked philopatry to their winter quarters. The Great Northern Diver in Galicia occurs in open coastal waters as well as in the sheltered rías, and is particularly concentrated off exposed sandy coastlines where groups of up to 30 have been recorded: the Baldaio coast in the north-west is a notable haunt (de Souza *et al.* 2010).

Gijón Bay, Asturias, formerly attracted some of the largest concentrations recorded in Iberia. Notable concentrations in the 1990s included 44 on 19 January 1994 and 34 on 12 February 1996 (*Ardeola* 44: 243). However, dredging and port expansion work has greatly reduced the Bay's attractiveness to this and other divers and peak midwinter counts for the whole Asturian coast have not exceeded 20 birds since 2001 (*A. O. Asturias*). It also appears regularly at Santoña, where up to eight are usually present in winter: more occur occasionally, including an exceptional 21 on 6 December 2006 (*Ardeola* 54: 171). A few are reported annually in winter off the Basque coast, notably at Txingudi (*A. O. Gipuzkoa, A. O. Bizkaia*).

The Great Northern Diver is regarded as the commonest of the divers in Portugal but it is nonetheless rarely reported there and there were only 42 known records for the entire country up to 2010 (AP, Muchaxo *et al.* 2011). Nearly all records have involved single birds but there were six together off the Ria de Aveiro on 27 December 1997, following severe storms (AP). An exceptional observation of ten in the Sado estuary on 24 March 2012 is among 20 Portuguese records from 2010–12 (RBP).

A few individuals are reported in some winters on the Atlantic coasts of Andalucía, notably at the Odiel estuary, Huelva and in Cádiz Bay. A count of five in Cádiz Bay on 29 January 1983 (*Ardeola* 30: 115) is noteworthy. The Great Northern Diver is rare in the Strait and there are only two recent records from Gibraltar. It is also rare on the Costa del Sol, where only occasional individuals are usually reported: records of seven together in Almería Bay on 4 April 2009 and one flying west there on 9 May 2009 (*Ardeola* 56: 154), are unusual in terms of the number involved on the first occasion and the late dates.

A fairly regular winter presence in small numbers is reported from the Levant coast, particularly at the Albufera de Valencia, the Ebro delta, and Roses Bay, Girona. Arévalo (1887) noted it in severe winters at the Albufera de Valencia and in the Júcar estuary, Valencia. Bird report records total some 105 individuals in Catalonia during 1996–2009 and at least 27 in the Comunidad Valenciana during 1993–2009, about seven and two birds per year respectively. Most observations are of single birds but there have been up to five together off the Ebro delta and there were five off Pineda de Mar, Barcelona, during January–March 2008 and January–February 2009. Six appeared on the Albufera lake during a cold spell in winter 1984–85 (*Ardeola* 34: 275). A few occur at least occasionally in Murcia, where there were four on the Mar Menor in November 2003 (*A. O. Murcia 2003*: 2). The Great Northern Diver is a rare vagrant to the Balearic Islands: there are two records from Menorca (Muntaner & Congost 1984).

This is the most likely of the divers to appear inland and there are a number of records from widely-dispersed sites far from the sea, including in Extremadura (two records); Castilla-La Mancha (six); Navarra (four), Castilla y León (three); and Aragón and La Rioja (one record each). One site 50km inland, the large Ullibarri-Gamboa reservoir, Álava, has attracted 1–3 individuals with some regularity, with birds in February and December 1998, February 2000, January 2003, February 2004 and January 2009, some of which remained for a month or more.

In common with the other divers, Great Northern Divers tend to appear quite late in Iberia and many do not arrive until December or January, these presumably being birds that spend the autumn and early winter further north. Once arrived, many remain for some weeks at favoured locations. The monthly distribution in Asturias (Fig. 49) may be representative of the species' presence in Iberia. Records in the Mediterranean have a similar pattern: the extreme dates for observations in Catalonia during 1996–2009 (*A. O. Catalunya*) are 13 November and 14 April but the great majority are found from mid December to mid

March. However, Galician records show two peaks: from mid November to mid January and again from late March onwards suggesting the possibility of birds arriving from other wintering areas (de Souza *et al.* 2010). There are a number of records of stragglers, including rare individuals that have over-summered, having acquired their breeding plumage.

Figure 49. Records by month in coastal Asturias during 1996–2000, n = 167 (*A. O. Asturias*)

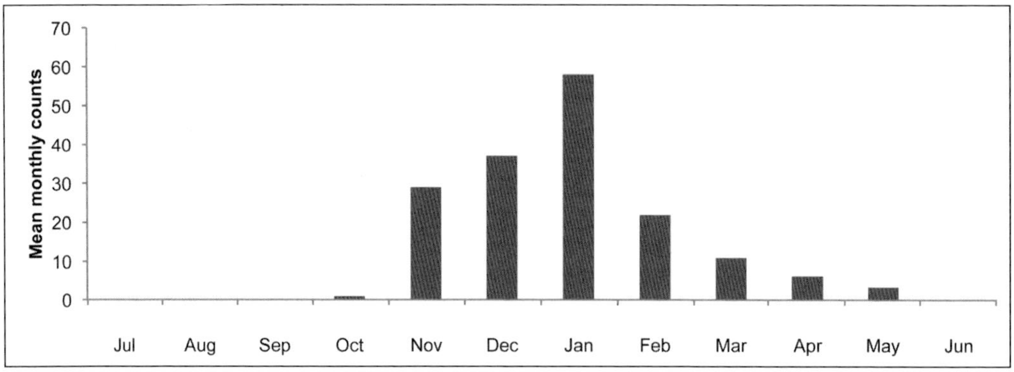

PIED-BILLED GREBE *Podilymbus podiceps*

(Zampullín Picogrueso, Merrgulhão-caçador)

Vagrant to Spain.

A rare vagrant to Europe, with records of 38 individuals from Britain, ten from Ireland and eight from France. There are also 21 accepted records from the Azores and one from the Canary Islands. Most first appear in autumn or winter and some become long-stayers or return repeatedly to the same sites.

The accepted Spanish records involved perhaps six different individuals. The earliest was one that remained at the Embalse de La Grajera, La Rioja, from at least February 2001 until January 2002. One was at the Laguna de Vixán, A Coruña, from July to September 2001 and what was probably the same bird was at the Laguna de Traba, some 65km to the north, from December 2001 to March 2002. One was at the Ría de Villaviciosa, Asturias, from December 2006 until 27 March 2007, when it moved a short way inland to the San Andrés reservoir, where it remained from 3 April to 3 May 2007. Subsequently one appeared at the Lagunas de Riocaldo, Lugo, and remained there from August 2007 until March 2011: it may have been the same individual that was in Asturias in 2006–07. One was well inland on a pool near Trujillo, Cáceres, on 13–15 May 2010. Another was at Sobrado dos Monxes A Coruña, from January to late March 2011.

LITTLE GREBE *Tachybaptus ruficollis*

(Zampullín Común, Mergulhão-pequeno)

***T. r. ruficollis*. A locally common resident throughout Spain and Portugal, and in Mallorca and Menorca in the Balearic Islands.**

The Little Grebe is a widespread resident throughout Iberia where it inhabits a wide range of freshwater bodies. It occurs virtually anywhere that offers plenty of submerged vegetation as well as adequate emergent plants for nesting. The breeding habitat includes quite small ponds as well as lakes, sluggish river backwaters, gravel pits, reservoirs and irrigation ponds and canals. It breeds in Mallorca and Menorca, and was first recorded nesting in Ibiza in 2007 (*Ardeola* 55: 290). In Portugal it is widespread south of the Tagus but absent from large areas further north, generally the higher elevations (PA2).

The size of the Iberian population is uncertain. The nesting population of Doñana alone has been said to number 8,000 pairs in some years (García *et al.* 2000). Nesting birds can be unobtrusive, making accurate censusing difficult (Gutiérrez & Figuerola 1997). However, at least several hundred pairs have been located in most of the autonomous regions, the highest counts outside Andalucía being 1,200–1,600 pairs in Castilla y León (Sanz-Zuasti & Velasco 1999), 500–600 pairs in Galicia (Fernández Cordeiro & Domínguez 1991) and at least 500 pairs in Catalonia: including 200 pairs in the Ebro delta and 80–90

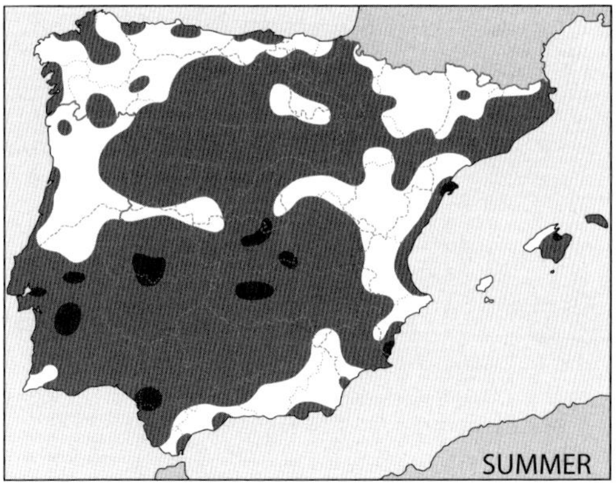

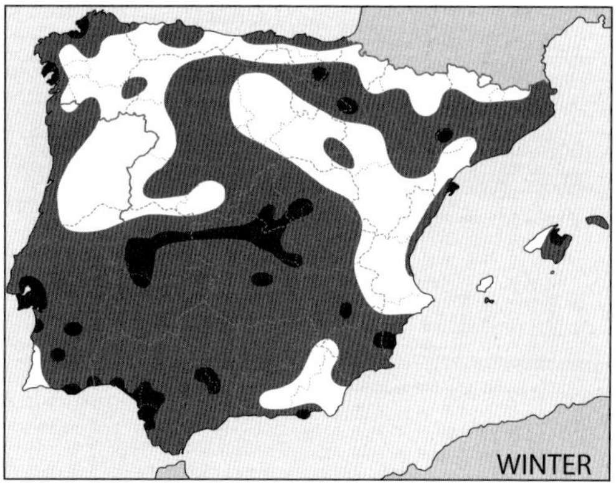

pairs in the Llobregat delta (SA1). The coastal wetlands of the Comunidad Valenciana have a good population, which fluctuates considerably from year to year according to water availability, as also happens in other hot regions: for example, there were 237 pairs in 1997 and 594 pairs in 2000, the latter including 453 pairs at El Hondo (SA2).

Breeding pairs are solitary although several nests may be found in fairly close proximity (Bernis 1945, Gutiérrez & Figuerola 1997). However, loose aggregations, sometimes of 100 or more, form immediately after the breeding season and at favoured locations in winter: there were 200 at the Azud de Riolobos, Salamanca, on 5 September 1998 and 395 at Brozas pool, Cáceres, on 17 August 2011 (*Ardeola* 45: 241, 58: 485). In Portugal, nearly 500 were at and around the Lagoa de Santo André in August–September 1993 and 1994, and 500 were at the Lagoa dos Salgados in August 2005 (AP). Many are resident but others abandon colder inland sites in winter and move to sheltered, often coastal locations. There they frequent coastal lagoons, saltpans and sheltered arms of estuaries but they do not normally occur on the open sea.

The Iberian birds are probably joined by others from further north in Europe in winter but there is a complete lack of foreign recoveries in Iberia. However, one ringed at Doñana in July 1987 was recovered at Settat, Morocco (448km) in July 1991. The January waterfowl counts in Spain during 1990–2009 found 5,192 birds on average (range 2,457– 7,387 birds; AAE). However, the actual wintering numbers are likely to be substantially higher. All regions showed a substantial increase in wintering numbers during 1990–2009. The Guadalquivir and Tagus wetlands together held 45% of the birds found during the January censuses: the highest site count then was 2,264 birds at Doñana in January 1993. Other mainland sites that have recorded high numbers in January include the Ebro delta (898), El Hondo (650), Santomera reservoir, Murcia (472) and Brozas pool (408). Substantial numbers have also been recorded in January at Balearic wetlands, notably at the Estany Pudent, Formentera (635), the Albufera de Mallorca (350) and the Albufera de Es Grau, Menorca (278) (AAE).

GREAT CRESTED GREBE *Podiceps cristatus*

(Somormujo Lavanco, Mergulhão-de-poupa)

***P. c. cristatus*. A locally common resident and partial migrant in Spain and eastern Portugal, more widespread in winter. Has bred Mallorca.**

This grebe breeds across much of Iberia wherever there are lakes, gravel pits and reservoirs offering fairly deep water and a supply of fish prey as well as marginal emergent vegetation for nesting. It is unsurprisingly lacking from clearly unsuitable terrain but it is also scarce and local north of the Cantabrian Range and in western Portugal (SA2, PA2). The breeding season is prolonged in the south, where nesting birds may be encountered in mid winter and chicks may be found at almost any time of year (Llandres & Urdiales 1990).

The species suffered intense persecution for its plumage throughout western Europe during the 19th century and its numbers only recovered during the 20th century. The Spanish population certainly seems to have been much smaller in the past and early accounts, such as Saunders (1871) and Arévalo (1887), suggest that it was much more local than is now the case. Chapman & Buck (1893) found it 'numerous' at Santa Olalla lagoon, Doñana, but Valverde (1960) was aware only of 6–12 pairs in the Marismas and a few more pairs in the lagoons. There must be little doubt that the species has increased as a result of the widespread proliferation of reservoirs during the 20th century, which have provided ideal nesting habitat in regions such as Extremadura where it was formerly lacking. Evidence for a larger nesting population was provided by the 2007 census of breeding waterbirds in Spain (Palomino & Molina 2009), which found a total of 4,025 birds in a sample of just 9% of Spanish

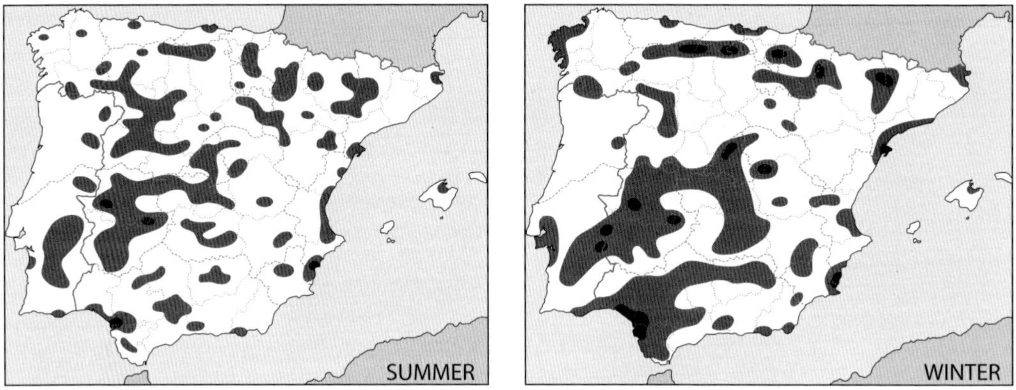

wetlands. This total allowed the total numbers present to be estimated at 44,100 individuals, with a minimum of 11,800. There are nesting concentrations in Doñana: where numbers fluctuate between 500 and 1,000 pairs; the reservoirs of Extremadura; the wetlands of La Mancha; the reservoirs of Castilla y León, including the Ebro reservoir in the Cantabrian mountains; the Ebro delta; the coastal wetlands of the Comunidad Valenciana, and the Ebro valley: notably at Gallocanta (SA2).

The Great Crested Grebe was rare in Portugal during the late 19th century and the first half of the 20th century (Tait 1887), when the large reservoirs did not exist. The first proved nesting in the country was at the Odivelas reservoir, Alentejo, in 1973. Since then the breeding population has continued to increase and is now widespread in the lowlands of the eastern half of the country, notably in Baixo and Alto Alentejo (AP, PA2).

Iberian birds may be fairly sedentary but ringing recoveries have shown that some young individuals may disperse widely. Six ringed as pulli or juveniles at Doñana travelled a long distance: three were found in Morocco, the farthest at Fez (351km); one reached the Mar Menor, Murcia (500km) and two reached the Haute-Garonne in southern France (880km). Many appear at coastal locations in winter, when the species is generally more widespread than during the breeding season. Some of these are probably migrants from further north in Europe and some cross Iberia to winter in North Africa, since concentrations of 1,000 birds or more occur locally in Morocco in winter (BM). The number of winter immigrants may be quite low since there is a lack of ringing recoveries and winter census results are also relatively low, given the size of the Iberian breeding population. The January waterfowl censuses in Spain during 1990–2009 found 5,179 birds on average (range 2,879–5,179; AAE). Most regions have shown a steady increase in wintering numbers during this period so that the mean population 2007–2010 was 7,120 individuals (range 5,171–9,568; SWA).

Post-breeding flocks, sometimes of several hundred individuals, form at favoured locations from late summer onwards. For example, 630 were on the Alange reservoir, Badajoz, in August 2012 and 700 were on the Sierra Brava reservoir, Cáceres, in August 2012 (*Ardeola* 59: 416). Some gatherings are in sheltered salt waters such as Santoña, Odiel and, especially, Cádiz bay. Birds begin to arrive in Cádiz bay from late September and numbers peak at up to 1,000 birds in mid-winter before they depart from March onwards (Tébar 1998); the largest January count there during 1990–2009 was 1,045 birds in 2001 (AAE). The Ebro delta holds large numbers in winter, 800+ in some years (AAE). Large reservoirs also often hold significant numbers: maximum January counts in Spain during 1990–2009 included 625 at the Ullibarri reservoir, Álava; 506 at the Yesa reservoir, Navarra; 505 at the Buendía reservoir, Cuenca; and 496 at the Ebro reservoir, Burgos/Cantabria (AAE). In January 2012 as many as 2,356 birds were found in Extremadura, including 665 at the Alange reservoir and 504 at La Serena reservoir (*Ardeola* 59: 171). Some Portuguese reservoirs, notably those at Santa Maria de Aguiar, Guarda; Alqueva, Beja, and Caia, Elvas/Campo Maior, similarly attract considerable numbers; for example, 769 were counted at Caia in February 2008 (AP).

The Great Crested Grebe was formerly a rare winter visitor to the Balearics (Clavell 2002) but several wintered at the Albufera de Mallorca in 1999–2000 and 2000–01 and two pairs nested there for the first time in 2002, followed by four pairs in 2003 and 2004 (*Ardeola* 52: 207, 423). These had increased to at least seven pairs in 2010 (*A. O. Balears 2010*: 112). Two pairs nested for the first time in Menorca in 2010, at the Albufera des Grau, and three pairs bred successfully there in 2011 (*Ardeola* 58: 485, 59: 416). The wintering population remains small but has clearly increased: the January counts of 2007–10 found 21 birds on average (range 13–28; SWA).

RED-NECKED GREBE *Podiceps grisegena*

(Somormujo Cuellirrojo, Mergulhão-de-faces-brancas)

***P. g. grisegena*. Almost annual in winter on northwestern and Biscay coasts. Rare elsewhere in Spain and not recorded from Portugal.**
***P. g. holboellii*. Vagrant to Spain.**

There are historical records from the Balearic Islands, including one in Alcudia Bay, Mallorca, in 1930 (Bernis *et al.* 1958, Ferrer *et al.* 1986). The species has yet to be identified in Portugal but it was first reported on the Spanish mainland at the Ebro delta on the rather late date of 2 May 1969 (*Ardeola* 15: 87). The first sighting from the Biscay coastlands, where most subsequent observations have occurred, was at Santoña on 6 October 1979 (*Ardeola* 28: 149). It was followed by one at Gijón, Asturias, in January 1982 (*Ardeola* 31: 141). The rarities committee considered the species during 1984–93 during which period the total number of records reached 41 (de Juana 2006). Together with subsequent records there is a sample of 156 records (197 individuals). Although 85% of observations were of single birds there are 14 of two together and 10 of three or more: including six off Betanzos, A Coruña, on 14 January 1989 (*Ardeola* 38: 150). The birds included up to 16 that were identified as adults, 34 first-winters and a juvenile, this last at Estarit, Girona, on 8 August 1996 (*A. O. Catalunya 1996*: 33).

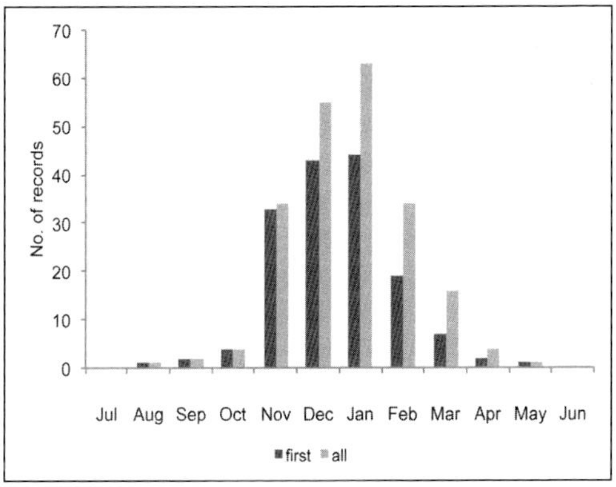

Figure 50. Records by month in Spain, 1930–2011 (n = 156)

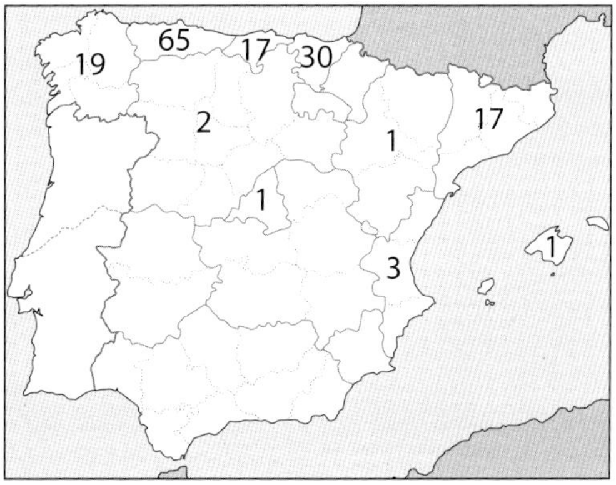

Figure 51. Records by region, 1930–2011 (n = 156)

Apart from the August juvenile, the extreme dates are 13 September and 2 May but 77% of first observations are during November–January (Fig. 50). Observations are concentrated on northern coasts, from Galicia to the Basque country (Fig. 51) where 84% of records and 87% of the individuals have occurred. Elsewhere there are 21 records from the Mediterranean, most of them from Catalonia, and just four have appeared far inland: at reservoirs in León, Valladolid, Zaragoza and Madrid. There have also been inland observations in the coastal regions of Galicia, Cantabria and the Basque country, at the A Rua-Petín, Ebro and Ullibarri reservoirs respectively. The great majority of Red-necked Grebes occur in the sheltered waters of bays, estuaries, rías and harbours.

There are two accepted records from Galicia of individuals that showed the characters of *P. g. holboellii*: an adult at Laxe on 28 January and 19 February 1984, and a probable first-winter at Ponteceso on 3–10 January 1987 (*Ardeola* 37: 109).

SLAVONIAN GREBE *Podiceps auritus*

(Zampullín Cuellirrojo, Mergulhão-de-penachos)

Small numbers occur annually on northern Spanish coasts but there are only two accepted records from Portugal. European breeding birds chiefly winter in northern coastal waters. Some 300–400 birds winter in France, mainly on Atlantic coasts south to Charente-Maritime.

The earliest Iberian record was in the far south, a bird obtained in the Strait of Gibraltar in October 1867 (Irby 1875). Another was captured in Santander harbour on 2nd December 1876 (Irby 1883). Bernis (1954a) suggested however that some of the earlier observations in Spain, including some claims of nesting, may have been due to confusion with the Black-necked Grebe. One was apparently obtained in the Balearic Islands in August 1969, an unlikely date (Nadal 1971). Some 20 reports from the 1970s and early 1980s (e.g. Lucientes 1971, Muntaner *et al.* 1974, Ferrer *et al.* 1986, Álvarez-Balbuena *et al.* 2000) include a surprising number of early observations in August and September and an equally surprising lack of observations from the Biscay coasts (de Juana 2006). The rarities committees considered the species from 1984 to 2004, accepting 71 records. Others have been reported subsequently but we have limited our analysis to formally accepted records, to avoid possibly misidentified birds. There are a minimum of 66, only two of which have been in Portugal: two in the Minho estuary on 1 November 1993 and one at the Óbidos lagoon, Leiria, from 7 January until 11 February 2006. A further five Portuguese observations of single birds occurred during 2010–12. The accepted records involve some 172 individuals, although some may have returned in successive winters.

Records are concentrated in winter (Fig. 52). The extreme dates are 28 September and 16 April but 74% of first observations have been during November–January. The majority are found on northern coasts, some 80% of the records being distributed between Galicia, Asturias, Cantabria and the Basque country (Fig. 53). Some 61% of observations are of single birds and a further 21% of two together. However, Santoña uniquely has attracted a small wintering population since 1994 and this exceeded 11 birds annually during 1998–2004, with a maximum count of 18 in January 2002. There were again 11 at Santoña in December 2010 (*Ardeola* 57: 221). Records elsewhere are much more irregular. There were reports from 11 different locations in 1995–96, when there was an influx into France (Duquet 1997), and from eight locations in 2010–11: with 26 birds in all, and groups of five at Ría de Ortigueira, Galicia, and the Ebro delta (*Ardeola* 58: 198). Nevertheless, Slavonian Grebes are usually only seen at 2–4 sites each winter, generally in estuaries, rías or bays, more rarely on reservoirs or lagoons. The furthest inland include single birds on Las Navas reservoir, Huesca, in January 1994, on a gravel pit near Madrid in February 2000, and on the Monteagudo de las Vicarías reservoir, Soria, in December 2008 (*Ardeola* 43: 105, 49: 144, 55: 290).

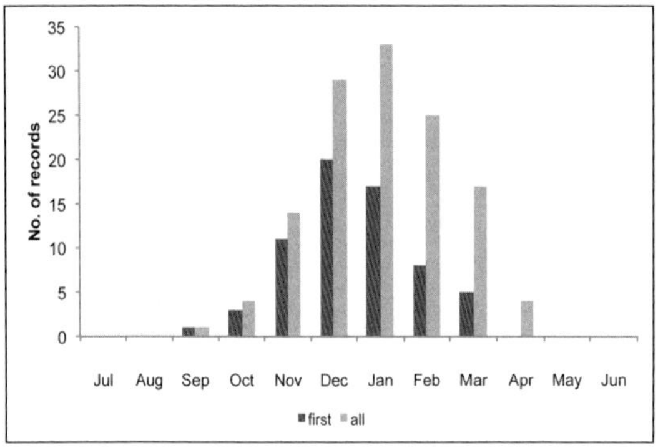

Figure 52. Records by month 1980–2004 (n = 65)

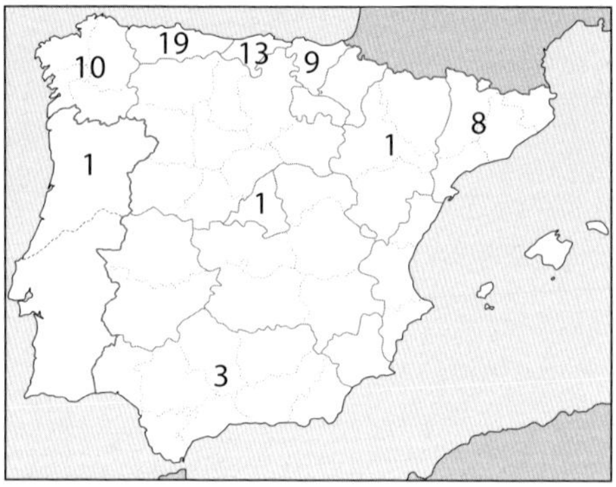

Figure 53. Records by region 1980–2004 (n = 65)

BLACK-NECKED GREBE *Podiceps nigricollis*

(Zampullín Cuellinegro, Cagarraz)

***P. n. nigricollis*. A scarce resident in Spain and has bred recently in Portugal. More widespread and locally numerous in winter, including in the Balearic Islands.**

The Black-necked Grebe has a highly fragmented breeding range in mainland Iberia, where the majority are found at Doñana and other wetlands of the Guadalquivir valley and Cádiz, the wetlands of La Mancha and at El Hondo, Alicante. Small numbers breed at other scattered locations, often sporadically: including Gallocanta; the Ebro delta; the Madrid gravel pits; the Laguna de Pitillas, Navarra, and the Monteagudo reservoir, Soria (SA2). Individuals were present in the breeding season at the Lagoa dos Salgados, Algarve, in 2001, 2002 and 2007, and one pair bred at the Albufeira de Poço da Rua, Évora, in 2007: the first recorded instance of nesting in Portugal (AP).

The Black-necked Grebe typically breeds in loose colonies, often in association with Whiskered Terns and Black-headed Gulls. It selects freshwater lakes, gravel pits, shallow arms of reservoirs, saltpans and riparian backwaters that combine open, shallow water with well-developed marginal vegetation. It settles to breed late in spring, which presumably allows it to establish whether or not a water body will persist through the season (SA2).

The breeding population fluctuates according to rainfall and there may be considerably more birds in wet years. Valverde (1960) reported colonies of up to 1,000 pairs in Doñana in wet years but he noted that none at all bred during droughts. The Doñana population has since been estimated to be as many as 2,000 pairs in some seasons (García *et al.* 2000) although recent prospecting suggests that the population in the National Park ranged from 383–445 pairs during 1998–2002, with a further

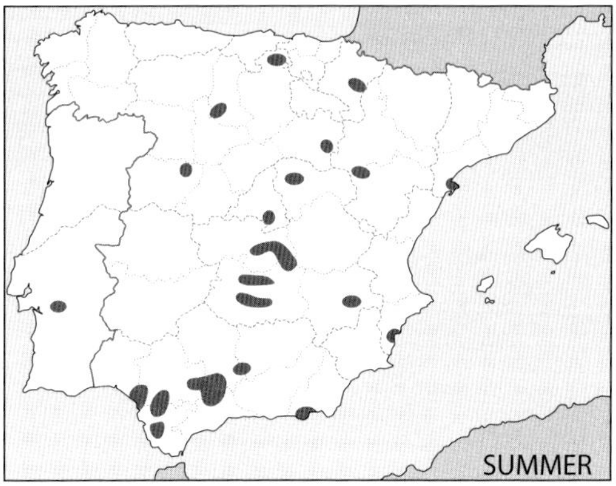

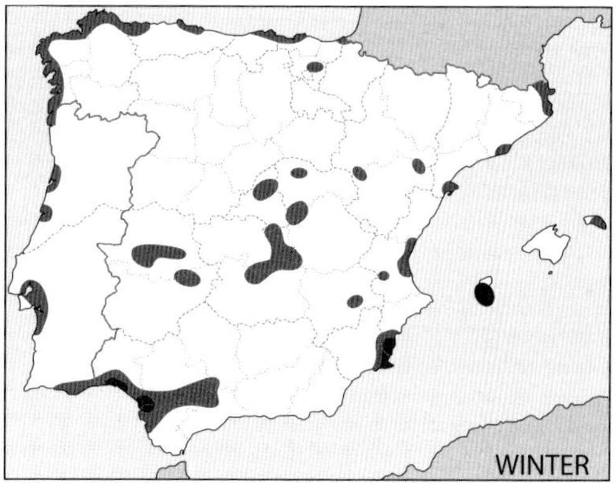

100 pairs in the park periphery (LR). Up to 500 pairs may nest in La Mancha in wet years, as in 1977, but many fewer do so normally (Velasco 1999). The breeding population at El Hondo was 115 pairs in 1998, 550 pairs in 1999, 300 pairs in 2000 and 287 pairs in 2009 (SA2; *A. O. C. Valenciana 2009*: 21). The El Hondo population highlights the almost total absence of breeding Black-necked Grebes elsewhere in the Comunidad Valenciana. Other Spanish breeding sites generally hold no more than twenty pairs and often fewer but numbers fluctuate locally. The total Spanish population in spring 2007 was some 4,000 birds (Palomino & Molina 2009).

Post-breeding flocks begin to form in July–August when many undertake moult migrations to favourable locations. These are predominantly coastal lagoons and saltpans, where the saline habitat supports quantities of the staple food item, the brine shrimp (Varo *et al.* 2011). The birds moult there between August and December, becoming flightless temporarily, and they may overwinter at the same sites or depart to winter quarters elsewhere. Iberian birds are joined during this period by an unknown number of migrants from elsewhere in Europe. Particularly important concentrations occur at the saltpans of the Marismas del Odiel and the Estany Pudent, Formentera, and well as on the Laguna de La Mata, Alicante, and at the Mar Menor (SWA).

Post-breeding flocks may attain considerable size early in the season. For example, 1,052 birds were at lagoons at the water purification works of Cabezo Beaza, Murcia, on 14 July 2008 (*Ardeola* 56: 347). Counts at the Estany Pudent have also tended to peak at 1,000+ birds in late summer in recent years, declining to a few hundred in winter (*A. O. Balears 2001, 2002, 2005; Ardeola* 50: 51, 52: 207, 53: 380). However, numbers typically peak later at other sites: for example, numbers at the Laguna de la Mata in 1985 built up steadily from zero in early June to peak at 3,000 birds on 9 October before falling back steadily to just 500 by the end of December (García & Calvo 1987). An autumn peak is also seen at Odiel, where some very large concentrations sometimes occur: 10,725 were counted in November 2002 (SWA). At Odiel numbers built up steadily during August and September to peak in late October at about 1,500 birds in 2008 and 2,500 in 2009, before declining through November to about 500 in 2008 and 700 in 2009: the decline corresponding with a sharp drop in brine shrimp abundance in both years (Varo *et al.* 2011).

Some of the Black-necked Grebes that leave their moulting locations perhaps move to Africa for the winter, although counts from Morocco and Mauritania have involved only modest numbers and few local concentrations of over 100 birds (BM, Isenmann *et al.* 2010). The average January count at Odiel during 1990–2009 was 501 birds, with a maximum of 1,286 (AAE), so a majority of the birds that moult there winter elsewhere. Some wintering Black-necked Grebes frequent coastal inlets and sheltered bays. Irby (1873) reported 'great numbers' in the Bay of Gibraltar in winter but only a few winter there now. They may be encountered along Mediterranean coasts, in the Balearic Islands and on the coasts of the Gulf of Cádiz as well as, further north, in the Rías of Galicia and more locally in Asturias and Cantabria. Northern coastal sites that held some numbers in winter during 1990–2009 include the Ría del Eo (up to 217 birds), Santander Bay (up to 107) and Santoña (up to 198) (AAE). A minority also winter well inland on lakes and reservoirs. The Spanish January counts of 1990–2009 found 4,970 birds (range 2,017–9,915; AAE), some 70% of them at the principal southwestern and Mediterranean coastal locations. The Spanish wintering population was largely stable until the mid-1990s but it increased markedly thereafter, with some more recent indications of decline towards the original numbers (AAE), although with local fluctuations that may be associated with hydrological conditions but probably also with brine shrimp availability (Varo *et al.* 2011).

The small Portuguese wintering population averaged 71 birds during 1992–95 and 109 birds during 2005–07, with a peak count of 172 in January 2005 (AP). Most are found at coastal wetlands south of the Tagus but a few occur in the northern coastlands. The Sado estuary and the Castro Marim reserve account for the largest numbers and occasionally attract gatherings of 100+: for example, 110 were at Castro Marim on 23 November 2008 (*Anu. Ornitol.* 6: 74).

Until recently the only foreign-ringed bird recovered in Iberia was one ringed at Lake Constance, Switzerland, in August that was recovered in January in northwestern Spain. However, over 10,000 moulting grebes at Odiel have since been marked and distant recoveries of 44 birds confirm that the wintering population includes some from further north and east in Europe: of these 39% were found in Italy/Southeastern Europe, 27% in Northern Europe, 20% in Central Europe and 14% in France/Belgium. Three individuals were found east of the Urals in Chelyabinsk, Russia (5,331km). Four Spanish-ringed Black-necked Grebes have also been recovered in Morocco.

WANDERING ALBATROSS *Diomedea exulans*

(Albatros Viajero, Albatroz-viajante)

Vagrant to Portugal.

The only substantial claim from Iberian waters is the record of an immature bird that followed a British warship some 50 miles south-west of Cape St Vincent, Algarve, on 18 October 1963 (Bourne 1967). There are only five records from European waters, including an immature male that was captured at Palermo, Sicily, in October 1957.

BLACK-BROWED ALBATROSS *Thalassarche melanophris*

(Albatros Ojeroso, Albatroz-de-sobrancelha)

Vagrant to Spain and Portugal.

The most frequent of the albatross species to reach the North Atlantic although it is still very rare. There are records of 27 individuals from Britain, 12 from Ireland, ten from France, three from Morocco and one from Italy.

There are seven accepted records from Spain and one from Portugal. They comprise an immature off Tina Mayor, Cantabria, on 3 July 1984; a dead adult found on the beach at Cangas do Morrazo, Pontevedra, on 26 December 1987; a possible adult off Conil de la Frontera, Cádiz, on 7 August 1999; an adult off A Coruña on 18 October 1999; an immature seen from Punta de la Chullera, Málaga, on 15 July 2002; an adult off Burela, Lugo, on 6 November 2011, and presumably the same bird off Muxia, A Coruña, on 10 November and off Cape Peñas on 29 November. The Portuguese record was of a probable adult seen at the Berlengas Islands in February 1983. There are two later Spanish records of single birds: at Cape Estaca de Bares on 21 April 2012 and off Punta La Vaca, Asturias, on 1 June 2013 (RBS). Four other observations of small albatrosses in Spanish waters are also most likely to have been of this species: two off Isla Villano, Vizcaya, on 7 October 1973; one at Vivero, Lugo, on 6–7 August 1984; one at Cudillero, Asturias, on 26 August 1985, and one off Punta de la Vaca on 9 July 1994.

NORTHERN FULMAR *Fulmarus glacialis*

(Fulmar Boreal, Pombalete do Norte)

***F. g. glacialis*. A scarce visitor to Atlantic coasts but virtually unknown in the Mediterranean.**

The Iberian Peninsula lies at the southern extreme of the non-breeding range of the Atlantic populations, the nearest nesting colonies being in Brittany. The Fulmar was virtually unknown in Iberia prior to the 1980s although Bernis (AMI) gives three winter recoveries from the Biscay coast of birds ringed as juveniles in Scotland. Some are now recorded annually but typically only very small numbers are present and the birds generally remain well offshore.

The great majority of records of live birds come from well-watched seabirding sites in Galicia and Asturias, notably Cape Estaca de Bares and Punta La Vaca, and there are relatively few reports from Cantabria and the Basque Country. The few records from Portugal, and the coasts of Huelva and Cádiz in southwestern Andalucía, are almost entirely of tideline corpses. There are no records from the Strait and we are only aware of one unpublished observation from Mediterranean Iberia: at Punta Calaburras, Málaga, in October 2005. There are at least three records from Mediterranean France and one from Italy (Bonaccorsi 2003, OF).

The temporal distribution of observations is confused by the high proportion recorded only as tideline corpses, which largely appear after severe weather in autumn and winter. No fewer than 20% of 335 individuals recorded during 1980–2010 were found dead or sick and a few had clearly died long before they were found. The five months from November to March account for 68% of records of dead or dying birds (79% of individuals) but only 19% of those of apparently healthy individuals. The records of live birds (Fig. 54) show a marked peak in late summer and a lesser one in March–April and it is clear that the early perception of the Fulmar as a rare winter visitor was incorrect: it is very scarce but some occur all year round.

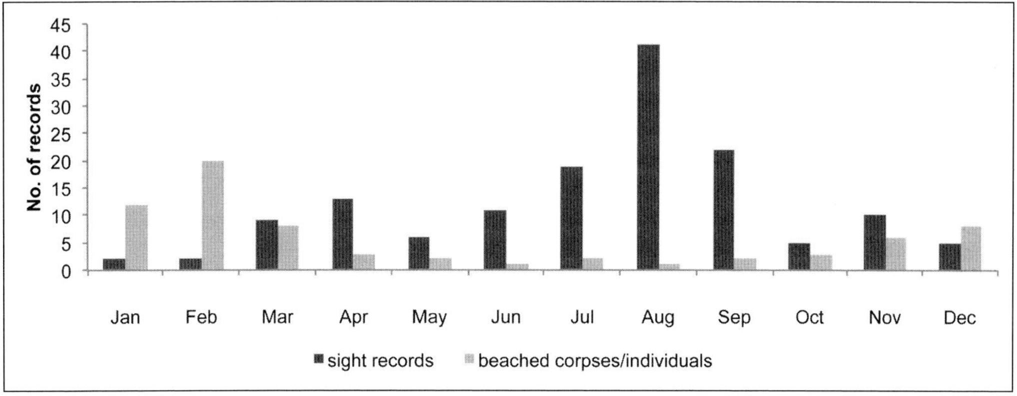

Figure 54. Monthly distribution of 145 sight records at sea and 68 tideline corpses or beached individuals reported in Iberia during 1980–2010.

The distribution pattern is also affected by rare influxes of storm-driven birds. By far the largest of these on record was seen in Asturias from 20–27 March 2007 when 90 birds were reported following north-northeasterly gales. Most were flying west and they included 73 that passed Punta La Vaca in three hours on 22 March (*Ardeola* 54: 172). Nearly all Fulmar records are sightings/findings of single birds and these observations of some numbers were unprecedented. The other noteworthy multiple occurrences have chiefly included dead or dying birds, such as 22 recovered on northern and northwestern coasts between November 2002 and August 2003 following the shipwreck of the tanker '*Prestige*' (*GIAM* 27: 3), 27 on the Biscay coast (nine alive) during January–March 2005 (*GIAM* 27: 3) and 13 (12 dead) in Cádiz Bay after westerly gales during 27 January–12 February 2009 (*Ardeola* 56: 154–155). The last of these incidents was exceptionally far south: the other four Andalucian records have all involved single birds.

Several individuals have been seen apparently prospecting coastal cliffs in Asturias and Galicia in spring (Paterson 1997) and one was seen flying over Isla Conejera, Suances, Cantabria, on 23 April 2007 (*Ardeola* 54: 390). One seen overflying the cliffs at Ponta da Atalaia, Faro, on 4 August 1997, and another seen at Cape Carvoeiro on 7 August 2010 (*Pardela* 11: 5; *Anu. Ornitol.* 8: 11) are the only accepted records of live birds in Portugal, although both the dates and the southerly locations are surprising. At least 17 had been found as tideline corpses on Portuguese coasts up to 2010 (AP, CPR 2010).

CAPE PETREL *Daption capense*

(Petrel Damero, Pintado)

Vagrant to Gibraltar.

One was seen at close quarters below Europa Point, Gibraltar, on 20 June 1979. There were ten European records of this southern petrel during 1844–1979, but most were considered of dubious origin (Haas & Crochet 2009).

BLACK-CAPPED PETREL *Pterodroma hasitata*

(Petrel Antillano, Freira das Antilhas)

Vagrant to Spain.

One was seen 180 miles north-west of Cape Finisterre, Galicia, on 30 April 2002 (Howell 2002, Ardeola 51: 521). It breeds in the Caribbean and occurs regularly in tropical and subtropical waters from Brazil to the USA. There have been two records for the Azores and one for Madeira (Haas 2012).

FEA'S / ZINO'S PETREL *Pterodroma feae / madeira*

(Petrel Gongón / Freira, Gon-gon / Freira)

Rare but recorded increasingly frequently off Atlantic coasts in autumn.

Individuals of this species-complex have been recorded on still rare but increasingly regular occasions from European Atlantic coasts. There were 45 accepted records of Fea's/Zino's Petrels from British waters up to 2010 and 73 from Ireland up to 2011, as well as three from France, three from the Azores and one from the Canary Islands. Two additional French records, in 2010, have been accepted as Fea's Petrel.

The *Pterodroma* petrels of the Macaronesian archipelagos have received detailed scrutiny in recent years, with the resulting discovery that although they are all but identical in appearance they clearly differ in their vocalisations (Zino *et al.* 2008). Fea's Petrel *P. f. feae* nests in the Cape Verde Islands and a larger and larger-billed subspecies, the Desertas Petrel *P. f. deserta* nests on Bugio island in the Desertas archipelago near Madeira. The Desertas Petrel also differs from both Fea's and Zino's Petrel in its morphology and vocalisations and it too may merit specific status (Robb *et al.* 2008, Jesus *et al.* 2009). Zino's Petrel breeds in the high mountains of Madeira. Fea's and Desertas Petrels are indistinguishable at sea (Shirihai *et al.* 2010) but the smaller, more delicately built and more slender-billed Zino's Petrel is separable from the other two under the most favourable circumstances.

All three forms are rare, with populations of the order of 500+ pairs (Fea's Petrel), 250+ pairs (Desertas Petrel) and 80 pairs (Zino's Petrel) (Shirihai *et al.* 2010). All three are pelagic when not breeding and geolocators are being employed to plot their movements. Preliminary results from 12 Zino's Petrels have revealed that the birds ranged widely in the northeastern Atlantic during their breeding season, April–September, reaching the waters off southwestern Ireland and also occurring off the Iberian west coast and the African west coast south to Senegal (Zino *et al.* 2011). They were not recorded in the Bay of Biscay or in the western approaches to the Strait of Gibraltar. The same birds ranged further west and south between late October and late March, when they were not recorded in European waters.

One was seen 130 miles west of Lisbon on 23 August 1992. Another was off Cape Estaca de Bares, Galicia, on 6 August 1995 and there were additional records there on 19 August 2008, 27 August 2009, 4 September 2010 and 1 November 2010, the September record involving two birds. Another was seen 200 miles west of Fisterra, A Coruña, on 13 August 2010. A significant number of Spanish records, also of birds seen off Galicia or from Cape Bares, were pending assessment at the time of writing, including at least 16 sightings in autumn 2011 and 14 in autumn 2012 (RBS). One of the latter has been specifically assigned to Fea's Petrel: an individual photographed 17 miles off Malpica de Bergantiños on 26 August 2012.

BULWER'S PETREL *Bulweria bulwerii*

(Petrel de Bulwer, Alma-negra)

Regularly present off Portugal in summer but has never been reported from land there. There are very few reliable observations from Spain.

Bulwer's Petrel nests in the Macaronesian islands, dispersing in the western and south Atlantic when not breeding (Bourne 1995). Records in Europe include one in Ireland, one in The Netherlands, four in France, two in Italy and one between Italy and Corsica (Flood 2009). There are also eight records from Morocco, mainly south of 32° N.

The oldest known observation in Iberian waters was one seen over the edge of the continental shelf near the Berlengas Islands in September 1964 (Bourne & Norris 1966). One was reported from Torremolinos, Málaga, on 7 February 1982, and an injured, probably shipborne bird was found in Vigo harbour on 1 June 1983 (*Ardeola* 30: 115, 33: 203). More recently there was the remarkable capture of a Bulwer's Petrel in mistnets set for waders at the Odiel Estuary on 7 June 2000; a tideline corpse was recovered at Isla Cristina, Huelva, on 25 October 2006, and one was seen 200 miles off Cape Finisterre on 25 August 2010.

The great majority of Iberian observations have been off Portugal since 1987, where birds have been seen mainly some dozens of miles from the coast, beyond the continental shelf, from 36°N south of the Gorringe bank north to 40°42'N off Cape Mondego. Most of the records result from 'pelagic' excursions by Colm Moore off southwestern Portugal. There were 39 accepted records by 2009, mainly of solitary birds, as is typical of the species. However, there were counts over some hours of 16 on 21 August 1999 and 17 on 23 August 2000 at the latitude of Sagres, Algarve (*Anu. Ornitol.* 1: 8). There are three subsequent Iberian records, one in Portugal and two in Spain (Flood 2009, RBS).

A recent datalogging study (Zino 2012) has confirmed the presence of Bulwer's Petrels from the Selvagem Grande colony, Madeira, off the Portuguese continental shelf but only during their breeding season, April–September. Accepted Iberian records fall between 6 April and 25 October, with a marked concentration in August that may be an artefact since it corresponds with the timing of the Moore trips. However, all the Moroccan records also fall in April–September. The observations thus presumably involve foraging birds from the Macaronesian islands.

CORY'S SHEARWATER *Calonectris diomedea*

(Pardela Cenicienta, Cagarra)

C. d. diomedea **(Scopoli's Shearwater). Breeds in the Balearic Islands and off eastern Spain. Common and sometimes numerous in Mediterranean waters. Chiefly present March–October.**
C. d. borealis **(Cory's Shearwater). Common and sometimes numerous in Atlantic waters and the Strait of Gibraltar. Breeds Portugal and Galicia. Chiefly present March–October.**

Cory's and Scopoli's Shearwaters differ significantly genetically and each has diagnostic vocalisations. They are increasingly regarded as distinct species (Gómez-Diaz *et al.* 2006, Robb *et al.* 2008, Sangster *et al.* 2012).

Scopoli's Shearwater

Scopoli's Shearwater has a significant breeding presence in the Balearic Islands; chiefly in Menorca, but also on islets off Cabrera, Ibiza, Mallorca and Formentera. There are also small colonies on the Columbretes islands off Castellón and the Islas Palomas off Murcia. Breeding is also suspected at Cape Tiñoso, Murcia (LR). The birds nest in caves and in holes in cliffs.

The size of the Spanish population is uncertain and various censuses and estimates give inconclusive results for the principal concentration in the Balearic Islands. These seem to number fewer than 10,000 pairs (SA2, LR), the great majority on the northern coast of Menorca. The Columbretes population is small: there were no more than 200 pairs until the 1990s since when numbers declined, to 50 pairs in 2009 (*A. O. C. Valenciana 2009*: 22). The Islas Palomas colony was put at 29 pairs in 1985–88.

It is often encountered foraging offshore in Mediterranean waters, generally between February and November. Small numbers are usually involved but large gatherings occur near the Balearic colonies and sometimes elsewhere, especially during passage periods. For example, 2,000 were off El Saler, Valencia, on the relatively early date of 13 February 1985 and passage rates of 288 birds/hour and 400 birds/hour were reported on 15 and 21 October 1989 respectively at Cape Cullera, Valencia (*Ardeola* 32: 409; *GIAM* 8: 7).

Most if not all leave the Mediterranean in winter and recent datalogging studies have suggested that they winter off West Africa in the waters of the Canarian current or in the Gulf of Guinea, as well as further south off Namibia (González-Solís *et al.* 2007, Reyes-González & González-Solís 2012). There are sometimes spectacular concentrations when the birds pass through the Strait of Gibraltar. The influx into the Mediterranean is chiefly in February and early March, with a return passage from mid October to mid November. Passage rates at the Strait may exceed 3,000 individuals per hour for long periods. However, the width of the Strait even at its narrowest point (14km) precludes monitoring the entire passage from shore. Many also clearly pass undetected at night since unabated passage is sometimes observed as darkness falls. Despite these difficulties, attempts have been made to estimate the size of the total passage through the Strait on the basis of sample counts. Tellería (1980, 1981), counting from Tarifa, calculated that well over 100,000 birds left the Mediterranean during the autumn passage in 1976. It was subsequently found that in autumn a very large flow of Scopoli's Shearwaters sometimes approaches the Strait from the south-east and passes close to Ceuta on the south shore. Counts there between mid October and mid November in 2005–07 found that the total passage may exceed 500,000 individuals (Navarrete 2008).

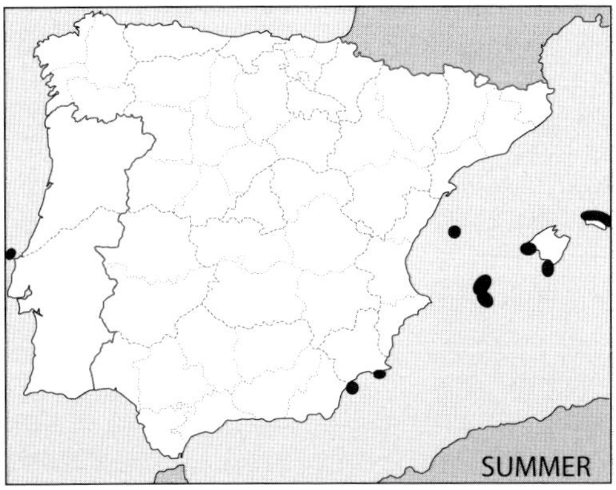

Some ringed at Mediterranean colonies have been recovered on the coasts of sub-Saharan Africa. There are three recoveries from Ghana, two from Senegal, one from Guinea Bissau and one from Angola. An adult ringed on the islet of Terreros, Almería, in July 2000 provided the most distant recovery: at Romansbaai, South Africa (8,300km) in April 2003; however, it may have been a Cory's Shearwater (see below). There is also a remarkable recovery of a bird that was ringed as a nestling at Ciutadella, Menorca, in September 1993 and recovered 11 years later on the far side of the Atlantic: at Little Tybee Island, Georgia, in September 2004. Ringing has revealed frequent inter-colony exchanges between the Balearic colonies and between these and the colonies on the Columbretes. One ringed as a nestling in

Corsica in August 1995 was found on the Columbretes in June 2002. Scopoli's Shearwaters from the Iberian colonies have also been recovered on the Mediterranean coasts of Morocco and Algeria, and there is one recovery from Tunisia, but there are none from France or Italy. Scopoli's Shearwaters presumably also occur at times off Iberian Atlantic coasts away from the Strait. Indeed a small colony has been found on the French Biscay coast of Gironde (Mays *et al.* 2006), the only ones known to nest outside the Mediterranean. However, the two forms can only be separated in the field under ideal conditions.

The Iberian breeding populations of both subspecies, but especially those of Scopoli's Shearwater, are regarded as 'Endangered' in Spain and 'Vulnerable' in Portugal (LR, LV). The main threat appears to be mortality due to longline fisheries. A study of the bycatch of the fishing fleet around the Columbretes islands, based on 129 fishing operations in 1998–99, found 22 dead seabirds including 18 'Cory's' Shearwaters (Belda & Sánchez 2001). A second study, that examined the bycatch from 2,587 fishing operations throughout Spanish Mediterranean waters during 2000–09, found 261 dead seabirds including 80 'Cory's' Shearwaters (García-Barcelona *et al.* 2010). The two subspecies were not distinguished in these studies but most victims are likely to have been *diomedea*. As elsewhere in the Mediterranean (Thibault 1995, Igual *et al.* 2006) the species seems also to suffer from the presence of introduced mammals, such as rats and rabbits, on the islets that hold the breeding colonies.

Cory's Shearwater

Very few birds of the Atlantic form breed in Iberian waters. The main colonies are instead in the Atlantic archipelagos, notably the Azores, Canaries and Madeira. However, there are small colonies on the Berlengas islands, Portugal. The Berlengas population is estimated at 700–800 pairs, largely on the basis of a thorough census on the island of Farilhão Grande, where 400–500 pairs were found in 2005 (AP). A small breeding population was discovered in Galicia in 2007: on the rocky islet of Coelleria, just south-east of Cape Estaca de Bares, and nesting was also confirmed on the Sisargas islands in 2009; the total Galician population that year being estimated at 30–40 pairs (*Ardeola* 57: 221). The Galician birds are the first to have been found nesting on the Spanish Atlantic coast and their biometrics confirm that they are of the Atlantic subspecies.

More surprisingly, a few pairs have been found nesting in Spanish Mediterranean waters on the Terreros islet, Almería (Gómez-Díaz *et al.* 2006). This is the only known 'pure' colony of *borealis* in the Mediterranean. However, individual *borealis* ringed as pulli in the Selvagens have been found within nesting colonies of *diomedea*; in the Columbretes islands in 1992 and 1993 (*Ardeola* 41: 91), on the Chafarinas islands near the Moroccan coast south of Málaga in June 2005 (*Ardeola* 52: 424) and on Linosa island, Sicily (Lo Valvo & Massa 1988). The Columbretes bird was a male and measurements suggested that its mate too was a *borealis* individual. Cory's Shearwaters are largely faithful to their natal colonies but these records show that exceptions occur. Breeding *borealis* have also been found in Corsica (Thibault & Bretagnolle 1998) and morphological evidence has identified several other individuals on the Columbretes (Martínez-Abraín *et al.* 2002a). A recent investigation has also found that as many as 22% of the breeding birds of the Chafarinas are actually *borealis* and that hybrid pairs are frequent there (Gómez-Díaz *et al.* 2006). There are also at least four recoveries during July–September of birds from colonies in the Canary Islands that had entered the Mediterranean: two from Girona and one each from Almería and the Columbretes islands. There is thus increasing evidence that the two forms are not as well segregated geographically when breeding as was once believed.

The movements of *Calonectris* shearwaters in Iberian waters are not fully understood. In particular, large numbers may be seen in the Strait from June to mid October, when they are also common in Málaga Bay. Observations from Gibraltar in summer suggest that most if not all these birds are Cory's Shearwaters and not Scopoli's Shearwaters. The birds are clearly foraging and may often be seen feeding in association with cetaceans and other seabirds off Gibraltar. Most such observations do not exceed a few hundred birds but there were exceptional concentrations of up to 1,600 in September–October 2004 (*GBR* 2004: 14). A few *Calonectris* shearwaters may be encountered occasionally in the Strait in December and January but their subspecific identify is unknown.

Cory's Shearwaters are also present off Atlantic and Biscay coasts where they occur chiefly during March–October. They winter in the south Atlantic and Iberian winter records are exceptional. The largest numbers occur in summer and autumn, when many feed in the Bay of Biscay. Significant counts have been made from key coastal watchpoints at this time, chiefly of birds coasting westwards, particularly past Cape Estaca de Bares. The mean annual post-breeding daytime passage there during 2004–08 was of the order of 100,000 birds (Sandoval *et al.* 2010) and many more must pass at night.. However, it is impossible to distinguish migrants from birds performing feeding movements, particularly early in the season, during June–August, when Cory's Shearwater may be the most abundant species recorded from the Cape (Sandoval *et al.* 2010). The maximum passage rate recorded was 1,573/hr on 19 August 2005 (Sandoval *et al.* 2009). Large numbers may also be seen from Portuguese headlands, at least occasionally: a count of 12,000 birds in a few hours during stormy weather on 19 September 1999 (AP) is particularly noteworthy.

A storm-driven bird found alive far inland in a city garden in Salamanca city on 18 October 2005 (*A. O. Salamanca 2004–2006*: 103) was highly exceptional.

GREAT SHEARWATER *Puffinus gravis*

(Pardela Capirotada, Pardela-de-barrete)

Regular and sometimes common in summer and autumn off Atlantic coasts and in the Bay of Biscay. Vagrant to the Mediterranean.

The Great Shearwater visits Iberian Atlantic waters during the summer and autumn but is relatively seldom observed from shore. The numbers involved vary considerably between years but sometimes involve thousands of birds that linger to feed in the Bay of Biscay. It is most often seen during August–November but some are reported in July from Galicia and Asturias and there have been exceptional north coast sightings in late May–June and December–January 2009 (*A. O. Galicia*, SWA). Five were in Cádiz Bay in December 2009 (SWA).

Observations from land are most often possible from headlands on the north and northwestern coasts, notably Cape Higuer, Punta La Vaca, Cape Peñas and Cape Estaca de Bares. Most birds are seen coasting west. Counts at the last of these sites reveal the high year to year variation in inshore occurrences: the highest count on record was 866 birds/hour on 29 September 1991 but recent seasonal totals here have ranged from just 85 in 2006 to 9,313 in 2011 (Sandoval *et al.* 2009, A. Sandoval pers. comm.). On the basis of sample counts, the mean annual post-breeding daytime passage at Cape Bares during 2004–08 was of the order of 6,000 birds (Sandoval *et al.* 2010).

There are few records from Portugal and southwestern Spain, none involving significant numbers (Paterson 1997, AP), although these areas are relatively underwatched. Many thousands move south in autumn off Portugal and clearly there is a potential for observations from land should inclement weather drive the birds inshore. A few individuals have been recorded in the Strait of Gibraltar in autumn, in August–November. There are also rare occurrences within the Iberian Mediterranean: at least 16 such records during 1960–2008, most of single birds, are listed in the *Ardeola* bird news and elsewhere, including several highly unseasonal records during January–April. Assuming that the identifications were correct, such birds would most probably be individuals that entered the Mediterranean through the Strait but failed to find their way out again. There is also a 19th century record for Sardinia and another for Algeria (Bonaccorsi 2003).

SOOTY SHEARWATER *Puffinus griseus*

(Pardela Sombría, Pardela-preta)

Common and sometimes numerous in summer and autumn on Atlantic and Biscayan coasts. Vagrant to the Mediterranean.

Large numbers pass southwards through Iberian waters in summer and autumn and many often linger to feed in the Bay of Biscay. They are readily observed from coastal headlands, particularly with onshore winds, and are more often visible to shore-based observers than the Great Shearwaters that occur at the same season. The first individuals generally appear off northern coasts in late July but there are several June records of single birds from Galicia (*A. O. Galicia*). Sooty Shearwaters become more numerous in August, peaking in abundance from mid September to mid October. Some are reported throughout November with stragglers to mid December (Sandoval *et al.* 2009). The largest reported counts have been at Cape Estaca de Bares, where the maximum recorded rate of westward passage was 766 birds/hour on 30 September 2000. Seasonal totals at Cape Bares may be high: 11,310 were counted in about 400 hours between July and December in 2008 and 35,867 during June–December 2011 (Sandoval *et al.* 2009, A. Sandoval pers. comm.). On the basis of sample counts, the mean annual post-breeding daytime passage at Cape Bares during 2004–08 was of the order of 55,000 birds (Sandoval *et al.* 2010). Other Biscayan headlands further to the east see lower numbers but these may still be substantial. For example, 3,265 were counted from Punta la Vaca from late June to December 1991 (*GIAM* 17: 4).

There are comparatively few reports from Portuguese coasts and southwestern Spain. However, large numbers pass south offshore in autumn and there is a potential for coastal influxes in stormy conditions. Thus 820 were counted from Cape Carvoeiro, Leiria, on 24 October 1999 during northwesterly gales (AP).

The Sooty Shearwater is reported from the Strait of Gibraltar in autumn with some regularity, although always in small numbers. There are 15 records (38 birds) from Gibraltar itself during the period 23 July–21 October, including an exceptional flock of 15 on 7 October 1979 (GONHS), as well as a few sightings from Ceuta and Tarifa in the same period and a record of two near Ceuta on 15 November 2008 (*Ardeola* 56: 155). It is rarely recorded within the Mediterranean but there are over 25 records up to 2008, some of them unseasonal spring sightings, in the *Ardeola* bird news and other sources. This raises the possibility of misidentification: some Balearic Shearwaters are very dark. However, some were reported by experienced observers and are doubtless authentic. There are also a number of records from other Mediterranean countries, including France, Italy, Malta, Tunisia and Algeria (Bonaccorsi 2003).

In addition to the Iberian Mediterranean records, a small number of observations from the northern and western coasts suggest that some individuals may remain in Iberian waters throughout the winter and spring. The few May and early December records may have involved early and late 'autumn' migrants respectively but there were 96 off Corrubedo, A Coruña, on 3 February 2002; one at Cape Bares on 29 March 2002 and, in 2008, two there on 13 February, two on 23 March and one on 29 March; six at Cape Higuer, Gipuzkoa, in January 1990; and single birds at Getaria, Gipuzkoa, on 30 December 2000 and 2 January 1998 (*A. O. Galicia 2002*: 42; *A. O. Gipuzkoa 1990*: 7, *1998*: 36, *1999–2000*: 54). Portuguese records include one 20 nautical miles west of Odeceixe, Faro, on 29 January 2008 and one 14 miles south of Faro on 19 April 2009 (*Anu. Ornitol.* 7: 74, 8: 56). There are also two March records (4 birds) from Gibraltar.

MANX SHEARWATER *Puffinus puffinus*

(Pardela Pichoneta, Fura-bucho do Atlântico)

Common and sometimes abundant in Atlantic and Biscayan waters. Vagrant to the Mediterranean.

The Manx Shearwater was long regarded as polytypic, including the Balearic Shearwater *P. (p.) mauretanicus* and the Levantine Shearwater *P. (p.) yelkouan* as subspecies. The two latter forms are now each accorded specific status. However, a consequence of the earlier situation is that some pre-1990 records, especially references to 'Manx Shearwaters' in the Mediterranean, cannot be confidently assigned to this species.

The Manx Shearwater occurs in Atlantic and Biscayan waters all year round, although very few are present in winter and it is generally scarce in spring and summer. It is common on southward passage in autumn, from July to December (SWA), and very large numbers are recorded on northern and northwestern coasts in September and early October in some years, particularly when northwesterly gales divert southbound birds into the Bay of Biscay. The spring passage extends from February to May but is less apparent from land. Autumn passage is most marked at Cape Estaca de Bares, where exceptionally large numbers were recorded in September and October 2008: 41,871 birds were counted in about 400 hours between June and December and record high westward passage rates of 1,682 birds/hour and 1,344 birds/hour were observed on 2 October and 7 September respectively (Sandoval *et al.* 2009). An even larger count was made in 2011: 135,804 Manx Shearwaters were counted from the Cape during the second half of the year (A. Sandoval pers. comm.). On the basis of sample counts, the mean annual post-breeding daytime passage at Cape Bares during 2004–08 was of the order of over 100,000 birds (Sandoval *et al.* 2010).

Numbers counted elsewhere on the north coast are much smaller but considerable numbers are still observed in some autumns. An observation of 91 on 15 February 1997 at Cape Silleiro, Pontevedra (*GIAM* 23: 17) is a high count for a late winter date: it may have involved early migrants but it is also the case that some Manx Shearwaters may be found in Iberian Atlantic waters throughout the winter. Some 60 were recorded between late December and early February during the 2007–10 winter atlas fieldwork.

There are comparatively few reports from Portuguese coasts, probably due partly to a lack of regular observation. Nevertheless, 1,250 were counted flying south from Cape Carvoeiro on 19 October 1999 following onshore gales (AP), and such movements are likely to occur regularly under the appropriate conditions.

There are also very few unambiguous records of Manx Shearwaters from southern Iberia or from the Strait of Gibraltar and all involve observations of single birds. One was reported from Tarifa on 20 May 2008 (*Ardeola* 55: 291). There are no records at all from Gibraltar itself. Records within the Iberian Mediterranean are very rare: they comprise single birds in Málaga Bay on 6 November 1990 and off the Albufera, Valencia, on 14 July 1990 (*Ardeola* 37: 326; *A. O. C. Valenciana 1990*: 11). Elsewhere in the Mediterranean there is a recovery near Montpellier, France, of a bird ringed on Lundy Island, England, as well as a capture in Malta on 6 September 1995 and an old 'possible' record from Italy (Bonaccorsi 2003).

Bernis (AMI) notes 37 recoveries of British-ringed birds in Iberia, 36 of them ringed on Skokholm island, Wales. Some 20 additional recoveries have also been of birds from Irish Sea colonies. Nearly all were recovered from the Biscay coast and adjacent waters but there is one recovery off western Galicia and four off Portugal (AP). Most recoveries fall during the spring (25) or autumn (10) passage periods.

LEVANTINE SHEARWATER *Puffinus yelkouan*

(Pardela Mediterránea, Fura-bucho do Mediterrâneo)

Breeds locally in Menorca, Balearic Islands. Regular and sometimes numerous off Catalonia but generally scarce elsewhere in Iberian Mediterranean waters, where formerly more abundant.

The historical status of the Levantine (Yelkouan) Shearwater in Iberia is obscured by the fact that both it and the Balearic Shearwater were formerly regarded as subspecies of the Manx Shearwater. More recently, it was retained for a time as a subspecies of the Balearic Shearwater when the former received specific status (Bourne *et al.* 1988) but it is now treated as a species in its own right (Sangster *et al.* 2002a). These recent taxonomic changes mean that it is often impossible to determine to which species the older accounts in the literature refer.

A small breeding colony was discovered at the Mola de Maó in Menorca as recently as 2001 (SA2) but there is fossil evidence that the species has bred there since at least Neolithic times (Gutiérrez 2004). This is the only known Iberian colony of a species that nests elsewhere from the Gulf of Marseilles eastwards to the eastern Mediterranean. The Menorcan colony numbered a minimum of 100–150 pairs (SEO/Birdlife 2001). Nesting elsewhere in Menorca has been suspected but not proved. It is subject to all the threats affecting the Balearic Shearwater (*q.v.*) and is clearly highly vulnerable as an Iberian breeding species in view of its small and localised population.

The genetic identity of the Menorcan breeding birds has not yet been fully established. An initial and small-scale analysis of mitochondrial DNA (Genovart *et al.* 2005) suggested that pure Balearic and pure Levantine Shearwaters were nesting sympatrically. However, subsequent and more extensive investigations that also examined nuclear DNA (Genovart *et al.* 2011, 2012) have suggested that although the Menorcan birds are phenotypically similar to Levantine Shearwaters they average somewhat smaller than other populations of that species and there is significant evidence of gene flow from Balearic Shearwaters.

Munn (1930) obtained a specimen of this form in the Balearics in 1928 and considered that both it and the Balearic Shearwater were present in the archipelago. The presence of the Levantine Shearwater in mainland Iberian waters was perhaps first reported by Bourne (1965), who observed a raft of 1,000 moulting shearwaters in the Bay of Gibraltar on 11 September 1964 that he identified as this species. They were accompanied by a few Balearic Shearwaters. A summer and autumn presence of 'Manx' shearwaters was then a familiar phenomenon at Gibraltar and had been noted in autumn there previously by Irby (1875) and Tuke (1953). Considerable numbers arrived in late summer and formed large rafts especially in the Bay of Gibraltar and off the northern shores of the Strait. The birds dispersed to feed and a continuous stream of shearwaters could often be seen returning to the Bay from the east at Europa Point, Gibraltar, in the afternoons and evenings, from late June to September (pers. obs., GONHS records). By September the birds were in heavy moult. Up to 6,000 birds comprised this concentration and photographic evidence confirmed that most were Levantine Shearwaters (GONHS records). The post-breeding flocks remained at Gibraltar until the late 1970s but thereafter disappeared quite abruptly. Larger counts towards the end of this period included a raft of 1,300 on 24 June 1976. Since then only very small numbers of Levantine Shearwaters – seldom reaching double figures – have been seen at or near Gibraltar, in late summer, autumn and winter. The abandonment of the area as a moult site coincided with a marked expansion in marine traffic and the establishment of long-term anchorages for large ships both in the Bay and east of the Rock of Gibraltar.

In more recent years Levantine Shearwaters have been found to occur regularly in some numbers off Catalonia, south to Tarragona, where they may be observed in the company of the more abundant Balearic Shearwaters throughout most of the year but particularly in winter (Gutiérrez 2004). The largest numbers are seen off Girona in winter: in Roses, Ter and Palamós bays and especially around Cape Creus, where the waters supply a high density of shoaling sardines and anchovies. Peak numbers off Cape Creus, estimated at 7,500–8,500 birds, are generally in December–January, declining in February as the birds depart to nest, with large numbers also recorded at times in November (SWA, CWA). For example, over 12,000 were counted in six hours from Cape Creus on 1 November 2008 (*A. O. Catalunya 2008*: 43). Some occur further south off the Comunidad Valenciana but only small numbers have been reported there in recent years: the regional bird report gives only six records for 2009, involving 46 individuals in total (*A. O. C. Valenciana 2009*: 22). Small numbers also occur outside the breeding season around the Balearic Islands and south to Murcia and a few are then seen in Andalucian waters as far as the Strait of Gibraltar.

The occurrence of Levantine Shearwaters in the Atlantic is hard to establish conclusively due to identification problems and the species is not yet included in the Portuguese List. There are convincing recent records from Galicia during 1999–2003. They include seven records (eight birds) during late May–August (Gutiérrez 2004). Single birds thought to be this species were also seen at Cape Estaca de Bares on 11 and 25 July 2009 (Sandoval *et al.* 2010). There is also a record of 20 seen from Punta La Vaca, Asturias, on 20 October 2003 (*Ardeola* 50: 340). The species is not a rarity in Spain and so such records are only subject to local evaluation. A few other records from French Biscayan coasts and the English Channel are under review.

BALEARIC SHEARWATER *Puffinus mauretanicus*

(Pardela Balear, Fura-bucho das Baleares)

Locally numerous as an endemic breeding species confined to the Balearic archipelago. Common and sometimes numerous seasonally in all Iberian waters.

The Balearic Shearwater is an Iberian endemic species although it was formerly regarded as a subspecies of the Manx Shearwater. It is apparently restricted as a breeding species to the Balearic archipelago where its total nesting range is less than 10km^2 (Ruiz & Martí 2004, LR). The first colonies were discovered by Ticehurst & Whistler (1930) on Ibiza and Formentera and by Munn (1934, 1936) on islets off Mallorca. About half the population nests on Formentera, the remainder being distributed between Ibiza, western and southern Mallorca, Cabrera and Menorca, and their associated rocky islets (SA2). Nests are in sea caves and coastal rock crevices, although some have been found up to 500m inland (Oro *et al.* 2009).

The breeding sites may be visited as early as November, exceptionally earlier, but the breeding season is largely from February to June. At that time the birds forage in continental shelf waters, particularly those shallower than 200m, notably in Palma Bay and along the Levant coast, especially off the Ebro delta and Cape La Nao, Alicante (Oro *et al.* 2009). Satellite-tracking has revealed that nesting adults routinely cover great distances when foraging and have mean home ranges of nearly 40,000 km^2 that may extend as far as Algeria (Ruiz & Martí 2004). Laying of the single egg is in March. Adults abandon their young prior to fledging and depart on their migration, followed by the juveniles when they leave the nests in late June or early July. A large proportion of the population does not nest every year and non-breeders may be observed near the nesting islands in spring and summer.

The size of the population is not known with certainty but there are clear indications that it is declining. Estimates have ranged from 2,083–4,414 pairs in 1999 (SA2) to only 1,200–1,800 pairs just two years later in 2001 (*Ardeola* 52: 207). The population was estimated at 2,000–2,400 pairs in 2005, distributed between Formentera (< 1,000 pairs), Mallorca (350–500), Menorca (100–175), Cabrera (50–100) and Ibiza (200–300) (Rodríguez-Molina & McMinn-Grivé, 2005). A reappraisal of breeding numbers in 2009 put the total population at 3,193 pairs but it was emphasised that the increase is more apparent than real and arises from more thorough prospection of the colonies, and from a revision of the criteria used to arrive at indirect estimates (Carboneras 2012). However, observations from coastal watchpoints (see below) indicate there is certainly a substantial, additional, non-breeding population, presumably including many immature birds, so that the total global population of the Balearic Shearwater may be of the order of 20,000–30,000 individuals (Birdlife International 2012b).

Capture-recapture data at particular colonies points to an annual decline recently of 7.4%, with adult survival at 78%: a low figure for a shearwater (Oro *et al.* 2004). In particular, 60% of the colonies on Cabrera have disappeared during the last few decades and a 2004 survey of Formentera – where *c.* 50% of the global population breeds– did not record any breeding birds at all in 30 suitable caves visited in a single cliff. Population viability analysis has estimated global extinction in about 40 years in the present circumstances (Oro *et al.* 2004). There are thus reasons for concern for regarding the long-term survival of the species and conservation measures have been put in place, including the designation of all the breeding sites as SPAs. The decline is due to mortality occasioned by fishery activities (bycatch), food shortages in the breeding season resulting from a reduced availability of discards from trawlers as well as overfishing, disturbance of the nesting sites and predation at the nests by rats and feral cats, among other factors (Ruiz & Martí 2004, Oro *et al.* 2009, BirdLife International 2012). The Balearic Shearwater is thus listed as 'Critically endangered' in Spain. Two recovery plans, in 1998–2001 and 2003–08, have been put in place by the regional government of the Balearic Islands. The implemented measures include the declaration of colony sites as reserves, predator control, prevention of disturbance by humans and establishment of marine reserves (Arcos 2011).

Much of the population leaves the Mediterranean after the breeding season. The early departure of the adults, in May, gives rise to the curious circumstance that the post-breeding movements begin in spring and the pre-breeding ones in autumn. Westward passage is apparent at the Strait of Gibraltar from May to August. The Migres programme counted over 18,000 birds flying west through the Strait of Gibraltar in May–July 2008 (Birdlife International 2012). Northward migration becomes apparent off Portuguese coasts in June. There are records of flocks of 5,000+ moulting off the Tagus estuary in July and August (AP) and 1,682 were off San Jacinto, Aveiro, on 19 August 2010 (*Anu. Ornitol.* 8: 57). The majority follow the Iberian Atlantic coasts to the Bay of Biscay, where they feed and moult during the summer and early autumn. A proportion travel even further north at this time, to British waters. Birds ringed at colonies in Mallorca have produced ten recoveries on the French Biscay coast, as well as two on the Portuguese west coast, three on the Algerian coast and one in northwestern Italy.

Southbound passage is evident on Biscay and Portuguese coasts especially during September–November. Large numbers have been counted from major headlands here. In total 3,777 birds were counted moving west at Cape Estaca de Bares in 209 hours during September–November 2008, with a record peak rate of 150 birds/hour noted on 5 October (Sandoval *et al.* 2009). On the basis of sample counts, the mean annual post-breeding daytime passage at Cape Bares during 2004–08 was of the order of 28,000 birds (Sandoval *et al.* 2010). However, a count from Cape Raso, Lisboa, of nearly 11,000 birds flying south in just two hours on 29 September 1990, is exceptional (AP). Such large counts, from both Atlantic and Mediterranean waters (see below), further emphasise that the censuses of breeding pairs at the known nesting colonies do not reveal the actual size of the global population.

The return to the Mediterranean is also apparent at the Strait and occurs from September to January. Most birds return to winter off the Levant coasts but they display only limited fidelity to winter quarters and these may have shifted in recent years. The principal winter quarters were off the Comunidad Valenciana until the 1980s but the major concentrations from the early 1990s and until very recently have been off Catalonia (Gutiérrez & Figuerola 1995): there January counts increased from zero in 1983 to a peak of 7,080 birds in 1993. There have since been some very large winter counts off the Comunidad Valenciana once more, including a record 16,420 flying north off Cape Cullera on 12 December 2009 (*Ardeola* 58: 486). There was also a raft of some 8,000 off the Albufera de Valencia on 31 January 2010 and a similar number flew south in three hours in the same general region on January 2011 (*Ardeola* 58: 486).

The movements of Balearic Shearwaters are complex and not fully understood. It is clear that some remain in the Mediterranean all year round and it has been suggested that these are mainly adults (Martí & Ruiz 2001). Some occur off Atlantic coasts of Morocco in autumn and others remain in Iberian Atlantic waters in winter. One ringed as a chick in Mallorca in June 1997 was recovered in January 1998 on the Spanish Biscay coast at Cudillero. Observations at Cape Bares in 2008 showed a presence throughout the year: the birds were scarcest in April–May and most abundant in September–October, when there was a clear westward passage out of the Bay of Biscay, presumably of birds heading back to the Mediterranean (Sandoval *et al.* 2009).

BAROLO'S SHEARWATER *Puffinus baroli*

(Pardela Chica, Pintainho)

Occurs off Atlantic coasts, mainly in late summer and autumn. Recorded in Spain, Portugal and Gibraltar.

Barolo's Shearwater is the endemic small shearwater of the Azorean, Madeiran and Canarian archipelagoes, with a total global population of just 3,000–4,400 pairs (Brooke 2004). It was formerly included as a race of the Little Shearwater *P. assimilis*. It is present in the northeastern Atlantic all year round but its movements are poorly known. Individuals are reported with some frequency off the Atlantic coasts of Britain, Ireland, France, Iberia and north-west Africa, and more rarely in the North Sea and Mediterranean. There are records of 62 individuals from Britain, 22 from Ireland, 74 from France and four from Italy. Most records off western Europe are during August–October, with a lesser peak in April–June. The same occurs in Morocco, from where there are three August records from the Strait and 13 from the Atlantic coast: mainly during September–November (BM).

The at least occasional presence of Barolo's Shearwater in Iberian waters has long been assumed (Bernis 1954a, 1966a) but the earliest record of which we are aware was off Torremolinos, Málaga, on 2 November 1984. One was then seen from Gibraltar on 16 August 1985 and another from Cape Raso, Lisboa, on 11 October 1987. Records have since become more frequent, especially due to the Colm Moore's summer pelagic trips off southern Portugal since the late 1980s. There were 33 accepted Iberian records of the species by 2010: 25 from Portugal, four from Gibraltar and five from Spain. This shearwater tends to be solitary and 73% of observations have been of single individuals. Very few have been seen from land but observations become more frequent far offshore. Thus a total of 21 Barolo's Shearwaters were counted on 30 August 1994 over the Gorringe Bank, south-west of Portugal. There are further unsubmitted observations from Algarve and one was heard on Berlenga island in April 1996 (AP). Others are regularly reported from the Bay of Biscay, where seven were seen in an hour from the Bilbao–Portsmouth ferry on 23 August 1999 (*Ardeola* 48: 119).

Observations in Iberian waters are largely during July–October, as elsewhere in Europe and in Morocco. This period matches the non-breeding season of the species, in contrast to the Iberian records of both Bulwer's and Zino's Petrels (q.v.) which are mainly during their breeding seasons. There is, however, a recent record of a tide-line corpse in Cádiz on 2 January 2010 (*Ardeola* 50: 367). Observations are chiefly in the south-west, although the pattern may be biased by the large number of records that resulted from Moore's trips. There are three records from the Moroccan side of the Strait, in addition to the four from Gibraltar, but only two of the Spanish records have been within the Mediterranean proper, and these were very near the Strait: the Torremolinos record of 1984 and one at Sotogrande, Cádiz, on 12 May 1986.

WILSON'S STORM-PETREL *Oceanites oceanicus*

(Paíño de Wilson, Casquilho)

***O. o. oceanicus/exasperatus*. Regular off Iberian Atlantic coasts in summer and autumn but seldom seen from land. Both subspecies probably occur but they are indistinguishable in the field.**

Its breeding season extends from November to mid-April, following which the birds ascend all the oceans to reach the northern hemisphere, where they remain from May to September. In the Atlantic they are then especially found over the Gulf Stream off North America, but there are indications that they travel clockwise and return south further east, off European coasts. Observations in European waters have increased considerably in recent years due to the regular excursions in summer to find pelagic seabirds. The species was withdrawn from the British rarities list in 2006, by when some 408 individuals had been recorded. In Ireland there are records of 239 individuals. However, there are only 27 records of single birds from France, and there are very few records from the Mediterranean, the furthest east being from Sardinia (Bonaccorsi 2003). There are also relatively few observations from Morocco. In contrast, groups of several hundred birds are regularly found off Mauritania, in March–April and September–October (Isenmann *et al.* 2010).

The earliest Iberian records date from the late 19th century and comprise several specimens in Lisbon Museum that were obtained off Cascais by the King, Dom Carlos (Tait 1924), as well as one obtained in Mediterranean waters off Andalucía (Arévalo 1887). Thereafter Wilson's Storm-petrel was seldom reported until the 1980s, when three tideline corpses were recovered: in Cantabria in October 1984, in Galicia in September 1987, and in western Andalucía in June 1988. However, by then observations at sea had indicated that it could sometimes be numerous in Iberian Atlantic waters. Thus Bourne and Norris (1966) observed flocks of up to 50 off southwestern Iberia in September 1964, and there were counts in late July 1985 of 38 off Cape Trafalgar, 32 south of Cape St Vincent and at least 656 off Cape Finisterre (Van den Berg *et al.* 1991). The summer 'pelagic trips' to see seabirds off southern Portugal, by Colm Moore from 1987, subsequently led to a rapid increase in records. There were 20 accepted records from Portugal up to 1997, when the species was withdrawn from the rarities list, and a further 14 have been published thereafter in *Pardela* and the *Anuário Ornitológico*. There were also 53 accepted records (273 birds) from Spain up to 2011.

An analysis of just the 64 accepted records up to 2009, which refer to at least 676 individuals, shows a notable summer peak that is not entirely due to the considerable number of August observations generated by the Moore trips (Fig. 55). Records from British waters peak from the last week of July until the third week in August and there are few September records (Flood & Fisher 2010). September records are relatively frequent off Iberia and there are four October records, two of them on 24 October. One was reported eight miles off Cádiz on 10 November 2011 (RBS). There are only two spring records: 30 seen from Cape Raso, Lisboa, on 24 May 1980, and one off Aveiro on 20 May 2008 (*Ardeola* 36: 113, *Anu. Ornitol.* 7: 74-75). The rarity of such observations accords with the idea of a clockwise circulation in northern latitudes, even though, as stated, spring birds are commonly seen at the latitude of Mauritania.

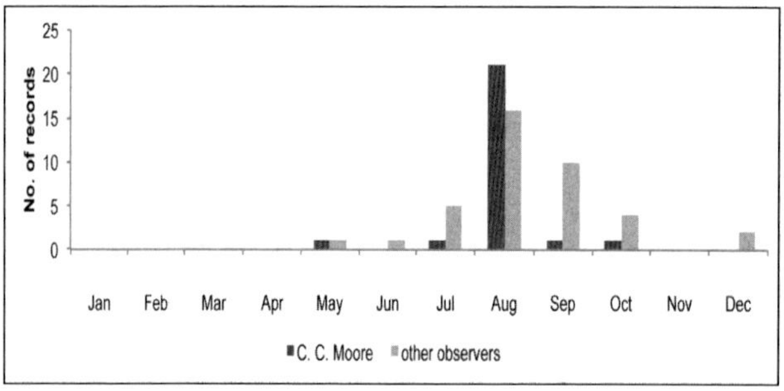

Figure 55. Records by month, 1984–2009, comparing those resulting from the Colm C. Moore excursions (n = 25) with the remainder (n = 39)ww

Figure 56. Records by region, 1984–2009 (n = 64)

Wilson's Storm-petrels are mostly observed off the Iberian west coast (Fig. 56) and there are relatively few reports from the Bay of Biscay, which accords with the paucity of records from French waters. The sole recent record from the Iberian Mediterranean was of one near Cape Palos, Murcia, on 25 July 2003. Accepted Iberian observations include few records of 20+ individuals and about half have involved single birds. Moore found 100+ off Porto on 21 August 1987, 80–100 off Lisboa on 21 August 1999 and 240 off Cape Espichel on 26 August 1999. More recent observations include 150+ individuals in the western Strait of Gibraltar near Tarifa over several days in August 2011 and 130 off the Costa da Morte, Galicia, on 3 September 2011 (RBS).The recent upsurge in records has led to the species having been withdrawn from the list of species considered by the Spanish rarities committee.

WHITE-FACED STORM-PETREL *Pelagodroma marina*

(Paíño Pechialbo, Calca-mar)

May occur fairly regularly off Atlantic coasts in summer but always likely to be uncommon given its essential scarcity. Recorded in Portugal.

The nearest colonies to Iberia are in the Salvage and Canary Islands, where the subspecies involved is *P. m. hypoleuca*. The Cape Verde islands have colonies of *P. m. eadesi*. Birds have been reported on rare occasions off Morocco in February, May and June, and off Mauritania in January, April and December, but they may occur there regularly. There are several records from the Atlantic coast of the United States and one from Britain.

The first Iberian record was of several in August 1962 off central Portugal at 40°N, 9°52′W (Buckley & Wurster 1970). Six more recent observations, of a total of 16 individuals, result from the pelagic seabirding trips made in successive Augusts by Colm Moore. Most have been south-west of Cape St Vincent but up to six were seen 40 miles off Sines on 26 August 1996 and one bird was 100 miles west of Lisboa on 20 August 1994. There are not yet any records from Spanish waters.

EUROPEAN STORM-PETREL *Hydrobates pelagicus*

(Paíño Común, Alma-de-mestre)

***H. p. pelagicus*. A local breeder off northern coasts. Largely a summer visitor.**
***H. p. melitensis*. A local breeder off eastern coasts and in the Balearic Islands. Largely a summer visitor.**

There are mainly small colonies on rat-free islets and islands off the coasts of Galicia, Asturias, Cantabria and the Basque Country, but apparently not on the Berlengas islands, which are inhabited by the Madeiran Storm-petrel, nor anywhere else in Portugal. Its Iberian stronghold in the Mediterranean is the Balearic archipelago. However, there are also colonies off Almería, Murcia and the Comunidad Valenciana and indications of small-scale nesting on some mainland seacliffs on the Costa Brava, Catalonia (Gutiérrez *et al.* 2007).

The Spanish population was put at 1,700–2,000 pairs by SA1 but fuller prospecting since then has shown that there at least 4,000 pairs and probably as many as 7,000 (SA2 & LR). These are divided between 44–59 Atlantic colonies totalling 1,175–2,590 pairs, and 15–19 Mediterranean colonies totalling 3,830–5,310 pairs. The Atlantic colonies are on the Cíes, Ons and Farallón islands and islets off the Biscay coast. Most (75%) of those in the Mediterranean are on islets off the Balearics, especially Mallorca and Ibiza, with others on the Columbretes; Benidorm island; Isla Grossa and Cueva de Lobos islands, Murcia; and Terreros islet, Almería. Some sites have been censused more thoroughly than others and more colonies are likely to be discovered.

Population trends are largely unknown although there are documented cases of local extinction following the arrival of rats on nesting islets. In the Comunidad Valenciana the small colonies of the Columbretes and Isla Mitjana have shown some recent decline whereas the Tabarca island colony, and the large colony on Benidorm islet (400–600 pairs in 2009), have remained stable (*A. O. C. Valenciana 2009*: 23–24). Predation by Yellow-legged Gulls is also a problem locally (Mínguez 1994). Nestboxes have been provided at key colonies in Galicia and Asturias and on Benidorm, Alicante, with some success: at the last of these 29% of nestboxes placed in 1996 were occupied in 2001 (SA2). Natural nest sites are in rocky crevices and holes, often inside caves. Laying at the Benidorm colony occurs from mid April to early July, but chiefly in May (Mínguez 1994). Ringing has revealed that inter-colony movements are quite frequent, generally between nearby colonies, as occurs in the Balearic archipelago and between colonies off Alicante, Murcia and Almería, but at least occasionally also between archipelagos. One ringed as an adult in Mallorca in June 2004 was found in Sardinia in August 2008 and one ringed as a chick in Sicily in July 2005 was on the Medes Islands, Girona, in June 2006.

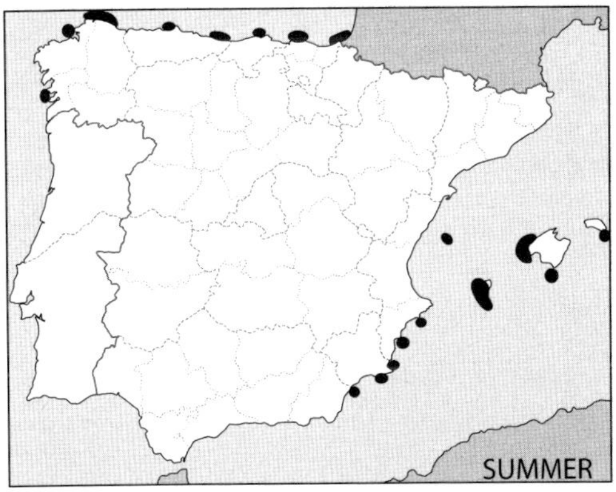

The European Storm-petrel may be encountered offshore in all Iberian waters, chiefly during May–November, but is rarely observed from shore and then usually only in very small numbers. However, 50 arrived following a fishing boat in the Ría de Aviles, Asturias, on 30 May 2007 and no fewer than 200 were at the same site on 1 June (*Ardeola* 54: 391): this exceptional observation hints at unusual feeding conditions. Indeed, 2007 saw an also exceptional total of 394 at Cape Estaca de Bares, including a record passage rate of 47 birds/hour on 6 July: this compares with whole-season counts there of just 12 birds in 2008 and 23 in 2011 (A. Sandoval pers. comm).

Birds are also observed with some regularity from Portuguese headlands, notably Cape Carvoeiro and Cape Raso, especially during foggy conditions (AP). For example, in 2010, 220 were counted from Cape Raso on 11 June and 113 were just off the estuary at Aveiro on 12 June (*Anu. Ornitol.* 8: 57). Some thousands have been mist-netted and ringed on the south Portuguese coast, especially in June, by using voice playback to attract birds at night. The capture near Lagos, Algarve, of some individuals with brood patches during the breeding season suggests that some may nest on offshore islets in the region, but this has yet to be confirmed. In total 23 individuals ringed in the British Isles have been recovered in Portugal and there are 150 distant recoveries of birds ringed in Portugal, including 100 from the British Isles, 33 from Norway and single birds from the Faroe Islands and Iceland (AP). Voice playback thus clearly attracts many birds that are in northward transit through Portuguese waters, where captures begin in May, peak in June and largely cease by the end of July. In contrast, similar luring at Cape Silleiro and Cape Mañón in Galicia has produced numerous recoveries mainly from the Basque Country and Brittany: especially on Banneg island, but just a few recoveries in more northerly regions, from Cornwall and Shetland. The European Storm-petrel is present in small numbers in the Strait of Gibraltar in summer and it may nest on rocky islets within the Strait or in the Alborán basin but any colonies there remain undiscovered. However, five were reported from Alborán island on the late date of 28 November 2009 (*Ardeola* 57: 221).

Most of the population winters in the south Atlantic but there are winter month records of some numbers from Atlantic waters. The post-breeding passage period extends to late December and some winter records may involve late migrants rather than overwinterers. Such observations include 50 at Cape Peñas, Asturias, on 21 December 1989 and 37 at Cape Higuer, Gipuzkoa, on 25 December 1989 (*GIAM* 9: 5) as well as sightings of up to 14 off Cape Raso, Cascais, on three dates in December 2010 (*Anu. Ornitol.* 8: 57). Also, 100+, as well as 40+ Leach's Storm-petrels, were close inshore at Hondarribia, Gipuzkoa, on 4 January 1997 after stormy weather (*Ardeola* 45: 118). Some very large counts of European Storm-petrels, estimated at several tens of thousands, have been made in November and December in offshore waters of the Gulf of Cádiz (Arcos *et al.*, 2009; J. M. Arcos pers. comm.) but it is not yet known whether any remain there in winter proper. Similarly, the numbers of European Storm-petrels in the Strait of Gibraltar are known to increase during October–November, when many birds have been seen flying eastwards, into the Mediterranean (Hashmi & Fliege 1994), but once again it is unknown whether or not these return later to winter in Atlantic waters. Winter month records from the Mediterranean are very scarce and the species appears to be largely a summer visitor to the Iberian Mediterranean on present evidence.

Occasional storm-driven birds have been found well inland during November–January. They include two at Fregenal de La Sierra, Badajoz, and single birds found in Sevilla city; in Madrid; at the Calanda reservoir, Teruel, and at Echávarri-Viña, Álava (*Ardeola* 36: 232, 49: 174, 57: 221; *A. O. Aragón 2004–2007*: 208).

LEACH'S STORM-PETREL *Oceanodroma leucorhoa*

(Paíño Boreal, Paínho-de-cauda-forcada)

***O. l. leucorhoa*. Present in winter in the Atlantic off Iberia. 'Wrecks' of storm-driven individuals occur occasionally and sometimes involve large numbers. Very rare in the Mediterranean.**

Leach's Storm-petrel has large colonies in the northern Atlantic, principally in Iceland. Most probably winter in the south Atlantic but large numbers clearly remain north of the equator, including in Iberian latitudes. They are most often observed from Atlantic and Biscay coasts in November–January, but some are present from September to early March. Prolonged stormy weather may result in hundreds, occasionally thousands, appearing onshore. Notable examples include: 500 alive off Doñana and hundreds dead on nearby Cádiz beaches in late December 1981; 625 counted in two hours off Gijón Bay on 21 November 1989, and 7,800 flying west off Gipuzkoa on 1 December 2009 (*Ardeola* 29: 177, 37: 326, 57: 222). In such circumstances the birds appear weak and many indeed are washed up dead or dying on the beaches. Such events are typically in mid winter but some have been as early as mid September. Wrecks of this type have long been noted in Iberia: they were reported from Portuguese coasts by Paulino d'Oliveira (1896) and Reis Júnior (1931).

Wrecks in southern Iberia bring birds into the Strait of Gibraltar, sometimes in significant numbers, as on 11 November 1995, when 200+ were seen inshore off Tarifa beach (*GIAM* 21: 6). However, very few ever enter the Mediterranean. There is a 19th century record from the Alborán basin (Irby 1895) and two from Catalonia: at the Aiguamolls de l'Empordà on 21 December 1980 (Sargatal & del Hoyo 1989) and off Badalona on 7 January 1984 (Carboneras 1987). There are also three records of single birds from the Balearic Islands on 11 June 1924, 28 August 1925 and 16 July 1926 (Munn 1928) that Paterson (1997) considered could probably be discounted as errors of identification. However, three recent unseasonal records from the Alborán basin, one of them supported by good photographic evidence, show that Leach's Petrels may occur in the Mediterranean other than in winter: they comprise three seen two miles off Cape Pinos, Málaga, on 8 July 2009; four off Rincón de La Victoria, Málaga, on 29 September 2009, and three five miles east of Alborán island on 7 April 2010 (*Ardeola* 58: 486–487).

Storm-driven Leach's Storm-petrels occasionally appear far inland. For example, there are three records of single birds from Ciudad Real in Castilla-La Mancha: on 14 December 1996, in November 1997, and on 30 December 1980 (*Ardeola* 44: 243, 46: 150; Jiménez *et al.* 1992). Similarly there are records of single birds picked up 100km from the sea at Cazalla de la Sierra, Sevilla, on 8 March 1982; in the streets of Mértola, Baixo Alentejo, on 8 November 1997; in flight at Sevilla city on 24 December 2010, and found dead at Miajadas, Cáceres, on 21 January 2013. The influx reported on northern coasts in late 2009 also produced the first four records (seven individuals) for Navarra, well inland in the Pyrenean region, all of birds found dead or in poor condition (*Ardeola* 30: 116, 57: 222; *Pardela* 11: 6; J. Prieta, pers. comm.)

SWINHOE'S STORM-PETREL *Oceanodroma monorhis*

(Paíño de Swinhoe, Painho-de-swinhoe)

Very rare. Recorded in Spain and Portugal.

The surprising capture of a Swinhoe's Petrel in the Salvage Islands in 1983 was the precursor of a series of other observations and captures of this Asiatic species in the north Atlantic that may imply the existence of an as yet undiscovered colony in the region (Bretagnolle *et al.* 1991, Cubitt *et al.* 1992). By 2008 there were 16 records from Europe: five in Britain, four in Norway, two in Ireland, one in France and four from Iberia, as well as five in the Salvage Islands –between the Canaries and Madeira– and three off North Carolina, USA (Flood 2009). Five of the captured birds had incubation patches and four individuals were recaptured at the same sites in successive years, generally within European Storm-petrel colonies. Another adult with an incubation patch was captured in the Canaries in June 2009 (Gutiérrez & Lorenzo 2011). Nearly all records have been between late June and late August.

There are three accepted Iberian records, all of individuals captured on land: on Benidorm islet, Alicante, on 13 July 1994; on Cabrera island in the Balearics on 13 August 1997 and at Punta de Almadena, Faro, on 27 June 1998. The first two were found in burrows at European Storm-petrel colonies. The third was mistnetted when a tape lure was being used to attract European Storm-petrels. The Benidorm and Faro birds both had incubation patches. There is a later sight-record: one at sea south-west of Lisbon on 19 August 2008 (Flood 2009) and another was caught on Benidorm islet in June 2013 (RBS). The Spanish records were from Mediterranean waters: one at Genoa in August 1991 is the only other Mediterranean record (*Riv. ital. Orn.* 65: 63–68).

MADEIRAN STORM-PETREL *Oceanodroma castro*

(Paíño de Madeira, Roquinho)

Breeds on the Berlengas Islands, Portugal. Rarely reported elsewhere in Iberia.

The band-rumped, squarish-tailed, storm-petrels of the northeastern Atlantic archipelagoes were once all considered to be Madeiran Storm-petrels but recent investigations of their DNA, vocalisations, biometrics and breeding seasonality have revealed the existence of at least two and possibly four distinct species (Robb *et al.* 2008). These are nonetheless too similar to be separable in the field, so that birds seen at sea cannot safely be identified as other than 'band-rumped' petrels. 'Band-rumped' storm-petrels have very rarely been identified in European waters, other than in Portugal, but there are at least ten records from the waters off the Moroccan Atlantic coasts, where pelagic trips have reported up to 34 birds: off Agadir on 24 October 1999 (BM, Bergier *et al.* 2011).

The Madeiran Storm-petrel was discovered breeding on the Farilhões islets in the Berlengas archipelago in 1981 (Teixeira & Moore 1983). These islets are some five miles north-northwest of Berlenga island, which lies a similar distance from Cape Carvoeiro, on the Portuguese west coast. The birds involved are of a larger form of band-rumped petrel that also nests on the Azores, the Canary and Salvage Islands and on the Madeiran archipelago: it may come to be separated as a distinct species, 'Grant's Storm-petrel' (Robb *et al.* 2008). The Portuguese colony, the sole one known in Europe, was initially estimated at 50 pairs but subsequently at 200–400 pairs (Hagemeijer & Blair 1997). More recently it was put at just 250 individuals, although on the basis of just one capture-recapture exercise (PA2). The population is constrained by predatory rats, so that the birds can only nest in burrows and crevices on islets and sheer cliffs (Granadeiro *et al.* 1998). The laying period extends from at least late

August to December, with a marked peak in October, while chicks fledge mostly in January-February (Granadeiro *et al.* 1998).

'Madeiran' storm-petrels have rarely been recorded anywhere else in Iberia. The earliest record was of two tideline corpses found at Punta Umbría, Huelva, in early November 1951 after a fierce storm (*Ardeola* 1: 128). Storm-driven birds were also found inland in Badajoz city on 15 February 1970 and at San Bartolomé de la Torre, Huelva, on 3 January 1982 (*Ardeola* 15: 101, 29: 177). There have since been 11 accepted records from Spain and three from Portugal. Four were birds that were attracted into mist nets by tape lures: two at Cape Silleiro, Galicia, in July 1994 and two at Punta de Almadena, Faro, on 10 July 1998 and 11 June 2008 (*Ardeola* 43: 106; *Pardela* 11: 6–7; *Anu. Ornitol.* 7: 11). One was seen at sea 220 km west of Sagres, Faro, on 20 August 1996 (*Pardela* 8: 6) and one was reported inshore at Denia, Alicante, on 30 June 1999 (Koppejan 2001). Another was found in a building in A Coruña city, Galicia, on 24 October 2002 (*Ardeola* 53: 170). A tideline corpse was found on a Cádiz beach on 26 October 2006 (*Ardeola* 55: 266). An oceanographic survey off Galicia in August 2010 produced six further records totalling 17 birds (*Ardeola* 59: 368–369) and two observations of single birds were made in the same region in October 2011 and October 2012 (RBS). All but the Denia bird have occurred off Atlantic coasts between Galicia and Andalucía and most records have been during July–October.

RED-BILLED TROPICBIRD — *Phaethon aethereus*

(Rabijunco Etéreo, Rabijunco)

Vagrant to Portugal.

A subadult was seen 87 miles off Cape Sardão, Beja, at 37°37'N, 10°38'W, on 13 August 1988 (Moore 1990). A tropicbird of undetermined species was reported from Cape St Vincent on 30 May 1977 (AP). *P. a. mesonauta* breeds in the Caribbean, the Cape Verdes and off Senegal. It has nested recently in the Azores (Furness & Monteiro 1995) and the Canaries: one pair on El Hierro during 2007–2010 (*Ardeola* 59: 369) and is the most likely to occur in Iberian waters. It reaches Mauritanian waters more or less regularly and a tropicbird species was seen off Cape Spartel, west of Tangier, in April 1995 (Géroudet 1965). There have been three recent records in British waters: on 7 June 2001 and on 29 March and 21 April 2002.

RED-FOOTED BOOBY — *Sula sula*

(Piquero Patirrojo, Atobá-de-pes-vermelhos)

Vagrant to Spain.

The nominate race nests in the Caribbean and SW Atlantic. There are eight records from the Cape Verdes and one each from the Canary Islands, France and Spain. A subadult was photographed off Estepona, Málaga, on 11 August 2010, and what was presumably the same individual was found at Cape Creus, Girona, on 3 December 2010 (*Ardeola* 59: 370). It moved next day to the nearby harbour of Estartit but was found there in a poor state on 15 December, having swallowed three fish-hooks, and died in care soon afterwards (Armada 2011). The Spanish record was followed by the appearance of a near-adult some 60km inland in France, at the Lac du Sainte-Croix, Var, where it was present during the first half of July 2011 (Reeber *et al.* 2012).

MASKED BOOBY — *Sula dactylatra*

(Piquero Enmascarado, Atobá-grande)

Vagrant to Spain.

An adult was off Sotogrande, Cádiz, on 10 October 1985, and another or the same bird was seen 80km away at Torremolinos, Málaga, on 14 December 1985. The third record was the obliging adult that boarded the Portsmouth/Bilbao ferry in French waters of the Bay of Biscay on 3 September 2003 and remained there until the vessel was docking at Bilbao the following morning. Birds from the Antilles regularly travel north along the coasts of the eastern USA. There are single records from Morocco, the Cape Verde Islands, the Azores and Israel (Haas 2012).

BROWN BOOBY *Sula leucogaster*

(Piquero Pardo, Alcatraz-pardo)

Vagrant to Spain and Portugal.

In Andalucía, an adult was reported from the Málaga coast at Benalmádena on 28 May 1983; an immature was at nearby Torremolinos on 9 September 1986, and a possible adult was seen off Huelva on 22 July 2010. In Portugal, an adult was present in Algarve at Piedade, Lagos, between 27 July and 20 August 1996, and an individual was seen from Cape Raso, Lisboa, on 17 April 2010. Three further records have been from Galicia: a dead bird was picked up on a beach in Pontevedra on 10 December 2002 in the aftermath of the *Prestige* oil spill; a possible adult was seen off Cape Estaca de Bares on 12 September 2003, and an adult was at Cape Bares on 13–16 August 2011. Five of the eight listed records were in July–September.

The nominate form inhabits the Gulf of Mexico, the Caribbean and the tropical Atlantic, the closest breeding sites to Iberia being in the Cape Verdes. Birds perhaps originating from there occur along the Mauritanian coast in small numbers. There are five records from the Moroccan Atlantic coast, the most northerly at Salé. There are also historical or unsubmitted records from the Azores and Canary Islands, as well as an accepted record from the Azores and another from Madeira. There are two records from the Italian Mediterranean, on 1 May 1996 in Tuscany and on 17 March 2003 at Messina.

CAPE GANNET *Morus capensis*

(Alcatraz de El Cabo, Alcatraz do Cabo)

There are no definite Iberian records but the species is included since it has figured in the Iberian literature.

Paterson & Riddiford (1990) have reviewed occurrences of apparent Cape Gannets and list ten records of single birds from Spanish Mediterranean waters during 1986–90, mainly adults showing the all-black tail and secondaries of the species. They concluded that such individuals are possibly Northern Gannets in an atypical plumage but that there remains a chance that some Cape Gannets reach European waters. A ringed first-year bird that was recovered at 21°40′ N, north of Cap Blanc on the Western Sahara/Mauritanian border (Crawford *et al.* 1983) appears to be the most reliable record of this species in the Western Palearctic. There are six unconfirmed records of a total of ten adult and four immature Cape Gannets seen between Cap Beddouza and Cap Juby, Morocco, during 1987–91 (Hillcoat *et al.* 1998).

NORTHERN GANNET *Morus bassanus*

(Alcatraz Atlántico, Alcatraz do Norte)

Common on passage and in winter in both Atlantic and Mediterranean waters. Some non-breeders remain in summer.

The global population increased massively during the 20th century (HBW) and there is little doubt that the Gannet is now more numerous and widespread in Iberian waters than formerly. However, the historical data is too sparse to enable any precise assessment of any change in abundance to be made. Nevertheless, Irby (1875) described it as uncommon in the Strait; Arévalo (1887) said that it was scarce in Spanish waters, where it appeared in cold winters, but that many occurred at Gibraltar in some years; Munn (1948a) described it as only a rare straggler to the Balearics, and Bernis (AMI) considered it scarce in the Mediterranean.

It is now a very common and locally abundant migrant along Atlantic and Biscayan coasts and large numbers remain offshore there in winter. Large numbers also enter the Mediterranean and the Gannet is common offshore in the Alboran basin and along the Levant coast, although it remains comparatively scarce in Balearic waters. Small numbers of non-breeders remain in the Mediterranean and in the Strait of Gibraltar in summer. These are chiefly but not entirely immature birds. Non-breeders are relatively common off Atlantic coasts and particularly in the Bay of Biscay in summer. Adults also comprise a large proportion of birds seen off Biscay coasts in summer.

The post-breeding movements are conspicuous all along the Biscay coasts and down the Atlantic seaboard. Within the Bay of Biscay they occur on a much larger scale than is seen in spring. Birds enter the Bay from the north and then turn west along the Spanish coast. This passage occurs chiefly during September–November. The first juveniles appear in mid August and they comprise a relatively high proportion of the passage early in the season, in September and the first half of October. The peak passage of subadults and adults is in October–November, when they predominate. The autumn westward movements are particularly evident at Cape Estaca de Bares where they are sometimes spectacular. For example, 128,483 birds were counted during September–November in 209 hours of observation in 2008 (614 birds/hour) (Sandoval *et al.* 2009): the total number of birds involved was clearly much higher since a watch was kept for only 40 days and counting is impossible at night, when some passage probably occurs. On the basis of sample counts, the mean annual post-breeding daytime passage during 2004–08 was of the order of 785,000 birds (Sandoval *et al.* 2010). It is thus evident that a very significant proportion of the Northeastern Atlantic Gannet population (900,000–930,000 individuals: Birdlife International 2004) passes Cape Bares each autumn. Mean passage rates during June–December were 207 birds/hour in 2009, 347 birds/hour in 2008, 210 birds/hour in 2007, 206 birds/hour in 2006 and 464 birds/hour in 2005, indicating considerable year to year variation. Peak rates are much greater: the record at Cape Bares is 6,468 birds/hour on 31 October 2004 and there were 4,608 birds/hour on 5 November 2005 and 4,275 birds/hour on 25 November 2008 (Sandoval *et al.* 2009, 2010). The maximum rates are thus observed late in the season, when the passage is dominated by adult birds.

Southward passage is also conspicuous in autumn down the Atlantic coasts of Galicia and Portugal and off the western approaches to the Strait of Gibraltar. Large numbers have been reported off Estremadura, Portugal: for example, there were 7,000 off Cape Raso on 30 October 1988 (Paterson 1997). Some 12,400 passed southeastwards at Cape St Vincent in November–December 1992, at rates of 120–780 birds/hour, and 38,024 were counted there in September–November 1993, at mean rates of 205 birds/hour in September and *c.* 430 birds/hour in October–November (Walker 1994). The relative proportion of adults increased as the season progressed.

A very large proportion of Gannets continue their southward migration along the African Atlantic coast, the passage along the Moroccan coast being described as 'huge' (BM). Large numbers winter in Atlantic waters off north Africa and the wintering range extends south to Mauritania and Senegal, with a few reaching the Gulf of Guinea (Wernham *et al.* 2002).

Significant numbers pass through the Strait of Gibraltar and winter in the western Mediterranean, many of them in Iberian waters (Garcia 1973, de Juana & Paterson 1986). The passage into the Mediterranean was estimated at 20,000–24,000 birds by Hashmi (1993). The first migrants at the Strait are a few juveniles that appear in late August, followed by large numbers of juveniles in September and an influx of subadult and adult birds in October–November (Cortes *et al.* 1980, Finlayson 1992).

Gatherings of several hundred or more may form off Biscayan and Atlantic coasts and in the Strait of Gibraltar in response to fish shoals. Feeding assemblages of a few thousand birds are not infrequently reported off Cape da Roca and Cape Raso, Lisboa, during October–January (AP). Mid-winter influxes of hundreds of birds into the Strait from the Atlantic are sometimes observed, generally in response to westerly gales, and these may soon be followed by return movements westwards (Finlayson 1992, pers. obs.). Wintering Gannets are also common off Iberian Mediterranean coasts north to Catalonia, although an annual mean decline of 1.1% was noted in Catalan waters during 2003–08 (CWA). However, they remain comparatively uncommon around the Balearic archipelago. Munn (1948a) gives only four records for the Balearics and Mayol (1981) considered it scarce there. Since then Gannets have become more frequent but still scarce: maximum reported counts include 21 off Mallorca on 6 January 1999 and 25 off Formentera on 14 February 2001 (*Ardeola* 49: 174, 50: 152).

The northward spring passage is noted as early as late January at the Strait, when adults begin to leave the Mediterranean. The exodus is prolonged. Adults predominate in February–April and a smaller passage chiefly of immature non-breeders continues through May and into June (Finlayson & Cortes 1984). The spring passage is also noticeable at Cape St Vincent, where passage rates are highest in January–February but sharply decline thereafter. Here in both 1993 and 1994 passage began in mid January and peak rates were seen in late January and during the first half of February (Walker 1994), with maximum rates often exceeding 1,000 birds/hour and reaching 2,918 birds/hour on 27 January 1994. Here too adults predominated during the earlier passage: in 1994 they comprised 98.1% of birds in January (n = 24,357), 92.3% in February (n = 19,209) but only 59.7% in March (n = 6,558).

Further north a largely eastward spring passage into the Bay of Biscay is detectable from late December to March: in 2008, 2,609 birds were counted at Cape Bares in sample counts during January–March, with a maximum passage rate of 171 birds/hour on 31 January (Sandoval *et al.* 2009). This movement is far smaller than the corresponding westward autumn passage and indicates that the flow of Gannets ascending the Iberian Atlantic coast in spring largely continues northwards without diverting into the southern Bay of Biscay.

Large numbers of Gannets have been ringed in Europe, chiefly as pulli at nesting colonies in Britain. Bernis (AMI) summarises 131 recoveries that are fairly evenly spread around Biscay and Atlantic coasts as far south as the Strait of Gibraltar but they include only four in the Mediterranean. Tellería *et al.* (1999) give a further 108 recoveries of which 17 were in the Mediterranean. About half of all of these and some 300 subsequent recoveries are of birds ringed at colonies in Scotland: notably the Bass Rock and Ailsa Craig, and at Grassholm in Wales; with others from the Channel Islands, the Faroes, Ireland, Norway (30), Denmark (1) and Iceland (8) (SMA, AP). There is also a Portuguese recovery of a bird that had been ringed in Newfoundland (Dennis 1987) and another there of one ringed in Germany (AMI).

GREAT CORMORANT *Phalacrocorax carbo*

(Cormorán Grande, Corvo-marinho-comum)

***P. c. sinensis*. Common and locally abundant in winter and on passage. Has recently established nesting colonies in Spain and has bred in Portugal.**
***P. c. carbo*. Common on Atlantic coasts in winter.**

The Great Cormorant has changed from being a scarce winter visitor to Iberia prior to the 1960s to becoming common and locally very numerous in winter across the entire region. Even more recently it has become a breeding species, with a scattering of colonies in Spain and a clear tendency to establish more.

Munn (1921, 1924) claimed that it bred both in Mallorca and Menorca but later recognised (Munn 1930) that it was just a winter visitor there and that only the Shag nested in the islands. The first known instance of nesting in Iberia was nonetheless in the Mediterranean, on the Medes islands, Girona, where one pair nested successfully in 1973 and adults were seen at a nest in 1978 and 1982 (de Juana 1984, Ferrer *et al.* 1986). The nearest Mediterranean colonies at that time were in Sardinia. There was subsequently a failed attempt at the Santillana reservoir, Madrid, in 1989, followed by successful nesting by one pair at the Navalcán reservoir, Toledo, in 1991 (*Ardeola* 36: 232–233, 38: 327–328). Since then, and most particularly since 2000, there has been an exponential increase in the breeding population although still at a limited number of sites, almost all of them inland (Fig. 57).

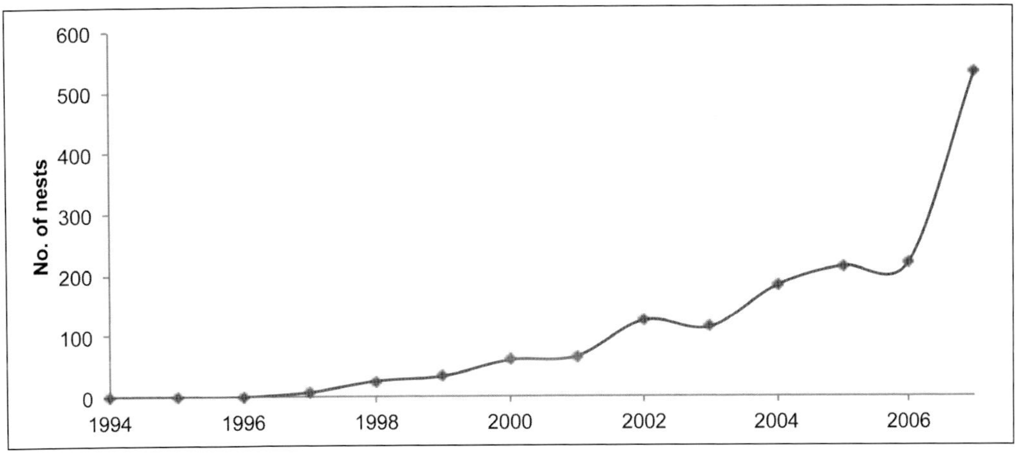

Figure 57. The increase in the Great Cormorant breeding population in Spain, 1994–2007 (Palomino & Molina 2009)

A census of the Spanish breeding population in 2007 found 532 pairs at 11 sites (Palomino & Molina 2009). The contiguous regions of Castilla-La Mancha (216 pairs), Extremadura (130 pairs) and Andalucía (118 pairs) held 87% of the total, these within the Tagus, Guadiana and Guadalquivir river basins. There have since been further increases in the numbers present at established colonies and new sites have been occupied. In 2010 breeding was confirmed for the first time in the respective provinces at small colonies at the Aracena reservoir, Huelva; the Linares reservoir, Sevilla, and on the river Duero at Toro, Zamora (*Ardeola* 57: 520). By 2011 La Serena reservoir, Badajoz/Córdoba, alone had a breeding population of 341 pairs (*Ardeola* 58: 487) and numbers there virtually doubled to 619 pairs in 2012 (*Ardeola* 59: 417). However, Cormorant colonies have been characterised by their instability and a number of sites have been occupied for just one or a few years and have then been abandoned, and others have been used sporadically. For example, 10–12 pairs nested at the Gasset reservoir, Ciudad Real, during 1998–2001 but not subsequently (Palomino & Molina 2009).

In some cases the birds have moved on as a result of persecution. Cormorants sometimes clash with fishing interests, especially at inland sites, and such conflicts are likely to increase as the population expands. A breeding attempt by 20 pairs at the Albufera de Adra, Almería, in 2004 (*Ardeola* 51: 543) apparently failed because of intervention by fish farmers and the birds have never returned. The great majority of colonies have been established at reservoirs that also attract large wintering populations. This has also been the case in Portugal where the first recorded instance of breeding was in 2007 at the Alqueva reservoir: three pairs nested there in 2008 (AP). Colonies are generally established on waterside trees. Large numbers of non-breeders are now often present around the colonies in summer and nest-counting is the only reliable way to assess breeding numbers (Palomino & Molina 2009).

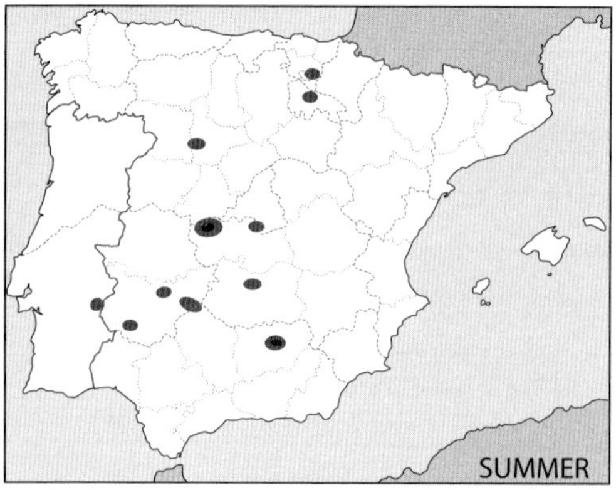

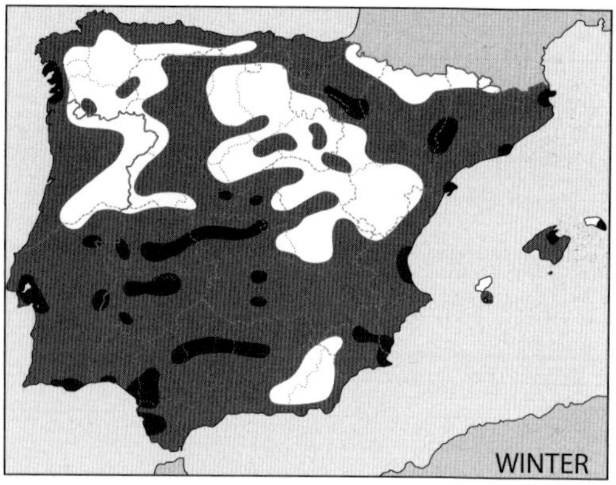

All the Iberian breeding birds are of the race *sinensis*, with the apparent exception of a single nominate race pair that nested at the Ullibarri reservoir, Álava, in 1994 and 1998 (*Ardeola* 46: 150). The records show a wide dispersion of colonies and a rapid expansion of numbers at some sites that can only be accounted for by immigration. The Spanish population is clearly largely comprised of wintering birds from elsewhere that have opted to stay and nest. In fact, three individuals originating from French colonies have been identified at the Castronuño reservoir, Valladolid (Palomino & Molina 2009). The colonisation of Iberia by *P. c. sinensis* is only a recent stage of a process of breeding range expansion that has been underway in western and southern Europe for some time. The subspecies first nested in France in the 1950s but there were 46–47 colonies (4,100 pairs) in 2006, although only six of them in the south (OF).

The Iberian Peninsula has only become an important wintering area for Great Cormorants since the 1990s. Bernis (1966a) considered them to be frequent winter visitors but the great majority of records at that time were from coastal locations. Bernis (AMI) analysed 39 Iberian recoveries of *P. c. carbo*, 37 of which originated from the British Isles and all of which were recovered on northern and western Atlantic coasts, south to Algarve. There were then also 67 recoveries of *sinensis*, nearly all from the Low Countries, and these were recovered equally from Atlantic and Mediterranean coasts: they included three inland recoveries in Spain. Nineteen of 20 Portuguese recoveries given by Teixeira (1984) were *P. c. carbo*.

More recent recoveries indicate that the wintering population chiefly originates from northwestern Europe (Campos & Lekuona 1994, Ibarra & Martín 1996, Pérez-Tris 2000). They confirm that *P. c. carbo* occurs almost entirely on Biscay and Portuguese coasts and is rare inland or in the Mediterranean. However, *P. c. sinensis* occurs on all coasts on passage and also crosses the Peninsula from the western Pyrenees south-westwards. Wintering *carbo* are chiefly from Irish Sea colonies in Ireland, Scotland and Wales, with a few from England and northwestern France. Birds from Britain and Ireland comprise 45% of 857 Iberian recoveries of extraregional origin (SMA, AP). The bulk of the wintering population, comprised by *sinensis*, is now largely from Denmark or Fennoscandia (29% of all recoveries); or from Germany, France, the Low Countries or the recently established colonies in Switzerland (24%); with a few from the Baltic States, Poland or Russia. Up to the 1960s virtually all *sinensis* recovered in Iberia originated from the Naardermeer colony in The Netherlands but they now come from a much wider region (AMI, Pérez-Tris 2000).

The wintering population of *P. c. carbo* remains almost entirely coastal in distribution but *P. c. sinensis* increasingly winters inland (Pérez-Tris 2000). Wintering inland is a relatively recent phenomenon in Iberia. The first count of any numbers was of 126 at the Valdecañas reservoir, Cáceres, in October 1968 (Fernández-Cruz 1969). Known inland occurrences prior to that date were summarised by Bernis (1969a) and they amounted to just 14 records (73 birds) during the period October 1953 to January 1968. Inland wintering was well established by the 1980s and expanding rapidly, very likely largely due to the proliferation of reservoirs throughout Spain and in Portugal.

National censuses of winter roosts in Spain (Table 10) show an approximate doubling of the population between 1996 and 2003 although with considerable inter-regional variation. That of the Comunidad Valenciana – the vast majority, if not all, being *P. c. sinensis* – increased spectacularly by over 500% and those of Andalucía and Murcia increased by nearly twice the national average of 92.4%. Some 2,000 were found at Los Alcázares, on the Mar Menor, on 9 December 2009, nearly twice as many as were censused in Murcia in January 2003 (*Ardeola* 57: 222). The change has been very apparent around Gibraltar, where the Cormorant was rare until the 1990s, since when a regular total wintering population of several hundred birds has frequented the Bay of Gibraltar, the harbour at Algeciras, the Guadiaro estuary and the Rock of Gibraltar itself (GONHS).

Table 10. Minimum numbers of Great Cormorants wintering in the Spanish regions (after Hidalgo 1998, del Moral & de Souza 2004).

Region	January 1996	January 2003	% change
Andalucía	4,218	11,962	+183.6
Aragón	2,194	4,743	+116.2
Asturias	1,395	1,428	+2.4
Balearic Islands	272	472	+73.5
Basque Country	1,099	1,233	+12.2
Cantabria	605	737	+21.8
Castilla y León	1,993	3,299	+65.5
Castilla-La Mancha	2,120	4,220	+99.1
Catalonia	4,709	8,316	+76.6
Comunidad Valenciana	1,923	11,711	+509.0
Extremadura	8,656	12,265	+41.7
Galicia	3,461	3,325*	-
La Rioja	583	155	-73.4
Madrid	819	1,740	+112.5
Murcia	365	1,077	+195.1
Navarra	1,333	2,102	+57.7
TOTAL	35,745	68,785	+92.4

* Incomplete count

Increases elsewhere have been more modest, especially in the north. In general, Extremadura, the Comunidad Valenciana, Andalucía and Catalonia together held 54.5% of the Spanish January population in 1996 and this increased to 64.3% by 2003, largely due to the increased importance of the Comunidad Valenciana fraction. The Spanish waterfowl counts of 1990–2009 confirmed the marked increase in wintering Cormorants throughout this period, a trend that was apparent in most regions but again with indications of relative stability in the north and in the Duero basin and perhaps a more general stabilisation in wintering numbers in recent years (AAE). The Spanish wintering population is concentrated at several key sites, the Albufera de Valencia holding by far the largest number: there were 8,603 there in January 2003. The remaining principal sites are associated with the valleys and estuaries of the Ebro, Guadalquivir and Guadiana rivers, although none of these held more than 2,000 in January 2003 (del Moral & de Souza 2004).

The numbers of wintering Cormorants in Portugal have increased in parallel to those in Spain, although no complete census data is yet available. The population in the early 1980s was estimated at about 1,000 individuals, mainly at the Tagus and Sado estuaries. It increased to 8,000–10,000 birds by the early 1990s and has since exceeded 12,000 birds (AP). The Ria de Aveiro and the Tagus, Sado and Formosa estuaries attract some numbers: often 1,000–2,000 birds in total and sometimes more, but inland reservoir sites are now also important: nearly 2,000 have been counted at the Alqueva reservoir.

Wintering birds begin to arrive in September but the main influx is the second half of November (SWA). The wintering population is at a maximum during November–February. Wintering birds generally roost communally and may fly over 30km twice daily between their roosts and feeding locations. Some 70% of roosts are at reservoirs or on riverbanks and 60% are in trees, although those wintering on the coast favour rocky islets offshore (del Moral & de Souza 2004). A study of the diet of wintering cormorants in central Spain (Martín Fernández *et al.* 1997) showed a clear predominance of Common Carp (52% of items), followed by barbel species (17%) and an alien catfish, the Black Bullhead (12%). It is noteworthy that salmonids were not a principal prey item since concern about the possible impact of cormorants on salmon and trout stocks has resulted in local persecution of the birds.The immigrants birds begin to return to their breeding areas from late February onwards. Most have left by the end of April but small numbers of non-breeders remain in Iberia all year round (AI1).

EUROPEAN SHAG *Phalacrocorax aristotelis*

(Cormorán Moñudo, Galheta)

***P. a. aristotelis*. Largely resident on Atlantic coasts of northern Spain and Portugal.**

***P. a. desmarestii*. Largely resident on Mediterranean coasts and in the Balearic Islands.**

The Shag is a locally common breeding species on rocky coasts and islets around the Iberian Peninsula and in the Balearics. It is an early nester, beginning as early as January on the Atlantic coasts and from mid December in the Mediterranean. Shag colonies are usually on sea-cliffs, often on islets, and nests are typically placed on ledges, in caves or holes, or under boulders. Shags always fish at sea, often in loose groups. The Atlantic subspecies has colonies all along the Biscay coasts, from the Basque Country west through Cantabria and Asturias to Galicia. It is especially common in western Galicia, having its largest colonies in the Cíes islands and on Ons island (SA2). Further south there are scattered colonies in Portugal (PA2), the largest (82–109 pairs) being in the Berlengas islands. The remaining Portuguese colonies, totalling fewer than 40 pairs, occur on rocky coastlines in the south-west between Cape Carvoeiro and southwestern Algarve.

There are no breeding Shags off southwestern Spain, where the habitat is in any case largely unsuitable. The Mediterranean subspecies has an outpost colony in Gibraltar and is also found at a number of sites on the Spanish Mediterranean coast north to Catalonia. Its principal Iberian populations are in the Balearic Islands. The first national census of breeding Shags in Spain was conducted in 2006 and 2007: there were 3,764 pairs in Spain, with another 60 pairs probably present (Table 11).

Table 11. Regional distribution of breeding Shags in the Iberian Peninsula. Sources: Álvarez & Velando 2007 (Spain), PA2 (Portugal), GONHS (Gibraltar).

Region	Population (pairs)	Trend
P. a. aristotelis		
Basque Country	105	=
Cantabria	73	=
Asturias	235	= ↓
Galicia	1,264	↓
Portugal (2002)	106–146	=
P. a. desmarestii		
Gibraltar (2011)	4–5	=
Almería	4	↓
Murcia	0	
Comunidad Valenciana	25	↑
Catalonia	41	↑
Balearic Islands	2, 017	↑

P. a. aristotelis numbered 1,667 pairs in Spain in 2006–07, some three-quarters of them in Galicia. There were also 100–150 pairs nesting in Portugal. The largest colonies, on Cies and Ons islands, exceeded 300 pairs but colonies on the Biscay coasts generally ranged from 5–20 pairs.

P. a. desmarestii numbered 2,087 pairs in Spain in 2006–07. There were also 4–5 pairs in Gibraltar, the westernmost colony of this race, where the population remains stable. Some 96% occur in the Balearics, including 1,179 pairs in Mallorca and smaller populations on all the other main islands: 242 in Menorca, 224 in Formentera, 157 in Ibiza and 152 in Cabrera. The other Iberian colonies probably number fewer than 80 pairs in total, chiefly distributed between Cape Gata, Almería; the Columbretes Islands, Castellón, and the Medes islands and Girona coast. Two pairs nested on Isla Grossa, Murcia, in 2009, and a total of 13 pairs nested on the Alicante coast the same year, at the Peñón de Ifach, Benidorm island and the Montgò coast (*Ardeola* 57: 520; *A. O. C. Valenciana 2009*: 26).

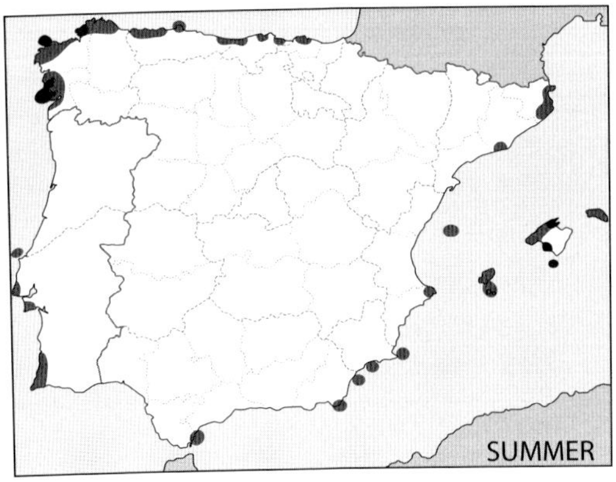

The fortunes of the Shag in Iberia have fluctuated considerably. Many were hunted, apparently for sport, in the Balearic Islands and in Galicia especially, prior to the key sites receiving protected status in the 1980s. At that time the Iberian Atlantic population was estimated at some 1,125 pairs (Bárcena *et al.* 1984). This was followed by a significant recovery in numbers in Galicia but the population there is again declining, perhaps by as much as 10% per year, due to losses of birds caught in fishing nets (Velando & Freire 2002). The species also suffered from the shipwreck of the tanker 'Prestige' (Martínez-Abraín *et al.* 2006). The Portuguese population is at least stable and may be increasing (AP). The principal populations of *desmarestii* have fared better and the Balearic and east coast populations are increasing. The Balearic population was estimated at 645–1,045 pairs in the 1980s, when there were also a few pairs on the Columbretes and Medes islands and on sea cliffs in Almería and Gibraltar (de Juana 1984). The increases have been most evident in Ibiza and Formentera, where adults and chicks were traditionally taken for food (Mayol 1977b) and the total breeding population was only 25–50 pairs in the early 1980s (de Juana 1984) but was 381 pairs in 2006–07. In contrast, the Menorcan population, which was not persecuted in the past, seems to be stable or to have increased only slightly.

Shags are relatively sedentary. Ringed birds have travelled from Galicia to Asturias and from Asturias to Cantabria, and four from Balearic colonies have been known to reach the Levant and Catalan coasts, but none of these birds moved further than 300km from their natal colonies. Single birds from Mallorca have also been found on the coasts of southern France and Algeria. However, a chick ringed on the Pantorgas Islands, Asturias, in April 2004 was seen in the Netherlands in November–December 2007 (Álvarez 2009a) and one ringed at the same colony in April 2006 was found 719km away at Aznalcázar, Sevilla, in May 2006. There are also a few recoveries of foreign-ringed immature birds in Iberia, including a Corsican bird in Mallorca, three birds from the south-west of the British Isles in northern Spain and, most surprisingly, a Welsh bird found on the Levant coast, i.e. within the range of *P. a. desmarestii* (AI1; Wernham *et al.* 2002).

GREAT WHITE PELICAN *Pelecanus onocrotalus*

(Pelícano Común, Pelicano-branco)

Vagrant to Spain and Portugal.

It was formerly more widespread in Europe. The remaining birds occupy the Danube colonies during April–October and winter mainly in east Africa (Izhaki *et al.* 2002). South of Iberia there is a large colony at the Banc d'Arguin, Mauritania, and large numbers also breed at Sahel wetlands. There are four records of single birds from Morocco. Vagrants have been reported from most North African and European countries but many records are attributed to escapes since many are kept in bird collections (Crivelli *et al.* 1997). Hence, in France only the 19th century records were accepted to be of wild birds although there are 96 records, of 113 individuals, from the 20th century and up to 2009, which are concentrated in summer and autumn. A recent analysis of 216 European records has shown that these are significantly most frequent from sites that are closer to the nearest colonies. Moreover, records occur more often in years when the Greek colonies experience poor breeding conditions, indicating that the birds disperse further when this happens. These patterns are indicative of wild birds, rather than fugitives from captivity, reaching other European countries (Jiguet *et al.* 2008).

There are undated references to the occasional occurrence of the White Pelican in Portugal during the 19th century (AP) but the only specific Iberian record then was the capture of an individual at the Albufera de Valencia in 1856 (Arévalo 1887, Dies *et al.* 1999). Another may have been obtained in Mallorca in 1773 (Bernis *et al.* 1958). Reliable 20th century records include individuals that were killed at the Albufera de Mallorca in June 1952 and in Girona in September 1977, as well as one seen flying towards the Strait at Tarifa on 27 August 1972, one killed in the Marismas del Guadalquivir a couple of days later and one captured at Medina del Campo, Valladolid, in April 1973, although the possibility of escapes was raised with respect to the last three records (Motjé 1988; *Ardeola* 1: 24, 19: 7–8). Further records were not until the 1990s. At least four White Pelicans appeared at various locations in northern Spain: in Catalonia, Aragón, Navarra and Asturias, during summer 1990. Their appearance coincided with a small 'invasion' in France, where 14 records resulted, and elsewhere in Europe, suggesting an influx of wild birds from the east (Zwaaneveld & Ebbels 1994, de Juana 2006). By 2009 there were 28 accepted records from Spain and nine from Portugal. They included six records in 2006 and 11 in 2007. Up to four adults frequented the fish farms at Veta la Palma, Doñana, during 1999–2006.

We here consider 34 records (39 birds), excluding obvious duplications. The age of 30 individuals was reported and 80% were adults, a similar proportion to that reported from France, Hungary and Poland (Jiguet *et al.* 2008). There are records throughout the year (Fig. 58) but there is a peak in August, as also noted in France. Observations have predominantly been in the south-west (Fig. 59) which may suggest arrivals from Africa. They include three records of southbound birds at the Strait: in addition to the 1972 observation at Tarifa, one flew south with White Storks on 18 August 1996 and one arrived from Spain near Tangier on 11 September 2003 (*Go-South Bulletin* 7: 1–14). However, a marked adult is known to have had a different, eastern, origin. This bird wintered near Lleida city from October 2008 to February 2009, where it was always in the company of White Storks and often visited a rubbish dump with them. It was previously tracked through southern Germany, Switzerland and southern France and stopped at the Aiguamolls de l'Empordà (Palomares 2009). A further accepted record is of an adult at Don Benito, Badajoz, on 21 October 2011.

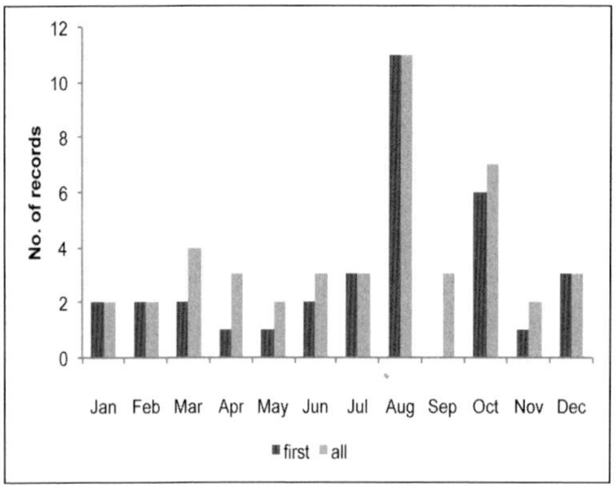

Figure 58. Records by month, 1952–2008 (n = 34)

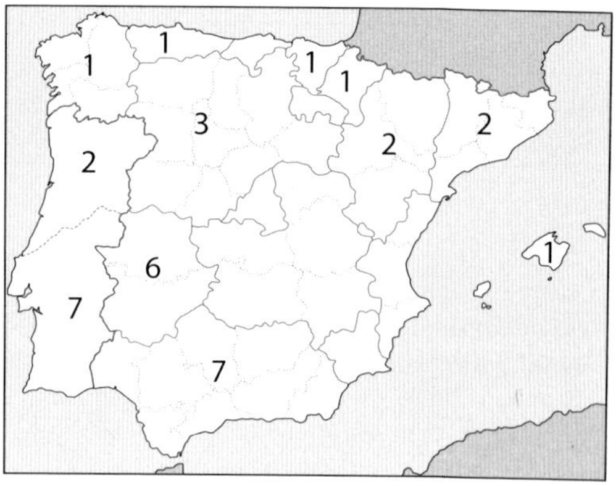

Figure 59. Records by region, 1952–2008 (n = 34)

PINK-BACKED PELICAN *Pelecanus rufescens*

(Pelícano Rosado, Pelicano-cinzento)

Recorded in Spain and Portugal. Possible vagrant.

It is a rains migrant in the Sahel, moving north for the wet season during June–September, when it also regularly reaches Lake Nasser in southern Egypt. Occurrences in Europe have usually been attributed to escapes. There are 22 records from France, but the species is retained in Category D there since a free-flying colony at a zoo near Narbonne may partly account for the high number of French records and perhaps for some of those in Iberia (OF). Nevertheless, an analysis of records from Europe and Israel (Jiguet *et al.* 2008) has found a highly positive correlation between occurrences and annual rainfall in the Sahel: high rainfall allows birds to reach further north in the Sahel so the finding implies that at least some of the birds recorded in the Palearctic originate from Africa during wet years there.

The earliest Iberian observation was an adult at Las Cañas reservoir, Navarra, during 1–30 August 1990. The same year saw two birds in France and individuals in The Netherlands, Germany, Austria and Poland, all between September and 1 December, as well as unusually large numbers at Lake Nasser: where there were 64 in June (*British Birds* 84: 2). The implication is that there may have been an arrival of wild birds in 1990.

The 1990 bird from Las Cañas probably accounts for several records in the Ebro valley in 1991. Thereafter there have been eight accepted records from Spain and one from Portugal, all falling within just four years: in 1998, an adult in November at the Aiguamolls del Empordà, Girona, and one in December in Mallorca and Cabrera; in 1999, an adult during September–October at the Santillana reservoir, Madrid, and another possible adult at Gallur, Zaragoza, in October; in 2001, one in April at the Celemín reservoir, Cádiz, and another in October at the Mondego estuary, Coimbra; and, in 2004, an immature at the Santillana reservoir in January, an individual at the Valdesalor reservoir, Cáceres, in May and an adult in June–July at Salburua, Álava. More recently one was reported off the coast at Roses, Girona, in June 2012 (RBS).

The Iberian records are well dispersed across the year but they show a tendency to concentrate during August–October.

This pattern also occurs with the French records and those from elsewhere in Europe, which show a September peak that may be related to the movements of the Pink-backed Pelican in the Sahel. Nevertheless, the occurrence of wild birds in Europe, including Iberia, has yet to be demonstrated unambiguously.

MAGNIFICENT FRIGATEBIRD *Fregata magnificens*

(Rabihorcado Magnífico, Rabiforcado)

Vagrant to Spain.

The only Iberian records of confirmed Magnificent Frigatebirds were of a subadult female at Torremolinos, Málaga, on 22 October 2000, and a female off Santander on 8 November 2008. In addition, male frigatebird sp. were reported from A Coruña on 22 May 1983; and from Baiona, Pontevedra on 29 September 1985.

Frigatebirds are occasionally transported by storms along eastern North American coasts and across the ocean to Europe. The Magnificent Frigatebird is by far the most likely to occur in the North Atlantic, although an Ascension Frigatebird was found in Scotland in 1953. There are two records of the Magnificent Frigatebird from Britain as well as a further five identified as frigatebird sp. There are two old records of the present species from France and four post-1990 records of frigatebird sp., one of them in the Mediterranean. Two frigatebird sp. appeared in the Canary Islands in November 2005 in the aftermath of Hurricane Wilma, which also resulted in records from the Azores, Britain and France. A frigatebird sp. at Souss-Massa on 10 April 2009 is the sole record from Morocco.

GREAT BITTERN *Botaurus stellaris*

(Avetoro Común, Abetouro-comum)

***B. s. stellaris*. A very rare and local breeding species in mainland Spain. Has bred in Portugal and the Balearic Islands. Widespread but scarce in winter at major wetlands throughout mainland Iberia and in the Balearics.**

The Great Bittern has never been abundant in Iberia but it declined to near-extinction during the 20th century and is now present as a regular breeding species at only a handful of locations in Spain. They comprise Doñana, the Aiguamolls de l'Empordà and, especially, wetlands of the upper Ebro valley in Navarra and Aragón. The last region has until recently been its Iberian stronghold, with 12–17 territorial males recorded early in the 21st century (LR), distributed between wetlands at Cinco Villas, Las Bardenas Reales and the Pitillas and Sariñena lagoons. However, Doñana has supported a significant and apparently increasing breeding population in recent years. Territorial males are recorded occasionally at a diversity of other wetlands, with recent records from Galicia, the Comunidad Valenciana, Madrid and Castilla y León. Breeding sites generally offer extensive reedbeds and stable water levels in regions of fresh water or low salinity. However booming males are also reported from wetlands smaller than two hectares in the Ebro valley, where the birds can also seek food in nearby ricefields and other wetlands. Most breeding locations are below 200m but nesting has been recorded as high as 900m, at the Laguna de Gallocanta.

The Spanish heron census of spring 2011 found 40 territories (Garrido & del Moral 2012), including 26 at Doñana. There were also five in Navarra: including three at Pitillas; four in Aragón: three of them at the recently restored Laguna del Cañizar at Villarquemado; two each in Castilla-La Mancha and Catalonia, and one in Mallorca at the Albufera. The Spanish population had earlier been estimated at 29–30 territories in 1991 (Blanco & González 1992): including ten in the Ebro delta, six at Doñana and five at the Aiguamolls de l'Empordà. Assessments in 2003–04 (SA2) put the population at no more than 25 territorial males. Some recent modest increase is therefore apparent.

The history of the Spanish population is one of accelerating decline throughout the 20th century followed by recent small increases locally. The original decline was largely due to hunting and habitat loss: often through drainage, water shortages and reedbed destruction by burning and grazing (LR). The species was already quite local by the 1960s, breeding only within large expanses of marshland (F. Bernis: AMI). Earlier writers point to a relative abundance, for example at Doñana (Saunders 1871, Arévalo 1887, Valverde 1960) and at the Albufera de Mallorca (Munn (1926), as well as to its high vulnerability to snipe-hunters in winter (Chapman & Buck 1893). The occasional droughts during the second half of the 20th century, particularly that of 1992–94, caused further declines and local extinctions. There has only been limited recovery during the early 21st century, partly due to specific efforts to protect the species. Inadequate management of reedbeds and water levels may be now preventing a faster recovery (LR), while variable, rainfall-dependent, water levels may account for the large local variations in numbers, especially at Doñana.

The Great Bittern was also once well distributed as a breeding species in Portuguese wetlands but had become rare by the 1940s (Tait 1887, AP). Since then there has only been one confirmed breeding record, at the Sado estuary in 1987 (*Airo* 1

(2): supplement). The 2008 Portuguese breeding bird atlas (PA2) found three during the breeding season and it may be that occasional nesting still occurs.

It is evident that it was also a locally numerous breeding species at the Albufera de Mallorca. Munn (1921) described it as a common resident, whose booming could be heard day and night for most of the year. It declined to extinction by 1970 but has bred irregularly since 1990: there were 10 territories in 1996 but only 2–3 in 2001(Vicens 2004).

The Great Bittern may be more easily encountered in Iberia in winter, when birds occur with some regularity at the principal wetland sites and also at ricefields, reservoirs, gravel pits, rivers and wet meadows throughout the Peninsula and also at Balearic wetlands. The Spanish winter atlas estimated the presence of some 24–84 birds in 2007–10 (SWA), while the 2010–11 census of wintering herons found at least 35, almost half of them in Navarra (Garrido & del Moral 2012). Portuguese records are chiefly from reedbeds at coastal wetlands along the entire coastline. Most records involve single birds but small concentrations occur in particularly favoured areas. For example, there were five in the Corredor Verde del Guadiamar, Doñana, on 13 January 2009 (*Ardeola* 56: 155). Tait (1887) flushed 11 together at Leça da Palmeira, on the coast north of Porto, in winter 1873 (and shot eight of them). It has been suggested that the introduced Red Swamp Crayfish is an important prey item in winter and attracts Bitterns to quite small wetlands in the Ebro valley (LR).

Iberian breeding birds are thought to be resident or dispersive over short distances but individuals from elsewhere in Europe enter Iberia and the Balearic Islands to winter and a very small number continue across the Peninsula to winter in Morocco (BM). Bernis (AMI) records three recoveries in March of birds ringed as nestlings: one from southern Germany in Navarra and two from southern Sweden in Cantabria and Valencia, and there has since been a recovery of one from Switzerland (SWA).

AMERICAN BITTERN *Botaurus lentiginosus*

(Avetoro Lentiginoso, Abetouro-americano)

Vagrant to Spain.

One was obtained in January 1961, near Roses, Cataluña (Maluquer 1964b). There have since been three accepted records, all from the Galician coast: at the Laguna de Traba on 17 and 30 October 1982; at A Ínsua on 10 October 1998, and at the Brañas de Laíño, Dodro, on 23–24 September 2006. These records fit the pattern of the occurrences in Britain, where nearly all have been during September–March peaking in October–November. There are 41 records from Britain up to 2010, but none from either France or Morocco. There are also at least 16 records from the Azores (Le Grand 1983) and two from the Canary Islands.

LITTLE BITTERN *Ixobrychus minutus*

(Avetorillo Común, Garçote-comum)

***I. m. minutus*. Widespread and locally common in mainland Iberia. Also nests in Mallorca. Largely a summer visitor but some remain in winter.**

It is widespread as a breeding species throughout the Mediterranean climatic region but most abundant in Spain: in the Guadalquivir, Guadiana, Tagus and Ebro valleys and in the wetlands of the Levant coast. It also nests at the Albufera de Mallorca in the Balearics. It breeds very locally in the Eurosiberian region in Galicia, Cantabria and the Basque Country, but only sporadically in Asturias (SA2). It is widespread but uncommon in Portugal, with scattered populations in western and southern wetlands, and inland mainly in the southern half of the country (PA2). Breeding birds occur in loose colonies or as isolated pairs in the emergent or waterside vegetation of a broad range of wetland types, including lagoons, reservoirs and riversides.

The Spanish population was thought to be in severe decline due to habitat loss (SA1) but there are no indications of recent trends. It may be considerably larger than the 1,900–2,300 pairs estimated by SA1 in particularly favourable years. The Ebro delta is a major stronghold, where a partial census gave an estimate of 533–902 pairs (Fouces & Estrada 1992). The mean population at the Albufera de Valencia was 250 pairs (range 124–358) during 1988–2000 (SEO/BirdLife-EOA 2000). In the Balearics it was known to breed only at the Albufera de Mallorca, where it was reported by Jordans (1924–25). A notable rise in numbers was detected during the 1990s and early 2000s, although breeding was not proved (Vicens 2004). A minimum of 30 pairs is more recently estimated there (*A. O. Balears 2010*: 118). The first known nesting in Menorca was in 2011 when a pair bred at Es Mercadal (*Ardeola* 59: 418).

The Portuguese population has not been systematically surveyed. It has shown local declines, with disappearances from

some formerly occupied coastal sites, but there has also been some apparent range expansion at inland localities, so no clear overall trend emerges (PA2).

Most individuals winter in tropical west Africa from where they return in April–May. The southward passage is in August–September. Some birds from the small populations of other western European countries occur on passage. Bernis (AMI) records recoveries in northeastern Spain during passage periods of one nestling ringed in Belgium and two ringed in The Netherlands and there is a more recent recovery from western Galicia of one ringed in Italy. A juvenile ringed in Málaga in June 1986 was found at Agadir, Morocco, in late 1987 and another juvenile ringed at Doñana in June 1998 was recovered in Mali in January 2004. There has been a subsequent recovery of a Spanish-ringed bird from Mali and another from the Ivory Coast. Migratory movements are seldom apparent and most passage is probably nocturnal. A record of 40 circling and then departing out to sea from the Río Antas estuary, Almería, on 11 September 1996 (*Ardeola* 44: 244) is very unusual.

There is a very small wintering population chiefly at coastal wetlands: notably at Doñana, the Albufera de Valencia and in Almería, with occasional winter records inland (SWA). However, the 52 individuals detected by the winter 2010–11 heron census in Spain were quite widely spread: 16 in Extremadura, 16 in Castilla-La Mancha, eight in the Comunidad Valenciana, five in Andalucía, four in Madrid, two in the Balearics and one in Catalonia (Garrido & del Moral 2012). A bird ringed at the Guadalquivir estuary in August 1971 was in central Portugal at Castelo Branco in December 1974. Most winter sightings are of single individuals but birds are present all year round at the Albufera de Mallorca, where up to 16 were counted in November 2002 (*A. O. Balears 2002*: 122).

BLACK-CROWNED NIGHT HERON *Nycticorax nycticorax*

(Martinete Común, Goraz)

***N. n. nycticorax*. Widespread and locally common in mainland Spain south of the Cantabrian mountains. A local breeder in west-central Portugal. Introduced to Mallorca. Partial migrant, with rapidly increasing wintering populations.**

The Night Heron has a fragmented range in mainland Spain throughout the Mediterranean climatic region, with scattered colonies along the major rivers and their tributaries, and around associated reservoirs, lagoons and gravel pits, as well as in the Mediterranean coastal wetlands: notably at the Ebro delta, the Albufera de Valencia and the Santa Pola saltpans. It was an abundant nester during the late 19th century at Doñana (Saunders 1871, Irby 1875) and the Albufera de Valencia (Arévalo 1887), and both sites retain their importance for Night Herons today. In particular, Doñana has recently supported nearly 60% of the Spanish breeding population (Garrido & del Moral 2012). The small Portuguese population is concentrated in the Tagus valley with some also nesting in Coimbra and Alentejo. There is also a long-established colony of 10–15 pairs at Lisbon zoo (PA2). The Night Heron was introduced to the Albufera de Mallorca, where 54 birds from Barcelona zoo were released during 1991–93 and where a small breeding population has nested since 1993 (Vicens 2004).

The nocturnal habits of Night Herons make them easy to overlook and they are perhaps most apparent at any time of year when they leave their daytime roosts in the evening, often in sizable flocks. Nesting colonies are generally in riparian or waterside trees; notably willows, poplars, elms and alders, and sometimes in eucalyptus, but reedbeds and tamarisks may be used where no suitable trees are available. They often nest in mixed heronies alongside egrets and sometimes White Storks or Spoonbills. Roosts are also usually in dense, shady stands of trees alongside fresh water.

The Spanish population was censused in 2011, when over 5,000 pairs were found (Garrido & del Moral 2012), considerably more than previously estimated. Some 59% were in Andalucia, where 2,500 pairs were breeding at the mixed heronries or *'pajareras'* of Doñana and 300 at another three colonies also within the park periphery. A further 16% were in Catalonia: with 603 pairs at the Ebro delta, 169 at the Medas Islands colony on the Costa Brava and 57 at a colony on the river Ter. Elsewhere in the Mediterranean area there were also 328 pairs in the Comunidad Valenciana: most of them at the Albufera de Valencia (148) and at Santa Pola (135), and 50 pairs were found at the Albufera de Mallorca. Inland populations were mostly along the Guadalquivir, Guadiana, Tagus, Douro and Ebro rivers, where the largest colonies included 80 pairs in Badajoz, Extremadura,

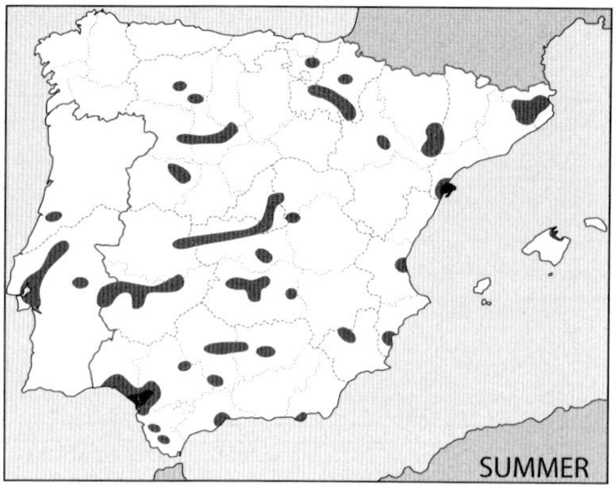

and 70 pairs at the Embalse de Azután, Castilla-La Mancha.

Former estimates of the Spanish breeding population were 2,175–2,211 pairs in 1990 (Fernández-Cruz *et al.* 1992), 1,480–2,210 pairs in 1997 (SA1) and 2,167–4,430 pairs in 2003 (SA2). The overall increase has been most evident at Doñana and at the major Mediterranean coastal wetlands, where rice cultivation has greatly expanded: a link that has also been seen in Italy and France (Fasola *et al.* 1996, 2010). Despite fluctuations occasioned by interannual variation in water levels, the Doñana population increased from 88 pairs in 1986 and 625 in 1988 (Máñez 2001), to 2,583 pairs in 2010 and 3,137 in 2011. Similarly, the population of Catalonia rose from 438–650 pairs in 1994–2001 (CA2) to 900 pairs in 2011. In contrast, the 2011 census revealed that inland populations have declined markedly, reflecting a possible population shift to the rice-growing areas of the south and east. For example, the breeding population of Navarra fell from 218–500 pairs during 1991–99 to 57 in 2011; numbers in Aragón declined from 70–200 pairs during the 1990s to just ten pairs; the 462 pairs in Castilla y León in 1988 declined to 138, and the 200 pairs in Madrid in the 1990s declined to 50 pairs (Garrido & del Moral 2012).

The Portuguese census in 2000 found 134–149 pairs (PA2), about half of them at the Paul do Boquilobo, in the Tagus valley. That colony held 60–80 pairs in 2005. The census indicates a marked decline in Portugal, especially in Alentejo: where there were at least 157 pairs in 1990 but only 11 by 2007 (PA2), following the loss of significant colonies when the Alqueva reservoir was constructed (AP). Nearly all the population was in west-central Portugal south to the Sado estuary at the time of the 2008 atlas.

Most Iberian Night Herons winter in sub-Saharan Africa but there is a sizable wintering population, chiefly in the lower Guadalquivir valley and at Mediterranean coastal wetlands. Wintering in Iberia appears to be a relatively recent phenomenon and has been an increasing trend since the early 1990s, when only small numbers remained to winter, at Doñana (SWA). The Spanish census of roosting herons in January 2011 found some 2,800 Night Herons (Garrido & del Moral *et al.* 2012). Some 52% were in Andalucía, where the largest roosts were at Doñana: at the Brazo del Este (290 birds) and the Cañada de los Pájaros (250). A further 23% were in Catalonia, where two roosts in the Ebro delta held 360 and 100 birds. There were also 349 at the Albufera de Valencia and smaller numbers elsewhere: in Castilla-La Mancha (151), Extremadura (52) and Mallorca (125). The wintering numbers in Andalucía appear to have been quite stable since the early 1990s and the 2011 count of 1,456 birds compares with the 1,426 birds counted in total at seven roosts there in winter 1992–93 (Sarasa *et al.* 1993) and the 1,190 at a roost in the Brazo del Este in November 2001 (*Ardeola* 50: 152). Wintering numbers elsewhere have evidently increased, especially in Catalonia, where wintering was first detected in 1993 (CWA). Earlier counts at other locations include 168 in winter 1994–95 on the Tagus in Toledo and 200 in winter 2001–02 at the Albufera de Valencia ((*Ardeola* 42: 212, 50: 152). Only a few individuals are known to winter in Portugal (AP).

Some birds from elsewhere in western Europe, notably from the sizable French population, cross Iberia on passage and some may remain to winter. Passage is nocturnal and birds are seldom seen on the move. Bernis (AMI) notes recoveries in Iberia of 19 French-ringed birds as well as single Dutch-ringed and Hungarian-ringed individuals, all of them ringed as pulli and recovered during passage periods. All these recoveries, and a further 17 from France, were from the eastern half of Spain or the lower Guadalquivir valley, with the exception of two in Portugal. There are also recent Spanish recoveries of one ringed in Croatia and another on the Russian shore of the Caspian Sea. Birds ringed as nestlings in Spain have been recovered in Italy (3), France (3), Malta (1), Hungary (1) and Bulgaria (1) and have also produced 22 recoveries in north Africa: 16 in Morocco and three each in Algeria and Tunisia, and nine in sub-Saharan Africa: seven in Mali and one each in Senegal and Sierra Leone. Occasional records from outside the breeding range in northern Iberia; including Galicia, Asturias and Cantabria, may involve migrants or individuals on post-breeding dispersal. With respect to the latter, ringing at Doñana has shown that many move north-west after fledging, with recoveries as far as northern Portugal (Bernis 1996a).

SQUACCO HERON *Ardeola ralloides*

(Garcilla Cangrejera, Papa-ratos-comum)

A local and generally scarce breeding species in Spain, chiefly in the south and east, and at the Albufera de Mallorca. A rare and irregular breeder in Portugal. Partial migrant.

The Squacco Heron is only at all common at the major wetlands of southern and eastern Spain, in particular at the Ebro delta, the Albufera de Valencia, El Hondo and the Santa Pola saltpans, and Doñana. A few nest, often irregularly, at other wetlands in Andalucía, Extremadura, Castilla-La Mancha, Madrid and Aragón, and exceptionally elsewhere. The Albufera de Mallorca was colonised in 1977, when 3–4 pairs bred: 11 pairs nested there in 2010 (*Ardeola* 44: 244; *A.O. Balears 2010*: 120). It is rare and irregular as a breeding species in Portugal, where the population has probably never exceeded 15 pairs in recent decades (PA2).

It frequents freshwater marshes and lagoons offering abundant aquatic vegetation, as well as ricefields. Nests are often in reeds on the periphery of mixed heronries. The 2011 Spanish census of breeding herons found 82% of birds nesting in reedbeds or reedmace, the remainder particularly favouring tamarisks; and 91% of colonies were shared with other herons and allies (Garrido & del Moral 2012).

The 2011 census found 2,076 breeding pairs, half of them at the Ebro delta, nearly a third in the Comunidad Valenciana and 15% in Andalucía, together with tiny populations in Castilla-La Mancha, Extremadura, Mallorca and Aragón (Garrido & del Moral 2012). Earlier counts have found smaller numbers in Spain. For example, a census in 1986 found only 204 pairs in Spain (Fernández-Alcázar & Fernández-Cruz 1991) but what was perhaps a more thorough survey in 1990 found 822 pairs (Fernández-Cruz *et al.* 1992): comprising 356 pairs in Catalonia, 346 pairs in Andalucía, 105 pairs in Valencia, eight pairs in the Guadiana valley and seven pairs in Alicante-Murcia. Since then the Spanish population has increased and generally continues to do so, although numbers fluctuate from year to year according to rainfall and the consequent state of the wetlands. Thus the Doñana population ranged from 20 to 430 pairs during 1996–2008 (Santoro *et al.* 2010). The 2003 atlas survey estimated the Spanish population at 850–1,100 pairs. The first reported nesting in Aragón was at a heronry near Belver de Cinca in 1997 (*Ardeola* 45: 118). The first known nesting in Extremadura was in 2000, when a pair nested along the Guadiana river in Badajoz and another was at the Arrocampo reservoir: 2–3 pairs have nested at the reservoir since then (*Quercus* 186: 10).

In Portugal the scant historical data available suggests that it was already rare during the late 19th and early 20th centuries (Tait 1887, PA2). The fieldwork for the 2008 atlas during 1997–2005 confirmed nesting at only two sites: the Paul do Boquilobo, Santarém, a traditional location, and the Ribera de Moinhos, Setúbal, with probable nesting also at the Sado estuary and at Vilamoura on the Algarve coast. One pair nested at the Alqueva reservoir in 2009 (*Anu. Ornitol.* 8: 58).

Most individuals winter in sub-Saharan Africa. Birds ringed as nestlings or juveniles in Spain, largely at Doñana or the Albufera de Valencia, have produced five recoveries in Morocco and one each in Algeria and Tunisia, as well as five in sub-Saharan Africa: two in Sierra Leone and one each in Mauritania, Mali and Ghana. Their movements are largely imperceptible but arrivals are in late March–May and departures in August–October. Some migrants from the small French population probably cross the Peninsula on passage: two birds from the Camargue have been found in Spain and a juvenile ringed at Doñana has also been recovered in the Camargue. There are occasional observations at wetlands north of the breeding range as far as and including Galicia, Asturias, Cantabria, the Basque Country and Navarra. They usually involve single birds but there were seven at the Trasona reservoir, Asturias, on 15 May 1979 and six at the Marismas de Alday, Cantabria, on 18 May 2009 (*Ardeola* 28: 150, 56: 156).

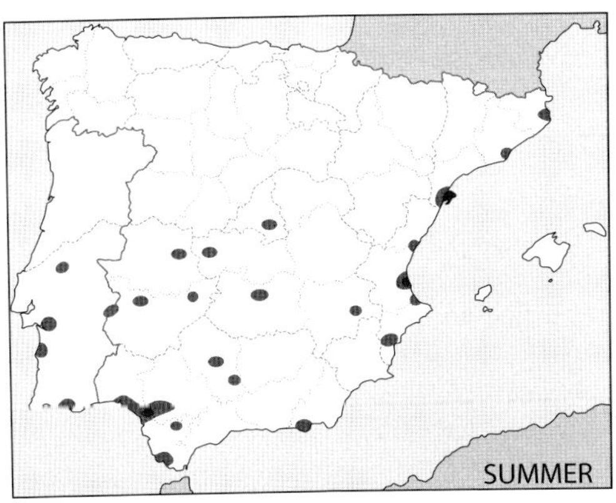

Some are recorded in winter, generally at coastal wetlands and in the lower Guadalquivir ricefields, but occasionally further inland, particularly in the Guadiana and Tagus valleys. Most observations are of single birds or just a few individuals but there was an exceptional count of 26 at La Puebla del Río, Sevilla, on 9 January 1998 (*Ardeola* 45: 242). Overwintering has been an increasing trend since the 1990s but the wintering population twenty years later is not thought to exceed 100 birds (SWA). The January 2011 census of roosting herons found only 24 birds in Andalucía, four in the Comunidad Valenciana and three in Extremadura (Garrido & del Moral 2012).

CATTLE EGRET *Bubulcus ibis*

(Garcilla Bueyera, Carraceiro)

***B. i. ibis*. Abundant in southwestern Iberia and increasingly common locally in the east and north. Introduced to the Balearic Islands. Mainly resident but dispersive.**

Cattle Egrets accompanying herds of cattle are emblematic of the pastures of southwestern Iberia, the heartland of their regional distribution. There the birds are common in western Andalucía: in the lowlands of Huelva and Cádiz and in the Guadalquivir valley, as well as in the river valleys and lowlands of Extremadura; Toledo and Ciudad Real in Castilla-La Mancha and in Madrid. They are also common in the southern half of Portugal. Scattered and tiny populations in Castilla y León are recent outposts of this western population centre. There is also a long-standing presence on the Mediterranean coast, with notable concentrations in Valencia and in the Ebro delta, from where the species has recently extended its range up the Ebro valley north to the Basque Country. The Cantabrian Range has traditionally marked its northern limits but there are now breeding populations in Cantabria and eastern Asturias, both at least partly derived from introduced birds (García-Fernández 2000). The Cattle Egret remains rare or absent as a breeding species in the north-west, the northern and eastern meseta and the arid south-east.

Cattle Egrets are less tied to waterside habitats than other heron species and they range over a diversity of open terrain, particularly outside the breeding season, although always at low altitudes. They characteristically frequent dry pastures and steppe habitats where livestock are grazed and where they accompany herds to seize the insects that they disturb. They also pluck invertebrates, principally flies but also perhaps ticks, from cows or horses. Cattle Egrets also forage in dry herbaceous habitats in the absence of livestock, seeking the large grasshoppers and other insects that may abound there. They feed in similar habitats in the lusher growth around the margins of wetlands and in ricefields, taking invertebrates, frogs and small fish. Flocks often accompany tractors that are pulling ploughs, cutting hay or harvesting – and thus unearthing or disturbing potential prey items. A relatively recent trend has been their affinity for municipal rubbish dumps, which attract flocks sometimes numbering several thousand that feed on organic discards.

The birds are more closely associated with water when nesting and many colonies, although not all, are in riparian trees such as tamarisks, willows, poplars and alders, often on islands. Such sites provide a measure of protection from disturbance and predators. Others nest in reedbeds. The 2011 breeding heron census in Spain found that 70% of colonies and nests were in trees or tamarisks and the remainder in reedbeds, reservoirs being especially favoured (Garrido & del Moral 2012). Several well known colonies use shrubs on sheltered sea cliffs and offshore rocky islets, notably in Alentejo and Algarve (PA2). A famous colony inhabited the sandstone seacliff of the Tajo de Barbate, on the Strait of Gibraltar, but this site has recently been abandoned. A small colony was found on Terreros islet, off the coast of Almería, in 1995, with 45 nests in June 1996 (*Ardeola* 43: 241). The colonies often also attract Little Egrets, other herons and White Storks, but are always numerically dominated by the Cattle Egrets. Colony size may range from a few pairs to typically several hundred and sometimes several thousand nests.

It is believed that the Cattle Egret was originally a colonist from North Africa but its presence in Iberia is long-standing. There is evidence that it occurred both in southwestern Spain and in the Mediterranean coastlands during the 16th and 17th centuries but its range then contracted (Guardiola 1999). By the late 19th century it was only common in western Andalucia, at and near Doñana, and it was then considered rare elsewhere in Spain, for example in Málaga and Valencia (Saunders 1871, Arévalo 1887, Bernis 1944, Riddell 1944). It was also common, and apparently nesting, in southern Portugal in the late 19th and early 20th centuries but it was known only as a summer visitor there, although a year-long presence was established by the 1940s (Tait 1887, AP).

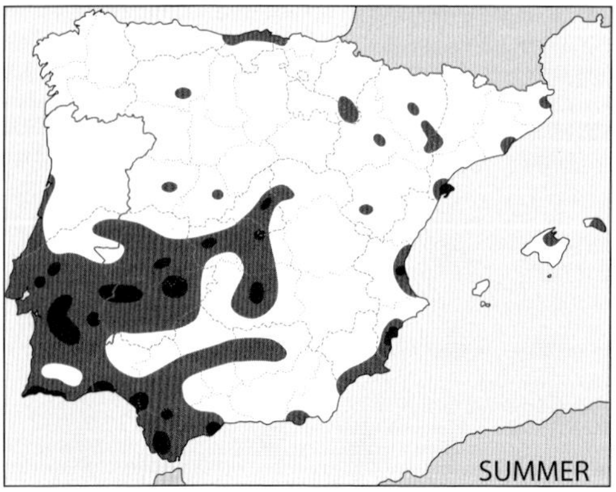

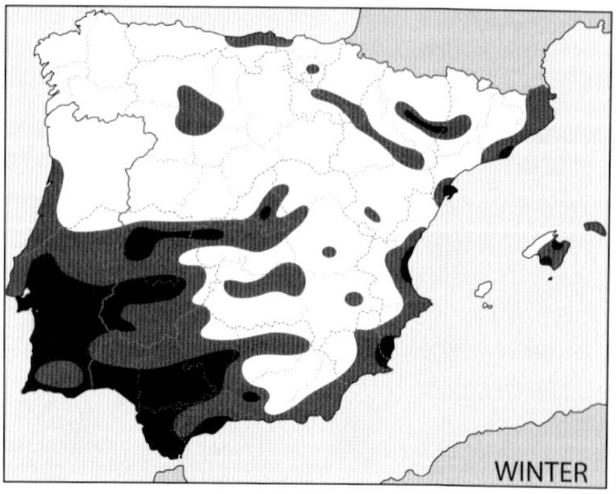

By 1960 there were already indications of increase in Andalucía, but the regular Spanish range, year-round, was still restricted to Cádiz, Sevilla and Huelva provinces (Bernis 1961a). There were however occasional records elsewhere: in Galicia, the northern Meseta, Catalonia, the Balearics and southern France. Range expansion was evident by the mid-1960s. First breeding at the Albufera de Valencia was in 1964 (Pechuán 1965) and at El Hondo, Alicante, in 1965 (Martorell 1966). Thereafter the Cattle Egret spread northwards up the Mediterranean coast, where it first nested at the Ebro delta in 1977 (Ferrer *et al.* 1986). It later ascended the Ebro valley where first nesting occurred in turn in Aragón in 1987, in Navarra in 1992, and in Álava in 1999 (Sampietro *et al.* 1998; *Ardeola* 40: 88, 46: 306). It also expanded its range along the Guadalquivir, Guadiana, Tagus and Duero valleys (Fernández-Alcázar & Fernández-Cruz 1991). In Extremadura there were six known nesting colonies, totalling over 5,400 pairs, by the early 1970s (Fernández-Cruz 1975) and first nesting in Castilla y León was in 1993: in the Duero basin at Coreses, Zamora.

The Spanish population was put at nearly 70,000 pairs in 1989 (Fernández-Cruz *et al.* 1993) after the long period of increase during the course of the 20th century. This total seems to have marked a peak in abundance and later counts and estimates have signalled a considerable decline. Between 1989 and 1991 the population was reduced to some 49,000 pairs. This rapid fall coincided with a severe drought in 1990 that brought about a (temporary) reduction in rice cultivation in Extremadura. Breeding numbers seem to have remained fairly stable until at least 2002 but the 2011 Spanish heronries census found a further decline, to around 40,000 pairs (Garrido & del Moral 2012). Most of the Spanish breeding population in 2011 (72%) was in the south-west, with 24.5% on the Mediterranean coastlands from Catalonia to Murcia and very small numbers elsewhere (Table 12). Breeding Cattle Egrets were only absent from Galicia and Asturias.

The results of the 2011 census imply an annual reduction in breeding numbers of about 2% since the late 1980s. However, numerical changes have differed markedly between regions and the steepest declines have been in the Guadiana valley in Extremadura, from 16,415 pairs in 1989 to 5,725 pairs in 1997, with some recovery to 6,655 pairs in 2011 (Garrido & del Moral 2012). The overall Andalucian population has also declined from some 20,000 pairs in 2000 to 12,000 in 2011. A large decline has also been seen in Catalonia, from 6,900 pairs in 1998: 5,577 at the Ebro delta and 1,182 on the Medes islands, to 4,825 pairs in 2011: 3,877 in the delta. The very small population of Castilla y León has been stable post-2000 but that of Castilla-La Mancha has increased substantially. In 2011 only ten nesting locations supported over 1,000 pairs. Seven were in the south-west and three on the Levant coast.

Table 12. Regional distribution of breeding Cattle Egrets in Spain in spring 2011 (Garrido & del Moral 2012).

Region	Population (pairs)	% of population	Density (pairs/100km^2)
Andalucía	12,202	30.7	14.0
Extremadura	8,320	20.9	20.0
Castilla-La Mancha	7,532	18.9	9.5
Catalonia	4,825	12.1	15.0
Comunidad Valenciana	3,652	9.2	15.7
Murcia	1,351	3.4	11.9
Madrid	800	2.0	10.0
Cantabria	600	1.5	11.3
Balearic Islands	155	0.4	3.1
Navarra	150	0.4	1.4
Aragón	84	0.2	0.2
Castilla y León	63	0.2	0.1
Basque Country	36	0.1	0.5
Total	39,770		8.0

The Cattle Egret was rare in northern Spain until the 1990s, with only occasional records of individuals or small groups from Galicia, Asturias, Cantabria, the Basque Country and the northern Ebro valley. There was an established presence in coastal Cantabria by the late 1990s, with 110 birds counted at a winter roost in Santander bay in January 1998, but at least some of the population there may originate from escapes from a nearby colony at Santillana zoo (*Ardeola* 45: 242). Birds began to appear regularly in eastern Asturias and Galicia at about the same time and attempted breeding was reported from the Basque Country in 1999, with 25 birds and a number of empty nests found at the Ullibarri reservoir in April (*Ardeola* 46: 150, 305–306). Nesting colonies were established on rocky islets just off the coast, with 147 birds present on the Isla del Castrón de Santiusti, Llanes, in 2001 – although that colony did not endure – and 350 pairs on the Isla de San Pedro, Noja, in 2013 (*Ardeola* 53: 197, 60: 514). The species has consolidated its presence on the north coast, so that a census in January 2013 found 6,900 birds at roosts in Cantabria (*Ardeola* 60: 514).

The Cattle Egret was formerly no more than a rare visitor to the Balearic islands (Munn 1921, Bernis *et al.* 1958, Ferrer *et al.* 1986) but one pair nested at the Albufera de Mallorca in 1997, following the release during 1991–93 of 20 birds originating from Barcelona zoo (*Ardeola* 44: 224, Vicens 2004), and there were 55 pairs by 2003 (*Ardeola* 50: 152, 52: 208). First breeding in Menorca was in 2006 when 30 birds were counted at a colony near Ciutadella (*Ardeola* 55: 136).

With the exception of two small colonies in Beira Baixa (Castelo Branco), all the Portuguese breeding birds are in the southern half of Portugal, mostly in Ribatejo, Alentejo and Algarve. This distribution is similar to that existing in the 1980s, although with some northward expansion. The population in Portugal showed no clear trend during 2004–09 (ACP).

Cattle Egrets forage fairly close to their colonies when nesting, disappearing from many of their regular feeding areas between April and early August. They disperse much more widely thereafter throughout the breeding range and somewhat beyond. There are small-scale movements across the Strait of Gibraltar (BG, Tellería 1981, Pineau & Giraud-Audine 1976). Some clearly travel further and there were irruptive arrivals in France of Spanish birds in the early 1980s and from 1992 onwards (OF). Most recoveries of birds ringed in Spain are local but during 1973–2010 there were 128 recoveries in Morocco, four in Algeria and one in Melilla, as well as three in France. There is also a clearly exceptional record involving a chick ringed at Barbate, Cádiz, in May 1986 that was found dead at Akersberg, Sweden (3,122km), in July 1987.

Table 13. Regional distribution of wintering Cattle Egrets in Spain in January 2011 (Garrido & del Moral 2012).

Region	No, of birds	%
Andalucía	95,303	51.9
Extremadura	31,162	17.0
Comunidad Valenciana	13,650	7.4
Madrid	12,584	6.9
Catalonia	10,175	5.5
Castilla-La Mancha	5,162	2.8
Cantabria	4,229	2.3
Murcia	3,908	2.1
Islas Baleares	3,712	2.0
Navarra	1,370	0.7
Aragón	1,174	0.6
Castilla y León	623	0.3
Asturias	376	0.2
Basque Country	203	0.1
Total	183,631	

Cattle Egrets congregate in large roosts outside the breeding season, commuting between these and their feeding grounds. Roosts are invariably in sheltered stands of trees, often along watercourses but sometimes in large parks and gardens in urban areas. Roosts may hold from a few dozen birds to several thousand, exceptionally as many as 25,000 (SA1). The wintering birds are largely found within the breeding range, comprising the southern half of the Peninsula, the Mediterranean coastlands and the Ebro depression, as well as the Balearic Islands (SWA). The January 2011 Spanish census of roosting herons found some 184,000 Cattle Egrets distributed between at least 311 roosts (Garrido & del Moral 2012). Over 50% of the birds were in

Andalucía but there were also large concentrations in Extremadura, the Comunidad Valenciana and Madrid (Table 13). Similar winter censuses in 1991–92 and early 1993 found 110,000–126,000 Cattle Egrets in Spain (Fernández-Cruz & Farinha 1992, Sarasa *et al.* 1993), so the 2011 count suggests a population increase of between 50% and 60%. A Portuguese census in 1993 found 46,407 birds.

The south-west quadrant held 78% of the wintering population, but 72% of the breeding birds, and only 15% were on the Mediterranean littoral where about 25% of breeding birds were found, an imbalance that raises questions regarding the supposed geographical separation of the two populations. Galicia was the only region without wintering birds. Fifteen roosts contained over 2,500 birds. The largest of all, with over 7,000 birds, was in Madrid and the 12,584 birds found wintering in that region greatly exceeds the 800 pairs that nested there. The attraction of Madrid seems to be the large landfill rubbish dumps of the city and most of the major roosts both there and elsewhere are near large cities, and are sometimes located at the rubbish dumps themselves, as seen in Sevilla and Córdoba. The wintering population in Andalucía was also far larger than the breeding population there, perhaps because the fluctuating water levels of Doñana may be a limiting factor on nesting activity there.

There is a marked discrepancy between the known breeding and wintering populations. Whereas the breeding population apparently declined from some 70,000 pairs to 40,000 pairs between the early 1990s and 2011, the wintering population increased from 126,000 birds to 184,000 over the same period. Counting errors are possible but seem unlikely since both colonies and roosts are conspicuous and well known. A possibility is that the relatively recent provision of landfill rubbish dumps is sustaining a substantial non-breeding population. Suitable nesting locations are not necessarily, or even often, near such dumps so birds that feed there in summer may not be able to nest that season. Conversely, the breeding population may be limited by dependence on natural resources such as the large insects on the meadows or the crayfish in the rice fields, and they may then also be subject to such adverse factors as agricultural intensification and fluctuations in water levels.

WESTERN REEF EGRET *Egretta gularis*

(Garceta Dimorfa, Garça-dos-recifes)

Rare. Recorded in Spain and Portugal.

The nominate subspecies breeds on the West African coast north to the Banc d'Arguin, Mauritania. Frequent observations of this race in Iberia and France seem to be due to natural arrivals. Such individuals have occasionally formed mixed pairs with Little Egrets that have given rise to a confusing variety of hybrids of intermediate appearance (Fábián & Sterbetz 1965, Grussu 1993, Dubois & Yésou 1995, Dies *et al.* 2001). There are accepted records of 40 individuals in France during 1981–2011, most of them clearly of the nominate race that appeared in the Mediterranean coastlands, mainly in the Camargue, between May and September. There are also 24 records from the Moroccan Atlantic coast up to 2010, 16 of them during March–April and four in September–December. Four apparently pure-bred birds and four hybrids have been reported in the Canary Islands.

The earliest Iberian observations of 'dark egrets' were all at Doñana. One was obtained at a heronry on 26 June 1956, although it was later thought that it might have been a melanistic Little Egret (*Ardeola* 3: 91–92, Bernis 1956b, Bernis 1971). Another incubating bird was photographed there in early June 1960 (*Ardeola* 6: 375–376). Later an adult was in the Marismas on 14 June 1971 (Hiraldo 1971b) and one was present from April to June 1980. The latter or a different bird was also seen in April 1981, when it was caught and identified as probably a Little Egret on biometric grounds (*Ardeola* 30: 116). Dark egrets were then seen in Cataluña, at the Ebro delta in May 1979 and at the Aiguamolls del Empordà in May 1976 and May 1984, although the possibility of melanistic Little Egrets was also raised here (Ferrer *et al.* 1986, Martínez Vilalta & Motis 1989). The existence of melanistic Little Egrets has never been proved but the possibility led to a prolonged review of the records by the Spanish rarities committee. In 2000 the committee published the details of 18 records of *gularis* individuals, four of Western Reef Egrets of indeterminate subspecies and seven of Reef Egret x Little Egret hybrids (*Ardeola* 49: 146–149). One of the apparently pure-bred birds seems to have appeared annually at the Albufera de Valencia during 1988–91 where it paired with a Little Egret during at least two seasons, giving rise to hybrid young that were grey or showed a mixture of white and dark feathers (Dies *et al.* 2001).

By 2010 there were 35 accepted records of apparently pure-bred birds, and 69 of presumed hybrids, from Spain. There are also six accepted records from Portugal, the earliest in 1995, one of which may have been a hybrid. If we accept the records that precede the establishment of the rarities committees, there are at least 43 Iberian records of pure Reef Egrets and a further 74 records of hybrid individuals. The age was noted for 23 of the pure birds, 16 of which were adults and seven were identified as definite or probable first-summers.

Occurrences are concentrated in spring and summer (Fig. 60), over 80% of the first observation dates falling during April–August. Observations from southern France and Morocco show a similar pattern, which supports the hypothesis that all these birds are seasonal visitors from West Africa (de Juana 2006), although the relative scarcity of autumn records from Iberia and Morocco is noteworthy.

Spanish records of hybrids show a very different pattern with records spread fairly evenly throughout the year (de Juana 2006).

Pure Western Reef Egrets are chiefly recorded in southern Iberia and along the east coast (Fig. 61), suggesting that a migratory corridor extends from Mauritania to southern France. There is only one record from the Balearics: at Salines, Mallorca, on 17 May 1997. There are also just two well inland: an adult in a heronry at the Borbollón reservoir, Cáceres, during June–August 1986, and another at the Alfajarín ricefields, Zaragoza, on 24 May 1993, although very recent records include at least three inland observations: in La Rioja, Navarra and Castilla-La Mancha (*Ardeola* 60: 453, RBS). Hybrid egrets are also mainly encountered in the south and east: 35% in Andalucía, 41% in the Comunidad Valenciana and 15% in Cataluña.

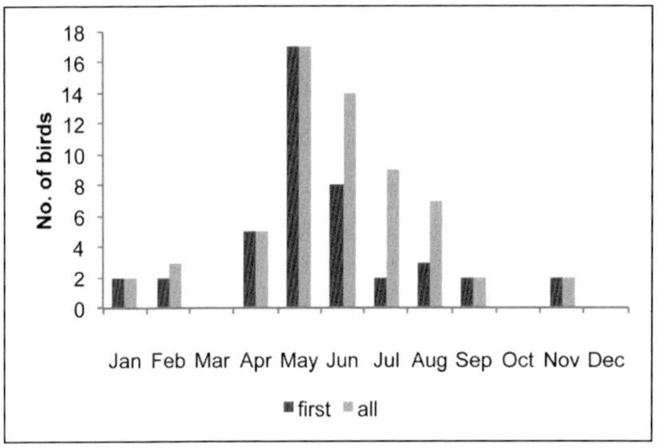

Figure 60. Records by months of apparently pure-bred Western Reef Egrets, 1976–2010 (n = 43)

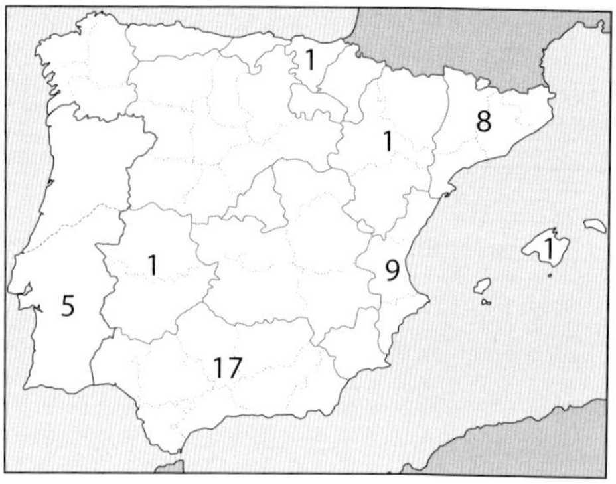

Figure 61. Records by regions of apparently pure-bred Western Reef Egrets, 1976–2010 (n = 43)

LITTLE EGRET *Egretta garzetta*

(Garceta Común, Garça-branca-pequena)

***E. g. garzetta*. A common breeding species in southwestern and eastern Iberia and increasingly so elsewhere. Resident or partial migrant.**

The Little Egret is closely associated with aquatic habitats, unlike the Cattle Egret, but the Iberian breeding distribution of the two species is quite similar. It has been making up lost ground since the low point of the 19th century, when white herons were persecuted for their plumes by the millinery trade. Saunders (1871) found it abundant at Doñana where Noble (1902) said that it was common locally, but Arévalo (1887) considered it uncommon in Spain and noted that just 'some' pairs nested at the Albufera de Valencia. Ticehurst & Whistler (1930) were unable to find any at the Ebro delta in late April and commented on the general absence of herons there, which they attributed to persecution.

By the mid 20th century it bred at scattered colonies in southwestern Spain and southern Portugal, with a separate population inhabiting the coastal wetlands of eastern Spain, but the total population then was still modest and very locally distributed. The earliest known breeding in Castilla y León was in 1953 (Bernis & Valverde 1954a) and this was followed by first nesting in Castilla-La Mancha in 1961 (Club Alcyon 1961) and in Catalonia in 1973 (Muntaner *et al.* 1983). The total Spanish population in the early 1970s was around 2,600 pairs; 2,000 of them nesting at Doñana and 400 in the Comunidad Valenciana: at the Albufera (375 pairs) and El Hondo (25), the remainder being found at a few locations in Catalonia (Ebro delta), Aragón, Extremadura, Castilla-La Mancha and Castilla y León (Fernández-Cruz 1975).

The 1997 Spanish atlas found inland colonies along the Ebro, Duero and Guadiana valleys (SA1). First nesting in Navarra occurred in 1983 (Elósegui 1985). The 2003 atlas showed further expansion with the colonisation of north coast wetlands in Cantabria and the Basque Country (SA2). The first reported nesting on the Biscayan coast was in 1998, when a colony of 17 nests was found on the islet of Ízaro, Vizcaya, although a breeding season presence on the Isla del Castrón de Santiusti, Llanes, Asturias, was known from the early 1990s (*Ardeola* 46: 150, 48: 138). Breeding in Cantabria was first noted in 2003 when at least nine pairs nested on San Pedro islet, Noja and 1–2 pairs nested in the Ría de Tina Mayor (*Ardeola* 50: 341). There were 22 nests at Noja in 2007 and 50 in 2009 (*Ardeola* 54: 391, 56: 349). The first recorded successful nesting in Galicia was in 2009, when a few pairs raised some young on an islet in the Ría de Arousa; a further attempt by eight pairs in 2010 failed (*Ardeola* 58: 200–201).

Breeding in the Balearics was first reported in 1996, when a few pairs nested at the Albufera de Mallorca, with 20–30 pairs there in 1997 (*Ardeola* 44: 244). This 'colonisation' followed the release of 25 birds originating from Barcelona zoo during 1991–93, at the same time as Night Herons and Cattle Egrets were also introduced (Vicens 2004). The population increased to 60

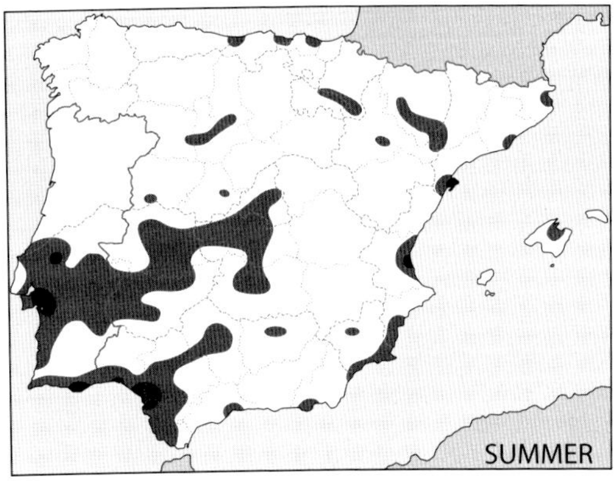

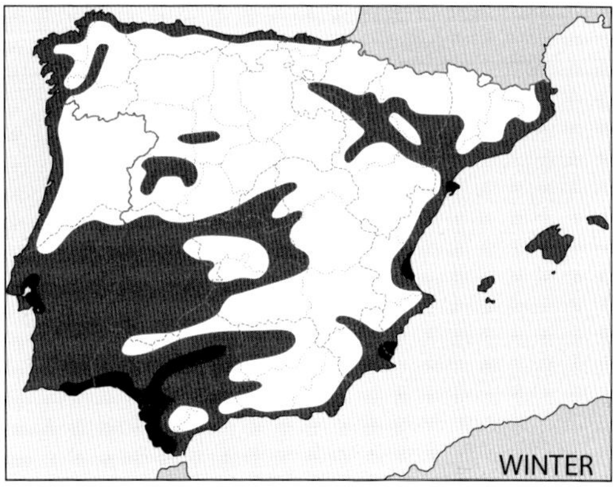

pairs in 2003 (*Ardeola* 50: 152, 52: 208). Previously the Little Egret had been described in turn as a 'very rare straggler' to the Balearics (Munn 1928), as a not uncommon visitor (Bernis *et al.* 1958) and as scarce but regular on passage and in winter with a few non-breeders remaining in summer (Ferrer *et al.* 1986).

The expansion of the Iberian breeding range has accompanied – and indeed lagged behind – northward range expansion in western Europe, through France (OF) and into the British Isles (Brown & Grice 2005). However, the spread in Iberia has been linked to the expansion of rice cultivation in the east and in the Guadalquivir, Guadiana and Tagus valleys, which has provided a large increase in suitable habitat. The impression also is that the species was formerly much more of a summer visitor whereas perhaps most of the population is now resident, although there is little historical data that could confirm this. Irby (1875) considered that it was a summer visitor to western Andalucía, arriving mainly from mid-April, with some remaining as late as November, and other historical references, including Saunders (1871) and Arévalo (1887) do not mention wintering birds. Small scale wintering was noted by Bernis (AMI) by the 1960s.

Iberian Little Egrets frequent a wide range of habitats when breeding: notably freshwater and salt marshes, riversides, lakes and reservoirs, saltpans and ricefields. They nest colonially, often alongside Cattle Egrets and other herons and allied species, typically in riparian or waterside trees and shrubs, but also in reedbeds. Some nest locally on rocky coastal islets, often with Cattle Egrets, as seen on Terreros islet, Almería and Algarve (*Ardeola*. 43: 241, PA2). Both breeding and wintering birds forage for aquatic vertebrates and invertebrates in shallow waters, typically in rivers and marshes but locally also in coastal pools on rocky shores.

The breeding population is very unevenly distributed. Recent censuses found an increase of some 25% in the Spanish breeding population, from an estimated 7,600 pairs in 1989 (Fernández-Cruz *et al.* 1992) to at least 10,400 pairs by the time of the 2003 atlas. The 2011 Spanish census found 9,347 pairs, somewhat fewer than in the 2003 estimate but probably explicable in terms of the fluctuations in breeding numbers that arise from variation in water levels, especially at Doñana. The Doñana area colonies totalled fewer than 3,500 pairs in 2011 although there were 6,000 pairs there in 1998 and over 5,000 in 2001, but only 25 pairs in the drought year of 1999 and 160 in 2005, another dry year. The 2011 results indicate relative stability in the population of the Comunidad Valenciana (1,995 pairs) but large recent increases in Catalonia (2,502 pairs) and Extremadura (424 pairs), as well as in the small populations of Murcia (233 pairs) and Castilla-La Mancha (134 pairs). The overall distribution shows a breeding presence in every Spanish region but over half the birds are in the Mediterranean littoral regions, 47% in south-west quadrant and very few elsewhere. The largest breeding concentrations in 2011 were at the Ebro delta (2,411 pairs), the '*Pajareras*' in Doñana biological reserve (1,880) and the Albufera de Valencia (1,849): no other locations held over 300 pairs (Garrido & del Moral 2012).

The Portuguese breeding range is almost entirely in the southern half of the country, with concentrations at the Paul do Boquilobo, Santarém; Montemor-o-Novo, Évora, and coastal Algarve (PA2). The population was put at 1,300 pairs in 1997 (Farinha & Costa 1999), although Dias (1991) estimated that were 3,000–5,000 pairs. A more recent census found 2,188–2,578 pairs in Portugal during 2000–04 (PA2) and concluded that no clear trends are apparent since some of the earlier counts may have overestimated the population.

Little Egrets are more widespread outside the breeding season, and some move south to winter in Africa. A northward post-breeding movement takes some birds to the coastal areas of the north and west, where some remain to winter. The birds are usually widely but thinly distributed but local concentrations occur; for example, there were 400 at a pool near Hondarribia airport, Gipuzkoa, on 15 October 1992, and 175 at the same site on 24 August 1993, and 300 at Plaiaundi, Gipuzkoa, on 13 September 1998 (*Ardeola* 44: 244, 49: 176). In Portugal there were some 300 at the Carvalhal ricefields, Setúbal, on 30

September 2009 (*Anu. Ornitol.* 8: 59). Valverde & Weickert (1956) found that birds ringed as pulli in Doñana moved chiefly north-west during July–September, throughout western Andalucía and Portugal, although a few reached as far afield as Italy, Madeira, the Canary Islands and Morocco. More remarkable are two transatlantic recoveries of individuals ringed as pulli in Doñana: one ringed in July 1956 was recovered in Trinidad in January 1957 and one ringed in July 1962 was recovered in Martinique in October of the same year (AMI).

Since the 1960s there have been ten recoveries in France, three in Italy and one in Poland of birds ringed in Spain. Six birds ringed at the Albufera de Valencia moved to the Ebro delta (4) and the Camargue (2). Thirty ringed in southern France, two from northern Italy and single birds from Serbia and Morocco have been found in Spain. There are also two recent recoveries involving the recently established English population: a chick ringed in Kent in May 2005 was found at Irún, Gipuzkoa, the following October and a chick ringed in Norfolk in May 2009 was recovered at Oleiros, A Coruña, in January 2010.

Ringing recoveries and observations of wing-tagged birds (Bartolomé *et al.* 1996) have shown that some birds from both eastern and western Iberia coincide in northwestern Africa in winter, although these two subpopulations are otherwise discrete and the eastern birds range north to southern France. There are 29 recoveries of Spanish-ringed pulli from North Africa: 24 in Morocco and five in Algeria, and six from sub-Saharan Africa: two each from Mauritania and Senegal and one each from Gambia and Guinea Bissau. Only small-scale movements are observed, including occasional passage of small flocks across the Strait of Gibraltar mainly in August–September and March–April.

The major wintering concentrations in Iberia are in the eastern and southwestern coastal areas and the lower Guadalquivir valley, with small numbers inland. Birds concentrate at winter roosts of up to several hundred individuals, dispersing to feeding areas during the day. The Spanish winter population was estimated at 10,000 birds by Sarasa *et al.* (1993) and is substantially higher in some years at least. The January waterbird censuses in Spain during 1990–2009 found 9,384 Little Egrets on average (range 3,176–18,505) and the population increased markedly during this period in all regions (AAE). The Albufera de Valencia alone held 5,300 Little Egrets in January 2009 (*A. O. C. Valenciana 2009*: 30). In addition to those at the Albufera, the largest wintering concentrations in Spain during 1990–2009 were at the Ebro delta (up to 5,822 birds), Doñana (up to 6,538), the Cádiz Bay marismas (up to 1,203) and other coastal wetlands of the south-west and Levant coasts, the remaining birds occurring inland along the major river valleys and on the Biscay and Galician coasts (AAE). The January Balearic population averaged 230 birds during this period, most of them at the Albufera de Mallorca (AAE). The January 2011 census of heron roosts provides a good snapshot of the recent distribution of wintering Little Egrets in Spain. They totalled 16,960 individuals, at the upper end of the range of recent winter counts. Wintering birds were similarly distributed to the breeding population, with about 49% in the southwestern quadrant and about 47% on the Mediterranean littoral. The largest roosts were at the Albufera de Valencia (3,425 birds), the Sevilla sector of Doñana (2,401 birds), the Ebro delta (1,703) and the Capitán ricefields, Sevilla (1,600): all four are locations offering ready access to ricefields. No other roosts exceeded 500 birds and most were much smaller (Garrido & del Moral 2012).

The wintering population includes a proportion of birds from elsewhere in western Europe, principally from France (AMI). However, the Little Egret winters in large numbers in France, where it was formerly a summer visitor, with a mid-December population of around 20,000 birds in the early 21st century, declining to 8,000–10,000 by mid January, presumably due to late displacement further south (OF).

GREAT WHITE EGRET *Egretta alba*

(Garceta Grande, Garça-branca-grande)

***E. a. alba*. A scarce and local breeder in mainland Spain. Increasingly widespread and numerous in winter throughout the region.**

This species declined severely in Europe during the 19th and early 20th centuries, largely as a result of the commercial demand for ornamental *aigrette* plumes, but has been recovering there since the 1970s (BWP, BE). Its principal European populations are in southern Russia, Ukraine and Hungary but there are also hundreds of breeding pairs in Romania and Austria. The breeding range has extended to western Europe since the 1970s, with first breeding recorded in The Netherlands in 1978, Italy in 1992, France in 1994 and Spain in 1997. It was formerly only a vagrant to Morocco but it has appeared there increasingly often since the mid 1980s and there were 86 records (159+ birds) up to 2012 (Bergier *et al.* 2013).

A few Spanish records date from the late 19th and early 20th centuries: from Doñana (Saunders 1871), the Ebro delta (A. Brehm *in* Maluquer 1971), the Albufera de Valencia (Arévalo 1887), the Laguna de La Janda (Irby 1895) and Mallorca (Munn 1925), but there is then a gap of 71 years until the next recorded occurrences in Iberia: in 1964 in the Balearic Islands and in 1965 on the Peninsula (de Juana & Ferrer 1996). The first recorded occurrences in Portugal were in 1971 and 1988 (AP). Since then records have increased virtually exponentially. A review of Spanish records to 1993, by when the Great White Egret could be regarded as a regular winterer, found 119 reports (156 birds), only eight of which were during 1964–1974 (de Juana & Ferrer 1996). These were widely spread geographically but with major concentrations at the Catalan wetlands: notably the Ebro delta, in Mallorca and in the Comunidad Valenciana. Although 72% were in October–March, some were reported in all months, which suggested future breeding. This occurred for the first time in 1997, on the Isla de Buda in the Ebro delta, where a pair fledged young (Martínez Vilalta *et al.* 1998). A downy chick was found at Doñana in 1998 and there were signs of nesting at the Albufera de Valencia in 2001 (SA2).

There was a particular upsurge in records during the 1990s, when there were 117 published reports in *Ardeola* and eight in *Pardela* (Fig. 62). Records outside the wintering period also increased during the same period: those between March and August comprised 12% of records during 1991–95, 23% in 1996–99 and 31% in 2000. First observations during the 1990s were principally during September– December (69%) and peak counts were in November–December (Fig. 63).

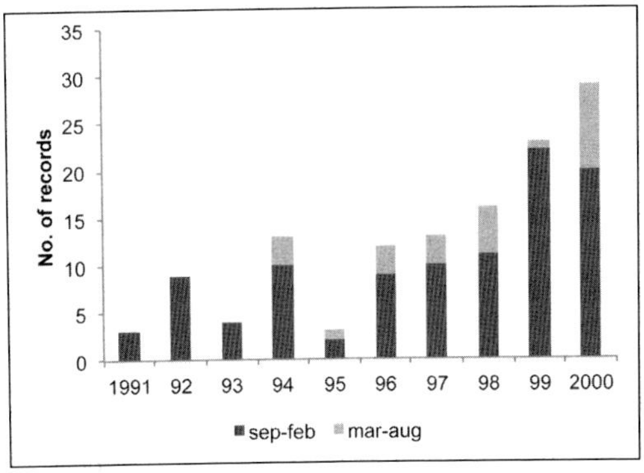

Figure 62. Annual records in Spain, 1991–2000 (n = 125)

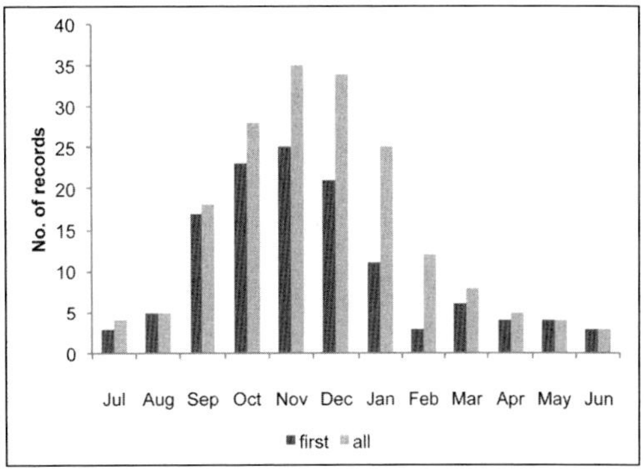

Figure 63. Records by month, 1991–2000 (n = 125)

The increased presence in Iberia has continued apace post-2000 and clearly many birds are no longer reported from areas where they are now a common sight. The 2011 census of breeding herons in Spain found 53 pairs (Garrido & del Moral 2012). They comprised 28 pairs distributed between five locations at Doñana, 11 at the Ebro delta, seven at the Tablas de Daimiel national park, and scattered pairs at six other sites: the Aiguamolls de l'Empordà; the Laguna del Cañizar, Teruel; the Arrocampo reservoir, Cáceres (two pairs); the Azután reservoir, Toledo; the El Vicario reservoir, Ciudad Real, and on the Guadiana river in Ciudad Real (Garrido & del Moral 2012). None bred at the Albufera de Valencia in 2011, however. Nesting at El Vicario reservoir dates from 1999 (Velasco *et al.* 2007), but first nesting at Azután was in 2009 and at Arrocampo in 2010 (*Ardeola* 56: 156, 58: 489). A pair nested at the Sariñena lake, Huesca, in 2010 (*Heraldo de Aragón* 11 July 2010). About half of the 2011 nests were in reedbeds or reedmace, the remainder using tamarisks or other trees, virtually always within mixed heronries.

The January 2011 counts of roosting herons found 1,459 winterers in Spain (Garrido & del Moral 2012). The Ebro delta is the principal wintering region: there were 468 birds there in January 2011 and the delta generally accommodates 80–90% of the wintering population of Catalonia and over half of that of Spain. A further 300–400 were in Doñana and other wetlands of western Andalucía, *c.* 310 within the Ebro basin, 86 at the Albufera de Valencia, 60+ along the middle reaches of the Guadiana and lesser numbers inland along the Tagus, Duero and Guadalquivir and at other coastal wetlands in Andalucía and in the Balearics. In Portugal 60 birds were censused wintering in 2007, the main sites being the Lagoa de Santo André, the Lagoa dos Patos, the Sado estuary and the Paúl de Barroca, Alcochete (AP).

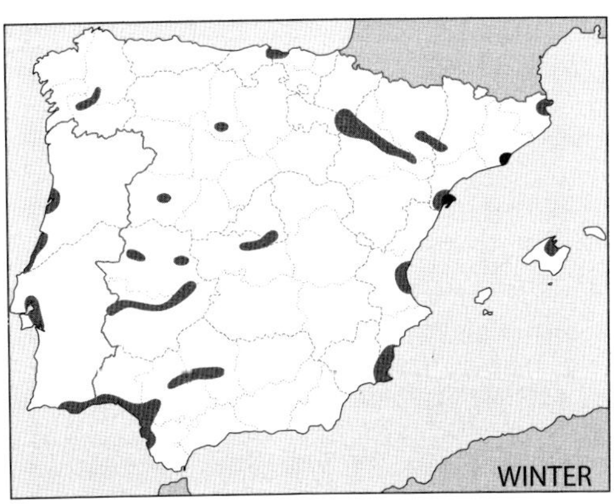

Notable recent concentrations of winterers have included 190 birds at the Albufera de Valencia in January 2012, as well as 56 at Irún, Gipuzkoa; 42 at the Ricobayo reservoir, Zamora; 29 at Casatejada, Cáceres; 26 at Figueira da Foz, Coimbra, and 21 at Villaviciosa, Asturias (*Ardeola* 58: 489, 59: 419; *Anu. Ornitol.* 7: 76). They occur at a broad range of wetland types that includes reservoirs, lakes, rivers, estuaries and saltpans as well as on irrigated crops, notably ricefields. One at the Mondego estuary that had been colour-ringed as a nestling at the Lac du Grand Lieu, France (AP) provides the only available indication of the origin of the wintering birds.

GREY HERON

Ardea cinerea

(Garza Real, Garça-real-comum)

***A. c. cinerea*. Increasingly widespread as a breeding species. Common on passage and in winter throughout the Peninsula and in the Balearic Islands.**

The Grey Heron was an extremely local breeding species in Iberia until the last quarter of the 20th century. Saunders (1871) said that it bred in the Guadalquivir estuary and Irby (1895) noted some resident pairs near the former Laguna de la Janda, Cádiz. Bernis & Valverde (1954a) knew of only three Spanish colonies, in Andalucía and Castilla y León, in 1953, totalling just 60 pairs, but Campos & Fernández-Cruz (2006) were aware of four colonies in the early 1950s, totalling 168 pairs. Breeding numbers at Doñana began to increase in the mid 20th century, from just six pairs in 1952 to over 200 in 1961 (Bernis 1961a) and 335 in 1963 (Fernández Cruz 1975). Away from Andalucía and Castilla y León, breeding was first noted in Extremadura in 1957 (Bernis *et al.* 1958); in Portugal near Évora, Alentejo, in the early 1970s (Vicente 1974); in Catalonia at the Ebro delta in 1975 or 1976 (CA1); in the Comunidad Valenciana at the Albufera in 1984 (*Ardeola* 32: 410); in Navarra at the Laguna de Pitillas in 1990 (*Ardeola* 38: 327–350); in Murcia in 1990 (SA2); in La Rioja at La Grajera reservoir, in 1994 (Campos & Fernández-Cruz 2006); in the Basque Country at Lekeitio and Areatza, in 1996 (Zuberogoitia & Torres 1998); in Almería in 1998; in Jaén in 1999; in Granada and Asturias in 2001 (SA2); in Galicia at Sandiás, Ourense, in 2002 (*Ardeola* 50: 342); and in Madrid at Algete gravel pit in 2004 (*A. O. Madrid 2004*: 192). Breeding in Cantabria was initiated in the 1990s by free-flying birds from Santillana zoo, with first nesting inland at the Ebro reservoir in 2003 (*Ardeola* 50: 341–342). The free-flying colony at Barcelona zoo, which numbered 128 pairs in 2011, is noteworthy (CA1, Garrido & del Moral 2012).

In general, the breeding population has increased greatly (Fig. 64) and spread extensively since 1980, perhaps due in part to the legal protection afforded to the species since 1975 (SA1, Campos & Fernández-Cruz 2006). A 1986 census in Spain found 786 pairs, distributed between western Andalucía (228 pairs); the Duero valley (304 pairs) and the Tagus valley (165 pairs) in western Castilla y León, and the Albufera de Valencia (89 pairs) (Fernández-Alcázar & Fernández-Cruz 1991). By 2000 the Spanish population numbered 4,790 pairs and 75 colonies (Campos & Fernández-Cruz 2006) and there were over 5,000 pairs during 2000–02 (SA2). Local censuses emphasise the rapid increases in numbers. For example, the population of the Albufera de Valencia went from 320 to 1,095 pairs during 1994–2000 and that of Doñana increased even more rapidly, from 495 pairs to 1,281 pairs during 1997–2000. There have also been some local declines, however: only 525 pairs were found at the Albufera in 2009 (*A. O. C. Valenciana 2009*: 32); but the breeding populations of aquatic species in Iberia are prone to fluctuate.

The most recent census of breeding Grey Herons, in 2011, found 6,994 pairs in 241 colonies, although some nested in isolation. They nested in every region except Asturias. The northern coastal regions generally accommodate small populations and inland regions, notably Extremadura and Castilla y León are relatively more important to this species than to other herons. In general some 65% nested within the western river basins.

Breeding Grey Herons are more widely distributed than other heron species, which tend to be concentrated at Doñana, the Albufera de Valencia and the Ebro delta. In 2011 the first two of these locations held the two largest concentrations of Grey Herons: 795 pairs at the Albufera and 768 pairs at the '*pajareras*' of the Doñana biological reserve. However, each of these only accounted for little more than 10% of the Spanish population. Only a further ten widely distributed colonies comprised 100 pairs or more, the largest of these being 320 at La Grajera reservoir. The Ebro delta, with 77 pairs or 1% of the population is relatively unimportant as a breeding location (Garrido & del Moral 2012).

Clavell (2002) reported rare instances of breeding in Mallorca, where Munn (1921) considered that some nested on the

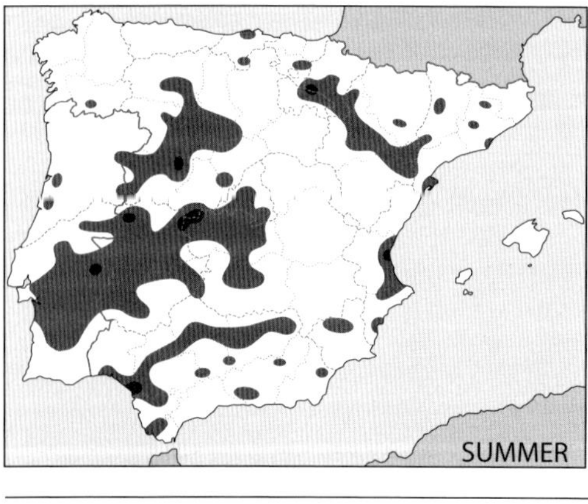

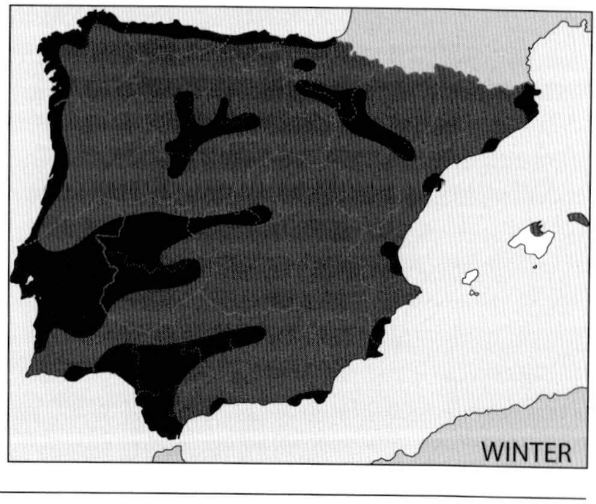

cliffs of the Alcudia peninsula and probably within the Albufera, although his account is unconvincing. There is no current breeding presence in the Balearic Islands.

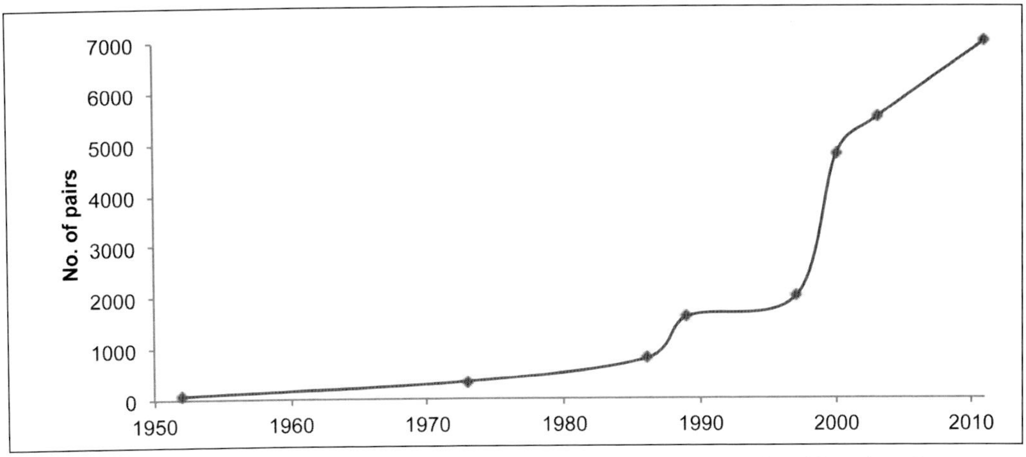

Figure 64. The increase in the Spanish Grey Heron population during 1952–2011 (SA2, Garrido & del Moral 2012).

The Portuguese population has also increased during recent decades. It was put at 300 pairs by Dias (1991) but a national census during 2000–04 found 1,414 pairs (PA2). The breeding range has expanded northwards but the majority remain south of the Tagus and chiefly in Alentejo. Elsewhere there are colonies in the Ria de Aveiro, on the Douro Internacional and in Baixo Mondego, but non-breeders occur widely throughout the country in the breeding season (PA2).

Grey Herons are versatile birds that hunt fish, amphibians and large invertebrates in a broad range of both freshwater and saltwater habitats: including estuaries, saltpans, ricefields, rivers, large and small lakes and ponds, reservoirs and irrigation channels. They will also attend fish farms, where they are obviously unwelcome and may face persecution. They typically nest in waterside trees, often alongside White Storks and other herons, but there are large reedbed colonies where suitable trees are lacking, as at the Albufera de Valencia.

Grey Herons are much more widespread and numerous throughout mainland Iberia, and also in the Balearic Islands, outside the nesting season, when large numbers enter the Peninsula from elsewhere in Europe and the breeding birds disperse more widely. Wintering birds are most abundant at the principal coastal wetlands and within the major river valleys, and local concentrations occur. Nevertheless, the wintering population is very widely dispersed and birds occur throughout Iberia, at wetlands of all sizes. The 1990–2009 January counts in Spain found 12,354 Grey Herons on average (range 4,436–18,283 ; AAE), but many birds will have been missed at small sites that were not surveyed. The January counts have shown a steady increase in wintering numbers since at least 1980. The largest concentrations found during the January surveys were at the Ebro delta (up to 5,383 birds), Doñana (up to 2,684), the Albufera de Valencia (up to 2,397) and the Cádiz Bay marismas (up to 2,201), each of which accommodated over 5% of the wintering Spanish population on average. No other sites held more than a few hundred Grey Herons and most of them supported much fewer (AAE).

The combined results of the January 2010 waterbird census (AAE) and the January 2011 heron roost census put the Spanish wintering population at 16,862 Grey Herons (Garrido & del Moral 2012), more than half of them in the Mediterranean littoral regions and with notable concentrations in Andalucía and the Ebro valley (Table GH2). Once again the largest numbers were at the Ebro delta (4,243 birds), the Albufera de Valencia (2,058), Doñana (1,025) and Cádiz Bay (792), and there were also 710 at the Marjal del Moro and 530 on the Río Tormes, Salamanca.

Campos *et al.* (2001) have analysed Iberian recoveries of foreign-ringed birds that were ringed as pulli. Some birds from the whole of Europe occur in Iberia or cross the region to winter in Africa. However, although recoveries of those from western Europe are evenly dispersed throughout the region, those from eastern Europe are principally recovered on the Mediterranean coast. It may be significant that there has been an ongoing decline in the proportion of recoveries of Grey Herons of Scandinavian origin: in Spain these amounted to 32.7% of all recoveries up to the mid 1960s (AMI), but only 10.5% of those during the latter part of the 20th century (AI1) and there were none at all during 2002–07. The suggestion is that northern Grey Herons may be wintering less far south than used to be the case, another instance of 'short-stopping' migration.

Campos *et al.* (2001) also considered movements within Iberia of birds ringed as pulli within the Peninsula. Dispersal patterns differed between regions. Those from colonies in western Spain dispersed to the western half of the Peninsula. Those from southern Spain settled relatively near their natal colonies or in Morocco, although one moved to Italy. However, those ringed in eastern Spain, at the Albufera de Valencia, moved throughout the Peninsula and, in a few cases, beyond: a chick ringed in May 1984 was recovered in Poland in September 1985 and another ringed in March 1986 was recovered in Senegal in September 1989. A Spanish bird has also been found in Mali. Another interesting recovery of a Spanish-bred bird is of a chick that was ringed at Castrejón reservoir, Toledo, in May 2007 and recovered in Tenerife, Canary Islands, the following August.

Small flocks of Grey Herons are recorded on passage at the Strait, chiefly during March–April and September–October (Pineau & Giraud-Audine 1979, BG, Tellería 1981) and along coasts elsewhere. A few hundred are seen entering Spain each autumn via the western Pyrenean passes, mainly during September and the first half of October. Passage is most evident in the form of changes in numbers present at wetlands, however, and birds seen on the move represent a small fraction of migrants. Many presumably move at night, unobserved.

PURPLE HERON *Ardea purpurea*

(Garza Imperial, Garça-imperial)

***A. p. purpurea*. Breeds at scattered wetlands in Spain, Portugal and Mallorca, where locally common. Principally a summer visitor.**

The Purple Heron is characteristic of all the major Iberian wetlands south of the Cantabrian Range and it also nests at smaller sites that offer reedbeds of some size. The principal concentrations in Spain are at Doñana, the Ebro delta, the Albufera de Valencia and the Albufera de Mallorca. It also nests widely throughout the Guadalquivir and Ebro basins, along the Tagus and Guadiana valleys, at coastal wetlands in the east and south-west and at the wetlands of La Mancha (SA2). A few pairs nest in the Biscay coastlands: at Santoña, the Ullibarri-Gamboa reservoir and the Lagunas de Laguardia, Álava. The distribution in Portugal is principally coastal, from the Minho estuary south to Algarve, with concentrations around the Ria de Aveiro and the Tagus estuary (PA2). Some also breed at a few inland sites in Portugal, chiefly in the southern half.

The Purple Heron is closely associated with reedbeds, where it often nests in small scattered colonies. Tree-nesting, other than in tamarisks at wetlands, is comparatively rare. The nesting habitat comprises large and small freshwater wetlands with dense peripheral vegetation, although birds may travel some distance to feed along rivers and irrigation channels and in saltpans and ricefields. The expansion of rice cultivation, especially in the Guadalquivir and Ebro estuaries and at the Albufera de Valencia, has provided them with much additional suitable habitat in recent years.

Both the 1997 and 2003 Spanish breeding bird atlases put the Spanish population at around 2,000 pairs. However, the spring 2011 census of Spanish heronries (Garrido & del Moral 2012) found 5,379 pairs, an increase of some 170% since 2003, in what may have been a particularly favourable year for the species. The regional distribution of the 2011 nesting population (Table 14) shows that the great majority, nearly 70%, were in Andalucía, most of them at Doñana, where the 3,659 pairs censused was a local record. Catalonia and the Comunidad Valenciana accounted for a further 15% of the population, with much smaller numbers elsewhere and only a very minor presence in the Biscay coastlands.

The Purple Heron has enjoyed a population recovery since the 1970s. It appears to have been relatively common in the past, at least in comparison with other herons, according to Saunders (1871), Arévalo (1887), Noble (1902), Witherby (1930) and others. The Spanish population seems to have declined strongly during the mid 20th century. Wetland drainage has been a serious cause of decline. The population of Castilla-La Mancha was 140 pairs in 1995–96 and 159 pairs in 2011 but would have been considerably greater in the heyday of the marshlands of La Mancha, now much reduced in extent. Most of the lost wetlands that were drained during the 20th century, notably La Janda and La Nava (now partly restored), held large numbers. Pesticides seem also to have had a serious effect at Doñana (Alberto & Pena 1981) and at the Ebro delta: where DDT application is thought to have reduced Purple Heron numbers from over 1,000 pairs in the 1960s to 50–60 in 1972–73 (CA1). Direct persecution has also been a problem locally, as at El Taray, Toledo (Bernis 1952). It seems likely that the Spanish population was less than 1,000 pairs by the early 1970s and the Purple Heron was listed as 'Vulnerable' in the Spanish Red Book.

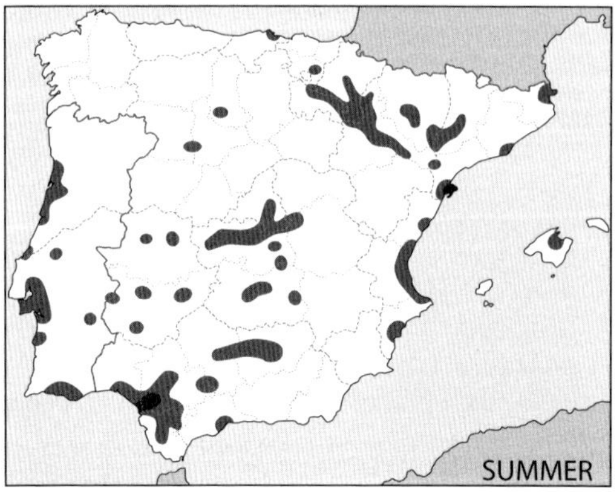

The numbers nesting at Doñana and in La Mancha especially may fluctuate considerably from year to year according to water availability. Thus the Doñana population ranged from 136 pairs to over 3,000 during 1996–2011 (Santoro *et al.* 2010, Garrido & del Moral 2012). The fortunes of the population of the Ebro delta have varied, declining from *c.* 1,000 pairs in the 1960s to just 60 pairs in 1973, but since recovering steadily to 577 pairs in 1998 and 527 pairs in 2011. The Comunidad Valenciana has a fluctuating population that numbered 191 pairs in 2000, a large proportion of them at the Albufera de Valencia where there were 59 pairs in 2002. The same region held 114 pairs in 2009 but the main colony (40 pairs) was then at Pego-Oliva (*A. O. C. Valenciana 2009*: 32). Extremadura held 70–80 pairs in 2002, chiefly at the Arrocampo reservoir, and these had increased to 105 pairs in 2011. A particularly noteworthy increase has been seen in Castilla y León, from just one or two pairs in 1995–97 to 134 pairs in 2011 (Garrido & del Moral 2012).

Table 14. Regional distribution of breeding Purple Herons in Spain in spring 2011 (Garrido & del Moral 2012).

Region	No, of pairs	%
Andalucía	3,748	69.7
Catalonia	702	13.1
Castilla-La Mancha	159	3.0
Castilla y León	147	2.7
C. Valenciana	119	2.2
Extremadura	105	2.0
Navarra	96	1.8
Aragón	81	1.5
Madrid	78	1.5
La Rioja	68	1.3
Islas Baleares	59	1.1
Basque Country	14	0.3
Galicia	2	<0.1
Cantabria	1	<0.1
Total	5,379	

Fluctuations have been noted in Navarra: where 96 pairs in 1989 had increased to 242 pairs in 1994 but declined to 96 pairs again in 2011. The population of the Balearic Islands has also declined recently: the Albufera de Mallorca held 60–65 pairs in 1991, 95–105 pairs in 2000, and up to 120 pairs in 2003 (*Ardeola* 52: 208, Vicens 2004) but the 2011 count found only 59 pairs in the archipelago. The tiny populations of Cantabria and the Basque Country are of recent origin but appear to be increasing and expanding northwards. Breeding in the northern coastlands was first recorded at the Marismas de Victoria, Noja, in 1998, when two pairs nested; there were four pairs there in 2007 and three in 2013 (*Ardeola* 47: 162, 60: 515). The first known breeding within the Basque Country was at Laguardia, where a pair nested in 1999 (*Ardeola* 46: 306).

The Portuguese population is known to fluctuate although there are indications of an overall increase since the 1990s (AP). A national census in 2000 found 319–399 pairs, including 155–188 pairs at the Ria de Aveiro and 60–75 pairs at the Tagus estuary. The Aveiro population was estimated at 200–250 pairs in 2004–05 (PA2). The national estimate has recently been revised upwards to 353–490 pairs (AP).

The great majority of Purple Herons winter in sub-Saharan Africa. There are 57 extra-regional recoveries of birds ringed in Spain, most of them marked as chicks. Of these, 31 were in North Africa, where many will have been in transit: 26 in Morocco, two each in Algeria and the Canary Islands and one in Tunisia, and 16 were found in sub-Saharan west Africa:

five in Mali, three each in Senegal and Liberia, two in Sierra Leone, and single birds in Cape Verde, Guinea Bissau and Guinea Conakry. There are also eight recoveries from France and a chick ringed at Doñana in June 1969 was recovered in Malta in June 1972.

Migrants enter the Peninsula during March–May and depart during late July–October. Migrating individuals and flocks of up to ten are occasionally seen at the Strait of Gibraltar (BG) and elsewhere, but visible passage is nowhere prominent and most movements are probably nocturnal and on a broad front. Observations in Portugal have shown that adults generally depart in late July–August with juveniles lingering into September–October (AP). Post-breeding concentrations are sometimes evident: there is a count of 114 flying to roost at the Paul do Taipal, Coimbra, on 10 July 2001 (AP). Some Purple Herons from other European countries cross Iberia on passage and there are recoveries of birds ringed in The Netherlands (58), France (25), Switzerland (6) and Italy (1). Purple Herons may appear at wetlands where they do not nest during post-breeding dispersal. They also occur widely but sparsely across the Peninsula during the passage periods, with occasional records from north coast wetlands in Galicia, Asturias, Cantabria and the Basque Country.

A few remain in winter, chiefly at Doñana, but individuals are reported occasionally elsewhere, generally at coastal sites. The wintering population is extremely small: for example, the January 2011 surveys found only three birds in Spain and no more than 20 are thought to occur in any year (Garrido & del Moral 2012). A count of five at the Caño del Guadiamar, Sevilla, on 21 December 2003 (*Ardeola* 51: 246) is noteworthy. A first-winter bird at La Granda reservoir, Asturias, on 31 January 2004 (*Ardeola* 53: 198) was unusually far north.

BLACK STORK *Ciconia nigra*

(Cigüeña Negra, Cegonha-preta)

Widespread and locally common as a breeding species in west-central Spain and eastern Portugal. More widespread on passage. Partial migrant.

This is another of the iconic species of the wilder parts of Iberia, where it is most often seen soaring over a rocky river gorge or hunting alongside a pristine river. The core breeding distribution is in west-central Iberia, where it is associated with the Duero, Tagus and Guadiana rivers and their tributaries, where these cross the wilder areas of western Castilla y León, western Castilla-La Mancha, Madrid, northern Andalucía and, especially, Extremadura, together with adjacent parts of eastern Portugal.

Over 60% of pairs nest on cliffs, in particular within riverine gorges but also on rocky outcrops in the sierras (Pizarro & Ferrero 1999, Cano 2012). Here they require large ledges, often under protective overhangs, to site their nests, which are generally occupied traditionally for many years. The remaining birds nest in trees, particularly within cork oak dehesas but also in pines. Such tree-nesters are generally harder to detect, and census, than the cliff-nesting birds. In addition to suitable nest sites, the breeding range typically offers large tracts of relatively undisturbed habitat including perennial watercourses for foraging. The birds hunt fish, amphibians or large invertebrates within large and small rivers and along the shores of lakes and reservoirs, as well as in damp meadows along watercourses. Such habitats are largely unavailable in the drier terrain of eastern and southernmost Spain but the birds' absence from the cooler regions of the north is harder to explain. Some recent range expansion northwards has, however, been seen in Portugal and western Castilla y León, with some nesting north to Sanabria. No breeding pairs were found in Madrid during the 1987 census (González 1987) but there were 12 pairs there in 2000 (SA2).

It is not numerous in Iberia but the population seems to be increasing slowly. Early accounts suggest that it was very scarce during the late 19th century and until the mid 20th century, with breeding records from the Montes de Toledo (Saunders 1871, Chapman & Buck 1893), Cáceres province (Witherby 1928, Gil Lletget 1945) and Coria del Río, in the Guadalquivir delta (Valverde 1960). There are also indications of possible breeding in Burgos province, near the Silos monastery, around 1950 (Valverde 1956). The earliest comprehensive census, in 1987, found *c.* 175 pairs (González 1987). By the early 1990s the number was put at 250 pairs (San Segundo 1992), with 300 pairs by the mid 1990s (Heredia & Soto-Largo 1996), although this latter increase has been attributed to more thorough prospection (SA2). However, the most recent assessment of the Spanish population during 1997–2002 found 387 pairs, including 322 definite and 65 possible records of nesting (LR, Cano *et al.* 2006).

Early records from Portugal are anecdotal and do not give any clear indication of former abundance. Tait (1887) referred to a specimen in the Lisbon Museum, obtained near Evora, as well as to 'many' in the Coimbra Museum from Alentejo, where he thought the species bred. Some decline and range contraction clearly occurred during the first half of the 20th century since published records refer to specific instances of nesting. For example, Coverley found two nests with eggs at Portel, Alentejo, in April 1938 and nesting occurred at the Sado estuary in the 1930s and early 1940s (AP). The first census in Portugal during 1995–97 found 83–96 pairs (Rosa *et al.* 2001), as well as several others whose territories were partly in Spain. The 2002–04

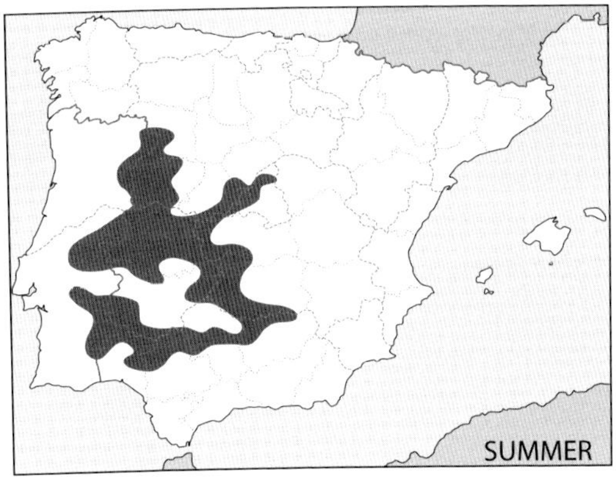

census of the Portuguese population found 97–115 pairs 46% being in the Tagus valley, 27% in the Duero and 24% in the Guadiana; the remainder were in the Mondego and Sado basins (PA2).

The total Iberian breeding population early in the 21st century was therefore about 500 pairs and is now almost certainly somewhat greater. Some 40% are in Extremadura, 23% in Portugal, 15% in Castilla y León, 13% in Andalucía, 6% in Castilla-La Mancha and 3% in Madrid. An assessment by Cano *et al.* (2006) considers the populations of Portugal and Extremadura to be largely stable, those of Castilla-La Mancha and Castilla y León to be increasing moderately and those of Andalucía and Madrid to be increasing strongly. The Iberian population is nonetheless only 4–7% of the European population (Cano *et al.* 2006).

Black Storks gather post-breeding at favoured locations within or near the breeding range in August–September (Moreno-Opo *et al.* 2009). San Segundo (1992) identified 98 such sites distributed between Extremadura, western Andalucía, southwestern Castilla y León, Castilla-La Mancha and Madrid. Satellite-tracking is currently yielding considerable information on the movements of European birds, including the frequency and duration of their stopovers (Chevalier *et al.* 2011). For example, a nestling from Madrid wandered around an ever-increasing area of the Tagus basin between late June and September, before descending to the Strait of Gibraltar; it arrived there on 24 September but aborted an attempt at crossing and wintered instead at Doñana (Cano 2012).

Iberian birds are joined by migrants from central Europe in September. The latter occur widely in eastern Spain especially as they cross the Peninsula, mainly in March–May and late August–early November (Madroño *et al.* 1992). Most arrive through the western Pyrenean passes, where annual counts of 430–600 (Urcun 2003, OF) have increased more recently to average 1,275 birds during 2009–11. The Pyrenean movements peak during the second half of September. These migrants presumably include most of the small French and Belgian populations as well as many birds from central Europe. There are two old autumn recoveries from Córdoba and Zaragoza (Witherby 1928) of birds ringed as pulli the same year in Denmark: where the Black Stork no longer nests, as well as 20+ more recent recoveries of birds ringed in Germany, Austria, Hungary, the Czech Republic, Slovakia, Poland, Belgium, Luxembourg, France and Switzerland (Cano 2012). One ringed as an adult in Sevilla in September 2007 was recovered in Upper Franconia, Germany, in June 2008 (CMA).

Black Storks occur as rare migrants in the Balearic Islands but a few have appeared there in most recent years, with records from the main islands and Cabrera: for example, there were seven in 2008 and six in 2009 (*A. O. Balears 2008*: 126, *2009*: 129).

Most Black Storks winter in sub-Saharan west Africa, crossing the Strait of Gibraltar principally from mid September to mid October, although a few pass as early as August and stragglers cross in November. There are recoveries or observations of birds ringed as chicks in Spain or Portugal from Mali (13), Burkina Faso (2), Sierra Leone, Ghana, Guinea Conakry, Mauritania, Senegal, Niger and Ivory Coast (LR), as well as five from Morocco, and radiotracked Iberian birds have also overwintered in Mauritania, Mali, Senegal and as far as 18°E in Chad (Cano & Tellería 2013, Cano *et al.* 2013). Black Storks return from Africa at the Strait during February–early May, most passing in March–early April.

Autumn counts at the Strait of Gibraltar have showed a marked increase. They totalled 200–400 birds during the 1970s (Bernis 1980) but there were 943 in 1993, 986 in 1994, 1,358 in 1995, 1,706 in 1996 and 2,296 in 1997 (San Segundo *et al.* 1994; BM). Spring counts show a similar trend but there is no data available for entire seasons. They cross the Strait singly or in flocks of up to 50, exceptionally over 100. Black Storks are smaller and lighter than White Storks, and they readily employ flapping flight and are apparently less dependent than White Storks on thermals and other upcurrents. Thus they are less tied to the shortest crossing point at Tarifa and cross on a broader front.

A significant and evidently increasing number of Black Storks winter in southwestern Iberia, chiefly in the ricefields of the Guadalquivir delta but also at the Orellana reservoir, Badajoz, and in the Tiétar valley in Cáceres, Ávila and Toledo (Cano 2006). Very few are reported from Portugal in winter (*Anu. Ornitol.* 8: 60). There are also occasional winter records elsewhere: for example, one in Oviedo, Asturias, on 6 January 2003 (*Ardeola* 52: 208) and two at the Albufera de Valencia in January 2009 (*A. O. C. Valenciana 2009*: 33). Winter groups of up to 25 occur in favoured localities but loose gatherings of over 100 occur in the Guadalquivir ricefields especially: for example, there were 142 at Isla Menor and El Trobal on 11 December 2004 (*Ardeola* 53: 381). The Red Swamp Crayfish is a key food item in the ricefields (Cano 2012). Observations of marked individuals have shown that wintering birds include individuals marked in Germany, Belgium, France, Luxembourg and the Czech Republic, as well as in Spain and Portugal (Cano *et al.* 2006).

WHITE STORK *Ciconia ciconia*

(Cigüeña Blanca, Cegonha-branca)

***C. c. ciconia*. An abundant breeding species in western and central Iberia, scarce in most of the north and east. Partial migrant.**

The White Stork is an unmissable element of the landscape across the greater part of mainland Iberia, where its nests seem to top every church and large edifice. It is now abundant as a breeding species throughout southern and eastern Portugal, western Andalucía, Extremadura, western Castilla-La Mancha, Madrid, Castilla y León and the Ebro valley. The marked contrast between its abundance on the western meseta in Madrid, Toledo and Ciudad Real and its almost total absence immediately to the east, in Guadalajara, Cuenca and Albacete, is striking. It has typically always been rare or absent in the wettest areas of the north but it has recently colonised parts of northwestern Portugal and the Biscay coastlands. It remains largely absent from the arid south-east, from mountainous regions: including most of the Cantabrian Range and Pyrenees, from areas of unbroken forest or scrub, and from the Balearic archipelago (Bernis 1981, Carrascal *et al.* 1993).

White Storks are most associated with open, lowland habitats with some element of water. They are thus typical of river valleys and floodplains, meadows and pastures, farmed areas and, especially, dehesas: open woodlands where scattered trees punctuate crops or pastures and where there are frequent dewponds and drinking pools for livestock. They patrol these habitats on foot in search of large invertebrates: notably the grasshoppers and locusts that often abound in pastures. Earthworms are often taken and the birds follow the plough to obtain these and other unearthed animals. Wetter areas provide amphibians and fish. Ricefields have also proved highly attractive to the species in recent years, including in winter, providing supplies of crayfish, and other animal prey. White Storks have also recently discovered that landfill rubbish tips can offer plentiful food and large numbers are attracted to such sites at some times of year.

The size of the Spanish White Stork population has been studied extensively for over 60 years, since the first national enquiry (for any bird species in Spain) was organised by Francisco Bernis in 1948. That census was conducted by sending questionnaires to the mayors of all Spanish villages, many of whom replied. The conclusion was that the 1948 population amounted to 14,503 pairs. Further postal censuses followed in 1957 and 1974, and a mixture of direct censusing and postal enquiry was used in 1984. Since then there have been two direct censuses, in 1994 and 2004, in which observers set out to locate and count all the nests in Spain. The results of the last census have been analysed by Molina & del Moral (2005), on whose report much of the following summary is based. Their account should be consulted for detailed province-by-province results. The results of the various censuses may not be closely comparable, given the different methodologies employed and the different degrees of reliability of each, but it is clear that the fortunes of the White Stork in Spain during the second half of the 20th century changed dramatically for the better (Fig. 65). The first four censuses show a decline from over 14,000 pairs to fewer than 7,000 pairs (Bernis 1981, Lázaro *et al.* 1986). Fortunately, a recovery was first noted by the early 1990s (Gómez-Manzaneque 1992, Martí 1999) and the White Stork has continued to increase ever since, probably beyond the 33,000+ pairs found in 2004.

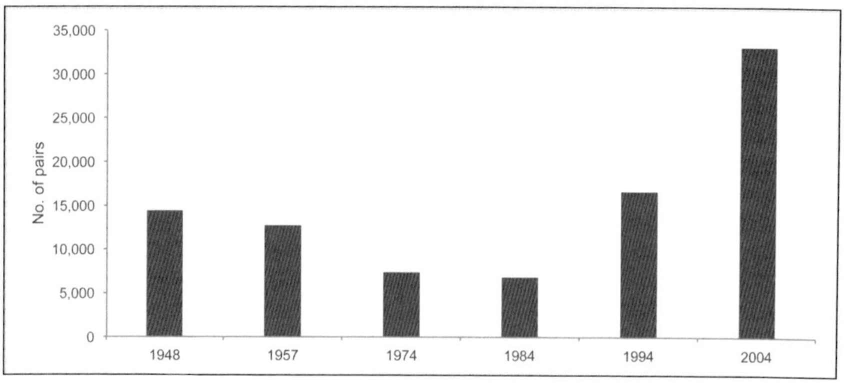

Figure 65. Results of successive censuses of the breeding population in Spain (Molina & del Moral 2005).

The broad distribution pattern has not varied greatly between censuses, with some exceptions. The great mass of the population breeds in Castilla y León and Extremadura, each of which accounted for over one third of the 2004 count, with (western) Andalucía holding a further 10% (Table 15). However, the small populations of the Ebro valley, Galicia, Cantabria and the Basque Country have showed relatively large increases from 1994 to 2004, although Asturias only accommodated a single pioneer pair in 2004. The population at least doubled between 1994 and 2004 in each region where the stork nests except in Extremadura, where the population was already high in 1994.

Table 15. Results by autonomous regions of the national censuses of breeding White Storks in Spain (Molina & del Moral 2005).

Region	1948	1957	1974	1984	1994	2004	%[1]	% increase[2]
Castilla y León	4,953	4,500	2,375	2,043	5,197	12,017	36.18	131
Extremadura	4,936	4,400	2,775	3,022	7,508	11,190	33.69	49
Andalucía	2,512	1,798	985	666	1,551	3,409	10.26	120
Castilla-La Mancha	1,333	1,200	779	525	1,005	2,315	6.97	130
Madrid	250	280	151	215	582	1,221	3.68	110
Aragón	140	164	117	145	339	1,205	3.63	256
Navarra	82	77	35	41	95	621	1.87	554
La Rioja	122	110	57	46	168	499	1.5	197
Cataluña	5	8	8	19	55	270	0.81	391
Galicia	26	16	6	4	56	256	0.77	357
Cantabria	114	126	49	26	79	177	0.53	124
Basque Country	30	22	6	1	8	36	0.11	350
Asturias	0	0	0	0	0	1	0	-
Murcia	0	0	0	0	0	0	0	0
Valencia	0	0	0	0	0	0	0	0
Balearic Islands	0	0	0	0	0	0	0	0
Total	14,503	12,701	7,343	6,753	16,643	33,217		

[1] % of 2004 population
[2] % increase from 1994 to 2004

The Portuguese population has fluctuated in parallel with that in Spain, for similar reasons (PA2). The first census in 1958–59 employed questionnaries and found 3,471 nests, which was thought to be a considerable underestimate. A census in 1984 indicated a marked decline and only 1,533 occupied nests were found in Portugal. By 1994 these had increased to at least 3,302 nests. The 2004 census found a further large increase, to 7,684 pairs (AP, PA2).

The breeding distribution of the White Stork may have been rather similar at the end of the 19th century and during the early 20th century, the bird being very common in the south, especially in Andalucía, and absent in most of the north, the east and the mountains. It nonetheless seems to have been largely absent then from the Ebro Valley: Boxberger (1921) did not see any in Zaragoza province and knew only a single nest in Navarra, at Tudela, whereas Ticehurst & Whistler (1925) found just a single individual during their visit to Navarra in May. The White Stork seemed also to be absent or very rare in northern Portugal: Tait (1887) did not see any close to Porto and Ticehurst & Whistler (1935) only saw one pair at Sabugal and another at Guarda. In central Spain the White Stork seemed to be much rarer than now: Witherby (1928) found it 'very locally distributed' in the area between Madrid, the foothills of the Guadarrama and Gredos mountains and at Guadalupe, Cáceres, where nowadays it abounds.

The White Stork has never nested in the Balearics and was seldom recorded there in the past. A specimen was apparently obtained in Menorca in May 1919 (Munn 1926). In Mallorca three were seen together on 4 May 1933 and one was at Alcudia for several weeks during April–May 1935 (Munn 1936). Records have become annual post-2000. Most are of single birds or very small numbers but some larger groups have occurred on spring passage in some years. There was something of an influx in 2009 when four flocks of ten or more birds were recorded (*A. O. Balears 2009*: 130), the largest being 18 at Sa Pobla, Mallorca, on 1 March. A flock of 17 at Sineu, Mallorca, on 4 April 2010 was also noteworthy (*A. O. Balears 2010*: 125–126).

Many of the most conspicuous White Stork nests are on buildings. Some of these: on castles, cathedrals and other ancient edifices are traditional sites that may have been continuously occupied for centuries. Others are on modern buildings, sometimes even on top of high-rise apartment blocks, as well as on a diversity of other artefacts, notably electricity pylons. Man-made purpose-built elevated platforms are also used in many areas. However, despite appearances, natural sites are still used in a high proportion of cases: 45% in 2004. Most such nests are on trees, often in holm oak dehesas and also in riparian formations and eucalyptus plantations. A small minority nest on rocks, especially in the boulder country of Extremadura: these rocky pinnacles are used as elevated platforms and are not equivalent to the cliff nests of many Black Storks. There are local instances of cliff-nesting White Storks in Extremadura, however (*Ardeola* 51: 545). There are also White Stork nests on coastal stacks and rocky islets in the south-west of Algarve, which are unique in Iberia (PA2). The use of electricity pylons is an increasing trend, as already noted in the 1984 census (Lázaro *et al.* 1986), with pairs switching to them from natural sites in trees. Presumably the height of pylons offers ease of accessibility and greater security from predation. The 2004 census found over 4,000 nests (12%) on pylons, 1,700 of them in Extremadura.

White Storks are highly sociable at all times and most of them nest in colonies, whether on trees, buildings or pylons. Only 14% of pairs in the 2004 census had isolated nests: at least 500m from the nearest neighbours. Colonies vary greatly in size from just a few pairs to a few dozen, and sometimes many more. The 2004 census found 33 colonies of over 70 pairs. By far the largest was the loose colony of the Dehesa de Abajo, Sevilla, with 382 tree nests in 2004, mostly on wild olives. The next largest was the 193 nests at the Ebro valley village of Alfaro, La Rioja, most of them on a single building, the church of La Colegiata de San Miguel.

Ringing recoveries or observations of birds marked outside Iberia indicate that birds from throughout western and west-central Europe travel to or through the Peninsula. There are over 8,500 such Iberian recoveries up to 2010, all but 15 of them from Spain. The great majority (98%) are of birds ringed in Germany, France, The Netherlands or Switzerland. The others include 94 from Belgium, 14 from Austria, 11 from the Czech Republic, eight from Denmark and just a few from Slovenia, Italy, Poland, Sweden, Lithuania, Hungary and the Kaliningrad enclave. Autumn counts of White Storks entering Iberia in autumn have been quite modest: they averaged some 2,000 birds at the western Pyrenean passes during 2009–11, with a few hundred also reported at Eyne in some years. The Pyrenean movements occur mainly in August–early September but some pass in July and there are stragglers into October.

White Storks ringed in Spain up to 2010 have given just 109 recoveries further north or east in Europe: 53 in France, 43 in Germany, eight in Swizerland, three in Italy and two in Hungary. There are, however, many recoveries of Spanish-ringed birds from North and sub-Saharan Africa, of individuals that were on passage or in their winter quarters. The 563 recoveries from North Africa are mainly from Morocco and the Western Sahara (509), the remainder being from Algeria (50) and Tunisia (3), as well as one from the Canary Islands. The 280+ recoveries from sub-Saharan Africa are mainly in the west, particularly in Mali (165): where wintering by White Storks has been well documented and in Mauritania (36), but there are also 26 from Nigeria, 13 from Ghana, 13 from Niger, 12 from Senegal, six from the Ivory Coast, five each from Chad and Burkina Faso and three from Guinea, as well as a few from ten other countries. The furthest travelled were one ringed as a chick at Barbate, Cádiz, in June 1968 that was recovered in Zimbabwe in March 1970; a chick ringed at El Gordo, Cáceres also in 1968 that was recovered in Zambia, and another chick ringed at Tarifa, Cádiz, in 1969 that was found in South Africa (8,100km). Birds ringed in Portugal have contributed a further 13 recoveries in Morocco, five in Mali and one in Nigeria.

The White Stork is generally regarded as a typical summer visitor to Europe, with winter quarters in sub-Saharan Africa. However, the timing of migration by Iberian storks is quite different from that of the central and eastern European populations. Formerly nearly all White Storks left Iberia in late summer, soon after the young had fledged, and were absent in October and November, although a few individuals remained in the south-west. The first returns to the nesting areas were relatively early, in December–January, most returning from February onwards (Bernis 1959). A pair may visit and repair the nest as early as December and may roost there, but laying is in March–April, earlier in the south. This migratory pattern allowed them to take advantage both of the wet season in sub-Saharan Africa and also the winter rainfall of Iberia. However, a significant 'wintering' population of White Storks, some of them of central European origin, has built up in Iberia since the early 1980s and these individuals do not migrate beyond Iberia at all (Máñez *et al.* 1994; Marchamalo 1995, 2002).

A winter month presence in Iberia is not new: Chapman (1888) saw them in southern Andalucía in winter, although these may have been birds that had already returned from Africa, but Bernis (AMI) makes no reference to wintering birds. However, the overwintering population in Spain was censused in 1995 and 2004, showing a large increase in numbers: from 7,594 to 31,229 birds. The censuses in both years were conducted in October to exclude the early returners from Africa (Molina & del Moral 2005). The wintering distribution shows close correspondence with the breeding range but the great majority wintered in the south-west. Together Andalucía and Extremadura held over 89% of the wintering population in 1995 and 85% of that in 2004. The latter census however showed an increase in wintering in the Ebro valley and the north-east.

The core wintering population is in the ricefields of the lower Guadalquivir, where gatherings of up to 9,000 birds have been reported (*Ardeola* 54: 174). Ricefields here, in Extremadura and elsewhere are a major winter habitat, providing storks with a supply of vertebrate and invertebrate food, including the abundant alien Red Swamp Crayfish. Major gatherings also occur both in winter and during passage periods at landfill rubbish dumps, many of which have large stork roosts in the vicinity. For example, there were 2,850 at the tip at Mérida on 28 September 2010 (*Ardeola* 57: 523). Some large concentrations, whether of wintering birds or early returning migrants, are also seen in Portugal: for example, in January 2009 there were 500 at a rubbish dump at Ferreira do Alentejo, Beja, and 830 at the Ria de Alvor, Faro (*Anu. Ornitol.* 8: 60). Observations of birds carrying field-readable marks, and radiotelemetric studies, have shown that the

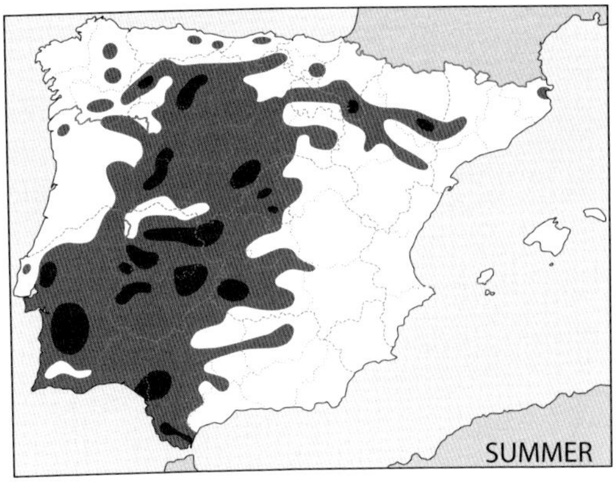

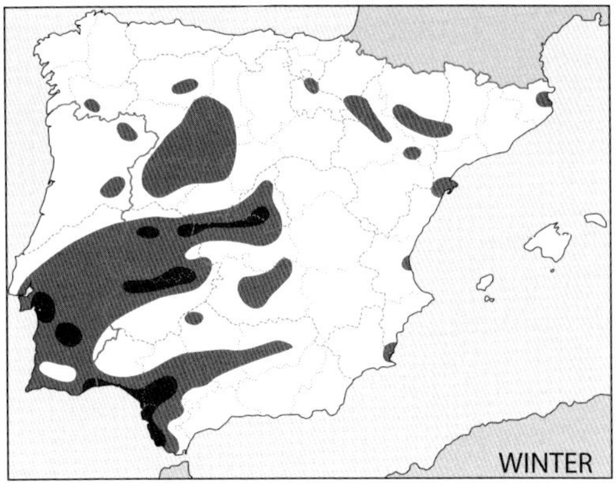

wintering Spanish population shows high site fidelity. It largely comprises adult birds: very few juveniles remain to winter. The wintering population is partly drawn from the Spanish population, including local birds, but also includes central European birds, principally from France, Germany and Holland.

Despite the increase in wintering numbers, migrants to and from sub-Saharan Africa still comprise the great majority of the populations involved. Migrants from western Europe cross Iberia diagonally during both passages and relatively few White Storks are usually reported from the Mediterranean or Atlantic coastlands. European White Storks enter and leave Africa at the Strait of Gibraltar, together with great numbers of Iberian birds. References to autumn and spring migration are largely meaningless when applied to White Storks, especially the Iberian birds. The southward exodus across the Strait begins early in July and peaks in late July–early August. The great majority have left by the end of August but some continue to pass south throughout September, these latter perhaps being most likely to be extra-Iberian birds. The return movements may begin as early as late October. Flocks of storks arrive back in Iberia from Africa throughout November–February and there is then a long-drawn-out passage of smaller numbers from March to early May. Observations of marked individuals have indicated that the earliest returners are old birds, that most of the Iberian breeding population returns in January, and that first-year and second-year birds are most evident from early March (Fernández-Cruz 2004)

The southward passage is concentrated in the central sector of the Strait, where the crossing is narrowest, with downwind drift as far east at Punta Carnero during westerlies and as far west as Punta Camarinal during easterlies, a front of some 36km. The northward passage is also concentrated at the central Tarifa narrows but wind drift means that arrivals are on a broader front. Seasonal counts of the southward passages of 1985 and 1993–99 inclusive (Fernández-Cruz 2004) show an increase during the earlier seasons that is consistent with the observed increases in the breeding populations. Fewer than 30,000 were counted during the 1985 season but from the mid-1990s around 100,000 birds were counted at the Strait each year. The number is likely have been much greater in more recent years, perhaps as many as 200,000 birds, given the size of the source populations in the early 21st century. However, Strait-wide coordinated counts, enterprises that demand a great deal of organisation and manpower over a very long period, were not carried out during 1999–2009.

The widespread decline in European White Storks during the mid 20th century seems to have been due to a combination of adverse factors, including habitat loss, exposure to pesticides and direct persecution, including hunting in the winter quarters (Chozas 1985, Biber *et al.* 1995, Schulz 1999). Climatic influences, notably the Sahel droughts of the 1970s and 1980s, are also likely to have been important. The Sahel rainfall peaks during June–September, including the period when White Storks are present in West Africa, and the effects of the rains include plagues of locusts and grasshoppers, on which the birds feed. A return to more normal rainfall patterns since the mid 1990s is likely have improved the birds' survival in their 'winter' quarters and the increased tendency to winter in Iberia may also have improved individual survival rates. There have also been large changes in land use in Iberia that have clearly contributed to the increase in White Storks. These especially include the recent and ongoing expansion of rice cultivation, providing extensive areas of new feeding habitat, and the proliferation of landfill rubbish dumps, which also provide large and reliable food supplies. Direct conservation measures, including the modification of pylons to reduce the incidence of electrocution, the provision of nesting platforms and the suspension of building works while storks are nesting, have all benefited the population (Molina & del Moral 2005). There have also been local introduction projects that account for small breeding populations in Girona, Tarragona and Málaga provinces.

MARABOU STORK

Leptoptilos crumeniferus

(Marabú Africano, Marabu)

Vagrant. Recorded in Spain (Category A) and Portugal (Category D).

The Marabou breeds at least occasionally at the Senegal delta and it occurs erratically at the Sahel wetlands in Mauritania, although there is only one record from the north there, at the Banc d'Arguin in March 1966 (Isenmann *et al.* 2010). Nonetheless, vagrants have been recorded in Israel and Morocco. In Morocco, one or two were found at the Chebeika estuary, between Tan-Tan and Tarfaya, in late 2001 and there were four there in late 2002 (Bergier 2004). There are also three older Moroccan records from Meknes and its vicinity during 1942–52: they included four or five birds feeding on refuse from an abattoir and two seen at a cow carcase (Bergier *et al.* 2006). There are recent observations of birds at large in Spain, Portugal and France. The 15 or so French records have hitherto been attributed to birds that have escaped from collections (OF).

The first Iberian observation was an adult at Vicién, Huesca, on 27 August 1988, and another was seen at the same site and various other locations in the same province during May 1989. Also in 1989 there was at least one at Doñana on various dates in April, September and October, and another at Ablitas, Navarra, on 13 September 1989. By 2010 there were at least 27 accepted records from Spain and seven from Portugal, involving up to 35 different individuals, although some duplication between sites and years cannot be excluded entirely.

There are records from most months, which would seem to favour the hypothesis that the birds are escapes. However, there is a notable concentration in summer–autumn and a relative scarcity in winter: some 62% of first observation dates are during July–October compared with just one during December–January. The pattern becomes even clearer if the French records are added, so that there also seems to be a second peak in spring (Fig. 66). Andalucía and southern Portugal account for more than half of the records (Fig. 67), which may suggest arrivals via the Strait. Some Marabous are known to have escaped from captivity in Spain (de Juana *et al.* 2003). Nevertheless, the significant number of records, and their temporal and spatial patterns, makes the possibility that some wild birds reach Iberia from Africa quite plausible.

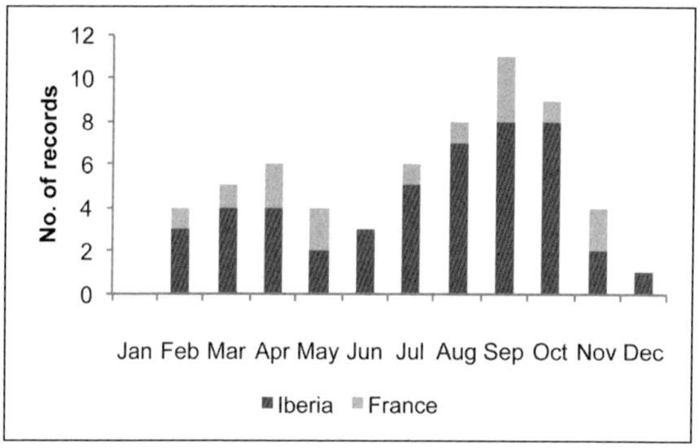

Figure 66. Combined records by month in Iberia (n = 34) and France (n = 14).

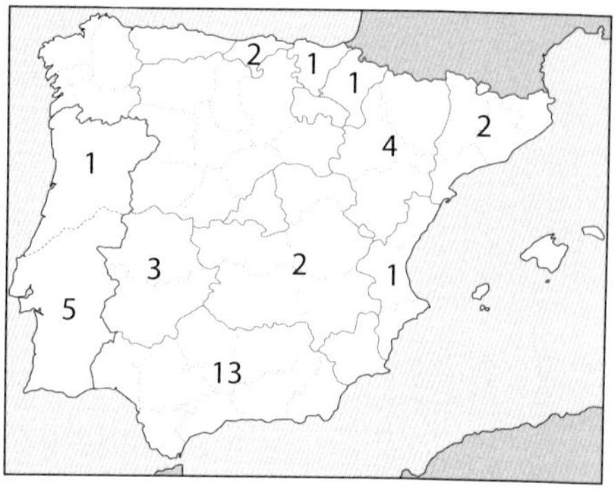

Figure 67. Records by region, 1988–2008 (n = 34)

GLOSSY IBIS

Plegadis falcinellus

(Morito Común, Ibis-preta)

Breeds locally in southern and eastern Spain. Has bred in Portugal. More widespread but generally very scarce on passage and in winter, chiefly along coasts.

The Glossy Ibis declined to extinction as a breeding species during the 20th century. This decline was part of its widespread demise in western Europe which saw its breeding range limited to the south-east, largely in Ukraine, Bulgaria and Romania, by the 1980s, probably due to a combination of habitat destruction, disturbance, the effects of pesticides and direct persecution (HBW). Since then it has recovered some of its former range and breeding was once again reported in France in 1988, Spain in 1993, Morocco in 1994, Portugal in 2005 and Tunisia in 2008.

The Glossy Ibis seems formerly to have been a regular nesting species in Spain. Irby (1875) referred to nesting near Vejer, Cádiz, at Soto Torero in the vicinity of the Laguna de La Janda. It nested in tamarisks at the Laguna de Santa Olalla, Doñana, from at least 1774 until 1899–1909 (Valverde 1960). Saunders (1871) found it abundant in the Marismas and considered that it undoubtedly bred there. Elsewhere in Spain it was noted only as a scarce passage migrant at the Albufera de Valencia and in winter in Málaga (Arévalo 1887). Thereafter there were failed attempts at breeding in the Marismas during 1930–35, 1940–42 and 1958 (Valverde 1960, LR). It may have bred at the Laguna de Las Madres, Huelva, prior to 1940 after which the colony was apparently exterminated when the waterbirds were used as a source of food by local people in the aftermath of the Civil War (Garrido 1996). It then ceased to breed in Spain until 1993, when two pairs nested at the Albufera de Valencia (Table GI1). Since then it has established colonies and bred in increasing numbers at Doñana, the Ebro delta and the Santa Pola saltpans, Alicante.

The Glossy Ibis is said to have nested in Portugal during the late 19th century, at the Guadiana estuary and perhaps at Pancas in the Tagus estuary (AP). Tait (1887) described it as a regular autumn migrant on the coast near Porto, appearing from late July. The only subsequent Portuguese breeding records were in 2005 in the Tagus valley, when at least three pairs nested at the Paul do Boquilobo, Santarém, and at least two did so at Escaroupim, Ribatejo, (AP, PA2) in a year when drought prevented nesting at Doñana: they did not return in 2006.

Glossy Ibises were recorded sporadically in Spain during the 1960s, 1970s and early 1980s. Thereafter observations became more frequent. A sample of 105 records in Spain from 1981–94: prior to the re-establishment of a breeding population, shows that most appeared in autumn, with almost 75% of first sightings in August–December, and a further 16% were during February–April, (Fig. 68). This pattern is similar to that noted earlier by Bernis (1969), and suggests that most birds seen before 1994 were passage migrants. Most were found in coastal regions. Of the 1981–94 sample 41% of records were in Andalucía, 18% in the Mediterranean from Catalonia to Murcia – including one in Mallorca – , and 16% on the Atlantic from Galicia to the Basque Country, with 25% in inland regions. These records totalled 351 individuals. Most (52%) were of single birds, or pairs or trios (26%), but 14% were of flocks of 10 or more birds, with a maximum of 24 together. A total of 48 individuals were observed at Portuguese coastal sites during 1985–93 (AI1). It has been suggested that these Iberian records involved birds of central European origin (Bernis 1969b) but the only concrete evidence is an old recovery of a bird ringed as a chick in Hungary and recovered at Sevilla in November (Gil Lletget 1945).

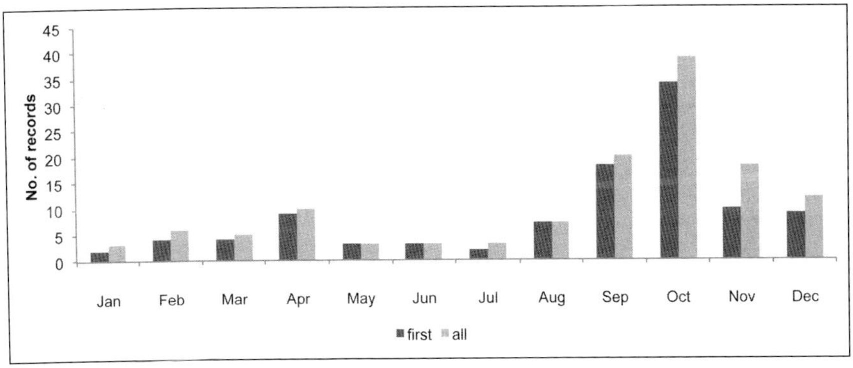

Figure 68. Records by month in Spain, 1981–1994 (n = 105).

The first of the more recent instances of nesting in Spain were at the Albufera de Valencia in 1993 and 1994, within mixed heronries in reedbeds. Two pairs raised three and five chicks respectively in 1993 but apparently only one pair nested in 1994, raising two young, and there were no more than isolated sightings at the Albufera during 1995 and 1996. However, 1996 saw four pairs nesting at the Ebro delta, raising four young in total and seven nests, containing 17 young, were found at Doñana in mid July during the same year (*Ardeola* 43: 242, 44: 245).

Individuals and small groups had been recorded all year round at Doñana from 1987 onwards (García *et al.* 1989) but the prolonged drought of 1991–95 seems to have prevented breeding then. The population there has expanded very rapidly indeed (Table 16). The four pairs that bred in 2005 did so at the Brazo del Este, on the periphery of Doñana, but all other nesting records from the Marismas during 1996–2007 have been within Doñana national park, where there were five separate colonies in 2007. The majority nest at the Lucio Cerrado Garrido, a system of permanent pools at the José Antonio Valverde Centre where the managed environment includes *Typha* beds, which are otherwise locally scarce (Santoro *et al.* 2010).

The Spanish population was censused in 2007, when over 96% of the breeding population was at Doñana. Annual counts of all the colonies since their establishment show their rapid expansion (Table 16), although hardly any pairs nested during the drought years of 1999 and 2005. Subsequent further increases have meant that some 5,300 pairs bred at Doñana alone in 2010 (Toral *et al.* 2012). Although the initial pairs in the Comunidad Valenciana in 1993 and 1994 were at the Albufera, thereafter the birds moved south to the Santa Pola saltpans (*Ardeola* 44: 245), where there were 30 pairs in 2009 as well as three pairs at El Hondo (*A. O. C. Valenciana 2009*: 35). However, at least three pairs bred at the Albufera in 2010 (*Ardeola* 58: 490). Several pairs were at a mixed heronry in Daimiel, Castilla-La Mancha, in the spring of 2011 and one of them bred successfully (*Ardeola* 59: 422). The 2012 season was also markedly affected by drought in western Andalucía, resulting in poor breeding and the establishment of several new small colonies away from the usual sites; for example, at least nine pairs nested in a mixed heronry at the Azud del Guadiana, Badajoz city, for the first known time in Extremadura (*Ardeola* 59: 422). Glossy Ibises nest colonially in the company of other waterbirds, notably herons and egrets, in reedbeds and other marsh vegetation or in low shrubs such as tamarisks. The principal colonies, at Doñana, are generally occupied in early April and the young fledge in mid June, but laying may be as late as mid July (LR).

Table 16. Population changes at Spanish Glossy Ibis colonies (Máñez *et al.* 2009).

Year	Doñana	Ebro delta	Comunidad Valenciana	TOTAL
1993	0	0	2	2
1994	0	0	1	1
1995	0	0	0	0
1996	7	4	0	11
1997	40	1	2	43
1998	86	6	1	93
1999	4	?	?	>4
2000	73	12	5	90
2001	199	19	?	>218
2002	400	23	?	>423
2003	673	26	?	>699
2004	1,125	51	12	1.188
2005	0	?	10	>10
2006	832	102	11	945
2007	3,643	119	15	3,777

The resurgent Spanish population has allowed for no fewer than 10,771 birds to be ringed, many with field-readable marks, up to 2009. These have resulted in numerous recoveries/observations outside Iberia, although many birds remain to winter near their natal colonies. In particular, an unprecedented arrival of Glossy Ibises in southwestern England, south Wales and southern Ireland in September 2009, involving at least 50 birds, included 12 juveniles that had been ringed at Doñana that spring (Gantlett & Millington 2009). Similar but smaller arrivals have occurred in the British Isles in subsequent years. Other findings of birds of Spanish origin include at least four each in France and the Czech Republic, three in Italy, two each in Germany and Hungary, and two even further to the east. These last comprise one ringed as a chick at Doñana in June 2004 and recovered on the island of Crete (2,827km) in May 2006, and another Doñana chick that was ringed in June 2006 and recovered in Odessa, Ukraine (3,117km) in August 2008. Doñana-ringed individuals have been found still further afield in recent years: in the Azores and in the Caribbean (SWA).

There are at least 28 recoveries of Spanish-ringed birds from Morocco and two from Algeria. In addition, there have been several recent sightings of small flocks crossing the Strait of Gibraltar and birds often occur on the ricefields at La Janda, in the hinterland of the Strait, during passage periods. Also single Glossy Ibises ringed as chicks in Algeria have been found in Portugal and at Doñana (Samraoui *et al.* 2011). One ringed in Italy was found at Doñana in early March 2010 and there are over 150 recoveries/observations of birds ringed at nesting colonies in the Camargue: 83% of them at the Ebro delta, 10% at Doñana but fewer than 1% in Portugal (SWA).

The Spanish wintering population is almost entirely confined to Doñana and the Ebro delta, which together accounted for 95–100% of the wintering birds during 1998–2010. The wintering numbers increased by over 30% per year during this period (Figure 69) so that the mean January count in Spain during 2008–10 was 4,715 birds (Máñez *et al.* 2009, AAE; SWA).

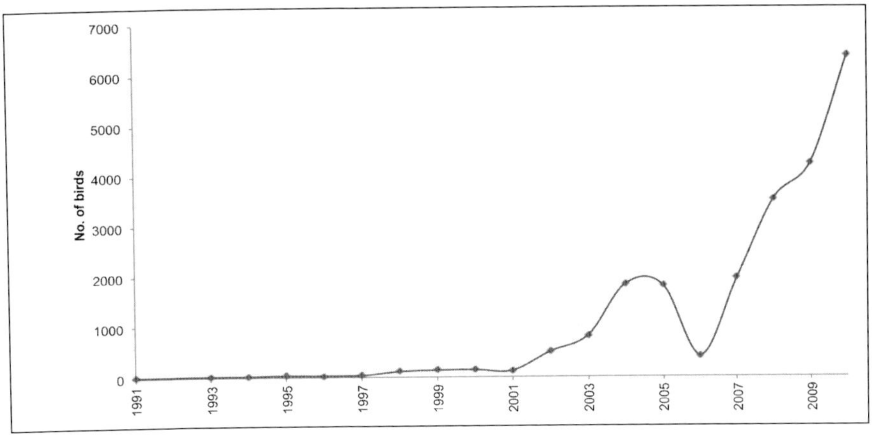

Figure 69. The increase in the Glossy Ibis wintering population in Spain during 1991–2010 according to January waterbird counts (SEO/BirdLife, 2011b).

Ricefields are especially important as a wintering habitat, and particularly during September–November (Toral *et al.* 2012). Indeed, it may be speculated that the increase in rice cultivation in Spain during the late 20th century has been a key factor in the recolonisation of Iberia by the species and may enable its spread to other sites in future. Although Glossy Ibises mainly feed on aquatic insects, principally Coleoptera and Odonata, during the breeding season (Macías *et al.* 2004), in winter they rely heavily on spilt rice in the flooded rice paddies (Toral *et al.* 2012). The ricefields of the Isla Menor in the Guadalquivir Delta, including those at the Brazo del Este, have attracted very large numbers in recent years. There was a record count there of *c.* 12,000 on 3 November 2011 (*Ardeola* 51: 546, 59: 176).

Wintering in Portugal has attained increasingly significant levels in recent years, with birds attracted to the ricefields of the Tagus and Sado estuaries, to the Lagoa de Santo André and to the Baixo Mondego valley. Prior to 2003 the largest flock recorded in the country was 21 birds but 95 wintered at the Mondego estuary in 2004–05 and nearly 200 birds in total were present then at Portuguese coastal locations. Thereafter wintering numbers have increased progressively so that 3,500 were present at the Tagus and Sado estuaries during winter 2009–10 (AP).

In Spain the resurgence in the population has been accompanied by only a modest increase in observations away from the breeding and wintering areas: 106 records (635 birds) during 2000–10. The temporal distribution of these records shows clear peaks corresponding to passage periods, with a marked concentration in September, and very few are in winter. Migrants are typically recorded at Mediterranean coastal wetlands on both the southern and eastern coasts but some may occur almost anywhere, including inland. The ricefields of Extremadura have attracted small groups during August–January in recent years: for example, there were 70 at a heron roost in the area on 11 November 2011 (*Ardeola* 59: 176). The Biscay coastlands have seen small influxes in some recent years: notably during January–May 2005, when Doñana suffered a severe drought. In 2012 a flock of 67 birds was recorded at Santoña in January and flocks of 20 or more were reported from several other coastal sites in Cantabria and Asturias during January–March (*Ardeola* 59: 176). Birds are also recorded occasionally in the Balearic Islands, chiefly at the Albufera de Mallorca in August and September and more rarely in spring or winter (*Ardeola* 50: 153). An old record of a flock of about ten at Alcudia, Mallorca, in early October 1933 (Munn 1936) is noteworthy.

NORTHERN BALD IBIS *Geronticus eremita*

(Ibis Eremita, Ibis-pelado)

Vagrant to Spain, where currently there is a reintroduction project.

The sole surviving viable wild population is in coastal southern Morocco, between Tamri and the Souss-Massa National Park, although reintroduction projects are underway in southern Germany, Italy and Spain. The Moroccan birds are now largely resident, although some disperse southwards along the coast after breeding. Further north one was taken near Tangier in the late 19th century (Irby 1895) and another in southern Spain in 1958.

The species is known from the fossil record of Spain and Gibraltar (Sánchez Marco 1996). It may also have survived in Iberia into historical times since medieval and Renassaince period treatises on falconry refer to a 'bald crow' that seems unlikely to have been anything else. A 16th century account by the Extremaduran native Luis de Zapata even gives specific locations where it occurred in Badajoz province: including Llerena, Villagarcía de la Torre and Medellín (Sánchez 2007).

An adult female was taken at Doñana on 11 July 1958 (Valverde 1959). This record clearly suggests a natural arrival from Morocco but the origin of two recently recorded individuals is less clear-cut. One was an adult that was present between Piedrahita and La Aldehuela, Ávila, from December 2004 to March 2005. The second was an immature that remained between Trujillo and Belén, Cáceres, from at least March to August in 2005, although it may already have been present since December. The close temporal and geographical coincidence of these two records, some 120km apart, and the fact that neither bird was ringed, is remarkable but the Spanish rarities committee regarded a natural origin as unlikely (Dies *et al.* 2007).

A reintroduction project is underway in southern Spain in the Barbate region, Cádiz (Quevedo *et al.* 2007; Proyecto Eremita: www.zoobotanicojerez.com). The first 28 birds were released in October 2005 and 215 had been liberated by 2010, all of them clearly marked with field-readable rings. A pair of captive origin produced young in both 2008 and 2009. In 2010 the male of this pair was replaced by the sole survivor of the chicks that fledged in 2008 (notice in *El País*–Cádiz, 11 June 2010): the pair nested unsuccessfully. However, seven pairs nested in 2011, eight pairs in 2012 and 15 in 2013: fledging 17 young (Ardeola 60: 515). By 2012 all the breeding birds comprised a small colony on the cliffs at La Barca de Vejer, Cádiz, quite close to the original release point. The 2012 free-flying flock numbered 66 individuals, eight of which had been born in the wild, but in total 216 birds were known either to have died or to have left the area (Proyecto Eremita 2012). Three ringed Bald Ibises at Utrera, Sevilla, on 10 December 2007 (GAE 2007) had clearly originated from this project, as had five marked birds in a group of eight that dispersed as far as southern Portugal, where they overwintered at the Lagoa dos Salgados, Silves, from October 2008 until at least March 2009 (AP, *Anu. Ornitol.* 8: 97).

SACRED IBIS *Threskiornis aethiopicus*

(Ibis Sagrado, Ibis-sagrado)

Recorded in Spain (Category C) and Portugal (Category E).

European observations are generally attributed to birds of captive origin, especially from the recently established feral population in France. These originated from a free-flying population that was established in a park at Morbihan, Brittany, from 1975 to 1980. The birds eventually established colonies at coastal locations both in Loire-Atlantique and in the Mediterranean. Both the Atlantic population, of some 5,000–6,000 birds, and the Mediterranean one, of over 300 individuals, disperse widely along coasts outside the breeding season.

Sacred Ibises have been found in Spain increasingly often as the French populations have grown. In Portugal three birds escaped in 1998 from a zoo at Montemor-o-Velho, Baixo Alentejo, and a pair of these bred locally and fledged three young: they were seen feeding in ricefields until October after which they disappeared (AP). There are at least four additional Portuguese records: at Barroca d'Alva, Setúbal, in January 2008, at Ria Formosa, Faro, in October 2011, at the Ria de Alvor, Faro, in March–April 2011, and at Quinta do Ludo, Faro, in November 2011.

The first sighting in Spain was in 1994 and the second in 1997, since when the Sacred Ibis has been reported annually. Observations were considered by the rarities committee until 2005, by when there were 21 records (Dies *et al.* 2007). The number of birds reported rose from eight in 2003 to 21 in 2008. Some observations may involve escapes or releases within Spain itself: Barcelona zoo has free-flying breeding birds (*Ardeola* 47: 144) and several that were released at El Retiro park, Málaga, in 2003 probably account for those seen during the following months at the Guadalhorce estuary nearby (GAE 2007).

Observations from Mallorca since 1999, especially from around the Albufera, may be due to one or two long-staying individuals. One was also found in Menorca in February 2006. Single birds made up 55% of a sample of 40 Spanish records (55 birds) and there are seven observations of pairs, one of three birds and two of groups of four: these last both in 2008. The 21 individuals for which ages were reported included 11 adults, four immatures and six juvenile/first-winters. Sightings are well distributed across the year but they tend to concentrate in winter, with a peak in first observations in December that could be due to birds arriving from France. Sacred Ibis sightings are widely spread but over 70% have been from coastal regions.

The Sacred Ibis has proved to be invasive in France, where it has done serious damage by preying on heronries and tern colonies (Yésou & Clergeau 2006). Some steps to control the French population began in 2007, when 340 birds were eliminated (OF). By 2011 the French Atlantic population had been reduced to 560–600 pairs and that of the Mediterranean was almost eradicated (Dubois *et al.* 2012). It would be wise similarly to prevent the establishment of the species in Iberia.

EURASIAN SPOONBILL *Platalea leucorodia*

(Espátula Común, Colhereiro)

***P. l. leucorodia*. A very local breeding species in southern Spain and Portugal. More widespread on passage and locally in winter. Partial migrant.**

This is another large aquatic species that has enjoyed a revival in western Europe after a long period of decline that reduced it to just two small populations, in the Netherlands and extreme southwestern Andalucía, by the mid 20th century (BirdLife International 2004). The western European population had increased to about 4,800 pairs and an estimated 19,000 individuals by 2007 (Triplet *et al.* 2008).

The Spoonbill has its principal Iberian nesting colonies in western Andalucía: in Huelva, Sevilla and Cádiz, with a few also nesting in Extremadura, Castilla y León and Castilla-La Mancha, as well as in the southern half of Portugal. The core populations in Andalucía fluctuated around 500 pairs in total during 1984–95 (de le Court & Feria 2009). Since then breeding numbers have tended to increase but there have been very large year-to-year fluctuations in relation to annual rainfall. The mean population from 1996 onwards has been around 1,400 pairs, with a maximum of 2,700 pairs in 2001 and minima of fewer than 300 pairs during the dry years of 1999 and 2005. The national census of 2007 found 1,614 pairs in Spain, 98.8% of which were in Andalucía (García *et al.* 2009). The main colonies were and remain those at Doñana, Odiel, the Marismas de Isla Cristina and Cádiz Bay.

There is some evidence that the Spoonbill nested in Portugal in the 17th century (Tait 1924) but it certainly ceased to do so by the 1970s and is only known to have resumed nesting in the country in 1988, when four pairs nested at the Paul de Boquilobo, Santarém. A census in 1996 estimated the Portuguese population at 43 pairs at four colonies: the Paul de Boquilobo; Pêro Peão, Evora; the Ria Formosa, Faro, and the Paul de Muge, Santarém (Farinha & Encarnação 1996). The population had increased to 92–99 pairs at ten sites by 2002 (PA2). The Portuguese nesting colonies are in the Tagus valley, central Alentejo and at the Ria Formosa (AP).

The Spoonbill has a long history of nesting in southwestern Andalucía. It is known to have nested in Doñana at the Santa Olalla and Taraje lagoons between 1765 and 1912 (Saunders 1871, Irby 1895, Valverde 1960). Irby also reported a nesting colony at the Laguna de la Janda, Cádiz, in the late 19th century (Irby 1895) and a small colony was found there in 1909 (Congreve 1943). There was also a colony at the Laguna de Las Madres, Huelva, from at least 1912 until 1961, when the lake was drained (Weickert, 1963). Intermittent but unsuccessful breeding attempts were reported at Doñana between 1909 and 1953 (Valverde 1960). Thereafter the species became better established in Doñana, where the '*pajarera*' in the Biological Reserve holds the longest-established colony. The Entremuros colony dates from 2000 and Las Nuevas colony is also recently established, from 2002 (García *et al.* 2009). Several other small colonies have been occupied intermittently within the Marismas and other small colonies have been found elsewhere in western Andalucía in recent years. Breeding at Odiel dates from at least 1960 but the region was not monitored until 1977, since when nesting has been continuous (Garrido 1996). The colony of La Covacha on Trocadero island in Cádiz bay was discovered in 1997 (*Ardeola* 45: 119). Nesting at Isla Cristina, Huelva, began around the same date, but the colony has been abandoned during some years, apparently because of disturbance.

Away from Andalucía, the first nesting attempt in Extremadura was in 1999 (*Ardeola* 47: 162, 49: 177). Since then nesting has occurred there at eight sites, most regularly at Casatejada and at the Montijo reservoir, and the Extremaduran populations increased from five pairs in 2000 to at least 55 pairs in 2013 (*Ardeola* 60: 515). The sole colony in Castilla y León, at Candeleda, Ávila, in the Tiétar valley, was discovered in 2006 (*Ardeola* 54: 175). In Castilla-La Mancha, there was a failed nesting attempt by three pairs at the Vega del Jabalón reservoir, Ciudad Real, in 2005 and one pair nested successfully at the Azután reservoir, Toledo, the same year (Velasco *et al.* 2007).

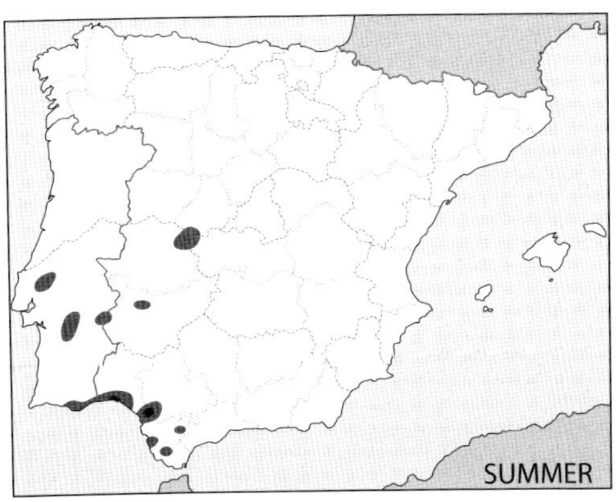

Spoonbills typically nest colonially and often in the company of herons, egrets and White Storks. The nests are often in trees: including species as diverse as eucalyptus, elms, poplars, willows, pines and cork oaks, but nests are built on the ground where trees are not available, notably at Odiel, where they are thus vulnerable to flooding. One or two pioneer pairs may settle in a heronry and may give rise to a future colony, as has been seen in the recent small-scale expansion of the breeding range inland in western Spain. The nesting habitats include marismas and the fringes of lakes and reservoirs. The birds feed in a diversity of freshwater, brackish and saltwater habitats, taking a wide range of animal prey, chiefly invertebrates. At Odiel the principal prey item is an introduced North American fish, the Mummichog, but within Doñana another North American alien, the Red Swamp Crayfish, is highly important in the diet (García *et al.* 2009).

Post-fledging dispersal takes a small minority of Spoonbills, some 2–4%, as far afield as northern Iberia, France and the Low Countries, prior to migrating south to West Africa (de le Court & Aguilera 1997). One ringed as a chick at Doñana in April 1994 was recovered in Malta in April 2003. Most remain close to the nesting areas and some notable concentrations are reported in late summer in Andalucía. For example, there were 1,000 at the saltpans of La Algaida, Sanlúcar de Barrameda, on 20 August 2000 (*Ardeola* 48: 139).

Eurasian Spoonbills from colonies in France and the Netherlands also occur in Iberia on passage. The autumn passage occurs from August to mid-October. Many regularly use the wetlands of the Biscay coast, notably the Marismas de Santoña and Txingudi as staging posts, and this has been a growing trend in recent years as continental populations have increased. For example, at Santoña a count of 109 on 14 September 1999 was then regarded as noteworthy but numbers peaked there at 602 on 21 September 2009 and some 1,900 were censused during the course of the 2010 autumn season, and 2,500 during September 2011 (*Ardeola* 48: 139, 56: 351, 57; 524; GRUSEC). Postnuptial migrants at Txingudi totalled 1,847 birds in 2009 and 2,036 in 2010 (*Ardeola* 57: 223 & 524). From there some follow the coast west to Galicia and then southwards along the Atlantic coast but many cross the central Peninsula southwestwards, appearing in small flocks at major wetlands across Castilla y Léon, western Castilla-La Mancha and Extremadura especially. At least 271 birds staged at the Azud de Riolobos, Salamanca, during autumn 2006 (Ceballos & Ramos 2007). Some may linger for several months at staging sites prior to crossing the Peninsula in autumn and there has been an increasing recent trend for overwintering at northern coastal sites, including Santoña where 43 overwintered in 2006–07, 57 were present throughout December 2007 and 62 remained in December 2009 (*Ardeola* 54: 392, 55: 136, 57: 223). Similarly, there were 157 at O Grove, Pontevedra, on 12 December 2009 (*Anu. Ornitol.* 7: 77).

Formerly relatively few were reported on Mediterranean coasts in autumn (Galarza 1986) but records of small numbers have become more frequent in recent years. For example, there were 38 at the Albufera de Valencia on 3 September 1997 and 41 on 20 September 2009 (*Ardeola* 46: 151; *A. O. C. Valenciana 2009*: 35), and up to 15 were at the Santa Pola saltpans in January–February 2009 (*A. O. C. Valenciana 2009*: 35).

Most Iberian birds, and migrants from other western European populations, spend the early winter in West Africa, especially at the Banc d'Arguin, Mauritania, and in the Senegal Delta. They travel via the wetlands of the Moroccan Atlantic coast. Immature birds remain in Africa for about three years until adulthood (de le Court & Aguilera 1997) but the adults begin to return to Europe in January, spending the rest of the winter in southern Iberia. There are 22 recoveries from sub-Saharan Africa of birds ringed in Spain: 13 from Senegal, eight from Mauritania and one from Guinea Bissau, as well as 28 from Morocco.

The regular international surveys of wetland birds in January have shown an increasing mid-winter presence of Spoonbills, including some from The Netherlands, in Spain since 1991. The Spanish January counts of 1990–2009 found 1,123 birds on average (range 92–3,200; AAE). Over 80% of them occur in southwestern Andalucía: at Doñana, the Marismas del Odiel, Cádiz Bay and nearby coastal wetlands. Elsewhere the Ría de Arousa in Galicia, Santoña and the Ebro delta have together held about 8% of the January population. However, it has been pointed out that January counts include birds that have already returned from Africa (de le Court & Rodríguez 2005). The monthly aerial surveys of Doñana indeed show much lower numbers of genuine overwinterers: with mean counts of 91 in November and 52 in December, rising to 221 in January (Rendón *et al.* 2008), but the November and December totals are also increasing.

Recent observations from Portugal indicate the increasing abundance there of birds in January, counts in that month finding an average of 214 birds in the country during 1993–97 but an average of 1,000 individuals during 2005–07 (AP). The largest numbers are on the Algarve coast: at the Formosa estuary and in Castro Marim, quite close to the colonies in Huelva. The Tagus and Sado estuaries and the Lagoa de Santo André also attract birds both on passage and in winter (AP). A count of 156 at Figuera da Foz on 27 January 2008 (*Anu. Ornitol.* 7: 77) is notable for the central Portuguese coast.

Spring passage through Iberia occurs from February to April, peaking in March (Galarza 1986). The spring movement is less concentrated than the autumn passage and more birds appear then at Mediterranean wetlands such as the Ebro delta. Some non-breeders remain in summer at the Biscayan staging sites, notably Santoña.

The Spoonbill occurs annually in the Balearic archipelago, particularly at the Albufera de Mallorca, but only in very small numbers. It is regarded as a rare migrant and winterer in the islands. Most records are of single birds, sometimes two together, so a flock five at the Salobrar de Campos on 20 October 2007 was exceptional (*A. O. Balears 2007*: 159).

Observations of marked individuals have shown a clear preponderance (80%) of birds from the Netherlands among migrants that cross the Peninsula, with the remainder being birds marked in Germany and France. Two ringed in Denmark have also been recovered in Spain. However, in 2009 there were surprising observations at Santoña in September of single birds that had been marked in Serbia and Hungary (*Ardeola* 57: 223). An earlier study of birds marked at the Odiel colonies (de le Court & Aguilera 1997) has shown that a small proportion of adults (2.4% of 596 birds ringed as young and 3.2% of 62 birds ringed as adults) are found during the breeding season in other European countries, in particular The Netherlands, France and

Portugal, suggesting that there is significant gene flow between the western European populations.

The Iberian population has recently shown encouraging trends. In Spain alone it increased by 5.5% annually during 1959–2007 (García *et al.* 2009). Nevertheless, the total population remains small and over 90% breed at just five colonies in Andalucía. Locally, birds are at risk from disturbance of colonies by humans and feral dogs, from flooding by high tides of nests placed on the ground and from disintegration of nesting trees at long-used sites. More widely, they are subject to environmental degradation of the feeding grounds both in Iberia and in Africa (LR, Zwarts *et al.* 2009).

GREATER FLAMINGO *Phoenicopterus roseus*

(Flamenco Común, Flamingo-comum)

Breeds very locally and sometimes irregularly at a few localities in southern and eastern Spain. Locally common in winter, especially in southwestern Spain and coastal Portugal. Uncommon to rare elsewhere in Iberia. Partial migrant and nomadic to some extent.

Iberian birds form part of a metapopulation that inhabits much of the Mediterranean basin and the Atlantic coasts of north-west Africa south to the Senegal river (Johnson & Cézilly 2007). There is considerable interchange between the colonies in this region, with birds switching from one location to another when conditions, essentially changes in water levels, demand. The two principal colonies in the western Mediterranean are in the Camargue and, in Spain, at the Laguna de Fuentedepiedra, Málaga. In an average year, Spain may hold some 31% of the birds breeding in the western Mediterranean, although that figure rose to 41% in 2007 when none nested at the Camargue (SA1, Rendón-Martos *et al.* 2009a). The concentration at Fuentedepiedra was even more exceptional in 1988, when it accommodated 53% of the breeding pairs in the western Mediterranean and 70% of the chicks fledged in the region (M. Rendón, pers. comm.).

Greater Flamingoes are a familiar feature of the saltpans, estuaries and shallow brackish lakes of southern and eastern Spain: in Andalucía, Catalonia, the Comunidad Valenciana, Murcia and southern Castilla-La Mancha. Occasional individuals and small groups are recorded in Spain outside these regions, most often in Extremadura and Madrid. They occur in similar habitats in southern Portugal and in the Balearic Islands but they do not breed there. Their principal haunts are coastal or some way inland but although many sites in these regions attract small numbers the really large concentrations are restricted to very few localities.

The first coordinated censuses of Greater Flamingo populations in Spain took place during 1985–89 (Fernández-Cruz *et al.* 1991). Since then the species has been monitored by regional authorities and considerable information is now available on the often marked numerical fluctuations at different sites and between different seasons. The species was censused particularly thoroughly throughout 2007 under the auspices of SEO/Birdlife and detailed information on the various colonies and wintering concentrations that year, together with a wider discussion of the species in Spain, is provided by Rendón-Martos *et al.* (2009b).

The Spanish breeding population during 1985–2007 ranged from 80 pairs (in 2005) to 34,197 pairs (in 2004) and the number of chicks fledged annually ranged from zero to 16,701. The fact that the maximum and minimum recorded populations occurred in successive years highlights the variability in breeding effort between seasons, which is largely due to water levels at the seasonal wetlands where nesting occurs. The key to whether it is or is not a 'good' year is the situation at Fuentedepiedra lake, which accommodates what is usually by far the principal colony in Spain. The 22,000 pairs in 2013 and 15,387 chicks reared in 1998 are site records. Fluctuations notwithstanding, however, the breeding population showed a significant upward trend during 1985–2007: by 4.2% annually at Fuentedepiedra (Rendón-Martos *et al.* 2009b).

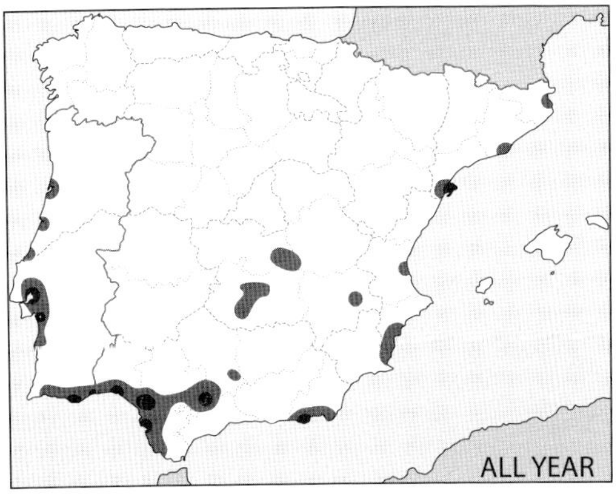

Fuentedepiedra lake has long been attractive to flamingoes (Saunders 1871) and breeding was reported there during some years in the late 19th century and early 20th century (Arévalo 1887, Bernis & Valverde 1954b). Boxberger (1931) counted 7,000–8,000 birds there in winter. Nevertheless, the first reported breeding colony in more recent times was found in 1963 (Valverde 1964b). The site is protected as a Nature Reserve. Fuentedepiedra is a shallow endorheic lake, 1.4m deep at its maximum, dependent on rainfall for its water. It is therefore highly affected by interannual fluctuations in precipitation and dries up completely by June or July in most years, refilling with the autumn rains from early October onwards in normal years. The basin is some 6.5km long and 2.5km wide, with a water surface area of over 1,350ha in wet seasons (Garcia & Paterson 2008). Water levels determine whether or not flamingoes breed in any given year: a water depth averaging at least 50cm is necessary for successful nesting and the birds do not nest at all if the lake is shallower than 30cm (Rendón-Martos *et al.* 2009b). Breeding occurred in 21 of the 28 years from 1985 to 2012 (M. Rendón, pers. comm.). Water levels are now usually maintained artificially to enable successful breeding, if the lake threatens to dry up after the chicks have hatched. The lake is highly salty and supported commercial saltpans until the 1950s. The remnants of the saltpan dykes are used as nest sites by the flamingos, as well as artificial islets that have been provided more recently. There are two other important breeding locations in Spain: Doñana and the Ebro delta. Breeding at the Marismas seems always to have been irregular (Valverde 1960), probably due to the interannual variation in water levels, as remarked by Irby (1875) and Chapman (1884).

Reproductive output is highly variable between sites and years (Table 17). The birds nest colonially but the breeding cycle is not closely synchronised, so that some eggs are still being incubated when the first young have already fledged. In Spain colony occupation begins in February but some pairs do not settle to breed until April or May. The single clutch, nearly always of one egg, is laid between late February and early June, but principally between mid March and mid May (BWP, Rendón-Martos *et al.* 2009b). Mean productivity in Spain during 1985–2007, expressed as total chicks fledged as a percentage of the number of pairs present, was 45.9% ± 28.1. Of the three main sites, breeding tends to be most erratic at Doñana, where low water levels made nesting impossible in 11 years during the period. Productivity at the Marismas is also much lower than at Fuentedepiedra and the Ebro delta (Table 17), largely because of the depredations of foxes and the abundant Wild Boar. In 2007 the 52 chicks that survived at the Marismas were birds that were taken into care before they were lost to predators along with the rest of the cohort (Rendón-Martos *et al.* 2009b). The individuals that breed at the Marismas are younger on average that those that nest at Fuentedepiedra, suggesting that they have been excluded from the principal location by older birds (Rendón *et al.* 2011).

Table 17. Greater Flamingo breeding populations and comparative productivity at Fuentedepiedra, Doñana and the Ebro delta during 1985–2007 (Rendón-Martos *et al.* 2009b)

Site	Fuentedepiedra	Guadalquivir	Ebro delta
Mean population (pairs)	10,237	1,901	770
Breeding attempts	235,452	43,719	17,715
Chicks fledged	132,389	4,994	8,937
Productivity (%)	56.2	11.4	50.4
No. of years when none bred	5	11	7
No. of years when breeding failed completely	0	3	3

Breeding away from the three main sites has been highly intermittent and on a comparatively modest scale. In the Comunidad Valenciana there are records of breeding at the Santa Pola saltpans in 1973 and 1975 (Ibáñez *et al.* 1974, Ferrer *et al.* 1986). During 1985–2007 there was attempted breeding there in 1992, 2001 and 2002, which was successful only in 2002 when 650 pairs fledged 220 young. Nesting by 650–1,000 pairs also occurred at El Hondo reservoir during 1996–98, successfully in 1997 and 1998, when 491 and 700 young fledged respectively.

Flamingoes often visit the lakes of Castilla-La Mancha but seldom nest there successfully. However, at the Laguna de Pétrola over 98 pairs nested in 1999, 232 pairs in 2000, 831 pairs in 2012 and 1,437 pairs in 2013, with numbers of young fledged each time (LR; *Ardeola* 59: 423, 60: 516). Successful nesting was achieved at Manjavacas lagoon in 2010 where a count on 7 August found 2,187 adults and a creche with 488 chicks (*Ardeola* 57: 524).

Unsuccessful breeding attempts have been observed in several other Spanish regions: in 2006 at the Salobrar de Campos, Mallorca; in 1979, 1982 and 1983 at the Cape Gata saltpans, Almería; and, in Murcia, at the San Pedro del Pinatar saltpans in 1994, 2000 and 2001 and at the Marchamalo saltpans in 1990. Most or all of the birds involved in these attempts were young and although nests were built these were later abandoned. The Marismas del Odiel, Huelva, often attract feeding flamingos, and breeding attempts were recorded there from 1989 onwards. Provision of artificial nests and later an artificial island at the Aragonesas saltpans led to nesting attempts there from 1989 onwards (Garrido 1996) but no eggs were laid for years. However, the birds were successful at last in 2008, when some 500 pairs raised 399 chicks on the artificial island (Rendón-Martos *et al.* 2008).

There has been no firm evidence of Flamingoes attempting to breed in Portugal until very recently. There are vague indications that they may have done so at the Guadiana estuary in the late 19th century, since King Carlos claimed to have received eggs from there, and there is also an unconfirmed and second-hand report of nesting at the Tagus estuary very early in the 20th century (AP). However, the increasing abundance of non-breeders in the country made it more likely that breeding attempts would follow. In 2010 there were two small-scale and highly unusual instances of attempted nesting at widely separated sites: two birds were seen incubating at the Lagoa dos Salgados, Faro, on 2 May and an egg was found at Vasa-Sacos, Santarém, on 30 April (*Anu. Ornitol.* 8: 61).

Nesting Flamingoes may commute considerable distances to feed when necessary. In particular, those at Fuentedepiedra regularly travel to and from Doñana (at night), a distance of some 140km each way (Rendón-Martos *et al.* 2000). After the breeding season the adults and fledged young disperse and join non-breeders at suitable wetlands in southern and eastern Iberia. Precisely where they go and when they do so depends on water levels at the various sites, which may differ widely between years. Hence flamingo movements tend to be complex and birds may be seen travelling in unpredictable directions. There are over 270 Iberian recoveries of birds ringed in France: mainly at the Camargue, as well as 11 from Italy: mainly Sardinia. Birds ringed in Spain have produced over 360 extra-Iberian recoveries: 64% in the Camargue as well as 14 in Italy and two in Greece. They also include 51 from North Africa: including 35 in Morocco/Western Sahara, 27 in Algeria, 23 in Tunisia, three in Libya, one in Egypt and one in the Canary Islands. Recoveries in sub-Saharan Africa comprise 14 from Senegal and six each from Mauritania and Guinea Bissau. The great majority of African recoveries are from the coastal wetlands.

January censuses of wintering Greater Flamingoes in Spain have found that some 60% on average are then found at Doñana and a further 25% at the Ebro delta, although the former site alone has been known to hold over 80% of the total. The January population in Spain has ranged from 4,236 birds in 1975 to 62,826 in 2007 (Rendón-Martos *et al.* 2009). The numbers wintering in Iberia have shown a highly significant increase during 1973–2007 although with year to year fluctuations according to climatic conditions, especially during 1973–87 (Johnson 1991). The increase in the wintering population of the Guadalquivir delta has been especially sustained by the construction of fish ponds at Veta La Palma during 1984–92, which now offers excellent habitat all year round irrespective of rainfall. The extensive areas of more or less permanently flooded saltpans at Bonanza/La Algaida have also enabled large numbers to remain there all year (Rendón *et al.* 2008).

Flamingoes were censused monthly throughout their breeding and wintering areas in Spain in 2007 (Rendón-Martos *et al.* 2009b). That year they were found in at least 84 locations distributed across 16 provinces in six autonomous regions, although Andalucía accounted for an average of 80% of the birds and the Ebro delta for a further 16%. The population peaked at 62,826 birds in January, remained at 40,000–50,000 until May and then reached a minimum of 28,080 birds in June, when many birds abandoned those wetlands that dried out and left the country. Numbers increased again with the beginning of the rains in September before declining to another low level of 29,307 in December. Such changes are not necessarily typical. Indeed 2007 was unusual in that no Greater Flamingos nested in the Camargue for the first time in 38 years (Béchet *et al.* 2009) and many of the birds from there moved to the Ebro delta and elsewhere in Spain then.

In addition to the Ebro delta, the Santa Pola saltpans are an important staging and wintering area for Greater Flamingoes in the coastal Levant. Peak counts there in 2009 were of 713 in January, 849 on 4 February and 2,412 on 6 October (*A. O. C. Valenciana 2009*: 36). A count of 829 at the Ibiza saltpans on 16 September 2011 was a record for the Balearic Islands (*Ardeola* 59: 424)

Greater Flamingoes seem to have been relatively uncommon in Portugal during most of the 20th century, occurring most often at the Tagus and Sado estuaries and at the Castro Marim reserve, Algarve (AP). They began to become more numerous and more widespread in the late 1980s when the wintering population of up to 250 birds increased to average 3,208 during 1992–95 and 5,092 during 2005–07, with maximum counts exceeding 7,500 birds (AP). A study at Castro Marim found that 125 of 189 marked birds had originated at Fuentepiedra, with 61 from Camargue and three from Sardinia (AP). Although flamingoes are often most numerous in Portugal in winter they may be seen there all year round and there are large and variable fluctuations in abundance. An influx of juveniles occurs in late summer or autumn, largely depending on whether or not the Fuentedepiedra colony has been successful. Large influxes of adults may occur as early as May when that colony has a poor year (AP). The flamingoes are drawn to the shallow waters of saltpans, ricefields and fish farms as well as coastal lagoons and estuaries. The most important sites are the Tagus and Sado estuaries, the Ria Formosa and the Castro Marim saltpans but they are also now found regularly at the Mondego, Arade and Alvor estuaries, the Lagoa dos Salgados, Faro, and the Lagoa de Santo André, Setúbal. The Ria de Aveiro, a relatively northerly site in the coastal north-west, has seen some large concentrations in recent years, including 575 on 31 January 2008 and 1,100 on 24 June 2008 (*Anu. Ornitol.* 7: 77). The Alentejo wetlands seem to attract peak numbers in summer and early autumn: for example, there were 432 at the Lagoa dos Patos on 10 August 2008 and 327 at the Albufeira dos Fartos, Évora, on 20 September 2008 (*Anu. Ornitol.* 7: 77).

Remarkably few birds wander north of the breeding range but preferred habitats such as saltpans are unavailable there. The Flamingo is unknown in the Biscay coastlands but there a few records from the upper Ebro valley, including one at La Grajera reservoir, La Rioja, in October 1991; single birds at the Embalse del Ebro, Burgos, in January and October 1972, and several in Navarra, where up to 24 birds were present in August 1998. There were ten published records for Madrid during 1998–2008, mainly of 1–2 individuals but including flocks of 40 in flight at San Martín de Valdeiglesias in March 2001 and at Villanueva del Pardillo in May 2005 (*A. O. Madrid 2001*: 164, *2005*: 165).

LESSER FLAMINGO

(Flamenco Enano, Flamingo-pequeno)

Phoenicopterus minor

Vagrant to Spain and has bred. Vagrant to Portugal (Category D).

By far the largest sub-population of the Lesser Flamingo inhabits the African Rift valley. Much closer to Iberia there are several thousand in west Africa: mainly at the Senegal delta. Two to three thousand non-breeders have been present all year further north, at the Banc d'Arguin, Mauritania, since the early 1970s. Apparent vagrants have occurred still further north, in Morocco and Iberia, but appearances elsewhere in Europe have generally been attributed to escapes. There are 41 accepted records from France, many of them relating to a pair that has nested in the Camargue since 1994; two pairs attempted to breed there in 2010. There are 13 records from Morocco, one from Algeria (Merzoug *et al.* 2010), one from Tunisia (Azafzaf *et al.* 2010) and two from the Canary Islands (*Ardeola* 51: 523, 55: 268).

Successful nesting occurred in Spain in 2007, 2011 and 2013, within Greater Flamingo colonies. Saunders (1871) referred to the probable sighting of an individual at Málaga, although the first certain record for Iberia and Europe was in early May 1966 at the Laguna de Calderón, Sevilla; what was perhaps the same bird then appeared in late May fairly close by, at Fuentedepiedra (*Ardeola* 12: 229; Guillou 1969). Six birds appeared at Fuentedepiedra on 21 June 1972, four of which were still there a month later (*Ardeola* 19: 14–15). One seen in Mallorca in 1988–89 was thought to be an escaped bird. The Spanish and Portuguese rarities committees have considered an increasing number of records since 1990, and other unsubmitted observations probably exist. By 2010 there were 42 accepted records for Spain and ten for Portugal, 72% of them during 2005–10.

The accepted records involved a minimum of 86 birds. Single birds were reported on 68% of occasions, and 23% of records involved two together, although the birds were nearly always accompanying Greater Flamingoes. Reports generally involve adult birds but a first-year bird was seen with an adult at La Vega del Jabalón reservoir, Ciudad Real, in October 2005. Lesser Flamingoes generally appear at lakes and saltpans. Most records have been in the south and east, although the lakes of La Mancha have also proved attractive (Fig. 70). There are records from all months (Fig. 71) but there is a spring peak in arrivals, with 51% of first dates falling in March–June. Regarding the accepted records from southern France up to 2009, 78% of first observations were during March–June and there were no new arrivals after July. There are winter records from both countries but they are a clear minority, particularly in France. Four adult birds were in Doñana on 14 January 2010 and at least one was still there the following October.

Iberian and southern French records fit geographical and temporal patterns that suggest the arrival of wild birds from Africa (de Juana 2006), probably accompanying Greater Flamingoes. Nevertheless, some escaped birds have certainly occurred. A ringed individual, of unknown origin, was seen at the Lagoa dos Salgados, Faro, in March 2007 (*Anu. Ornitol.* 6: 9). Also at least four birds that had been colour-ringed in captivity have appeared in Spain in recent years. Marked birds nonetheless make up too small a proportion to suggest that all the records involve escapes from captivity.

The first report of breeding was at Fuentedepiedra in 2007, where there were seven adults on 21 April, four of which were in attendance at two nests: a chick was seen there on 4 May (RBS). There were again seven birds at Fuentedepiedra in 2009 and a nest was built, but laying was not proved (RBS). Single pairs fledged a chick in 2011 at Manjavacas lagoon, Cuenca, and at Fuentedepiedra in 2013 (*Ardeola* 60: 454–455, RBS). Are we seeing the early stages of the colonisation of southern Europe by the Lesser Flamingo?

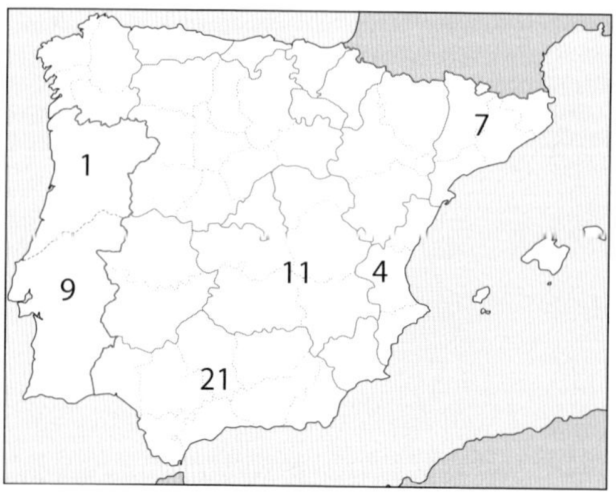

Figure 70. Records by region, 1966–2010 (n = 53)

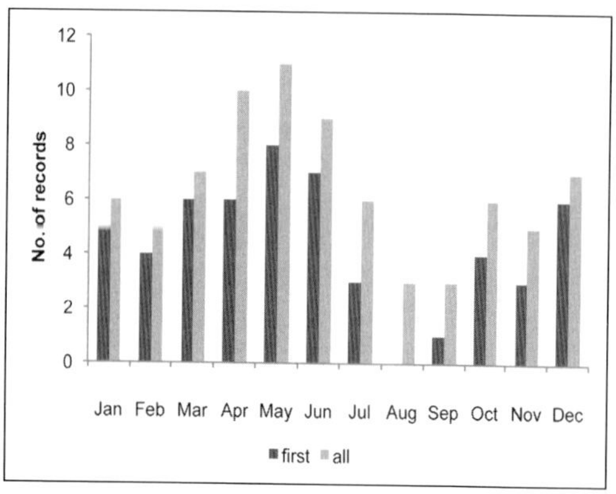

Figure 71. Records by month, 1966–2010 (n = 53w

EUROPEAN HONEY-BUZZARD *Pernis apivorus*

(Abejero Europeo, Bútio-vespeiro-ocidental)

Widespread but generally scarce as a breeding species, chiefly in northern areas. Common on passage in eastern and southern Iberia, when abundant at the Strait of Gibraltar.

The Honey-buzzard is relatively little known as a nesting species in Iberia but it is nonetheless widespread in some parts of the Peninsula, chiefly in the northern half, where it is most numerous in the cooler areas of the north-west, under Atlantic maritime influence. It thus inhabits the Cantabrian Range and the Biscay coastlands from Galicia west to the Pyrenees, ranging from sea level to some 1,600m (SA2). The largest regional population is in Galicia, followed by Castilla y León and Asturias. It also occurs throughout the Pyrenees east to Barcelona and Girona, although it is commoner in the western sector.

Further south it is widespread in the northern Iberian Range and across the Central Range, at 300–1,600m. Some breed in dehesas of Pyrenean Oaks and sometimes Lusitanian Oaks in western Castilla y León in Salamanca and Zamora (SA2). It is relatively common in northern Extremadura: in Pyrenean oakwoods of the southern foothills of the Sierra de Gredos and the Valle del Jerte, and has its southernmost regular outpost in Las Villuercas range, northeastern Cáceres. There have been occasional breeding reports from further south in Spain and the 2009–10 woodland raptor census (Palomino & Valls 2011) found some 20–30 territories in the southern Iberian Range, as well as a scattered breeding season presence in Andalucía and Murcia.

The Honey-buzzard is thinly distributed in Portugal, principally in the north-west, where it is relatively common in Peneda-Gerês National Park and in the sierras of the centre-north and north-east: in Trás-os-Montes, Beira Alta and Beira Baixa (PA2). There are no breeding records from Andorra although individuals have been observed there in July (ANA).

Breeding Honey-buzzards select areas offering deciduous woodlands with abundant clearings. They are especially associated with beechwoods and Pyrenean Oak woods, although some occur in a broad range of deciduous and coniferous woodlands (SA2). The farmland/woodland mosaics of the foothills of the Cantabrian Range and the western Pyrenees are especially favoured and Iberian breeding birds generally occur in areas where the spring rainfall exceeds 800mm (Palomino & Valls 2011).

The breeding population in Spain in the 1970s was estimated at 1,000–2,000 pairs (SA1). A revised estimate in 2003 suggested a population of 900–1,300 pairs, of which 80–90% were in the northern third of Spain (SA2). The 2009–10 census estimated the Spanish population at 1,850 pairs, about 60% of them in and north of the Cantabrian Range from Galicia to the Basque Country, and 28% in Castilla y León. Fewer than 8% were in the Pyrenees and north-east, from Navarra to Catalonia, which seems a low proportion, although this unobtrusive species is difficult to census.

The 2008 Portuguese Atlas confirmed nesting in 11 10km-squares, with probable nesting in 26 squares and possible nesting in a further 104 squares. The Portuguese population was estimated at 50–200 pairs by Birdlife International (2004).

The Honey-buzzard is best known in Iberia as a common migrant in the eastern half of the Peninsula, the Balearic Islands and in the south. Six ringed birds from Scandinavia, one from Finland and five from central Europe have been found in Spain. There are also five foreign recoveries of individuals ringed on passage in Spain: two from Morocco and one each from Sweden, France and Ivory Coast. The great majority travel via the Pyrenean passes in the north-east and the Strait of Gibraltar in the south. Among European broad-winged raptors it is one of the least dependent on thermals or upcurrents and hence is not as closely tied to short sea-crossings. It is also often on the move from first light and birds have been seen arriving at Gibraltar in spring an hour after sunset in May (pers. obs.), suggesting that some may move at night.

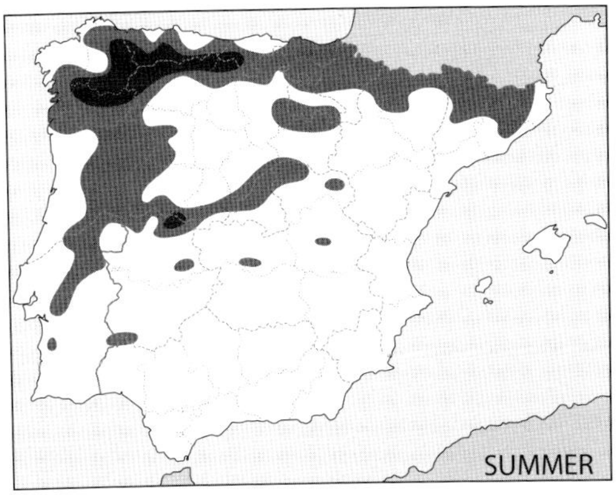

The spring passage at the Strait invariably peaks during the last few days of April and the first ten days of May, when over 75% of the birds pass. There is nonetheless a very prolonged arrival of presumed non-breeders that extends not just throughout June but also, on a very small scale, through July and exceptionally into August (Garcia & Bensusan 2006). The earliest known arrival at Gibraltar itself during 1964–2010 was on 4 April 1988 (*Alectoris* 8: 5) and there are very few records earlier than 20 April. The spring passage across the central Mediterranean, via the Sicilian narrows, also begins in late April, with few records before 20 April, and similarly peaks in early May (Corso 2001). Thus occasional claims of Iberian observations in February and March seem likely to involve misidentification.

The post-breeding passage follows a similar pattern and routing. Honey-buzzards enter Iberia via the Pyrenean passes from late July onwards, the movement peaking around 30 August (Urcun & Bried 1998b). Seasonal counts during 2009–11 at the three principal monitoring sites in the western Pyrenees averaged 26,700 birds and an additional 15,000 birds were counted on average in the eastern Pyrenees at Eyne during 2010–11. Southbound passage at the Strait begins in mid August and peaks during the last three days of the month and the first week of September. Very few pass later than the end of September but some stragglers occur to late October and very rarely into November.

The autumn passage at the Strait was of the order of 100,000 birds in the 1970s (Bernis 1980) and is still highly spectacular, but there are no recent coordinated Strait-wide counts available. The spring passage has never been fully counted but a recent comparison of rates of passage at Gibraltar itself during the spring seasons of 1967–72 and 1990–2004 has shown a significant decline of 46.5% (Bensusan *et al.* 2007). It is unclear whether this reflects actual population decline or whether a change in spring routing, with more birds returning to Europe via the central Mediterranean, is involved. In this context it is interesting to note that passage through the Balearic Islands can sometimes reach significant proportions. The Honey-buzzard usually occurs on passage in the Balearic archipelago in small numbers but 1,400 were counted at Pollença, Mallorca, on 4 May 2004 (*Ardeola* 52: 246). Passage dates in Mallorca in 2004 fell between 22 April and 30 June, and 29 August and 16 October, matching the usual passage periods at the Strait.

BLACK-SHOULDERED KITE *Elanus caeruleus*

(Elanio Común, Peneireiro-cinzento

***E. c. caeruleus*. Scarce but increasingly widespread in mainland Iberia. Mainly resident but dispersive.**

The sometimes elusive Black-shouldered Kite is now more widespread in Iberia than it has ever been, after a period of range expansion that began in the mid 20th century and accelerated rapidly from the 1970s onwards. Its elusiveness is due to its small size, its low population density within the immense areas of suitable habitat, its often crepuscular feeding habits and its secretive behaviour when nesting. This is countered by its habits of hovering when searching for its rodent prey and of sitting for long periods on exposed perches such as fenceposts and pylons.

The earliest Iberian records date from 1865 in Spain and 1867 in Portugal. Lilford (1865) was shown a recently killed specimen taken near Sevilla that April, one was seen near Vejer, Cádiz, on 10 May 1876 (Stark, in Irby 1879) and Chapman (1884) records a pair seen in April at Sanlúcar de Barrameda, Cádiz. Bernis (1961b) referred to two 19th century captures, in Cádiz and Granada. Professor Bocage showed a recently obtained specimen to Smith (1868), that was then the only known record for Portugal. Tait (1887) refers to two specimens in the Lisbon Museum that originated from Salvaterra, Santarém, and two more, from Montargil, Portalegre, in the museum at Coimbra. Its presence in Portugal during the late 19th century was known to Carlos de Bragança, who considered it very uncommon (AP). However, the regular presence of the species in Iberia was still in question as recently as the 1950s (Voous 1960). This encouraged England (1963) to search for it in Portugal. His enquiries established that it had nested in southern Portugal in 1944, when two nests were found, and that local people had some acquaintance with it. England succeeded in finding and photographing a nest in the same area in 1963. It seems therefore that the Black-shouldered Kite was a rare resident of southern Portugal from at least the 1940s onwards and it may be that its initial colonisation of Iberia began as early as the mid 19th century, although the early records may simply have involved individuals dispersing from Morocco. There are 19th century records from France (OF) so some vagrancy at least to western Europe existed then.

The Black-shouldered Kite was first found nesting in Spain as recently as 1973, in Toledo (*Ardeola* 19: 467–468), but by then there were widespread observations from western Andalucía, Extremadura and western Castilla y León (Ferrero & Onrubia 1998). Further confirmation of nesting in these regions followed from Salamanca (1975), Cáceres (1975) and Badajoz (1978) although nesting in Andalucía was not proved until 1986 in Huelva, followed by Córdoba (1991) and Cádiz (1992) (Ferrero & Onrubia 1998). Nesting birds were possibly overlooked in southwestern Spain earlier, as suggested by Collar (1978), but the first observations came shortly before the first recorded breeding attempts so that colonisation from Alentejo in the 1970s seems just as likely. Elsewhere, breeding was confirmed during the 1980s in León (1980), Ciudad Real (1984), Ávila (1985) and Valladolid (1985) and, much further afield, in Huesca, Aragón (1983).

Northward range expansion in Portugal by the 1980s was highlighted by Sacarrão & Soares(1985). It was also apparent in Spain by the 1990s that the species was spreading north and west from south-west Iberia and there followed breeding records from Palencia (1990), Zamora (1990) and Burgos (1994), as well as from Madrid and Segovia in central Spain and, in the north-east, in Lleida and probably also in Navarra and Soria. Eastward expansion was also noted in Castilla-La Mancha: into Guadalajara, Cuenca and Albacete, and in Andalucía: into Jaén, Málaga and Granada (Ferrero & Onrubia 1998, LR). The first

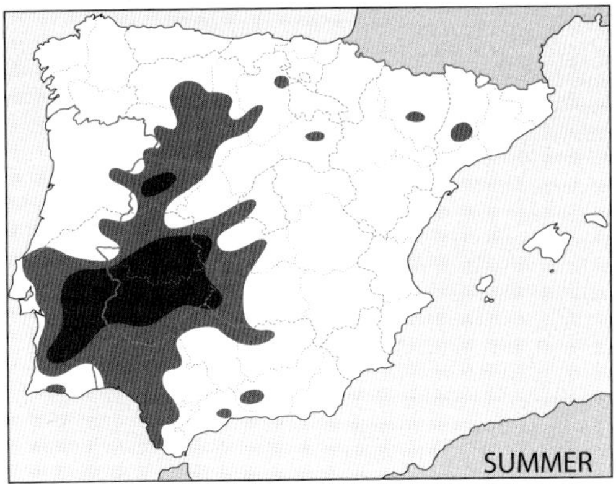

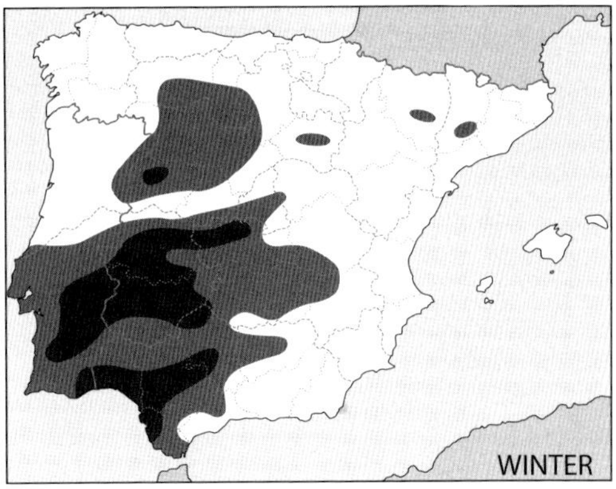

report of breeding in Catalonia was in 1998 (Clavell 2002) and in the Comunidad Valenciana a pair nested at Vilanova d'Alcolea, Castellón, in 2008 (*A. O. Castellón 2008*: 42). The first confirmation of breeding in Álava was in 2005 (*Ardeola* 52: 426). In 2010 the first nesting in Murcia occurred and a pair was found nesting in coastal Cantabria, the first recorded instance north of the Cantabrian range (*Ardeola* 57: 224, 524).

The post-2010 distribution is still clearly centred on southwestern Iberia where the range is largely continuous. In Portugal it takes in Alentejo and Ribatejo, with a presence also in Algarve and in the east of Beira Baixa, Beira Alta and Trás-os-Montes (PA2). In Spain the core range includes the Sierra Morena in Huelva and Sevilla, the Cádiz sierras, the whole of Extremadura, and also western Castilla y León in Salamanca, Ávila and Zamora, western Castilla-La Mancha in Toledo and Ciudad Real, and Madrid. Breeding in eastern and northern Iberia is widely reported but the range is fragmented and much potential habitat may yet remain unoccupied.

The size of the Iberian population remains largely conjectural. The Spanish population has been put as high as 1,000 pairs (SA1) but a more recent estimate is 350–800 pairs (LR). A breeding season survey in Andalucía in 2011 found a minimum of 177 pairs, with nesting birds in all provinces although 85% were shared between Sevilla, Cádiz and Córdoba (*Ardeola* 59: 424). The Portuguese population in the mid 1980s was estimated at 150–200 pairs (Rufino *et al.* 1985) and there is a later estimate of 300–1,000 pairs (BE), but that population too has never been rigorously assessed (AP).

The Black-shouldered Kite favours open country with scattered trees and an abundance of its preferred rodent prey. It nests in a wide range of tree species provided they are in suitable feeding habitat (Carbajo & Ferrero 1985, Silva & Beja 2001). The dehesas of southwestern Iberia are an ideal habitat provided that they are given over to crops, ideally cereals, rather than pasture: crop dehesas support many more rodents. Cereal dehesas have proliferated at the expense of grazing ones since the mid 20th century and this, together with an apparent increase in Common Voles across the northern meseta, may partly explain the ongoing increase and spread of the species (Balbontín *et al.* 2008, LR).

Black-shouldered Kites are perhaps easiest to observe outside the breeding season, at least locally, when some numbers gather at favoured locations in August–September and also in winter. As many as 65 have been reported in winter in the irrigated farmland of the Vegas Altas del Guadiana in Badajoz, in this case exploiting high densities of Mediterranean Pine Voles in alfalfa crops (LR). A census in western Badajoz in January 2004 found four roosts that totalled some 200 individuals (SWA). In Portugal, post-breeding roosts of several dozen birds are known from Ribatejo and Algarve and 82 were seen on 28 August 2004 in a small area of the humid flood plain at Ponta da Erva in the Tagus estuary (AP). Irrigated farmland has proved to be very attractive to them, especially in winter, when they seem to be most abundant in such habitats in the Duero, Tagus, Guadiana and Guadalquivir basins (SWA).

Some disperse after the breeding season, giving rise to observations all over Iberia. Wing-tagging and radiotracking studies in Extremadura have shown that females tend to disperse further than males, the longest movements being of some 300–400km (SWA). Some dispersed eastwards across the meseta and others moved north towards Galicia. The Galician bird reports give 18 records during 1993–2004, involving some 26 individuals, with annual occurrences since 1999. Most Galician records (14) fall between July and October with a few also in winter and the majority are from the east and south-east, in Lugo and Ourense provinces. There are at least five records from coastal Asturias, all of single birds and including one at Cape Peñas on 28 October 2007 (*Ardeola* 56: 157). Further east one was at Santurce, Bizkaia on 20 August 2009 (*Ardeola* 56: 351). There are a few records of single birds from both Navarra and La Rioja as well as more frequent but still scarce observations from all the provinces of Aragón and Catalonia. Reports from outside the known breeding range are generally more frequent in autumn and winter than during the breeding season, suggesting that post-breeding dispersal is a precursor to eventual settlement in a region (Fig. 72).

Figure 72. Records by month in Spanish regions where Black-shouldered Kites do not breed or did not then breed (n = 43).

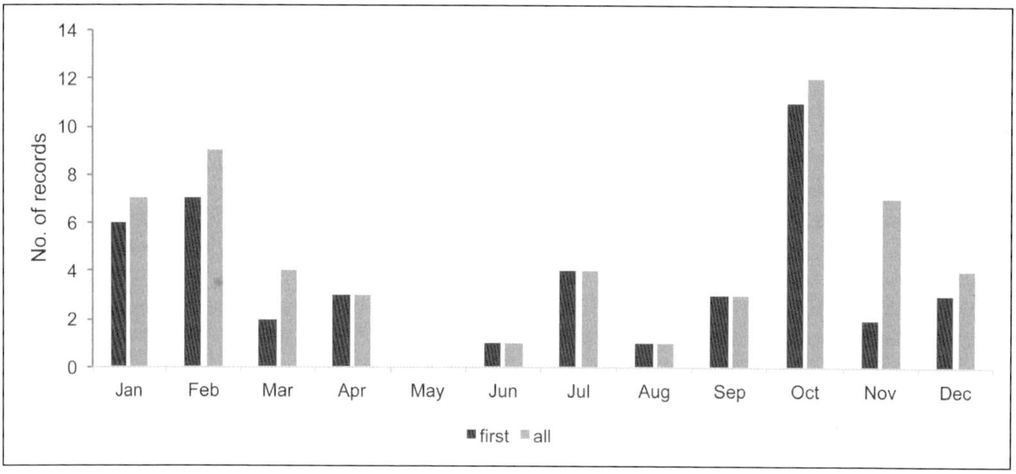

Black-shouldered Kites are also more widespread in Portugal outside the breeding season, occurring then in parts of Algarve and Beira Litoral where they do not nest, but they are still most abundant in winter within the breeding range (AP). Two Spanish-ringed birds have been recovered in Portugal (AP).

Noteworthy examples of post-fledging dispersal of birds ringed as pulli include one ringed in Badajoz in April 1995 that was recovered in Burgos, 381km north-east, the following November, and another ringed in Cádiz in May 2008 that was also found in Burgos (662km) in July 2009. A few individuals have been seen crossing the Strait of Gibraltar during the main raptor migration periods: there were at least 11 records during 1977–2009, seven of them at Gibraltar itself (GONHS). Although there is no suggestion of regular trans-Strait movement on any significant scale, it is clear that the Iberian and Moroccan populations remain in contact.

BLACK KITE *Milvus migrans*

(Milano Negro, Milhafre-preto

***M. m. migrans*. Widespread and locally abundant in the west and north of mainland Iberia but uncommon in the Balearic Islands. Largely a summer visitor and common on passage, when abundant at the Strait of Gibraltar.**

The Black Kite is characteristic of western Iberia, including most of Portugal, western Andalucía (Huelva, Sevilla and westernmost Cádiz), all of Extremadura, Madrid and western Castilla y Léon. In Spain within this region the largest nesting densities are in Cáceres, Salamanca and Zamora (Palomino & Valls 2011). Elsewhere in Spain it is common in the Ebro valley and increasingly so in the Biscay coastlands (SA2; ; LR; Palomino & Valls 2011). It is progressively more uncommon eastwards across Castilla y León, Castilla-La Mancha and eastern Andalucía, and scarce or absent along most of the Mediterranean coast, especially in the arid south-east. It does not nest in the Balearics but occurs there as a scarce or rare migrant: a few have summered in Menorca (*A. O. Balears 2010*: 262). It does not nest in Andorra but occurs there on passage.

The breeding range in Portugal has shown some expansion in the north and some contraction in the south, in Algarve (PA2). In Spain the Black Kite has become conspicuously more common in the breeding season in northern Cantabria in recent years, where it is now a familiar sight within the city of Santander. It has also become more widespread in Asturias and Galicia. By way of contrast, Irby (1883) spent seven weeks in Cantabria in May–June 1876 but saw only one individual, whereas Ticehurst & Whistler (1928) did not record any during two weeks in Galicia in May 1927.

Black Kites in Iberia are chiefly lowland birds, nesting below 1,000m. They are especially common in the major river valleys and some are clearly attracted to waterside habitats, including reservoirs, where they scavenge shorelines and snatch dead fish, and unwary live ones, from the water surface. They are associated with open broadleaved woodlands, notably grazing dehesas, but pines are important for nesting in Andalucía. Totally treeless areas are avoided in the breeding season. They are attracted in large numbers to municipal rubbish dumps – and often nest nearby. They often scavenge roadkills – and sometimes get killed themselves: Black Kite (and Red Kite) ringing recoveries due to roadkill are relatively frequent in comparison to those of other raptors (de Juana 1989). Numbers are also attracted to carrion, including the large carcases put out for vultures. The birds are sociable and often form loose colonies when nesting. Nests are usually in trees but sometimes also on riverside cliffs (Arroyo 1976).

The population includes a large proportion of non-breeders, both immature and adult. These flock at rubbish dumps and other reliable food supplies and also roost communally. The existence of a large contingent of non-breeders has been a complicating factor when censusing the breeding population and it is exacerbated by their tendency to be somewhat nomadic. Large communal roosts also occur ahead of the breeding season in February–March (see below) and post-breeding: in June–August. In summer large numbers join White Storks and Cattle Egrets to feed on the abundant grasshoppers that then swarm on the plains of La Serena, Badajoz, and other grassland areas. The numbers using a rubbish dump at Segovia in 2006 increased from around 60 in April to over 170 in May and June, over 300 in July and nearly 400 in August, before dropping sharply to fewer than 50 in September (Donázar 1992).

The Spanish population was censused for the first time in 2005 (Palomino 2006). The result was a population estimate of 10,295 breeding pairs in Spain. More than half of these were in Castilla y León (*c.* 3,700 pairs) and Extremadura (*c.* 3,000 pairs). In addition, the non-breeding adults and immature birds were estimated at *c.* 52% of the breeding season population, which therefore amounted to nearly 43,000 individuals. The Spanish breeding population had been estimated on four occasions prior to the 2005 census but the results are of unknown validity. Garzón (1977) suggested a population of 25,000 pairs. F. de Juana (1989) considered that there were then only 9,000 pairs. The second breeding atlas suggested at least 8,800 pairs (SA2 2003). Finally, the Spanish red book put the population at not more than 5,000 pairs (LR). These figures suggested a significant decline, that was attributed among other reasons to the impact of poison baits and agrochemicals that were implicated in declines of the Egyptian Vulture and Spanish Imperial Eagle, among others, in the 1970s and 1980s. More recent observations from both Doñana and Madrid suggested stable or slowly increasing breeding populations there (Palomino 2006).

Table 18. The breeding populations (pairs) of Black Kites in the Spanish regions in 2005 and 2009–10. After Palomino (2006) and Palomino & Valls (2011). In 1996–2001 Catalonia had a population of 30–50 pairs and Murcia 1–3 pairs (LR).

Region	2005	2009–10
Andalucía	1,149	1,470
Aragón	650	1,140
Asturias	48	230
Basque Country	411	340
Cantabria	283	280
Castilla y León	3,737	4,290
Castilla-La Mancha	297	1,190
Catalonia	No count	95
Comunidad Valenciana	13	82
Extremadura	2,995	2,310
Galicia	560	660
La Rioja	304	150
Madrid	258	330
Murcia	No count	34
Navarra	457	450
Total	10, 295	13,060

Black Kites were again censused in Spain in 2009–10 (Palomino & Valls 2011) and the population was found to have increased further, to 13,060 pairs, with Castilla y León and Extremadura once again together accounting for more than half (Table 18). The 2009–10 results are also noteworthy for an apparent enormous increase (by over 300%) in the population of Castilla-La Mancha since 2005. A large increase was also noted between the two periods in Aragón and the small east coast populations have apparently also increased since 2007. Only La Rioja shows a very large decline (about 50%) since 2007 but the apparent decline by 23% of the key population in Extremadura is also noteworthy. If the non-breeding fraction of the breeding season

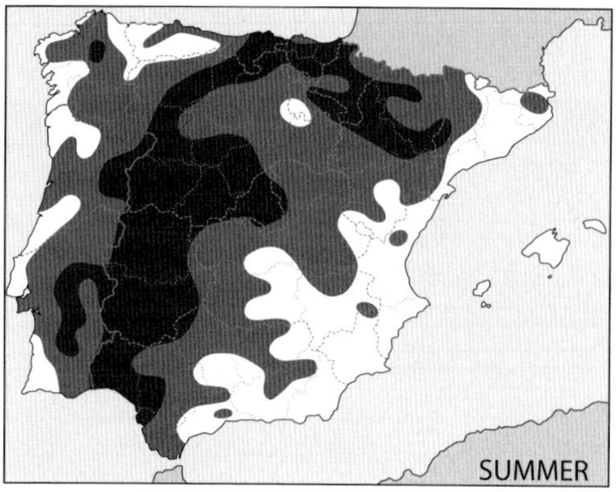

population is again taken as 52%, as in 2005, the total population in 2009–10 was *c.* 54,400 birds, an increase of about 27% since 2007. Such an increase seems very large for an interval of just 4–5 years and differences in methodology, and in the completeness of counts between different regions, could well account for some of the apparent change.

There is no equivalent census information available from Portugal but the Black Kite is widespread there except in the north-west, Estremadura, coastal Alentejo and Algarve, where the range is more fragmented (PA2). The Portuguese population was put at 800–1,600 pairs and considered stable by Birdlife International (BE). It showed no clear trend during 2004–09 (ACP). Earlier estimates put the population at 900–1,200 pairs in the early 1980s and 500–800 pairs in the mid 1990s (AP), a pattern also seen in Spain. The very large French population: 19,300–24,600 pairs (Thiollay & Bretagnolle 2004) is thought to be stable or increasing.

The Black Kite is a conspicuous migrant across the Peninsula, especially at the Strait of Gibraltar, where the Iberian birds are joined by most of those from France and some from Germany and Switzerland. Most ringing recoveries of extralimital origin are of birds ringed in Switzerland, Germany and France. Two Polish-ringed birds have been recovered in Spain but this is probably exceptional. The continental birds figure prominently at the western Pyrenean passes, notably at Organbidexka where the numbers passing have increased eightfold since the 1980s (OF). Black Kites cross the Pyrenean passes from early July onwards, the movement peaking in early August (Urcun & Bried 1998b). Seasonal counts during 2009–11 at the three principal monitoring sites in the western Pyrenees averaged *c.* 40,000 birds but comparatively few cross the eastern Pyrenees, where counts at Eyne during 1986–2011 averaged some 2,000 birds. There is also evidence of a long-term increase in the numbers crossing the Strait. Mean passage rates at Gibraltar in spring showed a highly significant increase from about 25 birds/hour during 1967–77 to 54 birds/hour during 1990–2004 (Bensusan *et al.* 2007), an increase of 119%. All the indications therefore are that the source populations were very healthy at the start of the 21st century.

Most Black Kites winter in sub-Saharan Africa. There are 39 recoveries of Spanish-ringed birds and one from Portugal from a region extending from Mauritania and Mali in the north to the coastal fringe between Senegal and Benin in the south: 13 were in Ghana. There are also 65 recoveries of Iberian-ringed birds found on passage in North Africa, mainly in Morocco but 11 in Algeria and one in Tunisia. The southward passage at the Strait starts early, at the beginning of July, with a pronounced peak during the first half of August and some further passage until mid September, after which very few pass. The northward passage is much more prolonged. Some early individuals arrive in January, or even in late December, but passage proper begins early in February. There is a clear peak during the first half of March, when most of the Iberian breeding population arrives. Arrivals continue through April, with a second peak of presumed non-breeders from mid April until mid-May, and a trickle of late birds right through June (Garcia & Bensusan 2006).

The numbers of Black Kites that cross the Strait are hard to establish. On the basis of the Iberian and French populations alone, and allowing a very conservative one juvenile per pair, the post-breeding population would be of the order of 105,000 birds to which an unknown but high number of non-breeders, perhaps 75,000 birds, also needs to be added. Thus well over 180,000 Black Kites cross the Strait southwards during July–September, making this the most abundant of all the raptors using the flyway.

Strong easterly winds especially often delay crossings during both seasons. Such conditions during July–August lead to enormous accumulations of kites that are waiting for the conditions to improve, which may take several days. The resulting rush when the birds are finally able to cross is one of the most spectacular avian sights in Iberia, termed by local birders a '*milanada*' or 'kitefest'. Thousands of birds may rush southwards in just a few hours at such times. On 2 August 2011 over 24,000 crossed the Strait between 11.00 and 18.00 hours (Mosquera 2011). Similar mass arrivals also occur in spring, when poor weather has caused migrants to accumulate on the African side of the Strait.

Some Black Kites are reported in Iberia in winter, defined as November–January, but birds at the extremes of this period could include late/early migrants respectively. Most records are from mainland Spain. Very few individuals are involved and these are often seen with Red Kites at municipal rubbish dumps. The winter atlas programme detected a minimum of 85 individuals in December or January over the three-year period, principally in the Guadalquivir valley, although others were found on the meseta, the lower Ebro valley and the Levant coast (SWA). In contrast, Black Kites were unusually numerous at Córdoba rubbish tip during winter 2011–12: 50 birds there on 3 December had increased to 237 by 17 December but the birds soon dispersed since there were only ten there on 16 January (*Ardeola* 59: 177). The early return of many Black Kites is exemplified by a count of over 1,000 at a rubbish dump roost at Puebla de Cazalla, Sevilla, on 16 February 2003 and of 6,000 at Córdoba rubbish tip on 26 February 2010 (*Ardeola* 51: 248, 57: 525).

RED KITE

Milvus milvus

(Milano Real, Milhafre-real)

***M. m. milvus*. Resident in mainland Iberia, chiefly in the west, and also in the Balearic Islands. More abundant in winter.**

The principal breeding areas extend from central and eastern Portugal, western Andalucía, Extremadura and western Castilla y León, eastwards across the central and northeastern Meseta to the Pyrenees, although the breeding distribution is far from uniform. Treeless regions are avoided, as are the arid regions of eastern Iberia and those influenced by Atlantic climate in coastal northern Spain and western Portugal. There is a small resident breeding population in the Balearic Islands in Mallorca: in the Serra de Tramuntana, and in Menorca.

Breeding Red Kites are characteristic of undulating landscapes offering a mixture of pastures, non-irrigated crops and small woods or open woodlands. Trees of a wide range of species provide nest sites. Dehesas are favoured, as are riparian woodlands. The birds flock to such anthropogenic food sources as municipal rubbish dumps and carrion dumps (Donázar 1992) and often nest nearby. Loose colonies occur in suitable areas. Red Kites are absent from the highest mountain areas but are often found in the foothills and valleys. The pinewoods of Hoyos del Espino, at 1,400m on the Sierra de Gredos in the Central Range, still hold a significant population, as was originally noted by Witherby in 1928 (pers. obs.).

A large-scale recent decline of both resident and wintering Iberian Red Kite populations, that seems largely attributable to deaths from eating poisoned baits illegally set out for mammalian 'vermin' (SA2, Cano *et al.* 2008), has given cause for concern. The Red Kite is also particularly susceptible to electrocution and it has suffered direct persecution from rabbit hunters in some regions (Villafuerte *et al.* 1997). However, it is still common and locally numerous in many areas and the most recent population census (Palomino & Valls 2011) suggests a widespread recovery in breeding numbers in Spain. The Spanish breeding and wintering populations were first censused in 1994 (Viñuela *et al.* 1999), and then again during 2004 (Cardiel 2006). The nesting population was again censused in 2009–10 (Palomino & Valls 2011).

Table 19. Breeding populations (pairs) of the Red Kite in Spain in 1994 and 2004. After Viñuela *et al.* 1999, Cardiel (2006) and Palomino & Valls (2011)

Region	1994	2004	2009–10
Andalucía	85–130		500
Aragón	391–537	253–263	120
Balearic Islands	41–48	27	34
Basque Country	35–37	11–16	26
Cantabria			3
Castilla-La Mancha	47–62	20–23	350
Castilla y León	1,739–1,935	1,102	1,780
Catalonia	10–15	32	44
Extremadura	650–910	250–314	870
Galicia			17
La Rioja			18
Madrid	65–70	36	140
Murcia			19
Navarra	270–300	263	65
TOTAL	3,333–4,054	1,994–2,176	3,980 (3,810–4,150)*

* 90% confidence limits

A comparison of the 1994 and 2004 results shows a widespread decline in the breeding population that averages 46% (Table 19). These declines were especially marked in the species' strongholds in Extremadura (-62%) and Castilla y León (-57.4%) and only the very small population of Catalonia showed an increase. Andalucía was not censused adequately in 2004 but

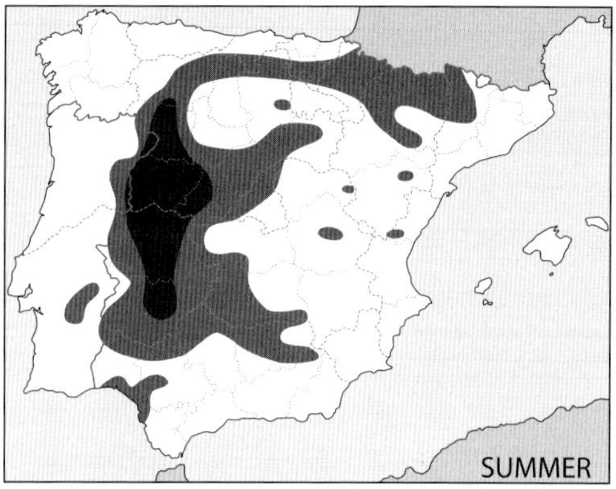

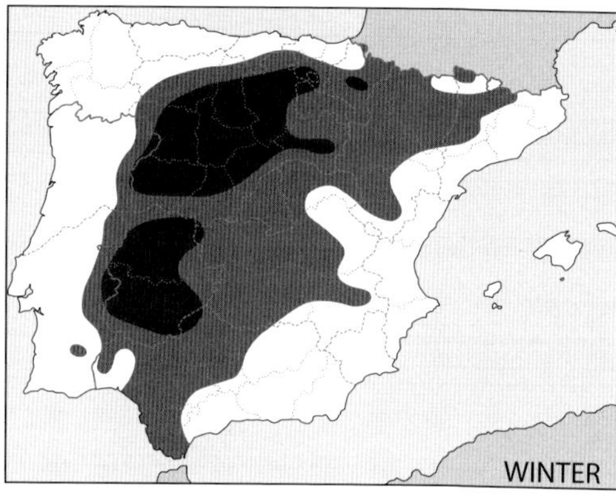

the population there too was known to have declined. However, the 2009–10 census suggests that the Red Kite had by then largely recovered its breeding population in Spain. Nonetheless, the 2009–10 figure of an estimated 500 breeding pairs in Andalucía contrasts with a regional census of just 39 pairs in 2009, most of them in and around Doñana (*Ardeola* 59: 424) and it may be prudent to consider the larger figure, which is an extrapolation from sample data, as provisional. The same caveat applies to the other Spanish regions (Blanco *et al.* 2012).

The Balearic population, in Mallorca and Menorca, was also censused in 2008 and showed a small increase since 2004 to 30 pairs in total, which fledged 14 young in Mallorca and 17 in Menorca (*Ardeola* 56: 351). This is a welcome improvement on the estimated 4–6 pairs that remained in 1999 following high mortality brought about by poisoning, electrocution and shooting, but is still well below the 135 pairs found during the late 1980s (Adrover *et al.* 2003, de Pablo & Pons 2003). A recovery plan for the species is in place on the islands and appears to be having some success.

The small Portuguese population has also seen recent decline. The Red Kite was considered common in central and southern Portugal especially during the late 19th and early 20th centuries but a marked decline was noted by the 1960s (Tait 1887, AP). The breeding population in the mid 1980s was estimated at 100–120 pairs. However, a census in 2001 only found 36–67 pairs and recent atlas work (PA2) has shown range contraction away from central Portugal, so that it is most numerous north of the Tagus and in the eastern regions bordering Extremadura and Castilla y León. The range is fragmented south of the Tagus and it does not breed in Algarve or in the western coastlands, although some occur there in winter.

There has also been a parallel change of similar magnitude in the wintering population in Spain. Iberian birds are largely resident. Hence the wintering population includes local birds as well as immigrants from central and northern Europe. The total population, *i.e.* residents plus immigrants, was put at 66,235–72,165 birds in 1994. In 2004 there were only 35,223–36,233 individuals, a mean decline of over 46%. The immigrant fraction has been estimated by deducting the resident birds from the total censused in winter, taking the residents to be the number of breeding pairs plus one third to include immatures (Table 20). Taking immigrants alone, the decline in the wintering population has been 45.8%. Castilla y León has seen an especially large decline (53.9%) in immigrant numbers. In contrast some northeastern regions: Navarra, the Basque Country and Catalonia, but not Aragón, have shown stable or increasing immigrant populations. Although actual decreases in Red Kite populations seem to underly the overall fall in wintering numbers, it may be that some of the apparent decline in Spain is due to fewer birds travelling as far south, as seems to occur in the Common Buzzard (q.v.). Wintering in France was first reported in 1964 and 5,800 wintered there in 2007 (OF). A similar pattern has also been reported in Switzerland, Germany and Sweden, all of which now support large wintering populations (de Juana *et al.* 1988).

Wintering Red Kites are most abundant within the Iberian breeding range but they also range more widely over open agricultural land, notably in the Duero and Ebro basins (SA1), returning to roost communally in copses or, sometimes, on electricity pylons. Very few occur in the Mediterranean coastlands. Winter roosts often exceed several hundred individuals and sometimes up to 1,000 birds may be present, using the same traditional sites in successive years (LR).

There are further indications that the wintering range in Spain has contracted northwards. Ringing recoveries in the southern half of Spain made up some 60% of the total during 1956–66 but only 26% during 1967–86 (de Juana *et al.* 1988). A decline in wintering numbers on the southern meseta was noted by Viñuela *et al.* (1999) and a more recent larger decline has been reported from the south-west: in Badajoz, Huelva and the lower Guadalquivir (SWA). Any range contraction may reflect a broader tendency to 'short stopping' but Viñuela *et al.* (1999) highlight the greater availability of food on the northern meseta offered by the periodic vole plaques and the large number of abbatoirs and pig farms there.

Table 20. Regional distribution of immigrant wintering Red Kites in Spain in 1994 and 2004. After Cardiel (2006).

Region	1994	2004	% change
Andalucía	1,716–2,579	?	?
Aragón	4,464	3,082–3,397	-30.96
Balearic Is.	0–23?	23	
Basque Country	76–112	239–249*	Positive
Castilla y León	36,199–39,324	16,683	-53.91
Castilla-La Mancha	909–1,259	591–677	-34.98
Cantabria	130–150	117	-10
Catalonia	105–170	224	113.33
Extremadura	8,253–10,733	6,336–6,485	-23.23
La Rioja	410	207	-49.51
Madrid	40–105	152–392	280
Navarra	1,698–1,788	1,733	2.06
TOTAL	54,000–62,140	29,289–30,094	-45.76

* Incomplete data.

Many of the Red Kites that enter Iberia to winter are of German origin: German-ringed birds comprised 351 (84%) of 420 foreign-ringed birds recovered in Spain, but there are 33 recoveries of birds ringed in Switzerland, a few from France, Denmark, Sweden or the Czech republic and two of Polish-ringed birds. Fifteen of 17 foreign-ringed Red Kites recovered in Portugal were also from Germany. The populations of these countries are only partly migratory and, as noted, may be becoming more sedentary, but large numbers enter Spain from early September to mid-November across the Pyrenean passes, where the movement peaks in mid October (Urcun & Bried 1998a). Most cross the Pyrenees in the west, where the combined autumn counts from Organbidexka, Lindux and Lizarrieta averaged some 10,000 birds until 1990, before declining to just 2,000 in the early years of the present century and then increasing once more to average 9,454 birds during 2009–11. Only a few dozen cross on the eastern route, at Eyne. The return movements across the Pyrenees extend from late February to the end of May.

Very few cross the Strait of Gibraltar. Systematic autumn counts in the 1970s gave a maximum of 103 Red Kites, in 1974 (Bernis 1980). Such Strait-wide censuses have not been repeated since but there is some evidence that even fewer now cross the Strait since none at all have been seen at Gibraltar in some recent spring seasons, despite an increased watch intensity there. Most migrants at the Strait cross during September–November and March–April but there are records as early as late July southbound and in February and May northbound. There are three recoveries from Morocco of Red Kites ringed in Spain and also a recovery from Ghana of a putative Red Kite pullus ringed in Cantabria: the latter seems far more likely to have been a young Black Kite, ringed in error.

WHITE-TAILED EAGLE *Haliaeetus albicilla*

(Pigargo Europeo, Pigargo)

Vagrant to Spain.

The Western Palearctic breeding range contracted under persecution during the 19th and 20th centuries, although some ground in the north-west has since been recovered, partly following reintroductions. Adults in Europe tend to be sedentary but immature birds disperse more widely in winter. However, they are rarely seen in the south-west and only 20–25 individuals winter in France, chiefly in the north-east but also regularly in Les Landes and Camargue.

The species may once have nested in Iberia (Bernis 1995) but the available evidence is highly inconclusive. A record of three juveniles seen on Dragonera, Balearic Islands, by Homeyer (1862–63) is considered doubtful (Bernis *et al.* 1958). There is a reference to a nest containing a chick on cliffs in an unspecified location of the Strait of Gibraltar in May 1876 (Irby 1879) and an ancient nest-site that may have belonged to this species has been identified on the Chafarinas islands, in the south

Alborán basin (de Juana 2006). The White-tailed Eagle is likely to have visited Iberia more frequently in the past, when it was more widespread in Europe, but there are very few records from any period.

The late 19th century records include one taken in Cádiz (Saunders 1871), a female killed at Doñana in December 1898 (Chapman & Buck 1910) and one taken at the Albufera de Valencia (Arévalo 1887, Boscá 1916). In Portugal, the King, Dom Carlos de Bragança, shot a juvenile female at Cascais, Lisboa, on 4 October 1902 (Soares 1970). These early records were followed in the mid 20th century by the collection of single immatures at Xilxes, Castellón, in late 1942 (Gil Lletget 1945); at Santoña, Cantabria, in December 1944 (de la Lama 1959a) and at the Ebro delta in February 1953 (Palaus 1960). One ringed as chick in southern Sweden in June 1946 was recovered near the Albufera de Valencia on 16 October same year (España Cantos 1948). In addition, in 1973 a mounted specimen was found in private hands at Penhas de Saúde, Castelo Branco, the bird having been killed by a hunter some years earlier in the Serra da Estrela (*Ardeola* 20: 348–349). The low point in the species' fortunes in western Europe during the second half of the 20th century probably accounts for the accompanying gap in Iberian records during that time. Similarly the positive trends shown recently in several countries (BE) may explain the observations that have occurred in the region since 2001.

These recent records all involve single immature birds in winter. There were seven accepted records during 2001–11: at the Serra de Tramuntana, Mallorca, from December 2001 to February 2002; at Santoña from December 2003 to March 2004; at the Embalse de Las Cañas, Navarra, from December 2004 to February 2005; at Zugaire, Navarra, on 29 December 2005; at Puértolas, Huesca, February 2007; at the Aiguamolls del Empordà, Girona, during February and March 2011, and at Carcastillo, Navarra, from December 2011 to January 2012. Another bird was at several sites in Extremadura from December 2012 to at least January 2003 (RBS). All the records fall between October and March, as is typical in France (OF) and most have been in northeastern Spain, in a region extending from Cantabria to Catalonia. Rings indicated that the 2004–05 bird in Navarra was hatched in 2004 on the Swedish Baltic coast (*Ardeola* 53: 172).

LAMMERGEIER *Gypaetus barbatus*

(Quebrantahuesos, Brita-ossos)

***G. b. barbatus*. A scarce resident in the Pyrenees and the subject of reintroduction projects in the Cantabrian Range and Sierras de Cazorla and Segura. Very rare elsewhere.**

The recovery of the Iberian population of the Lammergeier from its nadir in the 1980s is a notable and ongoing success story for conservation. The 2010 population in the Spanish Pyrenees was in the region of 250 mature individuals, plus a similar number of juvenile or subadult individuals. It is also by far the largest population in Europe and it is increasing significantly. The Pyrenean population inhabits both the high central axis and the flanking pre-Pyrenean sierras.

The Lammergeier was much more widespread until the end of the 19th century. It inhabited all the main mountain ranges in addition to the Pyrenees, nesting across the Basque mountains and the Cantabrian Range into Portugal; in the Iberian, Central and Betic Ranges; the Montes de Toledo and the Sierra Morena (Hiraldo *et al.* 1979). Lilford (1866) and Castellarnau (1877) saw it in the Sierra de Guadarrama, and Chapman & Buck (1910) in the Sierra de Gredos. Irby (1883) also found it common in the Picos de Europa, around Potes, where Chapman & Buck (1893) mention the Deva gorge as very attractive to it. In the south Lilford (1865) reported that pairs could be found throughout the Serranía de Ronda and Lynes (1912) saw a nest there near Grazalema in March 1910. Chapman & Buck (1893) refer to a nest on the Sierra Bermeja, Málaga, on the Mediterranean coast, and there was also a nest at the Desfiladero de los Gaitanes, Málaga (Saunders 1869). Saunders (1871) found it to be especially common in the Sierra Nevada, where Lynes (1912) also often saw it in spring 1910. It disappeared progressively from most of these in the face of relentless persecution. The collection of eggs and specimens for scientific collections is also likely to have hastened its decline: Hiraldo *et al.* (1979) found no fewer than 35 clutches in diverse European museums that had been taken from Andalucía alone. By the 1950s the only extra-Pyrenean locations in Spain in which the Lammergeier remained were the north of Burgos, where Valverde (1956) saw a pair with a nest at the Pancorbo gorge in April 1954, and in the Sierra de Cazorla, Jaén (Cano & Valverde 1959). It disappeared from the latter around 1987 (Máñez 2001). Its

historical presence in Portugal is poorly documented but King Carlos remarked on its presence in the north-east, in the Serra do Marão, during the late 19th century, and he was responsible for the collection of two specimens in the Guadiana valley in 1888 (AP). There have been no Portuguese records since then.

The Lammergeier is strongly associated with rocky mountain ranges, where it nests in inaccessible caves on remote cliff faces at 600–2,300m above sea level (Heredia & Heredia 1991, Margalida & Heredia 2005). It patrols the high tops and open mountain sides in search of carrion. At the highest altitudes this often takes the form of chamois and marmots, with sheep, cattle and horses predominating elsewhere (Margalida *et al.* 2009). As is well known, the Lammergeier is a specialist consumer of bones, which it may smash to pieces by dropping them from a height on to rocks, before swallowing the fragments. It visits carcases with other vultures and readily attends 'vulture restaurants' and carrion dumps where available.

The population suffers losses due to poisoning, collisions with power lines and illegal hunting (LR, Margalida *et al.* 2008) but these have diminished sufficiently to allow a progressive recovery in numbers (Fig. 73). The Spanish population was reduced to some 40 pairs by the late 1980s but has since increased progressively. In 2010 there were 105 breeding pairs in Spain plus another 12 occupied territories (Margalida 2011). In addition, there was one pair in Andorra and 35 pairs in the French Pyrenees. The distinction between territories and breeding pairs arises because some pairs may settle for several years before breeding and some may not nest every year. It is also a feature of the Pyrenean population that up to 15% of territories are occupied by polyandrous trios i.e. two males and one female, instead of pairs, which may be an indicator of habitat saturation (Donázar 1990). A 'foursome' has also been reported from Aragón (Gómez de Segura *et al.* 2012).

Figure 73. The increase in the Lammergeier breeding population in Spain, 1986–2011 (after SEO/Birdlife 2010 and A. Margalida, pers. comm.)

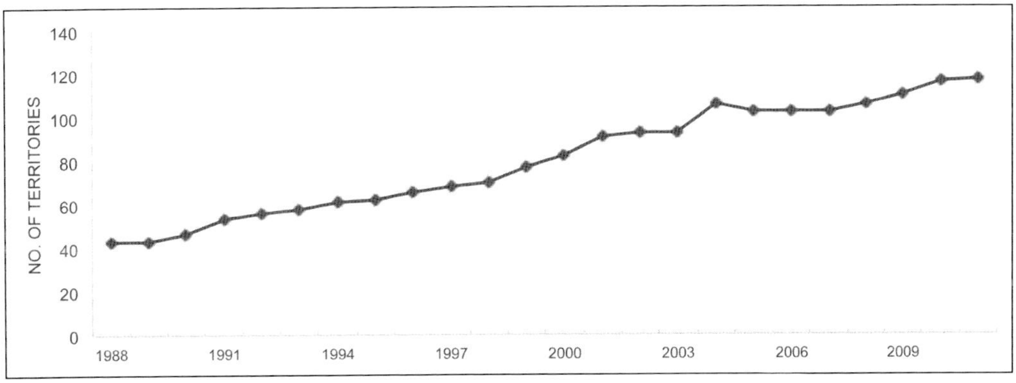

The great majority inhabit the Pyrenees. There the core population is in Aragón, where there were 53 pairs in 2001, followed by Catalonia and Navarra, with respectively 30 and seven pairs at that time (SA2). There are also a few pairs in the Basque Mountains, within Navarra, some 40km from the nearest Pyrenean birds (SA2, LR). The Spanish territories in 2010 comprised 71 in Aragón, 38 in Catalonia and eight in Navarra (Margalida 2011). The Spanish population has low productivity. A Pyrenean study population (n = 115 pairs) averaged 0.43 fledgings per breeding pair per year during 1992–99 and productivity declined from 0.57 in 1992 to 0.31 in 1999, possibly due to a decline in the mean age of the population and an increase in breeding density (Margalida *et al.* 2003, Carrete *et al.* 2006). Nevertheless, the survival rates of juveniles and subadults are high enough to have allowed the population to increase at a steady rate, averaging 5.6% during 1995–2002 (Heredia & Razin 1999).

Established adults are sedentary but immature birds and non-territorial 'floaters' tend to move to the central sector of the southern side of the Pyrenees in winter, returning to the northern side in the central and western Pyrenees in summer (LR). A few individuals, especially young birds, disperse much further, and they are being observed away from the Pyrenees more often as the population has grown. Known extra-Pyrenean sightings amounted to just four during 1960–80 but there were 14 in the 1980s and 87 during 1990–2002 (Heredia & Margalida 2003). These last comprised 60 in the Cantabrian Range, 17 in the Iberian Range, seven in the Cordillera Penibética and three in the Central Range. An immature bird was at Guadalupe, Cáceres, on 31 May 2012 (*Ardeola* 59: 425). There are several relatively recent records from both sides of the Strait of Gibraltar and it is quite probable that the occasional individual crosses the Strait. On the southern shore two were seen at Jebel Musa on 6 March 1971 and an adult was at Tangier on 27 December 1972 (Pineau & Giraud-Audine 1979). On the Iberian side an immature bird was over Gibraltar on 28 August 1998 and one was seen near Tarifa on 6 August 2005 (GONHS, *A. O. Cádiz 2004–2008*: 55): we are also aware of several unpublished records from the area.

Far-ranging birds hold out some possibility of natural recolonisation of lost range but this is countered by their strong tendency to return to their natal area to breed (Margalida & Heredia 2005). Any possible natural range expansion is thus likely to be extremely slow (Hiraldo *et al.* 1979). Steps have therefore been taken to reintroduce birds to their former haunts. A captive breeding station, the Centro de Cría Guadalentín, was established in 1996 at Cazorla in Andalucía (Simón 2003). Since then a total of 39 chicks have been reared and 23 juveniles had been released in the area up to 2012, mostly at the Sierra de Segura (www.gypaetus.org). Some of the individuals released at Cazorla/Segura have travelled long distances and have been tracked by a combination of satellite telemetry and observation of wing tags: ranging through Andalucian sierras: notably the Sierra Nevada, but also as far as the Pyrenees and Cantabrian Range. A parallel reintroduction programme in the Cantabrian Range began with the release of two juvenile females, obtained from Aragón, in the Picos de Europa in July 2010 (Aja 2010).

EGYPTIAN VULTURE *Neophron percnopterus*

(Alimoche Común, Britango)

***N. p. percnopterus*. Widespread and locally common in Spain, eastern Portugal and in the Balearic Islands. Chiefly a summer visitor.**

The Egyptian Vulture is typical of broken country, nesting on ledges and, particularly, in small south-facing caves on rock faces of all kinds (Ceballos & Donázar 1988), from sea level up to around 1,800m. Its Spanish breeding range is markedly discontinuous, so that a census in 2008 (del Moral 2009a) found 50% of the population within just seven provinces, five of them in the north-east: Burgos, Navarra, Huesca, Zaragoza and Guadalajara, and two in the west: Cáceres and Salamanca. The nesting distribution is most continuous in the north and north-east, taking in the Cantabrian Range from Asturias east to and including the Pyrenees, and extending throughout the Iberian Range and the upper and central Ebro valley. A second centre of population is in west-central Spain, including Zamora and Salamanca in Castilla y León, the whole of Extremadura, and the western Sierra Morena in Castilla-La Mancha. Smaller subpopulations are in Andalucía in the Cádiz ranges and the Sierras de Cazorla and Segura, Jaén (LR). It is largely absent when nesting from the Mediterranean coastlands, the Guadalquivir valley, central and northeastern Castilla-La Mancha and the agricultural meseta in Castilla y Léon, and has only a token presence in Galicia. The Egyptian Vulture breeds in eastern Portugal, principally in the north-east along the Duero in Trás-os-Montes but also in east-central Beira Baixa: in Portalegre and Castelo Branco districts (PA2). It also occurs in the Balearic Islands, almost entirely in Menorca but with one pair in Mallorca.

Egyptian Vultures are notoriously catholic in diet. One or more may often be found attending carcases alongside large vultures. Many are attracted to carrion dumps and rubbish dumps: up to 200 have been counted at roosts near rubbish dumps in the Ebro valley (Donázar *et al.* 1994). Unfortunately, their habits make them vulnerable to consuming poison baits and some recent local declines have been attributed to this (Cortés-Avizanda *et al.* 2009). At least 294 poisoned Egyptian Vultures were found in Spain during 1990–2007, largely victims of illegal baits intended for predator control in hunting areas (Hernández & Margalida 2009).

The Spanish population was censused in 1987 (Perea *et al.* 1990), 2000 (del Moral & Martí 2002) and 2008 (del Moral 2009a) (Table 21). Despite impressions of decline, each census has found more pairs than its predecessor and not all of these can be attributed to better prospection. The global figures nonetheless mask significant local variations, both increases and declines. It also seems clear that the Egyptian Vulture was much more numerous in the more distant past: Saunders (1871) noticed 'hundreds' around cattle pens in the south Spanish lowlands, an improbable occurrence today. More recently, Bernis (in Bijleveld 1974) referred to a large decline during the first half of the 20th century and he estimated the Spanish population to be 600–1,000 pairs in the early 1960s.

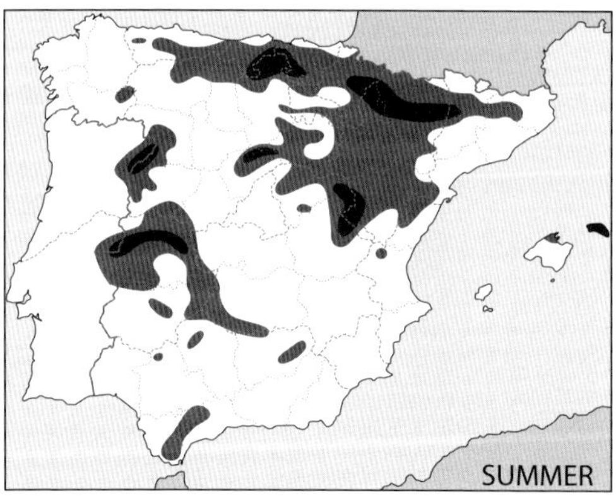

The 2008 results, when compared with earlier ones, show populations increasing to various extents in Asturias, Cantabria, Castilla y León, Castilla-La Mancha, Catalonia, the Comunidad Valenciana, Extremadura and the Basque Country. The species has returned to Galicia, but is absent in Madrid and Murcia, although often observed in both regions. The Madrid population numbered 8–10 pairs in the mid 20th century but the last nesting pair disappeared in 1988 (Ortega *et al.* 2006). The census results suggest that the population in Aragón has been largely stable since 1987, but there have certainly been substantial local declines there: only some 90 pairs remained in 2005 in a well-studied area of the Ebro Valley that held 190 pairs in 1979 (Grande *et al.* 2010). There have also been declines in Navarra, La Rioja and, in particular, in Andalucía. The situation in Andalucía gives most cause for concern since the region has a lot of potentially ideal habitat, including the mountains of the Betic Range and the Sierra Morena. Nevertheless, 48 Egyptian Vulture territories have been lost there since the 1987 census, 21 of them since 2000. As a general rule, declines everywhere have been more severe in agricultural areas than where livestock is reared (LR). Although the censuses indicate a population increase of around 12%, closer monitoring of 433 pairs has shown a 26% decline in productivity, a 16% fall in breeding success and a 6% decline in fledging success between 2000 and 2008 (del Moral 2009a), so that there are clearly negative underlying trends that may reverse the observed gains.

The Balearic population inhabits coastal and inland cliffs. One pair nests on a coastal cliff in northern Mallorca but the remaining 50 or so pairs are in Menorca, 87% in the west of the island (del Moral 2009a). The Mallorcan population became extinct in the late 1970s, apparently through poisoning, but the island was recolonised in 1993. Poisoning seems also to have been a major factor in the 26% decline seen in Menorca between 1988 and 2000 but the population has since recovered and has been stable since 2004: it may have reached the carrying capacity of the island.

The Portuguese population was censused in 2000 when at least 83 pairs were found, rising to 130 pairs if the pairs that occur along the Douro Internacional, along the north-east border with Spain, are included (del Moral & Martí 2002). There is historical evidence that the Egyptian Vulture was much more numerous and widespread in Portugal during the late 19th and early 20th centuries, when it even nested on sea cliffs in the south-west, but it declined progressively during the second half of the 20th century, especially in the 1980s and early 1990s (Palma 1985, AP). The breeding range has contracted since the first Portuguese breeding atlas survey and it no longer breeds in the Guadiana basin (PA2). However, numbers are apparently stable in the core areas.

Table 21. Minimum numbers of breeding pairs of Egyptian Vultures in Spain in 1987, 2000 and 2008. (del Moral 2009a).

Region	1987	2000	2008
Andalucía	81	50	33
Aragón	264	251	260
Asturias	48	50	55
Basque Country	20	37	45
Cantabria	25	40	46
Castilla y León	365	377	380
Castilla-La Mancha	78	101	151
Catalonia	29	34	66
Comunidad Valenciana	3	5	14
Extremadura	108	166	167
Galicia	0	0	2
La Rioja	27	25	18
Madrid	1	0	0
Murcia	0	0	0
Navarra	141	130	127
Balearic Is.	37	32	46
TOTALS	1,227	1,298	1,410

The Egyptian Vulture is largely a summer visitor, although the Balearic population is resident (de Pablo 2000) and some winter regularly in Extremadura and in the lower Guadalquivir, and occasionally elsewhere (SWA). Only small numbers of winterers are involved but the habit is not new since small-scale wintering in southern Spain was noticed by Saunders (1871). They include individuals of all ages. Winter roosts near El Rocío, Doñana, held up to 20–30 birds in the 1990s but no more than five have appeared subsequently (SWA). A pig farm in central Cáceres has been a regular winter haunt since it was opened in the mid 1980s (*Ardeola* 56: 157, 57: 224–225): the numbers present there each year vary considerably but there were a record 49 birds in winter 2012–13, some 60% of them adults (J. Prieta, pers. comm.).

Migrants converge on the Strait of Gibraltar during both passage periods. The first arrivals there in spring are in February but there is a protracted passage period that extends until May, with stragglers in June. The breeding adults arrive chiefly during late February and March, the later movements chiefly involving immature birds. The southbound passage is in August, peaking during the second half of the month and in the first half of September, with a few early individuals in late July and stragglers in October. Spring migrants arrive individually or in small groups but flocks of 30 or more are often seen in autumn, these typically including a mixture of adults, subadults and juveniles. It is likely that such flocks derive from the assemblages of Egyptian Vultures that form after the breeding season at favoured sites such as rubbish dumps, from where they presumably depart together. A study of 14 such gatherings in the Ebro valley (Donázar *et al.* 1994) found that numbers peaked at up to 200 birds in late July and early August, juveniles and other immature birds comprising 35–50% of those present.

Given the known size of the Iberian and French populations, and assuming that the wintering population is very small, the post-breeding passage at the Strait should now involve over 4,000 birds, comprising some 3,000 adults and an additional 30% immature birds: this being the ratio of adults to immatures recorded by Bernis (1980). There are single recoveries of Spanish-ringed birds from Morocco and Mauritania and one ringed in Portugal was recovered in Senegal. Recent radiotracking studies have also followed a number of individuals of Spanish origin, and two from Provence, France, to wintering grounds in southern and northeastern Mauritania (Meyburg *et al.* 2004, García-Ripollés *et al.* 2010). Very few Egyptian Vultures breed north of Iberia. The French population numbered 88 pairs in 2010, comprising 68 pairs in the French Pyrenees and 20 pairs in northeastern France (Dubois *et al.* 2012). Thus only very small numbers cross the Pyrenees, where some autumn passage occurs in August–September (Urcun & Bried 1998b). There are nonetheless nine recoveries of French-ringed birds, from the Pyrenees and south France, in Spain.

AFRICAN WHITE-BACKED VULTURE *Gyps africanus*

(Buitre Dorsiblanco Africano, Grifo-africano)

Vagrant to Spain. Recorded in Portugal (Category D).

One was at Cape St. Vincent, Faro, on 14 October 2006: this record has been assigned to Category D in Portugal (Matias *et al.* 2011). There are also three accepted Spanish records of immature birds found in the area of the Strait of Gibraltar: at Tarifa on 7 September 2008 and on 25 June 2009, and near Algeciras on 19 September 2011. The recent regular appearances of Rüppell's Vultures in southern Iberia raised the possibility that the White-backed Vultures too may have been genuine vagrants and the species has been admitted to Category A in Spain, but the possibility of escaped birds remains. One picked up at San Fulgencio, Alicante, on 26 November 2005 was assumed to be of captive origin (*Noticiario GAE 2003–2005*: 13). This species moves north in West Africa for the wet season in July–September and has reached 18°N in Mauritania. One was obtained in Cuneo, Italy, in June 1933.

EURASIAN GRIFFON VULTURE *Gyps fulvus*

(Buitre Leonado, Grifo-comum)

***G. f. fulvus*. Widespread and locally numerous in mainland Spain and Portugal and a recent immigrant to the Balearic Islands. Adults are mainly resident but some immature birds winter south to west Africa.**

The fortunes of Iberian Griffon Vultures are strongly tied to human activity and probably first took an upturn in the distant past when the Mediterranean forests were cut down and replaced by open country, much of it given over to grazing herds of domestic animals. Much more recently the Griffon has benefited from the carcases put out for the vultures at carrion dumps, as well as the ready availability of edible scraps at municipal landfill rubbish dumps. There were more Griffon Vultures in Iberia at the start of the 21st century than ever recorded before and very probably more than there have ever been. The Iberian populations are undoubtedly of the highest global significance and they far outnumber those of France: about 1,000 pairs (Dubois *et al.* 2012) and Morocco: where the Griffon is practically extinct as a breeding species (BM).

Griffon Vultures are sociable birds that establish colonies on the cliffs of river gorges, escarpments and mountains, nesting as

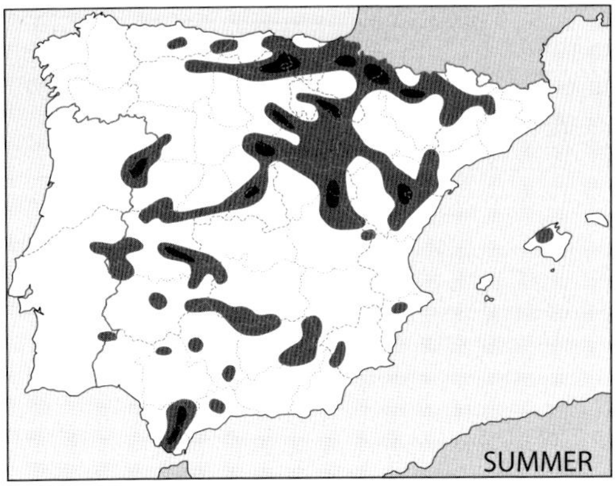

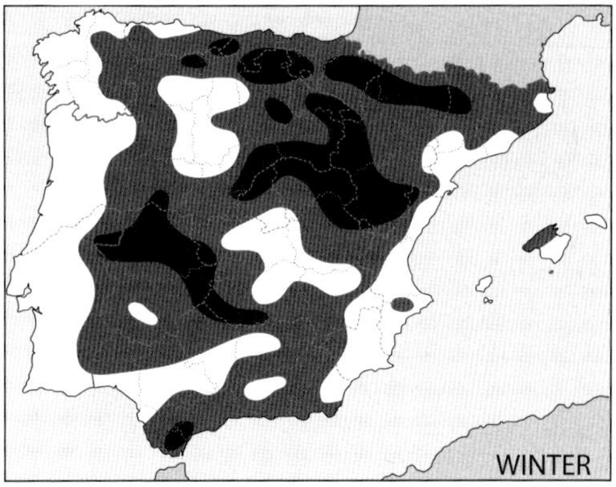

high as 2,000m in the Sierra de Guadarrama, Madrid, or in the Pyrenees around Jaca (SA2, del Moral 2009b). The great majority nest in such situations but some have been found taking over the tree-nests of other large raptors, including those of the Black Vulture and Spanish Imperial Eagle, perhaps to the detriment of these species. A few nest, very locally, on coastal cliffs, notably at Monte Candina in easternmost Cantabria. The birds range near and far from their colonies and may be found feeding on carcases on the plains, distant from any nesting areas, at any time of year. Immature, non-breeding birds are especially prone to wander great distances, roosting nightly on cliffs, in trees or even, if necessary, on the ground.

The Spanish population has been the subject of four national censuses: in 1979 (SEO 1981), 1989 (Arroyo *et al.* 1990a), 1999 (del Moral & Martí 2001) and 2008 (del Moral 2009b) which have shown a massive increase in the breeding population during the period spanned (Figure 74). The rate of increase has been spectacular, especially for such a large bird that raises only one chick at a time. In percentage terms the increases between censuses were 229% (1979–89), 131% (1989–99) and 42% (1999–2008), indicating a continuing enlargement of the population but with indications of stabilisation. However, coverage improved very significantly between the first and most recent censuses and some of the earlier increases may not have been quite as large as the data suggests. Prior to the first census, Valverde & Bernis (1960) estimated the Spanish population at 5,000–8,000 birds and Garzón (1977) considered that there were some 5,000 pairs.

This increase has principally involved enlargement and extension of existing colonies. However, since 2000 the nesting range has expanded westwards in Cantabria, eastwards in Barcelona and south-eastwards in Cuenca, Castellón and Murcia, and breeding has been recorded in Valladolid, Gerona, Albacete and Alicante. The Griffon Vulture nests in all the principal mountain systems, as well as in the Duero and Tagus gorges in the west and indeed anywhere that suitable cliffs are available. It is thus scarce in the Sierra Morena, whose rolling hills offer scant nesting opportunities.

Figure 74. The minimum number of breeding pairs recorded by the four Spanish national censuses of Griffon Vultures during 1979–2008 (after del Moral 2009b).

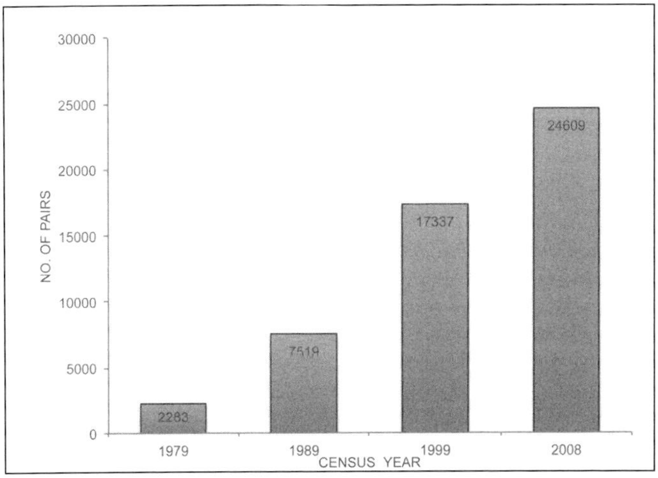

The regional distribution of the breeding population is very uneven (Table 22), showing the importance of Castilla y León and Aragón, and to a lesser extent Andalucía, Castilla-La Mancha and Navarra, which together hold 78% of the breeding birds. The range is most uniform in the northeastern quadrant of Iberia, bounded approximately by, and including, the Cantabrian Range and Pyrenees in the north, the provinces of Girona and Castellón in the east, Guadalajara and Teruel in the south and Segovia and Burgos in the west. This predominantly limestone region is rich in ravines and cliffs and has always been the core of the Iberian distribution. In 2008 it held 76% of the total Spanish breeding population and 1,030 of the 1,560 known colonies. Most of the remaining Spanish birds, 23% in 2008, breed in the south-west, in a roughly triangular area taking in the Duero gorges in Salamanca and Zamora in the west and including Extremadura: particularly Cáceres province, and much of Andalucía east to Murcia. Nevertheless, the only Spanish provinces without any breeding Griffon Vultures in 2008 were Huelva, Valencia and the four provinces of Galicia, although the small populations in Alicante and Almería provinces: 17 and 21 pairs respectively in 2008, derive originally from reintroduction projects. However, prospecting birds have been visiting potential nest sites in Galicia recently and that region too seems likely to have a nesting population before long (del Moral 2009b).

The Portuguese population is comparatively small and is based largely on the riparian gorges of the Duero, Tagus and, locally, the Guadiana. It too has increased, largely in parallel with that in adjacent parts of Spain. There were 102–189 pairs in 1989, 155–159 pairs in 1996 and 267–272 pairs in 1999, after which numbers are thought to have stabilised (PA2). The current distribution has shown some minor westward expansion as the population increased but it is still much more limited than in the historical past. In the late 19th and early 20th centuries the Griffon Vulture occupied all the principal mountain ranges north of the Tagus and it was widespread throughout most of southern Portugal. It declined during the mid 20th century, in some regions due to poisoning by baits intended to control wolves (AP). It does not nest in Andorra, although foraging birds may be seen there (ANA).

Table 22. Regional distribution of breeding Griffon Vultures in Spain in 2008 (after Del Moral 2009b)

Region	Breeding population (pairs)		% of total population
	Minimum	Maximum	
Castilla y León	5,965	6,062	24
Aragón	5,174	5,174	21
Andalucía	2,978	3,037	12
Navarra	2,783	2,783	11
Castilla-La Mancha	2,410	2,501	10
Extremadura	1,560	1,943	6
Catalonia	939	1,115	4
Basque Country	805	805	3
La Rioja	639	707	3
Cantabria	443	467	2
Madrid	454	461	2
Comunidad Valenciana	253	255	1
Asturias	151	176	1
Murcia	55	55	<1
TOTAL	24,609	25,541	

The 2008 Spanish census found 1,566 colonies plus an additional 225 solitary pairs. There were 39 colonies of more than 90 pairs, which together harboured 24% of the Spanish breeding population. The three largest were in Castilla y León: those of the Hoces del Río Duratón (566 pairs) and the Hoces del Río Riaza (402 pairs), both in Segovia, and the Arribes del Río Huebra (349 pairs) in Salamanca. Nevertheless, small colonies were relatively more numerous in the 2008 census than in earlier surveys and 560 new colonies were established between 1999 and 2008 (del Moral 2009b).

The increase in the breeding population was at first paralleled by an increase in nesting productivity, from 0.65 young fledged per nest in 1989 to 0.69 in 1999, but this declined to 0.62 in 2008, perhaps due to reduced food availability in certain regions. Such local productivity declines are sometimes attributable to increased competition for food, a density-dependent effect, as has been reported for Navarra where the population increased from 242 pairs in 1969 to 1,395 pairs in 1994 (Fernández *et al.* 1996). Productivity reduction accompanying population increase has also been reported from the Hoces del Riaza and in Castellón province, but has not been seen in other areas where Griffon Vultures nest at high density, such as in La Rioja or Cádiz (Camiña 2004a,b; del Moral 2009b). The negative impact of the EU regulations on carcase disposal has also been implicated in some productivity declines but, in practice, the regulations were ignored across most of Spain (Camiña 2004b) and the population continued to increase while the restrictions were in place. There is a sizable non-breeding population, largely comprised by immature birds (Donázar 1993). Previous investigations have suggested that there are an additional 1.1–1.3 non-breeders per breeding pair and hence the total number of Griffon Vultures in Spain in 2008 may have been as high as 91,500–95,000 individuals (del Moral 2009b).

The population was clearly suppressed during the 19th century, and until at least the 1960s, by the then antipathy to raptors of all kinds that resulted in direct persecution, as well as deaths from eating poisoned baits. At least 841 poisoned Griffon Vultures were reported during 1990–2005 (Cano *et al.* 2008) and this cause of death must have been much more serious prior to 1984, when the use of strychnine was prohibited (Bijleveld 1974). More enlightened attitudes since then have allowed the birds to take advantage of the abundant food and widespread nesting sites available in large parts of Iberia, which they have clearly done to impressive effect. They nonetheless remain vulnerable to changes in human activities, as shown by local problems caused by the temporary ban on leaving livestock carcases in the countryside. The European Union regulations on carcase disposal were relaxed in 2011, with autonomous regions made responsible for designating areas where the practice could resume, and this process began in 2012. The regular and widespread presence of vultures at municipal rubbish dumps is also a new phenomenon that may reflect increased intraspecific competition for food as the population has grown.

The proliferation of wind farms has also caused problems for Griffon Vultures in certain regions, notably in the north-east and at the Strait of Gibraltar. There were over 13,000 wind turbines in Spain by 2007 (Tellería 2009) and their literal impact on vultures has been monitored in some key areas. The large and extensive wind farms at the Strait around Tarifa were found to kill 0.19 birds per turbine per year (Barrios & Rodriguez 2004): in an area where thousands accumulate during their southward migration especially. Elsewhere, a single wind farm in Navarra accounted for 409 Griffons in a single year (Lekuona 2001).

Adult Griffon Vultures patrol large home ranges but are essentially resident. Immature birds are prone to wander over much larger distances. Many migrate across the Strait of Gibraltar, to winter in northern and sub-Saharan West Africa, with observations from Morocco, Algeria, Mauritania, Senegal and Niger, although the location of the principal winter quarters is not yet known in any detail. This passage has long been noticed but it has increased very greatly in magnitude in more recent years, as the source population has burgeoned. The peak movements are relatively late, from mid October to mid November (Griesinger 1998, pers. obs.). However, very large numbers of Griffon Vultures of all ages do not leave the Peninsula. For example, winter counts in Andalucía between 2007–08 and 2009–10 found some 14,000 birds in the region each winter, concentrated in the Betic Range and the Sierra Morena (*Ardeola* 59: 425).

The spring passage across the Strait is extended. A few arrive in February and March but most pass during the second half of April, throughout May (Fig. 75), and also well into June (Garcia & Bensusan 2006), i.e. long after the breeding cycle has begun. However, most if not all of the migrants are immature birds. There are no complete Strait-wide counts of raptors arriving in spring but an extended partial count during January–June 2009 recorded 5,280 Griffons arriving in the central part of the Strait (A. Onrubia, pers. comm.) and an additional 636 birds arrived at Gibraltar itself.

Figure 75. Distribution by approximately ten-day periods of Griffon Vulture arrivals in Iberia at the Strait of Gibraltar during 2008–12 combined (n = 16,673). birds. Passage continued throughout June but was incompletely monitored. (Fundación Migres, GONHS)

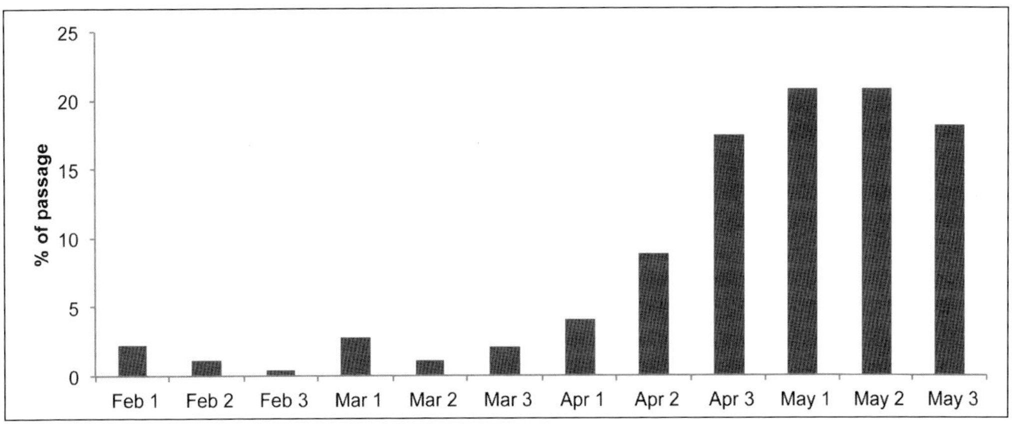

Griffon Vultures are ponderous fliers, when unable to soar or glide, and only attempt to cross the Strait when winds are light or otherwise favourable. Large numbers may build up in the hinterland at peak migration periods while the birds wait for suitable crossing conditions. Bildstein *et al.* (2009) concluded that the Strait is a significant barrier to the birds, to the extent that the difficulty of the crossing might come to limit population growth. We consider that this far overstates the challenge that the birds confront. Not only is there no indication that population numbers are being held back by losses on migration and in winter but also thousands of birds do cross every season. Bildstein *et al.* suggest that the width of the Strait at its narrowest point, 14km, is close to the maximum that Griffon Vultures are capable of crossing regularly. Whether or not this is the case, the Griffon is nevertheless capable of much longer sea crossings when the need arises. The best demonstration of this is the autumn 2008 influx into the Balearic Islands (see below). However, some birds do get into difficulties, often when they are accosted by the Yellow-legged Gulls that nest abundantly around the Strait: a few end up crashing into the sea, where they drown (*GBR* 2005: 17–18).

A study combining observations of birds carrying field-readable marks and the use of telemetry was conducted during 1990–93 (Griesinger 1998), on a total of 31 juvenile birds marked in northern Spain. It was found that 90% of juveniles left the breeding colonies during the second half of October, to appear in southern Spain two to three weeks later. Griesinger recorded *c.* 4,000 birds crossing the Strait in 1992 and 1993, on just a few days between 10 October and 10 November, and considered that at least 30% of the Spanish juvenile population migrates to Africa. There are 81 foreign recoveries of birds ringed in Spain up to 2010 and these include 21 from North Africa: 13 in Morocco, seven in Algeria and one in Tunisia; and 11 from sub-Saharan Africa: four in Senegal, two in Mauritania and single birds in Gambia and Mali.

Little is known in detail about the wider movements of Iberian Griffon Vultures but there is certainly a degree of nomadism, again probably involving young birds mainly. There are recoveries in France of 24 birds ringed in Spain and one ringed in Portugal, and at least 70 Spanish recoveries of French-ringed birds and of single birds ringed in Italy and Croatia. A regular movement of non-breeders from Spain into France has been reported across the eastern Pyrenees during April– September, the birds travelling as far as Provence, the southern Massif Central and the southern Alps, and a southwards movement of French-origin birds into Spain also occurs in autumn (Terrasse 2006). Flocks believed to include birds of Spanish origin have appeared erratically as far afield as Switzerland, Germany, Holland and Belgium (Terrasse 2006) and four Spanish-ringed birds have been found in Denmark. A Griffon Vulture that was ringed as a pullus in Segovia in May 2004 was recovered in May 2006 at Ciechanow, Poland, 2,179 km to the north-east. Another bird of Spanish origin was at Dabrowice, Poland, in July 2011 (*Birding World* 24: 279).

Griffon Vultures are more widespread in southern Portugal during the post-breeding dispersal and migration period, particularly in Algarve in October and November. Here hundreds sometimes wander as far as the Sagres peninsula, in the extreme south-west. A count of 1,900 at Vila do Bispo, Faro, on 25 October 2005 (*Anu. Ornitol.* 5: 45) is especially noteworthy. Such flocks are undoubtedly mainly comprised by juvenile and immature individuals of Spanish origin, and probably some French birds, some of which may eventually return eastwards and cross the Strait. Elsewhere in Portugal small numbers may appear almost anywhere, again chiefly in autumn (AP).

The Griffon Vulture was very rarely reported in the Balearic Islands until recently. It was recorded in Mallorca by Homeyer (1862) and Southern (1958). Much later, one arrived in Mallorca in 1983 and remained there until at least 2002 (Clavell 2002). At least eight more were recorded in the archipelago during 2002–07. These observations were eclipsed by an unprecedented influx in 2008 (*Ardeola* 56: 351–352). The distance between the mainland and the nearest point of the archipelago is the 80km between Alicante and Ibiza. However, the initial observations were in Menorca, a sea crossing from Spain of at least 200km if they flew there directly. In Menorca the first birds were 30 at Mahón on 31 October. The following day numbers had increased to 68 at Ciutadella. The birds then crossed to Mallorca, 38km away, where two groups totalling 25 birds were seen on 6 November at Escorca. The peak count in Mallorca was 51 on 7 December at Moncaire. This arrival occurred during the peak southward migration period of the species and is thought to have been brought about by a powerful southwesterly gale, with some indications that the birds were actually displaced from as far away as Murcia (Muntaner 2012). Some birds, including a flock of 14, were seen leaving the archipelago in April 2009, by flying south-west towards Alicante from Ibiza (*Ardeola* 57: 525), but others remained on the islands, including 53 in the Serra de Tramuntana, Mallorca, in mid November 2010 (*Ardeola* 58: 493). The 2008 influx may yet lead to the establishment of a breeding population and indeed the first reports of nesting came in 2012, when up to five incubating birds were found on cliffs in the centre-south of the Sierra de Tramuntana and another nest was located at a coastal site in Mallorca (Muntaner 2012). A Griffon Vulture found dead in Formentera on 2 February 2011 had been ringed at El Torcal, Málaga, in December 2010 (*Ardeola* 59: 425) and so had clearly travelled to the islands in mid winter.

RÜPPELL'S VULTURE *Gyps rueppellii*

(Buitre Moteado, Grifo-pedrês)

A rare but regular non-breeding visitor, recorded in Spain, Portugal and Gibraltar.

This tropical vulture is not known to nest any closer to Iberia than the Sahel in eastern Mauritania and in Senegal. It has become scarce in West Africa in recent years and is regarded as severely threatened there (Thiollay 2006). Its current status as a rare but regular non-breeding visitor to the Iberian peninsula has therefore been a great surprise. Those that occur in Iberia are generally immature individuals that perhaps attached themselves to the large flocks of Griffon Vultures that return to Iberia after wintering in West Africa (de Juana 1999, 2006; Gutiérrez 2003, Forsman 2005).

The earliest accepted record for Iberia was of a subadult that was photographed by the Cedillo reservoir, Cáceres, in May and June 1992, following reported sightings of a Rüppell's Vulture around the Sierra de San Pedro since April 1990. Another subadult was at Doñana in August 1992 and a second-year bird was captured in Portugal in October 1993: at Ribeira de Nisa, Portalegre. By 2009 there were over 40 accepted records (involving at least 59 birds) for Spain, 13 for Portugal (15 birds) and one for Gibraltar (Dies *et al.* 2011, Jara *et al.* 2010, Bensusan & Perez 2007). These certainly underestimate the number of occurrences since many records are unsubmitted, particularly from the vicinity of the Strait, where the birds are most frequently seen: photographic evidence suggested that at least nine individuals were present in Cádiz province in September 2010, with up to four in view at once at Tarifa. The 2011 report of the Spanish rarities committee accepts eight records for that year and an additional 16 from earlier years.

There is unequivocal evidence that Rüppell's Vultures cross the Strait of Gibraltar. Individuals have been seen arriving on the north coast from the south on several occasions, for example at Tarifa on 28 May 2003 and 10 June 2005, and at Gibraltar on 23 July 2006. A fluid accumulation of Griffon Vultures on the south shore at Ceuta between 27 May and 17 June 2006, where the birds waited for weather conditions to permit a crossing to Spain, included five Rüppell's Vultures: two juveniles and three immatures (*Ardeola* 55: 268; Cambelo 2007). Single immature birds were also reported from Jebel Musa, in the central region of the south shore of the Strait, on 18 June 2002 and 24 August 2006 (Bergier *et al.* 2010). Conversely, five individuals that had lingered for some time around the Tarifa area in autumn 1998 were not seen after 19 October, a day on which *c.* 1,000 Griffons crossed to Africa.

Allowing for possible duplications, we here consider 41 accepted records in Spain, 12 in Portugal and one in Gibraltar up to 2009 (Fig. 76). The most northerly were in Castilla y León, at the Hoces del Riaza, although more recently accepted observations include an immature at Viniegra de Arriba, La Rioja, in September 2007, and a bird at Teruel rubbish dump, Aragón, in November 2011. The records up to 2009 involved at least 59 individuals, although observations of two or more birds together have all been near the Strait. Only two have been identified as juveniles or first-years: at Gibraltar on 23 July 2006 and at Castro Verde on 18 September 2005 (AP). The remainder include 38 that were reported as 'immatures', five or six subadults and seven or eight adults, these last away from the Strait region. The number of adults may have been fewer since observations from several years of an adult in the frontier region of Cáceres/Castelo Branco may all have involved the same bird. An adult was found on a nest in this region in February 1999, within a Griffon Vulture colony on the Tagus, and it may have been paired with one of the Griffons: the outcome of the nest is unknown. An adult photographed in the Sierra de San Pedro, Cáceres, in January 2009 is thought to have been a Griffon x Rüppell's hybrid (*Ardeola* 58: 453, *Quercus* 317: 43).

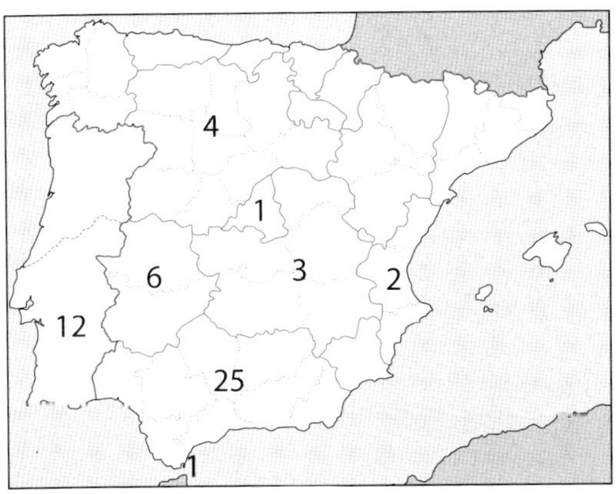

Figure 76. Records by region up to 2009 (n = 54).

There are observations from all months but they are less frequent in winter, suggesting that many individuals arrive in spring and return to Africa in autumn. Records from the northern hinterland of the Strait tend to peak during September–October, including the period when there is a large exodus of immature Iberian Griffons to Africa. Both Griffon and Rüppell's Vultures only attempt the crossing when conditions are favourable and it is not unusual for birds to linger for a week or more in the area before departing, which they generally do around Tarifa, where the Strait is narrowest. Hence autumn records here tend to outnumber spring records considerably: birds arriving in spring keep flying north and offer only a brief opportunity for detection. Spring arrivals are also on a broader front, so that it is much easier for individuals to pass unnoticed.

LAPPET-FACED VULTURE *Torgos tracheliotus*

(Buitre Orejudo, Abutre-real)

Vagrant to Spain.

It formerly bred in the western Sahara and perhaps also in mountainous regions of Morocco, Algeria and Tunisia. It may still breed in Mauritania. Vagrants have reached Syria, Jordan, Libya, France and Spain on very rare occasions. In France one was captured on Le Crau, Bouches-du-Rhône, in the 19th century.

The Spanish record, apparently reliable, was of two that were seen in the Pyrenean border region, in the Valle de Arán near La Picada pass, on 17 September 1940. They were at the carcase of a horse in the company of Griffon and Black Vultures (Mayaud 1941; *Ardeola* 12: 19–20).

EURASIAN BLACK VULTURE *Aegypius monachus*

(Buitre Negro, Abutre-preto

A locally common breeding species in southwestern Spain with a marginal presence in adjacent Portugal. Also breeds Mallorca. Very scarce or rare elsewhere.

The Black Vulture is once again becoming a familiar sight within its Iberian range, which holds a significant proportion of the global population, after a long period of decline during the 20th century.The current mainland breeding range is almost entirely restricted to the southwestern quadrant; including parts of Castilla y León, Extremadura, Castilla-La Mancha, Madrid and Andalucía, and with a marginal presence in adjacent parts of Portugal. There is also a colony in Mallorca and a reintroduction project is underway in the Catalan Pyrenees. The great majority nest in loose colonies. Foraging birds and non-breeders range outside the periphery of the breeding range, becoming progressively rarer with distance from it.

The Spanish population was censused in 2006 (de la Puente *et al.* 2007) and then numbered at least 1,845 pairs. These comprised 35 colonies and just five isolated pairs. The estimated total breeding population, taking into account that not all colonies were fully censused, was 2,440 pairs. Breeding birds were found in 13 provinces within six autonomous regions but the breeding range comprised just 3,651 km^2. Extremadura alone held 46% of the population. The five largest colonies, together holding 55% of the breeding population, were in the Sierra de San Pedro, Cáceres (336 pairs); Monfragüe, Cáceres (287 pairs); Cabañeros, Ciudad Real (165 pairs); the Alcudia valley, Ciudad Real (129 pairs), and the Sierra Pelada, Huelva (96 pairs).

Nesting Black Vultures select often steep slopes covered with open Mediterranean forest or scrub with scattered tall trees to provide nest sites. The population is strongly associated with the mountain ranges of southwestern Spain: the Sierras de Guadarrama, de Gredos, de Granadilla and de Gata, and Las Hurdes, in the Central Range; Monfragüe, the Sierra de San Pedro, Montes de Toledo and the Sierras de Canalizos and Alcudia; and, within the Sierra Morena of Andalucía, the Sierra Norte de Sevilla and the Sierras de Hornachuelos, Andújar and Pelada. Two thirds of the nests are located within or near the Tagus basin, with a further 25% in the Guadiana watershed and the remainder on the northern side of the Guadalquivir valley. Colonies are in protected or otherwise largely undisturbed areas, often difficult of access and on private land. Many pairs have alternative nest platforms and the 2006 census found 3,083 such nests, 1.7 per pair on average. Over 70% of

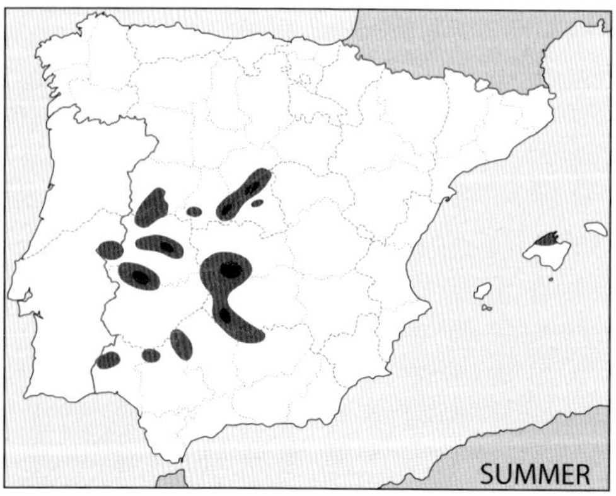

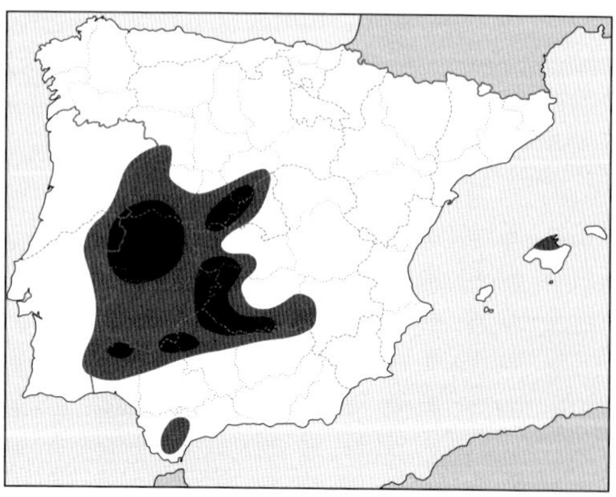

nests are on Holm and Cork Oaks but there are many on pines, especially at higher elevations: up to 1,250–1,900m in the subalpine pine forests of the Sierras de Guadarrama and Gredos.

The very small Portuguese breeding population is essentially an extension of adjacent colonies in Spain, inhabiting frontier districts of the centre and south, from Beira Alta south to Baixo Alentejo (AP). It may number as few as 1–3 pairs although survey work for the 2008 atlas confirmed breeding in one 10km-square and noted probable breeding in nine others. The Black Vulture was more widespread in Portugal until the mid 20th century but was more or less extinct as a breeding bird during the 1970s, 1980s and early 1990s (AP, PA2).

The Iberian population is of critical importance to the species in the Western Palearctic, representing about 90% of the European total (de la Puente *et al.* 2007). Recent trends have been healthy and encouraging, given that the species declined very severely during the 19th and early 20th centuries. The historical literature gives the impression that the Black Vulture was still common in central Spain during the late 19th century (Lilford 1866, Castellarnau 1877). However, there is no firm evidence that it then nested either in Andalucía or the Pyrenees: claims that it was common in the Sierra Nevada and elsewhere in Granada and in the Sierra Morena (del Moral & de la Puente 2010) are attributed to López-Seoane and other 19th century sources that we cannot regard as reliable. It was already very rare by the early 20th century, when the only well known colonies were in Cáceres (Skinner 1925) and Ávila (Witherby 1928). These too were subjected to heavy exploitation by egg collectors and specimen hunting on behalf of museum collections elsewhere in Europe. A notorious inhabitant of the Sierra de Gredos, known as 'El Pompa', is known to have harvested no fewer than 200 eggs there during 1928–39 (Bernis 1966a). There is also apparently reliable testimonial evidence from local people of former breeding to the north-east of the current range, in pines at Tardelcuende, Soria, until 1949; in a mature juniper wood at San Esteban de Gormaz, Soria, until the mid 1950s; and in a small cork oak wood in the Sierra de Vicor, Zaragoza, perhaps until 1960 (*Ardeola* 37: 333, 34: 282–282; Aragüés & Lucientes 1980).

The decrease was also brought about by losses to poison baits, which particularly affect the Black Vulture because of its propensity to forage for small carrion items such as rabbits. Bernis (AMI) thought that there may have been a little over 200 pairs in Spain in the 1960s. In 1973 the population was estimated at just 206 pairs (Hiraldo 1974) but a count in 1986 found an apparent increase to 365 pairs (González 1990). Numbers have continued to increase since then (Fig. 77), although to different extents in different regions, and some lost ground has been made up.

Figure 77. The increase in the Spanish breeding population of Black Vultures, 1973–2006 (de la Puente *et al.* 2007).

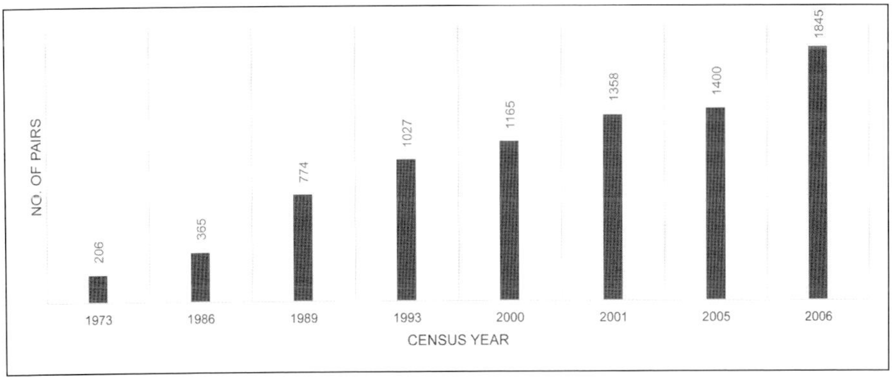

Mallorca boasts the only insular population of the species, in the Serra de Tramuntana. Despite its long history (Homeyer 1862–63, Munn 1921, Mayol 1977a, Tewes 2004) it is vulnerable on account of its small size. In 1982 there were only 20 birds and just two breeding pairs (LR). Releases from a captive breeding programme increased the population to 50 birds by 1992. By October 2008 there were 123 ± 14 birds and 15 pairs nested that year, fledging nine young in total (*Ardeola* 56: 352). A record 16 pairs nested in 2011 and fledged ten young (*Ardeola* 59: 425). The birds occupy a single colony, nesting on pines on coastal cliffs in the north of the island.

There have been claims that the Black Vulture may once have bred in the Pyrenees in the distant past (Saunders 1884, Mayaud 1936) but there is no clear evidence that this was the case. It was certainly absent from there for over a century prior to the start of an introduction project in Catalonia in 2007. This used chicks translocated from Extremadura and 38 were released at the Serras de Boumort and de Alinyá during 2007–12. Several pairs settled to nest in the area and the first chick flew in June 2010 (http://es.blackvulture-pyrenees.org/). Birds from a reintroduction programme in France, where 22 pairs bred in 2010 (Dubois *et al.* 2012) have also been observed in the area, and several released birds have moved to France. In addition, one ringed as a chick in France was recovered in its third year in Málaga and another nestling ringed in Madrid in 2004 was found 2,000km away near Paris in June 2005 (del Moral & de la Puente 2010).

Adult Black Vultures are largely sedentary, although they may range for up to 60km from their colonies when foraging. Radio-tracking and marking studies of nestlings from the Sierra de Guadarrama colonies have shown that whereas some juveniles remain fairly close to the natal sites, roosting communally and visiting carrion dumps, others disperse as far as 600km (del Moral & de la Puente 2010). It is largely immature birds that regularly attend carcases some distance from the nearest breeding colonies. For example, they are frequently observed in Doñana: about 100km south of the nearest colony at the Sierra Pelada in the north of Huelva province, and in eastern Portugal: where very few breed. Wandering birds are occasionally recorded far from the breeding range, most often in the northern Iberian Range: perhaps a former breeding area, in the Cantabrian Range and in the Pyrenees. Records from these regions are largely in spring and summer, peaking in May (Fig. 78), suggesting that non-breeders and immature individuals are mainly involved. There are hardly any during December–February. Reports are generally of single birds so records of six together at Peña Trevinca, Ourense, in July 2004, five at a horse carcase at Cangas del Narcea, Asturias, in May 2000 and up to six at Mecerreyes, Burgos, in July-November 2006 (*A. O. Galicia 2004*: 57; *A. O. Asturias 2000*: 88–89; *A. O. Burgos 2006*) are noteworthy.

Figure 78. Records by month of Black Vulture observations in northern Spain, from Galicia to Aragón, during 1980–2011 (n = 131 records)

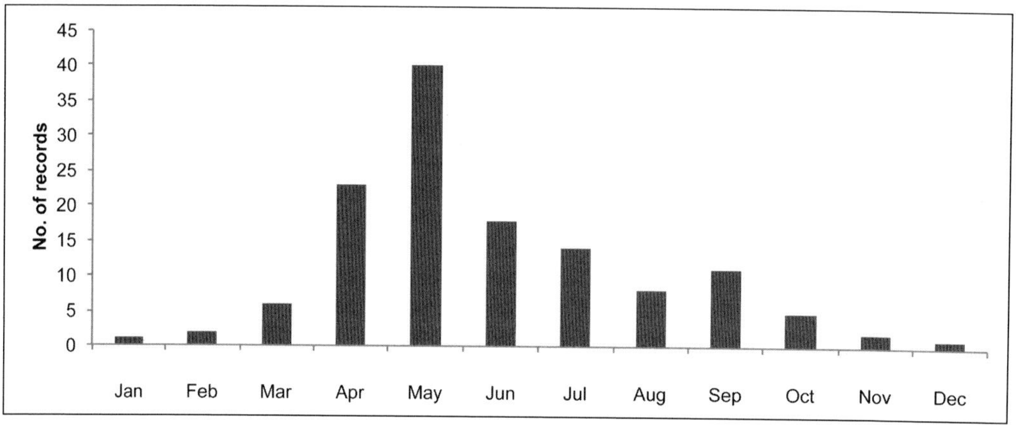

Black Vultures become progressively rarer eastwards across southern Spain and they are seldom recorded in the arid south-east: for example, there are only five records of single birds from Murcia (*Ardeola* 56: 352, 57: 525). One or two are recorded in winter from time to time around the Strait of Gibraltar (Bernis 1980, EG pers. obs.) and a few cross the Strait during migration periods, probably erratically. There are 22 records of individuals arriving at Gibraltar in spring during 1972–2012, apparently from Morocco. The species is regarded as accidental in Morocco, although it once nested in the north (BM). There are nonetheless two interesting and perhaps surprising recoveries from sub-Saharan Africa: a juvenile ringed in Cáceres in November 1994 was found 3,171km away in Mali in January 1995 and one ringed as a pullus in Madrid in June 2004 was in Senegal, 3,201km to the south, the following January. In addition, one was photographed at a carcase at Mbour, Senegal, in February 2007 (Talamelli 2007).

The current situation of the Black Vulture in Iberia is promising but the susceptibility of the birds to poison baits and disturbance potentially threatens their future. The latter has certainly affected colonies in the past and such activities as logging, cork-harvesting and other forestry works need to be timed sensitively to avoid the breeding season (Donázar *et al.* 2002, Moreno-Opo 2007). The birds may also be affected by the large declines of rabbits that resulted from the combined effects of myxomatosis and the haemorrhagic virus, and from the closure of carrion dumps – although the latter situation seems to have been resolved. A study in the Sierra de San Pedro found that Black Vultures there switched their diet from rabbits and other small carcases in the 1970s to one dominated by sheep, wild and domestic swine and Red Deer during 1998–2000, in response to the decline in rabbits and the increased availability of sheep and game carcases (Costillo *et al.* 2007).

SHORT-TOED EAGLE *Circaetus gallicus*

(Culebrera Europea, Águia-cobreira)

Widespread and locally common as a breeding species across much of mainland Iberia. Largely a summer visitor but a few winter in the south. A very scarce migrant in the Balearic Islands.

The breeding range extends across most of mainland Iberia but the Short-toed Eagle is largely absent as a nesting species from the mainly treeless areas of the arid south-east in Almería, the Guadalquivir valley and large tracts of both mesetas and southern Extremadura (SA2). It is thinly distributed in Galicia and the Biscay coastlands, as well as in coastal regions of northern Portugal (PA2), where the climate is cooler and strongly influenced by the Atlantic.

Breeding birds are strongly associated with woodland, usually nesting within extensive stands of trees. However, they require open terrain nearby for hunting for their preferred reptilian prey, chiefly large snakes (Ontiveros 2011). The typical nesting habitat is within undulating or hilly areas, where large woods form part of a mosaic that also includes open scrub, pastures, woodland clearings or cultivated land. The Short-toed Eagle shows some preference for Mediterranean pinewoods but also readily nests within corkwoods, holm oak woods and other deciduous woodlands, from sea level to 1,400m (SA2), and to 1,800m in the Sierra Nevada (Garzón & Henares 2012). An association with holm oak woods has been especially emphasised by recent studies (Palomino & Valls 2011).

The Spanish breeding population was estimated at 3,000 pairs by Garzón (1977), 1,700–2,100 pairs by F. de Juana (1989) and at least 2,772 pairs by Mañosa (SA2). However, a census in 2009–10 (Palomino & Valls 2011) found a much larger population: over 10,000 pairs, nesting in all regions but with over 60% inhabiting Andalucia, Extremadura and the meseta, and a further 13% in Aragón. In support of this, some Spanish regions have reported recent increases: for example, a clear rise in numbers and a 56% increase in the breeding range since the 1980s has been reported for Catalonia (CA2). The Portuguese population was estimated at 250–300 pairs by Palma *et al.* (1999) and seems likely have increased subsequently since the breeding range has expanded, especially north of the Duero (PA2). The post-2000 population in France is 2,400–2,900 pairs (OF).

Most of the Iberian population winters in sub-Saharan Africa. There are single recoveries of Spanish-ringed birds from Morocco and Mali, as well as one recovered in France, and seven French-ringed birds have been recovered in Spain. Three satellite-tracked birds migrated from France to Africa through Spain, two of them to Mali and the third to Niger (Meyburg *et al.* 1996, 1998, Cavallin 2007). Similarly, two satellite-tracked birds marked as chicks in Spain wintered in Mali, close to the Niger delta (Pavón *et al.* 2010). The birds enter and leave Iberia across the Strait of Gibraltar. This route is also used by a significant proportion, probably most, of the population that inhabits southern and central France. However, autumn counts at the migration watch points in the French Pyrenees: at Organbidexka, Lindux and Lizarrieta in the west and Eyne and the Plateau de Beille in the east, have yielded relatively modest counts of no more than one thousand birds per season in total, most of them using the more eastern crossing points (Urcun & Bried 1998b). The Pyrenean passage extends from early September to early October, peaking in late September. A few cross the central Mediterranean at the Sicilian narrows but most of the the several hundred pairs of Short-toed Eagles that nest in Italy migrate to and from their breeding range via the Iberian peninsula, a much longer route but one that offers the shortest possible sea crossing (Agostini *et al.* 2002a,b; Agostini & Mellone 2008).

The spring passage at the Strait of Gibraltar is extremely prolonged (Fig. 79). The dark-breasted breeding adults, arrive from early February into April, with a marked peak in early March. Birds continue to arrive throughout April–May, with some in June and stragglers even in July. These later birds are clearly non-breeders and they are generally very pale below, with the dark breast markings poorly developed or absent, a characteristic of immature individuals (Campora & Cattaneo 2005). The annual combined spring totals for birds recorded by GONHS and Programa Migres arriving on the north shore of the Strait during 2008–12 averaged 9,670 birds but ranged from 7,832 to 11,178.

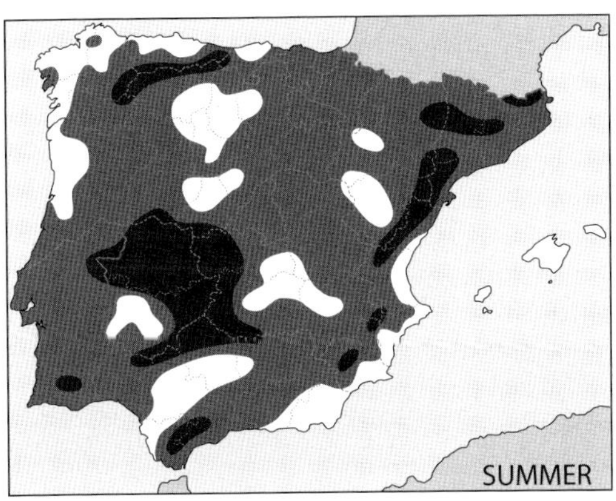

The southward passage at the Strait is largely between August and mid October, with stragglers into November. The peak movement is during the second half of September. The Strait-wide counts of departing birds in the 1970s recorded between 2,947 birds (in 1977) and 8,784 birds (in 1972) in total (Bernis 1980). Full recent counts are not available but on the basis of the reported breeding population sizes, totalling over 13,000 pairs, and allowing for half of these pairs fledging a chick, the southward passage across the Strait would be expected to be some 32,000 individuals. This figure is far in excess of any seasonal count at the Strait so far. However, the 2009–10 Spanish estimates may not be reliable (Blanco *et al.* 2012).

Figure 79. Distribution by approximately ten-day periods of Short-toed Eagle arrivals in Iberia at the Strait of Gibraltar during 2008–12 combined (n = 48,288 birds) (Fundación Migres, GONHS)

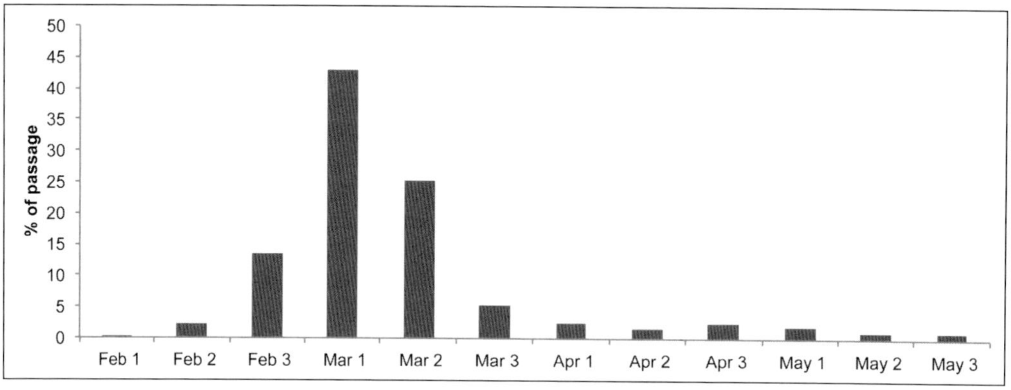

The Short-toed Eagle is known in the Balearic Islands as a rare migrant (Clavell 2002). However, there are a number of recent records from Mallorca, Menorca and Ibiza, including one in winter, on 13 December 2002, at Ciutadella, Menorca (*Ardeola* 52: 212). There was an influx into Mallorca during both passage periods in 2010, with five records in spring between 28 March and 21 April and 11 records (22 birds) between 5 July and 13 November, including seven together at Calvià on 17 July (*Ardeola* 58: 493).

Some Short-toed Eagles remain in warmer regions of Iberia in winter. Most of the very few December and January records have been of individuals found in eastern and southern coastal provinces, especially in Murcia and Alicante (Martínez & Sánchez-Zapata 1999), and Andalucía, with several from Gibraltar (*GBR 2008*: 20, *2009*: 19). Nonetheless, the 2007–10 Spanish winter atlas survey found only six birds in total in December or January (SWA). There were 18 mid December–January records during 1990–2004 from Portugal, only three of them north of the Tagus, with eight in Algarve, five in Alentejo and two in Estremadura (AP). Fourteen records were of single birds, two of two birds and one of three. The Portuguese winter records show some preference there for calcareous coastal regions.

The Short-toed Eagle has a favourable conservation status in Iberia but it is nonetheless quite susceptible to electrocution on account of its habit of perching on electricity pylons (Ferrer *et al.* 1991, Mañosa 2001) and it is also known to be vulnerable to collision with wind turbines (Ontiveros 2011). However, its predilection for reptiles means that it is not affected by poison baits.

BATELEUR — *Terathopius ecaudatus*

(Águila Volatinera, Águia-pintada)

A first-year bird arrived with other raptors at Carnero point, on the northern shore of the Strait of Gibraltar on 5 April 2012 (RBS). It was seen again in the same area two days later. Adult Bateleurs are largely resident but young birds tend to wander extensively (HBW). Individuals have reached Tunisia, Egypt, Israel, the Arabian Peninsula, Iraq and Cyprus on rare occasions.

WESTERN MARSH HARRIER — *Circus aeruginosus*

(Aguilucho Lagunero Occidental, Águia-sapeira)

***C. a. aeruginosus*. Widespread and locally common as a resident breeding species in mainland Spain and Portugal and in Mallorca. Common on passage and in winter. Reports of the north-west African subspecies *C. a. harterti* (J. L. González 1991, BWP) seem unsubstantiated.**

The Marsh Harrier was clearly numerous in Spain in the 19th century (Lilford 1865, Saunders 1871, Arévalo 1887) but widespread persecution, large-scale drainage of wetlands across the Peninsula and very possibly also the effects of pesticides, reduced the Iberian populations to a precarious remnant by the 1970s. Its numbers have since recovered and are continuing to increase.

The breeding range is linked to lowland wetlands and so is closely associated with the Ebro, Duero, Tagus and Guadiana valleys within Spain. It also nests in Doñana and at other wetlands of the Huelva coast. In Portugal the main concentrations are at coastal estuaries, notably those of the Sado and Tagus, and the Ria de Aveiro (PA2). In contrast, it is largely absent from the Mediterranean coastal wetlands, including the Ebro delta, and the only significant populations there are at the Aiguamolls de l'Empordà and the Albufera de Mallorca. The Marsh Harrier nests very locally and irregularly in Galicia and does not breed in Asturias or Cantabria. During the breeding season it favours wetlands large and small in proximity to open country, and especially to cereal crops, over which to hunt (J. L. González 1991). It is an opportunist, taking mammals as large as rabbits, waterfowl and other birds, reptiles, amphibians and large insects, as well as carrion occasionally. Nests are typically within reedbeds and other emergent wetland vegetation but nesting in cereal crops has become increasingly common in some areas.

A series of estimates and counts of the Spanish breeding population show a clear pattern of recovery from a low point in abundance, although differences in methodology mean that the figures are not strictly comparable (Table 23). The earliest estimate, for the 1980s, was of 274–329 pairs (J. L. González 1991) but a census in 1990 found 481–522 pairs (Martínez *et al.* 1994). The fieldwork for the 2003 atlas suggested a population of 817–851 pairs (SA2). A census in 2006 found 1,149–1,494 pairs (Molina & Martínez 2008).

The Marsh Harrier was common in Portugal during the late 19th century (Tait 1887) but had already declined seriously by the 1930s due to persecution (AP). That population also began to recover its numbers late in the 20th century, increasing from 38–49 pairs during 1990–94 to 70–75 pairs in 1998, with indications of some further increase since then (PA2).

Table 23. Regional distribution of the Spanish Marsh Harrier breeding population in 1990 and 2006 (Molina & Martínez 2008).

Region	Breeding population 1990 (pairs)	Breeding population 2006 (pairs)		% of minimum population
		Minimum	Maximum	
Andalucía	70–84	139	160	12.10
Aragón	69	98	228	2.53
Basque Country		7	7	0.61
Castilla y León	109–127	247	307	21.50
Castilla-La Mancha	135–141	237	309	20.63
Catalonia	13	50	50	4.35
Comunidad Valenciana	0	1	1	0.09
Extremadura	0–1	86	97	7.48
Galicia	1	0	1	0.00
La Rioja		56	79	4.87
Madrid	19–21	61	69	5.31
Navarra	55	153	168	13.32
Balearic Islands	10	14	18	1.22
TOTAL	481–522	1,149	1,494	

The 2006 survey confirms the discontinuity of the breeding distribution in Spain. The middle reaches of the Ebro: in Álava, La Rioja, Navarra, Aragón and Lleida have the greatest concentration of nesting birds, with at least 354 pairs (31% of the Spanish population) distributed between its diverse irrigation lagoons and reservoirs. The population increase in this area has been especially impressive: the Marsh Harrier became extinct in La Rioja during the 1980s and there were just three pairs there in 1993 (Gámez & Gutiérrez 1995), whereas it was considered very scarce in Navarra by the early 1980s (Elósegui 1985). Notable breeding locations in 2006 were the Laguna de Pitillas, Navarra (18 pairs); the Laguna de la Veguilla, Ciudad Real (16–17); the Utxesa reservoir, Lleida (16); the Riberas de Castronuño, Valladolid (15), and the Balsa de Pulguer and environs, Navarra (15)

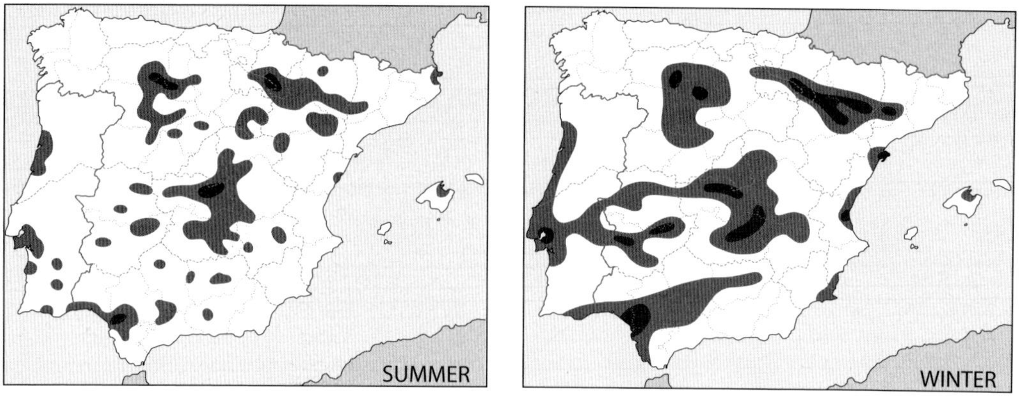

Much of the southern population is at Doñana, where numbers vary considerably between years according to rainfall: 2006 was a poor year there (Molina & Martínez 2008) but 130–150 pairs occur in average years (García *et al.* 2000).

The continued virtual absence of breeding Marsh Harriers from most of the Mediterranean coastal wetlands seems surprising in view of the resurgence in numbers elsewhere. Such extensive wetlands as the Ebro delta and the Albufera de Valencia once supported significant populations. It was described as abundant at the Ebro delta by Ticehurst & Whistler (1930), Westernhagen (1956) and Maluquer & Pons (1961), and Arévalo (1887) obtained chicks at the Albufera and said that it was quite numerous in marshy areas of the region. A sudden decline at the Ebro delta was reported by Maluquer (1971), who attributed it to hunting but most of all to applications of agricultural pesticides, including DDT. Norman (1971) saw only two during a visit to the delta in spring 1969 and Vaughan (1971) was surprised to see none there at all during a week in August that same year. The first Catalan breeding bird atlas found only two pairs in the entire delta. Breeding at the Albufera is thought to have occurred until the 1970s but not subsequently: a pair at the Prat de Cabanes, Castellón, was the only one in the Comunidad Valenciana in 2006 (Molina & Martínez 2008). There is evidence that the delta harriers are affected by lead poisoning from pellets within dead or injured waterfowl that hunters have not retrieved (Mateo *et al.* 1999). Disturbance by duck hunters, and shooting during the duck hunting season, may also be significant factors: nearly half the ringing recoveries of Marsh Harriers involved shot birds (F. de Juana 1989).

The 2006 breeding survey found that the great majority of nests (84.0%) were in marshland vegetation but 10.9% were in cereal crops and the remainder were mainly at small wetlands surrounded by pastures or crops. The tendency to nest in cereals showed an increasing trend, especially in the Guadiana valley in Badajoz and in Toledo, which together accounted for 60 (43%) of the 141 instances recorded.

The Iberian birds are thought to be largely resident but large numbers of Marsh Harriers from northern Europe winter in the Peninsula and Balearic Islands and many must also cross the region on passage to and from African winter quarters.

The Spanish wintering population was also censused in 2006, chiefly by counting at communal roosts (Molina & Martínez 2008). Such winter roosts are in reedbeds and other wetland vegetation or in adjacent crops. The 2006 census found 5,601–5,919 birds in Spain, many more than are present in summer. The winter distribution differs in major respects from the breeding one. Although the principal river basins hold many birds the Mediterranean coastal wetlands accommodate large numbers in winter and Doñana has a particularly significant wintering population: 1,633 were there in January 2012 (*Ardeola* 60: 517). Hardly any winter in the north-west or in the Biscay coastlands, however. The winter habitat includes wetlands, wet grasslands and arable land. Ricefields, which are unsuitable for nesting, gain particular importance as a hunting habitat in winter wherever they are found. The Portuguese wintering population was censused in January 1999 when 347 individuals were counted at roosts, 165 of them at the Tagus estuary (AP).

The 2006 survey found that over 25% of the Spanish wintering population was in Andalucía. There the Brazo del Este, in the periphery of Doñana, was the most important site, holding 443 birds in eight roosts, 384 of them at a single roost. Other sites that attracted very significant concentrations included the Ebro delta (292), the Albufera de Valencia (253), the Laguna de La Nava (256), the Laguna de Sariñena, Huesca (137), the Carrizal de la Torre, Navarra (128) and the Laguna Grande de Berciano, León (126). Such large gatherings are exceptional and only 11 roosts (3.6%) held over 100 birds, most accommodating just 1–10 birds. The Albufera de Valencia roost held 309 birds in January 2009 (Tirado 2011), perhaps indicating a continuing wider increase in wintering numbers. The Balearic Islands held up to 100 birds in winter 2006, most of them in Mallorca and 90% of them at the Albufera there. An earlier census of wintering birds, in 1992 (Jubete *et al.* 1995), found 1,713 individuals in Spain; far fewer than in 2006. The recent increase in wintering numbers, by some 225%, reflects known increases in the breeding populations in most European countries.

Very little is known of the movements of Iberian-bred birds but some adults and young birds from central Spain have been found in Doñana in winter (J. L. González 1991) and two ringed as chicks in Palencia were recovered over 800km to the south in Morocco. Birds ringed in Spain have also produced two recoveries from Sweden and one each from France, The Netherlands, Germany, Poland and Switzerland. An analysis of 51 Spanish winter recoveries of foreign-ringed birds ringed as pulli shows that 21 (41.2%) originated from Germany, 17 (33.3%) from Fennoscandia and nine (17.6%) from the Netherlands, with one recovery each of birds from Poland, France, Belgium and the Czech Republic. Another 147 Spanish recoveries of extraregional origin show a similar pattern. There are also two recoveries from Spain and one from Portugal of birds ringed as pulli in Britain. In addition, there are five recoveries during the breeding season of foreign-ringed birds less than three years old, suggesting that some non-breeders oversummer in their winter quarters.

Satellite telemetry studies of birds marked in Sweden have revealed that most of those that travel to Africa do so via the Iberian peninsula. Counts of migrants at the Pyrenean passes have detected only modest numbers. Autumn totals at Organbidexka during 1981–2001 have ranged between 70 and 350 birds a season (www.migraction.net) but they have tended to increase in more recent seasons and averaged nearly 1,000 birds in total at the three western recording sites during 2009–2011. Counts in the eastern Pyrenees at Eyne have varied widely, ranging from 150 to 600+ birds per season during 1987–2011. Marsh Harriers cross the Pyrenees from late August into November but with the peak movements during the second half of September (Urcun & Bried 1998b).

The spring passage tends to be further west than the autumn movements: when many cross the Alborán basin (Strandberg *et al.* 2008, Klaassen *et al.* 2010). Migrating harriers are less tied to short sea crossings than many other raptors but modest numbers cross the Strait of Gibraltar during both seasons. The limited available data from incomplete counts (GONHS, Programa Migres pers. comm.) suggests that fewer than 1,000 Marsh Harriers cross the Strait during either season. The spring passage at the Strait is apparent as early as February and continues through March–April into May. Males predominate in March and early April, followed by the females in April–May (BSG). Some fly south at the Strait as early as August but the autumn passage is largely in September–October, with stragglers into November, with no evident segregation of the sexes at that season (BSG).

HEN HARRIER *Circus cyaneus*

(Aguilucho Pálido, Tartaranhão-cinzento)

Widespread but scarce as a breeding species in northern Iberia, nesting more locally south to Castilla-La Mancha. More widely distributed in winter throughout the Peninsula and in the Balearic Islands.

The Hen Harrier has a markedly northern breeding distribution in mainland Iberia, where it is characteristic of the cooler, more humid eurosiberian zone, but there are local breeding populations south to central Spain. The core population extends continuously from Galicia and northernmost Portugal in the west across northern Castilla y León, Asturias, Cantabria and the Basque Country to La Rioja and Navarra in the east. Further south there are isolated populations in southern Castilla y León, Madrid, Castilla-La Mancha and in the north of Aragón and Catalonia. Nesting occurs occasionally in Extremadura.

The breeding habitat extends from sea level locally in the north to 1,800m. The birds favour open areas of the Cantabrian Range and Pyrenees, the Basque Mountains and the northern Iberian Range and their foothills, as well as the coastal ranges of Asturias and Galicia, nesting in montane heaths and low scrub: typically in gorse, and foraging over pastures, montane grasslands and arable land. Further south they also nest to an increasing extent in cereal crops and, occasionally, in reedbeds.

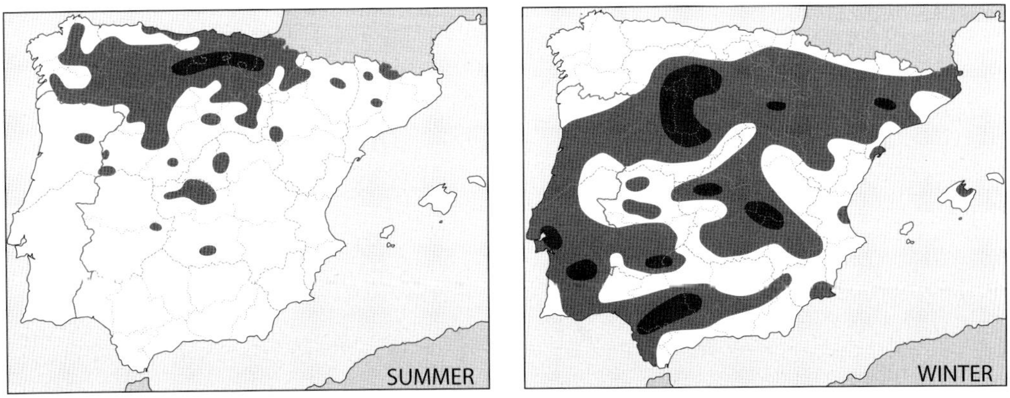

The Spanish breeding population was partially censused in 2006 during a survey of the Montagu's Harrier population (Arroyo & García 2007). The Hen Harrier population was assessed at some 1,100 pairs. About 54% were in Castilla y León and La Rioja, 28% in the northern fringe of the Peninsula from Galicia to the Basque Country, 10% on the southern meseta in Castilla-La Mancha and Madrid and 7% in the Pyrenean region, from Navarra to Catalonia. The 2006 figures compare with 837–999 pairs in the early 1990s (Pinilla *et al.* 1994) and an incomplete estimate of at least 798 pairs in the 2003 atlas (SA2), suggesting a moderate increase in breeding numbers in Spain. The 2006 survey showed a marked increase in the numbers of pairs that nest in cereal crops, however. Of 369 pairs where the nest location was noted, 51% were in cereals and 45% in natural vegetation. Most of the latter (67%) were in woody scrub, with 13% in grasslands and 2% at wetlands or other natural sites. F. de Juana (1989) considered that it bred only occasionally in cereals and it is clear that a major shift in breeding habitat has occurred since then, although the Hen Harrier remains much more reliant on natural habitats than Montagu's Harrier. Nests in cereals are prone to destruction when the crop is harvested but Hen Harrier nests are less vulnerable than those of Montagu's Harrier since the former nest earlier in the season (Arroyo 1994).

The breeding presence in south-central Spain may reflect recent range expansion or may have been overlooked in the past (Pinilla *et al.* 1994). The former is considered more likely and seems certain to have occurred in Madrid, Guadalajara and Toledo at least (Castaño *et al.* 1995, Arroyo & García 1999). In Castilla-La Mancha nesting was first recorded in Guadalajara in 1978, in Toledo in 1983 and in southernmost Ciudad Real in 1988 (*Ardeola* 28: 155, 32: 413, 36: 243). First nesting in southern Castilla y León, in Segovia, was recorded in a cereal field in 1993 and the earliest indications of nesting in Extremadura were also in 1993 (*Ardeola* 41: 96). Pinilla *et al.* (1994) suggested that there were 7–10 pairs in Extremadura, chiefly in the pseudosteppe areas of eastern Badajoz. The 2006 census did not find any in Extremadura so possibly it only nests there irregularly. However, isolated pairs are easily overlooked.

The small Portuguese nesting population, 10–20 pairs, seems to be declining and there are indications of some range contraction (Palma *et al.* 1999, AP, PA2). The breeding birds are mainly or entirely in the sierras of the northern border regions in Minho and Trás-os-Montes but breeding occurred in the early 20th century in the northwestern coastlands at the Ria de Aveiro, in the lower Mondego valley and, at least occasionally, south to eastern Évora and Beja (Palma 1985, PA2).

In winter Hen Harriers may appear in suitable habitat anywhere in the Peninsula and in the Balearic Islands, although they are thinly spread. They are often encountered hunting over arable land, as well over pastures, wetlands and ricefields. The birds flock to communal roosts in the evenings, often at wetlands in the company of Marsh Harriers. An important winter roost occurs at the Laguna de La Nava, Palencia, where 2006 was a vole year and 180 Hen Harriers (and 270 Marsh Harriers) were counted in November, a local record (*Ardeola* 57: 225, Molina & Martínez 2008). An exceptional concentration such as this suggests that the winter population is quite mobile and able to take advantage of local food concentrations. It would not be surprising if future, complete, winter censuses were to find several thousand Hen Harriers in Iberia as a whole: a population of over 3,000 birds has been suggested (SWA).

The Hen Harrier is a scarce winterer and passage migrant in the Balearic Islands (*A. O. Balears 2010*: 262). Observations there occur between mid August and early May with peaks of presumed migrants in November–December and March–April.

The very few available ringing recoveries indicate that the wintering population includes some Hen Harriers from France (11 recoveries), Britain, Germany and Finland, as well as presumably the Iberian birds themselves. Migrants enter Iberia on a broad front mainly across the western Pyrenees: the three western watchpoints recorded an average of 332 birds during 2009–11. Passage extends from early August to mid-November, with a poorly-marked peak in mid-October (Urcun & Bried 1998b). Wintering birds leave Iberia during late February–May, the majority probably in March–April.

Some Hen Harriers cross Iberia to winter in North Africa but the numbers involved are certainly small. The Strait-wide counts in autumn in the 1970s recorded a maximum of 114 birds (in 1974). Migrants cross the Strait in autumn during September–October, with some passage continuing through November. There are exceptional records of migrants in late August. The return passage extends from February into May, peaking in late March and early April. Males pass earlier than females in spring (BSG).

PALLID HARRIER — *Circus macrourus*

(Aguilucho Papialbo, Tartaranhão-pálido)

Very rare but increasingly frequent in Spain on passage and in winter. Recorded in Gibraltar. Vagrant to Portugal.

The winter quarters of the Pallid Harrier extend westwards across the Sahel to Senegal and Mauritania. The southward passage lies further east than the return movement (Corso & Cardelli 2004), which brings birds regularly through Tunisia and Italy. Spring counts at the Strait of Messina suggest that this passage is modest (Corso 2001). The Pallid Harrier is very uncommon further west but records include 58 in Britain and five in Ireland. There are accepted records of 182 individuals in France, where occurrences have been recently more frequent: there were 22 in 2008 alone. There are also 32 records (38 birds) for Morocco.

The Pallid Harrier is rare in Iberia but it has been recorded increasingly often in recent years. The earliest specific record was the capture of a bird in Sevilla in 1872, although Irby (1895) described it as 'not uncommon in spring' near the Strait. One taken in Menorca prior to 1911 and observations of single birds in Mallorca in April 1924 and May 1927 (Munn 1926, Bernis *et al.* 1958) are the earliest 20th century records. Subsequently a female was taken at Térmens, Lleida, in September 1950, and another female was found dead at the Llobregat delta in April 1981 (Ferrer *et al.* 1986). A male in Salamanca in March 1982 (*Ardeola* 29: 181) also antedated the Rarities Committee. There are no accepted records from Portugal up to 2010 but at least eight were reported there during 2011–12. There are seven records from Gibraltar.

The 45 records up to 2010, including three from Gibraltar, included 32 males, seven females and seven juveniles/first winters. The unmistakable adult males are much less likely to be overlooked than the females and immatures, some of which no doubt pass unnoticed. The Iberian records have a clear spring peak, with 62% of records falling within the space of about a month, between 18 March and 24 April. Autumn records are much fewer but 20% of records fall between 8 August and 28 September (Fig. 80).

There have recently been a number of winter records in Spain. A first-winter bird overwintered between Almenara and Xilxes, Castellón, from at least 5 December 2005 and 2 March 2006. What was very probably the same individual, by then recognisable as a second-year female, wintered in the same area between November 2006 and February 2007 (*Ardeola* 55: 269). An adult male was at Quinto, Zaragoza, in January 2006 (*Ardeola* 56: 321). Both 2010 and, especially, late 2011 saw unprecedented numbers of Pallid Harriers in Iberia, including several that overwintered. The latter influx was matched by exceptional numbers in other western European countries, including Britain and Ireland, in autumn, some of which overwintered (*Birding World* 24: 513, 25: 31; Forsman & Erterius 2012). Post-2010 Spanish records total at least 23 (24 birds) in 2011 and at least 16 in 2012 (*Ardeola* 60: 456, RBS). There was also a further spate of spring records in 2013, involving some 41 birds in Spain during March–May, most of them in April (RBS), as well as three from Gibraltar in April. Nearly all were found in the Levant, Andalucía and the Balearic Islands, as is usual (Fig. 81), but in April 2010 there were different single birds in Asturias at Carreño and at Cape Peñas (*Ardeola* 59: 373). There are at least four autumn records from Catalonia and five from Navarra, suggesting that birds enter the Peninsula then at both ends of the Pyrenees.There were three in Portugal in October–November 2011, a further three in March–April 2012 and two in November–December 2012 (RBP).

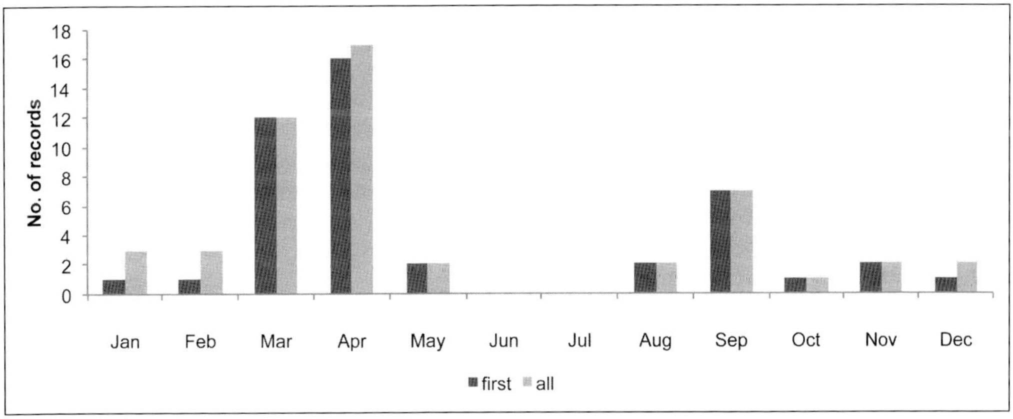

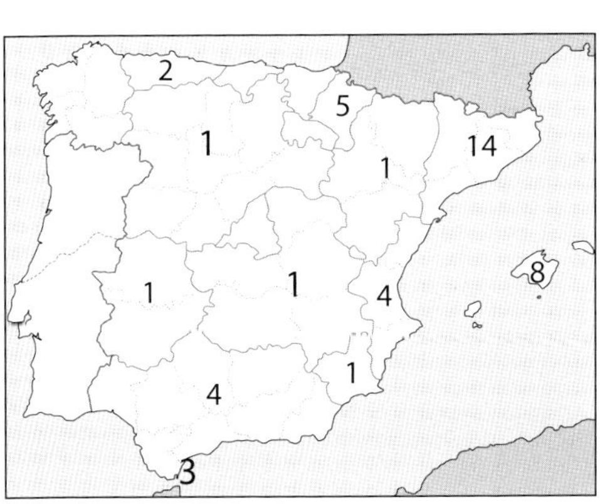

Figure 80. Records by month, 1925–2010 (n = 45).

Figure 81. Records by region, 1925–2010 (n = 45).

A juvenile seen at La Janda in January–February 2011 may have overwintered there (*Ardeola* 60: 456–457). The autumn 2011/winter 2011–12 observations in Spain comprised up to nine records in November, seven in December and six in January, most of them in Andalucía, including two seen repeatedly in the La Janda area: an adult male from September to January, and an adult female from December to January (RBS).

MONTAGU'S HARRIER *Circus pygargus*

(Aguilucho Cenizo, Águia-caçadeira)

Widespread and locally common as a breeding species in mainland Spain and Portugal. Occasionally nests in the Balearic Islands. Also a common passage migrant.

The sight of a Montagu's Harrier quartering a field is typical of agricultural regions throughout Iberia, in particular where cereal crops, the preferred nesting habitat, are grown. Often there will be several birds since the harriers frequently nest in loose colonies: 87% of birds in Madrid bred in such aggregations, comprising seven pairs on average and a maximum of 16, with inter-nest distances as short as 30m (Arroyo & García 1999). The breeding range extends across the greater part of mainland Iberia, avoiding high mountains and forests. The principal gaps in the distribution are the western seaboard of Portugal, the Biscay lowlands north of the Cantabrian Range, much of the arid south-east and most of the Mediterranean coastal regions.

A census of the Spanish breeding population in 2006 found 6,096–7,389 pairs (Arroyo & García 2007). Over 70% were shared between Andalucía: in particular the western provinces and the Guadalquivir valley, Extremadura and Castilla y León. Previous surveys of Montagu's Harriers in Spain have been incomplete and used different methodologies in different areas, and so are not entirely comparable with the 2006 survey. Ferrero (1995) estimated that there were 3,600–4,600 pairs. The 2003 breeding bird atlas survey found a minimum of 4,900 pairs (SA2). Taking these results together it would appear that the Spanish population is at least stable but an overall increase seems probable. There have been marked increases in the north-east, in the Comunidad Valenciana and Catalonia, reversing large declines there. There were none breeding in the Comunidad Valenciana by the early 1990s but there were over 160 pairs in Castellón alone by 2006 (Limiñana *et al.* 2006). Similarly, there were only two pairs in Catalonia by the late 1980s (CA2) but at least 119 pairs by 2006. The population of Castilla y León has also shown a steady increase from around 1,000 pairs in 1995 to 1,800 in 2002 and 2,300 in 2006, possibly in response to seasonal increases in vole populations but perhaps involving displacements of birds from elsewhere, such as from Madrid and Castilla-La Mancha (Arroyo & García 2007).

Montagu's Harrier regularly occurs as a scarce migrant in the Balearic Islands. It has bred occasionally in Mallorca, most recently in 2001, 2003 and 2004 (*A. O. Balears 2010*: 262). The first known instance of breeding on the islands was at the Salobrar de Campos, Mallorca, in 1967, and nesting occurred at saltpans in Ibiza in 1977 (Cayuela & Mayol 2002).

The Portuguese population was estimated at 1,000–1,300 pairs in the early 1980s and 900–1,200 pairs in the mid 1990s (AP). The population is largely concentrated in two large tracts of eastern Portugal (PA2). The principal nucleus is in the cereal croplands of the south-east, in Castro Verde and Campo Maior in Alentejo. The second centre of population is in the north-east, in Trás-os-Montes and Beira Alta, here nesting particularly in upland scrub, as occurs in adjacent parts of Spain. There is a scattered presence elsewhere in Portugal but there has been range contraction in the coastal regions, attributed to habitat loss due to irrigation and afforestation.

The great majority (80–90%) of Iberian Montagu's Harriers nest in cereal crops. Nesting in other, natural, low cover is characteristic of those in the north-west: in Galicia, Asturias and northwestern Castilla y León, and also in southern Tarragona and Castellón on the Mediterranean coast. Nests in cereals are at grave risk when the crop is harvested. Local conservation groups and regional administrations work hard to locate and save endangered nests and their efforts are clearly important in maintaining the current healthy populations of the farming regions (Arroyo *et al.* 2002, Castaño 1997).

Western European Montagu's Harriers winter in sub-Saharan Africa, particularly in the Niger basin. There are 27 foreign recoveries of birds ringed as pulli in Spain, including 22 in Morocco and single birds from Mauritania, Mali, Gambia and Guinea Bissau, as well as one from France. Six satellite-tracked birds from Castellón wintered in a narrow belt of the Sahel: in southern Mauritania and nearby areas of eastern Mali and Senegal (Limiñana *et al.* 2007). The movements of western European Montagu's Harriers to and from winter quarters are probably on a broad front, with some concentration at the Sicilian narrows and the Strait of Gibraltar. The earliest spring arrivals at the Strait are in early March but passage proper does not begin until mid-March and continues into May, peaking in April. Males generally arrive earlier than females. Montagu's Harriers arrive in central Spain from late March and most clutches are laid between mid April and mid May, earlier in the south than in the north. A recent study of Montagu's Harriers from Castellón followed by satellite telemetry (Limiñana *et al.*

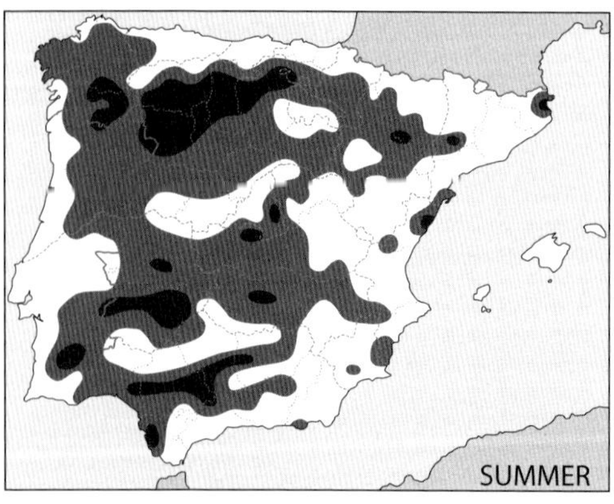

2008) has revealed post-breeding movements that take the birds further north and/or to higher elevations, where their preferred orthopteran prey is still abundant in late summer. Such movements recall those performed by other 'steppe' species such as bustards, Quails and Lesser Kestrels.

The return movements at the Strait are in August–September, peaking early in September, with a few early individuals – often juveniles – passing in late July and some passage extending into October. Melanistic individuals are often seen on passage: they amounted to 4.7% of all birds in 1976 and 3.0% in 1977 (Bernis 1980). There is some evidence from the Portuguese breeding population that melanistic birds are relatively frequent in the cooler, more humid uplands of the north but that they are relatively rare in the dry croplands of the south (AP). In Spain melanistic birds are also relatively frequent in the humid north-west, in Galicia (pers. obs.).

Analyses of ringing recoveries and observations at the above sites led García & Arroyo (1998) to suggest that there may be a loop migration, in which proportionately more fly south through Iberia in autumn and return north through the central Mediterranean in spring. The actual numbers seen at the Strait of Gibraltar during either season number a few hundred birds, only a fraction of the Iberian population alone, however, suggesting that many cross the Sea of Alborán and avoid the Strait altogether. One Dutch female tracked by satellite from The Netherlands to Morocco crossed the Alborán basin departing near Almería (Trierweiler *et al.* 2007) and eight from Castellón province crossed closer to Gibraltar, although some possibly not at the Strait itself (Limiñana *et al.* 2007).

Some Montagu's Harriers from other western European populations cross the eastern peninsula and the Balearic Islands on passage (García & Arroyo 1998) but, once again, relatively few are seen on the move through the Pyrenean passes. There are at least 34 Iberian recoveries of birds ringed elsewhere, generally as pulli: 22 from France, five from Germany, three from The Netherlands and two each from Britain and Sweden. Autumn seasonal counts at the western Pyrenean watchpoints averaged 276 birds in total during 2009–11, with fewer than 100 usually counted annually in the east, at Eyne. The passage is concentrated in a six-week period between mid August and the end of September, peaking in late August (Urcun & Bried 2008b).

Wintering birds are extremely rare. Very late or winter records include a male at the Azud de Riolobos, Salamanca, on 1 December 2001, a male at Alcañizo, Toledo, on 21 November 2004, a female at Doñana on several dates in January 2012 and two observations of birds of unspecified sex at the Saladares del Guadalentín, Murcia: one on 22 December 2006 and four on 27 December 2006 (*Ardeola* 49: 183, 53: 200, 55: 138, 59: 178). One at Castro Verde, Beja, on 13 February 2008 and a male at Quinta das Flores, Faro, on 26 February 2009 (*Anu. Ornitol.* 7: 78, 8: 63) were also very unseasonal. The recent occurrences of overwintering Pallid Harriers in Spain raise the possibility that at least some wintering harriers have been misidentified: any putative unseasonal Montagu's Harriers merit close scrutiny.

NORTHERN GOSHAWK *Accipiter gentilis*

(Azor Común, Açor-comum)

***A. g. gentilis*. A widespread resident in mainland Iberia, more abundant in the north.**

The ever-elusive Goshawk is a frequent presence in the forested areas of mainland Iberia, especially in the northern half of the Peninsula. Nevertheless, the 2009–10 census of breeding forest raptors in Spain found it to be especially abundant in west-central Spain: in Madrid, Ávila and Toledo, as well as in adjacent parts of eastern Extremadura (Palomino & Valls 2011). It nests in all the Spanish regions but its distribution is much more fragmented south of the Central Range, where it occurs mainly in the Sierra Morena and the Betic Range, although it is known to have nested in the pinewoods of Doñana since 1989 (Llandres & Urdiales 1990). In Portugal it is almost entirely absent south of the Tagus (PA2).

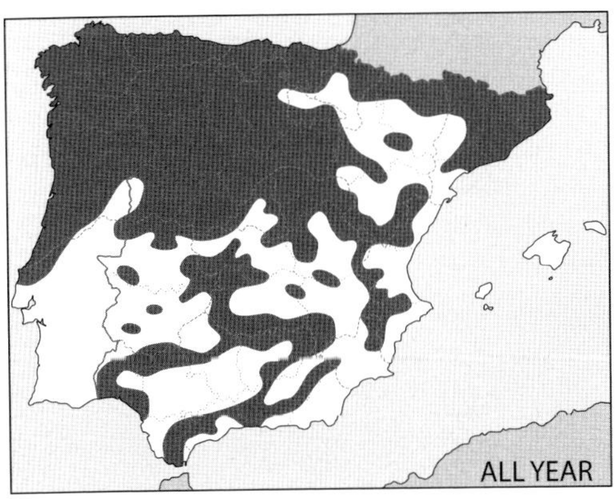

Nesting birds occur as high as 2,000m in pinewoods of the eastern Sierra Nevada (Garzón & Henares 2012). It nests in forests, woods and stands of tall trees, both deciduous and coniferous, and including poplar and eucalyptus plantations. It is widespread in the farmlands of Castilla y León, nesting in riparian woodland and in isolated copses and plantations. The ideal habitat combines woodland for nesting and roosting and open areas in which to surprise prey on the ground. A recent study in the Guadarrama foothills, Madrid, in an area offering a mosaic of pine and oak stands and riparian woodland intermixed with cultivation, found a breeding density of some six pairs/100 km^2: of 29 nests found, 27 were in pines and two in poplars (Alcobendas 2009). The Goshawk is a powerful predator, taking mammals as large as hares and birds as large as crows; pigeons and, locally, rabbits figuring prominently in the diet in Iberia (Mañosa 1994, Petronilho & Vingada 2002).

The Spanish population was estimated at 2,300–3,000 pairs by F. de Juana (1989). However, survey work for the 2003 breeding bird atlas put it some 5,000 pairs (SA2), including particularly large sub-populations in the north-west in Galicia (800–1,400 pairs), Castilla y León (800–1,100), Asturias (250–300) and Cantabria (150–175). There were also good numbers in Catalonia (250–300 pairs) and Aragón (150 pairs) in the north-east. The 2009–10 Spanish census of breeding woodland raptors (Palomino & Valls 2011) estimated the population to be 11,350 pairs, a remarkably large figure since the estimated breeding population for France is just 4,500–6,500 pairs. However, it is extremely difficult to say whether any evident local population increases are real or the result of more thorough prospecting. The 2009–10 census estimated that 22% of the population is in Castilla y León, 18% in Castilla-La Mancha, 12% in Andalucía and 11% in Aragón, which is surprising since it is relatively easier to meet with it in the pine-covered mountains of the north.

The Goshawk is extremely rare in the Balearic archipelago: two have been seen in Menorca, at Son Bernadiin January 2006 and at Maó in December 2012 (*Ardeola* 55: 295, 60: 518).

The Portuguese population estimates of 200–300 pairs in the mid 1990s and of 50–200 pairs by Birdlife International (2004) are based on limited information (AP). The 2008 breeding atlas recorded confirmed or probable nesting in 56 10km-squares, and possible nesting in a further 101 squares (PA2). The 2008 atlas found some shift in the population with an increase in western coastal areas and some decline inland. There are also 6–8 pairs in Andorra (BE).

The Spanish population seems to be stable although local fluctuations have been recorded. For example, an 18% range reduction has been reported in Catalonia since the 1980s (CA2). Conversely, the population of the Rias Baixas in Galicia expanded rapidly to 22 pairs during the 1980s as birds exploited the expansion of eucalyptus plantations and an abundance of domestic pigeons there, but declined to 7–10 pairs by the early years of 21st century as the maturing plantations were disturbed by foresters and pigeons also became scarcer (SA2).

The Goshawk is also known to be susceptible to electrocution, predation by Eagle Owls, death from trichomoniasis caught from infected pigeons and the robbing of chicks from nests by falconers, as well as killing by pigeon fanciers and illegal hunting (SA2). There is no doubt that its Iberian populations were depressed by persecution in the past, as occurred elsewhere in Europe (Bijleveld 1974). There are no useful historical references to its abundance in Spain but Bernis (in 1966) described it as very scarce and local and estimated that there were only a few hundred individuals in Spain (in Bijleveld 1974). Similarly, Garzón (1974) considered it to be scarce and local in west-central Spain, where the most recent census suggests that it is now common. A combination of legal protection and widespread afforestation can only have helped the species to increase.

Iberian Goshawks are largely sedentary. Twenty recoveries of Spanish-ringed birds showed a mean displacement of 37km, within a range of 0–140km (AI1). The farthest displacement involved one ringed as a chick in Galicia and recovered in northern Portugal three years later. There are no recoveries of foreign-ringed birds but a few individuals are recorded annually on passage through the Pyrenean passes (OF). There are annual records of apparent migrants at the Strait of Gibraltar but again very few birds are involved in any one year. Some Goshawks disperse outside the breeding range in winter. For example, the species was a rare winter visitor to Doñana prior to the establishment of a breeding population there.

EURASIAN SPARROWHAWK *Accipiter nisus*

(Gavilán Común, Gavião comum)

***A. n. nisus*. A widespread resident in the northern third of the Peninsula and present but more patchily distributed elsewhere. Also common on passage and in winter. It does not breed in the Balearic Islands.**

Breeding Sparrowhawks are quite uniformly distributed from northern Portugal and Galicia eastwards through the Cantabrian Range and the Biscayan coastlands to and throughout the Pyrenees as far the Mediterranean. They also inhabit riparian woodlands, copses and plantations in agricultural land throughout Castilla y León and the woods and forests of the Iberian Range, the Central Range, the Sierra Morena and the Betic Range. Sparrowhawks, like Goshawks, are very scarce in parts of the Ebro depression, much of Castilla-La Mancha, the dehesas of Extremadura, the arid south-east and the Guadalquivir valley. They may have once nested in Menorca but if so became extinct there around 1970 (Muntaner 1979). They are found in all types of woodlands, with some preference for conifers (SA1), nesting from sea level to at least 1,600m, and to over 2,000m in the Sierra Nevada (Garzón & Henares 2012), typically selecting denser stands than used by Goshawks. They favour habitat mosaics that include clearings and fields with hedgerows, where they can surprise their quarry, and also often occur close to towns and villages, which offer good opportunities for hunting small birds.

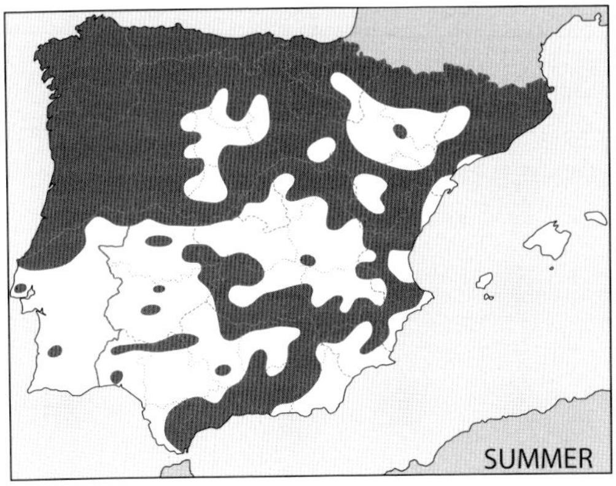

The Spanish population was estimated at 3,000–8,000 pairs by F. de Juana (1989). The 2003 atlas fieldwork suggested that there were somewhat more, perhaps 6,000–10,000 pairs (SA2), with the highest concentrations in Galicia (2,000–3,000 pairs), Castilla y León (800–1,000) and Cantabria (300–400 pairs). However, the 2009–10 Spanish census of breeding woodland raptors (Palomino & Valls 2011) estimated the population at 13,810 pairs, over 30% of them in Andalucía; 25% across the meseta, including Madrid & Extremadura, 18% in the Cantabrian region and 12–13% each in the Pyrenees/Ebro valley and the Levant regions: in total a much higher figure that may however reflect differences in methodology and fuller prospection. The 2009–10 estimate is not high given the extent of the potential breeding areas. The French breeding population post-2000 is some 30,000–50,000 pairs (OF).

It is likely that the Sparrowhawk was scarcer in the past. It is frequently mentioned in historical sources, and some authors regarded it as common, but in the 1960s Bernis (AMI) still described it as generally very scarce and sporadic except in the Cantabrian/Atlantic region and in some inland mountains, but nowhere common. Our subjective impression is that it is now seen much more frequently than in the 1970s and 1980s and very probably some population recovery has occurred, as has happened elsewhere in Europe since the banning of organochlorine pesticides (Hagemeijer & Blair 1997).

The Portuguese population was estimated at 200–300 pairs by Palma *et al.* (1999) but put at 300–1,500 pairs in 2002 by Birdlife International (2004). The great majority nest north of the Tagus, where they are widespread and seem to have increased in coastal regions (PA2). In southern Portugal they are clearly absent from the largely treeless regions of Baixo Alentejo and Beira Baixa but they have a scattered distribution elsewhere that has not been fully described. Some 15–20 pairs occur in Andorra (BE).

Iberian birds are thought to be resident (AMI) but they are joined during passage periods and in winter by migrants from further north in Europe. Sparrowhawks are most widespread in winter, when they may be found in regions where they do not usually nest, including the Balearics (SWA, AP). The migrants originate chiefly from north-central Europe with some also from Scandinavia and the Low Countries. Díaz *et al.* (1996: AI1) note 31 recoveries of foreign-ringed birds, chiefly marked as pulli, which include 12 from Germany, nine from the Low Countries, five from Fennoscandia/Baltic States, two from Switzerland and single birds from the Czech Republic, Poland and Italy. Subsequent recoveries show similar origins. Most recoveries are during September–February and many have been found in northeastern Spain, although they are widely scattered. Eighteen birds ringed in Spain, none as pulli, have been recovered abroad: nine in France, two each in Switzerland, Belgium and Finland, and single birds in Denmark, Germany and Italy.

Figure 82. Distribution by approximately ten-day periods of Sparrowhawk arrivals in Iberia at the Strait of Gibraltar during 2008–12 combined (n = 6,511 birds). (Fundación Migres, GONHS)

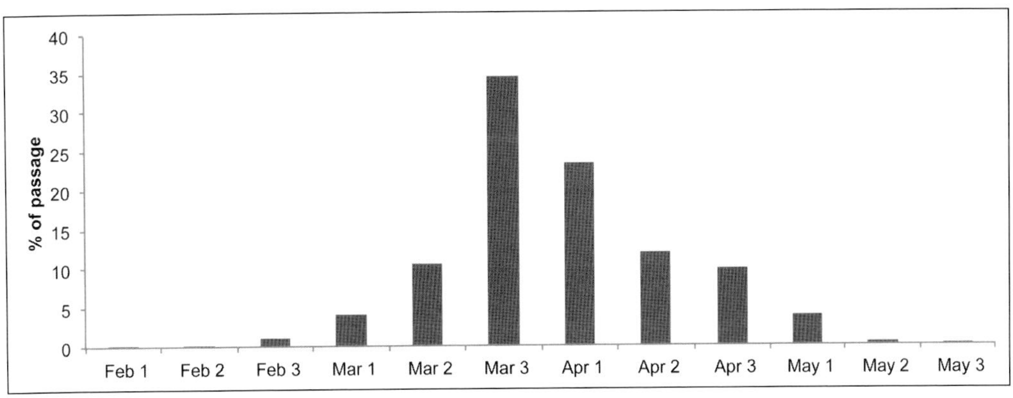

Sparrowhawk passage is evident at the Pyrenean passes although the numbers counted are never very large. Annual autumn counts at the three western Pyrenean watchpoints recorded an average of 1,158 birds in total during 2009–11 and there was a mean count of 672 birds at Eyne during the same period. The autumn passage through the Pyrenees occurs from early August until mid November, peaking during the second half of September (Urcun & Bried 1998b). The return passage is in March–May but has been little studied at the Pyrenees.

Some of those that enter the Peninsula cross the Strait of Gibraltar to winter in North Africa. The southward passage at the Strait also includes a few early individuals in August but most pass during September–October, with stragglers in November. The passage peaks during late September and early October, at much the same time as in the Pyrenees, which led Bernis (1980) to speculate whether in fact much of the trans-Strait movement might in fact involve Iberian birds after all. The matter remains unresolved.

Spring passage across the Strait is protracted, with some birds passing from late February until mid May and most arriving between late March and mid April (Figure 82). A striking long-term trend during both the spring and autumn movements has been a significant and continuing increase in the numbers of birds involved. This change has most clearly been revealed by the long-term spring counts at Gibraltar itself. The mean rate of spring passage during 1967–77 was 1.06 birds/hour but it increased significantly to 4.27 birds/hour during 1990–2004 (Bensusan *et al.* 2007). Total annual spring counts at Gibraltar and the central Strait region combined averaged 1,303 birds during 2008–12 (range 1,139–1,514; GONHS, Programa Migres). There is every indication that at least a couple of thousand Sparrowhawks cross the Strait during both passages in recent seasons. The reasons for the observed increase in the passage volume are unclear but they may reflect the known recovery in the European populations following the pesticide-linked declines during the early and middle years of the 20th century.

COMMON BUZZARD *Buteo buteo*

(Busardo Ratonero, Águia-de-asa-redonda)

***B. b. buteo*. A very common and widespread resident in most of mainland Iberia. Has bred recently in Mallorca. Also common on passage and as a winter visitor throughout the region.**
***B. b. vulpinus* (Steppe Buzzard). A rare passage migrant in Spain.**

The Common Buzzard is a familiar sight across the greater part of the Peninsula, including most of Spain and all of Portugal. It may be found nesting wherever there is a mosaic of woodland and pastures or arable land, or where tall trees in hedgerows provide nesting sites in farming areas. It is highly opportunistic and takes a wide range of small vertebrates and invertebrates as available (Tapia 2010). Typically it hunts small mammals and large invertebrates, especially orthopterans, in open country, often scrutinising likely areas from posts and other exposed perches, or by hovering. Outside the breeding season it may often be found in treeless areas, where it takes advantage of fenceposts, pylons and any natural perches.

In Spain breeding birds are only really absent from much of the Levant coastlands and the arid south-east, as well as from areas of the Guadalquivir valley and the southeastern meseta. They are most abundant in the Cantabrian region, from Galicia east to the Basque Country, where breeding densities of 8–10 pairs/km^2 were reported in 2009–10. Densities elsewhere in Spain range from 4.5–7.5 pairs/km^2, with relatively high populations in Navarra, Extremadura and Andalucía: where there are large expanses of very suitable habitat. Nests are nearly always placed in tall trees but some nest on cliff faces. Breeding birds nest from sea level to at least 1,600m, and to nearly 2,000m in the Sierra Nevada (Garzón & Henares 2012). They often hunt over grasslands above the treeline in mountainous areas.

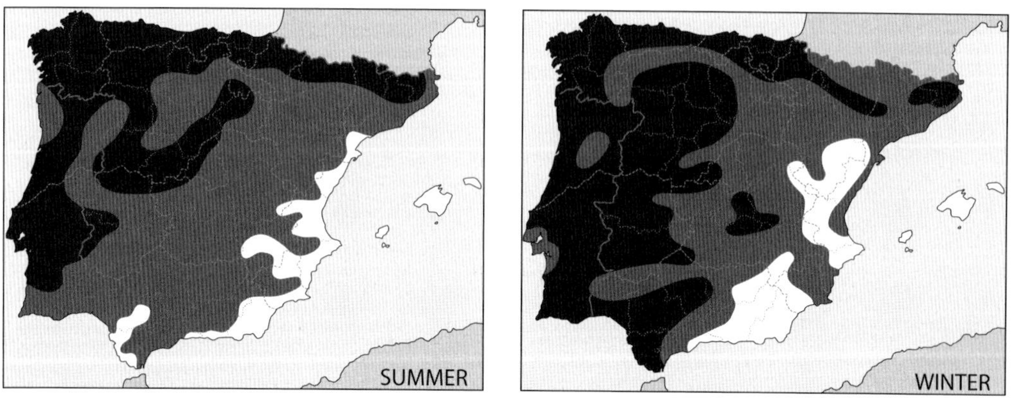

The Spanish breeding population was estimated at 5,000–5,500 pairs by F. de Juana (1989) but the survey work for the 2003 atlas found many more: at least 13,000–18,000 pairs. This higher figure was thought to reflect more thorough prospection rather than any real population increase. Nevertheless, it is highly probable that the populations in Spain increased during the course of the 20th century, as has occurred across much of Western Europe. Bernis (1966: AMI) considered it widespread in forested areas throughout the Peninsula, but nowhere common, and the Common Buzzard may have been among the most persecuted raptors in Spain before its legal protection in the late 1960s (de Juana 1989, Balbás 2003). The Spanish census of woodland raptors in 2009–10 estimated an even larger breeding population of some 29,820 pairs, which is of a comparable order of magnitude with that of other large European countries (Palomino & Valls 2011).

The highest densities in Portugal are in the south-centre in Alto Alentejo and in the western coastal regions of Ribatejo and Estremadura (PA2) and it is generally scarce in Algarve (AP). There is no recent census data but Palma *et al.* (1999) estimated that there were 2,000–4,000 pairs and Birdlife International (BE 2004) give an estimate of 1,500–3,000 pairs for 2002. The Portuguese population showed no clear trend during 2004–09 (ACP). The breeding population of Andorra is 20–30 pairs (BE 2004).

Until very recently the Common Buzzard was regarded as just a rare winterer and a scarce migrant in the Balearic Islands. The wintering numbers post-2000 averaged 4–9 birds in Mallorca and at least four in Menorca, but there was an exceptional influx during winter 2010–11, when there were up to 43 in Mallorca and 25 or more in Menorca (Ventoso & Muntaner 2011). There were indications of nesting in Mallorca from 2006 onwards and one or two pairs apparently bred there in 2010 near Esporles (*A. O. Balears 2010*: 135); breeding was confirmed in 2011 at Valldemossa, where a pair fledged two young (*Ardeola* 59: 426).

Iberian birds have been assumed to be largely sedentary but post-fledging dispersal can nonetheless be far-reaching in some cases, since at least eight birds ringed as pulli travelled over 200km within Spain, the longest displacement being 620km: from Granada to Tarragona. Only one, ringed as a pullus at Doñana on 5 June 1974, has been recovered abroad: near Port Etienne, Mauritania (2,100km), on 12 January 1975 (*Ardeola* 29: 70) – which is probably quite exceptional if the initial identification was correct.

Significant numbers of Common Buzzards from northern Europe join the Iberian population in winter, these being only a fraction of the much larger numbers that remain to winter in France and Germany (AMI, de Juana *et al.* 1988). Most (35) of 52 recoveries of foreign-ringed birds recovered in Iberia originate from Fennoscandia and the Baltic region, with nine from Germany, three each from Switzerland and France and one from the Czech Republic. Sixteen birds ringed in Spain: many or most of them full-grown, have been recovered abroad: eight in France, five in Germany and one each in Switzerland, Poland and Austria. Immigrant Buzzards appear to winter largely in northeastern Spain: in the Basque Country and Ebro valley, and it is thought that the large concentrations of wintering Common Buzzards found in Galicia, the northern meseta and Extremadura largely involve Iberian birds that descend to lower levels in winter (de Juana *et al.* 1988).

Common Buzzards may be seen entering Iberia through the Pyrenean passes from mid August until November, most passing during the second half of September and in October. The numbers recorded since systematic counting began in the early 1980s have not been very large, combined seasonal totals at the main passes during 1981–94 ranging up to 650 birds (Urcun & Bried 1998b) and they averaged only 341 birds at the three western watchpoints during 2009–11 and 70 birds at Eyne during the same period.

Migrant Common Buzzards are also recorded at the Strait of Gibraltar, passing south during September–November, with a peak in October, and returning from February to early May, chiefly during March–April. However, in recent years there has been a dramatic and highly significant decline in the numbers that cross the Strait. The coordinated autumn counts of the 1970s (Bernis 1980) detected a total of 7,315 migrants during the combined seasons of 1972, 1974, 1976 and 1977, showing that a few thousand individuals then used the crossing. Long-term observations from Gibraltar in spring have shown an unequivocal pattern of decline. Hourly passage rates declined significantly from 2.50 birds/hour during 1967–77 to 0.50 birds/hour during 1990–2004 (Bensusan *et al.* 2007). Total spring counts at Gibraltar and the central Strait region averaged just 57 birds during 2008–12 (GONHS, Programa Migres). The sharp decline in the trans-Strait movements is clearly not due to any decline in European or Iberian Common Buzzard populations, since these have either been stable or have increased (BE). The alternative would seem to be that migrants of either northern continental or even Iberian origin are short-stopping and increasingly remain to winter in Europe.

Some individuals of the eastern subspecies, *B. b. vulpinus* (Steppe Buzzard), occur in Spain on passage, with occasional records south to the Strait. This form is identified at the Pyrenean passes in autumn more or less annually and represents about 1.3% of the passage at Orgambidexka (Urcun & Bried 1998b). Bernis (1980) recorded 61 birds in 1972, 28 in 1976 and 16 in 1977 on southward passage at the Strait, although he queried whether the birds were genuinely *vulpinus* or merely *vulpinus*-like individuals of the nominate race.

LONG-LEGGED BUZZARD *Buteo rufinus*

(Busardo Moro, Bútio-mourisco)

Birds of the North African race *B. r. cirtensis* have been recorded increasingly often in southern Spain and Gibraltar. A pair first nested in Spain, in the vicinity of the Strait, in 2009. Vagrant to Portugal.

The nominate race, which breeds in the Balkans and Central Asia and winters mainly in the Middle East and N Africa, is known as a vagrant in central and eastern Europe, where it has been recorded mainly during August–October, although it has nested recently in both Hungary and Serbia. There are records of at least 22 from France, where one or two have occurred with some regularity in Camargue since 1995, although at least some may have been escaped birds.

The earliest Iberian record was of one captured in Málaga in February 1873 (Arévalo 1887) although this and some more recent claims (*Ardeola* 5: 143–148, 13: 247–248,, 16: 257) have been queried as perhaps due to confusion with the Steppe Buzzard *B. buteo vulpinus* (Bernis & García-Rúa 1974, AI1). However, a 19th century specimen obtained at La Almoraima, Cádiz, in June 1882 and currently in the British Museum collection, indeed appears to be *B. r. cirtensis* (Garcia 2004).

More recently, CR-SEO has accepted 32 Spanish records (of 37 individuals) from 1995–2011 (*Ardeola* 60: 457), over 70% of them from Andalucía. One at Gibraltar in June 1966 was only a 'probable' (Lathbury 1970, Cortes *et al.* 1980) but there have been seven accepted records (eight birds) there during 1985–2011. Although there are no accepted records from Portugal, there have been four recent sightings (RBP): single birds near Castro Verde, Beja, on 25 April 2010; at Sagres, Faro, on 22 November 2010; at Lagos, Faro, on 23 September 2011, and at Monte Horizonte, Setúbal, on 26 May 2012. In general, the increase in Iberian records has been pronounced, from only six during the 1990s to 32+ during 2000–11. They have included 23 birds aged as adult or sub-adult and 11 juvenile or immature individuals.

There are records from all months although with a peak in September (Fig. 83). Two individuals, an adult and a juvenile that appeared in Gibraltar on 2 September 2005 were seen intermittently there into the winter (*GBR* 5: 19; 6: 20). A wintering bird, at La Lantejuela, Sevilla, was present there from October 2001 but its wild status has been queried since it carried a ring (*Ardeola* 50: 134). Another adult overwintered at Los Barrios, Cádiz, each year from 2008–09 until at least 2011–12 (*Ardeola* 58: 454, RBS). An adult was at Jerez de la Frontera, Cádiz, from December 2011 to March 2012, and another was at Sucina, Murcia, on 31 January 2011. The concentration of records in southernmost Iberia (Fig. 84), and the presence of breeding birds in the Spanish hinterland of the Strait, reinforces the perception that the Iberian Long-legged Buzzards originate from Morocco, arriving on post-breeding dispersion, and individuals are seen crossing the Strait annually. Only one accepted record was assigned to the nominate subspecies: a possible juvenile at the Embalse del Voltoya, Ávila, on 5 September 1998 (*Ardeola* 47: 147).

The first Iberian breeding record was in 2009 near Tarifa, only some 20km from the nearest breeding birds on the Moroccan shore of the Strait of Gibraltar. The pair, both of which showed the characteristics of *B. r. cirtensis*, fledged two young around 20 July (Elorriaga & Muñoz 2010, *Ardeola* 58: 454). In 2010, however, the female in this territory paired with a male Common Buzzard and they raised three hybrid chicks that were ringed at the nest in June and successfully fledged in July (*Ardeola* 59: 373–374). Since then there have been reports of up to six breeding pairs in the hinterland of the Strait (A. Onrubia, pers. comm.).

Figure 83. Records by month, 1958–2010 (n = 41).

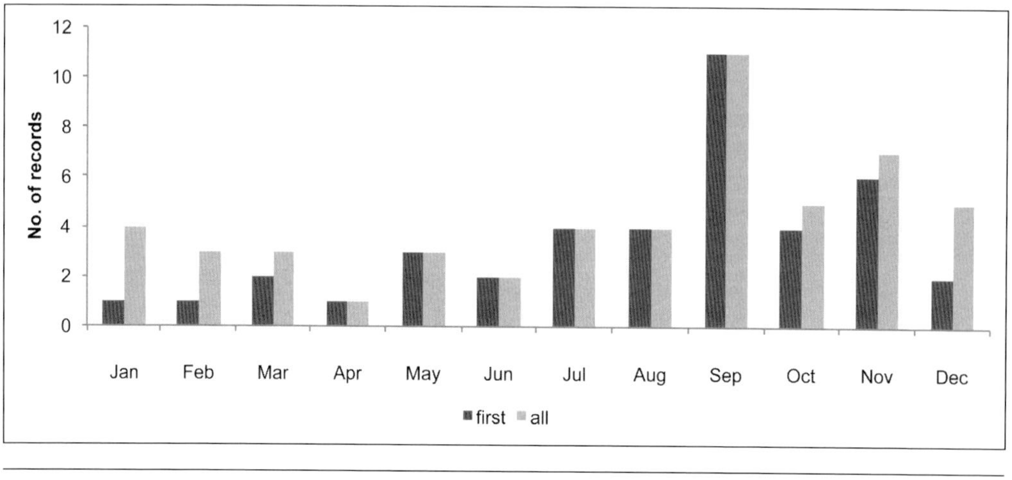

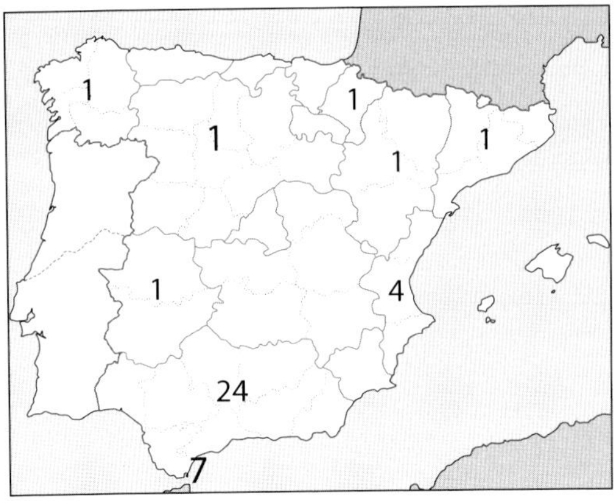

Figure 84. Records by region, 1958–2010 (n = 41.)

The resident buzzards of northernmost Morocco, in particular those of the Tangier peninsula, may represent an intergrade between *cirtensis* Long-legged Buzzards and Common Buzzards (Rodríguez *et al.* 2013). Some of these apparent hybrids, which these authors have termed 'Gibraltar Buzzards', combine such *cirtensis* features as rufous underwing coverts and a rufous tail with an overall browner appearance than is typical of *cirtensis* and sometimes a pale breast-band as in the Common Buzzard. Some quite closely resemble Steppe Buzzards *Buteo buteo vulpinus*. The status and distribution of breeding buzzards in the vicinity of both sides the Strait requires closer attention but it seems clear that Long-legged-type buzzards have become increasingly frequent on the Iberian shore over recent decades.

ROUGH-LEGGED BUZZARD *Buteo lagopus*

(Busardo Calzado, Bútio-calçado)

Vagrant to Spain.

Wintering birds range south to central and eastern Europe and a few reach the British Isles. They are rare in France, with 450 accepted records during 1981–2012, nearly all from the north-east. There are three accepted records from the Azores and one from Madeira but none from the Maghreb.

There are quite a number of published references to Rough-legged Buzzards in Spain but few survive close scrutiny, most of the records involving confusion with pale Common Buzzards, Long-legged Buzzards (see photograph in *Ardeola* 13: 247) or even Booted Eagles (AMI).

The Spanish rarities committees examined 30 claims of this species during 1984–2003 but only accepted five (de Juana 2006). Two were long-stayers in Girona province: at Pals from 12 November 1989 to 21 March 1990, and at the Aiguamolls de l'Empordà from 19 October 1998 to 16 March 1999. An adult was at the Llobregat delta from 5–22 March 1990, and one was at the same location from 13 January 1991 into February. One was at the Ebro delta on 17 January 1996. There have since been two more accepted records: a first-winter bird in Tarragona from 18 December 2010 to 9 March 2011, and an adult or subadult female near the Aiguamolls de l'Empordà on 4–14 March 2011. All accepted records have been of single birds in Catalonia between October and March.

LESSER SPOTTED EAGLE *Aquila pomarina*

(Águila pomerana, Águia-da-pomerânia)

Very rare migrant, recorded in Spain and Gibraltar.

Iberia lies far west of the principal breeding, passage and wintering ranges of this species. However, there is a small wintering presence in West Africa, in Cameroon, Nigeria and Mali (Thiollay 1977) and the species is regularly recorded on passage at Cap Bon, Tunisia (Dejonghe 1980). It is also known as a vagrant from several western European countries including France: from where there are 93 accepted records for 1981–2011 and where a pair bred in the far east during at least 2004–10 (Dubois *et al.* 2012). There are two accepted records for Morocco, in April 1996 and September 2007, the latter near the Strait. An unprecedented breeding attempt occurred in a Catalonian pine forest in 2011 but was unsuccessful. The male at least returned to the site in 2012 (Bosch & Meyburg 2012).

The only known certain occurrence in Spain until the 1990s was one taken at 'Sevilla' in 1889 and preserved as a specimen in the Sevilla University museum, labelled '*Aquila naevia*' (Herrera & Hidalgo 1971). There are also several records of '*Aquila naevia*' in the historical Spanish literature: most of them were probably Greater Spotted Eagles but some could have been Lesser Spotted (Editor's note in *Ardeola* 15: 134). Two late 19th century specimens from Portugal were identified as this species before they were destroyed in the Bocage Museum fire in 1978 (Soares 1970, AP).

The first accepted recent record for Spain was an immature bird found in Menorca during 10–24 September 1997 (Capó & Pons 2002). Since then there have been a further four accepted records from Spain and five from Gibraltar, all of single birds. The sightings at Gibraltar were on 18 April 2001, 17 September 2001 and 6 October 2002, followed by two separate individuals that arrived from the south on 1 April 2010. The Spanish records were at Doñana on 18 March 2003, in Menorca on 9 October 2004, and at Tarifa, on 29 September 2008 and on 12 November 2010. The latter was a juvenile that had been marked at Western Pomerania, Germany. Post-2010 there have been at least six more Spanish records, one from Gibraltar on 8 April 2013 and two from Portugal, both from the Sagres area (RBS, GONHS, RBP). In addition, a satellite-tracked individual that was the offspring of a male Lesser Spotted Eagle and a female hybrid Lesser Spotted x Greater Spotted Eagle was tracked to Iberia from northeastern Germany and crossed the Strait southwards from Tarifa on 28 October 2008 (Ramírez 2009). A second satellite-tracked German bird reached the Strait on 1 October 2009 but continued westwards into Portugal to a point 55km north-east of Lisbon on 7 October before returning to the Strait, where it crossed on 9 October (Onrubia *et al.* 2011). Two young Latvian Lesser Spotted Eagles, translocated into Germany, also migrated via Spain in 2009: one crossed the Strait on 9 October and the other apparently drowned near Melilla, after heading south for five hours across the Alborán basin from Cape Gata (Onrubia *et al.* 2011).

The Lesser Spotted Eagle may well prove to be a more frequent migrant across the Peninsula than the accepted records so far suggest. Observers at the Strait monitoring raptor migration under the auspices of Programa Migres have reported 47 additional observations during 1998–2009, from one to eight per year (Onrubia *et al.* 2011). Unfortunately none of these have been submitted to the Spanish rarities committee and so they have to be regarded as provisional, in particular in the case of this species, whose identification is not always straightforward. It may well be that the Lesser Spotted Eagle is or has become a very scarce but regular migrant across the Strait but the clarification of its Iberian status really requires the independent evaluation of all records.

GREATER SPOTTED EAGLE — *Aquila clanga*

(Águila Moteada, Águia-malhada)

A very rare but locally regular winter visitor to southern and eastern Spain and western Portugal, and a very rare migrant at the Strait of Gibraltar.

It is somewhat more frequent in France, with accepted records of some 140 individuals during 1981–2011, many of them from Camargue, where up to ten are present in some winters. Separating this species and the Lesser Spotted Eagle in the field is not always straightforward and the situation is complicated by the existence of hybrids between them, arising from frequent mixed pairings in some areas where their ranges overlap (Väli & Löhmus 2004). Nevertheless, the two differ markedly in phenology in southern Europe, where the Greater Spotted Eagle is chiefly a winter visitor and the Lesser Spotted Eagle occurs mainly on passage or in summer. Hence, in France most Greater Spotted Eagles occur during November–March and most Lesser Spotted Eagles are recorded during April–October (OF).

There are Spanish records from the 19th century, from the Albufera de Valencia (Saunders 1871, Arévalo 1887), the Guadalhorce estuary, Málaga (Arévalo 1887) and Doñana, this last including a specimen in the Lilford collection obtained there in October (Irby 1895). A specimen taken in Portugal in the 19th century is preserved in the collections of the Coimbra museum, the whereabouts of another 19th century specimen from the collection of King Dom Carlos are unknown, and a third 19th century Portuguese specimen was destroyed in the Bocage Museum fire: although it may have been a Lesser Spotted Eagle (Tait 1924, Soares 1970, AP).

More recently, one was taken alive at Doñana in February 1958 (*Ardeola* 10: 62) and another was seen there in March 1967 (Voous & Bijleveld 1968). Two were at the Ebro delta in April 1974 (Martínez Vilalta & Motis 1989) and one wintered at the Aiguamolls de l'Empordà during 1981–82 and 1982–83 (Sargatal & del Hoyo 1989). There are 22 accepted records from Spain during 1984–2010, as well as a record of an apparent hybrid juvenile, at El Rocío, Doñana, on 27 December 2001 (*Ardeola* 51: 525). A further half-dozen Spanish records were rejected since the Lesser Spotted Eagle was not definitely excluded, although the dates suggest that most were probably Greater Spotted Eagles. There are also nine accepted records for Portugal, in 1998–2004, most of them at the Tagus estuary (AP), and two for Gibraltar (*GBR* 4: 20).

All records have involved single birds with one exception, when two were seen together. They have included at least nine juveniles/first-winters, nine immatures or second/third year birds, and 13 adults or sub-adults. Winter records predominate (Fig. 85): although the extreme dates are 18 September and 14 May, 89% of first observations have been during October–March. Wintering birds are chiefly reported from the major coastal wetlands, from the Aiguamolls de l'Empordà and Ebro Delta in the east to the Tagus estuary in the west (Fig. 86). Favoured locations include El Hondo, Alicante, and Doñana, two sites from where there are a number of non-submitted records that suggest a small wintering presence there (García *et al.* 2000, Clavell 2002). Up to four different birds: one adult, two immatures and one juvenile, were reported from El Hondo in winter 2010–11 and there were also up to four there in winter 2011–12 and winter 2012–13 (RBS).

There is some evidence that at least occasional individuals cross the Strait to North Africa. There are at least four records of apparent migrants from the north side of the Strait: at Algeciras on 5 October 1999; at Jimena de la Frontera, Cádiz, on 18 September 2000; one arriving at Gibraltar, apparently from Morocco, on 10 March 1987, and one soaring over Gibraltar on 18 September 2004. In addition, one was recorded arriving at Cape Ceres, on the south shore of the Strait, on 22 September 1974 (Pineau 1975). In conjunction with the French records, these observations suggest an increasingly thin autumn migration down the Iberian east coast (de Juana 2006).

Figure 85. Records by month, 1958– 2010 (n = 38).

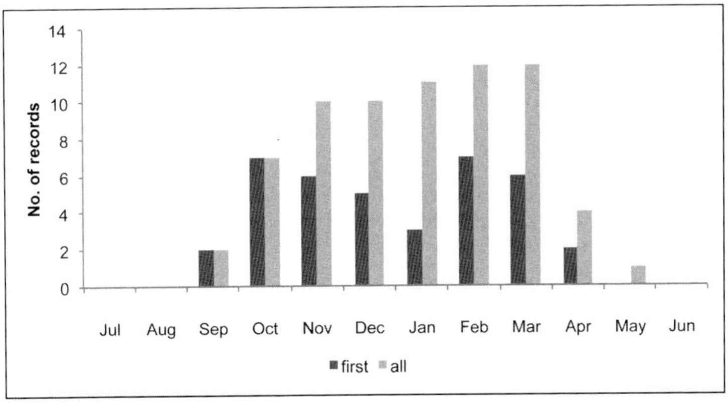

Figure 86. Records by region, 1958– 2010 (n = 38).

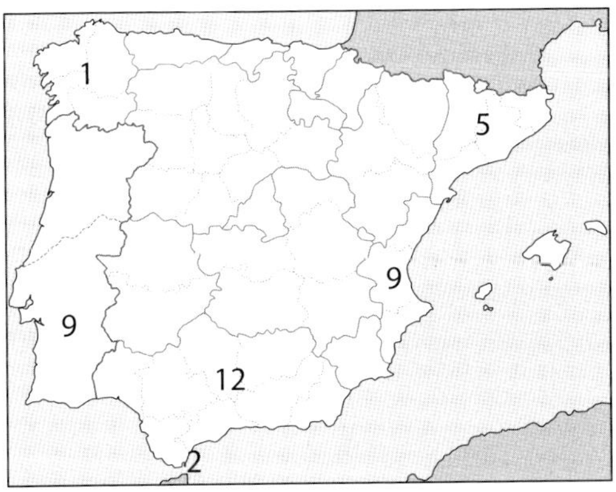

Satellite tracking of a juvenile Greater Spotted Eagle marked in Estonia largely corroborates other observations (Reeber *et al.* 2010; *Ardeola* 59: 375-376, 60: 458). This bird, known as 'Tõnn', began its first migration on 25 September 2008. It travelled successively across Latvia, Lithuania, Poland, Germany and France, to arrive in Catalonia on 16 November. Thereafter it wandered around eastern Spain until it finally settled at El Hondo, where it remained from late December 2008 until 13 April 2009. It was back in Estonia by 18 May. The following autumn it returned to El Hondo, rather more purposefully this time, since it arrived in late September, and remained there until early April 2010. Thereafter it returned annually to El Hondo until at least winter 2013–14.

TAWNY EAGLE *Aquila rapax*

(Águila rapaz, Águia-fulva)

Vagrant to Spain.

Irby (1895) refers to a specimen in the Lilford collection that was obtained near Málaga in winter 1877. Two observations from Spain antedate the Spanish rarities committee: a doubtful record of an adult at the Aiguamolls de l'Empordà on 16 May (Wallace & Sage 1969) and a young bird at Trebujena, Cádiz, on 15 November 1981 (*Ardeola* 29: 181). The breeding range of *A. r. belisarius* extends to the Souss valley and adjacent parts of Morocco. It was formerly much more widespread in Morocco, where it wandered occasionally north to the Strait (BM), but it is now rare.

STEPPE EAGLE *Aquila nipalensis*

(Águila esteparia, Águia-das-estepes)

Vagrant to Spain.

An apparently northbound adult was seen and photographed at Tarifa, in the Strait of Gibraltar, on 6 May 2008 (*Ardeola* 57: 493–494). Subsequently, an adult was photographed at Peguerinos, Ávila, in January 2013 (RBS). This species is a very rare vagrant to western Europe; three individuals had been recorded in France up to 2008.

SPANISH IMPERIAL EAGLE *Aquila adalberti*

Águila Imperial Ibérica, Águia-imperial-ibérica)

A rare Iberian endemic species that breeds only in southwestern and central Spain and marginally in adjacent parts of Portugal. Resident, but young birds disperse more widely. Monotypic.

The Spanish Imperial Eagle has formerly often been regarded as a subspecies of *Aquila heliaca*, the latter now separated as the Eastern Imperial Eagle. The two are clearly closely related. An early comparison of their mitochondrial DNA, using the cytochrome b gene, put their divergence at about one million years ago (Seibold *et al.* 1996). However, a more recent study of elements of their nuclear DNA has suggested that their divergence is only some thousands of years old, dating from the Holocene or the extreme end of the Pleistocene (Martínez-Cruz & Godoy 2007). The latter finding accords with paleontological evidence that 'imperial eagles' were absent from Iberia during the Pleistocene and only appeared there in the Holocene (González 2008). Martínez-Cruz & Godoy (2007) also found that asymmetrical gene flow between *heliaca* and *adalberti* has continued post-divergence, with one male *heliaca* per generation introgressing into the *adalberti* population, and they recommend that the two be regarded as allospecies. Both eagles are morphologically and ecologically similar but the Spanish Imperial Eagle differs in adult plumage in being slightly darker than its eastern counterpart and in having conspicuous white leading edges to the wings and much larger white scapular patches. Immature birds, especially juveniles, are strikingly different. Juvenile Spanish Imperial Eagles have predominantly rufous, unstreaked bodies and wing coverts, these being sand-coloured and heavily streaked in juvenile Eastern Imperial Eagles.

The Spanish Imperial Eagle is one of the rarest raptors in the world, classified as 'Vulnerable' (BirdLife International 2012c) on account of its very small population. It is endemic to the Iberian Peninsula and Morocco but it ceased to nest in the latter early in the 20th century (BM) and declined to extinction in Portugal by the late 1970s or early 1980s (PA2). Its nadir was during the 1960s and 1970s when the global population was about 50 pairs. Since then there has been an ongoing recovery in numbers in Spain (Figure 87), where the population reached 327 pairs in 2011 (González 2012b) and 365 pairs in 2012 (Grupo de Trabajo del Águila Imperial Ibérica: provisional figure). This led to a recolonisation of adjacent areas of eastern Portugal in 2003 and at least two pairs bred there in 2005, increasing to four confirmed pairs plus two probable pairs in 2008 (AP) and seven pairs holding territories in 2011 (González 2012b). The Spanish Imperial Eagle does not now nest in Morocco although breeding was suspected in the Rif mountains in 1977 and an adult pair was observed at Chechaouen there in 1991 (Fouarge 1992). A report of nesting in Morocco in 1995 (BM) is unsubstantiated.

Figure 87. The Spanish Imperial Eagle breeding population, 1967–2012 (after SEO/Birdlife 2010 and Grupo de Trabajo del Águila Imperial Ibérica)

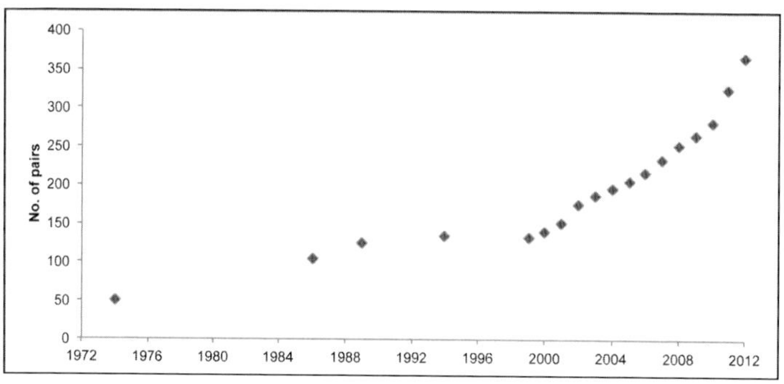

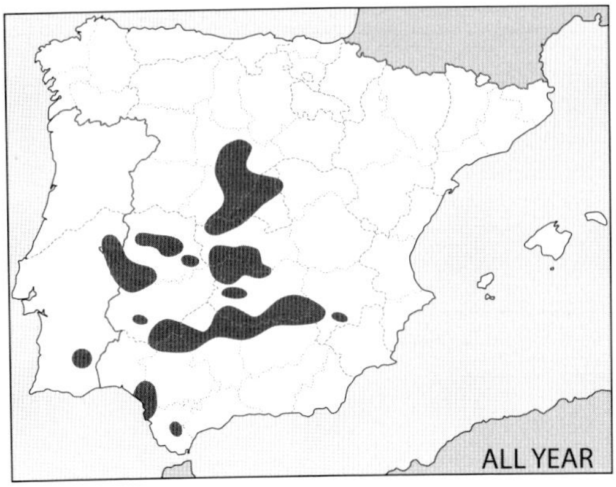

This is a large and powerfully built eagle, although smaller and less heavy than the Golden Eagle (BWP). Unlike that species it inhabits Mediterranean forests, as well as montane pinewoods in the Central Range, and it always nests in trees, building a large stick nest in the crown of a suitable oak or pine. The principal prey is rabbits, and the periodical decline of these in recent years has been a serious problem for the species (Ferrer & Negro 2004, Sánchez *et al.* 2009). It was noted at Doñana that aquatic birds became a very significant component of the diet during the 1950s, when myxomatosis reduced rabbit populations, and again in the 1990s, when rabbit haemorrhagic disease severely affected preferred prey availability (Delibes 1978, L. M. González 1991). An established pair may have several nest sites in its territory that it will re-use. Nest sites avoid areas of human occupation or frequent transit, such as tarmacked roads, although pairs that include subadult birds are more likely to nest in areas of significant human influence (González *et al.* 1992). The clutch size can be unusually large for such a big eagle, 1–4 eggs, mean 2.2 (Margalida *et al.* 2007). Also unusual is the relatively weak tendency to Cainism among the brood and pairs may fledge two or three chicks, when food is abundant. In other circumstances sibling rivalry leads to the death of the younger chicks and on average it accounts for the loss of some 26% of chicks annually (L. M. González 1991). Mean productivity is therefore *c.* 1.2 young per pair per year. First breeding occurs when the birds are three or four years old.

The Iberian breeding range may have extended throughout most of the Peninsula, with the exceptions of western Portugal, the Biscay coastlands, the Cantabrian Range, the Pyrenees and Catalonia, until the early 20th century (González *et al.* 1989). It was still regarded as quite frequent in southern Portugal at that time (Catry 1999). Thereafter the eagle retreated in the face of persecution until by the 1950s and 1960s it was confined to just four areas in the southwestern quadrant of the Peninsula: the Sierra de Guadarrama, the Tagus valley in Extremadura and the adjacent fringes of Portugal, the Monte de El Pardo in Madrid and Doñana (SA2). Then the total population was estimated to be just 30–50 pairs by Valverde (1960). The first census, in 1970–74 (Garzón 1977), found 39 pairs and estimated that the total population was around 50 pairs. Numbers increased once more from the 1980s onwards, with a temporary setback in the late 1990s that seems largely to have been caused by poisoning. By 2003 the population was increasing by 4% per year, the annual increase in the number of breeding pairs having risen from 3.6 pairs/year during 1986–95 to 11 pairs/year during 1999–2002 (SA2).

The breeding range during the first decade of the 21st century extended across sectors of five autonomous Spanish regions (and adjacent areas in Portugal). In 2002 these were Castilla-La Mancha (50 pairs), Extremadura (38), Andalucía (42), Madrid (24) and Castilla y León (18–20). In 2011 the total nesting population was 330 pairs, seven of them in Portugal. The Spanish population in 2011 comprised 37 pairs in Madrid, 119 in Castilla-La Mancha, 48 in Extremadura, 48 in Castilla y León and 71 in Andalucía (González 2012a). Within these regions it is possible to identify three subpopulations. The northernmost of these comprises the Sierras de Guadarrama-Gredos and the Alberche-Tiétar valleys; the Sierras de Monfragüe-Llanos de Trujillo-Alcántara reservoir, the Sierras de Coria, and the Sierra de San Pedro. Further south is the subpopulation of the Montes de Toledo; the Tierra de Barros; the Sierras de Almadén-Guadiana; the Sierra Madrona and eastern Sierra Morena, and the western Sierra Morena. The third subpopulation is the relatively isolated southernmost outpost at Doñana. The first two subpopulations have shown some range expansion in recent years and are increasing locally. In particular, there has been some northward spread of the breeding range in Castilla y León where there were 22 pairs in Segovia and four in Valladolid in 2012, which compares with nine in Segovia in 1999 and the first instance of breeding in Valladolid by a single pair in 2011 (Grupo de Trabajo del Águila Imperial). The Doñana population, however, is in decline, having decreased from 15 pairs in the 1990s to just 7–8 since 1999 (LR), showing high adult and juvenile mortality and very low productivity, as well as reduced levels of genetic diversity (Ferrer & Calderón 1990, Martínez-Cruz 2011).

Adults are largely resident, although some are reported well away from the breeding range outside the breeding season. Juveniles, however, show marked post-fledging dispersal: although most return eventually to their general natal areas when old enough to breed. Both juvenile males and juvenile females settle to nest about 100km from their nests of origin (González *et al.* 2006). Radio-tracking during the 1990s of 87 juveniles, 50 of them from Doñana, has allowed a number of important wintering areas to be identified (LR). These are: south-west Madrid and the neighbouring Tiétar-Tagus valleys ; the Campo de Montiel in La Mancha; Hornachos, Llerena and Azuaga (Badajoz) with northern Sevilla and north-west Córdoba; the Andévalo in Huelva; the region of Isla Mayor, Gibalbín-Medina Sidonia in Cádiz-Sevilla; Fresno de Torote,Valdeolmos, Villar del Olmo and Ambite (Madrid); the Valle del Tamuja (Cáceres); Logrosán–Zorita, the south-east of the Sierra de San Pedro and Cornalvo (Cáceres/Badajoz) and the

Tierras de Alhama, Loja and Temple in south-west Granada. The Doñana birds especially disperse to the Andévalo: in the Sierra Morena in Huelva, and also to the nearby Marismas, and south-east as far as La Janda, near the Strait of Gibraltar.

Irregular dispersion also takes individuals to far-flung corners of the Peninsula, with recent records of both immature and non-breeding adult birds from Almería, Albacete, Cuenca, Huesca, La Rioja, Lleida, Navarra, Palencia, Salamanca, Zaragoza, Zamora and Valladolid (Sampietro *et al.* 1998; *Ardeola* 57: 526, 58: 493; Regional bird reports). One ringed as a nestling in Huelva in June 2004 was found 369km away in Toledo in October 2004. A record of a juvenile in Galicia: at Porto do Son, A Coruña, in mid October 2010 was highly exceptional but was followed by a sighting of a second-year bird at Covelo, Pontevedra, in July 2012 (*Ardeola* 58: 202, 59: 426). A few stray as far as southern France, with 11 records to 2005: five of them from the 19th century but including four records of immatures in the French Pyrenees and one in Camargue since 1999 (Duchateau 2007).

A few individuals, the great majority of them immature birds, cross the Strait of Gibraltar and winter in North Africa. Three pulli from Huelva were subsequently recovered in Morocco, having travelled up to 646km from the natal sites (*Ecología* 19: 232). A Doñana bird has reached Libya (Calderón *et al.* 1988). There is also an exceptional record of a juvenile that was tracked from its nest in Extremadura to Senegal, returning to breed in its natal region three years later (LR).

The Spanish Imperial Eagle is the flagship species of the Iberian avifauna and its future survival and success will be an acid test of the effectiveness of conservation measures in Spain in particular (González *et al.* 2007; Ortega *et al.* 2009, BirdLife International 2012c). The indications are promising given that it is the focus of intense efforts to protect, sustain and expand the population, both numerically and geographically. The measures taken include legal protection, obtaining the collaboration of key landowners, the establishment of a captive fall-back population, public education via specific information centres, ongoing campaigns against poison baits and hazardous electricity pylons, reinforcement of rabbit populations and the reintroduction of the species to suitable sites. A National Strategy for the Conservation of the Spanish Imperial Eagle is in place in Spain and there are also regional Recovery Plans in Castilla y León, Castilla-La Mancha and Extremadura, although not yet in Andalucía or Madrid, the other two key regions for the species. SEO/Birdlife has also established a stewardship programme, 'Taking Flight' ('Alzando el Vuelo', http://www.aguilaimperial.org) to monitor and sustain the conservation and population recovery of the species, involving municipalities, landowners and schools and focusing on habitat management, species conservation, raising awareness and disseminating information. Two population viability analyses (PVAs) have been published, the first of which (Madero & Ferrer 2002) had predicted the extinction of the Spanish Imperial Eagle within 200 years. The second PVA (Ortega *et al.* 2009) has a much more promising outlook, based on the positive trends seen during more recent years: it considers that there is nil probability of extinction provided the current demographic parameters do not change. An EU action plan for the Spanish Imperial Eagle (Sánchez *et al.* 2008) has the removal of the species from the list of globally threatened species by 2018 at its objective, which will be met when there is a stable or increasing population of at least 1,000 mature individuals. The 2012 population of at least 367 pairs (Grupo de Trabajo del Águila Imperial) shows that the target is likely to be met comfortably on present evidence and the eagle has already been classed as 'Vulnerable' rather than 'Critically Endangered' by the IUCN.

The threats that resulted in the demise of the Spanish Imperial Eagle and that have set back its recovery have been very closely investigated. The International Action Plan (Sánchez *et al.* 2008) identifies three threat categories as critical: the scarcity of key prey species, largely due to the massive decline in rabbit numbers occasioned by myxomatosis and haemorrhagic pneumonia; electrocution on power pylons, and poisoning. Habitat alteration, mainly through the construction of infrastructures including roads, power lines and wind farms, is also considered to be of 'high' importance. Human disturbance during the breeding season is regarded to be of 'medium' importance. Chemical contamination, lead poisoning by ingesting lead shot in wildfowl prey, nest robbery and nest collapse, are of 'low' importance. Some direct persecution by illegal shooting occurs but the significance of this factor is unknown.

A 16-year study (1998–2004) of 267 instances of non-natural mortality showed that electrocution (47.7%) and poisoning (30.7%) were the most frequent causes of non-natural mortality. The latter was the more frequent in dispersal areas and (hence) affected subadults particularly. Conversely, illegal poison baits set for vermin were the principal cause of death within the breeding range and affected adults more frequently. Electrocution rates have declined recently despite the ongoing expansion of the power transmission network, largely as a result of changes in pylon design and the modification of existing dangerous pylons (López-López *et al.* 2011), such steps of course benefiting a diversity of other species as well as the Spanish Imperial Eagle. The campaign against the placement of poison baits has especially been orchestrated by SEO/Birdlife, who have succeeded in gaining the cooperation of land owners in key breeding areas, but poisoning continues to be an important major threat to the species.

A project to reintroduce the Spanish Imperial Eagle in three locations in the La Janda area, Cádiz, began in 2002, with a view to reinforcing and extending the southern subpopulation (Madero & Ferrer 2002). A population reinforcement programme also began in 2005 at Doñana. Between 2002 and 2010, a total of 73 young Spanish Imperial Eagles were released at these locations. As a result, a released male bred successfully with a non-reintroduced female and two young were reared in 2010, at a site in Cádiz. This was the first successful breeding event recorded in the province since 1960. An additional territorial pair also settled in the Cádiz release area in 2010 and four more translocated individuals have made breeding attempts within other subpopulations since 2006 (Muriel *et al.* 2011). The La Janda programme has nonetheless been criticised by SEO/Birdlife and others since the releases were not preceded by the correction of some dangerous pylons in the region: which accounted for a number of the birds, because it was not integrated within the National Recovery Plan and because released birds were apparently collected from the wild in an unregulated manner. Moreover, the releases took place in a region where young Spanish Imperial Eagles regularly winter, raising the possibility that natural recolonisation of the area would follow more favourable habitat management there. Captive breeding of Spanish Imperial Eagles was finally achieved in 2011 (Blanco *et al.* 2011) but, as with the La Janda programme, such an expensive measure seems to be unnecessary given the strong positive responses of the wild population to recent conservation measures.

EASTERN IMPERIAL EAGLE *Aquila heliaca*

(Águila Imperial Oriental, Águia-imperial-oriental)

Vagrant to Spain.

A long-dead bird found at the Sierra de Escalona, Alicante, in March 2010 had been ringed in June 2005 near Bratislava, Slovakia (*Ardeola* 60: 459). Two earlier records precede the establishment of the Spanish rarities committee: a 'probable', at Lizoain, Navarra, on 12 October 1967, and an adult at Bárcena Mayor, Cantabria, on 17 February 1982 (*Ardeola* 15: 133, 29: 181). This partly migratory eagle has been recorded as a vagrant in several western European countries, including France, where six were recorded during 1981–2008.

GOLDEN EAGLE *Aquila chrysaetos*

(Águila Real, Águia-real)

***A. c. homeyeri*. A widespread resident throughout most of Spain and eastern Portugal, but absent from the Balearic archipelago. Somewhat smaller and darker than the nominate. Those inhabiting the Pyrenees are sometimes assigned to *A. c. chrysaetos*.**

The Golden Eagle is characteristic of most of the mountainous regions of the Peninsula. It also inhabits the Ebro valley: where a large proportion nest in trees, and riverine gorges that offer cliff-nesting opportunities, for example along the Duero and Tagus on the Spanish/Portuguese border. In Spain it is absent from large tracts of the meseta, the coastal lowlands and the Guadalquivir valley and it has only a marginal presence in Galicia and in Cádiz province. In Portugal most of the population is in the north-east in Trás-os-Montes, with smaller concentrations along the Tagus bordering Extremadura and along the Guadiana in the south-east. Saunders (1871) claimed to have seen the Golden Eagle in Mallorca, and Munn (1921) 'frequently' saw large eagles there that he considered were Golden Eagles. It does not occur there now.

The Golden Eagle hunts over open country, whether in the lowlands or on the high tops above the treeline, and so it avoids areas of continuous forest. It kills a wide range of vertebrate prey as available, including rabbits, hares, young deer and Chamois; pigeons, partridges and other sizable birds, and large lizards occasionally. Rabbits are the most important prey in Mediterranean areas (Delibes *et al.* 1975, Fernández & Purroy 1990). It also regularly takes carrion, often of dead ungulates. Treeless plains are unsuitable during the breeding season but non-breeders and others often occur in agricultural regions in winter especially. Nests may generally be found between 160m and 2,150m above sea level, the mean being 950m (Arroyo *et al.* 1990). Some nest even higher in the Sierra Nevada, where eyries may be found at 2,500m (Garzón & Henares 2012). The great majority nest on cliffs, sometimes quite modest ones where the birds are undisturbed, but a recent survey found 17% nesting in trees, mainly in large pines or oaks (del Moral 2009c). Nests in firs, junipers, beeches, wild olives and even eucalyptus are also reported (Arroyo *et al.* 1990).

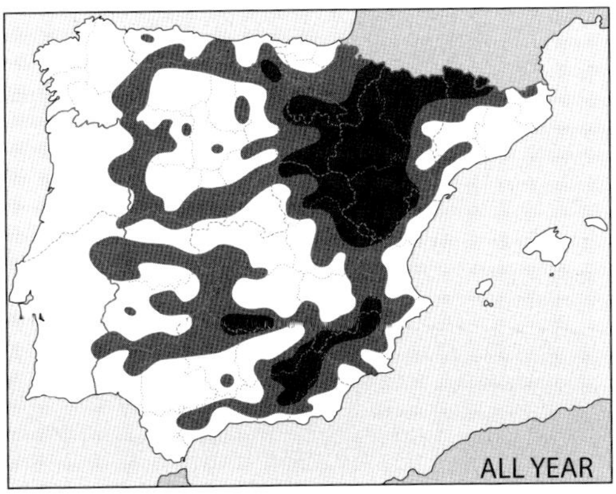

A survey of the Spanish population in 2008 found 1,553–1,769 pairs (del Moral 2009c). The majority (70%) were in the extensive regions of Andalucía, Aragón, Castilla-La Mancha and Castilla y León although the highest densities were found in La Rioja and Navarra. An analysis of breeding performance of a sample of the pairs surveyed in 2008 found productivity to be 0.80 and overall breeding success to be 1.07 young per pair, with 1.15 young fledged per successful pair. Reproductive output is strongly affected by Cainism in this species, in which more than one chick survives only at nests where the food supply is high. A positive relationship has been found between breeding success and the frequency of rabbits in the diet so the reduced availability of rabbits may account for recent productivity being somewhat lower than in an earlier study, when 1.31 young fledged per successful pair (Arroyo *et al.* 1990).

The Portuguese population numbered 53–61 pairs during 2002–04 (PA2), including 38–44 pairs in the north-east in Trás-os-Montes, seven pairs in east-central Portugal in the Tagus frontier area and six in the south-east along the Guadiana in Baixo Alentejo. The population is thought to be fairly stable, having increased from 20–40 pairs in the 1980s (AP), but the small population in the Peneda-Gêres National Park in the north has recently declined from four pairs to just one (PA2). A few nest in Andorra: three pairs during 1999–2001 (BE).

The Golden Eagle has always been the particular target of persecution by shepherds and farmers and this undoubtedly depressed the Iberian populations during the 20th century, as elsewhere in Europe (Bijleveld 1974). Iberian numbers probably reached their lowest point during the 1960s and 1970s (Garzón 1975, F. de Juana 1989) but Arroyo *et al.* (1990) found *c.* 1,300 pairs in Spain in the late 1980s, by when some recovery was already apparent. Since then numbers have been stable or have increased in most regions, a process accompanied by reoccupation of some former traditional nesting areas. Only 10% of nests were in trees in the earlier survey, compared with 17% in the 2008 survey, which may reflect a reduction in persecution and the reoccupation of non-mountainous areas. The Spanish population may have increased by 20% since the mid 1990s (del Moral 2009c). Golden Eagles remain subject to illegal persecution in Iberia and electrocution on poorly-designed pylons has accounted for quite a few birds. In addition, some traditional nest sites have suffered from habitat modification and accidental disturbance by human activities such as rock climbing.

Adults are sedentary but young birds disperse for varying distances, often wintering in agricultural areas and wetlands in the lowlands. The few recoveries of Iberian-ringed birds, and recent studies of 13 satellite-tracked juveniles from nests in the Comunidad Valenciana (Soutullo *et al.* 2006a, b), suggest that these movements are relatively short. Juvenile females ranged further from nests than juvenile males, mean maximum dispersal distances of the satellite-tracked birds being about 140km and 70km respectively. The farthest ringing recoveries on record were 265km and 276km from the ringing sites. There is no indication of any passage through the Pyrenean passes nor across the Strait of Gibraltar, where only occasional vagrants are recorded.

BOOTED EAGLE *Aquila pennata*

(Águila Calzada, Águia-calçada)

Widespread as a breeding species across much of Spain and Portugal, and in Mallorca and Menorca. Principally a trans-Saharan migrant but some remain in winter. Balearic birds are resident.

This small eagle is a familiar sight over or near open woodland across the greater part of the Peninsula and it is the most numerous raptor in large areas of the centre-west. The breeding distribution in Spain is most uniform in the west, becoming more fragmented across much of Castilla-La Mancha and the north-east. It is very scarce or absent in the eastern Mediterranean coastal provinces, from Almería north to Girona, as well as in much of the Guadalquivir depression. It is also uncommon or absent in northern and western Galicia, the Biscay coastlands and the northern side of the Cantabrian range (SA2). The distribution in Portugal includes the north-east, most of the centre and also the south but it is largely absent from the Atlantic coastlands and the north-west generally (PA2). It breeds in the Balearic Islands, chiefly in the Serra de Tramuntana in Mallorca and more widely in Menorca (Viada & de Pablo 2009).

The breeding habitat includes woodlands of diverse types, notably Mediterranean pinewoods and dehesas of cork and holm oaks, but also other deciduous and coniferous stands, as well as riparian woods and eucalyptus, chestnut and poplar plantations. Nesting birds have been reported as high as 1,900m in the Sierra Nevada, where they often hunt over open terrain even higher up (Garzón & Henares 2012), but the species is more typical of lower elevations. It occurs in farming areas provided that some trees are available for nesting. In general, it requires a mixture of tree cover alternating with open expanses of low scrub, pasture or arable land. Here it hunts a diversity of prey, chiefly mammals as large as rabbits, birds up to the size of partridges and often Ocellated Lizards and other reptiles (García-Dios 2009). Birds often hunt over towns and villages, snatching pigeons from the rooftops.

Nests are almost always in trees and those used in previous seasons may be re-used. Cliff-nesting is seen occasionally on the mainland but is the norm in the Balearic Islands. A number of Spanish studies have put productivity at 0.70–1.88 young per

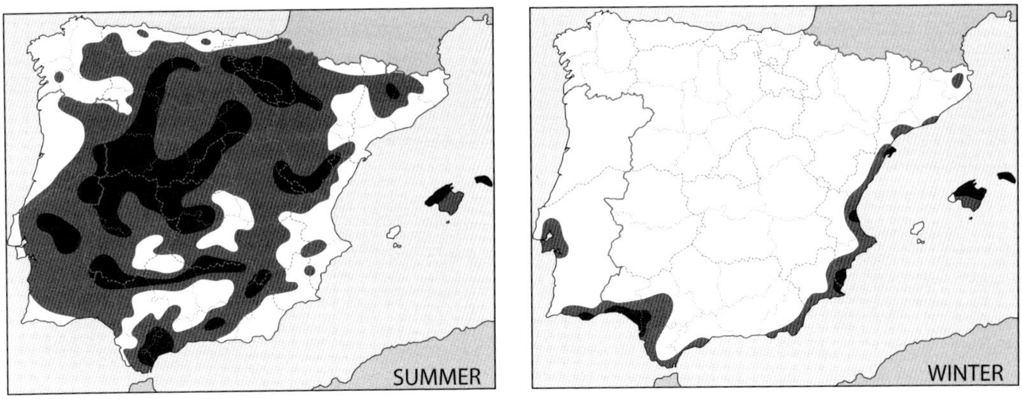

occupied territory (García Dios 2009). A recent and perhaps surprising discovery has been the existence of nesting trios, which may make up as many as 25% of nesting units in Madrid (Martínez *et al.* 2005, Díaz 2006).

The Spanish Booted Eagle population was censused in 2009–10, when 18,490 pairs were estimated by extrapolation from sample counts (Palomino & Valls 2011). The Balearic population was counted directly. Earlier estimates suggested that there were around 3,000 pairs (F. de Juana 1989) and at least 2,905 pairs (SA2), but neither of these assessments can be considered to have been comprehensive. Local indications are that the population is increasing in many regions, including in the Balearic Islands, and undoubtedly this is another raptor species that has benefited from the ongoing expansion of forested areas in Iberia. There has been some recent expansion in the breeding range, particularly in Catalonia (CA2), but the highest densities remain in the centre-west: in Extremadura, Andalucía and Castilla y León. Spring migration counts at the Strait of Gibraltar also strongly indicate an increase in numbers. Indications of an upward trend were already apparent in the early 1970s, when densities of up to 12 breeding pairs in 8km^2 were found in southwestern Spain (Garzón 1974). However, the Booted Eagle was already considered common in the more distant past, at least in the centre and south (Lilford 1865, Saunders 1871, Castellarnau 1877, Witherby 1928).

The size of the Portuguese population is uncertain, with a rough estimate of 300–1,000 pairs in 2002 (BE). The 2008 Portuguese atlas has shown marked range expansion, however, especially north of the Duero (PA2), suggesting an increase in numbers. Together Spain and Portugal account for the great majority of western European Booted Eagles, the only others being the French population of 400–650 pairs (OF).

Most European Booted Eagles winter in sub-Saharan Africa and French, Spanish and Portuguese birds converge at the Strait of Gibraltar in spring and autumn. The Balearic birds are resident. There are 24 foreign recoveries of birds ringed in Spain up to 2010: two of them from Italy, one from France, 14 from North Africa: in Morocco (11) and Algeria (3), and seven in sub-Saharan Africa: two each in Mauritania and Mali and single recoveries in Nigeria, Burkina Faso and Togo. The wintering range of six Spanish adults tracked by satellite in 2011–12 took in parts of Mauritania, Mali, Niger, Nigeria, Guinea and Sierra Leone (SEO/BirdLife, www.migraciondeaves.org).

Figure 88. Distribution by approximately ten-day periods of Booted Eagle arrivals in Iberia at the Strait of Gibraltar during 2008–12 combined (n = 37,632 birds). (Fundación Migres, GONHS)

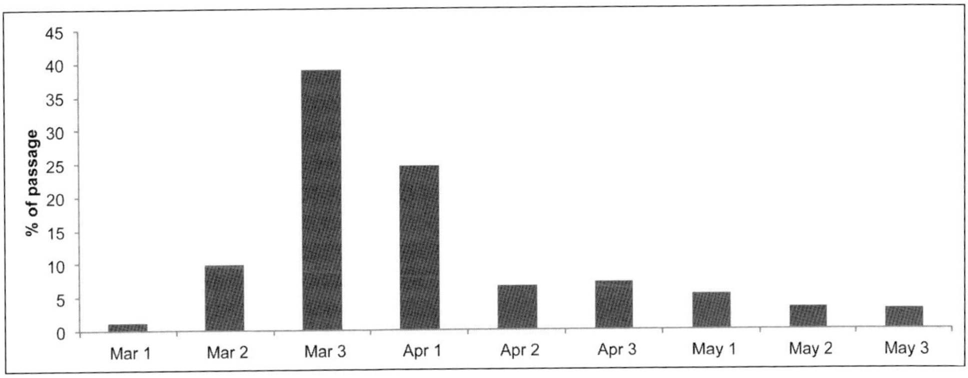

Passage through the Pyrenean passes is on a modest scale, although the size of the current French population implies that the total passage is now larger. Counts averaged 219 birds in total at the three western watchpoints during 2009–11 and just 19 at Eyne. The autumn passage extends from late August to early October, peaking in mid to late September (Urcun & Bried 1998b), but the spring passage has not been monitored. In contrast, the Booted Eagle is one of the commonest of the migrant raptors that cross the Strait of Gibraltar. The spring passage there begins in mid March, and extends throughout April–May, with stragglers in June. Peak arrivals are in late March and early April (Fig. 88). The mean rate of passage at Gibraltar in spring during 1990–2004 was 4.55 birds/hour and represents a highly significant increase over the rate of 1.58 birds/hour observed during 1967–77 (Bensusan *et al.* 2007), suggesting an increase in the populations involved. Total spring counts at Gibraltar and the central Strait region combined during 2008–12 averaged 7,546 birds (range 6,296–10,113; GONHS, Programa Migres).

The southward migration extends from August to early November but peaks during the second half of September until early October. Booted Eagles give every impression of being powerful, confident fliers and indeed many arrivals at Gibraltar in spring, especially in the later, warmer months, occur at considerable altitude. Nevertheless, they are particularly dilatory about embarking on the over-sea crossing, so accumulations of migrants occur on the north side of the Strait in autumn, especially when strong easterlies are blowing. These sometimes reach impressive proportions, with temporary flocks of hundreds of birds forming, often accompanied by numbers of Short-toed Eagles. An exceptionally large gathering of about 1,000 occurred during easterly wind conditions at Gibraltar around noon on 21 September 2003 (*GBR 2003*: 20). The numbers counted crossing the Strait in recent years are much higher than in the 1970s but they have not approached the theoretical totals that are implied by the population estimates of the source populations. These would be of the order of 40,000 adult birds and an unknown number of juveniles: perhaps another 20,000 birds, assuming that most cross the Strait, which is almost certainly the case, and that the population assessments prove to be reliable, which has yet to be established (Blanco *et al.* 2012). Dark morph birds comprise 17–21% of the migrants at the Strait (Bernis1980, BSG).

Some young Booted Eagles appear to disperse for considerable distances within Iberia before migrating to Africa. There are three same-year recoveries of birds ringed as nestlings in western Andalucía all of which moved north-east after fledging, to Barcelona, Tarragona and northern Italy (AI1). More recently, movements of this kind have been reported not only in eastern Spain but also continuing eastwards. An exceptional influx into Italy in autumn 1985 included some Spanish-ringed birds (Premuda & Baghino 2004). A strong northward movement was seen from 2–9 October 2004 in coastal Castellón in northern Valencia, with 629 birds counted, nearly all of them heading north-east. These observations were followed progressively later in October by sightings of birds travelling through France and down the Italian peninsula (Premuda *et al.* 2007). Some similar movement on a much smaller scale was detected in 2005 and 2006, with 83 and 18 birds respectively reported from Castellón. The 2004 passage was thus clearly exceptional but it may be that a new, circuitous, migration route, partly involving Iberian birds, is developing, taking birds to winter quarters in southern Italy and/or into Africa via the Sicilian narrows.

Another apparently increasing phenomenon is overwintering of Booted Eagles in Iberia. This was first reported by Chapman & Buck (1893), who mention that some wintered near Málaga. More recent reports came from Doñana, where 15–20 birds were thought to be wintering by the mid 1980s (García *et al.* 1989). Martínez & Sánchez-Zapata (1999) analysed published observations of at least 134 individuals found wintering in Spain during 1979–96. The great majority of such records were from eastern and southern coastal locations, many of them wetland sites, notably in Alicante, Murcia, Málaga (at the Guadalhorce estuary especially) and Doñana, with a scattering of inland records from Extremadura and Castilla y Léon and one at Santander on the north coast. Reports in *Ardeola* of Booted Eagles recorded during December–February from 1997–2010 total at least 138 individuals, again chiefly at eastern and southern coastal sites but with clearly increased numbers at these locations in some years. For example, over 50 wintered in Alicante in winter 2004–2005 (*Ardeola* 53: 383). As previously, there were also scattered observations, usually of single birds, at inland sites, widely dispersed across the meseta, Extremadura and the Ebro valley, with at least six January reports from Galicia, including one of two together at Ferrol in January 2001 (*A. O. Galicia 2001*: 75). A few individuals have wintered in and around Gibraltar annually since 2000. More recent reports show that 10–20 winter in coastal Catalonia, largely at the Ebro delta and on the outskirts of Tarragona, and this population is increasing (CWA). There is also a recent and increasing wintering presence in the Camargue and elsewhere in southern France (OF). Wintering birds are a regular sight in the coastal regions of the Comunidad Valenciana, notably at the Albufera de Valencia; the Marjal de Almenara, and around Elche, with reports of up to ten together in 2009 (*A. O. C. Valenciana 2010*: 46–47). There is also a small but regular wintering population in Portugal, where the birds are especially associated with coastal wetlands and those in the Tagus and Sado valleys (Costa 1994, AP): a count of eight at Quinta do Ludo, Faro, on 16 January 2008 (*Anu. Ornitol.* 7: 79) is notable. In general, it seems apparent that the Iberian wintering population *c.* 2010 numbered no more than a few hundred individuals, still very much a minority of the western European population as a whole but probably an increasing fraction.

BONELLI'S EAGLE *Aquila fasciata*

(Águila Perdicera, Águia-perdigueira)

***A. f. fasciata*. A scarce resident in mainland Spain and Portugal. Vagrant to the Balearic Islands.**

Bonelli's Eagle is a characteristic inhabitant of broken country, especially limestone formations, across the warmer regions of the Peninsula. Its strongholds are in the east, south and south-west of Spain and the east and south of Portugal. It has never had more than a marginal presence in the north-west in Galicia, or in the Biscay lowlands, and it has always been scarce in the Cantabrian Range: where Irby (1883) found it nesting near Potes in the Picos de Europa in 1876; as well as in the western Pyrenees and the northern meseta. However, the breeding range has contracted away from all these northern areas in recent decades. It is an opportunistic predator that readily switches to whatever prey is most available although it selects rabbits where these are available (Moleón *et al.* 2009). Despite its Spanish and Portuguese names – which mean 'partridge (hunting) eagle' – it usually takes more pigeons than partridges, domestic pigeons accounting for a high proportion of the diet in southern Portugal (Palma *et al.* 2006). Pigeons, Jackdaws and other medium-sized birds are often caught in flight. The Ocellated Lizard is another frequent prey item in some areas but carrion is seldom eaten.

Nearly all Bonelli's Eagles nest on cliffs but a few nest on trees, especially in southern and western Spain and southern Portugal (LR, Palma *et al.* 1999). Nesting on electricity pylons has also been recorded on rare occasions. Nests occur from sea level to 1,600m, but the mean nesting altitude is below 400m (del Moral 2006). The breeding territory typically has a number of nest sites that are selected from each year. The clutch of 1–2 eggs, rarely three, is laid very early in the more southern, core populations: in mid-January in some cases but in mid-February on average. Laying is later in the northern parts of the range, generally in early March but sometimes not until early April (Arroyo *et al.* 1995).

Iberia accommodates over 80% of the European nesting birds (BE). The Spanish population has been monitored to different extents in different regions for some time, most recently in a national census of breeding birds in 2005 (del Moral 2006), which found 733–768 pairs. In 2005, 82% were in just four regions: Andalucía (44%), Extremadura (12.3%), the Comunidad Valenciana (12.3%) and Castilla-La Mancha (11.6%) (Table 24), with a concentration in the Mediterranean coastal provinces of Cádiz, Málaga, Granada, Almería, Murcia, Alicante, Valencia, Castellón and Tarragona, which together held 52% of the total Spanish population. A further 32% were in the inland southern provinces of Cáceres, Badajoz, Córdoba, Jaén, Albacete and Ciudad Real. There were few breeding pairs in Castilla y León and the Ebro valley and none at all in Galicia, Asturias and Cantabria, where a few bred until relatively recently (del Moral 2006).

The 1987 census was incomplete, and so probably was an early estimate of 500 pairs in the early 1970s (Garzón 1977), but the total Spanish population was largely stable during 1990–2005. However, the national totals mask opposing regional trends: breeding numbers declined markedly in the north during this period but were stable or increasing in the south and east (Table 24). The most serious recorded declines have been in Catalonia: where there were up to 87 pairs in the 1970s, Aragón, Castilla y León and Murcia, with some recent recovery in the first two of these. All or most of the most northern outpost pairs, in the Cantabrian Range, have also disappeared. The three pairs in Navarra in 2005 are the survivors of a population of eight pairs and the two in Madrid compare with up to 15 pairs there in the 1970s (del Moral 2006), so these regions too have shown clear declines. The apparent increases in Andalucía and Extremadura are often attributed to improved coverage by recent surveys and these sub-populations are thought to be stable, although some traditional sites have been abandoned, and new ones established, in recent years (del Moral 2006). Nevertheless, the trends in the two best-monitored Andalucian provinces: Cádiz and Granada, are clearly positive: together they held 69 pairs in 1990 and 96 in 2005 (14% of the Spanish population). A breeding season census in Andalucía in 2009 showed a further increase in the regional population, to 334 pairs (plus 14 probable pairs): 75 pairs in Almería, 41 in Cádiz, 39 in Córdoba, 54 in Granada, just one in Huelva, 34 in Jaén and 18 in Sevilla (*Ardeola* 59: 426). Most Bonelli's Eagles in Spain are in coastal or pre-littoral ranges notably in the Betic range and in the east, from Almería to Girona. Inland important sites are chiefly on low ranges and in riverine gorges.

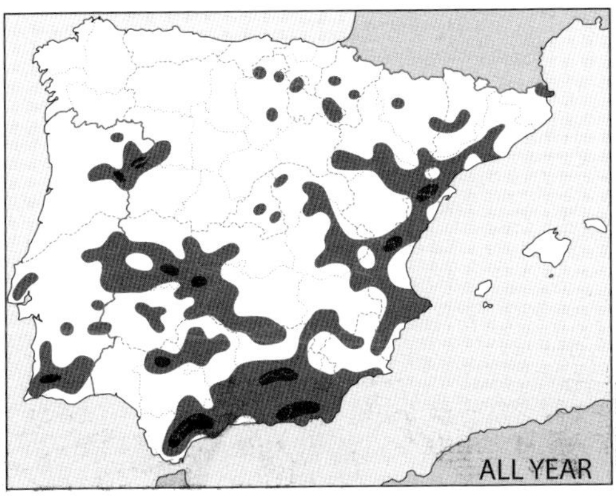

The Portuguese population was estimated at 92–99 pairs during 1999–2005 (PA2), these principally in the east and south. The largest concentration, with over 30 pairs, is in the north-east, in Trás-os-Montes and the Duero gorges. The species is well distributed in Algarve and the south-west Alentejo, with *c.* 30 pairs, and there are smaller numbers in

Beira Baixa and the Guadiana valley/Castro Verde. A few nest in the extreme west in Estremadura. The population appears to be healthy overall and increasing in the south.

An analysis of breeding performance of a sample of the pairs surveyed in 2005 found mean productivity of the Spanish population to be 0.92 and overall breeding success to be 1.17 young per pair, with 1.43 young fledged per successful pair (del Moral 2006). Reproductive success varied significantly between regions, being highest in the core populations of the south and lower in the north and east: 1.41 in Andalucía/ Murcia (n = 277), 1.22 in Extremadura/ Castilla-La Mancha (n = 162), 1.19 in the Comunidad Valenciana/ Catalonia (n = 136), 1.12 in the Ebro Valley (n = 37), and just 0.80 in Castilla y León (n = 14). Early studies have also highlighted marked regional differences in breeding output and these are apparently climate-related, associated particularly with mean annual temperatures (Ontiveros & Pleguezuelos 2003).

The declines in some of the Spanish regional populations have been attributed to a number of factors but especially to direct persecution by hunters and pigeon racers: which especially affects nesting adults, electrocution on poorly designed pylons: which mainly kills young birds, and disturbance of nesting cliffs by human activities such as forestry, road building and climbing: which can result in territory abandonment (Real *et al.* 2001, LR, Ontiveros *et al.* 2004). However, recent negative trends are only evident in northern areas, where the above threats are certainly not more acute than in the east and south, but where both population densities and breeding success are much lower. Lower densities and lower breeding success imply poorer habitat quality and it is probably this that ultimately underlies the observed population trends (Carrascal & Seoane 2009), as well as the viability of populations in such regions as Castilla y León, that perhaps already act as sinks within the population dynamics of the Iberian metapopulation (Real & Mañosa 1997).

Table 24. Regional distribution of breeding Bonelli's Eagles in Spain during the five surveys conducted between 1987 & 2005. Figures are the minimum numbers of pairs found. The maximum estimated total numbers of pairs are also given. After del Moral (2006).

NORTH	1987	1990	2000	2003	2005
Aragón	43	42	30	30	31
Asturias	3	0	0	0	0
Basque Country	1	0	1	1	1
Cantabria	1	1	0	0	0
Castilla y León	34	40	18	17	14
La Rioja	5	4	4	5	6
Navarra	4	3	2	3	3
North Total	91	90	55	56	55
SOUTH & EAST	**1987**	**1990**	**2000**	**2003**	**2005**
Andalucía	172	238	270	323	321
Castilla-La Mancha	72	88	76	80	85
Catalonia	54	62	58	66	65
Comunidad Valenciana	109	99	81	107	93
Extremadura	62	74	84	93	90
Madrid	3	2	2	2	2
Murcia	33	24	24	20	22
South & East Total	505	587	595	691	678
Minimum Spanish TOTAL	596	677	650	747	733
Maximum estimate	695	752	713	782	768

An important factor that affects habitat quality seems to be the expansion of forest cover, due both to abandonment of marginal cultivation and pastures and to large scale afforestation with pines, which reduces effective hunting areas (LR, Carrete *et al.* 2002, Ontiveros *et al.* 2005). A second major factor is the decline in favoured prey species, notably rabbits, which is also linked to forest expansion but is also a consequence of rabbit diseases. A positive relationship between breeding success and the proportion of rabbits in the diet has been demonstrated in Provence, southern France (Cheylan 1981) and falls in rabbit numbers have been shown to be important in the population decline in Castilla y León (Fernández *et al.* 1998). The importance of rabbits in the diet of Bonelli's Eagles in Spain, Portugal and France fell by a third after the outbreak of rabbit haemorrhagic disease in 1988 (Moleón *et al.* 2009). Demographic models show that reducing the mortality of both adults and subadults is of the utmost importance in maintaining population stability, outweighing improvments in reproductive output (Real & Mañosa 1997, Carrete *et al.* 2002). Displacement by Golden Eagles has sometimes been suggested to explain the local disappearance of Bonelli's Eagles but the available evidence seems ambiguous (Jordano 1981a, Gil-Sánchez *et al.* 2004, Carrete *et al.* 2005, Martínez *et al.* 2008). However, Golden Eagles are known to occupy former territories of Bonelli's Eagles (Fernández & Insausti 1990) and there is a case of a sub-adult Golden Eagle killing and eating an adult male Bonelli's Eagle (Bosch *et al.* 2007).

Adult Bonelli's Eagles are largely resident, although some may wander some distance outside the breeding season. Juvenile and immature birds disperse most widely (Real & Mañosa 2001, Cadahía *et al.* 2005, Balbontín & Ferrer 2009). Certain areas have been identified that attract young birds regularly, these generally being places offering a high abundance of such prey as rabbits or partridges as well as an absence of territorial adults (Mañosa *et al.* 1998). In Spain they include the Ebro depression in Lleida; the Sierra de Escalona in Murcia/Alicante; the lowlands of Toledo, Albacete and southwestern Madrid; the dehesas of Trujillo in Cáceres; the plains of La Serena in Badajoz and the depression of La Janda in Cádiz. Some juveniles travel 400km or more from their natal sites. Ringing recoveries include three juveniles from Catalonia that travelled to France: one of these was ringed in Tarragona in April 1993 and recovered quite exceptionally 1,033km away near the Belgian border, the following February. There are also seven recoveries in Spain of birds ringed in southern France. The Spanish population occasionally exchanges individuals with the remnant population in southeastern France, which numbered 30 pairs in 2010 (Dubois *et al.* 2012).

Dispersion may take a few individuals across the Strait of Gibraltar, where occasional birds are seen among migrant raptors, but there is no regular passage there. Wandering birds are probably also responsible for occasional sightings in the Balearic archipelago, where Bonelli's Eagle does not breed. Ferrer *et al.* (1986) describe it as accidental in the Balearics and Clavell (2002) gives eight records (nine birds) during 1987–99: six in Mallorca, one in Menorca and one in Ibiza, including four birds that were apparently adults. However, it probably was a scarce breeding resident in Mallorca in the past, where Munn (1921) recorded a breeding pair on the Alcudia peninsula and reported that Witherby found also found a pair at Lluc, in the Serra de Tramuntana, in 1919. A project to reintroduce the species to Mallorca is under way (Mayol & Viada 2011).

OSPREY *Pandion haliaetus*

(Águila Pescadora, Águia-pesqueira)

***P. h. haliaetus*. A scarce resident in the Balearic archipelago. Formerly bred sparsely on mainland coasts but extinct in mainland Iberia until very recently. Widespread on passage and winters locally in small numbers.**

The Osprey is an increasingly frequent sight at estuaries and inland wetlands throughout Iberia, where migrants often break their journeys for long periods or, locally, remain to winter. However, its history as a breeding species in Iberia is one of progressive decline to extinction in Portugal and mainland Spain, with only the Balearic Islands retaining a nesting population at the start of the 21st century. Since then reintroduction projects and natural recolonisation have resulted in nesting in southern Spain that it is hoped marks the beginning of wider re-establishment in mainland Iberia. The Osprey remains by far the rarest of the Iberian nesting raptors, with a small and vulnerable population (LR). The traditional Iberian population of coastal nesting Ospreys forms part of the small western Mediterranean metapopulation of fewer than 100 pairs (Thibault *et al.* 1996).

The breeding population is closely monitored. A census in 2008 (Triay & Siverio 2008) found 15 pairs in the Balearics: eight in northern/western Mallorca, four in northern Menorca and three in Cabrera, and one in Andalucía at the Marismas del Odiel, Huelva: this last the result of a reintroduction programme. The Balearic birds are resident when adult and recruitment is from locally-bred juveniles, so that the population is reproductively isolated (Triay 1995). First breeding is at an age of 3–5 years. Young birds generally build new nests or rebuild old abandoned ones when they first breed and thereafter they use their old nests traditionally.

The present Balearic distribution is a remnant of a larger historical presence there (González *et al.* 1992, Triay & Siverio 2008). The total population in the mid 20th century was 35–40 pairs and the birds then nested more widely in Mallorca and Menorca, as well as on Dragonera, Formentera and Ibiza. A sharp decline began when the islands began to attract mass tourism from the 1960s onwards, a process which resulted in former haunts being disturbed or built-up. Direct persecution for taxidermy also accounted for significant numbers. As a result, the Balearic population reached a low ebb of 5–8 pairs by 1980, having become extinct on Dragonera and in Formentera by the mid or late 1960s and in Ibiza in 1975 (González *et al.* 1992). Thereafter protection resulting in reduced persecution allowed numbers to recover during the 1980s and 1990s, peaking at 16–18 pairs in 1999 and with 13–16 pairs thereafter. The population is still far from secure, however, and high adult mortality due to electrocution on pylons has been identified as a problem in Menorca especially.

Balearic birds nest almost invariably on sea cliffs and on rocky stacks. The nesting of a pair on an Aleppo Pine in Mallorca in 2008 was unprecedented. The Ospreys that used to breed in mainland Iberia were also mainly non-migratory cliff-nesters, as is typical of the species throughout the Mediterranean basin. The species is not known ever to have been more than scarce, with a fragmented range principally occupying those parts of the Iberian littoral that offered sea cliffs and access to sheltered waters for fishing. Nesting on trees, generally coastal pines, occurred at least locally in Portugal, however, most recently at Albufeira, Faro, around 1960 (Palma 2001).

There are references to nesting Ospreys in Portugal, both on cliffs and on pines, from as early as 1616 (Palma 2001, AP). Nesting at Gibraltar in 1776 was reported by the Reverend John White (Irby 1875). The Rock of Gibraltar was a regular breeding site, where a nest site on the southeastern sea cliffs was inhabited throughout the 19th century (Reid 1871–1874, Irby 1895) and until about 1933 (Rait-Kerr 1934b), with a second pair said to be present in some years (Verner 1909). Reid saw the local pair in January 1874, confirming their non-migratory status. The loss of the species from Gibraltar is almost certainly due to indirect disturbance.

Other pairs on southern and eastern Iberian coasts survived longer. The cliffs at Barbate, Cádiz, were a traditional haunt, with two or three pairs nesting there until the 1950s (Verner 1909, Westernhagen 1962, Alonso 1983). Elsewhere in mainland Spain breeding Ospreys are known to have nested on the eastern Spanish coast in Gandía, Valencia, during the late 19th century, and in the Empordà area, Girona, until the 1960s (Ferrer *et al.* 1986). Reports of former nesting on Biscayan coasts, for example at the Ría de Ribadesella, Asturias (LR) cannot be confirmed, although the habitat seems suitable and supports migrants and winterers. In Portugal, the rocky shoreline of Algarve offers numerous breeding sites and may once have accommodated a significant number of pairs. Three pairs still remained near Cape St Vincent in 1978 but they were reduced to one pair by 1992. The female of this pair died in 1997 but the male remained on territory until 2002. It attracted a new female briefly in 2000 and 2001 but no breeding occurred. The nest has since been used by White Storks and the Osprey is extinct as a breeding species in Portugal (PA2).

The enormous proliferation of reservoirs within Iberia since the mid 20th century has created an abundance of potential nesting habitat that may well enable the establishment of an inland breeding population, using trees or other tall structures as nest sites. There have been a number of sporadic instances of nesting or attempted nesting in such circumstances in recent years. They include attempts at the Beniarrés reservoir, Alicante, in 1981 (Urios *et al.* 1991); in Huesca, Aragón: at the El Grado reservoir in 1992 and at the Joaquín Costa and Barasona reservoirs in 2005; at Palmones, Cádiz, in 2005, and at the Barbate reservoir, Cádiz, also in 2005 (Alcántara 2005, Ferrer & Oliveros 2005, *Ardeola* 52: 212, 427). The birds involved are almost certainly derived from the rapidly increasing populations of central and northern Europe rather than from the western Mediterranean population, which is strongly tied to rocky sea coasts.

The Barbate reservoir and the Marismas de Odiel are jointly the locations of an attempt to introduce the species, using wild-hatched chicks originating from nests in Germany, Finland and Scotland. In total, 129 juveniles were raised in semi-captivity and 'hacked' into the wild during 2003–09, 73 of them at the Barbate reservoir and 56 at Odiel (Muriel *et al.* 2010). Remarkably, though, a pair of wild Ospreys settled to breed in 2005 at the Guadalcacín reservoir in Cádiz, some 20km from the Barbate release site and before any of the released birds attempted to nest (Muriel *et al.* 2006). The first nesting attempt by released young was at Odiel in 2008, but no eggs were laid. The same pair returned in 2009 and this time they succeeded in raising three young. At the same time, a pair thought to comprise the resident male and a new female bred successfully at the Guadalcacín reservoir site, fledging two young (Muriel *et al.* 2010). These recent instances of nesting all involved trees or tree-like nest sites, as is typical of more northern European populations, and it remains to be seen whether a long-term population increase will eventually result in reoccupation of the traditional coastal cliff sites. In 2013 four pairs nested in Cádiz and three in Huelva, fledging 15 young in total (*Ardeola* 60: 518).

Migrant Ospreys from northwestern Europe regularly cross mainland Iberia and Balearic Islands on passage and individuals often linger for some time at favoured estuaries, saltpans, reservoirs and lakes, especially in autumn (Österlof 1977, Hake *et al.* 2001). Iberian recoveries/observations of foreign-ringed birds comprise at least 119 from Fennoscandia, 113 from Germany, 30 from Britain and ten from France. As many as 45% of these records were from Spain during 2002–09, reflecting the continuing increases in European populations. Although the Balearic adults are resident their young disperse more widely, returning when mature. Ringing recoveries suggested that juveniles from Mediterranean populations did not travel beyond the Mediterranean basin in winter but radio-tracking of three juveniles from Menorca found that they travelled further: one moved to Morocco south of the Rif mountains, another to coastal Mauritania and the third to inland Algeria, after having first visited Mallorca and the south of France (Triay 2002).

Osprey movements are on a relatively broad front and there is no marked concentration of migrants at the Strait of Gibraltar, many birds presumably overflying the Mediterranean. Nevertheless, a number use the Pyrenean passes in autumn: during 2009–11 they averaged 361 birds in total at the three western watchpoints but just 32 were counted on average in the east at

Eyne. Some do cross the Strait but the total passage there is unlikely to be more than 200 birds per season since it is unusual to see more than a few individuals on any one day. The passage periods at Gibraltar are from mid-February to May and from mid-August to October, peaking in late March/early April and September respectively. Satellite-tracking has also revealed that some Osprey migration occurs at night, and not only when birds find themselves over the sea and have no other option (Wernham *et al.* 2002). One that crossed the Strait of Gibraltar northwards in the evening had reached central Spain by the following dawn (Rutland Water osprey website: www.ospreys.org.uk).

Some Ospreys winter in Iberia, in addition to the resident birds, and they seem to be doing so in increasing numbers, although the total winter population is small (Triay & Siverio 2008). Up to 200 winter in Andalucía, most of them in Huelva and Cádiz (Sayago 2008). The principal sites here include Odiel and Cádiz bay, with a regular presence in Cádiz also at the Palmones estuary and the Arcos, Bornos and Barbate reservoirs. Elsewhere, wintering birds are also regular at the Ebro delta, at El Hondo and in the inlets of the Biscay coast, including the Ría de Villaviciosa and Santoña. Individuals have sometimes been reported in winter at major reservoirs inland, especially in Extremadura but also widely elsewhere (Sanz 1997, pers. obs.). Several dozen winter at both coastal and inland locations in Portugal, the principal sites being the Tagus and Sado estuaries, the Paul do Boquilobo, the Caia reservoir, the Alvor estuary and the Ria Formosa (AP).

Observations of marked individuals have confirmed that the same birds return to the same wintering sites in successive years. For example, most of 115 wintering birds that were marked at and near Odiel during 1996–2001 were recaptured or resighted in subsequent winters, some individuals returning for 12 or more years (Sayago 2002a, Gacio & Sayago 2011). Only seven were of Nordic origin but there have been 27 from Germany, six from Scotland, four from France, two from Menorca and one from Italy, which suggests a leap-frog migration pattern in which Ospreys of most northerly origin travel furthest south (Gacio & Sayago 2011). There are also occasional reports of summering non-breeders from widely distributed locations, although individuals seen before mid June may still be on passage. Sayago (2002a) noted oversummering by just two or three immature or subadult birds on the Huelva coast.

LESSER KESTREL *Falco naumanni*

(Cernícalo Primilla, Francelho)

Locally common as a breeding species in central and southwestern Iberia and in the Ebro valley. Chiefly a summer visitor. A rare migrant in the Balearic Islands.

The Lesser Kestrel is a characteristic element of the rooftops in old-style towns and villages. It is typically gregarious and nests in loose colonies, using holes in the walls of historical churches and castles, whether ruined or in use, and the spaces beneath roof tiles. There are also a number of well-known colonies at traditional cliff sites, such as Arcos de la Frontera and La Barca de Vejer, in Cádiz, and the North Face of the Rock of Gibraltar, but they accommodate only a minor part of the population, although presumably were its ancestral habitat. Colonies may number several hundred pairs but are usually smaller and isolated pairs occur occasionally. The birds hunt over open country; notably pastures, fallows and young crops in non-irrigated farmland, for their prey: chiefly orthopterans, coleopterans and other arthropods but also small reptiles, birds and mammals (Franco & Andrada 1977, Rodríguez *et al.* 2010).

The core breeding range is within the thermo- and mesomediterranean climatic zones in agricultural areas of central and southwestern Spain (SA2), extending into southeastern Portugal (Rocha 2008). Some 32% of the Spanish population in 2004, then estimated at 12,000 pairs, was in Andalucía, with a further 31% in Extremadura, 13% in Castilla-La Mancha, 12% in Castilla y León and 9% in Aragón (Tella 2004). The Portuguese population, which is almost entirely restricted to the Alentejo, numbered *c.* 480 pairs in 2005 (PA2).

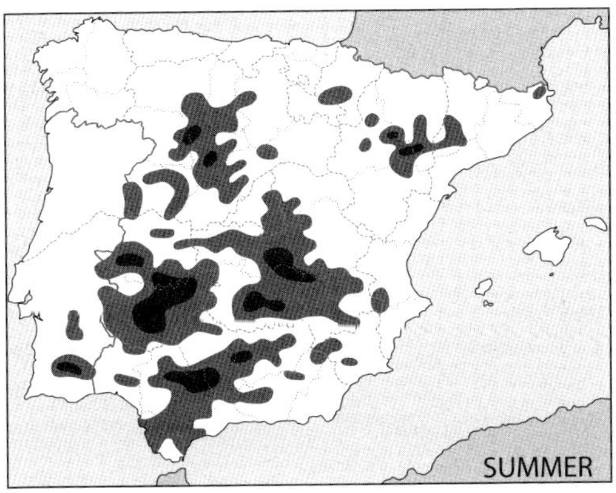

Historical references show that the Lesser Kestrel was once far more numerous. Irby (1895) said that 'vast numbers' bred at Gibraltar, a far cry from the 15 or so pairs that have nested there since the 1970s (BG). Lilford (1865) saw up to 300–400 'kestrels' on the wing at once at Coria del Río, Sevilla, in April 1864, and in April 1865 he saw many in the city of Madrid, where there are none today (Lilford 1866, Bernis 1974). It was abundant in Portugal south of the Tagus during the 19th century and was still locally 'as numerous as the swallows' in Évora in the 1930s (Reis Júnior 1931). The Spanish population was estimated to be of the order of 100,000 pairs in the 1960s (F. Bernis in Bijleveld 1974). Thereafter a marked decline was noted, not just within Iberia but throughout the European range of the species where numbers probably fell by over 95% (Biber 1996, BirdLife International 2012). Garzón (1977) still estimated the Spanish population at 20,000–50,000 pairs in the early 1970s but a national survey there in 1989 put the numbers as low as 4,293–5,089 pairs (González & Merino 1990), although its methodology has been criticised.

A catastrophic decline was clearly evident in Iberia by the 1970s and its coincidence with that of many other raptors in Europe suggests that it may have been caused by organochlorine pesticides, such as DDT, that were then in general use (Bijleveld 1977, BWP, González & Merino 1990). An analysis of unhatched eggs collected in southern Spain in 1988–91 detected contaminant levels below those known to affect reproduction (Negro *et al.* 1993), but DDT had already been banned in Spain in 1977. A study in Cáceres in 1987 found very low hatching success, just 62% (n = 61) and significant eggshell thinning, which again implies contamination with pesticides (Bijlsma *et al.* 1990). By way of comparison, recent hatching success in Los Monegros, Aragón, is 91.8% (Serrano *et al.* 2005) and in Castilla-La Mancha it increased from 78% in 1991 to 91% in 2003 (Ortego *et al.* 2010).

The Iberian populations have been recovering since the late 1980s but remain well below their historical levels. Regional censuses estimated the Spanish population at at least 12,000 pairs in 2004 (LR) and at over 14,000 pairs in 2010 (Íñigo & Barov 2010). Although there may have been some local declines, the recent general trends are rather positive, with significant increases in most of the main breeding areas (Ortego 2010). For example, the population in Aragón grew from some 100 pairs in 1988 to over 1,000 pairs in 2002, an annual increase rate of 15% (Alcántara 2004). In central Castilla-La Mancha the population increased by 42% during 1993–2000 and by a further 54% during 2000–05 (Ortego *et al.* 2007). The 1,605 pairs censused in Sevilla province in 2004–05, which comprised 63% of the Andalucian population, represented a 29% increase over the previous decade (Garrido 2008). Subsequent breeding season censuses of the Andalucian population have shown a continuing increase in the region: from 4,431 pairs in 2007 to 5,077 (in 687 colonies) in 2011 (*Ardeola* 59: 426–427). There has also been recent range expansion, with natural re-establishment in Burgos, Lleida and Teruel, and also in Murcia, where the Lesser Kestrel became extinct around 1979 but where 80 pairs were found in 2010 (Hernández 2010). The species is also benefiting from local conservation plans and from reintroduction projects in Catalonia, La Rioja and the Comunidad Valenciana (SA2), and from targeted conservation projects in Baixo Alentejo (PA2). The 480 pairs estimated in Portugal in 2005 may be compared with 270–272 pairs in 2001 and 155–158 pairs in 1996 (Rocha *et al.* 2002).

The Lesser Kestrel is vulnerable to a number of threats that have contributed variously to the reported declines (LR). In addition to the likely impact of organochlorine pesticides, perhaps the principal one is the loss of foraging habitat as an outcome of agricultural intensification that has often involved conversion of dry croplands to irrigated farmland, and through scrub encroachment on abandoned farmland (Donázar *et al.* 1993, Tella *et al.* 1998, Franco & Sutherland 2004, de Frutos *et al.* 2010). The birds have been affected locally by declines in large insects resulting from the use of insecticides to control 'plagues' of grasshoppers and locusts (Bonal & Aparicio 2008, Rodríguez & Bustamante 2008). Some colonies have also been lost when historic buildings have been refurbished or, in contrast, when ruins reach such a state of disintegration that they are uninhabitable even by kestrels. Conversely, the provision of nest boxes and even of purpose-built small buildings, '*primillares*', specifically designed to accommodate Lesser Kestrel colonies, has often proved to be an effective conservation measure (Catry *et al.* 2009). There is also the likelihood that the quality of the birds' winter habitats in sub-Saharan Africa has declined, although any such changes are unquantified.

The breeding colonies may be occupied early in the year, especially where there is a wintering population, from late February onwards. Laying is not usually until late April or early May. The colonies are abandoned in July and at least a proportion of the birds then disperse north to late summer quarters in northern Spain, notably in the northern meseta and the Ebro valley, where they form concentrations and communal roosts (Bustamante & Negro 1994, García 2000, de Frutos & Olea 2008). The median dispersal distance, chiefly northward, of 28 Spanish-ringed juveniles recovered before September of the same year was 210km (range 41–578km), the distance travelled being independent of the ringing latitude (Olea 2001). Two adult females marked with geolocators at Castro Verde, Portugal, also performed pre-migratory movements to Spain in summer, one travelling to the north-west and the other to the north-east (Catry *et al.* 2010b). In recent years there has also been a significant and unprecedented late summer presence in Cantabria, where up to 400, mainly juveniles, first roosted at Aloños from 8 August to 14 September 2005 (*Ardeola* 52: 427). Similar but smaller gatherings were reported from Asturias in September 2010 (*Ardeola* 57: 527). Records from the far north-west remain exceptional so a gathering of 33 at Cospeito Lugo in September 2013 is noteworthy (*Ardeola* 60: 518). Lesser Kestrels of both Spanish and Portuguese origin have also been found to gather together with some French breeders in the foothills of the French Pyrenees and in the Massif Central, with 200–600 counted at roosts in late August (OF). All these birds may be seeking feeding grounds in the less arid sectors of the Peninsula and beyond. Others, however, form premigratory gatherings within the nesting range. Such gatherings attained very large proportions locally in Extremadura in 2011, which may reflect a recovery in breeding numbers and productivity: up to 1,000 were at a roost in La Vera in mid August and 2,950 were counted at a roost in central Badajoz on 30 August, with 3,500–4,000 estimated to be present (*Ardeola* 58: 494).

Most Lesser Kestrels leave Iberia in autumn to winter in sub-Saharan Africa. Broad front southward passage across the Alborán basin has been confirmed in a study of satellite-tracked individuals and the same study also detected nocturnal migration, although most movements were diurnal (Limiñana *et al.* 2012). Visible passage at the Strait is negligible and no more than a few hundred individuals are counted there in either season (Bernis 1980, Programa Migres and GONHS records). Passage periods at the Strait are chiefly September–October in autumn and mid February–April in spring. Very few nest in France and hence passage through north-east Spain and the Balearic Islands is scarcely perceptible. Observations from Menorca of 15 at Maó on 22 April 2005 and 11 at Es Mercadal on 26 April 2010 (*A. O. Balears* 2005: 166, 2010: 138) are the largest published counts from the Balearics for that decade.

There have been extremely few foreign recoveries of birds ringed in Spain: just 18 in France, one in Italy, seven in North Africa: in Morocco (6) and Algeria (1) and two in sub-Saharan Africa: in Mauritania and Senegal. Four individuals marked with geolocators in Huelva province wintered near the Senegal River, on the border between Mauritania and Senegal (Rodríguez *et al.* 2009) and eight similarly marked in Portugal wintered in Senegal, Mauritania or Mali (Catry *et al.* 2010b), as did a further five Spanish individuals that were tracked by satellite (Limiñana *et al.* 2012). A roost of over 28,600 individuals was found in Senegal in January 2007 and most of these are likely to have been of Iberian origin (LPO 2007).

Some Lesser Kestrels may be found during November–January in southwestern Iberia, chiefly in the Guadalquivir valley and Cádiz and Málaga provinces, but also in Extremadura, and locally in very small numbers in Castilla y León, the Ebro valley and elsewhere (Saunders 1871, Andrada & Franco 1975, Negro *et al.* 1991, SWA). It is unclear whether all are genuine winterers or whether they include birds that have returned very early from Africa but genuine overwintering seems to occur on a small scale. Bernis (1980) noted 40 at La Barca de Vejer on 29 December 1960 and up to 50 roosted at the Gibraltar colony in December and January in the early 1970s (BG). Up to 40 wintered in Cáceres city during winter 1997–98 (*A. O. Extremadura* 1: 168). Winter observations have tended to be of birds arriving at roosts at known colony sites in the late evening, or departing from there at first light.

COMMON KESTREL — *Falco tinnunculus*

(Cernícalo Vulgar, Peneireiro)

***F. t. tinnunculus*. A widespread and common breeding species. Migrants from northern and central Europe also occur on passage and in winter.**

This versatile species is a widespread and often common resident throughout the Iberian Peninsula and the Balearic archipelago. In Spain the highest concentrations are distributed between Castilla y León; an area of the north-east taking in much of the Basque Country, Navarra and La Rioja; and west-central Andalucía, notably in Sevilla province. Conversely it is relatively scarce in the north-west in Galicia and Asturias, and across much of Castilla-La Mancha (SA2). In Portugal it is most abundant in coastal Estremadura and the south-west coastal areas in Algarve, being very scarce in the coastal north-west and in central areas of Beira Alta and Beira Baixa (PA2).

The Common Kestrel readily adapts to a very wide diversity of habitats, only requiring open country in which to hunt and an availability of nest sites. It occurs from sea level to above the treeline in all the major mountain ranges. It is especially common in agricultural areas but also frequents pastures and open woodlands such as dehesas. Only densely forested areas and extensive tracts of treeless open country are avoided, although Common Kestrels frequently hunt in the latter and will nest there if nestboxes are available on pylons (Avilés *et al.* 2001, Fargallo *et al.* 2001). Many now nest in towns and cities, where they may attain high densities: 35 pairs/100km^2 were estimated in the metropolitan area of Barcelona, where a colony of some

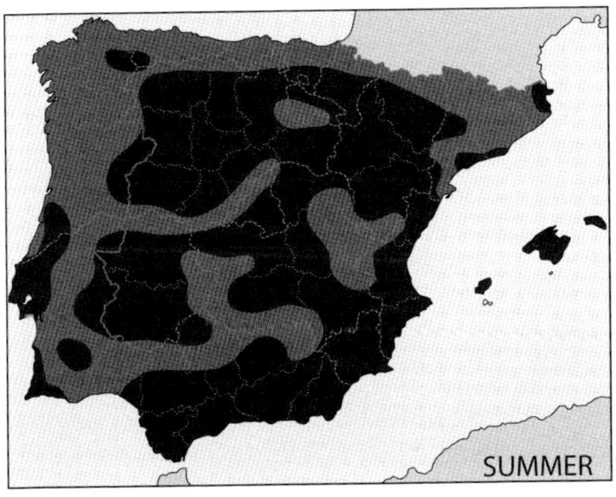

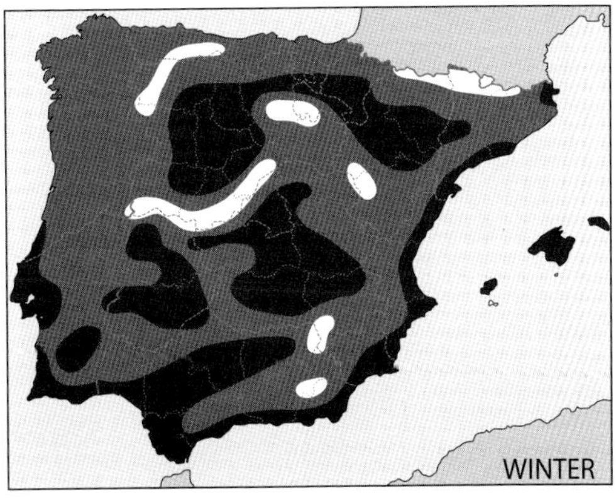

20 pairs occupies the sea-cliffs at Montjuïc, close to the harbour (Durany *et al.* 2003).

In Iberia a large proportion nest on cliffs, whether coastal, riverine or montane. Others use cavities in trees and buildings or the abandoned nests of corvids and other large birds. Common Kestrels may nest in small colonies and also form loose flocks where feeding opportunities are good, especially in late summer. Groups of up to 20 are common but a gathering of 200 at La Albuera, Badajoz on 30 August 1997, at a sunflower field during harvesting (*Ardeola* 44: 251), was more unusual although probably not too exceptional. The diet in Spain is based on voles and other rodents, but includes birds, insects, reptiles and even bats, with insects and reptiles being comparatively more important in the centre and south (Fargallo *et al.* 2009, Martínez-Padilla 2012).

The Spanish population was estimated at up to 30,000 pairs by Garzón (1977) and at 20,000–25,000 pairs by Purroy *et al.* (1997). The 2003 Spanish Atlas found at least 17,600 pairs but coverage was incomplete. At least 5,000 pairs and perhaps as many as 9,000 may breed in Castilla y León alone (Sanz-Zuasti & Velasco 1999). Numbers are likely to have declined during the heyday of organochlorine pesticides, as occurred elsewhere in western Europe, with a subsequent recovery since the late 1980s (Bijleveld 1974, Garzón 1974, SA1). A population increase has been apparent at least in Catalonia, where an estimated 400–1,000 pairs were estimated by Muntaner *et al.* (1983: CA1) and 2,912–4,676 pairs were present 20 years later (CA2).

The Common Kestrel is abundant and widespread in the Balearic Islands (Mayol 1978, SWA). There are over 1,000 pairs in Mallorca and smaller numbers in Menorca, Ibiza and Formentera (*A. O. Balears* 2010: 262).

The Portuguese population has been estimated at 1,000–3,000 pairs (BE) and may have shown some recent decline in the north-east (PA2). It is common in Andorra, with a population of 35–90 pairs estimated during 1999–2001; Eagle Clarke (1889) found it common there, nesting on the crags. It may have nested in Gibraltar during the late 19th century and early 20th century (BG) but was then certainly absent there until the 1990s, since when some ten pairs have become resident .

Iberian Common Kestrels have been regarded as largely resident (AI1) although two ringed as juveniles have been recovered further north in late summer: one travelled 326km from Málaga to Ciudad Real and the other moved 400km northwest from Toledo to León. Nevertheless, many appear to desert the highest and coldest areas in winter (de Juana *et al.* 1988, Palomino 2011), and there are a number of remarkable ringing recoveries of birds that were ringed as pulli in Spain: three travelled from 400–620 km to Morocco, one ringed in El Espinar, Segovia, in July 1972 was recovered 2,600km away on a ship off the western Sahara the following October, one travelled 2,900km from Doñana to Cape Verde and one from Alcázar de San Juan, Ciudad Real, reached Guinea-Bissau, a journey of at least 3,292km. Such recoveries hint at a hitherto unsuspected southward migratory tendency among at least some young Iberian birds, none of which have yet been recovered elsewhere in Europe.

Many Common Kestrels from northern and central Europe winter in the Peninsula, especially in the Levant, the Guadalquivir valley, Navarra and the Basque Country (de Juana *et al.* 1988). Wintering birds select more open and cultivated landscapes, as well as lower elevations, than are used in the breeding season (Deán 1996, Sunyer & Viñuela 1994, SWA). Densities of up to 6–8 individuals/km^2 then occur in some croplands. Winter recoveries of Common Kestrels ringed as pulli in other European countries comprise 16 from The Netherlands, 12 from Switzerland, nine from Germany, six from Sweden, five from Britain and two from France (AMI, de Juana *et al.* 1988). The greater part of 319 recoveries of Common Kestrels ringed outside Iberia are of birds ringed in the Low Countries (27%), Fennoscandia (21%) or Germany (21%), with most of the remainder from Britain, France or Switzerland and a few from Denmark, Austria or the Czech Republic; but many of these correspond to passage periods rather than winter.

Autumn passage is more apparent at the western and eastern ends of the Pyrenees than across the central sector. Out of 49 recoveries in Spain in August–October, 16 were in the Basque Country or Navarra: these mostly of Dutch origin, and 21 were found in Catalonia or the Comunidad Valenciana: these mostly originating from Switzerland and Germany (de Juana *et al.* 1998). The southward passage through the Pyrenees occurs from mid August to early November, peaking between mid September and early October (Urcun & Bried 1998b). However, the numbers recorded on passage are very modest: they averaged 784 birds at the three western Pyrenean watchpoints during 2009–11 and 192 at Eyne during the same period. The return passage is chiefly in March–April but has been poorly monitored.

Passage occurs across the Peninsula on a broad front and there are no significant concentrations at the Strait of Gibraltar, although small numbers are recorded there during both passages. The southward passage at the Strait extends from September into November and returning birds are recorded from February to early May. In recent years a number of Common Kestrels have been seen arriving at Gibraltar from Morocco in July and August (pers. obs.) and these could be North African birds engaged on post-breeding dispersal, as also seen in Long-legged Buzzards and Lanner Falcons.

RED-FOOTED FALCON *Falco vespertinus*

(Cernícalo Patirrojo, Falcão-vespertino)

A rare passage migrant, chiefly in eastern and northeastern Spain and in the Balearic archipelago. Most frequent in spring when irruptions have been recorded, rarer in autumn and occasional in winter. Vagrant to Portugal and Gibraltar.

The Red-footed Falcon usually breeds no closer to Iberia than Hungary (BE). It winters in southeastern Africa and some cross the central and eastern Mediterranean on passage. The species has long been reported in Iberia as a rare migrant. Lilford (1865) reported seeing one in Andalucía, King Carlos of Portugal may have hunted one in November 1884 (AP) and there is a report of one at Tarragona in May 1930 (Ticehurst & Whistler 1930). However, it has appeared annually since the 1980s and was almost certainly often overlooked previously, when observer density was much lower.

Iberia lies at the extreme western fringe of the migration routes and generally only small numbers occur on passage and then chiefly in the Levant coastal regions and the Balearic archipelago, where Ferrer *et al.* (1986) considered the species to be a very scarce migrant, regular in spring and sporadic in autumn. Red-footed Falcons are progressively rarer further west, although there are records from every Spanish region. The species is a vagrant in Portugal, with only two records validated by the CPR: single birds at Castro Verde on 5 September 2004 and at Cape St Vincent on 16 September 2004. There are also at least ten older Portuguese observations, all of them south of the Tagus, the most recent being single birds at Portimão on 2 May 1990 and 12 October 1990 (*Airo* 2: 54, AP). There is a more recent record of a juvenile at Sagres, Algarve, on 28 September 2011 (RBP). There are extremely few records from the well-watched Strait of Gibraltar and only one from Gibraltar itself: a male on 21 April 1975 (BG).

A search of the Spanish regional bird reports has allowed us to compile 268 records from 1996–2007 (excluding two influxes described below): 152 from Catalonia, 67 in the Balearics, 19 in the Comunidad Valenciana, 15 in Aragón and just a further 15 in other Spanish regions. They total 458 individuals, the mean number of birds per record being 2.51 in the Balearics, 1.49 in Catalonia, 1.46 in Aragón and 1.42 in the Comunidad Valenciana. Records elsewhere largely involve single birds. There are some records all year round, including a few in winter, but 78% of the records in our sample (81% of individuals) were in April–June: 63% in May, and 16% of records (13% of individuals) in August–October (Fig. 89). All November–March records were of single birds. Similarly, Ferrer *et al.* (1986) noted that 55% of 74 records were in May and 14% in August–October. Red-footed Falcons return north from Africa by a more westerly route than they take in autumn, occurring regularly in West Africa and in North Africa west to Algeria (BWP), which accounts for the preponderance of spring occurrences in Iberia and western Europe generally. Birds occur in open country and often in the vicinity of wetlands.

Figure 89. Records by month of Red-footed Falcons in Spain, 1976–2007 (n = 268 records, 458 birds). The 1992 influx is excluded (see text).

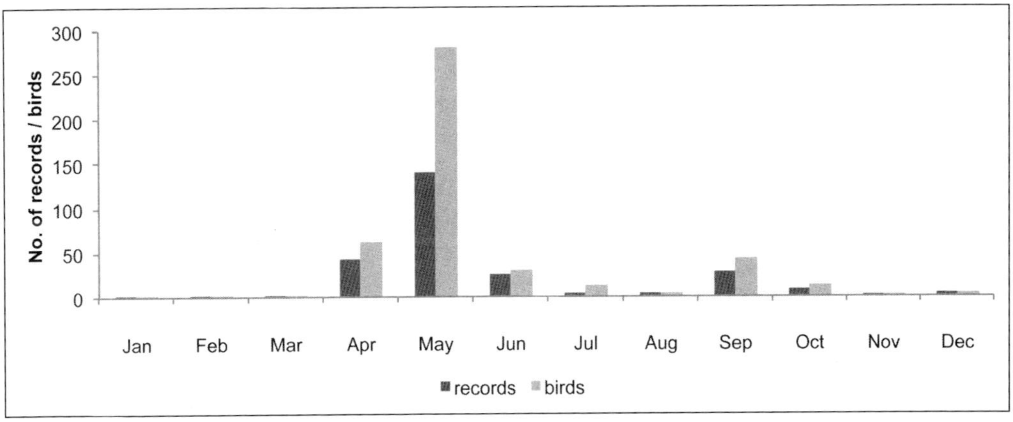

Two exceptional spring influxes involved considerable numbers. They presumably occurred when weather conditions produced a westward drift of migrants crossing the Mediterranean. The first was in 1992 and this was noted very largely on the island of Menorca, where at least 128 were recorded during 9–16 May (*Ardeola* 41: 96). The largest gathering was of 68 at Son Moscard on 13 May. The very local nature of this influx is shown by there having been only two reports from other Balearic islands that season. and it barely affected the mainland, where there were just six observations of single birds. The 2008 influx was much larger and was noted over a much wider area. It coincided with exceptionally heavy rainfall in the north-east and the Balearic Islands. The largest numbers were reported in Catalonia and the Balearics, with good numbers also in the Comunidad Valenciana, La Rioja and eastern parts of Castilla-La Mancha and Castilla y León. Relatively few were seen in the south-east, with some in Murcia and none at all in eastern Andalucía, suggesting that the influx was largely from the east and from the Comunidad Valenciana northwards. A few individuals nonetheless occurred as far afield as Asturias (2) and Cádiz (5). A minimum of 764 individuals was reported but it is likely that well over 1,000 birds were involved. Of the 764, 56 were during 28–30 April, 635 in May and 73 in June. Small flocks were present at many sites and particularly notable accumulations included peak counts in the Balearic Islands of at least 87 on 12 May and 79 on May 16 as well as site counts of 24 at Cistella, Girona, on 7 May; 27 at Delfià, Girona, on 18 May, and 40 at Plan de Sió, Lleida, on 24 May (*Ardeola* 55: 295–296, 56: 158–161, 56: 253). This influx was also the largest ever recorded in France, where some 4,300 were counted, most of them in Corsica (Dubois *et al.* 2012). A marked individual observed in the Balearics during the 2008 event had been ringed as a chick in southern Hungary (1,500km) on 8 July 2007.

MERLIN *Falco columbarius*

(Esmerejón, Esmerilhão)

***F. c. aesalon*. Scarce but widespread in mainland Spain in winter and a scarce passage migrant. Scarce in winter in Portugal. Rare in the Balearic Islands on passage and in winter.**
***F. c. subaesalon*. There is one ringing recovery in Spain.**

Iberia appears to be of limited importance as a wintering area for Merlins, which generally winter further north in Europe. However, they are not uncommon in open country, although thinly distributed, in Spain in winter although much scarcer then in Portugal or in the Balearic archipelago. The 40 available ringing recoveries suggest that most are of Fennoscandian origin (27 recoveries), with a minority from the British Isles (8), the Low Countries (3) and France (1). An Icelandic bird (*F. c. subaesalon*) has been found in Spain (Sunyer & Viñuela 1990). Two ringed in Spain have been recovered in Denmark.

The earliest arrivals are in September but peak passage occurs in October with stragglers in November. Small numbers – up to *c.*100 per season - are seen crossing the Pyrenees in autumn, especially at the western end (Urcun & Bried 1998b), the passage peaking in mid October. Autumn recoveries of ringed birds are concentrated in the Basque Country and Navarra (AMI, de Juana *et al.* 1988). They begin to depart from mid February but some remain as late as the end of April and there are several May records. One in Cuenca on 6 June 1992 (*Ardeola* 40: 93) and two July recoveries, one in Portugal and the other in southeastern Spain (Sunyer & Viñuela 1990), are exceptional. Extreme published sight record dates of 5 August (Ciudad Real in 2005) and 20 June (Comunidad Valenciana in 1995) are also exceptional, assuming the identifications were correct.

Some Merlins cross the Peninsula to winter in North Africa but the numbers involved are probably small; Thévenot *et al.* (BM) found only 81 certain winter records for Morocco. Very few are seen on passage at the Strait of Gibraltar, with records chiefly during September–November and March–April, but migrants occur there annually. The autumn passage is more apparent than the spring return, especially in eastern Iberia, where 48% of records fall during September–November. Records in central regions especially are chiefly in winter

Wintering birds are especially frequent in the northern half of the Peninsula, in the supra- and mesomediterranean climatic regions (SWA). The majority are reported from the northern meseta in Castilla y León but there are also wintering concentrations in the southern meseta and the Ebro and Guadalquivir valleys. The birds are especially associated with cereal croplands: habitats that attract large numbers of wintering larks, pipits and finches. Others frequent the margins of natural wetlands, which also attract abundant wintering passerines that often roost in reedbeds. Comparatively few winter in the Biscay or Mediterranean coastlands. The Merlin is widely reported in Portugal in winter, often at estuaries but also in open farmland, in dunes, montane scrublands and around wetlands, but very few individuals are involved. The cereal croplands of Alentejo attract birds regularly (AP).

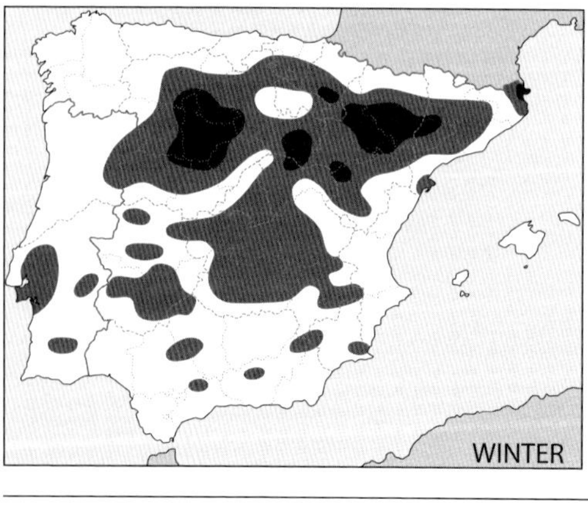

It was considered to be accidental on passage in the Balearic Islands by Ferrer *et al.* (1986) but Clavell (2002) gives 22 records (23 birds) during 1986–96 and there were at least 33 records (38 birds) during 2001–10 (*A. O. Balears*), chiefly in October–November and in winter. Some have remained into April but one at Pollensa on 18 May 2008 was exceptionally tardy. The Albufera de Mallorca is a favoured site.

EURASIAN HOBBY — *Falco subbuteo*

(Alcotán Europeo, Ógea)

***F. s. subbuteo*. Breeds widely in the northern half of the Peninsula, more locally in the south. A summer visitor, widespread on passage. Has bred in the Balearic Islands, where usually only a scarce migrant.**

The Hobby is a widespread summer visitor to mainland Iberia, wintering in sub-Saharan Africa. It is locally common in the northern half of the Peninsula, which accommodates some three-quarters of the Spanish population. It is distributed more or less continuously in Castilla y León, Galicia, La Rioja, Navarra and the Basque Country and it is also common in central and northern Asturias and Cantabria, northern Madrid and across much of Aragón and Catalonia. Within this range it is absent from higher elevations, nesting from sea level to 1,800m (SA2, LR). The breeding range is highly fragmented in southern Iberia, where the Hobby is thinly spread across Extremadura, Andalucía, Castilla-La Mancha, northern Murcia and the Comunidad Valenciana. There are local concentrations here, notably within the pinewoods of Doñana.

The Portuguese range is largely restricted to the north and centre of the country, where it occurs in similar habitats as in Spain (PA2). It is usually only recorded in the Balearic Islands on passage but attempted breeding occurred in Mallorca during 1988–90 (GOB 1997) and a pair bred in Menorca in 2003 (de Pablo & Capó 2004).

The Hobby requires open country for hunting and trees for nesting. It commonly uses old Carrion Crow nests and its relative scarcity in the south may be associated with the rarity of that species there, requiring it to use the disused nests of other corvids or other raptors. Within the croplands of the meseta it uses copses, especially pinewoods, and riparian woodlands for nesting. Those in the Biscay coastlands use heathland habitats and woodland edges bordering farmland. Pinewoods are again the preferred nesting habitat in the Mediterranean regions. Wetlands attract Hobbies, both for foraging during the breeding season and during passage periods, dragonflies being a frequent prey item. Large flying insects figure prominently in the adult diet but small birds are often taken to young in the nest. Bird prey ranges from hirundines and sparrows to larks on the meseta and small waders at wetlands (Palacín 2010).

The Spanish population was estimated at a minimum of 900–1,300 pairs by F. de Juana (1989) and a maximum of 1,600 pairs by Palacín (1990). The 2003 breeding bird atlas found rather more birds and estimated the population at 2,300–3,000 pairs. Most recently, the 2009–10 census of Spanish woodland raptors found an estimated 4,410 pairs, a substantially largely figure than previously expected but one that probably reflects more careful prospection, at least partly. The Portuguese population is thought to number 100–500 pairs. One or two pairs may breed in Andorra, perhaps irregularly (BWP).

Migrants cross the Peninsula on passage but their movements are inconspicuous. Ringing recoveries within Spain include at least four ringed as pulli in the Low Countries, two from Britain and one from Germany. There are three recoveries in Portugal of birds ringed in France (AP). Two pulli ringed in Aragón were recovered the following autumn in Algeria and Morocco, presumably when on passage there. Southward passage across the Pyrenees begins in mid August but the peak movements are from late September to early October, with stragglers into November (Urcun & Bried 1998b). Only modest numbers have been recorded there, up to 250 per season in 2009–11, most of them crossing the western passes. Birds presumably cross the Peninsula on a broad front and relatively few are recorded at the Strait of Gibraltar during either passage: in total they averaged just 70 birds (range 44–154) at Gibraltar and the central Strait sector during spring 2008–12 (GONHS, Programa Migres.). The timing of passage is similar to that seen at the Pyrenees, peaking in late September and early October. Spring movements at the Strait occur in April–May, peaking in late April and early May, with stragglers into early June. A two-year satellite tracking study of an adult Hobby (Meyburg *et al.* 2011) showed that it migrated across the western and central Mediterranean on different routes in different seasons and that it sometimes covered long distances at night. Unless this bird was somehow untypical this helps to explain why relatively few are seen at the usual migration bottlenecks. Nocturnal migration by Hobbies has been reported elsewhere (Strandberg *et al.* 2009) and may not be unusual.

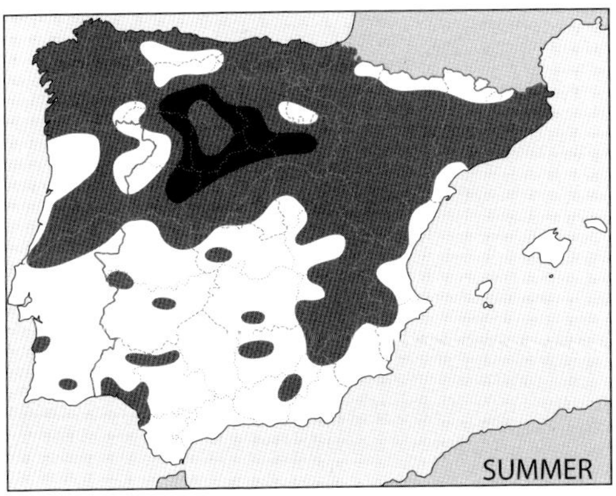

A number of published records of Hobbies seen in December, February and March are highly exceptional. They include single birds in Spain at Foz, Lugo, on 25 December 1988 and at Villanueva de la Vera, Cáceres, on 2 February 2004 (*Ardeola* 37: 336, 51: 549), and in Portugal at Ponte de Sor on 9 January 1991, at Mora on 28 December 1991 and at Mourão on 13 December 1994 and into January 1995 (*Airo* 2: 20, 3: 36; AP).

ELEONORA'S FALCON *Falco eleonorae*

(Halcón de Eleonora, Falcão-da-rainha)

A summer visitor, nesting colonially on the Columbretes islands and in the western Balearics. Migrants and non-breeders occur regularly in mainland Iberia.

Eleonora's Falcon is one of the most remarkable elements of the Iberian avifauna: a dimorphic colonial breeder that nests late in the summer on offshore islands from where it hunts over the sea, often cooperatively, for southbound migrant passerines to feed its young. It is a long-distance migrant that spends the northern winter chiefly in Madagascar. The highly fragmented breeding range extends from the Canary Islands and the coasts of Atlantic Morocco in the west, across the Mediterranean basin to Cyprus and the Turkish islands in the east but its core is in Crete and the Greek islands, which held 12,000 pairs of the global population of *c.* 15,000 pairs censused in 2007 (Papaconstantinou 2007).

The Iberian breeding population is confined to the Columbretes islands and the Balearic archipelago, and there are no certain records of mainland breeding, although the Rev. John White noted that 'a species of Hobby' was nesting on the Rock of Gibraltar in 1776 and fossils attributed to this species have been found in Gibraltar (BG). Eleonora's Falcon has been known to nest in Mallorca since at least 1861 (Homeyer 1862–63), in the Columbretes since 1885 (Brú 1913) and in the Pytiusas (western Balearics) since 1927 (Von Jordans 1928). The Columbretes/Balearic birds constitute a subpopulation that is some distance from the next nearest colonies: in Sardinia, Corsica and coastal Algeria.

Eleonora's Falcon nests colonially on often-inaccessible sea cliffs and is consequently difficult to census. The Balearic Islands were censused in 2004 and a minimum of 629 pairs (1,258 individuals) were found, distributed between Mallorca (27 colonies), Ibiza (8 colonies) and the islets of Cabrera, Dragonera, Tagomago, Es Vedrà and Vedranell. It is something of a mystery that none nest in Menorca, where there are apparently suitable cliffs. There were 56 pairs on the Columbretes in 2007. The total minimum Iberian population is thus of the order of 685 pairs (del Moral 2008).

The Iberian populations have shown significant and ongoing increases, almost certainly due to the protection now afforded to the breeding colonies. It was not always so and the chicks, which accumulate quantities of fat prior to fledging, were once locally considered a culinary delicacy (Mayol 1977b). Census methods have differed between years but there were fewer than 20 pairs in the Columbretes islands prior to the 1980s and only 254 pairs were counted in the Balearics in 1976 (del Moral 2008). The population of Cabrera, a National Park since 2002, increased from just nine pairs in 1974 to at least 36 in 2004 (LR).

Eleonora's Falcons return to their breeding sites from late April and during May (Mayol 1977c) but the clutch is not laid until late July or early August. The nest is a scrape on a sheltered cliff ledge. The young hatch in late August or early September and are fed on southbound migrant birds, chiefly passerines, for 35–40 days, fledging in October. Fledging success has been assessed at 2.1 young per successful pair in both the Columbretes and Balearics, annual productivity being 1.55 young per pair in the Balearic Islands (del Moral 2008). The last individuals leave the breeding sites in November.

Eleonora's Falcons do not breed in mainland Iberia but they are seen there with some, perhaps increasing, regularity. It has long been suggested that those from the central and western parts of the breeding range travel to and from Madagascar via the southern edge of the Mediterranean basin and the Red Sea coast (Walter 1979). However, satellite tracking of the autumn passage of a total of 11 individuals (five adults, two sub-adults and four juveniles) marked on the Balearics and Columbretes has shown that these birds travel much more directly, crossing the Sahara desert and Central Africa, regularly flying by night as well as by day (López-López *et al.* 2010; Mellone *et al.* 2012). Nevertheless, at least two individuals from the Atlantic colonies have been reported in Iberia on passage. A female ringed on Alegranza, Canary Islands, in September 2007 was observed in Valladolid, Castilla y León, in July 2008 before being seen again on Alegranza in August 2008 and yet again in Valladolid in 28 June 2009 (*Ardeola* 57: 226–227). Also, one ringed as a pullus on Mogador island, off western Morocco, in September 1960 was recovered the following September in Ávila, central Spain. Another ringing recovery of interest is of a bird ringed as a pullus in Crete in October 1988 that was recovered in Guadalajara in August 1989, indicating that some long-distance movement of individuals can occur within the breeding range. There are four foreign recoveries of birds ringed in September on the Columbretes: 474 km south in Algeria the following November, in Madagascar (8,163km) the following January, on Réunion Island (8,840km) the following November and in the Seychelles (7,447km) in December.

Mainland observations fall into two categories. There are records from Mediterranean and Portuguese coasts that largely correspond with migration periods, but there are also recent records from inland sites in Spain that fall mainly within June–July, a period immediately preceding the breeding season and a time when some colony sites, especially those on the smaller islands such as the Columbretes, offer limited hunting opportunities (Fig. 90). It is apparent that adults first visit their colonies in April–May but they soon move to the mainland to feed on flying insects and acquire breeding condition.

Figure 90. Records by month of Eleonora's Falcons in coastal and inland provinces in mainland Iberia (n = 521 coastal birds and 159 inland birds). Percentage of total number of individuals in each category recorded is shown.

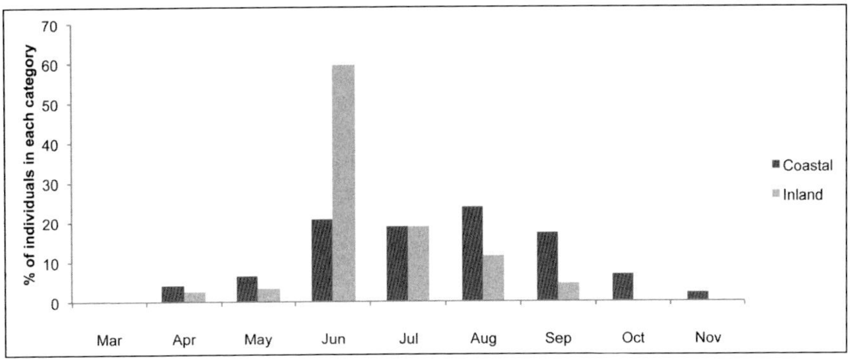

Figure 90 separates sightings of Eleonora's Falcons within regions that have a coast (e.g. Catalonia) from those seen in regions that are totally inland (e.g. Madrid). For simplicity no distinction has been made between birds seen on the coasts themselves and those seen further inland within coastal regions. Under these criteria, all 63 Portuguese observations are coastal: 40 (64%) of them are in Algarve in the extreme south and south-west and there are none from the northern half of the country. To these may be added 104 birds seen in Gibraltar and 354 seen in Andalucía, the Comunidad Valenciana and Catalonia. The coastal records fall within the period 25 March and 24 November, excluding an exceptional bird in Murcia on 11 January 2006 (*Ardeola* 55: 139). Three records of single birds from Asturias in August 2013 are also highly exceptional (*Ardeola* 60: 519).

The observations in Portugal, western Andalucía and Gibraltar peak in April (Gibraltar) and May (Portugal) with a much larger August–September peak in both areas, and these perhaps involve birds travelling to and from the Canary Island and Moroccan Atlantic colonies. The August–September peak is early for the return passage in this species so perhaps it involves non-breeders or failed breeders, but if so it is puzzling that it is not followed by more sightings of breeding adults and juveniles in October. The most likely explanation, as shown by the satellite-tracked Iberian birds, is that many or most birds from the Atlantic colonies fly directly across Africa from their breeding colonies, missing the Mediterranean altogether. Most coastal observations from Portugal, Gibraltar and Andalucía involve single birds or, more rarely, pairs. The exceptions are a flock of five at Gibraltar on 14 April 1975 (BG), six together near Tarifa on 10 October 1976 (Bernis 1980), and five at Tarifa beach on 29 October 2007 (*A. O. Cádiz 2004–08*: 66). Larger gatherings have recently been reported from inland areas in Catalonia, that fit a pattern that is comparable with the observations from the meseta described below. The most noteworthy have been in the Solsona/Olius area, where there were up to 40 present between 21 June and 1 July 2000, 24 in June–July 2002 and 25 on 11 July 2003 (*A. O. Catalunya*)

The occurrence of numbers of Eleonora's Falcons far inland in mainland Spain has only recently come to light and there is no way of knowing whether it was overlooked in the past or is a recent development, perhaps associated with population increases in the Balearic and Columbretes colonies. However, Lilford (1866) saw three on 3 June 1865 near Segovia, between San Ildefonso and Valsaín, forerunners of the more recent observations. The birds appear in groups of up to 20 that typically select a particular locality and remain there for some weeks, roosting communally in trees at night. This phenomenon has been reported from sites as far apart as Cuenca, Madrid and the southern slopes of the Sierra de Gredos in Ávila (*Ardeola* 37: 336, 41: 197, 51: 549, 52: 428). An observation of 40 at Portillo, Valladolid, on 20 June 2011 is particularly noteworthy (*Ardeola* 58: 495). There are similar observations of smaller numbers from León, Zamora, Valladolid, Soria, La Rioja, the interior of the Basque Country, Huesca, Zaragoza, Guadalajara, Badajoz and Toledo, showing widespread occurrence right across the meseta, with most records during June–August and thus preceding the onset of breeding. The age composition of the inland groups in Spain has not always been recorded but both adult and immature birds have been identified.

LANNER FALCON *Falco biarmicus*

(Halcón Borní, Alfaneque)

Occurs regularly in the south, where there are increasingly frequent records of undoubtedly wild individuals from Gibraltar and the region of the Strait in particular.

Some observations elsewhere have been discounted as involving escaped falconers' birds. Thus there are no accepted records from Portugal although there have been a number of recent observations there (AP). *F. b. erlangeri* breeds in north-west Africa, from Mauritania to Morocco and Tunisia, and formerly bred in southern Spain. There is a relatively numerous breeding population in Morocco, of some 1,000 pairs (BM). *F. b. feldeggii* breeds in southern Italy and the Balkans, east to Armenia and Azerbaijan, occurring occasionally as a vagrant elsewhere in Europe; there are 55 records of Lanner Falcons from France up to 2010.

It seems certain that the Lanner Falcon formerly bred in Spain in the vicinity of the Marismas del Guadalquivir, where clutches were found in old nests of other raptors during the late 19th century (Arévalo 1887, Irby 1895, Kumerloeve 1963, McGowan & Massa 1990, *Ardeola* 20: 349). One was caught alive in this region in 1954 and kept captive at Jerez de la Frontera, where it was seen by José Antonio Valverde and Mauricio González (*Ardeola* 2: 193). Other historical records from peninsular Spain comprise one taken at Barcelona and another seen at Girona in 1964 (Fuset 1913, König & König 1973). There are also several early observations from the Balearic Islands, including an instance of attempted breeding in Menorca in 1971 (Muntaner & Congost 1984, Ferrer *et al.*, 1986). In Portugal, single birds were in the collections of the Porto museum and the Bocage Museum, Lisbon, but the latter, whose species-identity was in question, was lost in the fire that destroyed the ornithological collection there (Tait 1924, Soares 1970, AP).

The Lanner Falcon has been reported increasingly frequently in recent decades, particularly near the Strait of Gibraltar where it is seen annually. The first observations at Gibraltar itself were in May and September 1969 (BG) and there were at least 73 records (75 birds) there by the end of 2012 (GONHS). The earliest reported elsewhere on the northern shore of the Strait was at Tarifa on 2 August 1973 (*Ardeola* 20: 349). The Spanish rarities committee accepted 13 records during 1992–2010 (Dies *et al.* 2011) but rejected numerous others in which the identification was uncertain or escape from captivity was suspected. The accepted Spanish records include seven from Cádiz, five of them from Tarifa, which accords with the comparatively high frequency of observations at Gibraltar. The other records comprise two from Extremadura: at Cáceres and Cabeza del Buey, three from Catalonia and one from the Comunidad Valenciana. The four records from the peninsular north-east were all in the vicinity of wetlands: at the Aiguamolls de l'Empordà, the Llobregat delta, the Ebro delta and the Albufera de Valencia. The last of these is the only known Iberian record of *feldeggii* (*Ardeola* 55: 269). Those from the Aiguamolls and Cáceres, and one of the records from Cádiz, as well as most if not all the records from Gibraltar are considered to have been of the North African race *erlangeri*. The two are readily separable in adult plumage at least, the former being the darkest race and the latter the palest (Forsman 1999, Ferguson- Lees & Christie 2001).

The Gibraltar records to 2010, together with the accepted Spanish records and the 1973 Tarifa record, show a bimodal temporal distribution, 54% being during August–October and 40% during March–June (Fig. 91). Most records from France, however, have been during September–February. In Spain, there are only two midwinter observations, from Catalonia: an adult at the Ebro delta on 19 January 1996, and a first-winter that was at the Aiguamolls de l'Empordà from at least 2 December 2001 until 2 January 2002. In addition, an adult remained in Gibraltar from 18 December 2011 into 2012. A juvenile at Gibraltar from at least 23 May to 24 July 2001 was another long-stayer. The 38 individuals of known age reported up to 2010 comprised 24 juveniles/first-winters (63%), six older immature birds and eight adults. Some of the Gibraltar observations in late summer are of juveniles dispersing northwards from Morocco. However, the bimodal pattern of the records as a whole hints at the existence of an as-yet-undiscovered migratory Iberian breeding population.

Figure 91. Records by month in Spain and Gibraltar, 1974–2010 (n = 67).

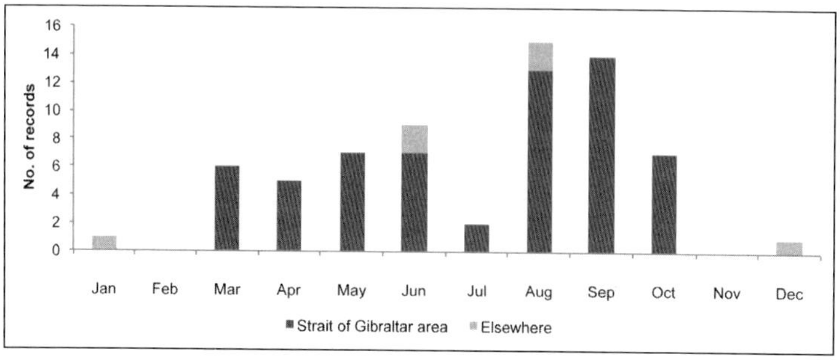

SAKER FALCON *Falco cherrug*

(Halcón Sacre, Sacre)

Vagrant to Spain, Portugal and Gibraltar.

The Saker is generally regarded as a vagrant in western Europe but it is popular with falconers so escaped birds may mask the occurrence of wild individuals. There are 15 apparently reliable records from France, two of them from the French/Spanish border: at Eine, Roussillon, on 4 September 1993, and at the Organbidexka col in the Atlantic Pyrenees on 27 September 1994 (OF, Reeber *et al.* 2011). There is a doubtful 19th century record for Morocco and a more recent occurrence of a juvenile that was captured 200km from Casablanca in May 1963 (Mayaud 1970).

The earliest Iberian observation of a possibly wild bird was one seen near the Strait on 10 April 2002 (*Ardeola* 53: 174). There were three observations in Gibraltar in autumn 2003: a female without jesses flew south on 25 July, a bird flew south with other raptors on 16 August, and one was observed on 8 September (*GBR* 3: 21).

The provenance of all these birds cannot be known for certain but there has been recent proof that individuals of natural origin may reach Iberia (Dies *et al.* 2011, Jara *et al.* 2010). A radiotracked juvenile female, named 'Piros', hatched in western Hungary in 2009, entered Spain via the Catalan Pyrenees on 30 July. It travelled up the Ebro valley and then remained on the outskirts of Valladolid for around six weeks. During this time it made a brief visit to eastern Portugal on 30 August, returning to Spain the following day. It began travelling south on 14 September and crossed the Strait five days later. By 29 September it had reached Nouadhibou, on the Mauritanian coast – where unfortunately she was found dead in early October. 'Piros' provides the sole certain record of a wild Saker in Spain and Portugal. However, its occurrence in Mauritania is not unprecedented since small numbers of Sakers have been recorded there and in northern Senegal (Isenmann *et al.* 2010). Small-scale passage through Iberia may therefore be less infrequent than once believed.

GYR FALCON *Falco rusticolus*

(Halcón Gerifalte, Gerifalte)

Vagrant to Spain. Recorded in Portugal (Category D).

There are four Portuguese records but the possibility that all four were escaped birds was raised and the species is in Category D of the Portuguese list (AP). They comprise one killed in coastal dunes at Vila do Conde, on 3 May 1928, one seen in the same region on 26 April 1959 and one seen near Alcácer do Sal on 2 March 1986 (AP). More recently a white phase individual was seen and photographed at Odemira, Beja, on 15 and 22 March 1991 (*Ardeola* 40: 184). One taken in the Balearic Islands at Petra, Mallorca, on 1 March 1960 is believed to have been an escaped falconer's bird (Ferrer *et al.* 1986). The first accepted (Category A) record for Spain was a first-winter female that was on the Biscay coast at Tapia de Casariego, Asturias, from 29 December 2009 to 21 March 2010 (*Ardeola* 59: 376). This last bird occurred at a more likely time of year and its arrival coincided with a cold spell and strong northerly winds. The nearest breeding populations to Iberia are in Greenland, Iceland, Fennoscandia and Russia and few range south of 60°N in Europe. There are 387 records for Britain, 124 for Ireland, and 22 for France.

PEREGRINE FALCON *Falco peregrinus*

(Halcón Peregrino, Falcão-peregrino)

***F. p. brookei*. A widespread resident in mainland Iberia and the Balearic islands.**
***F. p. peregrinus*. Breeds in the French Pyrenees and presumably intergrades with *brookei* in that region. A winter visitor from western and central Europe, probably scarce.**
***F. p. calidus*. A rare winter visitor from arctic Eurasia.**

The Peregrine is very widely distributed within Iberia with a healthy population in many regions (Table 25). In Spain it is only really absent as a breeding species from the extensive open country of the meseta and southern Extremadura, the gentle hills of the western Sierra Morena and the Guadalquivir valley (SA2). The range in Portugal is more discontinuous: in the south the distribution is predominantly coastal and includes the south-west Algarve, the Alentejo coast and the coast between Setúbal and Peniche; in the north it breeds mainly in the Peneda-Gerês National Park and in the north-east along the Duero gorges (PA2). It breeds throughout the Balearics and also in Andorra and Gibraltar.

It is commonest in areas that offer cliff sites for nesting and so is widespread in all mountain ranges, in river gorges, quarries and on the sea cliffs of all rocky coastlines. In Spain a 2008 survey (del Moral & Molina 2009) found that 95.2% nested on such sites but a small minority (3.1%) nested on disused corvid nests on pylons, these especially in Valladolid and Córdoba provinces. A small but increasing proportion (1.4%) use man-made structures, including tall viaduct supports and coastal Moorish watch-towers: as in Doñana, but especially high-rise city buildings; with nesting pairs present in Barcelona, Madrid, Salamanca and Burgos, among other cities. Urban nesting has also been reported in Portugal, in Lisbon (AP). Tree-nesting is very rare but two pairs used such sites in Andalucía in 2008 (del Moral & Molina 2009).

Table 25. Regional distribution of breeding Peregrine Falcons in Iberia. Figures are numbers of breeding pairs. (del Moral & Molina 2009, PA2, Birdlife International 2004, GONHS)

REGION	Definite	Probable	Estimated	Pairs/100km^2
Andalucía	284	16	300	0.3
Asturias	70	86	156	0.7
Aragón	293	26	319	0.5
Basque Country	101	0	101	1.4
Castilla y León	375	28	403	0.4
Castilla-La Mancha	181	19	200	0.3
Cantabria	101	0	101	1.9
Catalonia	268	28	296	0.8
Comunidad Valenciana	180	17	197	0.8
Extremadura	37	7	44	0.1
Galicia	80	6	86	0.3
La Rioja	40	0	40	0.8
Madrid	28	0	28	0.4
Murcia	116	18	134	1.0
Navarra	74	0	74	0.8
Mainland Spain 2008	2,228	251	2,479	0.5
Balearic Islands 2008	89	91	180	1.8
Portugal 1999–2005	70	0	100	0.1
Gibraltar 2008	7	0	7	–
Andorra 1998	3	0	6	1.6
IBERIAN TOTAL	2,397	342	2,772	

The 2008 survey found 2,228–2,479 pairs of Peregrine Falcons in Spain. The very large regions of Castilla y León, Aragón, Andalucía and Catalonia together hold nearly 50% of the population but the Peregrine occurs at high density in Cantabria, the Balearic Islands, the Basque Country and Murcia. The Balearic birds nest on all the main islands and many of their islets, inhabiting inland sites as well as coastal cliffs. The total Iberian population is probably in the region of 3,000 pairs, given that not all the Spanish regions received complete coverage in 2008, and is the largest in Europe (BE).

There are historical references to declines in the Iberian Peregrine population during the 20th century (Bijleveld 1974, Garzón 1977, Heredia *et al.* 1988) but little information on population sizes has been available until recently. It seems clear, however, that the impact of agricultural intensification, especially the effects of pesticide use, on Peregrine populations during the 1950s and 1960s was much lower in Iberia than in many other parts of Europe.

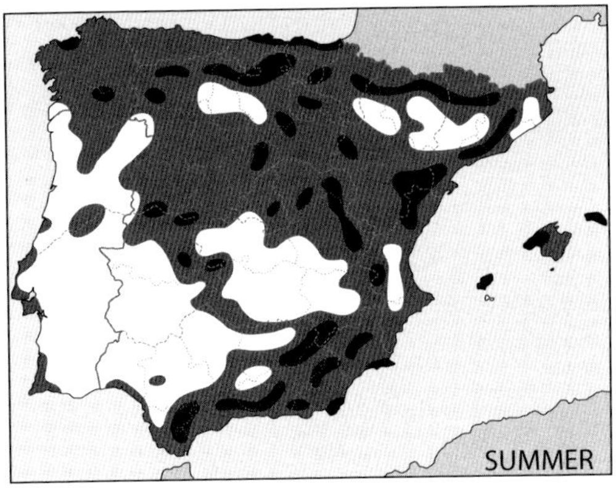

An early estimate (Garzón 1977) put the population at some 2,000 pairs. An incomplete assessment in 1986 (Heredia *et al.* 1988) estimated 1,685–1,751 pairs. A more recent survey based largely on bibliographical searches and requests to ornithologists and public bodies for unpublished information (Gainzarain *et al.* 2002) put the Spanish total at 2,435–2,743 pairs, a comparable result to that of the more direct census of 2008. The two latter surveys certainly reflect genuine population increases in some regions, although an unknown proportion of the increase probably results from improved coverage. Trends on regional and provincial scales suggest that the population is stable or slowly increasing in Andalucía and Asturias; stable in Cantabria and Extremadura; increasing in Catalonia, the Comunidad Valenciana, Galicia, the Balearics, La Rioja and Navarra; and declining in Madrid, Murcia and in some provinces of Castilla y León, Castilla-La Mancha and the Basque Country (del Moral & Molina 2009). No information was available for Aragón. The overall impression is of a largely stable Peregrine population, albeit with regional variations.

The Portuguese population is still small (79–100 pairs) but considered stable in the north, increasing in central regions and increasing slowly in coastal areas south of Cape Carvoeiro (PA2). It may have declined to some 30 pairs during the 1970s and 1980s, but the population has been poorly monitored (AP). Smith (1868) was informed that the species was rare in Portugal so its present relative scarcity in that country may be of long-standing. The Gibraltar population of up to seven pairs (in 6.5km^2) represents the highest known density in Iberia but the birds there occupy cliffs around most of the periphery of the Rock, a location that sees a high volume of both landbird and seabird passage for much of the year. The Gibraltar Peregrines often hunt over the sea, often at some distance from land. A comparable situation occurs on the Balearic islet of Cabrera (13.18km^2), where 12 pairs nested in 2005 and 11 in 2008 (*Ardeola* 53: 283; del Moral & Molina 2009).

Iberian Peregrines are subject to similar threats that have affected populations elsewhere, including reduced productivity attributable to pesticide accumulation in some northern regions and in Madrid (SA2). However, both the Portuguese and Spanish populations have suffered seriously in the past from the depredations of nest raiders: collecting eggs or young for falconry, and pigeon fanciers who kill both adults and young. These latter threats have diminished in recent years, partly due to the guarding of susceptible nests. An altogether different problem is posed by the impact of other raptors on nesting Peregrines. Pairs are known to be displaced sometimes from existing or potential nest sites on cliffs by eagles but Eagle Owls are the main hazard, both to adult and young Peregrines that are killed at their nests or roosts. The depredations have become more widespread as the owl population has increased and this factor is considered the chief cause for some local declines, such as the dramatic reduction from 30 pairs to two pairs on the river Tagus cliffs in Toledo (del Moral & Molina 2009).

F. p. brookei is largely sedentary although juveniles may disperse some distance (Zuberogoitia *et al.* 2009). There are nine recoveries of Spanish birds ringed as pulli that travelled over 100km, including two birds from northeastern Spain that were recovered in France. One ringed in Tarragona in June 2005 was at Caserta e Napoli, Italy (1,094km) a month later. In addition there is a remarkable recovery of a pullus ringed at Doñana in April 1997 that was found a year later at L'viv, Ukraine (2,618km), far outside the normal range of the subspecies (*Ecología* 13: 352).

Some Peregrines of northern races enter the Peninsula to winter and some may continue south to North Africa. The scale of this movement is unknown and certainly very few are seen on passage across the Pyrenees or at the Strait of Gibraltar, although distinguishing migrants from residents is difficult in both regions. The great majority of immigrants are of the nominate subspecies, as are those foreign birds ringed as pulli that have been recovered in Iberia, which include at least 11 from Finland, five from Sweden and three from Germany; and a bird from Northern Ireland recovered in south-central Portugal (Wernham *et al.* 2002) Individuals of *F. p. peregrinus* favour wetlands and steppe areas throughout Catalonia in winter (Clavell 2002) and probably elsewhere.

Large, pale Peregrines that are perhaps of the race *calidus* are very scarce but some are probably overlooked. Clavell (2002) gives six records (seven birds) for Catalonia during 1988–99, with extreme dates of 7 October and 15 April: all were at coastal wetlands, including four birds at the Ebro delta. A female whose dimensions strongly indicate *calidus* was captured and ringed in Mallorca in August 2004 and then recaptured during the winters of 2005 and 2006, which it spent at the airport, using the airfield as its hunting territory from which it excluded other falcons; what may have been the same bird returned in winter 2007 (*Ardeola* 55: 140, 55: 296). Similar single females were at the Albufera in Mallorca in January 2006 and September 2009 (*Ardeola* 55: 140, 57: 528). A very large individual at the Laguna de La Nava, Palencia, in January–February 1999, and another at the Laguna de Boada in October 2001 (*A. O. Palencia 1998–2001*: 84–85) were thought to be of this race, as was one that was at the

Guadalhorce estuary on several dates in January 2010 (*Ardeola* 57: 528). In addition, a breeding female was satellite-tracked in 1994 from its nesting range in the Kola peninsula, Russia, to a wintering area in Sevilla between Osuna and Marchena, where it arrived on 10 October after flying 4,609km in 26 days (Ganusevich *et al.* 2004). A northern Peregrine that was on the Berlengas islands during 8–14 November 2010 was considered to be *calidus/tundrius* (RBP).

WATER RAIL *Rallus aquaticus*

(Rascón Europeo, Frango-d'água-comum)

***R. a. aquaticus*. A widespread resident in mainland Iberia and in the Balearic Islands. The local birds are joined by numbers of continental migrants in winter.**

The Water Rail occurs throughout Iberia wherever suitable wetland habitat is available. It frequents dense emergent vegetation but principally occurs in sedge beds and reedbeds at freshwater lakes and marshes and along river courses. In Spain there are good populations in coastal Galicia, the Biscay and Levant coastal wetlands, Castilla y León, Castilla-La Mancha, Extremadura and the Ebro depression. It is especially common in the western Andalucían wetlands, especially those of the lower Guadalquivir and Doñana (SA2). It also nests in Mallorca and Menorca. The Portuguese distribution is predominantly coastal, with some birds inland chiefly south of the Tagus (PA2). The expansion of rice cultivation has led to some recent local range expansion in Extremadura, the Guadalquivir and Ebro valleys and the Levant coastlands. Virtually all breed at relatively low altitudes, below 900m (SA1).

Extrapolation from a few local estimates put the Spanish population in the early 1990s within the wide range of 11,400–68,000 pairs (SA1). There may also be a further 500–2,000 pairs in Portugal (BE 2004). In Spain, Castilla y León alone may hold a minimum of 1,500 breeding pairs (Sanz-Zuasti & Velasco 1999), and the wintering population in Catalonia has been put at 1,812–2,111 individuals (CWA). At least 200 pairs were estimated to breed at the Albufera de Mallorca in 2010 (*A. O. Balears 2010*: 140). Population trends are also uncertain but the loss of many important wetlands during the mid 20th century undoubtedly also caused local extinctions.

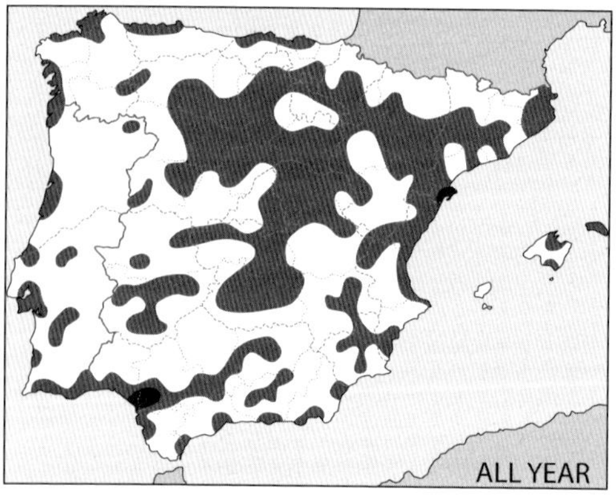

Iberian Water Rails are thought to be largely resident but local movements must occur during drought years when southern lakes especially may become uninhabitable. The furthest travelled among 62 recoveries of birds ringed in Spain during 1973–2009 moved 90km within Tarragona province (*Ecología* 11: 337). The winter distribution is largely similar to that during the breeding season (SWA) although some then occur in such habitats as salt marshes and rice paddies. The Iberian birds are joined in winter by some immigrants from further north in Europe and some also cross the Peninsula to winter in Morocco (BM). The size of the immigrant population is unknown but is probably small since there are only three recoveries of foreign-ringed birds in Spain: two from Hungary and one from Belgium, and none in Portugal (AP).

SPOTTED CRAKE *Porzana porzana*

(Polluela Pintoja, Franga-d'água-malhada)

A very scarce and local breeding species in mainland Spain. Also a scarce passage migrant and very scarce winterer in mainland Iberia and the Balearic Islands.

The past breeding status in Spain has been inferred from very little concrete information: chiefly from sightings, captures or individuals heard calling in suitable habitat during the breeding season. Nevertheless, it was probably more abundant in both Spain and Portugal in the 19th century than is now the case: in Portugal a nest with eggs was found near Coimbra in 1866 (AP). Breeding has been reported from Aragón, Castilla-La Mancha, Catalonia, the Comunidad Valenciana, the Basque Country, Galicia and Menorca, and has been assumed to occur at Doñana, although Valverde (1960) considered that 'extremely doubtful' (SA2). The very few recent confirmed breeding records include an adult seen with chicks at the Río Mijares estuary, Castellón, in June 1992; the finding of an abandoned nest at the laguna de Navaridas, Álava, in June 1993; an adult seen with a chick at La Furta reservoir, Asturias, on 24 July 2000, and a week-old chick captured at the Brazo del Este, Sevilla, on 11 August 2000 (*Ardeola* 41: 96–97, *A. O. Asturias* 7: 98, *A. O. Sevilla 2000–2005*: 75).

The population estimate of 750–2,400 pairs suggested by Purroy *et al.* (1997) was clearly wildly optimistic. Fieldwork for the 2003 atlas put the minimum population at 53 pairs but breeding was only proved for one pair (SA2). Possibly no more than a modest number nest in any one year but detecting this elusive species, let alone confirming nesting, is extremely difficult and some very probably nest unnoticed.

Spotted Crakes favour freshwater wetlands that offer dense stands of emergent vegetation such as sedges, rushes and irises, together with shallow, clear water and expanses of wet mud. Migrants and wintering birds occur at a wider range of habitats, including reservoirs and ricefields. The majority of observations probably involve scarce European migrants travelling to and from winter quarters in sub-Saharan Africa. Saunders (1871), Arévalo (1887), Irby (1895) and Munn (1921) variously refer to its abundance in Andalucía, Murcia and the Comunidad Valenciana, especially on passage and in winter, giving the impression that it was once much commoner. Indeed, Irby states that it was commoner than the Moorhen in Andalucía. Large declines have been reported in most European countries, especially in the west (BE), and the French population may only number a few hundred widely scattered pairs (OF).

The spring passage extends from February to April or early May but has a clear peak in March. Southbound birds may appear from mid July but most are reported during August–October. Autumn records are much scarcer than spring ones in nearly all parts of Iberia with the exception of the Biscay coastlands, where the two passage peaks are similar (Fig. 92). The Levant coastal wetlands and the Balearic wetlands account for a high proportion of observations. Well monitored sites record some numbers during the course of both seasons. This is most evident at the Ebro delta where seasonal totals in 2006 exceeded 60 individuals in autumn and 50 in spring (*A. O. Catalunya 2006*: 77). Many other records also come from well-watched sites, notably including Doñana, the Brazo del Este and other locations in the Guadalquivir estuary; the Albufera de Valencia and the Albufera de Mallorca. The Spotted Crake is rarely reported in Portugal and there are only some three dozen records for the country (AP). Nearly all recent Portuguese records were in the south and south-west: in Lisboa, Santarém, Setúbal, Beja and Faro. There are single recoveries in Spain of birds ringed in The Netherlands and Germany and one from The Netherlands has been found in Portugal in February (AMI, AP).

Figure 92. Monthly distribution (%) of observations comparing the northern coastal regions (Galicia, Asturias, Cantabria and the Basque Country: n = 70 records) with the rest of the Peninsula (n = 674 records).

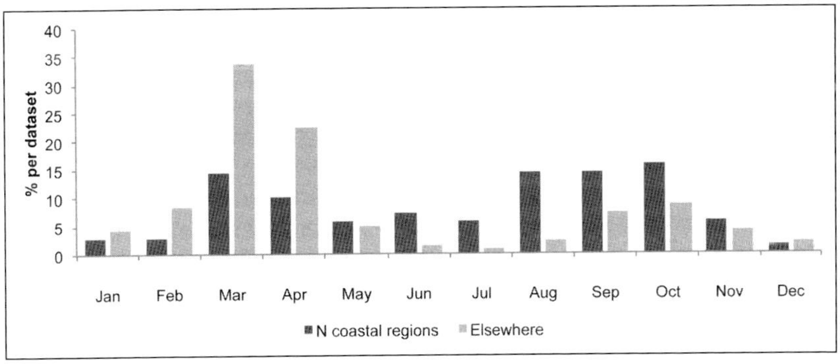

Most recent observations of Spotted Crakes are of single birds, some of which linger at particularly sites for some time, especially and inevitably in early autumn when the birds become flightless during the post-breeding moult. Very few remain to winter in Iberia: 17 in total were recorded during the 2007–10 Spanish winter atlas survey but only one was in mid winter (SWA).

SORA CRAKE *Porzana carolina*

(Polluela Sora, Franga-d'água-americana)

Vagrant to Spain.

The skin of an adult that was collected in the Marismas del Guadalquivir on 30 December 1975 is in the collection of the Estación Biológica de Doñana. One was seen at the Laguna de Traba, A Coruña, on 10 April 1988 (García *et al.* 1989; *Ardeola* 39: 77, 50: 134). This is a very rare vagrant to the Western Palearctic, where the records include three in Ireland, six in Britain, two in France, one in Morocco and one in Madeira.

LITTLE CRAKE
(Polluela Bastarda, Franga-d'água-bastarda)

Porzana parva

A rare migrant in Spain, where it has bred exceptionally. Very rare in Spain in winter. Vagrant to Portugal.

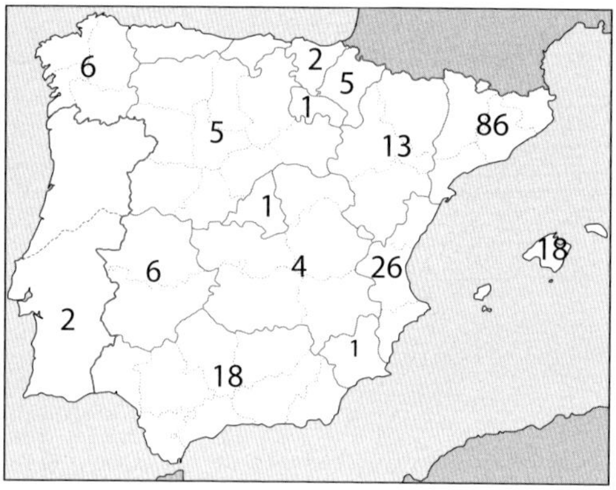

Figure 93. Records by regions, 1971–2011 (n = 194).

The Little Crake is uncommon in Iberia, where it is known chiefly as a rare passage migrant and occasional winter resident in mainland Spain and in the Balearic Islands. It has bred in Spain on very rare occasions but there is no regular breeding population. There are two old records for Portugal: an undated specimen obtained prior to 1931 – and since lost in the Bocage museum fire – and one found dead on 22 May 1903 at the Río Ave, now conserved at the Porto University Museum (Matias *et al.* 2007). Other Portuguese records comprise a recent undated specimen from the Paul do Boquilobo (AP); and single juveniles at Quinta do Lago, Faro, during 9 August–11 September 2008 and on 20 September 2009 (CPR 2008–09). A male was at Quinta do Lago during 27 March–19 May 2012 (RBP).

There are few historical references to this species: Saunders (1871) considered it "not uncommon" in southern Spain but Irby (1895) did not record any on the Spanish side of the Strait of Gibraltar, nor did Munn (1921 *et seq.*) find it in the Balearic Islands. Gil Lletget (1945) reported the existence in the Madrid Museum of specimens originating from Vigo, Badajoz, the Albufera de Valencia and Aranjuez. From the 1960s there is an interesting series of 16 individuals captured by hunters in Mallorca and taken to a local taxidermist (Nadal 1965, 1966, 1971). Thereafter, the Little Crake was assumed to be a regular, if scarce, breeding species largely on the basis of calls heard in apparently suitable nesting habitat, for example at the Laguna del Taray, Castilla-La Mancha, during 2–5 May 1963 (Bernis 1964b), although this type of evidence has been considered insufficient at the Ebro Delta (Ferrer *et al.* 1986) and Doñana (Llandres & Urdiales 1990). Valverde (1960) regarded its possible nesting in Doñana to be 'more than doubtful'. In all, there are just two confirmed breeding records in Spain: a downy chick found in a Barn Owl nest at Doñana in the early 1970s (Fjeldså 1975) and a chick found at the Llobregat delta in 1995 (A. *O. Catalunya 1998*: 116). The estimate of 50–100 pairs by Purroy *et al.* (1997) is wildly optimistic and there are no grounds to consider the Little Crake to be other than an extremely occasional breeder in Spain (de Juana 2003).

Figure 94. Records by month, 1971–2011 (n = 194).

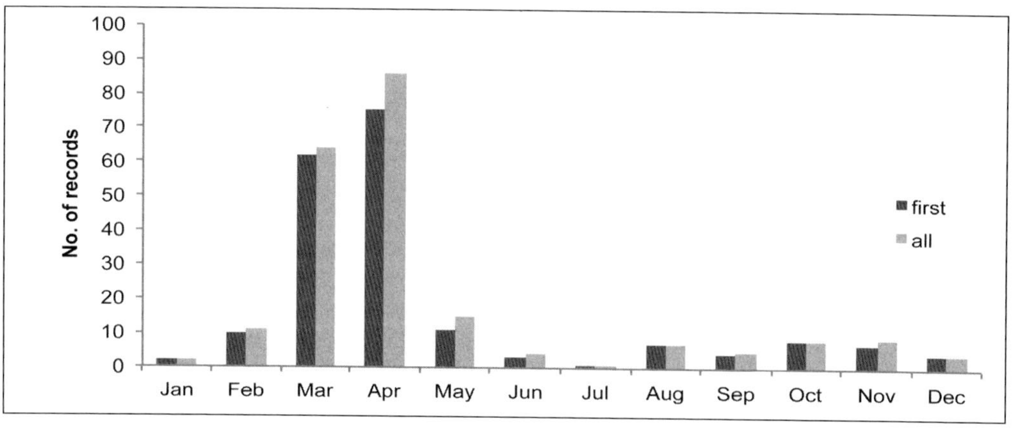

The core breeding distribution extends from central Europe eastwards across Asia. It has only a marginal presence in France, where breeding has not been proved in recent years (OF) and an average of only 1.8 singing males per year were recorded in 2002–10 (*Ornithos* 18: 160). Iberia thus lies well to the west of any likely migration routes and indeed most records are in easternmost Spain. In a sample of 194 published records from 1971–2011, excluding those that were only heard or found dead (and thus not easily dated), some 44% were in Catalonia and 67% were in Catalonia, the Comunidad Valenciana and the Balearic Islands combined (Fig. 93). The Little Crake is no more than scarce or rare even in these regions: for example, only seven records were accepted in the Balearics during 1997–2010 (*A. O. Balears 2010*: 79).

Most records are of migrants travelling to and from their winter quarters in sub-Saharan Africa, although there are some observations from all months (Fig. 94). The majority (81%) appear on spring passage, during late February–May, especially in March and early April. The prolonged return passage extends from August to December, with most in October–November. Winter records, from mid-December to mid-February, are very rare: de Juana (2003) found only four, so wintering in Iberia must be seen as exceptional. A further four individuals were recorded during the 2007–10 Spanish winter atlas survey period, all at Levant coastal wetlands (SWA).

BAILLON'S CRAKE *Porzana pusilla*

(Polluela Chica, Franga-d'água-pequena)

***P. p. intermedia*. A very scarce and local breeding species in mainland Spain. It bred formerly in Portugal but seems no longer to do so. Chiefly recorded on passage. Very rare in winter. A very rare migrant in the Balearics.**

Baillon's Crake is a very scarce summer visitor to Iberia, wintering in sub-Saharan Africa. Since the mid 20th century it has been recorded nesting at a limited number of locations in Spain: in Castilla y León, Castilla-La Mancha, southwestern Andalucía and the Levant coastal wetlands. There are also records of probable or confirmed nesting from Galicia; Asturias: confirmed at the Ría de Villaviciosa during 1984–86 (Álvarez-Balbuena *et al.* 2000); the Basque Country and the Ebro valley but any breeding presence in these regions is probably irregular (AI1, LR, *Ardeola* 60: 521).

It is evidently commonest and most regularly observed in and around Doñana but the nesting population there has probably not been large in recent years. Many clearly nested in the Marismas in the past since Noble (1902) found 11 nests with eggs, as well as many others that had been trampled or destroyed by predators, in a short period in May 1901. Valverde (1960) considered that thousands bred there when water levels were favourable, but that few did so in dry years. The preferred nesting habitat was the flooded *Salicornia* marshes, where nests were typically within *Salicornia* clumps and under a canopy of *Scirpus* stems that the birds constructed: nests were often near Whiskered Tern colonies and may have benefited from the ability of the terns to drive away raptors. A record of two females and two downy young that were captured on ricefields at Isla Mayor, Sevilla, on 11 October 1998 (*A. O. Sevilla 2000–2005*: 76) indicates occasional late breeding. Breeding in the Guadalquivir ricefields appears to occur regularly, nesting starting in June when the rice crop has grown tall enough (Muñoz Fernández 2008). The Spanish population in the early 1970s was estimated at 3,000–5,000 pairs by Glutz von Blotzheim *et al.* (1973), probably on the basis of Valverde's observations at Doñana. This may have been little more than a guess. The recent Spanish population may not exceed 100 pairs in any one year.

Figure 95. Records by month, 1970–2009 (n = 146).

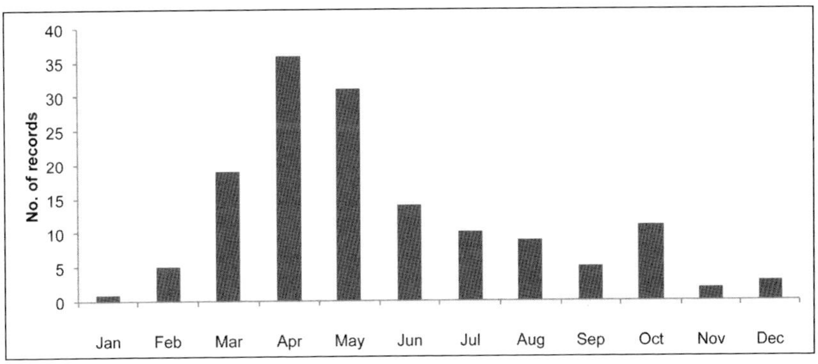

Baillon's Crake also nested in Portugal formerly and was regarded as locally common there during the late 19th century (AP): Tait (1887) found two nests on the coast north of Aveiro in June 1882. However, it declined to apparent extinction during the 20th century (Catry 1999) and is now very seldom recorded there at all. It was only reported at two sites during the fieldwork for the first Portuguese breeding atlas, and at only one site in the 2008 atlas (PA2), and systematic searches of apparently suitable sites during 1987–99 proved fruitless. However, there were at least three records in March–April 2012 (RBP): at Quinta do Lago, Faro; at the Alverca saltpans, Lisboa, and at Amoreira, Leiria. Only rare passage migrants are reported in the Balearic Islands, where just three records were accepted during 1991–2010 (*A. O. Balears* 2010: 80).

Baillon's Crake is extremely rare as a breeding species anywhere else in western Europe, including France: where it was once relatively common but where it does not now breed every year (OF, Dubois *et al.* 2012), so no significant passage would be expected across the Peninsula. Migrants, presumably of Iberian origin, are not infrequently seen at southern and eastern wetlands where they have seldom been known to breed. Iberian observations are generally between late February and early October and midwinter records are rare (Fig. 95). Spring records are relatively numerous but thereafter any breeding pairs are less easily detected as aquatic vegetation thickens and the birds become less vocal as the season progresses. A bird at Cartagena, Murcia, on 12 December 2008 was one of only three recorded during the 2007–10 Spanish winter atlas survey (SWA). Wintering birds may have been considerably less rare in the past, when Baillon's Crake seems to have been more abundant overall in Iberia, since Irby (1895) considered it frequent in some Cádiz wetlands during October–February.

STRIPED CRAKE *Porzana marginalis*

(Polluela Culirroja, Franga-d'água-raiada)

Vagrant to Spain.

A female was found within the village of Las Navas de la Concepción, Sevilla, on 12 December 2010. This intra-African rains migrant has been found in Algeria in January 1867, in Libya in February 1970 and in Morocco in November 1989 (Heim de Balsac & Mayaud 1962, BWP). There are also records for Italy: a female at Livorno in January 1997 (*Riv. ital. Orn.* 67: 190–191) and for Malta: birds collected in March 1991 and April 2004 (Haas 2012). The winter dates of the records recall the occurrence pattern of Allen's Gallinule.

CORN CRAKE *Crex crex*

(Guión de Codornices, Codornizão)

A scarce passage migrant and very infrequent winterer in Spain. Also recorded on passage in the Balearic Islands. Seldom reported in Portugal.

The Corn Crake is not easily encountered in Iberia. It occurs as a migrant, travelling between its remaining breeding sites in northwestern Europe and its sub-Saharan winter quarters, although a few may winter around the Mediterranean basin, including Spain. It is seldom detected in the field since it frequents dense herbaceous cover. A high proportion of records are of birds flushed, and often shot, by quail hunters, with dogs. The evocative Spanish name means 'Quail guide' and was given to the bird by hunters who associated it with an influx of quails, of which this 'superquail' might appear to be the leader. Other recent records are often of dead individuals that have flown into buildings and other structures during their nocturnal movements. Claims of a small breeding population in the cereal fields of Lleida (Chacón *et al.* 1985a) have not survived close examination and are discounted by E. de Juana (SA1).

Figure 96. Records by month in Spanish bird reports, 1981–2008 (n = 61 records, 68 birds).

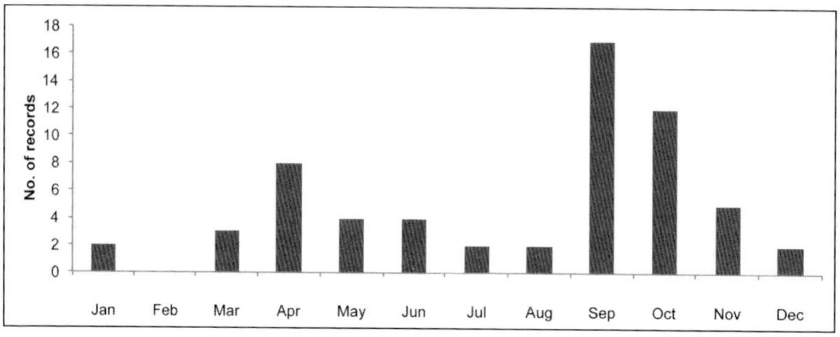

The European populations declined drastically during the 20th century, chiefly due to agricultural intensification (BE). The species was relatively frequently recorded in Spain during the 19th century: for example, 35 were hunted in the Columbretes Islands, Castellón, between 16 April and 2 May 1886 (Brú 1913). Arévalo (1887) described it as common on passage in Valencia during April and again in September–October, and stated that it wintered around coastal wetlands in Málaga. One ringed in Scotland in June 1932 was recovered in Girona on 2 September the same year (Gil Lletget 1945). It was also well known on passage in Portugal during the 19th century but was already becoming scarce there during the early 20th century (AP). There are comparatively few recent Iberian records, principally from southern and eastern Spain, where passage occurs in March–April and late August–mid November, peaking in September (Fig. 96). Corn Crake records have been considered by the Spanish Rarities Committee since 2006; just seven records were accepted up to 2010. There are only seven recent records from Portugal during 1985–1998: single birds in April and May and five individuals between 15 September and 19 November (AP). Subsequently, one was found dead at Benavente, Santarém, on 9 November 2008.

The circumstances of the observations that comprise Figure 96 confirm that many birds are only discovered when they come to grief. Of the 68 individuals involved, at least 12 were killed by hunters in autumn, another 12 were found dead or injured, one was caught by a dog and three were identified from remains within Eleonora's Falcon nests on the Columbretes islands. The prominence of the autumn peak may therefore be largely due to its coincidence with the autumn hunting season.

COMMON MOORHEN *Gallinula chloropus*

(Gallineta Común, Galinha-d'água-comum)

***G. c. chloropus*. A common resident on the Peninsula and in the Balearic Islands. Iberian birds are joined by some extra-regional migrants in winter.**

The Moorhen inhabits freshwater habitats throughout the region, generally below 1,200m (SA2). It may be found at the margins of reservoirs and lakes; in ponds and ricefields and along rivers, canals and irrigation ditches, always where some dense marginal or emergent plant cover is available. Birds often feed on wet grasslands and other open areas near such wetlands. In Doñana during winter they feed mostly on spilt rice, reed seeds, algae and other plants (Amat & Soriguer 1984). The Moorhen is thus least abundant in the arid south-east and most abundant in irrigated areas, especially where there are extensive ricefields: as in the Ebro delta, the Albufera de Valencia, the periphery of Doñana and the Vegas del Guadiana in Extremadura. In this last region 1,325 individuals were counted feeding on duckweed along a 1.5km stretch of the Guadiana river at Medellín on 3 January 1995 (*Ardeola* 43: 247).

The Spanish breeding population was estimated at 90,000–120,000 pairs by Purroy *et al.* (1997). A national sample census during the breeding season in 2007 (Palomino & Molina 2009) suggested a total population size of 116,800 individuals: 38% of them in Andalucía, 16% in the Duero basin, 16% in the Guadiana basin, 9% in the Comunidad Valenciana, Murcia and eastern Castilla-La Mancha, 8% in the Ebro basin: including all of Catalonia, 7% in the Tagus basin, 3% in Galicia and the Biscay coastlands, and 2% in the Balearic Islands. This may be an underestimate, however, since regional populations have been put at over 12,000 pairs in Castilla y León (Sanz-Zuasti & Velasco 1999), over 15,000 pairs in Sevilla (Chiclana *et al.* 2002) and 12,784–30,772 pairs in Catalonia (CA2). The Spanish population has been thought to be increasing (SA2) but the available information is perhaps insufficient to establish any definite trends (Palomino & Molina 2009). The Portuguese population has been estimated to be within the range of 5,000–50,000 pairs (BE 2004), with some recent increase and range expansion in the north-east (PA2).

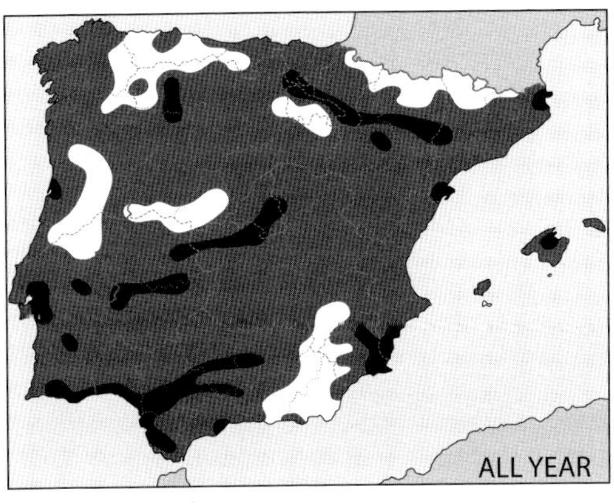

Iberian Moorhens are thought to be largely resident, although local movements occur in response to seasonal desiccation and reflooding of ephemeral wetlands. The longest recorded displacements involved two full-grown birds that travelled 155km from Jaén to Sevilla. An unknown number of birds from northern and central Europe enter Iberia to winter and some continue to North Africa. The 32 Iberian recoveries of foreign-ringed individuals comprise a few ringed mainly on passage or in winter in France, Belgium, the Netherlands, Germany, Poland and Denmark. Most recoveries are in coastal areas during November–January (AMI).

LESSER MOORHEN *Gallinula angulata*

(Gallineta Chica, Galinha-d'água-pequena)

Vagrant to Spain.

A female was picked up at Algeciras, Cádiz, on 10 March 2003. It was found near the harbour in poor condition and soon died. The possibility of an assisted shipborne passage arises but, conversely, several aquatic Afrotropical species have been found to reach Iberia, North Africa and the Canary Islands. The species is widespread in sub-Saharan Africa, where many individuals are rains migrants. A male collected in Madeira on 26 January 1895 had previously been identified as a Black Crake *Amaurornis flavirostris* (Matias 2009).

ALLEN'S GALLINULE *Porphyrio alleni*

(Calamoncillo Africano, Camão-pequeno)

Vagrant to Spain, Portugal and Gibraltar.

Iberia provided the first known occurrence of Allen's Gallinule in Europe when one was obtained in Murcia in 1854 (Guirao 1859). Six further records followed during the 1970s: at Monforte, Portalegre, in May 1973; in Murcia in December 1975; in Málaga in December 1975; at Santa Fe, Granada, in January 1976; in Málaga in December 1977, and at La Puebla del Río, Sevilla, in January 1978 (Calderón 1976, Amat 1979b, Garrido *et al.* 1985, AP). Between 1990 and 2010 there were six accepted records from Spain, two from Portugal and one from Gibraltar. There are subsequent records of one found dead at Palma airport, Mallorca, on 1 December 2011 (*Ardeola* 60: 460) and a first-winter seen at Motril, Granada, on 25 January 2012 (RBS). Three-quarters were in December–January (Fig. 97), a pattern also noted elsewhere in Europe (Hudson 1974, Yésou 1991), where there are records from as far north as Britain, Denmark and Finland. These months correspond with the height of the dry season in sub-Saharan Africa north of the equator, when many wetlands dry up and aquatic species have to find suitable habitat elsewhere. Iberian records have mainly been in the south (Fig. 98). The age of ten individuals was reported: eight were first-years and two were adults.

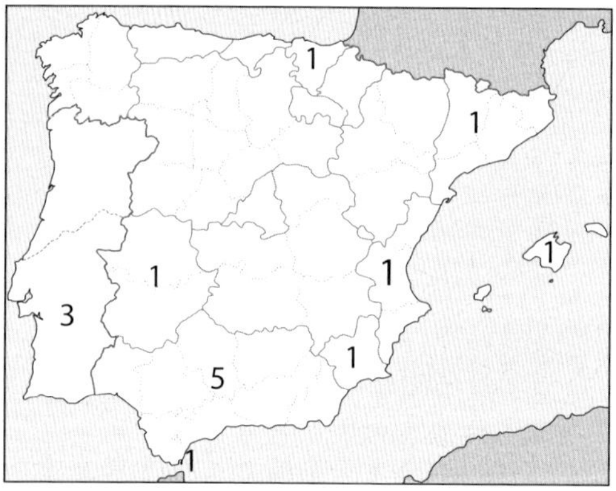

Figure 98. Records by region, 1973–2010 (n = 15)

Figure 97. Records by month, 1973–2010 (n = 15)

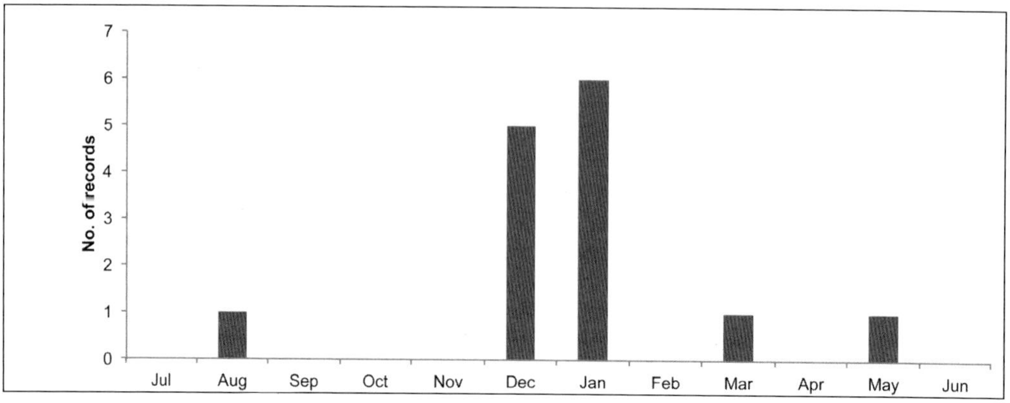

AMERICAN PURPLE GALLINULE *Porphyrio martinica*

(Calamoncillo Americano, Camão-americano)

A moribund first-winter individual that was found in Monsanto Park, Lisbon, on 7 November 2013 is the sole Iberian record (RBP).

PURPLE SWAMPHEN *Porphyrio porphyrio*

(Calamón Común, Camão-comum)

***P. p. porphyrio*. Resident and locally common in southern, central and eastern Iberia and in the Balearic Islands.**

The fortunes of the Purple Swamphen in Iberia have varied greatly during the past 150 years but it has enjoyed a widespread population increase and range expansion during recent decades, although with sometimes considerable numerical fluctuations at particular sites. The Swamphen was probably still widespread and locally common in western, southern and eastern Iberia during the early 19th century but it was reported as almost extinct in Valencia and declining in Mallorca by Saunders (1871). It was known from the Ebro delta in the 16th century (Maluquer Sostres 1971). Arévalo (1887) noted that it had been persecuted to extinction in Valencia but that it lingered on in Murcia, Granada and Málaga. It may actually have survived at the Albufera de Valencia until about 1916 (Maluquer Sostres 1971) and it is said to have been present on the Castellón coast, between Oropesa and Torreblanca until 1930–31 (von Jordans, 1933).

It declined to extinction in the Balearic Islands and most mainland sites by the mid 20th century. It was then restricted to just Doñana: where it was scarce and very local, and a small colony on the Guadalquivir river upstream of Sevilla, with a foothold presence in Portugal at the Ria Formosa, Faro (Valverde 1960, Bernis 1966c, Sánchez-Lafuente *et al.* 1992, Pacheco & McGregor 2004, AP). Occasional records from outside these areas pointed to its dispersive capacity: it was reported at the Ebro delta in 1963 (Westernhagen & Pons 1966) and rare individuals were reported from the Balearic Islands; in Menorca in 1941 (Moll 1957), in Ibiza in 1961 (Palaus 1962) and, more recently, in Formentera (Wijk 1989).

Numbers remained at a low ebb until the early 1980s, when a recovery in its core area led to recolonisation of many former haunts. The recovery was assisted by local reintroductions: at the Aiguamolls de l'Empordà in 1989, at the Albufera de Mallorca in 1991 and at the Albufera de Valencia in 1995 (Clavell 2002). For example, 28 were released in the Albufera de Mallorca in 1991 and there were *c.* 200 pairs there by 1999 (*Ardeola* 49: 184). Colonisation of other sites in Mallorca, and also in Menorca, stem from this introduction. Recolonisation of some other sites in eastern Spain, such as the Ebro delta, is thought to have been largely natural (Clavell 2002).

The upturn in its fortunes followed a great reduction in hunting pressure – it is now a protected species in Iberia – and the proliferation of new artificial wetlands, including reservoirs and ricefields, that proved to its liking (Sánchez-Lafuente 2010). It may also be that the severe droughts in southern Spain in the late 1970s and early 1980s provoked dispersion from Doñana that led birds to become re-established further afield. Natural range expansion in Spain continued during the 1990s and post-2000, with colonisation or recolonisation of wetlands in Huelva, Cádiz, Málaga, Almería and Granada; the Guadalquivir valley east to the Jaén reservoirs, and, in the east, in coastal Murcia, the Comunidad Valenciana and Catalonia. It also became established in Madrid; Castilla-La Mancha in Toledo, Ciudad Real and Guadalajara; Extremadura: at the Arrocampo-Almaraz reservoir in Cáceres and along the Guadiana river in Badajoz, and at the Laguna de Sariñena, Huesca. By 2010 this species was a prominent feature at more Iberian wetlands than at any time in the past century, and there is considerable promise of further range expansion.

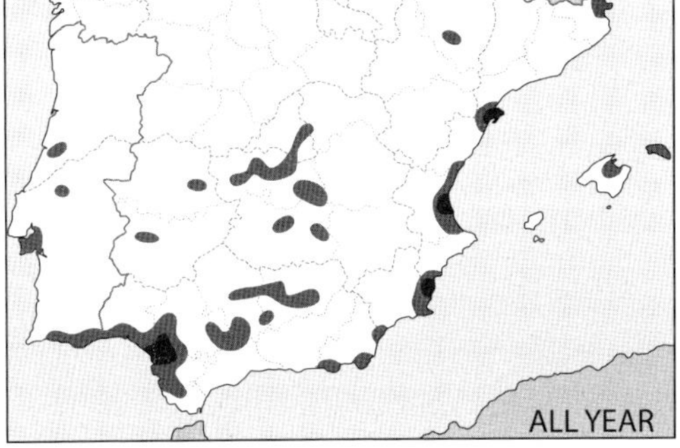

The Spanish population had reached an estimated 3,000–5,000 pairs by the mid 1990s (SA1) and 6,400–6,700 pairs by the end of the 20th century, these including some 5,000 pairs at Doñana (SA2). However, a national sample census during the breeding season in 2007 found only an estimated 5,500 individuals, with a notable decline at Doñana. That decline followed a severe drought in Andalucía in 2005 and poor breeding success locally thereafter. Illegal persecution by rice farmers is also known to be a factor in the decline. The population in Sevilla declined by some 50% during 2003–06 (Muñoz-Fernández 2008). Thus far fewer Swamphens were present in 2007 than had been expected (Palomino & Molina 2009).

There is considerable evidence of the long-standing presence and former abundance of the Swamphen in Portugal. Indeed, it figures in 2nd century Roman mosaics there and a fixed market price for it was listed in 1253 (AP). It was well distributed and locally common in western and southern Portugal during the late 19th and early 20th centuries. Tait (1887) knew it as common in the Mondego valley. However, it declined progressively thereafter, perhaps due to uncontrolled hunting (AP). The Portuguese population was only 5–10 pairs in the early 1990s but increased to over 100 pairs by 2005 and is continuing to do so (Pacheco & McGregor 2004, PA2). The Swamphen has recovered most of its former Portuguese range and spread to new sites, occurring chiefly at coastal wetlands and along the lower reaches of the major rivers, from Algarve to Baixo Mondego, with an isolated, introduced population at Elvas, Alentejo (PA2). Further north, a female with two young was seen at Águeda, Aveiro, in June 2008 (*Anu. Ornitol.* 7: 79).

The Swamphen occurs at freshwater wetlands with dense emergent vegetation, especially stands of reedmace, whose rhizomes are a favoured food. It occurs at lakes and ponds, along slow-flowing canals and rivers, in reedy drainage ditches, in reedbeds on reservoir margins and in marshes. Many feed in ricefields, locally causing damage that has led to illegal persecution in some areas. Particularly large concentrations have been seen at the Guadalquivir ricefields in Sevilla in winter, where birds often forage on rice stubble: for example there were *c.* 2,000 at the Brazo del Este on 24 December 2003 (*Ardeola* 51: 249).

The Iberian populations are essentially resident but their considerable dispersive ability brings birds to new locations which may sometimes be colonised. For example, a series of observations at the Llobregat delta culminated in nesting there in 2001 (Clavell 2002). Wandering birds have been recorded in Ibiza and Formentera, Navarra, Galicia, Asturias and Gibraltar (Clavell 2002, *A. O. Navarra 1997*: 117, *Ardeola* 37: 337, L. Carrera, pers. comm., *GBR* 2004: 22). A ringed immature travelled 253km from Sevilla to Cáceres but clearly those found in Galicia and Asturias must have travelled perhaps twice as far. Two birds ringed in Girona have also been found some 120km to the north in Hérault, France. A number of observations in France far beyond the known breeding sites, as far as north the English Channel, also confirm the dispersive tendency of some individuals (OF).

COMMON COOT *Fulica atra*

(Focha Común, Galeirão-comum)

***F. a. atra*. A widespread and locally common resident in mainland Iberia. It also nests in Mallorca and Menorca. Migrants from northern Europe occur on passage and in winter.**

The Common Coot is widely distributed across the region. Nesting birds require sheltered areas of open water, with a belt of emergent vegetation for nesting and, especially, quantities of submerged macrophytes that provide food, both directly and through supporting aquatic invertebrates that are also consumed. They find such conditions at a diversity of freshwater and brackish lakes, reservoirs, riparian backwaters, gravel pits, ornamental ponds, marismas and saltpans, generally below 1,000m but occasionally higher: to 1,200m at the Lago de Covadonga, Asturias, in the Cantabrian Range, for example (SA1).

The Spanish population was estimated at 12,300–18,700 pairs in the mid 1990s (SA1) and over 20,000 pairs by SA2 (2003). A national sample census during the breeding season in 2007 (Palomino & Molina 2009) estimated a population size of 112,500 individuals. The methodology employed in these assessments is different and the results are not necessarily indicative of a real population increase. Nevertheless, it is clear that it is a relatively abundant breeding bird with a fluctuating but probably increasing population.

The Ebro basin held nearly 30% of the Spanish breeding population in 2007 including 6,000 birds at the Ebro delta. In Andalucía there were 4,000 at Doñana, although in 2007 the water levels there were lower than usual; in good years the breeding population may be substantially higher: some 10,000–15,000 pairs (García *et al.* 2000). Indeed, professional collectors once harvested hundreds of thousands of coot eggs in the Marismas annually (Valverde 1960). El Hondo reservoir, in the Comunidad Valenciana, held 800 birds, but there were 2,800 there in 1984 (Ferrer *et al.* 1986). In the Tagus basin, the El Taray and Dehesa de Monreal lagoons, both in Toledo, held 1,300–1,700 birds each. In the Balearic Islands, the Albufera de Mallorca was the principal site with 620 birds in 2007. There is insufficient data to determine any trends in the breeding population but local fluctuations occur, especially when birds abandon Doñana and other wetlands during drought years (Palomino & Molina 2009). The Albufera de Valencia used to be another important site, with over 1,000 breeding pairs, before heavy eutrophication resulted in the disappearance of submerged macrophytes (Urios *et al.* 1991). The Portuguese population has not been censused but has been estimated at 200–2,000 pairs (Birdlife International 2004). The lower figure is certainly far too small and the population is substantial, and has increased in recent years (PA2).

Many Iberian Coots may be largely resident but those that inhabit seasonal wetlands may disperse over considerable distances. Some very large post-breeding concentrations have been reported: for example there were *c.* 20,000 at the Laguna

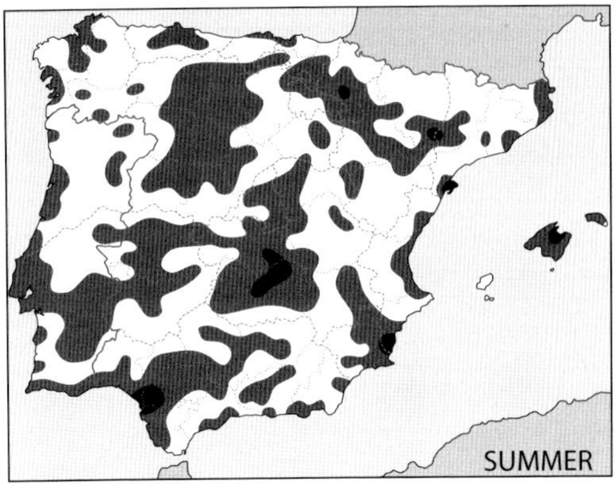

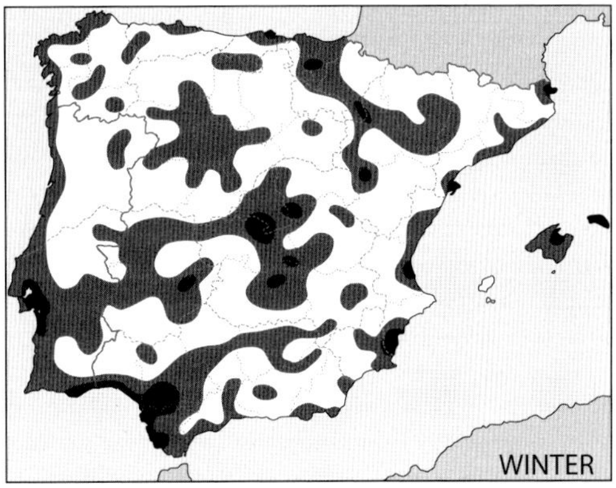

de Medina, Cádiz, on 24 August 1991 and at least 28,000 there on 17 September 2003 (*Ardeola* 38: 337, 50: 346). These very probably originated from Doñana which often dries up in summer. The mean distance travelled within the Peninsula by 42 birds ringed as pulli was 211km (AI1) but three juveniles ringed at Doñana in June–July have been recovered surprisingly far away: in the Shetland Islands, Scotland (2,590km); at Hortobagy-Halasto, Hungary (2,539km), and at Doneck, Ukraine (3,700km) (*Ardeola* 29: 74; *Ecología* 5: 296, 7: 325). Out of a further 94 recoveries of birds ringed in Spain as pulli or as juveniles, most of them at Doñana, 72 were found in Morocco, 14 in France, two each in Algeria, Belgium and Italy, and one each in The Netherlands, Denmark, Germany and Poland. Older ringed birds, also mostly from Doñana, have produced 67 recoveries in Morocco, 11 in France, two in Germany and one each in Denmark and The Netherlands. The preponderance of ringing recoveries of Spanish birds in Morocco is due to the large-scale exodus that occurs from the Marismas del Guadalquivir, including Doñana, in dry years, when some of the sizable breeding population moves to permanent wetlands on the Moroccan Atlantic coast, notably the Merja Zerga (BM).

Some Common Coots from central and northern Europe enter Iberia to winter but evidently very few cross the Peninsula to North Africa. Birds ringed abroad as pulli and recovered in Iberia up to 1995 comprised five from The Netherlands, four from the Baltic countries, two each from France and Sweden, and one each from Denmark, Poland and the Czech Republic (AMI, AI1). The total number of Coots of foreign origin of all ages recovered in Spain by 2010 is 150. These are very low totals given the importance of the Common Coot as a quarry species in some parts of Spain and the implication is that the winter influx is not on a large scale. Wintering foreign birds in Spain are present during September–April, 96% of recoveries being in October–March (AMI, n = 51). They are most numerous by far in coastal regions of the east and north-east: the coastlands between Girona and Alicante, together with the Balearic Islands, accounted for 72 of 75 recoveries of foreign-ringed birds found in Spain, with just single recoveries in Extremadura, Andalucía and Portugal (AMI, AI1). Recoveries in Portugal comprise 67 birds ringed in Spain and single birds from Britain and The Netherlands (AP).

The mean wintering population in 1978–89 in Spain was some 64,000 birds and it averaged 4,700 birds in Portugal during 1987–89 (AI1). The Portuguese wintering population increased to an average of 7,700 during 1992–1995 and 14,200 during 2005–07 (AP). This followed a progressive decline in Portugal during the 20th century due to excessive hunting. Numbers in Spain have also tended to increase considerably in most regions. During 1990–2009 the Spanish January population averaged 86,065 birds (range 57,402 –126,806 birds; AAE), but many winterers at small wetlands would have been missed. The Spanish Winter Atlas 2007–10 estimated 76,000–140,000 individuals, with some 30,000 at the Ebro delta and 28,000 at Doñana (SWA). Increases in wintering numbers may be partly due to the increased availability of suitable habitat in the form of reservoirs.

The numbers present in winter at particular sites may vary considerably between years and some very large counts are sometimes made. For example, in January 1989 there were 54,760 birds in Doñana, 17,097 in the Ebro Delta and 15,391 at the Laguna de Gallocanta (Troya & Bernués 1990). The Ebro delta has held up to 30,960 birds in January 1990–2009, with an average count there of 18,362 birds: nearly 17% of the January population in Spain (AAE). The wintering numbers at both the Ebro delta and the Albufera de Valencia fell sharply from the early 1970s to the 1980s due to eutrophication (Ferrer *et al.* 1986). There has since been a recovery at the delta, but not at the Albufera de Valencia: which is still heavily contaminated by fertiliser run-off from the adjacent ricefields as well as urban sewage (Giménez & Benavente 2005). The Albufera was formerly a Common Coot stronghold and during the 19th century 'immense' numbers were hunted there in winter and sold in the markets (Arévalo 1887) Many thousands were still being hunted there every year during the early decades of the 20th century, especially in the hunting reserves of Cullera and Silla, but a sharp decline then followed, probably due to overexploitation (Bernis 1964). In Portugal the relatively new Alqueva reservoir attracts large numbers and the Lagoa de Santo André has recently seen concentrations that are noteworthy for the country, including 6,084 in early October 1991 and 21,000 birds on 11 December 2007 (AP).

AMERICAN COOT *Fulica americana*

(Focha Americana, Galeirão-americano)

Vagrant to Spain and Portugal.

The first was in Portugal at Ludo, Faro, on 17 September 1992 (*Ardeola* 41: 108). Single birds were at Txingudi, Gipuzkoa, during 15–30 November 1999; at Sanxenxo, Pontevedra, from 5 January until March 2003, and at Valdoviño, A Coruña, from 19 January to 11 April 2009. All four were at Atlantic or Biscay coastal locations and at least two were first-years. The Iberian dates fit the wider European pattern, where records are during September–April with spring and autumn peaks that correspond with migration (de Juana 2006). The few European records include six from Britain, two from Ireland and 15 from the Azores.

RED-KNOBBED COOT *Fulica cristata*

(Focha Cornuda, Galeirão-de-crista)

A rare breeding species in southwestern Spain and recently introduced to the Levant coast and Balearic Islands. Vagrant to Portugal. Resident but dispersive.

The Palearctic range in the 19th century included parts of Andalucía, La Mancha, Aragón, Catalonia and the Levant in Spain, western and southern Portugal and also the Maghreb countries, with vagrants straying as far as southern France, Sardinia, Sicily, Italy and Malta (BWP, LR, AP). The principal Iberian location was Doñana, where Valverde (1960) considered that one in ten breeding coot pairs may have been this species. It began to decline during the first half of the 20th century, disappearing from Portugal and from central and eastern Spain, doubtless in part due to habitat loss, when some key sites were drained, as well as excessive hunting: especially in Valencia (Bernis 1964). By the 1980s it was confined to Doñana and the Cádiz lagoons and it was scarcely observed at all during the early years of that decade, when water levels at its preferred nesting sites were exceptionally low (LR). Since then the population has increased slowly but the numbers present and, especially, nesting in any one year fluctuate considerably. In general the birds breed only in exceptionally wet years (SA2).

In Portugal the Red-knobbed Coot was reported during the 19th century and early 20th century at a diversity of coastal locations including, from north to south, the Porto area, the Barrinha de Esmoriz, the Ria de Aveiro, Baixo Mondego (where nesting was confirmed), Ribatejo, Alentejo and Algarve (Catry 1999). Many from the Ria de Aveiro were offered for sale in Porto in October and November 1882 (Tait 1887). There have been very few Portuguese records since the mid 20th century: only five during 1987–1996 (AP). Occurrences have become more frequent in very recent years and there are at least 29 records from 2007–12: most of single birds but including a group of up to five in the Tagus estuary at the Ribeira das Enguias, Alcochete, during February–March 2012 (CPR 2008–09, RBP). Two Spanish-ringed birds have been found in Portugal: one ringed at Sevilla in February 1993 was at Bragança (457km) in October 1994 and a first-year bird in Córdoba in March 2002 was recovered in Coimbra (492km) in September 2003. The fieldwork for the second Portuguese breeding atlas found two birds, possibly paired, at each of two sites in south-central Portugal and a single bird in Algarve: raising the possibility that occasional nesting in the country now occurs (PA2). One of the individuals originated from a Spanish release scheme.

The ecological requirements of the Red-knobbed Coot in Europe have been little studied. However, its relationship with the very similar Common Coot has received some attention (Varo 2007, Varo & Amat 2008). These two species very often, if not always, share the same wetlands and are likely to be potential competitors. Studies in Spain suggest that although both favour calm freshwater wetlands with plenty of both emergent and submerged vegetation, the Red-knobbed Coot is less given to feeding by diving in open water and instead feeds largely on submerged macrophytes by pecking from the surface at the more nutritious apices of the plants (Fernández Palacios & Raya 1991). High habitat quality appears to be relatively more important to the Red-knobbed Coot than to the Common

Coot: at shared sites in Spain Red-knobbed Coots need to spend more time foraging than the Common Coots to meet their daily energy requirements (Varo & Amat 2008). This latter study also found that Red-knobbed Coots in Spain spent more time foraging than those in Morocco, which suggests that the Spanish sites are of lower average quality for this species.

The North African population is now confined to Morocco where it is rapidly declining and estimated at only 200–300 pairs, although mid-winter counts in the 1980s and 1990s included gatherings of up to 3,000 (BM). Movement between Iberia and Morocco would help to account for sudden increases in numbers in Spain in some wet years. However, the small size of the Moroccan population now makes such potential interchanges less likely to be significant.

The Spanish population was very small indeed during the 1980s and 1990s. No more than ten birds were seen, at eight lakes in western Andalucía, during 1985–1989 and breeding was only proved at three sites. There was no evidence of any breeding at all in Doñana in the early 1990s, when the species was apparently confined to the Cádiz lagoons and numbered 10–25 pairs. There were no nesting records even there during 1992–1995, when drought conditions returned, but 5–10 pairs nested from 1996 onwards, at the Laguna de Ratosa, Málaga, and the Lagunas de Espera, Cádiz (Ballesteros *et al.* 2008). A pair attempted to breed away from Andalucía at El Hondo, Alicante, in 1993: perhaps involving birds displaced from Andalucía.

The Spanish population has been repeatedly reinforced by the release of captive-bred birds. A captive-breeding programme was initiated at the Cañada de los Pájaros, Sevilla, in 1992, and released 1,000 individuals during 1992–2013. Initial releases were at the Cañada de los Pájaros itself, as well as at Doñana, the Lagunas de Espera, Odiel, and the Laguna del Portil, Huelva: all within the then-existing breeding range. Subsequent releases established birds further east in Andalucía: at the Balsa de Villargordo in Jaén, and in Almería (Ortega *et al.* 2007).

A second breeding programme released 662 individuals in eastern Iberia, where the species was long extinct, during 1999–2007. Initial releases were at the Albufera de Valencia followed by other coastal wetlands in Valencia and Alicante and then at the Llobregat Delta, Barcelona, in 2003 and in the Balearic Islands at the Albufera de Mallorca from 2004 and the Albufereta de Mallorca in 2007. Released birds everywhere generally carried white neck collars, which made them unusually easy to identify. Eleven marked individuals released in the Levant were seen in Doñana during 2002–07, indicating that there may be significant interchange between the Andalucían and eastern sub-populations (Ballesteros *et al.* 2008). Birds ringed in Sevilla have been recovered as far afield as Almería (317km) and Tarragona (625km), as well as in Portugal.

The Red-knobbed Coot breeding population has continued to fluctuate, despite these releases, but has also generally increased (Fig. 99). The increases have occurred in years when water levels were favourable, as in 2001–04 and 2007, and the declines in dry years, as in 2005 and 2006.

Figure 99. Fluctuations in the Red-knobbed Coot breeding population in Spain, 1990–2007 (Ballesteros *et al.* 2008)

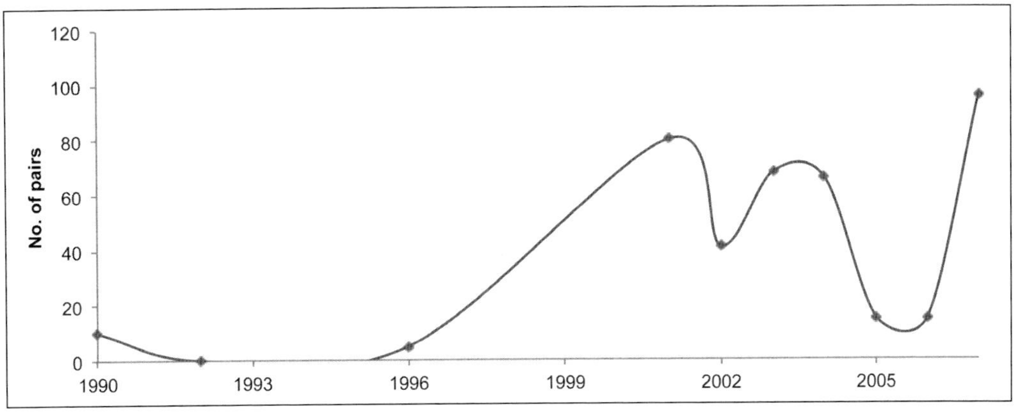

The species was closely monitored in 2007 and the distribution of the breeding, non-breeding and wintering populations throughout that year is known in considerable detail (Ballesteros *et al.* 2008). Breeding pairs totalled 96, 80% of them in Andalucía and the remainder – of captive origin - in the Comunidad Valenciana, Catalonia and Mallorca (Table 26). This was the largest total breeding population since at least the mid 20th century but, in view of the large number of captive birds released since 1992 it is still relatively modest and suggests that environmental factors, such as habitat availability, are limiting any real upturn in numbers. Many traditional nesting sites in Andalucía apparently had no breeding birds at all in 2007.

Table 26. Distribution of the Red-knobbed Coot breeding population in Spain in 2007 (Ballesteros *et al.* 2008)

PROVINCE/REGION	Sites	No. of pairs	No. of chicks
Almería	Graveras de Cerrillos	1	3
Cádiz	Laguna Salada de Zorrilla	1	3
Huelva/Sevilla	Doñana	64	2
Huelva	Marismas del Odiel	1	7
Jaén	Balsa de Villargordo, Laguna Grande	9	60
Sevilla	Laguna Capellanías Grande	1	?
ANDALUCÍA		77	75+
CATALONIA	Ebro delta	5	?
Castellón	Marjal de Almenara	1	2
Valencia	Albufera, Marjales de Pego-Oliva & Xeresa-Xeraco	8	24
COM. VALENCIANA		9	26
BALEARIC IS.	Albufera de Mallorca	5	13
TOTAL		96	114+

The Red-knobbed Coot was more widely distributed outside the breeding season in 2007 (Table 27) when it was recorded at various times at 33 wetlands in Andalucía: in every province except Granada; at eight wetlands in the Comunidad Valenciana spread between all three provinces, at the Ebro delta in Catalonia and at two sites in Mallorca. Similarly, the January waterfowl counts in Spain during 1990–2009 found birds at 55 sites during that period, with an average annual count of 58 individuals (range zero to 237 ; AAE): 97% of records came from the breeding range in Andalucía, the Levant and the Balearic islands. The relatively low total in April 2007 is attributed to the greater difficulty of locating birds when they are nesting, although this is not an easy species to census at any time since separating the two coot species can be difficult or impossible at long range.

Table 27. Regional Red-knobbed Coot census results in Spain in 2007 (Ballesteros *et al.* 2008)

Region	Jan	Apr	Jun	Sep	Nov
Andalucía	159	65	104	110	80
Catalonia	1		10		
Com. Valenciana	31	33	30	7	24
Balearic Islands	8	8	18	15	12
TOTAL	199	106	162	132	116

The occasional, rare, appearances of birds outside the breeding range: in Extremadura, Castilla-La Mancha, Madrid, Aragón, Galicia and Asturias, confirm that individuals may travel long distances. An observation of up to five at the Laguna de Pétrola, Albacete, from 31 August to 31 October 2011 (*Ardeola* 58: 497) is noteworthy. Such extralimital records include a white-collared bird, probably the same returning individual, at the Laguna de Louro, A Coruña, in autumn 2003 and autumn 2004 (*A. O. Galicia 2003*: 87, *2004*: 65). The sole record for Asturias was of a bird that remained at the Laguna de Ercina in the Picos de Europa from May 1992 until June 1993 (*A. O. Asturias 1993*: 69).

The future of the Red-knobbed Coot in Iberia is far from certain. It has suffered from habitat loss, followed by a widespread decline in habitat quality. The latter chiefly involves a reduction in the availability of submerged macrophytes, as a result of such diverse influences as trampling by livestock and destruction by alien animals, notably carp and the American Signal Crayfish. The Red-knobbed Coot is still also a frequent victim of 'accidental' shooting in winter, when the Common Coot is an important quarry species in some regions (Martínez-Abraín *et al.* 2007). Hunters do not, and probably cannot, distinguish the two and hunting still accounts for a significant number of Red-knobbed Coots, a loss that their tiny population cannot afford.

SMALL BUTTONQUAIL *Turnix sylvaticus*

(Torillo Andaluz, Toirão-do-mato)

***T. s. sylvaticus*. An extremely rare resident of southwestern coastal Spain and possibly southern Portugal. Critically endangered.**

It has been many years since the old name 'Andalusian Hemipode' had any real meaning since the bird is now extremely seldom recorded and may be on the verge of disappearance from the Iberian peninsula. The Small Buttonquail also has a precarious foothold in coastal Morocco (Gutiérrez Expósito *et al.* 2011) but otherwise it is widely distributed and locally common in sub-Saharan Africa, the Indian sub-continent and much of south-east Asia. In the 19th century the European range included Sicily and a wide area of southern Portugal and southwestern Spain, and in North Africa it occurred in Algeria and Tunisia as well as Morocco. It was last recorded in Sicily around 1920 (Violani & Massa 1993). Specimens were collected in the Aveiro region in north-central coastal Portugal in the late 19th century (Tait 1887) but it is not known whether they bred there or had wandered from the south. The most recent record in Portugal seems to have been in June 1973, when two were flushed south of Lisbon (BWP.) There is also anecdotal and unconfirmed information that some survived in coastal scrub near Cape St Vincent into the 1980s (Catry 1999, AP).

Table 28. Published records of Small Buttonquails in Spain post-1950. Three possibly doubtful records are marked ?

Date	Record	Locality	Source
1958	1 caught in a quail net.	Jerez de la Frontera, Cádiz	Mountfort 1958a
16 Oct 1952	1 hunted	Huelva, marisma	Garrido 1996
9 Nov 1952	1 seen	NW of Odiel estuary	Garrido 1996
12 Jan 1953	1 hunted	Huelva, marisma	Garrido 1996
February 1955	1 hunted	La Janda, Cádiz	Trigo de Yarto 1960
1956	1 hunted	Jerez de la Frontera, Cádiz	Gutiérrez Expósito *et al.* 2011
1958	Several seen	Jerez de la Frontera, Cádiz	Gutiérrez Expósito *et al.* 2011
1958	Present	San Juan del Puerto, Huelva	Garrido 1996
1960	Clutch taken	Utrera, Sevilla	Gutiérrez Expósito *et al.* 2011
20 Oct 1978	1	Barbate, Cádiz	Montoya 2000
Early 1980s	Several hunted	El Rocío, Doñana. Skins preserved.	Garrido & Gutiérrez 2010
6 May 1986	1	Coto del Rey, Doñana	Llandres & Urdiales 1990
Autumn 1987	1	El Acebuche, Doñana	Llandres & Urdiales 1990
Spring 1988	2	El Acebuche, Doñana	Llandres & Urdiales 1990
1989	1 heard	Doñana	Gutiérrez Expósito *et al.* 2011
1992	hunted	Near Odiel estuary	Garrido 1996
5 Jun 1992	1 heard	Coto del Rey, Doñana	*Ardeola* 40: 94
April 1994	1 flushed	Cabo de Gata, Almería (?)	Gutiérrez Expósito *et al.* 2011
18 May 1995	1 heard	Chiclana, Cádiz	Montoya 2000
21 Nov 1995	1 seen	La Alcolea, Huelva	Garrido 1996
17 Apr 2002	heard	10km north of Cáceres (?)	*Ardeola* 50: 346
22 Apr 2002	1 flushed	Cabo de Gata, Almería (?)	*Ardeola* 50: 346
2000–01	Several hunted	Near Doñana and in SW Cádiz	SA2
Spring 2002	heard	Doñana	SA2

In Spain the bird was known from western Andalucía, in the provinces of Huelva, Cádiz, Málaga and Sevilla. There it chiefly inhabited coastal areas, in particular the large expanses of Dwarf Palm *'palmito'* scrub that were widespread and characteristic of wide tracts of what are now known as the Costa de La Luz and the (western) Costa del Sol. The buttonquail was considered to be a typical member of the avifauna of the hinterland of the Strait of Gibraltar by Irby (1895), Saunders (1871) found it abundant near Algeciras and not uncommon near Málaga, and Arévalo (1887) described it as an abundant resident of the Málaga coast west to Gibraltar. However, historical references to its former occurrence in Catalonia are unsubstantiated (Gordo *et al.* 2006).

The buttonquail was still locally common in southwestern Iberia at the end of the 19th century and probably also well into the 20th century (Martínez Gámez 1906). Many records were actually of birds shot by quail hunters, whose dogs put up both species and, most importantly, found them in the first place. The Small Buttonquail is otherwise notoriously difficult to flush and so is very hard to detect in the field, although the mournful hooting call of the courting female is diagnostic. Specific records in Spain since the mid 20th century are few and far between (Table 28) and indeed the species was considered to be virtually extinct there by 1966 (AMI).

The decline to near extinction of the Small Buttonquail in Iberia occurred almost unnoticed, probably because this was a bird with a restricted geographical range that inhabited little-visited habitats and was hardly ever seen except by hunters. Hunting quail is now prohibited within the buttonquail's former haunts but it is unlikely that hunting was the principal reason for its original demise. The hunting ban also means that records of its continuing presence are even less likely to occur. The chief reason for the decline is usually attributed to habitat loss (LR, AP) and indeed extensive tracts of palmito and other low scrub on coastal sandy formations were lost during the 20th century, to tourism infrastructure such as hotels, beach resorts and golf courses, and to agriculture and afforestation. Nevertheless, large expanses of apparently suitable habitat still remain and, since 2005, these have been the object of intensive campaigns to find and conserve the species, with zero success to date (Gutiérrez Expósito *et al.* 2011). It is still possible that relict populations will be detected and the species saved from local extinction, and there are suggestions that a reintroduction of captive-bred individuals could be attempted. However, the situation is perhaps beyond recovery and the Small Buttonquail may be fated to become 'extinct in Europe' very soon.

COMMON CRANE — *Grus grus*

(Grulla Común, Grou-comum)

Locally abundant in winter, chiefly in southwestern and central Iberia, and on passage across a broad swathe of central and northeastern Spain. Occasional on passage and in winter in the Balearic Islands.

The Common Crane once nested locally in Iberia, and may have been relatively widespread in suitable habitat across the region several centuries ago (Bernis 1960a). It may have nested in Portugal in the late 19th century in the lower Guadiana valley and at Pancas in the Tagus estuary (AP). In Spain nesting was reported from Doñana (Saunders 1871) and possibly the Laguna de La Nava, also in the late 19th century, and – finally – at the Laguna de La Janda, Cádiz, until 1954 (F. Bernis: AMI). La Janda once held a nesting population of 30–40 pairs (Irby 1895) but its drainage in the 1960s meant the extinction of the Crane as an Iberian breeding species.

The Peninsula has always been much more important to Cranes in winter, when a major fraction of the northern European population congregates in parts of central and southwestern Spain, and in southeastern Portugal. Migrating Cranes enter Iberia across the western half of the Pyrenees, in a concentrated passage that begins from mid October and peaks in early November. However, there is recent evidence that wintering Cranes are arriving in Spain earlier in autumn, passage across western passes having advanced by 20 days over a 30-year period, perhaps reflecting an advance in the breeding season resulting from increasing mean temperatures in northern Europe (Filippi-Codaccioni *et al.* 2011). The peak passage has been in October in more recent seasons, around 17 October on average but later in some years: 28 October at Organbidexka in 2012, for example (www.migraction.net). The combined passage at Organbidexka, Lindux and Lizarrieta in 2009–2011 averaged 80,068 birds (www.migraction.net) but it varied widely between seasons: from 21,711 in 2010 to 175,239 birds in 2011, most probably due to the effects of wind and weather on the routes taken. Since the 1980s the migrants have used the Laguna de Gallocanta as a staging post; roosting at the lake itself and feeding on the surrounding farmland. Most of them continue south-westwards across the Meseta, within a broad band of territory that extends from Gallocanta and the Laguna de El Hito, Cuenca, in the east to the Lagunas de Villafáfila, Zamora, and the Sierra de Gata, Cáceres/Salamanca in the west. However, in recent years, significant numbers have remained at Gallocanta to winter. The Guadiana valley in Extremadura holds most of the wintering population but some numbers continue into southeastern Portugal and south to western Andalucía. There is what is now a modest passage across the Strait of Gibraltar to wintering areas in Morocco.

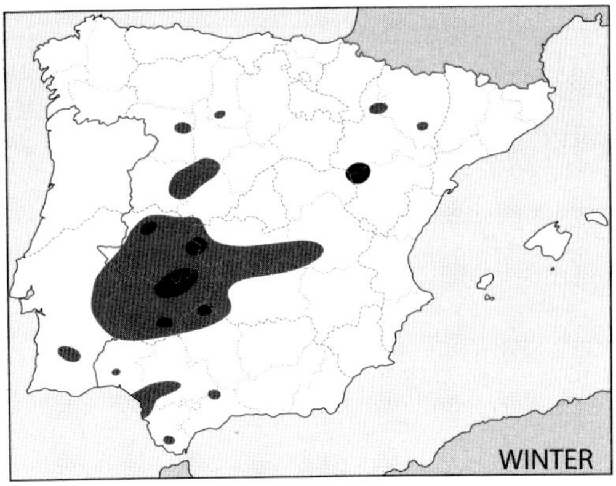

Birds begin to withdraw from the wintering areas as early as late January but passage is chiefly in late February and early March and, once again, Gallocanta is a key staging area where the flocks rest for up to a week before crossing the Pyrenees: 70,000 were counted there on 26 February 2012 (*Ardeola* 59: 180). Another very important spring staging area close to the Pyrenees is La Sotonera reservoir, Huesca, where some 50,000 birds were counted on 28 February 2012 (*Ardeola* 59: 180). The return route across the Pyrenean passes is across the west-central sector of the range, further east than in autumn.

The Spanish wintering population has been censused nationally on several occasions and there are regional counts in key areas. The most recent nationwide census was in December 2007. Cranes are censused as they return to their communal roosts in the evenings. They spend the day foraging in farmland, being especially attracted to the remnants of cereal, maize and rice crops and then to newly sown cereal fields. They also flock to the Iberian Holm Oak dehesas where they consume quantities of acorns. The diet is opportunistic and also includes a diversity of other plant material, such as the bulbs of Friar's Cowl and Sand Crocus, and some animal material in the form of snails, earthworms and arthropods (Fernández-Cruz 1981, Avilés *et al.* 2002, Bautista 2009). The vast majority of crane roosts are at wetlands, in sheltered areas alongside lakes and reservoirs, and sometimes on islands in these where available, and also at seasonal pools and in ricefields. No fewer than 91% of roosts in Extremadura during winter 2007–08 were at artificial wetlands, including reservoirs and ricefields (Prieta & del Moral 2008).

The December 2007 counts found 151,243 wintering Cranes in Spain (Prieta & del Moral 2008). More than half (53%) were in Extremadura, where the species is widespread in the dehesas but is particularly numerous in the irrigated croplands of the Guadiana valley: the 60,566 birds found in the Guadiana basin in 2007–08 make this the most important wintering area for the Common Crane in Spain and Europe (Sánchez *et al.* 1993, Prieta & del Moral 2008). The highest numbers there were in the vicinity of the village of Navalvillar de Pela: over 33,000 birds. A further 45% of the wintering population was shared between Castilla-La Mancha, Aragón and Andalucía but in these regions the species is much more locally distributed: notably at Doñana; Belalcázar, Córdoba; La Janda, Cádiz, and Fuentedepiedra, Málaga, in Andalucía; at Gallocanta in Aragón and around the wetlands of central Castilla-La Mancha. Elsewhere there is a relatively small wintering population at the wetlands of Castilla y León, chiefly in Salamanca, but only an insignificant presence in a few other autonomous regions. No wintering Cranes at all were found during the 2007 census in Galicia, Asturias, Basque Country, La Rioja, Navarra, Catalonia, Murcia or Madrid, although large numbers regularly occur on passage in Navarra and Madrid, often passing over the city itself in the latter case. Common Cranes have been recorded in all parts of mainland Iberia and in the Balearic Islands on passage and in winter but their presence outside their regular haunts is irregular and small-scale. Summering individuals are very rare but not unknown: for example, in 2010 five were found in summer in Extremadura, including three together at the Orellana reservoir, Badajoz, and three remained at Gallocanta (*Ardeola* 57: 530).

The wintering population in Portugal is largely confined to the Guadiana valley. In 1988–1992 up to 3,142 birds were found: chiefly at Campo Maior, Mourão, Évora and Castro Verde, sharing roosts with birds in Extremadura at the first two locations (Almeida 1996). Numbers here have apparently been fairly stable until very recently, with a maximum of 2,873 birds counted in winter 2000–01, but up to 6,000 were present in December 2007 (Prieta & del Moral 2008). The wintering range extended further west in Portugal, to Ribatejo, until the mid 20th century but only occasional birds now appear there, or indeed elsewhere in the country (AP).

The exact locations used by the wintering birds include many traditional sites, especially the roosts themselves and the flight paths used to reach and leave them, but food availability determines where the birds spend the day and this may change not only between years but at different stages of the same winter. Adult Cranes tend to be faithful to their winter quarters and they are accompanied there by their young of the year. Roost sizes vary greatly. The smaller concentrations are of a few hundred birds, and sometimes fewer than 100, but larger gatherings attract 1,000 and more. The largest roost of all in December 2007 was at Gallocanta, with 22,391 birds. Gallocanta has become an important wintering location as well as a staging post since the early 1980s, when just a few thousand birds – and sometimes a few hundred – spent the winter there. Wintering numbers at Gallocanta peaked at 26,000 birds during winter 2008–09 (Pelayo *et al.* 2009). The next largest roost in December 2007 was at the Casas de Hitos estate, Cáceres/Badajoz, with 11,325 birds, followed by the Rosarito reservoir, Ávila/Toledo, with 6,912 (Prieta & del Moral 2008).

Successive estimates, extrapolations from regional counts and full national censuses reveal the dramatic increase in the wintering population in Spain since the mid 20th century. The earliest, and imprecise, estimate resulted from an enquiry during winter 1957–58 by Bernis (1960a) who concluded that they numbered not fewer than 4,500 birds in Spain and 5,000 in the entire Peninsula. Four subsequent national censuses found progressively increasing numbers: at least a ten-fold increase over some 50 years (Figure 100).

Figure 100. The wintering population of the Common Crane in Spain found by national censuses in winter 1979–80, winter 1987–88, January 2004 and December 2007 (Fernández-Cruz 1981, Alonso & Alonso 1990a, de la Cruz & Montoya 2004, Prieta & del Moral 2008)

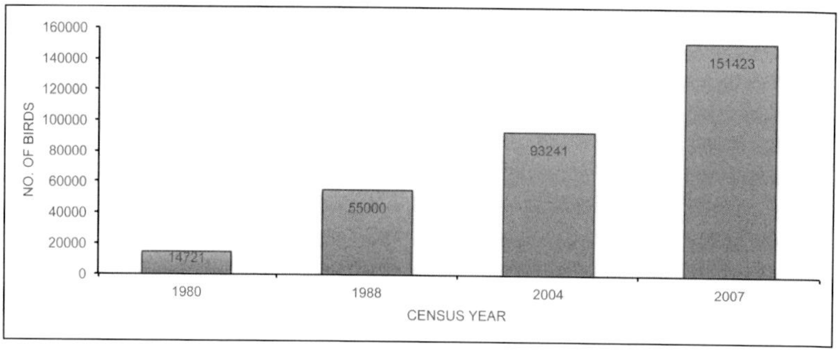

The wintering Cranes chiefly originate from western European populations that follow a western flyway through France and Iberia and sometimes into North Africa. The countries of origin of 605 marked birds observed in Spain were Norway/Sweden (42%), Germany (40%), Finland (11%), Estonia (6%) and Poland (0.2%) (SWA). The very large recent increases in the breeding populations in these countries largely account for the increased numbers wintering in Iberia. However, the European wintering distribution has also altered and there is a growing tendency for numbers of birds to remain in Germany and France in winter, where few did so before the 1980s. In 2007 there were some 50,000 birds wintering in France and up to 7,000 in Germany (Prieta & del Moral 2008), a change that may in part be due to protection from hunting, greater availability of food in winter and climatic amelioration. The wintering distribution within Spain has also altered since the mid 20th century. Bernis (1960a) found small numbers of wintering birds at 150 sites but in 2007 there were 60 extensive and much larger concentrations. The proportion of the Spanish population that remains in eastern Spain, in Castilla-La Mancha and Aragón has also changed: from 13% in 1980, to 35% in 1988, 42% in 2004 and 47% in 2007, largely as a result of young birds cutting short their ancestral journeys (Alonso *et al.* 2000, 2008). Should the tendency for wintering in France and Germany increase it is likely that the Iberian wintering numbers will show a corresponding decline, at least in some, milder, winters. Nevertheless, at least 101,000 Cranes were found in Extremadura in December 2012, a 20%+ increase over the equivalent 2007 total (*Ardeola* 60:171).

A further change has been an apparent reduction in the numbers that cross the Strait of Gibraltar to winter in Morocco. Early in the 20th century these may have comprised the majority of the birds that entered Iberia (Sánchez *et al.* 1998) but the Moroccan wintering population now varies within the range of 500–2,500 birds (BM). Passage across the Strait is chiefly in the western and central sectors and the Common Crane is only occasionally recorded at Gibraltar itself. Most southward passage across the Strait is in November, with a return movement from February to mid March, but some small-scale movement occurs in both directions throughout the winter as some birds switch between wintering locations on either side as the winter progresses (BM, pers. obs.).

SANDHILL CRANE *Grus canadensis*

(Grulla canadiense, Grou-americano)

Vagrant to Spain.

An adult was accompanying Common Cranes in the irrigation area of Vegas Altas del Guadiana, between Torviscal and Palazuelo, Badajoz, on 4 November 2011. It was subsequently relocated nearby on 2 January 2012 (RBS). What may have been the same individual was seen and photographed with Common Cranes at Gallocanta on 24 February 2013, where it remained until 26 February, when there was a large northward departure of cranes (RBS). In autumn 2011 there were also sightings of single birds in Finland, Estonia and Britain, all of which may have involved the same individual (but see Hudson *et al.* 2012). Up to 2011 there were just six Western Palearctic records: in Britain (2), Ireland, The Netherlands, Faeroes and Azores (Haas 2012).

DEMOISELLE CRANE *Grus virgo*

(Grulla Damisela, Grou-pequeño)

Vagrant to Spain and Portugal.

Rarely found in westernmost Europe and those that appear there may often originate from collections. It is listed as a Category D species in France, from where there were 15 records to 2009, eight of them post-1981, all of unknown origin, but one seen in October–November 2010 is thought to have been wild (Reeber *et al.* 2013). Some nested in Morocco until the mid 1980s, on the high plateaus of the Middle Atlas, but the species is now probably extinct there (BM).

It was described as a passage migrant and winter visitor to Andalucía in the late 19th century (Saunders 1871, Arévalo 1877), when it was also reported on passage in northern Morocco, raising the possibility that there was once regular passage across the Strait (BSG, BM). It may also have bred locally in Spain in the 19th century, at the Marismas del Guadalquivir (Irby 1895) and in the lower Guadiana valley in eastern Portugal (AP), although the only definite Portuguese record from the time is a specimen collected at the Guadiana by King Carlos in March 1893 (Tait 1924). Breeding in the Guadiana valley was also reported from Extremadura, where apparently chicks were captured around 1923–24 (Valverde 1992). Elsewhere three were obtained in the Balearic Islands, in Mallorca, during the 18th century (Gil Lletget 1945) and bones have been found in a medieval midden in Madrid (Hernández-Carrasquilla 1993).

The Demoiselle Crane is now only a very rare vagrant, with three accepted records from Spain and one from Portugal. An adult was with Common Cranes at Navalvillar de Pela, Badajoz, from 4–16 February 1996, and what is likely to have been the same bird was then seen at Gallocanta, Zaragoza, on 7–8 March, still with Common Cranes. One was accompanying Common Cranes at various locations between Ataquines and San Vicente del Palacio, Valladolid, from 11 January to 24 February 2004. A rather confiding individual that was at the Cavado estuary, Braga, from 6 November to late December 2004, is the only recent observation from Portugal but it was suspected to be an escaped bird.

LITTLE BUSTARD *Tetrax tetrax*

(Sisón Común, Sisão)

Widespread and locally common in mainland Iberia, but fast declining. Largely resident in the south, partly migratory in the north. Vagrant to the Balearic Islands.

The Little Bustard is highly characteristic of the open landscapes of Iberia. It was once very numerous indeed. Smith (1868) related that it was so abundant in Portugal that it was extremely cheap to buy in the poultry markets and that it was constantly being served at table as 'pheasant'. Arévalo (1887) also remarked that it was abundant in central and southern Spain, forming big flocks in autumn.The core of its current Spanish distribution comprises the southern meseta: Castilla-La Mancha and Madrid, and Extremadura. There are smaller and more scattered populations in Castilla y León, the Ebro valley and Andalucia, as well as a now-relict presence in La Rioja, Murcia and Galicia. None breed in the Levant or Biscay coastlands, or in the Balearic Islands (de Juana & Martínez, 1996, SA2). The Portuguese population breeds almost entirely in Alentejo but there are nesting birds along the international border between Miranda do Douro and Castelo Branco, and isolated groups in Algarve and at the Tagus estuary (PA2). The Little Bustard was formerly more widespread as a nesting species, especially in Castilla y León, but it has suffered recent range contraction, and a widespread decline in numbers, largely due to agricultural intensification. It was also certainly once more abundant and widespread in Portugal but suffered progressive decline and range contraction during the 20th century (AP).

Little Bustards are most abundant in the breeding season in habitat mosaics, favouring areas offering a mix of long- and short-term fallows, rough field margins, pastures and leguminous crops. They are comparatively scarce in extensive cereal monocultures as well as on ploughed land or in low scrub (LR). In general, traditionally farmed areas are much more attractive to them than intensively exploited areas (Martínez & Tapia 2002, Delgado & Moreira 2010). In winter they chiefly select fallows and alfalfa fields: they may spend most of the day in the latter and even roost there (Silva *et al.* 2004, LR).

The size of the Iberian breeding population has been estimated on several occasions on the basis of local counts. It is difficult to census and breeding season counts depend on assessing local densities of the relatively conspicuous displaying males in spring. The species employs a dispersed lek mating system, in which males form loose aggregations of territories, to which females are attracted. Simple extrapolation of the densities at these leks across the totality of the extent of breeding habitat is likely to result in overestimation (García de la Morena *et al.* 2006). Mean densities of 24 males/km^2 at La Serena, Extremadura, and 26 males/km^2 at Castro Verde, Alentejo, have been recorded in these core areas in the past (Moreira & Leitão 1996, de Juana 2009).

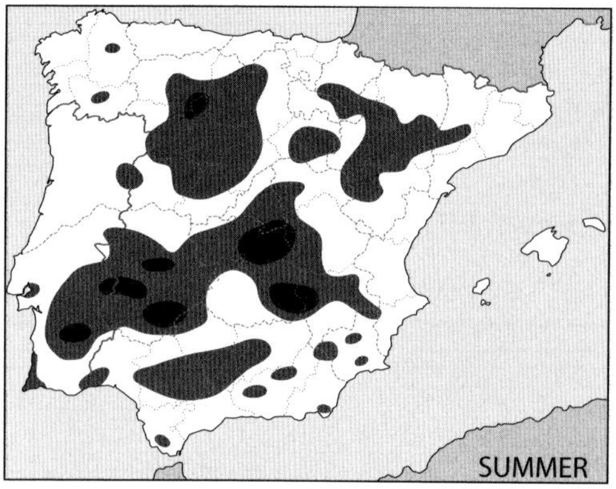

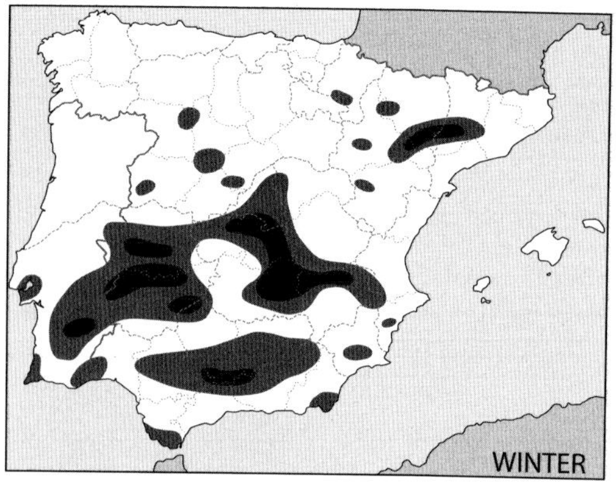

The Spanish breeding population was initially estimated at 100,000–200,000 breeding males on the basis of preliminary data obtained in 1993 and 1994 (de Juana & Martínez 1996, 2001). Within ten years the population was thought to number no more than 50,000–100,000 breeding males (SA2). The first nationwide census in Spain was attempted in 2005 (García de la Morena *et al.* 2006) and gave a national estimate of the order of 60,000 breeding males. The sex ratio in this species has been found to be skewed in favour of males, at approximately 1.4:1, giving a total breeding population in 2005 of around 105,000 individuals. However, the 2005 census of wintering birds (see below) found only 16,429–35,929 birds in Spain, strongly suggesting that the breeding census methodology seriously overestimates the nesting population. Data from closely monitored populations in Castilla y León and Aragón has allowed some correction to be made to the census results, giving a best estimate (95% confidence limits) for the 2005 breeding population of 26,900–44,800 males and 43,000–71,700 individuals (García de la Morena *et al.* 2006).

Irrespective of the accuracy of the figures, the 2005 survey revealed the nesting distribution of the species in Spain in detail. Castilla-La Mancha (46%) and Extremadura (21%) together held nearly 70% of the breeding males estimated, followed by Castilla y León (11%), Andalucía (5–10%) and Aragón (7%). The highest breeding male density was in Ciudad Real (3.3 birds/km^2) and Toledo (3.0 birds/km^2). A mean male density of 1–2/km^2 was found in a further eight provinces: Lleida, Zaragoza, Cuenca, Zamora, Badajoz, Cáceres, Madrid and Navarra. Spring 2005 saw much lower rainfall and higher temperatures than average and this may have affected the survey results on local or regional scales. During dry years the Little Bustards in Extremadura tend to vacate the driest areas, such as La Serena, but maintain or increase their usual densities in other more temperate locations at higher elevations, such as the Trujillo plains (de Juana & García 2005, own data). The population of Andalucía may have been underestimated in the 2005 survey, or may have increased subsequently, since a census of breeding males in spring 2007 found 5,259 birds: 54 in Almería, 192 in Cádiz, 913 in Córdoba, 492 in Granada, 82 in Huelva, 527 in Jaén, 249 in Málaga and 2,750 in Sevilla (*Ardeola* 59: 430).

The Portuguese breeding population in Alentejo was censused during 2003–06 and comprised 17,500 males (PA2). This approximates to 30,000 individuals in total, after applying the 1.4:1 sex ratio.

The winter distribution has a more southerly tendency than the breeding range, birds from northern Iberia moving to the southern Meseta and Extremdura, as well as parts of the Ebro and Guadalquivir valleys (Morales *et al.* 2002). In Portugal the Tagus estuary pastures have also attracted concentrations of up to several hundred birds in winter (Leitão & Costa 2001). The wintering population in Spain was also censused nationwide, between December 2005 and February 2006. A large proportion of the known wintering range was surveyed and 14,788 birds were found. The total population was estimated at 22,442 individuals (95% confidence limits 17,589–34,829 individuals) which is still thought to have been an underestimate since wintering birds were also found in areas where they were not previously known to occur in winter. Nearly 63% were on the Meseta Sur, largely in Castilla-La Mancha. A further 20% were in Extremadura, 13% in the Ebro valley – chiefly in Catalonia – and some 3% in Andalucía, with a minimal presence on the Meseta Norte and in Murcia. No winterers were located in Galicia or in the Levant coastlands. The wintering population also includes part of the residual French population. During 1997–2007 19 of 32 marked individuals from Poitou-Charentes were tracked to wintering areas in the southern meseta, Extremadura and coastal Portugal (Morales *et al.* 2002, Villers *et al.* 2010). There is also a recovery of a French-ringed bird from Cáceres (OF).

Little Bustard males take up their territories in late March and early April and the lekking period may continue until early June. Fledged young remain with the female into their first winter. The males largely abandon their territories by late June and move to post-breeding sites, generally locations offering plant and invertebrate food in summer and autumn. These are more humid areas, often at higher altitude or further north than the breeding sites themselves. Two adult males ringed in Talamanca de Jarama, Madrid, in May 2003 and May 2004, were recovered respectively in Soria in September 2003 and in Zaragoza in July 2005

(*Ecología* 22: 188). Females and juveniles may join the males once the young can fly. The Little Bustard is gregarious both post-breeding and in winter, when it forms flocks of hundreds of individuals and sometimes as many as three thousand. An exceptional gathering of 5,500 was recorded in the early 1980s in alfalfa crops at Herencia, Ciudad Real (Otero 1985, C. Otero pers. comm.). The post-breeding flocks move to the wintering areas in autumn and remain there until March, when all but the local nesters (if any) depart. Post-breeding dispersal often takes some individuals far from the breeding range. The Little Bustard is seldom reported from the Biscay coastlands but individuals, pairs and, rarely, groups of up to six have been recorded in coastal Galicia, Asturias and the Basque Country, nearly always in late summer and autumn. However, some large concentrations have been noted in inland Galicia: for example, there were up to 750 birds in A Limia in September 1984 (Bárcena *et al.* 1987).

The Spanish population undoubtedly declined markedly during the late 20th century and subsequently. Local studies have found declines in breeding numbers of up to 40% in Catalonia during 1994–2001 and c. 27% in Navarra during 1997–2001 (García de la Morena *et al.* 2004). The decline in Extremadura may have been even more severe, *c.* 75% during 1998–2008 (de Juana 2009). A further indication of decline is seen in the relatively small size of wintering flocks found in the 2005–06 survey. The largest of 436 flocks found comprised 1,100 birds and only 7% of flocks exceeded 100 individuals. The largest flock seen in Extremadura numbered 275 birds yet flocks of 2,000–3,000 birds were reported with some regularity in that region up to the 1990s (García de la Morena *et al.* 2006). More recently, flocks of 426 and 280 birds were found in Cáceres on 13 November 2009, at Trujillo and Zorita respectively, and 700 were at La Serena, Badajoz, on 29 January 2012 (*Ardeola* 57: 227,59: 181), but these concentrations were still relatively small.

Agricultural intensification has been blamed for the decline of the Little Bustard in other European regions and is certainly a key factor in Iberia. It affects both habitat extent and habitat quality through such processes as the reduction of fallows and uncultivated field margins, the expansion of irrigation and arboreal crops (such as olives and almonds), a reduction in invertebrate availability through application of insecticides and the loss of ground cover through overgrazing (LR). The species has also been badly affected by urban encroachment in some regions, as on the periphery of Madrid, and by abandonment of cultivation in less productive areas, which results in habitat loss to scrub encroachment (García de la Morena *et al.* 2006). Climatic change, which is predicted to reduce overall precipitation in Iberia, may represent a further long-term threat, especially in the already dry regions of Castilla-La Mancha, Extremadura and Algarve that currently hold the main populations (Osborne & Suárez-Seoane 2007, Delgado *et al.* 2009, Delgado & Moreira 2010). Spain alone now probably accommodates 50–70% of the global population of the Little Bustard, which elsewhere is now numerous only in southern Russia and Kazakhstan (HBW), so its conservation in Iberia must be regarded as a priority.

HOUBARA BUSTARD *Chlamydotis undulata*

(Avutarda Hubara, Hubara)

European records of vagrants as far west as Britain are considered to originate from the highly migratory central Asian populations (*C. u. macqueenii*). The nominate form breeds across North Africa from Mauritania and Saharan Morocco east to Egypt. There are only two undated 19th century records from Iberia: Saunders (1871) noted that he had examined two specimens killed in Andalucia and Arévalo (1887) said that one was obtained near Málaga city a few years earlier. A possible 19th century/early 20th occurrence in Portugal is even vaguer and dismissed as inconclusive (AP). Any Iberian occurrences are perhaps more likely to be of nominate race birds from North Africa than of *macqueenii* (AMI). The species is not formally admitted to the Spanish list.

GREAT BUSTARD *Otis tarda*

(Avutarda Común, Abetarda-grande)

***O. t. tarda*. Widespread and locally common in mainland Spain, chiefly on the meseta and in Extremadura, and in east-central Portugal.**

The Great Bustard is an icon of the avifauna of Iberia, which supports more than half of its global population. It is still common, and even increasing, in many areas, although the Iberian populations probably declined by some 30% during 1960–1980, with accompanying range contraction: the result of excessive hunting and changes in land use among other factors (LR). The Iberian populations are now fully protected.

The breeding distribution in the early 21st century is concentrated on the Spanish meseta, together with Extremadura and adjacent areas of southern Portugal. The largest and most continuous subpopulation is in Castilla y León, a region that

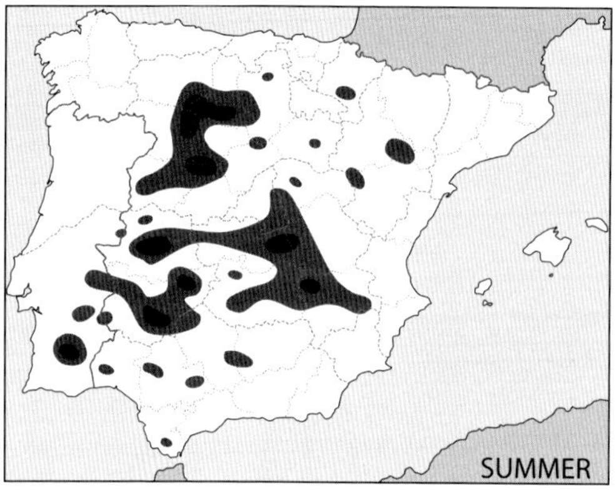

offers great expanses of cereal crops. The Great Bustard is widely distributed in the pseudosteppe habitats of central and eastern Castilla-La Mancha, and in open habitats throughout Extremadura. It is also relatively numerous in the croplands of Madrid. Elsewhere in Spain, it has a highly fragmented breeding range within Andalucía: chiefly in the Guadalquivir valley, and in the Ebro valley in Aragón and Navarra, which together account for a very small fraction of the population. The Portuguese birds are nearly all in the south-east, from southeastern Beira Baixa to the south of Alentejo: almost 80% are in Castro Verde, Beja (PA2).

The Great Bustard is characteristic of wide open spaces, as now occur especially in the cereal croplands of Iberia. These are level or undulating, near-treeless expanses, covering great areas. The bustard especially favours locations where the crops form part of a mosaic that also includes fallow areas, alfalfa and other leguminous crops and rough pastureland. Plants make up over 90% of its diet and these often include such crop species as alfalfa, vetches, chickpeas, wheat, sunflowers and grape vines, with beetles an important fraction of the animal diet (Alonso & Palacín 2009). The males establish their leks in spring, from early March to April, and court the females, who are entirely responsible for incubation and care of the young. The females nest in taller cover, including cereal crops. The fledged young accompany their mother for up to a year. The marked sexual size dimorphism of the Great Bustard is already evident from hatching and the young males, which require more food, show higher juvenile mortality than the females (Martín *et al.* 2007). Males do not breed until they are at least five years old but females will do so from the age of two or three. Reproductive output is low. Breeding success, expressed as the number of young found per female in September, has ranged from 0.08–0.20 in various studies, and also fluctuates between years: on average a female might expect to raise just one chick to maturity every ten years (Alonso & Palacín 2009).

There have been a number of attempts to estimate the size of the breeding populations since Trigo de Yarto (1971) put the Spanish population in 1969 at 16,000–17,000 birds. Palacios *et al.* (1975) used data from gamekeepers and partial counts to arrive at a Spanish population of 11,890 birds. Alonso & Alonso (1990b) attempted to reconcile data from different sources and suggested a population of 13,500–14,000 individuals in the 1980s. Further partial censuses in the early 1990s allowed Alonso & Alonso (1996) to suggest a mininum population then of 17,000 birds. A subsequent review of the global population (Palacín & Alonso 2008) has put the Spanish population at 27,500–30,000 individuals and there are about 1,400 in Portugal. The most recent assessment suggests that there were at least 29,400–34,300 individuals in Iberia in 2004–11 (Table 29). The Iberian populations account for at least 60% of the global population and are thus of the greatest conservation importance (Palacín & Alonso 2008).

Table 29. Regional distribution of Great Bustards as indicated by recent censuses (Source: www.proyectoavutarda.mncn.csic.es)

Region	Population	Year	Reliability	Alonso *et al.* (2005)
Castilla y León	14,000–16,000	2008–09	High	11,000
Extremadura	5,500–6,500	2002–03	Low	5,500–6,500
Castilla-La Mancha	5,000–6,500	2001–10	Medium	4,775 +
Portugal	1,900	2009	High	1,150*
Madrid	1,400–1,500	2011	High	1,300
Andalucía	350–360	2001–11	High	338
Aragón	115	2006	High	115
Navarra	50–60	2006	High	25
Murcia	6	2004	Low	6

(* Pinto *et al.* 2005)

Some provincial populations away from the core areas are very small and in danger of disappearance, notably the relict presences in Murcia and Valencia. Although hardly any birds were found in these two provinces during the surveys included in Table 29, there were a series of observations at Ayora, Valencia, in 2004, including a flock of up to 20 females and a young male reported between mid July and mid October (*Ardeola* 52: 214), whereas a flock of 15 including at least one displaying male was at the Estepas de Yecla, Murcia, on 24 February 2007 (*Ardeola* 54: 395). The isolated populations of the Guadalquivir and Ebro valleys are also vulnerable to extinction. Indeed, the Andalucían population is undoubtedly a pale remnant of large numbers that occurred in the 19th century and early 20th century: Saunders (1871) described it as generally distributed across southern Spain and especially abundant around Sevilla, Chapman (1884) refers to its abundance around Jerez and Irby (1895) also noted its universal presence in the farmlands of Cádiz and the lower Guadalquivir valley. Nevertheless, the results of recent censuses in Andalucía are encouraging: 444 individuals were found in 2011: 138 in Córdoba, 31 in Huelva, 80 in Jaén and 195 in Sevilla, and 471 in 2013 (*Ardeola* 59: 431, 60: 522). A promising increase in numbers has also been seen in Castilla y León, from 10,071 in 1998 to 14,025 in 2008, a 34% increase if only the area surveyed in both censuses is taken into account (Martín *et al.* 2012): this increase may have been fostered by the spread of unirrigated legumes in the region, which increased by 40% during 1998–2008.

The positive trends in the Iberian Great Bustard populations since the 1980s are undeniable. Some local increases however may have been at the expenses of more outlying locations, the birds tending to be concentrated in a smaller area (LR). This may culminate in eventual declines due to overpopulation. In Portugal too there has been a tendency towards concentration in a single high-quality area: Castro Verde (LV). Spanish locations with very large populations, over 1,000 individuals, include the Llanos de Cáceres and southern Badajoz in Extremadura: the latter mainly in the winter, as well as the eastern parts of Toledo and Madrid on the southern Meseta and, most of all, the Villafáfila reserve in Castilla y León: where some 2,500 Great Bustards occur in an area of 327 km^2.

Iberian birds are relatively sedentary. The females especially show strong attachment to their natal areas and hence colonisation or recolonisation of apparently suitable vacant habitat is seldom observed. Young males however tend to disperse further and begin breeding at leks 5–100km from their natal areas (Martín *et al.* 2008). Most males in central and southern Spain abandon the lek sites after the courtship season, in May or early June, and may then travel up to 250km to summering sites that are often in taller vegetation, including sunflower fields and areas with a scattering of small trees; such as young olive and almond groves and open dehesas. These are places offering a food supply and also some respite from the worst of the summer's heat, which is thought to affect the heavy males more than the females (Alonso *et al.* 2009). Those breeding in more southern areas travel further on average and to higher elevations. Summering areas are generally within the breeding range but some individuals may travel further, resulting in occasional records in Cantabria, Asturias and Galicia. These usually involve single birds but there were seven at Lalín, Pontevedra, on 12–13 September 2005 (*Ardeola* 53: 201).

Both sexes, and the young of the year, may travel usually short distances in autumn to wintering areas (Palacín 2007). Little more than half of the females in central Spain move, usually in a south-easterly direction, the mean distance travelled being just 50km. Some 85% of males have returned to their breeding localities by January, either directly from their post-breeding dispersal sites or after spending some time in winter aggregation areas. In contrast, migratory females usually leave their breeding areas in October–December and return during February–April.

The Great Bustard is gregarious outside the breeding season but flocks are most often segregated by sex (Alonso & Palacín 2009). Male-only and female/juvenile-only flocks occur in summer but mixed flocks sometimes occur in winter (Martínez 1988). Flocks of either or both sexes are generally up to 50 individuals, often fewer where the species is scarce, but some large gatherings are on record, notably *c.* 270 at Castro Verde, Beja, on 25 March 1997 (*Pardela* 7: 18)

The Great Bustard is vulnerable to a diversity of threats, which affect different subpopulations to different extents (LR). Habitat loss through replacement of cereal crops with irrigated land or their conversion to pasture is significant in Extremadura and the meseta. It is widely affected by the results of agricultural intensification, including the loss of fallows and crop mosaics and the application of herbicides and insecticides. Powerlines and fences are a hazard in many areas and disturbance and urban encroachment are problematic near cities and towns. The evident local abundance of these conspicuous birds has prompted some demands that hunting be reinstated: the effects of this would be potentially very serious should it ever be allowed.

EURASIAN OYSTERCATCHER

Haematopus ostralegus

(Ostrero Euroasiático, Ostraceiro-comum)

***H. o. ostralegus*. A scarce breeder on northern coasts and in the Ebro delta. Common on passage and in winter on Atlantic coasts, scarce in the Mediterranean and rare inland.**

The Oystercatcher has a marginal breeding presence in Iberia. The breeding population totalled 49–51 pairs in 2005–07: 34 at the Ebro delta, seven in Asturias, 6–7 in Galicia and 2–3 in Cantabria (Palomino & Molina 2009). None breed in Portugal although nesting was recorded at the Sado estuary in the 1930s (Rufino 1989a, AP).

Oystercatchers have nested in the Ebro delta since at least 1973 (SA1) and the population has fluctuated since, peaking at 40 pairs in 1997 but with only 23–30 pairs in the late 1990s (Palomino & Molina 2009). The delta birds are especially affected by predation by foxes and Yellow-legged Gulls, and disturbance by humans, and productivity is low. There were single breeding attempts at the Llobregat delta in 2008 and 2009, but the nests were lost to predators (*A. O. Catalunya 2008*: 86, *2009*: 105).

The few breeding pairs on the Atlantic seaboard are all on the Biscay coast, scattered on rocky islets just offshore, from the Farallón islands and other sites in Lugo east to Santa Marina islet in Santander Bay: twelve sites were used in 2007. Breeding in Galicia was reported as long ago as 1914, when Tait (1924) found eight pairs and a nest with two eggs on the Sisargas Islands in June. Tait also received three eggs obtained near A Coruña. There were no further reports of nesting in Galicia until 1982, since when the north coast population appears to have been fairly stable at 6–7 pairs (SA1). The population in Asturias has varied between five and ten pairs since 1981, when the first nests were found: productivity here is very low, from zero to 0.25 chicks per pair (LR). Between two and three pairs have nested in Santander Bay since 1990 (Palomino & Molina 2009).

The Oystercatcher is much more familiar as a migrant and winter resident, at least on Atlantic and Biscay coasts. Few linger on the Mediterranean coasts, with their low tidal range, but some occur on passage at coastal wetlands there. Arrivals begin in August and migrants are apparent at coastal wetlands on Biscay and Atlantic shores in September–October. A less conspicuous return movement is noted from March, peaking from mid-April to mid-May on the north coast (SA1, AP). A few individuals are reported annually at Mediterranean and Balearic wetlands, chiefly on passage. A flock of 75 at the Albufera de Mallorca on 17 August 2011 (*Ardeola* 59: 431) was a record count for the island. In addition, there is a small but regular presence of summering non-breeders in Galicia and Asturias (LR) and occasionally elsewhere. There are also occasional brief occurrences of migrant individuals far inland, usually involving one or two individuals that appear on passage briefly at wetland sites: a record of 25 at the Laguna de Pétrola, Albacete, on 30 August 1987 (*Ardeola* 38: 337) is quite exceptional.

The north coastal breeding birds appear to be resident but those of the Ebro Delta are migratory, perhaps wintering on African Atlantic coasts. Nevertheless, a few may be found in winter at the Delta where January counts averaged 11–12 birds during 1996–2009, with a maximum of 18 in January 1996 (*A. O. Catalunya*). Iberian ringing recoveries include birds from Britain, Ireland, the Low Countries and France and also a few from Scandinavia, one from Iceland and one ringed in Mauritania, nearly all of which have been found on Atlantic coasts. A chick ringed at the Po delta, northern Italy, was recovered three years later in Tarragona (*Ardeola* 19: 236). There are also single recoveries of Spanish-ringed birds from The Netherlands and Morocco.

The Iberian wintering population in Iberia is quite small and almost entirely limited to Atlantic and Biscay coasts. Prior to 1990 it averaged *c.* 2,700 birds in total (AI1), less than 0.3% of the western European population, with concentrations in Galicia (*c.*50%): especially in the Rías de Arousa and Ortigueira; on the Portuguese coast, chiefly at the Ria Formosa, Faro (40%) and Doñana (10%), with some individuals on the Biscay coasts and south from the Guadalquivir to the Strait of Gibraltar (AI1). The 1990–2009 January counts of wintering aquatic birds in Spain found 2,389 birds on average (range 1,412– 3,362), with the great majority (70%) once again in Galicia and the Biscay coastlands and virtually all the others in coastal western Andalucía (AAE). The largest concentrations recorded by the January surveys were at the Ría de Arousa (up to 750 birds), Cádiz Bay (up to 595) and Doñana (490). The Spanish winter atlas estimated 2,950–3,560 birds during 2007–10.

More recent winter counts in Portugal still average around 1,000 birds, although the Ria Formosa alone may hold 800–1,000 in some winters (AP). A census of non-estuarine coastal habitats in Portugal in winter 2009–10 found just 150 birds (AP).

BLACK-WINGED STILT

Himantopus himantopus

(Cigüeñuela Común, Pernilongo-comum)

***H. h. himantopus*. Widespread and locally numerous as a breeding species in much of peninsular Iberia and also in the Balearic Islands. Partial migrant, with a large and increasing winter population.**

This is one of the most widespread breeding waders, nesting regularly in all regions except Galicia, much of northern Portugal

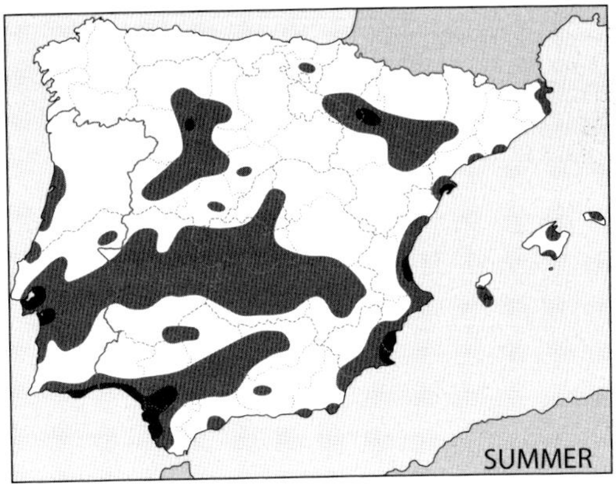

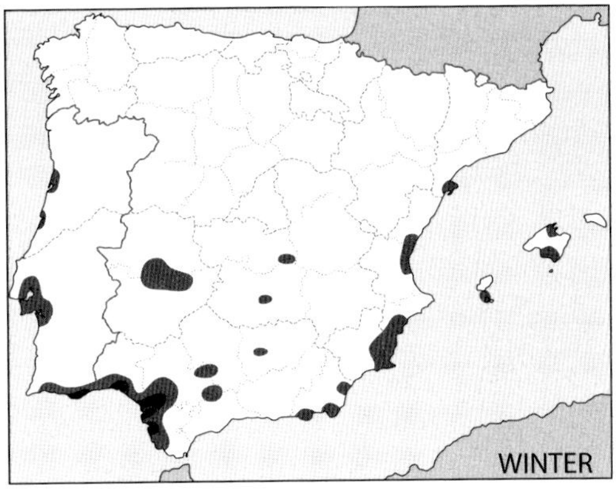

and north of the Cantabrian Range – although there are records of occasional nesting in all these areas. It was formerly regarded as a summer visitor (AMI), arriving in March–April and departing in September. It wintered largely in Sahelian wetlands, although Irby (1895) remarked that a few overwintered at Doñana in some years. Increasingly large numbers now winter in Iberia and these include some birds from elsewhere in western Europe (Rufino & Neves 1995).

The breeding range has traditionally comprised the key wetlands of southwestern and southern Portugal, southwestern and southern Andalucía, and the Levant coast, with a more scattered and local presence inland (Rufino & Neves 1991a, SA1). Increasing numbers have settled inland since the 1980s, taking advantage of increased habitat availability in the form of ricefields and new reservoirs, and the Stilt now breeds more widely in Spain across Extremadura, Castilla-La Mancha, Madrid and Castilla y León, and has also expanded its presence in the Guadalquivir and Ebro basins (SA2). A sizable population inhabits the Aragón ricefields: 728 nests were counted there in 1993 (Sánchez Sanz 1997). It was first reported breeding in La Rioja in 1988 and two pairs nested at Noja, Cantabria in 1989 (*Ardeola* 36: 246). Several pairs also nested in Galicia at A Ínsua inlet, A Coruña, in 1992 and 1993 (*Ardeola* 41: 97). Breeding has also been reported at several sites in the Basque Country: notably at Salburúa, where the population increased from 13 occupied nests in 1999 to 59 in 2000 (*Ardeola* 49: 185). It also nests widely at Balearic wetlands: in Mallorca, Menorca, Ibiza and Formentera, where some remain all year.

In Portugal the Stilt breeds widely south of the Tagus and also on the west coast at the estuarine lagoons of the Vouga and Mondego rivers (PA2). The principal Portuguese breeding sites are in the Tagus and Sado estuaries and at the Ria Formosa and Castro Marim reserve, Faro (AP).

The Black-winged Stilt is an opportunist that readily colonises a broad range of wetlands, both natural and artificial. At all seasons it chiefly requires stagnant or slow-moving shallow waters, which may take the form of marismas, marshes, lakes, ephemeral pools, large ponds, saltpans, ricefields, reservoir and gravel pit margins, and fish farms. Saltpans are especially important as a breeding habitat. Many nest as isolated pairs but the majority form loose colonies, whose nests are defended jointly against predators, which are mobbed noisily and incessantly.

The Spanish breeding population shows an increasing trend. In 1989 there were some 12,000 pairs (Martínez-Vilalta 1991) and these had apparently increased to 14,000–15,500 pairs by the late 1990s (Hortas *et al.* 2000). A sample census in 2007 revealed a further increase, after extrapolation, to an estimated 56,500 individuals, although with very wide margins of error (Palomino & Molina 2009). In 2007 the largest breeding concentrations were at the Ebro delta (>4,200 birds), Doñana (3,000) and the Albufera de Valencia (1,400). The Portuguese population was censused in 1990 when there were 1,202 pairs (AP). Numbers at particular sites may fluctuate very widely between years, largely due to variation in rainfall. Dry years in Doñana increase the likelihood that breeding colonies may appear elsewhere. Some sites have seen declines linked to habitat loss, notably in Cádiz Bay where some 1,400 pairs nested in the late 1980s but only 270 pairs bred on average during 1994–98, following the loss of much saltpan habitat (SA2).

As many as 22 Iberian sites are of international importance for the species, each holding at least 1% of this western European and north-west African subpopulation either regularly or in some recent years (Delany *et al.* 2009). They include Castro Marim, the Sado and Tagus estuaries, Pera marsh and the Ría de Faro in Portugal; eight sites in the lower Guadalquivir and Huelva/Cádiz coastlands including Doñana and its hinterland; the Ebro delta, the Albufera de Valencia, the Marjal del Moro, El Hondo and the Mijares estuary on the Levant coast; the Ebro valley ricefields in Aragón; the Laguna de El Hito, Cuenca; the Lagunas de Villafáfila and the Parque Natural del Sureste, Madrid.

The species shows some nomadic tendencies and considerable interchange has been reported between different breeding areas. Birds fledged in France have been found nesting in Portugal, birds from Portugal have been found in northwestern Italy and some Iberian birds have been found nesting within the Peninsula far from where they hatched (Delany *et al.* 2009). In addition to nomadism or dispersal there are regular migratory movements in which an unknown proportion of the population travels south in autumn to winter in northwestern or sub-Saharan Africa. Spanish-ringed birds have been recovered elsewhere in Europe: seven, including one each in Hungary and Albania; in Syria (1); in North Africa: three in Morocco and two each in Algeria and Tunisia, and in sub-Saharan Africa: seven in Mali, two each in Mauritania and Senegal, and one in Guinea Bissau. One ringed in Portugal has also been found in Hungary.

Migrants from other western European populations occur on passage and in winter. The populations involved are not large and so cannot account for more than a small proportion of the Iberian wintering population. There are Iberian recoveries of 13 birds from France, three from Italy and one from Tunisia. Migration is apparent from numerical fluctuations at Mediterranean coastal wetlands and at inland sites chiefly from March to April in spring and from late July to October in autumn, peaking in August (Castro 1993). Stilts are recorded occasionally at wetlands in Galicia and the Biscay coastlands, where they do not normally breed, most often during the spring passage period. The numbers involved are usually small but there was an exceptional presence at the Ría de Avilés in April–May 1989, peaking at 30 birds (Álvarez-Balbuena *et al.* 2000).

The great majority of the wintering population is in the south where thousands of individuals occur. Most birds abandon inland locations and the Stilt has usually been relatively scarce in winter on the Levant coast (Clavell 2002). The Iberian wintering population has shown marked recent increases (SWA). For comparison, censuses during 1978–80 found an average total of just 447 wintering Stilts, most of them in southwestern Andalucía (Alberto & Purroy 1981). However, the January counts in Spain during 1990–2009 found 6,159 birds on average (510–15,909 individuals; AAE): the vast majority of which (91%) were in coastal Andalucía, principally at the marismas, saltpans and ricefields of the lower Guadalquivir valley and the coastal wetlands of Huelva and Cádiz, with only 3% on the southern Levant coast and a further 5% divided between the Balearic Islands, Murcia and Extremadura. Larger numbers have tended to occur inland since the 1990s in at least some winters, notably in the ricefields of Extremadura: for example, there were 1,835 at the Vegas Altas ricefields on 14 January 2013 (*Ardeola* 60: 171).The mean wintering population in Portugal also increased from 120 in 1975–78 to 1,000+ in 1992–95 and 2005–07, with some wintering at west coast locations north of the Tagus, such as the Mondego estuary and the Ria de Aveiro in recent years (AP).

PIED AVOCET *Recurvirostra avosetta*

(Avoceta Común, Alfaiate-comum)

A common breeding species in coastal wetlands of southwestern Spain, the Levant and southern Portugal, and more locally inland. It has bred sporadically in the Balearic Islands. Widespread on passage and locally numerous in winter.

The Avocet breeds in some numbers at coastal locations in southern and eastern Iberia. More than half of the Spanish population usually nests in southwestern Andalucía: at the Guadalquivir estuary, the Marismas de Isla Cristina and Cádiz Bay, although breeding numbers fluctuate there and are low in dry years. The most important site by far is at the Veta la Palma fish-farms, which hold a very high proportion of the 4,000–7,000 pairs that usually nest in the Doñana area, probably due to their relatively constant water levels (García *et al.* 2000, SA2). Most of the remainder are at eastern coastal sites, notably the Cabo de Gata saltpans, Almería; the Mar Menor, Murcia; the Santa Pola saltpans, Alicante, and the Ebro delta (SA2). Nearly all Portuguese birds breed in coastal Algarve, with sporadic nesting also in the Tagus estuary (PA2). Some nest inland at scattered wetlands, chiefly in Castilla–La Mancha, but locally and sometimes irregularly also in Galicia, Castilla y León (Villafáfila and La Nava), Madrid, Aragón (Gallocanta) and Andalucía (Fuentedepiedra, upper Guadalquivir valley). The first breeding record from Extremadura was in 2006, when a nest with eggs was found at the Sierra Brava reservoir, Cáceres (*A. O. Extremadura 2004–2008*: 88). The Avocet is much less widespread as a breeding species inland than the Black-winged Stilt, with which it often associates.

The Spanish breeding population was estimated at 4,356–4,476 pairs in 1989 (Martínez-Vilalta 1991) and it apparently increased to 5,500–6,000 pairs during the 1990s (SA2). A census in 2007 found over 6,000 individuals at Doñana: in a year with low water levels, other significant concentrations being at the Ebro Delta (800), the Santa Pola saltpans (600) and, inland, at the Lagunas de Villafáfila (>400 birds) (Palomino & Molina 2009). Breeding Avocets occur at shallow coastal wetlands, often at man-made sites, such as saltpans and fish farms, but also at natural lagoons and marismas. Those inland chiefly frequent brackish lakes. The birds nest in loose colonies, often in areas with little or no plant cover.

The historical literature gives the impression that the Avocet was considerably scarcer as a breeding species in Iberia in the past (Saunders 1871, Arévalo 1887, Gil Lletget 1945). Noble (1902) said that it was 'abundant' in Doñana in spring but only mentions one nesting colony on a small island with 'a dozen or more nests'. Much later, although Valverde (1960) regarded the Avocet as a rather common but localised breeder in the Marismas, he reported 35 pairs as the maximum colony size there.

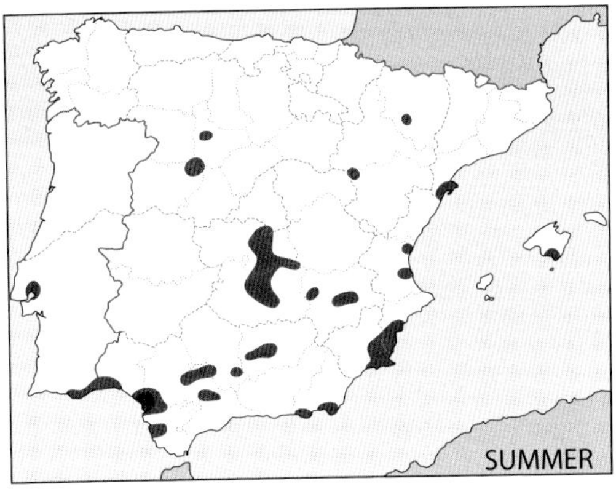

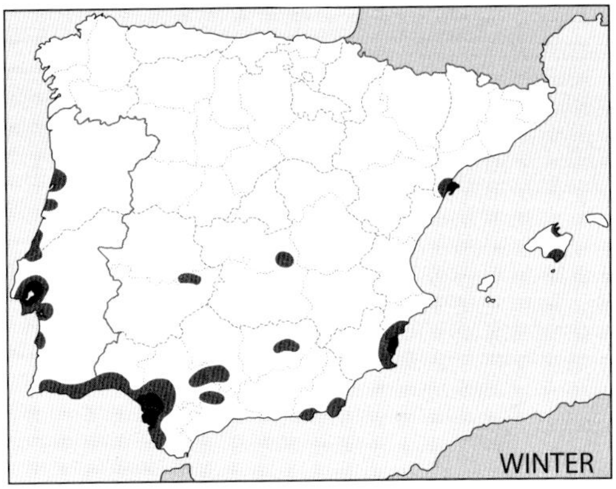

Ticehurst and Whistler (1930) found the Avocet 'strangely absent' during their spring visit to the Albufera de Valencia and the Ebro delta. There were only some 100 breeding pairs at the Ebro delta in 1973 compared with 400 pairs in both 1980 (Ferrer *et al.* 1986) and 2001 (CA2). Bernis (AMI) considered that only a 'few pairs or groups of pairs' bred in Iberia, at various coastal marshes and brackish lakes but mainly at Doñana.

The Portuguese breeding population during 2001–03 was estimated at 350–456 pairs (PA2), almost entirely confined to the saltpans of central and eastern Algarve. A few pairs have bred in recent years at the Tagus estuary and breeding has occurred sporadically elsewhere in the past, most recently at the Sado estuary, Setúbal.

Avocets breed somewhat irregularly in the Balearic archipelago (Ferrer *et al.* 1986, Clavell 2002), but have nested at the Salobrar de Campos, Mallorca, since 1998 (*Ardeola* 50: 159). Maximum counts at that site, during July–August, increased from 17 in 2002 to 260 in 2009 (*A. O. Balears*).

Iberian birds are joined in winter by migrants from France and northwestern Europe. There are at least 207 ringing recoveries in total of birds ringed in The Netherlands (41%), Denmark (27%), southern Sweden, Germany, France, Belgium, Italy and Britain. The great majority of recoveries are between October and March, peaking in November–January, from the principal wintering areas in the south-west (AMI, AI1). Migrants occur regularly in the Biscay coastlands, northern Portugal and other sites where they do not breed. Autumn passage occurs during August–October but is more apparent on Atlantic coasts than in the Mediterranean. Return movements are during March–May. Iberian birds are partly migratory and some may migrate over long distances: birds ringed in Spain, mostly as chicks at Doñana, have given 32 extra-regional recoveries from central Europe: including 24 from The Netherlands and one from as far east as Hungary, and four in Africa: in Morocco (2), Senegal (2,924km) and Guinea Conakry (3,179km) (*Ecología* 11: 338, 13: 354–355, 14: 314). Many of the recoveries in The Netherlands have been of adult birds during the breeding season, suggesting that some may move there to breed when water levels in the Andalucian wetlands are unsuitable.

The Iberian wintering population has increased in recent years. The censuses of wintering waders during 1978–80 found an average total of 3,240 wintering Avocets in Spain, most of them in southwestern Andalucía (Alberto & Purroy 1981). However, the Spanish January censuses of 1990–2009 found an average of 14,404 birds (range 5,525–28,519; AAE), 86% of them in Andalucía. Three Portuguese sites and 22 Spanish sites have been identified as supporting internationally important concentrations of Avocets, chiefly post-breeding and in winter (Delany *et al.* 2009). Nearly all the Spanish wintering sites form part of the complex of wetlands of the lower Guadalquivir together with the coastal wetlands of Huelva: including the Odiel estuary, and Cádiz Bay. A maximum winter count of 11,883 was recorded at Veta La Palma, Doñana, in 1983, and Doñana as a whole has held up to 23,093 birds in January during 1990–2009 (AAE). Other important Spanish sites that have each held post-breeding and wintering flocks of up to 1,500 birds include the Ebro delta, El Hondo, and the salt-pans at Cabo de Gata, Santa Pola and the Mar Menor. Small wintering groups have also been noted in Galicia and the Biscay coastlands, notably at Santoña, Txingudi Bay and at the Ría de Villaviciosa, since the mid 1980s (*Ardeola* 36: 246, 44: 252, 48: 142). Occasional observations of individuals inland in winter are also not unusual but a record of 75 at the Sierra Brava reservoir, Cáceres, in December 2005 (*Ardeola* 54: 177) is noteworthy.

The principal Portuguese wintering locations are the Sado and Tagus estuaries and the Ria Formosa, with a maximum non-breeding count of 13,661 birds at the Tagus estuary in 1989 (Delany *et al.* 2009). The Tagus estuary regularly holds about half of the wintering Avocets in Portugal and, in addition to the Sado estuary and Ria Formosa, the Ria de Aveiro, the Mondego estuary and Castro Marim also attract up to some hundreds of winterers (AP). The mean Portuguese wintering population during 1992–95 was 12,257 birds but there were only 8,633 during 2005–07 (AP).

STONE-CURLEW

(Alcaraván Común, Alcaravão-comum)

Burhinus oedicnemus

***B. o. oedicnemus*. Widespread in mainland Iberia. Partial migrant. Other western European populations occur on passage and in winter.**

***B. o. saharae*. Those resident in the Balearic Islands are sometimes attributed to this subspecies (AI1, Clavell 2002).**

The Stone-curlew is widely distributed in mainland Iberia outside the Eurosiberian zone, where it is largely absent. The core distribution in Spain includes the meseta, Extremadura, and the Guadalquivir and Ebro valleys (SA2). It is largely absent when nesting from the Mediterranean coastlands, except in the arid south-east in Murcia and Almería (SA2, LR). The distribution in Portugal is chiefly in the south and east (PA2). The Balearic population is widely distributed across all the main islands: Mallorca, Menorca, Ibiza, Formentera and Cabrera, the chief distribution gap there being the Tramuntana range in Mallorca.

It inhabits dry, fairly level, open country: including pseudosteppe pasturelands and dwarf shrub formations, sand dunes, semi-deserts, fallow areas and dry croplands: such as the early growth stages of cereal, maize, sunflower and potato cultivation. It also occurs in open vineyards and almond groves. The majority are thus found at relatively low levels but some breed locally at over 1,500m in hill pastures.

The Iberian Stone-curlew population is sizable and comprises perhaps half of the entire European population (BE). Stone-curlews are thinly spread in most countries but in Spain the Brozas–Membrío steppes, Cáceres; La Serena, Badajoz, and the Lleida steppes, Aragón, have large, internationally important breeding populations, each of over 1,500 pairs (Delany *et al.* 2009). The Spanish population was estimated at 22,000–30,000 pairs by Purroy *et al.* (1997) and 30,000–40,000 pairs by de Juana *et al.* (SA2). A recent estimate put the mean Spanish population in spring during 2004–06 at *c.*357,000 birds (ACRE). This total is about twice as high as the 77,000 pairs estimated for the whole of Europe by Birdlife International (2004) and so needs to be regarded with caution. The Portuguese population was estimated at 1,000–2,000 pairs in 2002 (BE).

The Stone-curlew is most abundant in Castilla-La Mancha, with high densities also found in Extremadura, Andalucía, Murcia and in the Ebro valley in Aragón. It is widely distributed across Castilla y León but seems to be relatively thinly spread there (SA2, ACRE). Densities of 2.1–3.7 birds/km^2 have been found in cereal crops and vineyards in meso- and thermomediterranean regions, and areas bordering wetlands seem to be especially favoured, with up to 12 birds/km^2 (ACRE). The Spanish population is thought to be in overall decline, largely because of habitat loss arising from changes in land use, and this may have amounted to a 20% population reduction during 1970–90 (SA1). There has also been some range contraction in Portugal: in Trás-os-Montes, the Alto Douro, the north of Beira Alta and in Estremadura and Ribatejo, attributed variously to afforestation, a decline in traditional pasturage and the spread of irrigated crops (PA2).

Many Iberian birds, including those on the Balearic Islands, appear to be sedentary. However, those nesting at higher elevations appear to move lower to winter and many in the northern meseta and the Ebro valley may migrate further south. The wintering distribution is largely within the warmer, low-lying agricultural regions of the southern half of Iberia, including the Guadalquivir and Guadiana valleys, northwestern Badajoz and the coasts of Alicante and Murcia (SWA). There is no data on the size of the winter population, other than the 2,570–3,402 individuals estimated for Catalonia (CWA).

At least 40 recoveries of foreign-ringed birds show that Iberian birds are joined on passage and in winter by Stone-curlews from England (32), France (6) and The Netherlands (2). Small numbers are detected on passage on both sides of the Strait of Gibraltar, in October–November and March–April (Pineau & Giraud-Audine 1979, BG, BSG). A few migrant or perhaps wintering individuals are also recorded from time to time in the northern coastlands, where none breed. A wintering flock at Santander

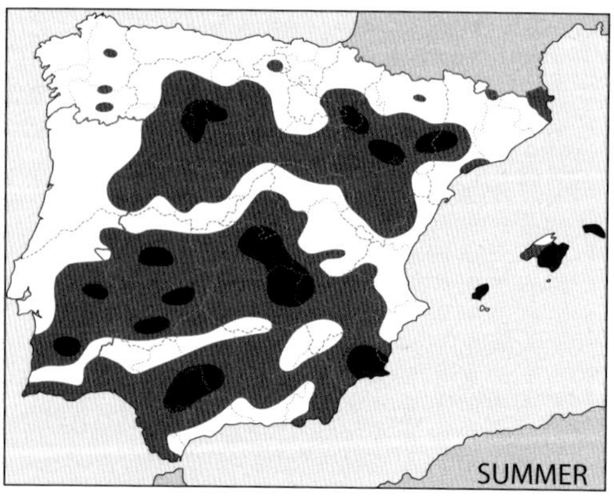

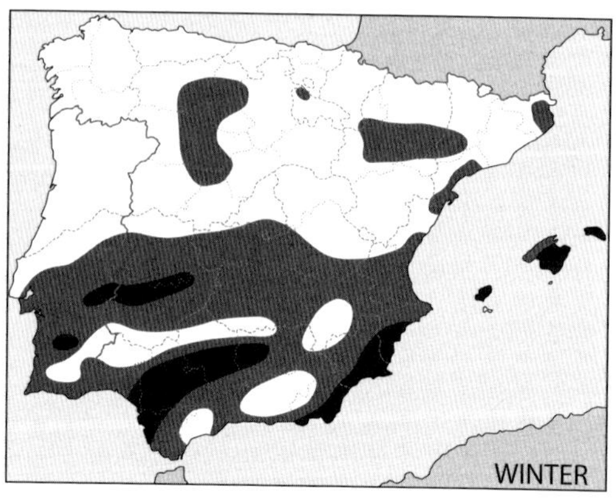

airport, that peaked at 50 birds in January 2008 was unprecedented (*Ardeola* 55: 142).

The Stone-curlew is gregarious outside the breeding season, from August onwards (Calvet *et al.* 2004). Flocks of up to 200 are not unusual and larger gatherings are sometimes reported: as many as 500 at La Victoria, Córdoba, in November/December 2006; 500 at Fuentedepiedra in November 2007, and 702 at Miralcamp, Lleida, in November 1991 (*Ardeola* 54: 177, 396; Calvet *et al.* 2004). Such flocks favour somewhat less open habitats than are used for nesting, at least during the day, when they may be found lying-up in open dehesas, vineyards, olive groves and broom scrub as well as in agricultural areas, very often irrigated ones. They are also seen commonly on the periphery of reservoirs, ponds and other wetlands (SWA). As ever, they are most active at night, when they forage for their food, which chiefly comprises large invertebrates (Amat 1986).

EGYPTIAN PLOVER *Pluvianus aegyptius*

(Pluvial, Ave-do-crocodilo)

There is a vague reference to a 19th century occurrence in Spain (Arévalo 1887). An adult was on a beach at L'Alfàs del Pi, Alicante, from 29 May to 12 June 2008 and later at a water purification plant at Cartagena, Murcia, where it remained from 18–26 June. Its wild origin has not been accepted (*Ardeola* 57: 495). Individuals have appeared in Jordan, Libya and the Canary Islands as well as in several European countries, although these last have tended to be attributed to escapes.

CREAM-COLOURED COURSER *Cursorius cursor*

(Corredor Sahariano, Corredeira)

Vagrant to Spain and Portugal. Has bred in Spain.

This courser breeds in the Canary Islands, North Africa and the Arabian peninsula, those of the Maghreb wintering south of the Sahara. It often wanders north of its breeding range, particularly during April–May and September–October (Isenmann & Moali 2000, BM, de Juana 2006). A considerable number have appeared in Europe although much less frequently recently than formerly. Thus in Britain only eight of 45 records have been post-1957. Similarly 16 of 25 records for France were in the 19th century and only one has been post-2000 (OF).

Single birds were obtained at Málaga on 28 August 1870 (Irby 1879, *Ardeola* 50: 135), Coria del Río, Sevilla, in 1883 (Irby 1895) and in Catalonia around 1888 (Fuset 1913). According to Arévalo (1887) this species tended to appear in summer on the beaches of Valencia, Málaga and Granada. There is a pre-1911 record from Menorca (Ferrer *et al.* 1986) and one was killed at Porto on 4 October 1928 (Reis Júnior 1931). Subsequent records have been much more recent. One was captured at Prats, Lleida, on 7 September 1965 (*Ardeola* 10: 64). Since 1984 there have been 19 accepted records from Spain and six in Portugal.

An unprecedented small invasion occurred in southern Iberia in spring 2001, giving rise to five occurrences between Algarve and Almería (*Ardeola* 50: 135–136). They included three or four breeding attempts in Almería: a probable nesting at Huércal-Overa and three confirmed instances in Tabernas. A flock of 11 in Tabernas on 28 May included a recently hatched

Figure 101. Records by month, 1870–2008. (n = 26)

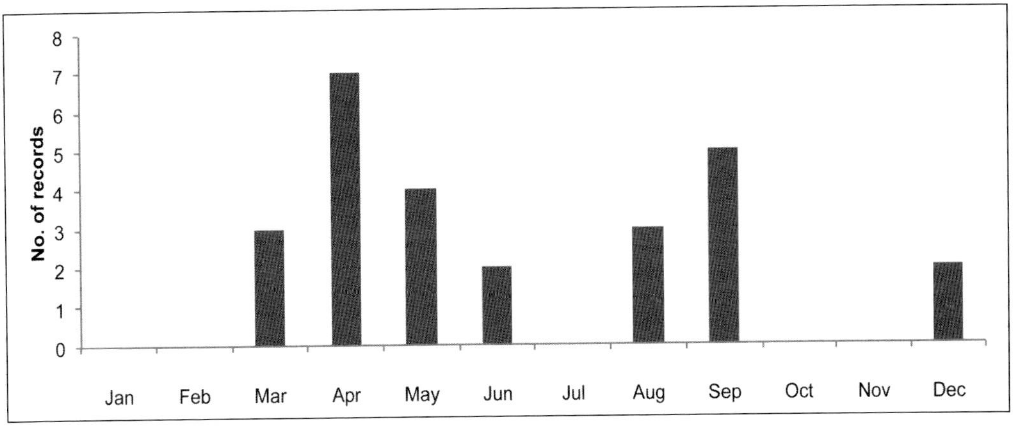

chick. Two nests were discovered there subsequently, in June–July: one failed and two chicks hatched in the other but their eventual fate is unknown. These were then the only known instances of Cream-coloured Coursers nesting in Europe (Gutiérrez 2001). Later in 2001 one was seen in coastal Galicia in September, at Ponteceso, A Coruña (*Ardeola* 50: 135). Two were present in suitable nesting habitat at Alpera, Albacete in May 2012, and in August two pairs were found to have nested successfully in the area, fledging at least two and one young respectively (RBS).

Breeding records apart, there are 26 Iberian records involving 28 individuals, including two records of two together and one group of three. Reported ages comprise 10–11 adults, one or two immatures and two juveniles: these last in August and September. The records are concentrated in the south and largely in spring (62%), with extreme dates 19 March and 9 June, and autumn, with extreme dates of 7 August and 25 September (Figs. 101, 102). A similar pattern is reported in the northern Maghreb but autumn records predominate in northern and central Europe and spring ones are most frequent in Sicily, Malta, Greece and Cyprus (de Juana 2006, Verducci *et al.* 2012).

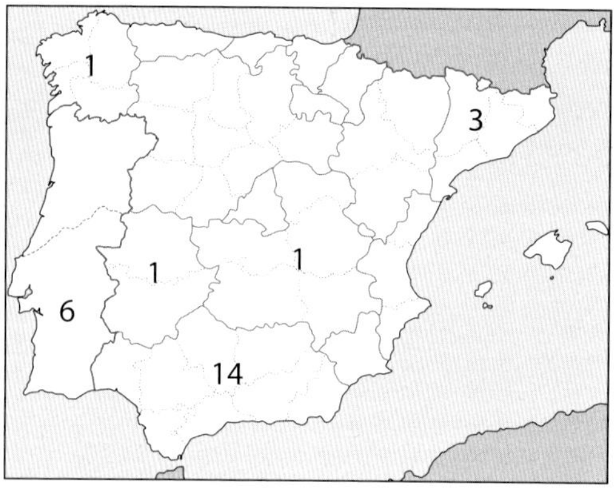

Figure 102. Records by region, 1870–2008 (n = 26)

Post-2010, single first-year birds were seen at three widely separated locations during 22–23 September 2011: on the Belén plains, Cáceres; at 1,200m in the Pyrenees at Auritz-Burguete, Navarra, and the Ría de Ortigueira, A Coruña. One was at Tarifa, Cádiz, in April 2012 (RBS). In Portugal, one was at the Ria de Alvor, Faro, in March–April 2013, and it or another was near Cape Raso, Lisboa, in May 2013 (Noticiario SPEA).

COLLARED PRATINCOLE *Glareola pratincola*

(Canastera Común, Perdiz-do-mar-comum)

***G. p. pratincola*. A summer visitor chiefly to southwestern Iberia and locally also in Castilla-La Mancha and the Levant coast. Uncommon or rare elsewhere.**

The Collared Pratincole is associated with the warmer regions of the Peninsula at the hottest times of year. It winters in sub-Saharan Africa. The chief breeding concentration is in southwestern Spain, including the Guadalquivir estuary and valley, Extremadura, coastal areas of Cádiz (Cádiz Bay and La Janda) and Huelva (Marismas del Odiel and Isla Cristina), and adjacent areas of Portugal in central and eastern Alentejo and eastern Algarve. There are also scattered populations across La Mancha and on the Levant coast, at the Ebro delta and coastal wetlands of the Comunidad Valenciana (SA1). It becomes progressively rarer away from the breeding range and is only a scarce migrant in the Balearic Islands (*A. O. Balears 2010*: 263).

The breeding habitat comprises low-lying, bare or sparsely vegetated land generally not far from water bodies. These last include lakes, reservoirs, ricefields and saltpans. The birds hawk in flocks for flying insects over such areas as well as over agricultural land. Much of their foraging activity is crepuscular. Many choose salt flats and dried-out pool beds on which to nest, which may make the nests vulnerable to flooding in wet spells. Others select newly tilled agricultural land or young crops: mostly cotton and sunflowers in Andalucía, where they may be at the mercy of further farming activities (Calvo & Furness 1995). The nest itself is no more than a scrape.

The Iberian population is not large but the Spanish birds alone now comprise over 75% of the population inhabiting south-west Europe and the Maghreb (Delany *et al.* 2009). There are often marked fluctuations in year-to-year abundance at particular sites. A national census in 1989 found 3,761–3,815 pairs: including 419–463 pairs in Doñana National Park, 2,500 elsewhere in Sevilla, 140 pairs in the Cádiz marismas, and 220 pairs on the plains of La Serena in Extremadura (Martínez-Vilalta 1991). The Spanish population in 1989–90 was at least 5,000 pairs, some 4,200 of them in Doñana (Calvo *et al.* 2003). More recently, a breeding season census in Andalucía in 2010 found 4,431 pairs in 250 colonies: 68 pairs in Almería, 632 in Cádiz, 52 in Córdoba, 924 in Huelva and 2,655 in Sevilla (*Ardeola* 59: 432).

Some 700–1,000 pairs breed in Extremadura, where the population has increased since the 1990s but is now considered stable. About half nest at reservoirs and the remainder do so in agricultural land, particularly in ricefields and fallow areas

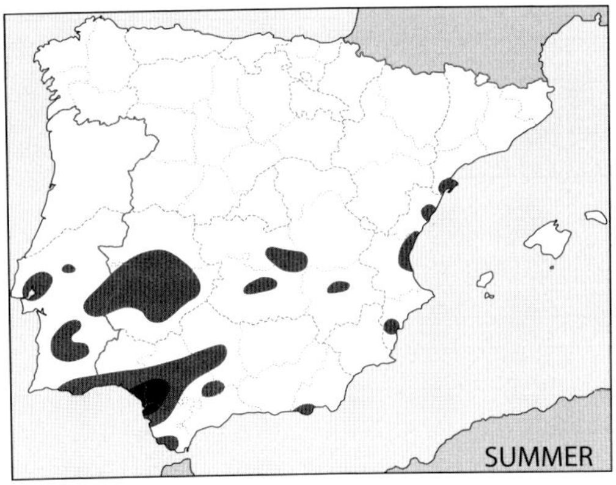

associated with wetlands (J. Prieta, pers. comm.). In particular, about half of the population is found within the Vegas Altas of the Guadiana river and on the plains and nearby reservoirs of La Serena (200 pairs), with notable colonies also at the Alange (50–100 pairs), Llerena (50–100), Los Canchales (30–50) and Guadiloba (25–50) reservoirs, among others, and smaller concentrations in steppe areas and ricefields elsewhere.

The Pratincole was once far more abundant in Doñana than it is now (Saunders 1871, Chapman & Buck 1893, Noble 1902). Valverde (1960) considered it even commoner than the Whiskered Tern, a species for which he estimated a population of over 50,000 pairs, and J. Vielliard (in Calvo 1993) estimated some 10,000 breeding pairs of Pratincoles there in 1962–65. Large interannual fluctuations in abundance occur in the Marismas: the nesting population during 1994–2000 ranged from 100 to 2,000 pairs (LR).

The Andalucian and Extremaduran populations together account for up to 90% of the Spanish breeding birds. Much smaller numbers occur in Castilla-La Mancha, chiefly in the Cigüela and Záncara valleys and in the Guadiana valley. The Toledo population ranged from 42 to 53 pairs during 1997–98 and that in Ciudad Real was 106–137 pairs during the same period (LR). The Pratincole also nests at least occasionally in Cuenca and Albacete. Its presence in Extremadura was first noted by Witherby (1928) who mentions eggs being sent to K. L. Skinner from the neighbourhood of Badajoz. For Castilla-La Mancha, Gil-Lletget (1945) refers to specimens in the Madrid Museum obtained in May–July at Daimiel, where Bernis saw the species in spring 1951 in addition to a colony of some 25 pairs at Quero, Toledo (Bernis 1952).

The chief concentrations on the east coast are at coastal wetlands. There the northernmost Iberian outpost is at the Ebro delta, where there were 75–85 pairs in 2001–02 and 72 pairs in 2009 (CA2, *A. O. Catalunya 2009*: 108). Other east coast sites that have supported colonies more or less regularly since the 1990s are the Marjal de Almenara and Prat de Cabanes-Torreblanca in Castellón; the Albufera, Marjal del Moro and Devesa del Saler in Valencia; and the Santa Pola saltpans and El Hondo in Alicante. There were at least 240 pairs in the Comunidad Valenciana in 2001 (Polo & Polo 2007). The breeding numbers on the Levant coast fluctuate as they do elsewhere but most colonies contain 30–60 pairs, often fewer, and none have exceeded 100 pairs until recently (LR): in 2009 there were 106 pairs at the Marjal de Almenara (*A. O. C. Valenciana 2009*: 59). The Collared Pratincole had a presence at the Mar Menor until the 1970s but was then extinct as a breeding species in Murcia until nine pairs attempted to nest at the Mar Menor in 2003 and six pairs nested near Cartagena in 2004; with poor results in both cases on account of disturbance (*Ardeola* 51: 250, 551).

The Portuguese breeding birds are chiefly in steppe habitats of central and eastern Alentejo, from Castro Verde and Ferreira do Alentejo north to Elvas, in the wetlands of eastern Algarve as at the Ria de Faro and Castro Marim, and in the Tagus estuary (PA2). The total population during 1986–95 was estimated at 315–550 pairs, including up to 150 pairs on the Tagus estuary pastures (Farrobo & Leitão 1997). The islands on the recently constructed Alqueva reservoir have provided secure nesting sites and 325 pairs nested there in 2006 (AP).

Collared Pratincoles begin to return to Iberia in late March but a few appear earlier in the month: for example, in 2009 three were at Ponta da Erva, Lisboa, on 14 March and one was at Quinta do Lago, Faro, on 17 March (*Anu. Ornitol.* 8: 67). The main arrivals are during the second half of April and into May. They depart between late July and the end of September, with stragglers remaining into October and, rarely, November. Post-breeding concentrations in July–August prior to migration often involve flocks of hundreds. Even larger gatherings occur in some years: for example 3,800, mostly juveniles, were at the Brazo del Este on 18 August 2007 (*Ardeola* 54: 396). Winter records, such as single birds at Santarém on 7 January 1996 and at the Brazo del Este, Sevilla, on 17 December 2009 (*Pardela.* 1: 19, *Ardeola* 57: 228) are highly exceptional. There are single recoveries of birds ringed as chicks in Spain from Senegal and Morocco.

Very few Collared Pratincoles are ever reported at any distance from the breeding range but there are occasional records of individuals or very small groups appearing on the northern Meseta, in Palencia; up the Ebro valley, in Navarra and the Basque Country; and in the Balearic Islands, most often in spring. Observations north of the Cantabrian Range are exceptional: for example there were only seven records from Asturias during 1900–97 (*A. O. Asturias* 4). Records from northern Portugal are also extremely unusual but there was one at the Minho estuary in April 1990 (*Airo* 2: 55).

BLACK-WINGED PRATINCOLE *Glareola nordmanni*

(Canastera Alinegra, Perdiz-do-mar-d'asa-preta)

Vagrant to Spain.

There are three accepted records of single birds, all from Mediterranean coastal wetlands during passage periods: at the Albufera de Mallorca on 25 April 1991 and on 9 May 2005, and at the Albufera de Valencia from 29 August to 19 September 2007. There are records of 35 individuals in Britain and 22 in France. A high proportion of the latter are on the Mediterranean coast, chiefly in Camargue.

LITTLE RINGED PLOVER *Charadrius dubius*

(Chorlitejo Chico, Borrelho-pequeno-de-coleira)

***C. d. curonicus*. A widespread breeding species in Peninsular Iberia and the Balearic archipelago. Chiefly a summer visitor, wintering in sub-Saharan Africa.**

The Little Ringed Plover is the most widespread of the Iberian breeding waders, although it is usually found at low densities. It nests in all regions of the Peninsula but is especially numerous in Spain in Andalucía, Extremadura, Castilla y León, western Castilla-La Mancha, Catalonia and Aragón, and also in southern and eastern Portugal. It also breeds in the Balearic Islands: in Mallorca, Menorca and Ibiza. It is scarcest in northwestern Portugal and Galicia and also north of the Cantabrian Range, as well as in the more arid regions of the meseta and southeastern Spain (SA2, PA2). Breeding Little Ringed Plovers are characteristic of inland rivers and lakes. They select areas of exposed riverbank, and gravel and sand banks, as well as the pebbly or sandy margins of reservoirs and gravel pits. Fast-flowing rivers and streams are avoided and so the species is absent from mountain ranges at any altitude. During a sample census in Spain in 2007 34% of birds were found along rivers, 21% at reservoirs larger than 20ha and 43% at small pools and farm reservoirs (Palomino & Molina 2009).

The size of the Iberian populations is not known with any precision. Available estimates during 1997–2003 put the Spanish population at 1,600–2,300 pairs (SA1), 2,500–3,000 pairs (Hortas *et al.* 2000) and at least 4,227 pairs (SA2). A sample census in 2007 (Palomino & Molina 2009) found 3,670 individuals, more than half of them in Andalucía, but there were 100+ at the Albuferas de Valencia and Mallorca; 110 at the Navalcán reservoir, Toledo; 80 at the Rosarito reservoir, Toledo and 80 at the Llobregat delta, Barcelona.

The Portuguese population has not been censused but the Little Ringed Plover is relatively common in Alentejo and within the Tagus, Sado and Guadiana basins. It is also present in the Sabor and Tuela basins in the north-east and along the lower reaches of rivers along the west coast (PA2). An estimate of 1,000–5,000 pairs is suggested for Portugal by Birdlife International (2004).

Nearly all European Little Ringed Plovers winter in sub-Saharan west Africa. Migrants from further north occur on passage, when small flocks appear at coastal and inland wetlands, including such habitats as saltpans where the species does not breed. The few foreign-ringed birds recovered in Iberia include individuals from the Low Countries (8), France (5), Britain (3), Germany (2), Denmark, Finland and Sierra Leone. There are single recoveries from Morocco and Senegal of birds ringed as chicks in Spain. Spring arrivals and passage across the Peninsula begin in March and continue during April and May. The return passage begins in August and is chiefly in September–October. Small flocks and some larger gatherings are

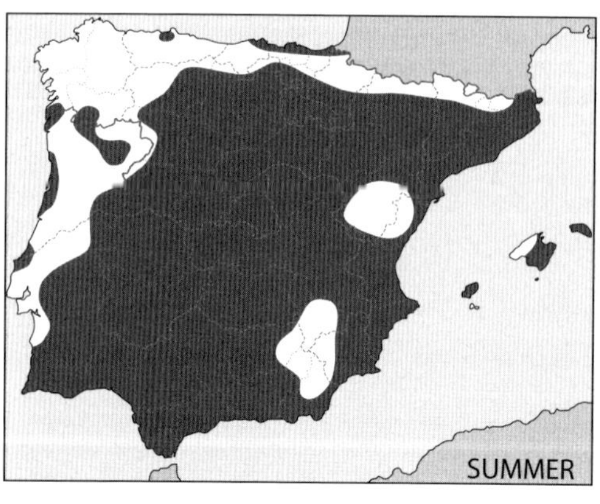

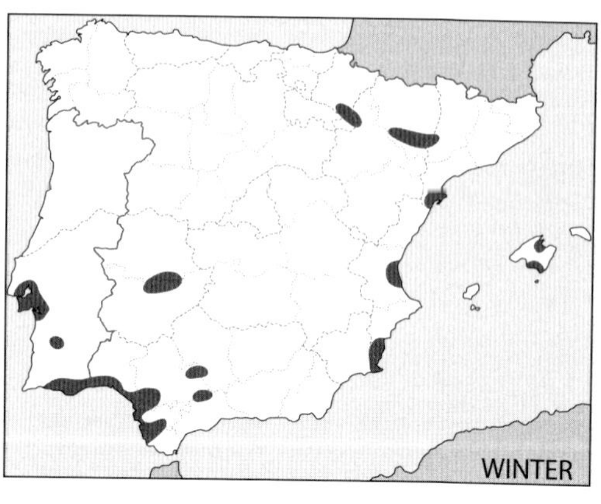

observed on passage and post-breeding: 371 in the Guadiana ricefields, Badajoz, on 20 July 2010 (*Ardeola* 57: 530) was an unusually large concentration.

Accounts prior to the mid 20th century make no mention of wintering but a small wintering population has since been detected, chiefly at estuaries and lakes in southwestern Andalucía and Extremadura, with some also at Levant coastal sites and in Mallorca (SWA). Alberto & Velasco (1988) suggested that the wintering birds might average no more than 100 birds in total. More recent observations suggest that this may now be an underestimate, or perhaps that the wintering population is increasing. The Spanish January counts of wintering waders detected an average of 243 Little Ringed Plovers in Spain during 1990–2009, although the annual totals ranged from just one bird to 1,241 (AAE). Some three-quarters of winter records came from the Andalucian coastal wetlands, particularly from Doñana, but regular wintering has also been reported from the ricefields of Extremadura; the Plana de Lleida, Catalonia, and the Albufera de Valencia: where there were at least 100 in January 2009 (*A. O. C. Valenciana 2009*, p. 60). There is also a count of 47 at Castro Marim, Faro, in January 1992 (AP)

COMMON RINGED PLOVER *Charadrius hiaticula*

(Chorlitejo Grande, Borrelho-grande-de-coleira)

***C. h. hiaticula*. Common on passage and also in winter, when locally abundant on Atlantic coasts.**
***C. h. tundrae*. Thought to occur on passage (Delany *et al.* 2009).**

The Common Ringed Plover is a familiar sight on Iberian Atlantic coasts during its passage periods and in winter. It also occurs on Mediterranean shores but is much less numerous there. Passage and wintering birds chiefly frequent sandy coasts and, especially, estuarine mudflats, as well as the shorelines of coastal lagoons, saltpans and ricefields. Migrants are regularly recorded on passage at inland wetlands, generally on sandy lake margins and usually in small numbers. This species does not breed any closer to Iberia than Brittany. Reports of nesting in Girona in 1969 (AI1); in Mallorca in 1989 (Clavell 2002) and the Ebro delta in 1993 (*Ardeola* 41: 98), remain unconfirmed and seem improbable. However, summering non-breeding individuals are not unusual in favoured locations.

Autumn passage begins in July, but is chiefly during August–October, extending into November (Tellería 1981, Galarza 1984, Castro 1993). Spring passage is apparent in March in the south but chiefly occurs during April–May, with stragglers into June. Small flocks appear on all coasts but large concentrations are generally only seen at a limited number of Atlantic coastal sites where the bulk of the wintering population also remains. Comparatively few are reported inland but small flocks may appear on passage at lakes, ricefields and reservoirs across the peninsular interior. There are also occasional records of wintering and summering individuals at inland sites. Most spring passage inland peaks in May and autumn birds occur from late August into October, peaking in September.

There are at least 161 Iberian recoveries of Common Ringed Plovers of which 71% were marked in Fennoscandia, and the remainder mainly in Britain, Germany or Denmark, with a few in Iceland (3), the Low Countries, Switzerland, Poland, Italy and the Czech Republic. One ringed in Spain was recovered in Senegal. Individuals ringed in winter in Iberia have also been recovered in Greenland (one) and Iceland (three), showing that at least some individuals of the putative race '*psammodroma*' may winter in Iberia and that not all descend to west Africa, as suggested by Delany *et al.* (2009). The countries of origin of most foreign-ringed birds confirm that the majority of those that occur in Iberia are of the nominate form. The race *tundrae* has seldom been identified and would usually pass unnoticed. However, three males believed to be of this form were at the Ría de Avilés, Asturias, on 15 June 1995 (*A. O. Asturias 2003*): the late date is compatible with migrants heading for Siberia.

Five Portuguese sites and six Spanish sites have been identified as internationally important for the species (Delany *et al.* 2009) since they sometimes hold at least 1% (730 birds) of the global population of *hiaticula* during passage periods and/or in winter. The Portuguese sites are the estuaries of the Minho/Miño, Sado and Tagus and the Rias de Aveiro and Faro (Formosa). The Spanish sites are concentrated around the mouth of the Guadalquivir but also include the Marismas de Isla Cristina and Cádiz bay. Some notable autumn passage counts have been made at some of these locations: including 5,280 birds at the Minho estuary in

2002, 4,334 at the Ria de Faro in 1995, 5,000 at Isla Cristina in 1996, and 2,593 in Cádiz Bay in 1996 (Delany *et al.* 2009). An even larger count was made in spring at the Marisma de Hinojos, Doñana, on 1 May 2001, when over 10,000 birds were present (*A. O. Doñana* 1: 110).

The total wintering population in the 1980s was estimated at around 4,500 birds in Spain (Alberto & Velasco 1988) and 2,500 in Portugal (Rufino 1989). The annual Spanish shorebird counts in January during 1990–2009 averaged 4,628 birds (range 929–12,646) with a moderate increase observed during this period (AAE). About 65% of the Spanish wintering population occurred at the coastal wetlands around the Gulf of Cádiz. Cádiz Bay has seen the largest January concentrations, averaging 2,011 birds during 1990–2009 with a maximum total of 6,533 birds (AAE). A further 20% of the wintering population was found in the Galician rías.

More recent Portuguese winter counts have found larger numbers than in the 1980s, although these may be at least partly result from more thorough prospection: the principal wetlands held an average of 5,447 wintering Common Ringed Plovers during 1992–95 and 4,974 during 2003–05 (AP). In addition a minority of birds winter on sandy coasts in Portugal; they totalled 281 in winter 2009–10 (Catry *et al.* 2010b).

The northern estuaries: in Galicia, Asturias, Cantabria and the Basque Country, including the Marismas de Santoña, hold some concentrations of Common Ringed Plovers during passage periods but only a minority remain there in winter. January counts in the Biscay coastlands during 1990–2009 found an average of 97 birds in Cantabria (maximum 234), 51 in Asturias (maximum 145) and 51 in the Basque Country (maximum 90): the principal location in this region has been Santoña, where up to 150 have been counted in January. Comparatively small numbers are also found at Mediterranean wetlands, both on passage and in winter. The largest winter concentration there is at the Ebro delta where January counts during 1996–2009 have averaged around 336 birds, ranging from 74 to 604 (*A. O. Catalunya*), with an average annual increase of 5% (CWA).

SEMIPALMATED PLOVER *Charadrius semipalmatus*

(Chorlitejo semipalmeado, Borrelho-de-bando)

Vagrant to Spain.

The four accepted records are all from the Galician coast at the Ensenada da Ínsua or the nearbyLaguna de Traba: an adult on 17 July 1988; a first-summer from 13 May to 15 August 1990 (Rabuñal 1990); a juvenile on 4 and 10 September 1999, and a juvenile on 4 October 2004. The species is unrecorded in France and there are only two records from Britain. However, there are accepted records of 52 individuals from the Azores and one from Madeira.

KILLDEER *Charadrius vociferus*

(Chorlitejo Culirrojo, Borrelho-de-coleira-dupla)

Vagrant to Spain and Portugal.

All Iberian records have been of single birds on the Atlantic coast. Four are from Galicia: at the Budiño lagoon from 8–18 September 1988 (Costas 1990); Baiona on 14 November 1997; the Cecebre reservoir from 24 November 2007 to 6 February 2008, and A Ínsua on 26 January and 9 February 2008. The sole accepted Portuguese record was at Azambuja, Lisboa, on 28 February 1998. There is second Portuguese record, one at Sacavém, Lisboa, in October 2011 (RBP). They included two adults, a possible adult and a first-winter. European records include 55 individuals in Britain, 20 in Ireland and ten in France; and a number from the Azores.

KITTLITZ'S PLOVER *Charadrius pecuarius*

(Chorlitejo Pecuario, Borrelho-do-gado)

Vagrant to Spain.

A male, probably an adult, was at the Laguna de Villacañas, Toledo, from 12–17 March 2001. This African species has been found in Israel, Cyprus, Greece, Morocco, Norway and France.

KENTISH PLOVER

Charadrius alexandrinus

(Chorlitejo Patinegro, Borrelho-de-coleira-interrompida)

***C. a. alexandrinus*. A common breeding species on sandy western, southern and eastern coasts in the Peninsula but largely absent from the Biscay coastlands. Also common in the Balearic Islands. A few nest very locally inland. Partial migrant. Birds from elsewhere in western Europe occur on passage and in winter.**

Kentish Plovers are common on sandy beaches around most of the Iberian Peninsula, and in the Balearic Islands, although they only nest sporadically in the Biscay coastlands east of Galicia. A small minority nest inland, chiefly in La Mancha (SA2). Beach nesters had to give a lot of ground during the 20th century when mass tourism led to coastal development on a very large scale, especially within the Mediterranean, and a considerable decline in the breeding population has occurred since the 1960s. Considerable damage to nests and breeding habitat is also done where amenity beaches are 'cleaned' regularly of flotsam, as has been reported especially in the Comunidad Valenciana (LR). The birds nest above the strandline on beaches, often in the first line of dunes if any. Others nest around saltpans and salt marshes, as well as in sparsely vegetated salt flats and coastal grasslands. Inland Kentish Plovers typically nest on the sandy margins of brackish lagoons but some also nest on reservoir shores or islands and, in recent years, at the Guadiana ricefields in Extremadura (Masero & Santiago-Quesada 2008).

The birds often nest in loose colonies and are typically double brooded. The earliest clutches are generally laid in April or early May and the second clutches in late May or June (Amat 2012). Many nests are lost to predators so replacement clutches are also frequent: the predators are most often Yellow-legged Gulls but Gull-billed Terns have been a problem for plovers nesting at the Laguna de Fuentedepiedra (Fraga & Amat 1996) and Carrion Crows affect those nesting on Galician beaches (Domínguez & Vidal 2003). The Spanish breeding population was estimated at 5,000–6,000 pairs (SA1: 1997), a minimum of 2,565 pairs (SA2: 2003) and 4,072–4,245 pairs (Palomino & Molina 2009, Table 30).

Table 30. Regional distribution of breeding Kentish Plovers in Spain in 2007 (2004 in Castilla-La Mancha). Figures are minima. After Palomino & Molina (2009).

Region	Pairs	Region	Pairs
Catalonia	1,651	Galicia	79
Andalucía	1,261	Extremadura	5
Comunidad Valenciana	524	Aragón	8
Castilla-La Mancha	158	Castilla y León	1
Murcia	149	Balearic Islands	236
		TOTAL	4,072

In 2007 over 60% of the Spanish breeding population was located within Andalucía and Catalonia, with a further 12% in the Comunidad Valenciana. Some 6% were in the Balearic archipelago. The only breeding Kentish Plovers in northern Spain were in Galicia. The species formerly bred locally in Asturias, until 1980–84 at Cape Peñas (Noval 2001) but no longer does so. Inland sites, excluding Doñana where the habitat is essentially coastal, accounted for just 5% of the mainland population. The

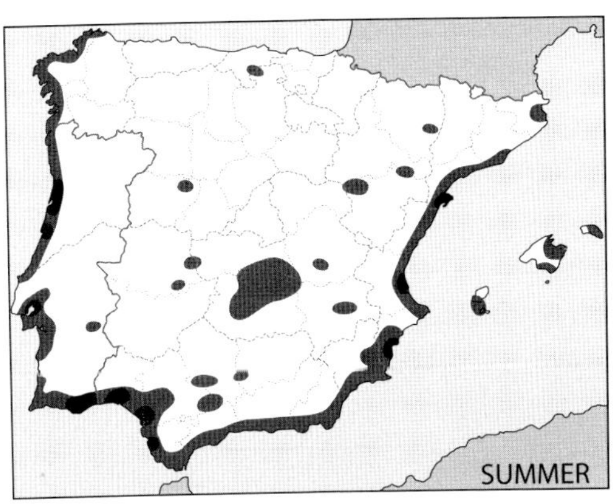

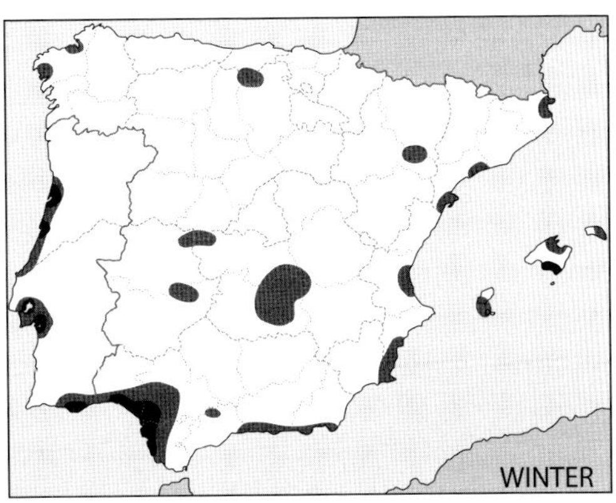

majority of the inland birds were at the lagoons of Castilla-La Mancha but Fuentedepiedra also had a significant population for an inland site, 27 pairs, although many fewer than the 100 pairs found there during 1991–93 (LR).

Population trends in Spain vary between regions although the principal tendency is a decline in numbers. Trends are obscured by between-year fluctuations associated with rainfall, particularly at inland sites – including Fuentedepiedra – that may dry up in some years, and at Doñana, where only a few pairs may nest in dry years but 550–2,000 pairs occur in wetter seasons (Hortas 2001, García *et al.* 2000). The 2007 census coincided with a poor year at Doñana, with only 124 pairs found in the National Park. Cádiz Bay is one area where a clear decline has been seen, from 770 pairs in 1991 to 420 in 1998 (LR). Since 2007 there have been some indications of an increase in Extremadura, where a minimum of 47 pairs nested at the Valdecañas reservoir in 2011 (*Ardeola* 58: 498).

The population in Catalonia is stable at the principal regional breeding site: the Ebro Delta, but has fluctuated at the two other major sites, the Llobregat Delta and the Aiguamolls de l'Empordà, with further recent decline at the Llobregat (Palomino & Molina 2009). Breeding numbers in the Comunidad Valenciana in 2007 were much lower than the 1,000 pairs estimated by Figuerola *et al.* (LR). The principal sites there are the Santa Pola saltpans, El Hondo and the Lagunas de La Mata y Torrevieja, all in Alicante, and the Albufera de Valencia. The regional population declined by some 50% since the early 1990s (Palomino & Molina 2009), most markedly in Castellón. In the Balearics the Kentish Plover was regarded as very common in Mallorca by Munn (1921, 1948b). More recently, its population is considered stable in Mallorca, where 80% nested in 2007 at two sites: the Salobrar de Campos and the Albufera, and also in Formentera, but the small populations in Ibiza and Menorca are declining (Garcias 1996, Palomino & Molina 2009).The ten sites that held the most breeding pairs in 2007 together accommodated about 65% of the Spanish population (Table 31), the Ebro delta alone having 36% of the Spanish breeding pairs.

SITE	Pairs
Ebro delta	1,575
Cádiz bay	434
Santa Pola saltpans	164
Marismas del Guadalquivir	124
Marismas del Odiel	96
Cape Gata saltpans	92
Lagunas de La Mata – Torrevieja	81
Salobrar de Campos (Mallorca)	80
Marismas de Isla Cristina	76
Albufera de Valencia	67
TOTAL	2,789

Table 31. The ten principal breeding sites of the Kentish Plover in Spain in 2007. Figures are the minimum number of pairs censused (After Palomino & Molina 2009).

The Portuguese population occupies almost the entire coastline as well as littoral wetlands and saltpans (PA2). The distribution is interrupted in rocky areas and where coastal developments are most intensive, as between Peniche and Setúbal. A few nest at inland wetlands, such as the Alentejo ricefields and the Alqueva reservoir. The breeding numbers have not been censused but the estimated population is sizable, within the range of 1,500–5,000 pairs (BE).

Iberian Kentish Plovers are resident or partly migratory. Bernis (AMI) considered the species to be chiefly a summer visitor but Chapman & Buck (1893) stated that many wintered in Spain and Munn (1921) noted that it was present all year in Mallorca, where numbers increased in spring. Significant numbers have certainly been present in winter since at least the 1970s and the wintering population may have been underestimated previously. The January wader counts in Spain during 1990–2009 found 4,685 birds on average (range 751–9,205; AAE) and probably reflect the fact that many birds winter on sandy coastlines that are difficult to census thoroughly. However, most wintering birds are to be found in saltpans, ricefields and intertidal mudflats (Velasco & Alberto 1994). The wintering numbers in Spain are thought to have increased considerably during 1980–2009 (AAE). The principal wintering areas in Spain are in coastal Andalucía, with significant numbers also wintering at the Ebro delta, the southern Levant coastal wetlands and in the Balearic Islands. In Portugal the Ria de Faro regularly holds some two-thirds of the Portuguese wintering population.

The largest reported January concentrations in Spain during 1990–2009 were at Doñana (up to 4,762 birds) and in Cádiz Bay (up to 2,260) but estuaries and sandy beaches from the Strait to the mouth of the Guadiana all supported gatherings of up to several hundred individuals (AAE). Only 35–40 pairs nest at the Marismas del Odiel but up to 500 birds winter there (Gacio & Sayago 2011). The Ebro delta has a significant wintering population in some winters: January counts there during 1996–2009 averaged some 400 individuals but between-year variation was large, ranging from 23 to 1,046 birds (*A. O. Catalunya*). The wintering population in the Balearic Islands during 1990–2009 averaged 467 birds in January (range 39–831; AAE), the Salobrar de Campos, Mallorca, and the Ibiza and Formentera saltpans regularly attracting most birds. Small numbers have also been found inland in winter in Castilla–La Mancha, Castilla y León, Aragón and Extremadura: a count of 321 at Nava Grande lagoon, Malagón, Ciudad Real, on 20 December 2001 was probably exceptional (*A. O. Ciudad Real 1995–2001*: 194).

The Portuguese wintering population averaged 840 birds during 1975–78 but was 3,800 birds in 1992–95 and 3,344 birds in 2005–07 (AP). In addition to the Ria de Faro, the Tagus, Sado, Mondego and Minho estuaries and the Ria de Aveiro are also important wintering sites (AP). A census of non-estuarine coasts in Portugal in 2009–10 found 578 birds (Catry *et al.* 2010b).

At least some Iberian birds are dispersive outside the breeding season. For example, colour-marked individuals from the Laguna de Fuentedepiedra have been found in coastal Andalucía from Huelva to Almería, at other lagoons in Málaga and Sevilla, and in Morocco (Amat 2012). However, those that nest further north in Europe are largely migratory and some clearly cross Iberia on passage with an unknown number remaining to winter. Passage has been reported from March to early June in spring and in September–October in autumn (AI1). Some very large counts have been made in coastal Portugal during the southward passage period, notably 4,314 at the Ria Formosa in August 1992 and 3,046 at the Tagus estuary on 13 August 1995 (AP). There are Iberian recoveries of birds ringed in Germany (17), the Netherlands (14), France (11) the Baltic States and Italy, as well as one from Ukraine, and eight foreign recoveries of full-grown birds ringed in Spain: from Belgium (3), Germany (2), France (2), Morocco and Senegal.

The movements themselves are chiefly apparent from numerical fluctuations at coastal sites, as well as records from areas where the species does not otherwise occur. For example, migrants are recorded on the coasts of Asturias in very small numbers, in April and early May and during August–October (Noval 2001). Similarly, 40 at ricefields at Santa Amalia, Badajoz, on 7 March 2010, and 62 at the Campo Lugar-Palazuelo ricefields, Cáceres/Badajoz, on 7 October 2010 (*Ardeola* 57: 228, 58: 498) are likely to have been on overland passage.

LESSER SAND PLOVER *Charadrius mongolus*

(Chorlitejo Mongol Chico, Borrelho-pequeno-de-colar-ruivo)

Vagrant to Spain and Portugal.

There are two Iberian records. The first was an *atrifrons* group individual at the Lagunas de Alcázar de San Juan, Ciudad Real, on 21 June 1981 (*Ardeola* 32: 140). More recently a *mongolus* group male was at Mitrena, Setúbal, from 12–15 August 2003. A *mongolus* group bird appeared on La Palma, Canary Islands, only 12 days after the latter record (*Ardeola* 53: 195). The very rare western European records are most often in summer. Five have been found in Britain: three apparently of the *mongolus* group and two of the *atrifrons* group. Six found in France have including at least one *atrifrons* group bird.

GREATER SAND PLOVER *Charadrius leschenaultii*

(Chorlitejo Mongol Grande, Borrelho-grande-de-colar-ruivo)

Vagrant to Spain.

The three accepted records were all of single birds on typical midsummer dates. Birds showing the characters of *C. l. columbinus* have appeared twice at the Ebro delta: an adult from 25–29 August 1993, and a juvenile from 11–13 August 1996 (Figuerola & Bertolero 1996a). A male showing the characters of *leschenaultii/crassirostris* was on the Atlantic coast, at Ponteceso, A Coruña, on 31 July 2007. This rare vagrant to western Europe is somewhat more frequent than the Lesser Sand Plover. There are records of 15 individuals from Britain and six from France. One has also been found in Morocco, in March 2006.

EURASIAN DOTTEREL *Charadrius morinellus*

(Chorlito Carambolo, Borrelho-ruivo)

A sporadic breeding species in the Catalan Pyrenees and in Andorra. Scarce but widespread on passage and also locally in winter in the south.

The breeding range extends from Scotland eastwards across the Arctic tundra of the Palearctic, with a few small outpost populations on mountain tops further south. These include a tiny population that nests in the eastern Pyrenees, at least during some years. A breeding presence in the Pyrenees had long been suspected but the first nests were found there as recently as 1982 in France (Lescouret & Gériard 1982) and 1994 in Catalonia (SA1). Earlier reports of nesting in the Pyrenees in Girona in 1956 and, away from this region, in the Picos de Europa in 1978 (AI1) are no more than sightings of two and four birds there respectively in late summer that possibly were just migrants (CA1, Lucio & Purroy 1985).

The known breeding range is centred on the Puigpedròs and Puigmal massifs, Girona, on the border with France, as well as on the nearby French massif of the Puig de Carlit (LR). All these peaks exceed 2,900m and Dotterels have been found here in the breeding season above 2,400m. The breeding sites are on very sparsely vegetated alpine plateaus. Birds have also been found in adjacent Andorra in the breeding season although there are no definite records of nesting there (Crozier & Argelich 1993). Nearly all Pyrenean breeding records, and all but one of those in Spain, have been during the 1990s. No more than five breeding attempts have been reported in the Catalan Pyrenees in any one year and the only instance of definite breeding anywhere in the Pyrenees since 1999 was in 2003, at a previously unknown site (CA2, OF).

The Dotterel is better known as a scarce passage migrant. Records of passage birds are widespread but usually involve very small numbers. The passage periods are well-defined (Fig. 103): from mid March to early May in spring, peaking in April, and from mid August to mid October, in autumn, peaking in September. All Portuguese records, with the exception of two winter observations, have been in autumn (AP). In general autumn records are much more frequent, accounting for some 80% of observations, and small flocks often then remain at particular sites for short periods. Most passage birds are reported from lowland sites, where they favour open, barren habitats notably including ploughed and fallow farmland, pastures and steppe areas (Lucio & Purroy 1985). There are occasional reports at or near wetlands and on coastal grasslands. A minority of reports involve usually single birds or very small groups found on mountain summits, with observations from the Pyrenees, the Cantabrian Range, the Iberian Range and the Sierra Nevada. Flocks of up to 30 sometimes appear at 2,300–2,700m on the Sierra Nevada in September, and sometimes as early as July, where they may remain for several weeks before resuming their migration (Garzón & Henares 2012). The probability of detection is much higher, however, at lowland sites.

Figure 103. Distribution of records by half-month periods (n = 140). Records are assigned to the date of first observation.

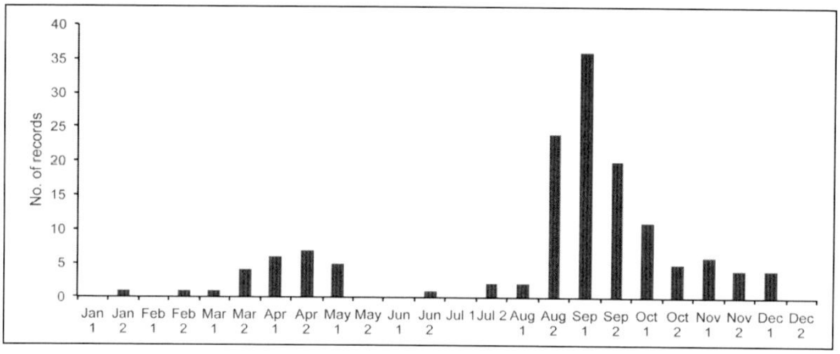

Dotterels are widely distributed on passage and may appear almost anywhere, although they are most seldom reported from northern Portugal, Galicia and the Biscay coastlands. Mean trip sizes are largest in Castilla-La Mancha (21.9 birds, n = 17) and Extremadura (14.6, n = 5), where birds are reported from farmland and steppes, suggesting that perhaps the majority of migrants cross the central part of the Peninsula. There are records of much larger aggregations, notably at a known staging area on the outskirts of Albacete city where Dotterels occur with some regularity during August–October and where peak numbers ranged from 126–300 birds during 2010–12 (*Ardeola* 57: 531, 58: 498, 59: 433). There are fairly frequent records from southern Portugal, eastern Spain and the Balearic Islands, but mean trip sizes here are quite small: (Portugal 3.0 birds, n = 21; Comunidad Valenciana 3.1, n =17, Balearic Islands 2.9, n = 13) and many sightings are of single birds. A flock of 11 at Valldemosa, Mallorca, on 14 September 2010 (*A. O. Balears 2010*: 80) was exceptional for the Balearics.

Trips are often reported fairly regularly at a limited number of staging sites, where birds occur in successive years at about the same period. Such places include Belchite, Aragón; Las Almoladeras, Almería; Los Lances beach, Cádiz; Doñana; and the farmland at Villacañas, Toledo, at Alcázar de San Juan, Ciudad Real, and near Albacete (see above). Other such places are likely to be to be found in the vast farmlands of the meseta especially. In Portugal the more regular sightings come from the Sagres peninsula and Cape Espichel (AP).

The winter quarters of western European Dotterels are in North Africa and those that cross the Peninsula chiefly winter in the uplands of Morocco. A winter presence has been reported from a few sites in Iberia, including Doñana; the steppes of La Serena, Badajoz and, especially, the steppes of Las Almoladeras, Almería. Records from the first two of these sites have been early in the winter and these were probably birds that subsequently continued south to Morocco later in the season. However, a small flock has overwintered at Las Almoladeras since at least 2003; the numbers involved vary but 40 were present in December 2011 (*Ardeola* 59: 181). Las Almoladeras is an area of dwarf shrub steppe near sea level in a region that is also particularly mild in winter. A further recent instance of over-wintering in Spain is also noteworthy for being in the north-east: 13 remained at Alfès, Lleida, from 2 December 2007 to 2 April 2008 (*A. O. Catalunya 2008*: 103).

PACIFIC GOLDEN PLOVER *Pluvialis fulva*

(Chorlito Dorado Siberiano, Tarambola-dourada-siberiana)

Vagrant to Spain and Portugal.

One taken at Málaga in 1878 was included in Lord Lilford's collection (Irby 1895). The second Iberian record was a bird in breeding plumage killed by a hunter at the Guadalhorce estuary on 17 August 1980 (*Ardeola* 44: 127). There have since been eight accepted records from Spain and two from Portugal, all of single birds. They included three first-year birds, two second/third-years and three identified as adults. Most have been in summer, as happens elsewhere in Europe, with extreme dates of 22 July and 9 September. They have generally been short-stayers but one remained from 23 August to 6 September. An exceptional long-stayer, an adult, was reported from the Ebro delta from 28 August 2013 until at least February 2014 (RBS). Nearly all have been seen on coasts: in deltas, marshes, lagoons or saltpans. Mediterranean coastal records (six) predominate over Atlantic ones (four), unlike in the American Golden Plover. The plumage details of the first Portuguese bird, at the Tagus estuary in August 2005, identified it as the same individual that had been seen in Ireland a week earlier (Moore *et al.* 2007). There are records of 77 individuals from Britain, 12 from Ireland and 22 from France. There are none from Morocco but there are two historical records from Algeria and a recent record from the Azores.

AMERICAN GOLDEN PLOVER *Pluvialis dominica*

(Chorlito Dorado Americano, Batuiruçu)

Vagrant to Spain and Portugal.

This plover is reported relatively frequently in Europe. There were 277 individuals recorded in Britain by 2005 and there are records of another 198 in Ireland and 54 in France. There are also 14 accepted records from the Azores, one from Madeira, 19 from the Canaries and at least five from the Cape Verdes (Hazevoet 1999). There are at least 20 records from coastal west Africa, from Morocco to Togo (Schmaljohann & Thoma 2005), and others further south.

The Iberian records are relatively recent. There are 24 accepted records from Spain during 1984–2010 and 12 from Portugal, all of single birds except for one of two together in Galicia in spring 2009 (*Ardeola* 58: 457). Three-quarters were on Atlantic coasts, from Galicia to Andalucía (Fig. 104). They have a major peak in autumn, extreme dates being from 9 September to 3 November, with a lesser spring peak, when records have been between 20 March and 24 May (Fig. 105). The autumn birds have included 17 first-year birds and eight adults. However, the spring records have included seven first-summer individuals but no adults. Adults migrate earlier than juveniles, as is frequent among waders, and indeed adults have been seen in Iberia between 9 September and 7 October and first-winters from 19 September to 3 November. There is one instance of overwintering: a first-year bird remained at the Ensenada da Ínsua, A Coruña, from January to May 1984. There are also three summer records: a probably immature bird at the Ria de Alvor, Faro, on 16–17 July 2000; an adult at the Lagoa dos Salgados, Silves, 17 July 2009, and an indeterminate bird at Gozón, Asturias, on 1–11 July 2009. A further 11 Spanish and eight Portuguese records during 2011–12 include a dispersed influx of at least nine birds in October 2012 (*Ardeola* 60: 462, RBS). One in Mallorca in May 2013 (RBS) would be the first Balearic record.

Iberian spring records make up a quarter of the total, a much higher proportion than observed in Britain (9%; Dymond *et al.* 1989) or France (only one spring record; OF). However, in the Canary Islands there have been four during March–May and 14 in September–December; in the Cape Verdes there is one October record and four during January–April, and all four records from Morocco have been during March–June. It may be that some individuals winter in Africa and return to North America in spring along a more westerly route than they followed in autumn, as happens with the seasonal movements of the species in the Americas (de Juana 2006).

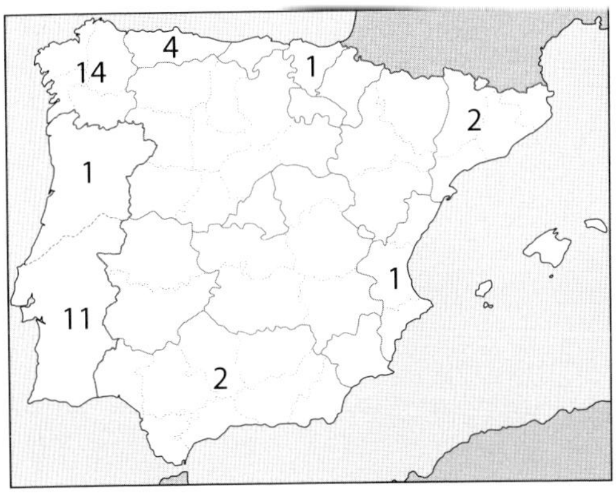

Figure 104. Records by region, 1983–2010 (n = 36)

Figure 105. Records by month, 1983–2010 (n = 36)

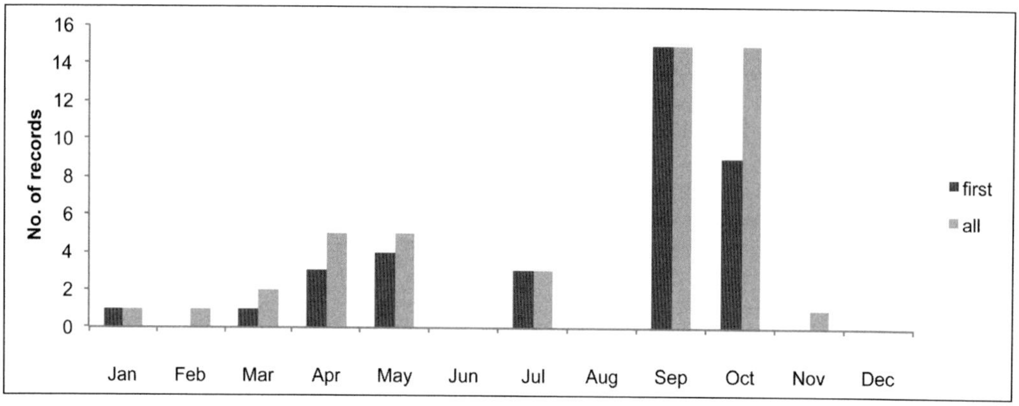

EUROPEAN GOLDEN PLOVER *Pluvialis apricaria*

(Chorlito Dorado Europeo, Tarambola-dourada-comum)

***P. a. altifrons*. Common on passage and in winter, especially in southwestern Iberia.**
***P. a. apricaria*. Present on passage and in winter, especially in northern and northwestern Iberia.**

A few Golden Plovers appear in late September on northern coasts but more widespread arrivals begin in October–November, with the largest influxes later in the winter. They are most abundant during December–February, numbers peaking in January (Leitão & Peris 2003, Calvet *et al.* 2004). They depart during February–April, with a few stragglers remaining into May (AMI). Occasional individuals, some perhaps winged by hunters, have been reported in summer. Some cross the Peninsula to winter in North Africa but the majority remain in Iberia, where they are most abundant in the south-west: in Alentejo, Extremadura and in western Andalucía: principally in the Guadalquivir valley and the Cádiz lowlands (AMI, AI1). They are more thinly distributed elsewhere and largely absent from much of the northern meseta where winter conditions are often harsh, although some may remain there in favourable localities: for example, up to 4,000 were at Villafáfila on 6 February 2004 (*Ardeola* 51: 551).

Flocks winter on pastures, steppe grasslands and arable land, often with Lapwings. In coastal areas they also frequent littoral grasslands and the fringes of marshes and wetlands. They are protected in Spain, but not in Portugal: where 31,000 are hunted annually on average (Béchet 2009). As a result, a considerable number of ringing recoveries are available to identify their countries of origin and/or the areas that they pass through on route to Iberia. They include at least 314 birds ringed in The Netherlands, 36 from Iceland, 20 from Belgium, 15 from Scandinavia, ten from Britain, two each from Finland, the Faeroe Islands, France and Italy, and single birds from Switzerland and Latvia. Many, including those from the Low Countries, were ringed on passage or in winter and so are of indeterminate origin but they include recoveries of birds ringed in Iceland (*altifrons*) and as pulli in Britain (*apricaria*), confirming the occurrence of both subspecies. *P. a. apricaria* most often occurs when harsh weather displaces the birds southwards from their usual winter quarters in northwestern Europe (Delany *et al.* 2009).

The Iberian wintering and passage populations are of the order of several hundred thousand birds (SWA), more when severe winters displace birds south from France and Britain. Castellarnau (1877) noted that they were more abundant during cold and wet winters, and Gil Lletget (1945) remarked that large numbers were sold in the Madrid markets in some winters. The largest flocks are regularly seen in southern Portugal and southwestern Spain. The principal Portuguese wintering areas, in Alentejo, Ribatejo and the Tagus estuary, attract very large numbers in some winters. January counts there in 2000–05 ranged from 44,500–454,000 birds and averaged some 200,000, 90% of the Portuguese wintering population (Leitão 2005). The large fluctuations in numbers in this region reflect both the incidence of cold spells further north, notably in Britain and The Netherlands, and also the occurrence of drought years in the wintering areas (Leitão & Peris 2004).

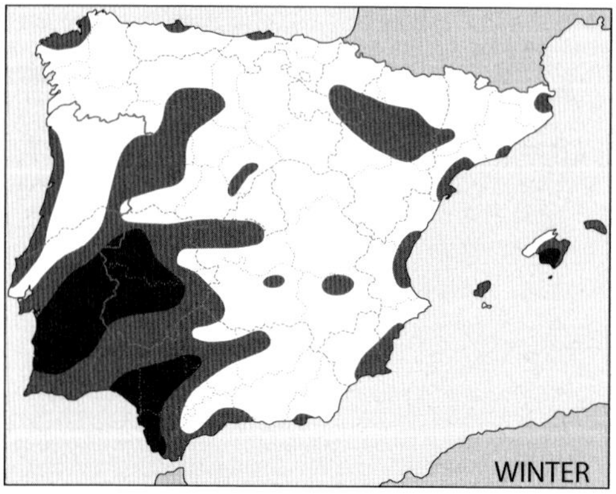

During the Spanish Winter Atlas fieldwork in 2007–10, some 49% of all Golden Plovers were found in Extremadura, with 21% in Andalucía and 8% in Castilla-La Mancha. The Ebro Delta sees the largest winter concentrations on the eastern Iberian littoral: January counts here during 1996–2009 averaged *c.*3,000 birds, and 5,839 were censused in 2009 (*A. O. Catalunya*). Other eastern wetlands that regularly held wintering concentrations during the same period included the Aiguamolls de l'Empordà, Llobregat delta, El Hondo and the Albufera de Valencia but these seldom exceeded a few hundred birds : the 1,763 at the Albufera de Valencia on 14 January 2009 was a site record (*A. O. C. Valenciana 2009*: 62). Counts in Mallorca of 1,800 at the Salobrar de Campos on 26 January 1999 and 2,400 at Vilafranca on 17 January 2010 (*A. O. Balears 1999*: 75, *Ardeola* 58: 498) are exceptional for the Balearic Islands.

GREY PLOVER — *Pluvialis squatarola*

(Chorlito Gris, Tarambola-cinzenta)

Common on passage and in winter on Atlantic and Biscayan coasts, scarcer in the Mediterranean and in the Balearic Islands. Occasional inland.

Significant numbers occur on passage and in winter at all the principal wetlands of the Atlantic and Biscay coasts. Comparatively few occur at Mediterranean wetlands, where only the Ebro delta regularly attracts a large passage and wintering population. They favour the intertidal zone, feeding on exposed mudflats and sand at low tide, which accounts for their relative scarcity in the Mediterranean. The largest concentrations are in the rías or estuaries of the north, west and south-west but some also occur on the sandy shores of coastal lagoons, on coastal grasslands and also in saltpans and ricefields.

Migrants begin to arrive from their Siberian breeding grounds in August but the principal arrivals are during October–November. Large numbers continue south to winter on the coasts of North-west and West Africa. Spring peaks in abundance at the key Atlantic sites are noted in February–March and then again in April–May, perhaps corresponding to birds that have wintered in Iberia and in Africa respectively (Rufino & Araújo 1987, Domínguez & Rabuñal 1989, Hortas 1990). Some late migrants occur in June and there are always a few summering non-breeders at key coastal locations: up to 400 were at Ria Formosa in June–July 1992 (Encarnação 1992). There are a few Iberian recoveries of birds ringed on passage in Germany (4), Britain (2), The Netherlands, France and Poland, as well as one from arctic Russia, and one ringed in Portugal was found in France. The mean Iberian wintering population in the 1980s was estimated at 5,800 birds (Alberto & Velasco 1988). In Spain the January counts of 1990–2009 found 6,258 birds on average (range 2,158–to 11,071; AAE). The total Spanish wintering population increased by about 3.5% per year during 1980–2009 (AAE).

Portuguese sites probably hold the major fraction of the Iberian wintering population, together supporting 5,000–10,000 birds in recent decades. However, the numbers using the Tagus estuary have shown a tendency to decline, reflected in the mean national wintering population averaging 8,941 birds in 1992–95 but only 4,882 in 2005–07 (AP). Three Iberian sites have been identified to be of international importance for the species (Delany *et al.* 2009): the Tagus estuary, Ria Formosa and Cádiz Bay. Other noteworthy passage and/or wintering sites include Santoña; the Ría de Villaviciosa, Asturias; the Ría de Arousa and other Galician Rías; the Miño estuary; the Ria de Aveiro, the Sado estuary, the Marismas del Odiel, Doñana and, in the Mediterranean, the Ebro delta. All of these sites regularly hold up to a few hundred birds and some attract many more. The Spanish wintering population is concentrated in the south-west, on the coasts of Huelva and Cádiz: where the largest January counts of Grey Plovers have been at Cádiz Bay (up to 5,883 birds) and Doñana (up to 1,593), and in Galicia and the Biscay coastlands: where the largest counts have been at the Ría de Arousa (up to 1,918 birds). Andalucía and Galicia together held some 70% of wintering Grey Plovers during 1990–2009 (AAE). The Ebro delta wintering population, by far the largest in Mediterranean Iberia, numbered some 600 birds in January during 1978–80 (Alberto & Purroy 1981) and has averaged somewhat higher more recently: 1,066 birds during 1996–2009 (range 631–1,709) (*A. O. Catalunya*, CWA).

Inland sightings are widespread but generally involve just single birds or very small groups, exceptionally flocks of 20–40, seen on the shores of wetlands or in the Guadiana and Ebro ricefields. They appear to be more frequent during the spring passage and may mainly involve immature birds (AMI). Winter records also occur inland.

SOCIABLE LAPWING *Vanellus gregarius*

(Avefría Sociable, Abibe-sociável)

Very rare. Recorded in Spain and Portugal, mainly in winter.

This formerly numerous but now 'Critically Endangered' species is still recorded in Europe with some frequency, mainly in autumn and spring. However, the Iberian records suggest that there may be a wintering area in the peninsular south-west (de Juana 1991, 2011). There are 119 records for France, but only one for Morocco, in December 1972 (Giraud-Audine & Pineau 1973) and another from the Canary Islands, in February 1992 (*Ardeola* 41: 110).

There are 19th century reports from Iberia. One was on sale at a Cádiz market in February 1868 (Saunders 1871) and there are references to winter occurrences in Málaga (Arévalo 1887). Much later, one was taken at Jerez de la Frontera, Cádiz, in January 1972; one was seen at Roses, Girona, in April 1979, and other was at Doñana during January–February 1984 (Heredia & Máñez 1985; *Ardeola* 28: 156, 29: 184). The rarities committees accepted 42 records from peninsular Spain and one from the Balearic Islands and 10 from Portugal up to 2010, all of single birds that were generally accompanying Northern Lapwings. They included at least five adults and 12 first-year birds. There have been ten subsequent Spanish records of single birds during 2011–12 (*Ardeola* 60: 462, RBS) as well as three Portuguese records during 2010–11.

Winter observations predominate (Fig. 106) unlike elsewhere in Europe where hardly any are in winter. French records, for example, peak in March and October–November (Baeta 2004). Sociable Lapwings in Iberia occur mainly in the north-east, in Catalonia, and in the south-west (Fig. 107). However, the occurrences are chiefly in autumn and spring in the north-east, and in winter in the south-west, where some 60% are in December–February (Fig. 108). Southwestern Iberia has a relatively mild winter climate and it attracts large concentrations of Northern Lapwings, with which the Sociable Lapwing associates. The density of observers in the south-west is still relatively low and they are generally drawn by the wetland sites that attract a large range of species. Tilled land and pastures attract relatively little attention so it seems likely that quite a few Sociable Lapwings could winter there unnoticed.

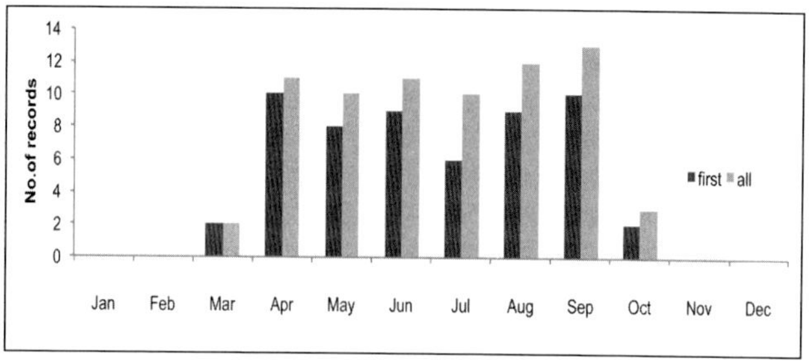

Figure 106. Records by month, 1868–2010 (n = 56)

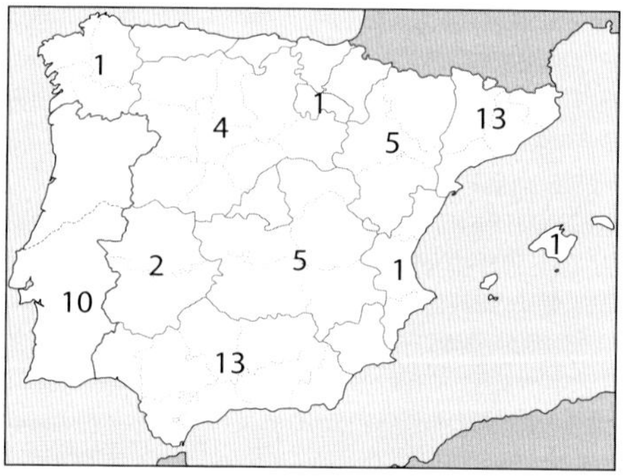

Figure 107. Records by region, 1868–2010 (n = 56)

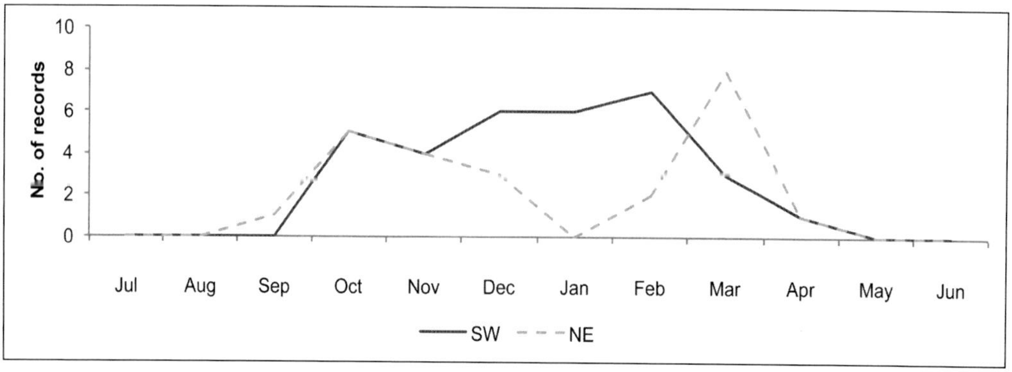

Figure 108. Distribution of first dates of records within the northeastern and southwestern halves of Iberia, to either side of an imaginary boundary line passing through Madrid (n = 24 NE and 33 SW records)

WHITE-TAILED LAPWING *Vanellus leucurus*

(Avefría Coliblanca, Abibe-de-cauda-branca)

Vagrant to Spain.

One was at Zeluán beach, Avilés, Asturias, on 21 September 2003. This Asian species is seldom found in western Europe, although it has bred at least occasionally at the Volga and Danube deltas (Helbig 1985, Kiss *et al.* 2001). There are six records from Britain and five from France. One was seen in Morocco in December 1983 (BM) and one was reported in the Canary Islands in November 1978 (*Ardeola* 45: 105).

NORTHERN LAPWING *Vanellus vanellus*

(Avefría Europea, Abibe-comum)

A scarce breeding species in Spain and an irregular breeder in Portugal. Widespread and often abundant throughout the region on passage and in winter.

The Northern Lapwing has a small breeding population in Spain with its largest concentrations in the Duero basin, La Mancha and Doñana. Some also nest in Galicia, northern Málaga, Extremadura, Madrid and the upper Ebro valley (SA2). A very few pairs have also nested in recent years in coastal Castellón at the Marjal de Almenara and the Prat de Cabanes (*A. O. C. Valenciana* 2009). Most breeding sites are inland, at or near wetlands. There the nesting birds favour lake and gravel pit margins, damp pastures, crops and fallow areas (SA2). The recent Spanish breeding distribution is much broader than was formerly the case. Breeding birds were only reported, in some numbers, during the 19th century from Doñana (Saunders 1871, Chapman & Buck 1893). However, Castellarnau (1877) saw several pairs in spring in meadows at Mozoncillo, Segovia, and wondered whether they were nesting there: the locality is within the current breeding area in Castilla y León. Elsewhere, Bernis (1952) found nesting birds in La Mancha, at Quero and Daimiel, and was informed that they had bred at the latter site for at least 40 years.

The Lapwing was first proved to have bred in Portugal very recently, during 1984–87 in Mértola and in 1993 in Algarve (AP, PA2: J. Petronilho). Only occasional scattered pairs have since bred, chiefly in Beira Litoral, Ribatejo and Alentejo (PA2), although five pairs nested at the Paul do Boquilobo in 2002 (AP).

The Spanish breeding population was estimated at 1,130–1,354 pairs during 1992–95 (SA1) and *c.*1,600 pairs post-2000 (SA2). The nesting population varies substantially between years according to rainfall, however. For example, only a few tens of pairs nest in Doñana during dry years but 500–800 pairs may breed when it is wet (García *et al.* 2000). Similarly, the Tierra de Campos, Valladolid, may support 500 pairs in years of favourable rainfall but only 12 pairs bred in the dry year of 2001. Post-breeding concentrations may be sizable in favoured nesting areas: for example, there were 610 at Santa Cristina del Páramo, León, on 21 August 2003 (*Ardeola* 52: 214). The Spanish population is regarded as stable overall, although recent declines have been reported in some areas: including Galicia, Gallocanta, the Laguna de La Nava and locally in Madrid (Pérez-Aranda *et al.* 2003). There is virtually no information on the movements of Iberian-bred birds but one ringed as a chick at Quero, Toledo, in May 1964 was recovered near Oran, Algeria, in January 1969 (*Ardeola* 16: 75). However, the earliest arrivals in Portugal are noted from July (AP) and some appear at Extremaduran wetlands as early as June (SWA).

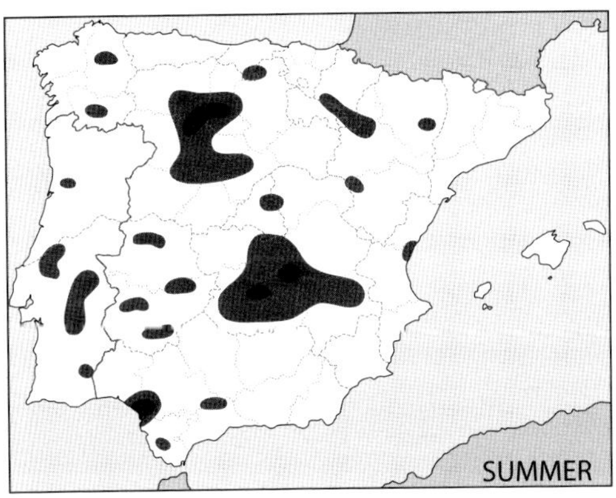

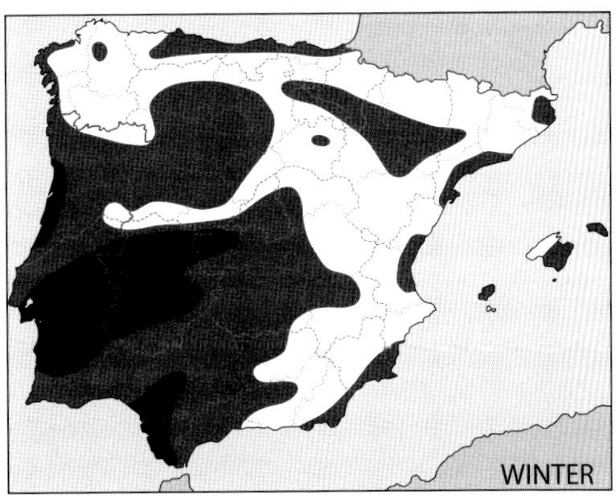

These movements are likely to involve Iberian birds in post-breeding dispersal and perhaps 'summering' non-breeders (Noval 1975, Asensio 1992, Calvet *et al.* 2004.).

The Lapwing is far more widespread in Iberia on passage and, especially, in winter. The Spanish vernacular name, *'avefría'*, marks its close association with cold spells, which may trigger conspicuous southward movements as well as sometimes massive influxes into Iberia from northwestern Europe. Migrants and winterers occur throughout the Peninsula and also in the Balearic Islands but the largest wintering concentrations are in Alentejo, Extremadura and western Andalucía (Leitão & Peris 2003, SWA). Extremadura held 29% of the Lapwings found during the Spanish winter atlas fieldwork in 2007–10. Wintering Lapwings may be found in a wide diversity of open habitats, with some preference for irrigated farmlands: including alfalfa fields and ricefields, and also damp grasslands, often around lakes and reservoirs.

The Lapwing has been ringed on a massive scale, especially in the Netherlands, and its status as a quarry species in Iberia has yielded abundant ringing recoveries there. These indicate that the majority originate from northwestern and northern Europe. Those from western countries, such as Britain and the Netherlands, tend to be recovered in western Iberia, from the Biscay coasts to Portugal, whereas those originating from central or eastern Europe tend to occur along Mediterranean coasts, and eventually often continue to North-west Africa (AMI, AI1). The furthest travelled include one ringed as an adult at Ryazan, Russia, in May 1961 and reported from Badajoz (*c.* 3,900km) in February 1963 (AMI) and one ringed as a chick near Moscow in May 1999 that was recovered at Durango, Vizcaya (3,198km) in December 2001 (*Ecología* 17: 254).

Migrant Lapwings generally spend only two to three months in Iberia. First arrivals may be in November but the majority arrive late, in December–January, and most have left by March. As many as 86% of the Iberian recoveries of birds ringed abroad are during December–February and 98% are in November–March (AMI, n = 929). Several hundred thousand are present each year (SWA), with very many more arriving when cold snaps affect western Europe. Some large concentrations occur in favoured areas in winter. Lapwings tend to be under-recorded in open country but during winter 1986–87 there were an estimated 65,000–70,000 birds on the Plana de Lleida in the Ebro Valley (Canut *et al.* 1986) and some 250,000 were on the steppes of La Serena, Badajoz: at densities of 27.9 birds/10ha (de Juana 1988). The Catalan winter population in 2006–09 was estimated at 46,000–67,000 birds, with concentrations in the lower Ter valley, the Aiguamolls de l'Empordà and the Ebro delta (CWA). Very large numbers often winter in Portugal. The estimated numbers found in the Tagus estuary, Lisboa region and Alentejo in January in 2000, 2001 and 2002 totalled 370,000, 559,000 and 203,000 individuals respectively (Leitão & Peris 2003).

Many of the Lapwings that enter the Peninsula may continue south to North Africa but again the numbers doing so very much depend on the severity of the winter further north. The regular movements are often inconspicuous and may involve nocturnal migration but the more precipitate responses to cold snaps often result in conspicuous daytime passages. These are sometimes evident in the vicinity of the Strait of Gibraltar where large numbers are sometimes seen from December to early February (BG). A particularly striking example occurred in 1965 when an estimated 40,000 birds passed south over a watchpoint near Jerez de la Frontera during 4–5 January; with similar large southward movements also visible in the distance (del Junco 1966). A count of 3,600 was made at Gibraltar on 5 January 1965 (pers. obs.) but there the birds were moving north.

GREAT KNOT *Calidris tenuirostris*

(Correlimos Grande, Seixoeirea-grande)

Vagrant to Spain.

The sole Iberian record, an adult at the Ebro delta on 7 April 1979, was the first for the Western Palearctic (Martínez-Vilalta & Motis 1985). The other European occurrences have been during July–November. This is an extremely rare vagrant to Europe, with only three records from Britain and none from France. One was recorded in Morocco in August 1980.

RED KNOT *Calidris canutus*

(Correlimos Gordo, Seixoeira-comum)

C. c. islandica. **Some winter south to Iberia.**

C. c. canutus. **Small numbers occur on passage and in winter in Iberia.**

The Red Knot is abundant on passage and in winter in the estuaries of northwestern Europe and very large numbers winter on African Atlantic coasts from Morocco south to South Africa but chiefly: over 90%, at the Banc d'Arguin, Mauritania, and in Guinea Bissau (Isenmann *et al.* 2010). It covers prodigious distances with non-stop flights, staging at a limited number

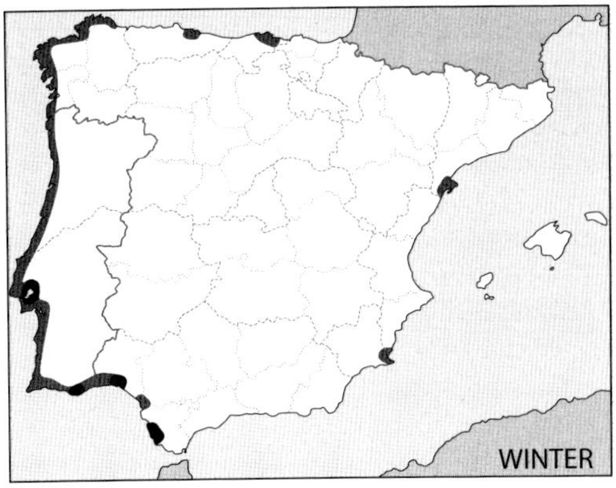

of traditional sites none of which are in Iberia (Delany *et al.* 2009). There is presumably a very large passage of *C. c. canutus* across Iberia, probably along and off the Atlantic seaboard, but few birds ever interrupt their journey in Spain or Portugal unless severe weather intervenes. The Tagus estuary may then serve as an 'emergency' stopover site.

Red Knots are seen annually on Iberian Atlantic coasts, including those of the Bay of Biscay. They favour the mudflats of the Atlantic estuaries and rías (Moreira 1999a, Masero *et al.* 2000) but seldom occur in the great flocks that typify similar habitats in northwestern Europe. Some sizable concentrations do occur irregularly in Portugal and northwestern Spain: for example, there were 3,337 at the Ria Formosa on 21 May 1985 (Rufino & Araújo 1987), but these are a negligible fraction of the very large populations involved. Migrants are generally more numerous during the spring passage, which tends to be late in the season, from mid April through to June. Only small numbers are ever recorded on Mediterranean coasts or in the Balearics.

The autumn passage, during August–October, is even less apparent. A few non-breeders remain in Iberia in summer, when up to 40 have been counted in the Marismas del Odiel (Garrido 1996). There are only a few Iberian recoveries of foreign-ringed birds, all of them ringed when on passage: they include 13 from Scandinavia, eight from Germany, six from Britain and a few ringed in Iceland, the Low Countries, France and Poland. One ringed in South Africa was found on the Huelva coast. Knots ringed in Portugal have been recovered in Guinea Bissau and South Africa and one ringed in Spain, at the Marismas del Odiel in September 1999, was found at the Banc d'Arguin in January 2000 (EBD, *Ecología* 15: 400).

The wintering population is variable and often very sparse. The average Spanish winter population in 2008–10 has been estimated at little over 300 birds, most of them in western Andalucía in Cádiz Bay (180 birds) and at the Marismas del Odiel (45), with smaller numbers at Doñana, the Ría de Arousa, the Ensenada da Ínsua and the Ebro Delta, this last being the only Mediterranean locality (SWA). The Spanish wintering population had earlier been estimated at 200 individuals by Alberto & Velasco (1988) and only 100–150 individuals were considered to comprise the entire wintering population of the Huelva coast from the Guadiana estuary to Doñana (Garrido 1996). Nevertheless, much larger winter counts have been made on occasion at Atlantic estuaries. The January wetland censuses of 1978–80 (Alberto & Purroy 1981) found 400 in the Ría de Arousa, in 1978 and *c.*1,000 there in 1979; and there were also *c.*1,250 at Santoña in 1978. In January 1969, there were 2,000+ at Santoña; 3,000 at the Ría de Arousa, and 5,000 at the Ría de Ortigueira, A Coruña (Fournier & Fournier 1972). Very few winter on Mediterranean coasts: the January wader censuses at the Ebro delta during 1996–2009 average just 13 birds but have ranged from zero in several years to 49 in 2003 (*A. O. Catalunya*, CWA). The 1978–80 censuses only found 108 Red Knots at Mediterranean wetlands: all of them in 1979 and the majority at the Mar Menor.

The Portuguese wintering population in recent decades has always been fewer than 1,000 birds: it averaged 455 during 1992–95 and 89 during 2005–07, with just a single bird in the entire country in 2005 (AP). The Tagus estuary and Ria Formosa are the principal wintering locations (AP). There have also been larger counts there in the past: a census of Portuguese estuaries in January 1969 found 14,000 birds at the Ría de Aveiro, 8,000 at the Tagus estuary and 2,000+ at the Sado estuary (Pagezy & Trotignon 1972).

An unknown number of Red Knots overfly the Peninsula, the evidence being the occasional records of individuals at wetlands far inland. Most sightings involve just a few birds or, less commonly, groups of up to ten. Occasional observations of groups of up to 40 hint at the at least occasional existence of a cross-Peninsular passage of some size. An exceptional 156 were at Villafáfila, Zamora, on 12 May 2002 (*Ardeola* 50: 348). Inland records tend to be in May and very few are reported inland in autumn.

SANDERLING *Calidris alba*

(Correlimos Tridáctilo, Pilrito-das-praias)

Widespread and locally common along coasts on passage and in winter, with some non-breeders remaining in summer. Occasional inland.

Flocks frequent sandy shores and estuaries throughout the region, although they are much more numerous on Atlantic coasts than in the Mediterranean. They may also be found around saltpans and on the sandy shores of coastal wetlands. Small numbers of non-breeders remain in summer in many areas. Sanderlings found in Iberia may originate from as far west

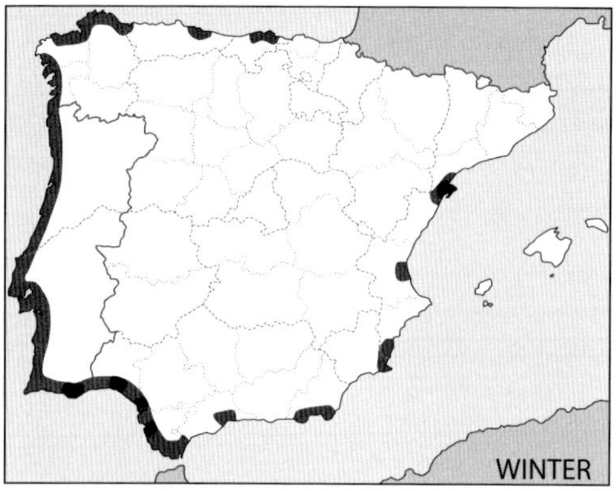

as northeastern Canada and Greenland and as far east as western Siberia, from where birds migrate along the East Atlantic flyway to winter especially in northwestern Europe and on African Atlantic coasts (Delany *et al.* 2009). There are Iberian recoveries of birds ringed in transit in Britain (13), Norway (9), Germany (3), Sweden and The Netherlands, and there are single recoveries from The Netherlands and Iceland of birds ringed in Portugal.

Sanderlings are locally numerous on the Atlantic and Biscay coasts, where significant concentrations are reported. Southward movements are evident from late July to November, peaking in August–September, and spring passage is in March–May (AI1, BSG, AP). The Galician coasts sometimes hold 1,000+ birds during passage periods and are included among European sites of International Importance for the Sanderling (Delany *et al.* 2009). Flocks of up to several hundred also occur commonly in Asturias (Noval 2001). Sanderlings also occur on passage on sandy Mediterranean beaches and at coastal wetlands there but generally only in small numbers. They are very uncommon in the Balearics: the January counts of 1990–2009 found only four birds on average, with 71 in the best year (AAE).

The wintering population is apparently not large, just a few thousand birds. However, it is probably underestimated given that there has never been a complete census of the entire sandy coastline. Wintering numbers in Spain were estimated at 2,100 birds by Alberto & Velasco (1988): with concentrations in Cádiz Bay; the Rias de Arousa and Ortigueira in Galicia and the Marismas del Odiel, as well as at the Albufera de Valencia and the Ebro delta in the Mediterranean. The January censuses in Spain of 1990–2009 found 2,246 birds on average (range 542–5,135; AAE). However, the Huelva coast population alone was put at 1,500–2,000 birds by Garrido (1996). During 1990–2009 sites in Andalucía: in particular Doñana (up to 1,293 birds) and Cádiz Bay (up to 945) held about 44% of the wintering Sanderlings, with Galicia and Catalonia: principally the Ebro delta, each holding about 19% of the population (AAE). Winter censuses in Catalonia during 1996–2009 found an average of 659 birds at the Ebro delta: although there were only 87 in 1996 and 1,176 in 2005 (*A. O. Catalunya*). A count of 100 at Torrevieja, Alicante, on 13 December 2009 (*A. O. C. Valenciana 2009*) was noteworthy for the Comunidad Valenciana. The Spanish wintering population apparently increased markedly post-1980, by 7.9% during 1990–2009 (AAE). Bernis (1966: AMI) referred to there being only a very few wintering birds, at a limited number of locations.

Censuses at the principal Portuguese coastal wetlands found an average of 530 wintering Sanderlings during 1992–95 and 1,040 during 2005–07, a large proportion of them being at the Ria Formosa (AP). As always expected, larger numbers of Sanderlings were found wintering on Portuguese sandy beaches by the 'Proyecto Arenaria' census of winter 2009–10: which counted 2,817 birds (Catry *et al.* 2010b). In general, the wintering distribution appears to have a southern bias, with the largest numbers on Atlantic shorelines of southern Portugal and southwestern Spain. Conversely, very few winter in the Biscay coastlands (Noval 2001, SWA).

There are occasional records of a few individuals or flocks of up to 20 at ricefields and lakeshores far inland in Spain and Portugal. A flock of 32 at Orgaz, Toledo, on 30 October 1998 is noteworthy (*A. O. Toledo 2001*: 267).

SEMIPALMATED SANDPIPER *Calidris pusilla*

(Correlimos semipalmeado, Pilrito-rasteirinho)

Vagrant to Spain and Portugal.

The earliest Iberian record was a juvenile that was ringed at Quinta da Rocha, Faro, on 18 October 1989 (*Ardeola* 41: 110). There have since been 11 accepted records of single birds in Spain up to 2008. In all there have been nine juveniles and two adults as well as a June record of a first-summer/adult bird. The majority have been in the north-west: eight in Galicia and two in Asturias. The sole exceptions are the Faro record and one on the Mediterranean coast at the Llobregat delta from 7–9 June 1997. The latter was also the only one not found in autumn. Iberian records peak in October, later than the September peaks noted in the British Isles and in France (Dymond *et al.* 1989, OF), although the Iberian sample is still small. Records of this rare vagrant to Europe have become much more frequent in recent years. There are observations of at least 115 individuals in Britain, 173 in Ireland and 28 in France. There were five records from the Azores up to 1979 (Le Grand 1983) but a further 32 have since been accepted: including one of a flock of eight juveniles (*Ardeola* 39: 78). There are four accepted records from the Canary Islands and two from Morocco, and it has also been reported in Madeira and Mauritania.

WESTERN SANDPIPER *Calidris mauri*

(Correlimos de Alaska, Pilrito-miúdo)

Vagrant to Spain.

The two Iberian records have been in Galicia: a juvenile at the Ensenada da Ínsua, A Coruña, on 8 September 1979, and an adult at the Laguna de Traba on 20–23 September 1999 (*Ardeola* 39: 78, 48: 128). A very rare vagrant to Europe, with only eight records from Britain, four from Ireland and five from France. Four accepted records from the Azores include a flock of six in October 1978.

LITTLE STINT *Calidris minuta*

(Correlimos Menudo, Pilrito-pequeno)

Widespread and common on passage, both at coastal wetlands and inland. Common in winter at coastal wetlands in the south-west and the Levant.

This Palearctic species is a common migrant at both coastal and inland wetlands throughout the Peninsula and in the Balearic Islands, although the numbers involved vary considerably between years. The autumn passage, from late July to October, generally involves larger numbers than are seen in spring, in April– June (AI1, Garrido 1996, AP). However, the opposite may apply on the Mediterranean coast: there are many more spring than autumn records at the Tordera estuary, Catalonia (Cordero-Tapia & López de Vilar 1985) and a more easterly routing of spring migration in Europe has been suggested elsewhere (BWP). Most Little Stints winter in sub-Saharan Africa, but some numbers remain locally on southern and eastern Iberian coasts, with relatively few wintering inland or in the north or north-west. About half of 55 Iberian recoveries are of birds ringed on passage in Fennoscandia, the others having been ringed in Germany, Denmark, France, England, Belgium, Poland, the Czech Republic, Italy or Croatia. Three ringed in Ukraine have been found in Spain. An individual ringed in Portugal has been recovered in Senegal and Spanish-ringed birds have been found in France (3), Finland (2), Italy, Tunisia and Ghana.

At all seasons the Little Stint frequents estuarine mudflats, saltpans, the shores of coastal and inland lagoons and ricefields, generally where there is very shallow water. Some inland sites attract notable concentrations during passage periods, again especially in autumn. For example, there were 400 at the Laguna de Boada, Palencia, on 30 September 2001 and 200 there on 7 October 2001 (*A. O. Palencia 1998–2001*: 103); and 450 were at the Finisterre reservoir, Toledo, on 23 October 2000 (*A. O. Toledo 2001*: 268). Spring counts inland seldom exceed 100 birds but there were 400 at the Finisterre reservoir on 10 March 2000 (*A. O. Toledo 2001*: 268).

The coastal wetlands of Huelva, including the Isla Cristina, Odiel and Guadalquivir marshes, support concentrations of international importance, as does the Ebro delta (Delany *et al.* 2009). Peak counts in these areas comprise over 10,000 birds at Doñana and over 7,000 birds at the Ebro delta. These high counts are associated chiefly with post-breeding migration: probably fewer than 500 Little Stints winter in the Huelva wetlands in some years but up to 15,000 are recorded there on autumn passage (Garrido 1996). Nevertheless, the numbers wintering at Doñana in some seasons can be very considerable: over 16,000 were censused there in January 2000 (*A. O. Doñana* 1: 113). The Ebro delta is also one of the principal European wintering sites of the Little Stint. The 1978–80 January counts (Alberto & Purroy 1981) found 1,190 birds at the Ebro delta and 300 at the Santa Pola saltpans, Alicante, in 1978. January censuses at the delta during 1996–2009 averaged 4,500 birds but ranged from 586 in 1996 to over 7,000 in both 1999 and 2009 (*A. O. Catalunya*, CWA).

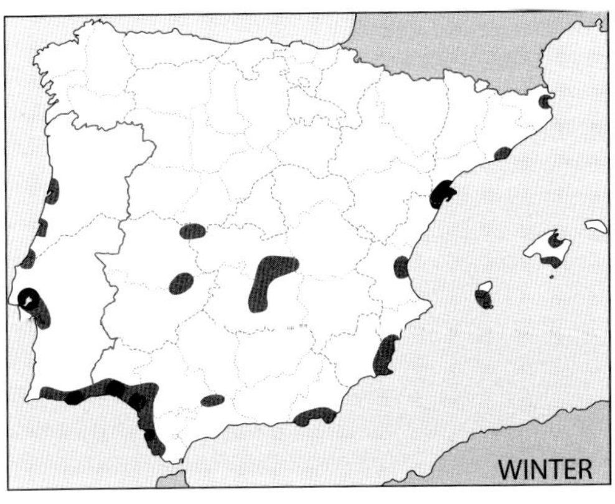

Bernis (1966: AMI) considered that very few Little Stints wintered in Iberia. However, the Spanish wintering population was estimated at *c.*1,400 birds by Alberto & Velasco (1988), who highlighted the importance of the Ebro delta and the wetlands of La Mancha. The wintering numbers have since been much larger and the January counts in Spain during 1990–2009 found 8,525 birds on average, although with marked inter-annual fluctuations, so that annual totals ranged from just 693 birds to 21,099 (AAE). A 9.5% increase in Spanish wintering numbers was noted during this latter period.

Clearly, by far the largest wintering numbers in Spain are at the Ebro delta and in southwestern Andalucía. Some numbers also occur in winter in the Balearics, where January counts during 1990–2009 averaged 205 birds, with a maximum of 680, the largest gatherings occurring at the Salobrar de Campos, Mallorca (AAE). The inland wetlands of the meseta do sometimes support some numbers in winter but this is likely to be a variable phenomenon: there were 43 at Alcázar de San Juan, Ciudad Real, on 30 December 1988 and these had increased to 150 birds a week later (*Ardeola* 36: 247). Other sizable day counts inland in December include 230 at the Laguna de Hoya Osilla, Albacete, in 1998; 320 at the Finisterre reservoir, Toledo, in 1999 and 290 at the same site in 2000 (*A. O. Albacete 1997–1998*: 101; *A. O. Toledo 2001*: 268). The January censuses of selected Spanish wetlands during 1978–80 (Alberto & Purroy 1981) found 350 at Santoña in 1978 and 200 there in 1980. No comparable concentrations have occurred on the Biscay coasts or in Galicia in more recent years, when only a few individuals have been detected by the January counts (AAE). Midwinter counts in Portugal have ranged from an average of 520 birds during 1975–78 to 1,615 during 1992–95 and 609 in 2005–07 (AP).

TEMMINCK'S STINT — *Calidris temminckii*

(Correlimos de Temminck, Pilrito de Temminck)

A scarce but regular migrant, most frequent in the east. Some remain in winter, mainly in the south.

The westernmost populations, in Scandinavia and northeastern Russia, winter principally in sub-Saharan west Africa (Delany *et al.* 2009), some birds migrating across Iberia. Temminck's Stint was seldom reported in Iberia prior to the 1980s and it was almost certainly overlooked at a time when there were relatively few observers capable of identifying it. It is now known to be a regular migrant that occurs annually in small numbers, mainly on passage but, locally, also in winter. A high proportion of records come from a limited range of well watched sites and are often submitted by the same experienced observers, so it may be commoner than observations suggest. There are a few Iberian recoveries of birds ringed on passage through central and western Europe: in Belgium, Britain, the Czech Republic, Finland, Germany and Sweden, as well as single British and Polish recoveries of birds ringed in Spain.

Temminck's Stint is more closely linked with freshwater and inland locations than the Little Stint, occurring on the margins of lakes, ponds and reservoirs and in marshes, ricefields and damp grasslands. The only coastal sites where it may be found are the more sheltered inlets in estuaries and salt marshes and at saltpans, and it avoids feeding on open mudflats. Autumn records somewhat outnumber spring sightings and winter records are in a minority. Records show a late spring peak in May, corresponding to northward passage, with return movements during August–October, with some early individuals recorded in July (Figure TS1).

Most records are of single birds (62% of 214 records) or two together (20%) and only a few (5%) involve counts of ten or more birds, the largest on record being a flock of 17 at the Brazo del Este on 1 August 2008 and another of 25 birds at the Ebro delta on 27 August 1998 (*Ardeola* 55: 298, *A. O. Catalunya 1998*: 133).

There are records from all regions except northern Portugal. There are also only six records for Galicia, suggesting therefore

Figure 109. Records by month (n = 214).

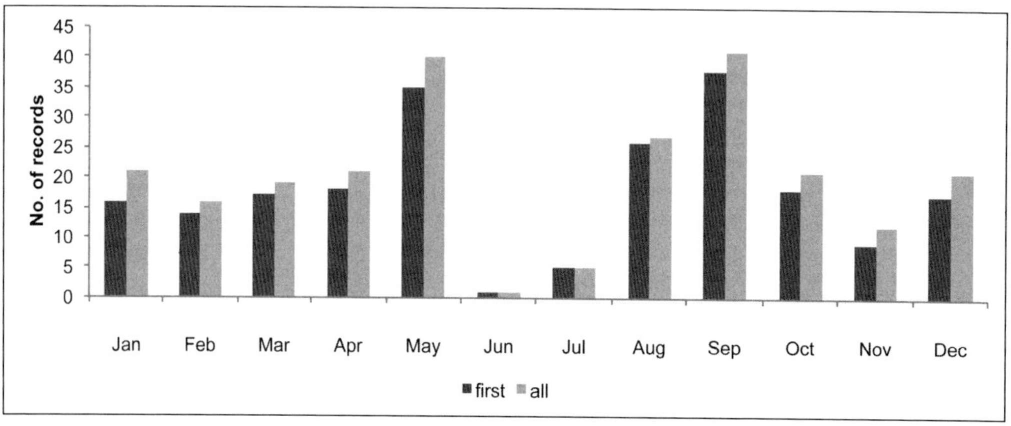

that it is uncommon in the peninsular north-west. It is also comparatively rare in Portugal generally but much commoner on the Mediterranean littoral, especially in Catalonia, the Balearics and the Comunidad Valenciana. It has been suggested that Temminck's Stint is commoner on eastern Iberian coasts in spring and more frequent inland in autumn (Velasco 1992a, AI1) but later records do not corroborate this: those from the inland regions of Aragón, Castilla y León, Castilla-La Mancha, La Rioja, Madrid and Navarra total 27 in spring (March–May) and 19 in autumn (August–October). Records from eastern Spain predominate in both spring and autumn but reports from western Iberia: including Galicia, Asturias, Extremadura and Portugal are approximately twice as frequent in autumn than in spring: 32 records in August–October and 17 in March–May. A predominance of autumn records in Portugal has previously been noted (AP). Winter records are relatively scarce in northern Iberia and relatively frequent in the east and, especially, in the south.

A number of sites stand out for the relative frequency with which Temminck's Stint has been recorded there, although sample sizes are low. It is apparent that wintering birds occur with some regularity in the lower Guadalquivir, particularly at the Brazo del Este, and also in the ricefields of the Ebro delta, at the Albufera de Valencia, the Albufera de Mallorca, El Hondo, Alicante, and the Lagoa dos Patos, Beja. These sites also hold birds on passage but migrants are also regular at wetlands of southern Portugal: including the Tagus and Sado estuaries, the Castro Marim reserve and the Ria Formosa; the coastal wetlands of Almería and the Levant: including the Llobregat Delta; the lagoons of La Mancha and the northern meseta, and at coastal wetlands of the north coast, including the Ría de Avilés and the Ría de Villaviciosa.

LEAST SANDPIPER *Calidris minutilla*

(Correlimos Menudillo, Pilrito-anão)

Vagrant to Spain.

Two records are from A Coruña, Galicia: single juveniles at the Ensenada da Ínsua on 11 October 1987 and at Malpica de Bergantiños from 6–8 October 1999. An adult was at the Ebro delta on 19–20 August 1996. Records of this very rare vagrant to Europe include 36 from Britain, 12 from Ireland, 14 from France, 28 from the Azores and one from the Canary islands.

WHITE-RUMPED SANDPIPER *Calidris fuscicollis*

(Correlimos Culiblanco, Pilrito-de-sobre-branco)

Vagrant to Spain and Portugal.

The first Spanish record was not until 1975: at the Albufera de Mallorca on 29 September (*Ardeola* 22: 136) and the first Portuguese record was in 1999: a juvenile at Peniche, Leiria, on 24–25 October. Accepted records to 2010 total 19 from Spain and three in Portugal. All have involved single birds, including 11 adults and seven reported as juveniles/first-winters. Records peak in autumn and there are very few at other times of year (Fig. 110). First observation dates range from 4 August to 18 October for adults and 11

Figure 110. Records by month, 1975–2010 (n = 23)

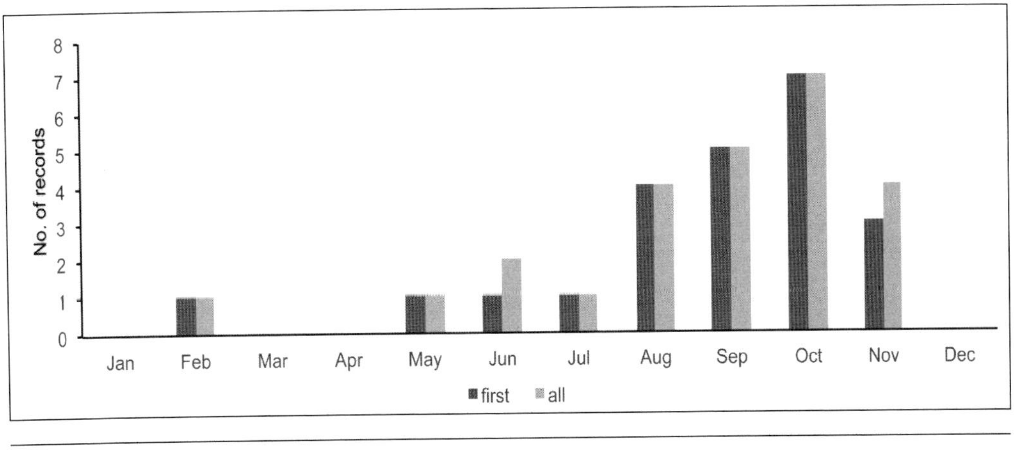

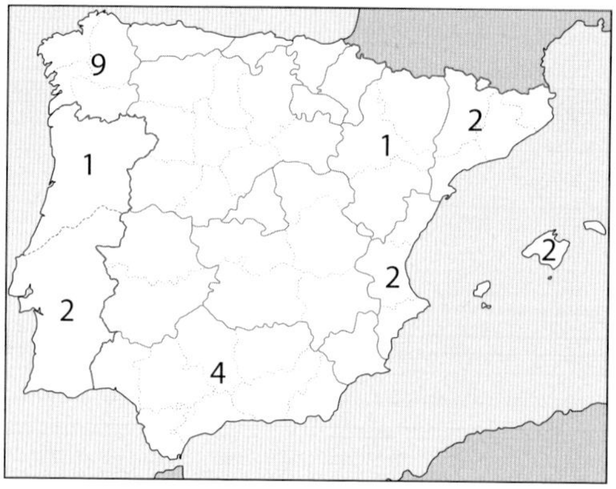

Figure 111. Records by region, 1975–2010 (n = 23)

October to 22 November for young birds. Most observations are on the Atlantic seaboard (Fig. 111) but the few non-autumn records are all from the Mediterranean: single birds in February in Málaga; in May–June at the Ebro delta; in June in Valencia harbour, and in mid July at the Albufera de Valencia. An adult in ricefields at Fraga, Huesca, on 16 and 21 October 2001 is the sole inland record. There were a further ten Spanish records in 2011–12 (*Ardeola* 60: 463, RBS).

The White-rumped Sandpiper is found in Europe relatively frequently. By 2005 428 had been recorded in Britain and there are records of 300 in Ireland and 53 in France. There are over 50 records for the Azores up to 2006: including one of a flock of 48, one from Madeira and 29 from the Canary Islands. There are, however, only two Moroccan records: in September 1999 and April 2012.

BAIRD'S SANDPIPER *Calidris bairdii*

(Correlimos de Baird, Pilrito-de-bico-fino)

Vagrant to Spain and Portugal.

An adult at the Ensenada da Ínsua, A Coruña, on 20 and 21 August 1988, was the first Iberian record. By 2010 there were a further eight accepted records from Spain and one from Portugal. All involved single birds; four adults and six juveniles/first-winters. All have been in autumn, between 19 August and 1 November for adults and 28 September to 14 December for immature birds. Records in the British Isles and France are also mainly in autumn, with a clear peak in September (Dymond *et al.* 1989, OF). All but two observations have been in Galicia, the exceptions being one at Ludo, Faro, on 14 December 2004, and one at the Albufera de Valencia during 19–27 August 2008. Two subsequent records in 2011 include the first during the winter: at the Ebro delta from 18 January to 19 February, and the first from the Balearics: in Palma de Mallorca on 8–12 October (*Ardeola* 60: 463).

Baird's Sandpiper occurs relatively frequently in Europe. There are records of at least 238 individuals in Britain, 132 in Ireland and 53 in France. There are also six accepted records from the Azores, one from Madeira and three from the Canary Islands. There are none from Morocco but it has been recorded in Senegambia and Namibia.

PECTORAL SANDPIPER *Calidris melanotos*

(Correlimos Pectoral, Pilrito-de-colete)

Rare but recorded increasingly frequently on passage in Spain and Portugal.

The Pectoral Sandpiper has previously been regarded as a transatlantic vagrant to Europe but the frequency with which it appears suggests that some occur there on passage to as yet unknown winter quarters in Africa and that a proportion of those seen in the Western Palearctic may originate from the Siberian population (Lees & Gilroy 2004). Nearly 2,000 individuals were recorded in Britain alone during 1968–2002 and some 409 individuals were found in France up to 2005: it was withdrawn from the respective rarities lists in 1963 and 2005. Accepted records from the Atlantic Islands include 21 from the Azores up to 2005, five from Madeira and 41 from the Canary Islands. There are nine records from Morocco and a number of others from widespread locations in Africa (Urban *et al.* 1986).

The earliest Iberian record was an individual affected by botulism that was found at Doñana on 19 September 1973 (*Ardeola* 22: 135). The Prat de Llobregat museum, Barcelona, has a specimen that was taken in the early 1970s (Clavell 2002). Two were observed in Asturias in October 1979 (*Ardeola* 28: 157) and one was taken there in September 1981 (Álvarez-Balbuena *et al.* 2000). The Pectoral Sandpiper has been reported annually in Spain since 1980. There are 245 accepted Spanish records from 1980–2010 and a further 25 from Portugal. Records have become increasingly frequent (Fig. 112) probably at least partly due to the increase in experienced observers. There were some 85 records from Spain during 2011–12: including at least 42 in September 2011 and 20 in September 2012 (*Ardeola* 60: 463–465, RBS), and at least seven Portuguese records from the same period (RBP).

Figure 112. The increased frequency of records during 1971–2010

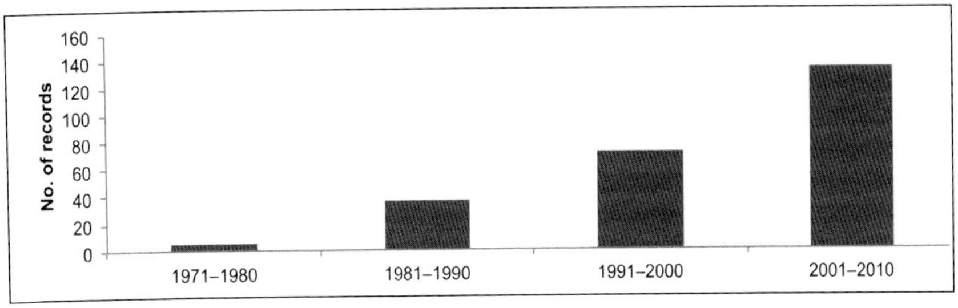

The 248 Iberian records up to 2010 involved 296 individuals in total. Although 85% of records were of single birds, 12% were of two together and there are five records of three or more, all of them since 2000. The mean number of birds per record increased from 1.11 up to 2000 (n=114) to 1.26 during 2001–10 (n= 135). Up to six juveniles were recorded together at the Albufera de Valencia during 12–22 September 2010. Autumn birds up to 2010 included 24 identified as adults and 211 juveniles.

Records show a pronounced peak in autumn and a much smaller one in spring (Fig. 113). One in Mallorca on 8 March 2004 and another at the Ebro delta on 20 February 2008 that had apparently been there since early January are exceptions. The adult autumn passage precedes that of the juveniles, adults having been found between 11 July and 22 September and juveniles from 16 August to 24 October: 17% of adults appeared in July, 44% in August and 39% in September. For juveniles, 7% were in August, 69% in September and 24% in October. Iberian records tend to be slightly later than observations in France and the British Isles, with a higher proportion of October sightings. They are correspondingly earlier than those for the Canary Islands, where the extreme dates for juveniles are 7 September and 8 November (de Juana 2006). This pattern suggests southward passage across the region.

Figure 113. Records by month, 1973–2010 (n = 248)

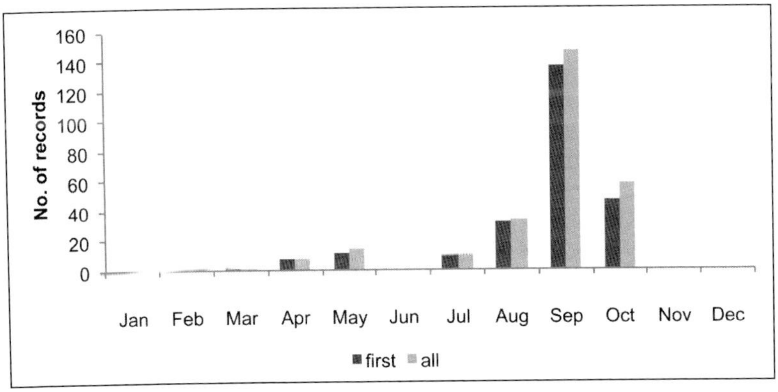

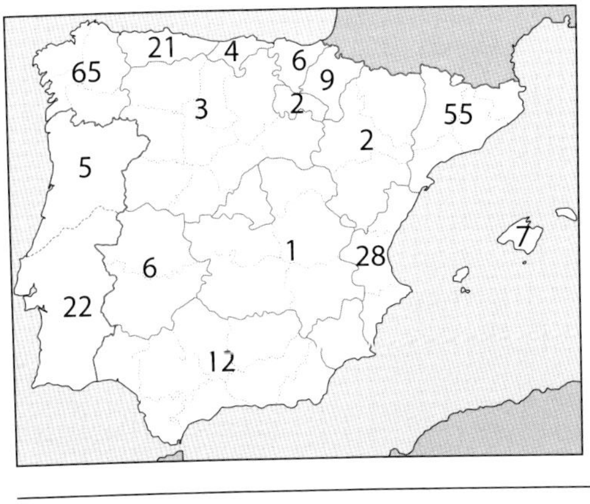

The geographical distribution of occurrences shows clusters in both the north-west and north-east: some 35% of the records are in Galicia or Asturias, and a further 33% are in Catalonia or the Comunidad Valenciana. In contrast there are relatively few records in the south or inland (Fig. 114). Spring records comprise a much higher proportion of observations in the north-east than in the north-west: 13.2% in the former but only 4.7% in the latter. A tendency for spring records to become more frequent further east has also been noted elsewhere in Europe (Alström *et al.* 1991).

Figure 114. Records by region, 1973–2010 (n = 248)

SHARP-TAILED SANDPIPER *Calidris acuminata*

(Correlimos Acuminado, Pilrito-acuminado)

Vagrant to Spain and Portugal.

The records comprise an adult at the Albufera de Valencia from 1–3 September 2001; a juvenile at the Ría de Villaviciosa on 30 August 2005; an adult at the Ria de Aveiro on 21 March 2007, and an adult at the Albufera de Valencia from 29 September until 3 October 2010. This is a very rare Siberian vagrant to Europe, where the records include 31 in Britain, seven in Ireland and seven in France. One was on the Azores in August 2000 and another in Madeira in August–September 2012.

CURLEW SANDPIPER *Calidris ferruginea*

(Correlimos Zarapitín, Pilrito-de-bico-comprido)

Widespread on passage both along coasts and inland but most numerous in the south and east, particularly in autumn. Small numbers winter at southern and eastern coastal wetlands.

The westernmost breeding populations winter mainly in coastal west Africa and migrate across western Europe, including the Iberian Peninsula. The size of this movement is greatly affected by fluctuations in arctic lemming populations since Arctic Foxes and other predators destroy most nests in years when lemmings are scarce but take relatively few when lemmings are abundant (Delany *et al.* 2009). The numbers recorded in Iberia are certainly variable, both on passage and in winter.

The autumn passage, from late July to October, peaking in August and early September, is generally larger and marked by observations of small flocks at shallow inland wetlands and larger gatherings at estuaries, lagoons and saltpans on the coasts, particularly in eastern and southwestern Iberia. The largest counts on record from Portugal have included 930 at the Tagus estuary in late July 1981 and 568 at the Ria Formosa in September 1992 (AP). Flocks of over 500 birds occur at the Huelva wetlands in autumn, with counts of over 2,000 at the Marismas del Odiel (Garrido 1996). A study of autumn passage at a reserve area within the Ebro delta in 1992 and 1993 (Figuerola & Bertolero 1996b) found that adults precede juveniles, the former passing there in early to mid August and the latter peaking in mid-September. Males also passed significantly earlier than females in both study years. Peak counts exceeded 1,500 individuals and considerably more must have been present within the delta as a whole. Similarly, at Odiel, where some 5,000 were ringed during 1997–2010, adult males pass from late July to early August, adult females mainly in mid-August and juveniles from early September (Gacio & Sayago 2011). The Curlew Sandpiper is comparatively scarce in the Balearic Islands: a count of 170 at the Salobrar de Campos on 28 July 2011 was a record for Mallorca (*Ardeola* 59: 435).

The spring passage is relatively inconspicuous, particularly in central and western Iberia. It is already apparent at the Ria Formosa in February–March (AP) but more widely it occurs in April–June, peaking in May. A count of 793 individuals at the Ria Formosa in March 1992 (AP) is noteworthy. Peak counts at Odiel are *c.* 400 birds, many fewer than in autumn (Garrido 1996). Peak counts at Doñana have included over 300 at the Brazo del Este on 22 April 2004 and 330 at the Dehesa de Abajo on 9 May 2005 (*A. O. Sevilla 2000–2005*: 76). However, considerable numbers are reported in spring in some years at east coast sites: for example, 1,451 were counted at the Cerrillos saltpans, Almería, on 6 May 1995 (*Ardeola* 42: 222).

A small number remain in winter chiefly (over 95%) in the coastal wetlands of southwestern Andalucía. The Spanish January counts of 1990–2009 found 407 birds on average but numbers ranged from zero to 1,772 individuals (AAE), with the principal concentrations at Doñana (up to 1,770 birds) and in Cádiz Bay (up to 1,026). Very small numbers, averaging single figures, occur in winter at sites in the Levant wetlands, chiefly in Alicante and Murcia, and in Almería (AAE). There are only a few winter records from the Ebro delta (CWA). Winter counts in Portugal found an average of 129 birds during 1992–95 and 125 during 2003–05, mainly at the Tagus estuary and the Ria Formosa (AP). Winter records inland are rare but not unknown: for example, on the Guadiana ricefields, Badajoz, there were ten on 13 December 1998, one on 30 December 1999 and two on 12 January 2003 (*Ardeola* 46: 156; *A. O. Extremadura 2001–2003*: 205). A few summering non-breeders have been reported in Portugal (AP).

There are at least 69 Iberian recoveries of Curlew Sandpipers ringed on passage in northern and central Europe, east to Russia (2) and Ukraine (10), as well as Spanish recoveries of four ringed in Senegal and one ringed in Tunisia. Birds ringed in Iberia have produced 31 far-flung recoveries: including eight in Ukraine, five in Italy, and one each in Britain, Bulgaria and Greece, as well as three in Senegal and two in Guinea Bissau, the remainder being in northern Europe.

PURPLE SANDPIPER *Calidris maritima*

(Correlimos Oscuro, Pilrito-escuro)

Regular in winter in Galicia and on Biscayan coasts. A few winter further south, with occasional observations from all coasts and in the Balearic Islands.

The European wintering range is concentrated on western coasts, reaching its principal southern limit in northern Iberia, where a small population winters on the rocky coasts of Galicia, Asturias, Cantabria and the Basque Country. This numbers no more than a few hundred individuals in total that usually occur in small flocks, often alongside Ruddy Turnstones. Larger counts have included up to 234 in Gijón Bay, Asturias, during January 1993 and 80 at Cape Silleiro, Pontevedra, on 20 January 1999 (*A. O. Asturias 1993*: 73, *Ardeola* 49: 187). The largest numbers found during the Spanish winter atlas survey in 2007–10 were at the Ría de Guernica, Vizcaya, but the peak count of 33 in January 2010 was relatively modest (SWA). In general, recent observations suggest that the wintering population has declined since the 1990s. The numbers using Gijón Bay, the prime Iberian location for this species in the past, fell sharply from 1995 onwards after beach enlargement works there (*A. O. Asturias 1994–1995*: 136).

Small numbers, no more than a few dozen birds, occur with some regularity on western Portuguese coasts, notably in the north-west and on the Oeiras coast in the south of Lisboa (AP). Observations further south and in the Mediterranean are few and widely scattered. Nevertheless, some individuals are reported with some regularity from sites in Algarve, the Huelva coastlands, the Strait of Gibraltar, the Costa del Sol, the Levant coast, the Columbretes islands and the Balearics. Sometimes one or two birds are seen at particular sites for several successive winters, suggesting that the same individuals return to the same places. A group of six was at Punta Secreta on the Strait of Gibraltar on 8 January 2000 and five were at Fuengirola harbour, Málaga, on 12 January 2009 (*Ardeola* 47: 167, 56: 356). The Purple Sandpiper is almost always encountered on coastal rocks but there are records, perhaps of migrants, from other coastal habitats, such as saltpans and beaches. A juvenile found inland at the Llanos de Cáceres in September 1993 (Lozano 1993) is a highly unusual record.

Most Purple Sandpipers are observed between mid October and April but there are records from early September and early May and one was at Burriana, Castellón, on 25 July 2001 (*Aves C. Valenciana 2000–2001–2002*). There is a 1954 recovery from Galicia of a bird ringed as an adult in Iceland (AMI).

DUNLIN *Calidris alpina*

(Correlimos Común, Pilrito-de-peito-preto)

C. a. alpina, C. a. schinzii. **Common on passage and in winter.**
C. a arctica. **Occurs mainly on passage to and from W Africa.**
C. a. sakhalina. **May occur as a vagrant.**

Large or very large numbers of Dunlins occur on passage at all the major estuaries of the Peninsula and the same sites also hold large numbers in winter. Migrants also occur commonly inland and some also remain inland in winter, but the numbers involved are always modest when compared with the tens of thousands that gather on coasts. Dunlins frequent estuarine mudflats, saltpans and the shorelines of shallow coastal lagoons. Inland they also occur on the margins of lakes and reservoirs and also in ricefields.

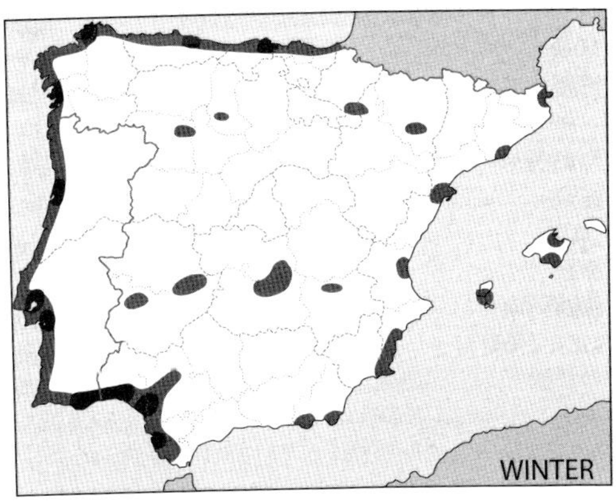

Migrants begin to arrive on northern coasts as early as mid July but principally during August–October (Galarza 1984). Adults precede juveniles, as in other waders. Passage in the centre and south is apparent from mid August to early November with the largest numbers present in September (Tellería 1981, BSG, Martín 2002). The return movements begin as early as February on the Huelva coast, when up to 20,000 birds have been counted in a single day (Garrido 1996), but more generally occur in April–May (AI1, Galarza 1984, Domínguez & Rabuñal 1989). Irby (1883) saw thousands on mudflats in Santander Bay during May 1876. Monthly counts at the Ría de Avilés, Asturias, in 1997 show spring passage from mid March to early June, with well-defined peaks of 1,000 birds in mid April and 3,000 birds in early May; the autumn passage there was in July–October with a peak of only 200 in mid September (*A. O. Asturias* 4: 111–113). A few non-breeders remain on coasts in summer.

At least 574 foreign-ringed Dunlins have been recovered in Iberia. The great majority were ringed as adults or fledged juveniles but at least three individuals were ringed as pulli in Iceland, Britain and Poland, within the breeding range of *schinzii*. The largest numbers of recoveries are of birds ringed in Fennoscandia (37% of recoveries) or Britain (33%) as well as in Germany (9%) or Poland (7%). There are some recoveries of birds ringed elsewhere in central Europe, east to Russia (2 birds) and Ukraine (1), and in Iceland (11). Birds captured on passage at the Odiel marshes are principally of the races *schinzii* and *alpina* (Garrido 1996), probably with a marked preponderance of the former (Cabot & Serrano 1984) but *arctica* has also been identified there. Studies in Portugal (AP) suggest that the majority of the wintering population there are *alpina* although some *schinzii* may winter in the south. *C. a. alpina* departs from Portugal largely in March and *schinzii* migrants, returning from west Africa, predominate in April–May, when small numbers of *arctica* also occur on passage. Conversely, *schinzii* reappears earlier in autumn, passing through Portugal from late July to September, being progressively replaced by the wintering population of *alpina*. Of 163 Dunlins ringed in Iberia and recovered abroad the majority were found in the British Isles (47%), with most of the remainder in northern Europe (19%) and central Europe (14%), as well as three in Iceland and two each in Italy and Ukraine. There are also six recoveries from Morocco and four from Mauritania.

The wintering population in Spain was estimated to be of the order of 28,000 birds by Alberto & Velasco (1988) and 45,000 by Martí & del Moral (2002). Some 55,000 were censused in winter on the Portuguese coast in 1989, half of them at the Tagus estuary (Rufino 1989b). The Spanish January counts of 1990–2009 found 65,259 birds on average but ranged from 13,040 to 124,776 birds, showing that the total Iberian wintering population exceeds 100,000 individuals in some years (AAE). There are indications, however, of a decline in the numbers that winter in Portugal. The Tagus estuary is the principal Portuguese wintering site, accounting for up to a third of the population, the remainder occurring mainly at the Ria Formosa, the Sado estuary and the Ria de Aveiro, with some hundreds also wintering at the Mondego and Minho estuaries and the Castro Marim saltpans (AP). The mean wintering population in Portugal declined from an average of 60,734 birds during 1992–95 to 33,605 birds during 2005–07 (AP).

The principal wintering concentrations in Spain are in the Andalucian coastal wetlands, at the Ebro delta and in the rías of Galicia and the Biscay coastlands, which together held over 95% of the birds counted in January during 1990–2009 (AAE). The major sites here (Table 32) saw some very large gatherings, although with marked interannual variation. Relatively few winter in the Comunidad Valenciana (up to 1,931 birds), Murcia (up to 1,328) or the Balearic Islands (up to 286). The Spanish wintering population increased by 6.5% during 1990–2009, mainly due to larger numbers remaining in the south-west (AAE).

Table 32. January counts of Dunlins at their six principal wintering sites in Spain during 1990–2009 (Source: AAE).

Site	January wintering numbers			% of national mean
	Minimum	Maximum	Mean	
Ebro delta	8,010	30,534	18,815	27.1
Doñana	570	63,025	17,008	24.5
Cádiz Bay	534	32,746	10,631	15.3
Ría de Arousa, Pontevedra	880	7,610	5,075	7.3
Marismas de Santoña	476	8,918	3,430	4.9
Marismas del Odiel	388	4,658	1,864	2.7

Earlier counts in Iberia found smaller wintering numbers than more recently reported. In January 1969, Fournier & Fournier (1972) censused fewer than 10,000 birds at the main wetlands of the north and north-west, the majority of them in the Galician rías. Similarly, Pagezy & Trotignon (1972) found *c.*15,000 birds in total at the Portuguese coastal wetlands in December 1968/January 1969, most of them at the Tagus and Sado estuaries. Alberto & Purroy (1981) found 19,500, 26,000 and 23,000 Dunlins at the principal Spanish wetlands during January censuses in 1978, 1979 and 1980 respectively, although Doñana was not censused in 1980. It is impossible to say whether the early counts are relatively low because numbers then were fewer or because the counts were incomplete. It is noteworthy, however, that Fournier & Fournier (1972) refer to hunting pressure at the Cantabrian estuaries being intense, which is not the case today, when most of the principal sites are reserves. A reduction in disturbance might well be reflected in an increase in wintering populations, which seems to have been the case at the Santoña. Nevertheless, the Asturian coastal wetlands have supported fewer Dunlins in winter in recent years following a period during which numbers increased. The January censuses for Asturias gave a mean total of just 110 birds during 1978–82 but this had increased to 1,121 during 1993–97 before declining to 697 birds during 2003–07, and there were only 385 counted in 2008 (COA).

Dunlins may often be seen in small numbers inland, especially during passage periods. The larger counts at inland wetlands have included 300 at Boada, Palencia, on 30 September 2001 and 417 at Pedro Muñoz, Ciudad Real, on 10 March 2007 (*A. O. Palencia 1998–2001*: 104, *A. O. Ciudad Real 2006–2007*: 172). The Guadiana ricefields in Extremadura have attracted larger numbers in recent winters: January counts there ranged from 2,500–4,500 birds during 2004–07 (Masero 2008).

STILT SANDPIPER *Calidris himantopus*

(Correlimos Zancolín, Pilrito-pernilongo)

Vagrant to Spain.

A moulting adult was at the Salinas de Levante, Mallorca, Balearic Islands, on 6–7 May 1983 (Catley 1983, Catley & Myers 1986). This is a very rare vagrant to Europe, where 29 individuals have been found in Britain, 14 in Ireland, and nine in France. One has also been reported in Morocco, in March–April 1996, and one was in the Azores in October 2007.

BROAD-BILLED SANDPIPER *Limicola falcinellus*

(Correlimos Falcinelo, Pilrito-de-bico-grosso)

Very rare. Recorded on passage in Spain and Portugal.

Migrant European Broad-billed Sandpipers occur regularly in Germany and are quite frequently recorded further west. There are records of 239 individuals in Britain, 24 in Ireland and 292 in France. There are also 15 records from Morocco, as well as observations in Mauritania, Mali and Nigeria.

The earliest Iberian record was one killed near Vila do Conde, Porto, on 8 October 1914 (Reis Júnior 1931). One ringed in Sweden was found on the Portuguese side of the Minho estuary on 6 November 1949 (*Ardeola* 1: 123), although some doubt has been expressed regarding this tardy record (Dierschke 1997). A further half-dozen records antedate the establishment of the rarities committees. The earliest of these was at the Santillana reservoir, Madrid, on 15 September 1957 (Corley-Smith 1958). The early records include one at Tavira, Faro, in September 1973 (*Ardeola* 20: 356–357). During 1985–2010, 28 records of single birds were accepted in Spain and two in Portugal : at Montenegro, Faro, in April 1997 and at Faralhão, Setúbal, in February 2009. Recently accepted records include one of three together at the Ebro delta on 8 May 2010. There were a further eight Spanish records of single birds during 2011–12 (*Ardeola* 60 : 466, RBS) as well as two 2012 records from Portugal (RBP), followed by at least eight on the Mediterranean coast in spring 2013 : including three together at the Albufera de Valencia on 9 May (RBS).

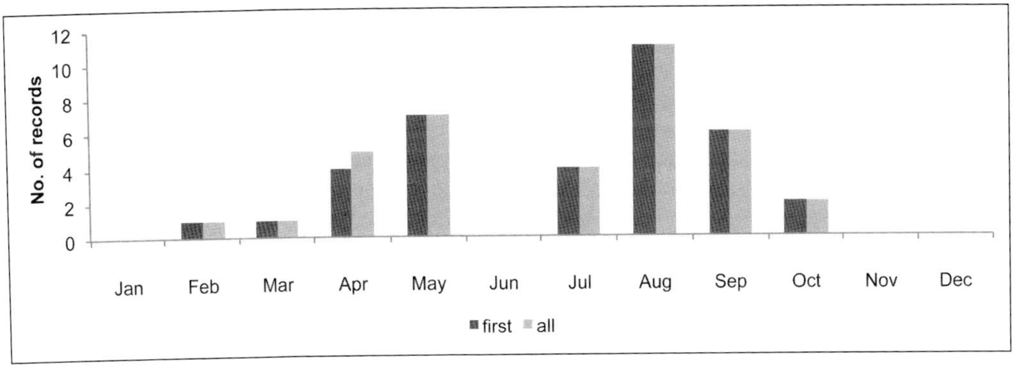

Figure 115. Records by month, 1957–2010 (n = 36)

Iberian records correspond with passage periods (Fig. 115). Extreme dates in spring are 23 February and 26 May. Autumn records fall between 12 July and 25 October. All but two: the Madrid September record and one at the Laguna de Manjavacas, Cuenca, from 30 March to 1 April 2007 (*Ardeola* 56: 325), have been at coastal locations (Fig. 116). Postnuptial records

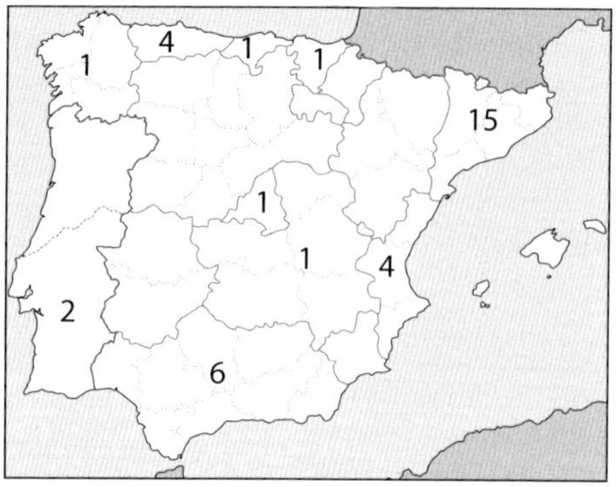

predominate in the Mediterranean, where there are 17 autumn records and three spring ones. In contrast, spring records are somewhat more frequent on Atlantic and Biscay coasts: nine in spring and five in autumn. Spring records also predominate on the European Atlantic seaboard, autumn records occurring further east (de Juana 2006).

Figure 116. Records by region, 1957–2010 (n = 36)

BUFF-BREASTED SANDPIPER *Tryngites subruficollis*

Correlimos Canelo, Pilrito-acanelado)

Vagrant to Spain and Portugal.

The first Iberian record was at Roses, Girona, on 26 August 1963 (*Ardeola* 10: 63). Four further records antedate the establishment of the rarities committees: at Lugo de Llanera, Asturias, in May 1975; at the Sabón reservoir, A Coruña, in September 1976; at the Guadalhorce estuary, Málaga, in September 1977, and at the Ría de Villaviciosa, Asturias, in May 1983 (Álvarez-Balbuena *et al.* 2000; *Ardeola* 22: 135, 24: 259, 50: 138). By 2010 there were 57 accepted records for Spain and nine for Portugal. The first Portuguese record was of a juvenile caught by ringers at the Ria de Alvor, Faro, in September 1990 (*Ardeola* 43: 111).

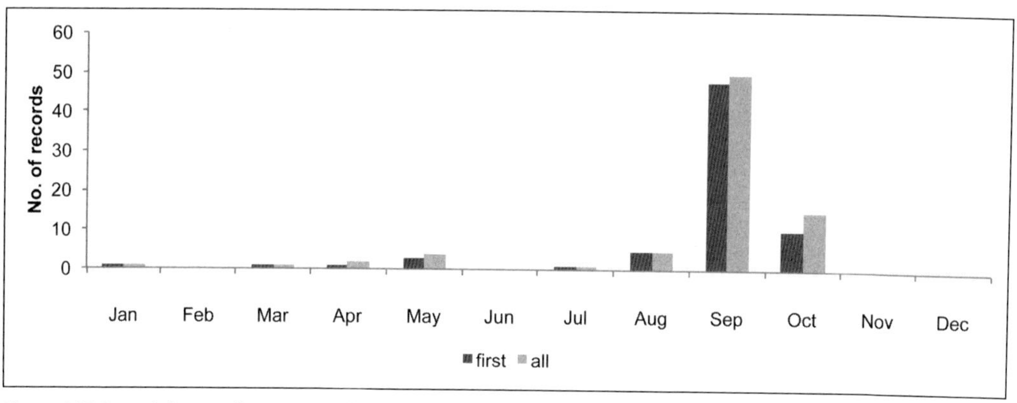

Figure 117. Records by month, 1965–2010 (n = 70)

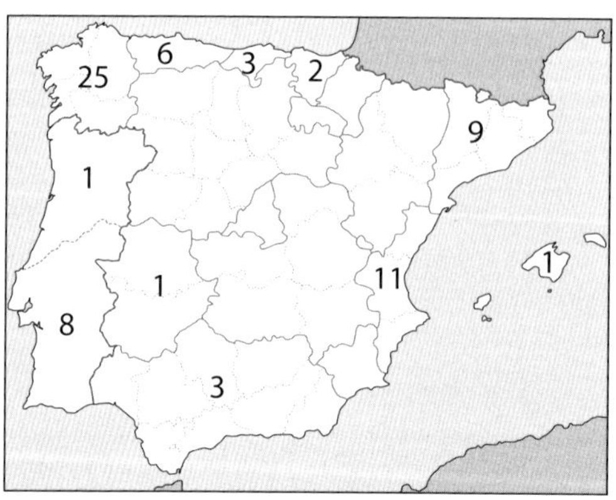

Figure 118. Records by region, 1965–2010 (n = 70)

Nearly all records have been of single birds but there are three of two birds together and two of groups of three. Only four adults were identified among 58 birds whose age was noted, the remainder (93%) being juveniles. Recent years have seen an upsurge in observations: there were up to 31 during September–October 2011, when up to 11 were simultaneously at the Ebro delta, and at least ten in autumn 2012 (*Ardeola* 60: 466–467, RBS, RBP).

Records peak markedly in September, as elsewhere in Europe (Fig. 117). Spring observations make up 7% of the total. The four adults were on relatively early dates: 13 July, 5 and 22 August, and 5 September, but juveniles have appeared between 28 August and 24 October. There is an untypical recent winter record: one at the Albufera de Valencia on 22 January 2008. Observations are predominantly from the Atlantic seaboard, especially in Galicia (Fig. 118). However, three of the five accepted spring records have been on the Mediterranean coasts; two in the Comunidad Valenciana and one in Catalonia, the remaining two being from Asturias. In addition, one was reported from Barcelona, in March 2013 (RBS). A juvenile at the Valdecañas reservoir, Cáceres, in October 2010 is the sole inland record.

This is among the more frequently recorded Nearctic waders in Europe, despite its relatively small global population. There were 629 individuals reported in Britain during 1986–2003, 478 have been found in Ireland and 256 in France. There are also 12 accepted records from the Azores, one from Madeira and eight from the Canary Islands, as well as five from Morocco.

RUFF *Philomachus pugnax*

(Combatiente, Combatente)

Widespread on passage. Winters locally, principally at Doñana.

The Ruff occurs commonly on passage throughout most of the region although it is scarce in Galicia and in the Biscay coastlands. There are 64 Iberian ringing recoveries; of birds from the Low Countries (24), Scandinavia (20), Finland, Britain, France and Germany, and one from Belarus, most of them ringed as adults or fully grown juveniles. There are also single recoveries from Morocco and Mali of Ruffs ringed at Doñana, as well as nine recoveries of Spanish-ringed birds elsewhere in Europe, one of them from Serbia. The western populations winter chiefly in the Senegal and Niger inundation zones in West Africa (Delany *et al.* 2009, Zwarts *et al.* 2009) and cross western Europe, including Iberia, on passage, although relatively few interrupt their journeys in the Peninsula. A small, although apparently increasing, proportion winter in western Europe, many of them in Iberia.

Sizable flocks occur locally on passage at marismas and in estuaries: for example, there were 2,200 at Las Nuevas, Doñana, on 4 April 2000 (*A. O. Doñana* 1: 115), but mudflats and the intertidal zone are avoided. The Ruff is also often encountered on passage in small flocks at inland wetlands, including shallow lakes and ricefields, as well as on damp pastures. Southward passage occurs from late July to October, mainly in August–September, with the return movements from February to May, peaking in April and with stragglers remaining into June (Garrido 1996, Martín 2004, AP). Early spring movements have been especially noted in Portugal, where the wintering population is small but where numbers have begun to increase in February or even late January in some years. The birds are often with migrant Black-tailed Godwits, which far outnumber them. Counts of nearly 1,000 at Campo Maior on 13 February 1991 and perhaps also the 250 at Palmela, Setúbal on 24 January 1999 were considered to involve early spring migrants (AP). Observations in the Huelva coastlands, including Odiel, show modest numbers on passage, with peaks of *c.*100 birds in April and *c.*200 in September–October (Garrido 1996).

January counts in Spain during 1990–2009 show that the wintering numbers remain fairly small: they averaged just 1,788 birds (range 258–4,483; AAE). The Spanish wintering population showed a small increase during 1980–2009 although the numbers present at the principal sites have fluctuated markedly during this period (AAE). Doñana (up to 3,843 birds) and the Ebro delta (up to 771 birds) together supported over 70% of wintering Ruffs. Smaller numbers winter inland at the wetlands of Castilla-La Mancha: notably Daimiel, and at a few sites in Castilla y León, notably Villafáfila. The Guadiana ricefields in Extremadura have attracted increasing numbers in recent winters. The Ruff is mainly granivorous outside the breeding season and rice may form a very important proportion of its diet (Tréca 1994). Specific winter counts at the Guadiana ricefields found 600 birds in 2004, 2,000 in 2006 and up to 3,000 in 2011; in an area where none wintered prior to 2002 (SWA). An increase in wintering numbers, probably associated with a lengthening of the flooding period in the rice paddies as an agri-environmental measure, has also been reported from the Ebro Delta, where January counts gave an average of 56 birds in 1977–85 and 375 in 2006–10 (Ferrer *et al.* 1986, CWA).

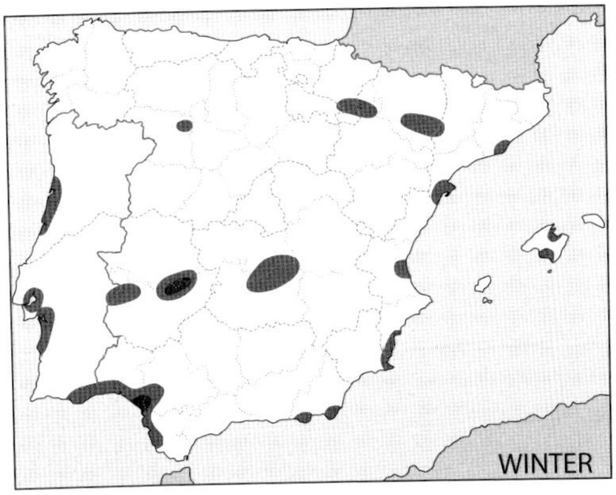

Mid-winter counts in Portugal averaged only 102 birds during 1992–95 and 199 during 2005–07, found largely at the Tagus and Sado estuaries, the Ria Formosa and Castro Marim (AP). Other Portuquese wetlands sometimes attract modest numbers in winter: for example there were 30 inland at ricefields at Alvito, Beja, on 28 December 2003 (*Anu. Ornitol.* 3: 35).

JACK SNIPE

Lymnocryptes minimus

(Agachadiza Chica, Narceja-galega)

Scarce but widespread in winter. Some cross Iberia to winter in North Africa.

The Jack Snipe is a winter visitor to wetlands throughout the Iberian Peninsula and in the Balearic Islands but its abundance and distribution are hard to judge given the difficulty of detecting it. Many observations, and a large number of ringing recoveries, come from snipe hunters whose dogs find birds that would otherwise go unnoticed. The Jack Snipe may occur anywhere that offers damp ground with sufficient plant cover to afford concealment. Such places include sedgebeds, reedbeds, the margins of freshwater and brackish pools, flooded grasslands and even muddy ditches. It often shares these habitats with the Common Snipe but is notoriously much less conspicuous. Chapman (1884) found 33 Jack Snipe (22%) among 150 snipe that he shot at Ovar, Aveiro, but G. Tait (AP) recorded that the snipe bag of 6,357 killed also at the Ria de Aveiro between 1912 and 1944 included only 212 Jack Snipes, 3.3% of the total.

Irby (1895) considered the Jack Snipe 'abundant' in the vicinity of the Strait of Gibraltar, specifically at the former Laguna de La Janda where he regularly hunted. The impression now is that it is not abundant anywhere in Iberia. A recent suggestion is that the entire wintering population in Spain numbers just a few thousand individuals (SWA). There have been regional estimates of 50–100 birds in Galicia (de Souza & Lorenzo 2003) and 542–678 in Catalonia (CWA). It is probably more frequent in the Spanish littoral regions than inland: for example, there are only 11 published records for Madrid during 1996–2008, despite the high density of observers there (*A. O. Madrid*). The Jack Snipe may be still relatively common in coastal Portugal, where a single hunter may bag up to 20–30 in a season (Pereira & Campos 2000). Noval (2001) considered that there were influxes into Asturias during cold snaps in winter.

Very few Jack Snipes are recorded outside October–April (Figure 119) and most observations fall between November and March. The sample examined included just two August, 16 September and four May records. Many winter in North Africa and across sub-Saharan West Africa and a proportion must cross the Peninsula on migration. There are 21 Iberian recoveries of birds ringed on passage in Germany (8), Scandinavia (6), Britain, France, the Low Countries, the Czech Republic or Switzerland (SMA, AP). In addition, birds ringed in Iberia have produced seven recoveries in France and one in Italy.

Figure 119. Monthly distribution of observations (n = 617).

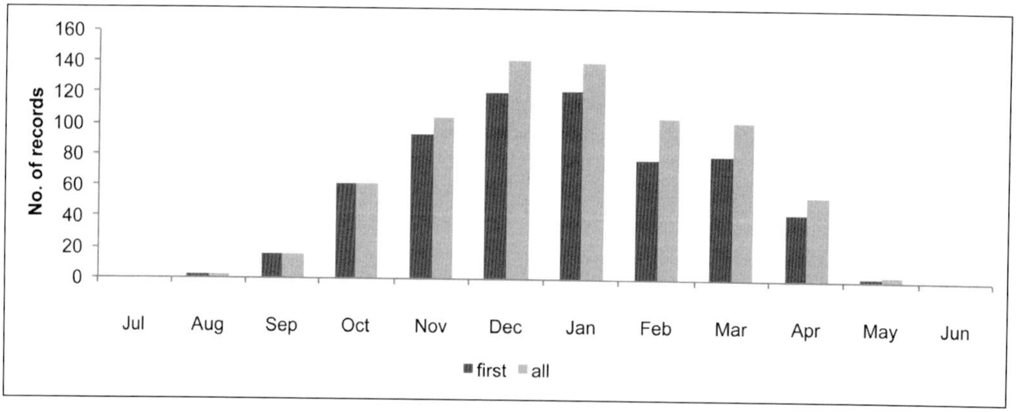

Most observations are of single birds or small, loose groups. Of 552 records in the Spanish bird reports, 73% were of single birds, a further 19% were of two together and just 6% were of 4–10 birds. The largest assemblages reported have been of up to 30 individuals: at the Aragón ricefields and the Llobregat and Ebro deltas, but these are exceptional. There is also an unparalleled observation of about 100 Jack Snipes, together with some 500 Common Snipes, at the Albufera de Valencia on 6 February 1998 (*Aves C. Valenciana 1998*). A sample of 55 records in Portugal gives an average of 4.5 birds per record, again suggesting higher mean densities there than in other Iberian regions; only 29% of the Portuguese records were of single birds.

COMMON SNIPE *Gallinago gallinago*

(Agachadiza Común, Narceja-comum)

***G. g. gallinago*. A very scarce breeding species in northernmost Portugal and adjacent Galicia and in the Central Range. Widespread and common on passage and in winter throughout the Peninsula and in the Balearic Islands.**
***G. g. faeroeensis*. Probably a scarce or very scarce winter visitor. There is one ringing recovery of an Icelandic bird in Portugal.**

The Common Snipe is at the southern limit of its western Palearctic nesting range in Iberia, where its populations, which have always been small and fragmented, have suffered recent declines (LR, LV). Both the Portuguese and Spanish populations inhabit montane habitats: in Spain at *c.* 850m in Galicia and at 1,100–1,700m in the Central Range, and in the uplands of northernmost Portugal. Nesting birds occur in bogs, wet heaths, sedgebeds and wet pastures, often where there is traditional grazing by livestock (LR).

The Common Snipe was known to breed in northernmost Portugal from the beginning of the 20th century (Reis Júnior 1931, Coverley 1932). The Portuguese population is in Trás-os-Montes where 100–150 pairs may have bred during the 1960s and until 1984–86 (Rufino & Neves 1991b) but these had declined to 50–100 pairs by 1989 and to just 8–10 pairs in 2006, on the Mourela plateau (PA2). The decline is attributed to drainage and deterioration of the montane habitats.

In Spain there is anecdotal evidence of regular breeding in Galicia, in the A Limia depression, Ourense, as well as around the headwaters of the Salas and Fírbeda rivers there, from at least the 1930s (Villarino *et al.* 2002). The A Limia population was estimated to be as large as 200–420 pairs during 1930–60. However, progressive land drainage reduced the available habitat drastically and numbers fell to an estimated 70–100 pairs by the late 1970s, 10–50 pairs during the 1980s and none at all by the early 1990s (Domínguez *et al.* 1987, Villarino *et al.* 2002). The upper Salas and Fírbeda basins supported an estimated 10–30 breeding pairs in the early 1990s and these two locations remain the only regular breeding locations in Galicia. The national census of breeding Common Snipes in Spain in 2009 found 9–13 pairs in Galicia, all of them in southernmost Ourense, adjacent to the remaining Portuguese population (Lorenzo & Planelles 2010).

The principal surviving Spanish, and Iberian, population is in the Central Range in Ávila and there is some evidence that breeding birds may also occur further east in Madrid and Segovia (Lorenzo & Planelles 2010). Breeding in the Central Range was first confirmed as recently as 1976, at Garganta del Villar (Castellanos 1976). The total population was estimated at 37–38 pairs by Hortas *et al.* (2000). The 2009 census found 25 pairs, all of them in Ávila province: six in the Sierra de Ávila, 10 in the Vega del Alberche, seven in the Sierra de la Paramera and two in the Sierra de Gredos, and the total population was estimated at 60–105 pairs.

It may have nested in Asturias prior to 1975 but there are no subsequent breeding records there (Álvarez-Balbuena *et al.* 2000, Noval 2001). Breeding has also been suspected occasionally in the Basque Country (Álvarez *et al.* 1985) and a nest that was later abandoned was found at San Telmo, Gipuzkoa, on 12 May 2002 (*Ardeola* 50: 348). A drumming male was seen in late April 2009 and again in May 2010 at the Laguna de Chozas de Arriba, León (*Ardeola* 58: 207). It remains quite possible that there remain undiscovered breeding locations or that sporadic breeding occurs in other areas. The known Iberian population in the early 21st century was nonetheless of the order of at least 50 pairs and estimated at 70–130 pairs. Restoration of the former A Limia lake would have a significant effect in boosting breeding numbers (LR).

The Peninsula is a major wintering area for Common Snipes originating from northwestern Europe and many cross the region to winter further south, in Morocco and also in sub-Saharan Africa (Delany *et al.* 2009). There are at least 659 Iberian recoveries of foreign-ringed Common Snipes: of birds ringed in Britain, the Low Countries, France, Germany, Scandinavia, the Baltic States, Poland, the former Czechoslovakia, Hungary and Switzerland, as well as single birds from Italy, Austria, Ukraine and Belarus, but hardly any of them involve birds ringed as pulli and many would have been on passage when captured. One ringed as an adult in Iceland in September was recovered in November in Alentejo (AMI). An analysis of 509 Iberian recoveries (Asensio & Carrascal 1987) identifies two migratory flows that terminate in the Peninsula. One begins in Fennoscandia and crosses Atlantic Europe to reach wintering grounds on Iberian Atlantic coasts and in the Biscay coastlands. The other crosses central Europe and enters Spain through southeastern France to

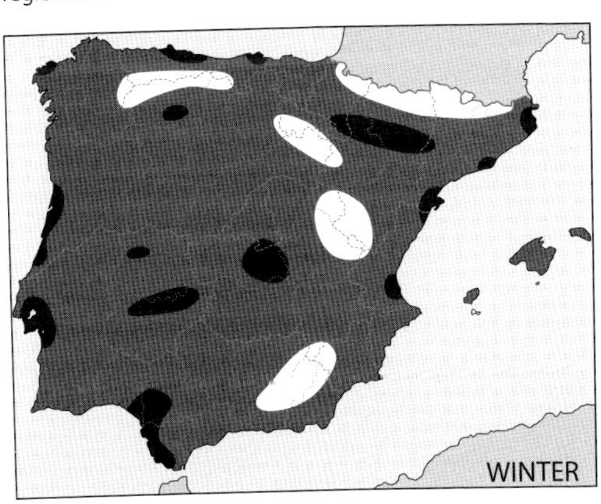

reach wintering areas in the northern half of the Levant coast. Common Snipes ringed in Spain during 1973–2009 have given 30 recoveries in France, five in Italy, four in Russia and one each in Finland, Sweden and Moldova. Ringing recoveries in Iberia are almost entirely during September–April, with a peak in December (Asensio & Carrascal 1987), although this may partly reflect hunting activity. Most migrants arrive during October–November and depart during February–March but some appear from late August and stragglers may linger throughout May.

Wintering and passage birds occur at a wide range of both coastal and inland wetlands, where they are often very common. They require damp ground, as may be found in marshes, lake and reservoir margins, wet pastures, flooded fields and ricefields. The wintering and transit populations are very hard to census and there are no national counts available but the numbers involved are certainly high. The numbers wintering at Doñana have been estimated at 10,000 birds (García *et al.* 1989), not an implausible number for such an extensive complex of wetlands, although a later estimate of 20,000–40,000 there (García *et al.* 2000) may be too generous.

A further 5,000 birds are estimated to winter in the Huelva coastlands outside Doñana, including Odiel (Garrido 1996). January counts at the Ebro delta found 1,142 birds in 1978, 3,500 in 1979 and 1,116 in 1980 (Alberto & Purroy 1981). The 1996–2009 January counts at Catalan wetlands found an average of 1,768 at the Ebro delta, the maximum count being 3,474 in January 2007. The Ebro delta is certainly a major wintering area but the other Catalan wetlands also support some numbers in winter: an average of 301 at the Llobregat delta and 563 at the Aiguamolls de l'Empordà in January during 1996–2009 (*A. O. Catalunya*). The Catalan winter population has recently been put at some 7,000–12,000 birds (CWA). The Albufera de Valencia and its nearby ricefields also hold many in winter: 629 were counted there in January 2009 (*A. O. C. Valenciana 2009*). Many also frequent the rías of Galicia and the Biscayan coast: notable concentrations reported there include 579 at Villaviciosa on 11 January 1997 and 215 at the Ría del Eo on 2 February 1997 (*A. O. Asturias* 4: 114–115). The flooded ricefields of Extremadura have proved attractive to the species and many also remain to winter there: 1,130 were counted at Galisteo, Cáceres, on 24 January 2004 (*Ardeola* 51: 552). The Common Snipe is generally widespread in Portugal in winter and is locally numerous at coastal wetlands, notably at the Tagus and Sado estuaries and the Ria Formosa, as well as at ricefields inland.

WILSON'S SNIPE *Gallinago delicata*

(Agachadiza de Wilson, Narceja de Wilson)

Vagrant to Portugal.

A first-year bird was killed by hunters at Vila Franca de Xira, Lisboa, on 30 December 2006. This Nearctic counterpart of the Common Snipe *G. gallinago* has formerly been regarded as a subspecies of that species and there are still few European records: seven in Britain and one in France. There are several from the Azores and a few seem to winter there regularly (Rodrigues & Gonçalves 2013).

GREAT SNIPE *Gallinago media*

(Agachadiza Real, Narceja-real)

Very rare. Recorded in Spain mainly on passage. Vagrant to Portugal.

The earliest Iberian records are of birds hunted near Benalup-Casas Viejas, Cádiz, in October 1868 and near Gibraltar in October 1871 (Irby 1895). Munn (1928) claimed to have flushed one near Santanyí, Mallorca, in April 1927. About 20 other 20th century records antedate the Spanish rarities committee (Ferrer *et al.* 1986, García-Rovés & García-Rovés 1989). There is also an old specimen record from Portugal of one killed at Ribatejo (Tait 1924). There are 21 accepted records from Spain and two from Portugal up to 2010. At least a further 31 Spanish records occurred during 2011–12 (*Ardeola* 60 : 467–468, RBS), most of them during an unprecedented influx in April 2011 in which at least 26 individuals were found, chiefly in Girona but with a few in the Ebro delta and two in the Balearic Islands. Five more in May 2013 were all in coastal Catalonia (RBS).

The 43+ pre-2011 Iberian records, nearly all of single birds, show clear peaks during passage periods, in April and September–October (Fig. 120), with extreme dates of 23 February and 22 May in spring and 15 August and 1 December in autumn. Spring records comprise 63% of the total, which rises to 79% for post 1986 records. This difference probably reflects the higher proportion among older records of birds killed by hunters during the autumn hunting season (de Juana 2006). All records are concentrated in the north-east and the Balearic Islands: Catalonia, the Balearics and the Comunidad Valenciana account for 70% of the total, and 79% of the post-1984 ones (Fig. 121).

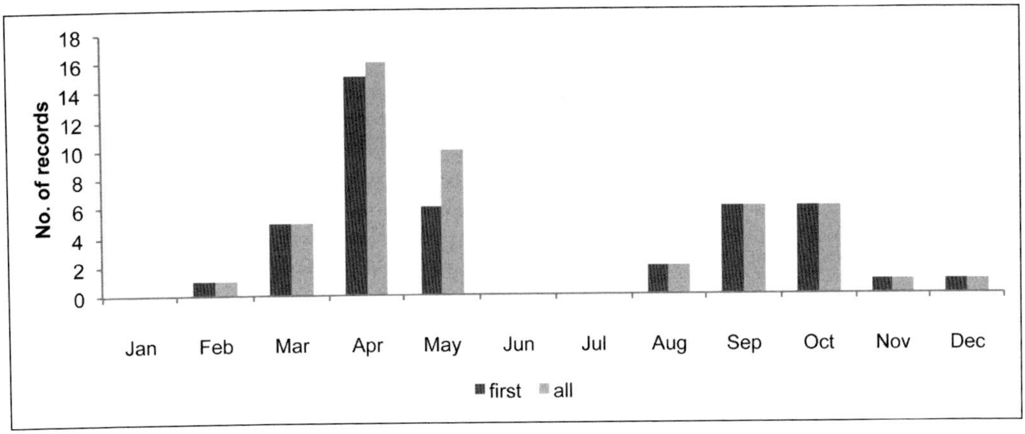

Figure 120. Records by month, 1986–2010 (n = 43)

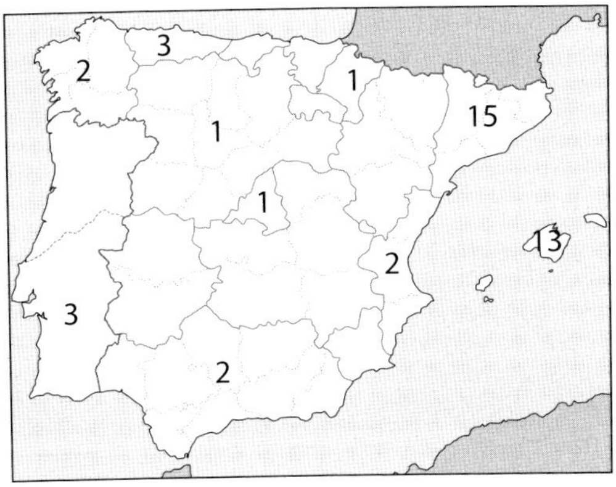

Figure 121. Records by region, 1986–2010 (n = 43)

Spring records make up the great majority of those from the north-east and Balearics (83%, n = 30) but only 15% of those found elsewhere in Iberia (n = 13). The great majority of British records are in autumn, so it may be that there is a loop migration: southwards through Britain and western Iberia in autumn, returning northwards in spring through eastern Spain and France: 65% of French records during 1994–2008 have been in spring (OF).

The Great Snipe was recorded more frequently in westernmost Europe in the past, when its breeding range extended as far west as Denmark, Germany and perhaps the Netherlands. For example, 532 individuals were reported in Britain pre-1950 but there have only been 163 thereafter. It occurred regularly in France in the distant past (Mayaud 1936) but there are only 54 accepted records since 1981. There are historical records from the Canary Islands and Madeira, and it is known as a rare migrant and winterer in Morocco, with two recent records.

LONG-BILLED DOWITCHER *Limnodromus scolopaceus*

(Agujeta Escolopácea, Maçarico-de-bico-comprido)

Vagrant to Spain and Portugal.

Regarded as conspecific with the Short-billed Dowitcher *Limnodromus griseus*, a much rarer vagrant to Europe, until 1951, and the identification criteria for separating the two in the field were unclear until quite recently. As a result many early records are indeterminate, although the great majority are likely to have been Long-billed Dowitchers. There are accepted records of 225 individuals of the present species from Britain, 115 from Ireland and 44 from France. There are also six accepted records from the Azores, two from the Canaries and eight from Morocco.

The earliest Iberian observations in Iberia were at the Ría de Villaviciosa, Asturias, in October 1969 and October 1970 (Noval 1986) but these and others that predate the rarities committees are only regarded as *Limnodromus* spp. The first accepted record of a Long-billed Dowitcher was of a juvenile at the Sabón reservoir, A Coruña, on 7 October 1975 (*Ardeola* 50: 138). Accepted records to 2010 total 18 in Spain and seven in Portugal, all but one of single birds. There was a further Spanish observation in 2011 and seven 2011–12 Portuguese records.

Autumn birds include 15 that were identified as juveniles and spring records include two adults and two second-year birds. The 19 autumn records were between 30 August and 30 December and the six spring ones fell between 17 February and 14 May (Fig. 122). Only one has been found inland (Fig. 123): at the Laguna de La Nava, Palencia, in April 1998 (*Ardeola* 47: 149). There are 13 autumn and three spring records from Atlantic coasts and six autumn and two spring records from the Mediterranean.

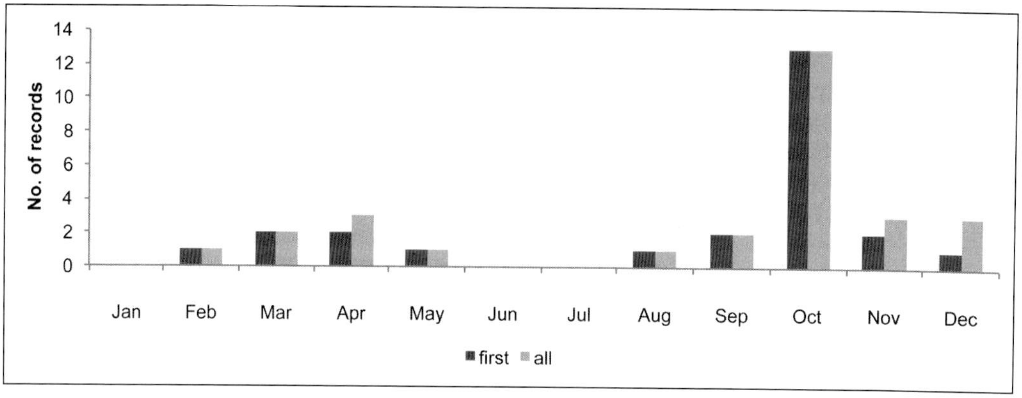

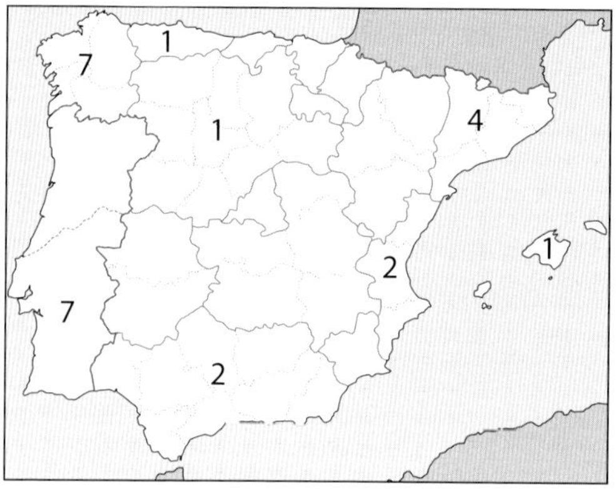

Figure 122. Records by month, 1975–2010 (n = 25)

Figure 123. Records by region, 1975–2010 (n = 25)

A further ten records from Spain and one from Portugal were only accepted as Long-billed/Short-billed Dowitcher. A record of three Short-billed Dowitchers at the Bonanza saltpans, Cádiz, in April 1964 (*Ardeola* 10: 63) is not considered reliable and there are no accepted Iberian records of that species. However, a putative Short-billed Dowitcher was at the Mar Menor saltpans, Murcia, during 9–26 March 2013 (RBS).

EURASIAN WOODCOCK *Scolopax rusticola*

(Chocha Perdiz, Galinhola-comum)

A resident breeding species in montane regions of northern Spain and Andorra. Widespread throughout the Peninsula and the Balearic Islands on passage and in winter, although still commoner in the north.

The Woodcock is a breeding resident of the montane forests of the cool, humid north, in the Eurosiberian zone, extending eastwards from Galicia across the Cantabrian Range and the Pyrenees to Catalonia. It also nests in the Iberian Range and, more locally, in parts of the Central Range. Breeding was first confirmed in Madrid in 2001, when a female with three chicks was at Montejo de la Sierra on 17 June (*Ardeola* 50: 160). Occasional observations in the breeding season outside the known nesting range suggest that scattered pairs may nest more widely in north-central Spain: for example, four birds were at Galende, Zamora, on 16 June 2005 (*Ardeola* 52: 430) and there are breeding season records from the Tiétar valley in northern Extremadura (Hidalgo & Rocha 2001). The preferred habitat, particularly in the Cantabrian Range, is mature beech and oak woodlands but mixed pine/deciduous woodlands also have nesting birds, particularly in the Pyrenees and Central Range (Lucio & Sáenz de Buruaga 2000, Onrubia 2003). The Spanish breeding population has been estimated at 3,600–4,000 pairs (Purroy *et al.* 1994), probably too conservatively (Lucio & Sáenz de Buruaga 2000). Some 15–30 pairs nest in Andorra (BE). The Woodcock does not breed in Portugal, although it may once have done so in the north (PA2). It may have bred sporadically in the Balearic Islands (SA2).

Breeding birds are thought to be resident, although some descend to lower levels in winter (AMI). However, large numbers of Woodcocks from northwestern Europe winter in Iberia and some continue south to North Africa. Cold spells in Europe may also lead to mid-winter influxes into Iberia, and from Iberia to Morocco (BM). The Iberian wintering population has been estimated at over 500,000 birds (SWA), presumably by extrapolation from hunting bags. Wintering Woodcocks suffer intense hunting pressure across Europe, where the annual bag is estimated at 2.6–3.7 million birds. Trained hunting dogs easily locate and flush roosting Woodcocks and much of what is known of their abundance and occurrence in Iberia comes from hunting records. The estimated bag for the whole Iberian peninsula in an average year is around 75,000 birds (Hepburn 1983), and

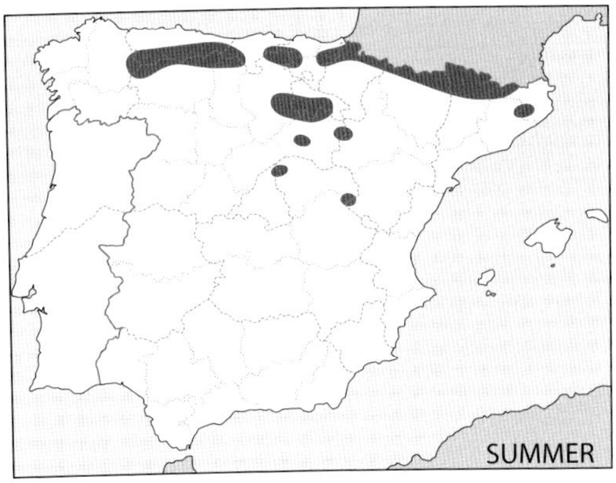

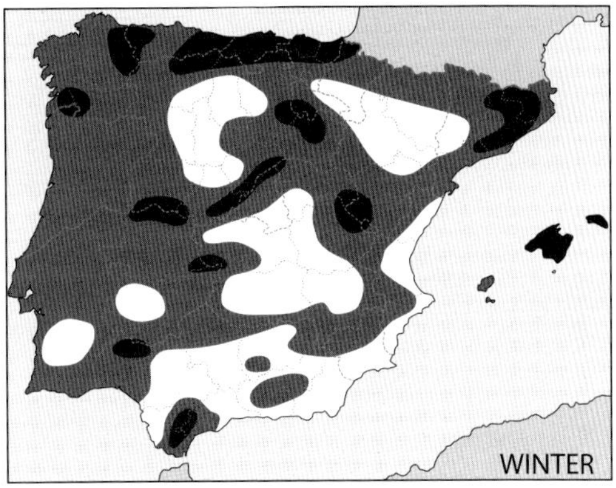

some 10,000 were killed annually in Catalonia during 2006–09 (CWA). Hunting pressure is locally high: for example, over 12,000 were shot in monitored hunting reserves in Asturias alone between October 1996 and January 1997 – at a density of 22.3 birds/1,000ha – most of them during a cold weather influx that January (*A. O. Asturias* 4: 115–116). A transect study using dogs, conducted outside the hunting season, together with an analysis of hunting records, shows that some Woodcocks arrive in Asturias and Álava in October but that the peak influx is during the second half of November; the spring passage peaks during the last week of February and the first week in March (Lucio & Sáenz de Buruaga 2000). Some birds may linger into April (Hidalgo & Rocha 2001).

At least 400 Woodcocks ringed both as adults or as pulli in the British Isles, France, Germany, the Low Countries, Scandinavia, Finland, the Baltic States, Russia, Belarus, the former Czechoslovakia and Hungary have been recovered in winter in Iberia, chiefly in the north. Thirteen birds ringed in Spain have been found abroad: eight in France, two in Russia and one each in Belarus, Austria and Italy. Most of the immigrants probably breed around the Baltic, from Sweden to Western Russia, with relatively few coming from more westerly countries, such as Norway or Britain, or from central Europe (Guzmán *et al.* 2011). There are at least 37 recoveries of birds ringed in Russia and in 2009 a satellite-tracked bird travelled 6,250km from Cantabria to central Siberia (Díez *et al.* 2010). Wintering birds are present throughout the Peninsula and also in the Balearic Islands but they are relatively scarce in the arid south-east and most abundant in the north and west. They inhabit damp woodland and scrub, including both deciduous and coniferous formations, provided that dense undergrowth is available.

BLACK-TAILED GODWIT *Limosa limosa*

(Aguja Colinegra, Milherango)

L. l. limosa. **Common on passage and in winter. A few pairs nest irregularly in Spain.**

L. l. islandica. **Common on passage and in winter.**

The Black-tailed Godwit is a common and locally abundant passage migrant and winter visitor at wetlands throughout the region. A few pairs of the nominate race also nest in some years but breeding records come from scattered wetlands in Spain and there is no established nesting population. The earliest known instance of breeding was in 1961 at the Ebro delta (Maluquer & Pons 1961), where a nest with eggs and possibly two other empty nests were found on the Isla de Buda. There were no further reports until 1981 when a nest was found at A Limia, Ourense (Bárcena & Domínguez 1995), followed by successful breeding at the Laguna de la Nava in 1990 and at the Laguna de Miguel Esteban, Toledo, in 1993 (*Ardeola* 37: 339, 42: 222). Since then there have been confirmed or suspected nesting attempts in Valencia, Lleida, Almería, Valladolid, Badajoz and La Rioja (SA2, LR). The most regular nesting site, however, is the Lagunas de Villafáfila, where breeding was first reported in 1997 and one or two pairs nested annually from then until 2004, although not in 2005–07 (Rodríguez & Palacios 2009).

Black-tailed Godwits are long-distance migrants that move characteristically early in the season, both in spring and in autumn. The earliest southbound migrants appear in Iberia in late June and July and passage continues through August and September into November. This autumn passage is as a rule much less evident than the spring one. Many birds are thought to fly from the Netherlands to sub-Saharan Africa directly, or with just a short stop in Morocco (Zwarts *et al.* 2009). Most of the best Iberian sites, often including Doñana, are dry in late summer and early autumn, and rice crops are still growing then. Some birds stop-over nevertheless at saltpans, for example in Cádiz Bay or at Odiel: concentrations of over 8,000 may occur in late July–August at the latter (Gacio & Sayago 2011).

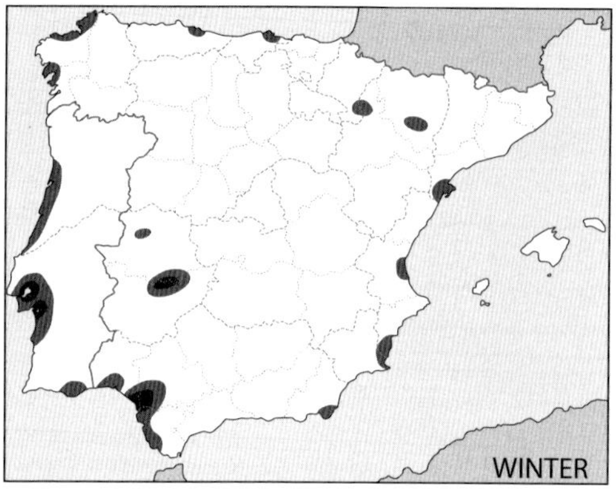

The return passage is already evident in Portugal in January and peaks during the second half of February. Staging duration is relatively brief: observations of colour-marked individuals give an average spring stay in Portugal of 24 days (Lourenço *et al.* 2010). The timing of spring migration at the Guadiana ricefields in Extremadura is similar and radio-tagging has shown stay durations of 35 days there by birds tagged in late January, declining to eight days by birds tagged in late February: almost all have left by early March (Masero *et al.* 2011). The early return of Black-tailed Godwits from Africa means that most of those seen in Iberia in mid winter are returning migrants rather than winter residents.

The Icelandic subspecies *L. l. islandica* winters chiefly in the British Isles but some follow Atlantic coasts to winter in Portugal, Spain and Morocco, where they arrive even earlier than *L. l. limosa*. The relative proportions of the two subspecies in Iberia in winter were formerly considered to be 85% *limosa* and 15% *islandica* (R. Rufino in Delany *et al.* 2009). However, large numbers of both subspecies have been colour-ringed in recent years and this has allowed the relative proportions to be reassessed: 30% are *islandica* and 70% *limosa*, although if this is a genuine change rather than the result of improved methodology it probably reflects the recent increase in the *islandica* and marked decrease in the *limosa* source populations (Gunnarsson *et al.* 2005, Alves *et al.* 2010). The decline in western populations of *limosa* may be up to 30% and those of the Netherlands have decreased by 50% since the 1980s (Thorup 2006).

Very large numbers of *limosa*, the majority of this subspecies, continue beyond Iberia to winter in sub-Saharan Africa (Delany *et al.* 2009, Zwarts *et al.* 2009). There are over 180 Iberian ringing recoveries of *limosa*, most of them (*c.* 80%) of birds ringed as pulli in The Netherlands, with a few from Belgium, Germany, Denmark and Sweden and one each from France and the Czech Republic. The great majority of recoveries have been in Doñana. There are also two Spanish recoveries of birds ringed in Britain that seem likely to have been *islandica* (AI1) and of another that was ringed in Iceland itself. Birds ringed in Spain have been recovered in The Netherlands (10), France (5) and Germany. Most migrating Black-tailed Godwits pass unnoticed but visible passage of godwit flocks is sometimes reported along coasts, for example in Asturias (Noval 2001) and at the Strait of Gibraltar, and flocks also appear at inland wetlands during passage periods.

Black-tailed Godwits are a familiar sight at freshwater and brackish wetlands, both inland and along coasts, as well as on coastal mudflats, in marismas and saltpans. They are especially attracted to ricefields. A gathering of over 2,600 birds in maize stubble near Valdivia, Badajoz, on 28 January 1989 (*Ardeola* 37: 338) was more unusual. In ricefields Black-tailed Godwits feed mainly on spilt rice left after harvest (Lourenço & Piersma 2008, Masero *et al.* 2009). Those in estuaries mainly consume bivalves, worms and snails (Moreira 1994) and quantities of chironomid larvae and pupae are taken in saltpans (Estrella & Masero 2010). Small numbers may be observed throughout the region but very large congregations occur at many of the major Iberian wetlands (Table BTG1). Three Portuguese sites are listed as of international importance for the species in winter and/or during passage periods (Delany *et al.* 2009): the Tagus estuary (maximum count 82,000 birds in February 1992: AP), the Sado estuary (20,030 birds in 1996) and the Ria de Faro (4,474 birds in 1995). No fewer than 23 Spanish sites are similarly designated, although 19 of these fall within Doñana. The remaining four sites are Cádiz Bay, the Extremaduran ricefields of the Guadiana valley, the Ebro delta and Santoña.

The mid-winter population in Iberia averaged some 64,000 birds in January counts during 1990–2001 (Alves *et al.* 2010) although, as noted above, most of those present in January are actually returning migrants from Africa. About 30,000 were in Portugal, 88% of them at the Tagus and Sado estuaries and ricefields. The remaining 34,000 were in Spain, over 70% of them in the Guadalquivir estuary. The mean January population in Spain during 1990–2009 was somewhat larger: 40,182 birds (range 10,557– 73,359) (Table 33), and showed a moderate increase (4.2%) during 1980–2009 (AAE). Given the decline in the western European breeding population, this might be due in part to a northward shift in the wintering range (Masero *et al.* 2011). Inter-annual fluctuations in mid-winter numbers in Iberia are considerable. Some sites have shown consistent falls in the numbers that use them. In particular the numbers wintering in Cádiz Bay declined from 4,250 birds in 1996 to 1,700 in 2002 (LR). Conversely, the expansion of rice cultivation in the Guadiana valley has seen enormous increases in the numbers found there from December onwards: 25,000 were present in February 2013 (Masero *et al.* 2009, *Ardeola* 60: 526). The two subspecies tend to occur in discrete flocks and colour-ringing has allowed the winter ecology of *islandica* and *limosa* to be compared (Alves *et al.* 2010), with the finding that at the Tagus estuary it is largely *limosa* that is attracted to ricefields whereas *islandica* is more reliant on estuarine mudflats. This segregation seems likely to occur more widely.

Table 33. The mean and maximum January populations in different sectors of Spain 1990–2009. Source: AAE.

Region	Mean	Maximum	Key sites
Andalucía	33,075	60,029	Doñana, Cádiz Bay, Huelva coastal wetlands, Cabo de Gata saltpans
Comunidad Valenciana & Murcia	486	1,244	Mar Menor, Santa Pola, El Hondo
Catalonia	3,388	4,797	Ebro delta
Galicia	456	862	Ría de Arousa
Castilla-La Mancha	70	340	Wetlands of La Mancha
Ebro Basin	2	20	Gallocanta
Biscay coastlands	286	665	Santoña
Extremadura	2,512	12,821	Vegas Altas, Palazuelos and Madrigalejo ricefields
Balearic Islands	2	12	
Castilla y León	5	81	

The mid-winter population of the Huelva wetlands west of Doñana was considered to be 5,000–6,000 birds by Garrido (1996), with much larger numbers occurring there on passage, over 25,000 in some autumns. Winter counts at the Ebro delta also tend to be lower than during passage periods, when up to 10,000 have been recorded (Delany *et al.* 2009). Significant numbers of Black-tailed Godwits occur at other east coast wetlands, especially on passage but also in winter. These sites include the Cabo de Gata saltpans, Almería; the Santa Pola saltpans, Alicante, and the Albufera de Valencia. For example, the Albufera held up to 3,500 birds during late February and early March 2002, 2,000 in February 2006 and some 3,000, 5% of them apparently *islandica* birds, in February and early March 2009 (*Ardeola* 50: 160, 54: 178; *A. O. C. Valenciana 2009*: 69). Black-tailed Godwits are uncommon on passage in the Balearic Islands and seldom recorded there in winter (AAE).

Comparatively few birds occur on the Biscay coasts but the numbers present in winter at Santoña have increased significantly in recent years; from 50 in January 1980 (Alberto & Purroy 1981) to a January average of 254 during 1990–2009, with a maximum count of 562 birds (AAE). The Santoña wintering birds are all *islandica* and the site is now of international importance (Navedo *et al.* 2007). The Galician rías also attract passage and wintering birds: January counts at the Ría de Arousa during 1990–2009 have averaged 370 birds, with a maximum count of 777 (AAE).

Black-tailed Godwits may be encountered at Iberian wetlands in summer and may then be locally numerous: up to 2,000 birds in the Huelva coastal wetlands, for example (Garrido 1996). The ricefields at Vila Franca de Xira in the Tagus estuary held nearly 3,000 birds on 23 June 2002 and there were 1,220 at Quinta da Vala, Faro, on 27 June 2009 (*Anu. Ornitol.* 2: 34, 8: 69). Some summer records may involve non-breeders but, since southward passage in this species is already underway in June, many birds seen at this time of year may be southbound migrants.

BAR-TAILED GODWIT *Limosa lapponica*

(Aguja Colipinta, Fuselo)

***L. l. lapponica*. Widespread and locally common in winter and, especially, on passage on Atlantic and Biscay coasts. Scarcer in the Mediterranean and uncommon inland.**

The Bar-tailed Godwits that nest from northern Europe east to western Siberia winter chiefly on European Atlantic and North Sea coasts, including the Atlantic and Biscay coastlands of Iberia. Bar-tailed Godwits tend to be most numerous on passage, although the largest concentrations in Portugal are seen in winter (AP). Migrants occur at all the significant estuaries of the Biscay coastlands and the Atlantic coasts of Portugal and Spain south to the Strait of Gibraltar. The autumn passage is mainly during August–October (Tellería 1981, Galarza 1984, AP) but a bimodal arrival pattern has been reported at the Marismas del Odiel, with an early arrival in July and early August and a second influx in October and November (Garrido 1996). The later birds will have interrupted their passage at the Wadden Sea, a key staging area during both passages, whereas those that pass in the summer months are early travellers to West Africa (Delany *et al.* 2009). Spring movements are principally during April–May but many migrants are then thought to travel to the Wadden Sea from West Africa in a single unbroken flight. However, sizable spring concentrations do occur: for example, there were 1,600 at Valdoviño, A Coruña, on 4 May 1998 (*Ardeola* 46:

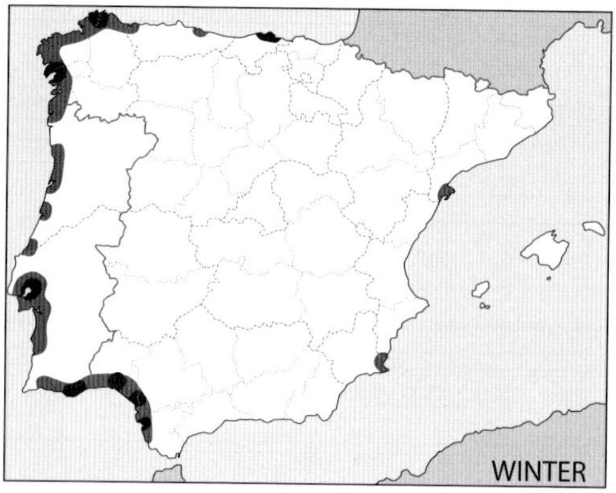

157). There are a few recoveries of birds ringed on passage in Britain, Norway, Denmark, Germany and The Netherlands. Two ringed in Portugal have been found in The Netherlands and there are single recoveries of Spanish-ringed birds from France and Britain.

Bar-tailed Godwits are less widespread in Iberia in winter but there are then significant concentrations at the Tagus estuary, the Ria de Faro, the Galician rías, and the Guadalquivir, Odiel and Santoña marismas. The first two of these sites are considered to be of international importance for the species, on the basis of of 1,500 birds at the Tagus estuary in 1996 and 4,218 birds at the Ría de Faro in 1995 (Delany *et al.* 2009). There is a much larger count from the Tagus estuary on record: 8,900 birds on 21 January 1996 (AP). Mean January totals of wintering Bar-tailed Godwits in Portugal were 4,101 birds during 1992–95 and 3,028 during 2002–05, although with large inter-year variation (AP).

Early censuses of Spanish wintering birds revealed modest numbers. Only 437 Bar-tailed Godwits were found during a national waterbird census in January 1973 (Araújo & García Rúa 1974), most of them in the Galician rías. The January censuses of 1978–80 (Alberto & Purroy 1981) found just ten birds in 1978, 431 in 1979 and 224 in 1980, the largest counts being in the Huelva marismas and Cádiz Bay. Nonetheless, the Spanish wintering population was estimated to be *c.*1,700 birds by Alberto & Velasco (1988). The January counts in Spain during 1990–2009 found a similar number on average, 1,769 birds (range 276–4,270), but averaged 3,800 during 2008–10 (AAE, SWA). The Spanish population is concentrated at a few sites on Atlantic and Biscay coasts, where some 90% of wintering birds occur. The principal site during 1990–2009 was Cádiz bay, where 715 birds occurred on average in January, with a maximum count of 2,703 birds. The other important wintering sites, in northern and western Spain, were the Galician rías: especially Arousa (up to 668 birds) and Ortigueira (up to 355), Santoña (up to 561), Odiel (up to 277) and Doñana (up to 975), although mean counts at these locations ranged from 100 to about 250 birds (AAE).

The Bar-tailed Godwit is much scarcer on Mediterranean coasts and in the Balearic islands but small numbers occur, both on passage and in winter, and there is a small but regular winter presence in the Ebro delta. The January counts at the delta during 1996–2009 found between 24 and 222 birds, with an average of 84 (*A. O. Catalunya*), although there were 975 birds there in 2000 (CWA). There are exceptional January counts from Murcia of 163 at the San Pedro del Pinatar saltpans and 136 at Las Encañizadas (AAE), although only about 20 normally winter there.

This species is strongly associated with muddy estuarine habitats but some also frequent saltpans. It is not attracted to ricefields, unlike the Black-tailed Godwit, although it may occur there on passage. Very few are recorded away from the coasts but some occasionally appear at inland wetlands on passage and, more rarely, in winter (SWA, AP). Most such records involve single birds observed in April, May and September but small groups sometimes occur. However, a confirmed record of two flocks of 25 and 200 birds seen arriving at and soon departing the Santillana reservoir, Madrid, on 1 May 2004 (*Ardeola* 52: 215) is exceptional. Small numbers of non-breeders remain on Iberian coasts in summer, at such sites as the Huelva coastlands (Garrido 1996) and the Ebro delta (Clavell 2002).

HUDSONIAN WHIMBREL

Numenius hudsonicus

(Zarapito de Hudson, Maçarico-galego-americano)

Vagrant to Spain.

There is an historical Spanish record: Lord Lilford (1873) obtained a specimen at Doñana on 3 May 1872. A long-stayer was seen at the Laguna de Baldaio, A Coruña, between 20 January and 22 December 2009, although it was apparently absent in May and June. One was seen from Punta La Vaca, Asturias, on 7 August 2013 (RBS). This Nearctic species has until recently been treated as a race of the Eurasian Whimbrel *N. phaeopus*. Seven have been recorded in Britain and three in Ireland, and there are also observations from the Azores, Madeira and the Cape Verde Islands.

EURASIAN WHIMBREL *Numenius phaeopus*

(Zarapito Trinador, Maçarico-galego)

N. p. phaeopus. **Common on passage on Atlantic and Biscayan coasts, scarce but regular in the Mediterranean. Small numbers winter on northern and western coasts.**

Whimbrels nesting from eastern Greenland and Iceland in the west and, particularly, from northern Britain to western Siberia in the east, cross Iberia on passage. There are Iberian ringing recoveries of at least four birds ringed in Iceland and two ringed in Britain, as well as 12 of birds ringed on passage in the Low Countries and one ringed in France. One ringed in Portugal was recovered in France and single individuals ringed in Spain were found in Britain and The Netherlands. Migrants are most common on the Biscay coasts and on Atlantic coasts south to the Strait of Gibraltar. They also occur regularly in small numbers on Mediterranean coasts, including in the Balearic Islands, where they are scarce, but are seldom seen inland. In Iberia they occur on both sandy and rocky shores as well as in estuaries, on mudflats and in marismas.

The Whimbrel is mainly recorded in modest numbers, probably because most migrants overfly or bypass the entire Peninsula without stopping when on passage. However, some concentrations are noted during passage periods. The southward passage is detectable from mid July to September on both northern and southwestern coasts, with peaks of up to 140 birds at Odiel in August, when up to 600 individuals may be present in the Huelva wetlands as a whole, usually in small groups along sandy and muddy shores (Garrido 1996). Even larger numbers have been recorded early in the southward passage period in southern Portugal, at the Tagus estuary and the Ria Formosa: for example, 1,361 were counted at the latter site in July 1993 (AP). Further north, Tait (1887) refers to large movements on the coast near Porto, where 'great numbers' arrived in August and could be heard on nocturnal passage; he also recorded numbers on passage there in spring in March and April, and saw large flocks at Esmoriz in May. Visible passage on a considerable scale was also reported from the Porto coast by Ticehurst and Whistler (1935), who saw a succession of flocks flying north all day on 1 May 1934.

Some visible passage is apparent in autumn on the Biscay coasts, when small flocks may be observed flying west. The visible autumn 2009 passage at Cape Bares was exceptional and 1,018 were seen in 275 hours between 18 July and 3 October, passage peaking at 29 birds/hr on 31 July (Sandoval *et al.* 2010). Coasting birds are also often seen heading south along the Portuguese west coast in August and September (AP). Spring passage is apparent in February in southwestern Spain and begins in March on the north coast, peaking generally in April with stragglers also in May (Domínguez & Rabuñal 1989, Garrido 1996, *A. O. Asturias* 2003). Some numbers are recorded on the Portuguese south coast, where the Alvor estuary in western Algarve is a favoured site at this time: there were 643 there in April 1997 (AP). An observation of 350 at Vélez-Málaga, Málaga on 14 April 2010 (*Ardeola* 57: 229) is a highly noteworthy count for a Mediterranean location. A few non-breeders remain around the Iberian coastline in summer.

Very few Whimbrels are seen inland but there are annual records of individuals and small flocks appearing at lagoons and ricefields across the Meseta and elsewhere, particularly during spring passage in April and early May. The largest of these flocks have included 27 at the Laguna del Cristo, Salamanca, on 25 April 1993 and 27 at Gallocanta on 20 April 1994 (*A. O. Salamanca 1924–2003*: 106, *A. O. Aragón 1993–94*: 51).

A small wintering population is present almost entirely on Atlantic and Biscay coasts. The January counts in Spain during 1990–2009 found only about 300 birds on average (range 100–1,000; AAE), with numbers tending to increase since 1980. During the 1990s wintering Whimbrels were mainly on northwestern and northern coasts, notably at the Ría de Arousa, Pontevedra; the Marismas de Santoña and, some way inland, at the Ebro reservoir (AAE). Since 2000 the southwestern coastlands have increased in relative importance as wintering sites, with some sizable counts from Cádiz bay (up to 451), Odiel (up to 298) and the Marismas del Río Piedras, Huelva (up to 139). A record of four at the Azud de Riolobos on 8 December 1998 is a rare winter occurrence far inland (*A. O. Salamanca 1924–2003*: 106).

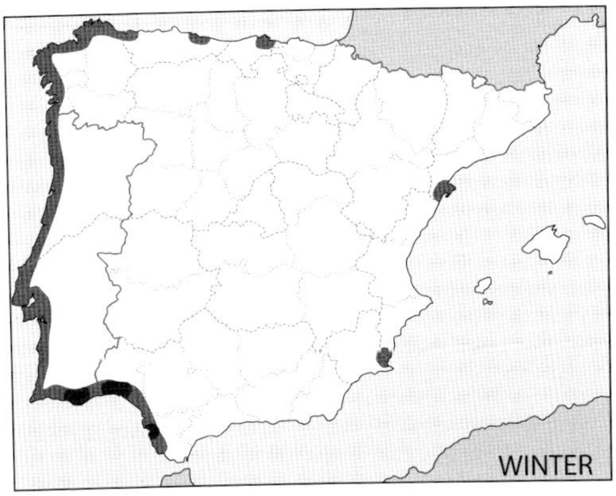

The numbers wintering in Portugal are also modest but show an increasing trend. Whereas fewer than ten were found annually during 1975–78, there was a mean mid-winter population of 192 during 1992–95 and 334 during 2005–07, the Ria de Faro being the principal site (AP). The winter survey of non-estuarine Portuguese coasts during 2009–10 found 86 Whimbrels (Catry *et al.* 2010b).

The Ebro delta is the only Mediterranean site to have held

some numbers of Whimbrels in winter, with 100 in January 1973 (Araújo & García 1974) and a similar number also in 1974 and 1975, but only a few have occurred in more recent years (CWA). January counts in Catalonia, the Comunidad Valenciana and Murcia during 1990–2009 have all averaged in single figures (AAE). Wintering individuals are exceptional in the Balearics (AAE).

SLENDER-BILLED CURLEW *Numenius tenuirostris*

(Zarapito Fino, Macariço-de-bico-fino)

Formerly a rare migrant and winterer in Spain. Vagrant to Portugal.

The Slender-billed Curlew declined rapidly to virtual or actual extinction during the second half of the 20th century. Over 600 were censused on the Moroccan Atlantic coast in 1964 (Blondel & Blondel 1964) but only a few individuals were found there during the 1980s and early 1990s, chiefly at the Merja Zerga (BM).

It was undoubtedly a scarce passage migrant and winterer in Iberia in the past, although there are few published records (Bernis 1966d, Gretton 1991). There are 19th century observations from the island of Cortegada, Pontevedra (Ríos 1850), the Albufera de Valencia, the Guadalhorce estuary and Fuentedepiedra (Arévalo 1887); Gallocanta (Saunders 1871) and Doñana: this last comprising a pair shot on 5 February 1898 (Chapman & Buck 1910). There are old specimens at the Ateneo de Mahón, Menorca (Munn 1924), and at the Lisbon museum: the latter is the only confirmed record from Portugal and it was obtained in the Tagus valley in the mid 19th century (Tait 1924, AP).

More recent observations include up to 11 seen at Doñana in January 1966 but two records of single birds from there in December 1990 were not accepted (*Ardeola* 11: 150, 42: 111). Other probable sightings have been on the Costa Brava, Girona, in August 1962; at Ses Salines, Mallorca, in late July 1977, and at Cape Trafalgar, Cádiz, in November 1979 and November–December 1980 (Gretton 1991).

EURASIAN CURLEW *Numenius arquata*

(Zarapito Común, Maçarico-real)

***N. a. arquata*. Common on passage and in winter on Biscay and Atlantic coasts, scarce in the Mediterranean and Balearic Islands. A few pairs nest in Galicia.**
***N. a. orientalis*. A rare winter visitor, reported from Huelva.**

The Eurasian Curlew has a marginal breeding presence in Iberia, where the few pairs that nest in Galicia represent a southwesterly outpost of the breeding range. Bernis (AMI) refers to possible nesting in Doñana but confirmed breeding outside Galicia is known only from Asturias during 1970–77, when several pairs regularly attempted to breed on the wet pastures bordering the Ría de Villaviciosa and a pair nested on the Monte Cubera nearby; one pair also nested on the Asturian side of the Ría del Eo (Noval 2001, LR). Most of these nesting attempts were unsuccessful, largely because the eggs were 'harvested' (Noval 1975). All definite breeding records since 1980 have come from Galicia (LR, Palomino & Molina 2009). Five pairs nested in 1983 in A Limia district, Ourense, and two pairs probably did so there in 1989: the area was formerly a traditional nesting site for the Curlew, around the Laguna de Antela before it was drained in the late 1950s (Villarino *et al.* 2002). One pair also nested on the Isla de Arousa, Pontevedra, in 1988. However, the Galician, and Iberian, breeding population has since been confined to Terra Chá district, Ourense, where up to four pairs nested during 1985–2007. The breeding area is mainly around Las Rozas airfield, Castro de Rei, where the birds nest in patches of heathland and peat bog. Other breeding attempts elsewhere have been in similar habitats as well as in damp grazing meadows and in fields of emerging crops. The nesting birds return to Terra Chá in late February or early March but only settle to breed in April. The tiny Iberian population is threatened by habitat loss and disturbance.

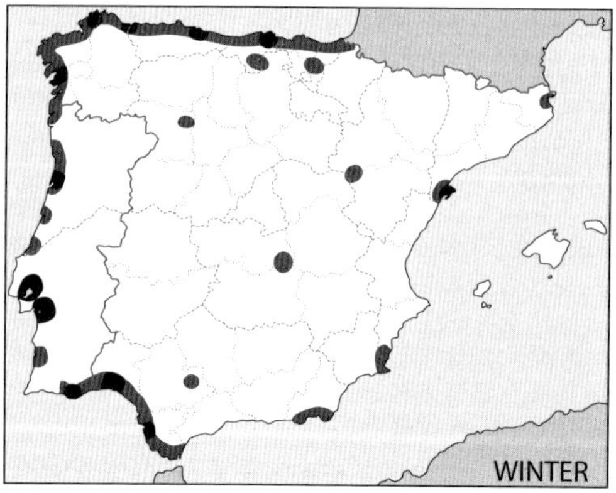

The westernmost European populations are only partly migratory but Curlews that nest in Fennoscandia, the Baltic and northwestern Russia winter south to Iberia, North Africa and Mauritania. Some 90 Iberian recoveries include individuals ringed as pulli in Britain, the Low Countries (40% of all recoveries), Germany, Denmark, Sweden and Finland, and a full grown juvenile ringed in Austria in May was recovered in October the same year in Cádiz. There are also single recoveries of birds ringed in France and Poland, one ringed in Portugal has been recovered in France and one ringed in Spain has been found in Britain.

Migrants begin to appear from July and passage continues until early November in the south. Many of these remain to winter on northern and northwestern coasts and others continue south to winter on western and southwestern Atlantic shores. Both passage and wintering birds occur in muddy estuaries, marismas, saltpans and ricefields, and on pastures and fields nearby. Small numbers are reported locally inland, both on passage and in winter. Wintering birds depart in March–April, at the same time as migrants returning from Africa, but some non-breeders remain on coasts all year round: several hundred do so in the Huelva coastlands (Garrido 1996) and a flock of 500 was at Sancti Petri, Cádiz bay, on 14 June 1999 (*Ardeola* 50: 348).

Wintering in Iberia may have increased in recent years (AAE). Earlier censuses and estimates of the wintering populations in Spain suggested totals of 2,365 birds (Araújo & García Rúa 1974), up to 1,800 birds (Alberto & Purroy 1981) and 3,000 birds (Alberto & Velasco 1988). These surveys found large numbers in Galicia but they gave inadequate coverage to the south-west. The Spanish January censuses of 1990–2009 found a mean wintering population of 4,143 birds (2,062–8,985; AAE), some 70% of which wintered at coastal sites in Galicia and the Biscay coastlands, with the remainder mainly at the southwestern coastal estuaries, although a few hundred occurred at the Ebro delta. Cold weather further north may produce influxes into the Peninsula: one such event in January 1997 resulted in a count of 2,584 birds in Asturias (*A. O. Asturias* 4: 119–121)

The Portuguese wintering population averaged 3,292 birds during 1992–95 and 2,280 during 2005–07 (AP). The Sado estuary is the principal wintering site for this species in Portugal and 1,000+ have been recorded there, although the numbers present have declined in recent years. Other important wintering areas, each of them holding several hundred birds, are the Ria de Aveiro, the Tagus estuary and the Ria Formosa, with lesser numbers also occurring at most other estuaries and also at coastal lagoons such as the Lagoa de Óbidos (AP).

The Curlew is comparatively scarce in the Mediterranean, both on the mainland coasts and in the Balearics, but small numbers winter in the Cabo de Gata saltpans, Almería; the Santa Pola saltpans, Alicante, and especially at the Ebro delta. During January 1996–2009 the Ebro delta held an average of 259 Curlews and there were 100 on average at the Aiguamolls de l'Empordà: maximum and minimum winter counts were 74 and 423 birds at the delta and 36 and 315 at the Aiguamolls (*A. O. Catalunya*).

Curlews are generally uncommon away from coasts but small flocks are occasionally reported inland in winter and on passage. An exceptional flock of 150 at the Urrunaga reservoir, Álava, on 16 January 1997 had increased to 500 by 26 January (*Ardeola* 44: 253): at the time of the large cold-weather influx of Curlews on the north coast described above. The presence of the eastern subspecies *N. a. orientalis* in Iberia is claimed by Garrido (1996) on the basis of birds handled at the Marismas del Odiel and a specimen obtained at Doñana that is currently in the scientific collection of the Doñana biological station. Intermediate individuals between both races have also been identified in the same area, where perhaps 10% of the wintering population is considered to show intermediate characteristics.

UPLAND SANDPIPER *Bartramia longicauda*

(Correlimos Batitú, Maçarico-do-campo)

Vagrant to Spain and Portugal.

One was found at Leça da Palmeira, Porto, on 12 October 1932 (Reis Júnior 1931). One obtained near Sevilla on 2 December 1988 had been ringed in Michigan, USA, on 5 July the same year (Dennis 1990). There are four more recent accepted records: at Ludo, Faro, on 22–23 September 1999; Navia, Asturias, from 30 October to 5 November 2005; Tavira, Faro, from 29 September to 20 October 2010 and Coaña, Asturias, on 21–24 September 2011. Subsequently, one was at Estaca de Bares in November 2012 (RBS). This is a rare vagrant to Europe, where 46 individuals have been recorded in Britain, 12 in Ireland and five in France. There are five accepted records from the Azores and one from the Canary Islands. One in Mauritania in May 1986 is the sole known occurrence in Africa.

TEREK SANDPIPER *Xenus cinereus*

(Andarríos del Terek, Maçarico-sovela)

A very rare migrant in Spain and Portugal. Has wintered.

The migration routes of the Terek Sandpiper lie well to the east in Europe. It is rarely observed in central Europe and occurs only sporadically in the west. There are records of 74 individuals in Britain, five in Ireland and 113 in France. There are three winter records from Morocco and another three in winter from the Banc d'Arguin, Mauritania.

A 19th century record from Málaga (Arévalo 1887) was the first from Iberia but no more were found until the 1970s and early 1980s, when individuals were found at the Marismas del Guadalquivir (*Ardeola* 22: 134, 29: 186), in Málaga (*Ardeola* 28: 157) and at the Ebro delta (Ferrer *et al.* 1986). There have been 54 accepted records from Spain and three from Portugal up to 2010. There were also a further three accepted Spanish records during 2011 as well as records of one in the Tagus estuary, Lisboa, in August 2011 and April 2012 (RBP). Most observations have involved solitary individuals but there are three records of two together and one of a group of three. Reported ages have comprised seven adults and nine first-years in autumn, and four adults and one second-year in spring.

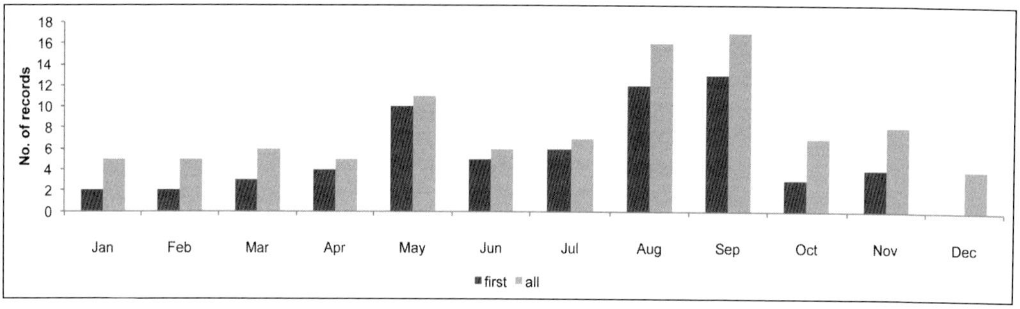

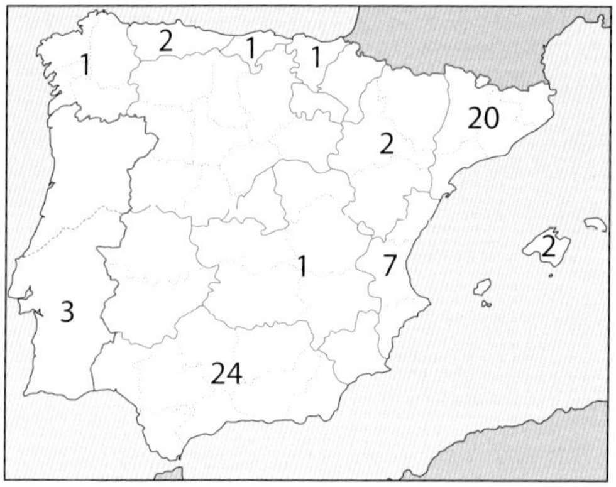

Figure 124. Records by month, 1974–2010 (n = 64)

Figure 125. Records by region, 1974–2010 (n = 64)

Records have occurred in all months but there is a spring peak in May and a larger autumn peak in August–September (Fig. 124). The earliest autumn adult has been on 16 July and the earliest juvenile on 27 July. Three first-winter individuals have overwintered at the Ebro delta: from November 1995 until April 1996; from August 1996 to February 1997, and from August 2000 to March 2001 (*Ardeola* 45: 107, 49: 159). There are also indications of overwintering at Sancti Petri and other locations in Cádiz bay during winter 2005–06 and 2006–07 by up to two birds, perhaps the same returning individuals. Most first observations (59%) have been during July–November, with 34% in March–June, although spring records tend to predominate in western Europe.

Terek Sandpipers appear chiefly at coastal locations in the south and east (Fig. 125). There are only three inland records: at Manjavacas lake, Cuenca, during 28–31 May 1988, at Bujaraloz, Zaragoza, on 20 April 2003 and at Quinto, Zaragoza, on 18–20 May 2010. The ratio of autumn to spring records is 2.4 in the Mediterranean (n = 41) and 1.0 on Atlantic coasts (n = 18). Similarly, in France, spring records predominate on Atlantic coasts and autumn ones in the Mediterranean.

COMMON SANDPIPER *Actitis hypoleucos*

(Andarríos Chico, Maçarico-das-rochas)

Widespread and locally common as a breeding species, particularly in montane regions, but absent from large areas of southern and eastern Spain. Common on passage throughout the Peninsula and Balearic Islands and some remain to winter.

This is indeed a common breeding species along the watercourses of the Pyrenees and Cantabrian Range and it is also widespread in much of northern and central Spain. Its breeding distribution is nonetheless incompletely known and many breeding season records may actually involve late spring or early autumn migrants (SA1, SA2). The Spanish breeding range is also more fragmented in the south and east, where nesting birds are largely confined to locations in the Montes de Toledo. Summering birds are reported from the Sierra Nevada, where they have been reported as high as 2,850m, but breeding is

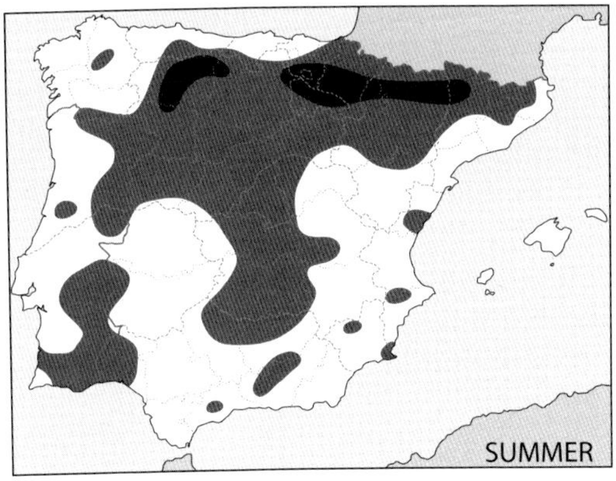

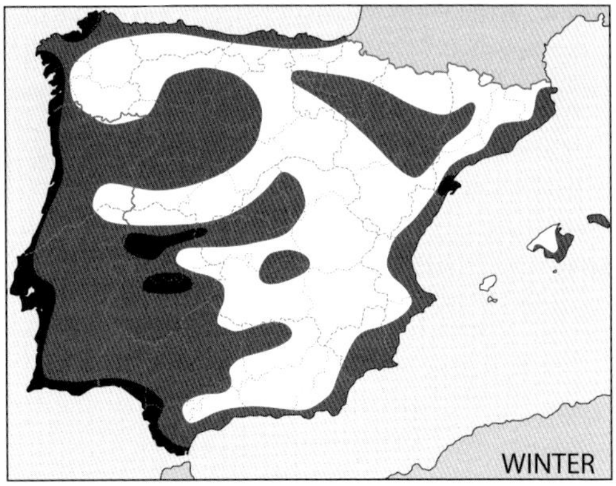

unconfirmed there (Garzón & Henares 2012). There are nonetheless occasional records of birds nesting at untypical sites in the south: for example, a pair bred successfully at Coín, Málaga, on a river gravelbank 200m above sea level (*Ardeola* 30: 119). The Portuguese breeding distribution, however, is mainly in the south and east, with few nesting in the low-lying Atlantic coastlands. In general, breeding birds are concentrated in the Eurosiberian and Supramediterranean climatic regions. Here they require streams and rivers that offer shingle beds and sand banks. Some also nest on the margins of reservoirs. The upper reaches of the Miño, Duero, Tagus and Ebro watersheds offer good nesting habitat but suitable locations are less frequent at lower elevations, where rivers and streams tend to be wider and more sluggish. The Common Sandpiper does not nest in the Balearic Islands. Historical references to breeding Common Sandpipers are very few (e. g. Tait 1887, Witherby 1928) and in the 1960s Bernis (AMI) still regarded the Spanish breeding population to be minimal. It is possible that there has been some subsequent increase in numbers but it may be more likely that the species was widely overlooked in the past, with nesting birds assumed to be migrants.

The Spanish breeding population has been estimated at 3,000–4,000 pairs, with the 2003 breeding bird atlas suggesting a figure closer to 3,000 pairs (SA2). The Portuguese population has been estimated at 250–2,000 pairs (BE 2004) but seems unlikely to approach the upper figure. Trends in the breeding population are uncertain but the Common Sandpiper has suffered from habitat loss and disturbance locally in Spain. The Portuguese breeding range has not altered greatly since the first breeding atlas was produced but numbers appear to have declined in parts of the north and centre of the country (PA2), so that the main breeding areas are in Beira Baixa, Alentejo and Algarve. It does not nest in Andorra but non-breeders are sometimes found there in summer (ANA).

The Common Sandpiper is frequently encountered on passage, both inland and on coasts, occurring at all kinds of wetlands, including lakes and reservoirs, estuaries, harbours and rocky seashores. The post-breeding passage is often more conspicuous and begins in July and continues until early October, peaking from mid August to mid September in most areas. In the Duero basin most adults pass from mid July to late August, and juveniles from late July to mid September (Balmori 2003). The spring passage is from March to May. The breeding birds may be resident locally but many descend from higher elevations to the coasts and some probably travel further: there is a Moroccan recovery of a breeding adult ringed in Spain. The breeding sites are generally occupied from May until July.

At least 157 Common Sandpipers ringed in Scandinavia (29%), Germany (27%), Britain, the Baltic States, Poland, Finland, the Low Countries, France, Switzerland and the former Czechoslovakia, many of them as migrants, have been recovered in Iberia on passage, especially in autumn, with some also found in winter. Birds ringed in Iberia: 27 in Spain and two in Portugal, have in turn been recovered in the above countries, with single recoveries also in Morocco and Senegal. The species is not highly sociable but small flocks, of up to 15–20 birds, are often observed during passage periods and larger concentrations are sometimes then reported. A count of 250+ in Txingudi bay, Guipuzcoa, on 14 July 1998 (*Ardeola* 49: 187) is exceptional and the early date may mean that at least some were Iberian birds that had just descended from their breeding sites in the mountains.

Most European breeders winter in sub-Saharan Africa but small numbers remain in western and southern Europe, including Iberia, in winter (Delany *et al.* 2009). Wintering birds are fairly solitary and are thinly and widely dispersed along sheltered coasts as well as at lakes and reservoirs inland, with some tendency to concentrate in the estuaries of the Biscay coast and Galicia, the Tagus and Sado estuaries, Doñana and other southwestern wetlands, the Ebro delta and, inland, the ricefields and other wetlands of Extremadura. None of these areas generally holds large numbers and the January counts of 1990–2009 found an average population in Spain of around 900 birds (range 250–1,400; AAE). The censuses themselves will have missed many individuals, perhaps the majority, at scattered sites along rivers and around gravel pits and reservoirs. Sizable winter concentrations in Spain occur in some years in the southwestern wetlands, particularly at Cádiz bay (up to 133 birds) and at

Doñana: where the January 2000 census located an exceptional 476 birds (*A. O. Doñana* 1: 122). The Portuguese wintering population has been conservatively estimated at up to 200 birds (AP) but as many as 320 were at a high tide roost in the Ria de Faro in January 1986 (Rufino & Neves 1987). The total wintering population of Catalonia has been estimated at 486–607 birds (CWA). There is also a winter presence in the Balearic Islands, where the 1990–2009 January counts found 45 birds on average and a maximum of 81: the Formentera and Ibiza saltpans have together held up to 47 birds in winter (AAE).

SPOTTED SANDPIPER *Actitis macularius*

(Andarrios Maculado, Maçarico-pintado)

Vagrant to Spain and Portugal.

The first Iberian record was an adult at the Ebro delta from 20–23 August 1979 (*Ardeola* 28: 157). There have since been 15 accepted records of single birds for Spain and two in Portugal. The autumn records include 12 juveniles and three adults. Two records involve overwintering: a first-winter bird was at the Ensenada da Ínsua, A Coruña, from 4 October 1986 to 1 May 1987 (*Ardeola* 36: 120), and one was at the Sado estuary, Setúbal, from 4 January to 28 February 1995 (*Pardela* 5: 10). In addition, a first-winter was at Gozón, Asturias, on 27–28 December 2009. All but two records have been in autumn, with first dates between 2 August and 28 October (Fig. 126). Autumn observations also predominate in the British Isles (Dymond *et al.* 1989) and France (OF). Most Iberian occurrences have been on northern coasts (Fig. 127) and only one has been inland: at Villafáfila on 6 April 2003 (*Ardeola* 52: 198). There were at least six Spanish records and one from Portugal during 2011–12 (*Ardeola* 60 : 468, RBS, RBP).

Figure 126. Records by month, 1979–2010 (n = 18)

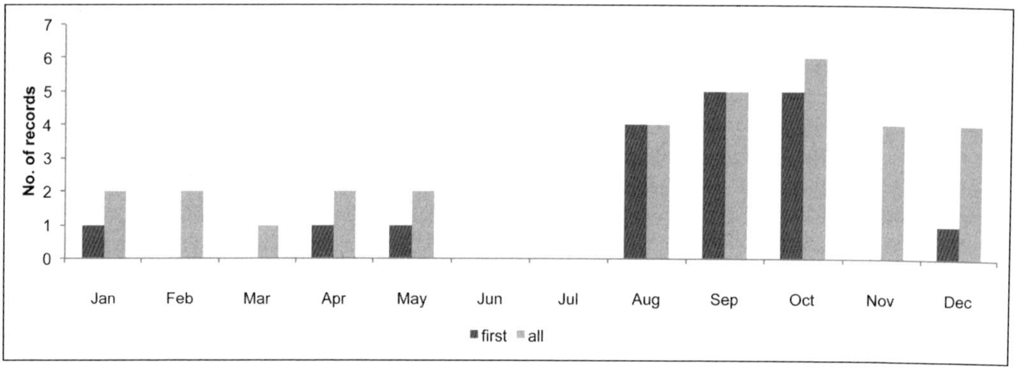

Figure 127. Records by region, 1979–2010 (n = 18)

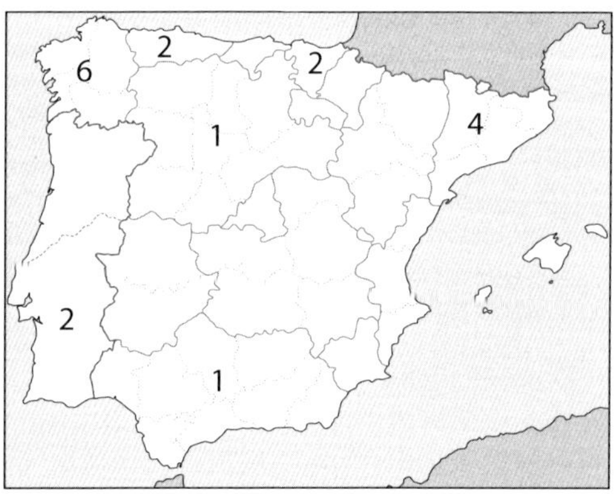

This Nearctic sandpiper would doubtless be recorded even more frequently if it were more easily distinguishable from the Common Sandpiper when in non-breeding plumage. Some three-quarters of records are from the British Isles where there have been 180 individuals in Britain and 37 in Ireland. There are only 19 records from France. It has been identified in the Azores since the 1960s (Le Grand 1983) and there are 36 accepted records up to 2006. There are also four from Madeira, ten from the Canary Islands, four from Morocco and one in Senegal.

GREEN SANDPIPER *Tringa ochropus*

(Andarríos Grande, Maçarico-bique-bique)

Widespread on passage and in winter throughout the region. Has bred Spain.

The Green Sandpiper is common at freshwater and brackish wetlands in mainland Iberia and the Balearic Islands. It may be found at almost any inland wetland, even at small farm ponds and temporary pools, and many frequent rivers, streams and irrigation ditches, both on passage and in winter. Seashores and mudflats are avoided but passage birds occur in wet pastures alongside the rías. It is not a sociable species and is seldom seen in any numbers, usually occurring individually or in small scattered groups. However, extensive wetlands such as ricefields may hold loose concentrations at times: for example, 95 were counted at the Albufera de Valencia on 21 August 2009 (*A. O. C. Valenciana 2009*: 73).

There is a single and quite exceptional confirmed record of a pair that nested in Spain in the Basque Country in 1996, at the Lertutxe reservoir, near Bilbao, Vizcaya (*Ardeola* 44: 254). The pair was observed in mid-May: the male being seen calling from tree tops, and again in July, then with four fledged young. There are reports of possible breeding in France in 1992 and 1995 (OF) but the nearest regular nesting locations to Iberia are in Scotland, Denmark and southern Germany (BWP).

The European populations winter in western Europe, the Mediterranean basin and North and West Africa (Delany *et al.* 2009). Southward passage begins in June, when the females leave the breeding areas, and is apparent in Iberia from mid June to early October (Calvet *et al.* 2004, AP). There are Iberian recoveries of single birds ringed as chicks in Norway and Sweden, as well as over 40 recoveries of full-grown birds ringed in Scandinavia (48%), Britain, Belgium, The Netherlands, Germany and France. Returning migrants are noted in March– April, when the winterers also depart.

Wintering birds are widely and thinly scattered across Iberia and accordingly difficult to census. In Spain the January wetland counts of 1990–2009 found 720 birds on average (range 26–1,800; AAE). Winter counts during 2008–10 averaged some 1,500 birds (SWA). Both these counts will certainly have missed large numbers that winter at minor and temporary inundations across the region. The Spanish wintering population probably comprises several thousand birds. The censused birds were especially associated with the Ebro and Tagus valleys, with concentrations also at Doñana and the Ebro delta, both of these offering extensive ricefields and other irrigated areas, where the birds are often seen along channels and ditches. Green Sandpipers are also relatively common in winter in Castilla-La Mancha and Extremadura, where the recent January censuses have found up to 420 and 157 birds respectively (AAE). In contrast very few individuals winter in the northern coastlands (AAE). The largest January counts in Spain have been at Doñana (up to 470), the Ebro delta (up to 295) and, which is surprising, at a reservoir at a sugarbeet factory between Linares and Baeza, Jaén (up to 240) (AAE).

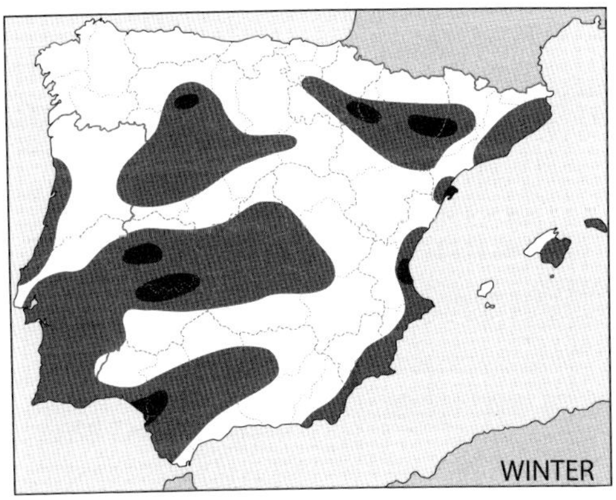

Winter counts in Spain have increased markedly since 1980, by over 12% per year (AAE), and this has been especially evident at the Ebro delta, where there were fewer than 30 during 1996–99 but over 100 since 2003, with 248 in 2008 and 295 in 2009 (*A. O. Catalunya*). The same January censuses have also found about 20 at the Llobregat delta, with over 40 present in some seasons. The wintering population of the whole of Catalonia has been put at 1,253–1,483 individuals (CWA). The Green Sandpiper is thinly distributed at Portuguese wetlands in winter and no large counts have been reported there. In general, however, few Iberian sites ever hold as many as 20 individuals in winter.

SOLITARY SANDPIPER *Tringa solitaria*

(Andarríos Solitario, Maçarico-solitario)

Vagrant to Spain and Portugal.

One was at the Ría de Villaviciosa, Asturias, on 15 and 17 September 1974 (Noval 1986). Subsequently, an adult was found at the Marismas de Alvor, Faro, on 26 June 1989, and a juvenile was at some pools at Cazanes, near Villaviciosa, on 8 October 2005. This is a very rare vagrant to Europe, with 36 records from Britain, five from Ireland, six records from France, six from the Azores and one from Madeira.

SPOTTED REDSHANK *Tringa erythropus*

(Archibebe Oscuro, Perna-vermelha-bastardo)

Widespread but scarce throughout the region on passage. Winters in the Guadalquivir estuary and locally in small numbers elsewhere.

The western populations winter mainly across sub-Saharan Africa, overflying western Europe on passage. There are a few Iberian recoveries of birds ringed on passage in Germany, Denmark, Belgium and Finland. A small minority winter in western Europe, including Iberia. Both migrant and wintering Spotted Redshanks occur widely at sheltered estuaries, lagoons, saltpans and ricefields. Migrants especially are often seen at inland wetlands, frequently at reservoirs, but seldom in more than small numbers. Counts of 40 at Gallocanta on 8 May 1991 and 35 at the Laguna de Sariñena on 27 February 1999 are noteworthy, and a count of 142 in the Vegas Altas ricefields on 20 October 2010 was a record for Extremadura (*A. O. Aragón 1991–92*: 44, *1999–2003*: 168; *Ardeola* 60: 173).

Southbound passage migrants may appear in Iberia as early as mid June and in July, these mainly being adult females that have laid their clutches and abandoned them to the care of the males, as is usual in Spotted Redshanks (BWP). The males and juveniles extend the migration period through August–October. Up to 2,000 may be present at peak periods in autumn in the Huelva coastlands (Garrido 1996). Spring passage in Iberia begins in March and is principally during April–May. A count of 526 at Castro Marim, Faro, on 15 April 1983 (AP) was high for Portugal.

The principal Iberian wintering site is Doñana, where a January count of 2,426 in 1993 remains the highest number recorded at any Iberian location during the winter surveys (AAE). These figures are much larger than the 300 estimated to winter in the whole of Spain by Velasco & Alberto (1993), or the 60 birds censused in Portugal in winter 1989 (Rufino 1989b). The Spanish January surveys of 1990–2009 found 791 birds on average (range 36–2,691; AAE). Doñana probably holds 85% of the Iberian wintering population in some years, with smaller concentrations regularly reported from the Ebro delta, the saltpans and ricefields of the Levant coast, the Huelva and Algarve coastal wetlands, the Tagus and Sado estuaries and the western rías of Galicia. Winter counts at these other sites generally average fewer than 20 individuals but are occasionally higher, with gatherings of about 100 birds reported from the Albufera de Valencia, the Marismas de Santa Pola and the Ría de Ortigueira (AAE). In particular, the Catalan January wader counts during 1996–09 have found that the Ebro delta has recently supported a larger wintering population, which averaged 344 during this period and ranged from 101 to 589 birds (*A. O. Catalunya*).

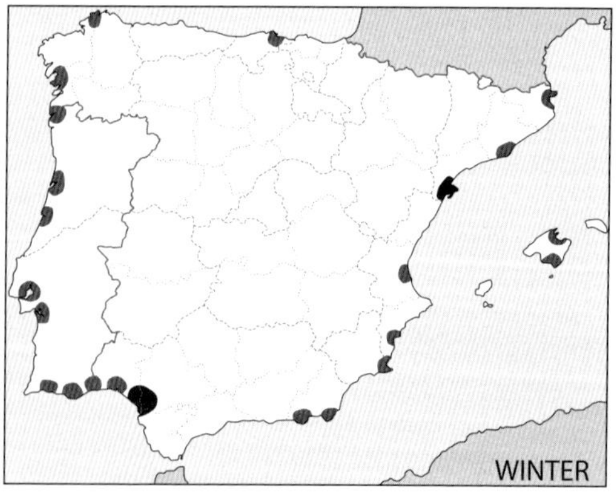

The numbers wintering in Portugal are variable but show a declining trend, averaging 59 birds during 1992–95 and 41 during 2005–07: in contrast 171 wintered on average at the Ria Formosa during 1975–78 (AP). Very few winter on the Biscay coasts but some individuals usually remain in the Ría de Villaviciosa, the Ría del Eo and at Santoña.

A few may also winter inland, notably at the Guadiana ricefields and the Llanos de Cáceres in Extremadura: *c.* 40 were recorded wintering regularly at the latter site (Acha & Velasco 1992). Elsewhere occasional individuals or small groups are reported in winter at wetlands across the meseta: for example, there were up to 14 at Villafranca de los Caballeros, Toledo, during winter 2004–05 (*A. O. Toledo 2002–2007*: 281).

GREATER YELLOWLEGS *Tringa melanoleuca*

(Archibebe Patigualdo Grande, Perna-amarela-grande)

Vagrant to Spain and Portugal.

One was taken at Punta Entinas-Punta Sabinar, Almería, on 13 August 1970 (López Martos 1991). Since then there have been eight accepted records for Spain and two for Portugal, all of single birds. Six occurred in autumn and three in spring. In autumn the extreme first dates were 10 July and 19 December but a juvenile overwintered between Valdoviño and Ortigueira, A Coruña: from at least 4 November 2000 to 13 May 2001, and a first-winter bird was at the Ensenada da Ínsua, A Coruña, between 18 January 2009 and 25 April 2010. Extreme first spring dates are 6 February and 16 April but a first-summer bird remained at the Salobrar de Campos, Mallorca, from 16 April to 29 June 1995 (Garcias 1997). All records have been coastal: five in Galicia, three in Andalucía, two in southern Portugal and one in Mallorca.

This is a very rare vagrant to Europe where 31 individuals have been found in Britain and 13 in Ireland, but only five in France. There are five records from the Azores, one from the Canary Islands: in November 1989, and one in Morocco: in November 2005.

GREENSHANK *Tringa nebularia*

(Archibebe Claro, Perna-verde-comum)

Widespread on passage throughout the Peninsula and in the Balearic Islands. Small numbers winter locally, mainly on coasts.

Western Greenshanks largely winter in Africa south of the Sahara but there is a small wintering population in coastal regions of western Europe, including Iberia. They are commonly seen on passage throughout the region, both at coastal and inland wetlands. A few individuals ringed in Scandinavia (10), Britain (10), the Low Countries (5) and Germany (1) have been recovered in Iberia and one ringed in Spain was recovered in France. At all seasons Greenshanks most often occur singly or in small, loose groups. They frequent tidal inlets, coastal mudflats, estuaries, coastal lagoons, saltpans and ricefields and, especially when on passage, lagoons, reservoirs, ephemeral pools and ponds inland.

Autumn passage is apparent in July both in northern and southern Iberia and continues until October (Barbosa 1997). The spring movements are in March– April in the south, extending into May in the north. Inland passage peaks in April and September (Calvet *et al.* 2004, Martín 2004, Arratíbel 2007). A few non-breeders remain in summer, especially at coastal sites such as the Ría de Villaviciosa. Migrants do not form large concentrations: peak passage counts at the Huelva coastlands are *c.*50 birds in April and *c.*100 in October (Garrido 1996). Counts of 112 at the Brazo del Este on 20 October 2005, 175 in the Ría de Avilés on 3 May 1997 and 129 at Alcochete saltpans, Setúbal on 23 August 2004 are noteworthy (*A. O. Asturias 1997*: 123, *A. O. Sevilla 2000–2005*: 86, *Anu. Ornitol.* 4: 33).

Wintering birds are mainly found at coastal wetlands in the north, west and south-west. Velasco & Alberto (1993) estimated the wintering population in Spain at *c.*400 birds: most of them at Doñana, the Galician rías and Cádiz Bay, and 44 were censused in Portugal in winter in 1989 (Rufino 1989b). More recent indications are that wintering numbers are somewhat larger; the Spanish January censuses of 1990–2009 found *c.*1,100 birds on average (range 200–2,300), and an increase of over 11% was reported during 1980–2009 (AAE). In particular, the numbers wintering on the northern coasts seem to have increased. The earlier censuses found only about ten birds at Santoña and 11 in Galicia in January 1969 (Fournier & Fournier 1972), and the January 1973 survey (Araújo & García Rúa 1974) found only nine Greenshanks in Galicia and none at all further east on the north coast. The January censuses of 1978–80 (Alberto & Purroy 1981) found up to 55 Greenshanks in Galicia, in 1980, but only five birds further east. The corresponding censuses during 1990–2009 found mean wintering populations of 363 birds in Galicia, 98 in Asturias, 45 in Cantabria and 36 in the Basque Country, the principal sites including the Ría de Ortigueira (up to 121), the Ría de Ribadeo (up to 105), the Ría de Villaviciosa (up to 96) and Santoña (up to 67) (AAE).

The earlier national censuses did not cover western Andalucía thoroughly but recent observations show that the principal wintering area in Spain extends from the Huelva coastlands to Cádiz bay, including Doñana. Garrido (1996) found 100–500 birds wintering in the Huelva coastlands, with considerable variation between years. More recently some 300 have been estimated to winter at Odiel (Gacio & Sayago 2011). The January censuses of 1990–2009 found up to 458 birds at Doñana and 319 in Cádiz Bay, but mean populations were around 150 and 100 birds respectively.

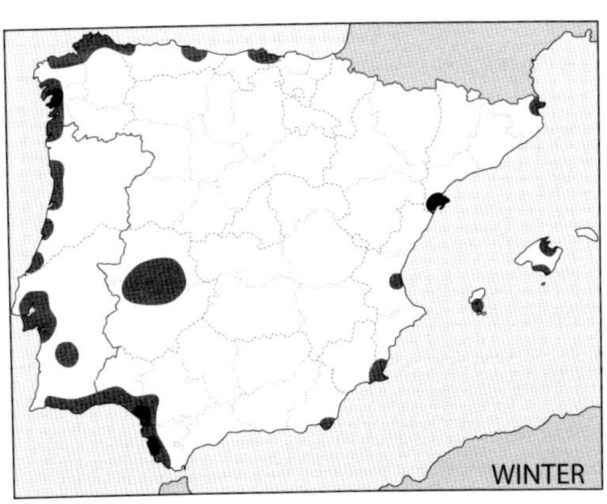

The Levant coastal wetlands hold relatively few Greenshanks in winter and the majority are usually at the Ebro delta. The January wader censuses of 1996–2009 found an average of 207 birds at the delta but wintering numbers increased there during the period from fewer than 70 birds up to 1998 to over 300 since 2006, with 435 counted in 2009 (*A.O Catalunya*, CWA). Some 40–50 have also been reported in recent winters recently at the Albufera de Valencia, (*Ardeola* 51: 250, 53: 203; *A. O. C. Valenciana 2009*: 73). Wintering numbers in the Balearic Islands averaged 20 birds in January during 1990–2009, with a maximum of 60 (AAE). The wintering population in Portugal averaged 168 birds during 1992–95 and 323 during 2005–07, mainly at the Ria Formosa and at the Tagus and Sado estuaries, with small numbers also at the Minho and Alvor estuaries and the Ria de Aveiro (AP).

Greenshanks may appear at almost any type of wetland inland, where they chiefly occur briefly when on passage, although there is a small but increasing winter presence in Extremadura and Castilla-La Mancha, where January counts during 1990–2009 have reached 167 and 46 birds respectively (AAE). Most inland records involve small numbers so April counts of 200 at the Azud de Riolobos, Salamanca, in both 2001 and 2003 (*A. O. Salamanca 1924–2003*: 109) are noteworthy.

LESSER YELLOWLEGS *Tringa flavipes*

(Archibebe Patigualdo Chico, Perna-amarela-pequeño)

Vagrant to Spain and Portugal.

This is one of the most frequent Nearctic waders to appear in Europe. There are observations of 328 individuals in Britain, 136 in Ireland and 50 in France. It is also relatively frequent on the Atlantic islands, with 26 accepted records for the Azores and 13 from the Canary Islands. There are seven records for Morocco, one from Mauritania and two from Senegal.

The first individual identified in Iberia was at the Sanlúcar de Barrameda saltpans, Cádiz, on 26 March 1967 (del Junco 1968). Another was at the Cabo de Gata saltpans, Almería, on 2 October 1981 (*Ardeola* 29: 192). Since 1984 there have been 52 accepted records for Spain and 18 for Portugal, as well as a further eight Spanish records and two Portuguese records up to 2012 (*Ardeola* 60: 469, RBS, RBP). One record involved two birds together but the remainder have all been solitary individuals. The autumn observations included 35 juvenile or first-winter birds but only seven adults. There are records from all months but most correspond with the spring and autumn passage periods (Fig. 128). An extreme long-stayer was at the Ensenada da Ínsua, A Coruña, from November 2003 to April 2005 (*Ardeola* 53: 177): only its arrival date is included in the Figure. This bird apart, the latest spring date has been 16 June and the earliest 'autumn' date 13 July. Adults precede juveniles and five of seven records of adults were between 13 July and 1 September, whereas juveniles occurred from 6 September but mainly in October. Spring records peak earlier in Iberia than in Britain or France: in April rather than May, and are later in autumn: in October rather than September, implying north/south passage within Europe (de Juana 2006).

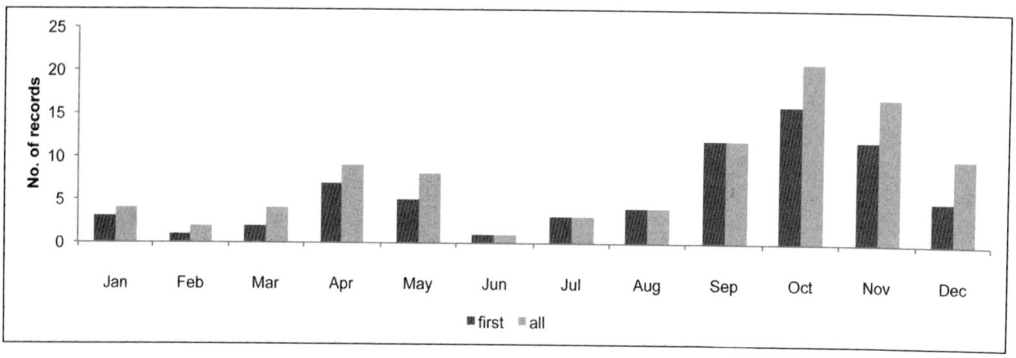

Figure 128. Records by month, 1967–2010 (n = 71)

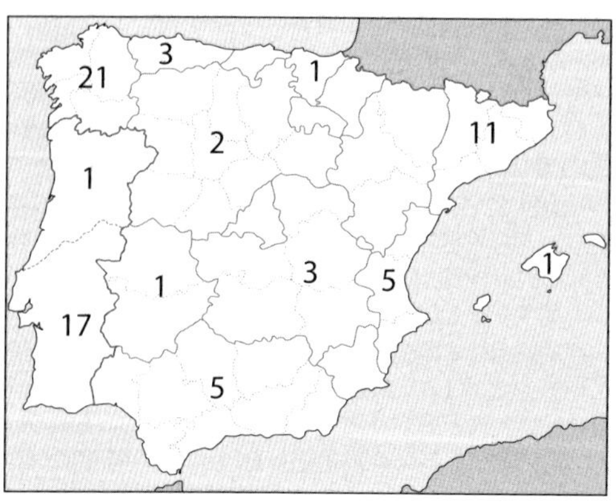

Iberian records are chiefly on the Atlantic seaboard and rarely inland (Fig. 129). Mediterranean coastal records are relatively more frequent in spring: there have been seven in spring and ten in autumn whereas in Galicia there have been two in spring and 19 in autumn. A similar pattern has been seen in France: up to 2008 there had been three in spring and two in autumn on the Mediterranean coast but 17 in autumn and seven in spring on the Atlantic coast. Spring records also make up a very low proportion of British records. It seems likely therefore that birds returning from Africa in spring follow a more easterly route than when flying south in autumn.

Figure 129. Records by region, 1967–2010 (n = 71)

WILLET *Tringa semipalmata*

(Playero Aliblanco, Maçarico-d'asa-branca)

Vagrant to Portugal.

The sole Iberian record is of an adult at the Ribeira das Enguias saltpans, Alcochete, Setúbal, on 29 April 2009. It showed characteristics of the nominate subspecies (Matias *et al.* 2011). The Willet is a very rare vagrant in Europe, known only from two records in the Azores, two in France and one each in Finland, Italy and Norway (Haas 2012).

MARSH SANDPIPER *Tringa stagnatilis*

(Archibebe Fino, Perna-verde-fino)

A scarce migrant in Spain, chiefly in the east and south, but only a vagrant to Portugal. Small numbers winter in southwestern Spain.

The western populations of the Marsh Sandpiper winter almost entirely in sub-Saharan Africa. The westernmost fringe of the passage includes eastern and southern Spain and the Balearic Islands, where the species is recorded annually in small numbers. It is progressively rarer further west in Iberia and is only recorded in Portugal as a vagrant: with just 16 records there, perhaps involving only 11 individuals, up to 2010 (AP, CPR 2010). There is also an additional Portuguese record in 2011 and about five there in 2012. A few winter in Spain, chiefly in Doñana, where they were first reported wintering by Hiraldo (1969). Passage and wintering birds are observed at marismas, estuaries, ricefields, saltpans and on the shores of both coastal and inland lagoons (Velasco 1992b).

The Marsh Sandpiper is mainly encountered in the coastal regions of the Levant: in particular at the Ebro and Llobregat deltas; the Tordera estuary, Girona; the Albufera de Valencia, and Santa Pola and El Hondo, Alicante; the Balearic Islands and in Andalucía: notably at Doñana and the saltpans of Almería, which together account for about 80% of published records. The remainder include a number of records from lagoons in La Mancha and a few in Extremadura, Castilla y León, Asturias and Galicia. It is commonest during spring passage, which extends from mid March to early May (Fig. 130). Autumn passage is prolonged and is in June–October. Most observations are of single birds (60%) or two together (19%) but small groups or loose assemblages of up to ten birds are occasionally seen. A flock of 26 at the Albufera de Mallorca on 10 July 2011 (*Ardeola* 59: 436) was exceptional.

Spring migrants predominate in eastern Spain: 57% of the sampled records were during March–May and 37% in July–October. For the whole of Spain Velasco (1992) found that 41% were in spring and 25% in autumn. However, the autumn passage may be relatively more noticeable in western Iberia: 13% of 39 records from Portugal, Extremadura, Galicia and Asturias, were during March–May, 56% in July–October and 31% in November–February.

The vast majority (96%) of winter records come from Doñana (Velasco 1992b). November–February records there during 1974–2006 averaged 3.8 birds per winter (García *et al.* 1989). Wintering in the Marismas may not even occur annually (SWA): during 2007–10 the maximum January count there was 12 birds: ten at Veta la Palma and two at Las Nuevas.

Figure 130. Observations in eastern Spain during 1988–2002 (n = 228: comprising 162 records from the Comunidad Valenciana and 66 from the Balearic Islands; the many records from Catalonia are not included since the bird reports do not publish all occurrences there).

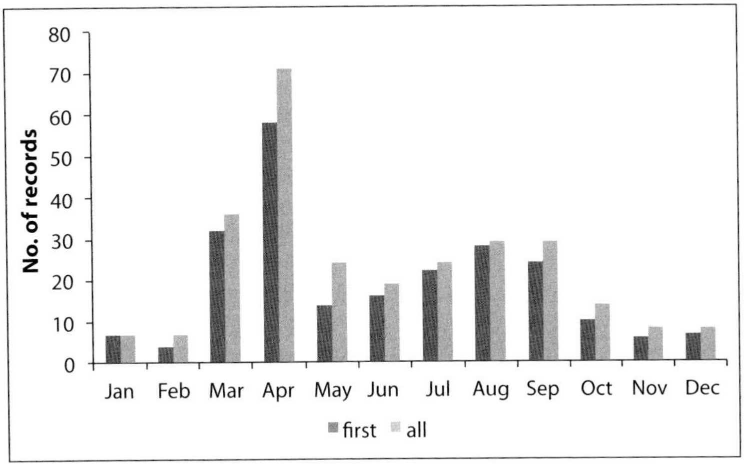

WOOD SANDPIPER *Tringa glareola*

(Andarríos Bastardo, Maçarico-de-dorso-malhado)

A scarce passage migrant throughout the region, most frequent in the east and south. Small numbers remain in winter.

The western Wood Sandpiper populations winter mainly in sub-Saharan Africa but passage across the Peninsula is relatively light. There are at least 45 Iberian recoveries of birds ringed on passage in Fennoscandia (42%), Denmark, Germany, the Low Countries, Switzerland and Poland. Five Spanish-ringed birds have been found outside the Peninsula: in Sweden, Finland, Germany, Italy and Senegal. In Iberia the birds occur at freshwater wetlands, ranging from ponds and flooded farmland, including ricefields, to the muddy margins of lagoons and marshland. Most records are relatively recent and many may have been overlooked in the past, although the expansion of rice cultivation may have attracted birds and perhaps contributed to the tendency for some to overwinter in the region.

Most occur on passage. Spring passage extends from mid March to May but most birds pass in April (Ferrer *et al.* 1986, Martín 2004). Returning individuals appear from late June but the main passage is during July–September, with a clear peak in August. In 1966 Bernis (AMI) stated that the Wood Sandpiper was frequent on passage but rarely wintered. Díaz *et al.* (AI1) also considered it to be only an accidental wintering species and could cite only 12 winter records for Iberia. Since then it has become clear that a small number winter regularly.

The Wood Sandpiper is much more gregarious than the Green Sandpiper and frequently forms small flocks, particularly during passage periods. These often comprise 3–20 birds but larger concentrations are reported occasionally: for example, there were 55 at the Lagoa de Santo André, Setúbal, on 8 April 1999 and 51 at the Alvor estuary, Faro, on 28 April 1994 (*Pardela* 10: 23, AP). The southward passage of 2010 was exceptionally evident at the Albufera de Valencia ricefields during 19–28 August, when gatherings of 100 to 340+ birds were observed at several locations (*Ardeola* 58: 500). Wintering birds are more likely to be seen in ones and twos but small flocks also occur.

Most Wood Sandpipers are found in southern and eastern Iberia, including southern Portugal, Andalucía and the Levant. They are very scarce in Extremadura (*A. O. Extremadura* 1998) but a flock of ten was at the Madrigalejo ricefields, Cáceres, on 28 January 2010 (*Ardeola* 57: 230). Passage birds occur at Daimiel with some regularity and there are occasional records elsewhere on the meseta. There were only 29 records from Madrid up to 2003 (Martín 2004). Only 38 individuals were reported in Asturias during 1993–97 (*A. O. Asturias 2003*: 124) and, in general, the Wood Sandpiper is not commonly reported from the Biscay coastlands or Galicia. The autumn passage is much more noticeable than the spring one (Ferrer *et al.* 1986), as also happens in France (OF). The winter status of the Wood Sandpiper in southwestern Andalucía, in particular Doñana, requires clarification. Estimates of up to 3,000 birds wintering in Doñana (García *et al.* 1979, 2000) have not been substantiated and seem very high. Within the Marismas, the Brazo del Este attracts birds both on passage and in winter and there were around 50 there during winter 2003–04 and winter 2004–05. Garrido (1996) considered that about 50 birds winter in the Huelva coastal wetlands but Gacio & Sayago (2011) consider it very scarce in winter in the Marismas del Odiel. However, the coastal south-west is probably the principal wintering area for the Wood Sandpiper in the region. Nevertheless, the January wader counts at the Ebro delta have shown a recent and perhaps new wintering presence there that may be of comparable importance. Only seven were counted at the delta in January 1996 but the mean wintering population during 1996–2009 was 93 birds; there were 211 there in January 2006 and 199 in January 2008 (*A. O. Catalunya*). The wintering population of Catalonia seems to have increased at an average annual rate of 29% during 1985–2009 (CWA).

Other winter records have come from diverse localities on the south and east coasts, and also inland, but most have involved very few individuals: a flock of 13 at the Albufera de Valencia on 23 January 2006 and 24 there in January 2009 (*Ardeola* 54: 178; *A. O. C. Valenciana 2009*: 74), and the Cáceres observation above, are exceptions. Only a few occur in Portugal in winter so a count of five at Barroca d'Alva, Alcochete, on 22 December 2007 (AP) is exceptional.

COMMON REDSHANK *Tringa totanus*

(Archibebe Común, Perna-vermelha-comum)

***T. t. totanus*. Breeds locally in the Peninsula and the Balearics. Widespread and common on passage and in winter.**
***T. t. robusta*. A scarce winter visitor.**

The Common Redshank breeds at a limited number of Iberian wetlands, both coastal and inland, including marismas, wet pastures, ricefields, saltflats and the margins of lagoons and rivers, generally nesting in short vegetation near water (LR). The majority breed at Doñana. There are also scattered, mainly small populations in the Ebro delta, Castilla y León and Castilla-La

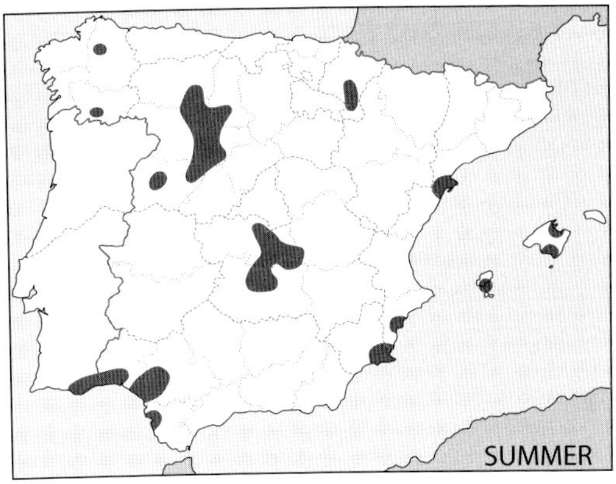

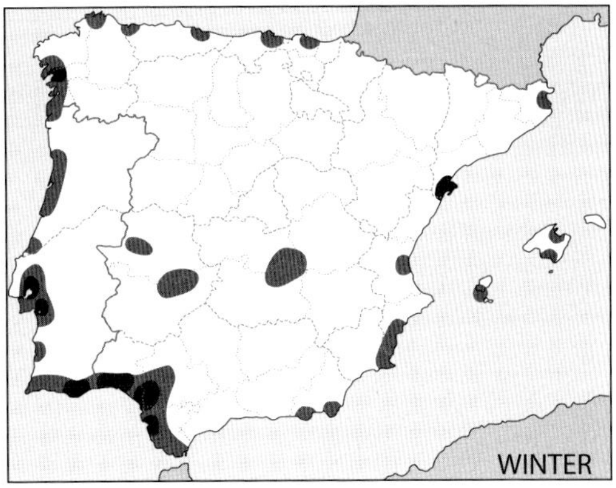

Mancha, and a few nesting pairs in Madrid, the Comunidad Valenciana and Murcia, as well as at least occasional nesters in Galicia, Aragón: at the Laguna de Gallocanta, Navarra and Extremadura. Small numbers nest in the Balearic Islands, mainly at the Salobrar de Campos, Mallorca, where there were *c.*20 pairs in 2000, but also at the Albufera and at the Ibiza saltpans. A few pairs also nest at coastal sites in Portugal, chiefly in Algarve in the Ria Formosa and at the Guadiana estuary, with at least occasional nesting at the Sado and Tagus estuaries and other west coast wetlands, as well as inland on rare occasions (PA2).

The Spanish breeding population has been estimated to be 650–800 pairs (Hortas *et al.* 2000) but it fluctuates considerably with rainfall, especially at Doñana where there may be over 1,000 pairs when water levels are ideal but perhaps just a dozen or so in drought years (LR). Valverde (1960) said it was very common in the Marismas, with scattered pairs throughout the region and large concentrations of nests on the Isla Mayor. Earlier authors (Saunders 1871, Chapman & Buck 1893, Noble 1902) also considered it to be an abundant breeding species in the Marismas. Bernis (AMI) estimated the 1960s population at 'many hundreds and perhaps thousands of pairs in some years'. García *et al.* (2000) suggested that there could be 4,000–8,000 pairs there in Doñana and Veta La Palma, while noting an alarming decline. However, Hortas *et al.* (2004) considered both the latter assessment and the suggestion of 50–200 pairs at Odiel and other Huelva wetlands (Garrido 1996) to be considerable overestimates. The mean breeding population in the Guadalquivir marismas may have been some 700 pairs during 1996–2001 (LR) and it is only a scarce breeder at Odiel (Gacio & Sayago 2011). Elsewhere in Andalucía a few pairs nest in Cádiz Bay (Tébar 1998). The Ebro Delta nesting population was put at 15–20 breeding pairs in the 1980s (CA1) but numbered around 100 pairs during 1992–2001 (Hortas *et al.* 2004). A thorough census there in 2001 found 124 breeding pairs (CA2). Only a few pairs nest at other wetlands of the Levant coast, south to Murcia, and similarly just occasional pairs nest in the northern coastlands, principally in Galicia.

Inland there are breeding birds at some of the lagoons of La Mancha, including El Hito and Manjavacas, Cuenca, and El Taray and Laguna Larga, Toledo (Palomino & Molina 2009), totalling perhaps 30–50 pairs (LR). Bernis (1952) found five pairs at El Taray and one or two at Daimiel in spring 1951. The Duero basin in Castilla y León has a breeding concentration of up to 50 pairs at Villafáfila, with probably fewer than 25 pairs in total nesting at other lagoons and wetlands in Palencia, Valladolid and Salamanca (Sanz-Zuasti & Velasco 1999).

The above information suggests a total Iberian nesting population of about 1,000 pairs. However, a national census in Spain in 2007 (Palomino & Molina 2009) found 1,100 individuals at selected wetlands and estimated by extrapolation that the total population could be around 5,600 individuals. It remains to be seen whether this relatively high estimate is supported by further monitoring. Water levels were low in 2007 at Doñana and other sites in Andalucía, and only 80 individuals were counted in that region. In addition to fluctuations associated with rainfall the nesting population has been affected by habitat loss and predation.

Although Iberian Redshanks are probably fairly sedentary (AMI) northern populations are largely migratory. The large Icelandic birds *robusta* winter chiefly in the British Isles and around the North Sea but this race has been recorded in Portugal as well as Morocco (Delany *et al.* 2009) so is at least an occasional winter visitor to Iberia. Large numbers of *totanus* winter in Iberia or cross the Peninsula to winter on West African coasts. Ringing recoveries are largely of birds ringed as pulli or fledged young in northwestern Europe, especially Denmark and southern Sweden but also in southern Finland, northern Germany, Belgium and The Netherlands (AMI). Birds ringed in The Netherlands comprise a high proportion of recent recoveries and there are some 40 controls of birds ringed there from the Marismas del Odiel alone (Gacio & Sayago 2011). Birds ringed in Iberia have produced over 100 recoveries outside the region, mainly in The Netherlands (64) and France (29) but also in Sweden, Denmark, Germany and Belgium. Four have also been recovered in Morocco and there is one sub-Saharan recovery: a young bird ringed at Aznalcázar, Sevilla, in late June 1964, was found in the Ivory Coast the following October (*Ardeola* 11: 33).

Migrants appear both at coastal sites and inland, at lakes, reservoirs and ricefields. They begin to arrive in Iberia in July and passage continues until October (Castro 1993, Barbosa 1997) although there may be midwinter influxes in cold winters. The southward passage is marked by influxes at Atlantic coastal sites, notably the Portuguese estuaries. For example, 5,185 were counted at the Tagus estuary in late July 1981 (AP).

Return passage is apparent during March–May. This passage, which peaks in April, tends to be much more evident inland than the autumn movements, as noted for example in Navarra (Deán 2001) and Madrid (Martín 2004). In 1997 the spring passage at the Ría de Avilés, Asturias, extended from 16 March to 13 June, with a maximum count of 361 birds on 3 May (*A. O. Asturias* 4). A count of 223 in the ricefields of the Albufera de Valencia on 31 March 2004 (*Ardeola* 52: 215) was noteworthy and 273 at Santoña on 7 May 2009 (*Ardeola* 56: 162) was a site record. Inland counts seldom involve more than small groups and a count of over 200 at the Azud de Riolobos on 18 April 2001 (*A. O. Salamanca 1924–2003*: 108) was quite exceptional.

The Spanish wintering population was estimated at 5,700 birds by Alberto & Velasco (1988). Early national censuses in Spain found just 773 in January 1973 (Araújo & García Rúa 1974) and 2,048 in January 1980 (Alberto & Purroy 1981). However, the January censuses of 1990–2009 found an average of *c.*4,800 birds in Spain (range 1,600– 9,000; AAE). The population is widely dispersed although overwhelmingly coastal in distribution, the largest January concentrations during 1990–2009 occurring in southwestern Andalucía: at Cádiz Bay (up to 3,398 birds), Doñana (up to 2,087) and Odiel (up to 1,221), as well as at the Ebro delta (up to 1,409) (AAE). Virtually all Iberian estuaries, saltpans and coastal lagoons support some Common Redshanks in winter but site totals of more than 200 individuals are exceptional and the numbers present are often much lower. Relatively few winter on the Biscay coasts but the extensive tidal inlets, rías and estuaries of western Galicia holds a larger wintering population: it averaged 562 birds during 1990–2009 with the largest gatherings seen at the Ría de Arousa (up to 762 birds) (AAE). Small numbers winter in the Balearics: the January counts of 1990–2009 found 94 birds on average, with a range of 35 to 163 (AAE). Hardly any winter inland although small numbers occur on the Extremaduran ricefields, where January counts during 1990–2009 found 44 birds on average, with a maximum of 315 (AAE). The Spanish wintering population showed a moderate increase of 3.9% during 1980–2009 (AAE), which has been more evident at sites in the north and north-west than elsewhere.

In Portugal 3,600 winterers were censused in 1989, half of them at the Tagus estuary (Rufino 1989b). The latter count is similar to the 3,500 found by Pagezy & Trotignon (1972) at the Portuguese coastal wetlands in winter 1968–69, when 3,000 were at the Tagus and Sado estuaries combined. National censuses in Portugal suggest that up to 6,000 birds may winter there: they averaged 5,410 individuals during 1992–95 although only 3,849 birds during 2005–07. The total Iberian wintering population is likely to be 8,000–10,000 birds on average although with considerable between-year variation.

RUDDY TURNSTONE *Arenaria interpres*

(Vuelvepiedras Común, Rola-do-mar)

***A. i. interpres*. Common on passage and in winter on Atlantic and Biscay coasts, scarcer and more localised in the Mediterranean and the Balearic Islands.**

Ruddy Turnstones that nest in northeastern Canada, Greenland and Iceland winter largely in northwestern Europe, with some also occurring south to Iberia. In contrast, those breeding in Fennoscandia and western Russia winter chiefly in West Africa, some occurring in Iberia on passage (Delany *et al.* 2009). There are at least three Iberian recoveries of birds ringed as pulli: in northeastern Canada, in Estonia and in Finland. There are also three recoveries of birds ringed as juveniles in June in southern Finland (AMI, AI1) and over 50 of birds of other ages ringed in Greenland, Iceland, Britain, Fennoscandia, the Baltic States, Poland, Russia, France, Belgium and Switzerland, as well as one that was ringed in Senegal. Two Turnstones ringed in Iberia have been recovered in Iceland and there are also a few recoveries from Belgium, Sweden, Finland, France, Holland and Britain.

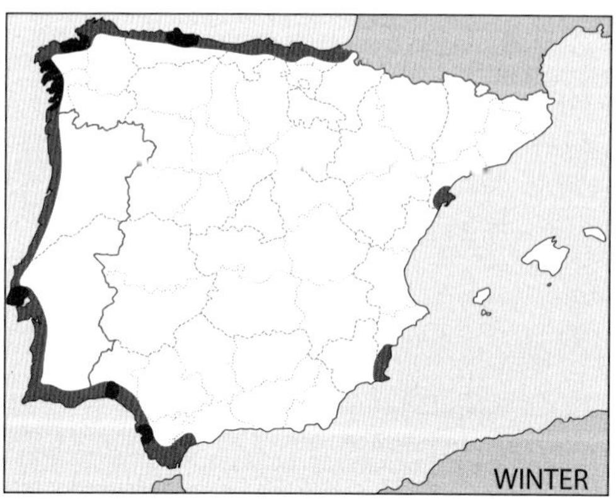

Turnstones are a familiar sight on Atlantic coasts where small flocks forage on rocky shores, along the strandline on beaches and in estuaries, on farmed mussel beds and in ports and harbours. They are less common in the Mediterranean but may still be encountered locally on Levant coasts, notably at the Ebro delta and the Mar Menor, as well as, more rarely, in the Balearics. Some hundreds of non-breeders summer on Portuguese coasts and locally in Spain, for example in Gijón and Cádiz bays.

Migrants appear on Iberian coasts mainly in August–September, with a return movement in April–May. A very small number occur on passage inland, more often in spring than in autumn (Fig. 131). The passage periods inland are clearly defined, the spring records falling mainly between 21 April and 10 May and the autumn records mainly in mid and late September. Birds found inland are nearly always on the shores of lakes and reservoirs across the meseta, with a few in the Ebro valley ricefields. Turnstones usually appear inland in ones and twos so observations of 15 at the Santillana reservoir, Madrid, on 1 May 2004) and 17 at the San Miguel de Cinca ricefields, Huesca, on 8 May 2004 (*Ardeola* 52: 215, *A. O. Aragón 2004–2007*: 279) are exceptional.

Figure 131. Monthly distribution of Ruddy Turnstones recorded inland (n = 67 records, 172 birds).

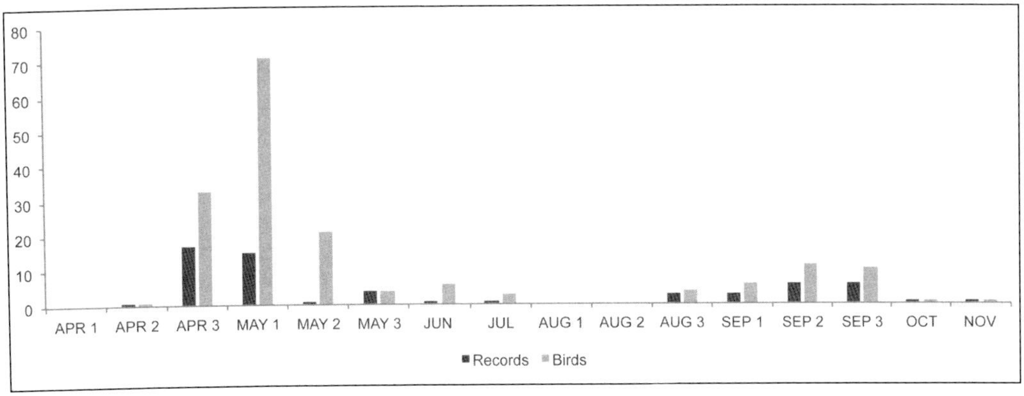

The wintering population is not large but the Turnstone is probably under-recorded by wader surveys since most of the birds are found away from the estuaries and marismas where most other wader species, and censuses, occur. The January censuses of 1990–2009 in Spain found around 1,000 birds in total, with a range of 250–2,000 individuals (AAE). The largest counts have been from rocky inlets and islands in Galicia (up to 1,000 birds in total), where gatherings of 200 or more may occur at favoured sites such as the western Rías, the O Grove peninsula and Cape Silliero. Similar gatherings also occur locally in Asturias, where up to 200–300 birds occur in Gijón Bay in some years (Noval 2001). January counts have shown some recent increases in the wintering population of Asturias, with maximum totals of 565 in 2006, 520 in 2007 and 437 in 2008.

Elsewhere in Spain there are notable winter concentrations in the south-west: in Cádiz bay (up to 456 birds) and at Odiel (up to 137) and other estuaries in coastal Huelva. Small numbers winter on the rocky coastline of the Strait of Gibraltar. January censuses have also found a small, variable but generally increasing wintering population at the Ebro delta (*A. O. Catalunya*). This averaged 105 birds during 1996–2009 but numbers have increased there from just ten and 18 birds in 1996 and 1997 respectively to peaks of 243 and 301 in 2006 and 2007 respectively. Very few winter elsewhere on the Mediterranean coasts, where the largest numbers are reported from the Mar Menor saltpans (up to 184 birds). They are also very uncommon in the Balearics, with no more than eight birds reported during the January counts of 1990–2009 (AAE).

Mean mid-winter counts at Portuguese coastal wetlands were 1,202 birds during 1992–95 and 703 during 2005–07 (AP). However, the winter count of non-estuarine wetlands in Portugal in 2009–10 found 1,989 Turnstones, on shorelines that would not have been covered by earlier surveys (Catry *et al.* 2010b). The coast to the west of Lisbon, at Estoril, holds over 200 birds. Bearing the more recent counts in mind, a wintering Portuguese population of up to 4,000 birds seems likely.

WILSON'S PHALAROPE *Phalaropus tricolor*

(Falaropo Tricolor, Pisa-n'água)

Vagrant to Spain and Portugal.

It was first found in Spain at Doñana on 28 May 1971 (*Ardeola* 19: 21). Another was at Zeluán, Asturias, in August 1983 (Noval 1986). There have subsequently been 18 records for Spain and five for Portugal by 2010, all of single birds. Nine were found during 1983–89 and 14 during 1998–2010. The eight-year gap in records may have coincided with a temporary decline in the source populations (Jehl 1999). There were also four Spanish and two Portuguese records during 2011–12 (*Ardeola* 60: 469, RBS, RBP).

Figure 132. Records by month, 1971–2010 (n = 25)

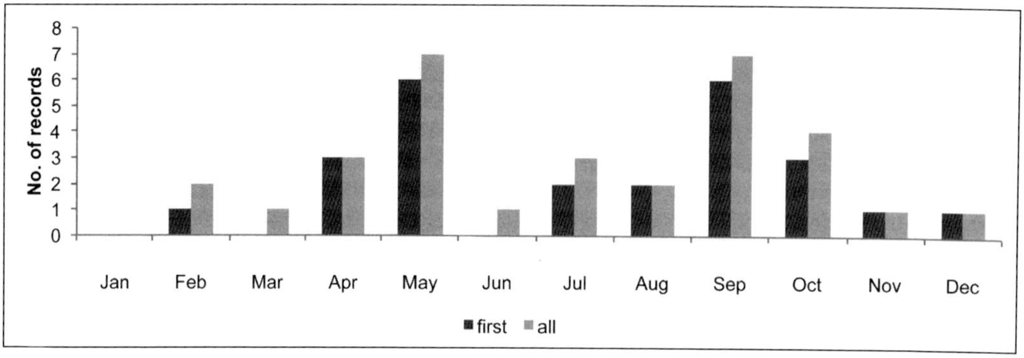

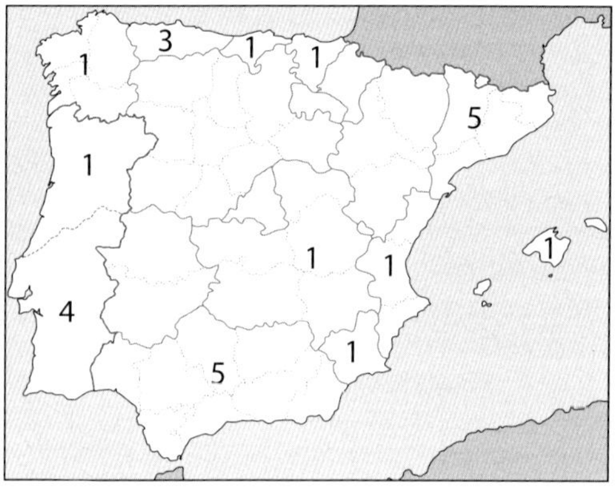

Figure 133. Records by region, 1971–2010 (n = 25)

In total by 2010 there were 15 post-breeding records, between 8 July and 1 December, and 10 spring ones, between 21 February and 28 May (Fig. 132). The proportion of spring records is greater than in France (OF) and much higher than in the British Isles (Dymond *et al.* 1989). The spring records included four in western Andalucía, three on the Mediterranean coast and two in southern Portugal. In all, 12 have been found in the Mediterranean, ten on the Atlantic seaboard and two inland (Fig. 133). The inland birds were at the Laguna de Manjavacas, Cuenca, in July 1989, and at Salburúa, Álava, in May 1999.

This Nearctic wader appears in Europe quite frequently. Some 238 have been found in Britain, 86 in Ireland and 80 in France. There are also four accepted records for the Canary Islands, four for the Azores and five for Morocco.

RED-NECKED PHALAROPE *Phalaropus lobatus*

(Falaropo Picofino, Falaropo-de-bico-fino)

A very scarce migrant, recorded in Spain and Portugal.

No major wintering areas are known in the Atlantic and those that breed in northern Europe and western Russia mainly winter in the Indian Ocean off the Arabian peninsula (Delany *et al.* 2009). Migrants from northern European populations thus travel well east of Iberia and only stragglers are recorded in the region.

A few are found annually, most often at saltpans and coastal lagoons in the south and east. The species has been included in the national rarity lists in both Portugal (since 2001) and Spain (since 2006). There are 35 accepted records during 2006–10 and there were at least 44 Spanish and four Portuguese records during 2011–12 (*Ardeola* 60: 469–470, RBS, RBP). However, many earlier published records exist and we have compiled up to 144 from the usual sources, excluding those not submitted to the rarities committees post-2001 in Portugal and post-2006 in Spain. In general, over 60% of records come from the Spanish east coast, especially from the Ebro delta, as well as from Santa Pola, La Mata lake and the saltpans at the Albufera de Valencia, Mar Menor, Cabo de Gata and Punta Entinas (Fig. 134). Very few have been found in the Balearic Islands, however. There are also regular observations in the south and south-west, especially at the saltpans of the Guadalquivir, Tagus and Sado estuaries. Occasional individuals appear at inland lagoons but there are very few records from the north coast. A southeastern predominance of records is also apparent in France, where a high proportion are from the Camargue, especially in spring (OF).

Observations show a spring peak in April–May with a more evident return passage in July–October (Fig. 135). Some 66% of the records are during July–November and just 30% during March–June. Similarly, in France fewer than 20% of records are during the prenuptial migration (OF). Many females are already on southward passage in July, having left one or more males to incubate the clutches they have laid and to care for the broods. Winter observations are rare but not unknown: some 4.5% of all records are in

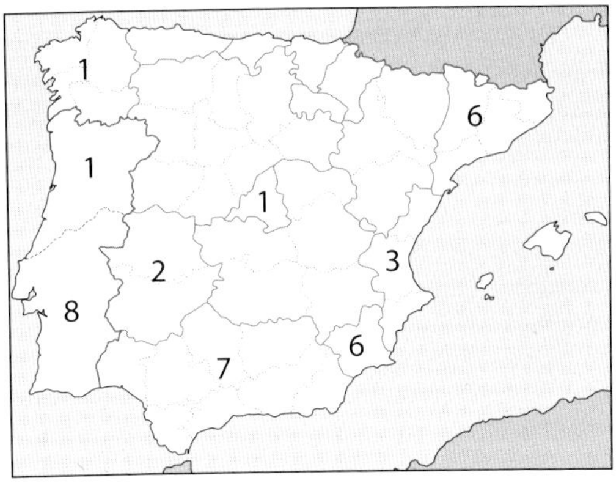

Figure. 134. Records by region, 2006–10 (n = 35)

December–February. The sample included eight December–February records of single birds during 1989–2010, one off Cape St Vincent, one from the Marismas del Odiel and the remainder from eastern Almería or the east coast.

Very few are recorded at any time: 83% of the records were of single birds, the remainder comprising 21 observations of two together and 14 of three to five birds. There were also eight at the Ebro delta during 22 September–11 October 1996 (A. *O. Catalunya 1996*: 137). As this last record shows, some individuals may linger for a week or more. Recently accepted autumn records include 15 of adults and 19 of juveniles or first-winter birds. Adults arrive earlier: in July–August there were ten adults and only two juveniles but during September–October there were five adults and 17 first-year birds.

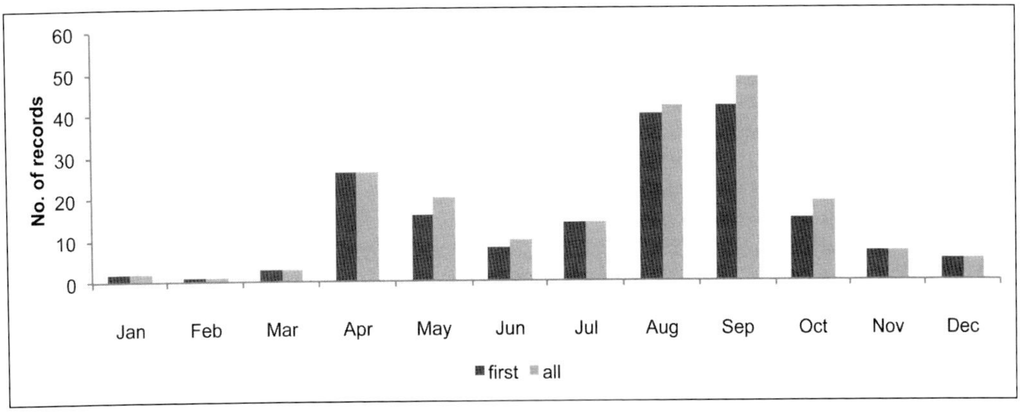

Figure. 135. Records by month, 1959–2010 (n = 179)

GREY PHALAROPE *Phalaropus fulicarius*

(Falaropo Picogrueso, Falaropo-de-bico-grosso)

Regular on passage off western and northwestern coasts, especially in autumn, and also recorded inshore on Atlantic and Biscayan coasts in winter, especially after stormy weather, when storm-driven birds may also occur inland and in the Mediterranean.

Very few breed in northern Europe: in Iceland, Bear Island and Svalbard, but these, and the very many that nest in the northern Nearctic, winter in the central and southern Atlantic (Delany *et al.* 2009). Adults leave the eastern Nearctic in late June/early July (females) and July/August (males) but juveniles may linger in Atlantic waters there until December (Delany *et al.* 2009). There are large southeasterly movements across the North Atlantic in autumn, normally far off the Iberian coasts. Nevertheless, the passage of Atlantic depressions may bring birds into Iberian waters and large numbers are sometimes recorded in autumn off Galicia and Asturias in the north-west and off western Portugal

A sample of 137 published records shows a peak in occurrence during October–January (84%) with only a few (5%) in August–September (Fig. 136). However, in a larger sample that also includes records from regional bird reports the number of individuals per record averages 27.7 during August–October but just 4.7 in November–February. It may well be, therefore, that the largest numbers of southbound migrants pass through Iberian latitudes early in the season, in August and September. The comparative scarcity of inshore observations in late summer seems likely to be due to the settled weather that prevails in summer ahead of the autumn storms. Nevertheless, significant numbers are reported in late summer in some years: for example, 1,100 Grey Phalaropes were observed off the Portuguese coast on 20 August 1996 (*Pardela* 8: 11) and 849 passed Cape Estaca de Bares in five hours on 21 August 2013 (*Ardeola* 60: 527).

Figure 136. Records by month in inshore waters (n = 137).

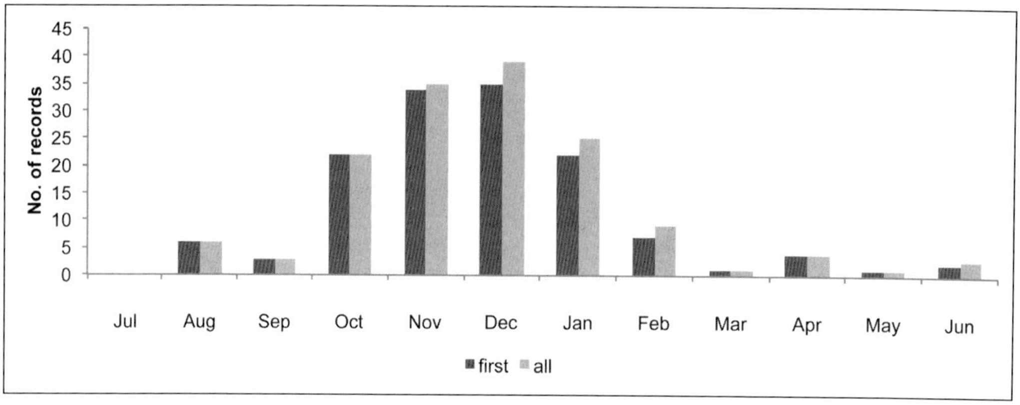

Most Spanish records are of birds seen from Galicia and Asturias: few are reported from further east on the Biscay coast, and these are generally in winter. Grey Phalaropes are mainly recorded from Biscayan and Atlantic coasts during October–February and the largest numbers generally occur in October. For example, 240 were seen by Walker (1994) on passage south off Cape St Vincent between 24 September and 24 October 1993. Significant passage was observed at Punta La Vaca in October 1982, with a peak count of 310 birds in 90 minutes, and 50 took refuge from stormy weather in the Ría de Avilés nearby later in the month (*Ardeola* 31: 143). In 2008, 253 migrants in total were seen from Cape Estaca de Bares between 13 August and 25 November, with a peak passage rate of 42 birds/hour on 28 October (Sandoval *et al.* 2009).

Most observations during November–February involve small numbers of storm-driven birds mainly seen taking shelter at coastal inlets, sometimes in harbours, or found on the coast itself or even inland. Dead birds are not infrequently found on beaches at this time and some are picked up inland occasionally. Some are recorded in winter every year but there are larger influxes, tantamount to small 'wrecks', in some years: notably in 1996, 2000–01 and 2005. An exceptional influx occurred in the north-west in late November 2009, occasioned by successive westerly gales that brought large numbers into the Bay of Biscay (SWA). Only 80 were seen at Cape Bares during August–October 2009 but 1,266 were counted in November, chiefly during the last ten days of the month: a passage rate of 159 birds/hour was seen on 29 November. The totals of birds on the move were eclipsed by counts of birds resting on the water: on 14 November some 500 were off Cape Ortegal, and several thousand were estimated to be present off Cape Bares during 26–29 November 2009 (Sandoval *et al.* 2010). There was also an influx in the south-west in November 1984, when 40 were sitting on the rocks of La Caleta, Cádiz, on 17 November (*Ardeola* 32: 415) and 32 in total were recorded off Gibraltar during 17–21 November (GONHS). Winter records are again most frequent on Atlantic and Biscayan coasts south to the Strait of Gibraltar but very few are recorded in the Mediterranean. There are no records from the Balearics (*A. O. Balears 2010*: 259). The presence of some numbers of Grey Phalaropes in Iberian waters in late autumn and winter could suggest that there is a regular winter presence in these latitudes, well north of the principal winter quarters. The Iberian birds in November–December may still be late migrants, however, since large numbers linger off the Atlantic coast of North America until early December (Delany *et al.* 2009).

Inland records are widely dispersed but show some tendency to decline from north-west to south-east. Most involve just single birds but occasionally two or three. An unparalleled exception is a sighting of up to 16 at the Balsa de Zolina, Navarra, from at least 30 November to 8 December 2009 (*Ardeola* 57: 230). There was also a significant influx into Castilla-La Mancha in 2000, when 27 birds in total were found at 18 wetlands in Ciudad Real, Toledo and Cuenca between 8 November and 21 December (Velasco 2002).

Grey Phalaropes are rarely seen in Iberia during their spring passage, which is generally in April–May (Delany *et al.* 2009), when the weather is again much less likely to be stormy than in winter. It is also very likely that northward passage in the Atlantic occurs further west, far from Iberian coasts. Certainly such storms as occur in Iberian waters in March or April do not result in springtime influxes. The *Ardeola* bird news summaries, 1980–2010, give just seven records of single. birds between 4 April and 17 June: four from Andalucía, two from the Mediterranean and just one from Galicia. There are no published spring records from Portugal but an unseasonal individual was at the Tagus estuary on 14 June 2009 (*Anu. Ornitol.* 8: 70).

POMARINE SKUA *Stercorarius pomarinus*

(Págalo Pomarino, Moleiro-do-árctico)

Migrants are regular on passage on Atlantic and Biscayan coasts but scarce in the Mediterranean, where most are recorded in spring. A few remain in winter.

Western Siberian and eastern Nearctic Pomarine Skuas winter largely at sea in the tropical Atlantic and off southern Africa. The southward passage is generally apparent in the Iberian north-west from August to October, where peak counts are usually in late September and early October. A few individuals appear as early as July and some passage continues until late November or early December. Autumn passage on the Biscay coasts comprises westward movement of southbound birds that have diverted into the Bay of Biscay. Observers at Cape Estaca de Bares recorded seasonal passage rates ranging from 1.7–4.2 birds/hour during the autumn passages of 2005–08 (Sandoval *et al.* 2009). On the basis of sample counts, the mean annual post-breeding daytime passage there during 2004–08 was of the order of 4,500 birds (Sandoval *et al.* 2010). The seabird movements at Cape Bares during the second half of 2011 were exceptional and included 4,434 Pomarine Skuas. Large movements are also apparent off Asturias, where observers at Punta La Vaca recorded a site maximum of 384 birds in 5½ hours on 19 October 1991 (*Ardeola* 40: 96). A record of 25 birds at Punta La Vaca on 2 December 1997 (C.O.A. 2003) is noteworthy for the relatively late date. The Pomarine Skua is also regular, although less numerous, further east in autumn, off Cantabria and the Basque Country (Paterson 1997).

Pomarine Skuas pass south off western Galicia and Portugal mainly during August–October, with peak numbers usually seen during adverse weather in October (Paterson 1997). Here, as elsewhere, peak observations do not necessarily correspond with the migration peaks themselves, but rather are dependent on the incidence of strong onshore winds, which are more likely late in the season than in late summer. Some numbers may thus be expected off the west Portuguese coast at any time from August to October. The passage of successive cold fronts accompanied by onshore gales in autumn 1999 enabled counts from Cape Carvoeiro, Peniche, of 215 on 19 September and 721 on 24 October (Moore 2000a). A count from the same area of 187 on 4 October 2010 is also noteworthy (*Anu. Ornitol.* 8: 70).

There are comparatively few observations from Algarve, Huelva–Cádiz, or within the Strait of Gibraltar, and these always involve just a few individuals. Some appear in the Strait from late July to November and Hashmi (in Finlayson 1992) observed a total of 32 in mid Strait between 18 September and 13 November 1987. Small numbers are also recorded in the Mediterranean in autumn, with records of individuals or groups of up to three birds off the Comunidad Valenciana and Catalonia chiefly in September and October but ranging from late August into December (Paterson 1997). Very few remain in Iberian waters in winter, and those seen in December may still be on passage, but there are occasional records of apparently wintering individuals from all coasts during December–February.

Spring passage is generally less apparent than the autumn movements. Most passage is in April and May but some apparent migrants are recorded in March. A few non-breeding individuals are also recorded off all coasts in summer. The Pomarine Skua is rarely seen at all from the Biscay coasts in spring but migrants are recorded off western Galicia and Portugal. A few individuals are also recorded in April and May, occasionally in March, in the Strait of Gibraltar.

There are also regular spring observations of small numbers off Mediterranean coasts and spring records predominate off the Levant coasts, where they include a group of 16 birds on 11 May 2002 off the Llobregat delta and another of 14 off the Ebro delta on 26 April 2008 (*A. O. Catalunya 2008*: 123). It has been suggested that the Pomarine Skua is commoner on spring passage off the Mediterranean coast than originally thought, on the basis of marine censuses off Valencia especially (Dies & Dies 1991). In recent years there have also been frequent records from Catalonia: 75 spring records during 2000–09 compared with ten during July–November in the same period. Extreme spring dates there were 1 March and 17 June but some 85% of the records were between April and mid-May (*A. O. Catalunya*). One at Cap de Ses Salines, Mallorca, on 29 March 1998 (*Ardeola* 46: 310) was the first record for the Balearics: there have been two subsequent observations at sea off the islands (*A. O. Balears* 2003: 99; *Ardeola* 56: 357). The regular occurrence of birds in the Mediterranean invites speculation that there is some overland passage across central Europe and that such birds are not just wanderers that have strayed inadvertently through the Strait of Gibraltar. However, there is little direct evidence of such movements since inland records of Pomarine Skuas in western Europe, including France, are uncommon, although there are regular overland movements from Siberia to the Baltic (BWP).

There are a few observations of Pomarine Skuas in estuaries or at coastal lagoons: for example, four were on the Laguna de Baldaio, A Coruña, on 25 August 1996 (*Ardeola* 44: 254), but inland records are almost unknown. A juvenile at the Laguna de Sariñena, Huesca, on 21 November 1999 (*Ardeola* 46: 310) provides a rare exception.

ARCTIC SKUA *Stercorarius parasiticus*

(Págalo Parásito, Moleiro-pequeno)

Common on passage on Atlantic and Biscay coasts, scarcer in the Mediterranean. Small numbers winter and some non-breeders occur in summer.

The Arctic Skua has a circumpolar arctic and subarctic breeding distribution, with some nesting as far south as northern Scotland. The birds migrate to and from pelagic winter quarters in the southern hemisphere. Migrants occur commonly off Iberian Atlantic coasts on passage, especially in autumn in the west and north-west, but the spring movements are less noticeable. A few remain in Iberian waters, including the Mediterranean, all year round. Several birds ringed at Scottish colonies have been recovered on northern and western Iberian coasts (Wernham *et al.* 2002) and there is also a Portuguese recovery of one ringed in Finland.

Autumn passage is during July–November but the largest numbers are seen in September–October in the north-west. As with all seabirds, the numbers seen from land are highly influenced by weather conditions. However, considerable numbers are recorded annually on passage off the coasts of Galicia and Asturias, with fewer off Cantabria and the Basque Country. The birds are moving out of the Bay of Biscay before following the western Iberian coast southwards. Passage rates at Cape Estaca de Bares have reached 64 birds/hour, on 28 September 1991, but average much lower: 1,380 birds were recorded in 410 hours during July–November 2008, a rate of c.1.5 birds/hour (Sandoval *et al.* 2009). On the basis of sample counts, the mean annual post-breeding daytime passage at Cape Bares during 2004–08 was of the order of 6,000 birds (Sandoval *et al.* 2010). In 2011 3,013 were counted there on post-breeding passage (A. Sandoval pers. comm.). At Punta La Vaca, Asturias, 1,268 migrants were counted between late July and late December 1991 and 250 passed west there in 5½ hours on 19 October 1991 (*Ardeola* 40: 96).

Southward passage off the Portuguese west coast is noted during August–November but peaks during the second half of September, when passage rates of 15–20 birds/hour are usual (AP). Onshore gales may sometimes result in much larger counts, such as the 1,350 counted flying south at Cape Carvoiero on 19 September 1999 (Moore 2000a). Such large counts, of this and other seabird species, point to the scale of the offshore movements that are normally not visible from land.

Small numbers are recorded in the Strait of Gibraltar chiefly between August and April but also occasionally in summer. Some also occur in the Alborán basin and off the Levant coast during passage periods and in winter (de Juana & Paterson 1986) but very few have been recorded in the Balearic Islands: Clavell (2002) gives four post-1985 records (five birds) from Mallorca, two in spring and two in autumn, but there have been no more up to 2010 (*A. O. Balears*). In general, Arctic Skuas are seldom numerous in the Strait and western Mediterranean but there is a regular presence, especially during passage periods, with some remaining in winter. Hashmi (in Finlayson 1992) observed a total of 65 in mid Strait between 18 September and 13 November 1987 and 54 were seen from Gibraltar in autumn 1992, mainly in September (GONHS). However, an exceptional influx of 100+ occurred at Tarifa during westerly gales on 11 November 1995 (GIAM 17: 6–7). The (modest) migratory peaks at Gibraltar presumably involve birds that enter the Strait from the Atlantic and then return westwards again, to continue further south or north according to season, as well as the few that winter in the Mediterranean.

The birds that occur off Iberian Mediterranean coasts during passage periods may also include some that have entered the region temporarily through the Strait as well as individuals that may remain to winter or summer in the region and perhaps a few that travel to or from the Mediterranean basin overland. In general, the occurrence of Arctic Skuas in this region is poorly understood. More are seen off the coasts of the Comunidad Valenciana and Catalonia during passage periods than in winter (Paterson 1996, Clavell 2002) so the movements appear to involve more than just the local wintering population. In the Comunidad Valenciana the autumn passage is reported from mid August into December, peaking in November (Dies & Dies 1991). In 1998 102 were counted in total on passage at Cape Cullera, Valencia, including 22 on 16 October (*Ardeola* 36: 249). Off Catalonia the autumn passage begins in mid-July and is at a maximum from mid August to mid October (Gutiérrez *et al.* 1995). In spring, passage extends from early March to early May, peaking from mid March to mid April, in the Comunidad Valenciana (Dies & Dies 1991, Gutiérrez *et al.* 1995). Further north, off Catalonia, the spring passage, which is much more noticeable than the autumn one, peaks in April but continues into June. There are counts of some numbers off the Levant coasts, including 54 birds recorded in 90 minutes off Premìa, Barcelona, on 29 April 1989 and 109 off l'Ametlla del Mar, Tarragona, on 13 May 1988 (Paterson 1997). There is also a record of 97 Arctic Skuas, including a flock of 94, on 11 July 1997 off Port-la-Nouvelle, on the French Mediterranean coast north of Catalonia (OF). These observations, involving much larger numbers than are usually detected at the Strait, not only show that the species may be more numerous at times in the Mediterranean than is generally thought but also suggest that there is southward passage off Levant coasts in summer/autumn and northward passage there in spring, which implies the existence of some overland passage across central Europe. However, Arctic Skuas seen in spring from Cape Creus, the most north-easterly Iberian headland, are usually flying south, as were, for example, 23 seen there between 4 March and 22 April 2001 (*A. O. Catalunya 2001*: 158).

Arctic Skuas are recorded inland in France annually, although they are rare in most years (OF), so perhaps a small but regular cross-country passage remains to be detected there. There are also rare records of birds found inland in Iberia. They include two in Aragón, two in Navarra, two in Extremadura and one in Jaén, Andalucía: five were in September and one in October (*Ardeola* 1981–2012, Paterson 1997). All involved single birds with the exception of two seen over the Lindux pass in the Navarran Pyrenees on 25 August 2012 (*Ardeola* 59: 437).

The Arctic Skua is the most frequent skua reported in winter off the Huelva coast (Garrido 1997) and in Málaga Bay (Paterson 1987), although it is very much scarcer than the Great Skua in winter at Gibraltar, where individuals are reported only occasionally. In Málaga Bay, Paterson (1987) recorded 306 sightings (not necessarily involving as many individuals) between September 1980 and December 1983, with records in all months but the majority from mid-September to December. The estimated wintering population off Catalonia during 2006–09 ranged from 31–59 individuals, most of them occurring between Barcelona and the Ebro Delta within the principal wintering area of the Mediterranean Gull, their principal victim in that region (CWA, Arcos 2000).

LONG-TAILED SKUA *Stercorarius longicaudus*

(Págalo Rabero, Moleiro-rabilongo)

Regular on autumn passage off Atlantic coasts, occasionally numerous. Rare elsewhere.

Long-tailed Skuas generally travel out of sight of land but a variable proportion appear inshore along Atlantic coasts, including in Iberia. They are often seen in some numbers in the British Isles but the species was still regarded as a rarity in France until 2005, by when there were records of 1,334 individuals. In contrast, there are only six records from the Azores, three from Madeira and two from the Canary Islands. There are only a dozen records for Morocco but the Long-tailed Skua appears in hundreds off Mauritania, where some remain in winter (Isenmann *et al.* 2010).

Irby (1873) noted that two immature birds had been obtained in the Strait of Gibraltar but otherwise the earliest report from Iberia seems to have been of one captured at Cullera, Valencia, in 1884 (Arévalo 1887). Much more recently, Bourne & Norris (1966) saw several off the Berlengas archipelago on 13 September 1964, and Huyskens & Maes (1971) saw six in total from Cape Estaca de Bares on 30 September 1967 and during 6–13 October 1969. There have since been 51 accepted records for Spain and 27 for Portugal up to 2010 and we here consider a sample of 82 records that involved 295 individuals, although there have been very many more records post-2010. Annual records since 1989 no doubt reflect the greatly increased interest in seawatching. The great majority of Iberian observations have been from Cape Bares in late summer and autumn, reflecting both its strategic location and the regular and increasing presence of observers there. However, some years, notably 1991, 1997, 2002, 2007 and 2009–11, have produced many more birds than others. This interannual variation is probably mainly due to meterological factors, notably the prevalence of conditions that drive birds inshore. However, Long-tailed Skua populations show fluctuations linked to the cyclical variation in lemming availability on the nesting grounds, and these would affect the numbers seen on passage.

Most records are in autumn, as happens in the British Isles and France, peaking during August–September (Fig. 137), and chiefly on the western and northwestern seaboards (Fig. 138). There are, however, six inland autumn records of single

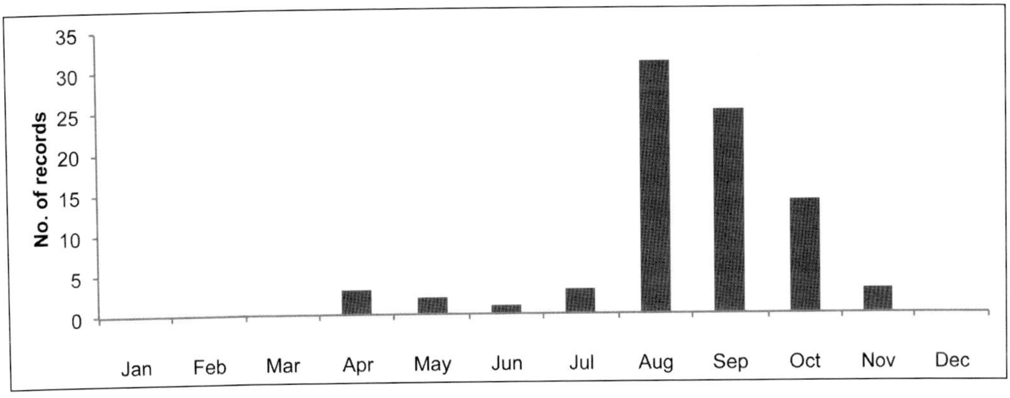

Figure 137. Records by month, 1967–2010 (n = 82)

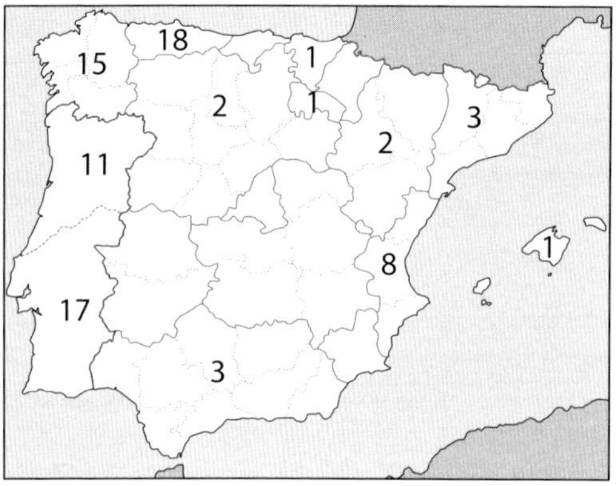

Figure 138. Records by region, 1967–2010 (n = 82)

birds, often involving sick or dead individuals: at Candasnos, Huesca, in September 1991; Igea, La Rioja, in August 1993; the Valle de Arán, Lleida, in July 1994; Santa Cristina del Páramo, León, in August 2001; the Azud de Riolobos, Salamanca, in August 2002, and Monreal de Ariza, Zaragoza, in September 2009.

The extreme autumn passage dates of the accepted records up to 2010 are 13 July and 26 November. There are only six spring records, that fall between 11 April and 5 June. All spring records have been of single birds but autumn records, especially in recent years, include flocks of migrants. Prior to 2009 the autumn birds included 25 adults and 89 aged as juveniles or immatures but the 2009 passage included 151 adults and only four young birds. Adults generally pass earlier in autumn: they comprised 50% of the total during July–August but only 24% during September–November. All five of the spring birds whose age was noted were adults. Four of the spring records have been in the Mediterranean, with one in the Strait on 11 April 2002 and another in southern Portugal, at Portimão, on 6 May 1999. The very limited spring passage in France is similarly chiefly noted in the Mediterranean, off Provence (Jordan *et al.* 2000, OF).

The 2010 passage was itself substantial in relation to earlier Iberian observations: 48 birds were recorded between 23 July and 7 November 2010 (RBS). The 2011 observations were altogether on a different scale. At least 2,287 individuals were counted between 19 July and 16 December. Of these 2,216 (97%) were counted from Cape Bares, where an unprecedented inshore influx of migrating Long-tailed Skuas was monitored daily, but there were also 52 seen from the Galician west coast, 11 from Asturias, five from Cantabria and three from the País Vasco. The monthly distribution of the records was July 1, August 233, September 1,992, October 33, November 26 and December 2. There was a later, unseasonal, observation of a first-winter off Burela, Lugo, on 8 January 2012 (RBS). The species was removed from the Spanish rarities list from 2013 although most of the recent records had not then been submitted to the rarities committee (*Ardeola* 60: 470) .

GREAT SKUA *Stercorarius skua*

(Págalo Grande, Alcaide)

Regular on passage and in winter off all coasts but more numerous in Atlantic waters. Some non-breeders remain in summer.

The Great Skua has a restricted breeding range that includes Iceland, the Faroes, Scotland, northern Norway and northwesternmost Russia. Most of the population winters in coastal waters of the northeastern and northwestern Atlantic (Magnusdottir *et al.* 2012). The principal wintering areas, especially of birds of Scottish origin, include the Atlantic and Biscay coasts of Iberia, although most generally keep well offshore. Nevertheless, large numbers are sometimes recorded from the key watchpoints on coastal headlands, both during passage periods and after stormy conditions in winter. Some numbers winter in the Mediterranean and regular movements occur in autumn and spring through the Strait of Gibraltar. Non-breeding birds may also be seen off Iberian coasts in summer. There are at least 96 Iberian recoveries of birds ringed as pulli in Scotland and at least 14 of birds ringed in Iceland: most Scottish-ringed birds recovered as adults in winter have been found in the Bay of Biscay and on Iberian coasts, including a minority within the western Mediterranean: immature birds are also recovered from these areas but some travel further south in the Atlantic.

The southward passage into Iberian waters begins in August but the principal movements are during September–December, when large numbers that have moved into the Bay of Biscay depart westwards. Maximum passage rates, of 30–50 birds/hour at Cape Estaca de Bares and Asturian headlands are usually seen between October and early December (SWA), although the exact timing of land-based observations is weather-dependent. In 2008 2,117 were counted at Cape Bares in 209 hours during September–November (Sandoval *et al.* 2009). A passage rate of 53 birds/hour on 29 October 2008 was then a site record: seasonal counts in recent years average 2–6 birds/hour but show a positive trend (SWA). On the basis of sample counts, the mean annual post-breeding daytime passage at Cape Bares during 2004–08 was of the order of 8,000 birds (Sandoval *et al.* 2010). In 2011 5,540 were counted on post-breeding passage at Cape Bares (A. Sandoval pers. comm.).

Such large movements have not often been detected off the Iberian west coast presumably because they normally occur well offshore but also because of a lack of systematic observation. However, in Estremadura on 9 December 1989 there was a southward movement of 600+ Great Skuas off Peniche and 48 were counted in 1½ hours on the same day at Cape Raso (Airo 2). A southward passage of 772 birds also occurred at Cape Carvoeiro on 27 December 2002; a further 580 were counted flying south there a few days later, on 6 January 2003 and 150 were counted at the same site on 6 December 2006 (*Anu. Ornitol.* 2: 35–37, 6: 60). These observations indicate the presence off central Portugal of large numbers of birds that were either late migrants or, more probably, winterers offshore. A count of 82 at Ponta da Atalaia on 12 November 1995 (*Pardela* 1: 19), is a significant record for Algarve.

Passage into the Mediterranean is apparent at the Strait in autumn during August–November. There are also occasional influxes into the Strait from the west in winter following the passage of cold fronts, when up to 40 have been counted in a single diagonal transect from Gibraltar to Tangier (Garcia 1973). This is the commonest skua in the Strait at all times of year, although the Arctic Skua is more numerous inshore off Huelva and the Costa del Sol. Paterson (1987) recorded only 87 sightings (v. 306 Arctic Skua sightings) in Málaga Bay between September 1980 and December 1983, and 80% of the Great Skuas were presumed non-breeders seen in June and July. By comparison, Hashmi (in Finlayson 1992) recorded 167 Great Skuas (v. 65 Arctic Skuas and 32 Pomarine Skuas) in mid Strait from 18 September–13 November 1987. It may be that the narrowness of the Strait forces birds to approach land more closely there than they do elsewhere. However, some inshore movements have recently been observed from the Málaga coast: 53 flew west at Punta Calaburras on 12 January 2008 and 15 flew east in one hour at Benajarafe on 28 September 2008 (*Ardeola* 55: 143, 299).

Great Skuas are regularly reported off Levant coasts both during passage periods and in winter. However, unlike the Arctic Skua, the Great Skua is most numerous in the region in winter and the passage movements thus seem to involve the wintering population. It is the commonest skua off Catalonia but is most abundant there from 5–13 miles offshore (Martínez-Vilalta *et al.* 1984). Numbers off Catalonia build up to a maximum during November–January. A wintering population averaging 40–60 birds has been reported at sea in the Gulf of Sant Jordi, Tarragona, on the north side of the Ebro delta, and up to 50 birds have been counted at Cape Creus, Girona (Paterson 1997). Further south, off the Comunidad Valenciana, the Great Skua is less common in winter but more obvious on passage, especially in spring, when easterly gales bring birds inshore. The spring passage is southward, presumably involving birds travelling towards the Strait and the Atlantic (Dies & Dies 1991), and occurs from February to late May, peaking in the second half of March.

The Great Skua is scarce in the Balearic archipelago. A few occur off the islands during October–May but the majority of observations there correspond with the spring passage period, in March–April. Records in the Balearic Bird Reports for 1993–2009 average about six per year. Single birds comprised 84% of sightings the remainder involving two or three individuals.

Spring passage is observed at the Strait in March–April, when birds are moving west to the Atlantic. Similar passage clearly occurs northwards up the west coasts of Portugal and Galicia, but these movements have seldom been reported. No significant spring passage is usually seen at Cape Bares or elsewhere on the Biscay coast, so presumably migrants normally continue north without entering the Bay of Biscay. However, bad weather may drive birds eastwards on occasion in spring: a strong westward passage was seen at Gijón Bay, Asturias, on 2 March 1989 after stormy weather, with 312 birds counted flying west in an hour (*Ardeola* 36: 249); the relatively early date may mean that wintering birds rather than migrants were involved.

Great Skuas are reported inland on rare occasions but more frequently than other skua species. For example, the Noticiarios in *Ardeola* give eight such occurrences during 1980–2007; of birds found in Castilla-La Mancha, Extremadura, Navarra, Aragón, Alicante and the Basque Country. All involved single individuals with the exception of three at the Gabriel y Galán reservoir, Cáceres, on 11 November 2007 (*Ardeola* 57: 231). Two records were in spring, in April and May, and the remainder in July–November.

SOUTH POLAR SKUA *Stercorarius maccormicki*

(Págalo Polar, Alcaide do Antárctico)

Three Spanish records were pending review in 2013 (*Ardeola* 60: 502, RBS): single birds, perhaps the same individual, seen from Cape Estaca de Bares on 13 August 2011 and 10 September 2011, and another bird there on 4 August 2012. There were also reports in 2013 of single birds off Cape Bares in August and off the Cádiz coast in September (RBS). The South Polar Skua is regularly found in summer off the coasts of North America, north to Greenland, but it has rarely been identified in the northeastern Atlantic, from where there is an 1889 record from the Faeroes, at least three from the Azores (Malling Olsen & Larsson 1997, Haas 2012) and one of three birds from the Canaries (*Ardeola* 60: 470–471).

BONAPARTE'S GULL *Chroicocephalus philadelphia*

(Gaviota de Bonaparte, Guincho-americano)

Vagrant to Spain and Portugal.

Bonaparte's Gull appears quite frequently in the British Isles, from where there are records of 178 individuals in Britain and 65 in Ireland. However, only 26 have been found in France. There are 16 records from the Azores, two from Madeira and one in the Canaries. There are also three records from Morocco and one from Senegal (Baillon & Dubois 1992).

The earliest known Iberian record was a first-winter bird at Corme, A Coruña, on 1 November 1982 (*Ardeola* 35: 172). The 29 accepted records for Spain and 12 for Portugal to 2010 are all of single birds. However, fewer individuals were probably involved since some tend to return to the same sites in successive years. For example, one wintered annually on beaches near Lisbon during 1990–99 (*Ardeola* 39: 79, *Anu. Ornitol.* 1: 24): pre-2010 Portuguese records total six if this individual is counted once only. Nine individuals were reported as first-years and 18 were adults, again counting the returning Portuguese individual only once. There was a further Portuguese record in 2011 and at least six Spanish records during 2011–13 (*Ardeola* 60: 472, RBS).

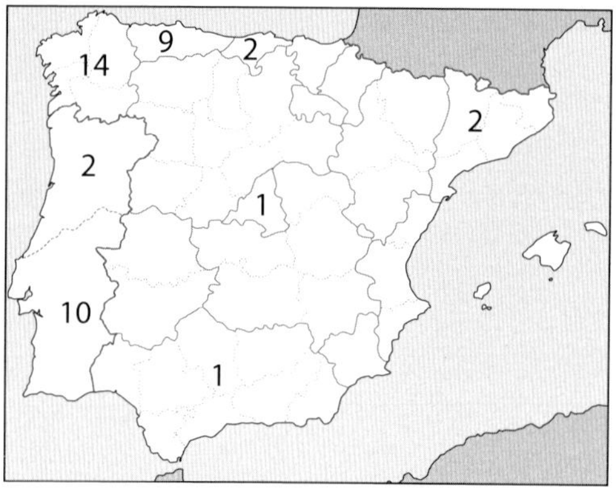

Figure 139. Records by region, 1982–2010 (n = 41)

Nearly all have been on Atlantic or Biscay coasts (Fig. 139), the exceptions being one on the Granada coast during 21–25 October 2004; another at Pinto refuse tip, Madrid, between 22 March and 18 April 2009; and one at Tarragona harbour, Catalonia, during 6–9 January 2009 and probably the same bird again from 16 December 2010 to 13 March 2011. Records have principally been in winter (Fig. 140) and the extreme dates are 21 October and 31 March, with the exception of two August records from Galicia in 2006. In contrast, records in Britain and France are spread throughout the year, with some instances of overwintering (Dymond *et al.* 1989, OF). It thus appears that those Bonaparte's Gulls that reach the western Palearctic perform north/south movements there (Hoogendoorn & Steinhaus 1990, de Juana 2006).

Figure 140. Records by month, 1982–2010 (n = 41)

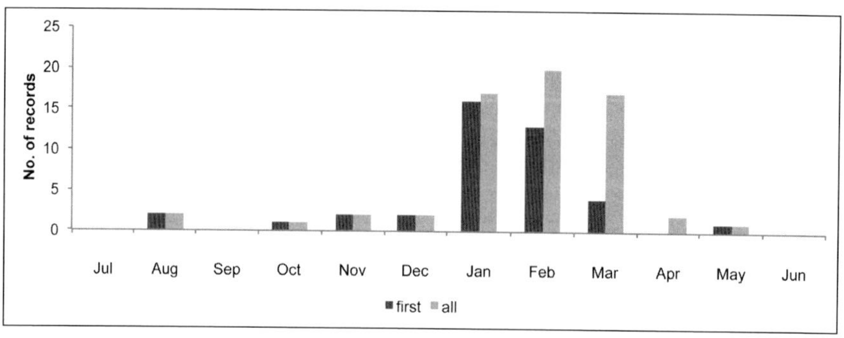

BLACK-HEADED GULL *Chroicocephalus ridibundus*

(Gaviota Reidora, Guincho-comum)

A local but increasing breeding species in mainland Iberia. Widespread and locally abundant on passage and in winter throughout the region, inland as well as on coasts.

The first known instance of breeding was in 1960 at the Ebro delta, when a colony of 25 pairs was detected (Maluquer 1960). Breeding then occurred at Fuentedepiedra in 1966, the Albufera de Valencia in 1967, Alcázar de San Juan in 1971, Gallocanta in 1973, the Laguna de Manjavacas in 1978 and increasingly widely thereafter (Cantos 2009). The Ebro delta remains the principal

Iberian nesting site but the Black-headed Gull has since become more or less established at some 40 sites in mainland Spain, nearly all of them in eastern and southern provinces (SA2). It also breeds locally in Portugal, where there is a small colony at the Mondego estuary (PA2). Colonies are always associated with wetlands, both coastal and inland, including estuaries, marismas, saltpans, lakes, gravel pits and reservoirs. Small islands in lakes and reservoirs are especially favoured. Some colonies are only occupied sporadically or may be abandoned during drought years. The numbers using particular sites may also vary widely between seasons: for example, La Lantejuela lagoons, Sevilla, had a colony of 30 pairs in 2007 but 250 pairs were reported there in May 2003 (*Ardeola* 50: 350).

The Spanish breeding population numbered at least 9,148–9,211 pairs in 2007 (Table 34). Early monitoring of breeding attempts was patchy but a first national census in 1993 found 3,077 pairs, 2,000 of them at the Ebro delta (Asensio & Cantos 1996). The Ebro delta colony grew to 3,778 pairs in 2000 and 4,008 in 2007 (*A. O. Catalunya 2000*: 159–160; Cantos 2009). Most colonies are on the east coast, in western Andalucía and in Castilla-La Mancha: where many of the lagoons of the La Mancha are occupied. Outside this region there were two small colonies in western Castilla y León, at the Lagunas de Boada and Villafáfila, as well as several in the Ebro valley: at Gallocanta, the Salada de Chiprana, the Saladas de Bujaraloz and the Estany d'Ivars, Lleida. Occasional breeding attempts by isolated pairs elsewhere: in Galicia, Madrid, Navarra and Extremadura may prove to be precursors of further range expansion. A possible indication of such a trend in Extremadura is offered by the observation of 31 birds, three of them incubating, at the Valdecañas reservoir on 20 June 2011, and of 20 pairs showing breeding behaviour at the Alange reservoir during May 2011 (*Ardeola* 58: 501).

Table 34. Regional distribution of breeding Black-headed Gulls in Spain in 2007. After Cantos (2009).

Region	Pairs	Sites
Catalonia	4,088	2
Andalucía	2,018	8
Castilla-La Mancha	1,312	19
Com. Valenciana	1,300	5

Region	Pairs	Sites
Murcia	122	1
Aragón	248	3
Castilla y León	60	2
TOTAL	9,148	40

Breeding season counts in 2009 showed a considerable regional increase to 1,732 pairs in the Comunidad Valenciana, including 1,081 pairs at the Albufera, 450 pairs at El Hondo, 150 pairs at La Mata-Torrevieja and 50 at Santa Pola (*A. O. C. Valenciana* 2009: 79).

The breeding presence in Portugal is minimal. The first recorded nesting was at the Sado estuary in 1995, when one pair bred, but the sole regular nesting site in Portugal is at the Mondego estuary, where two or three pairs have nested since 2000 (PA2). Occasional nesting is reported elsewhere: in 2009 nesting was confirmed at the Albufeira da Herdade, Évora, and three pairs nested at the Lagoa dos Salgados, Faro (*Anu. Ornitol.* 8: 71). Summering non-breeders are widespread in Portugal, as elsewhere, and both population increases and range expansion are also to be expected there.

Iberian birds disperse after nesting and are presumably among those that appear on coasts in July–August. Much larger numbers arrive to winter in the region from September onwards, especially in October and early November and a proportion continue south to winter in North Africa. Wintering birds depart from mid February to mid April but stragglers remain into May and some non-breeders are present in summer. Observations inland away from the breeding colonies are chiefly between July and April (Calvet *et al.* 2004, Cantos & Serrano 2011) but most recoveries of foreign-ringed birds are concentrated during

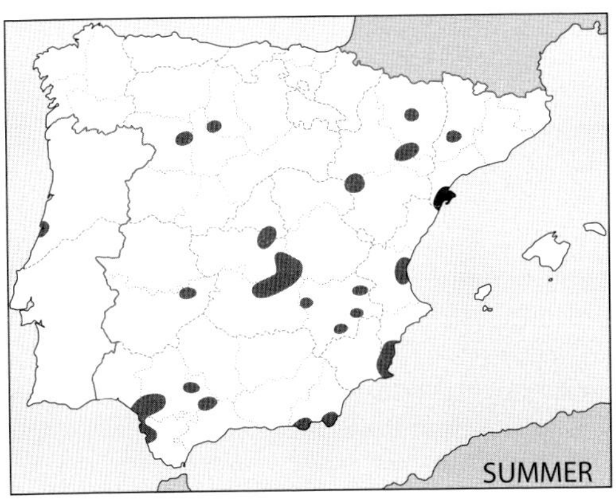

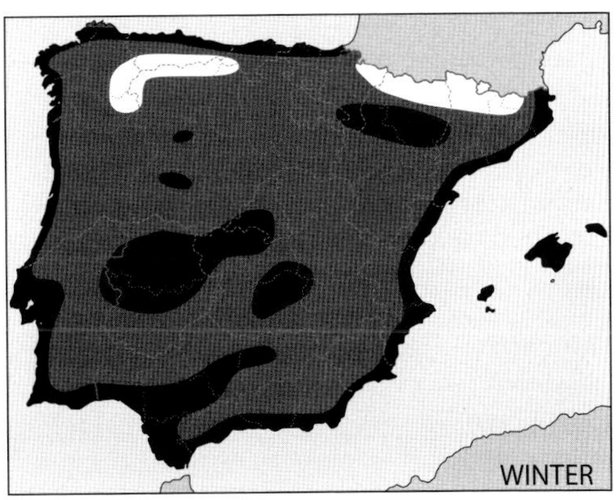

November–February (AMI 82% , n = 945). In one survey, first-winter birds made up 27% of the Madrid wintering population and 70% of 86 captured birds were females (Cantos *et al.* 1993).

There are Iberian recoveries of birds ringed in virtually all European countries east to western Russia, the Baltic States, Poland, the former Czechoslovakia and the former Yugoslavia, as well as three from Iceland. However, the majority of 3,396 recoveries originate from France/Belgium (36%), with 23% ringed in northern Europe and another 23% in central Europe and just 3% in the British Isles. Black-headed Gulls ringed in Iberia have been recovered mainly from the same areas but those ringed as chicks at colonies in Spain tend to move south after fledging and five of them have been recovered in Morocco. One ringed as a nestling at Alcázar de San Juan, Ciudad Real, in May 1972 was found in Aargub, Western Sahara (2,100km) the following October (*Ardeola* 20: 83).

The wintering population was censused throughout Iberia in January 1984 (Bermejo *et al.* 1986) and throughout Spain in January 2009 (Cantos 2009). A comparison of the results from the two counts (Table 35) is instructive, indicating not only an increase in total numbers but also a marked shift towards wintering inland. The latter has involved a much greater reduction in the numbers wintering on the Atlantic coasts, chiefly in the regions of northern Spain, than in the numbers that winter in the Mediterranean.

Table 35. The numbers of Black-headed Gulls wintering in Spain in January 1984 and January 2009. The Atlantic sector includes the Biscay coasts. Inland comprises only those provinces that lack a coastline. Since some coastal provinces are very large this means that the 'Inland' population is underestimated to some extent.

Period	Atlantic		Mediterranean		Inland		TOTAL
	No.	%	No.	%	No.	%	
Jan. 84	64,411	25.10	143,257	55.83	48,934	19.07	256,602
Jan. 09	33,495	10.14	138,335	41.87	158,562	47.99	330,392

The total Iberian wintering population in January 1984 was 299,625 birds (Bermejo *et al.* 1986). It included 43,023 birds in Portugal, the great majority of them (83%) at the Tagus/Sado estuaries. Censuses at the Tagus estuary in January 1981 and during winter 1981–82 found 58,180 and 42,570 Black-headed Gulls there respectively (AP). There are no recent national counts from Portugal that would reveal whether Black-headed Gulls have also tended to move away from Atlantic coasts there. However, reservoirs inland regularly attract thousands of roosting birds in winter (AP) and 5,700 were counted at the Lagoa dos Patos, Beja, in January 2008 (*Anu. Ornitol.* 7: 82). The census of non-estuarine coastal sites and beaches in Portugal in winter 2009–10 found only 1,302 Black-headed Gulls (Catry *et al.* 2010b).

The 2009 winter census found 330,392 Black-headed Gulls in Spain alone and it is likely that over 400,000 birds were present in the entire Iberian region that winter, an increase of some 33% since 1984 that can most probably be attributed to known increases in the source populations. On the basis of the January waterfowl counts the wintering population in Spain may have increased at an annual rate of 2.9% since the 1980s (AAE).

The regional distribution of the wintering population in Spain (Table 36) shows major concentrations in Catalonia, the Comunidad Valenciana, Madrid and Castilla-La Mancha, with lesser but still large numbers in Andalucía: chiefly in the west, and Extremadura. Comparatively few winter on the northern meseta or in the Biscay coastlands, although they are still widespread and common there. Local wintering numbers may vary considerably from year to year or even within a particular winter. For example, in Sevilla, the Isla Menor alone attracted 3,300 on 1 November 2005 (*A. O. Sevilla 2000–2005*: 89): 2.8 times the entire provincial population of January 2009.

Table 36. Regional distribution of wintering Black-headed Gulls in Spain in January 2009. After Cantos (2009).

Region	No. of birds
Catalonia	68,960
Madrid	64,070
Castilla-La Mancha	63,527
Comunidad Valenciana	52,221
Extremadura	24,619
Andalucía	23,135
Galicia	18,671

Region	No. of birds
Asturias	6,501
Castilla y León	3,799
Basque Country	3,548
Aragón	1,516
Balearic Islands	1,484
Murcia	992
Cantabria	477
TOTAL	330,392

The wintering population is well dispersed but there are notable local concentrations. That around Madrid city has been especially well monitored and approximately doubled between 1980 and 1990, to 52,210 birds (Gómez & de Juana 1984, Cantos & Asensio 1990). Further increases occurred thereafter but wintering numbers in Madrid have fluctuated markedly between years: 125,500 were counted during winter 2007–08 (Cantos & Serrano 2009) but the 2009 census found just 64,070 birds. The continuing increase has been associated with increased dumping of edible refuse at landfill sites, to which the birds flock daily from their roosts at the Santillana reservoir, El Porcal gravel pits and other large water bodies outside the city.

Wintering birds have traditionally been associated with ports and harbours: especially fishing ports, estuaries and coastal wetlands. They follow fishing boats together with other gulls and they may be seen feeding over coastal upwellings offshore, as often occurs in the Strait of Gibraltar. They also congregate at sewage outfalls. Inland the birds that are increasingly attracted to rubbish tips have taken advantage of the proliferation of reservoirs that offer secure roosting sites. Black-headed Gulls follow the plough and also frequent tilled land and ricefields.

GREY-HEADED GULL *Chroicocephalus cirrocephalus*

(Gaviota Cabecigrís, Gaivota-de-cabeça-cinza)

Vagrant to Spain and Gibraltar.

There are two accepted records: an adult at Doñana from 30 June to 15 August 1971 (*Ardeola* 19: 22–23), and a juvenile at Gibraltar on 17 August 1992 (British Birds 86: 284). An adult has been reported from the Albufera de Valencia in June 2013 (RBS). This is a rare vagrant to Israel, Jordan and the Maghreb. The latter include birds seen in Western Sahara in May 1947; in Morocco in May 1968, November 1988, October 2006 and May 2009; in Algeria in April 1981 and in Tunisia in July and August 1988,. There was one in the Canary Islands in February 2005. All Palearctic records are highly likely to involve birds of Afrotropical origin.

SLENDER-BILLED GULL *Chroicocephalus genei*

(Gaviota Picofina, Gaivota-de-bico-fino)

Breeds locally at a few coastal locations in Andalucía and the Levant. Partly migratory and occurs on coasts near the breeding areas on passage and winter. Rare in the Balearic Islands and elsewhere in Iberia.

The Slender-billed Gull is characteristic of coastal wetlands where it typically selects salty or brackish lagoons, fish ponds, saltpans and shallow bays, where it feeds on small fish and invertebrates. It largely avoids ricefields and freshwater lakes and is seldom encountered at sea. A breeding presence in Spain has been known since the late 19th century. Irby (1895) refers to a nest with eggs at the Guadalquivir estuary in May 1879 and Chapman & Buck (1910) found two nests in Doñana in 1883. Noble (1902) found a number of nests in the marismas and commented on how these were robbed systematically by egg-gatherers. Valverde (1960) considered that no more than six pairs were still breeding there but that there may have been several hundred pairs 50 years earlier. Five nests were found in 1960 and nine in 1962 (*Ardeola* 6: 383, 8: 275–276). A second breeding colony

Figure 141. Changes in the breeding population of Slender-billed Gulls in Spain 1910–2007. After Bertolero *et al.* (2008).

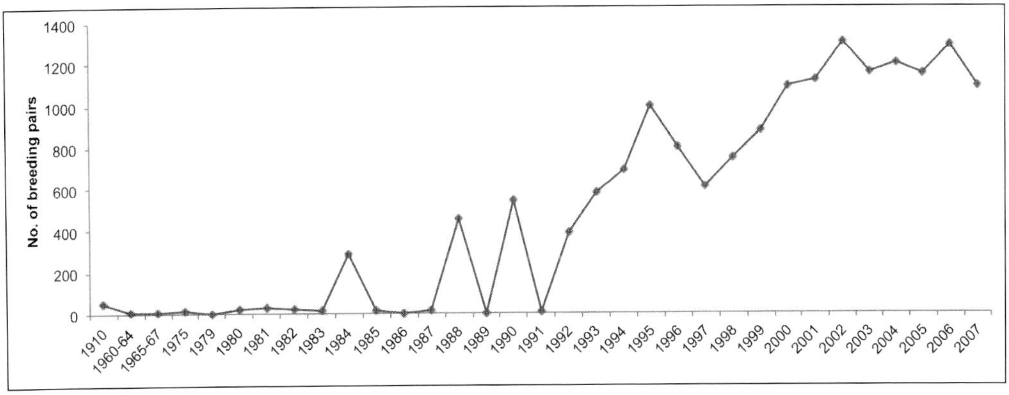

in Andalucía was thereafter discovered at the Laguna de Fuentedepiedra, with three nests in 1965 and 11 in 1967 (*Ardeola* 11: 154, 14: 167–174). Whereas the Andalucian population remained at low levels during the 1970s and 1980s, a new colony was formed at the Ebro delta in 1975 that rapidly increased: from 12 nests in 1975, to 24 pairs in 1980, 280 pairs in 1984 and 429 pairs in 1988, and the Comunidad Valenciana was colonised in 1991 (LR). Following steady increase during the 1990s, the Spanish population seems to have reached relative stability at 1,000–1,500 pairs since 2001 (Bertolero *et al.* 2008: Figure 141). The increase during the 1990s was paralleled by the population breeding at the Camargue, France, which reached 877 pairs in 2001 and which interchanges birds with the Ebro delta (LR).

Since 1998 the Slender-billed Gull has nested at only ten locations (Table 37), six in southern Andalucía: three of them in the Guadalquivir estuary; at the Ebro delta and at three major coastal wetlands of the Comunidad Valenciana. It is highly prone to re-siting its colonies within different parts of the breeding areas. It has used 24 different sites in the Ebro delta in a ten-year period although most recently it has nested at three sites: La Tancada lagoon, Punta de la Banya and Punta del Fangar, but not at all of them in any one year. The birds are also sensitive to disturbance, both by humans and by mammalian predators, and are prone to desert en masse if provoked. The breeding sites are on islets or slightly elevated ground at coastal wetlands, where flooding may curtail nesting in some seasons. The breeding season of 2008 was exceptionally wet and only eight pairs nested at the Ebro delta, where 242 pairs bred the previous year. In contrast, 2009 saw a notable increase in the breeding population of the Comunidad Valenciana to a total of 532 pairs: comprising 308 at the Lagunas de La Mata-Torrevieja and 224 pairs at the Albufera (*A. O. C. Valenciana 2009*: 79–80).

Table 37. Breeding sites and numbers of breeding pairs of Slender-billed Gulls in Spain 1998–2007. After Bertolero *et al.* (2008).

REGION/SITE	1998	1999	2000	2001	2002	2003	2004	2005	2006	2007	% of 2007 population
ANDALUCÍA											
Marismas del Guadalquivir	272	208	252	307	0	0	291	0	0	527	43
Veta La Palma, Doñana	0	0	0	0	300	332	48	281	522	47	4
Laguna de Fuentedepiedra	1	0	0	0	0	0	0	0	0	0	0
Cabo de Gata saltpans	0	0	0	0	30	50	0	0	0	0	0
Punta Entinas pools	0	0	0	0	0	0	0	135	225	131	11
Bonanza saltpans, Cádiz	0	0	0	0	0	82	102	0	44	0	0
CATALONIA											
Ebro delta	306	487	515	473	622	496	532	404	259	242	20
COM. VALENCIANA											
Albufera de Valencia	3	11	56	37	77	54	48	54	72	59	5
Santa Pola saltpans	162	170	169	0	247	120	153	0	15	0	0
Lagunas de La Mata-Torrevieja	6	0	98	304	30	103	131	414	423	214	18
TOTAL	750	876	1090	1121	1306	1237	1305	1288	1560	1220	

The Slender-billed Gull may be found in Spain all year round but a proportion of the population is migratory. Sizable counts are sometimes made near the breeding areas in Spain after the nesting season but numbers tend to decline through autumn and winter. For example, at the Santa Pola saltpans in 2009 there were 600 on 2 September, 400 on 23 October and 192 on 2 December (*A. O. C. Valenciana* 2009). However, the Lagunas de La Mata-Torrevieja, , the main breeding site in the Comunidad Valenciana in 2009, still held 599 birds on 4 December 2009 (*A. O. C. Valenciana* 2009: 80).

In general, migrants return to the breeding areas in March–April and depart in August–September. Small-scale passage has been reported at the Llobregat delta, Barcelona, chiefly during March–May and in August (Gutiérrez *et al.* 1995) and this may involve birds from the southern French colonies, where most are summer visitors (de Seynes 2011). Three birds ringed in France have been recovered in Iberia as well as three ringed in Ukraine.

Small numbers have wintered at the Ebro delta since the early 1980s (CWA) and locally elsewhere near other east coast colonies. January counts show that the wintering population at the delta grew from 70 birds in 1997 to average over 250 during 1998–2009, with 489 present in 2003 (*A. O. Catalunya*). Further south the Mar Menor and associated saltpans also sometimes hold noteworthy numbers in winter: for example there were 143 at San Pedro del Pinatar on 8 December 1990 (*Ardeola* 40: 97). Some have also been noted in winter in coastal areas near the Guadalquivir colonies, notably in Cádiz Bay, where there were 17 on 4 January 1998 and 21 on 28 December 1999, and 30–40 are estimated to winter at Odiel (*Ardeola* 45: 123, 47: 168; Gacio & Sayago 2011). Slender-billed Gulls are very seldom reported away from the Mediterranean coasts, the Strait and the Atlantic coasts of Cádiz and the south of Huelva. They are also rarely recorded in the Balearics although one or two are seen there on passage in some years.

The winter quarters of migrant Iberian birds are unknown but they seem likely to be in coastal West Africa, where the Slender-billed Gull winters from Mauritania to Guinea (Borrow & Demey 2001), with some also possibly in coastal North-west Africa, where small numbers winter on the Atlantic coasts of Morocco as well as on the Mediterranean shores of the Maghreb (BM, Malling Olsen 2003). There are 42 recoveries of birds ringed as pulli at Spanish colonies that largely indicate southward displacement after fledging: nine were found in France and there are five recoveries from Sardinia and one from Malta, but there are 27 from across North Africa: Morocco (6), Algeria (2), Tunisia (5), Libya (4), Egypt (1) and the Canary Islands (1). There are also eight recoveries from sub-Saharan West Africa: six from Mauritania and one each from Senegal and Gambia.

The first record from Portugal was of one killed at Figueira da Foz, Coimbra, on 10 May 1912 (Reis Júnior 1931) but all subsequent records have been in the south of the country. There were over 20 records, involving some 84 individuals, during 1983–2004, the great majority of them in Algarve, especially at the Castro Marim reserve (AP). Most of these records were between mid March and late May: a count of 32 at Castro Marim on 28 April 2000 being especially noteworthy (*Anu. Ornitol.* 1: 24). Some individuals occurred in summer and as late as October but there were few winter records there. A recent upsurge in records, nearly all of them at the Castro Marim and Tavira saltpans, Faro, has led to the removal of the species from the Portuguese rarities list since 2008 (CPR 2008–09). Counts at Castro Marim in 2010, peaked at 58 on 7 September and 73 on 26 September (*Anu. Ornitol.* 8: 71–72).

Rare occurrences in northern Spain include two from the Basque coast and eight records from coastal Asturias in April–November, (Paterson 1997, *A. O. Gipuzkoa 1999–2000*: 99, *Ardeola* 32:, 417, 56: 163). In Galicia, four were at Xunqueira beach, Pontevedra, on 10 May 1994 (*Ardeola* 44: 255). Inland occurrences are also exceptional. The generality of records outside the breeding range correspond with passage periods.

MEDITERRANEAN GULL *Larus melanocephalus*

(Gaviota Cabecinegra, Gaivota-de-cabeça-preta)

A very scarce, local but increasing breeding species in Spain. Common on passage and especially in winter, when abundant on Mediterranean coasts.

The Mediterranean Gull is widespread within the Mediterranean basin in winter but its core breeding distribution is centred on the Black Sea in Ukraine. However, it has increased greatly in numbers since the mid 20th century and expanded its nesting range westwards, so that it now nests in most European countries. The nearest large colonies to Iberia are in France, especially

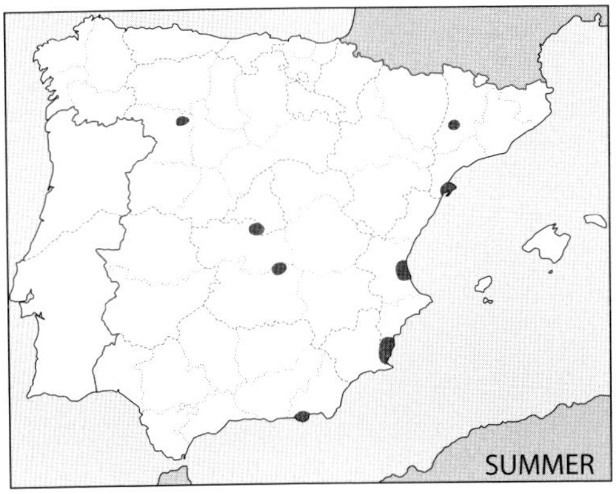

in the Camargue: where nesting was first recorded in 1965 and has been regular since 1981, with over 3,000 pairs present in 2006 (OF). An unsuccessful nesting attempt was made on the islet of Dragonera in the Balearic Islands in 1984 (*Ardeola* 32: 416) but the species has yet to breed successfully in the archipelago.

The first known successful nesting in mainland Spain was at the Ebro delta in 1988 (Martínez-Vilalta 1988) and up to three pairs have nested there irregularly since then. Chapman (1884) claimed to have found a gull's nest with eggs near which he shot an adult in Doñana in May, but the evidence given of a link between the bird and the eggs is inconclusive. The Mediterranean Gull has since been more successful at other Levant coastal sites, especially at the Albufera de Valencia, where 29 pairs nested in 2007 and 173 pairs in 2009, and it has also established a small breeding presence at a number of inland wetlands (Table MG1, *A. O. C. Valenciana 2009*: 78). The population was censused in 2007 (Molina 2009), when at least 52 pairs were found nesting in Spain. The majority were distributed between six sites on the east coast but single pairs bred at three inland sites and there were 3–5 pairs in Toledo. Nesting birds are often found within Black-headed Gull colonies but those that have attempted to breed in the Balearics have associated with Audouin's Gulls and those at the Ebro delta nest in a mixed colony of gulls and terns (Molina 2009). The nest sites are on bare or sparsely vegetated ground in saltpans, marismas, sandy beaches or lake margins.

The Spanish breeding population has been increasing rapidly since 2002 (Fig. 142) and the 2007 count was then the largest on record. However, a much larger number were found nesting in the Comunidad Valenciana in 2009: comprising 173 pairs at the Albufera, five pairs at El Hondo and two pairs at the Santa Pola saltpans (*A. O. C. Valenciana 2009*: 78). Inland, two pairs nested in the Black-headed Gull colony at the Laguna de Pétrola, Albacete, in 2011 and four pairs nested there in 2012 (*Ardeola* 59: 437). Non-breeding adults have been seen at other Black-headed Gull colonies, including at Doñana, and there is every indication that the still-small population will continue to increase.

Figure 142. The size of the Mediterranean Gull breeding population in Spain 1988–2007. After Molina (2009).

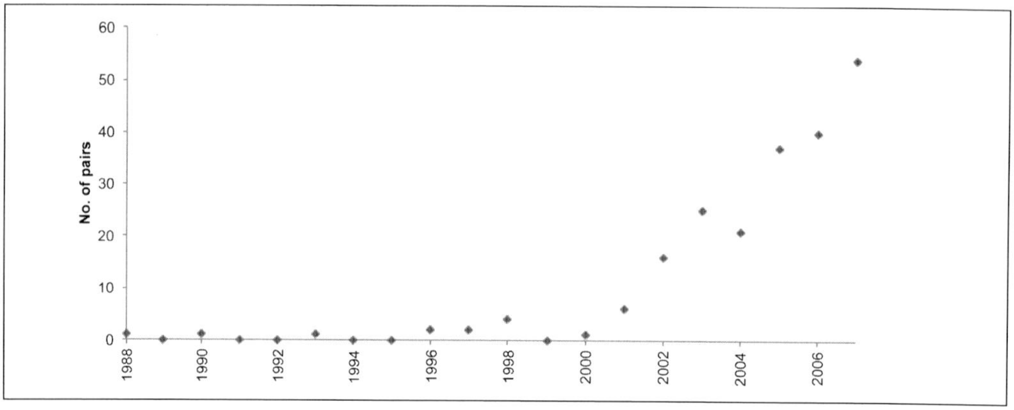

It is much more widespread and far more numerous on passage and, in particular, in winter. Mediterranean Gulls begin to appear in eastern Iberia from late June and there is a small westward movement at Gibraltar in July–August that includes adults, presumably failed or non-breeders (García 1973, García-Barcelona 2011). The principal arrivals on Mediterranean coasts are in October–November. Wintering birds begin to return east at least in March, some perhaps earlier, and few remain after the end of April. A small number of non-breeders remain in summer on all coasts, however.

A pan-Iberian census of wintering gulls and terns in January 1984 (Bermejo *et al.* 1986) confirmed what might be termed the traditional winter distribution of Mediterranean Gulls, prior to subsequent increases and range changes. There were 18,102 on the Mediterranean coasts: of these 58 were in Cádiz/Malaga and the remainder were all in the north-east, in Castellón (4,156), Tarragona (8,596), Barcelona (5,278) and Girona (14). In contrast, the census found just 177 birds on Atlantic coasts: the majority at the Tagus and Sado estuaries (53), in southwestern Portugal (50) and on the Cádiz Atlantic coast (73) and there were none on the Biscay coast, in Huelva or in south-eastern Algarve and only one bird in Galicia. There were no inland records.

The importance of the Levant coast for wintering Mediterranean Gulls was noticed by Isenmann (1976), who counted 9,000 in the Comunidad Valenciana and Catalonia, although it was already known that 26 of 41 ringing recoveries up to the mid 1960s were from this region (AMI). More extensive census work during winter 1976–77 and in January–February 1979 led Carrera *et al.* (1981a) to estimate the total wintering population to be c.15,000 birds, the great majority in Catalonia, and in Castellón province in the Comunidad Valenciana. The annual January counts of aquatic birds during 1990–2009 have confirmed the greatly increased abundance of Mediterranean Gulls in Spain: mean totals increased from just 277 during 1980–89 to 3,589 during 1990–99 and 9,008 during 2000–09 (AAE). Over 90% were on the northeastern Levant coast, from the Ebro delta northwards. Here the largest concentrations were at Cambrils harbour (up to 19,150 birds), the Ebro delta itself (up to 14,652), the Llobregat delta (up to 10,343) and Tarragona harbour (up to 4,110) (AAE). The harbours involved include fishing ports that attract large numbers of gulls of different species. A more recent survey (Cama *et al.* 2011) put the mean Spanish wintering population during 2005–09 at about 41,000 individuals – nearly half the known global population – the principal site being the northern Catalan coast in the Gulf of Sant Jordi, which extends from the Ebro delta north to Cambrils and Tarragona. Here the mean wintering numbers during this period averaged 17,000 birds but peaked at up to 45,000 during late winter and early spring.

In the 1980s and 1990s the wintering populations further south in Mediterranean Iberia, from Valencia to the Strait, were considered to be much smaller than in the northern Levant sector, with only small numbers present in Alicante, Murcia and on the Costa del Sol (Paterson 1997). Nevertheless, Isenmann (1976) counted 5,000 Mediterranean Gulls at Fuengirola, Málaga, in January 1975 and 4,000 were on flooded fields near the Río Guadalhorce, Málaga, in January 1989 (Paterson 1997). The January censuses of aquatic birds in Spain found only 714 birds on average in the whole of Andalucía during 1990–2009 and clearly overlooked large numbers that still occur then on the Málaga coast, at least in some years. In this area the fishing port of La Caleta de Vélez is highly attractive to gulls. There some 15,000 Mediterranean Gulls flew east on 2 February 2008 and 19,154 were counted in the same area on 16 February 2010, during easterly gales in both cases (*Ardeola* 57: 231). Large numbers have also been found raiding fruiting olive groves near the Málaga coast: up to 9,200 were counted at La Viñuela reservoir, Vélez-Málaga, 15km inland in January 2008 (*Ardeola* 55: 143), where they gathered after feeding on olives. Similar behaviour has also been reported from Tarragona, up to 10km inland from the Gulf of Sant Jordi (Carrera *et al.* 1981a, Cama *et al.* 2011). The olive-eating habit, and the attraction of Mediterranean Gulls to ricefields at harvest time (*Ardeola* 54: 178), further testify to the versatility of this species.

Very few winter in the Balearic Islands, and the recent January surveys found no more than two birds there (AAE). In general, the Mediterranean Gull is considered scarce in Mallorca and Ibiza and rare in Menorca, both in winter and on passage (*A. O. Balears 2010*: 264).

The Mediterranean Gull was scarce on Atlantic coasts until quite recently. A gull census of the Pontevedra coast, western Galicia, in December 1974 (Melendro & Rodriguez 1977) did not find any. It has since become regular and at times locally numerous in the north on passage and in winter. Noval (1986) considered it 'very occasional' in Asturias but 15 years later he described it as abundant on passage in autumn and a numerous winter visitor (Noval 2001). Only 13 Mediterranean Gulls were reported in central Asturias during 1983–86 (*Ardeola* 34: 286) but since then Gijón Bay has become a significant wintering area on the north coast and 300+ wintered there during the 1990s, with peak counts increasing to 628 in January 1997, and up to 726 thereafter (*Ardeola* 38: 340, 40: 97; *A. O. Asturias* 4; AAE). Further east, there has been a similar increase in the numbers found in the Basque Country, where Txingudi bay held 354 on 6 February 1999 (*Ardeola* 49: 188). Wintering numbers in Galicia are usually smaller: the January counts of 1990–2009 found 396 on average, but the Ría de Vigo and Cíes islands have together attracted as many as 3,944 birds (AAE).

In Spain Mediterranean Gulls are relatively scarce in the south-west, both in the Strait and on the coasts of Cádiz and Huelva (García *et al.* 1989, Garrido 1996, Tébar 1998). The largest numbers in the Strait are seen in winter, November–February, especially after periods of strong easterly winds, but counts of more than 50 birds are rare. However, Odiel has recently attracted concentrations of over 200 birds in winter and over 500 on passage (Gacio & Sayago 2011). There are occasional records of larger numbers inland in this region: 200 were at the Laguna de Medina on 3 January 1988 (GONHS).

The relative scarcity of Mediterranean Gulls in southwestern Spain contrasts with the increasing frequency of sizable wintering concentrations on Portuguese coasts, south to Algarve. Tait (1924) was only aware of four records for Portugal and it was considered a rarity there until the late 1960s, when it began to appear in increasing numbers (AP). The Mediterranean Gull has since become a regular wintering species on the Portuguese coasts, notably at the Tagus estuary and in Algarve. It was widespread by the late 1980s, when 660 were counted at Vila Nova de Milfontes, Odemira, on 1 March 1987 (AP). Significant recent winter counts include 2,200+ at Cape Raso, Cascais, in January 2005; 1,420 off Castro Marim, in December 2007; 1,800 at the Mira estuary, Odemira, in January 2008 and 4,000 at Largo da Guia, Cascais, in December 2008 (*Airo* 4: 79, *Anu. Ornitol.* 6: 61, 7: 82). The Portuguese wintering population south of Cape Roca during 2004–06 was estimated at 7,000–8,000 birds (Poot & Flamant 2006). Some Mediterranean Gulls are present on Portuguese shores as early as July but the largest numbers occur during October–February.

A few early migrants may also appear on Biscay coasts from late June but the principal passage there is in October–November. In autumn 2008 1,175 flew west at Cape Estaca de Bares in a concentrated movement chiefly between mid-October and mid-November (Sandoval *et al.* 2009). The equivalent counts in 2009 and 2011 found 619 and 1,721 birds respectively (Sandoval *et al.* 2010, A. Sandoval pers. comm.). In general, this species has become progressively commoner on passage at the Cape, where the vast majority are seen during the autumn passage period.

Very few occur far inland in Iberia, other than those that nest locally there, but occasional individuals are detected on the meseta, in Extremadura and in the Ebro valley, usually with other gulls at such sites as municipal rubbish tips. A count of 15 adults at the Valdecañas reservoir on 12 July 2011 was a record for Extremadura and recalls the summer passage of Black-headed Gulls across the region (*Ardeola* 58: 500). There were 122 records (250 birds) from Madrid during 1992–2007, with a marked increase during 2005–07: birds there occur from August to May, with peaks in December and March (Juan *et al.* 2007).

Bernis (1967: AMI) noted 41 Iberian recoveries, 36 on Mediterranean coasts and five in the Atlantic, all of birds ringed at nesting colonies in Ukraine. The recent establishment of the Mediterranean Gull as a breeding species in western Europe is reflected by the broader origins of marked birds found in the Peninsula. For example, 96 recoveries in Spain during 2002–09 involved birds ringed in 12 European countries other than Ukraine, from where just 21% originated. They included Italy (22%), the Low Countries (15%) and Greece (12%), as well as several birds each from France, Hungary, Poland, Serbia and Turkey and single birds from Romania, Britain and Germany. A recent survey of 1,125 colour-marked birds wintering at seven sites around Iberia during 2005–09 (Carboneras *et al.* 2010) has confirmed that different sub-populations tend to winter in discrete areas. Those birds wintering in northeastern Iberia are overwhelmingly (90%+) from the Mediterranean and Black Sea colonies. Those of Atlantic origin, from northwestern Europe, make up over 65% of winterers off Portugal and 73% of those found in Galicia. A perhaps surprising finding of this study was that 40% of 119 birds wintering in Málaga had originated from Atlantic colonies. Five colour-marked birds seen inland, in Madrid, originated from England, France, Belgium, the Czech Republic and Hungary: none were from eastern Europe (Juan *et al.* 2007).

LAUGHING GULL *Larus atricilla*

(Gaviota Guanaguanare, Guincho-americano)

Vagrant to Spain, Portugal and Gibraltar.

The Laughing Gull appears in Europe with some frequency. There are records of 193 individuals from Britain, 42 from Ireland and 41 from France. There are also 26 accepted records for the Azores, three from Madeira and seven in the Canary Islands, as well as six from Morocco, three from Mauritania and others from Senegal and Gambia.

Figure 143. Records by month, 1980–2008 (n = 50)

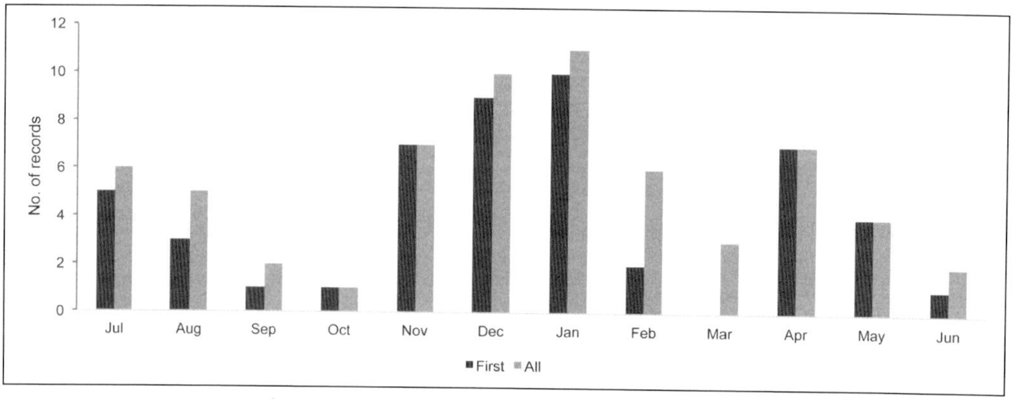

The earliest known Iberian occurrences were in 1980 when one was found at Baiona, Pontevedra, on 20 February, followed by another at the Ensenada da Ínsua from July to September (*Ardeola* 36: 120, 38: 158). Single birds were seen in 1981 at Fuengirola, Málaga, on 7 May (*Ardeola* 29: 186) and at Oporto on 6 June (Preiswerk 1981a). There were 41 accepted records for Spain, four from Portugal and three from Gibraltar by 2010. An influx into the northeastern Atlantic occurred in 2005 in the aftermath of Hurricane Wilma in November (Ahmad 2005) and over 40% of Iberian records are from 2005 or 2006. Nearly all have been of single birds but there are four of two together, all in 2005 or 2006. Recorded ages during October–March comprise 18 first-winters, eight second-winters and seven adults. During April–September there have been six first-year birds, four second-years and ten

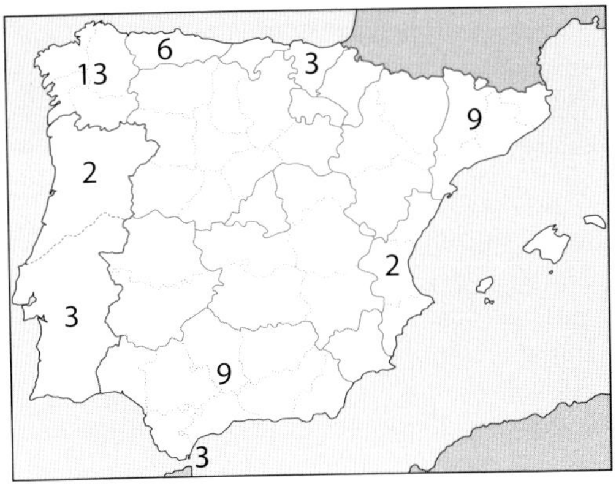

Figure 144. Records by region, 1980–2008 (n = 50)

adults. Five records from 2011–12: three from Spain and two from Portugal, include an exceptional inland record from Extremadura: an adult in breeding plumage at the Valdecañas reservoir on 14–20 June 2011 (*Ardeola* 60: 472, RBS).

The temporal distribution of Iberian records shows a concentration in winter, with half the first observation dates in November–January (Fig. 143), although this proportion falls to 37% if the exceptional 2005–06 influx is excluded. The proportion of winter records is still much higher than in Europe as a whole, where Laughing Gull records are more evenly distributed throughout the year (Dymond *et al.* 1989, Hoogendoorn & Steinhaus 1990). Possibly those that reach Europe tend to move south in winter, which also explains the late dates of most records in Morocco, the Canary Islands and Mauritania (de Juana 2006, Isenmann *et al.* 2010). Iberian occurrences are widespread (Fig. 144) but Atlantic coastal records predominate in autumn and winter and Mediterranean ones predominate in summer: there were 21 Atlantic records and eight Mediterranean ones during October–March and 12 Mediterranean ones and nine Atlantic records during April–September.

FRANKLIN'S GULL *Larus pipixcan*

(Gaviota Pipizcan, Gaivota-das-pradarias)

Vagrant to Spain and Portugal.

This is a rare vagrant to Europe where the records include 67 individuals in Britain, 11 in Ireland and 33 in France. There are a few from the Atlantic islands: four in the Azores; one in Madeira and one in the Canary Islands. There are also seven reports from Morocco, one from Mauritania and at least five from Senegal.

The earliest Iberian observation was an adult at the Guadalquivir delta ricefields at La Puebla del Río on 3 May 1978 (*Ardeola* 28: 158). There were 15 accepted records for Spain and four for Portugal by 2010: the first of the latter in 2002, all of single birds. Autumn–winter observations have included six first-winters, two second-winters and one second winter/adult. Spring–summer observations have included four first-summers and five adults. Adults predominate in European records as a whole (Hoogendoorn & Steinhaus 1990).

Figure 145. Records by month, 1978–2010 (n = 19)

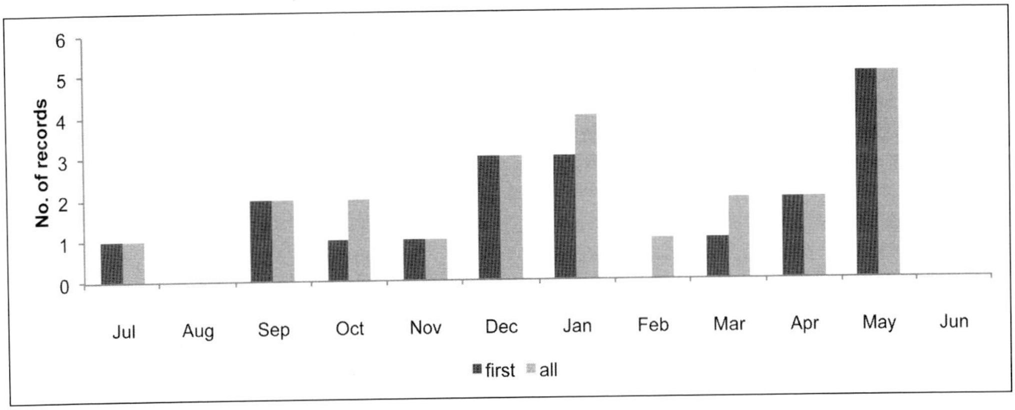

Iberian records are broadly distributed throughout the year (Fig. 145). They include three first dates in autumn (September–October), seven in winter (November–January), eight in spring (March–May) and only one in summer, in July: although there have been two more summer records subsequently: an adult on the A Coruña coast in June 2011 (*Ardeola* 60: 472) and an individual at the Sado estuary in July 2011 (RBP). Summer records predominate in the British Isles (Vinicombe & Cottridge 1996)

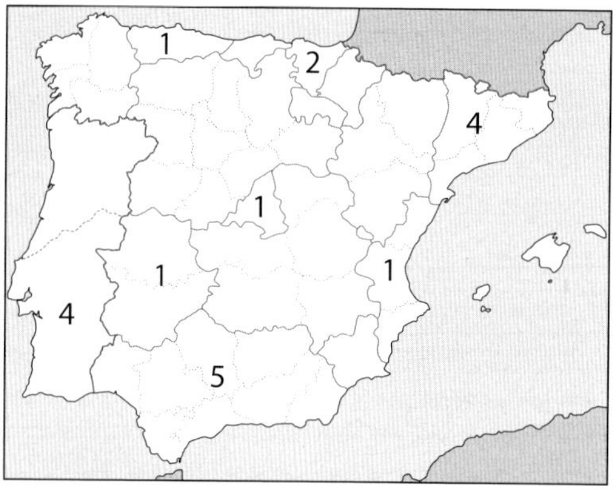

Figure 146. Records by region, 1978–2010 (n = 19)

whereas most in France are in winter (OF). Only two Iberian records suggest overwintering: first-winter birds were at the Llobregat delta from 26 December 1990 to 19 January 1991 and at Santurce harbour, Bizkaia, from 20 January to 19 March 2006. The concentration of Iberian records during passage periods may indicate the occurrence of birds moving to and from Africa: the four Moroccan records have first dates during November–February and the Mauritanian and Canary Islands records were in January. Iberian records show a tendency for southern locations, with half of them in Andalucía and southern Portugal (Fig. 146). The four from Catalonia contrast with the scarcity of observations from the French Mediterranean, from where there is only one accepted record up to 2010. There are two inland records: at Rivas-Vaciamadrid, Madrid, during January 2002, and at Malpartida de Cáceres, Cáceres, in December 2003.

AUDOUIN'S GULL *Larus audouinii*

(Gaviota de Audouin, Gaivota de Audouin)

A locally numerous breeding species on Mediterranean coasts and islands, and a scarce nester in southern Portugal. Common in Mediterranean waters and in the Strait, scarcer in southwestern Iberia and progressively rarer further north. Partial migrant.

The recent history of Audouin's Gull in Iberia is one of dramatic increases in population accompanied by colonisation of new western Mediterranean breeding sites that now accommodate the bulk of the global population. Until the mid 20th century this handsome gull was known as a rather scarce Mediterranean endemic that bred in small, scattered colonies, from the Balearic and Columbretes Islands, Corsica and Sardinia in the west to the Greek islands, Turkey, Cyprus and Lebanon in the east, as well as off the coasts of Tunisia, Algeria and Morocco. Information on its historical presence in Iberia is sketchy. Lilford found it at Alborán island in 1879 (Irby 1895). Saunders (1871) said that Natterer had observed the species at Tarifa and Arévalo (1887) mentions a specimen presumably obtained at Málaga. It was said to have occurred in the Balearic Islands by Saunders (1871), although he did not see it himself, and Munn (1943) found several nests there in a Yellow-legged Gull colony, most probably on Cabrera (Mayol 1978). It ceased to breed on Alborán at some time during the early 20th century and was then scarcely known in the region away from the small Balearic colonies.

The entire global population was estimated to be only 800–1,000 pairs by Makatsch (1969), including a then recently discovered colony of 400–500 pairs on the Chafarinas Islands, off the north-east Moroccan coast and part of the Spanish enclave of Melilla (Brosset & Olier 1966). Brosset (1959) had suggested that the species might nest on the Chafarinas and the size of the colony when this was confirmed suggests that it was established at least some years previous to 1966. By then also small numbers of Audouin's Gulls had been found wintering on the Moroccan Atlantic coast (Smith 1965). Audouin's Gulls also began to be recorded in the Strait of Gibraltar on post-breeding passage in 1967 and since then this has become a major and conspicuous movement.

The Iberian nesting colonies have been closely monitored during the past 20 years and much information is available on their numerical evolution and reproductive trends. The recent situation is given by Genovart *et al.* (2008) who summarise the Spanish population in 2007. Their account includes the Chafarinas, which are just outside our region, off North Africa in the south of the Alborán basin. However, we also consider that colony here since it is a key element in the western metapopulation and birds from there regularly occur on and off Iberian coasts. Its protection since the mid 1970s has certainly played an essential part in the global recovery of Audouin's Gull (de Juana & Varela 1993, Oro & Ruxton 2001).

The breeding population in 2007 (Table 38) was censused at almost 20,000 pairs: 17,575 pairs at Spanish Iberian colonies and 1,886 pairs in the Chafarinas. This total represented some 90% of the global population of the species. In addition, two small colonies have been established on the Algarve coast, the only known nesting colonies outside the Mediterranean. Breeding in Portugal first occurred at Castro Marim, where 12 pairs bred in 2001, and a second colony was established at the Ria Formosa in 2002. These Portuguese colonies held 41 pairs in 2005 (PA2).

The Ebro delta colony is by far the largest Audouin's Gull colony in the world and in 2007 it accommodated 65% of the global population. It was established as recently as 1981, when 36 pairs nested (Martínez & Carrera 1983), but since then it has grown by an average of 22% per year despite some notable fluctuations (Fig. 147). The initial rate of increase there has been shown to be due to large-scale immigration, thought to be largely from the Chafarinas (LR). Indeed, long-range dispersal has been a tendency of Audouin's Gull, perhaps only a recent one, that has enabled its rapid spread and colonisation of new sites in the western Mediterranean. The Ebro colony was the only one in Catalonia until 2009, when five pairs nested at the Llobregat delta, raising at least four chicks between them (GIAM 33: 9). This latter colony increased rapidly to 140 nests in 2010, 380 in 2011 and 546 in 2012: a further 87 pairs attempted to nest nearby in 2012 but the nests were lost during a storm (R. Gutiérrez, pers. comm.).

The Ebro Delta colony is at Punta de la Banya, at the southeastern extremity of the delta. This is a low-lying oval peninsula, 2,500ha in extent, whose only link to the mainland is along a narrow 5km-long sand bar. This has made it relatively easy to safeguard this protected area against human incursion and it also offers significant protection against mammalian predators (Oro *et al.* 1999). In 2007 the colony actually comprised 62 subcolonies, ranging in size from 1–1,102 pairs. The birds nest among the halophytic sand-dune vegetation that comprises the spit and on a large saltpan.

The Chafarinas Islands held the largest colony in the western Mediterranean until the Ebro colony was established. Numbers there increased from some 500 pairs in 1965 to 2,000 pairs during the 1980s, after which further increases led to a peak count of 4,300 pairs in 1992 (de Juana & Varela 1993). Since then the population has declined by about 3% per year, probably due to emigration to other sites, although the archipelago still houses the second-largest colony in the world. Predation by Yellow-legged Gulls was an acute problem there in the past (Bradley 1986) and was managed by a culling programme for some years, but this has long since been discontinued.

Table 38. Breeding colonies of Audouin's Gull in Spain and the Chafarinas Islands in 2007. After Genovart *et al.* (2008).

Region	Colony	No. of pairs	% of population
Andalucía	Alborán island	526	2.7
Catalonia	Ebro delta	14,177	72.8
Comunidad Valenciana	Albufera de Valencia	88	0.5
	Lagunas de La Mata-Torrevieja	450	2.3
	Columbretes islands	79	0.4
	Benidorm islet	1	<0.1
Balearic Islands	Isla del Aire islet, Menorca	125	0.6
	Porros islet, Menorca	37	0.2
	Cape Cavalleria, Menorca	2	<0.1
	Dragonera islet, Mallorca	100	0.5
	Mola de Andratx, Mallorca	30	0.2
	S'Estopa islet, Mallorca	3	<0.1
	Na Guardis island, Mallorca	233	1.2
	Cabrera island	39	0.2
	Cala Salada islet, Ibiza	245	1.3
	Sa Foradada, Ibiza	85	0.4
	S'Espartar islet, Ibiza	191	1.0
	Santa Eularia islet, Ibiza	24	0.1
	Espardell island, Formentera	360	1.8
Murcia	Isla Grosa	582	3
	Puerto Mayor	189	1
	Salinas de San Pedro	9	<0.1
Chafarinas islands	Chafarinas islands (NW Africa)	1,886	9.7
TOTAL		19,461	

Figure 147. The Audouin's Gull population of the Ebro delta, 1981–2007.

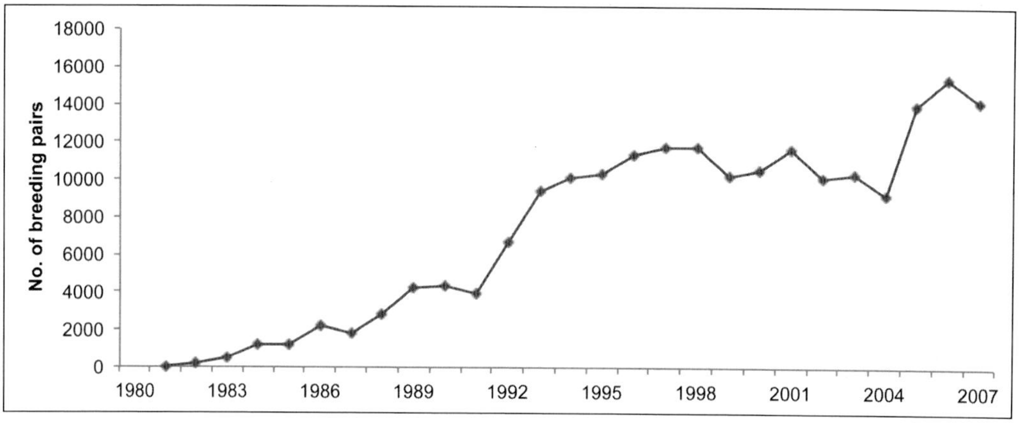

The Columbretes Islands, off Valencia, were colonised in 1974, when 45 pairs nested (Pechuán 1974). Numbers increased to 430 pairs in 1990 but thereafter they fluctuated wildly but largely declined, although they again peaked at 525 pairs in 1995 before decreasing to the lower numbers found recently: 79 pairs in 2007 and 120 pairs in 2009. This colony is strongly linked to the Ebro colony and the 1995 peak followed an influx of birds from there. The two mainland colonies of the Comunidad Valenciana: at the Albufera de Valencia and the La Mata-Torrevieja saltpans, Alicante, were founded respectively in 2003 and 2004. The Albufera colony was the result of an experimental introduction programme, initiated in 1992 with the release of captive-hatched juveniles from eggs originating from the Columbretes. The colony was established in an unexpected location, on a reedmace island within the Albufera lagoon (Martínez-Abraín *et al.* 2003), and it increased from seven pairs in 2005 to 88 in 2007 and 275 pairs in 2009. At La Mata-Torrevieja saltpans, some 30 adults and at least five nests were found in 2004 (*Ardeola* 51: 553); that colony then increased from 30 pairs in 2005 to 450 in 2007 and then increased sharply to a remarkable 2,826 in 2009 (*A. O. C. Valenciana* 2009: 81) and 3,823 pairs in 2010 (*Aves y naturaleza* 7: 27). Attempts to establish a fourth colony in the region by 'hacking' birds on to the Islet of Benidorm led to a few nesting attempts that failed due to predation by Yellow-legged Gulls (Oro *et al.* 2011). A fifth colony was discovered in 2009, when at least five pairs were found incubating at the Marjal de Almenara, Castellón, although all were unsuccessful (*A. O. C. Valenciana 2009*: 80).

The long-standing population in the Balearic archipelago (Mayol 1978, Muntaner 1998) is widely scattered on rocky coasts and, in particular, on rocky islets throughout the archipelago (Table AG1, Fig. AG2). There were 1,474 pairs distributed between thirteen colonies in 2007 but the gulls show a strong tendency to switch breeding sites between years (Oro & Muntaner 2000).

Isla Grossa, off Cape Palos, Murcia, first attracted Audouin's Gulls in 1989 when a few pairs were found there (LR). This colony increased to 300 pairs by 1993 and peaked at 1,200 pairs in 2001. The rapid increase was due to immigration from the Ebro delta colony. The colony has since declined as birds have moved to mainland colonies in Alicante and only 582 pairs were present in 2007. Two other colonies in Murcia are of very recent origin and both are on the mainland. A colony was found at Puerto Mayor, opposite Isla Grossa in 2006 and numbered 189 pairs in 2007. A small colony of nine pairs was also established at the San Pedro del Pinatar saltpans in 2007 (Genovart *et al.* 2008).

The only other Spanish colony is on Alborán island, some 90km south of Almería and 200km east of Gibraltar. There were none there in 1982 but birds returned in 1985 and four pairs nested in 1988 (Paracuellos & Nevado 1995). This population too has increased, to 200 pairs in 1999, 526 pairs in 2007 and 626 pairs in 2009. The increase accompanied the culling of Yellow-legged Gulls on the island, which reduced that population from over 400 pairs in the 1990s to about 100 pairs thereafter (Paracuellos & Nevado 2010).

The nesting cycle of Audouin's Gull begins in mid March when the birds settle at their colonies. Adult annual survival is high, c. 90%, but productivity is low, although it varies considerably both between colonies and years. For example, the main colony, at the Ebro delta, has produced 0.6–0.8 young per pair during most years, but it ranged from 0.2–1.1 during 1992–2007 and was practically zero in the catastrophic season of 2008, when 13,000 pairs fledged only 88 young between them, a consequence of both a commercial fishing moratorium and exceptionally heavy rain in May (*GIAM* 33: 2–7). Food supply is the principal factor determining breeding success everywhere but some variation has been due to predation, both by mammals and Yellow-legged Gulls. Productivity on Alborán island increased from 0.25 young/pair during 1990–2002 to c. 0.9 young per pair in 2006 following control of the Yellow-legged Gulls that also nest there (LR).

Audouin's Gull is a semi-pelagic species that specialises in the capture of sardines and anchovies, frequently at night, employing a shallow diving technique (Wallace 1969, Hoogendoorn & Mackrill 1987, Ruiz *et al.* 1996). Its remarkable increase in range and numbers seems to be the outcome of increasing reliance on accompanying fishing boats at sea combined with the availability of secure nesting sites near such fishing grounds. At least some of the observed variation in breeding success can be attributed to periodical changes in the fishing regime. Widespread breeding failure in 2009 is highly likely also to have been due to food shortage. Nevertheless, the species has shown itself to be surprisingly versatile when seeking food, especially outside the breeding season, taking not just other fish and squid at sea but also collecting grasshoppers and capturing migrant passerines in coastal Morocco near the Chafarinas, foraging for crayfish and water beetles in ricefields and – in Turkey - even raiding olive groves and peanut fields (de Juana *et al.* 1979, Witt *et al.* 1981, Paterson *et al.* 1992).

The movements of Audouin's Gulls, within Iberian waters at least, are well known and they have increased in magnitude and scope in recent years. The population increases have been accompanied by conspicuous post-breeding dispersal along Mediterranean coasts followed by large scale westward movements through the Strait of Gibraltar in late summer and autumn, to wintering grounds chiefly on the Atlantic coasts of Morocco, Mauritania and Senegal (de Juana *et al.* 1987, Oro & Martínez-Vilalta 1994a, Muntaner 1998, Sanpera *et al.* 2007, BM, Isenmann *et al.* 2010).

Birds ringed at the Spanish Iberian colonies (thus excluding the Chafarinas) have produced over 200 recoveries from North Africa, largely on Moroccan Atlantic coasts but including 14 from Algeria. There are also 59 recoveries from sub-Saharan West Africa: 45 from Senegal, 11 from Mauritania and three from Gambia. Those from the Chafarinas have produced numerous recoveries along the same coasts. Two probably exceptional long-distance recoveries within Europe of birds ringed at Spanish colonies are also noteworthy: one ringed as a nestling at the Ebro delta in June 1985 was recovered in Albania in August 1986 and another ringed on the Columbretes in June 1992 was reported from The Netherlands in May 1996. Conversely, very few birds ringed outside Iberia have been recovered in the region: five from Corsica and one each from Sardinia and Croatia.

The Mediterranean wintering population is almost entirely comprised by adults (de Juana *et al.* 1987) and these also predominate then in the Strait of Gibraltar. Winter observations on Spanish Mediterranean coasts indicate that usually only a small minority of the breeding population of the region remains there and most observations are of small numbers, but there have been more in some seasons. For example, no wintering birds were found in the Comunidad Valenciana in 1984 but 157 wintered at Torrevieja in 1991 and 1,900 were counted there in December 1995 (Jiménez & Carda 1997). At the same time the Columbretes were a regular wintering site where 1,000+ Audouin's Gulls were present in January during 1992–96: numbers there peaked annually in February, to a maximum of 2,600 in 1995, but thereafter declined considerably and there were only 173 birds in January 2010 (SWA). Colour rings show that the Spanish east coast wintering birds are mainly adults from the Ebro Delta, with a minority (23%) of Columbretes birds and just occasional records of birds from the Balearics or Chafarinas (Jiménez & Carda 1997). There is an unusual record of at least 200 birds that accompanied other gulls visiting the Pedrera reservoir, 14km inland from the Alicante coast, to bathe and drink on 5 January 1996 (*GIAM* 22: 5–6). January counts in Spain during 1990–2009 found 1,167 birds on average, with a maximum of 2,389, and they also indicated that almost the entire wintering population occurs on the Levant and southwestern coasts: the Torrevieja saltpans had the largest numbers found in those surveys, with up to 1,350 birds (AAE). A recent radiotracking study of 23 adult birds marked at Spanish colonies (Arcos *et al.* 2009) has confirmed the patterns established by visual observation. Some 70% spent at least part of the winter on African Atlantic coasts from Morocco to Senegal and most of the remainder wintered on the Moroccan Mediterranean coast, with only a couple of individuals spending part of the winter in Spain, on the Murcia coast.

The saltpans of Almería are a major staging area during the post-breeding migration: peak counts here have included 1,777 at the Cerrillos saltpans on 23 August 1991, 1,587 at Cape Gata on 16 August 1992, 2,976 at Cerrillos on 18 September 1992 and 1,000 at Punta Entinas on 20 August 2002 (*Ardeola* 38: 341, 41: 99,53: 287). The earliest post-breeding migrants at Gibraltar are immature non-breeders and these begin to appear from late June and predominate during July, although they make up

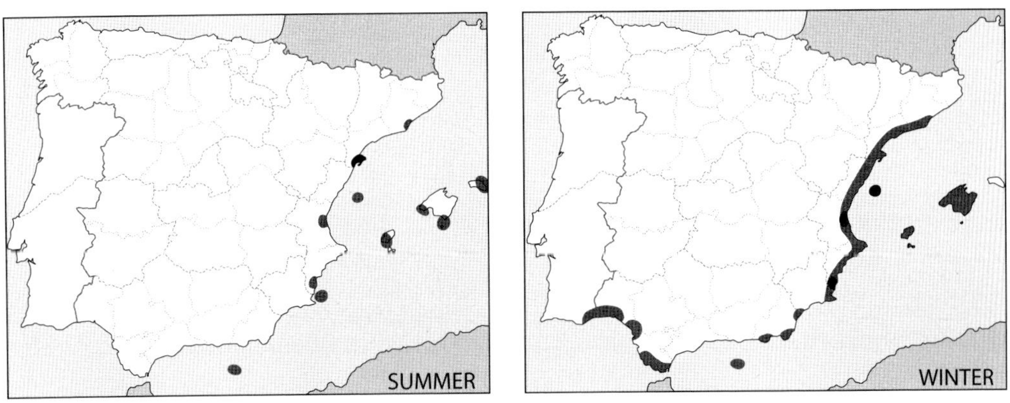

only a small proportion of the total passage. Adults, third-year birds and juveniles pass from late July to August, and juveniles predominate from late August to October (BSG).

Audouin's Gulls were still very scarce on the Atlantic coasts of southwestern Spain until the 1980s. They were first recorded in Doñana in November 1983 (García *et al.* 1989). Since then they have become frequent on the coasts of Cádiz and Huelva, especially in late summer and autumn, when flocks appear at such sites as Los Lances beach, Cádiz Bay and Odiel. Garrido (1996) then considered it frequent but scarce on the Huelva coast, both on passage and in winter, with a total wintering population that did not exceed 25–30 birds. It has since been recorded in increasingly large numbers, especially at the Odiel estuary: where recent peak counts have included 250 on 23 July 2000, 390 on 27 July 2002 and 860 on 2 August 2006. Increasingly large numbers remain in south-western Spain in winter instead of continuing south. Los Lances beach, is a regular wintering site: there were 40 there on 25 December 1987 and since then up to 200, mainly adults, have been present in December–January. More recently, 300–400 have overwintered at the Odiel estuary and over 1,000 may be found there on passage (Gacio & Sayago 2011).

Audouin's Gull was an extreme rarity in Portugal until the late 1980s, the only definite record being one taken near Cape St Vincent prior to 1914, although it may have nested on the Ilha da Baleeira, Sagres, in the 19th century (Tait 1924, *Airo* 10: 1–13). The first more recent Portuguese record was a third-year bird at Cape Raso, Cascais, on 30 October 1988 (*Airo* 2: 20) and this was followed by increasingly frequent records of small numbers of birds, nearly all of them juveniles in August and September, on the Algarve coasts. The species was considered by the Portuguese rarities committee during 1995–98, when 33 records were accepted. Such observations have continued post 1998, with juveniles especially appearing in small numbers along the entire Algarve coast in August–September, with a few coasting north to the beaches of Estremadura and exceptionally as far as Aveiro (*Anu. Ornitol.* 2: 35, 3: 37; RBP).

Large post-breeding concentrations have also been seen in eastern Algarve in recent years, especially at the Tavira saltpans, where there were 214 (adults) on 30 September 2004, 176 on 31 August 2005 and 322 on 4 August 2007 and at Castro Marim, where there was an exceptional gathering of 582 on 6 August 2010 (*Anu. Ornitol.* 4: 34, 5: 53, 6: 61, 8: 72). Gatherings of 300 at the Castro Marim saltpans on 28 June 2008, 214 at Tavira on 19 July 2008 and 240 also at Tavira on 16 July 2010 (*Anu. Ornitol.* 7: 82, 8: 72) were comparatively early and may have involved non-breeders. Audouin's Gulls are rarely reported from Portugal in winter, although a few remain then at the Ria Formosa (AP): the 2009–10 winter survey of non-estuarine coasts in Portugal found 15 individuals (Catry *et al.* 2010b). Here too the summer/autumn gatherings chiefly depart from late September onwards.

The return passage from the Atlantic through the Strait of Gibraltar extends from February to June but is clearly rapid and much less conspicuous than the autumn movements. Adults occur from February to early April but the later arrivals are chiefly of subadult or immature non-breeders. Spring passage is also less conspicuous than the autumn passage up the Moroccan Atlantic coast and may occur well offshore (BM).

A feature of the population increase has been the unexpected appearance of Audouins Gulls inland and on northern coasts. Small numbers are involved but the trend seems set to continue. There are at least 77 published records (190 birds) away from the regular coastal haunts during 1983–2009 (*GIAM* 24: 8–9, 28: 2–6; *Ardeola*, *Anuário Ornitológico*). They include a few from the north coast: four from Galicia, three in the Basque Country and at least one each in Asturias and Cantabria; three from Aragón, four from Castilla y León, 25 from Castilla-La Mancha, eight from Madrid, five from Andalucía, two from Portugal and 22 from Extremadura. Two ringed birds seen on the Biscay coasts, in Gipuzkoa, had originated from the Ebro delta (Gutiérrez & Guinart 2008). Most of the birds involved are juveniles on post-fledging migration (Fig. 148); spring observations are much scarcer and winter records are rare.

Figure 148. Monthly distribution of inland and northern coastal records 1983–2009 (n = 166 birds).

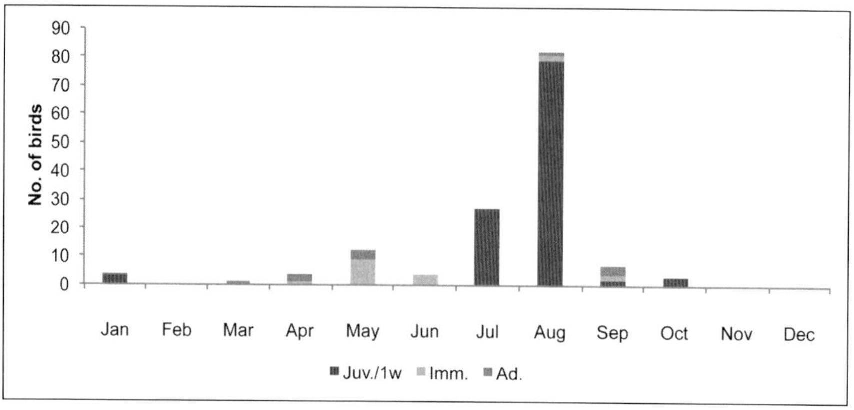

Sources: Marchamalo & Blanco 1998, Ocio & Sánchez 2007, *Ardeola*, *Anuário Ornitológico*.

The observations from Castilla-La Mancha are mainly of birds that appear in spring at the Lagunas de Alcázar de San Juan, Ciudad Real and post-breeding at the Finisterre reservoir, Toledo. The Madrid records principally involve juveniles found among other gulls at the city rubbish dumps from July onwards, some apparently lingering into the winter months (Juan *et al.* 2009). The sightings from Extremadura account for 63% of the minimum number of individuals recorded. They chiefly arise from what now seems to be a regular passage of juvenile Audouin's Gulls along the Guadiana valley in July and August, where they have been recorded annually since 2000 at Mérida and the nearby Los Canchales reservoir (*Ardeola* 51: 553). Ocio & Sánchez (2007) have suggested that these and other inland records involve birds that follow dispersing Black-headed Gulls and other gulls that habitually migrate along the Iberian river valleys.

RING-BILLED GULL *Larus delawarensis*

(Gaviota de Delaware, Gaivota-de-bico-riscado)

Very rare. Recorded in winter, mainly in northwest Spain and in Portugal. Recorded in Gibraltar.

The Ring-billed Gull was not recorded from the western Palearctic until the mid 20th century, when there were ringing recoveries in the Azores in 1945 and in Spain in 1951 and 1965 (Dennis 1981). Later, vagrant individuals were reported, chiefly in the British Isles from 1973 onwards (Vinicombe 1985, Hoogendoorn & Steinhaus 1990). Thereafter the number of European records increased very rapidly, in parallel with population increases in America. Thus by 2003 there had been records of at least 1,525 individuals in Britain. There were 365 records from France by 2005, involving some 200 individuals. There are also a considerable number of records from the Atlantic islands, including 21 from Azores, where it is no longer considered a rarity, and 17 from the Canary Islands. At least 52 have been found in Morocco, and one in Senegal, in October 1985 (Baillon & Dubois 1992). The temporal and spatial distribution of Ring-billed Gull records in Europe now corresponds to regular wintering rather than vagrancy (Malling Olsen & Larsson 2003, de Juana 2006, Vaslin 2010) although the numbers involved vary considerably from year to year.

The initial ringing recoveries in Iberia were all in Spain in January: at Vigo, Pontevedra, in 1951; Barbate, Cádiz, in 1965, and the Ebro delta in 1981, and all three had been ringed on the Canadian Great Lakes (Dennis 1981, 1987). The first sight record: of a first-summer near Lisbon in April 1981 (Moore 1983), was followed by an adult at Gibraltar in December 1983 (*Alectoris* 7: 53) and a second-winter at the Ebro delta in February 1984 (Llimona & Matheu 1986). Another adult was found in Gibraltar in March 1995. The Portuguese rarities committee considered the species until 2002, by when there had been 70 records, and the Spanish committee did so until 2005, by when there had been 176 records involving 235 individuals. We have considered these and other published records, from *Ardeola* and the *Anuário Ornitológico*, up to

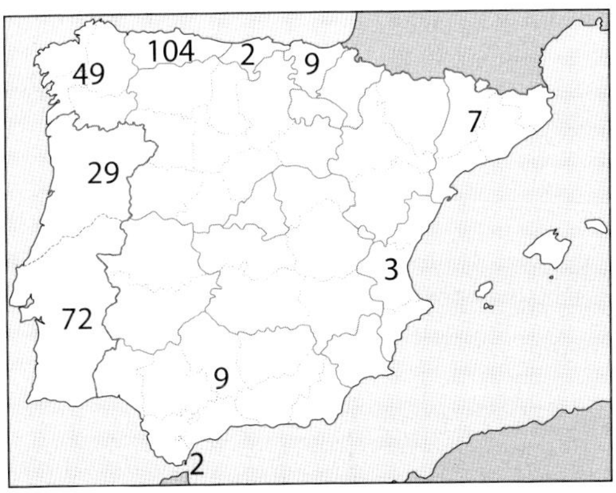

Figure 149. Records by region, 1981– 2007 (n = 286)

Figure 150. Records by month, 1981–2007 (n = 286).

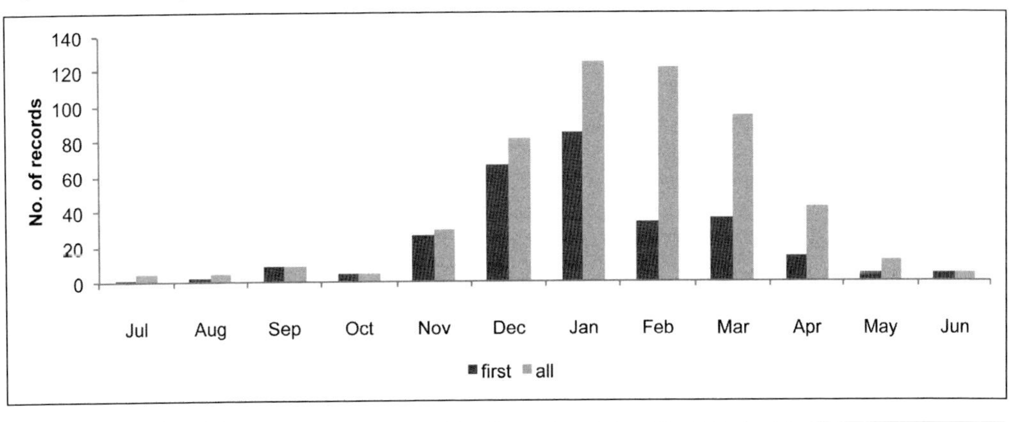

2007: giving a sample of 286 records. The number of individuals involved cannot be known precisely since some return to particular locations in successive years (García Sánchez 1993): they included individuals that wintered at Castropol, Asturias, from 1990 to 1998, and at Muxía, Galicia, from 1992 to 1998.

The Iberian data, as that from elsewhere in Europe (Malling Olsen & Larsson 2003), shows that the proportion of young birds is greater in some winters than in others. This was the case in 1989–90, 1991–92 and 2000–01, for example, and may indicate small irruptions. Such winters tend to be followed by a relative abundance of second-year birds the following winter, and of adults the winter after that, which supports the hypothesis that some birds return in successive winters (de Juana 2006). Immature birds tend to travel further south in Iberia. Ring-billed Gulls found in northern and northwestern Iberia during December–February comprised 53% first-winters, 11% second-winters and 36% adults (n = 181). The corresponding proportions in the south were 64% first-winters, 8% second winters and 28% adults (n = 60). Adults and immatures are equally frequent in the British Isles (Hoogendoorn & Steinhaus 1990) but only one of 21 individuals found in the Canaries was an adult.

Records are concentrated on northern and western coasts (Fig. 149). Asturias accounts for a good proportion of the total although the figures may be biased by returning birds and a high density of observers. In the 1980s and 1990s, 49% of records came from Asturias and 27% from southern Portugal, but post-2000 26% have been in Galicia, 23% in southern Portugal, 19% in Asturias and 15% in northern Portugal. There are only 14 records (5%) from the Mediterranean and four of those were in the Strait region.

There are records year-round but winter observations clearly predominate (Fig. 150): over half of first observation dates fall during December–January and 86% are during November–April. A similar pattern is seen in France, but not in Britain, where autumn and spring peaks suggest migratory movements (Dymond *et al.* 1989). In Iberia, adults arrive later on average and depart earlier, and the few summer records always involve immature birds. The earliest arrival date for a first-winter bird is 5 November.

In Spain more recent winters have produced nine birds in 2008–09 and 15 in 2009–10 (SWA). In Portugal there were reports of at least seven in winter 2008–09 and 12 in 2009–10. A new development has been the presence of three birds well inland: one that overwintered on the Río Guadiana in Badajoz in 2008–09; one at Alcázar de San Juan rubbish tip, Ciudad Real, during February–March 2009, and one at Pinto rubbish tip, Madrid, in February 2010.

COMMON GULL *Larus canus*

(Gaviota cana, Famego)

***L. c. canus*. Regular but scarce in winter in mainland Iberia and the Balearic Islands.**

The nominate race winters in large numbers in northwestern Europe but relatively few occur as far south as Iberia. It is commoner in some winters than in others since mid-winter influxes occur when cold weather affects the principal wintering areas further north (AMI). The wintering population in Spain has been estimated at 200–300 birds, although a 75% decline may have occurred between 1991 and 2010 (SWA). Up to the mid 1960s 54 (78%) of 69 ringing recoveries were on Biscay and Atlantic coasts, including those of Portugal and western Andalucía; five were from the Catalan coast and ten were widely scattered inland (AMI). Most Spanish recoveries are of first-winter birds ringed as pulli around the Baltic in Scandinavia, northern Germany, Finland and the Baltic States, as well as a few from the Low Countries, the British Isles and Italy, and found in Galicia or the Biscay coastlands.

The Common Gull is most often encountered on Biscay coasts and the largest numbers reported there are in Asturias, where the annual wintering population totalled 100+ birds in most years up to 1990s. However, there was an influx in January 1997, accompanying a cold spell further north, during which 868 were censused, with some numbers at most coastal inlets and reservoirs, including 127 in Gijón Bay and 116 at the Ría del Eo (*A. O. Asturias* 4: 141–143). As elsewhere, the numbers found in Asturias in winter have declined since the 1990s: the means of January censuses there fell from 257 during 1993–97 to in 2003–07, and only 38 were counted in 2008 (*A. O. Asturias 2008*).

It is annual in winter elsewhere on Biscay and northwestern coasts but there are comparatively few records from Cantabria and the Basque Country. Recent counts from Galicia suggest that 50–100 birds are present in the Rías in winter. The January 1997 influx saw over 100 in Galicia, including 25 at the Laguna de Valdoviño, 12 at Baldaio beach and 56 at the Cecebre reservoir, and also 21 at Umia–O Grove Bay (*Ardeola* 45: 249). The gathering at the reservoir may have been exceptional but the above and other sites have seen similar gatherings in other recent winters: for example, there were 49 at the Ría de Arousa in January 2002 (*A. O. Galicia 2002*: 81). There is no significant visible passage at Cape Estaca de Bares, where, for example, just seven birds were recorded during October–December 2008 (Sandoval 2009).

There are annual winter records in Portugal where it is reported with some frequency south to the Estoril coast. Most records involve just a few individuals but there are occasional midwinter reports of small flocks. For example, there were 16 at Aveiro on 24 December 2005 (*Anu. Ornitol.* 6: 62). Even fewer reach southwestern and southern Portugal, where the occasional records are generally of 1–3 birds. The 2009–10 winter survey of non-estuarine coasts in Portugal found 15 individuals (Catry *et al.* 2010b).

The Common Gull is extremely scarce or rare in winter on coasts of Huelva and Cádiz and in the Strait of Gibraltar (BSG, Garrido 1996, Gacio & Sayago 2011), where it is reported less than annually. Most records involve single individuals, rarely two together: a record of five at Zahara de Los Atunes on 13 December 1986 (*Ardeola* 34: 286) is exceptional. There are only four records for Gibraltar itself, the most recent being in 1990 (GONHS).

In the Mediterranean it is scarce but regular in Catalonia but less frequent in the Comunidad Valenciana and the Balearics. Munn (1941) saw one at Palma, Mallorca, in February 1939. Clavell (2002) gives 98 records (157 birds) for these regions during 1985–2000: 53 in Catalonia, 25 in the Comunidad Valenciana and ten in the Balearic Islands, but records have been more frequent in recent years, perhaps due to increased 'gull-awareness' among local birders. The regional bird reports give a further 169 records (211 birds) for the same regions during 2001–09. The wintering population in Catalonia during 2006–09 has been estimated at 38–48 individuals (CWA). In the south the Common Gull is almost unknown in Granada and rare in Málaga, where again most records are of single birds.

It is rarest of all in the south-east. A study of seabirds on the eastern coasts of Almería during 1985–97 did not detect the species there at all. Since then there is an unseasonal record of an adult at the Cerrillos saltpans in July and September 2002 and of a first-winter bird at the same site in September 2006 (*Ardeola* 51: 553, 53: 387). There were four records of single birds from Murcia during 2008–10.

First-winter birds greatly predominate in observations from Atlantic coasts. There were 1,456 birds observed in Asturias during the course of 1997, which included the January influx: where the age was noted, nearly all were first-winters and there were only 17 second-year birds reported and 12 adults (*A. O. Asturias 2003*). Similarly, the age was noted for 65 of 116 birds reported from Portugal during 1991–2007 and only three were adults. Adults make up a noticeably greater proportion of the records from the Mediterranean and from the south-west. Considering only those records in which the birds' age was recorded, there were 80 adults (32%) among 247 birds from the Levant during 1985–2010 and 15 adults (44%) among 34 birds from Andalucía and Gibraltar during 1984–2010.

The great majority are recorded during November–February, and many seem to arrive in January. They arrive in the principal wintering region in Asturias mainly during October–November and depart in March–April, with stragglers in May. There are nonetheless occasional records of summering individuals in all regions. The temporal distribution in the Mediterranean (Fig. 151) is broadly similar although there are indications that a few arrive earlier in autumn, in August–September, and linger longer in spring.

Figure 151. Records by month from Catalonia, the Comunidad Valenciana and the Balearic Islands during 1985–2010 (n = 202).

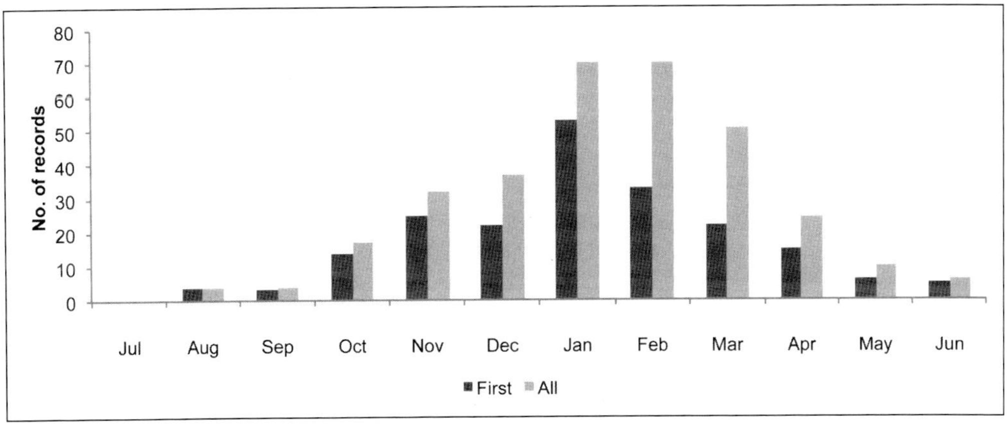

Common Gulls are often recorded on beaches, in ports and harbours and on lagoons and reservoirs near coasts, generally with other gull species. Few occur far inland but there are occasional records, usually of single birds in winter, at rubbish tips, lakes and reservoirs, from Madrid, Castilla-La Mancha, Castilla y León, Extremadura and the Ebro valley. An observation of 20–25 birds, mainly immatures, with Black-headed Gulls at Móstoles rubbish tip, Madrid, on 25 January 1985 (*Ardeola* 32: 417) was exceptional: they appeared during a cold spell but did not linger since only one was found the next day. Most of the 97 records for Madrid up to 2009 involve just one or two individuals, over 80% of them first-winters, that have appeared between October and May with peak occurrences in January (Juan *et al.* 2009). Regional bird reports show a similar and marked January peak in inland records elsewhere.

LESSER BLACK-BACKED GULL *Larus fuscus*

(Gaviota Sombría, Gaivota-d'asa-escura)

***L. f. graellsii*. A scarce breeder on Spanish Atlantic coasts, especially in Galicia, and on the Berlengas Islands, Portugal. Abundant on passage and in winter, both on coasts and inland.**
***L. f. intermedius*. A scarce breeder at the Ebro Delta. Common on passage and in winter, both on coasts and inland.**
***L. f. fuscus*. Baltic Gull. Rare on passage and in winter.**

The Lesser Black-backed Gull is a recent colonist of Iberia. The earliest known breeding birds were two pairs on the Sisargas islands, Galicia, in 1973 (*Ardeola 22*: 139). There were 12–15 pairs there in 1974 and, since then, breeding colonies have been established on other Galician islands that remain its Iberian stronghold. A few now also nest at coastal sites in Asturias, Cantabria and the Basque Country. Several pairs were discovered nesting on the Berlengas Islands, off Cape Carvoeiro, central Portugal, in 1977 (Teixeira 1983) and a tiny population has bred there since then (PA2) with sporadic nesting also reported from Ilha do Peseguiero, Setúbal, and the Ria Formosa (AP). Several small colonies have recently appeared on the southwestern Andalucian coast: at the Marismas del Odiel (ten pairs in 2007) and Isla Cristina (three pairs in 2007), and in the Cádiz Bay saltpans (six pairs in 2008). The only known breeding birds in Mediterranean Iberia have nested at the Ebro delta since 1981 (Ferrer & Martínez-Vilalta 1986) and the delta now supports the second-largest Iberian breeding population. However, whereas the birds at all other colonies, on Atlantic and Biscay coasts, are of the northwestern subspecies, *L. f. graellsii*, the Ebro birds are considered to be of the central European subspecies, *L. f. intermedius*, distinguished by its darker mantle and upperwings (SA2). The most recent assessment of the Iberian breeding population was in 2006–07 (Palomino & Molina 2009), which found 293 pairs at 21 colonies, including 191 pairs in Galicia and 75 pairs in Catalonia. In Iberia the Lesser Black-backed Gull is at the southwestern extremity of its European range and it seems unlikely to become at all numerous as a breeding species in the face of competition with the extremely abundant and increasing Yellow-legged Gull.

The pioneer colony on the Sisargas peaked at some 300 pairs in 1992 but then declined steadily to c.100 pairs in 2005. The second largest Galician colony, on the island of Sálvora, was discovered in 1979 and increased to 80 pairs in 2004 but declined slightly thereafter, apparently because of predation by American Mink. A colony on the Islas Cíes, off southern Galicia, was established in 1974 and peaked at 11 pairs in 1977 before declining to extinction in 2006, as the Yellow-legged Gull population there burgeoned. Small colonies or isolated pairs, both pure and mixed, have been found at Yellow-legged Gull colonies on islands all around the Galician coast, as well as on the coasts of Asturias, Cantabria and the Basque Country, but these involve only a tiny number of pairs in total and nesting at many sites is sporadic. The sole Cantabrian breeding record in 2006 was of a pair that nested on a rooftop in Santander. The Portuguese colony on the Berlengas comprises only a handful of pairs. It remains to be seen whether the newly established Andalucian colonies will endure or prosper and whether new breeding sites will appear. The Ebro population also increased rapidly from its initial establishment and it peaked at some 170 pairs in 1995 before declining to 75 pairs in 2007; the principal colony here is at Punta de la Banya (71 pairs in 2007) with a small number on the Isla de Buda (4 pairs). Mammalian carnivores have been implicated in that decline.

The Iberian peninsula is a major wintering region for the two western subspecies. Very large numbers also migrate along Atlantic coasts to and from wintering grounds on the Atlantic coasts of North-west and West Africa. The scope and extent of these movements has changed, however, since the mid 20th century, with increasing numbers now remaining to winter in northwestern Europe. These changes have occurred against a background of a continuing increase in the global population (BE). The Iberian wintering population was censused in January 1984 (Bermejo *et al.* 1986), when 50,028 birds were found in

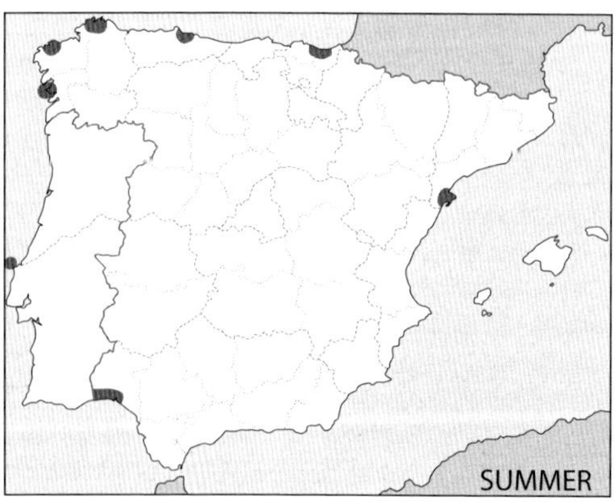

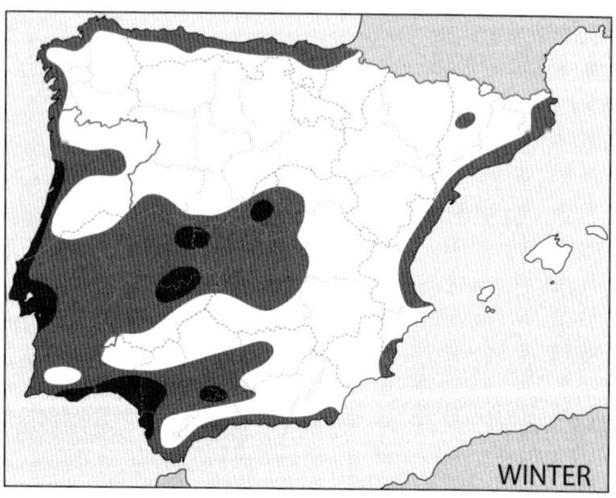

Table 39. Regional distribution of the wintering population of the Lesser Black-backed gull in Spain. January 2009. (After Molina 2009). No data was available from La Rioja and Navarra but few if any winter there.

Region	No. of birds	% of total
Andalucía	146,127	46.07
Madrid	66,814	21.07
Castilla-La Mancha	41,550	13.10
Extremadura	36,781	11.60
Comunidad Valenciana	6,905	2.18
Catalonia	6,174	1.95
Asturias	5,165	1.63
Galicia	4,708	1.48
Castilla y León	1,231	0.39
Cantabria	1,176	0.37
Basque Country	536	0.17
Murcia	6	<0.1
Balearic Islands	4	<0.1
(Gibraltar)	3	<0.1
Aragón	0	0.00
Total	317,180	

Spain and a further 48,880 in Portugal: the latter especially at the Tagus/Sado estuaries, which held 32,880 birds, 67.3% of the Portuguese total. A second national census in Spain in January 2009 located 310,180 birds, as well as 12,181 individuals, mostly first winters, that were either this species or Yellow-legged Gulls. In 2009 nearly half of the wintering birds were found in Andalucía and a further 45% were distributed between Madrid, Extremadura and Castilla-La Mancha (Table 39).

The population within those regions that held most birds in January 2009 was very unevenly distributed and just 12 sites held 70% of the national total. Five locations each held over 20,000 birds: Fuentedepiedra (44,000); the Santillana reservoir and El Porcal gravel pits, Madrid; the Castrejon reservoir, Toledo; and the Río Tinto marismas, Huelva. The reservoirs, gravel pits and lakes were used as roosts and drew birds from a wide area including, in the case of Fuentedepiedra, the Costa del Sol to the south. It is very likely that the distribution of the wintering population will continue to change as new feeding opportunities, such as new garbage tips, become available and previously used sites lose their attraction. An instance of this has been seen in Córdoba where the Pantano de Puentenuevo served as a roost for up to 20,000 birds that fed at Córdoba city rubbish tip, during winter 1987–88 (*Ardeola* 36: 251). This is one of the earliest recorded instances of large numbers wintering inland in Spain. In January 2009, in contrast, only 76 Lesser Black-backed Gulls were censused in Córdoba.

The Spanish wintering population increased by over 500% between 1984 and 2009, a massive increase but made even more striking by the equally dramatic regional redistribution of the birds that accompanied it (Table 40). In 1984 the birds were very largely confined to Atlantic coastal regions, with a modest proportion on Mediterranean coasts and a negligible presence inland. This was the same geographical distribution as that of the Iberian recoveries of birds ringed abroad up to the mid-1960s (AMI): only eight (3.2%) of 282 such recoveries were inland and just 15 (5.3%) were in the Mediterranean. By 2009 nearly two-thirds of the birds had moved inland, leaving a much smaller proportion on Atlantic coasts especially. The switch to inland sites, to take advantage of the rich pickings available at landfill sites especially, has a parallel in the Black-headed Gull (q.v.) but the change in Lesser Black-backed Gull distribution has been far more dramatic. It also contrasts markedly with the winter distribution of the Yellow-legged Gull, which remains overwhelmingly coastal.

Table 40. Coastal and inland distributions of wintering Lesser Black-backed Gulls in Spain in 1984 and 2009. Sources: Bermejo *et al*. 1986, Molina 2009.

Sector	January 1984		January 2009	
	No. of birds	%	No. of birds	%
Atlantic coasts	41,027	82.01	72,146	22.5
Mediterranean coasts	8,879	17.75	42,614	13.29
Inland	122	0.24	205,873	64.21
Total	50,028		320,633	

The Lesser Black-backed Gull continues to be widespread on all coasts in winter but wintering numbers in Galicia especially have declined dramatically. The 1984 census found 19,282 birds in Galicia, 38.5% of the Spanish wintering population, whereas only 4,708 birds (1.5%) were present there in January 2009. Major wintering concentrations have instead continued to develop at the Tagus estuary; the Portuguese coast: where 13,580 were censused in winter 2009–10 (Catry *et al.* 2010b); the Huelva and Cadiz coastlands: including the marismas of the Guadalquivir and their ricefields that extend into Sevilla province, and on the coasts of Málaga and Granada in Mediterranean Andalucía. The Lesser Black-backed Gull remains scarce in winter in the Balearic archipelago and on coasts north of Barcelona, but significant concentrations occur in Tarragona and Valencia especially: both of these provinces have ricefields that attract large numbers. Very few winter in Murcia or Almería.

The coastal winter distribution is substantially complementary to that of the Yellow-legged Gull, both in general and within each coastal sector, perhaps reflecting interspecific competition. As a rule, the Lesser Black-backed Gull is more abundant along low-lying and sandy coasts that are distant from the main Yellow-legged Gull colonies, which tend to be in areas with cliffs or on offshore islands and that continue to attract large numbers of Yellow-legged Gulls all winter. For example, the Lesser Black-backed Gull outnumbers the Yellow-legged Gull all along the western Portuguese coast and in Huelva, as well as, within the Mediterranean, on the beaches of Málaga, Granada and Castellón. In contrast, the Yellow-legged Gull is much more abundant along the Biscay coastlands, in Galicia, the western Algarve, eastern Cádiz including the Strait of Gibraltar, Murcia, Alicante and Valencia, and also in Catalonia and the Balearics (Fig. 152). Lesser Black-backed Gulls are almost absent in winter from some areas that have large Yellow-legged Gull populations: the 2009 census found only three Lesser Black-backs at Gibraltar, six in Murcia, nine in Girona and four in the Balearics.

Figure 152. Differences in abundance of Lesser Black-backed and Yellow-legged Gulls within different coastal sectors during the January 2009 census.

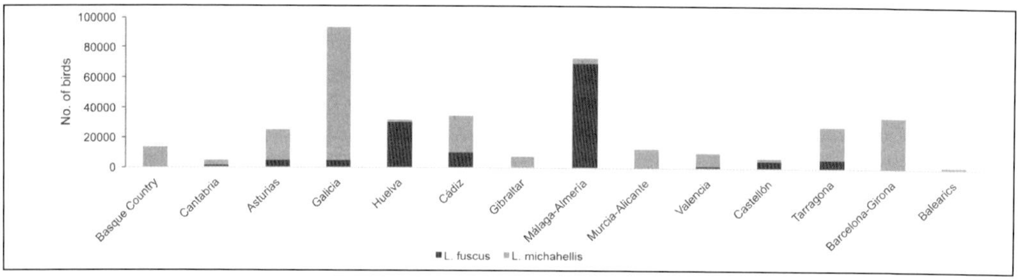

The massive inland wintering population is concentrated around Madrid; in Toledo and Ciudad Real in Castilla-La Mancha, and in Extremadura. Birds are thinly distributed or absent elsewhere. The big increases in numbers wintering in central Iberia are a very recent phenomenon. It began from 1989 onwards in Madrid and in 1990 in Extremadura (Traverso 2000, Cantos & Serrano 2009, Molina 2009). In Madrid, a marked increase was seen in winter 1996–97, when almost 4,000 birds were counted: since then the increase has been almost exponential, reaching 61,500 birds in 2008 and over 110,000 in 2013 (Serrano & Cantos 2013).

The results of ringing, marking and satellite-tracking studies (Pütz *et al.* 2007; Klaassen *et al.* 2012) show that a large proportion of the Iberian winterers originate from the British Isles and The Netherlands but there are also observations and recoveries of birds from France, Belgium, Germany, Denmark, Norway, Sweden, the Faeroes and Iceland. The relative abundance of *L. f. graellsii* and *L. f. intermedius* in Iberia in winter remains unclear, not least because these two forms are not sharply defined. However, a preponderance of '*graellsii*' is apparent on Atlantic coasts and '*intermedius*' is relatively more frequent in the Mediterranean, although '*graellsii*' still outnumbers it there. There is some evidence that birds of British origin are more frequent in Asturias and central Spain and that those of Dutch origin are commoner in Cádiz and Málaga (Álvarez Laó 2001). Similarly, *intermedius* ringed mainly in the Low Countries and southern Norway comprised three-quarters of colour-ringed individuals observed on the Málaga coast between October 2006 and May 2007 (*GIAM* 29: 2–7 2007) and most of those reported from Catalonia seem to originate from the Low Countries and southern Scandinavia (CWA). In a sample of 288 marked birds recorded mostly from Madrid and Badajoz, 58.8% were marked in the British Isles and 39.7% came from The Netherlands (Galván *et al.* 2003).

The globally scarce Baltic Gull *L. f. fuscus* winters in east Africa but there are a few recent observations of colour-marked individuals of known origin. Some earlier records of *fuscus* are unreliable: for example, the eight Iberian recoveries of Swedish-ringed birds assigned to *fuscus* by Bernis (AMI) originated from southwestern Sweden and so were probably *intermedius* birds. The Spanish ringing reports for 2002–10 record sightings of six marked birds of Finnish origin: at Palma de Mallorca in December and February, in Málaga harbour in December, in Galicia in April and at rubbish dumps in Madrid in October and December. There are also a recoveries of a Finnish-ringed bird from Esmoriz, Aveiro, on 10 September 2001 (AP) and a Swedish-ringed bird from Galicia. At least six observations of single birds in Spain during 2011–12 included one ringed in Poland, one from Norway and two from Finland.

Birds marked in the British Isles comprised some 85% of recoveries up to the 1960s (AMI) but only 38% of those found during 2002–07. At the same time, the proportion of Iberian recoveries of birds from the Low Countries increased from 0.7% during the first period to over 12% in the latter. There has also been a marked increase in Iberian recoveries of birds of Icelandic origin: there were none up to the mid 1960s but these accounted for over 12% during 2002–07. These changes seem to result both from a large recent increase in the numbers of Lesser Black-backed Gulls that overwinter in the British Isles and from huge increases in the breeding populations of The Netherlands and Iceland in particular.

Post-breeding movements are a conspicuous along Atlantic coasts in particular. The passage is prolonged but the early movements, in August–September, are small. Passage proper begins in October but peaks late in the season, during November and early December, even in the north. In general there is a large coasting passage westwards along the Biscay coasts and south along the Atlantic ones, with many thousands continuing south along the African Atlantic shoreline. A return passage on the same route begins in February and peaks in March, continuing through April.

HERRING GULL *Larus argentatus*

(Gaviota Argéntea, Gaivota-prateada)

***L. a. argenteus*. Scarce but regular in winter, chiefly on northern coasts.**

***L. a. argentatus*. Very scarce in winter.**

A recent analysis of both published and unpublished records (Álvarez Laó 2011) has done much to clarify the status, distribution and abundance of the Herring Gull in Spain at least. The northwestern European subspecies *argenteus* is largely sedentary or a short-distance migrant but small numbers reach Iberia, especially the northern coasts, in winter. The larger, darker, nominate race winters around the Baltic and North Seas south to northern France and the Low Countries; very few reach Iberia.The two subspecies intergrade around the North Sea producing some individuals of intermediate appearance (Malling Olsen 2003).

The Iberian distribution of the small wintering population is predominantly on the Biscay coasts, with increasingly fewer occurring further south on Atlantic coasts south to the Strait of Gibraltar. Rare individuals are also reported from Mediterranean coasts. There are at least 43 Spanish recoveries or observations of colour-ringed individuals: of birds ringed in Britain (23), France (9), The Netherlands (7), Denmark (2), Belgium (1) and Lithuania (1), chiefly from north and west coasts but including at least seven from Cádiz province and the area of the Strait and three from the Comunidad Valenciana. In addition there are 15 recoveries from Portugal: of birds ringed in Britain (7), France (5) and The Netherlands (3) (AMI, AP). An unknown proportion of all these recoveries may be unreliable since many birds have been ringed as pulli at mixed Lesser Black-backed/Herring Gull colonies, where young of the former may easily be misidentified as Herring Gulls but some are undoubtedly genuine: for example there is a Spanish recovery of a Herring Gull ringed as an adult in Shetland (Wernham *et al.* 2002).

Their occurrence on the Biscay coast has been well documented in Asturias, where a small wintering population of *argenteus* is mainly present from October to February (Fig. 153). Up to 40% are adults. A relatively large count in January 1997 corresponded with cold weather further north that displaced many other species into northern Iberia. Larger concentrations during that year included 41 in Gijón harbour and 24 at the Ría d'Aboño on 15 January; and 21 in Gijón harbour on 20 November and 19 there on 16 December (*A. O. Asturias* 4: 145–147). In general, 250–400 Herring Gulls are present in Asturias in mid January and the total mid-winter Spanish population is 500–700 birds (Álvarez-Laó 2011).

Figure 153. Monthly distribution of Herring Gulls wintering in Asturias during 1996–1998 (n = 844). Source: *A. O. Asturias* 3–5).

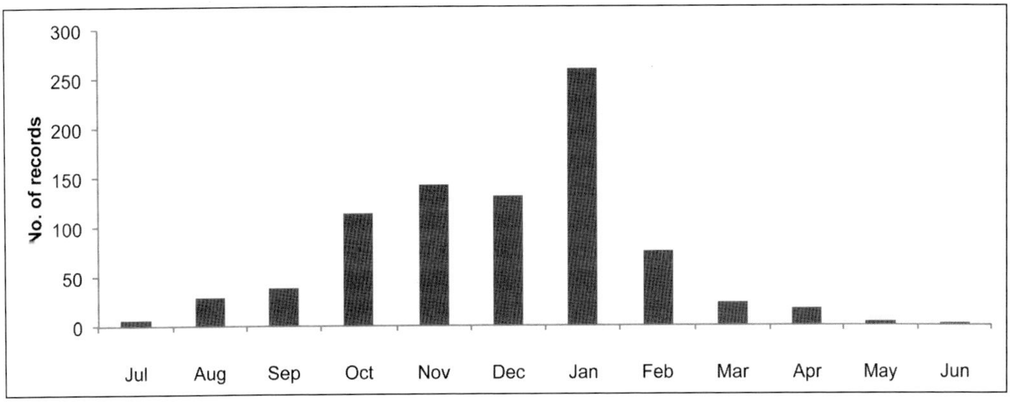

The northern race *L. a. argentatus* is a comparative rarity in Asturias, with only 12 definitely identified there prior to 1997, when there were at least nine, most of them part of the exceptional cold weather influx of other gulls. Up to five were present in Asturias during winter 2001/2002 but none appeared the following year (Álvarez-Laó 2011).

A few non-breeding immature Herring Gulls remain in Asturias, and probably elsewhere on northern and western coasts, in summer. Single adults were found apparently incubating nests in a Yellow-legged Gull colony at Pasaia, Gipuzkoa, both in 1993 and 1999 (Paterson 1997, *Ardeola* 49: 189): mixed pairings were likely.

Reliable records from western and southern Iberia are hard to find but the species has attracted very little attention until recently and it is almost certainly under-recorded. Thus the 'Noticiarios' in *Ardeola* and their Portuguese equivalents give just 23 records (33 birds) for Portuguese coasts and those of Atlantic Andalucía for 1991–2010, with extreme dates of 13 October and 7 February, except for an immature bird at the Marismas de Isla Cristina, Huelva, on 2 September. They comprised 25 adults, three second-years, four third-year birds and one first-winter individual, which suggests that young birds were being missed. At least a further 25 records (28 birds) from Portuguese coasts were reported during 2011–12 (RBP): these included at least six adults, five third-year birds, nine second-years, and eight first-winters.

The Herring Gull is uncommon in the Mediterranean. Clavell (2002) lists just eight confirmed records from the Spanish northeastern coasts and Balearic Islands during 1996–98, including four from the Ebro delta, two from the Barcelona coast and one each from Mallorca and Menorca. The Catalan Bird Reports list 60 records (63 birds) for 2000–09, many more than have been reported elsewhere in the Iberian Mediterranean. They comprised ten first-year birds, 26 older immatures and 27 adults. Five individuals were identified as *argenteus* and two as *argentatus*. Extreme dates were 14 August and 10 May but birds were generally present during October–March. About half the records came from well-watched Tarragona harbour, the remainder occurring at the Llobregat and Ebro deltas and other coastal sites. Some clearly returned in successive years: for example, a single adult was present every winter at Tarragona harbour from December 2001 until at least March 2009 (CWA). There are few recent records from the Comunidad Valenciana: just three during 2002–06 (*A. O. C. Valenciana 2000–2001–2002*, *Ardeola* 53: 204, 54: 179). Post-1998 records from the Balearics comprise eight recent winter observations of single immature birds from Palma harbour, Mallorca, during 2008–11: two were identified as *argentatus*. (*Ardeola* 56: 358, 57: 535–536, 58: 501–502).

A very recent development has been the occurrence of a few Herring Gulls inland, accompanying the large flocks of other gulls at rubbish tips and nearby roost sites. There were at least 25 records, involving a minimum of 11 birds, at Madrid tips, as well as one in Segovia, during 1991–2011 but mainly post-2008 (Juan *et al.* 2009). Some 75% were first-winters and the remainder were adults.

AMERICAN HERRING GULL *Larus smithsonianus*

(Gaviota Argéntea Americana, Gaivota-prateada-americana)

Vagrant to Spain and Portugal.

The American Herring Gull was first recorded in the western Palearctic in November 1937, when an individual ringed in New Brunswick landed on a ship 300 miles north-west of Cape Finisterre, Galicia (Dennis 1981). It was not reported again until one appeared in Ireland in 1986 but there have since been 90 Irish records, 27 in Britain and 12 in France, as well as seven from the Azores.

The first Iberian record was in Gijón Bay on 3 January 1991 (*Ardeola* 42: 118). The second was in southern Portugal, at Ludo, Faro, on 30–31 December 1992 *Ardeola* 42: 118). A meeting of gull enthusiasts at Foz do Douro, Porto, detected three separate individuals there during 30 March–1 April 2001 (Hoogendoorn *et al.* 2003). This record is serendipitous, since the birds coincided with some of the few people in Europe able to identify them, but the occurrence of three together raises the possibility that the American Herring Gull may be more frequent on the Portuguese coast than formerly thought, some perhaps being overlooked as European Herring Gulls (AP). There are three further accepted records from Spain: at Getxo, Bizkaia, from 28 January–19 March 2006; at Cariño, A Coruña, from 19–25 March 2007, and at Colmenar Viejo rubbish dump, Madrid, during 7–26 February 2010. One of the Foz do Douro birds and the Madrid individual were in second-winter plumage. An adult at Lires beach, A Coruña, from at least 28 December 2011 to 25 February 2012, returned there the following winter (*Ardeola* 60: 472–473). All other Iberian records have involved first-winters, with their distinctive very dark plumage, including one at Viana do Castelo in March 2012 (RBP). Most have been on Atlantic coasts and all between 30 December and 1 April.

CASPIAN GULL *Larus cachinnans*

(Gaviota del Caspio, Gaivota-do-cáspio)

Rare. Recorded in Spain in winter. Vagrant to Portugal (1).

The Caspian Gull is one of the 'yellow-legged gulls' that has recently attracted separate-species status. It has been found to occur commonly in winter in central and eastern Europe but it is rarer further west. It occurs regularly in small numbers in Britain, chiefly in summer and autumn. It was first identified in France in 1997 but at least 100 individuals are now thought to winter there, chiefly in the north. The former treatment of the Caspian Gull as a subspecies of the Yellow-legged Gull may have meant that it has been overlooked in the past. However, the growing number of gull enthusiasts in Iberia has no doubt facilitated the ongoing upsurge of records. The first Iberian observation, a first-summer bird at a rubbish tip in Beja (Baixo Alentejo) from 18–23 April 2006, remains the sole accepted Portuguese record. The Spanish rarities committee received reports of Caspian Gulls from 2005 onwards and had accepted 41 records up to 2010, about half of the submitted observations.

Figure 154. Records by month, 2005–2010 (n = 42)

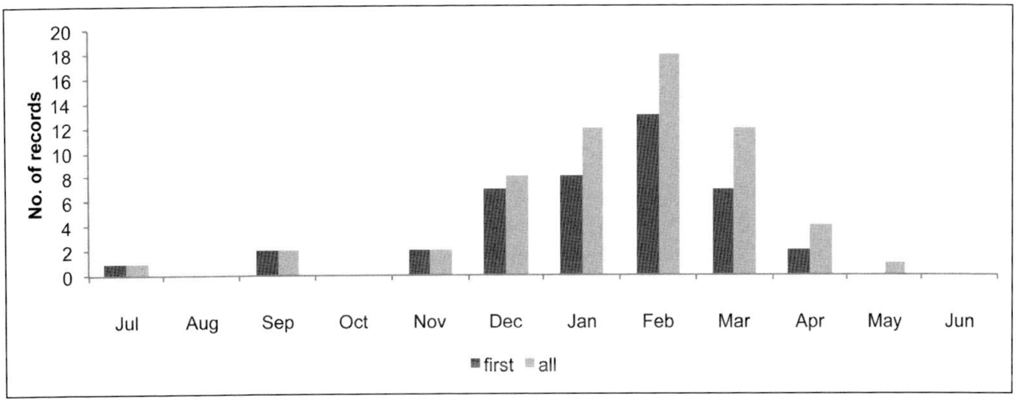

Allowing for possible duplications, there are thus at least 42 Iberian records, involving 49 individuals. None have been adults: they comprised 40 first-winters/first-summers, five second-winters/second-summers, two third-winters and one fourth-winter, this last being one that returned yearly to the Ría de Avilés, Asturias, during 2008–11 (*Ardeola* 59: 385–386). Observations peak in late winter (Fig. 154) and have tended to be concentrated on northern coasts (Fig. 155). Two were inland: at the Alcázar de San Juan rubbish tip, Ciudad Real, from 16 February to 1 March 2008 and at the Colmenar Viejo rubbish dump, Madrid, from 5 January to 14 February 2010. Most observations have been of single birds but there are three of two together and two of groups of three: at Vilanova i la Geltrú, Barcelona, in March 2006, and at Lago, Lugo, in January 2009. One at A Coruña, Galicia, had been colour-ringed as a nestling in Poland: it appeared as a first-winter in March 2006 and was seen again during January–March 2007 (*Ardeola* 58: 463).

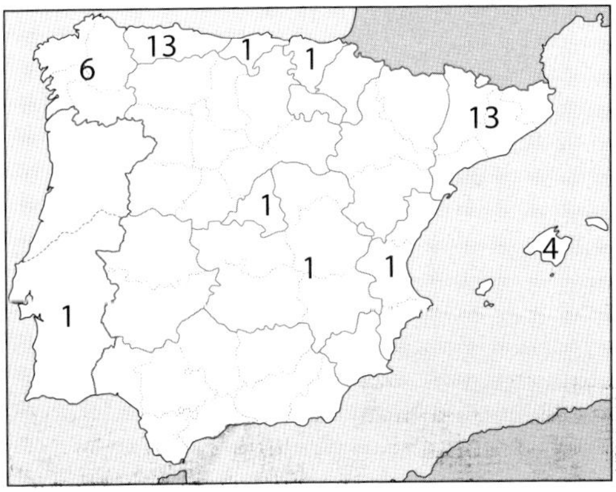

Later years have produced multiple sightings in Spain. For example, at least 27 were reported during winter 2010–11: 12 in Catalonia, 14 on the Biscay coasts or Galicia and one in Madrid. They included one ringed in Russia that appeared at Solius rubbish dump, Girona, and another six Polish-ringed birds: four at rubbish dumps in Girona and one each in Madrid and Lugo. Subsequent observations of at least 29 individuals in winter 2011–12 and 17 in winter 2012–13 have included more observations at the rubbish tips at Alcázar de San Juan and at Colmenar Viejo, Madrid, reflecting the attention paid by gull watchers to those sites.

Figure 155. Records by region, 2005–2010 (n = 42).

YELLOW-LEGGED GULL *Larus michahellis*

(Gaviota Patiamarilla, Gaivota-de-patas-amarelas)

***L. m. michahellis*. Very common and locally abundant on southwestern and Mediterranean coasts and in the Balearic Islands. A few nest inland.**

***L. m. lusitanius*. Very common and locally abundant on Atlantic and Biscay coasts, from northern Portugal to the Basque Country. A few nest inland.**

The Yellow-legged Gull, the 'common' gull of Iberia, has demonstrated a high degree of adaptability in recent years and has proved adept at exploiting the food supplies provided by human activities. The result has been a spectacular population increase since about the 1970s that is still continuing.

The Yellow-legged Gull has been 'split' from the Herring Gull but its relationship to the Caspian Gull and other species of the herring gull complex is far from clear. Moreover, there is considerable variation within the Yellow-legged Gulls of the western Palearctic, even within Iberia. The 'classic' Yellow-legged Gull *L. m. michahellis* inhabits the Strait of Gibraltar, Alborán island, the Mediterranean coasts of Spain and their offshore islands, and the Balearic archipelago. Very similar birds also nest on southwestern coasts, from the Strait north to central Portugal. Those of northern and northwestern coasts are smaller and shorter-legged than the Mediterranean birds and there are subtle plumage differences between the two forms in all age classes, as well as some genetic and vocal differences (Teyssèdre 1983, Carrera *et al.* 1987). They are sometimes assigned to a separate taxon '*lusitanius*' (Pons *et al.* 2004). About 20% of '*lusitanius*' adults have pink or partly pink legs (HBW) but genetic analyses show that any resemblance to the Herring Gull is probably due to convergence since the genetic affinities of the northern birds are closest to those of Mediterranean Yellow-legged Gulls (Pons *et al.* 2004). The two forms are compared by Malling Olsen (2003), who also notes that the birds in Galicia, and perhaps others on the Atlantic coast, have some affinities with *L. m. atlantis*.

The Iberian populations are at an all-time high following at least three decades of rapid increases in numbers. A national census in Spain in 2007–08 (Molina & Bermejo 2009) found over 116,000 pairs, to which may be added at least 16,000 pairs in Portugal (PA2). Gibraltar has a large nesting colony which numbered some 4,000 pairs in 2002 but this had decreased to fewer than 2,000 pairs by 2009 following ongoing culling (Cortes *et al.* 2005; Guillem 2009, 2010). The minimum breeding population of Iberia in 2007–08 was thus of the order of 135,000 pairs. There are also very large numbers of non-breeding, chiefly immature individuals. The breeding distribution is overwhelmingly coastal. It is most continuous in the north-west and north, where there is an abundance of sea cliffs, offshore stacks and islands that offer secure nesting sites. It is more fragmented in the south and within the Mediterranean, where similar habitats are scarcer but where large colonies have become established locally in such habitats as estuaries, saltpans and sandy beaches.

The regional distribution of the breeding population (Table 41) shows that some 77,000 pairs (60%) nested on Atlantic coasts and 53,000 pairs (40%) were in the Mediterranean. Galicia alone supported around one third of the breeding pairs in 2007–08. These largely inhabit the various islands of the Atlantic Isles National Park, as well as the adjacent rocky coastlines. Other major concentrations are in Catalonia: where the Ebro delta is the principal site, Murcia, the Balearics, Gibraltar and the Berlengas archipelago, Portugal. The Portuguese breeding range is predominantly in central and southern coastal regions, where rocky coasts are widely available (PA2). During 2007–08 breeding birds were found in all the Spanish autonomous regions except for La Rioja and Madrid, in which the species has yet to nest, and in Castilla y León and Extremadura, where it has nested previously.

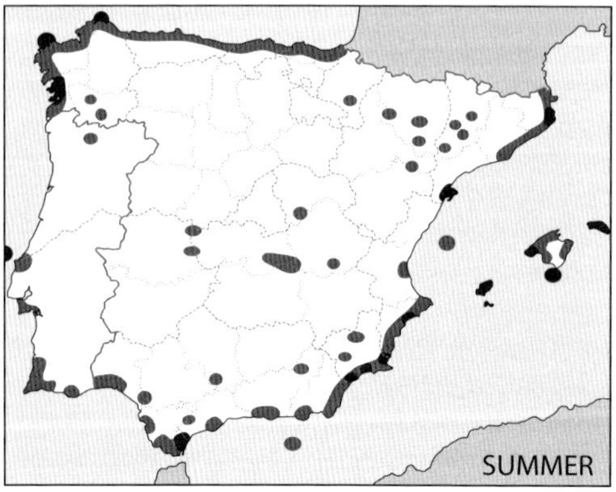

Table 41. Regional distribution of breeding Yellow-legged Gulls in Iberia in 2007–08.
Sources: Spain: Molina & Bermejo (2009); Portugal: Catry (2008); Gibraltar: Guillem 2009, 2010.

Region	No. of pairs	% of Iberian total
Galicia	45,901	35.2
Catalonia	18,966	14.5
Murcia	8,702	6.7
Andalucía	6,285	4.8
Basque Country	4,530	3.5
Asturias	4,237	3.2
Comunidad Valenciana	2,147	1.6
Cantabria	1,606	1.2
Aragón	68	0.1
Castilla-La Mancha	6	0.0
Navarra	1	0.0
Castilla y León	0	0.0
Extremadura	0	0.0
La Rioja	0	0.0
Madrid	0	0.0
Balearic Islands	18,000	13.8
Portugal (2002)	16,000	12.3
Gibraltar (2002)	4,000	3.1
IBERIAN TOTAL	130,449	

The 16 largest colonies, of 1,000+ pairs (Table 42), are mainly on islands, or quasi-islands such as the Rock of Gibraltar and the inaccessible sand spits of the Ebro delta. The very large dispersed colony of roof-nesting gulls in the city of Vigo, Pontevedra (see below), is a notable exception. The Berlengas Islands have long-held the largest concentration in Portugal. The Berlengas colonies increased from 1,300 pairs in 1974 to 3,750 pairs in 1981 after which the numbers present in the breeding season increased rapidly to 45,000 birds in 1994: control measures reduced the population to some 25,000 individuals by 2005 (AP).

The Vigo colony of 1,000+ pairs is the largest known urban colony in Iberia but roof-nesting is widespread at many coastal cities and towns. The recent census found a minimum of 4,532 roof-nesting pairs in Spain and there would certainly have been more were it not for the sustained local campaigns to remove nests. The largest rooftop colonies are all in the north. In addition to Vigo, in 2007–09 they included 934 pairs in A Coruña, 600 in Gijón, 390 in Ferrol and 313 in Santander. Rooftop nesting has also been reported in Portugal since 2000: in Viana do Castelo, Porto, Peniche, Cascais, Lisbon, Lagos and Portimão (AP). Mediterranean city colonies tend to be smaller: the larger ones known are in Málaga (100 pairs), Barcelona (50+) and Gibraltar (50+). Roof nesting in Barcelona was first reported in 1980 (García-Petit *et al.* 1986). The gulls favour such sites because they are inaccessible to predators and are often near garbage tips and other food sources. Gibraltar has a small number of pairs that nest in pine trees in the Botanic Gardens, building their usual nest in a major fork.

Inland breeding was first noted in Catalonia during the early 1980s (Ferrer *et al.* 1986) and in Aragón in the mid 1990s (Sampietro *et al.* 1998) but not until 1999 in Castilla-La Mancha (Velasco *et al.* 1999, Perea *et al.* 2004). It remains rare and the 2007–08 censuses found only 92–95 pairs inland in total. The majority were at just three sites: the Chiprana salt flats, Zaragoza (35 pairs); the Sotonera reservoir, Huesca (30) and the Negratín reservoir, Granada (12). The colony at the last of these sites had grown to 67 nests in April 2010 (*Ardeola* 57: 536). A further 16 Spanish sites, all of them beside lakes or reservoirs, held one or two breeding pairs in 2007–08. Four pairs nested at the Valdecañas reservoir, Extremadura, in 2011 and there were 58 birds there in June that year (*Ardeola* 58: 502, 59: 438). Inland breeding is also known locally in northern Portugal, where 4–7 pairs have bred since 2001 on islets in the Albufeira do Alto Rabagão, Montalegre (AP).

Table 42. Iberian Yellow-legged Gull colonies of more than 1,000 pairs in 2007–08. The large colonies on the Balearic Islands, notably on Cabrera, were not included in this survey.

Region	Site	No. of pairs
Galicia	Islas Cies	15,654
Catalonia	Ebro delta	9,744
Galicia	Islas Sisargas	7,500
Catalonia	Islas Medas	7,291
Galicia	Isla Sálvora	6,786
Galicia	Islas Ons–Onza	6,002
Portugal*	Islas Berlengas	5,000
Andalucía	Cádiz Bay saltpans	4,161
Gibraltar*	Gibraltar	4,000
Galicia	Isla Coelleira	2,500
Murcia	Isla Perdiguera	2,269
Murcia	Isla de Mazarrón	2,136
Murcia	Isla Grosa	1,272
Basque Country	Isla Ízaro	1,164
Galicia	Vionta–Noro–Erbosa	1,152
Galicia**	Vigo city	1,023

*in 2002, ** in 2009.

There were no complete national censuses prior to 2007–08, although in the early 1980s the Atlantic population was estimated at 50,000–55,000 pairs, 70% of the total, and that of the Mediterranean at some 15,000 pairs (Bárcena *et al.* 1984, de Juana 1984). Nevertheless, regional data reveals the magnitude of the 20th century population changes (Aguilar *et al.* 1994, Arcea 2005, IKT 2008) and Oro & Martínez-Abraín (2007) have demonstrated an annual growth rate in the breeding population of about 5% over the past 30 years, based on long series of data from a number of both Atlantic and Mediterranean colonies. The population of the Basque country increased from a mean of 1,728 pairs during the 1980s to a mean of 4,246 pairs post 2000 and the number of colonies increased from 12 to 28 during the same period (Arizaga *et al.* 2009b). There have also been large changes in the distribution of the population since the mid 20th century, with particularly large increases occurring in the Mediterranean. In the 1980s over half of the Iberian population was in Galicia (Bermejo & Carrera 1983), where 35% were found in 2007–08. The Ebro delta population had an estimated population of 500–600 pairs in the mid 1960s (Westernhagen & Pons 1966) and this fell to only 150 pairs in 1979 (Martínez-Vilalta in Carrera *et al.* 1981), but it accommodated well over 9,000 pairs in 2007–08. The Medes islands colony, Girona, comprised 3,000 pairs in 1961, 7,500 pairs in 1978 and over 14,000 pairs in 1990 (Carrera & Vilagrasa 1984; CA2), but there were only 7,300 in 2007–08 after a culling programme.

The Yellow-legged Gull was formerly scarce in the Balearics, not least because its eggs were habitually harvested by humans (Mayol 1977b). Munn (1921, 1924) said that it occurred in northeastern Mallorca and in Menorca as scattered pairs nesting along the coastline. Ticehurst and Whistler (1930) found a pair or two nesting here and there on sea cliffs in Ibiza, with one 'large' colony of 50–75 pairs in which the accessible nests had been robbed. The Balearic population had grown to 3,000–5,000 pairs by the mid 1970s (Araújo *et al.* 1977) and to an estimated 18,000 pairs in 2007–08.

A further useful case-history is provided by the colony at Gibraltar. Saunders (1871), Irby (1895), Jones (1900) and Verner (1909) mentioned seeing 'Herring' Gulls in winter but did not mention nesting. The earliest reference to nesting birds is by Rait-Kerr (1934c), although she did not estimate the numbers involved. Cortes *et al.* (1980) estimated that there were c. 600 pairs in Gibraltar in the late 1970s. A survey in spring 2002 found a minimum of 3,963 pairs and put the population, including non-breeders, at 11,000–13,000 individuals. Large-scale culling may since have had a significant effect since censuses in 2009 and 2010 found only 1,505 and 1,360 pairs respectively (Guillem 2009, 2010). The initial increase has been attributed to the establishment of a large municipal rubbish tip at Los Barrios, some 7km west of Gibraltar, to and from which the gulls commute frequently. Such tips have been shown to be important to nesting Yellow-legged Gulls elsewhere (Bosch *et al.* 1994, Sol *et al.* 1995, Ramos *et al.* 2009). However, much natural food is also available to the Gibraltar gulls in the Strait, in the form of shoals of anchovies and other fish, and of swimming crabs in spring.

Table 43. Regional distribution of wintering Yellow-legged Gulls in January 2009. Source: Molina & Bermejo 2009.

Region	No. of birds	% of total
Galicia	84,166	36.17
Catalonia	55,252	23.75
Andalucía	25,308	10.88
Asturias	20,119	8.65
Basque Country	13,058	5.61
Comunidad Valenciana	11,315	4.86
Murcia	10,591	4.55
Cantabria	3,426	1.47
Aragón	183	0.08
Castilla y León	82	0.04
Castilla-La Mancha	13	0.01
Extremadura	0	0.00
Madrid	0	0.00
Navarra	n.c.	
La Rioja	n.c.	
Balearic Islands	1,167	0.50
Portugal	No count	
Gibraltar	8,000	3.44
TOTAL	232,680	

Henslow's swimming crab has been found to be the second most important food item (after refuse) for adult Yellow-legged Gulls in Galicia (Munilla 1997a) and it is also a key item in Asturias (Álvarez & Méndez 1995). A study in the Basque Country (Arizaga *et al.* 2010) found that fish comprise 60% of the food biomass brought to chicks, 30% being refuse. These and other studies highlight the importance of natural food to this species despite its omnivorous habits. However, fishery discards account for much of the fish 'prey' taken (Ramos *et al.* 2009, Cama *et al.* 2012) and a fishing moratorium in the Spanish Mediterranean depressed its breeding success locally (Oro *et al.* 1995). Yellow-legged Gulls are notorious kleptoparasites of other seabirds, including Audouin's Gull (Oro & Martínez-Vilalta 1994b), and their increasing abundance at such key locations for breeding terns, waterfowl and waders as the Ebro delta, Veta La Palma in Doñana, the Albufera de Valencia and Cádiz bay raises concerns for such species, whose eggs and young are highly vulnerable to gull predation. The negative impact of the expansion of the Yellow-legged Gull in the Mediterranean basin has been highlighted by Vidal *et al.* (1998) who consider it to have become locally superabundant, with diverse and very severe effects on ecosystems locally.

The great majority of the population winters within Iberia. A census in January 1984 (Bermejo *et al.* 1986) found 137,380 in Spain and a further 3,700 in Portugal. A census of wintering birds in Spain and Gibraltar, but not Portugal, in January 2009 found 232,680 birds, an increase of 62.7% for Spain since 1984 (Table 43). The 2009 census did not cover Almería or Alborán, and the Balearic Islands received only partial coverage even though all hold significant numbers in winter: for example, 10,000 were counted at San Reus rubbish dump, Palma, Mallorca, on February 2002 (*Ardeola* 52: 216). Navarra and La Rioja were also not covered but these regions hold few if any wintering birds. A count of birds wintering at non-estuarine sites in Portugal during 2009–10 found 13,324 birds (Catry *et al.* 2010b). The largest concentrations in January 2009 were in the north-west in

Galicia; in Catalonia: over 21,000 at the Ebro delta, and in the south-west, including Gibraltar and Cádiz. The species was widely distributed elsewhere on all coasts, however.

A comparison between the January distribution of Yellow-legged Gulls in Spain in 1984 and 2009 (Table 44) shows that the increase in wintering numbers has been most marked in the Mediterranean, in parallel with the changes seen in the breeding population. Unlike the Lesser Black-backed Gull (*q.v.*) only negligible numbers winter inland and the Yellow-legged Gull thus remains an overwhelmingly coastal species, showing that of the two it is more reliant on natural food sources, and fish discards from trawlers, obtained in coastal waters. In 2009 the few inland wintering Yellow-legged Gulls were in Castilla-La Mancha, Castilla y León and Aragón and there were none in Madrid or Extremadura, the strongholds of wintering Lesser Black-backed Gulls.

Table 44. Coastal and inland distributions of wintering Yellow-legged Gulls in Spain in 1984 and 2009. Sources: Bermejo *et al.* 1986, Molina & Bermejo 2009.

Sector	January 1984		January 2009	
	No. of birds	%	No. of birds	%
Atlantic coasts	98,460	71.67	118,502	53.96
Mediterranean coasts	38,882	28.30	100,819	45.91
Inland	37	0.03	278	0.13
Total	137,379		219,599	

Inland sightings have been somewhat more frequent in Spain during June–October, rather than in winter, although they seldom involve more than two or three individuals. However, there were 30 adults at the Alarcón reservoir, Cuenca, in June 1999, and 13 birds at San Martín de la Vega, Madrid, in June 2001 (*Ardeola* 47: 168, 50: 162). An observation of 42, mainly second-years, at Villadangos del Páramo, León, on 26 April 1998 (*Ardeola* 46: 158) is an unusual inland spring record.

The Yellow-legged Gull often appears to be largely sedentary in Iberia but a proportion of some populations undertake complex movements, some of them apparently of recent origin. The northern populations, of the Iberian endemic form '*lusitanius*', have been found to be largely non-migratory and most recoveries of birds ringed as pulli in Galicia, Asturias, Cantabria and the Basque Country are within 50km of the ringing colony, although immatures move further than adults (Munilla 1997b, Arizaga *et al.* 2010). Those ringed in Galicia have produced at least 25 recoveries in Portugal: mostly in the north but some in the centre and one in Algarve, as well as one in Córdoba, Andalucía: only one has reached the Mediterranean and only two have reached Morocco (Arizaga *et al.* 2009c). There are however two surprising recoveries of birds ringed as pulli in Galicia: one ringed at the Sisargas in July 1989 was recovered on the Isle of Wight, England, in February 1992 and another ringed on Ons island on July 2000 was recovered at Den Helder, The Netherlands, in July 2001 (*Ecología* 7: 345,16: 365).

In contrast, the birds ringed in the Spanish Mediterranean in Catalonia: mainly on the Medes Islands, Girona, and at the Ebro Delta, are partly migratory and many have a northward post-breeding dispersion that takes them to the French shores of the Bay of Biscay (Carrera *et al.* 1981a, Carrera 1987). Those ringed in the Comunidad Valenciana: on the Columbretes Islands and on Benidorm island, show similar general movements that take them to southwestern France and onwards to the French and Spanish Biscay coastlands and as far as Galicia and northern Portugal (Martínez-Abraín *et al.* 2002b). Almost 200 controls of birds colour-marked on the Balearics were in northeastern Spain in Catalonia and Comunidad Valenciana (42%), within the Balearics themselves (30%), on the north and northwestern Iberian coasts from the Basque Country to northern Portugal (19%) and on French Atlantic coasts of France north to the English Channel (8%), and one reached western Andalucía (Rodríguez & Muntaner 2004). An individual that was marked on Dragonera islet in the Balearics in April 2009 was observed in England on 7 July (*Ardeola* 57: 536). The movements of Mediterranean Yellow-legged Gulls into the Atlantic appear to be largely restricted to the summer and early autumn, from July to November (Carrera 1987). Birds cross southwestern France to the Bay of Biscay early in the summer but from August onwards they may return to the Mediterranean or visit the southern coasts of the Bay, where a small minority reach Asturias, Galicia and northern Portugal. Most appear to be back in the Mediterranean by December. This pattern fits earlier observations from the French Atlantic littoral (Yésou 1985).

At least 168 Yellow-legged Gulls ringed in Spain have been found outside the Peninsula, mainly in France (123) but also in Britain (6), The Netherlands (5), Morocco (29) and Algeria (5). It has been suggested that seasonal food shortage in the western Mediterranean in summer triggers this mainly northward displacement, which is on a large scale. Carrera (1987) considered that some 50% of the juveniles from Catalonia travel to the Atlantic. Conspicuous post-breeding dispersal is seen at Gibraltar, where the entire large colony disappears in August and returns from September onwards: the destination of these birds is unknown. Paterson (1987) noted conspicuous westerly movement on the Málaga coast in June and July, declining in August, suggesting a displacement then of some birds from the Alborán basin to the Atlantic via the Strait.

Some Yellow-legged Gulls from elsewhere in the Mediterranean visit Iberia during their post-breeding dispersal. In particular immature birds that have been ringed at breeding colonies in Algeria have been found wintering or remaining in Spain during their early years (Samraoui *et al.* 2011). There are also 13 Iberian recoveries of birds ringed in France, six from Italy, three from Poland and one from Switzerland. Further studies are likely to reveal the extent and nature of interchange between the Mediterranean colonies.

THAYER'S GULL *Larus thayeri*

(Gaviota Esquimal, Gavià de Thayer)

Vagrant to Spain.

An adult was at San Cibrao, Cervo, Lugo, from at least 20 March to 8 April 2010. What was probably the same individual in immature plumages had already been recorded there in previous winters, in February 2008, March 2011 and March–April 2013 (RBS). Thayer's Gull rarely reaches the Atlantic coasts of southeastern Canada and the northeastern United States. It is a very rare vagrant to Europe: eight were found in Ireland during 1990–2011, and there are also records from Norway, Iceland and Denmark (Malling Olsen 2003).

ICELAND GULL *Larus glaucoides*

(Gaviota Groenlandesa, Gaivota-branca)

Occasional in northern Spain and Portugal, mainly in winter. Vagrant elsewhere in Iberia. Recorded in Gibraltar.

Iceland Gull breeds in Greenland (*L. g. glaucoides*) and north-eastern Canada (*L. g. kumlieni*), wintering in the North Atlantic region. It has been found more frequently in northwestern Europe in recent years but is still considered a rarity in France, with accepted records of 213 individuals by 2008, only two of them *kumlieni*. However, an irruption in 2009 may have brought several hundred into France and there are accepted records of at least 141 individuals that year (Dubois & Duquet 2009, Reeber *et al.* 2010). There are 13 accepted records from the Azores, one from Madeira and two from the Canary Islands, as well as nine from Morocco.

The earliest Iberian records are from Portugal: one taken at the Douro estuary in November 1912 and one seen at Leixões harbour, Porto, on 29 December 1914 (Tait 1924). The first Spanish records were not until 1977, at Cudillero, Asturias, and at A Coruña (García-Rovés & García-Rovés 1989, Grupo de Ornitoloxía da Delegación da Coruña da S. G. H. N. 1980). There were further records from Portugal in 1981, at Porto and Peniche (Preiswerk 1981b, Moore 1982). In all by 2010 there were 26 accepted records from Portugal, 88 from Spain and one from Gibraltar. Numerous records: at least 14 in Portugal and 28 in Spain, resulted from an influx in early 2009 that coincided with strong Atlantic gales. Prior to that there were 32 Iberian records during 1977–99 and 50 during 2000–08, including maxima of five during winter 1997–98 and nine in 2007–08. All pre-2009 observations involved single birds with the exception of a record of three in A Coruña in December 2006. In 2009 there were several records of two birds and five were seen together at a fish farm at Lago, Lugo, on 3 January.

The 2009 influx was in January, later than average: nearly a quarter of first dates pre-2009 were in December when there were very few in 2009 (Fig. 156). The influx records were also more southerly than in previous years: 36% were in southern Portugal and Andalucía, which accounted for only 10% of earlier records (Fig. 157). The 2009 influx also saw the first observation from Catalonia: a first-winter at Cambrils on 4–10 February. A first-winter bird at Talavera de la Reina, Toledo, on 31 December 2009 was the first inland Iberian record, although it preceded the influx. One of the 2009 arrivals at the Ría de Vigo, Pontevedra, remained in the general area from June 2009 until at least August 2011, by when it had attained third-summer plumage (*Ardeola* 60: 477).

Taking all pre-2010 records together, 83% of 120 individuals of known age were first-winters/first-summers and only 8% were adults: the same proportion of adults as in France (OF). The first adult seen in Iberia was in 2006 but there were two in 2008 and seven in 2009: including five *kumlieni*. There was only one prior accepted Iberian record of *kumlieni*: at Guetaria, Gipuzkoa, on 4 March 2008. Nine of the ten adults were single-day observations but many of the first-years were long-stayers: one was at A Coruña from at least 25 December 2002 until mid April 2003. Adults were only found between 25 January and 3 April but first-years were recorded between 7 November and 3 May.

Figure 156. Records by month, 1912–2008 (n = 82)

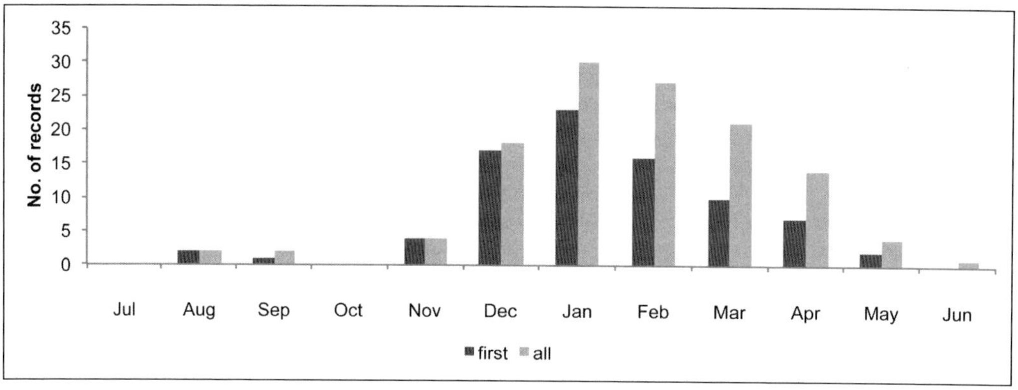

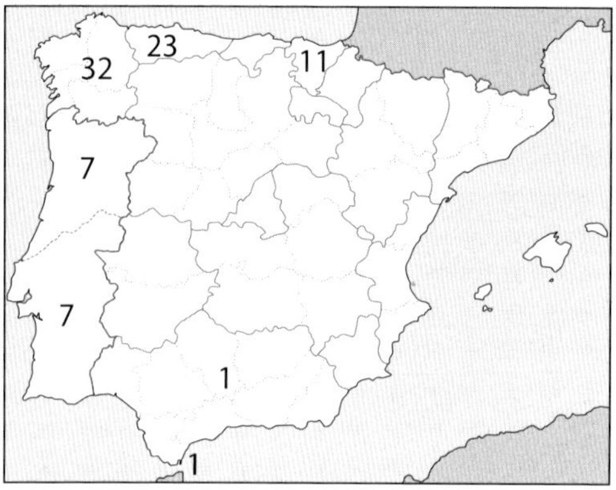

A further substantial influx during winter 2011–12 included at least 15 in Spain and six in Portugal (*Ardeola* 60: 477, RBS, RBP). Their arrival was clearly part of a much larger influx that occurred in the northeastern Atlantic, which brought unprecedented numbers to the British Isles and Faroes (Fray *et al.* 2012). At least three oversummering birds were located on the Biscay coast in July 2012 (RBS). Two of the 2011–12 Spanish birds were identified as *kumlieni* and two more of this subspecies were identified in January 2013 (RBS).

Figure 157. Records by region, 1912–2008 (n = 82)

GLAUCOUS GULL *Larus hyperboreus*

(Gavión Hiperbóreo, Gaivotão-branco)

Occasional in northern Spain and Portugal, mainly in winter. Vagrant elsewhere in Iberia.

This Arctic gull winters in Iceland and Norway, and south to the British Isles and North Sea coasts. Some 10–30 reach northern France, with occasional irruptions after winter storms (Malling Olsen & Larsson 2003, Dubois & Duquet 2009). There are also 21 records from the Azores, two from Madeira and five from the Canaries. There are 16 records from Morocco and two from Ceuta: in February 1998 and February 2009.

The first reliable Iberian records were in Galicia in the 1970s, the earliest being a first-winter bird at Finisterre on 28 March 1975 (Grupo de Ornitoloxía da Delegación da Coruña da SGHN 1980). There are early 20th century reports from Portugal (Reis Júnior 1931, Tait 1924) but the earliest definite observation there was of an adult at Caminha, Viana do Castelo, on 14 January 1988 (*Ardeola* 37: 122). Since then observations have accumulated quite rapidly and by 2010 there were 149 accepted records for Spain and 37 for Portugal. A presumed hybrid with a Herring Gull was present in Asturias during 2004–05 and another hybrid of unknown parentage was near Lisbon in January 2001.

A notable influx after severe storms in early 2009 resulted in some 50 Iberian records. These are treated separately here, as we have done with the parallel influx of Iceland Gulls. There were 139 Iberian records up to 2008 and a further six in 2010, involving 157 individuals in total: although some may have returned in different years. Some 93% of records were of single birds, with eight of two together and two groups of three. In 2009, however, single birds made up 78% of observations, there were six groups of two, one of four, one of five and two of six together, these last at Camariñas, A Coruña, on 1 February, and at Burela, Lugo, on 8 February. Maximum winter totals pre-2009 were 15 during 1995–96 and 11 in 2006–07. First-year birds predominate. They made up 64% of the autumn/winter totals excluding 2009, with 11% being adults and 25% of intermediate age (n = 108). In 2009 there were 19 adults, 32 first-winters and 20 of intermediate age.

The spatio-temporal distribution of the Glaucous Gull closely resembles that of the Iceland Gull, most occurring in winter in the coastal north-west (Figs. 158, 159). About 80% of first observations fall in December–March, although there are records in all months. The 2008–09 arrivals were comparatively late: three in December, 17 in January: most of them during the third week, and 22 in February. Adults generally arrive later and leave earlier. Adults have been seen between 20 October and 20 February, with the notable exception of one at Lepe, Huelva, on 7 May 2010, but first-winters have appeared as early as late September and many of them have remained well into the spring. One long-stayer, initially found as a second-summer bird, was seen intermittently in Asturias between 2nd May 2006 and 6 April 2008. Excluding 2009, some 75% of records were in Asturias/Galicia and only 10% were in Andalucía and southern Portugal combined. In winter 2008–09, however, there were 28 records in Asturias/Galicia and 12 in southern Iberia.

Figure 158. Records by month, 1975–2010, excluding the winter 2008–09 influx (n = 145)

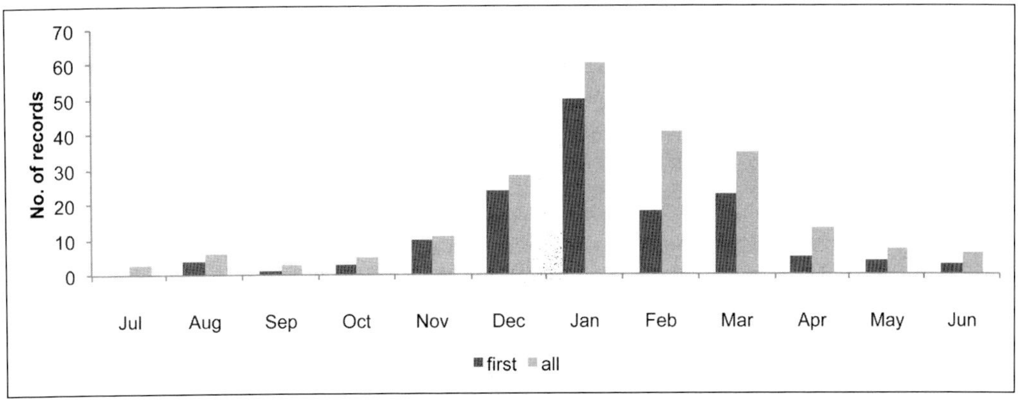

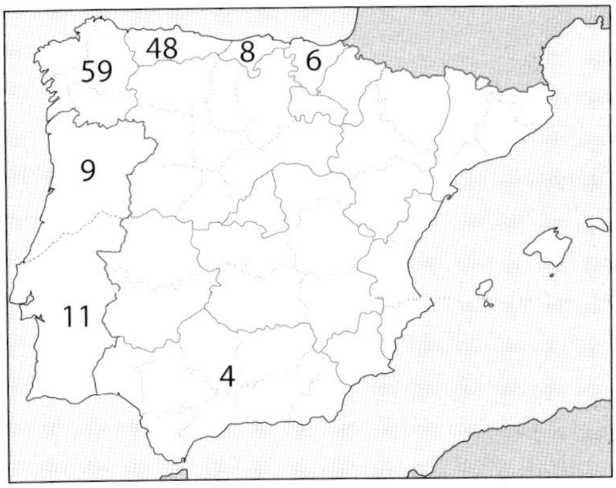

Figure 159. Records by region, 1975–2010, excluding the winter 2008–09 influx (n = 145).

There were also a considerable number of post-2010 records (RBS, RBP). The 2010–11 winter was unexceptional and two were reported in Portugal and about 14 in Spain: 12 of them in Galicia, one in the País Vasco and one at the exceptional southern location of Fuentedepiedra on 15 January (*Ardeola* 60: 477, RBS). The winter 2011–12 saw a small influx: five individuals were reported in Portugal and some 25 in Spain, all the latter from the Galician coast with the exceptions of single birds in Cantabria and the País Vasco. They included an apparent Glaucous Gull x Great Black-backed Gull hybrid (*Ardeola* 60: 477–478). A recently accepted record was in the centre of the Peninsula: at Colmenar Viejo rubbish tip, Madrid, in January 2010. A sizable influx occurred once again in December 2013 and January 2014, when some 50 individuals appeared, nearly all in Galicia (RBS).

KELP GULL *Larus dominicanus*

(Gaviota Cocinera, Gaivotão)

An adult was photographed at Vila do Conde in June 2013 (RBP). There were also reports of two other individuals on the Portuguese coast in 2013, at Peniche in May and Faro in August. This species has recently been identified breeding at the Khniffis lagoon in southern Morocco (Jönsson 2011).

GREAT BLACK-BACKED GULL *Larus marinus*

(Gavión Atlántico, Gaivotão-real)

Regular in small numbers in winter on Atlantic and Biscay coasts, particularly in the north and north-west, becoming progressively scarcer further south. Rare in the Mediterranean. Several pairs have bred recently in the north-west.

The Great Black-backed Gull is only at all common in the regions bordering the Bay of Biscay. Here at most a few hundred individuals occur in winter and a very small number remain in summer. A few winter on Atlantic coasts of Portugal, southwestern Andalucía and the Strait of Gibraltar, but they are very thinly spread and few sites report them annually. Nonetheless, there are three recoveries from the Portuguese west coast of birds ringed in Brittany (2) and Ireland (1) and a third-winter bird at Viana do Castelo harbour on 12 October 2004 had been ringed in Newfoundland in June 2002 (AP). Another individual from Newfoundland, a second-winter bird, was among the casualties of the Prestige oil spill in Galicia in December 2002 (García *et al.* 2003). At least three ringed in England, one from Sweden and one from northern Norway have also been found in northern Spain (SWA), and a Danish ringed individual was seen at Ter estuary, Girona, in September 2001 (*A. O. Catalunya 2001*: 167).The species is even rarer in the Mediterranean although there are records from all regions and from the Balearic Islands.

Although summering birds are not unusual in northern Iberia, the species is not known ever to have bred in the region prior to 2005, when two pairs, one of them with a chick, were found at a Yellow-legged Gull colony on the Farallons islands, Lugo (Bertolero *et al.* 2008). At least one pair was present there in 2006 and two pairs attempted to nest in 2007, when an adult was seen with two chicks. One pair nested on an islet at Ortigueira, A Coruña, in 2007. All nests were within Yellow-legged Gull colonies but none of the attempts during 2005–07 is known to have succeeded. An incubating pair was found at a Yellow-legged Gull colony in the Ría del Eo in Asturias in 2009, and another pair was near Tapia (*GIAM* 32:10). Up to five breeding pairs are reported to have bred on the Galician north coast in 2011 and a pair nested successfully at Cape Peñas in 2012 (R. Hevia pers. comm.; *Ardeola* 59: 439). Further east, a pair built a nest at Zierbena, Bizkaia, in 2012 but no eggs were laid (*Ardeola* 59: 439). Much further south, a male Great Black-backed x female Yellow-legged Gull pair laid eggs at a Yellow-legged Gull colony near Sintra, Lisboa, annually from 2004 to at least 2009 (Gonçalves 2009–10): three hybrid young may have resulted in 2004 but none are known to have fledged thereafter.

By far the largest numbers are reported from Galicia and Asturias in winter. The gull census of January 1984 (Bermejo *et al.* 1986) found 181 in Galicia: 149 of them in the Rias Bajas and 17 in Asturias. The high count in Galicia was considerably more than had been found there in previous years: winter counts during 1976–80 gave totals for the region of 14–65 birds (Bermejo *et al.* 1986). More recent observations suggest an increasing wintering population: the 2008–10 Spanish winter atlas estimated the mean population to be around 300 birds, over 80% in Galicia and nearly all the remainder on the Biscay coast.

Wintering birds are regular in Asturias in some numbers in the Ría del Eo, at Tapia de Casariego and in Gijón Bay, with a few individuals elsewhere. January counts from the region post 1990 have been around 150 birds in total (*A. O. Asturias 2008*). The Asturian January counts of 1996 and 1997 combined included 139 adults (46%), 98 first-winters (32%) and 68 (22%) of other ages (*A. O. Asturias 1996, 1997*). The Great Black-backed Gull is also regular in Cantabria and the Basque Country in winter but it is generally encountered singly or in very small groups.

Reports from Atlantic coasts south of Galicia seldom involve more than one or two individuals. The January 1984 gull census (Bermejo *et al.* 1986) found nine in Portugal: all at the Tagus/Sado estuaries; 16 on the Huelva coast and seven on the Atlantic coast of Cádiz. The 2009–10 survey of non-estuarine coasts in Portugal found 14 individuals (Catry *et al.* 2010b) and the Portuguese wintering population, including the estuaries, is now estimated to be a few dozen individuals (AP). Garrido (1996) considered that 20–50 winter regularly on the coasts of Huelva. The total numbers of published records suggest that this is a considerable overestimate and Gacio & Sayago (2011) note that whereas it is seen annually at Odiel, the largest count there has been five. Occasional individuals are reported from the Strait of Gibraltar, most often from Los Lances beach, but they are not even annual there

They are seldom reported from within the Mediterranean but there are occasional individuals in winter on the coasts of Málaga, Granada and Almería. The species is rare on the Levant coasts and in the Balearic Islands. Clavell (2002) gives only ten records of single birds for the period 1985–98: three in Catalonia, five from the Comunidad Valenciana and two in the Balearics. Ferrer *et al.* (1986) give a total of 25 pre-1985 records for these regions, not all of which may be reliable (Clavell 2002). Recent records up to 2010 include three from the Levant coasts and three from Palma harbour, Mallorca (*Ardeola* 53: 388, 55: 300, 57: 536).

The monthly distribution of records away from the north and west coasts (Fig. 160) shows that the majority are reported from November–February. However, some individuals appear there as early as August–September, even in the Mediterranean.

Figure 160. Monthly distribution of Great Black-backed Gulls reported from Portugal, Atlantic Andalucía and on Mediterranean coasts, 1985–2010 (n = 82)

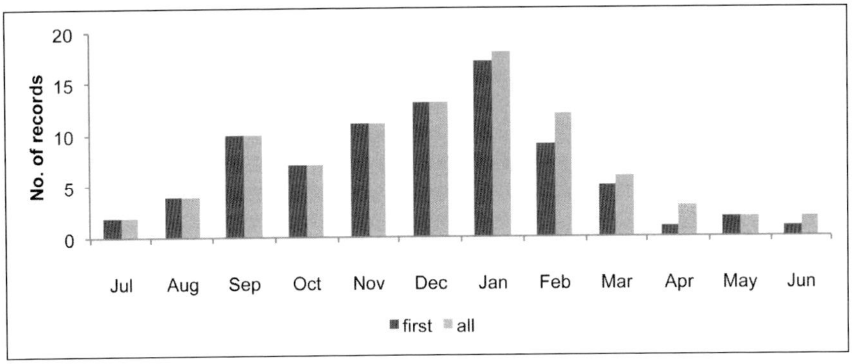

Inland records are rare but recent sightings include an adult at the Laguna Grande de Villafranca, Ciudad Real, in September 2002; a third-winter at La Mata reservoir, León, in April 2012, and at least ten records from rubbish dumps: including at least eight individuals in Madrid city between 1997 and 2012 (*Ardeola* 51: 553, 59: 186).

ROSS'S GULL *Rhodostethia rosea*

(Gaviota Rosada, Gaivota-rosada)

Vagrant to Spain.

There are three or four records of single adults. One at the Bidasoa estuary, Gipuzkoa, on 6 and 11 April 1994 (Gorospe 1996) was also reported from France on 7 April (OF). A Ross's Gull that appeared later the same month, on 17 and 18 April, at Zumaia, some 40km from the earlier sightings, was in nuptial plumage and thought to be a different individual. An even more remarkable occurrence was the finding of an adult in winter plumage at the Lagunas de Alcázar de San Juan, in the centre of the Peninsula, where it remained from 19 February to 1 March 2009. A first-winter bird was photographed in Pontevedra harbour in February 2014 (RBS). This high Arctic species rarely ventures very far south. There are records of 92 from Britain, 18 from Ireland and seven from France, but none from the Macaronesian islands or Morocco. There are, however, two surprising records from the Italian Mediterranean: birds captured in Sardinia in January 1906 and Campania in May 1997 (Bonaccorsi 2002).

BLACK-LEGGED KITTIWAKE *Rissa tridactyla*

(Gaviota tridáctila, Gaivota-tridáctila)

***R. t. tridactyla*. A very scarce breeding species in Galicia. Common and sometimes numerous off Biscay and Atlantic coasts but generally keeps well offshore. Usually scarce in the Mediterranean and exceptional inland, except when storm-driven.**

The two small Kittiwake colonies in Iberia are the southernmost of the species' range. A third colony, on the Berlengas islands, Portugal, was even further south but it was tiny and short-lived: three pairs bred in 1981 but they disappeared in 1983 (Teixeira 1983, AP). There is some anecdotal evidence from local fishermen that the Berlengas colony was first established during the 1970s.

The Iberian Kittiwakes nest on the Sisargas Islands and at Cape Vilán, in northwestern Galicia. In 2007 there were 11 pairs in the Sisargas and 10–12 at Cape Vilán (Bertolero *et al.* 2008). The colonies appear to be of quite recent origin. The Sisargas colony was discovered in 1975, when over 50 pairs nested at three points on Sisarga Grande (Rodríguez Silvar & Bermejo 1975, Bárcena *et al.* 1987). This colony increased to 152 pairs in 1981 but then decreased to a low point of three nests in 2001. A sharp decline was seen there between 1992, when there were still 142 pairs, and 1993, when just a few pairs nested, although 120 adults were present (Fig. 161). The cause of this decline is uncertain but it may have been partly due to the wrecking of the tanker Aegean Sea off A Coruña in December 1992, which led to widespread pollution by oil and dispersants (LR). Nesting on islets off Cape Vilán was first noticed in 1979, when there were 60 pairs, after which this colony too declined (Docampo &

Velando 1995). A second major oil pollution incident in the region, from the wreck of the tanker Prestige in November 2002 may also have had deleterious consequences for the population. However, the wreck of the Aegean Sea was not accompanied by any decline in the Cape Vilán population in 1993, nor in those of other breeding seabird species in the area.

Figure 161. Numerical changes in the Spanish breeding population of Kittiwakes (1979–2007). Source: Bertolero *et al.* (2008).

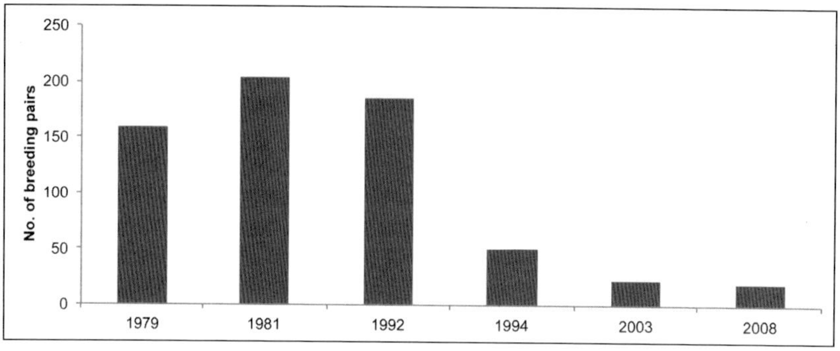

The Iberian Kittiwakes seem to form part of a meta-population that includes colonies in Brittany and in southern Britain. Individuals from Brittany and Wales have been found nesting in Galicia and juveniles from Spain have joined Breton colonies (Bertolero *et al.* 2008). The fate of the Iberian colonies may thus be influenced by recruitment from these other, much larger, subpopulations. The French population is in a comparatively healthy state, having increased from 785 pairs in 1970 to 5,820 in 2003 (OF).

Atlantic Kittiwakes disperse after breeding and lead a pelagic existence throughout the North Atlantic, chiefly north of 35°N, about the latitude of northern Morocco (Malling Olsen 2003, Frederiksen *et al.* 2011). Juveniles and immature birds disperse further than adults and non-breeders remain in the wintering range in summer. Many Kittiwakes occur in the Bay of Biscay and off the western Iberian coasts in winter especially but the numbers seen from land are highly variable. There are at least 23 Iberian recoveries of birds ringed in the British Isles as well as 17 from France and a few from Denmark, Norway and Russia: one from the Kola Peninsula. Three ringed in Spain have been recovered in France and one Spanish-ringed bird was found in Sidi Ifni, Morocco.

Some also winter in the Mediterranean every year, travelling via the Strait of Gibraltar, but sightings are less than annual at Gibraltar and infrequent on Mediterranean coasts so the numbers involved are normally small. Estimates of the wintering population from coastal seawatch points in Catalonia during 2006–09 ranged from 26 to 51 birds but a census in 2006 that included the continental shelf found 107 (CWA). Kittiwakes are most likely to be found in the Mediterranean from late October to March but offshore counts seldom reach double figures (Arcos & Albiol 1997, García-Barcelona 2002). There are a few records of individuals there in April–May. An adult at the Llobregat delta on 12 June 1984 (*Ardeola* 32: 418) may well have been a disorientated survivor from that winter's 'wreck'. However, a record of six adults in Palma Bay, Mallorca, on 26 September 2000 (*A. O. Balears 2000*: 115) is surprising.

The highest counts come from the Biscay coasts where large westerly movements are seen during November–March, generally following periods of strong or gale force westerlies that displace birds into the Bay of Biscay, from where they return westwards in due course. The numbers seen and the rates of passage thus vary greatly from year to year. For example, mean autumn passage rates at Cape Bares during 2005–08 ranged from 2.2 birds/hour in 2006 to 35.3 birds/hour in 2007 (Sandoval *et al.* 2009). However, gales on 12 November 2008 resulted in a movement of 1,451 birds/hour and a new site record of 1,635 birds/hour was set on 22 January 2009, during westward movements following the passage of the remnants of hurricane Klaus (see below). Very large counts are also on record from Punta La Vaca and Gijón Bay. Over 4,300 birds flew west past Gijón Bay in 90 minutes on 6 December 1988 (*Ardeola* 36: 252). Over 5,000 Kittiwakes were lingering off Punta La Vaca on 21 December 1992, as usual in the aftermath of westerly gales, and an exceptional westward passage rate of 2,696 birds/hour, was also reported there on 2 December 1997 (*GIAM* 17: 3–4; *A. O. Asturias 2003*). Such large numbers are not recorded further east along the north coast but significant counts are nonetheless made there occasionally.

Few comparable movements are on record for Portugal, which may partly be due to lack of coverage until recently. The largest counts there have been in winter: 1,200 at Cape St Vincent on 16 February 1986, and 3,100 flying south at Cape Carvoeiro on 27 December 2002, with a further 2,500 flying south at Cape Carvoeiro on 6 January 2003 (AP; *Anu. Ornitol.* 2: 36, 3: 38). Clearly here too Kittiwakes most often occur well offshore.

The passage of Atlantic storms can result in significant 'wrecks' of Kittiwakes on very infrequent, if not rare, occasions. These differ from the more or less annual influxes into northern coastal waters that often follow westerly gales in that the causal storms have been more prolonged and more severe, so that many birds are exhausted and in poor condition through having been unable to feed for several days. The result is that large numbers appear in Atlantic coastal waters, dead and dying birds are picked up on beaches and sometimes far inland, and sometimes large numbers enter the Mediterranean. Kittiwake wrecks will have often gone unnoticed in the past. However, Castellarnau (1877) found a dead individual at Valsaín, Segovia, well inland on the northern meseta, during winter 1875–76, and Valverde (1956) saw a dozen immature birds and several adults on the Río Pisuerga in Valladolid during 2–5 February 1954, shortly after severe storms had affected the Biscay coast. Similarly in 1955 one was found dead at Villahoz, Burgos, on 10 February (*Ardeola* 2: 185) and two were at the Casa de Campo lake in Madrid on 19 February (Díez Ponce de León & Sáez-Royuela 1955). Ferrer *et al.* (1986) also record the occurrence of Kittiwakes in the Pyrenees in 1954, and again in 1957. The latter coincided with a large influx into the Mediterranean from December 1956 to March 1957, when Brosset (1959) found numerous dead birds at the Moulouya estuary, on the south coast of the Alborán basin. A similar event may have occurred during winter 1978–79 when a storm drove 150 inshore at Doñana on 5 January 1979 and a further 150 were found dead on the beach there (García *et al.* 1989).

A widespread 'wreck' occurred early in 1984 when 1,455 were found dead on beaches in Spain and 1,334 in Portugal, chiefly in February (*Ardeola* 32: 418.). Virtually all the birds were on Atlantic coasts but 24 were found on Mediterranean beaches. However, Beaubrun (1985b) found c.1,000 dead birds on Moroccan Mediterranean coasts (and a further 1,500 between Tangier and Agadir) during January–March 1984, so clearly a major wreck affected the region and included a large influx into the Mediterranean through the Strait. One found in Valencia on 28 February 1984 had been ringed in Newfoundland and was the first Iberian recovery of a bird from a western Atlantic colony (Dennis 1987). The 1984 wreck was followed by an exodus from the Mediterranean. This peaked at Gibraltar on 5 March when 330 birds/hour returned west during a four-hour period (BSG). The 1984 event also produced a wider series of inland observations, most involving single birds (*Ardeola* 31: 144). A record of 25 at the Lagunazo de Moncayuelo, Zaragoza, on 6 November 1991 (*A. O. Aragón 1991–1992*: 47) probably also involved storm-driven birds.

The most dramatic and certainly the best-documented 'wreck' on record occurred in late January 2009. Ex-tropical storm 'Klaus' hit southern France and northern Spain on 24 January, when westerly gusts in northern Spain reached 194km/h. Numerous Kittiwakes were observed inshore and unprecedented numbers were driven inland, where many travelled right across northern Spain to Catalonia and the Mediterranean (*GIAM* 32: 8–9). Hundreds were recorded across Navarra, Aragón and inland Catalonia, most of them at reservoirs where they gathered (Aymí *et al.* 2011). At least 900 reached Catalonia, where the larger counts included 127 in Cambrils harbour on 30 January, over 90 at Riba-roja reservoir on 31 January and 107 passing Cape Creus on 7 February (*A. O. Catalunya 2009*: 146–148). Many were clearly in poor condition and large-scale mortality followed over several days. As many as 113 were found dead at the Sabiñánigo reservoir, Huesca, in the Pyrenean foothills on 1 February (*Ardeola* 56: 164–165). This 'wreck' was also apparent further south in Mediterranean Iberia, as well as in the Canary Islands and Morocco, although smaller numbers were involved there. Kittiwakes began to appear within Gibraltar harbour on 23 January and 120 were close inshore there on 24 January. They continued to be recorded through February by when a clear westward movement through the Strait became apparent, with a peak passage rate of 112 birds/hour on 24 February (*GBR 2009*: 24). This movement presumably involved a return of storm-driven birds to the Atlantic and is likely to have included individuals that had crossed northern Spain from the Bay of Biscay to Catalonia in January. Many Kittiwakes remained inshore off southern Atlantic coasts in February: over 1,000 were off the Odiel estuary on 7 February (*Ardeola* 56: 164). The 2009 event was also noted in Portugal, where some numbers of dead birds appeared on beaches but the largest reported counts of live birds were fairly modest: 300 at Cape St Vincent on 29 January, 170 at Cape Raso, Lisboa, on 29 January and 250 between Cascais and Oerias, Lisboa, on 5. A few live birds and some corpses were found far inland in Portugal during the same period, with records from Guarda, Portalegre and Santarém (*Anu. Ornitol.* 8: 74).

SABINE'S GULL *Xema sabini*

(Gaviota de Sabine, Gaivota-de-sabine)

Regular on autumn passage off Atlantic coasts. Vagrant to the Mediterranean.

Southbound Sabine's Gulls occur regularly off Palearctic Atlantic coasts in autumn but usually too far offshore to be visible from land unless stormy weather drives the birds inshore. They are regularly reported in variable numbers off southwestern Britain and off French Atlantic coasts. Some numbers too are reported on passage off Atlantic coastal Morocco. The species is a vagrant elsewhere in Europe and there are very few records from the Mediterranean (Bonaccorsi 2002).

The earliest specific records of Sabine's Gull in Iberian waters arose from observations at sea off the west coast in the mid-

1960s (Bourne 1965, Bourne & Norris 1966) and from a study of autumn seabird passage at Cape Estaca de Bares (Huyskens & Maes 1971). More observations followed from the Biscay coast during the 1970s (*Ardeola* 15: 142, 28: 158, 29: 186; García-Rovés & García-Rovés 1989). The species was considered by the Spanish rarities committee until 1993, by when some 220 individuals had been reported, chiefly due to regular monitoring of autumn seabird passage off Punta La Vaca, Asturias (de Juana 2006). Sabine's Gull remains on the Portuguese rarities list and 24 records had been accepted there up to 2009. There would doubtless be many more Portuguese records if more seawatching occurred in autumn from the western headlands. There are also two records of single birds from Gibraltar, on 7 November 1987 and 16 August 1992 (GONHS), but Sabine's Gull is a rarity there and in the Strait as a whole. In total, our analysis here considers 116 records involving over 1,000 birds, considering for Spain only those published in *Ardeola*.

About half of the observations are of single birds and the others are mainly of twos and threes or small flocks, but some larger concentrations are on record. They include 104 seen 20 miles north of Cape Bares on 29 September 1996 and a gathering of 300–400 off the Miño estuary on 15 November 1999. Some 55% in a sample of 182 birds were adults, the remainder immatures.

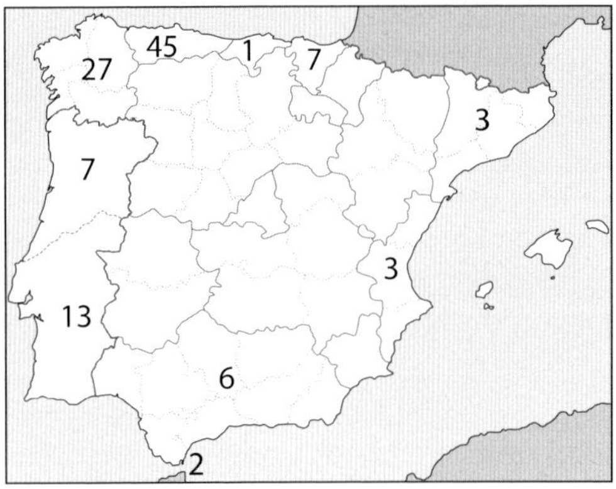

Figure 162. Records by region, 1967–2008 (n = 114)

The greater part of records come from the north-west, chiefly the key watchpoints in Galicia and Asturias (Fig. 162). There are many fewer from Cantabria and the País Vasco, probably because progressively fewer birds reach the eastern sector of the Bay of Biscay. Thus the 27 Galician records in the *Ardeola* sample involved at least 532 birds, a mean of nearly 20 birds per record, whereas the 45 Asturian records involved 132 birds, barely three per sighting. A similar pattern is shown by records in the regional bird reports for 1996–2000, when there were 28 Galician records and 51 Asturian records that averaged 18.3 and 4.3 birds/record respectively. The Gipuzkoa bird reports for this latter period give only six records, totalling 19 birds. In our general sample there are only eleven records from the Mediterranean and the sole inland record was not far from the sea: in the Guadalquivir delta near La Puebla del Río, Sevilla.

Figure 163. Records by month, 1967–2008 (n = 116)

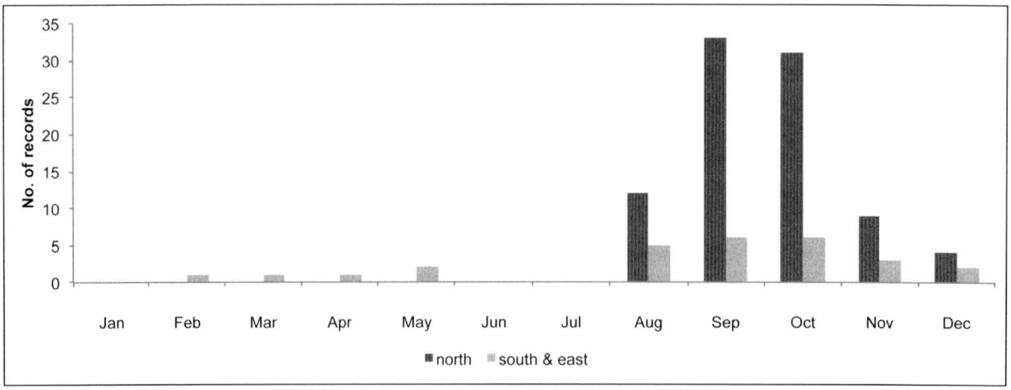

The seasonal distribution of records shows that the overwhelming majority are in autumn, as elsewhere in Europe, peaking during September–October (Fig. 163). Iberian passage dates peak somewhat later than in Britain, where the movement peaks in early September (Dymond *et al.* 1989). The Iberian passage dates range from 6 August to 30 December. Juveniles pass later than adults: they make up 17% of birds seen in August, 34% in September and 57% in October.

There are just five spring records: single birds at Adra, Almería, on 8 February 1998; at Cullera, Valencia, on 26 March 1988; at Cape Raso, Lisboa, on 24 May 1980; at sea north-west of Cape St. Vincent, on 19 April 2006, and at sea off Cape Roca, Lisboa, on 1 May 2008. The rarity of spring observations off western Europe suggests that the northward passage occurs further west in the Atlantic and that the birds involved during both passages are of Nearctic origin. Significant numbers are reported moving north in spring off southern Morocco (BM), from where they presumably head north-east, avoiding Europe altogether.

LITTLE GULL *Hydrocoloeus minutus*

(Gaviota Enana, Gaivota-pequena)

Widespread but generally scarce on all coasts on passage and in winter. Rarely seen inland.

The western populations winter chiefly at sea in western and southern European waters, including the Mediterranean basin. There are at least two Iberian winter recoveries of birds ringed at a nesting colony in Latvia and one from Finland (SWA). The birds are pelagic in winter and the largest numbers seen on Iberian coasts tend to occur after prolonged onshore gales. However, the patterns of occurrence of Little Gulls in Iberian waters are incompletely understood and may also have changed over recent decades.

Many Little Gulls from the Baltic colonies reach western European waters, wintering from the southern Irish Sea and western Channel approaches south to the Bay of Biscay and beyond (Wernham *et al.* 2002, OF). Influxes into French coastal waters in Brittany and the Bay of Biscay occur during stormy winter weather, suggesting a regular presence offshore there (OF). A series of monthly surveys off Punta Galea, Bizkaia, during 2001–03 found that wintering Little Gulls were present regularly during November–February beyond the continental shelf, some 15–20 miles off the Spanish coast (Ocio & Astigarraga 2008). The Little Gull is better known as a wintering species than as a migrant in many coastal areas but it appears irregularly inshore and generally only after prolonged stormy weather. The pan-Iberian census of wintering gulls and terns of January 1984 (Bermejo *et al.* 1986) found only 20 Little Gulls in total in the Mediterranean and a modest 272 on Atlantic coasts, 231 of them in Huelva at the Odiel and Tinto estuaries. However, there is marked between-year variation in the abundance of Little Gulls in coastal waters in winter, which undoubtedly results from the pelagic habits of the species, and January 1984 may have been a poor year in this respect.

It is possible that the Little Gull may have become less abundant as a wintering species in the western Mediterranean basin in recent years, since most larger counts there date from the 1970s and 1980s. For example, January counts at the Llobregat delta found up to 1,270 in 1977, 601 in 1979, 500 in 1980 and 841 in 1981 (Carrera *et al.* 1981a, Carrera 1983) and 1,000 birds were at the Ebro delta in February 1971 (Ferrer 1977). These counts are much larger than more recent ones: the January counts in Catalonia during 2006–09 found only 4–7 individuals, and the Little Gull is now regarded as a very scarce wintering species in Catalonia (CWA). In the Comunidad Valenciana there were 200 at the Turia estuary, Valencia, in January 1983 and 60 were there in January 1985, but here too recent winter counts have been much smaller in more recent years: in 2009 the largest counts were 19 at the Turia estuary and five in Valencia city harbour (*Ardeola* 32: 416, *A. O. C. Valenciana 2009*: 86).

The Little Gull is seen very irregularly in winter in the Strait of Gibraltar but influxes have occurred there following prolonged easterly gales during December–February. Gatherings of up to 200 birds have then appeared at Gibraltar itself but the birds disperse after a few days (GONHS). Such influxes are very far from annual and none at all were observed during 2001–10 (GBR). Similar variation is apparent in the numbers wintering in Atlantic waters off southwestern Andalucía. Garrido (1996) noted that the winter population of the Huelva coast west of Doñana varied between years from zero to over 500 birds. Recent records in this area include 190 in Doñana on 12 February 1996 and over 300 at the Marismas del Odiel during winter 2008–09 (*Ardeola* 44: 254, 45: 248; Gacio & Sayago 2011).

The Little Gull occurs regularly on passage on Biscay coasts in August–December and March–May but is generally scarce onshore (Paterson 1997, Noval 2001). However significant movements have been reported there in autumn, especially in November and chiefly in the aftermath of stormy weather, once again suggesting that larger numbers may be more frequent offshore off northwestern Iberia. For example, 206 were sheltering in Gijón Bay, Asturias, on 23 November 1991 following three days of storms and 850 flew west at Punta la Vaca nearby on 20 November 1992, most of them in a single flock (*Ardeola* 40: 97). Passage at Cape Estaca de Bares in 2008 was noted from late October to mid December, when 620 birds flew west, with a pronounced peak on 25 November when 493 were counted (Sandoval *et al.* 2009). In 2009 1,448 Little Gulls were seen from Cape Bares, most of them during the first ten days of December, but 171 were seen from mid January to mid February: once again suggesting a winter presence offshore. In 2011, 3,204 were counted at Cape Bares during the second half of the year (A. Sandoval pers. comm.). An influx during the spring passage period was noted at Txingudi Bay, Gipuzkoa, during 4–10 April 1994, when about 50 were present, again following a storm (*Ardeola* 42: 223).

Scarce migrants are seen regularly during both passage periods from western Iberian Atlantic coasts but there is no available information on a possible winter presence at sea here in winter. The Little Gull is generally scarce in Portugal, both on passage and in winter, although small numbers occur regularly. A count of 120–150 at the Douro estuary during 29 March–1 April 2001 (*Anu. Ornitol.* 1: 43) is the highest to date in the country: most records are of just a few individuals or flocks of up to 20. However, the passage counts in Galicia and the Biscay coastlands given above suggest that some numbers may occur regularly further off Portuguese coasts.

Spring migrants are very much more evident in southern Iberia and in Iberian Mediterranean waters in spring than in autumn. They are generally scarce then on the coasts of Huelva and Cádiz but larger numbers occur on spring passage in some years. For example over 100 were counted at Doñana on 13 May 1984 (*Ardeola* 31: 144). Some migrants are recorded at the Strait of Gibraltar from mid March to mid May and, much more rarely, during August–October (Garcia 1973, BSG). Very few birds are usually involved, which may indicate that the wintering range does not extend far south of the Peninsula: the Little Gull is considered uncommon to scarce in Morocco and there are no records of other than small groups there in winter (BM), although there is no available information from offshore waters. A record of 881 flying east off Gibraltar on 31 March 1990 (*Ardeola* 38: 341) is highly exceptional and followed easterly gales, suggesting that the birds were displaced from the western Mediterranean, but they could also have been on spring migration through the Strait. Small numbers: in flocks of up to 50 birds, have been seen flying north up the Moroccan Atlantic coast of the Tangier peninsula between mid March and early May (Pineau & Giraud-Audine 1976), which appears to indicate that there is also wintering presence off northwestern Africa.

In Málaga Bay some 90% of observations during 1981–83 were in spring, with a peak in sightings in April–May. Annual sightings during the three years varied considerably, totalling 103, 1,830 and 48 birds respectively. Adults comprised only 4.6% of sightings and were only seen in March–April (Paterson 1987). The Little Gull is markedly more common along the coasts of Catalonia and the Comunidad Valenciana in spring than in autumn. The Catalan bird reports (*A. O. Catalunya*) for 2000–09 note that some 53% of at sea observations were during March–May and just 12% in August–October. Autumn observations usually involve very few birds but spring counts are often of hundreds, with over 6,000 birds at L'Escala on 26 April 2002 (*A. O. Catalunya 2002*: 269). Northward passage of Little Gulls is often apparent from Costa Brava watchpoints during March–May, larger recent counts including 1,487 at Palamós on 5 April 2008 and 1,400 at Llançà on 7 April 2009 (*A. O. Catalunya 2008*: 127, 2009: 136). Significant movements have also been noticed at the Llobregat delta during February–May, peaking in mid-April at up to 1,000–2,000 birds/day (Gutiérrez *et al.* 1995). The Little Gull is also generally scarce on passage on the coasts of the Comunidad Valenciana but here again it is commoner in spring than in autumn. However, very few indeed are seen in the Balearic Islands and there are only six records in the regional bird reports 2001–2010. A similar migration pattern to that in eastern Spain is also reported in Mediterranean France, where the spring passage is pronounced in spring but poorly apparent in autumn (OF). Northward passage inland across eastern France is also seen in spring. It may be that there is a loop migration around Iberia in which many of the Little Gulls that migrate southwards off the Atlantic coasts of France and Iberia return northwards in spring off the eastern Iberian coasts (de Juana & Paterson 1986).

A few non-breeders are reported summering on coasts, for example in Málaga bay (Paterson 1997) and at the Llobregat delta (Gutiérrez *et al.* 1995). A second-year bird remained at the Ebro delta from June–September 2003 (*A. O. Catalunya 2002–2005*: 270). The Little Gull is rare inland but there are scattered records from many provinces, including sites in Castilla y León, Castilla-La Mancha, Madrid, Extremadura and the Ebro basin. Most observations involve single birds or very small numbers and the majority are observed during passage periods. There were at least 52 records in Madrid between 1954–2009, all between August and May with peaks in January and March, the latter probably reflecting spring passage (Juan *et al.* 2009). A count of 20 at the Laguna de Sariñena, Huesca, on 19 December 1988 (*Ardeola* 36: 250) is also noteworthy in view of the exceptional number of birds involved.

GULL-BILLED TERN — *Gelochelidon nilotica*

(Pagaza Piconegra, Tagaz)

***G. n. nilotica*. Breeds locally, often in large colonies, chiefly in southern and eastern mainland Spain and southeastern Portugal. A scarce migrant in the Balearic Islands.**

The Gull-billed Tern has an important breeding presence in Iberia, where the population of over 6,000 pairs constitutes 45–50% of the Western Palearctic population, following severe declines in eastern and central Europe in recent decades (BE, Sánchez Guzmán *et al.* 2004). In Spain it nests principally in Andalucía, Extremadura, Castilla-La Mancha and the Levant coastal wetlands. It is absent from most of the northern third of Iberia and from the Balearics, where it is only a rare migrant. In Portugal the breeding range has expanded in recent years and it occupies sites in eastern Beja and Évora, in the Guadiana valley, as well as in eastern Portalegre (PA2). It is a summer visitor, wintering in sub-Saharan Africa. Birds ringed in Spain have produced three recoveries from France and one from Italy, and seven in North Africa: Morocco (5) and Algeria (2), as well as eight in sub-Saharan Africa in January–February: from Guinea Bissau (4), Senegal (2), Sierra Leone (1) and Mali (1); the countries that accommodate most of the African wintering population (Zwarts *et al.* 2009).

It nests colonially, usually on islands in wetlands, both inland and on coasts. Colonies are generally at natural or seminatural lagoons, at coastal sites such as marismas and saltpans, and at reservoirs. The Gull-billed Tern is highly

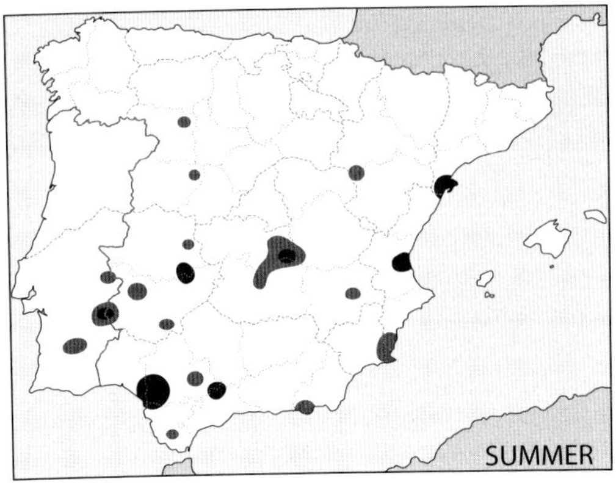

opportunistic with respect to diet but breeding birds principally forage over farmland, including ricefields and pastures, where they capture insect prey both in the air and on the ground. Their diet especially features large insects, such as locusts and grasshoppers, but includes a wide range of other animal prey that is captured as available. These items include earthworms, crustaceans, small fish, amphibians, lizards, small mammals and the eggs and chicks of other birds (Sánchez Guzmán & Fasola 2002). Their chief prey at both Doñana and the Albufera de Valencia is the Red Swamp Crayfish that inhabits the ricefields: it comprises some two-thirds of prey items at the latter site (Costa 1984, Dies *et al.* 2005). In fact, a large proportion of the Spanish population frequents the main rice-growing areas: Doñana, the Ebro Delta, the Albufera de Valencia and the Vegas Altas in Extremadura.

Gull-billed Terns are often seen during the breeding season at or around wetlands where they do not nest and readily switch to new sites when these become available. Their colonies fluctuate in numbers from year-to-year, especially those at lakes and reservoirs, where water levels may be highly variable between seasons. Such sites may only be occupied opportunistically and many lakes and reservoirs in La Mancha and in Extremadura support nesting birds in some years but not in others. Nesting in Extremadura is a relatively recent phenomenon that has resulted from the proliferation of reservoirs there, many of which offer islands that are ideal nesting locations. In this region the first colony established was at the Orellana reservoir, Badajoz, in 1989 (Sánchez & Sánchez 1991). Since then the population has been highly mobile and nesting has occurred at the Sierra Brava, Valdecañas, Alange, Los Canchales and Arroyo Conejo reservoirs as well as at Casas de Hito and the Charca de Gorbea (SA2). The Extremaduran population has grown from c.100 pairs initially to over 1,000 pairs in some years. In 2007 only Sierra Brava, Valdecañas and Alange had breeding birds and the great majority, 820 pairs, were at Sierra Brava, making it one of the largest colonies in Spain (Corbacho *et al.* 2009).

Breeding birds were well known at Doñana and the Albufera de Valencia in the late 19th century (Saunders 1871, Arévalo 1887) but the colonies clearly suffered from persistent egg harvesting by local people, as remarked by Noble (1902). Valverde (1960) estimated that there were then two colonies totalling just 200–500 breeding pairs in the Marismas. Another long-standing breeding colony known is that of Fuentedepiedra, where almost 700 nests were counted in 1967 (Studer-Thiersch & Studer-Thiersch 1969). At that time the Gull-billed Tern was still largely absent from the eastern and the central sectors of the Peninsula (Gil Lletget 1945), although Bernis (1952) found two birds at Daimiel and one at the Laguna de El Taray, Toledo, in June 1951. Much later, the Spanish population has shown an increasing trend since it was first estimated at c. 900 pairs in 1972 (Møller 1975), despite the often very large fluctuations in the numbers nesting at individual colonies. There were over 2,000 pairs by the late 1990s and 3,000–3,500 pairs by 2003 (Corbacho *et al.* 2009), with a rapid increase since then. The birds are nonetheless subject to disturbance by human activity and predators, to the effects of insectides used to control grasshopper 'plagues' and to the vagaries of rainfall.

In 1989 the Spanish population of some 2,000 pairs was mainly in Andalucía: 700 pairs at Doñana and 300 at Fuentedepiedra, with 400 in Extremadura. There were then also 170 pairs at the Ebro delta, 60 pairs at Gallocanta and 40 pairs at La Mancha wetlands (Martínez-Vilalta *et al.* 1991, Corbacho *et al.* 2009). Breeding at the Ebro delta was first noted in 1958 (Dorka 1960) but by 1980 there were only 54 pairs there (CA1). The first reports of nesting at Gallocanta were in 1973 (Aragüés *et al.* 1974). In 1989 the Gull-billed Tern was not yet present on the Mediterranean littoral south of the Ebro delta, where it first colonised the Albufera de Valencia in 1994 (Dies *et al.* 1999), nor on the northern Meseta, where breeding at the Lagunas de Villafáfila was first recorded in 1990 (Palacios & Rodríguez 1993). Since then a few pairs have also nested occasionally elsewhere in Castilla y León: at the Azud de Riolobos, Salamanca, in 2000 and 2001 (*A. O. Salamanca* 1924–2003: 112).

The Spanish population was censused in 2007 (Corbacho *et al.* 2009), when at least 5,764 pairs were located, the largest total on record. To these can be added a further 95 pairs that nested at Villafáfila (*Ardeola* 55: 145). The Portuguese population numbered over 500 pairs in 2005/2006 (PA2). The Spanish birds comprise six subpopulations: those of Doñana; the Extremaduran reservoirs; Fuentedepiedra; the La Mancha wetlands, the Ebro delta, and other Levant coastal wetlands, which together held 82% of the population in 2007 (Table 45). The remaining 18% breed in small scattered colonies in other regions: in 2007 these included 20 pairs at Gallocanta and 95 pairs at Villafáfila. A marked increase to 1,031 pairs was reported at the Albufera de Valencia in 2009 (*A. O. C. Valenciana 2009*: 87) and 1,100 pairs nested in Extremadura in 2013 (*Ardeola* 60: 530).

Table 45. Distribution of the Gull-billed Tern breeding population in Spain in 2007. After Corbacho *et al.* (2009).

Region	No. of pairs	No. of sites	%	Principal sites
Andalucía	2,419	9	41.3	Doñana, Fuentedepiedra
Aragón	20	1	0.3	Gallocanta
Castilla-La Mancha	732	9	12.5	Peñahueca, Manjavacas, Camino de Villafranca and Salicor lagoons
Castilla y León	95	1	1.6	Villafáfila
Catalonia	731	1	12.5	Ebro delta
Comunidad Valenciana	660	2	11.3	Albufera de Valencia
Extremadura	990	3	16.9	Valdecañas, Sierra Brava and Alange reservoirs
Murcia	212	1	3.6	San Pedro del Pinatar saltpans
TOTAL	5,859	27		

The Portuguese population has also grown rapidly and numbered over 500 pairs in 2005/2006 (PA2). The Gull-billed Tern nested at the Ria de Aveiro during the late 19th and early 20th centuries, and perhaps as recently as 1967 (Tait 1887, 1924; Paulino d'Oliveira 1896, AP). Perhaps sporadic nesting was also reported at the Sado estuary in 1937 (one nest found) and in the Mira region in 1947 (Coverley 1939, AP). Confirmed nesting was not known in Portugal thereafter until 1999, when 120 adults were present and at least ten pairs nested at the Herdade do Esporão reservoir, Évora (PA2), where 350 birds were counted in May 2008 (*Anu. Ornitol.* 7: 83), but breeding at that site is irregular. Since 2002 the principal breeding site has been the massive Alqueva reservoir on the Guadiana, which includes many small islands ideal for the species. The colony there grew from 105–130 pairs in 2003 to 575 pairs in 2005; but only 316 nested in 2009 (*Anu. Ornitol.* 6: 62, 8: 76). Breeding has also occurred in some years at other reservoirs: at the Barragem do Roxo, Beja, (150 pairs in 2000, AP) and at the Barragem do Caia, Portalegre (70–75 pairs in 2005) (*Anu. Ornitol.* 5: 54).

Gull-billed Terns mainly appear in Iberia in April–May, with some early arrivals in late March. They depart during August–September, with some lingering into October or even early November in the south. Some post-breeding dispersal is noted inland in summer and small numbers then occur regularly at the Tagus estuary, where none breed: a record of 29 there on 17 July 2005 (*Anu. Ornitol.* 5: 54) is typical.

There are Iberian recoveries of birds ringed in Denmark (8), France (2) and Ukraine (1) but western populations north of Iberia are small and any associated passage through the region is likely to be limited following the large-scale declines in France: to 300 pairs in 2007 (OF), and northwestern Europe. Nevertheless, a few migrant Gull-billed Terns are recorded flying west along the coast of Asturias in autumn more or less annually, and there are a few records at this season from Cantabria and the Basque Country (Paterson 1997; *Ardeola* 36: 252). Spring records in northern Spain are even rarer. Few migrants are ever recorded on passage along western or eastern coasts but small-scale movements are sometimes seen at the Strait of Gibraltar, with occasional records of individuals or small flocks moving east from late February to early June and west during late July–October (Finlayson & Cortes 1987), chiefly in April and August. Trans-Saharan passage occurs in Morocco (BM), at least in spring, so possibly movements tend to be on a broad front.

There are rare records of individuals in winter: for example, at the Odiel estuary on 9 December 1986 (Garrido 1996), at Villaviciosa, Asturias, on 25 December 1988 (Paterson 1997) at Ayamonte, Huelva, on 21 January 1991 (Garrido 1996) and at the Ria de Alvor, Faro, on 16 January 2008 (*Anu. Ornitol.* 7: 83), as well as at least seven records from Catalonia during 1996–2010, all in coastal Tarragona.

CASPIAN TERN *Hydroprogne caspia*

(Pagaza Piquirroja, Garajau-grande)

A scarce passage migrant. A small number winter on southwestern coasts and locally in the Mediterranean. Has bred at the Ebro delta.

The nearest breeding Caspian Terns to Iberia are in the Baltic and in Mauritania. However, a single pair nested successfully at the Ebro delta in 1988 (Martínez-Vilalta 1988). Western Palearctic birds chiefly winter in sub-Saharan Africa, especially in the Inner Niger delta (Zwarts *et al.* 2009) but a small number occur regularly in Iberia in winter, chiefly on southwestern coasts. A few also occur in winter in the Mediterranean, chiefly at the Ebro delta and the Santa Pola saltpans, Alicante. There are Iberian recoveries of birds ringed on the Baltic coasts of Finland (at least 17), Sweden (16) and Estonia (1), most of them found on the Huelva coast.

The total wintering population may not much exceed 100 individuals in most years. The January 1984 census of wintering gulls and terns (Bermejo *et al.* 1986) found 44 birds: one in Algarve and 43 on the Huelva coast. Subsequent observations confirm the regular presence of wintering Caspian Terns in these areas and also on the Atlantic coasts of Cádiz. The 2008–10 Spanish winter Atlas found 118 birds on average, 67% of them in Cádiz Bay and most of the remainder elsewhere in western Andalucían estuaries: just 1–8 were at the Ebro delta (SWA). Small numbers may be encountered at this time anywhere between the Tagus and Sado estuaries in southwestern Portugal and the Strait of Gibraltar but the birds are most commonly encountered in bays and estuaries and at nearby saltpans. They are usually in ones and twos, the latter often comprising an adult accompanied by a still-dependent juvenile. Larger groups occur where birds gather to rest or roost but flocks of more than ten are uncommon. The Ria Formosa, Faro, is the principal wintering location in Portugal and there is a record winter count of 80 birds there in 2006 (AP). Some 10–15 winter at the Marismas del Odiel, exceptionally up to 40 birds (Gacio & Sayago 2011). Favoured locations on the Cádiz coast include the Chiclana saltpans, Puerto Real and the Marismas de Sancti Petri, where peak counts of 20–30 birds have been recorded in January–February. Winter records from the Ebro delta and Santa Pola have usually involved just one or two individuals, occasionally three, but there were up to eight at the delta in January 2009 (*A. O. Catalunya 2009*: 149). Wintering there may be a recent development since the first winter records were in 1996–1997 (CWA).

Small-scale passage occurs along coasts. Spring and autumn records are equally frequent on the Mediterranean coast but autumn observations predominate in the north (Fig. 164). Some small-scale westward passage is reported from the Biscay coastlands during August–October. The majority of observations there are from Asturias but even there the species is less than annual: for example, 34 were reported in total during 1975–97, 20 of them in September 1994 (*A. O. Asturias* 4: 152–153, Paterson 1997). There are none from Portugal north of the Tagus (AP) but presumably there is some southward passage there in autumn.

Mediterranean coastal sites, such as the Albufera de Valencia and the Santa Pola saltpans regularly see migrants in ones and twos and occasionally in groups of up to five: nine at Santa Pola on 7 March 1990 was an exceptional local record (*A. O. C. Valenciana 1990*: 50). Similar numbers occur on passage at most Catalan wetlands but larger concentrations, of up to 30, are regularly reported from the Ebro delta, most often in September but also in March–April. The scarce records for Gibraltar show passage during March–May and late July–October, but most records there are in autumn: there were 28 records (48 birds) during 1976–2008 (GONHS). Mean flock sizes in Spain during both passage periods are very small: 1.7 in spring (n = 93) and 1.9 in autumn (n = 158). Single birds comprise 55% of observations, pairs 25%, trios 10% and flocks of 4–10 also 10% (n = 251). Adults made up two-thirds of 71 birds whose ages were recorded during August–October, the remainder being juveniles.

Figure 164. Monthly distribution of Caspian Tern records in northern and eastern Spain. North = the Biscay coastlands from the Basque Country to Galicia (n = 58). East = Catalonia, the Comunidad Valenciana and Murcia (n = 245)

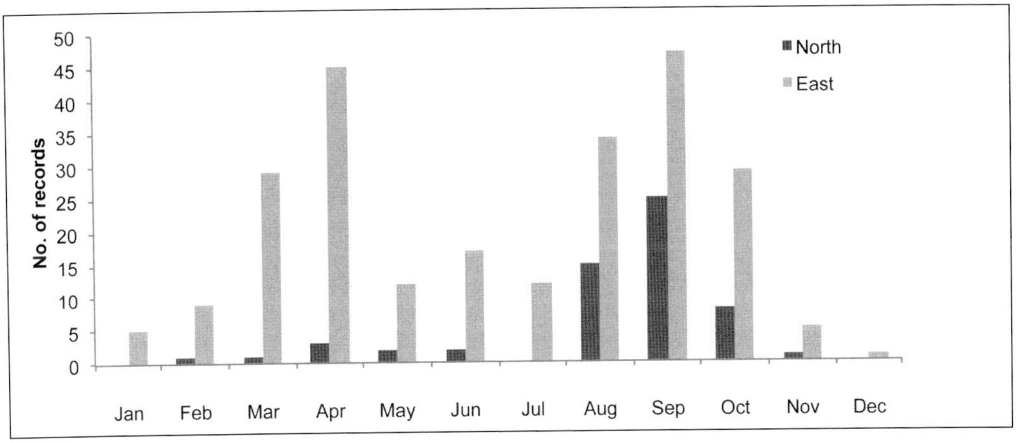

Very few Caspian Terns are seen in the Balearics, where occasional migrants occur less than annually. Ferrer *et al.* (1986) give six records of single birds, all from Mallorca. Clavell (2002) documents a further 16 records (17 birds) during 1986–99, one from Ibiza and all the rest from Mallorca: principally at the Albufera. Together with a further eleven observations (13 birds), all from Mallorca, during 2001–10 (*A. O. Balears*), they include seven April records, one in June, two in July, six in August, eleven in September and four in October.

Caspian Terns are rarely seen inland, although they may venture some way up major rivers. They occur at least occasionally as far up the Guadalquivir as Sevilla and may do so regularly. A very few individuals have been recorded much further from the sea. They include at least two records each from Aragón, Navarra and Madrid, three each from Castilla-La Mancha and Castilla y León, and nine from Extremadura. In Portugal there were two at the Albufeira do Caia, Elvas, on 17 September 2005 (*Anu. Ornitol.* 5: 54). All inland records have been during passage periods and involve birds seen at lakes and reservoirs.

A very small number of non-breeders are reported in June and July, again most often in the south-west. Records generally involve just one or two birds but there were eight at Castro Marim on 8 July 2007 and seven at Sanlúcar de Barrameda on 10 July 2011 (*Anu. Ornitol.* 6: 62, *Ardeola* 58: 503). The species is long-lived and first breeding is usually not before the fourth year. It has been assumed that summering non-breeders found in Iberia and Morocco are from European colonies but the possibility exists that at least some of them may have dispersed northward from Mauritania, as suggested by Thevenot *et al.* (BM).

ROYAL TERN *Sterna maxima*

(Charrán Real, Garajau-real)

Rare individuals reach southern Spain and Gibraltar. Vagrant to Portugal.

The Banc d'Arguin in Mauritania is the northernmost breeding location of *S. m. albidorsalis*, with some 15,000 pairs in 2004 (Isenmann *et al.* 2010). After breeding most disperse south but some hundreds move north, chiefly in August–October. Very few reach northernmost Morocco and most return south to winter. Some non-breeders remain on Moroccan coasts in June and early July, chiefly in the south (BM). Royal Terns of the nominate race also disperse north along the North American east coast: there are single recoveries of birds ringed in North America from Britain in 1979 and Spain in 1989 (Dennis 1994). Seven other records of Royal Terns from the British Isles and one from the Azores are also likely to have been of Nearctic origin. The sole record for France, in April 2005, was one identified as *albididorsalis* (Reeber *et al.* 2008).

The Royal Tern was not reported in Iberia until the 1970s, when there were records from Sanlúcar de Barrameda, Cádiz, on 25–27 April 1970; Isla Cristina, Huelva, on 4 July and 24 September 1977, and Estepona, Málaga, on 16 October 1979 (*Ardeola* 16: 262–263, 24: 260–261, 28: 186). These were followed by reports from Algeciras, Cádiz, on 19 September 1981; the Guadalhorce estuary, Málaga, on 20 September 1982, and Torremolinos, Málaga, on 14 July 1983 (*Ardeola* 29: 186, 30: 120). Since 1984 there have been 22 accepted records (27 birds) from Spain and two of single birds from Portugal. There are also 11 accepted records of single birds from Gibraltar during 1980–2006.

Figure 165. Records by month, 1970–2010; only first dates of observation are included (n = 41)

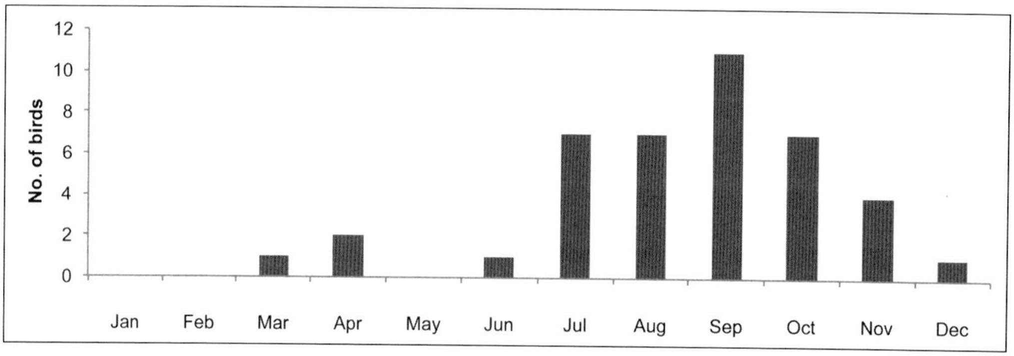

The 41 Iberian records involve 50 individuals. Some 80% of observations were of single birds but there are seven of two together and one record of three: at the Marismas del Odiel, Huelva, in July 2000. Twenty of those reported during July–November were adults and only one was a juvenile, seen with an adult at Tarifa beach in September 1989. All but three of the records have come from the southernmost coasts of the Peninsula, especially at or near the Strait (20 records). There are numerous additional observations from the southern shore of the Strait, of groups of up to eight during September–October (BM, Finlayson 1992). The geographical location of these records strongly indicates that the birds originate from the west African colonies. The temporal distribution of Royal Terns in Iberia also corresponds with the northward dispersal period of the west African birds: 88% of Iberian records fall between 4 July and 7 November (Fig. 165).

Although most of the Royal Terns found in Iberia are virtually certainly of the race *albididorsalis*, the American race *maxima* has reached the Peninsula at least once: a bird that was accidentally hooked by a fisherman at Roses, Girona, on 26 December 1989, had been ringed as a juvenile in Virginia, USA, the previous August. There are two additional records from Catalonia, both of unknown subspecies: at the Llobregat delta in July 1984 and at l'Ametlla de Mar in August 2010.

LESSER CRESTED TERN *Sterna bengalensis*

(Charrán Bengalí, Garajau-pequenho)

***S. b. emigrata*. Very scarce but regular on passage on Mediterranean coasts and in the Strait of Gibraltar. Rare in winter. Vagrant to Portugal and northern Spain. Has bred sporadically in eastern Spain.**

The Lesser Crested Tern has its core distribution in the Indian Ocean east to northern Australia. However *S. b. emigrata* breeds only in the south-central Mediterranean, on islands in the Gulf of Sirte off the Libyan coast, wintering mainly on western African coasts from Mauritania to Senegambia (HBW). There are records of 40 individuals in France up to 2011 (Reeber *et al.* 2012), these mainly in May and July–August, the latest being in September (OF). Passage to and from the Atlantic winter quarters occurs through the Strait of Gibraltar, where individuals and small flocks are regularly recorded on both the African and Iberian shores. In addition, a few individuals have nested in Spain within the large Sandwich Tern colonies at the Ebro delta and the Albufera de Valencia.

Breeding in Spain is a recent development but nesting has only been sporadic and most instances have involved one or two individuals that paired with Sandwich Terns. The first record was in 1979 at the Ebro delta and individuals were recorded at the Sandwich Tern colony there annually from then until at least 2002. One or two mixed pairs were the norm each year but there were one or two pure pairs in 2001 and 2002 (SA2). One or two mixed pairs were also reported at the Albufera de Valencia annually during 1994–98 and one of these fledged a hybrid juvenile in 1997 (Dies & Dies 1998, Dies *et al.* 1999). Individual Lesser Crested Terns are occasionally reported from Levant coasts from late April to August (Clavell 2002) and these seem likely to include the birds that have attached themselves to the Sandwich Tern colonies. The species is a rare vagrant to the Balearics: one was recorded at the Albufera de Mallorca on 16 August 1997.

Lesser Crested Terns are most likely to be encountered at the Strait of Gibraltar and on the shores of the Alborán basin but even here there are no records of large or even moderate numbers on the Iberian side. The first recorded observations were during the 19th century, when Saunders (1871) obtained a specimen shot off Gibraltar and Irby (1895) referred to the capture of two at Tarifa in April 1873. Nevertheless, the species was considered accidental in Iberia until quite recently (AMI) and Spanish records have been subject to review by the rarities committee since 2006: 23 records were accepted during 2006–10. This is surprising since the Libyan population numbers some 10,000 pairs (HBW), which would be expected to result in a more conspicuous presence at the Strait. It may be that an unexpectedly high proportion of the Mediterranean breeding birds remain there in winter, or travel to the Indian Ocean rather than the Atlantic. Observations at the Strait and its approaches indicate a concentrated and rapid spring passage that peaks in May and a much more conspicuous and prolonged autumn passage that builds up slowly from late July to September, peaking in October and extending through November (Fig 166).

Figure 166. Monthly distribution of observations from the northern shore of the Strait of Gibraltar and southern coasts of Andalucía in 1979–2010 (n = 68 records, 133 birds). Only accepted records have been included from 2006–10.

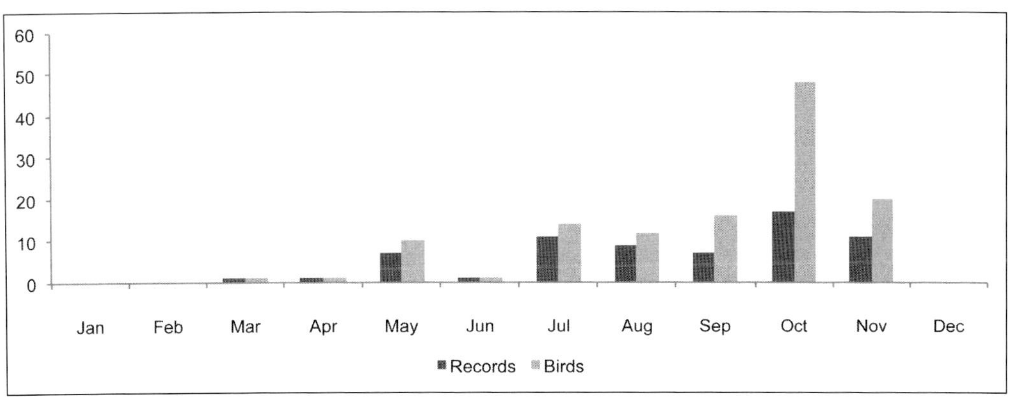

The available records suggest a sparse movement, certainly along the Iberian coastlines. The observations comprising Fig. 167 were 40 records of single birds, 17 of pairs, seven of 3–6 birds and only one record of more than six: 20 off Gibraltar in October 1979 (C. Perez, pers. comm). There were subsequent records from Spain of at least 22 birds during 2011 and 32 in 2012, most of them from or near the Strait, often at Los Lances beach. Occasional birds are reported from western Cadiz and Huelva coasts during passage periods, usually singly, but three were at Odiel on 7 July 2001 (*Ardeola* 50: 163). There are two accepted records from southern Portugal: an adult at Sagres on 10 October 2003 and an adult at the Lagoa dos Salgados on 27 October 2006, and a later record of an adult at Oeiras, Lisboa, on 23 August 2011 (RBP). There are also exceptional records of vagrants on the Spanish north coast: in the Basque Country, at Plaiundi on 21 October 1989; in Galicia, at the Ría de Foz on 17 May 2009, and one inland in Aragón, at the Barasona reservoir on 2 August 2006 (*GIAM* 8: 9; *Ardeola* 55: 278, 59: 388).

Some larger counts have been made from the southern coast of the Strait. Groups of 20–30 were present at Oued Smir, south of Ceuta and on the immediate south-east approaches to the Strait, during 4–8 September 1975 and 45 were at the same site on 4 October 1973 (Pineau & Giraud-Audine 1979). More recently, a flock of 62 resting on the shore at Ceuta on 21 September 2010 (*Ardeola* 57: 536–537) was much larger than any flock ever recorded in Iberia and also larger than any of the concentrations reported for Morocco (BM). One of the individuals in the Ceuta flock had been colour-ringed in Libya and was the second such bird recorded there. A Libyan-ringed bird was also recorded at Doñana in July 2011 (RBS). The 2010 Ceuta record was eclipsed by a count of 240 coasting at Ceuta on 5 October 2012 (*Ardeola* 59: 440). This last record is additional evidence that, in general, most Lesser Crested Terns seem to pass through the Strait by following the southern shore, which may account both for the relative lack of observations on the northern coast and for the overall sparsity of observations: the Iberian shore is much better-watched than the African one. Some winter on the southern shore of the Alborán basin in Morocco, apparently to an increasing extent (BM). Winter occurrences in Iberia are very rare but a first-winter bird was at Cape Cervera, Alicante, on 31 January 1991 (*Ardeola* 38: 342).

SANDWICH TERN *Sterna sandvicensis*

(Charrán Patinegro, Garajau-de-bico-preto)

***S. s. sandvicensis*. Nests very locally in Spain. Common on coasts on passage and in winter.**

The Sandwich Tern is best known in Iberia as a common migrant and moderately numerous winter visitor on coasts. However, it has become established as a regular breeding species since 1971, when four pairs nested at the Ebro delta. It may nonetheless have nested in Spain in the late 19th century since Saunders (1871) and Arévalo (1887) mention nesting at the Ebro delta and at the Mar Menor. The most recent assessment of the nesting population (Corbacho *et al.* 2009) found 3,796 pairs in 2007. The great majority (99%) comprised the large colonies of the Albufera de Valencia and the Ebro delta. The remainder nested at two small colonies in Alicante that may be seen as satellites of the Albufera colony, and at one site on the Biscay coast: Merón beach in easternmost Asturias, where seven pairs bred.

Figure 167. The increase in the Spanish breeding population of the Sandwich Tern 1971–2007. After CA2 and Corbacho *et al.* (2009).

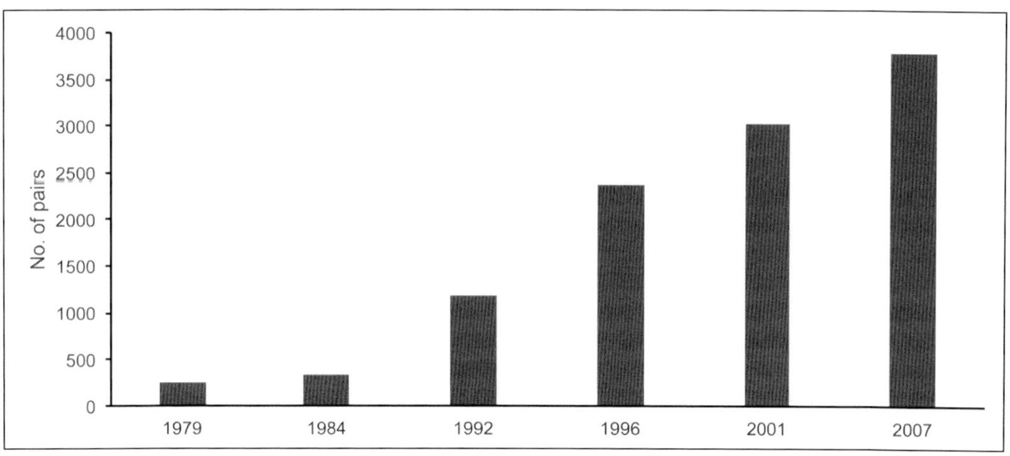

Breeding at the Ebro delta occurred in 1961 but the species only became established after 1971, since when the population grew rapidly to peak at 1,889 pairs in 1996 (CA1, Martínez Vilalta 1996). Nesting at the Albufera de Valencia was first reported in 1984 but regular breeding began in 1993, again with the nesting of four pioneer pairs (Dies *et al.* 1999). This colony now outnumbers the Ebro colony: the largest count to date there was 2,586 pairs in 2006. In general, the Spanish (and Iberian) breeding population has grown steadily since 1971 and is now of the order of 4,000 pairs, with numbers tending to stabilise (Fig. 167). The population increased by almost 25% annually during 1985–95 but by just 6% post-2000 (Corbacho *et al.* 2009). The Spanish Sandwich Terns undoubtedly benefit from having their major colonies in well protected areas but the limited number of sites inhabited makes them vulnerable.

There have been sporadic breeding attempts elsewhere, none of which has led to the establishment of an enduring colony. Three pairs nested at the Ría de Villaviciosa in 1971, fledging a single chick between them; repeat nesting in 1972 failed (Noval 2001). A colony of 15 pairs settled unsuccessfully at the Llobregat delta in 1988 and 23 pairs nested at the Santa Pola saltpans in 1994 (SA1): the latter colony also failed, on account of predation. Paterson (1997) notes unconfirmed reports of nesting attempts in Cádiz Bay and Almería. Breeding birds settled at two new sites in 2008, when there were at least 28 pairs at the Mar Menor saltpans, Murcia and at least 50 pairs at the Cabo de Gata saltpans, Almería (*Ardeola* 56: 167).

Spanish Sandwich Terns nest on bare or very sparsely vegetated islets in saltpans, coastal lagoons and marismas, as well as on sandy beaches. Suitable secure sites are limited and this may curtail further increases in numbers. The Ebro colony has proved to be highly mobile and has switched between a number of sites in different seasons. The two principal colonies are shared with Slender-billed and Black-headed Gulls, Common Terns and Avocets. The breeding birds are probably summer visitors rather than residents: the establishment of the large Albufera colony has not seen any increase in wintering numbers on the adjacent coast (LR).

Sandwich Terns ringed in Spain, many of them as chicks at the breeding colonies, have ranged widely. There are 52 recoveries from elsewhere in Europe, including 15 from France, 14 from Italy, ten from the Low Countries and a few from Denmark and Britain. One was found in Poland and one ringed as a chick at the Ebro delta in July 2003 was recovered in Albania (1,570km east) in March 2005. There are also three recoveries from Morocco and 11 from sub-Saharan Africa: from Senegal (8), Ghana (2) and Mauritania (1). In addition, one marked on passage at Odiel in 2003 was controlled in 2010 in South Africa (Gacio & Sayago 2011).

The majority of European birds winter on African Atlantic coasts, south to the Cape, and also around the Arabian peninsula (HBW) but many Black Sea birds winter in the Black and Mediterranean seas and some reach Iberia. Bernis (1967: AMI) notes 25 Iberian recoveries of birds ringed as pulli at Ukrainian Black Sea colonies: 14 were found on Spanish Mediterranean coasts but seven reached the Atlantic coasts of Andalucia and Algarve; two were in Leiria, central Portugal, and two were recovered in Galicia. In addition, there are at least ten recoveries in Morocco of birds ringed in Ukraine (BM). There is thus some evidence of migration from the Black Sea to the western Mediterranean and to the Atlantic beyond. Many and probably most of the Sandwich Terns that occur on Iberian coasts on passage and in winter come from northwestern Europe. The recoveries of birds from northwestern and central Europe analysed by Bernis included 56 found on Atlantic coasts and 14 in the Mediterranean. Autumn migration through the Strait of Gibraltar thus includes birds from northern colonies entering from the west and others from Mediterranean colonies flying east. In all, of some 380 recoveries to 2009, the great majority originated from Britain (46%), with 18% from southern Scandinavia and the Baltic, 12% each from Ukraine and the Low Countries, 9% from Germany and a few from France and Italy. One ringed in Mauritania has been recovered in Portugal.

Migrating Sandwich Terns are often observed from coastal watchpoints and large numbers tend to accumulate in sheltered bays, estuaries and saltpans, where they may linger for some time. The post-breeding passage is already noticeable in July and some large counts on Atlantic coasts from August indicate the existence of sizable early movements and/or the presence of many summering non-breeders. For example, in Portugal 900–1,000 were in the Tagus estuary at the Alcochete saltpans on 12 August 1990 and 420 were at Castro Marim, Faro, on 1 September 2006 (*Airo* 2: 55, *Anu. Ornitol.* 6: 62). The Sandwich Tern does not breed in Portugal and these birds were far from the nearest colonies, in eastern Spain.

Visible autumn migration at most coastal sites generally involves modest passage rates. Southbound birds occur in the Bay of Biscay, from where they head west to Galicia along the coasts of the Basque Country, Cantabria and Asturias. The largest counts have been made during well-watched seasons at Punta La Vaca, Asturias, and Cape Estaca de Bares. A seasonal count of 6,227 at Punta La Vaca during late June to December 1991 (*GIAM* 17: 3–4) is noteworthy but the heaviest visible autumn passage in Iberia is seen at Cape Bares: 13,314 were counted on post-breeding passage past the Cape in 2011. Here, as elsewhere, an unknown proportion of passage occurs too far offshore to be seen from land. Southward passage continues along the western coasts of Galicia and off western and southern Portugal and onwards to the African coast. No large counts of active passage are on record for western Iberian coasts which may indicate that most birds pass well offshore, although the region is under-watched. A minority coast eastwards along Algarve and southwestern Andalucía and enter the Mediterranean through the Strait. Passage at Gibraltar extends from August to early November, with maximum movements in September

(Finlayson & Cortes 1987). The migratory flow is bidirectional, as noted above, both at the Strait and along the Spanish Mediterranean coasts.

Return movements in spring tend to be less conspicuous than in autumn, especially in northern Iberia, including at Cape Bares. This may be because most birds following the west coast northwards then continue onwards without diverting into the Bay of Biscay. Movements at the Strait occur from late February to May, with a clear peak in April (Finlayson & Cortes 1987, Paterson 1997). The spring passage at the Cabo de Gata saltpans is more noticeable than the autumn one and extends from February to June, peaking in March–April (Castro 1993). Further north passage commences in mid March and again extends through May, peaking from mid to late April.

This is the only tern species that winters in Iberia in other than very small numbers. The majority occur on western and southwestern coasts and in the Mediterranean. The Biscay coastlands have a very small wintering population, with a few individuals found especially at sheltered inlets including the northern rías, Gijón and Santander bays and Santoña. Wintering birds in general favour sheltered locations, including bays, estuaries and harbours, as well as low-lying sandy coastlines (Vilagrasa *et al.* 1982, Fernández Cordeiro & Costas 1995). The wintering population is of the order of a few thousand individuals and may have increased since the mid 20th century. For example, censuses in Galicia during 1973–80 gave totals ranging from 7–57 birds but there were 264 in January 1984, all in the Rias Bajas on the west coast (Bermejo *et al.* 1986). The wintering population is concentrated at a number of key sites, with smaller numbers spread more thinly along intervening coasts and in estuaries and harbours. The 2009–10 winter survey of non-estuarine coasts in Portugal found 225 birds (Proyecto Arenaria). In the Atlantic the largest numbers occur at the Rías Bajas in Galicia, the Tagus and Sado estuaries (407 in 1984), the Algarve coast and the Huelva and Cádiz coasts: in particular the Marismas del Odiel and Cádiz Bay. Garrido (1996) estimated the wintering population of Odiel and the coastal wetlands of Huelva west of Doñana at 500–800 birds. The Spanish winter atlas put the total population during 2008–10 at some 1,000–2,000 birds, about half of them in the Gulf of Cádiz and 15–20% in Valencia (SWA). However, the wintering population in Catalonia alone has been recently estimated at 936–1,495 individuals (CWA).

The Strait of Gibraltar attracts up to 300 birds in winter, when a flock of up to 100 may often be seen resting at Los Lances beach, Tarifa. Wintering Sandwich Terns are present in small numbers on the Mediterranean coasts of Andalucía and Murcia. They are more numerous in the Comunidad Valenciana and Catalonia, with concentrations at the Santa Pola saltpans, the Albufera, the Ebro delta, and the harbours of Tarragona and Barcelona, all of which may hold 100 birds or more (Paterson 1997). The Albufera alone held 400 in January 1988 and 700 in total were found in Valencia in 1993 (Paterson 1997). Since 1996 the largest winter counts in Catalonia have been at the Aiguamolls de l'Empordà: 201 in 2001, and the Llobregat delta: 223 in January 1996, with January averages for these two sites during 1996–2009 of 71 and 65 birds respectively, although with considerable year-to-year variation. Many more winter in the Ebro delta but full counts from there are lacking (CWA). The western European and the Ukranian populations are equally represented among 23 Iberian ringing recoveries during December–February recorded by Bernis (AMI). Eight recoveries were on southwestern coasts from the Algarve to Cádiz and the remainder were in the Mediterranean.

The Sandwich Tern occurs regularly in the Balearic islands on passage and there is a small wintering population there. It is considered scarce both on passage and in winter, and is commoner in the western islands, the Pityusas, than in Mallorca and Menorca (*A. O Balears 2010*: 265). It is very rarely seen at any distance inland but there are records of single birds from Madrid, Castilla y León and Extremadura (2). The most remarkable inland record is probably an influx of over 40 at Loza reservoir, Navarra, on 9 April 1994 (*A. O. Navarra 1993–1994*: 118): apparently storm-driven from the Basque coast.

ELEGANT TERN *Sterna elegans*

(Charrán Elegante, Garajau-elegante)

Vagrant to Spain.

A species that breeds only on the Pacific coast of the New World and that is virtually unknown on the American Atlantic coast is an unlikely candidate for vagrancy to Europe. However, there have been a series of observations of at least nine different individuals from the French Atlantic coast since 1974 (Dubois *et al.* 1994, Reeber *et al.* 2011) and there are five reports from Ireland, the first in 1982 (Mullarney 1999, Irish Rare Bird Report 2005) and one from Denmark (*British Birds* 94: 139).

There is one accepted record from Spain: an adult at the Llobregat delta from 24 to 30 April 1993 (Gutiérrez & Muntané 1997, Gutiérrez 1998). There have been further sightings in Spain of yellow-billed terns showing characters of the Elegant Tern but their specific identity has not been established and the records remain under rarities committee scrutiny. One of these was an individual that appeared repeatedly during 2002–12, mainly at the Albufera de Valencia, but also north to the Ebro and Llobregat deltas and south to Cádiz Bay and the Marismas del Odiel (RBS). It was originally ringed at Odiel in October 2002, when it was identified as a Lesser Crested Tern, and was recaptured at the Albufera de Valencia in 2006, when it was marked with a yellow colour ring on the left leg. This bird, a male, paired with a female Sandwich Tern at the Albufera in 2006 and they fledged a hybrid chick. A metal-ringed bird, almost certainly the same individual, had previously paired with a Sandwich Tern at the Ebro delta in 2005, and this pair also fledged a hybrid chick. The colour-ringed male again paired with a Sandwich Tern at the Albufera in 2010 and 2011 but it paired with a second 'Elegant Tern' at the same location in 2009 and 2012. This 'pure' pair laid eggs in 2009 and fledged a chick in 2012. Post-2010 sightings of 'Elegant Terns' elsewhere in Spain, include two from Galicia: at Pontevedra in August 2011 and at Valdoviño in October 2012 (RBS).

ROSEATE TERN *Sterna dougallii*

(Charrán rosado, Gaivina-rosada)

***S. d. dougallii*. A very scarce passage migrant. Has bred Spain.**

The western European populations declined severely during the 1970s and 1980s since when they comprise some 700 pairs in Ireland: chiefly at the Rockabill colony, about 100 pairs in Britain and a few dozen in France. There were also around 1,000 pairs in the Azores in 2002 (PA2). European birds winter on west African coasts, notably off Ghana. However, their overall scarcity is probably the main reason why there are relatively few Iberian records, even though their movements must take them past the Iberian and Biscay coasts. The Roseate Tern is also rarely seen in Morocco (BM, Bergier *et al.* 2009) or in the Canary Islands (Martín & Lorenzo 2001, Gutiérrez *et al.* 2012b). It is rare in the Mediterranean on the coasts of France and Spain and is a vagrant to Italy, Malta and Tunisia (BWP). Nevertheless, it was reported to have bred in Tunisia during the late 19th century (Heim de Balsac & Mayaud 1962) and nesting in the Camargue was recorded in four years of the 20th century (Blondel & Isenmann 1981). There is also one breeding record from Spain, at the Ebro delta (Maluquer & Pons 1961).

There are a number of Iberian recoveries of birds ringed in Britain or Ireland, notwithstanding the scarcity of sight records. These have all been on Atlantic and Biscay coasts during September–October (Wernham *et al.* 2002). There are seven or eight such recoveries from Portugal, but only seven sight records there (AP). Recoveries in Spain comprise one in Asturias, two in Cádiz and one in Murcia. One marked as a chick at Rockabill was reported from El Puerto de Santa María, Cádiz, on 27 July–30 August 2010 (*Ardeola* 59: 388–389). In addition, one found dying at Odiel on 19 April 2003 had been ringed as an adult in the Azores in 2001 (*Ardeola* 52: 200).

Figure 168. Records by region, 1912–2010 (n = 72)

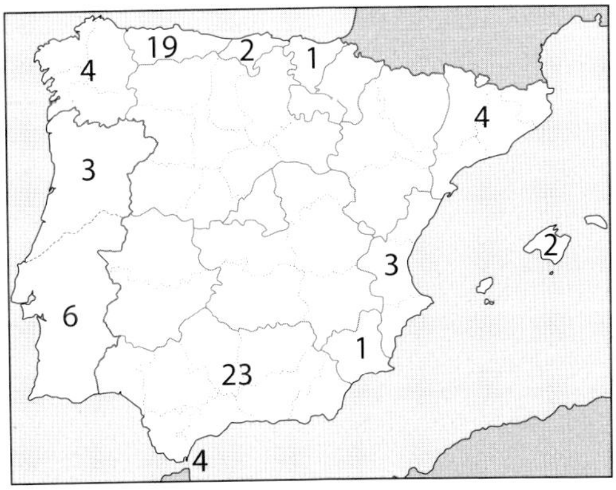

The earliest confirmed occurrence in Iberia was of one in northern Portugal at Vila do Conde in April 1912, after a storm (Reis Júnior 1931). The first recovery of a ringed bird, from the British Isles, was from Algarve in October 1958 (Tait 1960). The 1961 breeding record followed the detection of several adults in a common Tern colony at the Ebro delta, where they remained from mid May to mid July; a nest with eggs was found on 20 May but failed: however, a chick apparently of this species was found several days later (Maluquer & Pons 1961). Several migrants were reported off Atlantic coasts in September 1964 (Bourne & Norris 1966). Much later, stormy weather in autumn 1991 resulted in six observations from the Biscay coasts (*Ardeola* 38: 342, 40: 97, 41: 199).

Very few or none at all used to be seen in Iberia in most years so the Roseate Tern is currently included in the Iberian rarities lists: with seven records from Portugal during 2000–10, 30 in Spain during 2003–10 and four from Gibraltar during 1976–2007. We have considered a sample of 72 reliable sight records, excluding the breeding record, involving 112 individuals. Some 72% were of single birds and 21% involved two or three together. On 5 July 2008 there were at least seven different birds at El Puerto de Santa María. Where ages were recorded, all 13 reported in spring were adults, but the autumn records included 60 adults and seven juveniles. Most records have been from the Atlantic and Biscay coasts but there are a few from the Mediterranean (Fig. 168). They correspond with passage periods (Fig. 169) and fall between 3 April and 16 November. French Mediterranean records are chiefly in spring (OF) and nine out of ten records from the Spanish Mediterranean have been in April–July: it may be that some birds stray inadvertently into the Mediterranean during their northward passage in some years, which may lead to the rare instances of breeding at the Camargue and Ebro delta.

Recent years have seen an increase in observations, particularly during the post-breeding passage period in July–September, when small numbers have appeared among other terns on the beaches of Huelva and Cádiz and some passage has been apparent at Cape Estaca de Bares. There is also an unprecedented inland record: one at Villafáfila on 26 June 2011 (RBS). The autumn 2011 season was exceptional and at least 170 were recorded, most of them on passage west past Cape Bares, where 137 were counted on 18 September (RBS). In contrast, only 26 were reported in Spain in 2012, during July–October.

Figure 169. Records by month, 1912–2010 (n = 72: 62 from Atlantic/Alborán waters and 10 in the Mediterranean)

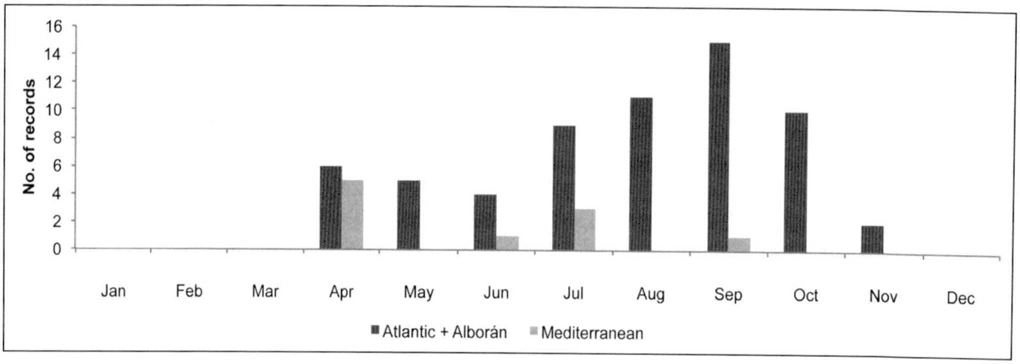

COMMON TERN

Sterna hirundo

(Charrán Común, Gaivina-comum)

***S. h. hirundo*. Breeds locally in Spain, chiefly at Mediterranean wetlands. Occasional pairs nest in Portugal. Widespread and common on coasts on passage, rare inland. Very small numbers occur in winter.**

Several thousand pairs breed at a limited number of sites in Spain but only sporadic breeding pairs occur in Portugal. The fortunes of the Iberian nesting birds have fluctuated markedly since the mid 20th century but have shown an overall decline more recently (Corbacho *et al.* 2009, LR) The Spanish breeding population was estimated at 5,800–11,000 pairs by SA1 (1997) and at up to 8,000 pairs by SA2 (2003). Breeding birds form colonies chiefly at coastal wetlands: on sandy islets in marismas, lagoons and saltpans, frequently with other terns, gulls and waders. All the principal breeding colonies are on the eastern Mediterranean coast and usually just two locations, the Ebro delta and the Albufera de Valencia, account for over 80% of the Iberian total.

Nesting at the Ebro delta was first documented in 1956 but seems likely to have occurred earlier, although Ticehurst & Whistler (1930) saw only a single bird there on 1 May. Some 1,500 pairs bred during the 1960s but the population declined to as few as 100 pairs during the 1970s. This decline may have resulted from DDT application on the ricefields, which apparently also extirpated the colony at the Albufera de Valencia (Docavo 1979). The Ebro delta population then recovered to reach 4,953 pairs in 1996, falling to 3,479 pairs in 1997 before reaching a record maximum of 5,495 pairs in 1998 (LR). However, the national census of breeding tern species in 2007 (Corbacho *et al.* 2009) coincided with the almost total failure of the Ebro colony and the total nesting population of Common Terns in Spain that year was 3,180 pairs, although it would have been closer to the usual average had the 2,000 pairs that initially settled at the Ebro not been displaced by flooding. In 2006 there were 2,000 pairs at the Ebro colony and the birds returned in 2008 (Corbacho *et al.* 2009).

The Ebro colony is the only site where nesting has been recorded in Catalonia. The Comunidad Valenciana, however, supported four colonies in 2007 and nesting has been recorded at additional sites in other years. The great majority nest at the Albufera, where breeding was reported in 1963 but ceased from 1973 until 1983 (Pechuán 1965, Dies *et al.* 1999). One or two hundred pairs nested at the Albufera in the later 1980s and numbers increased there to over 2,000 pairs from 1999 onwards. The Common Tern, together with the Sandwich Tern and various other larids and waders, has been a major beneficiary of the conservation work at the Racó de l'Olla, at the Albufera, which ensures the availability of a secure and protected nesting environment (Dies 2000).

The other colonies in the Comunidad Valenciana are of more recent origin. Nesting at the Santa Pola saltpans has occurred

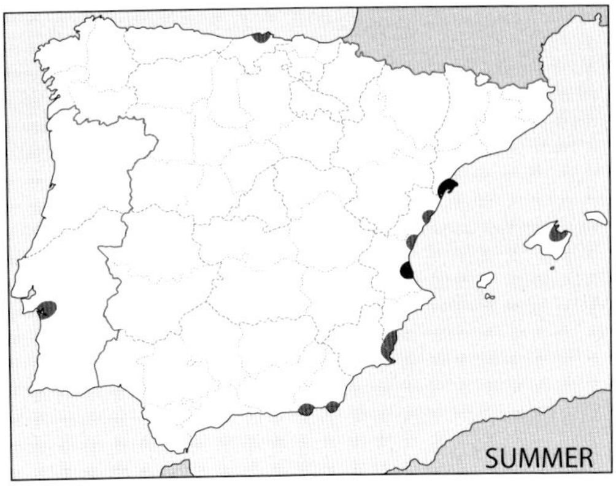

since 1978: 5–10 pairs bred there during 1978–84 since when numbers have increased considerably, peaking at 522 pairs in 2000. Nesting at the Lagunas de La Mata-Torrevieja began in 1988 and there were 200+ pairs there in 2000. The smaller colony at the Marjal del Moro was established in 1999. Nesting has also occurred at two other sites in Alicante: El Hondo and the La Pedrera reservoirs (LR, Corbacho *et al.* 2009).The colonies at the Mar Menor in Murcia also date from the late 1980s, although Saunders (1871) reported nesting there, without giving any details. Some 29 pairs nested in 1988 and there were 60 pairs by 1990. Numbers increased considerably post-2000 to 200–300 pairs.

The Common Tern has nested sporadically in some numbers at Doñana, where 350 pairs were censused in 1966 and where there were *c.* 100 pairs in the early 1980s (SA1) but only occasional pairs have been recorded subsequently. More regular breeding has occurred in eastern Andalucía, in Almería, where up to five pairs nested at the Cerrillos saltpans during 1988–93. Since then nesting in Almería has mainly been at the Punta Entinas-Sabinar and Cabo de Gata saltpans, with a fluctuating population that reached 86 pairs in both 2001 and 2007.

Breeding in the Balearics has been recorded only at the Albufera de Mallorca. A single pair laid eggs in 2001 but they were taken by Yellow-legged Gulls (Mas 2002). However three pairs nested in 2002 and 2003 and the colony numbered 8–10 pairs in 2008 and 18 pairs in 2010 *Ardeola* 52: 217, 56: 360; *A. O. Balears 2010*: 171). The Common Tern otherwise occurs annually in the archipelago on passage but is very uncommon and reports usually involve just a few individuals.

Away from the Mediterranean, the Common Tern was reported to have bred formerly in Asturias by Noval (1986). Subsequently, a small but growing colony has become established in Santander Bay, the only colony on the Biscay coast. A pair settled here on a partly-submerged wrecked boat in 1989 and up to three pairs were present by 1991. Conservation measures, including nest translocations to purpose-built gravel-covered rafts nearby, have since enabled the colony to consolidate and expand (Fernández-Calvo & González Sánchez 2008), reaching 33 pairs in 2007. In 2009 a pair nested on an artificial platform at Las Marismas Negras nearby (*Ardeola* 56: 360). Only 18 pairs nested at Santander Bay in 2010 but they fledged 30 young (*Ardeola* 57: 537).

There are no established colonies on the Iberian west coast. Two pairs nested at Comporta, near the Sado estuary, in 1937 (Coverley 1939) but other historical claims of nesting in Portugal are not fully substantiated (AP). One or two pairs nested annually, and unsuccessfully, at saltpans in the Tagus estuary during 1992–95 and in 1997 (AP, *Pardela* 9: 23). Between one and three pairs nested at the Sado estuary during 2000–08 (PA2, *Anu. Ornitol.* 6: 63 , 7: 83).

Common Terns very occasionally nest at wetlands inland. In 2007 the only such record was the pair in Aragón. Nesting by single pairs has previously been reported from the Laguna Larga de Villacañas, Toledo, in 2001; the Finisterre reservoir, Toledo, during 2001–05; La Cuerda del Pozo reservoir, Soria, during 1994–97 and 2002; the Velilla de San Antonio gravel pits, Madrid, in 1995 and 2000–2001; and the Valdecañas reservoir, Cáceres, in 2011 (LR; AOM, *Ardeola* 49: 190, 50: 163, 58: 503). The marked scarcity of inland breeding records is in sharp contrast to other western European countries.

The Common Tern is almost entirely a summer visitor to the western Palearctic, wintering all along the west African coastline from Mauritania southwards (Borrow & Demey 2001). Birds ringed as pulli at the Ebro delta and the Albufera de Valencia have been recovered from Mauritania to Ghana, chiefly in Senegal (AI1). In general, there are 38 recoveries of Spanish-ringed birds from the length of the West African seaboard south to Angola, Namibia and South Africa, 20 of them from Senegal. A more unusual recovery was of one ringed as a juvenile at Odiel in September 2001 that was recovered at San Salvador de Bahía, Brazil, in January 2002 (*Ecología* 17: 260). Most Iberian recoveries of foreign-ringed birds originate from countries bordering the North and Baltic seas and the great majority have been found on Atlantic or Biscayan coasts; only two of 77 recoveries analysed by Bernis (AMI) were on the Mediterranean coast. There are 17 recoveries of birds ringed in Italy. Over 80 birds ringed in Iberia have been recovered from northwestern Europe, mainly from the Low Countries. There are also three recoveries from Spain of birds ringed in Namibia as well as a Portuguese recovery of a bird ringed in South Africa.

Common Terns occur on all coasts on passage, sometimes in large numbers, but it is evident that much passage occurs out of sight from land. Flocks often appear at estuaries and on sandy beaches during passage periods, especially during the autumn passage, which is generally both more leisurely and more conspicuous than the spring movements.The post-breeding passage begins in July on Cantabrian and Mediterranean coasts but the principal movements everywhere are during August–mid October, with stragglers remaining through November. Southbound birds that have entered the Bay of Biscay coast westwards off the Iberian Biscay coasts. The largest counts are usually made from Cape Estaca de Bares, where migrants are logged as 'Comic' terns since most are too far off to be safely identified as either Common or Arctic Terns. In 2008 16,813 'Comic' terns passed Cape Bares towards the Atlantic in 410 hours between 14 July and 19 November, a mean passage rate of 41 birds/hour. Peak movements occurred during August and the first three weeks of September, attaining a maximum rate of 332 birds/hour on 13 September. On the basis of sample counts, the mean annual post-breeding daytime passage at Cape Bares during 2004–08 was of the order of 60,000 birds (Sandoval *et al.* 2010).

There are no comparable large counts available from the Iberian western and southwestern coasts, perhaps because the birds travel too far offshore but possibly also because of lack of coverage. However, flocks of 100–200 occur off the Huelva coast during July–October and the saltpans at El Puerto de Santa María in Cadiz bay sometimes hold large gatherings, including 2,200 on 10 August 2007 and 1,500 on 10 September 2008 (*Ardeola* 54: 399, 55: 300). Roosts at the Marismas del Odiel during passage periods sometimes also attract over 2,000 birds (Gacio & Sayago 2011).

Southward passage is also observable from Mediterranean coasts. Observation from Cape Cullera, Valencia, sometimes detects large numbers: for example, over 5,000 were counted flying south in 42 hours during autumn 1989, with a maximum rate of 263 birds/hour on 13 September (*A. O. C. Valenciana 1989*: 53.). Other Mediterranean coastal sites generally see much lighter passage and only small numbers are usually reported at the Strait of Gibraltar during most years.

Spring passage begins from late February in the south but is generally during March–April, with some passing in May. The return movements are usually even less noticeable from land than the southward ones, except when unusual weather conditions drive birds inshore. A count of 486 at Gibraltar on 31 March 1990 followed prolonged easterly gales (*Alectoris* 9: 64). Common Terns are scarce on passage along Biscayan coasts in spring and there is no large movement then at Cape Bares.

Virtually all passage is coastal and only tiny numbers are ever recorded inland, with occasional observations of one or two birds at wetlands in Extremadura, the meseta and Aragón. However, there was a small influx through Extremadura in spring 2012 when small groups appeared at several reservoirs, the largest counts being 11 at the Guadiloba reservoir on 26 April and 12 at the Portaje reservoir on the following day (*Ardeola* 59: 440). One recorded at an altitude of 1,440m in the Serra da Estrela: at the Albufeira de Vale do Rossim, on May 2008 (*Anu. Ornitol.* 7: 83) was also most unusual.

A few Common Terns overwinter, chiefly at sheltered sites on western and southwestern coasts but also in the Mediterranean and in the north. Recent observations have usually been of 1–3 birds. However, the census of January 1984 (Bermejo *et al.* 1986) found 35 in Galicia in the Rias Bajas, 23 on the Huelva coast, and seven in the Mediterranean in the Comunidad Valenciana. In contrast, there were only six December/January records from Catalonia during 1985–2009 (CWA). Eight first-winters were on Carcavelos beach, Oerias, on 11 January 1992: one of the few records of wintering in Portugal. There is also a record of 20+ at the Albufera de Valencia on 24 December 1963 (Érard & Vielliard 1966).

ARCTIC TERN *Sterna paradisaea*

(Charrán Ártico, Gaivina do Árctico)

Regular on passage off Atlantic and Biscayan coasts but very few are seen from land. Very rare in the Mediterranean and exceptional inland.

The Arctic Tern breeds regularly no nearer to Iberia than Britain, Ireland and the Netherlands. Reports of sporadic nesting in Spain at the Ebro delta (AMI, AI1) have never been substantiated.

It mainly winters at sea in the southern hemisphere. European populations, and those nesting in eastern North America, migrate south off eastern Atlantic coasts, the return passage being further west. Migrants are accordingly more numerous off European Atlantic coasts in autumn than in spring but, in both cases, the principal movements occur well out to sea. It was the most common tern seen by Bourne & Norris (1966) at sea off western Iberia during the first half of September 1964. It is seldom seen in Iberian Atlantic coastal waters in any numbers and it is a vagrant to the Mediterranean.

Arctic Terns are most often specifically identified on the coasts of Asturias and Galicia. For example, 100–400+ were identified on westward passage from Punta La Vaca, Asturias, in most autumns during 1993–2000 (*A. O. Asturias*). A count of 70 at the coastal Cecebre reservoir, A Coruña, on 21 September 2009 (*Ardeola* 51: 252) is high for a coastal inlet in Galicia: most records of birds that are not actively on the move usually involve few individuals. Arctic Terns are similarly most often identified on Portuguese coasts and in southwestern Andalucía in very small numbers: a record of three separate juveniles present at a large roost of other terns at the saltpans of el Puerto de Santa María, Cádiz, during 18–31 August 2010 (*Ardeola* 57: 538) is a typical observation.

Southbound birds occur from the second half of July to November, but an analysis of 51 records from Asturias and Galicia, involving a total of over 180 individuals that were identified at close quarters, shows that the largest numbers pass in August–September, with the spring passage peaking in May (Figure 170). As many as 47% of these records were of single birds and 27% involved two together, but some groups were seen, the largest being 45 birds in A Coruña harbour on 12 August 1997 (*A.*

O. Galicia 1997: 93). A count of 495 flying south in eight hours off Cape Carvoeiro on 19 September 1999 (Moore 2000a) was exceptional and occurred during onshore gales. There are winter records of single birds at Cádiz on 27 December 1977 and 9 January 1995 (*Ardeola* 24: 260, 42: 224).

Spring passage is from April to mid June. A record of 43 on passage 28 miles off Aveiro on 24 May 2008 (*Anu. Ornitol.* 7: 83) exemplifies what is likely to be a regular offshore movement there. A count of 408 in 30 minutes on passage off Baldaio, A. Coruña, on 9 April 2000, is more noteworthy (*A. O. Galicia 2000*: 41). Very few appear on Biscayan coasts in spring. At least eight Arctic Terns ringed in Britain and one from Finland have been recovered in Iberia.

Figure 170. Monthly distribution of Arctic Terns identified onshore on the coasts of Asturias (n = 28 records) and Galicia (n = 23 records) during 1993–2000.

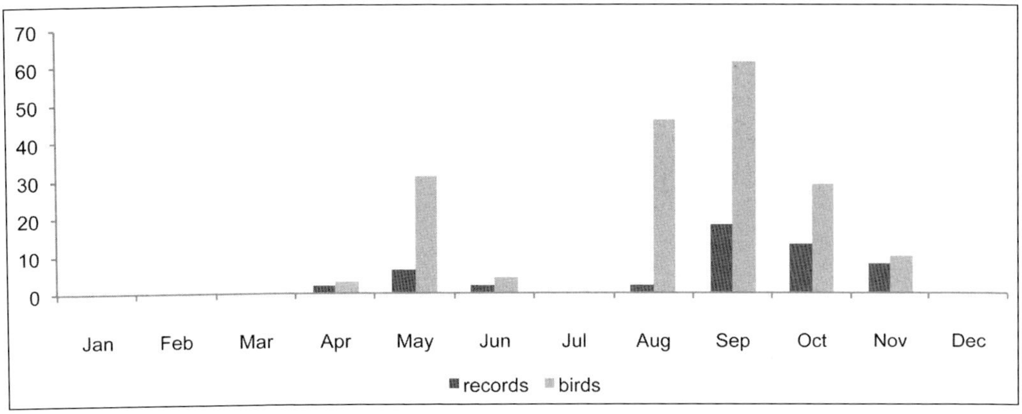

The Arctic Tern is very rarely identified in the Strait of Gibraltar and there is only one record of a single bird from Gibraltar itself: in October 1979 (BG). It is similarly rare in the Mediterranean and the reliability of some published records has been questioned (Clavell 2002) but there are a few credible records, some of oversummering individuals. For example, one was at the Racó de l'Olla ternery in the Albufera de Valencia on 6 July 2001, from 14–25 June 2002, on 21 July 2003 and 13 August 2004 (*Ardeola* 50: 163, 351; 52: 217); and two adults were at the Llobregat delta from 17–21 July 1984 (Gutiérrez *et al.* 1995).

Arctic Terns are very rarely recorded inland but there are two records of single birds from the Ebro valley: at Las Cañas reservoir, Navarra, on 10 October 1987 and at Calahorra, La Rioja, on 2 June 1990 (*Ardeola* 37: 341). Inland in Portugal five were reported on the River Duero at Castelo Branco on 17 May 2002 and one was at the Paul de Boquilobo on 16 June 2002 (*Anu. Ornitol.* 3: 39).

FORSTER'S TERN *Sterna forsteri*

(Charrán de Forster, Gaivina de Forster)

Vagrant to Spain and Portugal.

The first Iberian record was an adult in the northern sector of the Strait of Gibraltar on 27 October 1987 (Hashmi 1988, *Ardeola* 44: 133). There have since been three accepted records for Portugal and two more for Spain: an adult at the Castro Marim saltpans, Faro, on 31 December 1993 and 1 January 1994; a first-winter at Fuseta harbour, Olhão, Faro, on 12 February 2003; one at the Marismas del Odiel, Huelva, on 13 October 2003; a first-winter in the Ría de Camariñas, A Coruña, from 16 November 2003 until 5 January 2004, and a first-winter at Peniche, Leiria, on 12–15 November 2010. The records thus comprise three autumn observations and three winter ones, all from Atlantic coasts from Galicia south to the Strait. There were two more reports in 2012: one on the beach at Almonte, Doñana, in November 2012, and one at the Ensenada da Ínsua, A Coruña, in December 2012; and one in 2013: at Chipiona, Cádiz, in August–September (RBS). This is a very rare vagrant to Europe. Up to 35 have been recorded in Ireland and 20 in Britain. There are also five records for France and three for the Azores.

BRIDLED TERN *Onychoprion anaethetus*

(Charrán embridado, Gaivina-de-dorso-castanho)

Vagrant to Spain.

The sole Iberian record was one at the Ebro delta on 18 June 2008. Most European records have been in July–August. The Bridled Tern reaches European coasts very rarely. Records include 23 from Britain, nine from France and 11 from the Azores. *O. a. melanopterus* nests as close to Iberia as the Arguin bank, Mauritania, and Virginia island in the south of Western Sahara, but it has declined markedly in that region in recent years.

SOOTY TERN *Onychoprion fuscatus*

(Charrán Sombrío, Gaivina-de-dorso-preto)

Vagrant to Spain and Portugal.

The earliest reliable observation in Iberia was one photographed at El Rocío, Doñana, in May 1963 (*Ardeola* 10: 65–66). Much later an immature bird was at the Guadalhorce estuary, Málaga, on 27–28 August 1979 (*Ardeola* 28: 159). There have since been a further three accepted records from Spain and one from Portugal, all of single individuals: an immature at Laxe, A Coruña, on 20–21 June 1981; an adult at the Ebro delta on 24 July 1986; an adult at Gafanha do Carmo, Aveiro, on 18 August 1998, and an adult at Cabo Silleiro, Pontevedra, on 29 August 2001. There are thus four observations from the Atlantic and two from the Mediterranean, all during May–August. Increased observer effort in Iberia has not produced any corresponding increase in sightings, which accords with the recent decline in records from Britain and France (Vinicombe & Cottridge 1996, OF).

The Sooty Tern is very rarely found in Europe but there are records of 26 individuals from Britain, two from Ireland and 13 from France: these last including one in the Mediterranean. There are also four records from Italian waters. There is only one record from the Canary Islands, but there are several from Madeira and the Azores: including sporadic instances of breeding in the latter.

LITTLE TERN *Sternula albifrons*

(Charrancito Común, Chilreta)

S. a. albifrons. **Locally common, nesting colonially at coastal wetlands of the southern two-thirds of mainland Iberia, but absent as a breeding species from the Biscay coastlands and Balearics. Some breed inland in the south. Mainly a summer visitor, widespread on passage.**

The Little Tern is a common breeding species at coastal wetlands of southern and central Iberia. The largest colonies are in southwestern Spain: at the Guadiana, Odiel and Guadalquivir estuaries and in Cádiz Bay. There are also significant colonies at Mediterranean coastal wetlands, from southern Almería north to the Ebro delta. In Portugal, there are breeding birds at the Algarve wetlands and on the west coast north to the Ria de Aveiro. A small minority nest inland in Spain and eastern Portugal, most of them in the Guadiana and Tagus valleys in Extremadura and Castilla-La Mancha. However, the Little Tern does not breed in the Balearic Islands, where it is only a very scarce migrant. It seems to have nested in Asturias in the past (Noval 2001) but it has not done so for many years, nor does it breed elsewhere in the Biscay coastal regions.

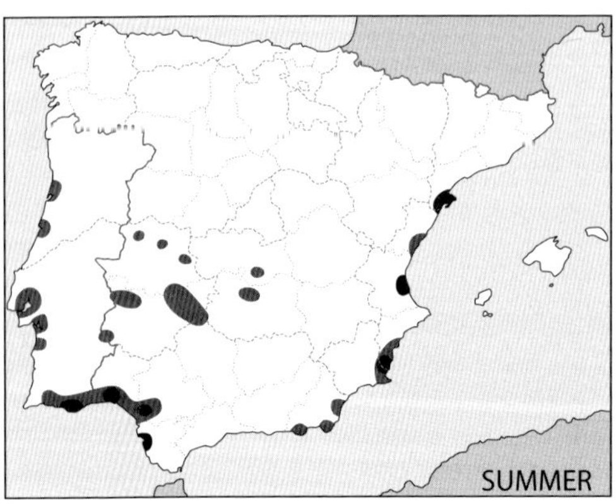

Breeding birds are especially associated with saltpans and the fringes of coastal lagoons, where colonies occur on sparsely-vegetated sandy or gravelly shores, sometimes on islets. Nesting on sandy beaches was the norm in Portugal prior to the 1990s but most of the population now nests at saltpans, following disturbance and destruction of the original colony sites (Catry *et al.* 2004). Nesting inland was traditionally on gravel spits and sand bars in rivers but the recent trend has been for nesting on the shoreline of reservoirs and lagoons, and on islands in these (Corbacho *et*

al. 2009). Little Terns mainly prey on small fish and crustaceans in both fresh and saline waters, especially in littoral channels and lagoons (Calado 1998, Bertolero *et al.* 2005).

The breeding population is substantial but given to considerable fluctuation, especially inland, where some nesting areas are only suitable in wetter years. The Spanish population was estimated at 2,500–3,000 pairs (Tucker & Heath 1992) but the first Spanish breeding bird atlas suggested 5,500–6,000 pairs (SA1). A census in Spain in 2007 found 3,575 pairs (Table 46), over 50% in coastal southwestern Andalucía. Here the Cádiz Bay complex was the principal breeding area that year, with 1,032 pairs in 15 colonies distributed within the saltpans, the natural park and sandy beaches. Cádiz Bay and the other major sites at coastal wetlands are traditionally occupied, although breeding numbers may fluctuate considerably from year to year. Thus, for example, Veta La Palma, in Doñana, accommodated 500–650 pairs in the dry years of 1994 and 1995 but there were 1,500–2,000 pairs in the wetter seasons of 1997 and 1998 (LR). The Cádiz Bay population underwent a serious decline during the 1990s, from over 1,000 pairs to fewer than 250 pairs, before recovering during the following decade (LR). In Huelva the beach colony of the Odiel spit, the Espigón, alone numbered 2,000 pairs; but 1,500–2,000 pairs was more usual during the 1980s and 1990s (Garrido 1996) after which the colony declined sharply due to human disturbance before moving to the dredging ponds nearby and increasing again to over 1,000 pairs (Gacio and Sayago 2011). Similar fluctuations are reported from most other major coastal breeding areas but the general trend, although uncertain, seems to indicate an overall increase in the Spanish Little Tern breeding population during 1970–90 that was followed by relative stability during 1990–2000 and then by a gradual decline (Corbacho *et al.* 2009). It is certainly sensitive to human disturbance of the breeding colonies, which occurs especially when these are on beaches (Calado 1996, Catry *et al.* 2004, CA2).

Table 46. Regional distribution and abundance of breeding Little Terns in Iberia in 2007. After Corbacho *et al.* 2009, PA2

Region	No. of pairs	%	Principal locations
Atlantic coastlands			
Andalucía (south-west)	2,198	51.4	Cádiz bay; marismas of the Guadalquivir, Odiel, Río Piedras and Isla Cristina.
Portugal*	707	16.5	Algarve saltpans; Lagoa de Santo André; Tagus, Sado, Alvor, Arade and Mondego estuaries; Ria de Aveiro
Mediterranean coastlands			
Andalucía (Almería)	156	3.6	Cabo de Gata saltpans, Punta Entinas-Sabinar, Cañada de Las Norias.
Catalonia	393	9.2	Ebro delta
Comunidad Valenciana	518	12.1	Santa Pola saltpans, Albufera de Valencia, Marjal del Moro, Lagunas de La Mata-Torrevieja
Murcia	235	5.5	Mar Menor: San Pedro del Pinatar saltpans
Inland			
Extremadura	69	1.6	Alange & La Serena reservoirs, Badajoz
Castilla-La Mancha	3	0.1	Navalcán & Rosarito reservoirs,Toledo
Andalucía (Inland)	1	0.0	La Serena reservoir in Córdoba
TOTAL	4,280		

* Figure for Portugal is from an estimate from 2003 of 707–734 pairs (Catry *et al.* 2004).

The Portuguese nesting population is also subject to considerable year-to-year fluctuations in numbers and the 2003 estimate of 700+ pairs is a high figure that has been attributed to thorough prospection and a good nesting year. The principal nesting site in Portugal is the Ria Formosa, which supports around 250 pairs (AP). Castro Marim at the Guadiana estuary had 51–82 pairs nesting in 2000 but 121–198 bred there in 2001 (AP). The Ria de Aveiro population has fluctuated between 35 and 100 pairs since the early 1980s and similar variation has been seen at other regular breeding sites such as the Tagus, Sado and Mondego estuaries and the Lagoa de Santo André. Inland nesting at reservoirs has also been noted in Portugal, for example at the Caia and Alqueva reservoirs, with up to 60 pairs nesting on islands at the former in some seasons (Venâncio *et al.* 2002, PA2).

The largest inland population in Iberia is in Extremadura, chiefly at reservoirs in the Guadiana valley in Badajoz, with smaller colonies at reservoirs on the Tagus in Cáceres. Together these have recently totalled up to 300 pairs (SA2) but breeding numbers are greatly influenced by water levels and 2007, with just 69 pairs in the region, was a very poor year. There were 100 pairs at La Serena reservoir in June 2012 (*Ardeola* 59: 440). Similar sites in Castilla-La Mancha attract small numbers but there has been a pronounced move away from nesting at the lakes of La Mancha, where water levels are especially variable, to reservoirs (LR). The first recorded instance of inland nesting in Iberia seems to be that by four pairs at the Laguna del Taray, Toledo, in 1951 (Bernis 1952). It was first seen nesting in Extremadura in 1963, on the Guadiana river at Villanueva de la Serena in 1963 (Pérez Chiscano 1965).

Little Terns generally begin to appear in Iberia in early April (Valverde 1960, Paterson 1997). Arrivals and spring passage continue through April–May, with stragglers in June. The return passage is already noticeable in late July in the north, when family parties sometimes linger in the rías of the Biscay coast (Noval 2001) but most pass during August–September with some also lingering through October and into November in the south. Numbers peak in southern Portugal, where many breed, in late July and early August and there is a notable count of 814 birds at the Ria Formosa in July 1994 (AP). There are also some large late summer counts on record from the sandy beaches at Doñana: including 351 on 31 August 2000 and 1,050 on 21 August 2001 (*A. O. Doñana 1*: 135).

There are 88 Iberian recoveries of Little Terns ringed mainly in Italy (21), Germany (19) and Britain (11), as well as a few from France, the Low Countries, southern Scandinavia and Poland, confirming the transit through the region of migrants from other western European populations. Three ringed in Spain have been recovered in Italy and there are single recoveries from Sweden and Morocco and two from Senegal, these last including one ringed as a chick at San Pedro del Pinatar, Murcia, in June 2001 and recovered the following January at the Sine-Saloum delta (*Ecología* 17: 261). The movements themselves are inconspicuous. Very few are seen actually on the move on Mediterranean coasts or at the Strait of Gibraltar but small-scale visible passage is seen on the Biscay coasts in autumn especially: recent autumn counts from Cape Bares total just 100–300 birds (Sandoval 2009, Sandoval *et al.* 2010).

There are a number of winter records, including one from the north coast at the Ría de Villaviciosa in January 1979 (Noval 2001). They include several recent observations of small groups of apparently wintering individuals from the Guadiana estuary and adjacent coasts of Algarve and western Andalucía. Groups of up to 30 have been recorded at the Ria Formosa in January (AP). A total of 11 at Sancti Petri, Cádiz, on 8 February 2003 (*Ardeola* 51: 252) also suggests wintering in the region. A few winter records come from the Levant coast. Seven individuals were seen from the coasts of the Comunidad Valenciana during the Spanish winter atlas surveys (SWA). However, a record of 11 at Malgrat de Mar, Barcelona, on 31 December 2002 (*A. O. Catalunya 2002–2005*: 297) is exceptional.

WHISKERED TERN *Chlidonias hybrida*

(Fumarel Cariblanco, Gaivina-dos-pauis)

***C. h. hybrida*. Nests commonly at some of the major Spanish wetlands and more widely, including in Portugal, in wet years; although not in the Balearic Islands. Widespread on passage. Chiefly a summer visitor.**

Hordes of Whiskered Terns once swarmed over Doñana, as noted by 19th century authors, including Saunders (1871), Arévalo (1887), and Irby (1895). Valverde (1960) estimated a breeding population in the Marismas of over 50,000 individuals in the late 1950s: he considered it the second most abundant nesting species there after the Collared Pratincole. Some numbers may then also have frequented smaller wetlands throughout Iberia but the evidence is scanty: for example, Bernis (1952) refers to 'a few' at the Laguna del Taray, Toledo, and at Daimiel in spring 1951. The breeding population remains centred on Doñana, with significant colonies also occurring at the Ebro delta, the lakes of La Mancha and the coastal wetlands of Valencia and Alicante. The total Spanish population now varies from year to year within a range of about 5,000–8,000 pairs, as nesting conditions permit. The Portuguese population is very small and breeding there is irregular. Whiskered Terns require shallow, clear, freshwater wetlands with short emergent vegetation to which they attach their floating nests. Such places include marismas and a diversity of lagoons. Suitable locations were formerly much more widespread (LR) but wetland drainage during the 20th century has meant that the current breeding range is quite limited.

The Spanish breeding population was censused in 2007 (Corbacho *et al.* 2009), which proved to be a moderately poor year away from the principal sites (Table 47). Many other sites, chiefly at lagoons on the meseta and in Extremadura, are used

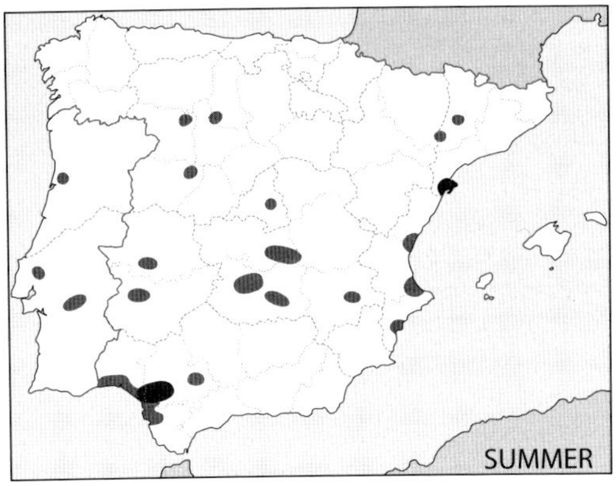

sporadically, generally when a wet winter has provided suitable conditions. The better known of such sites include the following but the list is far from comprehensive: in Andalucía: the Estero de Domingo Rubio, Huelva; in Castilla y Léon: the Lagunas de Villafáfila, Chozas and La Nava, and the Azud de Riolobos; in Castilla-La Mancha: the Vicario reservoir and the lagoons at Manjavacas, Pedro Muñoz, the Campo de Calatrava region and Pétrola; in Extremadura: the pools of the Llanos de Cáceres and Los Canchales reservoir; in Catalonia: the Aiguamolls de l'Empordà; in the Comunidad Valenciana: the Albufera de Valencia, Clot de Galvany and Laguna de La Mata (LR, Corbacho *et al.* 2009). Breeding has also occurred at Meco, Madrid, where there were 17 pairs in 2011 (*Ardeola* 58: 503). Up to 100 pairs bred regularly in Aragón at Gallocanta, until 1976, and some nested there at the Laguna de Sariñena in 1978 & 1979 (Aragüés & Lucientes 1980, Paterson 1997).

In 2007 the mean colony size was 230 pairs but ranged from two to over 1,500 pairs. However, just three large colonies: 1,569 pairs at Las Nuevas and 1,095 pairs at the Marisma de Hinojos; both in the Guadalquivir delta, together with 1,234 pairs at the Ebro delta, accounted for over 60% of the breeding birds in Spain. The concentration of the breeding birds at so few sites, and the dependence of breeding success on high quality wetlands that are of restricted availability, makes them vulnerable to predation and colonies have been known to suffer from attacks by a wide range of mammalian and avian predators (LR). Interspecific competition for nest sites with the increasing population of Black-headed Gulls may also cause problems in La Mancha (Paterson 1997).

The Spanish population is the largest in western Europe, where the species is otherwise numerous only in France (2,200–3,900 pairs; OF). A relatively recent estimate of 5,000–8,000 pairs comprising the Spanish breeding population (Blanco & González 1992) has been supported by all subsequent estimates, including the results of the 2007 census. Nevertheless, the breeding numbers at particular sites fluctuate greatly (Corbacho *et al.* 2009). For example, the population of Doñana during 1997–2002 ranged from nil during drought years to 5,300–6,300 pairs in the most favourable seasons. The Ebro delta population fluctuated between 100 and 800 pairs during the 1970s and 1980s but increased to 1,475 pairs in 1991 before declining slowly, to 1,234 in 2007. The Albufera de Valencia accommodated probably over 1,000 pairs in the 1960s and early 1970s but there were just some 20 pairs in 1977 and none have nested there since, although the location is used by foraging birds all summer; the decline was apparently due to eutrophication as well as to organochlorine insecticides, that were then widely used in the ricefields (Docavo 1979, Ferrer *et al.* 1986). The other colonies of the Comunidad Valenciana have numbered 310–1,740 pairs since the mid 1980s: El Hondo had the largest regional colony in 2009, 528 pairs (*A. O. C. Valenciana 2009*: p. 90): this colony numbered 500–1,200 pairs in the 1970s (Navarro Medina & Navarro García 1984) but there were only 200–300 pairs by the mid 1980s (Ferrer *et al.* 1986). The other principal breeding region, the lagoons of La Mancha, accommodated up to 1,000 pairs in total, spread between up to 20 lakes, but numbers there have declined considerably since, with none breeding in the driest years such as 1986 and 1987 (Jiménez García-Herrera *et al.* 1992, Paterson 1997) and only up to 500 pairs more recently.

Table 47. Distribution of breeding Whiskered Terns in Spain in 2007. Source: Corbacho *et al.* 2009.

Region	No. of pairs	No. of sites	%	Sites used in 2007
Andalucía	4,045	9	63.1	Doñana
Castilla y León	16	1	0.2	Laguna de Boada
Castilla-La Mancha	506	9	7.9	Tablas de Daimiel; El Taray, Pozo de La Puerta, Vadoancho, Navaseca, Navagrande and Alcázar de San Juan lagoons
Catalonia	1,292	3	20.2	Ebro delta, Ivars lagoon & Utxesa reservoir
Comunidad Valenciana	547	6	8.5	Pego–Oliva, El Moro, Xeresa-Xeraco and Almenara marshes; El Hondo & Santa Pola saltpans
TOTAL	6,406	28		

The very small Portuguese population does not seem ever to have exceeded 200 pairs (BE). Nesting in recent years has been reported from the Tagus and Mondego valleys and in Évora, as well as at the Ria de Aveiro: where 15 pairs bred in 1995 (PA2, AP). The Paul de Boquilobo, Santarém, has been a traditional breeding haunt, where up to 120 pairs nested in 1993 (Catry *et al.* 1997), but nesting there has been irregular in recent years. Four pairs nested at the Alqueva reservoir in 2006 and there were 38 birds and seven occupied nests at the Lagoa dos Patos, Beja, in July 2009 (*Anu. Ornitol.* 6: 63, 8: 77).

The great majority winter in sub-Saharan Africa. There are seven recoveries of birds ringed at Spanish colonies from sub-Saharan Africa: from Ghana (3), Mali (3) and Senegal. Birds ringed in Spain have also given recoveries in France (6), Italy and Hungary, and at least eight ringed in France and one from Switzerland have been recovered in Spain.

The Whiskered Tern is an unobtrusive migrant and is most often noticed on passage when small flocks appear at both inland and coastal wetlands. Small numbers are also seen occasionally on passage along coasts, sometimes accompanying Black Terns. The passage periods extend from mid March to May and from late July to October, with peaks in May and September. Early and late individuals may sometimes be seen in early March and early November respectively.

Passage birds are scarcest in Portugal, northwestern Spain and the Biscay coastlands. A gathering of 22 at the Tresona reservoir, Asturias, on 18 May 1999 (*A. O. Asturias 1999*: 114) was exceptional. Autumn observations are much more frequent than spring records on the Biscay coast, comprising some two-thirds of records, but flock sizes there are small, averaging 2.6 birds. In Portugal, non-breeding or migrant Whiskered Terns sometimes gather at the ricefields of Ponta da Erva, in the Tagus estuary, during May–July, sometimes in considerable numbers: 480 were counted there on 12 June 1999 (*Pardela* 12: 23) but fewer than 100 is more usual. Movements are most evident at Mediterranean coastal wetlands, where flocks appear at saltpans, ricefields and lagoons during passage periods, but very few are recorded on the coasts themselves. There is similarly no significant visible passage at the Strait of Gibraltar and the Whiskered Tern is rarely recorded at Gibraltar itself.

Perhaps 200–300 Whiskered Terns have been found to winter in southern France since the 1980s (OF) and possibly similar numbers also winter in Iberia, at least during some years. Wintering individuals have occasionally been reported from Mediterranean coastal wetlands since the 1960s (Erard & Vielliard 1966, Carrera *et al.* 1981b) and larger numbers have been found at the Ebro delta, with 100 on 11 January 1973 (Ferrer *et al.* 1986). The January censuses at the Ebro delta of 1996–2009 found about 150 wintering Whiskered Terns on average but counts there have ranged from 16 to 220 birds (*A. O. Catalunya*). Some numbers have also been found in winter in southwestern Andalucía, generally at Doñana in Sevilla. Wintering in the Guadalquivir delta was unknown in the 1970s (García *et al.* 1979) but recent counts there at the Caño del Guadiamar have included 95 on 12 January 2002, c. 100 on 2 January 2003 and 84 on 3 January 2004 (*Ardeola* 50: 163, 351, 51: 252). Annual totals during the Spanish winter atlas fieldwork in 2007–10 ranged from 134–289 individuals, with averages of 108 at the Ebro Delta, 65 at Doñana, 33 at other sites in Andalucía and 11 in the Comunidad Valenciana (SWA). Inland winter records of one at the Arroyo de La Luz, Cáceres, on 12 January 2002 and five at Daimiel on 5 December 2010 (*A. O. Extremadura 2001–2003*: 223; *Ardeola* 58: 503) are most exceptional. January observations in southern Iberia raise the possibility of early-returning migrants, and there is a record of a bird in breeding plumage arriving off the sea in Málaga in December (Paterson 1997). However, 22 at Isla Mayor, Sevilla, on 18 November 2006 (*Ardeola* 54: 179) were at various stages of moult so perhaps had settled to winter in the area. A few wintering individuals are also reported occasionally from coastal wetlands in southern Portugal, but there are no records of any numbers there (AP). Further observations are needed to establish the extent of wintering in Iberia, which may be a relatively new phenomenon (Carrera *et al.* 1981b).

BLACK TERN *Chlidonias niger*

(Fumarel Común, Gaivina-preta)

***C. n. niger*. Breeds sporadically in mainland Spain and has bred Portugal. Regular throughout on passage, when occasionally abundant. Rare in winter.**

The European populations have all declined since the 1970s (BWP, BE). Significant numbers nested in Spain until the mid 20th century, the majority at Doñana. Arévalo (1887) and Witherby (1930) also reported nesting at the Albufera de Valencia. Valverde (1960) considered that in the Marismas in the 1950s it was outnumbered by the Whiskered Tern by about 3–5 to 1, which would have given a breeding population there of the order of 5,000–8,000 pairs of Black Terns. Since then the population of the Marismas and elsewhere in Spain has declined inexorably and the Black Tern no longer nests regularly anywhere in Iberia.

The Black Tern has apparently closely similar, if not identical, breeding requirements to the Whiskered Tern and was commonly found nesting in mixed colonies with that species. It too requires shallow, clear waters with emergent vegetation on which to attach the floating nest. Both species feed on aquatic insects, amphibians and small fish and both also forage over open country near wetlands, capturing insects in the air and on the ground. Both too should be similarly susceptible to the effects of eutrophication and insecticide application. It is thus something of a mystery why the Black Tern should have declined so much more than the Whiskered Tern in Iberia, even though the populations of the latter also decreased greatly during the 20th century.

The Spanish population was estimated at 150 pairs by the late 1980s (Blanco & González 1992). During 1990–2003 it declined further and did not exceed 40 pairs in any year except 1996, when there were 33–44 pairs in Doñana and a further 26 pairs at the Brazo del Este: very few of which nested successfully in either area (SA2, LR). Breeding attempts in more recent years have been sporadic and the Spanish census of breeding terns of 2007 (Corbacho *et al.* 2009) did not find any nesting Black Terns at all. The Marismas of the Guadalquivir remained the principal breeding area post-1990 but the population there did not exceed 70 pairs in the most favourable year and probably none at all nested during drought periods. The lagoons of La Mancha had a fluctuating population that did not exceed 30 pairs during the same period (LR). The Black Tern has also nested sporadically at the Ebro delta, most recently in 1988 when five pairs bred. Breeding sites in the Comunidad Valenciana have included the Albufera de Valencia, the Marjal de Xeresa-Xeraco and El Hondo and Santa Pola saltpans. Sporadic nesting by occasional pairs is also known from Castilla y León and Extremadura: at Los Canchales reservoir, where at least two pairs nested successfully in 2005 (*Ardeola* 54: 179). There are several unconfirmed reports of nesting in Algarve but the only definite instance in Portugal was at the Ria Formosa in 1995, where one pair nested successfully (AP). Further nesting attempts in Iberia by small numbers of Black Terns seem likely to occur, especially at Doñana and at the wetlands of La Mancha and the Levant coast, but there are no signs that a regular population will become re-established in the foreseeable future.

The Old World Black Terns winter on west African coasts from Mauritania south to Namibia and South Africa (HBW). Passage is reported throughout the region but the largest numbers are seen along coasts. The presence and abundance of migrants is highly variable between years but there are clear indications that migrants are far less abundant than in the past, when the source populations were much larger. There are a number of Iberian recoveries of Black Terns ringed in Italy (16), The Netherlands (5), Poland (3), the Czech Republic and Russia, indicating that some birds from eastern Europe occur in the region on passage. There are also single recoveries from Algeria, Morocco and The Netherlands of birds ringed on passage in Spain. The passage periods are well defined. In spring the earliest birds may appear in the south in late March but the main movement occurs from mid April to mid May, with some arriving into June. The return passage begins in mid July, continues through August and peaks in September, with small numbers occurring through October and stragglers commonly seen in November.

The largest numbers in autumn have been reported on Mediterranean coasts and in the Strait. Black Terns also occur regularly in the Balearic Islands in autumn but are generally scarce there. Probably much passage occurs well offshore in the Mediterranean but large concentrations are reported at times along the shoreline and at estuaries and other coastal wetlands. For example, 3,000+ were at the Turia estuary, Valencia, on 29 August 1989 (*A. O. C. Valenciana 1989:* 54). Westward movement towards the Strait is sometimes marked in the Alborán basin, with passage rates of 500/hr reported from Málaga Bay in August (Paterson 1997). Large westward movements are similarly reported from the Strait in some years. For example, 3,303 were counted from Gibraltar in three hours on 31 August 1990 (GONHS). Such large counts at Gibraltar are very irregular and some seasons, particularly in recent years, have seen hardly any birds appearing there. Black Terns are often also numerous on the Atlantic coasts of Cádiz and Huelva in August and September although it is not clear whether they are birds that have continued coasting westwards after passing through the Strait or whether some or all of them arrive overland or after having followed the Portuguese Atlantic coast southwards.

The autumn passage on Atlantic and Biscay coasts appears to be much smaller than in the Mediterranean. There is a regular westward passage on the Asturian coast but the numbers involved appear to be modest. For example, in August–September annual totals during 2005–09 at Cape Estaca de Bares ranged from 242 to 1,011 birds; there were 734 in 2011 (Sandoval *et al.* 2009, 2010; A. Sandoval, pers. comm). Only small numbers have been found at inland wetlands in Spain in autumn in recent seasons.

It is widespread on autumn passage in Portugal but most observations involve small flocks at both coastal and inland wetlands giving the impression that it is usually comparatively scarce. There are no reports of significant southward passage off the west coast. Nevertheless, occasional influxes of up to 1,000 birds have been reported in August at the Tagus estuary (Paterson 1997): it is hard to account for these against the wider background of Portuguese records.

The spring passage is comparatively rapid and more concentrated than the autumn movements. Once again occurrences tend to be irregular but large numbers are seen at times. Influxes occur in some years on southwestern coasts, especially in Cádiz and Huelva, and also in the Strait: where the birds are usually flying east. Whereas autumn movements most often involve a leisurely succession of small groups, those seen in spring are more likely to be in large, compact flocks moving rapidly onwards. For example, Stead (1959) had two to three thousand in view at once moving east off Tarifa on 25 April 1958 and Nisbet *et al.* (1961) saw 'many thousands' arriving from the sea on the Doñana coast in May 1956. Such observations imply the existence of a large offshore passage northwards up the Moroccan coast (BM).

Small scale passage occurs widely in spring in Portugal and northern Spain but spring migrants often stage at coastal wetlands in southern and eastern Spain, where large numbers may be present briefly. For example, 2,000+ were at the Llobregat delta on 19 April 1991 (Gutiérrez *et al.* 1995) and gatherings of 1,500+ are frequent at the Cerrillos saltpans, Almería (Paterson 1997). Flocks are much more frequent at inland wetlands in eastern Spain in spring than in autumn and some large gatherings occur. For example, 1,800+ have been counted at Fuentepiedra in April (Paterson 1997) and 800 were at Gallocanta on 15 May 1992 (*A. O. Aragón 1991–92*: 47). The Black Tern is similarly more numerous in the Balearics in spring than in autumn. In general, there is evidence that the spring movements tend to occur further east in Iberia than the postnuptial passage, as elsewhere in Europe, and large numbers of northbound migrants may reach the Mediterranean by flying across the Sahara (Haverschmidt 1978).

A few Black Terns are widely reported in Iberia in November but December–January records are very scarce. They include records of one or two birds from Málaga, Sevilla, Almería, Gipuzkoa, Asturias, Pontevedra, Valencia and Coimbra (Paterson 1997, *Ardeola* 50: 163, 351; *Anu. Ornitol.* 6: 63). Records of 150 at the Albufera de Valencia on 26 December 1963 (Erard & Vielliard 1966) and six at Gavà, Barcelona, on 28 January 1974 (*Ardeola* 20: 315) are quite exceptional.

WHITE-WINGED BLACK TERN — *Chlidonias leucopterus*

(Fumarel Aliblanco, Gaivina-d'asa-branca)

Very scarce but annual on passage in mainland Spain and the Balearic Islands. Rare in Portugal.

The European and western Asian populations winter inland in sub-Saharan Africa west to Senegal, unlike the closely related Black Tern which has a maritime winter distribution. Large-scale movement occurs across the eastern Mediterranean and across the Sahara (HBW, BM). Iberia thus lies well to the west of the principal movements but a small number occur annually, the great majority in eastern and southern Spain: including eastern Andalucía and, in particular, the Levant coastal regions and Balearic Islands (Fig. 171). Saunders (1871) found it 'abundant' at the Albufera de Valencia in May but only once saw it at Doñana. Records become progressively fewer further west and the species is essentially rare in Portugal and northwestern Spain. In addition to some historical records, there are 30 published records (34 birds) from Portugal during 1987–2009 (AP, RBP). It is under-reported in those regions in which it is most common. In particular, Clavell (2002) comments that it is much more frequent in Catalonia, the Comunidad Valenciana and the Balearic Islands than published records suggest.

Figure 171. Records by region, 1978–2010 (n = 445 records). The relative importance of Catalonia and the Comunidad Valenciana is underestimated since not all records from these regions are published.

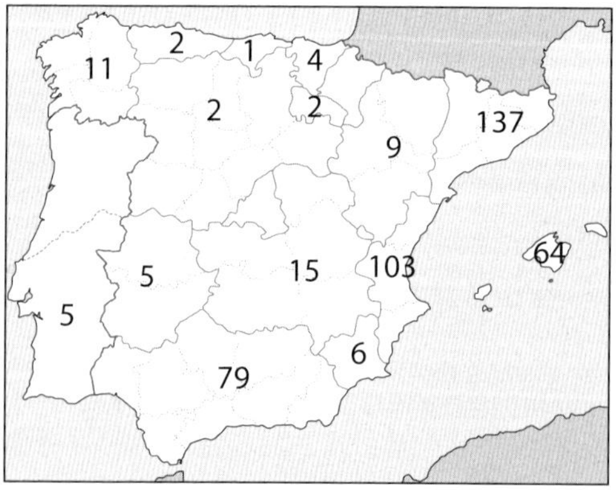

The majority are seen at saltpans, ricefields and at coastal and inland lakes and reservoirs, often with Black Terns. Many linger for short periods, especially in autumn. Most records are of single birds but 2–4 may coincide at the southern and eastern wetlands where the species is commonest. Larger groups are also found occasionally, the largest on record including 13 at the Laguna de Pedro Muñoz, Ciudad Real, on 1 May 1994; 20 at Los Monegros ricefields, Huesca, on 20 May 2011; and 37 on 17 May 1997 at the Albufera de Mallorca (*Ardeola* 41: 100, 58: 503; *A. O. Balears 1997*: 174).

The passage periods are clear-cut (Fig. 172). Spring passage is generally from mid April to the end of May but the great majority of spring records are in the first half of May. The autumn passage is more prolonged, from August to mid October. Spring and autumn records are equally frequent but autumn birds comprise some 60% of the individuals recorded. The spring passage has been reported to be larger than the 'extremely small' autumn passage in Morocco (BM). Nevertheless, it is quite obvious that whereas White-winged Black Terns are among the most unmistakable of terns when in their spring plumage, individuals in juvenile or winter plumage are much harder to separate from the far more abundant Black Terns at any distance. It is therefore likely that an unknown proportion pass unnoticed during the autumn passage.

Figure 172. Records by month, records in 1985–2010 (n = 445).

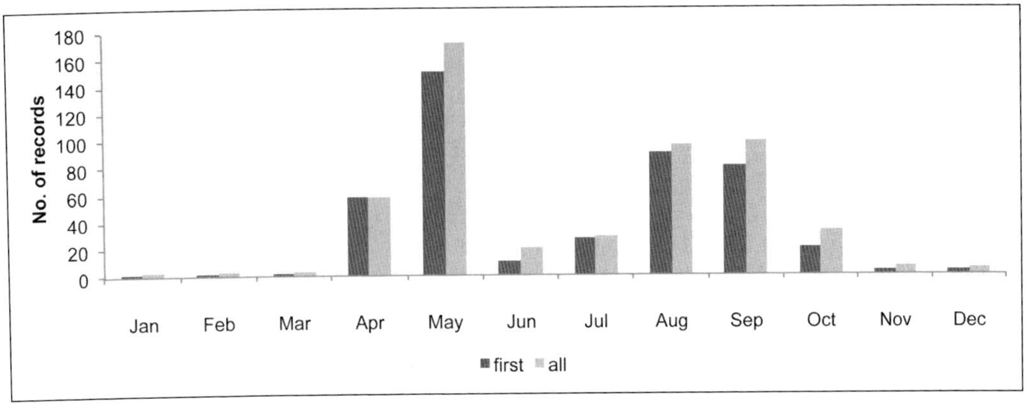

A few individuals have been recorded in winter on very rare occasions at eastern and southern locations. They include what was believed to be the same adult individual that overwintered at the Ria Formosa in 1991–92 and 1992–93 (AP). A first-winter bird remained at the Albufera de Valencia from 27 November 1994 until the end of March 1995, and one at the Cañada de Las Norias, Almería, on 30 January 2012 had apparently been there for several weeks (*Ardeola* 42: 225, 59: 186). There was one at Doñana on 27 January 2000 but there were three there on 27 February and five on 6 March, suggesting very early spring passage (*A. O. Doñana* 1: 138). Overwintering occurs at least occasionally in northern Morocco and wintering numbers have increased there since the 1990s (BM).

COMMON GUILLEMOT *Uria aalge*

(Arao Común, Airo-comum)

***U. a. albionis*. On the verge of extinction at its remaining Spanish nesting locations in Galicia. Extinct as a breeding species in Portugal. Common in winter on Biscay and Atlantic coasts, rare in the Mediterranean.**
***U. a. aalge*. Common in winter on Biscay and Atlantic coasts; probably very rare in the Mediterranean.**

The nominate race breeds in the northwestern Atlantic and east to the Baltic, wintering south to Iberia. The western European form has a more southerly breeding distribution, in Ireland, England, Wales and Brittany, with outpost populations in western Iberia. The Iberian population has undergone over 50 years of inexorable decline. A survey in Spain in 2006 and 2007 found only two breeding pairs and recorded a further four individuals: all at Cape Vilán, A Coruña (Alcalde & Docampo 2009). In Portugal an incubating bird and 26 individuals at the Berlengas Islands in 2002 constitutes the most recent breeding record (PA2).

The population early in the 20th century may have been of the order of 20,000 individuals (Munilla *et al.* 2007). It was apparently healthy until the mid 20th century, with small colonies on coastal islets and sea cliffs in the Basque Country, Cantabria and Asturias, and larger colonies in Galicia and on the Berlengas off central Portugal (Tait 1924, Bernis 1948 Alcalde & Docampo 2009). The north coast colonies from Asturias eastwards were the first to disappear, although a few may have nested in Asturias until about 1970 (Noval 2001). Nevertheless, the Spanish population around 1960 may have still numbered some 3,000 birds (Bárcena 1985), divided between at least eight colonies in Galicia: at Cape Estaca de Bares, Cape Ortegal-La Capelada, A Gaveira-Veigue islet in the Ría de Ares-Betanzos, the Sisargas islands, Cape Vilán, the island of Ons, Costa da Vela in Cangas do Morrazo, and on the Cíes islands (LR). The largest colony, with perhaps over 1,000 individuals, was on the Os Aguillons islets, off Cape Ortegal. There may also have been colonies on the Farallón and Ansarón islets, Lugo (Palomino & Molina 2009). However, the population was in steep decline, at an annual rate of 33.3% between 1960 and 1974, which slackened to 13.4% up to 1994. The decline coincided with increased commercial fishing in the region and the introduction of nylon filament nets, which are believed to have drowned many adults (Munilla *et al.* 2007). Only three colonies remained by 1982: at the Cíes and Sisargas islands and at Cape Vilán. Breeding ceased in the Cíes between 1988 and 1991 and was last reported on the Sisargas in 2004.

The population on the Berlengas was the southernmost breeding population of the species in the Atlantic, although nesting was confirmed further south, at Cape St Vincent, in May 1939 (AP). Lockley (1952) censused 12,000 individuals on the islands in 1939. Here too a rapid decline set in during the 1970s, from 1,683 birds in 1974, to 400 in 1977, 231 in 1979, 140 in 1983 and 51 pairs in 1985. Only 41 birds remained by 1997 and just 27 in 2002, since when no birds at all have bred there (Alcalde & Docampo 2009, PA2).

The decline to virtual extinction of the Iberian Guillemots has been attributed to a combination of losses of adults in fishing nets, declines in food availability: especially of shoals of anchovies, and mortality and ecosystem damage resulting from oil spills. The most infamous of the oil spillages was from the tanker Prestige which sank off Galicia in November 2002. Over 11,300 oiled Guillemots were among the many birds recovered on nearby coasts between then and February 2003 (García *et al.* 2003). Most of them were obviously winterers from further north but Iberian birds are likely to have been affected as well. The southern outpost nature of the Iberian Guillemots may mean that they have been more sensitive to broader ecological and environmental change, possibly linked to climate. Overall the patterns of population change in this subspecies are clearly complex and poorly understood but human influences are likely to be major factors (Arcos *et al.* 1995, Munilla *et al.* 2007).

Guillemots are regarded as dispersive rather than migratory in that many adults remain close to their breeding sites in winter (BWP). Some birds from western British colonies especially gradually move south to the Bay of Biscay where they occur from October to February, with some also descending the Iberian Atlantic coast as far south as the Strait of Gibraltar (Mead 1974, Wernham *et al.* 2002). Most corpses collected on northern Spanish beaches prove to be of first-winter birds: 76% in one study in Asturias and 66% in another in Galicia (Álvarez & Pajuelo 2004, Ramos *et al.* 2011). Iberian recoveries of foreign-ringed birds include at least 146 from Britain, 91 from Ireland and nine from France. They involve both *albionis* and the nominate race; 42% of 59 recoveries were ringed in Scotland, within the range of *aalge* (AI1). There is a recovery of a Welsh bird from Gibraltar (Mead 1974) but most recoveries are on Biscay coasts with a few south to Lisboa on the west coast. None were in the Mediterranean.

Guillemots occur regularly in winter off northern and western coasts but they are much scarcer within the Mediterranean than either the Razorbill or the Puffin, and accidental in Morocco (BM). They may commonly be encountered at sea in continental waters but relatively few are ever recorded from land and some of those are storm-driven weak birds or tideline corpses. Even well-watched headlands seldom report any considerable numbers. For example, only 128 were seen from Cape Estaca de Bares in 2008, chiefly in January and November–December, with extreme dates of 26 April and 4 October. The maximum passage rate on record for Cape Bares is just 36 birds/hour, on 12 December 2007 (Sandoval *et al.* 2009).

The numbers reaching Iberia, especially south of the Bay of Biscay, may vary between years although there is little available data. Something of an influx seems to have occurred in central Portugal in January 2003, when 280 flew south in three hours off Cape Carvoeiro on 6 January. The same day there were 22 in Baleal Bay, and these included at least eight *aalge*; nine dead birds were recovered at Alcobaça on 9 January (*Anu. Ornitol.* 3: 39).

Guillemots are rarely identified in the Strait but there are ten records for Gibraltar: comprising seven single birds, two groups of three and a flock of 12 (GONHS). All were observed between 14 December and 1 April except for one on 9 September 1974. There are very few records within the Mediterranean. One was collected in Valencia harbour on 10 January 1887 (Arévalo 1887). Clavell (2002) regarded the species as 'accidental' on the Levant coast and in the Balearics. Carboneras (1988) gave only two records off Barcelona. Most observations are of single birds but there were seven off Vinaroz, Castellón, on 23 February 1985 and five there on 24 March 1985 (*Ardeola* 32: 419). A surprising record of a flock of 30 flying north off the Llobregat delta, Barcelona, on the late date of 16 May 1982 (*Ardeola* 36: 253) seems hard to credit but there are three, more recent, equally unseasonal reports of single birds off Catalonia: in May 2007, June 1996 and August 2000.

A few non-breeders occur in summer, especially off Biscayan coasts and southwestern Galicia (Arcos *et al.* 1995). Twenty in total were counted off Castrillón, Asturias, on 6 June 1997 (*A. O. Asturias 2003*), although these could have been late migrants. Some enter the rías in late summer and early autumn. They include some very young individuals, thought to originate from Breton colonies, that have swum south during their post-fledging period.

BRÜNNICH'S GUILLEMOT *Uria lomvia*

(Arao de Brünnich, Airo-de-freio)

One collected in Galicia, in the Ría de Pontevedra, on 2 January 1942 (Iglesias 1952) is the sole Iberian record. It is in the possession of the University of Santiago de Compostela (E. Fernández de la Cigoña, *in litt.*). This Arctic species is a rare and sporadic visitor to the British Isles and other northern and central European countries. Over 40 have been recorded in Britain, but only three in France, in the north. There is a recent record from the Azores.

RAZORBILL

Alca torda

(Alca Común, Torda-mergulheira)

***A. t. islandica*. Common and sometimes numerous in winter on all coasts and by far the most likely auk to be encountered inshore in the Mediterranean.**

The subspecies *islandica* breeds as near to the region as the British Isles and the French Channel coast. A single recovery from Barcelona of one ringed in southern Sweden (AMI) would presumably have been of the more northerly nominate subspecies but there is no other evidence of the occurrence of this race in Iberia. Over 500 birds ringed as pulli in Britain (330) and Ireland (185) have been recovered in Iberia. One ringed in Iceland in June 1999 was controlled at the Odiel estuary in December the same year (Gacio & Sayago 2011). Most Iberian ringing recoveries are from Biscay and Atlantic coasts but a minority have been found on Spanish Mediterranean coasts north to Catalonia: of 259 recoveries considered in AMI and AI1, 41% were in northern Spain, 30% in Portugal, 15% in Atlantic Andalucía and 14% in the Mediterranean, where only one was in the Balearics. The Iberian recoveries involve both adult and immature birds but the latter account for a higher proportion of recoveries in the Mediterranean (Carboneras 1988). Most Iberian recoveries are during October–April, but some have been reported in every month of the year (AMI, AI1). The Razorbill is certainly more abundant off northern coasts than in the south and Mediterranean (AP, Carboneras 1988).

A few early migrants may appear on Atlantic coasts in September but most arrive from mid October to December. Coasting movements are often reported from headlands on the Biscay and Atlantic coasts and at the Strait of Gibraltar. These may sometimes involve considerable numbers, probably when wind and weather drive the birds closer to shore. For example, a westward passage of 950 birds/hour was reported at Cape Higuer, Gipuzkoa, on 26 November 2000 (*Ardeola* 49: 191) before an easterly gale, and 718 were counted flying south at Cape Carvoeiro, Leiria, on 27 December 2006 (*Anu. Ornitol.* 6: 63). On the basis of sample counts, the mean annual post-breeding daytime passage at Cape Bares during 2004–08 was of the order of 6,500 birds (Sandoval *et al.* 2010), a fairly modest figure.

Some thousands enter the Mediterranean through the Strait, where the earliest arrivals are in October but the majority pass during November–December. Carboneras (1988) estimated that 15,000–40,000 birds winter in the western Mediterranean, with the principal concentrations off Spain and within the Gulf of Lyons, France. Here they occur from mid-November to April, but peak abundance is during December–February. They depart during February–March, chiefly during early March, when large numbers are sometimes visible passing Gibraltar in a rapid succession of flocks. For example, 1,188 were counted flying west in three hours at Europa Point, Gibraltar on 2 March 1990 (*Ardeola* 38: 342). Very few remain in the Mediterranean by April but there are occasional records of individuals, perhaps oiled or sick birds, lingering into the summer (Carboneras 1988). Spring migrants return north off Atlantic coasts during February–March with smaller numbers reported in April (Paterson 1997). A few summering non-breeding individuals are also recorded on Biscay coasts in summer and more rarely on Atlantic coasts: there are two June records from western Portugal (AP).

Razorbills winter in coastal waters all around Iberia. In Catalonia, they occur throughout the continental shelf but they are more frequent there in shallow waters, including off beaches and in harbours (Carboneras 1987). Arévalo (1887) wrote of their abundance off the Valencia coast in February and March. They may concentrate off estuaries but seldom ascend any distance upriver. Large numbers have been reported off the Llobregat delta, Barcelona, where there were 1,500–2,000 on 12 January 1974 (Ferrer *et al.* 1986) and a peak count of 852 was made during winter 1987–88 (*Ardeola* 35: 309). Counts of over 500 were made at the Ebro delta in late December 1966 (Ferrer *et al.* 1986). They also frequent the Tagus and Sado estuaries (AP). Razorbills were badly affected by the wreck of the tanker Prestige in November 2002, after which 3,654 oiled birds were recovered on northwestern Spanish coasts and a further 240 in Portugal (García *et al.* 2003, AP). Accidental drowning in nylon fishing nets may also be an important cause of mortality (Bourne 1984, Teixeira 1986).

BLACK GUILLEMOT

Cepphus grylle

(Arao Aliblanco, Airo-d'asa-branca)

Vagrant to Spain.

There are four accepted records of single birds from the northwestern Spanish coasts, mostly in winter: at Sanxenxo, Pontevedra, on 27 December 1992; at Cabo Silleiro, Pontevedra, on 27 December 2001; at El Musel port, Gijón, Asturias, from at least 11 November 2004 to 7 February 2005, and at the Ría de Villaviciosa, Asturias, from 31 August to 5 September (Salvadores *et al.* 1994). Three were first-year birds. The most likely source of the very rare records in the region is probably the British Isles (*C. g. arcticus*). The Black Guillemot only occasionally reaches as far south as the Netherlands or France: where 71 individuals have been found, seldom south of Brittany.

LITTLE AUK *Alle alle*

(Mérgulo Atlántico, Torda-miúda)

***A. a. alle*. Very scarce in winter off Biscayan coasts and progressively rarer southwards along Atlantic coasts. Vagrant to Gibraltar and the Mediterranean.**

The Little Auk mainly winters very far north, on the edge of the pack ice, but a minority winter regularly as far south as northern Britain. Occurrences even further south involve mainly small numbers and may be associated with irruption years, when larger numbers than usual appear in northern European waters south to the Bay of Biscay and sometimes in Iberian Atlantic waters. The Little Auk is highly pelagic, feeding offshore on zooplankton and small fish, so its presence is often undetectable from land, except during 'wrecks' when large numbers of storm-driven birds may appear on coasts, many of them dead (Hope-Jones *et al.* 1984). Coastal records in Iberia similarly often follow bouts of stormy weather and many birds only appear as tideline corpses: 59 (23%) of 256 birds reported on Iberian Atlantic and Biscayan coasts during 1979–2009 were found dead. In addition, the sinking of the tanker Prestige off the Galician coast in November 2002 was followed by the appearance of 64 Little Auks on Spanish beaches between then and mid February 2003 (García *et al.* 2003) and a further three were found in Portugal (AP). This is another clear indication of the presence off northwestern Iberia of larger numbers than are ever seen from land, in at least some winters.

The first Iberian record was in January 1915, when a moribund bird was picked up on a beach in Vila do Conde, northern Portugal (Reis Júnior 1931). The first Spanish records were of birds captured at Segur de Calafell, Tarragona, in 1948 (*Ardeola* 5: 223–224) and in Santander bay in December 1954 (de La Lama 1955). Since then the Little Auk has been found to be a very scarce winter visitor to northern coasts but even here most birds remain far offshore. Ocio & Astigarraga (2008) found it regularly off Punta Galea, Bizkaia, in December–January 2001 and 2002, but always beyond the continental shelf at least 15 miles offshore: 75% were over 20 miles from land.

The great majority of records come from Galicia and the Biscay coasts but even here the species is very scarce. For example, the *Ardeola* Noticiarios and the respective regional bird reports give just 35 records (52 birds) for Galicia during 1982–2004 and 77 records for Asturias (124 birds) during 1979–2000. Reports from Cantabria and the Basque Country are less than annual. Only 13 were encountered during the Spanish winter atlas fieldwork in 2008–10 at six localities: three in the Basque Country, two in Galicia and one in Cantabria, and they included five birds on passage at Cape Higuer, Gipuzkoa, on 1 December 2009 (SWA). Very few have been reported in Portugal: just 12 records up to 2007, in addition to single historical records from 1915 and 1934 (AP). Further south there are at least eight records from the coasts of Atlantic Andalucía, involving 15 birds only one of which was alive, an exhausted individual that was found four kilometres inland at Sanlúcar de Barrameda on 17 December 1985 (*Ardeola* 5: 143–148): they include nine tideline corpses found after storms at Los Lances beach, Tarifa, on 2 January 1979 (*Ardeola* 29: 187).

Little Auks are almost entirely recorded in winter (Fig. 173), as would be expected, but records in November–December greatly outnumber those later in the winter. Coastal observations need not necessarily be an accurate reflection of the situation offshore but it may be that birds begin to leave Iberian waters in January. Spring records (of live birds) are unusual but in Galicia one was at Cariño on 30 March 2010, one was off O Grove on 24 April 1983, and two birds were off O Grove on 10 May 2000 (*Ardeola* 31: 145, 57: 234; *A. O. Galicia 2000*: 41). There is an additional, bizarre, April record that is not included in Fig. 173: of a flock of 30 flying 200m off the beach in Doñana on 2 April 1988 (*Ardeola* 35: 309).

Figure 173. Monthly distribution of live Little Auks in Galicia, Asturias, Cantabria and the Basque Country during 1979–2009 (n = 117 records, 190 birds).

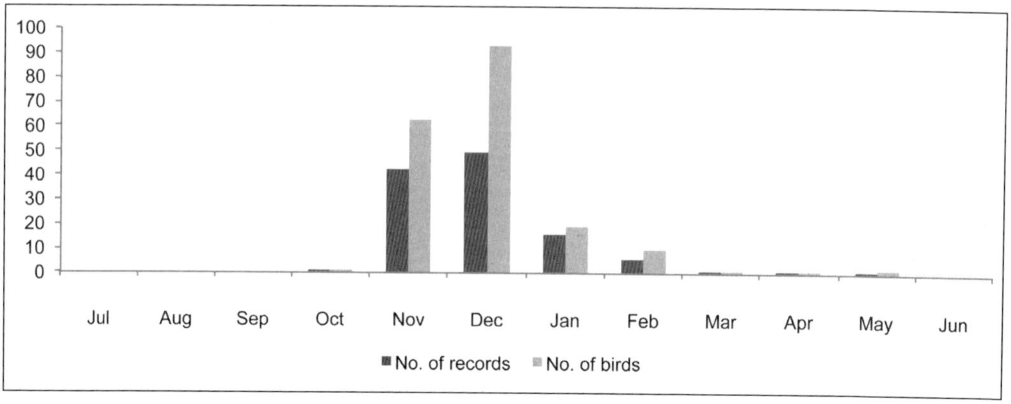

The sole record for Gibraltar was of one flying west on 27 March 1986. Very few enter the Mediterranean but there are two records there from Andalucía: single birds off Benalmádena on 23 February 1988 and off Motril from 20–22 November 2009 (Paterson 1997, *Ardeola* 57: 234). Arévalo (1887) also reported finding dead birds on the Málaga coast after storms. Further north there are at least eight records from Catalonia and one from the Columbretes islands (Paterson 1997, Clavell 2002). All were of single live birds except for three seen together in Roses bay, Girona, on 28 December 1975 (Sargatal & Llinàs 1978). The dates of some of these east coast records fall well outside the 'usual' winter period and it may be that the birds had tried to return north up the Spanish east coast. The late records comprise single birds at Estartit, Girona, on 12 May 1968 (Sargatal & Llinàs 1978), off Barcelona on 8 June 1985 (Carboneras 1988) and off the Columbretes islands on three dates during 3–14 May 1996 (*A. O. C. Valenciana 1995–1997*: 165). Three were photographed off the Llobregat Delta on the very early date of 19 October 2006 (*A. O. Catalunya 2008*: 14).

ATLANTIC PUFFIN *Fratercula arctica*

(Fraillecillo Atlántico, Papagaio-do-mar)

***F. a. grabae*. Common in winter off all coasts but seldom seen from land.**

The Puffin is pelagic outside the breeding season and so very seldom seen from land. Even coastal watchpoints seldom detect large movements, except at the Strait of Gibraltar. Many fewer ringed Puffins than Razorbills have been recovered in Iberia, no doubt partly because their oceanic habits make it less likely than dead birds will be found. There are at least 53 recoveries of ringed Puffins and all originated at colonies in the British Isles. All 13 ringed Puffins recovered in Morocco were also of British origin (BM). The breeding Puffins of Britain and Ireland, the Faroes, northwestern France, southern Sweden and southern Norway are of the subspecies *F. a. grabae*. The more northerly nominate subspecies has never been recorded in Iberia but some could reach the region in winter.

All Puffins abandon their breeding colonies and adjacent seas by the end of August (Wernham *et al.* 2002) but they are seldom noted in Iberian waters earlier than October and the main arrivals are in November–December. Chiefly small numbers are reported on autumn passage from headlands on Biscay and Atlantic coasts. For example, only 312 Puffins were reported from Cape Estaca de Bares in 2008 and just 68 in 2009, virtually all of them during January–March and mid October–December. Larger numbers are seen there in some seasons, the maximum movement recorded being 796 birds/hour on 12 December 2007 (Sandoval *et al.* 2009, 2010). A total of 2,426 Puffins were seen from the Cape during the well-watched season of 2011 (A. Sandoval, pers. comm.). The wreck of the Prestige off Galicia in November 2002 was followed by the finding of 3,584 oiled Puffins on northwestern Spanish coasts (SEO/Birdlife 2003) but only 27 were picked up in Portugal (AP); this confirms the presence of considerable numbers of Puffins off the north-west at that time.

Large numbers winter in the western Mediterranean, which they enter through the Strait of Gibraltar mainly in November–December, with some eastward arrivals noted into January (Garcia 1973). The wintering population there has been estimated at 35,000 individuals, on the basis of the frequency of ringing recoveries (Carboneras 1988). Dean Hashmi (In BSG) carried out transects within the Strait from August but saw no Puffins earlier than 9 November, with a peak arrival in 1986 on 17 December, when the passage rate reached 5,400 birds/hour. The influx into the Mediterranean occurs well offshore in mid-Strait and is seldom detectable from land, although Tellería (1981) was able to record an eastward flow of small groups off Tarifa from early October in 1977.

Puffins are also pelagic within the Mediterranean and are very rarely seen in coastal waters. Some numbers reach Balearic waters, at least in some years: for example, off Mallorca there were 110 on 23 February 1992; 250 on 17 January 1993 and 140 on 10 April 1993 (*A. O. Balears 1992*: 94, 1993: 89). Munn (1921, 1924) described it as common off the islands in winter, although seldom visible from land. Counts of 78 on passage past Cape Cervera, Alicante, on 23 March 2005 and 113 seen from Cape Palos, Murcia, in 75 minutes on 28 March 2009 (*Ardeola* 53: 388, *A. O. Murcia 2009-10*: 22) are noteworthy.

Puffins begin to leave the Mediterranean in February and passage peaks at the Strait in late March and early April, continuing through May and even into June (BSG). Unlike the autumn movements, the return to the Atlantic is often visible from Gibraltar itself, as a rapid succession of flocks. A passage at Gibraltar of 13,000 in five hours on 28 March 1985 (BSG) is the largest on record. Much larger numbers of Puffins than Razorbills winter in the Mediterranean: Puffins comprised over 92% of 25,936 auks that were specifically identified from Europa Point, Gibraltar, during 1980–86 (BSG).

Large-scale Puffin passage has seldom been observed from western Iberian coasts and presumably occurs at some distance from land, although more watching from prominent headlands would be likely to detect significant movements at times. A westward movement of 3,558 birds in four hours was counted at Ponta da Atalaia, Faro, on 24 March 2005 (*Anu. Ornitol.* 5: 55). Puffins are much scarcer on Biscay coasts in spring than in autumn, in common with most migrant seabirds. Inland records in Iberia are all but unknown but there is a remarkable record of one picked up alive 40km inland from the Málaga coast and 1,350m above sea level in the Sierra de Las Nieves on 11 December 2010 (*Ardeola* 59: 441).

Occasional Puffins are reported in Iberian waters in summer and these may be immature non-breeders. For example, in the Mediterranean five were off Peñíscola on 23 July 1966; four were identified at close quarters nine miles off Garraf, Barcelona, on 1 August 1985; one was off Cape La Nao on 26 June 1988; three were off Valencia on 16 July 1989 and four were seen there, on 4 June 1992, and one was off the Columbretes Islands on 28 June 2000 (*Ardeola* 12: 239, 33: 209; *A. O. C. Valenciana 1989*: 55). There is also a record of 17 near Cape Finisterre, A Coruña, on 27 July 1985 (Woutersen 1993).

BLACK-BELLIED SANDGROUSE *Pterocles orientalis*

(Ganga Ortega, Cortiçol-de-barriga-preta)

***P. o. orientalis*. Widespread but patchily distributed across central Spain and, locally, in adjacent parts of Portugal. Absent from the Balearic Islands.**

This sandgrouse chiefly occurs in central Spain and adjacent areas of southeastern Portugal. Its Iberian populations are the only ones in western Europe. It is absent from the Eurosiberian region and from all coastal lowlands except those in the dry south-east. It inhabits largely arid open country, selecting uncultivated land, long-term fallows, cereal crops and rough pastures (SA1). It avoids taller cover but is more tolerant of scattered trees and shrubs than the Pin-tailed Sandgrouse. It ranges from sea level in the arid south-east to over 1,300m in the Iberian Range, where it frequents the *páramos* or montane grasslands. An essential requirement, as with other sandgrouse, is the availability of water sources offering good all-round visibility, to which it flocks daily in summer to drink in the mornings and sometimes also at dusk (de Borbón *et al.* 1999, Ferns & Hinsley 1999). It is less gregarious than the Pin-tailed Sandgrouse and is usually encountered in pairs or very small groups during the breeding season and in flocks of up to 15 birds, more rarely up to 40, in winter. Nevertheless, larger concentrations do sometimes occur: for example in Ciudad Real in 2002 there were 210 at Granátula de Calatrava on 13 January, 250 at Alcázar de San Juan on 19 February and 160 at the latter site on 3 August (*A. O. Ciudad Real 2002–2003*: 286).

The Spanish population has a fragmented distribution, with subpopulations of varying importance inhabiting the northern and southern Meseta, Extremadura, the Ebro valley, the Iberian Range, the Guadalquivir valley, the eastern lowlands of Andalucía, and the arid south-east: in Almería and Murcia (Suárez *et al.* 2006a). The small Portuguese population is in Beira Baixa (Castelo Branco) and in the east of Alentejo (Campo Maior, Évora, Mourão, Moura, Castro Verde and Mértola) (PA2, AP).

Table 48. Minimum and maximum estimated numbers of Black-bellied Sandgrouse in the Spanish regions in 2005. After Suárez *et al.* 2006a.

Region	No. of individuals		% of mean national population
	Minimum	Maximum	
Aragón	2,000	2,500	26.9
Castilla y León	1,390	1,870	19.5
Extremadura	1,000	2,000	17.9
Castilla-La Mancha	1,000	1,500	14.9
Andalucía	800	1,020	10.9
Navarra	300	400	4.2
Murcia	200	250	2.7
La Rioja	70	130	1.2
Madrid	50	100	0.9
Comunidad Valenciana	54	83	0.8
Catalonia	10	20	0.2
Total	6,874	9,873	

Black-bellied Sandgrouse may be found within their breeding range all year round and thus appear to be largely sedentary. They are chiefly seed-eaters, consuming grain on cultivated land and mainly the seed of clovers and other leguminous plants

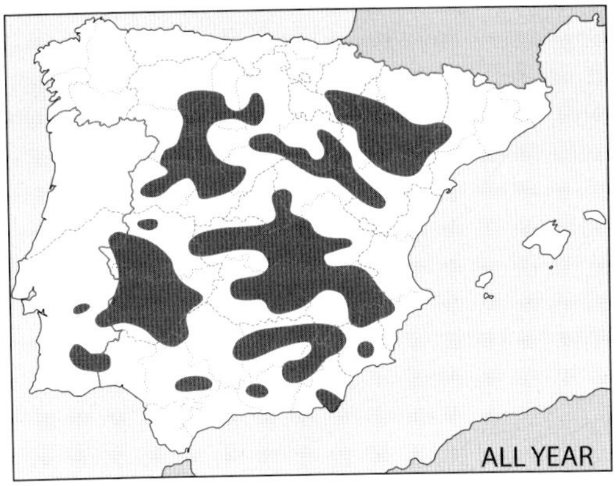

on pastures and fallows (Suárez *et al.* 1999b). In winter the majority favour agricultural land, whereas more natural habitats, such as pastures and long-term fallows, are selected when nesting (SWA). Some indications of numerical increases in winter in Castilla-La Mancha may be due to birds descending from the higher elevations of the Iberian Range (Suárez *et al.* 2006a). There it occurs at especially high densities in the Campo de Calatrava, Ciudad Real, where few remain to breed (SWA). They are very seldom recorded at any great distance from their regular haunts and, in particular, there are no records north of the Cantabrian Range. Recent observations away from the usual range include a flock of seven in the Sierra de Las Yeguas, Málaga, in January 2002; sightings of a few at Villena, Alicante, in February, March and June 2002, and one at Castro Marim, Faro, in October 2006 (*Ardeola* 50: 164, 51: 554; *Anu. Ornitol.* 6: 63).

Most Iberian sub-populations have shown worrying declines in recent decades. The most comprehensive survey, in spring 2005 (Suárez *et al.* 2006a), put the Spanish population within the range of 6,874 to 9,873 individuals, with the largest sub-populations occurring in Aragón: both in the páramos of Teruel and in the Ebro valley; in Castilla y León and Castilla-La Mancha: both on the meseta and in the páramos of Soria, Segovia and Guadalajara; in Extremadura and in Andalucía (Table 48). The Andalucían population is represented by small numbers (100–170 birds) in the farmlands of Huelva, Sevilla and Córdoba in the west, a population (550–650 birds) in the arid, stony lowlands of Granada and Jaén in the centre, and the Almería population (150–200 birds) that comprises part of the population of the arid south-east, together with the Murcia population. A breeding season census in 2010 found 803 birds in Andalucía (*Ardeola* 59: 441–442). Smaller populations occur within the Ebro valley in Navarra, La Rioja and Lleida. There is also a marginal presence in eastern Valencia. Historical references, notably Lilford (1886), Smith (1868), Arévalo (1887) and Gil Lletget (1945), indicate that it formerly occupied much the same areas of Spain and Portugal as those where it still occurs but that it was considerably more numerous.

Previous assessments of the Spanish population have differed very considerably. Purroy *et al.* (1997) gave an estimate of 54,000–100,000 individuals but these figures were apparently based more on 'intuition' than solid data. Estimates of 14,000–17,000 birds (Suárez *et al.* 1999a, c) and 10,000–15,000 birds (LR) have a firmer basis in local surveys. Birdlife International (2004) gave a broad estimate of 5,500–22,000 birds. Nevertheless, where comparative data exist all the indications are of significant decline and range contraction, the former perhaps amounting to 30% in Spain as a whole during 1974–94 and as much as 50% in the fringe populations in Catalonia and Navarra (LR). The species has declined sharply in Portugal and suffered extensive range contraction there during the 20th century. It has disappeared from Trás-os-Montes, Alto Douro and Beira Alta since the mid 1980s and the population in Beira Baixa (Castelo Branco) declined from hundreds of birds in the 1990s to fewer than 20 post-2000 (PA2). One of two population centres in Alentejo has also disappeared in recent years (AP). A census in April–May 2003 found 258 in Portugal (PA2).

The species has been severely affected by the problems that have diminished the quality and extent of pseudo-steppes and dry croplands throughout Iberia, mainly habitat loss to irrigation, scrub encroachment following land abandonment, planting of olive groves and vineyards, and loss of short-term fallows and uncultivated field margins (LR). Increased use of fertilisers, herbicides, seed dressings and insecticides has reduced habitat quality and may have affected drinking water sources. It is a long-lived species: up to ten years, which usually compensates for the very high rate of predation on nests: over 75% in studied areas in Spain (de Borbón *et al.* 1999), but these additional negative pressures are clearly having a serious impact.

PIN-TAILED SANDGROUSE *Pterocles alchata*

(Ganga Ibérica, Ganga-de-barriga-branca)

***P. a. alchata*. Widespread but patchily distributed across south-central Spain, with a marginal presence in Portugal. Mainly resident. Absent from the Balearic Islands.**

The nominate race is restricted to the Iberian Peninsula and southeastern France, and Spain accounts for almost the entire population. The French population has been declining since the early 20th century and numbered only 100–150 pairs post-2000, all of them in Crau (OF). The Portuguese population was considered to have declined to extinction by the mid 1980s but a tiny nucleus, no more than 5–8 pairs during 2002–05, was discovered in 2002 on the eastern border with Spain in Beira Baixa (PA2).

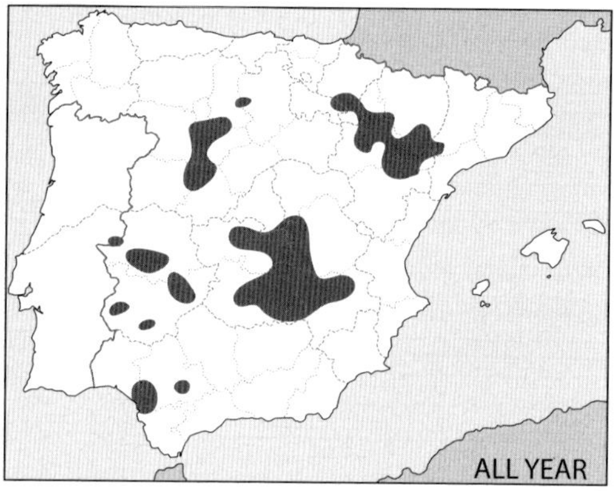

The Spanish population has a more restricted range than the Black-bellied Sandgrouse. It comprises five discrete sub-populations: in the central Ebro valley (from Lleida in the east through Aragón to southern Navarra and the east of La Rioja); in Castilla y León (mainly in La Armuña region of Ávila and adjacent areas of Salamanca and Valladolid, with a residual presence between Palencia and Burgos); in Castilla-La Mancha and southeastern Madrid; in Extremadura (chiefly on the plains of Llanos de Cáceres, Brozas-Membrío and La Serena); and in western Andalucía, chiefly in Doñana (SA2).

It is even more intolerant of tree or shrub cover than the Black-bellied Sandgrouse and may only be found in very open, semi-arid country comprised by cereal croplands, short-term fallows and rough pastures (Suárez *et al.* 1997, 1999a; Martín *et al.* 2010a). Those in the Guadalquivir delta also occur on the salt flats and pastures there. The distribution is more southerly in central Iberia than that of the Black-bellied Sandgrouse, and it is much scarcer than that species on the northern meseta. It does not occur on the páramos of the Iberian Range, also unlike the Black-bellied Sandgrouse, but breeds to 1,000m on the northern meseta. Breeding birds particularly select pastures and annual fallows and avoid cereal crops or dwarf-shrub steppes. They require well-exposed drinking water sources to which they fly daily in the mornings and, sometimes, in the afternoons, at least during the hottest months of the year (de Borbón *et al.* 1999). Several pairs may nest close together: at La Serena two nests were only 6m apart and within 50m of a third one. It has also been found nesting in small colonies at Doñana, often near Collared Pratincole colonies (Valverde 1960). The Pin-tailed Sandgrouse feeds on the seeds both of wild and cultivated plants, with some preference for legumes, although cereal grain may comprise most of its diet in summer (Casado *et al.* 1983, Suárez et al. 1999b).

The Spanish populations were censused in spring 2005 (Suárez *et al.* 2006a). This survey put the total numbers at 7,982–10,972 individuals, the great majority in Castilla-La Mancha (42%) and Aragón (32%), with significant numbers also in Extremadura (13%) and western Andalucía (5%) but only small or marginal concentrations elsewhere. Previous estimates have put the Spanish population much higher. Tucker & Heath (1994) suggested 10,000–22,000 birds and Suárez *et al.* (1999c) gave an estimate of 17,000–22,000, later revised to 20,000–40,000 (BE). However, it is thought unlikely that numbers have declined to the extent that some of these estimates might suggest. The 2005 survey was much more intensive and thorough than the earlier ones. The two sandgrouse species thus have Spanish breeding populations of a similar size although the Pin-tailed Sandgrouse population is more concentrated within its smaller breeding range.

The 2005 survey confirmed the numerical declines and range contraction experienced by at least some populations since at least the mid 20th century. These have been particularly evident in the most northerly reaches of the breeding range in the northern meseta, as well as in the Guadalquivir valley outside the Doñana region. There have also been at least local range contractions and numerical declines in Extremadura and possibly in Aragón, although overall the populations in the central Ebro valley and Castilla-La Mancha may have seen little or no recent change. It seems always to have been much rarer and much less widespread than the Black-bellied Sandgrouse in Portugal, as noted by Smith (1868). However, it bred formerly in eastern Trás-os-Montes, the Alto Douro and southern Alentejo, from where it has disappeared, and it may once have been abundant in the Guadiana valley (PA2, AP).

This sandgrouse is much more gregarious than the Black-bellied Sandgrouse, especially outside the breeding season when it often occurs in flocks of 200–300 birds and sometimes many more (de Borbón *et al.* 1999). Flocks of over 1,000 have been reported from Aragón, Castilla-La Mancha and Doñana outside the breeding season. For example, in Ciudad Real there were 2,000 at Ojos del Guadiana, Daimiel, on 5 December 2002 and 1,200 at the Campo de Criptana on 22 September 2003 (*A. O. Ciudad Real 2002–2003*: 286). The largest such flock recorded was of 4,000 birds at Herencia, Ciudad Real, in an expanse of irrigated alfalfa crops, where the birds were in the company of some 5,500 Little Bustards (Otero 1985, C. Otero pers. comm.). The association with Little Bustards is quite typical and apparently usually occurs in good feeding areas, such as alfalfa fields, in which the vegetation is taller than is usually acceptable to Pin-tailed Sandgrouse, which otherwise select sites offering good all-round visibility (de Borbón *et al.* 1999, Martín *et al.* 2010b). The inference is that the sandgrouse benefit from the presence of the much taller bustard 'lookouts'.

There are clearly some movements of varying extent outside the breeding season, as evidenced by regional numerical fluctuations, but these have not been properly investigated, although they seem to be confined to the Peninsula. A study of radiotagged individuals in La Mancha found that breeding and wintering areas were distinct and up to 20km apart (Martín 2011). Suárez *et al.* (2006a) have suggested that the populations of Castilla y León move to winter on the southern meseta, that those in Castilla-La Mancha concentrate locally in the most suitable habitat areas in winter, that the Guadalquivir delta populations may be augmented by immigrants from Extremadura in winter and that the Ebro valley may receive birds from more southerly regions also in winter. Non-irrigated croplands are the principal winter habitat, although many occur in vineyards and in the vicinity of wetlands during this season (SWA).

The species is very rarely recorded well outside the breeding range. However, it has wandered north to Asturias, where three were recorded during 1975–92 (*A. O. Asturias* 4). One near Miranda de Ebro, Burgos, on 30 May 2005 (*Ardeola* 52: 431) was marginally within the Basque Country. There are at least six records of single birds from the the Comunidad Valenciana and up to 18 were at the Estepas de Yecla, Murcia, in February–March 2007 (*Ardeola* 54: 399).

PALLAS'S SANDGROUSE *Syrrhaptes paradoxus*

(Ganga de Pallas, Ganga-das-estepes)

Vagrant to Spain.

Pallas's Sandgrouse is renowned for the far-reaching irruptions during the late 19th and early 20th centuries that took many birds vast distances from their usual range, some as far as western Europe. The largest such event was in 1888 when thousands were involved. Hundreds reached France that year, especially during May and June. The sole Iberian record was during the 1888 invasion, when one was obtained at La Devesa, a system of littoral sand dunes near Valencia city, on 9 June (AMI, Dies *et al.* 1999).

ROCK DOVE / FERAL PIGEON *Columba livia*

(Paloma Bravía, Pombo-das-rochas)

C. l. livia. **Ubiquitous and often abundant throughout the region. The great majority are feral pigeons living in close proximity to human habitation but wild populations also occur.**

Breeding population estimates. Spain: ACRE 6.6 million birds. **Gibraltar** 500 birds.

Wild Rock Doves still occupy their ancestral habitats in parts of Iberia, where they nest in colonies in holes in sea cliffs, riverine gorges and mountain escarpments. Elsewhere they have disappeared or been absorbed into the feral population. The largest wild-type populations occur inland in Andalucía and Extremadura but there are also nesting birds on sea cliffs throughout the region. Saunders (1871) and Arévalo (1887) described it as abundant in the sierras, particularly on the east coast and in Andalucía, and Munn (1921, 1924) refers to the abundant wild populations of Mallorca and Menorca, but wild type birds anywhere are now much more locally distributed than formerly. The Rock of Gibraltar had a large population in the 19th century and no doubt long before (Reid 1871–90, Irby 1875) and the species is abundantly represented there in subfossil cave deposits (BG). However, there are no longer any wild-type populations in Gibraltar.

Birds in other than wild-type plumage are not infrequently present among the 'wild' birds and it is uncertain whether there now remain any populations that are made up entirely of pure-bred wild birds. Wild-type birds also occur as a minority in and around cities, towns and villages, within the large flocks of phenotypically diverse 'feral' pigeons that live there. Rock Doves, feral or otherwise, flock to feed on agricultural land near their breeding sites. Grain must often comprise much of their diet and they certainly abound in the cereal croplands of the meseta, where they typically nest in village churches or in the traditional pigeon lofts that are scattered across the countryside. Those in human settlements of course also do much or all of their feeding by scrounging scraps discarded by people. Spanish city populations may be very large even though many municipalities now have pigeon control programmes: estimates include 43,000 individuals in Málaga and 180,000 in Barcelona (SA2, Senar & Sol 1991).

The Portuguese population has not been censused but the Rock Dove and feral Pigeon are widespread throughout and there are 'wild' populations in the Duero and Tagus gorges and on the rocky coasts of Algarve, for example. As elsewhere, Portuguese towns and cities have large populations: that of Lisbon was recently estimated at 153,000–182,000 individuals (AP).

STOCK DOVE

Columba oenas

(Paloma Zurita, Seixa-europeia)

***C. o. oenas*. Locally common as a breeding species in the north-east quadrant of the Peninsula but progressively rarer further west and south. Numbers are increased by migrants in winter, when the species is more widespread. A rare visitor to the Balearics.**

Breeding population estimates. Spain: SA1 36,000–54,400 pairs; ACRE 183,000 birds. **Portugal**: few.

The Stock Dove is widely distributed in Iberia but only at all common as a breeding species in the central and eastern sectors of the northern meseta and the Ebro valley (SA2 & LR). It becomes very much scarcer and more local further west and south in Spain and is now largely absent in the breeding season from Andalucía (Martín 2001) and is very uncommon or irregular in Castilla-La Mancha and Extremadura. It does not nest in the Pyrenees or the Cordillera Cántabrica, nor in the Biscay coastlands. It is only an accidental visitor to the Balearics (*A. O. Balears 2010*: 265), where Munn (1930) saw two at Alcudia, Mallorca, on 18 October 1927. There is a small breeding presence in extreme northeastern Portugal, in the Serra do Barroso and the Parque Natural de Montesinho in Trás-os-Montes, contiguous with Galicia where it also nests, but it does not breed regularly elsewhere in the country (PA2, AP). In essence, the Iberian breeding distribution is centred on parts of the supra- and mesomediterranean climatic zones.

The Stock Dove flocks to feed on farmland and in open woodland but requires cavities for nesting, in the form of holes in mature trees or in (usually) ruined buildings or cliffs. Nesting in rock cavities or in man-made structures is a particular feature of Iberian Stock Doves and is much less frequent elsewhere (Hagemeijer & Blair 1997). Suitable nesting habitats often take the form of open oak woodlands, juniper woods, pinewoods and riparian trees, as well as large urban parks: given the presence of holes large enough for nesting or the availabilty of suitable nest-boxes. Loose colonies form in favoured locations. Stock Doves nesting in banks and small ravines, where they may even make use of enlarged Bee-eater holes, can be found in the rather deforested areas of the central Ebro valley (Calvet *et al.* 2005). In many places the Stock Dove typically coexists with the Jackdaw, with which it may compete for nest holes (de Juana 1980). The availability of nest sites in trees has certainly been reduced by Dutch elm disease, since elm groves were a favourite breeding habitat (AMI), and also by the formerly widespread removal of trees from roadsides and hedges. Breeding in holes in buildings and cliffs may be a recent development: Bernis (1967: AMI) remarked that most Iberian Stock Doves nested in tree holes, with just a minority using cliff sites.

There is considerable evidence that the Stock Dove is a relatively recent colonist of the Iberian peninsula, having arrived during a period of westward range expansion that saw it becoming progressively more widespread in western Europe during the late 19th century (Voous 1960, Yeatman 1976, BWP). Within Iberia it is not mentioned by several late 19th century authors including Brehm (1857), Lilford (1865, 1866), Saunders (1871) or Chapman (1884, 1888) and a reference by Arévalo (1887) to its presence in southern and central Spain is not entirely convincing. However, Irby (1875) saw one near Gibraltar and he refers to a breeding population in northern Morocco, whose existence has since been confirmed (Heim de Balsac & Mayaud 1962, BM). The first published record of breeding in Iberia was by Witherby (1928), who found several nesting pairs in spring 1927 in the gardens of La Granja de San Ildefonso, Segovia: where it certainly did not nest during the late 19th century since Castellarnau, a highly competent observer, reported that it only occurred there in winter. Nevertheless, other authors did not find nesting birds during the 1920s in such regions as Navarra, Aragón or the Levant (Boxberger 1921, Ticehurst & Whistler 1925, 1930) where the Stock Dove is now common. Breeding was reported more widely during the 1940s, at El Pardo, Madrid, by Gil-Lletget (1943) and in Ledesma, Salamanca, and various sites in Madrid and Segovia, including the Casa de Campo park in

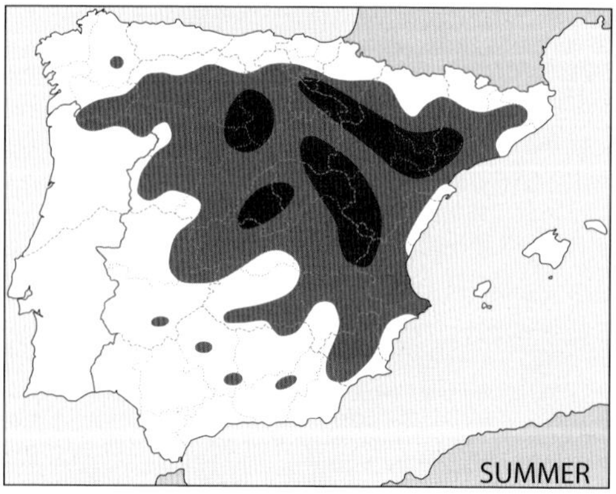

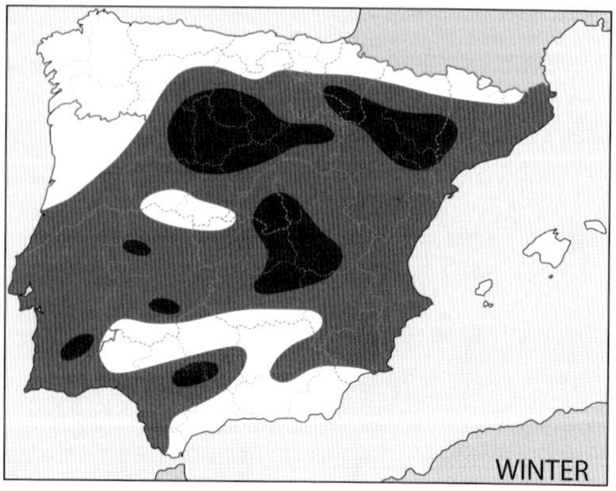

Madrid, by Bernis (1945). Bernis then described it as widespread but not numerous and he was surprised that it had apparently not been noticed earlier, although if the species was indeed a colonist that arrived in the 1920s there had been plenty of time for its numbers to have increased during the intervening period. An increase in breeding numbers in Spain certainly occurred during the 20th century but was poorly monitored. An example is provided nonetheless by observations in the gardens of Aranjuez, Madrid, where Witherby (1928) only managed to hear one bird whereas Moreno-Opo & Seoane (2004) were able to find a dense breeding population of some 50 pairs/10ha.

Changes in abundance have been reported in some regions since the first Spanish atlas survey of 1975–95 (SA1), the Stock Dove having apparently disappeared from Huelva and Cádiz in the extreme south-west and from Girona in the extreme north-east, but having apparently increased in Zaragoza, Teruel and Cuenca (LR). A significant 17% decline was seen on SACRE sites during 1998–2012 (Escandell 2013).

Large numbers of Stock Doves from central and northern Europe used to winter in Iberia, travelling in large flocks through the Pyrenean passes, often in the company of Wood Pigeons, many falling to the hunters who waylaid them there in autumn (AMI, Noval 1975, LR). The number of migrants through the western Pyrenees, however, has apparently declined greatly: at the Organbidexka pass, for example, some 5,000–10,000 birds were recorded annually in autumn during the early 1980s but 1,188 were counted on average during 2009–11, with 5,000+ annually at the three western passes combined during the same period. A similar decrease at the Echalar pass in Navarra has been evident (Purroy & Rodero 1984, LR). Counts provide limited information of the scale of the passage since most pigeons passing through the passes are recorded as 'pigeon species'. However, hunting bags at the Pyrenean passes declined by 82% between 1866–80 and 1969–83 and the proportion of Stock Doves among the total captures of pigeon species fell from 32% to just 4% between these two periods (LR). The autumn influx peaks in October but extends from September to November, with departures during February–April, peaking in March.

Ringed birds mainly of German, Finnish, Swedish and Swiss origin, and a few individuals from England, Belgium, Latvia, the Czech Republic and Austria, have been recovered in Iberia. Out of 102 nestlings ringed abroad and recovered in Spain, 58% were from Germany, 12% from Finland, 10% from Sweden and 10% from Switzerland (LR). Iberian Stock Doves appear to be sedentary but one ringed as a chick in Barcelona in June 1989 travelled to the Pyrénées-Atlantiques, France (256km) by the following October. There are also some indications that Stock Doves abandon higher elevations in autumn to winter in more favourable locations, at least in the northern half of Spain (SWA).

Wintering birds are more widely dispersed than the breeding population and may be found throughout most of Spain except for the coastal north-west and the Mediterranean coastlands. In Spain they are most abundant in the central Ebro depression, on the central meseta of Castilla y León, and within a large sector of the southern meseta, covering Madrid, Toledo, Cuenca and Albacete (SWA). Gil Lletget (1943, 1945) noted that Stock Doves were very common in the Holm Oak dehesas of Candeleda, Ávila, in some winters, particularly during winter 1941–42 when flocks of thousands appeared. The wintering distribution may have shifted northwards during the second half of the 20th century, however, since recoveries from Extremadura and the Alentejo comprised some 36% of recoveries up to the late 1960s but only 12% of those during the 1970s to early 1990s (AMI, AI1).

Stock Doves also occur more widely in Portugal in autumn and winter and there are records of autumn influxes around the Tagus and Sado estuaries, where the species does not winter, and of small scale passage at Cape Espichel, Setúbal, in October (AP). Stock Doves are usually encountered in winter feeding on farmland in small flocks. A flock of nearly 450 near Mourao, Évora, in late January 2003 (AP) is quite exceptional.

Some Stock Doves may continue south to Morocco but only very small-scale passage has ever been seen at the Strait of Gibraltar. Thiollay & Perthuis (1975) recorded a total of 67 migrants there in October 1974 and Tellería (1981) recorded 160 in total flying south at Tarifa during October-early November 1977.

WOOD PIGEON *Columba palumbus*

(Paloma Torcaz, Pombo-torcaz)

***C. p. palumbus*. A widespread breeding species throughout mainland Iberia and the Balearic archipelago. Millions of European immigrants winter in the Peninsula.**

Breeding population estimates. Spain: SA1 220,000 pairs; ACRE 9.5 million birds. **Portugal**: 10,000–100,000. **Andorra**: 60–120 pairs.

The Wood Pigeon is a common bird in suitable habitat throughout temperate and Mediterranean Europe, including Iberia. It also breeds commonly on the main Balearic Islands although it was first found to breed in Formentera only in 2001 (Costa 2002). It is well distributed and only absent when nesting from the most treeless and most arid regions, which include the coastlands

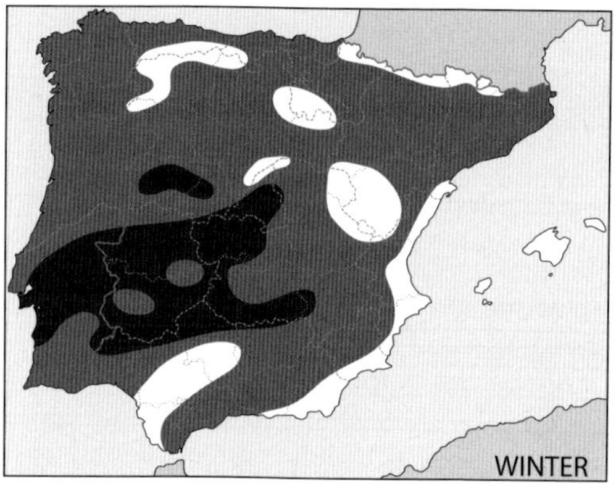

of Almería, much of the lower and central Guadalquivir valley and local tracts of Extremadura and the Ebro valley (SA2). For this reason, it is more abundant in the northern half of Portugal than in the south and it is very scarce in the west of Alentejo and Algarve. Nevertheless, it is also scarce or absent in the Atlantic sectors of the Basque Country and in Navarra. It may be found in farming areas provided that there are copses, shelter belts or riparian trees to provide nesting and roosting sites. The preferred habitats are decidous woodlands, especially oakwoods, but many also inhabit coniferous forests and areas of tall scrub.

Parks and gardens also have their breeding birds, as elsewhere in Europe, although in Iberia the Wood Pigeon only began nesting in close proximity to human habitation in relatively recent decades (AMI). Initial nesting in the urban parks of Madrid, where it is now abundant, seems to date from the early 1940s (Gil Lletget 1945) but the population there was still small during the 1970s and 1980s. For example, only 24 pairs were counted in 1976 in the 118ha park of El Retiro, Madrid, (Alonso & Purroy 1979). In other Spanish cities nesting in urban parks does not yet seem to be as widespread as in Madrid (SA2), whereas in Portugal it only occurs in sizeable green areas, as the 1,000ha Parque Florestal Monsanto in Lisbon (AP).

References in the historical literature strongly suggest that the Iberian Wood Pigeon breeding populations increased greatly during the latter part of the 20th century. Eagle Clarke (1889) considered it a rarity in the Pyrenees and did not find any in Andorra. Similarly, Lynes (1912) did not record it in the Sierra Nevada and Boxberger (1921) only gave two localities for Aragón: Ordesa in the Pyrenees and the Moncayo range in the Iberian Range. Witherby (1928), during his spring tour of central Spain in 1927, only encountered it in the royal gardens of Aranjuez; at Candeleda, Ávila, and on the slopes of the Gredos and Guadarrama sierras: he described the Wood Pigeon as 'fairly common' in Aranjuez but saw just a few elsewhere. The population density in 1927 even at Aranjuez was evidently far lower than is now the case: the 313ha gardens were estimated to accommodate 7,400–9,400 breeding pairs in 2002, at densities of 35–510 pairs/10 ha, and the wintering population there has been estimated at some 90,000–100,000 individuals, including flocks of up to 10,000 birds (Cano 2013). As recently as 1967 Bernis considered that the Spanish native population was scarce overall (AMI). Thereafter, the population is thought to have grown by at least 20% during 1970–90 (Tucker & Heath 1994), and it showed a significant 101% overall increase on SACRE sites during 1998–2012 (Escandell 2013). A 2002 estimate for Catalunya alone put the local population there at 200,000–275,000 pairs (CA2). The highest densities in Spain: 40–100birds/km^2, are in mesomediterranean habitats.

The Iberian populations are thought to be largely resident but those of northern and eastern Europe are largely migratory and several million winter in Iberia (AMI, Purroy *et al.* 1984, Purroy 1988). They enter especially by the western Pyrenean passes, where they are traditionally hunted from late September to mid or late November. The peak influx is during the second half of October. Smaller scale hunting at mountain passes also occurs in the northern Iberian Range, between La Rioja and Soria, and in the Guadarrama and Gredos ranges. A large proportion of the survivors continue to their preferred wintering areas in central and southern Portugal, western Andalucía and Extremadura (Bernis 1967). Very few appear to cross the Strait of Gibraltar (BG, Tellería 1981). The departure from Iberia, in February–March, has a more easterly orientation through the Pyrenees (SA1).

Wood Pigeons are widely distributed in winter but the oak dehesas of the south-west are especially attractive to them on account of the autumn acorn crop. Iberian Holm Oak woods are preferred but the birds also flock to Cork Oak and Lusitanian Oak woods. Large flocks of wintering Wood Pigeons also frequent farmlands, where they consume quantities of emergent crops. Great numbers are hunted in winter and bird scarers are widely employed to displace the birds. The annual bag in Spain has tentatively been put at one million birds (Purroy *et al.* 1984). Ringing recoveries indicate that the wintering population is largely derived from France, Germany, Scandinavia, Finland, the Baltic States and Switzerland.

The wintering population in Iberia, including the resident birds, was estimated at 6–7 million birds by Purroy (1988) and is evidently even larger now. The population that winters in the dehesas of Extremadura and the montados of southern Portugal has alone been estimated recently at some 2.5–3.0 million birds (Bea *et al.* 2003). The Wood Pigeons are most impressive when great flocks assemble to roost in selected woodland areas. Although some 80% of 210 roost sites in the main Extremadura–Portugal wintering area attracted fewer than 5,000 birds, in different winters 40–70% of the population roosted at a few locations each accommodating over 400,000 birds (Bea *et al.* 2003). Some of the most spectacular of such mass roosting sites are in the Sierra de San Pedro, Cáceres, and within the Herdade do Pinheiro and Mata de Valverde estates in Alcácer do Sal, Setúbal (Purroy *et al.* 1984, Elias *et al.* 1998, AP). In contrast, a count of 500 roosting at Capdepera, Mallorca, on 20 December 2003 (AOB, *Ardeola* 52: 217) is noteworthy for the Balearic Islands. Significant displacements may occur within the winter period but these may be irregular and are poorly documented. However, a trend for birds to move south during the course of the winter has been reported in Portugal (AP).

AFRICAN COLLARED DOVE *Streptopelia roseogrisea*

(Tórtola Rosigrís, Rola-rosada)

Breeds in sub-Saharan Africa and the Arabian peninsula. The domesticated form, the Barbary Dove '*S. risoria*', is in Category C in Spain.

Released birds established a small breeding population at the Devesa del Saler, near Valencia city, where at least ten pairs were counted in February 1986 (*Ardeola* 36: 255). Others were released around the same time at Denia, Alicante (SA1). The birds were initially misidentified as Eurasian Collared Doves, which are very similar and which reached the region around the same period (Dies *et al.* 1999). Conversely, the breeding bird atlas of the Comunidad Valenciana (Urios *et al.* 1991) assigned all observations of collared doves to the Barbary Dove, undoubtedly in error. Following the original releases, African Collared Doves spread elsewhere within the littoral zone of the Comunidad Valenciana so that by the early years of the 21st century there were an estimated 15–20 pairs in Valencia city and its neighbourhood, 10–20 in Denia and 4–6 birds in Guardamar de Segura, Alicante, although they were reported to be declining in the face of competition with the much more abundant Eurasian Collared Dove (SA2). Nevertheless, García & Català (2008) were still able to put the population at 80–100 birds.

EURASIAN COLLARED DOVE *Streptopelia decaocto*

(Tórtola Turca, Rola-turca)

***S. d. decaocto*. A widespread and increasing breeding resident throughout the Peninsula and the Balearic Islands.**

Breeding population estimates. Spain: ACRE 3.3 million birds. **Portugal**: 50,000++ pairs. **Gibraltar**: 50 pairs.

The dramatic westward range expansion since the 1930s of the Collared Dove from its original distribution centred on the Indian sub-continent brought it as far as northwestern Europe by the 1950s. Southwestern Europe was colonised more slowly (Bernis *et al.* 1985, Rocha & Hidalgo 2002a). The first record in the Iberian peninsula was a pioneering individual identified at Colunga, Asturias, in 1960 (*Ardeola* 16: 264). The next few observations were also in the Biscay coastlands, in Cantabria: single birds at Torrelavega in 1966 and 1968 and two there in 1971; and three together in Santander also in 1971 (*Ardeola* 16: 264–265). By 1973 the Collared Dove had spread west along the north coast to Galicia, where two were reported at Ferrol (*Ardeola* 20: 361). The first report of breeding was in Santander in 1974 (*Ardeola* 20: 361).

Portugal appears to have been colonised initially from the north, as a southward continuation of the progressive settlement of the Biscay coastlands, and the Collared Dove was first reported there in the city of Porto in 1974 (Santos Júnior 1979a). The 1980s saw the appearance of Collared Doves in the Mediterranean coastal regions, with the colonisation of Catalonia and the Comunidad Valenciana, and also inland along the Ebro Valley and in Castilla y León and Madrid (Rodríguez 2000, SA2, Pocino *et al.* 2005). The southward spread continued on the Portuguese west coast reaching Lisbon in the mid 1980s (PA1). Southern Iberia was colonised last, mainly during the 1990s, when birds became widespread in Algarve, Extremadura and Andalucía. Extremadura seems to have been colonised from the north-west, from Portugal (Rocha & Hidalgo 2002a). The first records in Gibraltar were in 1990 (GONHS). The earliest observations in the Balearic Islands were in Mallorca in 1995, with birds appearing in Menorca in 1997 (*Ardeola* 43: 252, Clavell 2002). The Collared Dove was not identified in Andorra until March 1997 (ANA) and may not yet be established there.

The colonisation of Iberia, as elsewhere in Europe, has been noteworthy both for its relative rapidity and for the speed of increase of local populations once the birds became established. The Spanish population was estimated to be 400 pairs in 1980 (Bárcena & Domínguez 1986), and 1,000+ pairs in the mid 1990s (SA1) but at least 36,500 pairs were detected by

1998–2001 (SA2). The current population is much larger. The SACRE surveys put the mean Spanish population in spring during 2004–06 at *c.* 3.3 million birds (ACRE). The population showed a tenfold increase in abundance on SACRE sites during 1998–2012 (Escandell 2013), by far the largest increase of any of the species monitored. This increase was seen in all regions but was most marked in Andalucía, where the population increased seventeen-fold over that period. The Portuguese population is now certainly larger than originally estimated and it too showed a marked increase during 2004–09 (ACP).

The success of the Collared Dove is undoubtedly associated with its close association with human settlements. It is very common in parks and gardens but is also attracted to farmsteads, where it feeds on spilt grain and animal fodder, sometimes in flocks of 100 or more: gatherings of 600 at a cattle farm at Massanassa, Valencia, in September 2009; at least 413 at an animal feed factory at Ciempozuelos, Madrid, in January 2007, and 500 at the sunflower seed depot at Salteras, Sevilla, in September 2006 (*A. O. C. Valenciana 2009*: 94, *A. O. Madrid 2007–2008*: 255, *Ardeola* 54: 179) are noteworthy. It also forages in farmland and orchards but is seldom found very far from human habitation, a habit that contributes to its being very much less a target of hunters than other pigeon species. The highest reported densities have been from towns, cities and their periphery, ranging from 22–64 birds/km^2 (ACRE). It is most abundant in coastal regions, both in Spain and Portugal (ACRE, PA2). The principal gaps in its range are in mountainous regions and extensively wooded or forested areas. Its success is also linked to its high reproductive output. A pair may nest three times a year and clutches have been found in every month, although principally between March and September.

Established birds appear to be sedentary but it is clear from its range expansion than young birds at least may be dispersive. Over 6,000 were ringed in Spain during 1973–2009 and at least 739 recoveries are available. Although 91% were within the ringing locality, and most of the remainder had not moved far, four travelled over 200 km: one ringed in Sevilla and recovered in the Algarve; two ringed in Huesca, Aragón, one of which travelled to the Basque Country and the other to Gard, France (408km), and one ringed in Navarra and recovered in Loire-Atlantique, France (536km). There is also a recovery of one that travelled from the Ebro valley to northern Germany.

EUROPEAN TURTLE DOVE *Streptopelia turtur*

(Tórtola Europea, Rola-brava)

***S. t. turtur*. Widespread summer visitor to Spain and Portugal and common on passage throughout the region.**
***S. t. arenicola*. Summer visitor to the Balearic Islands.**

Breeding population estimates. Spain: SA1 790,000–1 million pairs; ACRE 3.5 million individuals. **Portugal**: 10,000–100,000 pairs.

This is a common and characteristic species of Iberia although its populations have suffered severe declines there as elsewhere in Europe. Most populations winter in the African Sahel zone. It is a bird of farming landscapes and open woodlands, with a preference for habitat mosaics that offer a mix of arable land and weed-rich pastures with orchards, tall hedgerows, copses, clumps of tall scrub or riverine woodland. The dehesas of central and western Iberia are a major habitat, especially those where the trees are interplanted with cereal crops. It also favours olive groves, open pinewoods and heathlands. It is mainly encountered below 1,000m although it reaches higher altitudes in the warm south (SA2). It does not nest in very open country unless there is suitable tall woody vegetation available but flocks gather on farmland outside the breeding season, on such crops as sunflowers.

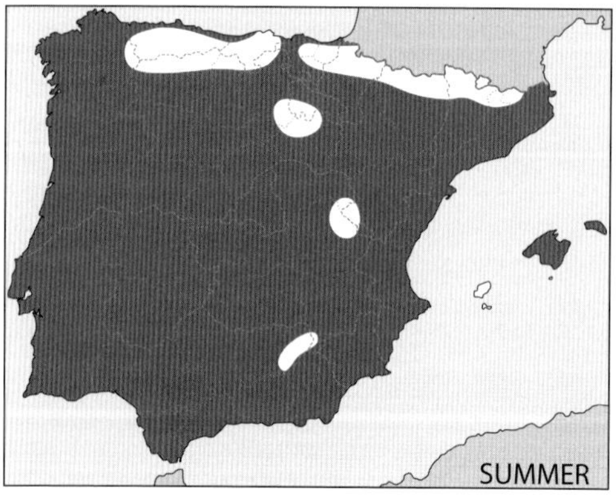

The Spanish breeding population has shown a marked decline dating from at least the 1970s. The population on SACRE sites showed a significant 29% decline during 1998–2012 (Escandell 2013). There is no numerical data on its historical abundance in the region but there is anecdotal information that it was far more numerous in the past. Saunders (1871) noted that it 'abounds throughout the country' and Irby (1875) found it 'in extreme abundance' in Andalucía, and Chapman (1886), Arévalo (1887) and Noble (1902) refer to great numbers on passage in spring. In 1967 Bernis (AMI) referred to great numbers occurring in western Iberia: from western Andalucía, Extremadura and the Alentejo to northern Portugal and Galicia.

The Portuguese population has not been censused but is also substantial and densities of 30–40 birds/km^2 have been reported in especially favoured areas in the north-east: in the Serra da Nogueira and Parque Natural de Montesinho (AP). Adjacent areas of Spain, in Galicia and western Castilla y León, are also particularly attractive to the Turtle Dove (LR). It becomes progressively scarcer from north to south across Portugal (PA2) but not in Spain, where Andalucía contained 31% of the breeding birds in the 2004–06 survey. Tait (1877) was able to describe it as abundant in summer throughout Portugal and the (numerically) dominant pigeon in the country, but that is certainly far from the case now. The Portuguese population too has declined and there is evidence, chiefly from hunters' bags, that this has been a long-term trend there throughout the 20th century (Santos Júnior 1979b, AP). The Portuguese population showed a further marked decline during 2004–09 (ACP). Hunters have also lamented the declines in their quarry in Spain, despite having accounted for 2.5 million there in 1989 alone (Purroy 1995, LR). However, such tallies do not distinguish between Iberian nesters and the many migrants that occur in the region on passage. The latter have also decreased substantially.

The earliest Turtle Doves may appear in southern Spain and Portugal in late March but passage proper begins in April. Spring passage is more evident in the eastern half of Iberia than the autumn movements, which tend to occur further west and are relatively more notable in Portugal, an indication also supported by the distribution of ringing recoveries (AMI, AP). The southward passage begins in August, although some Iberian birds may still have young in the nest then. For example, the breeding season in both Madrid and in Extremadura extends from early May to the second half of August (Peiró 1990, Rocha & Hidalgo 2002b). Most passage is in September, peaking in mid-month and continuing into October with a few stragglers occasionally reported in November. Flocks of up to several hundred birds are attracted to cereal and rapeseed stubble and harvested sunflower fields in autumn. A count of 482 roosting at Mira, Coimbra, on 27 September 2002 (*Anu. Ornitol.* 2: 38) is a notable gathering of presumed migrants.

Migrating Turtle Doves may be seen on the move in daytime, sometimes individually or otherwise in mainly small flocks, but much passage is nocturnal. There are Iberian recoveries of birds ringed in the Low Countries (56), the British Isles (32), France (23), and Germany (5), as well as at least one each from the Czech Republic, Switzerland, Italy and Russia, and two from Senegal. Birds ringed in Portugal have been recovered in France (24), Britain (6), Belgium (1) and Germany (1), and birds ringed in Spain, most of them on passage in the Balearic Islands, have been recovered in France (7), Italy (5) and Morocco (1).

There are very few records indeed during the winter months and those in early December or in February are likely to involve late or early migrants respectively. Midwinter sightings include single birds at the Lagoa de Santo André, Setúbal, on 27 December 2006; at Sant Feliu de Llobregat, Barcelona, on 5 January and 4 February 1992, and at Villaobispo de Regueras, León, on 27 January 1997, as well as two at Benavente, Santarém, on 7 January 2005 (*Anu. Ornitol.* 5: 55, 6: 64; *Ardeola* 40: 98, 44: 256).

The decline of the Turtle Dove in Iberia has diverse causes. It is heavily reliant on the seeds of a wide range of herbaceous plants: including *Chenopodium, Polygonum, Fumaria, Amaranthus* and *Papaver*, that were common as 'weeds' in agricultural areas but that have declined in the wake of agricultural intensification, notably the application of herbicides (BWP, Jiménez *et al.* 1992). In Extremadura, a significant relationship between the Turtle Dove decline and the progressive reduction of the regional extent of cereal cultivation has also been detected (Rocha & Hidalgo 2002b). Changes in land-use are thus implicated in the species' decline, alongside changes in habitat quality arising from deforestation in both the breeding and winter quarters (LR).

Excessive hunting pressure seems extremely likely to have exacerbated any existing problems (Purroy 1995). The start of the hunting season in mid-August means that some birds are killed when they still have young in their nests: during this time 2.7% nests in Extremadura contain chicks and some 13.7% of fledged young are not yet able yet to fly strongly (Rocha & Hidalgo 2002b): these proportions are considered to be much higher further north the Peninsula, where the breeding season is later (Peiró 1990, LR). It is also the case that the low reproductive rate of the beleaguered Turtle Dove does not permit intensive hunting pressure (Boutin 2001, Lutz 2007): its clutch size is just two eggs and although there may be several breeding attempts per season, the mean annual productivity is only 2.7 young per pair in northern Portugal and 2–3 in Extremadura, (Lutz 2007, Rocha & Hidalgo 2002b), far below the reproductive output of such other quarry species as ducks or partridges.

RUFOUS TURTLE DOVE *Streptopelia orientalis*

(Tórtola Oriental, Rola-oriental)

Vagrant to Spain.

Single birds were recorded on the Columbretes islands, Castellón, on 30 September 2001, and at the Laguna de Traba, A Coruña, from 30 October to 6 November 2005. Both were juveniles showing the characters of *S. o. meena*. Two migratory subspecies have occurred as rare vagrants in Europe: *S. o. meena* and *S. o. orientalis*. There are nine records from Britain and five from France. Most European records have been during September–December.

LAUGHING DOVE *Streptopelia senegalensis*

(Tórtola Senegalesa, Rola-dos-palmares)

Vagrant to Spain and Portugal.

S. s. phoenicophila expanded its breeding range during the 20th century, to reach the northern Maghreb. At least some of the considerable number of records from western Europe may involve true vagrancy but the species is retained in Category D in France, where there were ten records up to 2005. Nevertheless, the relatively frequent southern European records: from Spain, Italy, Malta and Greece, tend to be concentrated during passage periods, March–May and August–November, which may indicate movements from and to Africa (de Juana 2006). Observations of one on a ship in the Bay of Biscay in April 1966, and of a pair that came on board another ship between Corsica and Italy in May 1985, also suggest arrivals of wild birds.

Figure 174. Regional distribution of records, 1995–2009 (n = 18)

The earliest Iberian observations have been quite recent: on 26 March 1995 at Cartagena, Murcia; on 28 May 1995 at Laxe, A Coruña, and on 5–7 January 1996 at Parede, Lisboa (*Ardeola* 44: 133, 46: 141; *Pardela* 8: 14). However, by 2010 there were 16 accepted records from Spain, involving 18 individuals and two from Portugal. All but two, from Madrid and A Coruña, have been in southern Iberia, especially in Andalucía, and in the Mediterranean coastal regions (Fig. 174), which again suggests natural occurrence, as does the tendency of the records to show some concentration in spring and autumn: a pattern that is even more pronounced if the French observations are also included (Fig. 175). More recent Spanish records include single birds: at the Ebro delta in July 2012 and at Gibraleón, Huelva, in February 2013 (RBS).

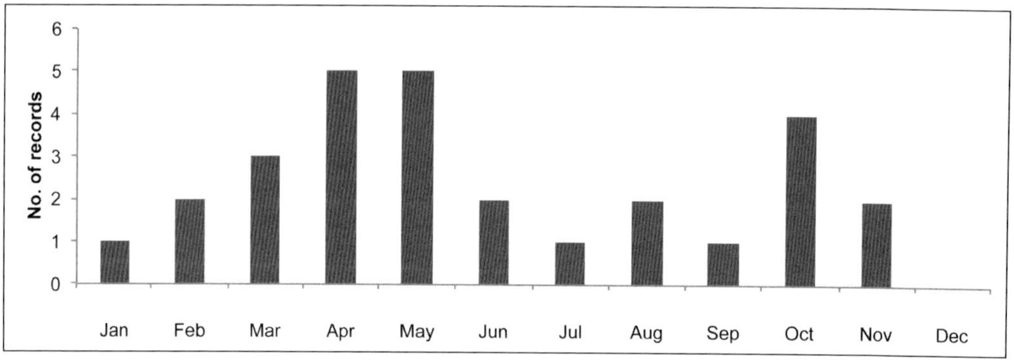

Figure 175. Records by month in Iberia and France combined, 1972– 2009.

ROSE-RINGED PARAKEET *Psittacula krameri*

(Cotorra de Kramer, Periquito-rabijunco)

Feral birds breed locally in mainland Spain, Portugal and the Balearic Islands. The subspecies involved are the Asian races *P. k. borealis* and *P. k. manillensis*, and hybrids between them. Included in Category C in Spain and Portugal.

This parakeet is frequently kept in captivity and the Iberian population derives from escapes or releases, which may still be occurring. In Spain the earliest encounter with the species in the wild seems to have been in Cáceres in 1970 (Pérez-Chiscano 1971) but breeding birds were first found in Barcelona and Málaga cities in the 1980s, followed by Valencia in the 1990s (SA2), since when they have become more widespread. Single birds or small groups, some of them perhaps recent escapes, may appear almost anywhere but regular breeding seems to be limited to the provinces of Barcelona, Valencia and Alicante on the Levant coast, to Málaga and Sevilla in Andalucía, and to Madrid, Asturias and the island of Mallorca. The

Spanish feral population was estimated at 213–254 pairs (SA2: 2003) but has increased significantly since then. The Catalan population, for example, was estimated at 80–100 individuals in 1999–2002 and 272–499 individuals in 2006–09, implying a 17% annual increase rate (CWA); it is still mainly restricted there to Barcelona and its periphery, where a roost of some 100 birds was found in the Ciutadella park in December 2009 (*A. O. Catalunya 2009*: 157). A flock of some 300 was in Sevilla city in June 2013 (*Ardeola* 60: 534).

The Rose-ringed Parakeet was first seen in the wild in Portugal in the late 1970s and by 1990 a roost of 20–30 birds had become established in the Jardim da Estrela, Lisbon (AP). This roost grew to 208 birds by November 2003, a large proportion of the national population that is estimated to be around 250 birds. Away from greater Lisbon the parakeet also nests on the central coast around Mira, Leiria (PA2) and there are occasional observations at scattered locations in Algarve and inland.

The dependence of the Rose-ringed Parakeet on large cavities, usually in trees, for nesting may be limiting its population expansion and it is certainly much less numerous in Spain than the Monk Parakeet, which builds its own nest. Most nests in Barcelona are in holes in palms and planes (CWA). It is often encountered in city parks but seems to be less adaptable than the Monk Parakeet when seeking food.

MONK PARAKEET *Myiopsitta monachus*

(Cotorra Argentina, Caturrita)

***M. m. monachus*. Feral birds breed locally in mainland Spain (Category C) and the Balearic Islands, but only sporadically in Portugal (Category E).**

Many escaped or released individuals have established an increasing feral population in Spain. It is most abundant on the eastern and southern Spanish coasts, in particular in the provinces of Barcelona, Valencia, Alicante and Málaga (SA2, GAE) as well as in Madrid. Isolated populations also occur in most other regions, nearly always in or around major cities. There are breeding populations in the Balearic Islands: in Mallorca, Menorca and Ibiza. It is occasionally reported in Portugal but there are no established populations there. However, sporadic breeding has been reported from Faro and Lisbon districts (AP).

The Monk Parakeet was first noticed in the wild in 1975, when two appeared in a Barcelona park (Batllori & Nos 1985), and it was also reported in Málaga during the late 1970s (Muñoz & Palomo 1996). It became widespread post 1985 and the Spanish population was estimated at 485–590 pairs during the 1990s (SA1) and has since increased to several thousand, greatly outnumbering the Rose-ringed Parakeet. The largest population is probably in Catalonia, where it increased from some 300 birds in 1991 to 2,600 in 2001 and 4,000 in 2006–09 (Clavell *et al.* 1991, Domènech *et al.* 2003, CWA). The earliest observations in Madrid city were in 1985 and there were at least 123 birds there in 1998 (Belinchón *et al.* 1999), some 1,300 in 2005 (Martín Pajares 2006) and a few thousand by 2012 (pers. obs.). The main colonies in Barcelona city are in La Ciutadella park and those of Madrid in the Casa de Campo park, both being the sites of the original population nuclei from which the species expanded. There were at least 305 pairs in Málaga province in 2002 (SA2), and 100 were counted at a breeding colony in Santa Ponça, Mallorca, in September 2010, despite control measures there (*A. O. Balears 2010*: 223). Further and considerable increases in range and population size are foreseeable, given the observed rates of increase and the availability of unoccupied potential habitat (Muñoz & Real 2006).

The conspicuous communal stick nests are most often placed high in trees of various species, with a marked preference for *Phoenix* palms in Barcelona and Cádiz, *Eucalyptus* in Málaga and *Cedrus* in Madrid. Others are on pylons and other artificial structures, and a small colony in Arganda, Madrid, nests within the bases of White Stork nests sited on posts. The breeding season extends from March to August.

The Monk Parakeet is highly adaptable and capable of exploiting a wide range of foods, gathered both from vegetation and on the ground. It consumes fruits, seeds, and buds, as well as insects and even bread. It is an agricultural pest in South America and has already caused significant damage in orchards and tomato crops in Spain (GAE), so it may not prove to be an entirely welcome addition to the avifauna. So far, however, it seems to be largely confined to urban environments, where it clearly selects parks and other tree-rich areas (Rodríguez-Pastor *et al.* 2012).

GREAT SPOTTED CUCKOO *Clamator glandarius*

(Críalo Europeo, Cuco-rabilongo)

Widespread when breeding in mainland Spain and Portugal, wintering in sub-Saharan Africa.

Breeding population estimates. Spain: SA1 55,000–64,500 territories; ACRE 278,000 birds. **Portugal**: 500–1,500 territories.

The Great Spotted Cuckoo is well-distributed in Iberia, occurring throughout mainland Spain and Portugal except north of the Cantabrian Range and along the west coast. It is also absent from higher elevations, above 1,200–1,400m. The Iberian breeding distribution is largely conditioned by two factors: the presence of its principal hosts, the Eurasian Magpie and the Carrion Crow, and the existence of a reliable source of its main food, large caterpillars (SA2). Iberia is its European stronghold and holds over 90% of the European population (BE).

It is a brood parasite of corvids. The Eurasian Magpie is the principal host, followed by the Carrion Crow. The Jackdaw and the Red-billed Chough are also sometimes used as hosts, particularly in treeless areas where Magpies are scarce (SA2), and there are records of successful parasitism of Raven nests (*Ardeola* 16: 267–268). The Azure-winged Magpie, however, is not parasitised regularly by the Great Spotted Cuckoo, despite assertions in the literature to the contrary. Valverde (1971) reported that the Azure-winged Magpie served as a host very occasionally but these observations, if correct, were certainly exceptional and no recent instances have been reported (Arias de Reyna 1998, Valencia *et al.* 2005).

The breeding distribution is thus closely associated with that of corvids, especially the Eurasian Magpie (*q.v.*). In particular, the absence of breeding Great Spotted Cuckoos from the lower Guadalquivir valley, the provinces of Cádiz and Málaga, the Costa del Sol, the Murcia lowlands, much of coastal Alicante and Valencia, and the Balearic Islands corresponds with the virtual or total absence there of the Magpie (and the Carrion Crow). Climatic and dietary considerations probably account for the absence of the Great Spotted Cuckoo from the Eurosiberian zone and the coastal areas bordering the Atlantic in Portugal. The Portuguese range is principally in the east, where it is more numerous south of the Tagus, and there is some evidence of recent range contraction north of the Tagus and in more western areas (PA2, AP).

The Great Spotted Cuckoo consumes a wide range of large invertebrates and also small vertebrates such as lizards (Soler 2008). The often abundant larvae of the Pine Processionary Moth are especially important in its diet and these are available in pinewoods from December to April, when the cuckoos return from Africa. The distribution of the moth is limited by winter temperatures and it is thus absent from the cooler, higher regions of the Peninsula. Areas that include suitable pinewoods are thus especially attractive to the cuckoo but it may be found in the diverse types of open country favoured by Magpies: dehesas and other open woodlands, orchards: especially almond groves, copses, tall hedgerows and waterside scrub (SA2). Nesting densities are generally low, as might be expected for a species with such specialist requirements, of the order of 2–2.5 birds/km^2 in Mediterranean farmland mosaics (ACRE) and up to 5 birds/km^2 in optimum Magpie habitat, as in the Guadix depression, Granada (SA2, Soler 2008).

Historical references to the species in Iberia are sparse and give the impression of scarcity in most regions, although it seems to have been relatively common around Madrid and at Aranjuez, Toledo, and elsewhere in western Castilla-La Mancha (Lilford 1866, Gil Lletget 1945), and Valverde (1956) found it common in the pinewoods of Segovia and Valladolid.

The wintering areas are in sub-Saharan Africa and a few satellite-tracked individuals marked in Spain have been found overwintering in Senegal (M. Soler, pers. comm.). There is also a ringing recovery from Gambia and three from Morocco of birds ringed as nestlings in Granada. The birds return to Iberia very early, from December to February, some as early as late November in the south. One at Orgaz, Toledo, on 6 November 1993 (*A. O. Toledo Revisión histórica 2001*: 283) was exceptionally early. The presence of Great Spotted Cuckoos during the winter months may have given rise to the impression that some winter in Iberia (e.g. AI1) but this does not seem to be the case. There were 79 observations in total during the Spanish winter atlas surveys, only five of which were before 23 December and only 12 were reported between 24 December and 20 January. However, there were 19 between 21 January and 3 February, and 43 (54%) appeared during 4–17 February (SWA). These observations thus confirm that there is a negligible winter presence and instead that birds recorded during the winter months have returned from their winter quarters (AMI, Herrera & Hidalgo 1974). The winter

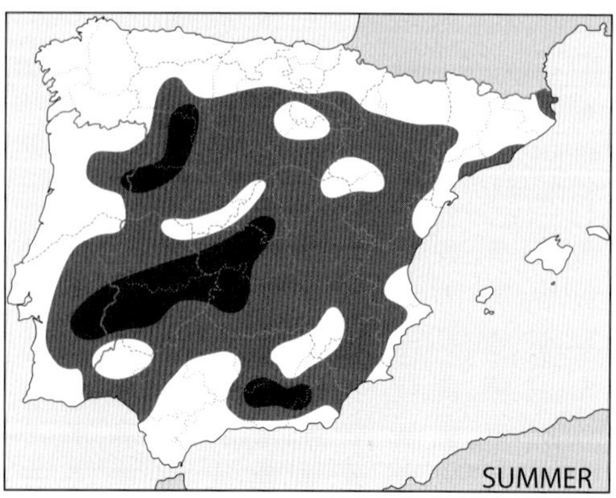

return to Iberia allows them to benefit from the processionary moth outbreaks before the Magpie breeding season begins, in March. Some cuckoo passage continues during March–April and may involve non-breeding, immature individuals.

Adult Great Spotted Cuckoos largely disappear soon after laying, in early June, and probably return to Africa in June and July. At that time some first move to more temperate areas, such as nearby mountain ranges where caterpillars still abound: in July the Sierra de Baza then receives some of the cuckoos that breed in the Guadix depression, for example (Soler 2008). The juveniles depart during August–September, with rare stragglers lingering into October (Soler *et al.* 1994). Passage is nocturnal but scattered migrants are often seen in small numbers in coastal pinewoods especially.

Records outside the breeding range are infrequent. It occurs in the Balearics only as a rare migrant. Single birds were found in Menorca in 1912 and Mallorca in 1941 (Munn 1943) and there were at least 15 on the islands during 1988–2010 (*Ardeola*, *A. O. Balears*). These latter records fall between January and August, eight of them during March–May. Some of these migrants may have belonged to the population of the French Mediterranean coast, which dates only from the 1940s but which numbered 300–600 territories post-2000 (OF).

Rare recent observations from northern Spain include at least 13 records (14 birds) from Asturias during 1975–2006 (Álvarez-Balbuena *et al.* 2000; *A. O. Asturias*) and a juvenile was at Cape Peñas in August 2012 (*Ardeola* 59: 442). Observations of two 'immatures' in the Serra do Candán, Pontevedra, in June 1995 and a single juvenile at Oímbra, Ourense, in July 2004 (*Ardeola* 51: 554, 52: 218) may have involved locally bred individuals. A juvenile found in the Sierra de Alpujata, Málaga, in July 2009, and another near Málaga city in July 2009 (*Ardeola* 56: 361), may be indications that the breeding range is now extending to the province since the recent increase in Eurasian Magpie records there.

COMMON CUCKOO *Cuculus canorus*

(Cuco Común, Cuco-cinzento)

***C. c. bangsi*. A common summer visitor to the Peninsula and the Balearic Islands. Summer visitor. Smaller and shorter-winged than *canorus*.**
***C. c. canorus*. Recorded on passage.**

Breeding population estimates. Spain: SA1 143,000–320,000 territories; ACRE 2.1 million birds. **Portugal**: 10,000–100,000 territories. **Andorra**: 50–150 territories.

The nominate race inhabits most of breeding range of the species, including the remainder of Europe. There is a Portuguese ringing recovery of a bird ringed in The Netherlands (AP), and another Dutch bird has been recovered in Morocco (BM). Two ringed in Britain and one in Belgium have also been found in Spain. It is apparent that much of the European population migrates along flyways east of Iberia (AMI, BWP). Even most British Cuckoos migrate to and from Africa through eastern France and Italy (Wernham *et al.* 2002) although three satellite-tracked individuals in 2011–12 crossed eastern Spain (British Trust for Ornithology: in www.bto.org). The Iberian and Maghreb populations are also summer visitors and are believed to winter in sub-Saharan west Africa, but their winter quarters are essentially unknown (AMI).

The race *bangsi*, occurs throughout the region in virtually all vegetated habitats, with local absences only on barren mountaintops, in built-up areas and in extensive crop monocultures (SA2, PA2). It appears commonest in open woodlands and farmland mosaics and scarcer in closed forests with little understorey. Deciduous woodlands, especially of oak species, are preferred to conifers. Many frequent scrub and heathland, including above the treeline in montane regions. The chosen habitats must supply the large invertebrates, notably caterpillars, but also grasshoppers and other large insects, that are their main diet. They are generally more frequent in the eurosiberian and supramediterranean climatic zones and scarcer in the thermomediterranean (AI1). SACRE surveys found the highest densities in Spain to be in the supramediterranean regions, with 8–13 birds/km² (ACRE). However, local studies have found higher concentrations in a range of habitats including 32 birds/km² in Sessile Oak woods in Navarra, 23 birds/km² in Pyrenean Oak woods in the Sierra Nevada, Granada, and 20 birds/km² in Iberian Holm Oak dehesas in Cáceres (SA1).

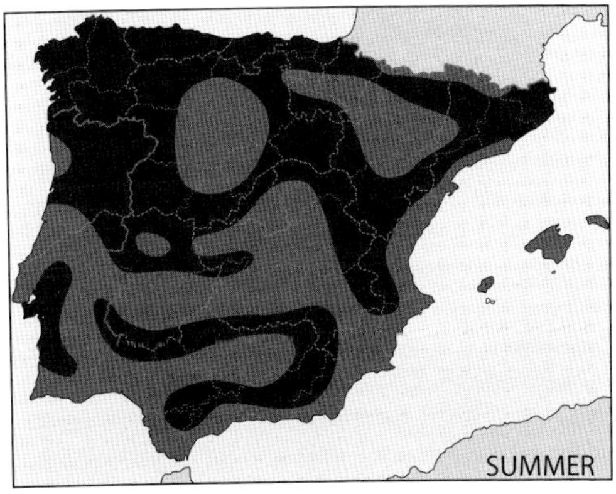

The Common Cuckoo parasitises a much wider range of host species than does the Great Spotted Cuckoo. Its hosts in Iberia include such typical Mediterranean species as Sardinian and Subalpine Warblers but also more widespread ones, such as Reed and Great Reed Warblers. However, it does not seem to parasitise the Iberian Azure-winged Magpie, whose sibling species is one of its main hosts in Japan (Davies 2000).

It is a nocturnal migrant and passage is inconspicuous but the males start singing as soon as they reach their nesting areas. The song is first heard in the south from early March but the main arrival is from mid March and in April (AMI). Earlier records, such as one heard at Rivas-Vaciamadrid on 11 February 2000 (*Ardeola* 49: 191), are unusual but in February 2010 single birds were reported from three widely separate locations in Portugal: at Castro Verde, Beja, on 2 February; at Idanha-a-Nova, Castelo Branco, on 7 February, and at Castelo Rodrigo, Guarda, on 14 February (*Anu. Ornitol.* 8: 77). There is a remarkable record of one found freshly dead at Lastras del Pozo, Segovia, on 11 January 2010 (*Ardeola* 57: 539). Singing males remain vocal throughout May and into June but they fall silent by the end of the month. The adults largely disappear from mid June onwards and are seldom observed later than the end of July, having presumably returned to Africa. The BTO satellite-tracking study has confirmed that the early disappearance of British adults from the breeding range is due to their departure for Africa in June–July, as long suspected, and it is likely that Iberian adults also depart at this time. Juveniles follow during August and into September and are rarely reported later. Single juveniles at the Guadalhorce estuary, Málaga, on 14 October 1997, and at the Paul do Boquilobo, Santarém, on 29 October 2003 (*Ardeola* 45: 320, *Anu. Ornitol.* 3: 40) were exceptionally late.

YELLOW-BILLED CUCKOO — *Coccyzus americanus*

(Cuclillo Piquigualdo, Papa-lagarta-norte-americano)

Vagrant to Spain.

An injured individual was picked up at Cala Millor, Mallorca, Balearic Islands, on 28 October 1994. This is a rare vagrant to Europe. There are records of 62 from Britain, ten from Ireland and 13+ from the Azores. Only two have been found in France and one in Morocco.

BARN OWL — *Tyto alba*

(Lechuza Común, Coruja-das-torres)

***T. a. alba*. A widespread resident throughout the Peninsula and the Balearic Islands.**
***T. a. guttata*. Occasional in winter in Spain and Portugal. Has bred.**

Breeding population estimates. Spain: SA1 50,400–90,500 pairs. **Portugal**: 2,000–6,000 pairs.

The nominate race is widely distributed throughout Iberia, being scarce or absent only in the arid south-east, within areas of unbroken forest and from elevations above *c.*1,000–1,200m, although it has been found in a village at 1,380m on the southern slopes of the Cantabrian Range (Alegre *et al.* 1989). The Barn Owl hunts in open country: over scrub, farmland, orchards and pastures, including montane grasslands. It is strongly associated with human habitation through its tendency to nest and

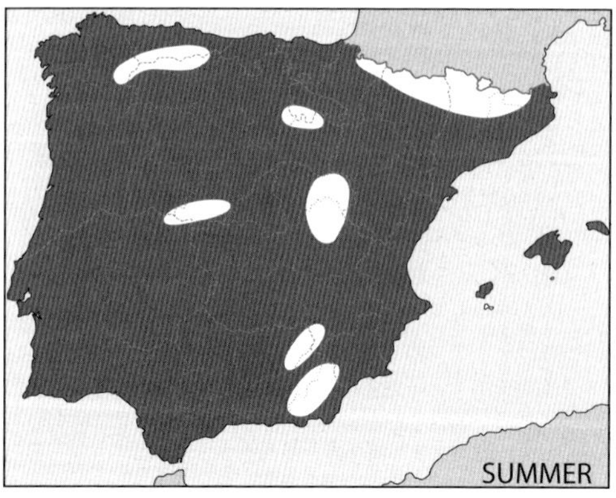

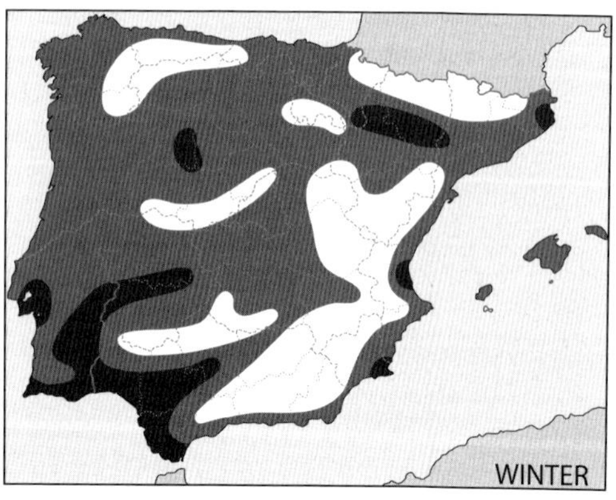

roost in buildings, especially in ruins, barns, granaries, castles and churches: anywhere that offers secluded holes or ledges. Holes in cliffs and old trees are also used widely for nesting and roosting.

The Barn Owl population is considerable but its numbers are not known with any precision. There are clear indications of population decline, however, at least in some areas in Spain. An especially marked decline has been reported from an agricultural area in the south-east of Madrid, where 40 breeding pairs were censused in 1983 but only nine remained in 1993 (Fajardo & Pividal 1994). A 69% decline has been reported from two areas of Valencia and Alicante during 1989–2000 following habitat modification (Martínez & Zuberogoitia 2004a) and a 25% reduction in occupied 10km-squares is apparent from a comparison of the 1984 and 2004 breeding bird atlases for Catalonia (CA2). The densest populations appear to be those of Extremadura, Cantabria and Castilla y León (Fajardo 1995). In Portugal some range contraction in western Algarve has been attributed to abandonment of farmland there (PA2). The Barn Owl has been affected by agricultural intensification, including the consequences of habitat modification and the use of rodenticides, as well as by direct and illegal persecution in some areas. Thousands, especially juveniles, die annually on roads: roadkill accounted for 36.5% of the known deaths in a study in central Spain (Fajardo 1990).

Micromammals are the predominant prey, especially mice: *Apodemus* and *Mus*, although the Common Vole is very important in the northern meseta, where its abundance periodically attains plague proportions. However, some owls take many birds at sparrow and starling roosts and others, especially in the south, consume many bats (Pérez-Barbería 1991). The breeding season is potentially long in Iberia and is linked to prey availability. Clutches have been found in every month (AP) but first clutches are usually in February or March, with second clutches later in the year if conditions allow. Second clutches and larger clutch sizes are particularly associated with Common Vole 'plagues' (Navarro *et al.* 1990). Examples of midwinter nesting include two broods of five well-grown young found in La Rioja on 5 December 1988 and 3 January 1989 and a young bird picked up in Huesca city, Aragón, on 7 January 1998 (*Ardeola* 36: 256, 45: 125).

Iberian Barn Owls are probably very sedentary although juveniles disperse further than adults, and in all directions (Martínez & López 1995). Most recoveries are within a 50km radius of the nest site but birds ringed as chicks have travelled from Cáceres to Setúbal (257km) and from Huesca to Murcia (441km). However, some birds are attracted after the breeding season to areas that presumably offer abundant prey. The highest reported densities in Spain in winter are in the Ebro, Guadiana and Guadalquivir valleys, as well as along the Mediterranean littoral, from Catalonia to Murcia (SWA). There they favour irrigated crops, especially rice paddies. Similarly, post-breeding concentrations of juveniles have been reported in Portugal during August–November in areas where few pairs nest, notably at the Tagus estuary, where a 40km nocturnal transect on 20 October 2004 detected 184 birds (AP).

Some individuals from central Europe occur in Iberia in winter but the size of any wintering population is unknown. There are 49 recoveries from throughout the region, mainly of birds ringed as nestlings in France, Germany, the Low Countries and Switzerland. One ringed in Britain has also been found in Spain. At least some of these are from the breeding range of *T. a. guttata* and individuals showing the characteristics of this dark-breasted race have been observed in Spain, with recent records of mixed pairs in Madrid and Vizcaya (Zuberogoitia & Campos 1999, J. A. Martínez *et al.* 2002).

EURASIAN SCOPS OWL *Otus scops*

(Autillo Europeo, Mocho-d'orelhas-comum)

***O. s. mallorcae*. A widespread breeding species throughout mainland Iberia and the Balearic islands. May intergrade with the nominate form in northern and central Spain (König & Weick 2008). Mainly a summer visitor.**
***O. s. scops*. Occurs on passage.**

Breeding population estimates. Spain: SA1 30,000–34,000 pairs. **Portugal**: 2,000–6,000 pairs. **Andorra**: 2–3 pairs.

The Scops Owl may be encountered almost throughout the Peninsula and on the principal Balearic islands, although it appears to be scarce in northern and eastern Galicia, southern Asturias and northern Castilla y León (SA2). It is also generally absent from the highest elevations, reaching its altitudinal limit at *c.*1,200m in La Rioja and *c.*1,400m in the Catalan Pyrenees (SA1), although nesting birds range as high as 2,500m in the Sierra Nevada (Garzón & Henares 2012). Its highest population densities are probably in the Mediterranean littoral region, including the Balearics.

The plaintive repetitive call of the male is a familiar sound in spring and summer in many towns and villages. It selects habitats that offer scattered trees alternating with open areas, as provided not only by parks and gardens but also olive, almond and citrus groves, dehesas and farmland mosaics with tall hedgerows or copses. Conversely, it avoids dense forest or woodland, as well as very open areas with little or no tall cover. It nests in cavities, especially in holes in mature trees, and readily takes to nestboxes. Some use in holes in walls: those of the old Pamplona fortress accommodate a sizable population (Berián 2008), as well as old woodpecker holes or abandoned Eurasian Magpie nests. It is mainly insectivorous, consuming

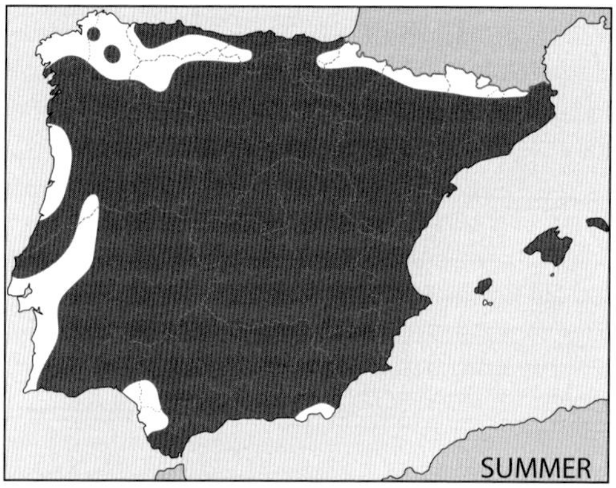

a wide range of large moths, beetles, crickets, cockroaches and other insects, as well as occasional small vertebrates.

Regional surveys have suggested that the SA1 population estimate, 30,000–34,000 pairs, may be too low (SA2). For example, the population of Catalonia alone has been put at 5,400–7,600 pairs (CA2). It is certainly the case that the Spanish population is a very important fraction of the European population, which has been estimated at 110,000–200,000 pairs (BE). The population trends in Spain are uncertain but there are indications of some decline (Escandell 2012). In Portugal it is most numerous in Trás-os-Montes, but some recent range contraction has been detected in the centre and north (AP). A few may nest in Andorra but regular breeding in Gibraltar ceased in the 1950s. The Scops Owl is vulnerable to insecticides, both directly and indirectly, and to any habitat changes, such as the removal of old trees, that result in the loss of nest sites.

Most birds leave the region from August onwards to winter in sub-Saharan Africa, but some Mediterranean populations are at least partly resident. Migrants are often reported in Gibraltar and the Strait region in late September–early October especially (Bensusan & Perez 2013, Tellería 1981). The return passage peaks from late March to mid April (Bensusan & Perez 2013). The region is probably overflown by some Scops Owls of the nominate race from further north, in particular from the French population. Individuals ringed in Spain have been recovered in France (2), Italy (2), Morocco (4) and Algeria, and there is a Spanish recovery of one from Italy.

The presence of wintering Scops Owls has often been confirmed in the Balearic Islands and southern and eastern Spain (SA1, Clavell 2002). Chapman & Buck (1893) said that some wintered in Andalucía, Boxberger (1921) mentions having heard it nightly in winter in Málaga from October onwards, and Munn (1921, 1924) found it resident in both Mallorca and Menorca. However, the winter atlas surveys found some winter presence in all the Spanish autonomous regions (SWA). The size of the wintering population is hard to assess since the birds generally pass unnoticed when they are not calling, making their presence apparent only in February when most males begin their songs. There are a few winter records from southern Portugal (AP), chiefly of birds that were heard but not seen.

EURASIAN EAGLE OWL *Bubo bubo*

(Búho Real, Bufo-real)

***B. b. hispanus*. A widespread resident in most of mainland Iberia but absent from the Balearic Islands.**
***B. b. bubo*. May intergrade with *B. b. hispanus* in the Pyrenees.**

Breeding population estimates. Spain: SA1 1,000+ pairs; SA2 2,345 pairs+. **Portugal**: 200–500 pairs. **Andorra**: 2–3 pairs. **Gibraltar** 1 pair.

The Eagle Owl is increasingly widespread across much of mainland Iberia where it is a typical member of the raptor community of rocky areas, nesting in suitable cavities or small caves on cliffs and steep slopes. However, some also inhabit wooded areas, nesting in tree hollows, in disused nests of other raptors or even on the ground: as happens, for example, in the Sierra Norte, Sevilla (Penteriani *et al.* 2005); in the Monte de El Pardo, Madrid (F. de Juana, pers. comm.), or in Doñana (Penteriani *et al.* 2012). Urban nesting has increasingly been reported, in Madrid and Bilbao for example (SA2, *Ardeola* 58: 504), with some nests located on large buildings. Breeding territories are generally in areas of open habitat, including pastures, scrub and woodland clearings, although those that inhabit denser forested areas may be underestimated since they are hard to find there. Territories occur from sea level to middle altitudes in mountainous areas, exceptionally up to 1,900m in the Sierra Nevada (SA1). It is an early nester. Calling and duetting between pair members begins in autumn, intensifies in winter and continues through to spring. Reported mean laying dates include 18 January in Murcia, 23 January in Alicante, 4 February in Catalonia and 18 February in Aragón (Serrano 2001).

In Spain it is most widespread and abundant in the sierras of Andalucía, Extremadura, the Centre and the Levant, although few occur in very open agricultural areas, such as the Guadalquivir valley. Territories are more widely spread in the farmlands of western Castilla y León, where birds have recently been found nesting in pine copses among the cereal fields there (SA2). The Eagle Owl

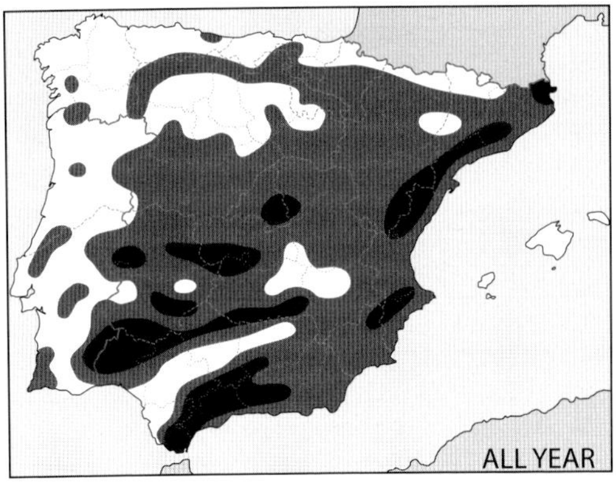

is scarcest in Spain in Galicia and in the Biscay coastlands but new territories have become established there in recent years, some of them resulting from deliberate reintroductions. The first known instance of nesting in Cantabria was near Ramales de la Victoria in 2009 (*Ardeola* 56: 168). It first bred in Doñana national park in 2006 and the population there numbered at least 19 pairs by 2008, although the initial colonists are thought to have escaped from a local rehabilitation centre (Penteriani *et al.* 2012).

In Portugal the Eagle Owl is most widespread in the east, in the more rugged interior, notably in Trás-os-Montes, Beira, Alentejo and northeastern Algarve. It is rare or absent in most of the western half, although its distribution is incompletely known (PA2, AP). There is evidence that it was formerly more widespread in western Portugal but that it was exterminated locally by persecution during the mid 20th century (AP). Elsewhere in Iberia, the species is unknown in the Balearic Islands but up to three pairs nest in Andorra (BE 2004) and at least one pair nests in Gibraltar: which was recolonised in 2005 (Garcia 2006).

The two Spanish breeding atlas surveys found progressively more birds but probably underestimate the current population. Regional surveys have shown increases in numbers but it is not yet possible to confirm a more general trend since there is a lack of historical data. There does seem to have been a reduction in direct persecution and nest robbing but mortality by electrocution or collision with wires continues to be a significant cause of death locally. The Eagle Owl is considered uncommon in Portugal, and there is no reliable estimate of its population size there, but the extent of the breeding range suggests a minimum of several hundred pairs.

The principal prey is the rabbit and areas with high rabbit densities and numerous nest sites may hold up to 35–40 breeding pairs/100km^2 (Sayago 2002b, Ortego & Calvo 2004, Penteriani & Delgado 2010, Pérez-García *et al.* 2012). In such favourable circumstances second broods may possibly occur (Martínez *et al.* 2003a,b, Ortego 2004) as well as broods of up to five nestlings (Pérez *et al.* 2010). The recurrent population crashes of this key dietary component due to myxomatosis initially and more recently to viral haemorrhagic disease, has posed problems not only for the Eagle Owl (Martínez & Calvo 2001, Martínez & Zuberogoitia 2001) but for many other Iberian predators. However, it is an opportunistic hunter and many other prey species are readily taken as available. They include rats: often sought at rubbish dumps, water voles, hares, hedgehogs, partridges and other birds, reptiles, amphibians and large invertebrates (Penteriani & Delgado 2010). Rats may be very important locally: when some Catalan rubbish dumps were closed down some 44% of the local pairs did not attempt to breed and those that did showed much reduced breeding success (CA2). Small carnivores, including cats and fox cubs, are often taken, as seen elsewhere in Europe (Herrera 1974a, Penteriani *et al.* 2012). The pair at Gibraltar consumes quantities of Yellow-legged Gulls. Other raptors, especially young birds taken from nests, may figure in its diet and its presence has been given as a negative factor affecting the Peregrine Falcon especially (del Moral 2001). There is conflicting evidence on whether or not predation on other raptors is inversely correlated with rabbit abundance (Tella & Mañosa 1993, Serrano 2000).

The Eagle Owl is sedentary and territories are often occupied traditionally for many years. Certain individuals, most probably young birds, disperse over some distance and these are likely to make the colonisation or recolonisation of new territories possible. There are 16 recoveries of juveniles in Spain during 1969–2009 found over 50km from where they were ringed, including one that moved from Mora de Toledo to Cardeña, Córdoba (165km) and another that moved to Buitrago de Lozoya, Madrid from Monfragüe, Cáceres (242km) (*Ecología* 3: 211, 21: 269). Most juvenile dispersal is probably much shorter. The mean distance travelled by 74 radiotracked individuals in the Sierra Norte, Sevilla, was only six kilometres (range 2–34km) (Penteriani & Delgado 2010).

LITTLE OWL *Athene noctua*

(Mochuelo Común, Mocho-galego)

A. n. vidalii. **A common and widespread resident throughout mainland Iberia but scarce in the Balearic Islands.**

Breeding population estimates. Spain: SA1 50,000–65,000 pairs. **Portugal**: 50,000+ pairs. **Gibraltar**: 10 pairs.

Its partly diurnal habits make the Little Owl a familiar roadside presence in Iberia, perched conspicuously on fence posts and pylons. It occurs almost anywhere that offers open country with at least a few trees and bushes. It avoids the higher

levels of the principal mountain ranges but some nest as high as 2,300m in favourable locations in the Sierra Nevada (Pleguezuelos 1992). It finds ideal habitat in olive, almond or carob groves, orchards and vineyards, on woodland edges and in farmland with hedges or patches of scrub. Even very arid zones are inhabited, with rocks providing perches. Many live around villages, where they often nest in holes in walls or under loose roof tiles. Others find nest sites in stone huts and shelters, ruins, holes in cliffs, among boulders, in hollow trees or even in rabbit burrows. The Little Owl is an opportunist hunter that captures a wide range of small vertebrates and large invertebrates, the latter being relatively more important in the south of the range (Máñez 1983).

It may be encountered throughout most of the Peninsula, including Gibraltar, but does not appear to nest in Andorra, although it has been recorded there (ANA). The Little Owl was regarded as an accidental winter visitor to the Balearic Islands (Ferrer *et al.* 1986) but there have been recent breeding records from both Mallorca and Menorca (SA2) and observations from Ibiza and Formentera (*Ardeola* 49: 191, 52: 218). The Balearic Bird Reports give only 56 records (59 birds) during 1987–2010 for the entire archipelago so it is clearly still extremely scarce there. The Little Owl is particularly vocal, and detectable, in February and March (Zuberogoitia *et al.* 2007). However, 52% of Balearic records were during October–January, peaking in December, and only 14% were in February–May, suggesting that many involved winter visitors.

The SA1 estimate, 50,000–65,000 pairs, is probably too low since the population of Catalonia alone, where the species does not seem to be particularly abundant, has been put at 8,890–14,448 breeding pairs (CA2). In general, there is no precise information available on the current size of the Spanish or Portuguese populations but the indications are that it has suffered a rapid and large-scale decline in Spain recently: a significant 37% decline was reported from 500 SACRE sites during 1998–2012 (Escandell 2013). This decline appears to be steeper in more intensively developed regions such as Catalonia, Madrid, the Basque Country and certain areas of the Comunidad Valenciana, whereas the populations in most of Extremadura, Andalucía and Castilla-La Mancha may be stable or just slightly decreasing (Fajardo 1995). In Catalonia, a comparison of the 1984 and 2004 breeding bird atlases shows a 21% drop in occupied 10km-squares (CA2). The Little Owl is known to be vulnerable to road traffic (Hernández 1988, Fajardo *et al.* 1998) and it rapidly disappears in the face of urbanisation: for example, 87 territories were censused in 1993 and just 14 in 2002 at Huerta de San Juan, Alicante, the decline being due mainly to habitat loss but probably also to direct mortality since 67 Little Owl carcasses were found on the roads there over a five-year period (Martínez & Zuberogoitia 2004b). Habitat loss also results from agricultural intensification and afforestation, especially with conifers, and Little Owls are also vulnerable to reduction in prey availability as a result of insecticide applications (SA2, Zabala *et al.* 2006). However, the Portuguese population showed no clear trend during 2004–09 (ACP).

The highest reported densities in Portugal are in the south: in eastern Alentejo and Algarve (PA2). In Spain the Little Owl is especially abundant in Andalucía, Murcia and the southern part of the Comunidad Valenciana; in olive groves and sunflower fields in Sevilla the population density ranges from 87 birds/km^2 in July to 17 birds/km^2 in winter (Fajardo *et al.* 1998). In contrast, densities are low in the north: only 272 territories were found in Bizkaia during an intensive census in 1992–96 (Zuberogoitia & Campos 1997).

The Iberian birds are apparently sedentary: 117 (82%) of 142 recoveries of birds ringed in Spain were at the ringing locality and only three had moved more than 50km: including a juvenile that moved 93km within Sevilla province. Nevertheless, the Spanish winter atlas observations suggest that some of the breeding birds of the Eurosiberian zone may leave that region in winter, a finding that requires confirmation. A January recovery in Mallorca of one ringed as an adult in Germany (AMI) shows that immigrants reach Iberia at least occasionally.

TAWNY OWL *Strix aluco*

(Cárabo Europeo, Coruja-do-mato)

***S. a. sylvatica*. A widespread resident in mainland Iberia. Absent from the Balearic Islands.**

Breeding population estimates. Spain: SA1 45,000–61,000 pairs. **Portugal**: 10,000–40,000 pairs. **Andorra**: 40–50 pairs. **Gibraltar** <5 pairs.

The Tawny Owl occurs in forests and woodlands throughout most of the region, being scarce or absent only in very open areas,

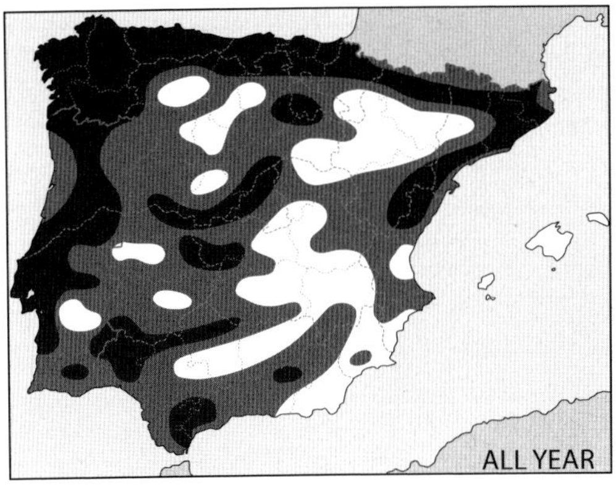

including large tracts of eastern Alentejo and Trás-os-Montes in Portugal and the Guadalquivir valley and much of the southern meseta especially, as well as the arid south-east, in Spain. It occurs wherever there are tall and mature trees: in forests and woodlands of all kinds, including dehesas and riparian treelines, and in parks and gardens including those in city centres. Beech forests seem particularly attractive to it (SWA). It is the most widespread owl in Andorra, although a shortage of mature deciduous trees may limit its abundance there (ANA). Mature trees provide cavities for nesting and roosting but some pairs use holes in walls or cliffs. Monterey Pine plantations in the Basque Country support dense populations (Zuberogoitia & Martínez, 2000). The breeding season is very lengthy in Iberia: first clutches appear as early as October and become progressively more frequent through the winter with peak laying occurring around 15 March (SWA, Zuberogoitia 2011).

The size of the Iberian population is not known with any precision. However, a census in the province of Bizkaia in the Basque Country found 1,704 territories there alone (Zuberogoitia & Campos 1997) giving a high density of 0.77 pairs/km^2. Population trends are uncertain but appear to be positive (Fajardo 1995). In Spain the NOCTUA programme showed an increasing trend during 1998–2001 but no significant change thereafter (Escandell 2012), as well as local range expansion. An example of the latter comes from Gibraltar, where the Tawny Owl was unknown until the 1990s but where several pairs have since become resident (GONHS). Similarly, the first records from the Sierra Nevada date from the 1990s but it is now established in mature forests and woodlands there (Garzón & Henares 2012). It is otherwise largely sedentary, with just short-range dispersal seen in juveniles: 50 of 73 recoveries during 1973–2007 were local and the three individuals that travelled further than 50km had been released from rehabilitation centres. There are no Iberian recoveries of birds of extra-regional origin.

LONG-EARED OWL *Asio otus*

(Búho Chico, Bufo-pequeno)

***A. o. otus*. Widespread throughout much of mainland Iberia and in the Balearics. Local populations are joined by winter visitors from northern Europe.**

Breeding population estimates. Spain: SA1 4,800–7,000 pairs. **Portugal**: 100–1,000 pairs. **Andorra**: 3–10 pairs.

This is a well distributed resident, but its secretive habits mean that its distribution and population are incompletely known and probably underestimated. It is easiest to find in winter, when it is most vocal on territory. The Long-eared Owl is associated with trees bordering open agricultural areas, in which it mainly hunts. It especially frequents isolated pinewoods but also occurs in riparian woodlands, copses and the edges of larger blocks of woodland and forest. Territories tend to form loose aggregations (Martínez & Zuberogoitia 2004c, Rodríguez *et al.* 2006). Most nests are in abandoned nests of corvids, especially Eurasian Magpies, although some nest in holes in trees or walls and even on the ground. Long-eared Owls in a Mediterranean coastal population have been found to use arboreal rat nests and accumulations of pine needles as nest sites in the absence of old raptor or corvid nests (García *et al.* 2011).

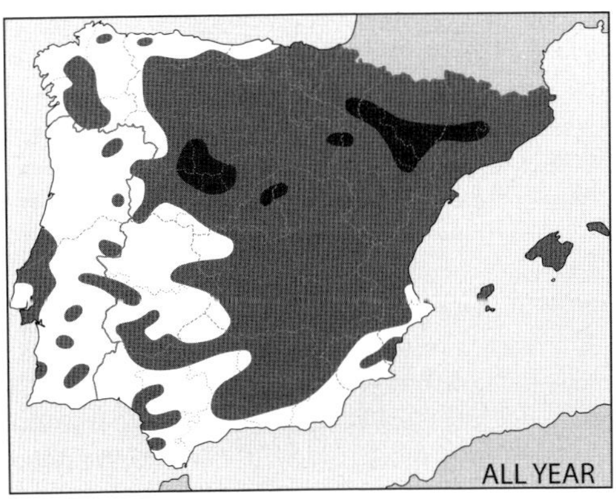

The known breeding range suggests that it is relatively common in Castilla y León, Madrid and the Ebro valley as well as in eastern Castilla-La Mancha and parts of eastern Andalucía, but that it is scarce or absent in the Biscay coastlands, Galicia, northwestern Portugal and southwestern Spain (SA2, PA2, SWA). This distribution pattern most probably indicates a concentration in those regions where voles and mice are most abundant (García & Cervera 2001,

Escala *et al.* 2009). Voles are most abundant on the northern meseta and Long-eared Owls in the Mediterranean coastlands of the Comunidad Valenciana have a more diverse diet that principally comprises mice but also includes birds and bats (García *et al.* 2011).

Preliminary information on population trends (NOCTUA) suggests a significant overall decline in Spain during 1998–2011 (Escandell 2012) although with local fluctuations that are probably associated with Common Vole cycles of abundance (Fajardo 1995). There is no census data for Portugal but the widespread if scattered breeding distribution there suggests the presence of a few hundred pairs, with perhaps some concentration in Estremadura (PA2).

The Long-eared Owl was formerly considered to be just a scarce winter visitor to the Balearic Islands (Bernis *et al.* 1958) and the first breeding records there date from the early 1970s (*Ardeola* 15: 148, 19: 25). It was only in 1993 that a systematic survey found it to be relatively common and widespread in Mallorca (Viada 1994). It also breeds in Cabrera, Ibiza and Formentera but only a sole breeding record has been reported from Menorca (Catchot 1998).

Iberian birds are probably fairly sedentary. Only five ringed as chicks have been recovered over 50km away from their natal nests, the most exceptional having travelled 342km from Torre de Juan Abad, Ciudad Real, to Jerez de la Frontera, Cádiz (*Ecología* 8: 333). Some migrants from northern Europe winter in the region: at least two birds ringed in southern Sweden, three from Switzerland and one from The Netherlands have been recovered in Spain (AMI, BWP), and there are single recoveries in Portugal of birds ringed in Germany and Belgium. An adult ringed in Spain at the Prat de Llobregat, Barcelona, in January 2007 was recovered the following April at La Spezia, northern Italy (703km). The size of the immigrant wintering population is unknown but is probably small given the scarcity of ringing recoveries.

Long-eared Owls may roost communally in favoured areas outside the breeding season. Such gatherings have been noted in westernmost Algarve, in September and October: where 13 were ringed in a small area of pinewood in 1996 and a roost of eight birds was seen in 2005 (AP). These observations are relatively early in autumn and so seem likely to involve Iberian birds but similar roosts later in the winter could also include immigrants. An unusual roost site has recently appeared in the village of Madrigal de las Altas Torres, Ávila, where the birds use quite small trees along a principal street (G. Sierra, pers. com.). A roost in pines inside the local cemetery of Corella, Navarra, has attracted up to 60 birds (Blanco 2007, Escala *et al.* 2009). It is very rarely recorded in Gibraltar but there are five records of single birds there, one in May and four in September–early November, suggesting small-scale passage.

SHORT-EARED OWL — *Asio flammeus*

(Búho Campestre, Coruja-do-nabal)

***A. f. flammeus*. Breeds very locally and irregularly in Spain. Has bred in Mallorca and possibly in northwestern Portugal. Widespread on passage and in winter.**

Instances of breeding by the Short-eared Owl were reported from the Ebro delta in 1962; the Salobrar de Campos, Mallorca, in 1976; and Murcia in 1990 (Maluquer 1971; *Ardeola* 24: 261, 38: 344), and was suspected on rare occasions in Barcelona, Castellón, Madrid, Huelva and Menorca (SA1). A more regular nesting population has become established in northern Spain since the early 1990s, although the numbers that nest each year are very variable.

The establishment of this regular population is associated with the spread of the Common Vole from the Cantabrian Mountains into the farmlands of the northern meseta during the 1980s and the subsequent increase in its populations. Since then the vole populations have fluctuated on a cyclical basis, reaching 'plague' proportions in 1983, 1988, 1989, 1993 and 1994, for example (SA1), and eventually leading to nesting by a considerable number of pairs of Short-eared Owls. The owl population reached *c.* 360 pairs in 1994, a peak vole year, most of them in the prime agricultural region of the Tierra de Campos: in the provinces of Palencia, Valladolid, Zamora and León, with at least three pairs in La Rioja and three pairs in the lower Ebro valley in Catalonia (Jubete *et al.* 1996, Onrubia & Jubete 1998). Since then there have also been a few, sporadic, breeding records elsewhere outside the core region: in Madrid in 1998, 2001 and possibly 2008; Toledo in 1998 and possibly in 2006; Salamanca in 2004; Lleida in 2003; La Rioja in 2000: where two pairs nested at altitudes

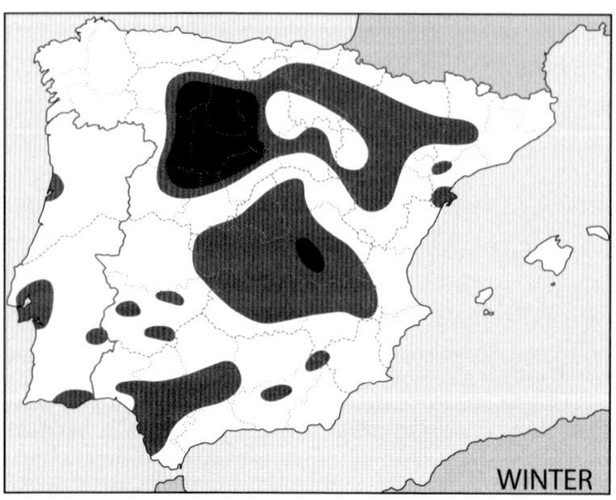

of 1,450m and 1,600m; Huesca in 2001 (two pairs) and possibly in 2003; Zaragoza in 2001 and possibly 2004; in Navarra and possibly also in Badajoz in 2000, and in Ciudad Real in 1997. These latter records are from the southern meseta and the Ebro valley. In addition, breeding has occurred occasionally in the north-west, where two pairs bred in Pontevedra in 2003 and up to two pairs were suspected to have nested in Portugal in 1997 and 1998, in the Serra do Gerês in the north-west (AP). The breeding population everywhere is highly variable from year to year.

The core population in 1993 and 1994 was closely studied (Onrubia & Jubete 1998). The great majority (71%) of 76 nesting territories were in cereal farmland: in the crops and in stubble fields, fallows and adjacent meadowland. Most of the remainder were in heathland, meadow and grassland mosaics. The study population was mainly 700–800m above sea level, exceptionally at 1,150m, well above the normal limit of 650m reported for central Europe. The population is vulnerable to agricultural practices conducted during the nesting season and 50% of nests located in Palencia in 1993 were destroyed by mowing, with stubble burning accounting for several broods during 1994 (LR). The effect of the application of rodenticides during vole plagues is unknown but likely to be negative.

Significant numbers of Short-eared Owls from northern Europe winter in Spain and Portugal, with some also in the Balearic Islands. A few ringed at nests in southern Sweden, Finland, Germany, Britain and The Netherlands have been recovered in Iberia (AMI, AI1) and one ringed in the Czech Republic on 3 November 1997 was found dead in Ciudad Real on 22 May 2001 (*A. O. Ciudad Real 1995–2001*: 200). There are single recoveries from France and Germany of birds ringed as adults in Spain in January and March. The wintering population varies but may comprise several thousand individuals in some years (Onrubia 2010). The birds begin to arrive in late September with most appearing during October, and they depart from March onwards, although some linger into May (Fig. 176). In Spain the wintering birds are most numerous in the central regions of both mesetas and in the Guadalquivir valley, and less frequent in the other major river valleys or in coastal regions (SWA). Inland Short-eared Owls range over fallows, stubble fields and uncultivated land but in coastal areas they are especially associated with marismas and wetland sites generally. The Short-eared Owl is rare and local in Portugal, where it winters chiefly at coastal wetlands, notably the Tagus and Sado estuaries, with some also occurring regularly at the Ria de Aveiro and the Ria Formosa (AP). It is often gregarious when roosting, forming groups of up to 20 individuals, sometimes more, often at wetland sites. There were 20 together at Doñana on 30 January 2012 and a daytime roost at Hernán Cortés, Badajoz, held 35 on 2 March 2012 (*Ardeola* 59: 187–188, 443).

Figure 176. Monthly distribution of records in the Spanish regional bird reports (n = 587)

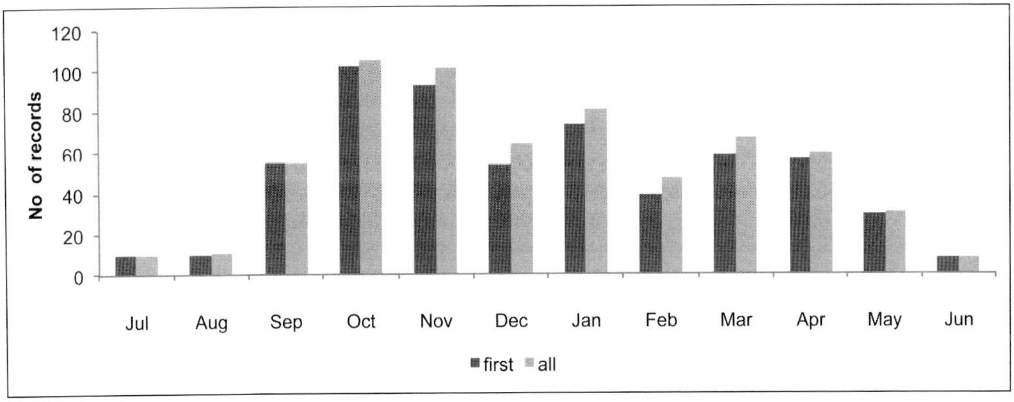

Individual birds are sometimes observed on passage, often along coasts. Records from the Biscay coastlands are generally in autumn, when individuals are not infrequently seen arriving from the sea: 75% of 57 records from Asturias were during September–November and only 12% were during March–May (*A. O. Asturias*). A large proportion of the birds seen on passage are along the Mediterranean littoral, including the Balearic islands (Asensio *et al.* 1992) and there the spring passage is much more evident: 35% of 71 records from the Balearics were during March–May compared with 42% during September–November (*A. O. Balears*). Winter records are comparatively prevalent inland and in Galicia. There are a few records of apparent single migrants at Gibraltar: three in April and one each in September and October (GONHS), but the Short-eared Owl is considered a rare passage migrant and winter visitor in Morocco (BM) so any trans-Strait movement is probably minor.

MARSH OWL *Asio capensis*

(Búho Moro, Coruja-moura)

Vagrant to Spain.

The isolated North-west African subspecies, *A. c. tingitanus*, was formerly relatively common in the Atlantic lowlands of Morocco but now survives as a small and endangered population, chiefly centred on the Lower Loukkos and the Merja Zerga. The nearest breeding birds to Iberia are some 50km south of Tangier (Bergier & Thévenot 1991). They are mainly sedentary but winter dispersal takes some outside the breeding range.

Most Iberian records are from southernmost Spain in the 19th century, when the owl was still quite common in Morocco and when clearly some occurred in winter at wetlands on the north side of the Strait. One was taken at Utrera, Sevilla, in November 1867 (Saunders 1871) and specimens were obtained at the former Laguna de La Janda, Cádiz: in October–November 1868, November 1870, December 1873 and during the winters of 1882–83 and 1893–94 (Irby 1895). In southern Portugal two were killed by the king, Dom Carlos, at Pancas in the Tagus estuary, one of them in December 1887 (Tait 1924, AP). Since then only one has been found in Iberia and it met the same sad fate as its 19th century predecessors: a first-winter individual was picked up with gunshot injuries at Jerez de la Frontera, Cádiz, on 10 December 1998.

BOREAL OWL *Aegolius funereus*

(Mochuelo Boreal, Mocho-funéreo)

***A. f. funereus*. A scarce resident in the Pyrenees: in Spain and Andorra, as well as France.**

Breeding population estimates. Spain: 180 pairs. **Andorra**: <20 pairs.

The Boreal (Tengmalm's) Owl was suspected to breed in the Pyrenees as long ago as the late 19th century (Vayreda 1883, Saunders 1884) but was not confirmed to do so until much later. It was first proved to be present in the French Pyrenees in 1962, but proof of occurrence in the Spanish Pyrenees was not forthcoming until 1983, the first nest being found there in 1987 (Alamany 1989, SA1). The species was known in Catalonia from the fossil record (Arribas 2004), however, and there is no reason to think that it is a recent colonist. It appears that the Pyrenean population is a post-glacial relic that is now *c.* 270km distant from the next closest population, in the Massif Central.

The habits and habitat of the Boreal Owl explain why it was overlooked for so long: it is almost entirely nocturnal and it inhabits old-growth forests at high altitudes. The territorial song of the male is given chiefly in winter and early spring, when the breeding forests are least accessible to humans. The Iberian range may not yet be fully known but it has been found to breed in the Pyrenean and Pre-Pyrenean mountains in Catalonia: chiefly in Lleida, but extending into Barcelona and Girona; in Huesca, Aragón, and on the eastern margin of Navarra. It is also well distributed in Andorra. Further prospection may well extend the known range, perhaps even to include the Cantabrian Range, from where a calling individual was reported at 1,200m in the Sierra de Gabaneda, on the south side of the Picos de Europa, on 20 August 1997 (*Ardeola* 46: 312). There is also a record of one in a mixed beech/oak wood at San Salvador de Cantamuda, Palencia, on 2 February 2008, at 1,500m in the southern approaches to the Cantabrian Range (SWA).

The Boreal Owl is generally found in coniferous forests at 1,700–2,300m. It descends locally to 800m in deciduous or mixed forests on the cooler northern side of the Pyrenees, in France, and this is also seen in Spain in the north-facing Valle de Arán. The preferred habitat in Catalonia and Andorra is Mountain Pine forest, where 94% of the population has been located, the remainder being in Scots Pine and Silver Fir (LR). It nests in tree cavities, especially Black Woodpecker holes, although it will resort to the smaller holes produced by other woodpeckers if these are the only ones available (Dalmau *et al.* 2001). It takes readily to nest boxes, which are being provided for it in its chosen forests, where they compensate for a certain shortage of natural cavities (López *et al.* 2010).

The total Pyrenean population may be of the order of 500–600 pairs. A survey in 2002 found 123 territories in Catalonia, 14 in Aragón and five in Navarra (SA2). More recent data has put the Catalan population at 144 pairs (CA2) and that of Aragón at 37 pairs (Lorente *et al.* 2007). The population in Andorra may comprise a further 17 pairs (Mariné & Dalmau 2000, Dalmau *et al.* 2001). Males are highly faithful to their territories and remain there all year round but females comprise a floating population and are attracted to those males that can best provide supplementary feeding (König & Weick 2008). The longest recorded displacement in Iberia is of a female ringed at its nest in Sahún, Huesca, in June 2005 and recovered 156km away at a new nest in Setcases, Girona in May 2006 (Gil *et al.* 2008). There are no confirmed observations in Iberia away from the Pyrenees, other than the two observations from the Cantabrian Range above and two winter records of calling birds at La Garrotxa, Girona, in the pre-Pyrenean region, in 2007 and 2008 (*A. O. Catalunya 2007*: 129, *2008*: 153). However, dispersing individuals are occasionally recorded well away from their nesting locations in France (OF) and the possibility that the Pyrenean population is not entirely isolated from other population nuclei cannot be discounted. On the contrary, it has been found to be genetically diverse, implying gene flow with other European populations (CWA).

EUROPEAN NIGHTJAR *Caprimulgus europaeus*

(Chotacabras Gris, Noitibó-cinzento)

***C. e. meridionalis*. Widespread in the northern half of the Peninsula and in the Balearic Islands, occurring locally at higher elevations in the south. Almost entirely a summer visitor but some remain in winter.**
***C. e. europaeus*. A passage migrant across the region but seldom identified.**

Breeding population estimates. Spain: SA1 82,000–112,000 pairs. **Portugal**: 1,000–10,000 pairs. **Andorra**: 20–50 pairs.

The breeding distributions of the two Iberian nightjar species are largely complementary, the European Nightjar nesting in the cooler north and locally at higher levels in the south, as well as in the Balearics: where the Red-necked Nightjar does not breed. It is well distributed in northern Spain except in the central Ebro valley and in the north-east of Castilla-La Mancha. Further south in Spain there are isolated populations, chiefly at higher elevations, notably in the Betic Range in Granada and Jaén, but locally also in the Sierra Morena in Córdoba and Huelva and in Extremadura. Breeding birds range to 800–1,000m on the páramos of the Iberian Range (AMI) and inhabit the open coniferous woodlands of the Sierra Nevada, at 800–2,000m (Garzón & Henares 2012). In Portugal it is widespread in the north and centre but scarce in Alentejo and absent from Algarve (PA2). In general, its distribution corresponds quite closely with the supramediterranean and eurosiberian climatic zones. It is associated with woodland edges, open woodlands and oak dehesas and open stands of Maritime Pines, as well as young pine plantations, low scrub and heathland, often on hills and plateaus. The Iberian breeding populations have never been censused with any exactitude. Long-term population trends in numbers remain unclear given the lack of adequate data on breeding numbers.

European Nightjars arrive from mid April onwards throughout May (AMI). The breeding sites remain occupied until August–September. Autumn passage at Gibraltar extends from early September into early November, peaking in mid October (Bensusan & Perez 2013). One at Vila Nova de Gaia, Porto, on 18 November 2002 (*Anu. Ornitol.* 2: 39) was very late for northern Portugal. The darker nominate race presumably crosses the region on passage, although it is very seldom identified as such. Single individuals ringed in the Czech Republic, Croatia and in England have been recovered in Spain and one ringed on passage in Formentera in May 2004 was recovered in England the following month. A full grown bird ringed in Tiana, Barcelona, in June 1984 was recovered in November 1985 in the Democratic Republic of Congo (4,500km).

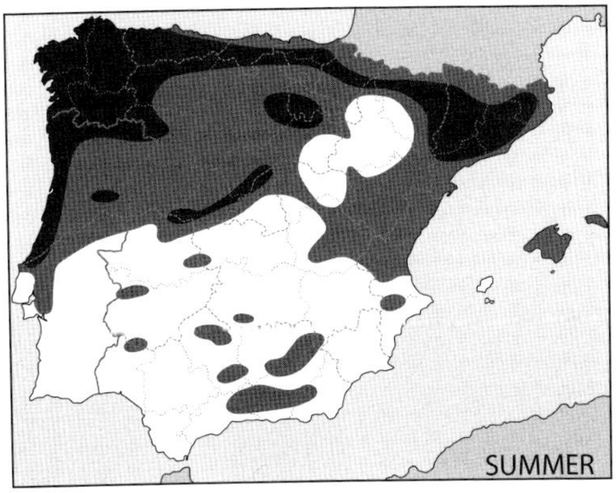

Winter records have been regarded as highly unusual until very recently. However, the winter atlas and Noctua programme fieldwork in Spain since 2009 has identified a very small but apparently regular presence in winter (SWA). There were some 45 records in total during the three winters up to 2009–10 that fall between 3 December and 15 February, well outside the migration periods of this species. They are well-dispersed, with records from 18 provinces, eight of them coastal but the remainder inland: including Jaén (6 records), Madrid (5) and Badajoz (5). This hitherto unsuspected behaviour has yet to be fully assessed, in terms of the numbers of birds involved, their origins and their winter distribution. Future trends may suggest whether European Nightjar wintering is a new phenomenon, which seems likely.

RED-NECKED NIGHTJAR *Caprimulgus ruficollis*

(Chotacabras Pardo, Noitibó-de-nuca-vermelha)

C. r. ruficollis. **Summer visitor, chiefly to low-lying areas of southern and eastern Spain and southern Portugal, local elsewhere. Absent from the Eurosiberian region and the Balearic Islands.**

Breeding population estimates. Spain: 20,000–100,000 pairs. **Portugal**: 7,000–11,000 pairs.

The Red-necked Nightjar is endemic as a breeding species to the Iberian Peninsula and the Maghreb, wintering in the south of the latter region and in sub-Saharan west Africa, where it has been recorded from Mauritania east to northern Ghana and northeastern Nigeria. Spain and Portugal together support the majority of the nominate race and an unknown but certainly substantial proportion of the global population. This large nightjar is associated with hot or warm dry regions at low altitudes, in contrast to the European Nightjar, that replaces it elsewhere in Iberia. The nesting distribution largely corresponds with the thermo- and mesomediterranean climatic regions. In Spain it is common in Andalucía, Extremadura, western Castilla-La Mancha, Madrid, the Levant coastlands and the Ebro valley, north to Navarra (SA2). In these regions it occurs from sea level to 800m, exceptionally higher, and so is absent from the upper levels of the Betic Range, the Sistema Central and the Iberian Range. However, it ranges as high as 1,100m in the Sierra Nevada and there it occurs locally alongside the European Nightjar, which replaces it at higher altitudes (Garzón & Henares 2012). It also breeds locally in western Castilla y León, north to Zamora, Valladolid and Burgos. It is absent from the supramediterranean and eurosiberian regions, and thus from much of eastern Castilla-La Mancha, the Pyrenees, the Cantabrian Range, Galicia and the Biscay coastlands. The Portuguese distribution shows a widespread presence south of the Tagus but it is absent from the north of the country except along its eastern boundary, from the Alto Duero to Trás-os-Montes (PA2).

The Red-necked Nightjar is a spring vagrant in the Balearic Islands: recent records there include 11 caught by ringers during 1994–2009, all of them between 21 April and 5 May (Clavell 2002, *Ardeola* 52: 432, 53: 389, 55: 146). It is extremely rarely reported north of its breeding range but one was found at Vitoria, Álava, on 15 September 1995 (*Ardeola* 44: 257).

The incessant 'kut-ock kut-ock' song of the territorial males can be heard in a diversity of habitats that provide open areas mixed with some tree cover. This species is more attracted to open woodlands than the European Nightjar. It is especially characteristic of Stone Pine, Aleppo Pine, Cork Oak and Iberian Holm Oak woodlands on level ground and with a well developed shrub layer. It is also found locally in olive groves and vineyards and areas where cultivated land forms a mosaic with patches of scrub with scattered trees. Birds often sit in the open at night on paths, tracks and roads, a habit which often leads to deaths from collisions with traffic. Such roadkills may be substantial: 250 are known to have been killed along a 20km stretch of road in Córdoba during 1986–90 (Aragonés 1996)

There are no reliable assessments of the size of the entire Iberian populations. The Catalan population has been estimated at 7,000–11,000 pairs (CA2). The Spanish population was regarded as stable during 1970–90 (BE) but the NOCTUA programme suggests a decline during 1998–2011 (Escandell 2012). Local decreases attributed to land-use changes have been reported from the Ebro valley (Woutersen & Plateeuw 1998, Calvet *et al.* 2004). The changes that are thought to be responsible largely involve the conversion of olive and almond groves and other Mediterranean crops to irrigated fields.

The Red-necked Nightjar arrives in April–May. Spring passage at Gibraltar peaks during mid April to mid May (Bensusan & Perez 2013). Birds depart in September–November (AMI, AI1). The autumn passage in Madrid, detected by observations at night on and near roads, extends from mid-August to mid-October, peaking in the second week of September (Ponce *et al.* 2008). At Gibraltar it peaks from late September into early October (Bensusan & Perez 2013).

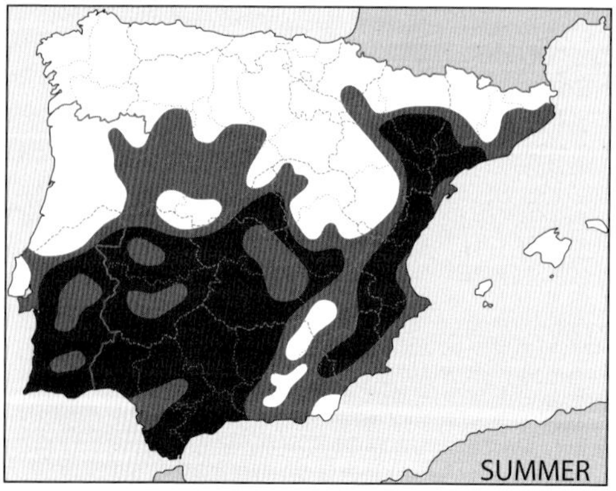

There are a few records of Red-necked Nightjars seen and/or heard singing in Spain during the winter months but these are too infrequent to suggest regular wintering, as occurs in southern Morocco (BM). Such observations include single birds heard singing in Murcia at the Sierra de Lavia on 17 January 1998 and the Sierra de Burete on 28 February 1998; in Cabañeros N. P. during winter 1999–2000; at Brunete, Madrid, on 23 January 2000, and at Horcajo de los Montes, Ciudad Real, on 12 December 2000 (*A. O. Madrid 2001*: 188; *Ardeola* 45: 125, 49: 191, 50: 164). There are also about ten winter records from Doñana (García *et al.* 2000). Recently, the winter atlas and Noctua programme fieldwork has produced two winter records from Granada and one each from Sevilla and Barcelona.

COMMON NIGHTHAWK *Chordeiles minor*

(Añapero Yanqui, Bacurau-norte-americano)

Vagrant to Spain.

The sole Iberian record is of an exhausted individual that was picked up at Puerto Real, Cádiz, on 4 November 2005, a month during which a diversity of Nearctic species reached Europe in the aftermath of hurricane Wilma. This American nightjar is a very rare autumn vagrant to Europe. There are some 22 records from Britain, two from Ireland and one from France. There are several observations from the Azores, four of them recently accepted. One was found dead in the Canary Islands in December 1972.

CHIMNEY SWIFT *Chaetura pelagica*

(Vencejo de Chimenea, Rabo-espinhoso)

Vagrant to Spain and Portugal.

Until recently this species was only known in Europe from five records in Britain. However, autumn influxes in 1999 and 2005 brought birds to several regions, including Iberia. The October 1999 influx, in the aftermath of hurricane Irene, brought five to Britain, four to Ireland, one to Sweden and four to Iberia: one bird at O Grove, Pontevedra, on the 23rd; one at the Lagoa dos Salgados, Faro, on the 26th, and two at the Guadiaro estuary, Cádiz, on the 27th. The passage of hurricane Wilma during the last few days of October and the first few days of November 2005 delivered another spate of records: there were 100+ in the Azores, two in Madeira, at least five in the Canary Islands, six in Britain, ten or more in Ireland, one in France and two in northwestern Spain, both on 30 October in Galicia: one bird at Baiona and two at Cape Estaca de Bares.

COMMON SWIFT *Apus apus*

(Vencejo Común, Andorinhão-preto)

***A. a. apus*. A very common, locally abundant, breeding species and passage migrant throughout the Peninsula and in the Balearics. Summer visitor.**

Breeding population estimates. Spain: SA1 450,000–600,000 pairs; ACRE 32.7 million birds. **Portugal**: 20,000–100,000 pairs. **Andorra**: 300–400 pairs. **Gibraltar**: 1,000–2,000 pairs.

The Common Swift is by far the most widely and uniformly distributed of the five breeding swift species. The great majority nest in holes or spaces in buildings, whether under the eaves, under loose tiles or in walls. Historic buildings, including churches, castles, bridges and city walls as well as dwellings, frequently offer many suitable nesting sites. Conversely, modern buildings that have been built with all possible cavities sealed against the elements and 'birds' are inimical to them. A minority still nest in holes in cliffs and nests in holes in trees are reported locally (SA1). The latter are generally within towns, often in palms. A notable concentration of tree-nesters has been reported from Badajoz city, where 36 pairs were found nesting in *Phoenix* palms, including a colony of 31 pairs in the Plaza de San Francisco, with up to four nests in a single tree (*Ardeola* 30: 121). Displaying birds are regularly seen over a Scots Pine forest at *c.*1,700m in the Sistema Central at Rascafría, Madrid and in this region active nests in woodpecker holes have been reported on Valsaín and Peguerinos (*Ardeola* 28: 160, 37: 343). Breeding birds may be found from sea level up to *c.* 2,000m in the highest villages of the region, and at 2,400m in the ski station on the Sierra Nevada (Garzón & Henares 2012).

The Iberian populations are very large. Bernis (1988) had suggested that the total Spanish population numbered at least 900,000 and probably one million birds, but by 1997 he was of the opinion that it comprised several million pairs (SA1). The population of Catalonia has been estimated at 203,000–290,000 pairs (CA2). More recently, however, and on the basis of the SACRE censuses, Carrascal & Palomino (2008) have estimated the Spanish breeding population during spring 2004–06 to number over 30 million individuals. This very large figure, based on extrapolation from local density samples, needs to be regarded with extreme caution since the entire European population has been put at 6.9–17.0 million pairs by BirdLife International (2004). A significant 26% decrease was seen in Spanish SACRE sites during 1988–2012 (Escandell 2013). The

Portuguese population has not been censused but is also very large, with higher concentrations found north of the Tagus, but large populations also nest in towns and cities elsewhere (PA2): the population showed no clear trend during 2004–09 (ACP).

The Common Swifts that nest in Iberia do not generally have to face the harsh conditions that quite often affect the populations breeding further north but unseasonal cold spells in spring sometimes cause problems and, at times, mortality. For example, inclement weather in June 1961 led to reports of dead swifts from Valencia, Alicante, Murcia, Almería and Granada and several dead birds were found in Salamanca on 3 July the same year, when temperatures dropped as low as 3°C (*Ardeola* 7: 259–260). Similarly, a prolonged cold, wet period in May 1971 led to large scale mortality at Alcalá de Henares, Madrid (*Ardeola* 16: 268).

Large flocks may be seen feeding over such locations as mountain ridges or water bodies, where flying insects may be concentrated. Very large numbers also accumulate at such sites during passage periods, especially when migration is impeded by cold, wet or otherwise unfavourable conditions. Thousands of birds may be involved at times, the estimated 50,000 over the Albufera de Mallorca on 4 April 2002 (*Ardeola* 52: 218) being a particularly noteworthy example.

The 'recent sightings' sections of Spanish and Portuguese ornithological publications (the Noticiarios Ornitológicos in *Ardeola* and the Portuguese *Anuario Ornitologico*) include a number of records of 'Common Swifts' recorded very early or very late in the year and even in mid-winter. We consider that the great majority of these records very probably arise from sometimes hasty misidentifications of Pallid Swifts, which do occur at such times, as commented by Bernis (1988a), Catry *et al.* (2010) and Prieta & Molina (SWA).

Common Swifts arrive at Gibraltar and elsewhere in the south from late March, progressively later in the north. Tait (1887) noted their arrival at Porto from 8–15 April. The first arrivals in Madrid are in mid April but sizeable numbers do not appear until late in the month and continue to build up through May and into June; the last local birds depart by mid August (Bernis 1988a). Fledged young probably depart for sub-Saharan Africa soon after they leave the nest and the bulk of the population flies south during August and early September, with stragglers in early October at the latest. Common Swifts, in common with other aerial insect feeders, often migrate overland, feeding as they go. Very large numbers congregate regularly on southward passage at the Strait of Gibraltar, between July and October, peaking from mid July to the first half August when the rate of passage may exceed 100,000 birds/hour (BSG). The spring passage there is much less concentrated. Common Swifts from other European countries cross Iberia on passage and there are Iberian recoveries of birds ringed in Britain (22), The Low Countries (8), France (3), Germany (3), Switzerland (4), the Czech Republic, Sweden and Finland (AMI, AI1, AP). Some 10,000 birds have been ringed in Spain but only one has been found in Africa, in the Central African Republic.

PALLID SWIFT *Apus pallidus*

(Vencejo Pálido, Andorinhão-pálido)

***A. p. brehmorum*. A common and locally numerous breeding species in southern and southwestern Iberia as well as on the Levant coasts and in the Balearic Islands, and locally elsewhere. Summer visitor.**

Breeding population estimates. Spain: SA2 40,000 pairs. **Portugal**: 5,000–20,000 pairs. **Gibraltar** >2,000 pairs.

The Iberian distribution of the Pallid Swift is probably incompletely known, especially inland, where it may have been overlooked in the past among the more abundant Common Swifts, although this assumption may mask any genuine recent northward range expansion. Historical references to the species in Iberia are very sparse. Saunders (1871) does not mention it at all but Arévalo (1887) said that it occurred in Andalucía where it outnumbered the Common Swift but had only been identified there quite recently due to confusion with the latter. Irby (1872, 1895) identified it at Gibraltar and north to Granada but implies that it was mainly scarce: although he saw many near Vejer de la Frontera, Cádiz, he was unable to find any Pallid Swifts among the hordes of Common Swifts in Algeciras – where it is now numerous. King Carlos was the only 19th century author to identify the species in Portugal, in coastal regions where it still occurs, and there were only sporadic Portuguese records until the fieldwork for the first breeding atlas (PA1) led to some clarification of its status there (AP).

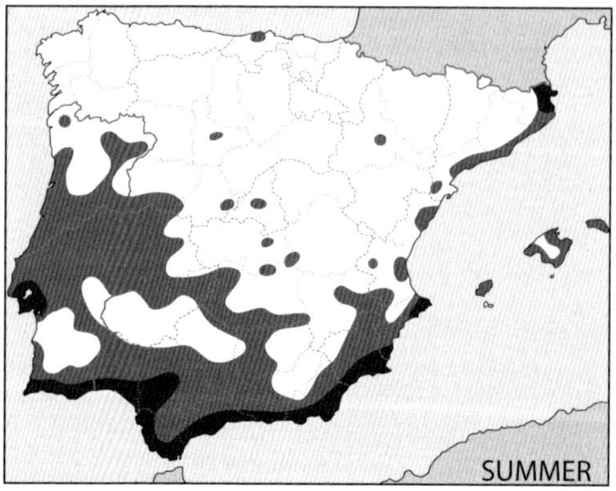

This swift is associated with the warmer regions of Iberia, in the thermo- and mesomediterranean zones. However, there is also a strong maritime association in its distribution, with large numbers nesting on not only in coastal towns and villages but also on sea cliffs throughout the Levant, the Balearic Islands, the Mediterranean and Atlantic coasts of Andalucía, and Algarve and Estremadura on the Portuguese west coast. Inland nesting, both at urban sites and more locally on cliffs, is common in the coastal provinces of Andalucía, including Sevilla, as well as more locally in Extremadura, Castilla-La Mancha, Murcia and Alicante. It is very uncommon relative to the extremely abundant Common Swift in Madrid and very few colonies have been identified there: Bernis (1988a) suggested that the latter outnumbered the Pallid Swift in the city by more than 2,000 to one. Inland nesting in Mallorca was first reported in 2010, when a small colony was found at Inca (*Ardeola* 58: 505). The second Portuguese breeding bird atlas (PA2) found the Pallid Swift to be widely distributed throughout the country, although it is more abundant there on coasts than inland and it is scarce and local north of the Douro (AP). There are important colonies in some Portuguese littoral towns and cities, including Lisbon, Setúbal and Faro.

There are also reports of breeding pairs from scattered locations further north in Spain, some of them perhaps reflecting ongoing colonisation. For example, a pair bred at the Castillo de Loarre, Huesca, in 1983 and one pair nested in Valladolid city in 2010, with other birds in the vicinity there (*Ardeola* 36: 256, 57: 539). A long-established colony exists in an old bridge on the river Ebro at Zaragoza and further pairs have recently been located in other buildings in the city, giving a total population of 16–20 pairs (Contel *et al.* 2010). A more remarkable development was the discovery of a nesting colony on the Biscay coast in Santander city. Nine or ten pairs were nesting in buildings around the Plaza del Cañadío in 2009 and the sites were reoccupied in 2010 and 2011 (Maroñas 2010). A small population at Biarritz, on the French Atlantic coast 182km east of Santander, has existed since at least 1990 (OF).

Until the recent colonisation of Santander there were relatively few records of Pallid Swifts in Iberia well to the north of the breeding range, although wandering individuals are easily overlooked. For example, only five birds had been recorded in Asturias up to 1997 and one at Txingudi, Gipuzkoa, on 8 August 1999 (*Ardeola* 49: 191) was the first record for the province. The Galician bird reports give 17 records (28 birds) during 1999–2004: including at least six over the Corrubedo dunes on 13 May 2000 and three at Sanxenxo on 30 September 2000. Twelve of 25 records from the northern Spanish coastlands from Galicia to the Basque Country were during April–May and 13 during July–December. The French population is small: 2,000–2,500 pairs, at least half of them in Corsica (OF), so there will be very little passage across northern Spain to and from locations further north. Overshoots in spring and dispersal in autumn may account for most northern Iberian records.

The Pallid Swift nests both in natural and artificial locations. Our impression is that it is more likely than the Common Swift to be found nesting on sea cliffs and this certainly seems to be the case in Gibraltar, where only the Pallid Swift nests in crevices on the sea cliffs, which hold 7% of the population (Finlayson 1979). Over 1,000 pairs nest on the sea cliffs of Cape La Nao in Alicante (Urios *et al.* 1991). In addition to cliffs, the Pallid Swift nests commonly on bridges, ruins, dams and buildings of all kinds, generally in small and often loose colonies. It nests alongside both the Common and Alpine Swifts in some of the river Guadiana bridges in Extremadura, as at Medellín and Mérida. Unlike the Common Swift it very rarely seems to nest in holes in trees (HBW). Most Pallid Swifts nest at relatively low elevations but there is a sizable colony at 1,200m in the village of Piornal in the Jerte valley, Cáceres (SA2). The Spanish population at least is likely to be considerably larger than current estimates suggest.

The interspecific relationship between the Common and Pallid Swifts has been closely studied in Gibraltar by Finlayson (1979, 1992). The Pallid Swift considerably outnumbers the Common Swift as a breeding species there, where the study population was estimated at 2,030 pairs of Pallid Swifts in 114 colonies and 1,110 pairs of Common Swifts in 56 colonies. Over 80% of Common Swift colonies were in pre-20th century edifices but only 23% of Pallid Swift colonies were in these older buildings, reflecting a species preference for the different types of nest sites offered by old and new constructions. The largest known Iberian colony, of some 8,000 birds was in a 23-storey building under construction in Sevilla that was left unfinished for many years during the 1970s and 1980s (Rodríguez de los Santos & Rubio 1986).

The principal differences between the two species are associated with the Pallid Swift being double-brooded and the Common Swift, as elsewhere, being single brooded (Rodríguez de los Santos & Rubio 1986, BSG). The Pallid Swift is present in Iberia for a much longer period than the Common Swift. It begins to appear at its southern colonies in February, with the bulk of the population arriving everywhere in March. It departs from September onwards but stragglers are regularly reported well into November. In Gibraltar the Pallid Swift arrives from mid February but the first clutches are not laid until the second half of April, peak laying being in mid May. The second clutches are started at the end of July. First brood young mainly fledge during the first week of July and those of the second broods in mid September, exceptionally into October. The last adult Pallid Swifts depart by mid November. There are seven recoveries from North Africa of birds ringed in Sevilla: five from Morocco and one each from Algeria and Tunisia (*Ardeola* 20: 87, 94–95).

There are occasional records during the winter months, usually of single birds seen in the warmer south and south-west, but there may be a small winter population in the city of Sevilla (Cuadrado *et al.* 1985). Observations of six at Loures, Lisboa, on 15 December 2001, four at the Laguna de La Mejorada, Sevilla, on 23 December 2004 and up to nine at the Alange reservoir,

Badajoz, during 4–8 December 2010 (*Anu. Ornitol.* 1: 44; *Ardeola* 52: 218, 57: 235–236) may have involved wintering birds or early returning migrants. Wintering has recently been confirmed at the last of these sites in Alange village, where 8–15 roosted in the church between December 2011 and February 2012 (*Ardeola* 59: 188). One at the Llobregat delta on 2 December 1988, five at Castelló d'Empúries, Girona, on 4 December 1993 and, especially, four at the Sabón reservoir, A Coruña, on 13 December 2003 (*Ardeola* 36: 256, 42: 227, *A. O. Galicia 2003*: 133) were exceptionally far north for the time of year. The Spanish winter atlas fieldwork produced only two winter sightings during 2008–10: in Cádiz on 18 January 2009 and in Badajoz on 24 January 2009 (SWA).

ALPINE SWIFT

Apus melba

(Vencejo Real,Andorinhão-real)

A. m. melba. **A summer visitor to mainland Iberia and the Balearic Islands, widespread as a breeding species in eastern Spain but with a fragmented range elsewhere. Widespread on passage.**

Breeding population estimates. Spain: SA1 5,700–8,000 pairs; SA2 10,000+ pairs. **Portugal**: 100–1,000 pairs. **Gibraltar**: 20 pairs.

The nesting range is discontinuous and becomes increasingly fragmented from east to west across the Peninsula. The birds seem especially to favour the limestone cliffs of the Levant and south-east, and they occur on sea cliffs from Gibraltar to Girona, as well as inland. Other substrates are also occupied, however, and there are good numbers in the Betic Range, the eastern Sierra Morena, the Iberian Range, the Ebro valley and the Pyrenees. Further west in Spain there are more scattered colonies in the Cantabrian Range, the upper Duero gorges on the Portuguese border, the sierras and Guadiana valley in Extremadura and, more continuously, on the sea cliffs of the Galician coasts (SA2). The absence of colonies in the Gredos and Guadarrama sierras in Central Spain is probably due to gneiss and granite predominating there, such rock lacking the many cavities that make limestone an ideal nesting substrate (Bernis 1952). Nesting birds range to 1,700m in the Betic Range and to 1,900m in the Pyrenees but this species prefers warm, often south-facing locations for nesting, exclusively so in the Pyrenees (SA1).

A small but perhaps increasing minority nest on man-made structures, including dams, bridges and buildings in towns and cities. Urban breeding takes place predominantly in Catalonia, where it is nonetheless a recent development. Nesting in buildings there was first reported at the Port Bou railway station in 1958 (Dorka 1960) and at the Montjuich castle (Palaus 1960). This Catalan urban population has very much increased and expanded during the last few decades. There are an estimated 6,000–10,800 pairs in the Barcelona metropolitan area alone, some 60% of the estimated regional population of the Alpine Swift (CA2). One of the best known colonies is at the Barcelona football club stadium. Significant increases in nesting in buildings were also reported from Switzerland and France during the 20th century (BWP, OF) but the tendency remains rare in Spanish regions other than Catalonia, although it is probably increasing. For example, the sole urban colony in La Rioja in the late 1970s was at Murillo de Río Leza but by 1990 there were colonies totalling 33–40 pairs in at least eight villages, all of them in the principal churches (de Juana 1980, *Ardeola* 37: 343–344). In Extremadura a colony in Plasencia cathedral dates from at least 1990 (*Ardeola* 37: 344) and there are others in castles, dams and, especially, bridges elsewhere in that region: up to eight bridge colonies, totalling 89–95 pairs were found in the early 1980s along the Guadiana and its tributaries, from Villanueva de la Serena to Olivenza (Pérez Chiscano *et al.* 1984). Nesting in the bridges at Mérida and Medellín was first reported as early as 1957 (Bernis *et al.* 1958). Bridge nesting in Extremadura may also be increasing: at Villanueva de la Serena there were six pairs in 1963, 17–20 pairs in the mid 1970s (Pérez Chiscano 1975) and 25 pairs in the early 1980s (Pérez Chiscano *et al.* 1984). In Andalucía a small colony in the main church at Campillos, Málaga, was first reported in 1990 (*Ardeola* 41: 1999). Bernis (1988) commented on the absence of Alpine Swifts in the towns of Castile and we are still unaware of any urban breeding there.

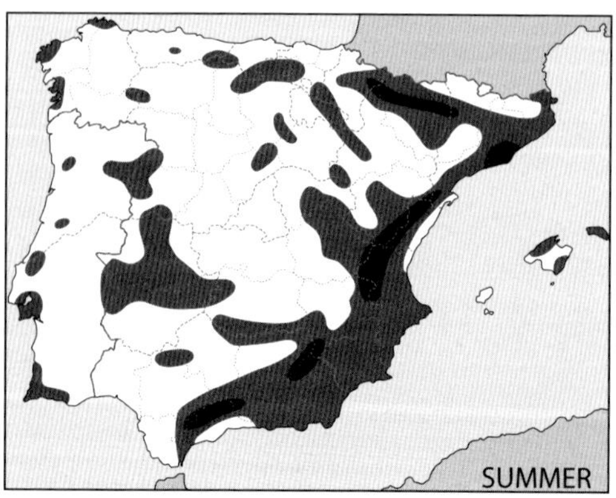

The Portuguese colonies are chiefly on coastal cliffs of the centre: for example at Nazaré, Cape da Roca and Cape Espichel, and in the south, notably in southwestern Alentejo and in Algarve, as well as inland along the Duero (PA2). Very few have been reported nesting on buildings in Portugal but some nest on a bridge across the Douro at Porto (PA2) and there is also a colony on a Roman bridge across the river Egres at Segura, Castelo Branco, on the Spanish border (AP). The species is very scarce and local in the Balearic Islands, where it breeds only in Mallorca and Menorca (SA2).

The published figures underestimate the Spanish population: a more recent survey has put the population of Catalonia alone at 10,000–18,000 pairs (CA2) so a Spanish population of 30,000–40,000 pairs may well be closer to reality. It is not known to nest in Andorra. There are small colonies in the cliffs of Gibraltar but the Alpine Swift was abundant there until the late 19th century (Reid 1871–90), when key colonies were destroyed by quarrying.

The historical literature conveys the impression that the Alpine Swift has increased and probably expanded its breeding range in Iberia relatively recently. F. Bernis commented on the scarcity of Iberian records as recently as 1946 and he considered that migrants then accounted for most observations. Purroy (1973a) censused 328 pairs across the Pyrenees, where the largest colony, 88 pairs, was at the Congosto de Ventamillo, Huesca. He also located 20 pairs at Ordesa on the same cliffs where Wallis (1895) found just one pair in May–June 1884. The present Pyrenean breeding population is reported to be much larger (Sampietro *et al.* 1998).

Alpine Swifts may appear anywhere in the region during March–October, both on passage and during their sometimes far-ranging forays from their nesting colonies. A few individuals are recorded as early as February in the south. For example, in Faro there were two at the Lagoa dos Salgados, Silves, on 19 February 2007 (*Anu. Ornitol.* 6: 64). Northern regions are occupied during April. The birds remain until September, with some migrants observed during October and stragglers into early November. Despite its long stay in the region, the species is single-brooded. Migratory movements are generally inconspicuous but flocks may appear during passage periods to feed or in response to unusual weather conditions. Perhaps the most noteworthy of such gatherings on record was the estimated 3,000 birds over the Laguna de Sariñena, Huesca, on 11 April 2004 (*Ardeola* 51: 253). Records of 340 on passage at the Serra de Monchique, Faro, on 19 March 2007; 500 near Almograve, Beja, on 6 October 2007 and 500 again at the Serra de Monchique on 10 October 2008 (*Anu. Ornitol.* 6: 64–65, 7: 84) are also noteworthy. The limited number of ringing recoveries suggests that most passage of birds of foreign origin takes place across eastern Spain: 12 of 18 recoveries of Swiss-ringed birds were on the Mediterranean coast, two others being found in the Basque Country and one inland in eastern Andalucía (AMI).

There is some concentration at the Strait of Gibraltar during passage periods. Some individuals may be seen among flocks of migrating Common Swifts there, especially in August. Occasionally, however, sizable compact flocks of Alpine Swifts are recorded on the move at Gibraltar: for example, 72 flew south there on 30 October 1981, a relatively late date, and 80 flew north on 4 March 2005 (GONHS). Such flocks are probably common generally on passage but may pass very high or on a broad front and thus tend to be missed. Tellería (1981) gives observations totalling 8,705 birds during the autumns of 1972–74 and 1977. He noted only a few birds in late July and during August, the main passage period extending from mid September to early October, with a final observation of a flock of 21 on 26 October 1977.

Alpine Swifts are rarely recorded during the winter months but there are a few records usually of single birds, for example: at Viana do Castelo in northwestern Portugal on 20 December 1996; at Vigo, Pontevedra, on 12 December 1996; at Cape Home, Pontevedra, on various dates during November and December 2003 and until 6 January 2004; at the Lagoa dos Salgados, Faro, on 17 January 2009 and at Quinta do Lago, Faro, on 23 December 2009 (*Pardela* 7: 19; *Anu. Ornitol.* 8: 78; *Ardeola* 44: 257, 51: 253).

WHITE-RUMPED SWIFT *Apus caffer*

(Vencejo Cafre, Andorinhão-cafre)

A scarce and local breeding species in southwestern Spain and eastern Portugal. Summer visitor.

Breeding population estimates. Spain: SA1 30–100 pairs; SA2 100–150 pairs. **Portugal**: 5 pairs.

The White-rumped Swift is a widespread Afrotropical species whose colonisation of the Iberian Peninsula in 1960s was a great surprise, given that it had never before been recorded in the Palearctic. Swifts with white rumps had first been reported in southern Spain in 1964 and they were found nesting in the Sierra de La Plata on the Strait of Gibraltar in 1966 (del Junco & González 1966, Allen & Brudenell-Bruce 1967). The initial assumption was that the anticipated crossing of the Strait by the Little Swift, a breeding species in Tangier since the 1950s, had come about. However, a photograph of one of the swifts leaving its nest on the Sierra de La Plata, that was published in *The Ibis* (Allen & Brudenell-Bruce 1967) and *British Birds* (Ferguson-Lees 1967) soon led to the correct identification (Benson *et al.* 1968, del Junco & González 1969). The birds were nesting in a small cave within the nest of a Red-rumped Swallow. A breeding population in the Moroccan High Atlas mountains was discovered shortly afterwards, in July 1968 (Chapman 1969).

The White-rumped Swift has extended its nesting range rather slowly since the initial colonisation. By 2003 nesting pairs had been found in Andalucía in Huelva, Sevilla, Málaga, Córdoba and Jaén, in addition to Cádiz, as well as in Extremadura: both in Badajoz and Cáceres, and on the southern fringes of Castilla-La Mancha in the Sierra Madrona (SA2). A pair nested at Los Navalucillos, Toledo in 2004 (*A. O. Toledo 2002–2007*: 294) and a recently fledged chick was picked up at Torrejón de Ardoz, Madrid, on 26 October 2006 (*A. O. Madrid 2006*: 152). Several individuals were also seen in 2006 during May–July in another area in southeastern Madrid (*A. O. Madrid 2006*: 152). The White-rumped Swift has also been seen in Salamanca, Granada and Almería, one was at the Almenara marsh, Castellón, on 22 July 2008 (*A. O. Castellón 2008*: 96) and there is a far-flung record from the Llobregat Delta, Barcelona, on 20 May 1997 (*Ardeola* 44: 257). One seen at Lespignan, Hérault, on 26 June 2010 was the first record for France (Reeber *et al.* 2012).

Nesting had been recorded in Portugal at four locations up to 2006 and there is a small population in the south-east of the country, in the lower Guadiana valley (AP). The earliest observations were of small groups in Algarve, one of adults and recently fledged juveniles at Ludo, Faro, in August 1983, and a flock of eight at Odelouca, Silves, on 11 April 1987 (*Airo* 2) but the first known nest was in 1995, at Mina de São Domingo, Mértola, in the south-east of Alentejo (Moore 1996). Up to three pairs bred there in 1996 and 1997, and nests were found in 2006 in the Alcoutim–Mértola area (AP). Elsewhere a nest was found in the eastern Algarve, some 40km south-southwest of Mina de São Domingo (Viney 2002) and breeding has been considered likely at Vidigueira, Beja, and Vila Velha de Ródão, Castelo Branco (AP).

The Spanish population is likely to have increased since it was last assessed but there is no recent data available. The highest known density is in Cádiz, where at least 30 pairs nest in the vicinity of the Strait and Los Alcornocales natural park nearby. In Iberia the breeding birds apparently always nest within nests constructed by Red-rumped Swallows or, exceptionally, in those of House or Crag Martins (SA2, *Ardeola* 32: 420). The nests used are most commonly sited under bridges or in culverts, but sometimes in disused buildings or in caves. The swifts evict the rightful owners and add lining material so that a Red-rumped Swallow nest that has been taken over usually has the opening ringed with feathers. Eviction of a pair of swifts by another pair, and even infanticide by other swifts, has been recorded (SA2), suggesting that the availability of nests to take over may be a limiting factor for the swift population at least in some areas. In Africa White-rumped Swifts also nest in rock crevices or in holes in buildings. A pair seen repeatedly entering a cavity under a roof tile in the village of Guijo de Santa Bárbara, Cáceres, on 7 August 2010 (*Ardeola* 57: 539) may have been nesting there independently of hirundines and this could be a more widespread habit than is currently known.

White-rumped Swifts are summer visitors that presumably winter in the Afrotropical region. They are regarded as late arrivers in spring, seldom reported earlier than April, with most arrivals in May (SA2). Two were seen arriving from the sea at Torremolinos, Málaga, with other migrating swifts on 25 March 1985 (*Ardeola* 32: 420). They are double-brooded and the breeding season extends until October. An active nest at Hornachuelos, Córdoba, on 26 October 1975, still held two fully-feathered chicks (*Ardeola* 24: 262). The scarcity of White-rumped Swifts, and their apparent intraspecific competition when nesting, means that they are only ever seen in small numbers, typically just one or two pairs. Small groups, presumably including fledged young as well as adults, appear late in the season, from August to October, often in the company of other swifts and hirundines feeding along ridges in the sierras of Extremadura and western Andalucía. A record of 12 together at Monfragüe on 12 September 2009 (*Ardeola* 56: 361) is a noteworthy example.

The White-rumped Swift is rarely seen in Iberia during the winter months, unlike the Little Swift which overwinters regularly. Observations attributed to the present species include sightings of one and then two at buildings in Almería from 18 January 1975 through February and March 1975; one at the Celemín reservoir, Cádiz, on 31 December 1975 and one at the Laguna de Tarelo, Cádiz, on 26 January 2007 (*Ardeola* 22: 147, 24: 262, 54: 180).

LITTLE SWIFT

Apus affinis

(Vencejo Moro, Andorinhão-pequeno)

***A. a. galilejensis*. One of the most recent African colonists of Iberia and still confined as a breeder to southwestern Spain, where it is at least partly resident. A rare vagrant elsewhere.**

Breeding population estimates. Spain: <30 pairs.

In the 1920s this swift extended no further north in Morocco than Marrakech (BM). However, it subsequently spread north to reach the Tangier peninsula at the Strait of Gibraltar in the early 1950s (Pineau & Giraud-Audine 1979). The colonisation of southern Spain was widely expected thereafter and was announced to have occurred when a small colony of swifts with white rumps was discovered at the Sierra de la Plata, Cádiz, in full view of Tangier and only 20km away (del Junco & González 1966, Allen & Brudenell-Bruce 1967). There was not a little surprise when these birds were actually revealed to be White-rumped Swifts (*q.v.*). Little Swifts did eventually settle to breed in Spain but the first proved nesting attempt was not until 2000. It is remarkable that these pioneer breeders also settled in the Sierra de la Plata, alongside the White-rumped Swifts that still breed there.

The first observation at the Sierra de La Plata was of one or two birds in May 1992 (*Ardeola* 41: 113) and up to five were there in spring 1996, when birds were seen entering a cave (*Ardeola* 45: 110; van der Vliet 1998) and where a pair were found feeding young in July 2000 (*Ardeola* 49: 161). The site has remained occupied ever since, with a maximum count of 13 on 7 October 2010 (RBS).

Little Swifts have been found nesting elsewhere in southernmost Andalucía since 2000. A colony of three nests was found at Chipiona in 2004 (*Ardeola* 53: 181) and had grown to seven nests in 2009 (*Ardeola* 58: 469): up to 24 individuals were counted there in March 2011 (RBS). Up to six were present at a site in the Sierra Norte, Sevilla, in spring 2006 (*Ardeola* 55: 278) and one nest was located. Nesting at this second site by at least one pair has also been reported subsequently (*Ardeola* 56: 330, 58: 469, 59: 389). There have also been recent reports during the breeding season of birds seen in the town of Zahara de Los Atunes: just west of the Sierra de La Plata, as well as over the Sierra de Retín nearby and over Barbate harbour to the west (*Ardeola* 58: 469, RBS).

The pioneer colonists at the Sierra de la Plata took over the nest of a House Martin but elsewhere the birds seem to have built their own nests: the usual untidy balls of straw and feathers, in cavities on buildings. The breeding season in Spain has not been fully studied but probably extends from March to July, as in Tangier. The species is sometimes double-brooded. It seems likely that the breeding range will expand and Little Swifts should be looked for in towns and villages in southern Spain and Portugal.

The Little Swift is rarely observed outside its breeding range and it remains on the rarities committee list for Portugal, although it was withdrawn from the Spanish rarities list from 2013. There were only 30 accepted records up to 2009 of individuals seen away from the nesting sites; 18 in Spain, one in Gibraltar and 11 from Portugal: these last all since 1995 and from coastal regions south of Lisbon. The earliest observation in Spain involved eight at Torremolinos on 4 June 1981, two of which were repairing a House Martin nest (*Ardeola* 32: 143). Only one accepted Spanish record has been outside Andalucía: one was seen at Torredembarra, Tarragona, on 22 February 2004, but there are post-2010 records of one at Cape Estaca de Bares in November 2011, at the Albufera de Mallorca in April 2012 and at Alange, Badajoz, in July 2013, as well as three records (four birds) from coastal Murcia in July and September 2013 (RBS, *Ardeola* 60: 535). Observations away from the breeding range are distributed throughout most of the year but with a pronounced spring peak (Fig. 177).

Figure 177. Monthly distribution of Iberian records away from the breeding range, 1981–2009 (n = 30)

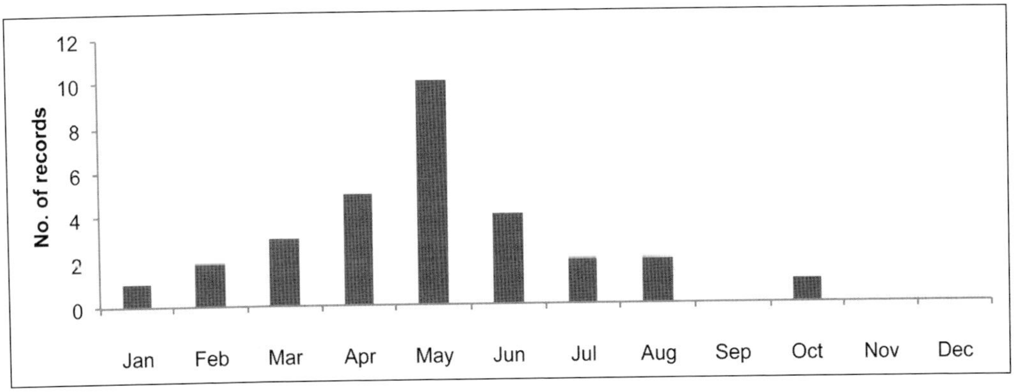

This is probably the only one of the five Iberian swift species that winters habitually within its breeding range. Twenty or more are regularly seen in December–January around the mouth of the Guadalquivir, in the vicinity of the Chipiona colony (F. Chiclana, pers. comm.) and seem likely to include the local breeding birds. Systematic observation of other colony sites in winter is lacking but no birds have been reported there at that season.

COMMON KINGFISHER *Alcedo atthis*

(Martín Pescador Común, Guarda-rios-comum)

***A. a. ispida*. A widespread breeder in northern, central and western Spain and in Portugal. Also a winter visitor from western Europe.**
***A. a. atthis*. Resident in southern and eastern Spain.**

Breeding population estimates. Spain: SA1 7,800–9,500 pairs. **Portugal**: 2,000–10,000 pairs.

The Kingfisher is widespread and locally common in suitable habitats throughout mainland Iberia. It is scarcest during the breeding season in La Mancha as well as in the arid south-east, from where it is largely absent. In Portugal it is commoner in the south than in the more rugged north, where the terrain is locally unsuitable (PA2, AP). Its presence is inevitably linked with the extent and persistence of river flows so that local increases in densities have been noted during wetter seasons (CA2). When nesting it requires still or gently flowing freshwaters with an abundance of small fish, as well as waterside vegetation offering shady over-water perches for fishing and sandbanks for excavating the nest tunnel. It is then especially associated with rivers, streams and canals rather than with more exposed wetlands such as lakes and reservoirs. Outside the breeding season it occurs commonly in a much broader range of waterside habitats, including large water bodies, irrigation canals, brackish lagoons and saltpans, rocky sea coasts, estuaries and harbours.

The SA1 assessment, 7,800–9,500 pairs, may have underestimated the Spanish breeding population: the Catalan population alone has been estimated at some 1,000–1,400 pairs (CA2). A survey in 2007 (Palomino & Molina 2009) put the population at *c.*48,400 individuals, although within such wide confidence limits, the minimum being only 5,400 birds, that its findings must be regarded as tentative. The Duero and the Andalucian basins together accounted for 47% of the estimated population in the 2007 survey. Earlier accounts (SA1, LR) have suggested declines of up to 20% due to reductions in water quality and nesting site availability and a decline has been reported from SACRE plots during 1998–2010, although the sample was small (Estrada 2013). Some local declines have been reported recently in Portugal (AP).

The species is much more widely distributed outside the breeding season, occurring on passage and in winter in many coastal regions, including at Gibraltar and in the Balearics: where it is common at the Albufera de Mallorca (SWA). Some of the birds may be dispersing from breeding sites inland, where harsh conditions may prevail in winter. In particular, the large population of the Duero basin in Spain appears to desert the region in winter (SWA). However, many Iberian Kingfishers are clearly fairly sedentary: only 12 of 45 birds ringed in Spain during the breeding season travelled further than 30km (Martín & Pérez 1990). Other passage and wintering birds are migrants from elsewhere in Europe. There are at least 76 Iberian recoveries, mainly from littoral areas, of birds ringed in France (26), Belgium (7), Germany (16), the Czech Republic (12), Switzerland (6), Italy (6), Slovakia, Poland and Britain (Martín & Pérez 1990, AP), although not all of them were ringed during the nesting season. There are also a few recoveries of birds ringed in Spain from France (2), Germany (3), the Czech Republic, Poland and Italy, as

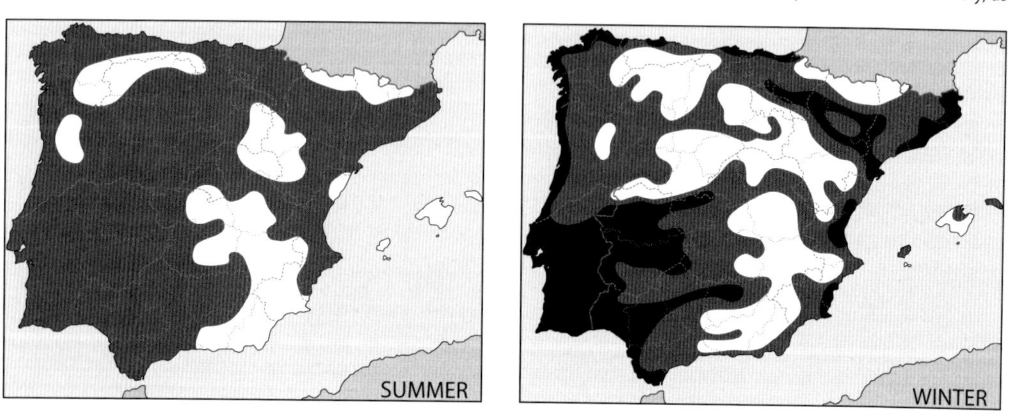

well as two from Morocco. Most recoveries of immigrants are of juvenile or immature individuals found between August and February (Martín & Pérez 1990). Wintering birds are especially associated with the Ebro, Tagus, Guadiana and Guadalquivir basins, and are also relatively common in the north-east of Catalonia, the Levant and Biscay coasts and western Galicia (SWA).

BELTED KINGFISHER *Megaceryle alcyon*

(Martín Gigante Norteamericano, Guarda-rios-cintado)

Vagrant to Spain.

The sole Iberian record, and the second for continental Europe, was at the Mar Menor, Murcia, where one remained from 16 November 1999 to 9 March 2010. A very rare vagrant to Europe, with six records from Britain, five from the Azores and one from the Netherlands. European records are mainly in October but several have then overwintered.

EUROPEAN BEE-EATER *Merops apiaster*

(Abejaruco Europeo, Abelharuco-comum)

A widespread breeding species in mainland Iberia and in the Balearics. Common and locally abundant on passage.

Breeding population estimates. Spain: SA1 23,000–30,000 pairs; SA2 >100,000 pairs; ACRE 6.0 million birds. **Portugal**: 5,000–25,000 pairs.

In Iberia the Bee-eater nests throughout the Mediterranean climatic region, which it virtually defines by its presence. Thus it occurs throughout the greater part of the Peninsula and on all the main Balearic Islands. It is absent as a breeder from the Eurosiberian zone: including the Biscay coastlands, all but southeasternmost Galicia, coastal regions of northern and central Portugal south to the Tagus estuary and all except the southern foothills of the Pyrenees and the Cantabrian Range. A pair that nested near San Esteban de las Cruces, Asturias, in 1980 and 1981 (*Ardeola* 31: 146) was exceptionally far north. Bee-eaters also avoid higher elevations but nesting birds have been found as high as 1,500m in the Sierra Nevada (Garzón & Henares 2012) and 1,800m in Madrid (AI1). They are seldom recorded north of the breeding range and never in any numbers. They are only sporadically observed in the Biscay coastlands, where recent records include single birds at Cape Higuer in May 1999; near Gorliz lighthouse in April 2008, and at Llanes in June 2010 (*Ardeola* 49: 191, 56: 362, 57: 540).

The Bee-eater occurs in all but densely forested regions, nesting in open habitats of all kinds provided suitable nesting sites are available. It typically nests in colonies that may number just a few pairs or ten or more, exceptionally 100 or even 200 pairs: although such very large colonies have reportedly become rarer in Spain in recent decades (SA2). The nest is at the end of a tunnel that the birds excavate in sandbanks, claybanks, established dunes, quarries, road cuttings and riverbanks. They will also tunnel into the ground where the substrate allows. Colonies may be found in areas of scrub or open woodlands, farmland, pastures, olive groves and orchards, often in the vicinity of wetlands. The birds regularly forage for bees, wasps and other large flying insects from perches on overhead cables, fences and bare branches, as well as high in the air. Pairs are often assisted in feeding their young by nest helpers (Fry 1984), generally the young from the previous year, so not all the individuals at a colony are breeding birds.

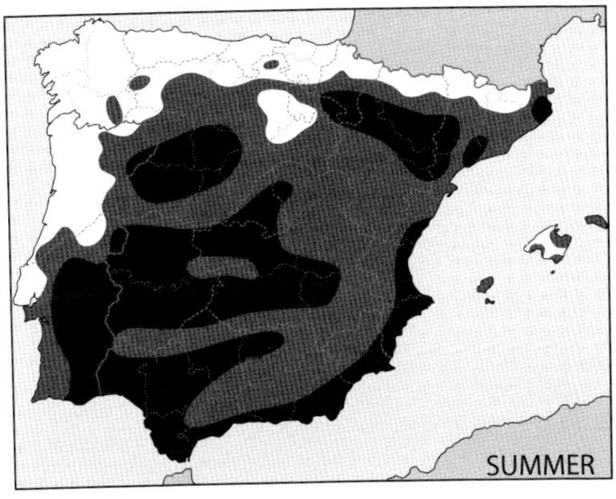

The ACRE assessment of the Spanish population at 6.0 million individuals suggests a nesting population of 2–3 million pairs, after allowing for helpers. The highest densities were found in the hottest regions: in the thermo- and mesomediterranean, where they ranged from 23–44 birds/km^2. Earlier assessments may have substantially underestimated this very widely distributed species but extrapolation from sample counts is not always straightforward for colonial-nesters and in our opinion the ACRE estimate must be viewed with considerable caution. The Catalan population has been put at some 22,000–30,000 pairs (CA2). The population was thought to be largely stable overall during 1998–2012 (Escandell 2013). Only 24 pairs were found in Mallorca in 2010 but there were over 100 in Menorca in 2012 (Adrover 2011, *Ardeola* 60: 535). In Portugal the Bee-eater is most abundant in the south and east, especially in Alentejo (PA2).

The Bee-eater may have been even more numerous in the past, at least in southern Iberia. Lilford (1866) found that it outnumbered all other species at Aranjuez, Madrid, in May 1865 and Saunders (1871) describes how they were brought by the sackful to the markets of Córdoba and Sevilla. Noble (1902) found them to be still quite common in Andalucía despite snaring, shooting and netting, but was informed by the local people that they were not nearly as abundant as formerly. On the other hand, there are indications of northward spread in the breeding distribution during the 20th century. The Bee-eater was already present in the Duero valley in Portugal during the 1920s (Tait 1924) but both Boxberger (1921), for Aragón, and Ticehurst & Whistler (1925), for Navarra, gave only localities alongside the river Ebro, whereas the Bee-eater now ranges commonly to the Pyrenees in those regions. In Central Spain Witherby (1928) only found it in Aranjuez, Madrid, and Candeleda, Ávila, and he commented that he was unable to find it anywhere in the Gredos and Guadarrama sierras. Earlier, Castellarnau (1877) did not note it at Segovia.

Bernis (1946) found a few nest-holes in the La Maragatería region, León, but since the local peasants were unfamiliar with what is a conspicuous and striking bird it seems likely to have then been a recent arrival there. Range expansion in the Duero basin is widely acknowledged (Valverde 1953a, AMI, de Juana 1980) and northward spread is also evident in Navarra when the range shown by first breeding bird atlas there (Elósegui 1985) and the recent distribution (SA2) are compared. Marked range expansion is known to have occurred in the Comunidad Valenciana and Catalonia since the 1920s, but principally during the 1950s. The Osona area in Catalonia was not occupied until 1985 (Ferrer *et al.* 1986, Aymerich *et al.* 1991, CA2). The present Iberian breeding range was essentially complete by the 1990s. Its northern limits are marked by the dry regions of inland Galicia, the plateau country south of the Cantabrian range in León, the upper Ebro Valley and the Pyrenean foothills.

The first Bee-eaters arrive in southern Iberia in mid March but the main arrivals at the Strait of Gibraltar occur between mid April and mid May, with stragglers passing in early June (BSG). After breeding, some Bee-eaters regularly move to feed in montane regions, as noted by Bernis in the Guadarrama and Gredos mountains during July–September (AMI). The southward passage at the Strait begins in early August but the peak movement is during the first half of September, after which the numbers passing soon decline until the last few pass in early October (BSG). Large numbers clearly concentrate at the Strait, as also occurs with such other aerial insect feeders as swifts and hirundines, producing spectacular movements at peak periods when a succession of often clamorous flocks may be seen. Estimates of 150,000–250,000 birds crossing the entire Strait region during the southward passage were made from local counts of around 35,000 birds in both 1974 (López Gordo 1975) and 1977 (Tellería 1979). Such estimates clearly amount to only a small proportion of the total passage if the Spanish population is indeed anywhere near as large as suggested. However, many birds probably migrate on a broader front and an unknown proportion travel at night (Irby 1895, BSG, pers. obs.), when they give themselves away by their unmistakable calls. Some of the birds that cross Iberia on passage are likely to originate from the relatively small but expanding population in France, which numbered 6,000–10,000 pairs post-2000 (OF). There are three Spanish recoveries of birds ringed at the Camargue and one ringed in Germany was recaptured on passage in Formentera. Birds ringed in Spain have given five recoveries in Morocco, one in France and one in Germany.

Bee-eaters are very rarely found in Iberia in winter. A juvenile at the Guadalhorce estuary, Málaga, on 25 October 1998 (*Ardeola* 46: 160) was already very late but Cape Espichel, on the Portuguese west coast has produced some even tardier birds, including two on 1 November 1997, one on 6 December 1997, and one on 4 December 2001 (*Pardela* 7: 19, 9: 23; *Anu. Ornitol.* 1: 44). Also in Portugal, single birds at San João da Venda, Faro, on 13 February 2000; at Odelouca, Faro, on 5 February 2002; at Esperança, Portalegre, on 4 February 2008; and at Gavião, Santarém, on 27 January 2008 (*Pardela* 6: 23, 14: 23; *Anu. Ornitol.* 7: 84) may have been extraordinarily early spring migrants.

BLUE-CHEEKED BEE-EATER *Merops persicus*

(Abejaruco Persa, Abelharuco-de-faces-azuis)

Vagrant to Spain and Gibraltar.

M. p. chrysocercus is an uncommon summer visitor to southern Morocco. It is a relatively rare vagrant to Europe that is most often recorded in southern countries such as Italy, Greece and Malta. There are ten records from Britain and seven from France. Many of the European records could involve the eastern, nominate subspecies (Ebels & van der Laan 1994) although overshooting migrant *chrysocercus* may be the most likely to occur in Iberia.

The earliest Iberian observation was one flying south at Gibraltar on 9 September 1973 (BG). Subsequently there was an adult at the Llobregat delta, Barcelona, on 15 June 1989 (Gutiérrez 1991); one at Zaragoza on 25 April 1997; two or three at Mijas, Málaga, during 22–26 April 1999, and an adult on Cabrera, Balearic Islands, from 11–17 April 2004. There is a more recent record of two together at Delfià, Girona, on 24–26 August 2012 (RBS).

EUROPEAN ROLLER

Coracias garrulus

(Carraca Europea, Rolieiro-comum)

C. g. garrulus. **Summer visitor to western, central and southeastern Spain and eastern Portugal, and locally elsewhere.**

Breeding population estimates. Spain: SA1 4,000–10,000 pairs. **Portugal**: 100 pairs.

The breeding range is fragmented, with the largest numbers nesting in Extremadura, the Guadalquivir valley, the arid south-east and the central Ebro valley, with others breeding at scattered locations in western Castilla y León, Madrid, Castilla-La Mancha and the southern Levant (Avilés 1999, SA2). The small Portuguese population occurs in east-central and southeastern regions (PA2). The breeding sites are generally at low elevations and away from Atlantic coasts.

The breeding habitats of the Roller are in open woodlands, in regions of traditional farmland mosaics offering fallow areas and low-intensity grazing, and in pseudosteppe areas. Trees provide both nest sites and hunting perches, which probably accounts for the positive selection of oak dehesas over treeless grasslands (Avilés & Costillo 1998). Sizable holes for nesting are essential and these may be found in walls, bridges and other isolated buildings as well as in trees. Nestboxes on powerline posts in Extremadura, in particularly at the steppes of Brozas–Membrío, the Llanos de Cáceres and La Serena, sustain large numbers, and allow them to occur at high density in treeless regions from where they would otherwise be absent (Avilés *et al.* 2000). Nestboxes have also proved effective in other regions: for example, boxes attached to trees, walls and pylons sustain a dense population at Hernán Valle, Granada (Rodríguez *et al.* 2011). Conversely, a lack of suitable holes may be a limiting factor in otherwise suitable habitats.

The Spanish population has declined in some areas, perhaps by over 20% during 1970–90 (SA1, SA2). The available evidence on population trends in Spain is largely from local surveys and the overall trend remains uncertain (LR). Nevertheless, there have been some significant falls in breeding numbers locally. For example, the population of the Alberche valley, Madrid, declined from 56 pairs in 1969 to nil in 1984 (LR). Numbers have increased in parts of Extremadura as a result of the successful nestbox programes there but it remains to be demonstrated whether or not the high take-up of nest boxes has accompanied a shift from natural nest sites elsewhere (Avilés & Sánchez 1998, Avilés *et al.* 1999, LR). The historical literature also gives the impression that a generalised and possibly very significant population decline has occurred in Spain: the present scarcity and patchy distribution of the Roller in many areas is certainly at odds with earlier descriptions that described it as common and locally abundant (Saunders 1871, Arévalo 1887, Chapman & Buck 1893, Irby 1895, Gil Lletget 1945).

The Portuguese population is restricted to the Castro Verde area in Baixo Alentejo, its national stronghold, and to the east of Alto Alentejo and Beira Baixa (Marques *et al.* 2005, PA2). It seems always to have been scarce and local in the country (Smith 1868, Tait 1924, AP).

Rollers are very rarely observed north of the breeding range. There are at least four records each from Galicia and the Basque Country. A juvenile was at Santoña, Cantabria, on 6 August 1999 (*Ardeola* 47: 169). Seven birds were recorded during 1900–96 in Asturias (*A. O. Asturias* 4). In northernmost Portugal there was one at the Minho estuary on 23 August 1990 and 'several' were observed at Montalegre on 15 August 1989 and 27 August 1991, with one there on 27 July 1990 (*Airo* 2: 21, *Ardeola* 40: 99). The timing of many of these northern records, as well as others from the north of Castilla y León, suggests northward post-breeding dispersal in July–August.

Iberian Rollers were assumed to winter in the Sahel regions of West Africa but in Mauritania at least the species only occurs there on passage (Isenmann *et al.* 2010). Recent tracking of four adults marked in Spain: in Lleida, Madrid, Ciudad Real and Granada, has revealed that they travelled much further east and south than expected, two of them reaching Namibia and the other two Botswana by the following December (www.migraciondeaves.org). They depart from Iberia from early July but chiefly in August–early September. Very few are recorded later than mid September so one at Sagres on 4 November 1988 (AP) was exceptional. They return to their nesting range between mid April and early May, with stragglers arriving to the end of May. Saunders (1871) claimed to have seen a small flock near Jaén on 13 March, and Chapman (1884) noted several near Jerez, Cádiz, about 26 March.

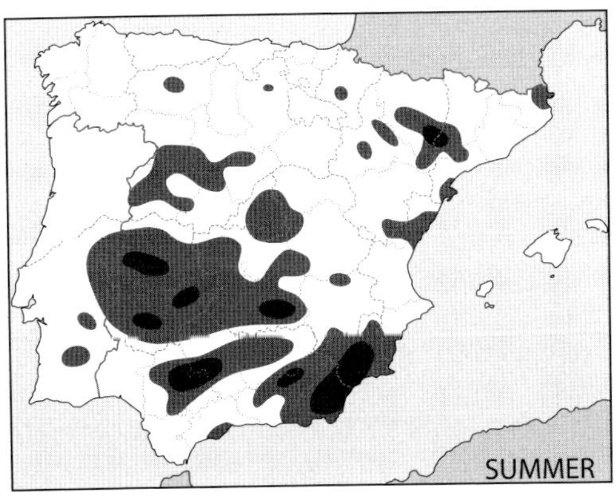

The Roller is recorded in the Balearic Islands as a rare and irregular passage migrant. The Regional Bird reports give 92 records (95 birds) during 1985–2010. Most records (74 birds:

78%) have been in spring, chiefly in May, with extreme dates of 11 April and 24 June, the remainder occurring between 11 July and 27 September. Exceptions include a bird that remained on Menorca from at least 26 August to 2 October 2003 (*A. O. Balears 2003*: 99) and a record from Cabrera on 9 March 1998 (*A. O. Balears 1998*: 99). Two or three are reported almost annually but there were at least nine in spring 2002, when strong westerly winds prevailed (AOB, *Ardeola* 52: 218), and ten occurred in spring 2010.

HOOPOE *Upupa epops*

(Abubilla, Poupa-comum)

***U. e. epops*. Widespread and common throughout most of mainland Iberia and in the Balearic Islands. Largely migratory but many winter in the south-west and in the Mediterranean coastlands.**

Breeding population estimates. Spain: SA1 497,000–710,000 pairs; ACRE 1.9 million birds. **Portugal**: 10,000–100,000 pairs. **Andorra**: 0–2 pairs.

The Hoopoe breeds throughout most of the Iberian Peninsula and in the Balearics, although generally below 1,000m. It is scarcest in the Biscay coastlands, where only a few scattered pairs occur. Most winter in sub-Saharan west Africa but a sizable wintering population occurs in the south-west, the Mediterranean coastlands and in the Balearics. It is a bird of warm, dry regions, hence its scarcity on the northern coastal strip of Spain and in the mountains, although nests have been found as high as 2,100m in the Sierra Nevada (AMI, Garzón & Henares 2012). It breeds throughout Portugal but is most abundant south of the Tagus, whereas in the northern half of the country it is much less frequent in coastal regions and becomes progressively more abundant inland (PA2). It does not nest in Gibraltar but does so occasionally in Andorra (ANA).

Open woodlands, such as oak dehesas, are a typical habitat but it is also characteristic of other open habitats with at least scattered trees: including lowland pinewoods, olive groves, woodland edges and parks. It also nests in much more open landscapes, including the arid lands of the south-east and in regions of both dry and irrigated crop cultivation. Nests are typically in holes in trees but it often also uses holes in drystone walls, buildings: especially ruins, and within piles of rocks or boulders (SA1, SA2).

The Iberian population is a substantial fraction of the European breeding population (BE). The highest densities in Spain, ranging from 5.8–11.1 birds/km^2, are in Iberian Holm and Cork Oak dehesas, olive groves and grasslands. The Spanish population was thought to be largely stable overall during 1998–2012 and the Portuguese population showed no clear trend during 2004–09 (Escandell 2013, ACP).

The Hoopoe is a partial migrant. There is a sizable wintering population in Spain, numbering at least tens of thousands of birds (AMI, SWA). The core wintering areas include the peninsular south-west: most of Extremadura, Huelva, the Sierra Morena and the north-west of Toledo, as well as the adjacent sectors of both mesetas and the Guadalquivir valley. There is also a contiguous wintering population in Portugal, largely south of the Tagus and away from the Atlantic coast (AP). A second core area includes the Ebro depression and the Levant coastlands from Almería to Girona, but especially in Alicante and Valencia. Conversely, the Hoopoe is very scarce or entirely absent in winter from the rest of the northern third of the Peninsula. Some remain all year in the Balearics, where Munn (1921, 1924) noted a few wintering in Mallorca and Menorca, but others occur there on passage or as summer visitors. Mallorca now has a large wintering population (SWA).

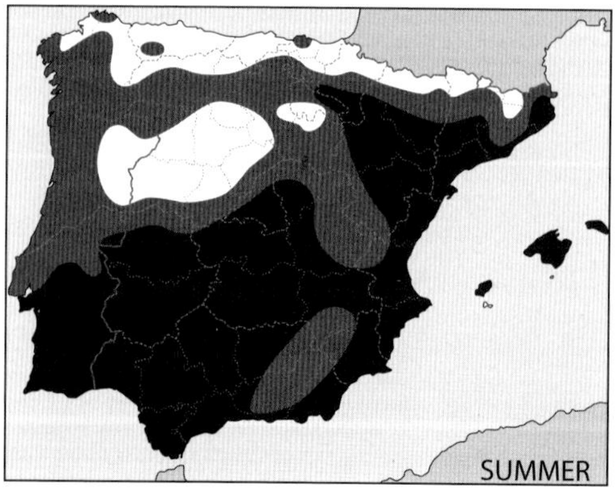

SUMMER

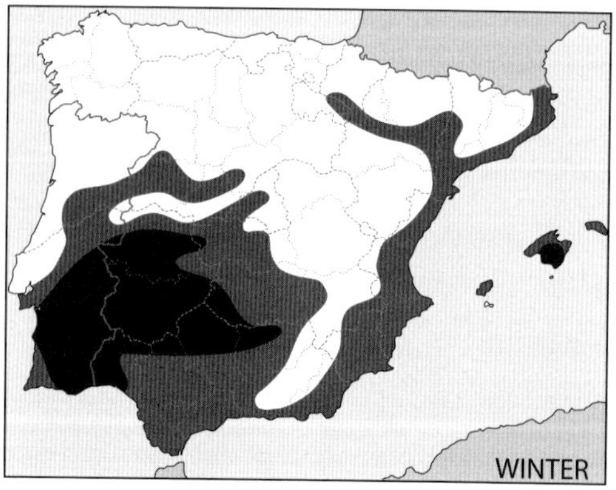

WINTER

Wintering in Iberia is not new and was noted in Valencia by Arévalo (1887) and in southern and eastern Spain by Gil Lletget (1945), but it appears to be increasingly frequent. Bernis (AMI: 1970) provided a detailed map of the wintering range that has since evidently expanded. The AMI map does not include the northern Meseta, the Ebro valley, most of Catalonia nor, in Central Spain, La Mancha and Madrid. At that time there were just a few February and November records from these regions that could have involved migrants. In Catalonia, for example, there were just two records in December–January, both in the south, whereas the present wintering population has been put at some 40,000 individuals, around 20% of the estimated breeding population (CWA). This Catalan wintering population still seems to be increasing: at a mean rate of 19% per winter between 2002–03 and 2008–09. The historical literature also suggests that wintering in Portugal was formerly less frequent than at present (AP). Wintering Hoopoes occur at highest densities in agricultural mosaics and in cork oak or Holm Oak dehesas, up to 1.07 birds/10ha in Cáceres (AI1), but birds may be encountered in a wide range of open habitats, although higher elevations are avoided in all regions (SWA). There are at least two records of early nesting from Andalucía: a nest with eggs was found in Almería on 21 January 1984 and a nest in Málaga on 3 March 1984 held naked young (*Ardeola* 32: 421, 36: 257).

Hoopoes nesting in more northern parts of Iberia are migratory and some of them probably winter in sub-Saharan Africa, although others may remain in the south of the Peninsula. There is insufficient data to assess the degree of residency of the southern Iberian breeders. There are 11 recoveries of Spanish-ringed birds in Morocco and two were found in France. Birds from other populations in western and central Europe occur on passage and some of these too may remain in Iberia in winter. Individuals ringed in Switzerland (3), Germany and France have been recovered on passage in Spain.

Migrant Hoopoes are especially obvious along coasts, where small groups may occur. A count of 175–200 seen feeding along an irrigation ditch at Fuente Obejuna, Córdoba, on 17 August 1994 (*Ardeola* 41: 200) is an extraordinary record, presumably the result of an unusual availability of food in damp ground at the height of the summer drought, in an area where the Hoopoe is very common. The first arrivals in the south are in early February, with the majority occurring in March–early April (AMI). Some passage is detected at Doñana and at Gibraltar into May. The return passage extends from late July to September, peaking in August and early September (AMI, Tellería 1981, BSG).

EURASIAN WRYNECK *Jynx torquilla*

(Torcecuello Euroasiático, Torcicolo)

***J. t. torquilla*. Breeds locally throughout most of mainland Iberia and in the Balearics, most widespread in northeastern Spain. Partial migrant.**
***J. t. tschusii*. May occur on passage since it has been identified in Morocco (BM).**

Breeding population estimates. Spain: SA1 46,000–53,000 pairs; ACRE 126,000 birds. **Portugal**: 1,000–5,000 pairs. **Andorra**: 2–5 pairs.

The Wryneck is a widespread breeding species throughout most of the Palearctic region. Most European birds are long-distance migrants, wintering in sub-Saharan Africa, but a minority winter around the Mediterranean, including parts of Iberia. The Wryneck is widely if thinly distributed in the northern third of Iberia and in the Levant, except in much of the Ebro depression, the Biscay coastlands and Galicia, where it is scarce. The range is scattered elsewhere in Spain, with local concentrations in Andalucía in the Sierra Morena and the Betic Range (SA2). However, the Wryneck is not easily detected where it nests at low density and it may go unrecorded in many areas. It is widespread in Portugal, with concentrations in Trás-os-Montes and Beira Alta, as well as in Ribatejo, Alentejo, Algarve and the north of Estremadura (PA2). It also nests in Mallorca and Ibiza and at least sometimes in Menorca. Those in Mallorca, which are apparently resident, are comparatively short-winged (Garcias 2011).

They are found in a diversity of open areas with scattered trees, wherever there is a ready supply of the ants and ant larvae that are its principal food. Many favour orchard crops, from apple orchards in the north to citrus, almond and olive groves in the Levant and south. In Andalucía they also occur in oak dehesas, although many apparently suitable similar areas, in Extremadura for example, remain unoccupied. Farmland with well-developed hedgerows and riparian woodland are also favoured. Nesting sites must provide suitable holes for nesting, generally in trees, although cavities in abandoned buildings and nestboxes are also used (SA1, SA2). Most nests in northern Spain are in old woodpecker holes, especially those of the Green Woodpecker (González *et al*. 2002). Most Wrynecks nest at altitudes below 1,200–1,300m, locally up to 1,600m in the Catalan Pyrenees (CA1): the population of Catalonia is estimated at some 4,800–7,300 pairs (CA2).

Densities of 1–2 birds/km^2 have been found in oakwoods, riparian woods and Mediterranean orchards (ACRE). The population on SACRE sites showed no overall trend during 1998–2010 but there was a decline in the north countered by an apparent increase in Andalucía (Escandell 2011). Local declines have been reported since the 1990s in Galicia, Asturias,

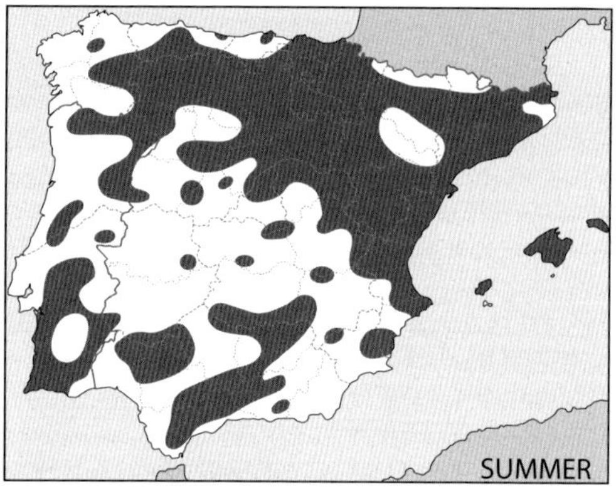

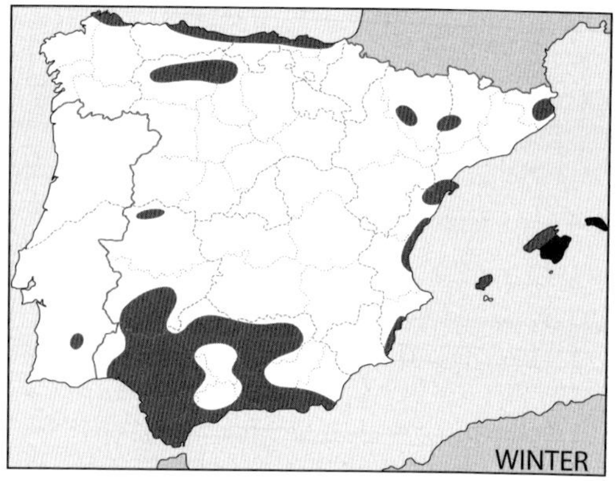

Cantabria, the Basque Country, and western Castilla y León, whereas the population has been stable or increasing in Aragón, Madrid, the Comunidad Valenciana, Andalucía and Mallorca (LR).

There is considerable evidence that the breeding range in southern Iberia expanded during the 20th century and continues to do so. There are no references to breeding birds in Andalucía in the historical literature. Irby (1833, 1895) met with it near Santander in May but only recorded it occasionally on passage in the Cádiz sierras in March and September. Saunders (1871) described it as abundant in spring and autumn in southern Spain but the only nest he found was in the north of his survey area, at Aranjuez, Madrid. Chapman & Buck (1893), who were based in Andalucía, noted explicitly that the Wryneck bred in Castile. More recently, Bernis (AMI: 1970) stated that the Wryneck only bred in the northern half of Iberia. Nesting in Andalucía was first proved as recently as 1975, in the western Sierra Morena (Herrera & Alcaide 1977). A comparison of the breeding distribution in the two Spanish breeding bird atlases clearly shows range expansion in the south, which may not simply be due to improved prospection. In SA1 there are just two centres of distribution in Andalucía, in northern Huelva and in the mountains of Cádiz and Málaga, with only a few records elsewhere in the region. SA2 shows an expansion of the Andalucian breeding range to Sevilla, Córdoba, Granada and Jaén, as well as into southern Ciudad Real. It will be interesting to see whether further spread occurs in Andalucía, into such areas as northern Córdoba where there were still very few occupied squares in SA2.

Western European Wrynecks, including most of those that breed in Iberia, are migratory, wintering in sub-Saharan Africa. Some from northern and central Europe overfly the region on passage and there are Iberian recoveries of birds ringed in Finland (4), Sweden (3) and Germany (3), as well as in Belgium, Switzerland, Austria, Croatia and Italy (AMI, AI1, AP, González *et al.* 2002). There is a German recovery in May of a bird ringed in Algarve in November and five Spanish-ringed birds have been recovered outside the peninsula: in Italy, the Czech Republic, Germany, Morocco and Egypt. The last of these was a bird that was ringed on spring passage in the Balearics, in May 1998, and that was recovered that September on the Egyptian coast (*Ecología* 13: 380). Spring passage at Gibraltar extends from late February into May, with most birds recorded in March–early April; the return passage there extends from September into November, peaking in October (BG, BSG). The spring passage on Cabrera and Dragonera in the Balearic islands extends from mid-March to early May, with a very marked peak in mid-April; the less marked postnuptial passage there extends from late August to early November (Garcias 2011). The first arrivals in northern Spain are in late March or the first days of April (González *et al.* 2002).

The wintering of the Wryneck in Iberia was formerly considered to be only sporadic (SA1), and some early authors (e.g. Irby 1875, Tait 1887) knew it only as a summer visitor, but Arévalo (1887) said it was a regular winter visitor to coastal Valencia. Recent observations confirm the presence of a regular wintering population in Andalucía, southern Castilla-La Mancha, the Levant coastlands, the Balearic Islands and, in Portugal, in Algarve and Alentejo, with a smaller winter presence in parts of the Ebro Valley, the northern Meseta and the Biscay coastlands (SWA). It is difficult to determine whether wintering is a recent habit or was overlooked in the past, although birds were reported in January by Stenhouse (1921) and Boxberger (1931), from near Gibraltar and in Málaga respectively. Bernis (AMI: 1970) noted that it wintered in very small numbers in the Balearics and in all the Mediterranean provinces, with rare winter records inland both in Spain and in Portugal. A recent increase in both in the number and the geographical spread of wintering birds seems nevertheless possible. It has been suggested that the winter presence of the Wryneck may be associated with that of the invasive and now often abundant Argentine ant, which appears to be more active in winter months than native ant species (CWA). The size of the wintering population is unknown but it is probably modest. In Catalonia it was put at 181–215 individuals during 2006–09 (CWA). Wintering individuals are often reported from scrub habitats in the vicinity of wetlands.

GREEN WOODPECKER *Picus viridis*

(Pito Real, Peto-real)

***P. v. sharpei*. A widespread resident in most of mainland Iberia: in Spain, Portugal and Andorra, but absent from the Balearics.**

Breeding population estimates. Spain: SA1 66,000–96,000 pairs; ACRE 636,000 birds. **Portugal**: 5,000–50,000 pairs. **Andorra**: 50–100 pairs.

The Iberian Green Woodpecker is endemic to Iberia and the whole of the Pyrenees, extending to the northern foothills on the French side (OF). Adults lack the black area that surrounds the eye in the nominate race and thus look rather pale-headed. The malar stripe of the male is almost entirely red, with only a little black on the lower margin. The laughing 'yaffle', given by both sexes, is higher pitched, less mellow sounding than in the nominate race. Recent assessment of the genetic variability of Green Woodpecker populations (Perktas *et al.* 2011), based on analyses of mitochondrial DNA, has provided clear evidence that the Iberian birds form a genetically uniform group distinct from populations elsewhere in Europe, and from Levaillant's Woodpecker *Picus vaillantii* in North Africa, and that they comprise a phylogenetic species *P. sharpei* (Saunders1872). However, some interbreeding with the nominate race is known from southern France and agreement regarding the taxonomic status of the Iberian form is pending further studies (Pons *et al.* 2011).

The Iberian distribution extends over most of the region. It is particularly abundant in the Eurosiberian and supramediterranean zones. Conversely, it is uncommon in much of the thermomediterranean, in particular being scarce or absent from the Guadalquivir depression and along southern and southeastern coasts. It is also scarce or absent in large areas of Extremadura and the southern Meseta. All Iberian populations are extremely sedentary so occurrences away from the breeding range are very infrequent. Thus it is only a rare vagrant to Gibraltar, with six records of single birds. It does not occur on the Balearics, except possibly as a rare vagrant to Mallorca (*A. O. Balears 2010*: 266).

This is the most widespread of all the Iberian woodpeckers and it is relatively catholic in its habitat preferences, requiring open or fairly open areas with some sizable trees to provide nest holes. The types of tree cover selected are very varied although there are regional differences in their apparent preferences. Thus the highest densities in the Comunidad Valenciana are in poplar plantations; in Madrid in ashwoods, riparian woodlands and Iberian Holm Oak woodlands, and in Ávila in pine, Pyrenean Oak and Holm Oak woodlands (SA2). It is characteristic of riparian woodlands, woodland and forest edges, open deciduous woodlands, parks and gardens, and orchards. Deciduous woodlands are preferred to conifers in the north but open pinewoods are selected in the south, including in central and southern Portugal (AP). Breeding birds may be found from sea level to the treeline, reaching over 2,000m in Mountain and Scots Pine forests (Bernis 1955, Ferrer *et al.* 1986). Breeding is known from some treeless areas such as the arid zones of Guadix, Granada, and in Almería, where it uses abandoned nest-holes of Bee-eaters or Rollers (Manrique 1993).

The ACRE estimate of the Spanish population of some 300,000 pairs may be compared with the 0.5–1 million pairs in France, where the Green Woodpecker is much more evenly distributed (OF). The Catalan population alone has been put at some 39,000–62,000 pairs (CA2). The highest densities found in Spain were in diverse habitats in the Eurosiberian, supramediterranean and mesomediterranean regions, ranging from 2.7–3.9 birds/km² but up to 5.3 birds/km² in the vicinity of mesomediterranean wetlands. The population on SACRE sites showed a significant decline during 1998–2012 that was chiefly evident in central regions (Escandell 2011, 2013). The Portuguese population showed no clear trend during 2004–09 (ACP) but there may have been local increases recently in Baixo Alentejo and Algarve, where pine plantations on abandoned farmland have provided new habitat (PA2).

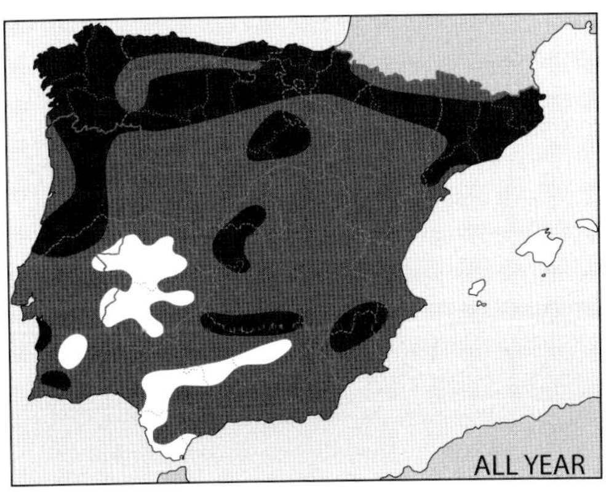

BLACK WOODPECKER *Dryocopus martius*

(Picamaderos Negro, Pica-pau-preto)

***D. m. martius*. Resident in Spain: in the Pyrenees and Cantabrian Range, and in Andorra.**

Breeding population estimates. Spain: SA2 1,000–1,500 pairs. **Andorra**: 25–30 pairs.

The Iberian population is an isolated outpost of an extensive breeding range that extends throughout much of Eurasia. There are two subpopulations, the principal one in the Pyrenees and a smaller and largely separate subpopulation in the Cantabrian Range. In Spain the Pyrenean population ranges from the east of Gipuzkoa in the Basque Country in the west, eastwards across Navarra and Aragón as far as the north-west of Girona in Catalonia. It also nests in Andorra. In the Cantabrian Range it occurs from the extreme east of Galicia, across Asturias and northernmost Castilla y León, to western Cantabria.

It occupies a wide range of mature forest and woodland types, provided that large trees are available (SA1, SA2). The massive oval nest hole is excavated in the trunk of a tall tree, usually a living one, often many metres from the ground. The Pyrenean birds occur chiefly in forests of spruce, beech and Scots Pine, and in Black Pines at lower elevations. Those in the Cantabrian Range are found in pure Beech forests or in mixed stands of Beech, Birch and Sessile Oak.

The Spanish Pyrenean population numbered 730–1,100 pairs at the start of the 21st century: including 300–600 pairs in Catalonia, 280 pairs in Aragón, 150–200 pairs in Navarra and just 1–2 pairs in the Basque Country (SA2). This is not a large number for such an extensive region and the birds are thinly spread and sometimes elusive, not least because their home ranges are often large, averaging 300–400ha. There were an estimated 280–320 pairs in the Cantabrian Range during the same period: 130 pairs in Castilla y León, 100 pairs in Asturias and 50–90 pairs in Cantabria (SA2). Subsequent fieldwork in Asturias found around 180 pairs there (Sánchez Corominas *et al.* 2009).

The range in France and elsewhere in western Europe has expanded in recent decades (OF) and a similar trend has been noted both in the Pyrenees and the Cantabrian Range. A breeding pair in the Valle de Carranza in 1999 represented an extension of the range from Cantabria to the west of the Basque Country (SA2). Range expansion in Asturias since the 1990s has brought it into the coastal mountains in the Sierra de Cuera (J. A. González Morales, pers. comm.). There have been recent observations in northern Burgos, in 2011 and 2012 in the Mena and Losa valleys and at San Zadornil (*A. O. Burgos*). Range expansion has also been seen in Catalonia (Camprodón *et al.* 2007) where the recent winter atlas estimated the population at 768–965 individuals (CWA). However, there are historical records suggesting former breeding in the northern Iberian Range and in the Gredos and Guadarrama mountains (Bernaldo de Quirós 1956, Trigo de Yarto 1960), where the species is now almost certainly absent, although one was seen at Canencia in spring 1991 and another was observed between the Somosierra and Navafría passes on 15 July 2007 (AI1, *Ardeola* 54: 401).

Records elsewhere in Spain indicate that dispersing birds may sometimes travel considerable distances but these occurrences are very rare indeed. The most far-flung of such records include one shot on the banks of the Tagus at Talavera de La Reina, Toledo, in late August 1982 (*Ardeola* 31: 146) and two observations in Málaga: at Cartájima on 15 August 1986 and in the Sierra de Las Nieves on 9 August 1990 (*Ardeola* 40: 100), on dates that suggest post-breeding dispersal.

The Black Woodpecker is not formally admitted to the Portuguese List. There is one published record: one seen at at Porto Chão in Gerês region on 22 August 1993 (Pimenta & Santarém 1996), as well as a number of unpublished records, but none of these is considered sufficiently well documented (AP). However, it is not unlikely that the occasional individual, most probably a dispersing juvenile, reaches northeastern Portugal.

GREAT SPOTTED WOODPECKER *Dendrocopos major*

(Pico Picapinos, Pica-pau-malhado)

D. m. hispanus. **Widespread throughout much of mainland Iberia. Vagrant to the Balearics.**

D. m. pinetorum. **Vagrant to the Balearics.**

Breeding population estimates. Spain: SA1 143,000–190,000 pairs; ACRE 702,000 birds. **Portugal**: 10,000–100,000 pairs. **Andorra**: 200–300 pairs.

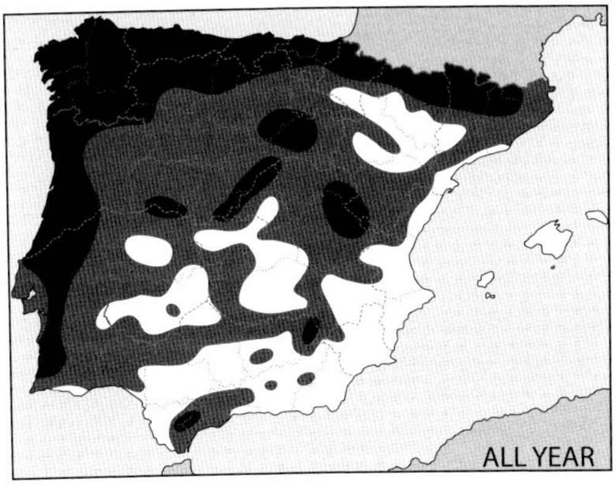

The endemic Iberian race is darker than *D. m. pinetorum*, the widespread race of most of continental Europe but the interface between the two subspecies within or north of the Pyrenees does not appear to have been defined. Irby (1879) remarked that Andalucian specimens sometimes resembled the Moroccan race *mauritanus* in having red breast markings, although less well defined than in the latter form. There are recent observations, including photographs published online, of such individuals from the Serranía de Ronda, Málaga, so it appears that southern Iberian birds may show intermediate characteristics.

It may be found wherever there are stands of suitably large trees; in woodlands, forests, shelter belts, riparian woods, copses, pine and poplar plantations, large parks and gardens or orchards. The highest densities have been recorded in northern Sessile Oak and montane Scots Pine woods, whereas eucalyptus plantations are avoided everywhere (AI1, SA2, Camprodon *et al.* 2007). It ranges from sea level to over 2,000m in the Pyrenean pine forests. Its range is largely continuous in the northern half of Spain and but it is more fragmented in the south, where it is largely confined to the woodlands of the Sierra Morena and the Betic Range and lowland oak dehesas (SA2). It occurs in in most of Portugal but is most abundant in the western half (PA2).

The principal gaps in the breeding range are in the arid south-east and other largely treeless regions: such as eastern Alentejo, parts of Extremadura and the southern meseta, the Ebro depression and the coastlands of the Comunidad Valenciana. Some recent range expansion has been seen in the Filabres range, Almería, where a nesting pair in 2005 was the first recorded instance of breeding in the province and where a population of 15–20 pairs was detected during 2006–08, these benefiting from recent afforestation with conifers (*Ardeola* 55: 302). A record of two on elms at the Moratalla estuary, Murcia, in May 2007 (*Ardeola* 55: 146) may be a further indication of southeastwards range expansion. Some range expansion has also been seen in certain northern areas, including Burgos and La Rioja (de Juana 1980, SA2) and Catalonia (CA2).

The species appears to be highly sedentary and none of 93 recoveries of birds ringed in Spain were at any distance from the ringing site. However, individuals of both *pinetorum* and *hispanus* are claimed to have occurred in the Balearics as vagrants (Ferrer *et al.* 1986). A male, race unknown, was in pines on Formentera in August 1981 (*Ardeola* 29: 187).

The Catalan regional population has been put at some 14,000–29,000 pairs (CA2) and at 87,000–135,000 individuals in winter (CWA). An average annual increase of 5% was detected in Catalonia during 2002–08 (CWA). The highest densities in Spain are in the north: in oak, spruce or pine woods and forests in the Eurosiberian and Supramediterranean regions, where they range from 4.0–8.4 birds/km^2 (ACRE). The population showed a significant increase on SACRE sites during 1998–2012 (Escandell 2013). The Portuguese population showed no clear trend during 2004–09 (ACP).

MIDDLE SPOTTED WOODPECKER *Dendrocopos medius*

(Pico Mediano, Pica-pau-médio)

D. m. medius. **A scarce resident in the Cordillera Cantábrica, the Basque mountains and in a few points of the Pyrenees.**

Breeding population estimates. Spain: SA1 650–680 pairs; SA2 1,045–1,205 pairs.

The Iberian birds have sometimes been assigned to a separate subspecies, *D. m. lilianae* (Witherby 1922), but this does not have general support (HBW). This subspecies was described from birds collected at Potes, Cantabria, where the Middle Spotted

Woodpecker had already been detected in 1876 (Irby 1883), but various other references in the historical Spanish literature can only be attributed to identification errors, probably involving juvenile Great Spotted Woodpeckers (AMI, Reig 2004b, Salas *et al.* 2005). The Iberian populations are small and isolated.

The Spanish birds occur in two main population centres, with a number of smaller sub-populations. There is a western population centred on the Cantabrian Range in and around the Picos de Europa, taking in parts of Asturias, southwestern Cantabria, León and Palencia (Purroy *et al.* 1984, Robles & Olea 2003). A breeding presence has also been reported in western Asturias; in the region of Somiedo and Degaña, as well as in the Sierra de Los Ancares in easternmost Galicia: although this last subpopulation may be among several small groups that may have recently become extinct (García Fernández *et al.* 2002, Sánchez Corominas *et al.* 2009).

A second sizable population is located in southern Álava: notably in Izki forest (Arambarri & Rodríguez 1996), and adjacent sectors of Navarra and Burgos. This was detected relatively recently (*Ardeola* 41: 100–101). Smaller subpopulations, none of them thought to exceed ten pairs, occur or have occurred in eastern Álava, Gipuzkoa and Navarra (SA2, LR). They include a breeding presence at Irati, in the Navarran Pyrenees, which is contiguous with a larger population on the French side of the range. Another French subpopulation extends into the Valle de Arán, which is the easternmost point of the Iberian breeding range (*Ardeola* 37: 344–345). Recent censuses there (Camprodón *et al.* 2007) have suggested that the population is some 31–52 pairs, considerably more than the ten pairs that were originally estimated.

Most Iberian birds inhabit Pyrenean Oak or Pedunculate Oak woods, with others occurring in mixed woods of Sessile and Lusitanian Oak, beech and pine (Camprodón *et al.* 2007). In León Middle Spotted Woodpeckers occur at similar densities both in mature woods and dehesas of Pyrenean Oak: there they select large diameter trees for foraging and decaying trees with holes for nesting, younger forests evidently being avoided (Robles *et al.* 2007).

The SA2 survey found 480–685 pairs in the Cantabrian Range and 365–520 pairs in the Basque mountains (SA2, 2003). These estimates may have been conservative since a recent census in Cantabria alone found 700 pairs (Fombellida *et al.* 2009). However, only 10–15 pairs are estimated to occur in neighbouring Asturias, all of them in a relatively small area in the south-east (Sánchez Corominas *et al.* 2009).

The Middle Spotted Woodpecker is highly sedentary and has not been reported at any distance from its native woods. Its breeding range may have extended into Portugal until the mid 20th century but there are no entirely reliable records there and it is not currently admitted to the Portuguese List (AP). The most recently claimed Portuguese record was in the Peneda-Gerês National Park in September 1992 (Pimenta & Santarém 1996).

LESSER SPOTTED WOODPECKER *Dendrocopos minor*

(Pico Menor, Pica-pau-galego)

***D. m. buturlini*. A widespread but often very local resident in mainland Spain and Portugal, absent from the Balearic Islands.**

Breeding population estimates. Spain: SA1 1,500–2,200 pairs; SA2 2,971–4,044 pairs. **Portugal**: <1,000 pairs.

The Iberian distribution may not yet be fully known. It is clearly absent from the arid south-east and from the Balearics but elsewhere it is often very local and elusive. Atlas surveys have identified three distribution centres in Spain where the species is present at relatively high density. These are in the north: including the Basque Country, La Rioja, Navarra and the upper Ebro valley; in the north-east in Catalonia and in the centre-west on the slopes of the Sistema Central in Salamanca, Ávila and Cáceres.

Less extensive sub-populations include those in Asturias; in the Guadarrama mountains in Madrid; in isolated regions of Extremadura including Las Villuercas, Sierra de San Pedro and the corkwoods of southwestern Badajoz, and in the Sierra de Aracena in Huelva. The surveys for the two Spanish breeding atlases produced a few observations in Galicia, western Andalucía, Castilla-La Mancha and in the interior of Valencia, but these were very sparse indeed. Fuller prospection may yet find more birds in some of these regions. For example, a search for the species in the Sierra Norte, Sevilla, located

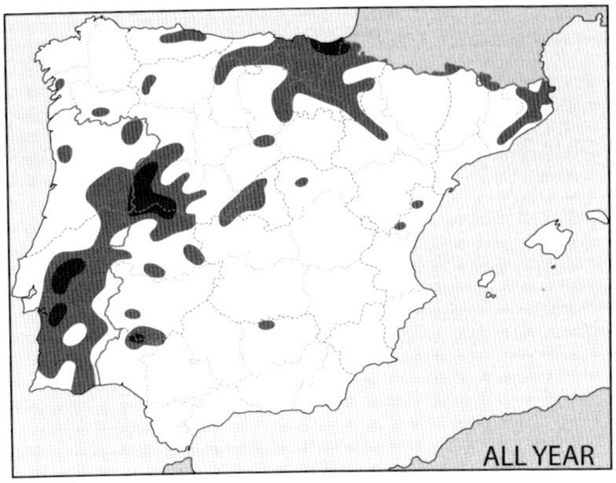

at least seven territories in an area of Cork Oak and mixed dehesa and riparian woodland where the Lesser Spotted Woodpecker was not previously known (*Ardeola* 52: 219). The Portuguese distribution is also incompletely known but there it is commonest south of the Tagus, especially in the Sado valley, western Alentejo and central Algarve. It ranges south to the Serra de Monchique, Algarve, where Ticehurst & Whistler (1933) obtained three specimens. It is much scarcer further north, occurring largely in scattered locations in the east, in Beira Baixa, Beira Alta, Alto Douro and Trás-os-Montes (PA2).

Improved coverage has also been suggested to account for the ongoing expansion of the known breeding range in Spain (SA2). In our opinion, however, there has been actual and relatively well documented spread in several areas, at least in Catalonia, the Ebro Valley, Madrid and Toledo, and probably elsewhere as well. The first Catalan breeding bird atlas found only five occupied 10km-squares but 43 were occupied at the time of the second atlas, 20 years later and some of the new records came from previously well surveyed areas in the north-east, in the basins of the Fluvià, Ter, Tordera and Besòs rivers (Ribas 2000), where the Lesser Spotted Woodpecker was found occupying either riparian forests or recent commercial poplar plantations (Romero & Gálvez 2001). Ferrer *et al.* (1986) considered it to be very scarce and local in Catalonia, with breeding unconfirmed south of the Pyrenees. By the early 2000s the Catalan population was put at 150–500 pairs and the winter atlas (CWA) suggests a further increase in numbers during 2002–09. Similar range expansion is apparent in the Ebro Valley when the 1997 and 2003 Spanish breeding bird atlases are compared, with the species appearing only in SA2 in some parts of La Rioja, Navarra and Aragón that had been relatively well covered earlier (e.g. Iribarren 1969, Purroy 1974, Aragüés & Lucientes 1980, de Juana 1980, Elósegui 1985). The new localities there have been found progressively further east along the Ebro river and tributaries. Additional occupied localities have been reported from Aragón in recent years (*A. O. Aragón 2004–2007*: 303–304). In like manner, in central Spain the number and spread of known occupied localities in Madrid and Toledo provinces has increased quite clearly in recent years (del Moral *et al.* 2002, *A. O. Madrid*, López de Carrión 2009). Nevertheless, the historical Spanish literature noted the presence of the Lesser Spotted Woodpecker in its current strongholds in Salamanca and Cáceres (Bernis 1945, Gil Lletget 1945, Abs 1958) but also from some localities where it is certainly long since absent: the gardens of Aranjuez and the Casa de Campo park in Madrid, and the Coto Doñana in Andalucía (Lilford 1866, Chapman 1888, Jourdain 1919, Witherby 1928, Bernis 1945).

It may be found in a wide range of deciduous woodlands but generally avoids conifers. It occurs in all types of oakwoods, including the dehesas in the west and south-west, as well as in Sweet Chestnuts in the Sierra de Gredos and the western Sierra Morena. It is often found in riparian woods: in alders, poplars and ash, and it also frequents poplar plantations, especially within the Mediterranean climate region. Its recent spread in the Ebro basin may be associated with both the recovery of riparian forests and the local proliferation of poplar plantations. The higher densities, of some 4 territories/km^2, have been reported from mature Pyrenean Oak woods on north-facing slopes in Cáceres and Salamanca (Serradilla & Calvo 2001, Prieta & Molina 2003). It ranges from sea level to about 1,200m in the eastern Pyrenees and 1,300m in the Sierra de Gredos, but more generally it breeds below 800m (SA2, CA2). It is strictly sedentary but short-range juvenile dispersal has been reported from Catalonia (CWA).

The low detectability and low population density make it difficult to assess the size of the Iberian populations. Most of the birds found in the SA2 survey were in just two provinces, Salamanca and Cáceres so the Spanish population is likely to number at least 5,000 pairs. Regional surveys in some areas at various times during 1996–2003 have found 25–35 pairs in Asturias, 5–15 pairs in Cantabria, 156–166 pairs in the Basque Country, 341–417 pairs in Navarra, 65 pairs in La Rioja, 20–30 pairs in Aragón, 117–500 pairs in Catalonia, 1,663–1,758 in Castilla y León, 500–1,000 pairs in Extremadura and at least six pairs in Madrid (SA2). More recent estimates are of 700–900 individuals in Catalonia in winter (CWA), 30–40 pairs in Aragón (*A. O. Aragón 2004–2007*: 304) and 25–50 pairs in Asturias (Sánchez Corominas *et al.* 2009). The SACRE scheme in Spain reported a significant population increase during 1988–2011 (Escandell 2012).

WHITE-BACKED WOODPECKER *Dendrocopos leucotos*

(Pico Dorsiblanco, Pica-pau-de-dorso-branco)

***D. l. lilfordi*. A rare resident in the western Pyrenees.**

Breeding population estimates. Spain: LR 60–70 pairs..

The relict and highly isolated Pyrenean population was first detected in the Pyrénées Atlantiques, France, in 1851 and was not proved to nest there until 1958 (OF). Its occurrence in the Spanish versant was not reported until 1968 (Bernis & Iribarren 1968) and it was first proved to breed in 1977, at Irati, Navarra (Senosiain 1978). An intensive survey during 1992–93 (Fernández *et al.* 1994) established that the Spanish breeding range extended along the French border from the Ansó valley, in the extreme west of Aragón, westwards in Navarra through the Roncal valley, Irati, Burguete-Roncesvalles to Quinto Real, with an isolated population further west at Bértiz. Earlier reported presence further east in the Huesca Pyrenees, at Ordesa (*Ardeola* 13: 258, Purroy 1972) and at the Selva de Oza and Aragüés del Puerto (Purroy 1972) could not be corroborated. The known breeding range remains as determined in the early 1990s although there may have been some westward expansion and, conversely, the marginal presence in Huesca may have disappeared: there were just three pairs there in 1992 but only one was found in 2000–02 (LR). The core population seems to be that of the Irati forests, with a mean density of 0.56 pairs/km^2, although this varies greatly with habitat quality.

The White-backed Woodpecker inhabits the oldest and best preserved beech forests of the western high Pyrenees although it also occurs locally there in beech-fir, beech-fir-pine and beech-pedunculate oak woodlands, but beech always predominates at such sites (Fernández *et al.* 1994, Fernández & Azkona 1996). The altitudinal range is from 400m at Bértiz to 1,700m at Larra in the Valle de Roncal, but over 60% of known sites are at 800–1,200m (Fernández *et al.* 1994). The majority of breeding locations are on northeast-facing slopes, which is where the most suitable beech forests occur.

A larger subpopulation on the French side of the border: 260–320 pairs in 2000 (OF), is largely contiguous with the Spanish population of 60–70 pairs (Fernández *et al.* 1994, LR). The conservation of this species demands the preservation of the mature beech forests that it requires. These forests need to offer a diversity of tree sizes and also an abundance of dead wood (LR, Garmendia *et al.* 2006).

FORK-TAILED FLYCATCHER *Tyrannus savana*

(Tijereta Sabanera, Tesourinha-do-campo)

Vagrant to Spain.

The nominate race breeds from Brazil to Argentina and winters as far north as Colombia and Venezuela, being an annual vagrant in the United States and Canada, especially along the Atlantic coasts, with some 120 records up to 1999 (Lockwood 1999). The sole Palearctic record is of a bird seen at El Rocío, Doñana, on 19 October 2002 (*Ardeola* 54: 433). It has been accepted as a natural occurrence not only because Fork-tailed Flycatcher records peak in North America in October but also because the observation coincided with the aftermath of two powerful Atlantic hurricanes: Lili and Kyle.

BAR-TAILED DESERT LARK *Ammomanes cinctura*

(Terrera Colinegra, Calhandra-das-dunas)

Vagrant to Spain.

Single birds have been found in Mediterranean Spain: at Cape Ses Salines, Santanyí, Mallorca, on 24 and 26 March 1994; at

Roses, Girona, on 28 April 2005, and on the Columbretes Islands, Castellón, on 9 May 2008. This lark is relatively common in southern Morocco where it is largely resident although given to dispersive movements, particularly in spring. Vagrants have been recorded in Italy, Malta and Cyprus and there are two accepted records for the Canary Islands.

DUPONT'S LARK *Chersophilus duponti*

(Alondra Ricotí, Calhandra de Dupont)

***C. d. duponti*. A scarce and local resident, chiefly in eastern Spain. Extinct in Portugal.**

Breeding population estimate. Spain: 2,800 pairs.

This is a scarce species, with a restricted fragmented range chiefly in the eastern half of Spain. It occupies – or has recently occupied – some 50–70 localities that have a total surface area of some 1,500km^2 (LR; Suárez 2010). All are in the Mediterranean bioclimatic region, chiefly in the meso- and supramediterranean, with an outpost population in the thermomediterranean in the arid south-east. The Spanish population is also the sole European population: elsewhere it now occurs only in North Africa, from Morocco to Egypt. The principal populations are believed to be in Morocco: including some 15,400 males (García *et al.* 2008).

Dupont's Lark may have bred in Portugal in the 19th century, when five were collected at Alfeite, Setúbal, during 1886–88 (Tait 1924). These specimens, that included a recently fledged juvenile, were lost in the fire that destroyed the Bocage Museum in 1978. An undated late 19th century sight record at Ponta da Erva in the Tagus estuary is the only other Portuguese record (AP). It probably also bred during the 19th century in steppic areas of Provence, southern France, where the type specimen was obtained in 1820 and where at least three others were captured subsequently, two in 1896 and one in 1915 (OF).

It is also likely to have bred in Spain during the 19th century but early records there were usually attributed to dispersal or vagrancy from North Africa. Francisco de los Ríos obtained at least two near Málaga during the late 19th century, to which reference is made by Irby (1879), Arévalo (1887) and others, and these remain in the collections of the British Natural History Museum (Suárez 2010). It was presumably these skins that were exhibited by Howard Saunders at a meeting of the Zoological Society of London on 1 May 1877 (*P.Z.S.* 1877: 368). Castellarnau (1877) claimed to have identified an individual between San Ildefonso and Segovia city in December 1874. Navas (1918) mentions a specimen from Cariñena, Zaragoza, that was examined by Hartert.

During the first half of the 20th century the species was well known to the peasants of the Cogolludo-Atienza region of southern meseta uplands, who thought that it was present all year round (F. Bernis: AMI). They knew it under the onomatopoeic name of '*ricotí*'. The first definite proof of breeding in Spain did not however come until 7 June 1967, when an adult was captured on a nest at Bellver de Cinca, Huesca (*Ardeola* 23: 236). EdJ heard the distinctive song of this lark at Níjar, Almería, in spring 1973 (*La Garcilla* 126: 8) and saw an individual near Arancón, Soria, in April 1977 (de Juana 1980). Many more nesting localities, outlining the Spanish breeding range of Dupont's Lark, had been identified by the early 1980s (Suárez *et al.* 1982, Aragüés & Herranz 1983).

The core population is restricted to two general areas: the páramos of the Iberian Range, mostly in Soria and Teruel, and the steppes of the Ebro valley: mainly at Los Monegros and Belchite in Aragón and locally in Navarra. The latter area formerly extended from La Rioja in the west: where a bird carrying food was at Alcandre in May 1986 (*Ardeola* 34: 288), to Catalonia in the east, where a small population at Alfés became extinct in 2005 (*A. O. Catalunya 2006*: 139). There are outpost populations elsewhere, notably in southeastern Andalucía: in the Sierra de Gádor and Las Amoladeras, Almería, and at El Padul, Granada, and in Castilla y León: at the Hoces del Río Duratón and Hoces del Riaza in Segovia and also in Zamora, in the far west. The Zamoran population, near the Portuguese border, was detected in spring 1993 (*Ardeola* 41:101) but may well have been overlooked earlier. Prospection in Almería in 2008 and 2009 has located a formerly unknown population of at least nine territories at Sorbas, as well as singing birds in Tabernas and at a diversity of new sites in the Sierra de Gádor (*Ardeola* 57: 236), but the population of Las Amoladeras, formerly the principal locality in the province, declined from

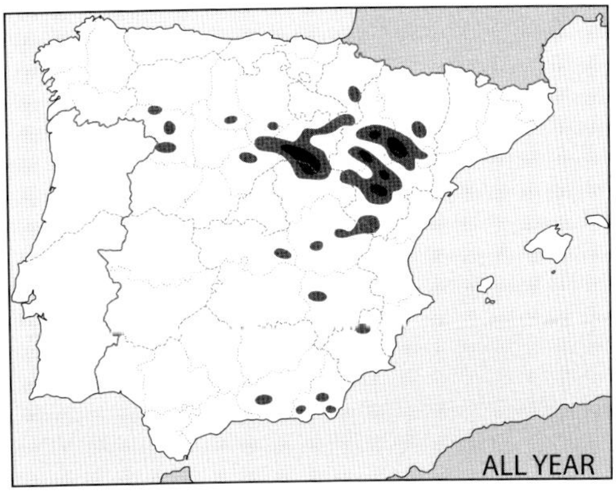

at least 48 males in 1988 to just 3–4 more recently (Suárez 2010). The very small populations of the Baza depression, Granada, and the Sierra de Espuña, Murcia, were no longer thought to exist (LR, Suárez 2010) but three singing males were located in the Sierra de Baza in 2011 (*Ardeola* 59: 445). The breeding sites generally range from sea level to over 1,500m but some 70% of the population inhabits locations at altitudes of 1,000–1,400m (Garza & Suárez 1988). A singing male at 1,950m on the Sierra de Baza in September 2008 (*Ardeola* 57: 236) was exceptionally high up.

The nesting habitat is sparse low scrub in flat or gently sloping terrain, featuring tussocks or cushions of thymes, lavenders, *Genista* spp., *Helianthemum* spp., esparto and other grasses, and halophytic vegetation, usually 20–40cm high. Pastures and cultivation are generally avoided, at least in the breeding season. It has been claimed that some join other larks on cereal stubbles in winter but this has been questioned by Suárez & Garza (1989). In fact, a radiotelemetric study (Suárez *et al.* 2008) found that although birds did not generally winter within cereal stubbles they were attracted to the interface between such agricultural land and the adjacent low scrub which they inhabited. The breeding season extends from mid-March to July.

The Spanish population was estimated to be of the order of 13,000 individuals by Garza & Suárez (1990), on the basis of transect surveys. More intensive investigations, employing actual mapping of occupied territories, showed that this was probably a major overestimation and that the total population was as few as 1,900 territories (Garza *et al.* 2003) or even 1,300 (Tella *et al.* 2005). Most recently, a thorough national census conducted during 2004–07 and additional survey work in 2008 (Íñigo *et al.* 2008, Suárez 2010) established the population at *c.* 4,500 singing males, which is equivalent to approximately 2,800 pairs when the known sex ratio of the species is applied (Table 49). Some 46% of the population was within Aragón, 32% in Castilla y León and 16% in Castilla-La Mancha, with Navarra, the Comunidad Valenciana, Murcia and Andalucía each holding fewer than 2% of the birds. Annual breeding season censuses in Andalucía during 2007–10 found between 20 and 34 singing males (*Ardeola* 59: 445).

Table 49. The extent of the breeding range, numbers of subpopulations and estimated minimum numbers of territorial males by geographical regions (Suárez 2010).

Region (Provinces)	Area (km²)	No. of subpopulations	Estimated minimum no. of territories	Density (males/km²)
Central Castilla y León (Burgos & Palencia)	10	1	18	1.8
Western Castilla y León (Zamora)	176	7	72	0.4
Ebro valley (Navarra, Huesca, Zaragoza & Teruel)	2,431	14	710	0.3
Iberian Range páramos (Segovia, Soria, Guadalajara, Zaragoza, Teruel & Valencia)	11,156	27	3,064	0.3
Southern Meseta (Cuenca, Toledo, Ciudad Real, Albacete & Murcia)	661	10	167	0.3
Eastern Andalucía (Granada & Almería)	247	4	26	0.1
Total	14,681	53	4,057	0.3

Birds are even more elusive in winter but seem to be largely resident. Most remain in their breeding areas but those of the high páramos may descend to lower altitudes when the heights are snowbound (Suárez & Garza 1989, Suárez *et al.* 2006b). A few disperse further and account for occasional records elsewhere but, in general, there is no evidence of any regular migratory activity. An early instance of such sporadic winter movements was reported from Torredembarra, on the Catalan coast, where several birds were seen and four collected on the beach dunes during 3–28 February 1963, following snowfalls and intense cold inland (*Ardeola* 13: 259). Other recent winter occurrences far from known breeding sites include single birds at Carmona, Sevilla, in January 2002 and at the Albufera de Valencia in January 2010 (*Ardeola* 49: 192, 57: 236).

Dupont's Lark has suffered a clear decline since the end of the 19th century that seems to have accelerated during the past 20 years, even in some of its core areas in the Ebro depression and in the páramos of Castilla y León. A third of known subpopulations are possibly extinct and over half of the remainder are declining. Localities where the species has recently become extinct include the northernmost ones: at the Páramo de los Franceses, Palencia, and the Páramo de Masa, Burgos, as well as many other peripheral sites (Suárez 2010). The surviving populations generally occur at low to moderate densities, ranging from about 0.5–8 birds/km² (LR, Suárez 2010).

It is principally threatened by habitat loss. Large areas of scrub have been lost to cultivation or afforestation, especially in key areas in Aragón and Castilla y León, and others have become overgrown as a result of reduced grazing by sheep, leading to the extinction of small subpopulations and the decline of others (Seoane *et al.* 2005, Tella *et al.* 2005, Laiolo & Tella 2006, Suárez 2010). The construction of windfarms, new roads and other infrastructures has also contributed to habitat loss and fragmentation. The coastal population in Almería has similarly suffered from the proliferation of plastic greenhouses there. The small size of the total population – very small indeed for a passerine – and its fragmented distribution, make it vulnerable to losses of any kind. Many nests are lost to predators: up to 84% in one study (Herranz *et al.* 1994), which can affect the viability of marginal populations especially. Fire has recently been shown to be a threat to its populations. Although fire can help to prevent scrub encroachment it also displaces breeding birds, which may lead to the loss of isolated subpopulations especially. The population in Almería has been severely reduced by recent fires (C. Pérez Granados, pers. comm.). Dupont's Lark is accordingly classed as 'Endangered' in Spain and consequently in Europe. At the very least, a comprehensive network of protected areas should be established to maintain and manage its main population centres.

CALANDRA LARK *Melanocorypha calandra*

(Calandria común, Calhandra-real)

***M. c. calandra*. An abundant resident in much of Mediterranean Spain. More local in Portugal, chiefly in the east. A rare vagrant to the Balearics.**

Breeding population estimates. Spain: SA1 1.03–3.40 million pairs; ACRE 8.5 million birds. **Portugal**: 1,000–10,000 pairs.

This is one of the defining species of the Iberian steppes (Valverde 1958). It is widespread throughout the Mediterranean climatic region but absent from the Eurosiberian. It inhabits open country across both Mesetas and in the Guadalquivir and Ebro valleys. It is also present near southwestern coasts in Cádiz & Huelva but is largely absent from the Mediterranean coastlands, where there is little suitable habitat (SA2). In Portugal it is mainly encountered in the east, on the interior plains of Alentejo and in the upland grasslands of Beira Baixa and Beira Alta, but small numbers occur in the Sado and Tagus valleys in the west (PA2).

This is a bird of treeless open country that provides ample low herbaceous cover. It thus avoids the arid semideserts of the south-east, as well as areas invaded by scrub or with taller plant cover. The principal habitats are pastures and cereal croplands (Martínez & de Juana 1999; Delgado & Moreira 2000). In the latter it is particularly attracted to fallow areas, whereas it tends to avoid cereal crops lacking fallows, probably because they become unsuitable as the crops grow taller and denser during the season. In Extremadura, densities in cereal areas increased in dry years when the crops were comparatively stunted and the fields had numerous clearings (de Juana & García 2005). It also occurs in salt-steppe habitats in the Ebro valley, often at high densities: 20–25 birds/10ha at Belchite and Alcañiz (Hernández & Pelayo 1987) and, more locally, in Almería and Murcia. Irrigated croplands are avoided (Díaz *et al.* 1993, Calvet *et al.* 2004). It mainly occurs at lower altitudes but many occupy cereal steppes at *c.* 1,000m on the northern Meseta and some occur locally to 1,400–1,600m, for example at Béjar, Salamanca (Carnero & Peris 1988).

The highest densities are in mesomediterranean pastures (101 birds/km^2) as well as in cereal-growing areas of the meso- and supramediterranean (46–92 birds/km^2) (ACRE). Breeding Calandra Larks may be regarded as semi-colonial, since their territories are small and tend to be clumped (e.g. Calvet *et al.* 2004). The Calandra Lark showed a significant decline on SACRE plots during 1998–2012 (Escandell 2013). The Portuguese population is also thought to be declining.

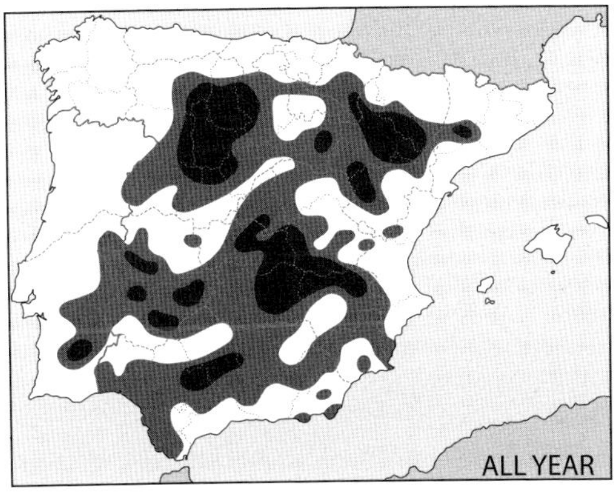

It is sensitive to agricultural intensification, in the form of the reduction or loss of fallow areas in cereal steppes, the spread of irrigated farmland, the loss of farmland to scrub encroachment after land abandonment and the conversion of arable land into olive groves, vineyards or arboreal croplands. In Iberia such changes have been particularly evident in Portugal, where the Calandra Lark has disappeared from Estremadura, part of Ribatejo and Algarve since the 1980s (PA2). It was clearly much more widespread in Portugal in the past, when it nested abundantly on the outskirts of Lisbon in the 19th century and was even more abundant in that region in winter, large numbers being killed for human consumption. It also occurred in Cascais and commonly in Trás-os-Montes during the first half of the 20th century (AP). Large numbers were formerly kept as cagebirds in villages and towns in Portugal, where Smith (1868) said that dozens could be counted along a single street. They were also popular captives, kept for their song, in Spain (Suárez *et al.* 2009).

Many populations are entirely resident but some northern birds winter further south in the Peninsula (AI2, SWA). There is very little indication of any significant passage at the Strait of Gibraltar: Tellería (1981) saw only 26 apparent migrants there during the autumns of 1976 and 1977 and there are only three records from Gibraltar itself (GONHS). Records north of the breeding range are also highly exceptional: one was killed by a hunter in Asturias in January 1902 (Álvarez-Balbuena *et al.* 2000) and one was at Vigo from at least February to April 1996 (*A. O. Galicia 1996*: 105).

Vagrancy to the Balearic Islands is also extremely rare. There were only seven records during 1957–82 and two thereafter: an individual at the Albufera de Mallorca in April 2008 and a group of four at María de la Salut, Mallorca, in April 2011 (*A. O. Balears 2008*: 90, *Ardeola* 59: 445).

Calandra Larks form flocks soon after the breeding season, sometimes as early as June. For example, there were flocks of 450 at Villalpando, Zamora, on 29 June 2005 and 350 at Aznalcázar, Sevilla, on 7 June 2007 (*Ardeola* 52: 433, 54: 401). Gatherings of several hundred, or even a few thousand, are not unusual. Flocks may wander extensively in farmland during the winter, concentrating particularly in fallow or set-aside areas and on newly-sown land, as well as on rough pastures, always avoiding areas with any significant presence of woody vegetation (SWA). Breeding starts early and the song period may already be underway in February.

WHITE-WINGED LARK — *Melanocorypha leucoptera*

(Calandria Aliblanca, Calhandra-ruiva)

Vagrant to Spain.

The sole Iberian record, of one taken at El Maresme, Barcelona, in October or November 1974 (Cordero-Tapia 1983), antedates the Spanish rarities committee and is not part of the official Spanish list. This lark is a rare vagrant to central and western Europe, with several records from Italy but none from France.

GREATER SHORT-TOED LARK — *Calandrella brachydactyla*

(Terrera Común, Calhandrinha-galucha)

***C. b. brachydactyla*. Common in mediterranean regions of Spain and Portugal and in the Balearics. Summer visitor.**

Breeding population estimates. Spain: SA1 2.2–2.6 million pairs; ACRE 2.6 million birds. **Portugal**: 2,000–20,000 pairs.

This is a common species of open country within the mediterranean climatic region. It extends only very locally into the eurosiberian zone, in southern Galicia (SA2). In general, it is more abundant in the thermomediterranean, and in the warmest sectors of the mesomediterranean, than elsewhere (AI2). It nests on all the principal Balearic islands where it is widespread if not abundant (*A. O. Balears 2010*: 266).

The breeding habitat is cereal croplands, pastures and steppes. However, it avoids growing crops and selects fallows, unsown ploughed fields and areas of set-aside in the cereal-growing regions (Tellería *et al.* 1988, Delgado & Moreira 2000). Some occur in coastal dune vegetation, and in vineyards and orchards that offer plenty of open ground. It most often occurs at low to moderate altitudes, up to 1,300–1,500m, but exceptionally it has been found nesting at 2,100m in the Sierra Nevada (Garzón & Henares 2012). The highest reported densities are in semi-arid level areas of the Ebro valley (LR).

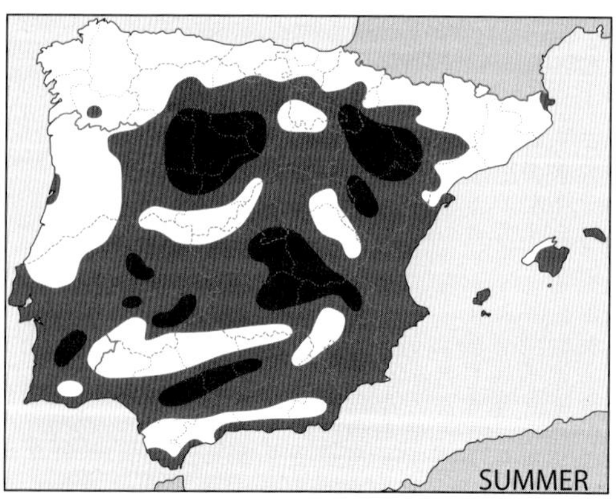

The Spanish population on SACRE sites showed mixed fortunes during 1998–2010, being stable overall but showing a significant decline in the north and an increase in the south (Escandell 2011). Elsewhere, a marked decline has been reported from Catalonia, where there were some 16,000–22,000 pairs in 1994–95 but only 680–1,080 pairs in 2004 (CA2). The decline in Catalonia has been accompanied by major range contraction there, the species having virtually disappeared from Girona and Barcelona, most of Tarragona outside the Ebro Delta and large parts of Lleida. Range contraction is also apparent in the north of León and Palencia in Castilla y León when the distributions in the 1997 and 2003 breeding bird atlases are compared. The Portuguese population is also thought to be decreasing; it is generally scarce here, the highest densities occurring in cereal growing areas in Baixo Alentejo and Algarve (PA2). The breeding distribution in Portugal has shown some contraction since the 1980s, particularly at some littoral sites, in the interior of the north and centre and also south of the Tagus (PA2).

The historical literature gives the impression that it was formerly much more numerous in both Spain and Portugal and that its populations have since suffered large-scale decline. Most 19th and early 20th century authors (e.g. Saunders 1871, Arévalo 1887, Chapman & Buck 1893, Boxberger 1921, Tait 1924) described it as 'abundant' or 'very abundant' and it was still regarded as very common during the 1940s and 1950s (Gil Lletget 1945, Mountfort 1954, Valverde 1958). It is probably a victim of the large-scale landscape modification associated with agricultural changes, mainly the reduction or loss of fallow areas in cereal steppes, as well as the decline in extensive grazing and spread of reafforestation. However, and unlike other larks, factors affecting it in its winter quarters and on passage may also be involved. Its decline in Iberia gives cause for concern.

The Greater Short-toed Lark is the sole summer visitor among the eight lark species that breed in Iberia. Its winter quarters are in Sahel belt of sub-Saharan Africa. Arrivals in southern Spain are from mid March to May, chiefly in April–early May (BSG, Suárez *et al.* 2009). Birds may gather in small flocks prior to the post-breeding migration: 120 at Vila do Bispo, Algarve, on 28 August 2008 and 300 there on 5 September 2009 (*Anu. Ornitol.* 7: 85, 8: 79) are noteworthy concentrations. Departures occur from mid August and continue to mid October, most leaving in September.

Records from the northern coastlands are rare. There were four May records of single birds in coastal Galicia during 1993–2004 and two were seen there at Baiona in October 2002 (*A. O. Galicia*). There were only nine records from Asturias during 1954–2006 (Álvarez-Balbuena *et al.* 2000; *A. O. Asturias 1996*: 134).The great majority of the northern records have been during passage periods: ten were in April–May, six in August–October and one in January.

Observations in winter are very rare. However, one was at Ludo, Algarve, on 22 December 1988, one was remarkably far north at Baldaio, A Coruña, on 20 January 2001 and one was on the Columbretes Islands, Castellón, on 5 January 2008 (*Airo* 2; *A. O. Galicia 2001*: 135; *A. O. Castellón 2008*: 102). Occasional wintering has been claimed in Alicante in recent years and there is an intriguing record of 40 in a mixed flock with other larks there at Villena on 15 February 2009 (*A. O. C. Valenciana 2009*: 106).

LESSER SHORT-TOED LARK *Calandrella rufescens*

(Terrera Marismeña, Calhandrinha-das-marismas)

***C. r. apetzii*. A local resident in eastern and southwestern Spain and southeasternmost Portugal. Vagrant to the Balearic Islands.**

Breeding population estimates. Spain: SA1 230,000–260,000 pairs. **Portugal**: PA2 70–76 pairs.

The form *C. r. apetzii,* one of the greyer and more heavily streaked races, is the only one to occur in Europe. It is endemic to the Iberian peninsula where it is typical of steppe habitats in the thermo- and mesomediterranean. Its range is highly fragmented but there are four principal population centres: in the Ebro valley; the arid south-east: including the Hoya de Baza and other inland sites in Granada and coastal and inland sites in Almería, Murcia and Alicante; the littoral regions of the south-west: including southeastern Algarve, Doñana and other coastal areas in Huelva and Cádiz; and the east of Castilla-La Mancha (LR). There are also small and isolated populations on the Levant coast in Valencia, Castellón and the Ebro delta. The delta population was discovered in 1929 (Witherby 1930) but those inland in the Ebro valley were only detected during the late 1970s, both in Aragón and Navarra (Aragüés & Lucientes 1980, de Juana 1980). The Castilla-La Mancha population has been a much more recent discovery: although there had been several winter observations there, breeding was not confirmed until 2001 (Velasco *et al.* 2002).

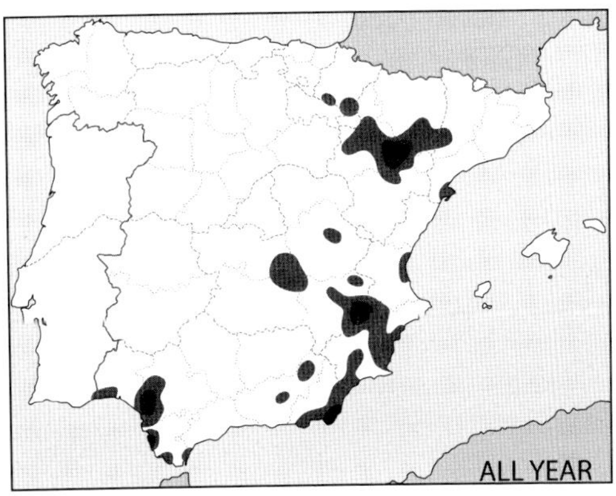

The habitats occupied differ between regions but all comprise low scrub formations on sandy, clay, dry mud or pebbly substrates. This species is strongly linked with largely natural vegetation and hence is absent from farmland when breeding, although it may flock there in winter, particularly in the Ebro valley. However, it occurs at low densities in vineyards in La Mancha (Velasco *et al.* 2002). It may be found alongside the Greater Short-toed Lark but the latter tends to favour sites with more grass and less scrub, i.e. those where the plant cover is lower and there is more bare ground (Suárez *et al.* 2002). Unlike the Greater Short-toed Lark, it frequently perches on the vegetation. The Ebro population breeds on salt flats, arid pastures and in areas of *Ononis*, *Gypsophila*, *Helianthemum* and *Rosmarinus* scrub, chiefly at elevations below 450m. Some, however, occupy areas of rosemary and Kermes Oak scrub on stony hillsides and summits in adjacent parts of the Iberian Range up to 950m. The birds in Castilla-La Mancha occupy salt flats, sometimes on the periphery of wetlands. Those inhabiting coastal areas, such as the marismas of the Guadalquivir and the coastlands of Almería, Murcia, Alicante, Valencia and the Ebro delta, occur in *Salicornia* scrub as well as in dunes, the margins of saltpans and sandy areas with sparse plant cover.

The Spanish population has shown some decline in numbers and geographical range since the 1990s. The principal causes are related to habitat loss. This species has been affected by the loss of coastal saltflats to plastic greenhouses, particularly in southeastern Spain. Birds can occur at high densities in favoured areas, 20–50 birds/10ha in the breeding season (Hernández & Pelayo 1987). Higher densities have been reported from the Hoya de Baza, Granada, with 34–89 birds/10ha in spring and 62–145 birds/10ha in winter (LR). The nests are typically clumped, forming loose colonies (Cañadas *et al.* 1982).

The entire Portuguese population is in or near the Castro Marim reserve at the Guadiana estuary in the extreme south-east, where its occurrence was noted by Tait (1924) and Themido (1933). There is little evidence that the Lesser Short-toed Lark was formerly any more widespread in Portugal although there are unconfirmable claims that it nested at the Tagus and Sado estuaries in the 19th century (AP).

Lesser Short-toed Larks are sedentary and remain in their nesting areas in winter or travel to forage on nearby margins of agricultural land, as well as in stubbles and fallow areas (AI2, Suárez & Sáez-Royuela 1987). Those in La Mancha winter in the vicinity of wetlands, remaining there from September until mid April (Velasco *et al.* 2002). Records away from the breeding areas are rare. Perhaps the most noteworthy was the appearance of 40–50 birds at the Santillana reservoir, Madrid, in December 1998: the first record for the region but followed by 30–35 at the same site in December 2003 (*A. O. Madrid 1998*: 174, *2003*: 167). In Extremadura one was at Santa Amalia on 2 July 1977 and there were five at Madrigalejo on 23 February 2003 (*Ardeola* 30: 122, *A. O. Extremadura 2001–2003*: 245). Single birds were seen in Asturias near Oviedo on 16 September 1974 and in Galicia at Cape Estaca de Bares on 5 October 1974 (*Ardeola* 23: 236). Two were present at Montejo de la Vega, Segovia, on 3 June 1992 (*Ardeola* 43: 255). One was found on the Columbretes Islands on 27 January 2008 (*A. O. Castellón 2008*: 102).

There are very few records for the Balearic Islands. The most recent were at the Albufera de Mallorca: six birds on 1 December 1988, one on 9 December 1990 and two on 20 April 1993 (*A. O. Balears 1988*: 38, *1990*: 62, *1993*: 108). There is an earlier record from Menorca, at Salinas de Addaia on 15 March 1975 (Muntaner & Congost 1984).

It does not occur as a regular migrant at the Strait of Gibraltar but one was seen west of Tarifa on 9 September 1957 (Bruhn & Jeffrey 1958) and there are two records from Gibraltar itself: four birds on 8 April 1969 and three on 8 April 1988 (GONHS).

CRESTED LARK *Galerida cristata*

(Cogujada Común, Cotovia-de-poupa)

***G. c. pallida*. An abundant resident in mainland mediterranean climatic zones. Paler above than the nominate race and less distinctly streaked below. Those from southern Spain are smaller and red-brown, tending towards the northwestern Moroccan race *G. c. kleinschmidti* (BWP).**
***G. c. cristata*. The mainly scarce resident birds in the eurosiberian climatic zone: from northern Portugal to the margins of the Pyrenees, have been assigned to this subspecies (BWP).**

Breeding population estimates. Spain: SA1 400,000–1 million pairs; ACRE 31.5 million birds. **Portugal**: 10,000–100,000 pairs.

G. c. pallida is endemic to Iberia where it is a common resident of the thermo-, meso- and supramediterranean regions of mainland Spain and Portugal (SA2, PA2). Birds resembling the nominate race are present, but scarce, in the eurosiberian region, both inland and on the coast. There are no records of either race from the Balearic Islands.

The species occupies a diversity of habitats, typically those managed by human activity. It may be found in all types of arable land, notably in fallow and set-aside areas in cereal-growing regions but also in irrigated croplands, vineyards, dehesas incorporating cereal crops and the outskirts of towns and villages. Pairs are often seen foraging along roads but the Crested Lark is not otherwise gregarious and seldom forms flocks of any size, even in winter. Scrub and woodlands are avoided and it does not generally occur much above 1,000m, selecting level ground and valleys in hilly or mountainous regions. It occurs in the foothills of the Sierra Nevada below 1,300m, being replaced at higher elevations by the Thekla Lark (Garzón & Henares 2012). It is particularly abundant in the major river valleys of western Iberia, especially within the

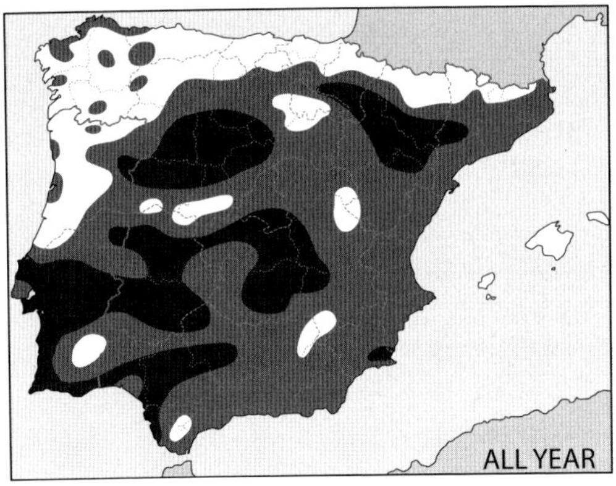

Guadalquivir depression but also in the Guadiana, Tagus and Duero valleys. In the east it is most numerous in the Ebro depression and also in Murcia (SWA).

The ACRE estimate of the Spanish population seems very high and must be regarded as provisional. The regional population of Catalonia has been reliably assessed at 94,400–199,700 pairs (CA2). The highest reported densities are in agricultural habitats in the thermo- and mesomediterranean, averaging 4.7 birds/10ha both in cereal croplands and in irrigated crops (SA2). The population is reported to have declined by about 6% during 1998–2012 (Escandell 2013).

It seems probable that an earlier historical decline of significant proportions affected the Spanish populations during the 20th century. The Crested Lark was considered 'very abundant' (Arévalo 1887), 'one of the most abundant birds in Spain' (Chapman & Buck 1893) or 'exceedingly common in the cornlands' (Noble 1902). By 1971, Bernis (AMI) was of the opinion that a decrease had taken place in at least some regions during the preceding 20–30 years. A large-scale decline in Crested Lark populations has certainly been apparent since the 1930s in the rest of western Europe, including France (HBW, BE, OF).

In Portugal there are some indications of decline and range contraction in the north-west since the Crested Lark occurred in the Porto district in the mid 20th century but it is now very rare there. It has also largely disappeared from the Minho valley since the 1980s (AP). Nevertheless, recent Portuguese census results have been more promising, showing a marked increase during 2004–09 (ACP).

Iberian Crested Larks are highly sedentary and nearly all ringing recoveries are local, although one travelled 370km from Murcia to Madrid (SMA). Some seasonal changes in abundance have been detected locally and there are indications that some marginal areas in Catalonia, largely on the Pyrenean fringes, are abandoned in winter (AI2, CWA), but little is known about any such movements. Occasional Crested Larks from elsewhere in Europe may enter the Peninsula to winter, but there is no firm evidence of this.

THEKLA LARK *Galerida theklae*

(Cogujada Montesina, Cotovia-escura)

G. t. theklae. **A common resident throughout mediterranean Spain, as well as in eastern and southern Portugal and in the Balearic Islands.**

Breeding population estimates. Spain: SA1 1.4–1.5 million pairs; ACRE 4.0 million birds. **Portugal**: 50,000–500,000 pairs.

The Thekla Lark is widespread throughout the thermo- and mesomediterranean and, to a lesser extent, the supramediterranean regions of the Iberian Peninsula, including most of the mountain ranges, but it is almost completely absent from the eurosiberian region (SA2). It may thus be encountered only to the south of the Cantabrian Range and the Pyrenees. It is especially frequent in the arid south-east: in Almería, Murcia and Alicante, as well as in the mountains of the east and south. Conversely, it is relatively rare in the northern half of the Peninsula, where it is often confined to especially rugged and dry localities. In Portugal it is most abundant in the drier east, with its greatest concentration in the far east of Trás-os-Montes, southeastern Baixo Alentejo and northeastern Algarve and it is almost entirely absent from the western half of the country north of the Taqus (PA2). Gaps in its broader Iberian distribution correspond to areas of unsuitable habitat, notably in the Guadalquivir valley, where it is replaced by the Crested Lark. It is also very common in the Balearics, where the Crested Lark is entirely absent, inhabiting all the principal islands.

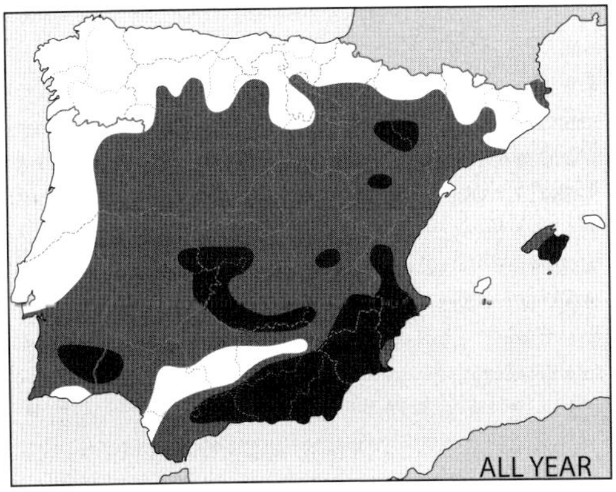

The Thekla Lark occupies different habitats from the

Crested Lark (Abs 1963; *Ardeola* 10: 81–82; Santos & Suárez 2005). In particular, it is a bird of Mediterranean scrub and hillsides covered with open bushy vegetation, notably Cistus, Kermes Oak, broom and esparto grass. It also occurs in open pinewoods, young pine plantations, dehesas, bushy steppe areas and in coastal sand dune habitats. All these combine areas of open ground and tall herbaceous or woody vegetation, on which the Thekla Lark often perches. The dehesas in which it occurs are those used for grazing: it avoids the cereal dehesas favoured by the Crested Lark. It is also absent from the agricultural habitats occupied by the latter. It frequently ranges much higher than the Crested Lark, reaching 1,400–1,500m in central Iberia and up to 2,500m in the Sierra Nevada (Garzón & Henares 2012).

It is largely sedentary and occupies its nesting habitats all year round, although those at higher altitudes may descend lower in winter. One ringed bird moved over 700km from the Ebro valley to the vicinity of the Strait but other Iberian recoveries have been local (SMA). *Galerida* larks recorded at Gibraltar, where none breed, are now all considered to be Thekla Larks: they occur there as winter visitors, a few individuals being present from October–March.

WOOD LARK *Lullula arborea*

(Alondra Totovía, Cotovia-dos-bosques)

***L. a. arborea*. A widespread resident or partial migrant, and winter visitor, in the northern half of the Peninsula. A rare vagrant in the Balearic Islands.**
***L. a. pallida*. A widespread resident or partial migrant in the southern half of the Peninsula.**

Breeding population estimates. Spain: SA1 560,000–1.3 million pairs; ACRE 2.2 million birds. **Portugal**: 50,000–500,000 pairs. **Andorra**: 20–75 pairs.

The Wood Lark is widespread throughout most of mediterranean Iberia, the Pyrenees and the whole of Portugal. It is also common in western and southern Galicia but is very local or absent across most of the northern side of the Cantabrian Range. The treeless areas of the Guadalquivir and central Ebro valleys also comprise significant gaps in its distribution (SA2, PA2). The nominate subspecies has a more northerly distribution than *L. a. pallida*. However, the boundary between the two is obscure and the subspecific status of Iberian birds is not agreed (AP, SA2).

The principal breeding habitat is open woodland, both coniferous or broadleaved, adjacent to pastureland. It also occupies plantations, dehesas, juniper woodland and tall scrub interspersed with clearings, as well as olive and almond groves locally. Dense forest, cultivated land and open country are avoided when nesting. However, flocks may forage on farmland in winter, generally close to woodland. Birds occupy a wide altitudinal range, from sea level to over 2,000m (AI2): up to 2,400m in the high pinewoods of the Sierra Nevada (Garzón & Henares 2012).

In Spain, the population on SACRE sites showed a 15% increase during 1998–2012 (Escandell 2013), which was most marked in Andalucía. The highest densities have been reported from the supramediterranean: in sessile oakwoods (16 birds/km^2), pasturelands (15 birds/km^2), matorrals (13–14 birds/km^2) and Holm Oak woodlands (12 birds/km^2) (ACRE). The Portuguese population showed no clear trend during 2004–09 (ACP).

Iberian Wood Larks are probably largely sedentary but altitudinal movements occur in which higher levels are abandoned in winter (Santos *et al.* 1983, Sánchez 1991, AI2, SWA). The Andorran population is absent between October and mid February (ANA). Some birds from elsewhere in Europe enter the Peninsula in winter and the occurrence of Wood Larks in the Biscay coastlands in winter may involve immigrants (SWA). There are single Spanish recoveries of birds ringed in Belgium, Switzerland and the former Czechoslovakia (SMA). One ringed as an adult at Montejo de Tiermes, Soria, in May 1987 was recovered at El Gastor, Cádiz (535km) the following November (*Ecología* 3: 213). Any passage is generally inconspicuous but migrants have been noted on the Asturian coast in September (Noval 1986). Few are recorded at Gibraltar itself but Tellería (1981) noted some passage across the Strait at Tarifa in autumn, in the second half of October and the first half of November, with 38 birds in 1976 and 133 in 1977. Returning migrants have been detected there in February–April (GONHS, Suárez *et al.* 2009).

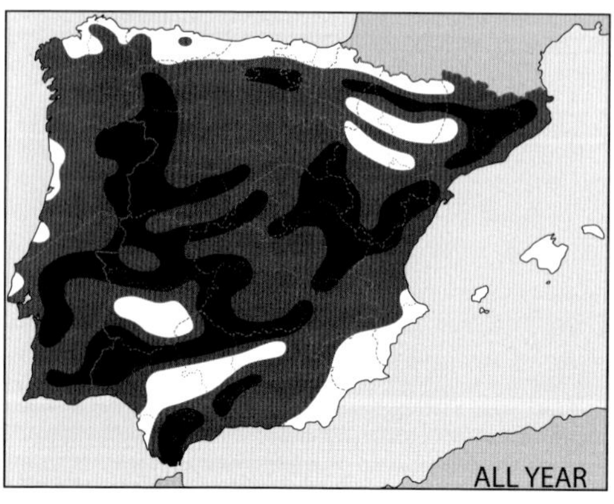

There are only a few recent records from the Balearics: a

flock seen off Ibiza in 1969 (Ferrer *et al.* 1986) and single birds in Mallorca in October 1995 (Clavell 2002); on Cabrera in October 2007 and October 2009, and on Dragonera in November 2009 (*Ardeola* 55: 303, 57: 540). However, Munn (1921, 1928) said that it occurred occasionally in Mallorca on passage and in winter and he reported an influx there during a cold spell in January 1926.

EURASIAN SKY LARK *Alauda arvensis*

(Alondra Común, Laverca-comum)

***A. a. guillelmi*. Common in northern Portugal and northwestern Spain.**

***A. a. sierrae*. Locally common in southern Portugal and in central and southern Spain.**

***A. a. cantarella*. Common in northeastern Spain.**

***A. a. arvensis*. An abundant winter visitor to Spain and Portugal. Also winters in the Balearic Islands.**

Breeding population estimates. Spain: SA1 2–6 million pairs; ACRE 3.1 million birds. **Portugal**: 1,000–10,000 pairs. **Andorra**: 500–750 pairs.

The Sky Lark is common as a breeding species in the northern half of Iberia but becomes progressively scarcer and more local in the south (SA2, PA2, Suárez *et al.* 2003). It is widespread throughout the eurosiberian zone, and in the supramediterranean across Castilla y León. It is only a local breeder in the southern half of Iberia, where confirmed breeding is chiefly reported from the higher elevations of the main mountain ranges. The Sky Lark does not usually breed in the hot lowlands of the thermomediterranean and is thus mostly absent from the coastal areas of the south and east.

It breeds commonly in pastures and mixed farms down to sea-level in the eurosiberian region and on western (and possibly southern) Portuguese coasts, (SA2, AP). It also nests on high level pastures with scattered scrub in the Cantabrian Range and in the Pyrenees, ascending to the subalpine zone: in Andorra it is found on montane moorland at 2,100–2,700m (ANA). However, it may also be found nesting in halophytic scrub both inland: as in the Ebro valley and Castilla-La Mancha, and in littoral areas: such as at the Tagus estuary, the wetlands of the Comunidad Valenciana and the Ebro delta (de Juana 1980, SA2, AP). Sky Larks in the supramediterranean of the northern meseta inhabit all manner of open land, including cereal crops and pastures and, especially, the páramos or high grasslands, up to about 2,000m. There are also breeding Sky Larks in croplands in Castilla-La Mancha, notably in Guadalajara and Cuenca, although they occur much more locally there than in Castilla y León and generally at higher elevations. Sky Larks (*A. a. sierrae*) are characteristic of the highest summits of the Betic Range in Andalucia: in the Sierra Nevada they occur at 900m in the cereal growing areas of the foothills but they range from 1,500–3,200m (Garzón & Henares 2012).

The highest densities in Spain are reported from the supramediterranean, in cereal-growing areas (39 birds/km^2) and in other open habitats including matorrals, field mosaics and pastures (14–28 birds/km^2) (ACRE). In Spain a significant decline was reported on SACRE sites during 1998–2012 (Escandell 2013). There are also indications of range contraction in Portugal since the 1980s, with a withdrawal from coastal areas (PA2).

Iberian Sky Larks flock in farmland in winter but generally move relatively modest distances. However, they are joined by very large numbers of birds of the nominate race. These enter through the Pyrenean passes and along northern coasts in September–December but the chief influxes are in October–early November. The wintering birds depart during February–March. The great majority of 98 Spanish recoveries of immigrants from further north have been in the Basque country/La

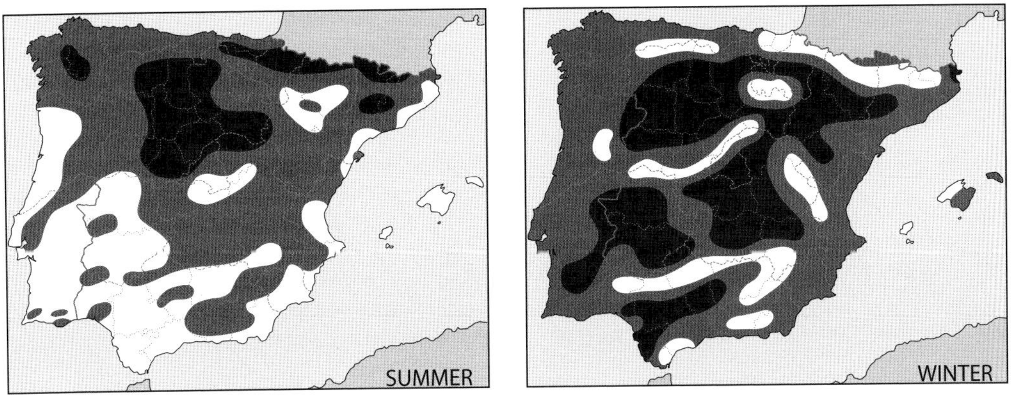

Rioja/Navarra and in the Peninsular south-west, and there are very few from eastern Spain (SMA). Ringing recoveries are chiefly of birds ringed in Belgium and France, largely on passage there (AMI, SMA), with a few also from Germany, the Netherlands, the Czech Republic, Sweden, Denmark, Finland and the Baltic States. There is also a Portuguese recovery in November of a bird ringed as a nestling in southeastern Scotland the same year (Wernham *et al.* 2002). One ringed at Milagro, Navarra, in December 1963 was recovered in Pskov, Russia, in July 1965 (*Ardeola* 13: 76), and one ringed at Tarifa in October 1992 was reported from Morocco in November 1993.

Diurnal Sky Lark passage is often conspicuous at the Strait of Gibraltar during the second half of October and the first half of November, when large numbers may be seen crossing to North Africa. Many others pass at night (Suárez *et al.* 2009). The return passage is less noticeable but occurs mostly from mid January to March, peaking in February. However, whether or not any Iberian-bred birds migrate to North Africa is unknown.

The total Iberian wintering population is unknown but is certainly of the order of millions of birds and was probably even larger in the past. Tait (1897) recorded 'immense numbers' in the vicinity of Porto from mid October and in winter. Both Saunders (1871) and Arévalo (1887) also considered the Sky Lark 'very abundant' in Spain and its former wintering numbers in Iberia must certainly have been huge prior to the massive declines suffered by the breeding populations in northern and central western Europe during the 20th century (BE, HBW). The largest winter concentrations occur in the major river valleys, notably those of the Duero, Tagus and Guadalquivir, as well as on the plains of La Mancha and southern Extremadura, and the Sky Lark is then comparatively scarce in Galicia, the Biscay coastlands and along the Mediterranean littoral (SWA). The largest wintering concentrations are seen on dry croplands, such as stubble fields, where recent densities in Spain have averaged 700 birds/km^2, much higher than in irrigated crops or pastures where mean winter densities are 30–60 birds/km^2 (SWA). However, a wide-ranging study in central and southern Spain has highlighted dry pastures as the principal wintering habitat there, with Sky Lark densities being 10–20 times greater than in any other habitat (Suárez *et al.* 2003). Some numbers winter in the Balearic Islands, notably in Mallorca and Minorca (SWA).

HORNED LARK *Eremophila alpestris*

(Alondra Cornuda, Calhandra-cornuda)

Vagrant to Spain.

A Swedish-ringed bird was reported from Almonte, Huelva, in October 1950 (Sáez-Royuela 1952), although some doubt has been attached to that record (Zink 1975). There are four accepted records of single birds. Three are from coastal Asturias: at El Musel harbour, Gijón, on 12 October 1994; at the Ría de Villaviciosa from 3 February to 16 March 2005, and at Bañugues beach, Gozón, on 21 October 2007. The fourth one is from the Pyrenees: at Baztán, Navarra, on 17 January 2010. One was at Cape Creus, Girona, in early May 2013 (RBS). European Horned Larks nesting in Scandinavia largely winter on the shores of the North Sea, where some hundreds regularly reach England. Very few winter in France, usually in the extreme north, although there are three records from the Camargue (OF).

SAND MARTIN *Riparia riparia*

(Avión Zapador, Andorinha-das-barreiras)

***R. r. riparia.* A widespread but local summer visitor to mainland Iberia. Common on passage throughout the Peninsula and Balearics.**

Breeding population estimates. Spain: SA1 540,000–750,000 pairs. **Portugal**: 10,000–100,000 pairs.

The Sand Martin nests throughout Spain and Portugal although it is most numerous in the northern half of Spain and the western half of Portugal (SA2, PA2). Its breeding colonies are concentrated within the major river basins, especially those of the Duero and Ebro. The birds frequently flock to feed over wetlands and irrigated crops, ricefields in particular (PA2). Nesting is mainly below 600m (AMI) but colonies have been found much higher where suitable sites permit: to 1,620m in the Cantabrian Range in León (García & Álvarez 2000). It does not nest in the Balearics but may have done so in Mallorca in the past (Ferrer *et al.* 1986).

Colonies are generally in sand or clay banks and so are mainly along water courses. However, the birds take advantage of other nesting opportunities, often artificial ones, and many nest in gravel workings in Madrid, Extremadura and elsewhere,

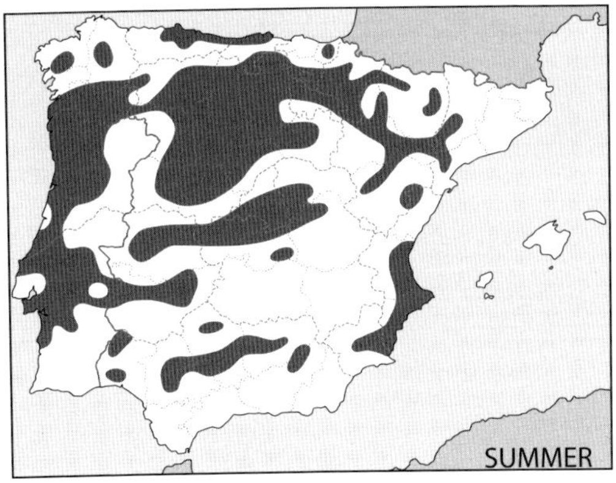

as well as in temporary sand heaps and holes in walls. Colony sizes vary from just a few pairs to several hundred. Two colonies at La Granja d'Escarp, Lleida, totalled almost 1,300 pairs in 1999 (Calvet *et al.* 2004). Juveniles and adults congregate with other hirundines after the breeding season at often large communal roosts, usually in reedbeds, prior to their southward migration.

The available regional counts: of 3,000 pairs at the most in Extremadura (de Lope *et al.* 1987), 3,000–5,000 pairs in Burgos (Román *et al.* 1996), some 800 pairs in Ourense (González & Villarino 1997) and around 2,700–4,500 pairs in Catalonia (CA2) suggest that the total population may be smaller than suggested by SA1. The population shows considerable fluctuations in size between years and these are often conspicuous locally, when they result from colonies being abandoned through human disturbance, as often occurs at gravel workings, or from new ones being established. SACRE (1998–2012) results show no clear recent trends in Spain. In Portugal there has been considerable range expansion in central regions and the coastal north since the 1980s, probably accompanied by a population increase (PA2).

The Sand Martin winters in the Sahel zone of sub-Saharan Africa. There are 16 recoveries from Senegal of birds ringed in Spain, as well as two there from Portugal, and six birds ringed in Senegal have been found in Spain (SMA, AP). There are also six recoveries in total of Spanish-ringed birds from Mauritania, Mali and Ghana, as well as single birds found in Algeria and Tunisia (SMA). A Portuguese-ringed bird was found in Ghana. The breeding population returns from mid-February to April, although few appear at inland sites before mid-March. There are occasional observations in December–January, especially in the south-west. Chapman (1888) saw a few near Vejer, Cádiz, in December 1887. Recent similar records include six at Ferreira do Alentejo on 17 January 2002; four at the Dehesa de Abajo, Sevilla on 27 January 2007, and four again at the Dehesa de Abajo on 30 December 2007 (*Anu. Ornitol.* 2: 40; *Ardeola* 54: 180, 55: 146), as well as a number of records of 1–2 birds. These may be early returning migrants rather than true winterers (SWA, AP). However, one of two ringed at La Puebla del Río, Sevilla, on 19 January 2012 was in active moult, suggesting possible local wintering (*Ardeola* 59: 445). Records of Sand Martins during the winter months are nonetheless rarer than those of other migrant hirundines.

Most of those that cross the Peninsula on passage originate from northwestern Europe, mainly Britain. As many as 73% of 372 extraregional birds recovered in Spain had been ringed in Britain, with 24% from France and the Low Countries and only a few individuals from elsewhere: Germany, Ireland, Norway, Denmark, Italy and Lithuania (AMI, Asensio *et al.* 1991, SMA). Recoveries of birds ringed in Spain show a similar pattern (SMA). Only four foreign-ringed birds have been recovered in Portugal, however, suggesting that most trans-Iberian passage occurs further east (AP). Migrant Sand Martins returning further north continue to cross the Peninsula throughout May and into early June: for example, over half the birds recovered in the British Isles were ringed in Spain in those months. It is also apparent that birds of foreign origin may settle to breed in Iberia in later years, showing that considerable interchange may occur between widely-separated colonies.

The autumn passage extends from August to October, with stragglers in November. Asensio *et al.* (1991) suggest on the basis of wing-length differences that most Iberian birds depart in August, those captured in September–October originating from elsewhere in Europe. Gatherings in July 2009 of 1,000 at the Albufera de Valencia and 1,000 over the Vila Franca de Xira ricefields in the Tagus estuary were presumably pre-migratory concentrations (*A. O. C. Valenciana 2009*: 107, *Anu. Ornitol.* 8: 79). Ringing recoveries show that autumn migrants enter the Peninsula at either end of the Pyrenees. The western flow is dominated by birds of British origin (Mead & Harrison 1979), which account for 75% of recoveries from this sector of the Pyrenees, but British and western French birds still comprise 35% of those using the eastern entry point (Asensio *et al.* 1991). In the south, Sand Martin passage from the Peninsula is clearly on a broad front and there is no great concentration at the Strait of Gibraltar during either season: it is the scarcest of the migrant hirundines there. Tellería (1981) counted only 793 birds in August–November 1977 at Tarifa, with a clear peak in passage at the end of September.

CRAG MARTIN *Ptyonoprogne rupestris*

(Avión Roquero, Andorinha-das-rochas)

A common resident or partial migrant throughout most of mainland Iberia and in Mallorca and Ibiza. A common winter visitor to other areas.

Breeding population estimates. Spain: SA1 84,000–100,000 pairs. **Portugal**: 5,000–50,000 pairs. **Andorra**: 3,000–4,000 pairs.

The Crag Martin nests throughout most of the Iberian Peninsula, except within the large tracts of the Meseta that lack suitable habitat. It is also largely absent as a breeding species from the Atlantic coastlands of Galicia and Portugal; from southwestern Andalucía, including most of the Guadalquivir valley and much of the Levant coast. Elsewhere it may be found nesting wherever there are suitable cliffs, gorges, cuttings or caves, or their man-made equivalents: such as ruins, dams, castles and bridges. It is seldom found nesting on inhabited human dwellings although it does so locally (*Ardeola* 7: 263–264, 9: 147–148, 19: 27–28, 20: 368–369; de Juana 1980). Some nest locally in sheltered positions on sea cliffs. They are especially common in the limestone mountains of northern and eastern Spain, which offer abundant nest sites. Breeding birds may be found at considerable altitudes frequently at over 2,000m and up to 3,000m in the Sierra Nevada (Pleguezuelos 1992, Garzón & Henares 2012). Everywhere they nest solitarily or in loose colonies of just a few scattered pairs. In the Balearic Islands they only breed in Mallorca and Ibiza, but they occur more widely in winter, including in Menorca (*A. O. Balears 2010*: 266).

Regional estimates in Spain include 7,200 pairs in Galicia, over 10,000 pairs in Burgos and 17,000–24,000 pairs in Catalonia (Guitián *et al.* 1980, Román *et al.* 1996, CA2). SACRE results show no significant trend (Escandell 2013). The Crag Martin is described as 'plentiful' in Andorra where the population is known to have increased since the mid 20th century, accompanying a switch from natural nest sites to nesting on buildings, as the latter became increasingly available (Eagle Clarke 1889, ANA).

Large numbers winter in Iberia. The populations of the Iberian periphery tend to be resident but those nesting at higher altitudes and in the colder central parts of the peninsula occur there during March– September and move to the coastal regions or south to Africa to winter. Wintering birds congregate in sheltered valleys in the northern and western coastal areas and in the warmest southern sectors of the Peninsula (Bernis 1961c, AP, SWA). They may often be seen sunning themselves on roofs and cliff faces and feeding low over sheltered areas or wetlands where flying insects may be found, often in small flocks but sometimes in considerable numbers: some 3,000 were counted over the Marjal de Almenara, Castellón, in January 2009 (*A. O. C. Valenciana 2009*: 108). A gathering of 1,000 at the Mata Nacional do Barão de São João, Algarve, in November 2008 (*Anu. Ornitol.* 7: 85) is also noteworthy. They roost at night in caves, sometimes in large numbers. A winter roost at Gorham's Cave, a sea cave in the south-east of the Rock of Gibraltar, was first noted by John White in 1770 (Holt-White 1901). Some 2,000–3,000 birds, exceptionally up to 4,000, were counted using this roost during the 1960s and 1970s (Elkins & Etheridge 1974, Cortes & Finlayson 1987, BSG). The birds spent the day feeding some kilometres away over the Spanish hinterland. Coastal erosion resulted in the progressive abandonment of this roost during the 1980s and only a few hundred birds have roosted in other Gibraltar caves since the 1990s. Roosting on buildings has also been reported from coastal areas in winter: for example about 1,500 were counted on waterfront apartments at Benicàssim, Castellón, in January 2009 (*A. O. C. Valenciana 2009*: 108). Wintering birds begin to arrive in Gibraltar in late September, with the majority arriving in October. They begin to depart in February but mainly do so in March and only a few stragglers remain by early April (BSG). Those wintering in Castellón and in littoral regions of central and southern Portugal also mainly arrive in October and depart in March (AP, Tirado 2011).

Some Crag Martins winter in Africa, chiefly in the north, although some have been noted as far south as Senegal (BWP).

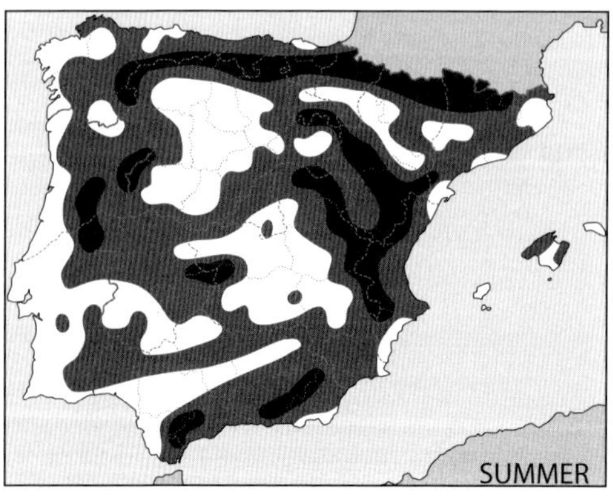

Migrants accompany other hirundines at the Strait of Gibraltar in October–early November but are not seen in large numbers. Tellería (1981) only counted 670 during this period in 1977. The Crag Martins wintering in Iberia clearly include some birds from other countries further north, although many of these winter in the Mediterranean coastal regions of France and Italy. Birds ringed at the winter roost in Gibraltar have been recovered in the breeding season in eastern Spain, the Balearics, the French Atlantic Pyrenees and the Italian Alps (BSG). Five ringed in Mediterranean coastal Spain in winter have been recovered in the breeding season outside the Peninsula: two in the French Pyrenees, two in Italy and one in Tyrol, Austria (SMA).

BARN SWALLOW *Hirundo rustica*

(Golondrina Común, Andorinha-das-chaminés)

H. r. rustica. **An abundant summer visitor throughout mainland Iberia and the Balearic Islands. Abundant on passage. Some occur in the south in winter.**

Breeding population estimates. Spain: SA1 783,000–812,000 pairs; ACRE 29.4 million birds. **Portugal**: 0.5–2.0 million pairs. **Andorra**: 15–20 pairs.

The Iberian Peninsula, with its warm climate, immense expanses of farmland, high levels of stock raising, numerous small towns and villages and an all too evident abundance of flying insects, offers a highly favourable environment for the Barn Swallow. This familiar species is accordingly present virtually anywhere in the region, from sea level to about 1,500m (AI2, SA2, PA2), and over 2,000m in the Sierra Nevada (Garzón & Henares 2012). Nearly all pairs nest on buildings; typically in farms, villages and small towns. Small colonies tend to form in suitable areas. There are still occasional birds that use the presumed ancestral habitat, nesting in caves (AMI), and nesting in trees has exceptionally occurred in Spain (*Ardeola* 15: 148, 38: 436). Few nest in the larger cities (Bernis 1988), perhaps because there are fewer insects there but also because these lack expanses of open ground for the low-level foraging typical of the species. In general, the Barn Swallow is more abundant where livestock is raised.

The ACRE estimate of the Spanish population is vastly greater than the SA1 assessment: future studies will no doubt establish the magnitude of the population more precisely in due course. However, the population on SACRE sites decreased significantly during 1998–2012 (Escandell 2013). The highest densities, 165–275 birds/km^2, are reported from villages and towns, and their peripheries, within the mediterranean climatic region (ACRE). Breeding season densities are especially high in Extremadura (92 birds/km^2) and the Mediterranean coastlands: in Catalonia, the Comunidad Valenciana and Murcia (74–82 birds/km^2). Conversely, they are relatively low in the humid northern coastlands: in Cantabria, Galicia, the Basque Country and Asturias (11–35 birds/km^2) (ACRE).

The Portuguese population shows no clear overall trends (AP, ACP). The small population of Andorra may be of relatively recent origin since Clarke (1889) was unable to find any there at that time. The few pairs that bred in Gibraltar last did so in the early 1970s (BG).

Iberian Barn Swallows winter in tropical West Africa, chiefly around the Gulf of Guinea, where there are ringing recoveries from Ghana, Ivory Coast, Benin, Togo, Cape Verde and, especially, Nigeria (de Lope 1984, SMA, AP). Of 406 recoveries of birds ringed in Iberia (including through-migrants) – 396 in Spain and ten in Portugal - 254 were in other European countries, 57 in North Africa: 86% in Morocco, and 95 in sub-Saharan Africa south to South Africa: 54% in Nigeria (SMA). Great numbers of birds from other West European countries cross the Peninsula twice-yearly on passage (Zwarts *et al.* 2009). Ringing recoveries from Iberia are chiefly of birds originating in the westernmost European countries: France, the British Isles, Belgium, Germany and the Netherlands, with some also from Scandinavia. Foreign recoveries of Iberian-ringed birds are also predominantly from this region but they also include a few from as far east as Hungary (3), the Czech Republic, Slovakia, Croatia and Greece. Very large roosts of hirundines, mainly Barn Swallows, form after the breeding season and during the southward migration period. The largest roosts are in reedbeds and these may hold many thousands of individuals. Such a roost at the Laguna de Encanyssada in the Ebro delta on 3 October 2012 was estimated to number 1.5 million birds (*Ardeola* 60: 181). Tall, uniform crops, such as maize, are also sometimes used as roosts by migrants (EG, pers. obs).

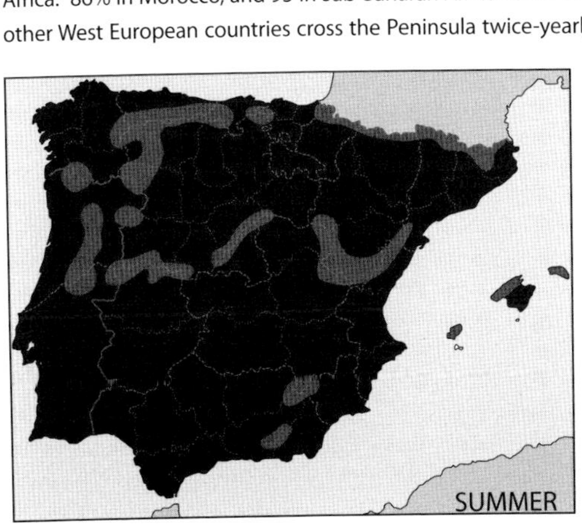

The Barn Swallow is described as a summer visitor to Iberia, and indeed to Europe. However, some may be seen in parts of Iberia all year round, even in December and January, chiefly in the thermomediterranean regions of the south, where they have been recorded in some numbers in winter. Winter records are generally from the vicinity of wetlands, including ricefields (SWA). Fieldworkers for the Spanish winter atlas recorded 390, 290 and 940 individuals in December–January in Spain during the three winters of the project, the lowest count coinciding with the coldest winter, in 2008–09 (SWA).

Winter records of single Barn Swallows, or of groups of up to 4–5 birds, are quite frequent in Andalucía and Extremadura, and also in Mallorca in some years, and there are winter reports from as far north as A Coruña and Logroño. A count of 15 at Santiago de Compostela in late January 2002 (*A. O. Galicia 2002*: 119) is particularly noteworthy for such a northerly location. Observations of flocks in December–January are most frequent in the south-west, generally at wetlands, but these seldom number more than a few dozen individuals. Thus a record of 200 at Doñana on 20 December 2010 (*Ardeola* 58: 214) is noteworthy and one of 800–1,000 at Puebla del Río, Sevilla, on 13 December 1997 (*Ardeola* 45: 251) is exceptional.

Winter month observations are not new: Arévalo (1887) noted that some appeared in Málaga in late January during mild winters, Tait (1887) saw some near Porto – where they usually arrived in early March– on 13 January 1878, and Chapman (1888) saw Barn Swallows on several occasions near Vejer, Cádiz, in December 1887 and winter 1888. Nevertheless, it has yet to be established whether any individual Barn Swallows spend the whole winter in Iberia. It seems more likely that most or all of those seen in late December and January are actually individuals that have returned early to the Peninsula, having already migrated to and from more southern winter quarters, where they moulted. Since some Iberian birds migrate south in July– August they have plenty of time to moult and return by December. Recent information from southern Iberia (A. Onrubia, pers. comm.) hints that the situation may be even more complex than suspected. None of several Barn Swallows captured in southern Spain in January 2009 and January 2010 were in moult but one of five ringed at La Puebla del Río, Sevilla, on 19 January 2012 was in active moult (*Ardeola* 59: 445) and others captured in northern Morocco in January were in full moult. Moreover, up to 20% of birds captured in Cádiz in September were in full moult and at least four were captured replacing the outermost primary, suggesting that they had more or less completed their post-breeding moult by then.

The passage of Barn Swallows across Iberia is protracted, in any event. Migrants can be seen arriving from Africa at the Strait of Gibraltar from late January and many nests in Andalucía are reoccupied during February. Passage at Gibraltar continues throughout March–May, with occasional stragglers into early June. The later birds are presumably those travelling to northern Europe. Migration through the Peninsula is most apparent on the eastern and southern coasts in autumn since many follow these coastlines south to the Strait of Gibraltar, where they cross to Morocco (AMI, AI2). The return passage at the Strait in spring is also marked but gives the impression of being less concentrated (BSG).

Some first broods of southern Iberian Barn Swallows may fledge as early as late March or early April and second and third broods are normal. The latest broods may not fledge until late October (de Lope 1983). It is not surprising therefore that southward passage at the Strait is quite evident from mid-July: there is even one record of a flock of 15 juveniles flying south at Gibraltar on 21 June 1969 (EG, pers. obs). The southward passage continues throughout August and noticeably increases in volume during September, the largest numbers passing during late September–first half of October. Very large numbers have been counted at the Strait in autumn, although still only a fraction of the source populations. Tellería (1981) recorded over 150,000 birds during August–November 1977, with peak passage in late September and the first half of October. Passage declines sharply after mid-October but there are stragglers through November and, occasionally, into December.

RED-RUMPED SWALLOW *Cecropis daurica*

(Golondrina dáurica, Andorinha-dáurica)

***C. d. rufula*. Widespread and locally common in the southern two-thirds of Iberia, very local elsewhere. Rare in the Balearic Islands. Summer visitor.**

Breeding population estimates. Spain: SA1 9,600–13,200 pairs; SA2 22,400+ pairs; ACRE 3.8 million birds. **Portugal**: 10,000–100,000 pairs.

The Red-rumped Swallow is now such a constant feature of mediterranean Iberia that it might be assumed to be an established long-standing component of the avifauna. In fact it is a relatively recent colonist from North Africa. Arévalo (1887) refers to one having been obtained in Valencia and noted that it had been recorded at Casarabonela, Málaga, perhaps nesting there. The first confirmed report of nesting in Iberia was however by Stenhouse (1921), who saw one at La Janda in Cádiz province, Andalucía, in April 1919 and found a pair collecting mud in the same general area on 2 May 1920: he shot the male (Jourdain 1936–1937). Verner claimed to have seen one in the La Janda area some years earlier (Stenhouse 1921). Thereafter a number of nests were

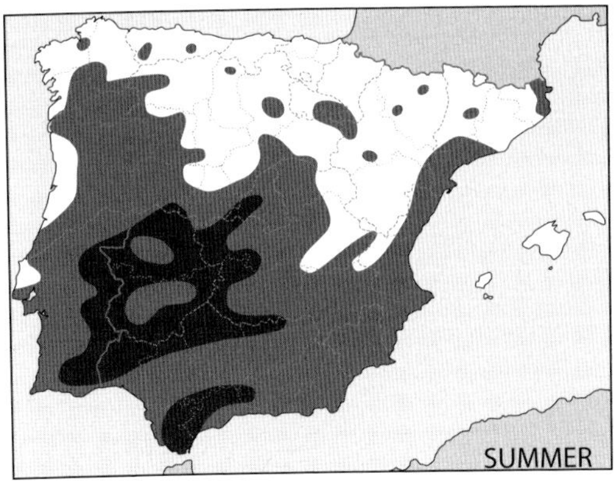

found in the Almoraima, Cádiz, in 1938 (Bridges 1938). Further reports seem to have been lacking until the early 1950s, by which time birds were established in the southern half of Spain as far north as Ciudad Real (Bernis 1952). Breeding birds were found at La Pedriza, Madrid, in 1953 (Bernis 1954b). The first confirmed breeding record in Portugal was in 1951 between Portel and Amiera, Évora, near the Spanish border (*Ardeola* 3: 305) but the Red-rumped Swallow only began to become widespread in the country during the 1960s (AP). By 1959 it was already breeding in the Duero region both in Portugal: near Mogadouro and at Barca d'Alva (Santos Júnior 1959) and in Spain: at the Arribes del Duero and reportedly also at Ricobayo, Zamora (Bernis 1960b). At the same time it had reached as far north on the Mediterranean littoral as Catalonia, where five nests were found under a bridge at Tossa de Mar, Girona (Mertens & Stehle 1960).

Range expansion in Spain was widely apparent during the 1960s and 1970s (de Lope 1981) and is still progressing, although slowly. The first nests were found in Aragón in 1974, in La Rioja in 1979 , in Galicia in 1987, in Asturias in 1999 and in Palencia in 2000 (*Ardeola* 28: 161, 35: 312, *A. O. Asturias 1999*: 135, *A. O. Palencia 1998–2001*: 142; de Juana 1980). The Basque Country and Cantabria remain the only two regions where breeding has not yet been proved, although there are several observations there (*A. O. Cantabria*, *Ardeola* 59: 188). Nevertheless, the Red-rumped Swallow remains absent from most of the northern humid belt, across the Cantabrian Range and the Pyrenees, and its range is still highly discontinuous in the northern meseta, the eastern part of the southern meseta, the Ebro valley and the coastlands of the Comunidad Valenciana and Catalonia (SA2). It is a scarce migrant in the Balearics (*A. O. Balears 2010*: 266) but there are three breeding records from Mallorca: a nest with chicks was found at Andratx in 2007 and again in 2011, and a pair nested at Manacor in 2012 (*A. O. Balears 2007*: 85–89; *Ardeola* 59: 446, 60: 536).

Red-rumped Swallows nest on cliffs and in gorges, under overhangs or in caves. They readily adopt man-made, cave-like structures, bridges and culverts in particular, as well as 'rocky' edifices such as ruins and castles, but seldom nest on occupied buildings. Nests are in areas that offer some open country nearby for foraging, notably all kinds of agricultural land. They can be found from sea level up to about 1,500m, and to 1,600m in the Sierra Nevada (Garzón & Henares 2012). They are not gregarious when nesting and pairs occur solitarily. Nonetheless, small flocks are often observed during passage periods and larger gatherings may occur after breeding. A count of 218 at the Serra de Monchique, Algarve, in June 2008 (*Anu. Ornitol.* 7: 85) was exceptional.

Population increase has undoubtedly accompanied range expansion but there is virtually no detailed local data available. However, in Catalonia the 50 pairs found at the time of the first breeding bird atlas in the early 1980s had increased to 217–361 pairs 20 years later (CA1, CA2). It is inconceivable that both the earlier national estimates and the vastly higher ACRE assessment could be explained by population growth so the actual size of the existing Spanish population must be regarded as unconfirmed. The population on SACRE sites showed no significant trend during 1998–2011 although there were indications of a continuing increase in numbers (Escandell 2012). By far the largest fraction of the Spanish population is in Andalucía (67%), followed by Extremadura (11%) and the Comunidad Valenciana (7%). The highest densities are reported from the thermomediterranean, in the vicinity of Cork Oak and Holm Oak woodlands (30 birds/km^2), olive groves (23 birds/km^2) and matorrals and farmland (15–20 birds/km^2) (ACRE). The Portuguese population is still expanding its range (AP), although it showed no clear trend during 2004–09 (ACP).

Red-rumped Swallows, like Barn Swallows and House Martins, begin to return from their winter quarters very early in the year. The breeding season in Extremadura extends from April to September (de Lope 1980). Passage is readily observable at the Strait of Gibraltar from early February and continues into May, with a peak in April (BG). The return passage is from mid August to October, peaking in late September and the first half of October, with some stragglers passing during November. The species winters in sub-Saharan Africa but the precise winter quarters are unknown: the sole recovery of an Iberian-ringed bird was from southern Morocco.

Some are reported each year in December and January, chiefly in southern Spain, but seldom in any numbers. Records of 100 on 2 December 1997 and 20 on 10 January 1998 in Sevilla (*Ardeola* 45: 251) are exceptional. A count of 12 over a lake at Loulé, Faro, on 28 December 2003 (*Anu. Ornitol.* 3: 42) is still noteworthy. There is also an observation of 3–4 at the Cecebre reservoir, A Coruña, on 20 January 2002 (*A. O. Galicia 2002*: 97). Most such birds seem likely to be early returning migrants rather than genuine winterers, as discussed for the Barn Swallow (q.v.). Nevertheless, a small group attempted to overwinter in the main square of Trujillo, Extremadura, where the 20 birds present on 20 November 2011 had dwindled to just a sole sickly individual by early February 2012 (*Ardeola* 59: 188).

HOUSE MARTIN

Delichon urbicum

(Avión Común, Andorinha-dos-beirais)

***D. u. meridionale*. An abundant summer visitor. Smaller than the nominate race but not otherwise distinguishable.**
***D. u. urbicum*. Abundant passage migrant.**

Breeding population estimates. Spain: SA1 2.14–2.16 million pairs; ACRE 8.7 million birds. **Portugal**: 100,000–1 million pairs. **Andorra**: 3,000–5,000 pairs. Gibraltar <10 pairs.

The House Martin is extremely numerous in Iberia, nesting throughout the region from sea level to over 2,500m (SA2, PA2). It is highly gregarious and nests are built clustered together. Quite small village houses may harbour 20–30 nests and large colonies, as occur for example on some dams in Extremadura, may have 600 nests or more. There is a count of 1,520 nests, not all of them necessarily occupied, under the arches of the bridge over the Guadiana at Medellín, Badajoz, on 16 May 1991 (*Ardeola* 38: 346). Birds nest on all kinds of edifices, within cities as well as in towns, villages and on isolated buildings. Unfortunately, some colonies are illegally destroyed by building owners, including local administrations. Many Iberian birds still nest locally in natural sites, in caves and under overhangs in gorges and on cliffs. Cliff-nesting was the norm in Andorra in the late 19th century (Clarke 1889) but all Andorran birds now nest on edifices (ANA).

The largest regional subpopulations in Spain are those of Extremadura and Andalucía. The highest densities are reported from villages and towns, and their peripheries, in the Mediterranean climatic zones (76–138 birds/km^2). SACRE results have shown a small decline in Spain during 1998–2012 (Escandell 2013). The Portuguese population showed a moderate increase during 2004–09 (ACP).

House Martins cross the Peninsula on migration in very large numbers but their movements are generally less obvious than those of the Barn Swallow, perhaps because they tend to fly higher and are less disposed to follow coastlines. Nevertheless, many are observed crossing the Strait of Gibraltar during both seasons. The spring passage is principally from early February until May, with an extended peak from mid-March to mid-May (BG, Onrubia *et al.* 2009). A few early migrants are seen arriving from Africa at the Strait in December–January and some stragglers may still be seen flying north in June. The return passage is from late July until November, with a peak in October.

There are only some 20 ringing recoveries within the Peninsula: of birds ringed in France, the British Isles, Germany, The Netherlands and Belgium. Only four birds ringed in Spain have been recovered abroad: in France, Ceuta and Morocco (2). Very large gatherings of House Martins are reported infrequently: examples include 25,000 at the Azud de Riolobos, Salamanca, on 11 September 2001 and 20,000 at Doñana, on 15 April 2006 (*Ardeola* 49: 192, 54: 180). The winter quarters of Iberian birds are unknown but House Martins in general winter in tropical and southern Africa, where they lead a highly aerial existence comparable to that of the Common Swift, seldom being observed there in large numbers (Wernham *et al.* 2002).

House Martins are regularly observed in the Peninsula in December–January, especially in the south-west, although there are occasional winter sightings from central Iberia and in the Balearics. Most records are of small groups, more rarely flocks of up to 30, so the count of 73 at the Dehesa de Abajo on 3 January 2008 (*Ardeola* 55: 147) is noteworthy. The Spanish winter atlas found a total of *c.* 600 individuals during the three seasons of the project, although only 38% were in December–January, the remainder being recorded during the first half of February, when most were likely to have been newly arrived migrants (SWA). There were also ten December–January records from as far north as Galicia during 2002–04, including one group of at least 13 at the Cecebre reservoir, A Coruña, on 27 December 2004 (*A. O. Galicia 2004*: 101). These winter occurrences, as with those of other hirundines, most probably involve Iberian birds that have returned early to their nesting grounds. Some nest sites in Extremadura and western Andalucía are already visited by birds in December–January but laying is unlikely before late February at the earliest. A well-studied colony in Badajoz proved to be triple-brooded but the laying dates of the first clutches ranged from the beginning of March to the end of May (Pajuelo *et al.* 1992). House Martins further north in Iberia do not start laying until much later, in early May in parts of Catalonia (Mestre 1980).

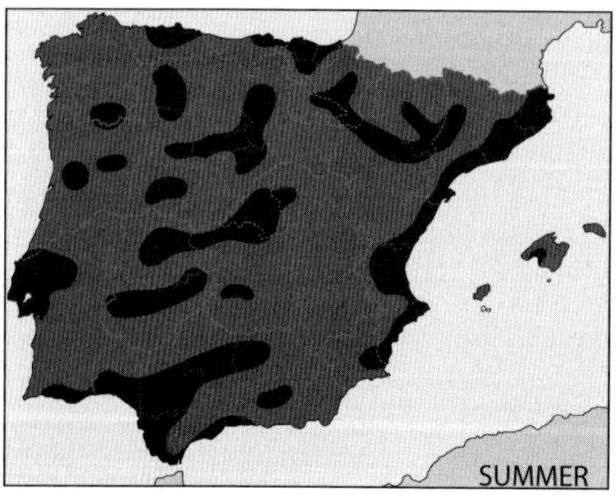

RICHARD'S PIPIT *Anthus richardi*

(Bisbita de Richard, Petinha-de-richard)

A very scarce migrant and wintering species.

The nominate form nests from southern Siberia and northeastern Kazakhstan south to Lake Baikal and winters mainly in the Indian subcontinent, but small numbers winter locally further west: in Europe, the Middle East and northwestern Africa. It has become clear that it occurs regularly on passage and in winter in many European countries, including those in Iberia (Gilroy & Lees 2003, Grussu & Biondi 2004). In France, for example, it is now known to winter regularly in very small numbers in the Camargue and occasionally elsewhere. Italy has an estimated winter population that averages at least 52 birds (Biondi *et al.* 2004). It was formerly regarded as a vagrant to Morocco but it now appears to be a scarce but regular winter visitor to the Atlantic coast. There are also four accepted records from the Canary Islands.

Numerous Iberian records, some dating back to the 19th century (Saunders 1871, Arévalo 1887, Irby 1895, Tait 1924, Bernis 1971a), indicate a regular presence of Richard's Pipits in coastal regions such as the Basque Country (Noval 1967), southern and central Portugal (Moore & Elias 1997), the Strait of Gibraltar (Tellería 1981, BSG) and Málaga (Garrido *et al.* 1985). The Spanish rarities committee considered the species until 2005, by when there had been 79 records (217 birds) from mainland Spain and the Balearic Islands. The Portuguese committee accepted 35 records to 2001, since when further Portuguese records have been published in the *Anuário Ornitológico*. There are also two accepted records of single birds for Gibraltar.

Figure 178. Regional distribution of records, 1989–2005 (n = 141)

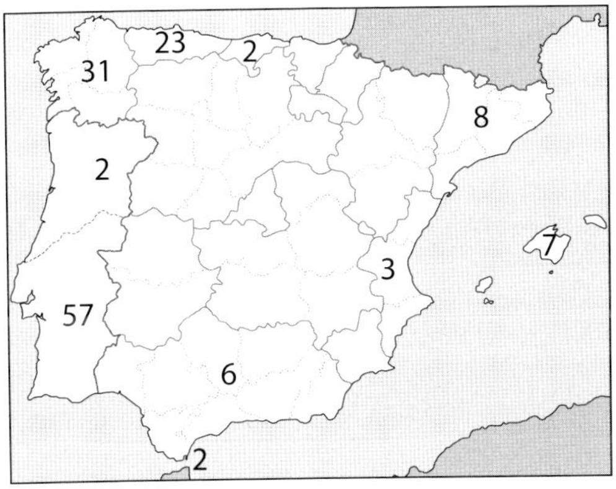

For purposes of analysis we have considered 141 pre-2005 records, involving 374 birds, excluding observations antedating the establishment of the rarities committees. Some 46% of observations were of single birds and 23% involved pairs but there were 22 instances of groups of five or more together: including five flocks of at least ten. The largest group on record falls outside the period sampled: there were 24 at Cabo Espichel, Setúbal, in October 2007 (*Anu. Ornitol.* 6: 66). The ages of the birds were seldom reported but first-year birds predominated where this was the case: there were 47 juveniles/first-winters but only 11 adults reported during October–January.

Observations are widespread from coastal locations around the Peninsula but are concentrated on the Atlantic coastlands of Asturias, Galicia and Portugal (Fig. 178). There are several from the Balearics. Inland records are virtually unknown: we are aware only of a record of two at Valdehornillos, Badajoz, in January 1977 (*Ardeola* 30: 122).

Figure 179. Monthly distribution of records, 1989–2005 (n = 141)

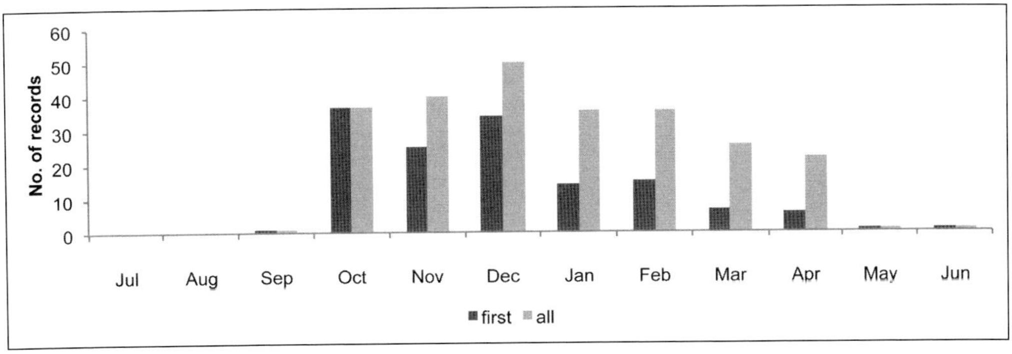

The temporal distribution of records indicates wintering, with 86% of new records in October–April, peaking in December (Fig. 179). Birds are generally at known wintering areas during October–April (de Juana 2006). An initial peak in new records in October presumably indicates arrivals and through-passage. However, there is no equivalent spring peak: five of the six April records of new birds are from the Mediterranean coast, which otherwise accounts for only 12.5% of Iberian observations. A comparison with neighbouring regions makes it clear that spring records become more frequent further east, suggesting a loop migration. Thus, although 97% of British records are in autumn (Dymond *et al.* 1989), as are 80% of those in France (OF), 17% of Italian records are in spring and only 4% are in autumn there (Biondi *et al.* 2004). Most Richard's Pipits found in Greece appear in April–May (Handrinos & Akriotis 1997). It thus seems apparent that the autumn passage is chiefly along western European coasts with the return spring passage occurring further east (Mitchell & Young 1997, Gilroy & Lees 2003, de Juana 2006).

The available records suggest that wintering numbers are very small, probably seldom exceeding 100 birds in total and with fewer in some years. Regular wintering is always at coastal locations, usually on pastures and rough grassland. Such sites include certain headlands in Asturias: such as Cape Peñas, Cape Vidio and Punta de la Romanela, and in Portugal, at Cape Espichel and Sagres. The vicinity of coastal wetlands also attracts birds to such places as the Lagunas de Traba and Vixán, in A Coruña; the Santo André and Salgados lagoons, in Setúbal and Faro respectively, and the Tagus and Sado estuaries. In the Mediterranean a wintering presence was formerly known at the Guadalhorce estuary, Málaga, and at the Albufera de Mallorca in some seasons.

BLYTH'S PIPIT *Anthus godlewskii*

(Bisbita Estepario, Petinha-de-Blyth)

Vagrant to Spain and Portugal.

In Portugal, an adult was at the Lagoas de Malhão, Beja, on 4–9 March 2003 and one was on Berlenga Island, Leiria, on 12–13 October 2010. In Spain, one was on coastal pastures at Tapia, Asturias, from 26 November to 12 December 2005. It is a very rare Asian vagrant to Europe, mainly in October or November, with 22 records for Britain and 11 for France.

TAWNY PIPIT *Anthus campestris*

(Bisbita Campestre, Petinha-dos-campos)

***A. c. campestris*. A summer visitor to large areas of Spain and Portugal, and to the Balearic Islands.**

Breeding population estimates. Spain: SA1 400,000–640,000 pairs; ACRE 895,000 birds. **Portugal**: 1,000–10,000 pairs.

This is the only pipit that breeds regularly in the southern half of the Iberian Peninsula but its range there is fragmented. It is predominantly a species of the supramediterranean and as such occurs across most of the northern meseta. It is largely absent from the eurosiberian zone, including the Pyrenees and most of the northern side of the Cantabrian Range, but there are small populations in southern parts of Galicia, Asturias and Cantabria, and along the Biscay coastlands. It is common in the mesomediterranean in the Ebro valley but more dispersed across the southern meseta and Extremadura. It is generally scarce or absent from Andalucía except from the upper levels of the Betic Range and eastern Sierra Morena. However, there is also a good population near sea level in Cádiz province, adjacent to the Strait of Gibraltar. The majority in the ACRE survey were found in Aragón (31.5%) and Castilla y León (24.4%). The Portuguese population avoids the Atlantic coastlands except in the south, where there are breeding birds in western Alentejo and western Algarve (PA2). They also breed in central eastern Portugal and in the north-east. Breeding birds occur on all the main Balearic Islands, being particularly evident in Menorca (SA2, *A. O. Balears 2010*: 266).

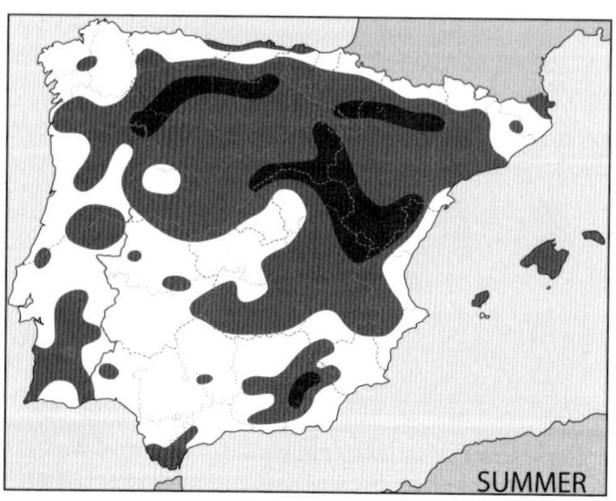

They are especially typical of the páramos and dry mountain areas of the supramediterranean. They also inhabit expanses of low scrub – of juniper, thyme or cushion-heath - and agricultural land, favouring set-aside and fallow areas in the latter (Tellería *et al.* 1988). Breeding birds occur on subalpine pastures to over 2,000m on the southern flanks of the Cantabrian Range and Central Ranges and reach 2,800m in the Sierra Nevada (Garzón & Henares 2012). Pastures grazed by sheep are especially attractive to them (de Juana 1980). Those of the Cádiz coast inhabit dry pastures with scattered thistles and other herbaceous plants. The highest densities recorded are from matorrals in the supra- and mesomediterranean (8–18 birds/km^2, SACRE).

The population showed no significant overall trend on SACRE plots during 1998–2012 (Escandell 2013). In Portugal there has been some range contraction in Algarve and Estremadura, perhaps in response to land use changes, although there may have been some range expansion in the north-east, in Beira Baixa and Beira Alta, and in the interior of Alentejo (PA2).

Tawny Pipits winter in the Sahel and the northern savanna belt of sub-Saharan Africa (Borrow and Demey 2001). Some return to the Peninsula in late March but most do so during April–May (BG). They depart from mid-August to mid-October, the peak passage in the south being in September (BG). Four at Cape Peñas, Asturias, on 22 November 1993 (*Ardeola* 41: 200) were very late. Winter records are exceptional but birds have been noted at the Lagoa de Santo André, Setúbal, in December, January & February (*Airo* 2: 21, *Airo* 4: 40; Moore & Elias 1997) and one was at Castro Marim, Faro, on 17 February 2009 (*Anu. Ornitol.* 8: 80). There are also records of at least three birds at Las Amoladeras, Almería, on 29 November 1996 and one at Boada de Campos, Palencia, on 4 December 2000 (*Ardeola* 46: 160, 49: 192).

Migrants are most apparent when small flocks settle along coasts but in general the passage is inconspicuous. Small numbers are reported annually on the north Iberian coast. Diurnal passage of small flocks is sometimes noted at Gibraltar but no large movements have been observed at the Strait. Individuals ringed in France and Switzerland have been recovered in autumn in Andalucía (AMI).

OLIVE-BACKED PIPIT *Anthus hodgsoni*

(Bisbita de Hodgson, Petinha de Hodgson)

Vagrant to Spain and Portugal.

The first accepted record was of one on the Formentor peninsula, Mallorca, on 10–11 October 1990. The other Spanish records comprise one ringed at Doñana on 13 December 2000, two seen at A Guarda, Pontevedra, during 16–25 December 2001, and four single birds on the islet of Cabrera: one during 19–26 October 2007, and the others on 4, 5 and 22 October 2008. Three of the Cabrera birds were caught and ringed and one was a first-winter individual. One ringed at Ria de Alvor, Faro, on 16 November 1994 is the sole accepted Portuguese record so far. Post-2010 there was a second Portuguese record: at Peniche, Leiria, in October 2011 (RBP), followed by six more Portuguese and three Spanish records of single birds in October–December 2012, when there seems to have been a small influx (RBS). *A. h. yunnanensis* is a frequent vagrant to northern and northwestern Europe, chiefly in autumn and especially in October, with 349 recorded in Britain and 18 in France. One was in Morocco in November 2007 and another in the Canary Islands in April 2011.

TREE PIPIT *Anthus trivialis*

(Bisbita Arbóreo, Petinha-das-árvores)

A. t. trivialis. **A common summer visitor, chiefly to northern Iberia. Common generally on passage.**

Breeding population estimates. Spain: SA1 300,000–400,000 pairs; ACRE 347,000 birds. **Portugal**: 50–100 pairs. **Andorra**: 750–1,000 pairs.

The Tree Pipit is characteristic of the eurosiberian climatic region and it breeds commonly in Galicia and the Biscay coastlands, and in the Cantabrian Range and Pyrenees. It also inhabits the supramediterranean in the northern Iberian Range, from where it has perhaps spread to the Sierra de Guadarrama and other points of the Central Range. The breeding range extends into northernmost Portugal in Minho, Tras-os-Montes and Alto Douro (PA2). It is generally found below 1,200m but it breeds to the treeline at 1,700m locally in the Cantabrian and Iberian Ranges. It does not nest in the Balearics but is a common migrant there.

The presence of nesting Tree Pipits in the Iberian Range was first confirmed relatively recently (de Juana 1980). It was later confirmed in the Central Range, at several points in Segovia, in 1976 and 1977 (*Ardeola* 23: 239) and thereafter in the Sierra de Ayllón and Sierra de Guadarrama. However, Castellarnau (1877) said that it was quite common from late March to October at some areas at San Ildefonso, Segovia, on the northern slopes of the Sierra de Guadarrama, nesting in grassy meadows, so it may well have been overlooked in the region during the intervening period. It was noted further west, in the Sierra de Gredos

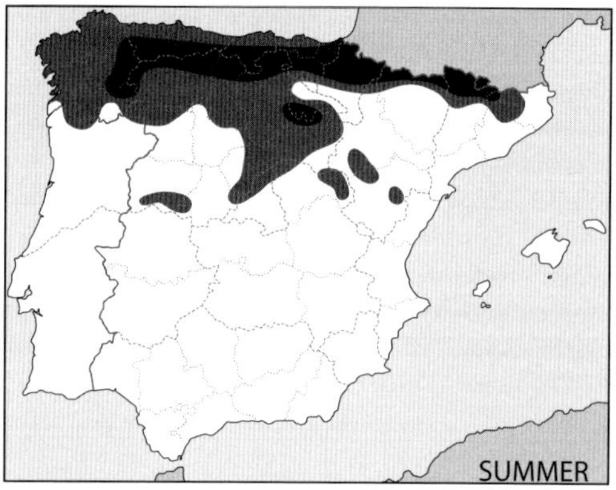

and the Sierra de Béjar, Salamanca, in the late 1980s (Carnero & Peris 1988) and has since spread further west to the Sierra de Gata. Over half of the birds in the ACRE survey were in Castilla y León.

It selects open woodlands and woodland clearings, with intervening low scrub, as well as heathland or pastures with scattered trees to serve as song posts. It seems particularly attracted to clearings in Pyrenean Oak and Scots Pine woodlands. It has also been noted in riparian poplars (de Juana 1980, AI2). The highest densities, 6–13 birds/km^2, are in eurosiberian matorrals and pastures with trees (ACRE). There are recent indications of some southward and westward range expansion in Portugal (PA2).

Tree Pipits from northern Europe cross the region on migration. There are a few recoveries of birds ringed in Belgium, the British Isles, France, the Netherlands, Norway and Sweden. One ringed in Hungary in June 1998 was recovered in Girona in April 2004 and there is a May recovery from Ivory Coast of one ringed in Mallorca the previous October (*Ecología* 21: 274). Spring arrivals in southern Iberia are evident from mid-March to May, with a peak passage in April. The return movements are from late August to October, with a peak in September and stragglers into November (AMI, BSG). One at Évora on 1 December 2008 (*Anu. Ornitol.* 7: 85) was exceptionally tardy. Visible passage is generally on a small scale but flocks of up to 60 have been recorded on occasion at Gibraltar (BG).

MEADOW PIPIT *Anthus pratensis*

(Bisbita Pratense, Petinha-dos-prados)

***A. p. pratensis*. A very scarce local breeding species in northern Spain. Abundant in winter and on passage throughout the Peninsula and in the Balearic Islands.**
***A. p. whistleri*. A winter visitor in small numbers.**

Breeding population estimates. Spain: SA2 50–100 pairs.

The Meadow Pipit has only a marginal presence as a breeding species in Iberia, where nesting was first proved in Asturias in 1972, and then reported again in 1981 (Noval 1986, Álvarez-Balbuena *et al.* 2000). Its spring presence in Cantabria was also reported by de La Lama (1976). Since then there are indications of a very small breeding population in Asturias, northernmost León and Cantabria, and also in Galicia, in pastures and heathland on the northern slopes of the Cantabrian Range (SA2). The breeding population may be of the order of 50–100 pairs (SA2) but its actual size and distribution have yet to be established fully. There is no confirmation of nesting in Andorra (ANA).

In contrast, the Meadow Pipit is one of the most abundant of the passerine species that winter in Iberia, and many pass through to and from North Africa. The wintering population is certainly of the order of many millions of individuals. There is an estimate of 1.25–2.25 million wintering birds in Catalonia alone, a region that comprises just 5.5% of the Iberian Peninsula (CWA). There are some 1,300 recoveries of birds ringed outside the region (SMA, AP) and these are chiefly of birds ringed in the British Isles, Belgium, France, Holland, Germany, Finland and Scandinavia with others from Iceland (20+), the Baltic States, Poland, Switzerland, Italy, the Czech Republic, Romania and Russia (23). Most of the many recoveries from Belgium, The Netherlands, Germany and France are of birds ringed on passage (AMI), so it appears that the breeding range of most of those wintering in Iberia extends from Iceland and the British Isles to Scandinavia and Finland. Although most British and Irish Meadow Pipits winter in Iberia (Wernham *et al.* 2002) the extent to which *A. p. whistleri* travels to the region is not known and there are only a few ringing recoveries to confirm its presence there. Birds ringed in Iberia have been recovered in the same regions of northern and central Europe and one has been found in Morocco.

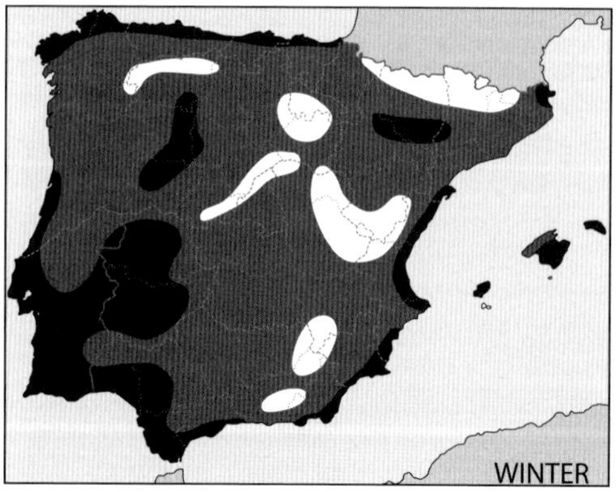

Wintering Meadow Pipits flock in humid pastures, grasslands and irrigated croplands across the Peninsula. They seem to favour the warmer sectors of the south-west, notably the Guadalquivir depression, and are also relatively numerous in the Duero, Tagus and Guadiana basins and on the plains of Extremadura (SWA). They are scarcer in the east but are still relatively numerous there in ricefields and other irrigated croplands. Many also winter in the Balearic Islands, especially in Mallorca. The wintering birds everywhere are generally encountered below 700m in areas with a mean winter temperature above 8.6°C (SWA). Densities as high as 332 birds/km^2 on irrigated croplands and 249 birds/km^2 on open country near wetlands, have been reported from Catalonia (CWA).

Migrating Meadow Pipits are often visible along coastlines especially and they concentrate to some extent at the Strait of Gibraltar. Tellería (1981) counted over 3,500 flying south towards Morocco at Tarifa in autumn 1977. The autumn passage at the Strait is chiefly in October–November, peaking in the second half of October. The return movement there extends from February to early April, peaking in March (BG).

RED-THROATED PIPIT *Anthus cervinus*

(Bisbita Gorgirrojo, Petinha-de-garganta-ruiva)

A very scarce migrant, chiefly on southern and eastern coasts and in the Balearics. Vagrant to Portugal and Gibraltar (3).

Its main passage routes across Europe are through the central and eastern Mediterranean, which accounts for the relative frequency of observations from the Levant and on the Balearic Islands in some years (Fig. 180). Published records from the Balearics may underestimate its frequency there (Clavell 2002) but it is still regarded as no more than a rare migrant in the archipelago (*A. O. Balears 2010*: 266). In the south, there are occasional records from the Guadalquivir estuary and the Strait of Gibraltar. It is rarely reported on northern coasts or inland: although an exceptional influx was noted between 21 April and 12 May 1992 at Gallocanta, where numbers peaked at 40 on the last date (*A. O. Aragón 1991–1992*: 52). It is a vagrant to Portugal, where the whole of the 20th century produced just 15 records, involving 27 birds, all but one at coastal sites (AP). Ten recent Portuguese records had been accepted up to 2010 and there were at least four more observations (six birds) there in 2012 (RBP).

Figure 180. Regional distribution in Spain. Three Gibraltar records are included (n = 252 records, 604 birds).

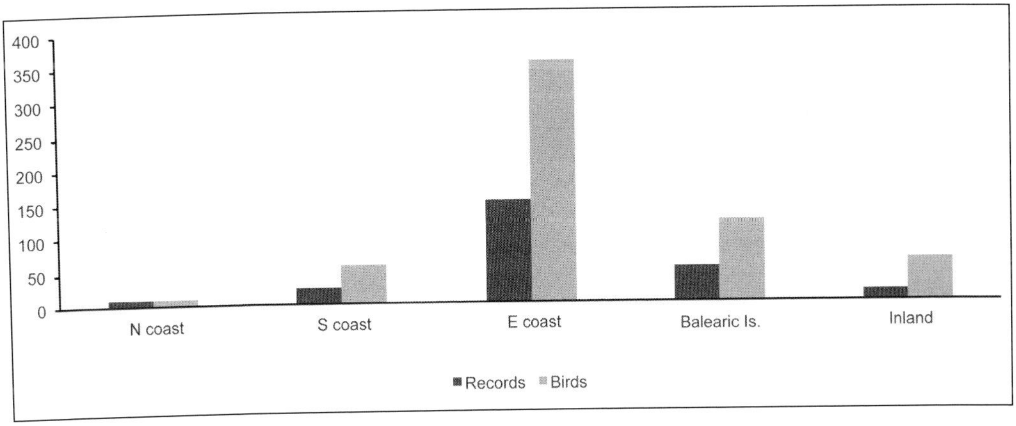

Red-throated Pipits are most often found in coastal scrub, saltpans and marismas. Records frequently come from well-watched sites, such as the Albufera de Mallorca, the Llobregat and Ebro deltas, the Aiguamolls de l'Empordà, the Albufera de Valencia, the Guadalquivir estuary and the Strait of Gibraltar, and it may be that some are missed elsewhere, especially inland. Spring observations, which clearly peak in April–early May, far outnumber autumn records, which peak in October (Fig. 181). A number of winter records suggest occasional wintering by some individuals, as is reported regularly in Morocco (BM). They include single December sightings from Asturias, Cantabria, Burgos, Alicante and Sevilla. Two were at the Cecebre reservoir, A Coruña, on 20 January 2002 and two were at the Albufera, Mallorca, on 9 February 2006 (*A. O. Galicia 2002*: 97, *Ardeola* 55: 147).

Figure 181. Records by month (n = 274).

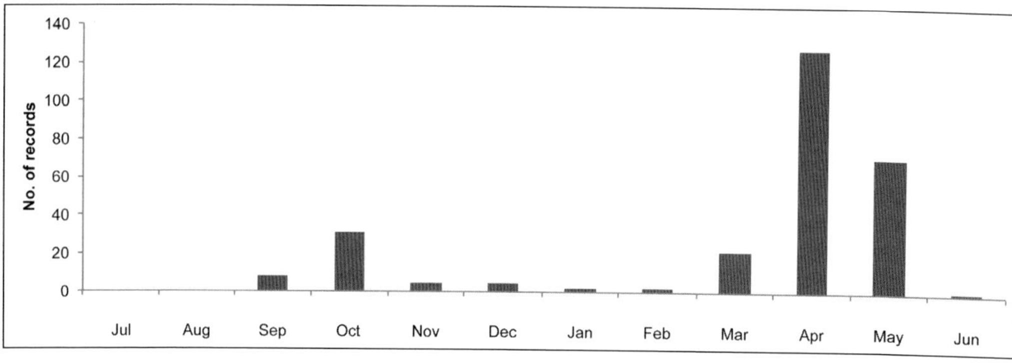

The 249 published records examined included 160 (64%) single individuals and a further 49 (20%) observations of 2–3 together. Flocks of up to ten are seen occasionally. The Gallocanta record of 40 is the largest known gathering seen in Iberia but other sizable concentrations have been noted at the Aiguamolls de l'Empordà especially: up to 19 in April–May 1996, 20+ on 27–29 April 2006, 30+ on 5 May 2008 and up to 11 in April–May 2009 (*A. O. Catalunya*).

BUFF-BELLIED PIPIT *Anthus rubescens*

(Bisbita Norteamericano, Petinha-fulva)

Vagrant to Spain.

One was in Bens park, A Coruña city, Galicia, during 13–30 September 2011. It is a rare Nearctic vagrant to the western Palearctic. Records include 27 in Britain, 18 in Ireland, two in France and six in the Azores.

WATER PIPIT *Anthus spinoletta*

(Bisbita Alpino, Petinha-ribeirinha)

***A. s. spinoletta*. A common breeding species on montane pastures in the northern half of Spain and in Andorra. A few nest in northernmost Portugal. Widespread in winter throughout the Peninsula and Balearic Islands.**

Breeding population estimates. Spain: SA1 16,000–32,000 pairs. **Portugal:** 5 pairs. **Andorra:** 2,000–3,000 pairs.

The Water Pipit breeds on many of the higher mountains in central and northern Iberia (SA2), where it is one of the most common passerines. It inhabits the Cantabrian Range and associated ranges, from Galicia west to the Basque mountains. It is

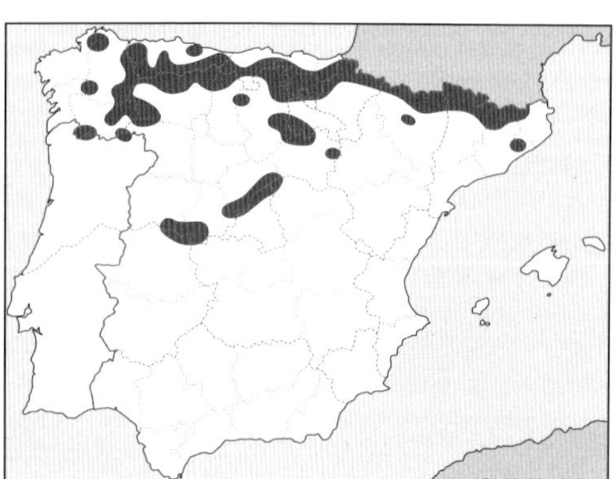

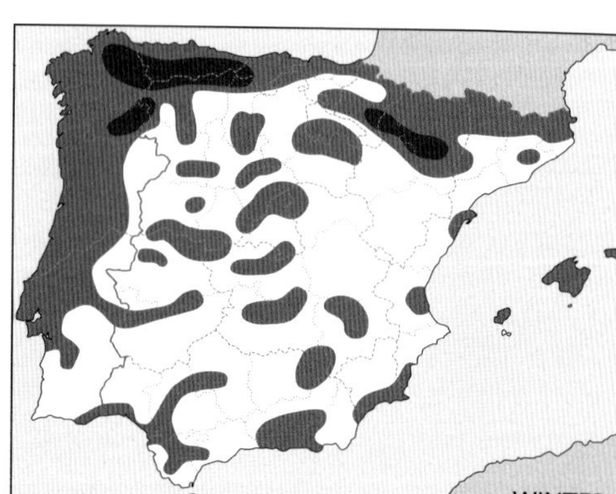

also common throughout the Pyrenees and in the northern Iberian Range (de Juana 1980). It nests less continuously across the Central Range, from the Sierra de Gredos west to the Sierra de Guadarrama. A few nest in northernmost Portugal in the Serra de Montesinho and probably in Peneda-Gerês National Park (AP). Breeding in southern Iberia in the Sierra Nevada, the highest of the Iberian summits, has never been confirmed but a singing male was present there at 2,600m on Pico Veleta on 26 June 2010 (*Ardeola* 57: 541).

This is a typical species of the high mountain pastures, often found near streams and in the vicinity of snowfields. It also frequents subalpine heaths and alpine juniper scrub. Most nest at considerable altitude, typically 1,400–2,600m above sea level, and some nest at over 3,000m in the Pyrenees (Antor & Pedrocchi 1989). Less typically, some are known to nest as low as 550m in similar habitat on the cool northern side of the Cantabrian Range, on the Serra de A Capelada in Galicia and on the Sierra de Cuera in Asturias (SA2). In general, birds nest at lower altitude in the north of their geographical range than in the south, and lower on the northern side of mountains than on the southern side, a familiar pattern among montane species.

The Iberian birds descend to middle levels or the lowlands at the onset of the first snows and return to the uplands as the snows recede, generally from late April onwards. The breeding season in the Catalan Pyrenees extends from late April to August (Ferrer *et al.* 1986). In winter they are joined by migrants from outside Iberia and Water Pipits may then be found throughout the Peninsula and also in the Balearic Islands (contrary to the statement in SWA). A few ringing recoveries, from France (4), Finland, Sweden, Switzerland (4) and the Czech Republic, show that some travel from the Alps and even the western Carpathians to winter in Iberia (SMA, AP). There are also two recoveries from the French Pyrenees of birds ringed in Spain, in Zaragoza and Málaga.

Wintering birds may be found in suitable habitats across Iberia but occur at higher densities in the north, often near the mountains where they breed (SWA). Many are found in coastal regions and the principal river valleys, on pastures, dunes, low scrub, ricefields, fresh and salt marshes, riversides and the shores of lakes and reservoirs. These locations all offer areas of low herbaceous cover, often in close association with water. Winterers on the Asturian coast are mainly present from mid October to February (Carrera 2003). They are generally encountered singly or in loose groups, although they may roost communally. Winter roosts of up to several hundred occur at the Brazo de la Torre, Sevilla (Chiclana *et al.* 2002) but most gatherings are much smaller. Some Water Pipits cross the Peninsula to winter in North Africa. In the south the southward passage is chiefly in October and early November, with a return movement in March and early April, but migrants are inconspicuous at both seasons (AI2, BSG).

ROCK PIPIT *Anthus petrosus*

(Bisbita costero, Petinha-marítima)

***A. p. littoralis*. A scarce winter visitor to Atlantic and Biscay coasts, rare or vagrant elsewhere.**
***A. p. petrosus*. May also occur as a scarce winter visitor.**

The Rock Pipit is most likely to be encountered on the Biscay coasts: in Galicia, Asturias, Cantabria and the Basque Country. Small numbers also occur on Portuguese coasts, more frequently in the northern half of the country, but there are records south to Algarve (AP). They are generally encountered singly or in very small loose groups, on beaches, rocky shores, in estuaries and coastal meadows. Most records are during October–March.

Some have been recorded some way inland in Spain, by reservoirs and other water bodies: for example, there are three recent records of single birds from Zamora: at Bengiles in October 2002, and at Villafáfilla in October 2004 and March 2006 (*Ardeola* 50: 165, 52: 220, 53: 208). Single birds were found in Salamanca: at Ciudad Rodrigo in November 2001 and at Alba de Tormes in October 2002 (*A. O. Salamanca 1924–2003*: 121). Such inland records away from the north coast are rare so an observation of four together at gravelpits at Suzana, Burgos, in October 2005 (*A. O. Burgos* 1: 241) was unusual.

Very few have been reported from the Mediterranean coasts. There are at least five records (six birds) for the Comunidad Valenciana, five records for Catalonia and three for the Balearic Islands, all since 1985: pre-1985 records for the Levant coast and Balearics are in question since at least some involve confusion with the Water Pipit (Clavell 2002; *Ardeola* 49: 192; regional bird reports). Single birds were found in coastal Murcia in March and November 2003 (*A. O. Murcia 2003*: 18). Further south single birds have been found in Gibraltar in February and at Doñana and Tarifa in March (Lathbury 1970; *Ardeola* 38: 346, 55: 304). Irby (1872) implied that it was regular in winter at the Palmones estuary, Cádiz: confusion with winter-plumaged Water Pipits seems likely. There are also observations from Sagres, the Alvor estuary, Monte Gordo and Vila Real de Santo Antonio on the south coast of Algarve (AP).

The two subspecies are hard to distinguish in winter but both are likely to be present. There are two recoveries of *littoralis* ringed on passage at Ottenby, Sweden: one in Asturias in late October and one in Beira Litoral in December (AMI).

YELLOW WAGTAIL *Motacilla flava*

(Lavandera Boyera, Alvéola-amarela)

***M. f. iberiae*. A widespread summer visitor to the Peninsula and Balearic Islands. Common on passage. Occasional in winter.**
***M. f. flava*. Very common on passage throughout the Peninsula and Balearic Islands.**
***M. f. flavissima*. Common on passage, mainly in the west. Scarce in the Balearic Islands.**
***M. f. thunbergii*. A scarce passage migrant, mainly in eastern Iberia and the Balearic Islands.**
***M. f. cinereocapilla*. An uncommon passage migrant, in eastern Iberia and the Balearic Islands.**
***M. f. feldegg*. Vagrant to Spain and Portugal.**

Breeding population estimates. Spain: SA1 70,000–240,000 pairs; ACRE 2.7 million birds. **Portugal**: 5,000–50,000 pairs.

Yellow Wagtails of the Iberian race are widespread and common, although their range is highly fragmented. They occur wherever there is suitable humid open habitat in the form of wet meadows, marismas, saltpans, fresh and salt marshes: often in estuaries, ricefields, irrigation channels and irrigated farmland. As a result they are more widespread in the northern half of the Peninsula, nesting across much of the northern meseta, in the valleys of Galicia and in the Atlantic and Biscay coastlands of northern Spain (SA2). They also breed in the western lowlands of northern Portugal as well as in the interior, in Trás-os-Montes and the Alto Douro (PA2). Nesting in dry crops, notably alfalfa, has been reported in Castilla y León (SA1).

Further south they occur at suitable points of the Levant coast, in saltpans and other coastal wetlands, notably in the Ebro Delta. They inhabit the few wetlands of the Costa del Sol but are much more abundant further west in the coastlands of southwestern Andalucía, especially in the Marismas and valley of the Guadalquivir. They are scarce in Extremadura and the southern meseta but occur there at the many scattered lagoons and other wetlands. In the Balearic Islands *iberiae* breeds in some numbers in Mallorca and Ibiza but is only a scarce nester in Formentera and does not breed in Menorca (*A. O. Balears 2010*: 266). Yellow Wagtails breed commonly in southern Portugal in Algarve, the lower Tagus and Sado valleys and in coastal Estremadura (PA2). The nesting habitats are generally at low altitudes but some nest at over 1,000m on meadows in the Central and Iberian Ranges (AI2). There is also a breeding population in the bogs and wet pastures of the Serra de Gerês, Trás-os-Montes, at up to 1,200m (AP).

Range expansion into some inland areas may have been favoured by the increase in irrigated crops there (SA1). The highest reported breeding densities in Spain are in the ricefields and irrigated croplands of the thermomediterranean (66–97 birds/km^2) but the breeding range is most continuous within the mesomediterranean. A significant increase was reported in SACRE sites during 1998–2012 (Escandell 2013).

Most Yellow Wagtails winter in sub-Saharan Africa. The Iberian nesting populations begin to arrive from late February in the south but chiefly in March–April, departing during August–October (AMI). Some individuals of *iberiae*, and also *flava*, are reported in winter from southern wetlands and, more rarely, elsewhere (Pérez-Tris & Asensio 1997a, SWA). There is a record of an individual of *thunbergi* from Bonanza, Cádiz, in February 2006, and one of *flavissima* at the Albufera de Valencia in January 2010 (*Ardeola* 53: 390, 57: 237). The small and well-watched estuary of the Guadalhorce, Málaga, produces regular winter sightings, generally of very small numbers, but there were 21 there on 28 December 1991 and 18 on 15 February 1992: two *flava* males were captured from the first group (*Ardeola* 40: 101). Records from Doñana in January 2009 of 16 on a muddy ricefield at Hato Blanco and 18 at Isla Mayor (*Ardeola* 56: 170, 58: 508) are also noteworthy although a few winter in the marismas annually. Winter records away from southern coastal regions are unusual: there was one at Miño, A Coruña, on 11 January 2004 (*A. O. Galicia 2004*: 103).

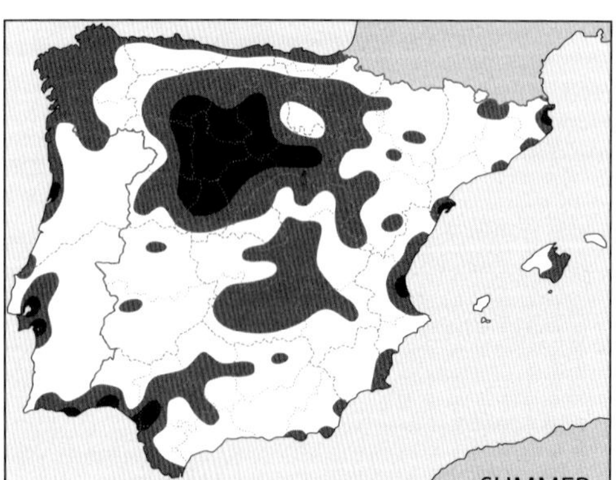

Observations and ringing recoveries (Pérez-Tris & Asensio 1997a, Cantos & Asensio 1989, SMA) show that central European birds (race *flava*) predominate on passage but *flavissima* of the British Isles and near continent is also often prominent, especially in the western half of the Peninsula, as is *thunbergi* of northern Europe in the eastern half and Balearics. All 52 autumn recoveries of *flavissima* were in the western half of Iberia, whereas most of *flava* and *thunbergi* were in the eastern half, and all of *cinereocapilla* were on Mediterranean coasts south to the Strait of Gibraltar. The pattern in spring seems to be less clear-cut but the available ringing data is more limited: 13 spring recoveries of *flavissima* were more or less equally divided between western and eastern Iberia.

There are a few recoveries of birds ringed in Poland (4), Italy (4), Croatia (1) and the Czech Republic (1) but the great majority of 371 recoveries in Spain have been have of birds ringed in Germany, Sweden, the British Isles, the Low Countries and France (SMA).

Individuals showing the characteristics of the Italian race *cinereocapilla* are generally scarce and occur chiefly in eastern Spain and the Balearics, although two males were reported at Galisteo, Cáceres, in August 2009 (*Ardeola* 57: 237). A pair that included a ringed Italian male of this subspecies nested in coastal marshes near Tarragona, Catalonia in April 1989 (Bartrolí 1991) and a pair in which the male at least was *cinereocapilla* nested in Mallorca in 2008 (Garcias 2009a). Bartrolí (1991) found that his study population near Tarragona showed intermediate characteristics between *iberiae* and *cinereocapilla*, as has also been noted in the Camargue. There is only one reported occurrence of *cinereocapilla* in Portugal; a male at Castro Marim, Algarve, in April 1999 (*Anu. Ornitol.* 2: 9, Moore 2000b).

The Black-headed Yellow Wagtail *M. f. feldegg* is a vagrant to the region. The earliest Iberian record, a male at Castro Marim, Faro, in April 1987, remains the sole Portuguese observation. There are eight accepted records from Spain, between 1989 and 2000, all of them of the easily identifiable adult males in spring. Five were in Catalonia, two in Andalucía and one in Galicia: this last during 14–27 June 1992. The other observations were all on Mediterranean coasts between 25 March and 3 May (de Juana 2006). There were at least a further three records from Valencia, four from Catalonia and one from Navarra during 2011–13 (RBS).

Over 76,000 birds were ringed in Spain alone during 1973–2009. Those that could be assigned to a particular race comprised 51% *flava*, 39% *iberiae*, 6% *flavissima*, 3% *thunbergi* and 1% *cinereocapilla* (n = 4,997; Oficina de Anillamiento). Birds ringed in Iberia have been recovered in northern and central Europe and in the British Isles (41: all but one of them ringed in Portugal) and there are also a few recoveries of birds ringed in Spain from Italy, Morocco (3), Algeria, the Canary Islands and Senegal (2). Migrants are most evident when flocks, sometimes of hundreds, stop to rest and feed in coastal wetlands and meadows. Spring migrants occur in the south from late February but the main passage is in March–May, with stragglers in June. Males of *iberiae*, and *cinereocapilla*, are the earliest to arrive (Palomino 2012). The return movements in the south are from mid August to late October, peaking in September and early October, with stragglers in November and into December (BG, BSG, Palomino 2012).

CITRINE WAGTAIL *Motacilla citreola*

(Lavandera Cetrina, Alvéola-citrina)

Vagrant to Spain and Portugal.

Westward breeding-range expansion during the last quarter of the 20th century has been accompanied by an increase in observations in western Europe. The first British record was in 1954 but there have since been at least 263 individuals there, as well as 27 in Ireland, 34 in France, three in Morocco, four in the Canary Islands and two in the Cape Verdes (Slack 2009).

The first reports from both Iberia and France were in 1987. The first Iberian records were a male in Mallorca during 14–18 April 1987 and a first-year at the Llobregat delta on 12 September 1987 (Hazevoet & Haafkens 1988; *Ardeola* 36: 121, 48: 131). By 2010 there were 47 accepted records for mainland Spain and the Balearics and two for Portugal, all of single birds with the exception of two together at the Albufera de Valencia from 12–21 September 2009. As many as 15 records were during 2004–08 and there were 14 in 2009–10, indicating an increased frequency in recent years. Post-2010 Spanish records numbered at least 12 in 2011 and seven in 2012 (*Ardeola* 60: 482–483, RBS).

Figure 182. Regional distribution of records, 1987–2010 (n = 49)

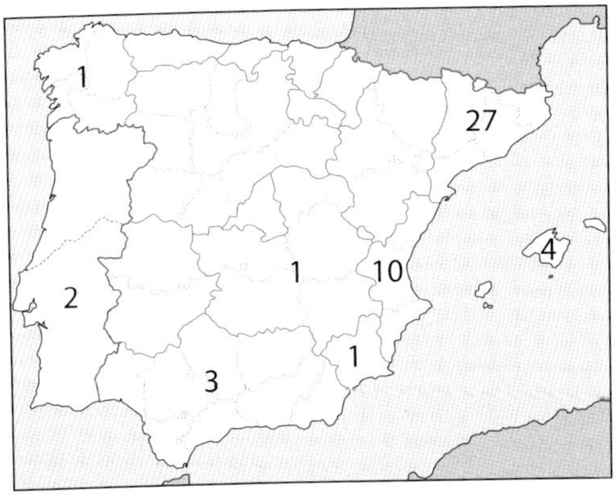

Accepted autumn observations up to 2010 included 29 juveniles/first-winters and eight adults. The ten spring birds, nine males and one female, included three adults and three second-years. All observations have been at coastal locations, except for one at Chinchilla de Montearagón, Albacete: between 30 March and 3 April 2009. They have mainly appeared at wetlands and predominantly in the east: with some 92% in the Mediterranean coastlands and 55% in Catalonia (Fig. 182). There are 37 autumn records and ten in spring (Fig. 183). One remained at the Aiguamolls de l'Empordà from 14 December 2006 until 16 March 2007 and a first-winter individual was at the Albufera de Valencia on 25 December 2010. We are unaware of any other midwinter observations in Europe, although one wintered in the Canary Islands in 2005–06 (*Ardeola* 54: 436, 55: 278) and two of the three Moroccan observations were in January (BM).

In autumn most birds were first recorded between 1 August and 16 October and the extreme spring dates are 22 February and 10 May. The frequency of autumn records (76%) is much lower than in Britain but considerably higher than in France (Slack 2009, OF). Up to 2010 there were 13 autumn and three spring records on French western coasts but 13 spring records and just one autumn record in eastern France. Nearly all records from central Europe are in spring (Mitchell & Young 1997). There may thus be a loop migration involving a small number of individuals that travel to Africa along western European coasts and return on a more easterly route.

Figure 183. Monthly distribution of records, 1987–2010 (n = 49)

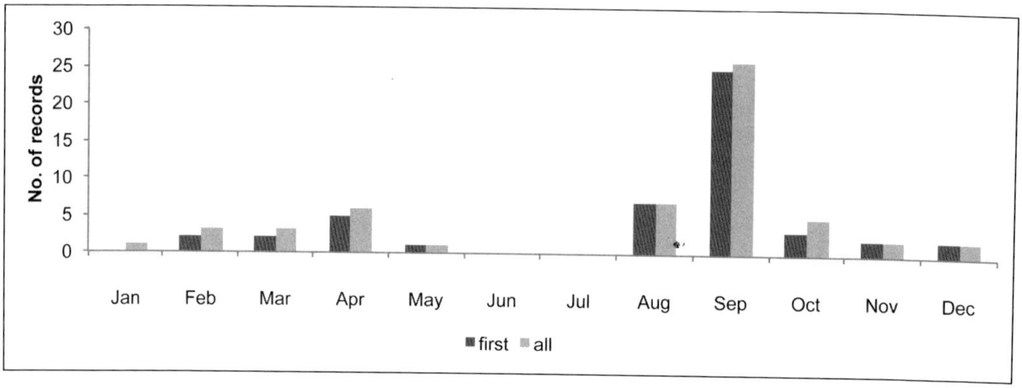

GREY WAGTAIL

Motacilla cinerea

(Lavandera Cascadeña, Alvéola-cinzenta)

***M. c. cinerea*. A widespread resident or partial migrant in mainland Iberia. Also a common migrant and winter visitor throughout the Peninsula and in the Balearic Islands.**

Breeding population estimates. Spain: SA1 13,500–17,000 pairs; ACRE 233,000 birds. **Portugal**: 5,000–50,000 pairs. **Andorra**: 70–100 pairs.

The Grey Wagtail nests wherever fast-flowing rivers and streams occur. It is therefore particularly widespread and numerous in all the major mountain ranges, ranging down to sea level on Biscay and Atlantic coasts and nesting to altitudes over 2,000m. Foraging birds range to over 3,000m in the Sierra Nevada in summer (Garzón & Henares 2012). It is widespread in Portugal but scarcer south of the Tagus (PA2). It does not breed in the Balearic Islands. In general, it favours stretches of watercourses with emergent rocks and stones or exposed gravel beds. A pair can often be found near a bridge, watermill or dam, the nest frequently being built in some cavity within the structure.

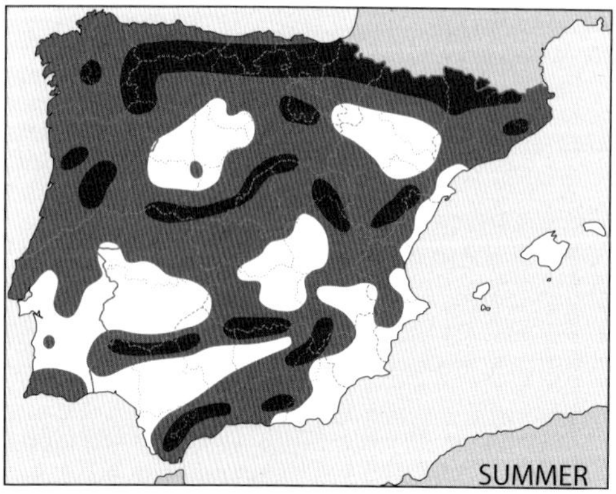

The SA1 assessment of the Spanish breeding population was clearly an underestimate. Catalonia alone is considered to accommodate some 7,800–11,800 pairs (CA2). The population on SACRE sites showed a significant 47% decline during 1998–2011 that was most evident in the north (Escandell 2011, 2012). The Portuguese population showed no clear trend during 2004–09 (ACP).

The Grey Wagtail ranges more widely in winter, then also often occurring some way from water, for example on lawns and paths in city parks and on rooftops in towns and villages, but it is also then drawn to temporary pools and inundated open fields, as well as to lake margins. Some individuals frequent sea coasts, especially around estuaries and in saltpans but also on rocky shores. The Spanish winter range includes some areas where the Grey Wagtail is absent or very rare during the breeding season, such as the eastern and southeastern littorals and the Balearic Islands (SWA).

Iberian Grey Wagtails are fairly sedentary but those nesting high in the mountains winter at lower levels (AMI). Many birds from other European countries enter the Peninsula and the Balearic archipelago, to winter there: mainly in coastal regions (Pérez-Tris & Asensio 1997b) or to continue further south to Africa, where some reach sub-Saharan West Africa. Autumn passage generally begins in September, peaking during the second half of October, with stragglers into November (Pérez-Tris & Asensio 1997b). Spring migrants appear in the Gibraltar region from mid February until April (BG). There are 80+ Iberian recoveries of birds ringed in Belgium, Germany (27), the Czech Republic (19), Austria, the British Isles, Switzerland, Holland, France, Denmark, Sweden, Finland, Poland and northern Italy (SMA, AP). Birds ringed in Iberia have also produced eight recoveries in northwestern and central Europe, and one in Morocco. Wintering birds are usually encountered singly but small numbers may gather in communal roosts at this season.

WHITE WAGTAIL *Motacilla alba*

(Lavandera Blanca, Alvéola-branca)

M. a. alba. **(White Wagtail). A common resident or partial migrant in most of mainland Iberia. Also abundant on passage and in winter throughout the Peninsula and in the Balearic Islands.**
M. a. yarrellii. **(Pied Wagtail). A scarce winter visitor.**
M. a. subpersonata. **(Moroccan Wagtail). Vagrant to Spain, Portugal and Gibraltar.**

Breeding population estimates. Spain: SA1 112,000–370,000 pairs; ACRE 4.5 million birds. **Portugal**: 10,000–100,000 pairs. **Andorra**: 100–200 pairs.

The White Wagtail breeds almost throughout the Peninsula but not in the Balearic Islands, although Munn (1921) saw a few summering individuals in Mallorca. The chief gaps in its range are in the south-west, where it is scarce or absent in southern Portugal in Baixo Alentejo and Algarve, the southwestern Spanish coastlands in Huelva and Cádiz, the Guadalquivir valley, most of southern Extremadura and parts of the southern meseta (SA2, PA2). Its absence there largely coincides with the thermomediterranean and southern mesomediterranean regions. Nevertheless, a pair nested in Gibraltar in 2004 (*GBR 2004*: 31). It may have been even scarcer in the south in the past since the late 19th century authors considered the White Wagtail to be just a winter visitor to Andalucía, Irby (1879) referring to an exceptional nest having been found near Álora. Subsequently Jourdain (1936–1937) noted that 'apparently a few birds breed in the (southern) sierras'. Similarly, in Portugal, Ticehurst & Whistler (1933) found the White Wagtail to be very common in the north but very scarce or absent south of Coimbra.

Breeding birds are ubiquitous in the eurosiberian region, particularly in farming areas around farms and villages, where they forage around livestock and in fields and paddocks. Some nest within towns and cities, where they can be seen feeding on lawns in parks and gardens. They are more closely associated with watercourses and mountain streams further south, in the Montes de Toledo, Sierra de Cáceres, Sierra Morena and the Betic Range (SA1). Nesting occurs to 1,200m in the Sierra Nevada (Pleguezuelos 1992) and to over 1,600m in the Sierra de Gredos (San Segundo 1990). White Wagtails may occur alongside Grey

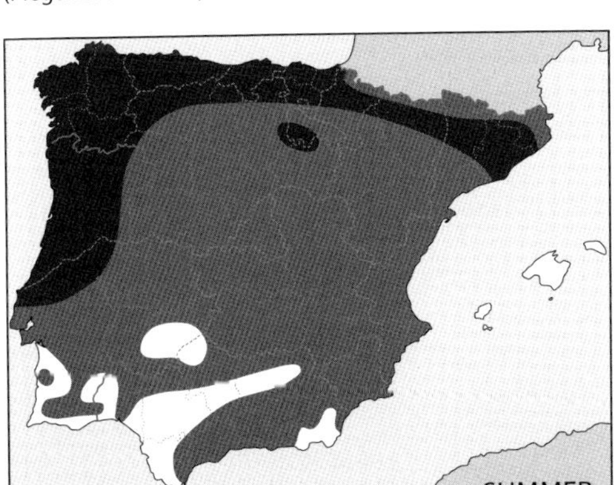

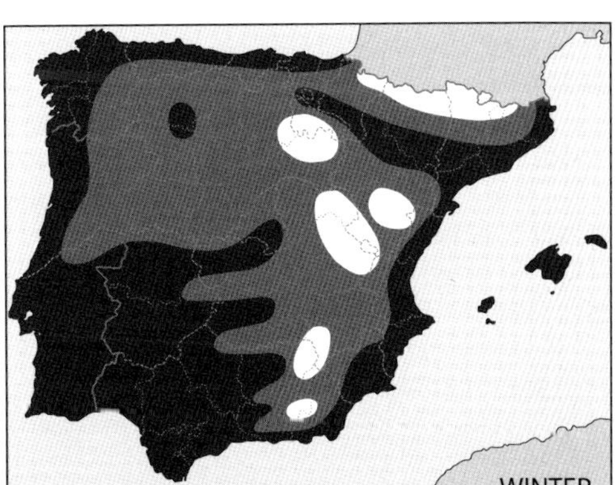

Wagtails along rocky streams, especially near and within mountain villages, but they are reportedly more tolerant of a degree of water pollution (Pleguezuelos 1992). Many nest in holes in bridges, walls and buildings.

The SA1 assessment was probably too conservative since Catalonia alone is estimated to support 39,000–71,000 pairs (CA2). However, it remains to be confirmed whether the population is as large as suggested by ACRE. The highest densities are reported from the eurosiberian coastlands of northern Spain, in villages and towns, and their vicinity (56–113 birds/km^2) and in pastures with trees (44–50 birds/km^2). The outskirts of urban areas of the supra- and mesomediterranean are also favoured (40–47 birds/km^2). A significant decline was reported on SACRE plots during 1998–2012 (Escandell 2013). The Portuguese population was estimated at 10,000–100,000 pairs (BE) and there are indications of some westward range expansion in the south since the 1980s (PA2); it showed no clear trend during 2004–09 (ACP). The Andorran population was estimated at 100–200 pairs (BE) and stable.

The Iberian birds are largely resident, with those at higher elevations descending lower in winter (AMI, AI2). Very large numbers of White Wagtails enter the Peninsula to winter. They are ubiquitous at this season, but are especially abundant in the warmer southern sectors of Iberia, especially in the Guadiana and Guadalquivir valleys, the coastal regions of the Comunidad Valenciana and Murcia and on the island of Mallorca (SWA). They then frequent ricefields and other croplands, damp pastures and the margins of permanent and seasonal water bodies. Densities of 25–100 birds/km^2 are reported from agricultural land in winter (SWA). White Wagtails are well known for their habit of congregating in large roosts in winter, sometimes in reedbeds but also often in densely vegetated trees and shrubs within towns and cities, even in well-lit streets and squares in city centres (Bernis 1988b). Such roosts often consist of hundreds or sometimes thousands of birds that flock to traditional sites that are used every year: 3,400 individuals were censused in January 1962 in La Cibeles square in Madrid city (Bernis 1988b) and a winter roost of nearly 100,000 birds is reported from Sevilla (García *et al.* 2000).

The principal wintering population arrives in October–early November but some individuals appear from late August. Iberian recoveries of birds ringed abroad fall during October–March, with a few in April, and peaking in November and January (AMI, AI2). Visible southward passage is noticeable at the Strait of Gibraltar from mid October to mid November, with stragglers into December (BG). Return movements at the Strait are seldom obvious but occur during February–April, chiefly in March.

Some Pied Wagtails, which breed only in the British Isles and near continent, are widely reported in winter, principally from northern, western and south-western coastal Iberia. One was trapped at La Línea, Cádiz, on the unusual date of 14 August 1958 (*Ardeola* 6: 392). Tellería *et al.* (AI2) reports 73 recoveries of birds ringed in Britain, all or most of which would be *yarrellii*, despite many British birds being sedentary (Wernham *et al.* 2002). It is readily observable among the abundant wintering White Wagtails at Doñana (Llandres & Urdiales 1990). Three roosts in Plasencia, Cáceres, in early February 2010 included a total of 11 *yarrellii* (*Ardeola* 57: 238). However, a report of 80 together, accompanying 20 *alba* birds, at Gozón, Asturias on 23 February 2002 (*Ardeola* 51: 555) is unusual. Even more surprising is a report of a female *yarrellii* seen apparently carrying food for nestlings also at Gozón, in August 2008 (*Ardeola* 56: 363). There are at least seven recoveries of *yarrellii* from Morocco, confirming that some travel even further south than Iberia.

The race *yarrellii* has been regarded as a rarity in eastern Iberia. Clavell (2002) gives only 13 records in total for Catalonia, the Comunidad Valenciana and the Balearics up to the year 2000. Subsequent observations suggest that it was probably overlooked in the past, unless there has been a change in its behaviour. They include at least 12 records from the Balearics, chiefly in Mallorca in November–February (*Ardeola* 52: 220, 53: 390, 55: 147, 304, 56: 363, 57: 542); three records from the Albufera de Valencia in February–March 2005 (*Ardeola* 53: 208) and 17 records from Catalonia during 2001–09: mainly from the Aiguamolls de l'Empordà and the Ebro Delta in winter (*A. O. Catalunya 2001*: 225).

The vast majority of the White Wagtails that winter within or cross the Peninsula or the Balearic archipelago are of the nominate form. Most recoveries are of birds ringed in Belgium, Germany, Switzerland and France but there are also a few from Iceland (3), the Czech republic, Scandinavia, Poland, the Baltic States and Austria. There is a clear tendency for birds from central and eastern Europe to winter on the eastern side of the Peninsula and in the Balearic Islands, in contrast to the pronounced westerly bias of *yarrellii* recoveries, which are largely from Portugal and western Andalucía, with some also in the Biscay coastlands (Wernham *et al.* 2002). There is only one recovery of a Spanish-ringed bird from Morocco. There is also a single recovery from Galicia of one ringed in Senegal (SMA) but clearly some numbers of *alba* cross the Peninsula to winter as far south as sub-Saharan West Africa.

The Moroccan Wagtail *M. a. subpersonata* has only recently been recorded in Iberia. There are at least four accepted records of single male birds: at Beja, Portugal on 13–14 July 1995; Gibraltar from 30 July to 7 August 2006; Tarifa on 24 March 2007, and Málaga harbour on 19–21 March 2009. It has recently expanded its range in Morocco northwards and occurs as close to Iberia as Larache on the Atlantic coast and also on the Moroccan Mediterranean coast (BM).

BOHEMIAN WAXWING

Bombycilla garrulus

(Ampelis Europeo, Tagarela-europeu)

Vagrant to Spain and Portugal.

The Waxwing is is renowned for its periodic irruptions into Western Europe in winter. At least eight such irruptions were recorded in the British Isles during the 19th century and there were some 24 during the 20th century (Wernham *et al.* 2002), with others subsequently, including during 2011–12 and 2012–13. Invasions apart, eastern France is at the regular western limit of the wintering range (OF). Waxwing invasions very rarely reach Iberia and only tiny numbers have been involved when they do. Most observations have been in the northernmost regions in winter.

There are probably reliable 19th century records of one at Valsaín, Segovia, and two in Cataluña, including one taken near Barcelona around 1850 (Fuset 1913, AMI, Ferrer *et al.* 1986). Two 20th century irruptions are known to have reached the Peninsula, in 1958–59 (Cornwallis 1961) and in 1965–66 (Érard 1967, Cornwallis & Townsend 1968). The former was only noted in the Basque Country where one was taken at Zumaia, Gipuzkoa, on 23 March 1959 (*Ardeola* 6: 392–393, 52: 201). The 1965 invasion took some as far as Portugal, where one was killed at Anadia, Aveiro (AMI) but most records came from the Biscay coastlands, with individuals collected at Castro Urdiales on 10 October, Santander on 22 November, Zarauz on 24 November, Camargo on 27 November, San Sebastián on 28 November and at Munguía in December (*Ardeola* 11: 159–161). Two were hunted in Girona, northern Catalonia during this period: one at Cartellà on 17 December and the other at Olot on 16 January (*Ardeola* 11: 161, 53: 182, 56: 332). Much later, one was captured at the Serra de Collserola, Barcelona, during winter 1999–2000 (*Ardeola* 59: 391).

The large irruption of autumn 2004, which brought as many as 100,000 Waxwings into France (Paul & Olioso 2006), resulted in three records in northern Spain: two birds were at Cedeira, A Coruña, on 1 November; four were at Ferrol, A Coruña, on 3 November, and three were at Llastres, Asturias, on 18 November. There is also a record of two at the Ulzama valley, Navarra, on 25 November 2005. Subsequently two birds were at Redilluera, León, on 6 January 2009, and one at Proaza, Asturias, on 10 December 2010. Post-2010 there were two more records from Asturias during winter 2010–11: two at Llena in December (RBS) and one at Avilés during 19 February–5 April 2011 (*Ardeola* 60: 483) as well as several from winter 2012–13, when there was a further small influx involving at least ten individuals, chiefly into Asturias and Cantabria but with some also recorded in Navarra, Lleida and as far south as Salamanca (RBS). They included a flock of five at Ramales de la Victoria, Cantabria, in January–February 2013 (RBS).

WHITE-THROATED DIPPER

Cinclus cinclus

(Mirlo-acuático Europeo, Melro-d'água-comum)

***C. c. cinclus*. Resident in northern and western Iberia, from the Pyrenees, Cantabrian Range and the mountains of northern and central Portugal, south in Spain to the Central Range. Black-bellied.**
***C. c. aquaticus*. Resident in eastern and southeastern Iberia, intergrading with *cinclus* in central Iberia. Rufous-bellied.**

Breeding population estimates. Spain: SA1 6,200–8,000 pairs. **Portugal**: 500–2,500 pairs. **Andorra**: 70–120 pairs.

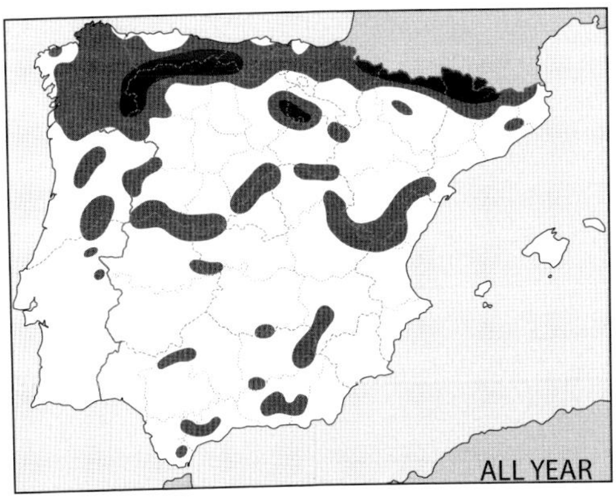

The Dipper requires fast-flowing, permanent rivers with a stony substrate and is thus far more widespread in the cooler humid areas of Iberia. It selects shallow, clear, neutral to alkaline streams with steep gradients (Peris *et al.* 1991). It ranges throughout the eurosiberian climatic region and is widespread and common there in the Pyrenees, the Cantabrian Range, the mountains of Galicia and adjacent sectors of northeastern Portugal, and their foothills. Further south it is largely confined to the mountains (SA2). It does not occur in the Balearic Islands.

It is common in the northern sector of the Iberian Range and across the Central Range. The distribution is more fragmented in the southern parts of the Iberian Range. It is also present in the Duero gorges on the Portuguese/Spanish border. Further south it has local, isolated populations in the central Portuguese uplands north of the Tagus; in Las Villuercas, Cáceres; in the Montes de Toledo, and at scattered points of the Sierra Morena and the Sierras of Cazorla and Segura. It is relatively widespread in the eastern Betic Range in the Sierra Nevada but much rarer and very local further west, in Málaga and Cádiz: where a nest found near Algeciras was probably exceptional (*Ardeola* 16: 272, SA2, PA2). Iberian Dippers often nest in holes in man-made edifices, mainly bridges: 60 of 73 nests in Navarra were in such artificial sites (Esteban & Campos 1995).

Breeding birds range down to sea level in the Biscay coastlands and may be found up to 2,000m in the Pyrenees, 2,100m in the Sierra de Gredos and to 2,450m in the Sierra Nevada (AI2, Garzón & Henares 2012). The Iberian birds are highly sedentary, although those nesting at high altitudes may descend lower in winter (AMI, Moreno-Rueda 2011). Short-range dispersal by juveniles has been recorded. The farthest known displacements have been of 74km in Valencia by a Dipper ringed in the province (*Ecología* 2: 394) and 54km by a juvenile ringed at Elizondo, Navarra, found also in that province, at Triabuenas (CMA). A Dipper on the Ribeira de Assêca at Tavira, Algarve, on 8 May 1997 (*Pardela* 7: 19), was far to the south of the normal range in Portugal. One ringed in Hecho, Huesca, in August 1976 was found at Arudy, Pyrénees-Atlantiques, France (43km) in May 1978 (*Ardeola* 29: 109) and another ringed bird crossed the Pyrenean watershed in the opposite direction. The only long-distance foreign recovery is of a bird from eastern Belgium that was found in Aragón (SMA).

Birds occur at densities of 1–15 pairs per 10km of those rivers that offer suitable habitat, the higher concentrations being in northern Iberia (Moreno-Rueda 2011). The Spanish population seems to be in gradual decline, partly as a result of pollution and acidification of watercourses. In particular, small range reductions have been reported from Navarra (Esteban & Campos 1998) and Catalonia (CWA). The Portuguese population is also considered to be declining. Mean densities of 8.4 pairs/10km are reported from Andorra (ANA).

NORTHERN WREN *Troglodytes troglodytes*

(Chochín Común, Carriça-comum)

***T. t. troglodytes*. An abundant resident in mainland Iberia.**

***T. t. kabylorum*. A common resident or partial migrant in Mallorca. Paler than *troglodytes*.**

Breeding population estimates. Spain: SA1 2.1–4.0 million pairs; ACRE 6.3 million birds. **Portugal**: 100,000–1 million pairs. **Andorra**: 500–700 pairs. **Gibraltar**: 100–150 pairs.

This adaptable bird is widespread throughout Iberia but its predilection for relatively humid areas results in some significant gaps in its distribution. In Spain it is largely absent from much of the Guadalquivir and Ebro valleys, the large tracts of crop-growing dehesas and steppelands of Extremadura and the southern meseta, the dry croplands of the central northern meseta and the arid south-east in eastern Almería, Murcia and southern Alicante (SA2). The main gaps in the Portuguese distribution are similarly in the croplands of Beira Baixa and Alentejo (PA2). In all of these it may still be encountered in riparian vegetation. In the Balearic Islands the Wren nests commonly in Mallorca and Ibiza but apparently it is only a scarce winterer and migrant in Menorca and is absent from Formentera (*A. O. Balears 2010*: 266).

The Wren occurs wherever there is dense low undergrowth, usually with some emergent shrub or tree cover. It is particularly characteristic of young plantations, often of exotic trees such as Monterey Pines or eucalyptus, so long as there is abundant undergrowth. In central Spain it is associated especially with woodlands of an Atlantic rather than a Mediterranean character, such as those of Ash, Pyrenean Oak and Scots Pine (SA2). Many inhabit parks, gardens and hedgerows. Wrens nest from sea level to over 2,000m in the Pyrenees and Central Range and to over 2,400m in the Sierra Nevada (AI2, Garzón & Henares 2012). Southern birds tend to frequent hilly and mountainous areas rather than plains and they prefer shady slopes to sunny ones.

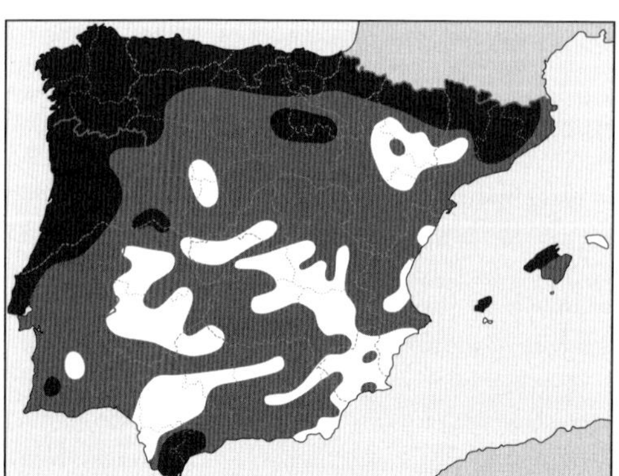

The largest regional populations found by the ACRE study were those of Galicia and Castilla y León (23% each). The Catalan population has been estimated at 314,000–431,000 pairs (CA2). In general, the highest densities are in the eurosiberian region and lowest in the thermomediterranean. The densest populations reported are, remarkably, in northern exotic pine and eucalyptus plantations (83–101 birds/km²). Other habitats in the eurosiberian zone also attract relatively high densities, in scrub, woodlands and farming mosaics (48–76 birds/km²). A significant 31% increase was reported on SACRE sites during 1998–2012 (Escandell 2013).

Many Wrens are entirely sedentary but others perform regular movements from and to high altitude territories that are abandoned in winter (AI2, SWA). The extent of this dispersal is not fully understood but the appearance of Wrens in winter in Menorca and in warm areas of the Peninsula where they do not breed suggests that some travel considerable distances. Such winter influxes have been noted in the Ebro valley at Lleida, and in parts of Almería and Murcia (SWA). Nevertheless, 1,260 of 1,286 recoveries of Spanish-ringed birds were at the original ringing sites and only one moved a significant distance. Some immigrants from northern Europe occur in winter. There are single Iberian recoveries of Wrens ringed in Belgium, Denmark and Sweden and a Spanish-ringed bird has been found in France (AMI, SWA). The Swedish bird was ringed as a chick in Gotland in June and was found the following December in Granada, a displacement of nearly 2,800km.

DUNNOCK *Prunella modularis*

(Acentor Común, Ferreirinha-comum)

***P. m. mabbotti*. A common resident in northern Iberia, very local further south.**
***P. m. modularis*. Common in winter and on passage throughout the Peninsula and in the Balearics.**

Breeding population estimates. Spain: SA1 400,000–1 million pairs; ACRE 1.8 million birds. **Portugal**: 5,000–50,000 pairs. **Andorra**: 5,000–7,000 pairs.

The Dunnock is widespread throughout the Eurosiberian region but it is almost entirely a montane species further south, occurring in the supra- and oromediterranean regions. It occurs down to sea level north of the Cantabrian Range, in Galicia and in northwestern Portugal. Elsewhere it is largely found only at high levels in mountains, above 1,000m and commonly above 2,000m (SA2). Breeding has been confirmed as high as 2,400m in Andorra (ANA). The populations of the northern lowlands may be found in scrubby habitats often with scattered trees, such as in heathlands, hedgerows, gorse clumps, woodland edges and in the understorey of open broad-leaved or coniferous woodlands. They extend above the treeline in subalpine scrub formations, notably in *Erica*, *Cytisus* and juniper heaths. The Portuguese breeding range extends south to the Serra da Estrela and the Dunnock does not breed south of the Tagus (PA2). A small population occurs in the mountains of Las Villuercas, Cáceres. A more isolated, sparse, southern outpost population, in broom and juniper scrub at 1,700–2,300m in the Sierra Nevada, was first detected as recently as 1998 (Garzón & Henares 2012). The densest populations reported are in habitats in the eurosiberian zone: in scrub above 500m and in pastures with trees above 750m (65 birds/km²). SACRE sites in Spain reported a significant decline during 1998–2012 (Escandell 2013). The Catalan breeding atlas put the regional population there at 51,000–92,000 pairs (CA2).

Many Iberian breeding populations are largely sedentary but those nesting at high altitudes descend lower for the winter (AMI, AI2) and may then disperse to coastal and southern regions of the Peninsula, although this has yet to be demonstrated. Numerous migrants of the nominate race enter Iberia to winter and then also occur in the Balearic Islands. They occur during October–April. There are over 140 recoveries of birds ringed across much of central and northern Europe: in Germany, France, Belgium, Scandinavia, Finland, the Baltic States, Poland, Finland, the former Czechoslovakia, Switzerland and Italy. Eighteen Spanish-ringed birds have been recovered in north-central Europe, six of them in Fennoscandia (SMA).

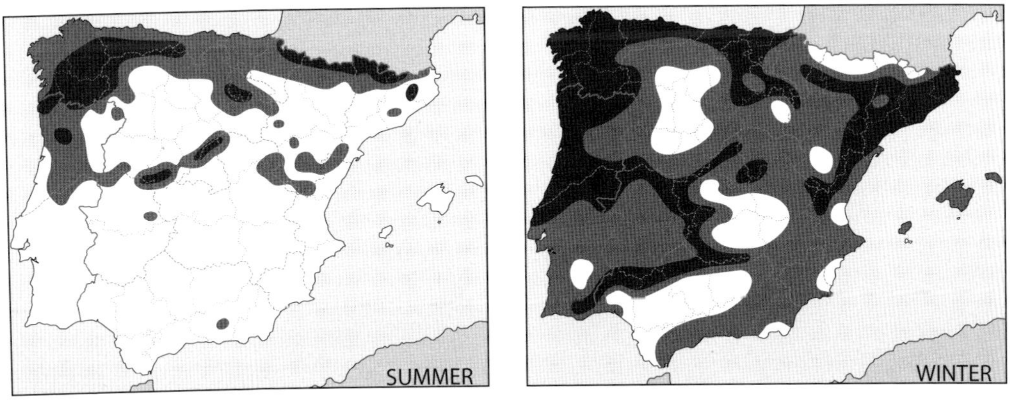

Wintering Dunnocks occur in dense scrub at low levels throughout much of Iberia in winter but ringing recoveries show an avoidance of the northern meseta and a concentration in the thermo- and mesomediterranean (AI2), although the Guadalquivir valley and the Cádiz coastlands are avoided. Seeds, especially of the Cistaceae, are the predominant item of their winter diet in southern Spain (Martínez-Cabello *et al.* 1991) and this would account for the attractiveness of the Mediterranean mountains, with their great tracts of Cistus scrub, to wintering Dunnocks. Nevertheless, the Spanish winter atlas found the highest densities to be in the north-west: in Galicia, western Asturias and the Montes de León, and, to a lesser extent, in parts of Catalonia and the northern Comunidad Valenciana (SWA). There is a concentration of recoveries of foreign-ringed birds in the latter area, but AI2 found none in the north-west, which suggests that the northern winterers may largely be of Iberian origin, most probably from the Cantabrian Range. The Iberian recoveries of birds ringed abroad fall during September March, but most are in November–February (AMI, AI2). Dunnocks are recorded in Gibraltar from mid October to March but are never more than scarce there (BG). Small numbers continue further south to winter in northern Morocco (BM).

ALPINE ACCENTOR *Prunella collaris*

(Acentor Alpino, Ferreirinha-serrana)

***P. c. collaris*. Common on the highest mountains in Spain and Andorra. More widespread in winter, then occurring in Portugal and occasionally in the Balearic Islands and Gibraltar.**

Breeding population estimates. Spain: SA1 17,000–20,000 pairs; **Andorra**: 200–300 pairs.

The Alpine Accentor is a defining species of the bird communities of the highest levels of the Iberian mountains. Breeding birds are normally found no lower than 1,800m in the Pyrenees, Cantabrian Range, the northern Iberian Range and the Central Range, with the Sierra Nevada population as a southern outpost. They nest at up to 2,500–2,800m, and at over 3,000m in the Pyrenees and Sierra Nevada (SA2). However, nesting at much lower elevations has been reported in the Basque Mountains (Álvarez *et al.* 1985): one pair there nested at just 700m in the Sierra de Amboto, Bizkaia in 1998 (*A. O. Bizkaia 1998–1999*: 81). Breeding birds occur above the treeline, often on scree slopes and other broken ground, and generally near alpine pastures or open scrub formations. They are particularly attracted to snowfields, where they forage for trapped insects (Martí *et al.* 1986, Martín-Vivaldi *et al.* 1995).

The SA1 assessment of the Spanish population may be a high figure since the Catalan population has been put at only 1,600–2,800 pairs (CA2) and that of Castilla y León at less than 1,000 pairs (Sanz-Zuasti & Velasco 1999). Breeding populations are regarded as stable. The wintering population is estimated at 30,000–36,000 individuals (SA1), possibly including at least some birds from the French Pyrenees and the western Alps (AMI).

Some remain at high altitudes all year round. Adult males in particular have been shown to remain at 2,000m in the Sierra de Gredos all winter (Martí *et al.* 1988) and similarly in the Sierra Nevada (Martín-Vivaldi *et al.* 1995), in both cases foraging around ski installations and other buildings. Females and juveniles tend to move further (Martí *et al.* 1989). Many disperse widely and descend to lower altitudes, even to sea-level. They may then appear in river gorges, rocky areas, cliffs and the summits of lower sierras, but also on large buildings such as castles. Winter reports come from much of Iberia (SWA), but a preference for the eastern half of the Peninsula, with its abundant limestone ranges, seems clear. In recent years there have been observations, sometimes annually, from sheltered spots in the central and southern Iberian Range, the Portuguese uplands, the Levant coastlands from

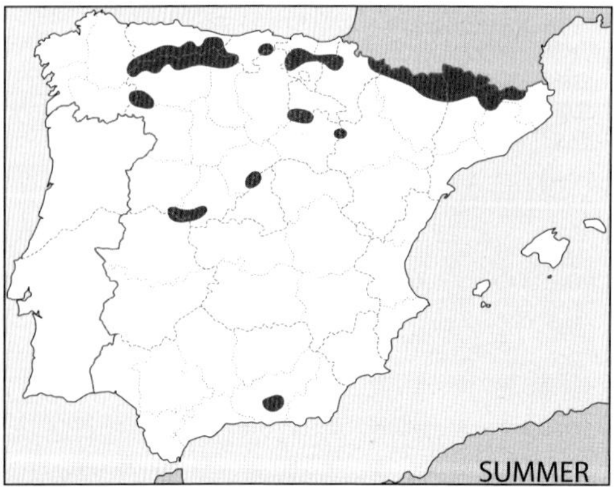

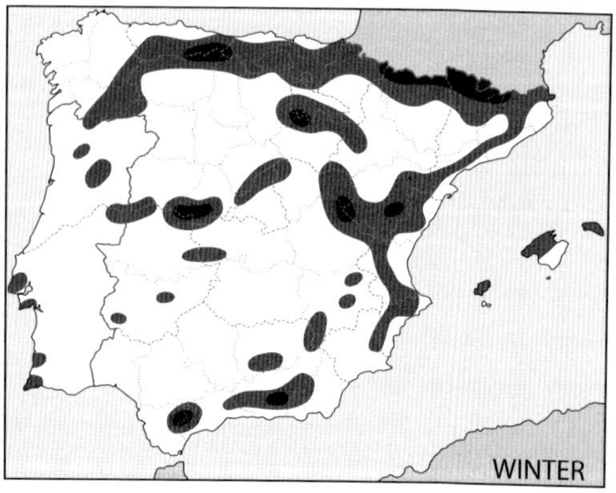

Barcelona south to Almería, the mountains of Extremadura, the Guadiana gorges on the Spanish/Portuguese border, the Sierra Morena; the Sierra de Cazorla, the Serranía de Ronda and Montes de Cádiz. The Spanish winter atlas similarly shows a high probability of occurrence then in the Pyrenees, Picos de Europa, the northern Iberian Range, the Sierra de Gredos and Sierra Nevada, but also in the southern Iberian Range: in the Alto Tajo region and Teruel mountains, and in the Serranía de Ronda.

Some are regularly reported at low levels in western and southern Portugal, for example in the Serra de Sintra and even on rooftops in Sintra town, as well as on rocky coasts and headlands, including the coast at Loulé, Cape Espichel, Cape Sardão and Cape St Vincent (AP). Wintering birds are seldom numerous locally, occurring singly or in loose groups of up to 5–6, but flocks of up to 40 have been reported occasionally. A count of 65 in two flocks at 800–950m on La Malena mountain, Morcín, Asturias, (*Ardeola* 57: 238) on 31 December 2009 is noteworthy. There was also a remarkable concentration of 57 at Marvão castle, Portalegre, on 7 January 2010 (*Anu. Ornitol.* 8: 80). Away from the breeding areas Alpine Accentors are mostly seen between late October and March, although there are also observations in September and April. A few appear in some years at Gibraltar, where they were first observed by Irby (1875), between late October and March, and some are then reported at low levels in northern Morocco (BM).

The Alpine Accentor occurs as a scarce migrant and winterer on the Balearic archipelago (*A. O. Balears 2010*: 266). The Balearic bird reports give 68 records during 1985–2010, involving at least 230 birds. Most are from Mallorca and Menorca but there are also records from Formentera, Ibiza, Cabrera and Dragonera. The origin of these birds remains unknown. Although 43 of the records are of just 1–2 birds the remainder include a number of observations of groups of ten or more, the largest count being 15 at Cúber on 28 December 1996 (*A. O. Balears 1996*: 100). The summits of the Serra de Tramuntana attract birds and probably support a small wintering population every year but some are recorded on cliffs and rocky terrain down to sea level. Most observations are during November–March with extreme dates for nearly all records of 23 October and 3 April. However, there are unseasonal records from Mallorca outside this period: 12 at Escorca on 18 September 1993, one at Pollença on 21 September 1996 and, most remarkably, ten on Puig Major (1,445m), the highest summit of the island, on 2 June 1990.

RUFOUS-TAILED SCRUB-ROBIN — *Cercotrichas galactotes*

(Alzacola Rojizo, Solitário-ruivo)

***C. g. galactotes*. A scarce local summer visitor to the southern third of Spain and adjacent areas of Portugal.**

Breeding population estimates. Spain: SA1 12,000–24,000 pairs; 2004 census 329,501 birds. **Portugal**: 500–5,000 pairs.

The Scrub-robin has a restricted distribution, chiefly in the thermomediterranean region but also in the mesomediterranean, where it is only present during the hottest part of the year (SA2, PA2). The population is concentrated in three main areas. The highest-density subpopulation occupies the Guadalquivir valley in Cádiz, Sevilla, southern Córdoba and western Jaén. A second subpopulation, that was formerly more important, is in the arid south-east in Almería, Murcia and Alicante, and is linked to the previous one by scattered populations in the lowlands of Cádiz and Málaga. The third subpopulation is north of the Sierra Morena, in the lowlands of Badajoz, extending into Ciudad Real in the east and Alentejo in the west. In addition, there are dispersed populations further north in Cáceres, north to Plasencia. There are occasional, apparently isolated, records of nesting in other regions, with recent instances in Madrid, Toledo, Valencia and Tarragona (SA2, Velasco 2009). The principal Portuguese population is in the Guadiana basin, in Alentejo and Algarve, with a more northern outpost in Beira Baixa.

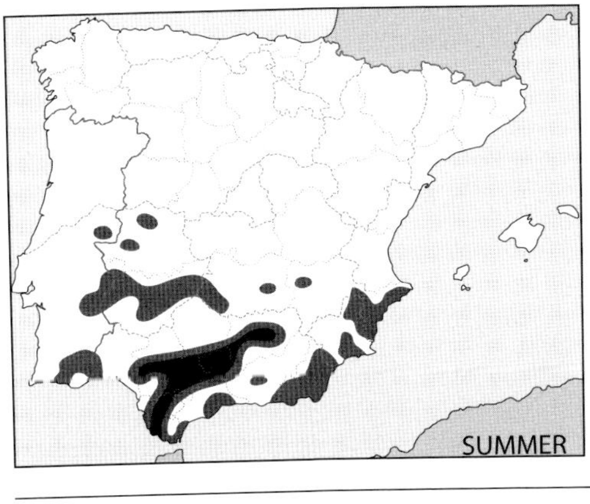

This species was familiar to 19th century authors who found it to be common in southern and eastern Spain. Saunders (1871) found it abundant in Andalucían vineyards, nesting in the cactus hedges. Arévalo (1887) described it as abundant in coastal areas of Andalucía, Valencia and Murcia. More recently, Gil Lletget (1945) also found it common in Extremadura, particularly in Badajoz, and stated that it occurred north to Candeleda, Ávila. He cites Abel Chapman as having found it north of Plasencia. Bernis (1952) found several pairs nesting at Oropesa, Toledo, in May 1951 and also observed it at Fuencaliente, Ciudad Real, in June 1951. The historical Spanish breeding range was thus very similar to the recent one, although the early references give the impression that it may have occurred at much higher densities than at present. For example, Chapman & Buck (1893) considered it to be one of the most frequent species in southern Spain, and Irby (1875) said that it was 'very plentiful' in the area of the Strait of Gibraltar.

The available estimates of the Spanish population (SA1, Seoane 2005) differ markedly. Irrespective of the population size, however, it is clear that the Scrub-robin declined considerably in Spain during the 1970s and 1980s, especially in Murcia, Alicante and the Andalucian coastlands, disappearing completely from some locations (López Iborra & Gil-Delgado 1988, Muñoz-Cobo 1990). The Scrub-robin may have been more widely distributed in the past in southern Portugal (AP) but there may have been some range expansion westwards in Alentejo in recent years (PA2). Population declines may be at least partly due to events in the winter quarters and on passage but the Scrub-robin has been affected on its breeding grounds by changes in land-use, including loss of ground to cultivation under plastic, and agricultural intensification, including use of herbicides and insecticides in vineyards and olive and citrus groves (LR). The creation of the Alqueva reservoir on the Guadiana in Alentejo may have displaced up to 200 pairs (AP).

Breeding birds occur where there are mixtures of dense low cover and open ground, interspersed with occasional taller shrubs or trees, quite often Carobs. They frequently forage on the ground and thus avoid continuous dense cover. The olive, citrus and almond groves and vineyards of southern Spain hold the bulk of the population (Cano 1960, SA2, Seoane 2005). They also occupy the ramblas of the south-east, clumps and hedges of Prickly Pear cactus and coastal Aleppo Pine plantations (López Iborra 1989), these last notably in Alicante. The Scrub-robin is normally encountered at low levels and is absent from the Betic Range and most of the Sierra Morena, but it reaches 1,100m at Guadix, Granada (Pleguezuelos 1992). It is a favoured host of the Common Cuckoo (LR, Álvarez 1994).

This is one of the latest summer visitors to return to Iberia from its winter quarters, in sub-Saharan Africa. Ringing in Sevilla has demonstrated a high degree of fidelity to its breeding areas (Domínguez & Cuadrado 1994) and three from that area have been recovered in Morocco during passage periods. There are records from the extreme south in late March but these are exceptional and passage is not normally evident at Gibraltar before mid-April. Most do not arrive until May and some do not appear until early June (BG). However, nests with eggs have been found in Alicante from May until early June (López Iborra & Gil-Delgado 1988). The breeding areas are abandoned from late July, chiefly in August and early September, although a few stragglers are recorded as late as October. Four winter records from the Comunidad Valenciana must be regarded as highly exceptional: they comprise single birds at Picassent on 9 February 1992, at Torrevieja on 14 December 1992, at Beniarrés on 25 January 1996 and at the Marjal del Moro on 29 December 1997 (*A. O. C. Valenciana 1992*: 95, *1996–1997*: 192).

The Scrub-robin is only a vagrant to the Balearic Islands, as in mainland areas north of the breeding range. Early observations were of two in Ibiza on 8 May 1930 and one on Formentera on 19 May 1930 (Ticehurst & Whistler 1930). There were 18 accepted records during 1989–2010 (*A. O. Balears 2010*: 85): often birds caught by ringers in Mallorca and Menorca and, especially, on the islet of Cabrera. Two of the Balearic records were in April but all the others were in May. There have been at least three occurrences on the Columbretes islands in late April–May (*A. O. C. Valenciana*) and several in Catalonia: dating back to 1955 (Ferrer *et al.* 1986) and with at least ten during 1986–2005 (Clavell 2002, *A. O. Catalunya*). Most of these records have clearly involved overshooting spring migrants. The Catalan records include an exceptionally early bird at the Aiguamolls de l'Empordà on 16 March 1996 (*A. O. Catalunya 1996*: 191) and single birds, perhaps the same individual, at the Llobregat delta on 12 August, 2 September and 16 September 1989 (Clavell 2002).

Elsewhere north of the breeding range there have been birds at Oviedo, Asturias, in August 1988 and August 1989 (*Ardeola* 37: 347); at Pendueles, Asturias, on the late date of 6 November 1995 (*A. O. Asturias 1996*: 144); and in Aragón at the Laguna de Sariñena, Huesca in May 1989 and near Zaragoza: in June 2005 (two) and July 2006 (*A. O. Aragón 1999–2003*: 193; *Ardeola* 52: 433, 54: 181).

EUROPEAN ROBIN *Erithacus rubecula*

(Petirrojo Europeo, Pisco-de-peito-ruivo)

***E. r. rubecula*. A widespread resident or partial migrant in the northern half of the Peninsula, more local in the south. Abundant on passage and in winter throughout the Peninsula and in the Balearics.**
***E. r. melophilus*. A scarce migrant and winter visitor.**

Breeding population estimates. Spain: SA1 1.2–3.0 million pairs; ACRE 9.9 million birds. **Portugal**: 50,000–500,000 pairs. **Andorra**: 700–1,000 pairs.

This is a common breeding species throughout the eurosiberian zone, where it occurs from sea level throughout the Atlantic and Biscay coastlands to the treeline in the mountains. It also occupies much of the supramediterranean, being present

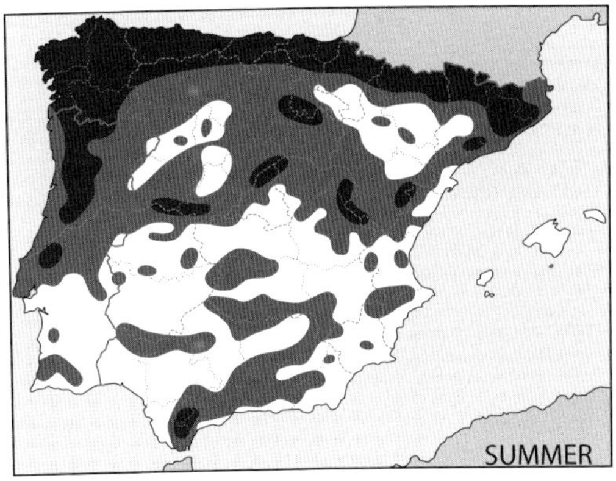

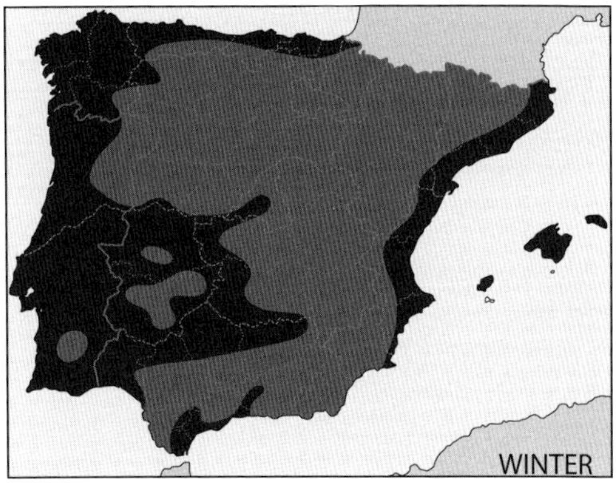

throughout the Iberian and Central Ranges and in much of the northern meseta. It is largely tied to relatively humid habitats and this is most evident further south, where it is progressively associated with mountains and, more locally, coastal areas. Thus it is present in the east coast mountains: from 800–1,500m in Valencia (Urios *et al.* 1991) but descending to the lowlands in Catalonia; in Las Villuercas in eastern Extremadura, the Montes de Toledo, the Sierra Morena and the Betic Range (SA2). It reaches sea level in the last of these in the humid wooded valleys of Cádiz and Málaga but in the east it ranges from 600m to 2,200m in the Sierra Nevada (Garzón & Henares 2012). In contrast it is largely absent from the Huelva coastlands, the Guadalquivir and central Ebro valleys, the arid south-east and central Levant coastlands, and much of Extremadura and the southern meseta, although some occur locally in these regions in riparian woodlands.

The Robin appears to have expanded its historical range in Spain. In the Madrid region, for example, it now breeds in mid-altitude riparian forests (Díaz *et al.* 1994), including the Casa de Campo park within the city, whereas formerly it was restricted to the Guadarrama and Somosierra mountains (Bernis 1952). During his visit to central Spain in May–June 1927, Witherby (1928) could only find it at San Ildefonso, Segovia; at the Navacerrada pass, Madrid, and in the Sierra de Gredos. Similarly, in their visit to Navarra, Ticehurst & Whistler (1925) recorded the Robin only as a shy inhabitant of the Pyrenean woodlands. Earlier, in Catalonia Saunders (1897) only found it in the Pyrenees, at Ribes de Freser, where it was 'chiefly an inhabitant of the higher wooded mountains'. Range expansion is still underway in Catalonia: a comparison between the first and second breeding bird atlases (CA1: 1983, CA2: 2004) shows a spread to coastal areas of Girona and Tarragona, with an overall 7% increase in occupied 10km-squares. The Robin is widespread and common in Portugal north of the Tagus but very local to the south, where it is largely restricted as a breeding species to upland areas, in particular the Serra de São Mamede as well as the Odemira region, Alentejo, and the Serra de Monchique, Algarve (PA2).

It was first reported nesting in the Balearics as recently as 2005, when a pair bred successfully at Palma, Mallorca (*A. O. Balears 2005*: 216). Successful breeding again occurred in Mallorca in 2009 and 2010 (*Ardeola* 57: 542, 58: 508). A progressive colonisation of the islands may be in progress, given the ongoing range expansion seen in Catalonia and southern France.

Some Robins occur in Iberian parks and gardens, and this may be an increasing trend, but the Robin is still predominantly a woodland species in the region. It inhabits the shrub layer and woodland floor in a broad range of forested habitats, whether broad-leaved, coniferous or mixed. It also occurs on woodland edges and in tall scrub formations, well-developed hedgerows and riparian growth. Many frequent eucalyptus plantations in the northern coastlands, where these typically have a dense native understorey.

The SA1 estimate of the Spanish breeding population may have been too conservative since 800,000–1 million pairs are estimated to breed in Catalonia alone (CA2). The population showed a significant overall increase on SACRE sites during 1998–2012 and this was especially notable in the south (Escandell 2011, 2013). The highest densities (81–106 birds/km^2) are reported from forested habitats in the eurosiberian region. The Portuguese and Andorran populations are at least stable.

Iberian Robins are partial migrants, deserting the colder regions of the Peninsula to winter in warmer areas, often in the south (AMI, AI2, SWA). They are joined by very large numbers of birds from further north in Europe, which disperse widely throughout the Peninsula but particularly settle in the thermo- and mesomediterranean regions. There may be a tendency for the immigrants to avoid areas densely occupied by local birds, as appears to be the case in the humid Cádiz woodlands (Tellería *et al.* 2001, Tellería & Pérez-Tris 2005). This may also occur in the northern and western Atlantic coastlands, where winter densities are very high but the number of recoveries of foreign-ringed birds is relatively low: in Galicia, for example, they amount to just 1% of the total compared with 36% in Andalucía, 15% in the Balearics and 10% in Catalonia (AI2).

Table 50. Recoveries of foreign-ringed Robins in Spain (n = 2,102). After Bueno (1998).

Country of ringing	Total recoveries in Spain	Natal origin confirmed
Scandinavia, Finland	801	90
Germany	339	51
Poland	237	11
Benelux	231	46
Switzerland	157	8
France	144	20
British Isles	63	6
Former USSR	58	7
Former Czechoslovakia	46	12
Italy	18	0
Austria, Hungary, former Yugoslavia	8	2

Wintering birds are present during September–April. Ringing recoveries suggest that 75% of foreign-ringed birds arrive during the last ten days of October and the first ten days of November (Bueno 1998). Autumn passage at a site in southern Madrid where the Robin does not breed extends from mid-September to early December with a very marked peak in October; the spring passage there peaks in March (Pinilla & de la Puente 2003). Robins arrive in the south chiefly from late October and most have departed by late March. The Robin is the second most frequently ringed species in Gibraltar (after the Blackcap) and the majority are caught there from late October into November: most are migrants that continue south to winter in North Africa, where they are abundant in Morocco from October to March (BM).

Ringing recoveries show that Iberia receives wintering and passage Robins from northern Europe in particular (Table 50). Some individuals from the weakly migratory population of the British Isles, subspecies *E. r. melophilus*, also occur (Wernham *et al.* 2002). Winter recoveries reveal that birds of Nordic origin tend to winter in southern Iberia whereas those from further south and east in Europe tend to remain on the Levant coasts, including in the Balearic Islands (Bueno 1998). Birds ringed in Spain to 2010, including through-migrants, have produced 152 recoveries in North Africa: 128 in Algeria, 23 in Morocco and one in Ceuta (SMA). The Algerian recoveries are mainly of birds that were ringed on passage in the Balearic Islands and on the Levant coast and most of them were found in the Algerian coastlands.

Wintering birds may be found in similar habitats to those used by the breeding population, chiefly woodlands with a well-developed shrub layer, but including the Mediterranean woodlands and taller scrub formations where the Robin does not breed. Fruits of a wide range of species, including Lentisc, Wild Olive and Myrtle, are an essential and often dominant component of the diet in winter (Herrera 1981, Jordano 1989). Retraps of ringed birds show that some Robins at least are faithful to the same wintering sites in successive years (Herrera & Rodriguez 1979, Finlayson 1980). There are no data on the total size or trends of the wintering population but local studies in the far south, in the Cádiz corkwoods, have suggested that wintering numbers may be declining there (SWA). Since the source populations are increasing it may be that some migrants are travelling less far south than used to be the case.

RUFOUS NIGHTINGALE *Luscinia megarhynchos*

(Ruiseñor Común, Rouxinol-comum)

L. m. megarhynchos. **An abundant summer visitor to the Peninsula and Balearics. Also abundant on passage.**

Breeding population estimates. Spain: SA1 0.45–1.70 million pairs; ACRE 10.7 million birds. **Portugal**: 100,000–1 million pairs. **Andorra**: 30–75 pairs.

The Nightingale is abundant throughout the Iberian Peninsula with the clear exceptions of coastal northwestern Portugal, most of Galicia, the Biscay coastlands and the northern flanks of the Cantabrian Range, in all of which it is a rare and local nester (SA2, PA2). Ticehurst & Whistler (1928) commented on its absence from Galicia. It is widely distributed throughout the

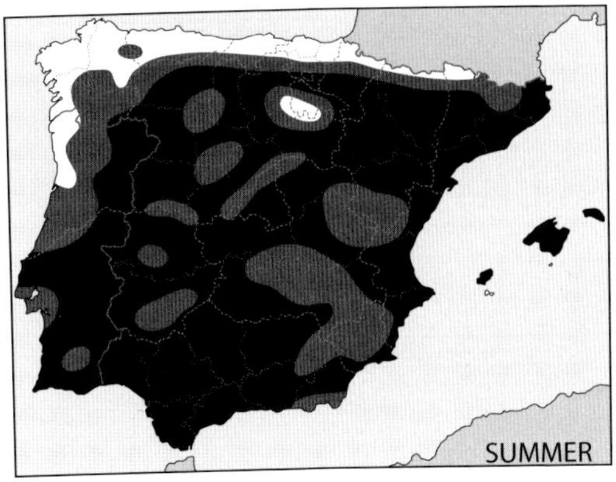

supra-, meso- and thermomediterranean climatic regions. It also nests commonly in the Balearic Islands.

It requires dense undergrowth, in the form of tall scrub or well-developed broad-leaved woodland and undergrowth. It may be found in untidy areas in cultivated land, including orchards, gardens and parks, and, especially in the tangled shrubbery of watercourses of all types and sizes: streams, rivers and irrigation channels, and in the peripheral scrub of lakes and marshes. It is very much associated with water in the warmest parts of the Peninsula and is very local there in arid zones, including significant areas of Almería, Murcia and Castilla-La Mancha. Few nest at higher altitudes than 1,000m, but there are some nesting pairs as high as 1,500m in the eastern Pyrenees (Ferrer *et al.* 1986) and Central Range (Carnero & Peris 1988), and to nearly 2,000m along the watercourses of the Sierra Nevada (Garzón & Henares 2012). Those in the northern Iberian Range occur up to 1,200m on the southern slopes but only reach 800m on the northern side (de Juana 1980).

The SA1 assessment was probably an underestimate since the Catalan population alone has been considered to number about one million pairs (CA2). The ACRE survey found 22% of the birds in Castilla y León, 20% in Catalonia, 19% in Andalucía, 13% in Aragón and 9.5% in Castilla-La Mancha. The population was regarded as highly stable during 1998–2006 but by 2012 the overall trend on SACRE sites was a significant increase (Escandell 2013). The highest reported densities (74–115 birds/km^2) are in riparian habitats and orchards of the supra- and mesomediterranean. The Portuguese and Andorran populations are considered stable.

The celebrated song is first heard in southern Iberia in mid March, when the first birds return from their winter quarters in the savannas and forests of West Africa. The earliest arrivals in the north are not usually before mid-April (de Juana 1980, Calvet *et al.* 2004). The spring passage peaks in April and continues throughout May and into June, the later birds very probably being migrants to more northerly countries. Few sing later than June but the southward departure is in August–September, with stragglers into October. Unseasonal records are very rare but one was ringed in Albacete on 22 November 1998; one was singing by the light of street lamps in Murcia city on 19 December 1987, and one was found at the Paul do Boquilobo, Santarém, on 21 January 2004 (*Ardeola* 35: 313, 46: 161; *Anu. Ornitol.* 4: 41).

There are few recoveries within Iberia of birds ringed outside the Peninsula. Bueno (1990) gives only 18 and they include eight ringed in Germany, four each in Belgium and France and two in the British Isles. There are three more recent recoveries of birds ringed in Switzerland, Slovenia and the Czech Republic (SMA). Eleven birds ringed in Iberia have been recovered in France and there are also single recoveries from Italy, the Czech republic, The Netherlands and Britain, as well as 17 from Morocco and one from Algeria (SMA).

BLUETHROAT *Luscinia svecica*

(Ruiseñor Pechiazul, Pisco-de-peito-azul)

***L. s. azuricollis.* Breeds locally in the mountains of central and northwestern Spain, exceptionally in Portugal. Summer visitor. Males tend to have unspotted throats or indistinct white spots. Larger and longer-winged than other races.**

***L. s. cyanecula.* A scarce passage migrant and local winter visitor. Males have white-spotted or unspotted throats.**

***L. s. namnetum.* A scarce passage migrant and local winter visitor. Smaller than *cyanecula* and greyer-brown above. Males are also white-spotted.**

***L. s. svecica.* Claimed to occur on passage in eastern Iberia and the Balearics but there are no conclusive records. Males have red-spotted throats.**

Breeding population estimates. Spain: SA1 9,000–12,800 pairs.

The Bluethroat has two principal separate breeding populations in the high mountains of central and northwestern Iberia. One occupies the Central Range, from the Sierra de Somosierra in the east through the Sierras de Guadarrama, Gredos and Béjar. The other is in the central and western Cantabrian Range and the Montes de León: in Ourense, Asturias, Zamora, Palencia and León. A third small population has been reported from the Puerto de Piqueras in La Rioja (SA2). There are only sporadic reports

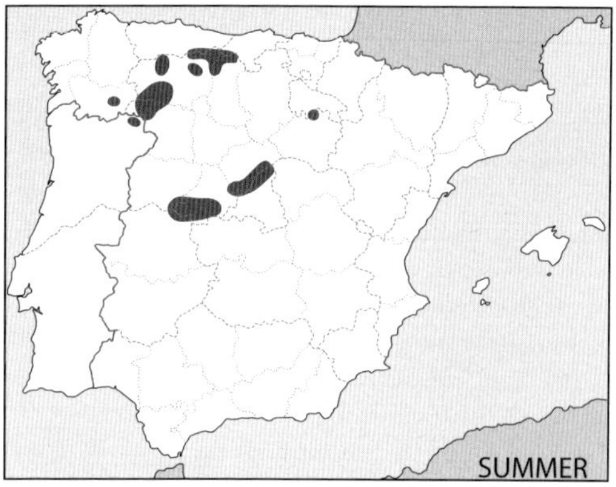

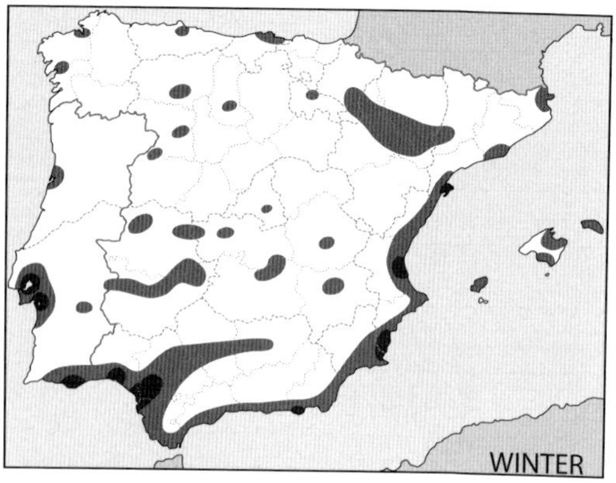

of nesting in Portugal: at the Serra da Estrela in 1992 and 1993, and the Serra de Montesinho in 2001 (AP). An exceptional instance of breeding from the Sierra Nevada, Granada, in 2010 (*Ardeola* 57: 542) has attracted reservations and is pending corroboration (J. Garzón, pers. comm.).

The Iberian birds are genetically distinguishable from those of western Europe and may be regarded as a separate subspecies, *L. s. azuricollis* (Johnsen *et al.* 2006). Corley Smith & Bernis (1956) draw attention to the extreme variability of Spanish Bluethroats with respect to the presence/absence and colour of the throat spot. Corley Smith (1959) noted that 80% of males in the Sierras de Guadarrama and Gredos lacked the throat spot and that only 5% showed a well-defined white one, most of the others having ill-defined spots. A breeding season study in the Central Range during 2003–09 (Campos *et al.* 2011) has confirmed this variability: about half of 185 captured males lacked a throat spot altogether, most of the remainder had a white spot and six individuals had a red spot – although five of these six were subadults. The Central Range *azuricollis* Bluethroats were also clearly longer-winged than *L. c. namnetum*, suggesting that the former are longer-distance migrants than the latter (Campos *et al.* 2011).

Iberian Bluethroats nest at high altitudes, above the treeline, from over 800m in the Montes de León to over 2,000m in the Sierra de Gredos. The birds of the northern population favour the warmer, south-facing slopes but those in the Central Range occur on the cooler, more humid, northern side. They inhabit montane scrub in both areas, especially of *Erica* and *Cytisus* species, selecting fairly open formations interspersed with low herbaceous growth. For example, in Madrid they select *Cytisus oromediterraneus* scrub 50–70cm tall interspersed with meadows and preferably near small water courses (Gómez-Manzaneque 1997). Those nesting at lower altitudes in the Montes de León, however, inhabit Cistus scrub in open Holm Oak woodland (*Ardeola* 42: 228, García *et al.* 2008). Breeding birds are commonly encountered in loose colonies of anything up to 25 pairs (SA2). Densities of up to 5 males/10ha have been reported in the most favoured areas.

The Bluethroat was first reported nesting in Iberia in June 1927 on the Sierra de Gredos (Witherby 1928). Chapman & Buck (1893) had earlier found a recently dead male there, after a stormy night in early May, but they assumed that it was just a migrant. Its presence in the Sierra de Gredos and also in the Sierra de Guadarrama was confirmed in the early 1950s by Corley Smith & Bernis (1956), who also searched suitable areas of the Cantabrian Range for the species but were unable to find it.

The breeding sites are abandoned in winter when most or all of the (*azuricollis*) population leaves the Peninsula to winter in the river basins of sub-Saharan West Africa. They are joined by Bluethroats from western and central Europe, specifically those of the French Atlantic coastal population (*L. c. namnetum*) and those nesting in the Low Countries (*L. c. cyanecula*). The great majority (81%) of 576 individuals ringed at the Laguna de La Nava during the autumn passages of 1999–2008 were *cyanecula*, the remaining 19% being *namnetum* (González Villalba *et al.* 2011). Migrants cross the Peninsula and the Balearic Islands during August–November, chiefly in September–October, and return during March–May, principally in April (BSG, AI2, Arizaga *et al.* 2011). A proportion remain to winter in Iberia, chiefly on the eastern, southern and western coasts, where they occupy wetlands and the marginal scrub around irrigation ditches, saltpans and estuaries. Favoured wintering sites include the Odiel and Guadalquivir estuaries, the Cádiz Bay saltpans, the Mar Menor, El Hondo, the Albufera de Valencia and the Ebro delta, but quite small sites often hold wintering birds (SWA). Some 300–500 have been estimated to winter in Catalonia (CWA). In the Balearics there is a small wintering population in Mallorca and some also winter in Menorca and Ibiza (*A. O. Balears 2010*: 267).

Some are also reported at northern and inland wetlands in winter and they occur regularly at such sites during passage periods. Bermejo & de la Puente (2004) describe regular wintering, as well as passage, at a site on the Río Jarama in Madrid region and consider that inland wintering in Spain has increased. Males predominated in the wintering population and showed a high degree of philopatry to the wintering site. The great majority of the 283 birds handled by Bermejo & de la Puente (2004) were also *cyanecula* but eight individuals of *namnetum* were handled, only one of which was in winter. However, up to 75% of the bluethroats found on passage on the Biscay coasts are *namnetum* (Arizaga *et al.* 2006, Quintana 2009) and this subspecies largely winters in Portugal, southwestern Spain and northwestern Africa (Zucca & Jiguet, 2002). The Iberian breeding birds are not readily distinguishable outside the breeding season from *cyanecula* individuals from central Europe and there is no ringing evidence to suggest whether or not any of the Iberian birds remain to winter in the Peninsula (Bermejo & de la Puente 2004). Wintering Bluethroats were seldom reported in Spain before the mid 20th century, although Arévalo (1887) said that they wintered in Andalucía and Valencia. They have been reported increasingly widely since the 1970s (SWA), so their winter presence may be a relatively new phenomenon, but it may also have been revealed by more thorough recent coverage, in particular the increase in ringing.

Recoveries of individuals ringed outside the Peninsula whose natal origin was known are chiefly of birds from Belgium (Bueno 1990). Recoveries in general are principally of birds ringed in France, the Low Countries and Germany, and these countries account for most recoveries of Iberian-ringed birds (Table 51). They show that birds generally enter the Peninsula at either end of the Pyrenees but predominantly in the east. They also show that adults predominate within the wintering population and that males outnumber females both during the postnuptial passage and in winter (Hernández *et al.* 2003). Ringing recoveries have shown a high degree of fidelity to winter quarters (Arizaga *et al.* 2011). The great majority of Iberian recoveries are from the Mediterranean coastlands, the south-west: especially the Guadalquivir estuary, and the Ebro valley (SMA).

Table 51. Iberian recoveries of foreign-ringed Bluethroats (n = 496) and foreign recoveries of Iberian-ringed Bluethroats(n = 155) (Sources: Bueno 1990, SMA, AP).

Country of ringing/recovery	Total recoveries of foreign-ringed birds	Total recoveries of Iberian-ringed birds
Belgium	231	24
France	139	38
Germany	32	31
Netherlands	75	39
Czech Republic	7	14
Switzerland	4	1
Others	8[1]	8[2]

[1] Morocco (2), British Isles, Hungary(3), Poland, Austria.

[2] Algeria (3), Switzerland, Slovenia, Hungary, Poland, Russia.

There are no definite observations or recent recoveries of the red-spotted Scandinavian race (*L. c. svecica*), which migrates southeastwards to Asian winter quarters. Corley Smith saw a presumed migrant Bluethroat with a red throat spot at El Pardo, Madrid, in May 1956 and at least one other among the breeding birds on the Sierra de Guadarrama in May 1956 (Corley Smith & Bernis 1956). They cite Drost (1935 *Anz. Orn. Ges. Bayern*) who notes two Iberian recoveries of birds said to be of this race: one in Braga, northern Portugal and the other at Casas Viejas (Benalup de Sidonia), Cádiz. Ticehurst & Whistler (1930) saw two birds, clearly migrants, that they assigned to this race in Ibiza in May 1930: one was a female identified by wing-length. A more recent claim is one captured in the Ebro Delta in May 1966 (Blondel 1966). There are also seven records, including individuals trapped for ringing, of red-spotted Bluethroats in Morocco (Thévenot *et al.* 2003). The possible occurrence of this subspecies on passage through southeastern Spain (Peiró 1997) has been contested on the grounds that misidentification is likely (Cortés *et al.* 2002). Probably none of the Spanish observations are conclusive, as Clavell (2002) remarked with respect to Catalonia and the Balearics.

SIBERIAN BLUE ROBIN

Larvivora cyane

(Ruiseñor Azul, Pisco-azul-siberiano)

Vagrant to Spain.

A first-winter male was ringed at the Ebro delta on 18 October 2000 (Bigas & Gutiérrez 2000). This is a very rare vagrant to Europe, with only three records in Britain, one in the Channel Islands and one in Spain, all of them in October.

RED-FLANKED BLUETAIL *Tarsiger cyanurus*

(Ruiseñor Coliazul, Pisco-de-flanco-ruivo)

Vagrant to Spain.

A female/first-winter was ringed at the Llobregat delta, Barcelona, on 17 November 1998; a first-winter male at the Devesa del Saler, Valencia, was first seen on 16 November 2005 and later ringed, but was found dead during a cold snap in January 2006; a first-winter was ringed at Aroche, Huelva, on 24 October 2010; a first-winter or female was at the Clot de Galvany, Alicante, on 1 December 2010, and a first-winter was at Sant Feliu de Buixalleu, Girona, from 26 December 2010 to 23 January 2011. The Huelva record was inland but the others were on the Mediterranean coast. All appeared late in the year and two lingered into January. More recently, one was photographed at Villar de Olalla, Cuenca, in March 2013 (RBS).

Western European records have been more frequent in recent years, perhaps due to the recent increase in the small Finnish population. There are 112 from Britain: 30 of them in 2010, two in Ireland and 12 in France, where the first was in 1993. Most appear in autumn, records peaking in October.

BLACK REDSTART *Phoenicurus ochruros*

(Colirrojo Tizón, Rabirruivo-comum)

***P. o. gibraltariensis*. A common resident in Spain north of the Duero river, and in Andorra. Common on passage and in winter throughout the Peninsula and in the Balearic Islands.**
***P. o. aterrimus*. A common resident in central and southern Spain and throughout Portugal. A typical male is jet black above and below, where *gibraltariensis* is dark grey. Included in *gibraltariensis* by HBW, the variation being somewhat clinal, but nonetheless distinctive.**
***P. o. phoenicuroides*. Vagrant to Spain (1).**

Breeding population estimates. Spain: SA1 400,000–900,000 pairs; ACRE 1.8 million birds. **Portugal**: 20,000–200,000 pairs. **Andorra**: 6,000–10,000 pairs.

The Black Redstart is widespread throughout the northern half of the Peninsula, where it ranges from sea level to high altitudes. Here it is only widely absent from the Ebro depression. It is almost entirely a montane species further south where it is common in the mountains of Cáceres, the Montes de Toledo, the eastern Sierra Morena and throughout the Betic Range and its outliers. All these southern areas are occupied by *P. o. aterrimus*. It does not breed in the southern lowlands in Spain nor in the Balearic Islands (SA2). It is widespread in Portugal north of the Tagus, and may be found nesting in the older quarters in Lisbon. However, further south in Portugal it is practically restricted to Ribatejo and the rocky southwestern coastline south of Cape Sines (PA2).

The Black Redstart breeds as high as 2,300m in the Sierra de Gredos (San Segundo 1990), to 2,700m in the eastern Pyrenees (Ferrer *et al.* 1986) and to 3,300m in the Sierra Nevada (Pleguezuelos 1992). It is strongly associated with rocky areas such as cliffs, scree slopes and mountain pastures with rocky outcrops. Some nest in open scrub and woodlands that are also interspersed with rocky stretches. It also nests readily on man-made equivalents: on ruins, bridges and large edifices, such as castles. It is a frequent nesting species among the rooftops of towns and villages throughout the northern half of the Peninsula

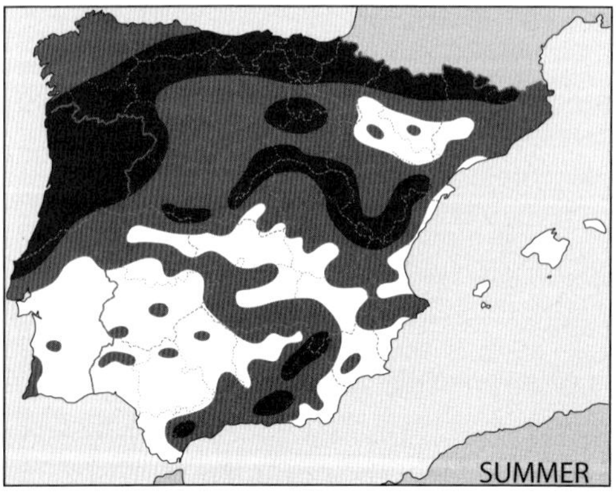

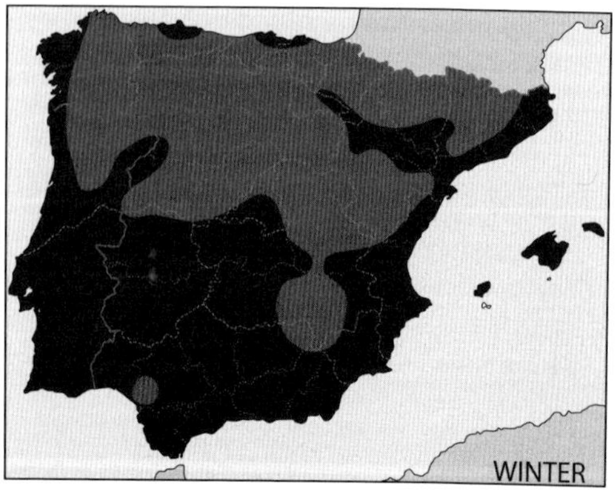

and this tendency towards urban nesting is tending to spread south. Urban nesting was already known during the 19th century from such locations as San Ildefonso, Segovia (Lilford 1866); Potes, Cantabria (Irby 1883) and Andorra la Velha, Andorra (Eagle Clarke 1889). However, in central and southern Iberia the Black Redstart was still largely restricted to high mountains during the mid 20th century (Gil Lletget 1945). A spread into lowland towns and villages has taken place since the 1960s and 1970s in at least the Castilian mesetas (Bernis 1988) and Catalonia (Ferrer *et al.* 1986, CA2).

The highest reported densities are in the villages and towns both of the Biscay lowlands and the supramediterranean region (23–56 birds/km²). The population of Catalonia alone numbers some 36,000–70,000 pairs (CA2). The Spanish population showed no overall significant trend during 1998–2012 (Escandell 2013) but a moderate increase was noted in Portugal during 2004–09 (ACP). It is regarded as the most widely distributed species in Andorra, with a stable population (ANA).

The Iberian birds are largely resident although those nesting at high altitudes descend lower in winter. However, the Iberian Peninsula together with the Balearic Islands is a major wintering area for very large numbers of *P. o. gibraltariensis* from further north in Europe. Wintering birds are present during September–April but the chief influx is in October–November and most depart in February–March (Bueno 1992b). The first wintering birds in the Madrid region are usually detected in mid October and the last ones depart in mid March (SWA). Ringing shows that the great majority of the wintering immigrants originate from central Europe (Table 52) and that they are mainly found in the south and east, and in the Balearics. The relative scarcity of immigrant Black Redstarts in the west may be illustrated by the fact that there are only 16 ringing recoveries of extra-regional birds from Portugal (AP) but 417 from Spain (SMA). As many as 39% of a sample of 180 Iberian winter recoveries of birds ringed outside the region were in Andalucía and a further 20% were in the Comunidad Valenciana, 12% in the Balearic Islands and 11% in Catalonia, whereas there were none at all in Galicia, Asturias, Navarra or La Rioja, and just one each in Cantabria and Castilla y León (Bueno 1992b). Wintering birds may be encountered at low levels, especially on rocky coasts, throughout the Peninsula as well as throughout the central and southern lowlands (SWA). Densities of 10–60 birds/km² are reported from favoured habitats (SWA). They occur in open country, frequently near rocky areas, often around buildings or bridges, including in towns and villages. Some may also be found in winter in cultivated areas including vineyards, almond groves, Holm Oak dehesas and irrigated crops. Individuals of *P. o. aterrimus* are also widespread in the southern lowlands in winter, having descended there from their montane breeding sites.

Table 52. Recoveries of foreign-ringed Black Redstarts in Iberia (after Bueno 1992b). N = 401.

Country of ringing	Total recoveries in Iberia to 1990 (%)	Natal origin confirmed
Germany	224 (55.9)	157
Belgium	71 (17.7)	50
Switzerland	47 (11.7)	24
France	42 (10.5)	26
Netherlands	8 (2.0)	6
Others*	9 (2.2)	2

* Denmark, British Isles, former Czechoslovakia.

Many cross the Peninsula to winter in North Africa and their movements are often evident around the Strait of Gibraltar, where the peak southward passage is from mid-October to mid-November, with a return movement from mid February to early April, peaking in March (BG). Birds ringed in Spain have produced seven recoveries from Morocco and 12 from Algeria, the latter largely of birds ringed on passage in the Balearic Islands (SMA).

There is one accepted record of the Asiatic subspecies *P. o. phoenicuroides*: a first-winter male trapped and ringed at Isla Grossa, Murcia, on 23 March 2011.

COMMON REDSTART *Phoenicurus phoenicurus*

(Colirrojo Real, Rabirruivo-de-testa-branca)

***P. p. phoenicurus*. Breeds locally in mainland Spain and Portugal. Summer visitor. Abundant on passage throughout the Peninsula and in the Balearics.**

Breeding population estimates. Spain: SA1 75,000–94,000 pairs; ACRE *c.* 188,000 birds. **Portugal**: 1,000–10,000 pairs.

The Common Redstart has a highly fragmented and perhaps relict breeding range in Iberia. The largest continuous area of distribution is in the Cantabrian Range and the Biscay lowlands, where it nests from sea level to over 1,500m in the mountains. Further south it is chiefly found in upland woodlands, principally at 800–1,200m but locally up to 1,800m. Here it is widespread in the northern Iberian Range between Soria and Burgos (de Juana 1980, Román *et al.* 1996) but it is much more local in the south of the range. It also inhabits the Central Range, the Sierra Morena in Huelva, the Sierras de Cazorla and Segura in Jaén, and points of the Serranía de Ronda (SA2). In addition there are numerous instances of occasional and sporadic nesting elsewhere. It is curiously uncommon and local in the Pyrenees, where it is not known to breed in Andorra (ANA), and it is absent when breeding from most lowland areas of the Peninsula and from the Balearic Islands. The Portuguese range is also highly fragmented. It occurs locally throughout the country, particularly in the north of Bragança district, Trás-os-Montes, Beira Alta and Beira Baixa, Ribatejo, western Alentejo and in the Algarve mountains (PA2, AP).

It is mainly associated with mature woodlands that offer scattered clearings and limited undergrowth. The association with large mature broadleaves is thought to be related to its requirement for nest holes and this may also account for its very patchy Iberian breeding distribution. Oakwoods are especially favoured but it also nests in conifers and in mixed woodlands. Some use holes in walls and in rock faces. The birds in the northern coastlands may be found nesting in orchards and gardens but those further south are especially associated with oak dehesas, although they will nest in Scots Pine forests if nestboxes are available there. Those in the Serranía de Ronda occur in stands of the Spanish Fir (SA2). The highest densities of breeding birds (5.6 birds/km^2) have been reported from pastures with trees above 750m in the Cantabrian Range.

The SA1 estimate of the Spanish population was later thought to be too high (SA2). Although the ACRE estimate is similar, some of the regional estimates offered by that survey are much higher than local knowledge would support: for example, the ACRE survey put the Catalan population at a minimum of 7,000 birds but the second Catalan Atlas (CA2) found only 5–20 pairs there. The actual population size thus remains to be confirmed. Similarly, the sample sizes available to the SACRE survey have been too small to enable any reliable trends in the population to be detected. Earlier assessments point to significant decreases and some breeding range contraction in Spain since the 1970s, matching serious declines elsewhere in Europe, that may be attributed to droughts in the Sahel wintering areas (LR). These declines were especially noticeable in the Biscay coastlands (Álvarez *et al.* 1985, Noval 1986, LR).

The Common Redstart is more widespread and obvious during its passage to and from its sub-Saharan winter quarters. The earliest arrivals in the south are in late March. The northward movements peak in April and continue until May, with stragglers in early June. The return movements are noted from late August, peaking in September–October, with stragglers into November (BG, Tellería 1981, Bueno 1992a). Ringing recoveries show that the spring passage is concentrated on the eastern side of the Peninsula and in the Balearic Islands, east of 4°W. In contrast, the autumn passage is principally west of 4°W (Bueno 1992a). Both of these trends are significant and hence the Common Redstart is relatively less abundant on spring passage in Portugal and western Spain, and on autumn passage in eastern Spain and the Balearics. Common Redstarts from all the western European populations cross Iberia on passage (Table 53). Recent recoveries also include individuals ringed in Hungary and Tunisia. Common Redstarts ringed in Spain, including in the Balearic Islands, have also produced 23 recoveries in North Africa: 16 in Algeria, six in Morocco and one in Tunisia (SMA), and one in Niger (*Ecología* 16: 367).

Table 53. Recoveries of foreign-ringed Common Redstarts (n = 569). After Bueno 1992a

Country of ringing	Total recoveries in Iberia to 1990 (%)	Natal origin confirmed
Germany	183 (32.1)	85
Scandinavia and Finland	130 (22.8)	43
British Isles	84 (14.7)	31
Low Countries	64 (11.3)	41
France	51 (9.0)	20
Switzerland	27 (4.8)	17
Poland	14 (2.5)	3
Others*	16 (2.8)	9

* Baltic States, Russia, former Czechoslovakia, Italy.

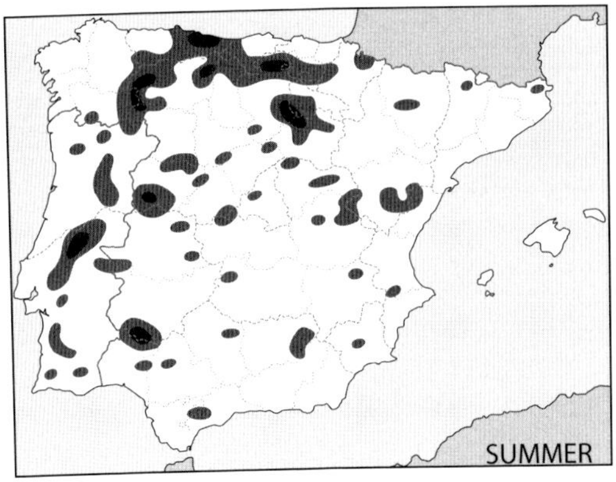

The recoveries considered by Bueno (1992a) include several in December–February. However, there are no unseasonal sight records of Common Redstarts in Iberia that cannot be attributed to late/early migrants. Munn (1921, 1924) described it as plentiful in Mallorcan orchards in winter and common in winter in Menorca but it is more than likely that he did not take care over distinguishing female and first winter Black Redstarts from this species. There are however midwinter records from Morocco and the species is considered an occasional winter visitor there (BM). In the absence of more concrete data we consider that the unseasonal ringing records may well have involved incorrectly reported rings or misidentifications.

MOUSSIER'S REDSTART *Phoenicurus moussieri*

(Colirrojo Diademado, Rabirruivo-mourisco)

Vagrant to Spain and Portugal.

Irby (1873) saw a male near Tarifa, Cádiz, in October. The five accepted records have been of the striking males: at the Ebro Delta on 7 April 1985; on the Málaga coast on 26 April 2000; at Cape St. Vincent, Faro, from 16 November 2006 to 14 January 2007; on Cabrera, Balearic Islands, during 19–23 October 2008, and at Grazalema, Cádiz, on 7 April 2010. Two further observations of males were made in 2012: at Punta de n'Amer, Mallorca, in January, and at Tarifa, Cádiz, in March, and another male was at Constantí, Tarragona, in March 2013 (RBS).

Moussier's Redstart nests as close to Iberia as the southern shore of the Strait of Gibraltar. It is largely resident but some dispersive and altitudinal movements occur. A few have appeared in Europe, largely in spring or overwintering, in Italy: six records (Janni & Fracasso 2011), Malta: 40 records (Fennech & Sammut 2013) and Greece, and one was found in Britain in April 1988 (Slack 2009).

WHINCHAT *Saxicola rubetra*

(Tarabilla Norteña, Cartaxo-nortenho)

Breeds locally in the northern third of the Peninsula. Summer visitor. Common on passage throughout the Peninsula and Balearic Islands.

Breeding population estimates. Spain: SA1 15,000–20,000 pairs; ACRE 154,000 birds. **Portugal**: 30+ pairs. **Andorra**: 75–200 pairs.

The breeding distribution is particularly associated with the supramediterranean region. It is most frequent in the northern third of Spain, with a marginal presence in northeasternmost Portugal, but its range is far from continuous there (SA2, PA2). It seems to avoid most of Galicia, the Biscay coastlands and northern Navarra, although some breed in southern Asturias and at scattered localities in Cantabria and the Basque Country. The Pyrenean population is largely in the east of the range, from eastern Huesca in the west through Lleida to western Girona in the east, taking in Andorra. The population of the Cantabrian Range is chiefly south of the watershed in León, Palencia and Burgos, with outposts in Valladolid and Salamanca in the south-west of the northern meseta, as well as extending to the western foothills of the northern Iberian Range. There is a southern outpost in the eastern Central Range, in Madrid and Segovia. A scattered population in southern Galicia extends into northernmost Portugal in Peneda-Gerês National Park and Montesinho Natural Park. Breeding has been reported exceptionally in the Serra de Tramuntana in Mallorca (Ferrer *et al.* 1986).

The montane populations favour pastures, including those above the treeline. Those in the eastern Pyrenees occur to 2,400m but are chiefly found at 1,000–2,000m (CA2). Some occur in pastures and wet meadows in valley bottoms. Areas of *Calluna* and Juniper heath are also used as well as fallows in the cereal croplands of the northern meseta. Overall, the Whinchat selects areas of tall grass, often not far from watercourses and having at least some shrubs, bushes or fences to serve as song-posts and hunting perches (de Juana 1980). The breeding of the Whinchat in Iberia was apparently first reported by Castellarnau

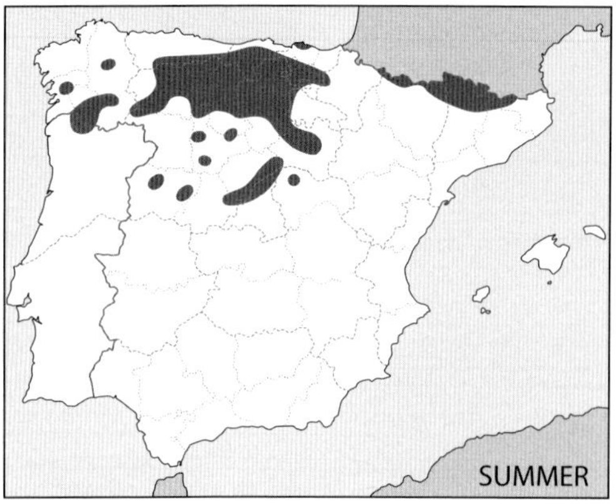

(1887), who considered it rather common in meadows in the northern foothills of the Sierra de Guadarrama, in Segovia. There were, however, no further breeding reports from this area until 1955 (*Ardeola* 4: 204). Sáez-Royuela (1958) reported breeding in Burgos in 1950–54.

The SA1 estimate was regarded as too high since Sanz-Zuasti & Velasco (1999) estimated just 6,000 pairs for Castilla y León, which accommodates the majority of the birds. Only 800–1,200 pairs are estimated to breed in Catalonia (CA2). In view of all these estimates, the ACRE figure seems implausible. In Portugal there are an estimated 20–30 pairs in Peneda-Gerês and a few more in Montesinho (AP).

The Whinchat occurs far more widely on passage when it may appear in all types of open country throughout the Peninsula. It is also a common migrant in the Balearic Islands. The spring passage commences in late March in the south and peaks in April–May, with some still arriving in early June. The return movement is from late August to mid November, peaking in September–October and with occasional stragglers into December (BG, Tellería 1981, Bueno 1991, Chiclana *et al.* 2002).

Bueno (1991) summarises 123 recoveries of foreign-ringed birds within Iberia. Only six of were in spring, which may mean that birds overfly the Peninsula rapidly during that season but in contrast remain for some time prior to crossing the Sahara in autumn. Conversely, BWP notes that there are many spring recoveries in North Africa and very few in autumn, suggesting that Whinchats cross the Sahara in a single flight from Iberia in autumn but frequently make landfall in North Africa in spring. The autumn recoveries are chiefly from Portugal (39%), the Basque Country/Navarra (15.4%) and Cádiz in Andalucía (12.1%), so it may be that most birds enter the Peninsula through the western Pyrenees and the Basque Country and thereafter head south-west. Birds from throughout the European range of the species have been recovered in Iberia but the great majority are from northern Europe. There are also two recoveries of Whinchats ringed in Italy and one ringed in Tunisia. One ringed in the Columbretes Islands has been recovered in Algeria and one ringed on Cabrera, Balearic Islands, in April was found in Morocco in October.

Occasional reports from the winter months seem to show that a very few individuals winter in the Peninsula, although the possibility of misidentified Stonechats arises in some cases and both December and February records seem likely often to involve very late or very early migrants respectively. Records in December–February are most frequent in the Levant region, from where the published bird reports include 14 winter records (15 birds) from the Comunidad Valenciana, four (five birds) from Catalonia, one (one bird) from Murcia and five (eight birds) from the Balearic Islands during these months. They include two that lingered for some time, both in Castellón during 1992–93: a female was at the Sitjar reservoir from 6 December until at least 30 January and a male was at Playa de Castelló from 15 December to 14 January (*A. O. C. Valenciana 1992*: 97). There are single January records from Andorra, Madrid and Ciudad Real and one was at the Valdoviño reservoir, A Coruña, from 23 January until 21 April 2001 (*A. O. Galicia 2001*: 147).

SIBERIAN STONECHAT *Saxicola maurus*

(Tarabilla Siberiana, Cartaxo-asiático)

Vagrant to Spain and Portugal.

The Portuguese observations were at the Lagoa de Santo André, Setúbal, in November 1993 and November 1995, and at Ponta da Atalaia, Faro, where a first-winter was present in October 1997. In Spain there was a male at the Ebro Delta in March 1997; a male at the Laguna de Traba, A Coruña, in April 1997; a possible female at the Ensenada da Ínsua, A Coruña, in September 1997; a male at Collserola, Barcelona, in March 1998; and a first-winter at Sant Adrià del Besos, Barcelona, in September 2010. Post 2010 there were records of one on the Berlengas Islands in October 2011, a first-year male at Cape Peñas in October 2012, a male at the Llobregat delta, Barcelona, in May 2013, and a male showing the characteristics of the Caspian Stonechat *S. m. variegatus* in the Ebro delta in January 2013 (RBP, RBS). Siberian Stonechats occur quite frequently in western Europe, chiefly in autumn in Finland, Scandinavia and the British Isles. At least 348 were identified in Britain up to 2011, and there have been seven in Ireland and 37 in France.

EUROPEAN STONECHAT *Saxicola rubicola*

(Tarabilla Común, Cartaxo-comum)

***S. r. rubicola*. An abundant resident and partial migrant throughout most of the Peninsula and in the Balearic Islands.**
***S. r. hibernans*. A common resident in northwestern Spain and on the Portuguese Atlantic coast. A common winter visitor from the French Atlantic coast and the British Isles. Darker and more extensively rufous below than *rubicola* but the two races are poorly defined, intergrade and are not safely separable in the field.**

Breeding population estimates. Spain: SA1 300,000–700,000 pairs; ACRE 7.7 million birds. **Portugal**: 25,000–250,000 pairs. **Andorra**: 50+ pairs.

The European Stonechat is a characteristic and often highly conspicuous bird of open landscapes throughout most of the Iberian peninsula and the Balearic archipelago. Breeding birds are only scarce or absent from much of the Guadalquivir valley and parts of the central Ebro depression, southeastern Castilla-La Mancha and the arid south-east in Murcia and Almería (SA2, PA2). Stonechats seem to favour regions under greater Atlantic influence and in general are much commoner in the north and west of the Peninsula than in the east and south. They range widely everywhere from sea level to considerable altitudes in the major mountain ranges: to 1,400m in the northern Iberian Range and the Navarran Pyrenees (de Juana 1980, Elósegui 1985), 1,800m in the Sierra de Gredos (Carnero & Peris 1988) and 2,500m in the Sierra Nevada (Garzón & Henares 2012). Breeding has exceptionally been confirmed as high as 2,000m in Andorra but most of the population there nests at 1,000–1,700m (ANA).

It may be found breeding in the full range of open habitats, requiring open ground and low shrubby vegetation, with taller bushes, fence posts or telephone wires to serve as perches from which to forage. It is common in both heathland and Mediterranean scrub; in untidy hedgerows, woodland edges and clearings, pastures and riparian woodlands in farming areas; on the margins of wetlands, especially in the southern meseta; in young conifer and broadleaf plantations; in coastal dune vegetation; in montane meadows and in juniper scrub above the treeline.

The population on SACRE sites showed a significant decline during 1998–2012 that affected all regions, following a now-familiar tendency among the passerine species of farmland habitats (Escandell 2013). The highest breeding densities (71 birds/km^2) have been reported from thermomediterranean grasslands and also from scrub in the supramediterranean and eurosiberian regions (48–56 birds/km^2). The Catalan breeding atlas put the regional population at 44,000–73,000 pairs (CA2). The Stonechat is widespread throughout Portugal but is most abundant in Estremadura, Ribatejo and Alentejo (PA2); that population showed no clear trend during 2004–09 (ACP). The Andorran population is thought to be increasing (ANA).

The breeding population abandons higher altitudes in winter and the Stonechat is then especially common in the thermomediterranean region, particularly in coastal Andalucia, the Guadalquivir valley, the southern Levant coastlands and the Balearic Islands (SWA). Many nonetheless winter on other coasts and also inland, especially in Extremadura, but the species is generally scarcer in winter than in the breeding season in the eurosiberian and supramediterranean regions. Some may migrate further south to winter in North Africa. Stonechats ringed in Iberia have been recovered in France (3) and Britain (2) and in coastal locations in Morocco (7) and Algeria (2) (SMA). One of the Moroccan recoveries was of a bird that was ringed as a chick at Somosierra, Madrid in July 1988 and found at Moulay Bousselham (661km) in January 1989 (*Ecología* 4: 274). Another ringed as a chick at Cubas, Cantabria in May 1990 was found that November at La Puebla del Río, Sevilla (711km) (*Ecología* 7: 353). These two recoveries confirm that some Iberian Stonechats are not sedentary and may winter south of their natal areas.

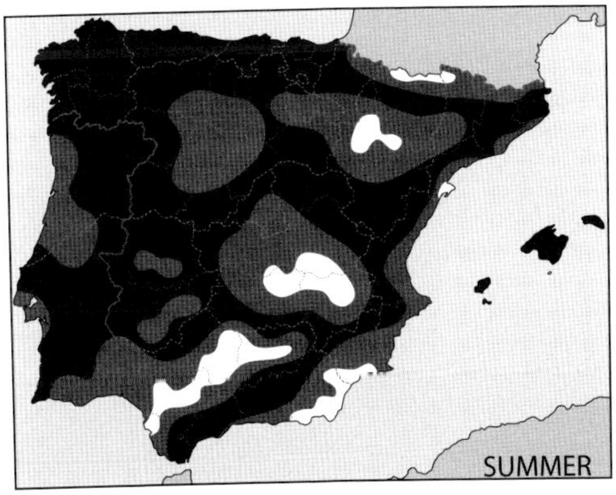

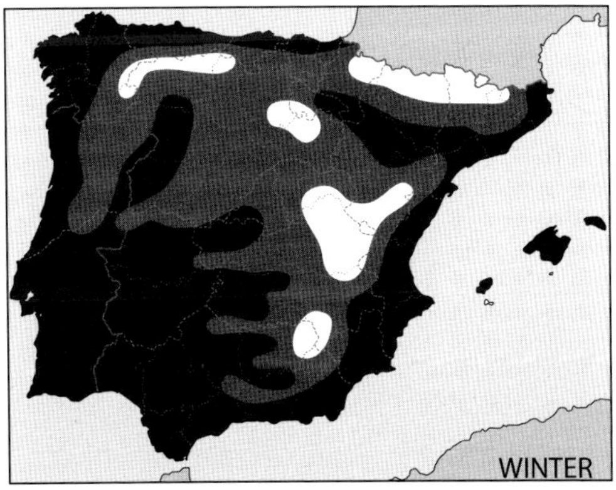

The Iberian birds are joined in winter by migrants from elsewhere in western Europe. The great majority (97%) of 123 Iberian recoveries, only one of which was in Portugal, comprised birds ringed in the Low Countries, France, Germany or the British Isles, with two from Switzerland and single recoveries also of birds from Norway and Sweden (Bueno 1991, SMA, AP). The recoveries show that these too winter especially on the warmer coasts and that many cross the Peninsula to North Africa. Most winter ringing recoveries are from the northeastern Mediterranean coastlands, including the Balearics (42%), and in Andalucía (33%), with very few inland and none at all in the Iberian north-west, from Asturias to central Portugal (Bueno 1991). The origins of foreign-ringed birds confirm that both *rubicola* and *hibernans* pass through or winter in Iberia. Migrants begin to enter Iberia in September and depart in March. The movements are generally inconspicuous and masked by the presence of resident birds. However, the species has not been reported nesting in Gibraltar since the late 19th century (Irby 1872) and there birds arrive from late September and depart in March; peak passage at Gibraltar is seen in October and February (BG).

ISABELLINE WHEATEAR *Oenanthe isabellina*

(Collalba Isabel, Chasco-isabel)

Vagrant to Spain.

Two observations that may be of 'possible' status antedate the Spanish rarities committee: one at Tarifa in September–October 1972 and another there in September 1974 (*Ardeola* 19: 29–30, 20: 378–379). Five accepted records comprise single birds at Cabrera island, Balearics, in October 2010; at Baldaio, A Coruña, in November 2010; at Muxía, A Coruña, in November 2010; at A Coruña city in October 2011, and at Ponteceso, A Coruña, in October 2011. In 2013 there were also single birds at Vilobí d'Onyar, Girona, in April and at Fisterra, A Coruña, in October (RBS).

Some winter as far west as Mauritania and they occur on passage in Tunisia, Algeria and southeastern Morocco. Nevertheless, the species is a relatively rare vagrant to western Europe, with 31 individuals found in Britain up to 2011 and 15 in France. There is also an accepted record for the Azores and another from the Canaries.

NORTHERN WHEATEAR *Oenanthe oenanthe*

(Collalba Gris, Chasco-cinzento)

***O. o. libanotica*. A common summer visitor to much of the northern half of Iberia, more local and largely montane in the southern half. Also breeds in the Balearics. Male paler above than nominate and cream or white below. Female often very similar to male, unlike in the nominate form (BWP).**
***O. o. oenanthe*. A common passage migrant throughout.**
***O. o. leucorhoa*. (Greenland Wheatear). A scarce but regular migrant throughout. Larger, longer-winged and darker than nominate.**
***O. o. seebohmi*. (Seeböhm's Wheatear). Vagrant to Gibraltar (1). Bib black in male, grey-black in female.**

Breeding population estimates. Spain: SA1 326,000–361,000 pairs; ACRE 2.4 million birds. **Portugal**: 500–5,000 pairs. **Andorra**: 2,500–4,000 pairs.

The Northern Wheatear is widespread in the eurosiberian region, where some nest down to sea level in the Biscay coastlands. However it is rather local on the northern side of the Cantabrian Range and also in Galicia. It is also widespread throughout the supramediterranean, in the northern meseta and in the Central and Iberian Ranges (SA2). It nests in the Balearic Islands, where the Black-eared Wheatear does not breed, in the Sierra de Tramuntana in Mallorca but down to 300m in Ibiza (Ferrer *et al.* 1986).

Elsewhere the distribution is essentially montane and the species becomes progressively restricted to higher altitudes the more southerly the location, as in the southern Iberian Range, the Levant coastal mountains and the higher levels of the Betic Range (SA2). Thus, in the north it breeds from 300m in the Ebro valley to over 2,300m in the northern Iberian Range (de Juana 1980) and to over 2,000m in the western Pyrenees (Elósegui 1985). It is largely found between 1,000m and 2,700m in the eastern Pyrenees (Ferrer *et al.* 1986). In the northern meseta it occurs from 700m in the cereal croplands, ascending to 2,300m in the Central Range (San Segundo 1990). It breeds above 850m in the Sierra Nevada, ranging to 3,400m (Pleguezuelos 1992, Garzón & Henares 2012). The breeding birds in the Sierra Nevada settle at progressively higher altitudes from April onwards as the snow melts, those that nest above 3,000m sometimes not laying until mid June (Garzón & Henares 2012). In general, this species is segregated by habitat

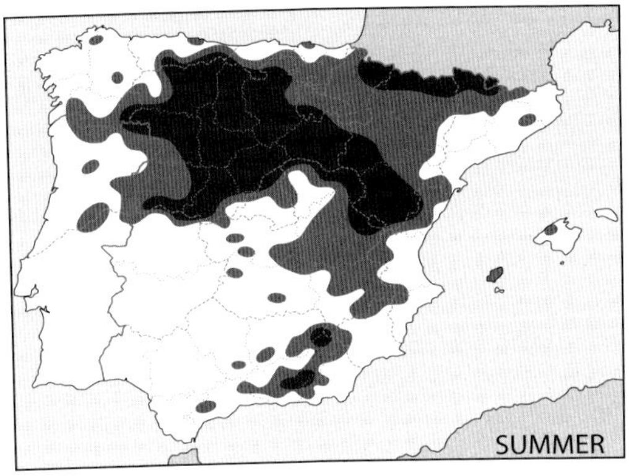

and by altitude from the Black-eared Wheatear, avoiding the drier areas favoured by the latter. However, it is also scarce or absent in the Atlantic coastlands. It is thus almost entirely absent as a breeding species from the south-west quadrant of Iberia, including all but the north-east of Portugal, where it occurs in the uplands south to the Serra da Estrela, generally above 800m (PA2). It nested in coastal northern Portugal until the mid 20th century (AP).

The Northern Wheatear is associated with grazed pastures, particularly where rocks and walls offer nest sites. Many occur on the alpine pastures as well as in high altitude expanses of dwarf shrubs near meadows, where grazing intensity may determine its relative abundance. It also breeds in the croplands of the northern Meseta and the Iberian Range, where there are mosaics of cereals, fallow areas and stony ground. In these areas it often nests in sheep pens, stone walls, ruins, stone piles and other man-made structures. A few nest in coastal dunes in the north and others in halophytic pastures in the Ebro valley, La Mancha and other low-lying areas. In Spain the population on SACRE sites showed a significant decline of 15% during 1998–2012 (Escandell 2013).

The species may be encountered in open country throughout the Peninsula and Balearics when on passage and it is sometimes numerous in falls of migrants on the coasts. The nominate race begins to arrive in February, with peak arrivals in March–early April, continuing into May. Its southward passage is from August to early November, peaking in September–October (Bueno 1991 AI2, AP). Individuals showing the characteristics of *leucorhoa* appear late during both passage periods. They are reported at Gibraltar in May and then again in late October–early November (BG). The nominate race winters in the Sahel zone of sub-Saharan Africa but the winter quarters of *leucorhoa* are centred in Senegambia (Borrow & Demey 2001).

Bueno (1991) maps 90 recoveries of foreign-ringed birds in Iberia, only seven of them in spring. Over 60% were found west of the 4°W meridian, which runs approximately through Madrid, and most of them were in western Andalucía. Spring recoveries were chiefly east of this line, but the sample size is small. The Basque Country, Navarra and La Rioja together also produced a sizable proportion of the total recoveries (18%). The recoveries include eight individuals identified as *leucorhoa*, six of them in autumn between late September and the end of October: six were from Iceland and two from Greenland. Most recoveries of birds ringed as nestlings originate in the British Isles (Table NW1, SMA), with a few from Scandinavia, Finland and Switzerland. Seven Northern Wheatears ringed in Spain have been recovered abroad: five in Morocco and one each in Algeria and Sweden (SMA).

The Northern Wheatear is occasionally reported in Iberia in winter, chiefly in the south-west. A few observations in early December may well have involved late migrants but recent records in late December and January include single birds at La Puebla del Río, Sevilla; Los Llanos aerodrome, Albacete; Doñana; the Vegas Altas del Guadiana, Badajoz, and Mallorca (*Ardeola* 45: 251, 50: 353, 53: 209; SWA). There were also two at Doñana on 27 December 2007 (*Ardeola* 55: 147).

There is only one known occurrence of the North African form, Seeböhm's Wheatear *O. (o.) seebohmi*, in Iberia and indeed in Europe: a male in Gibraltar on 28 April 2005 (*GBR* 2005: 33).

BLACK-EARED WHEATEAR *Oenanthe hispanica*

(Collalba Rubia, Chasco-ruivo)

***O. h. hispanica*. A common summer visitor to Mediterranean climatic regions of mainland Iberia. Migrants occur in the Balearic Islands.**
***O. h. melanoleuca*. Vagrant to Spain.**

Breeding population estimates. Spain: SA1 513,000–620,000 pairs; ACRE. **Portugal**: 1.1 million birds. **Andorra**: 1.1 million birds.

The Iberian breeding distribution is complementary to some extent to that of the Northern Wheatear. It is largely confined to the thermo-, meso- and supramediterranean regions, occurring down to sea level on Mediterranean coasts. The distribution is much more uniform in the eastern half of Iberia than in the west, probably reflecting its association with dry Mediterranean landscapes. Breeding birds nest to 600m in Portugal (PA2), to 700m in the western Pyrenees (Elósegui 1985) and to 1,000m in the Levant coastal mountains (Ferrer *et al.* 1986). Some breed higher locally: to 1,300m in the northern Iberian Range (de Juana 1980), to 1,600m in the Sierra de Gredos (San Segundo 1990) and to 2,000m on the southern slopes of the Sierra Nevada (Garzón & Henares 2012), but

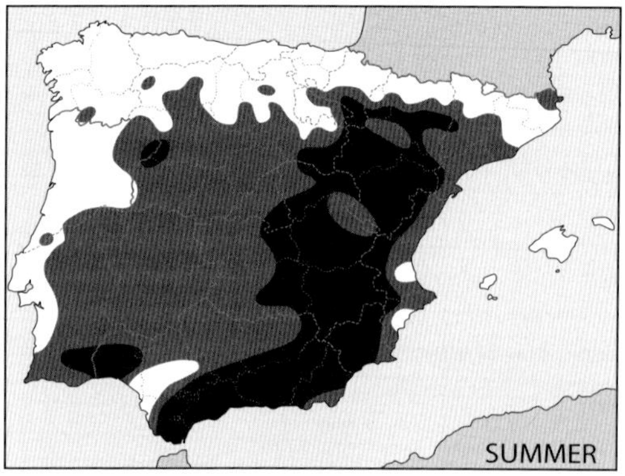

still at lower levels there than many Northern Wheatears. The two species are apparently mutually exclusive where they do coincide (Suárez 1987). The black-throated morph outnumbers the white-throated in most Iberian populations; typically 56–62% of males are black-throated (Suárez 1990).

The principal gaps in the breeding range are in the eurosiberian region. The Black-eared Wheatear is absent from the Cantabrian Range, Galicia and most of the Pyrenees: with the exception of occasional, perhaps sporadic breeding pairs in the south. Very few have been reported from the Biscay coastlands but in Asturias there was one at Somiedo in July 1991 and one at La Ballota in August 1997 (Álvarez-Balbuena *et al.* 2000, *A. O. Asturias 1997*: 190). It is also absent from the northern two-thirds of Portugal except in the east (PA2), and from much of the lower Guadalquivir valley: where the habitat may be unsuitable.

The Black-eared Wheatear occurs in a diversity of dry, open habitats with sparse vegetation. These include steppe areas, semideserts, stony pastures, patchy low scrub, fallow croplands and very open dehesas or other sparse woodlands. It also occurs in vineyards and olive groves. It selects more broken and scrubby terrain than does the Northern Wheatear. It avoids cereal crops and all types of tall or dense vegetation. It does not require stone walls or rocky areas for nesting, building instead in the shade of small shrubs or large stones (AI2).

A decline in the Iberian populations was noted as early as the mid 20th century (Mestre *et al.* 1987) and has continued since, although with local fluctuations. It is most apparent in parts of the Comunidad Valenciana and Catalonia (LR). Overall the population showed a significant 25% decline on SACRE plots during 1998–2012 (Escandell 2013). In Catalonia, a 20% reduction in the number of occupied 10km-squares over some 20 years is apparent when the findings of the first and second regional breeding bird atlases are compared (CA1: 1983, CA2: 2004). The population in Catalonia has recently been estimated at 17,000–27,000 pairs (CA2). A similar comparison between the first and second Spanish national atlases also suggests that the breeding range has contracted in some areas, mainly in the north (LR). The Portuguese population is also thought to be declining: the breeding range there has contracted since the 1980s in the west of Tras-os-Montes and it no longer nests in the coastal zone between Lisboa and Odemira (PA2, AP). Possible causes include habitat loss to afforestation, scrub encroachment after cessation of grazing, irrigation, urbanisation and loss of fallow areas to cultivation. Insecticide and herbicide use has been implicated locally. Factors affecting the winter quarters, notably droughts in the Sahel, may also be important (Mestre *et al.* 1987). In contrast, habitat loss to vegetational encroachment has been found to be reversed by scrub fires, which seem to benefit the species.

The Black-eared Wheatear winters in the Sahel zone of sub-Saharan Africa. The population begins to return in mid March, the peak arrival being in April, with some in May. One at Castro Verde, Beja on 27 February 2008 (*Anu. Ornitol.* 7: 86) was exceptionally early. Large numbers are sometimes seen in falls of migrants on southern coasts in spring. Breeding begins quite late in some areas, perhaps especially in those at higher elevations: in the páramos of the Iberian Range, clutches are mainly laid during the second half of May (Suárez 1977). The departure is normally less conspicuous and occurs chiefly in August–early September, with stragglers in October. Very few nest north of Iberia and there are no recoveries of foreign-ringed birds. A female ringed in Alicante in August 1989 was recovered at Nador, Morocco, in September 1991 (*Ecología* 6: 260). Only a few migrants occur in the Balearics, where the species does not nest. Winter records are exceptional. A female was reported at Puebla de Montalbán, Toledo, on 28 December 1980 (*Ardeola* 29: 188).

The eastern race *O. h. melanoleuca* has only once definitely been recorded in Iberia; a male ringed at Cabrera, Balearics, on 6 May 1994 (*Ardeola* 44: 135–136).

DESERT WHEATEAR — *Oenanthe deserti*

(Collalba Desértica, Chasco-do-deserto)

Vagrant to Spain, Portugal and Gibraltar.

Vagrants frequently appear in Europe, most often in autumn or early winter. At least 126 have been recorded in Britain, eight in Ireland and 35 in France. There are 12 accepted records from the Canary Islands.

The first Iberian observation was offered only as a 'possible', despite having an adequate description: it was near Lleida, Cataluña, on 25 September 1971 (*Ardeola* 16: 274–275). By 2010 there were nine accepted records from Spain since 1985, one

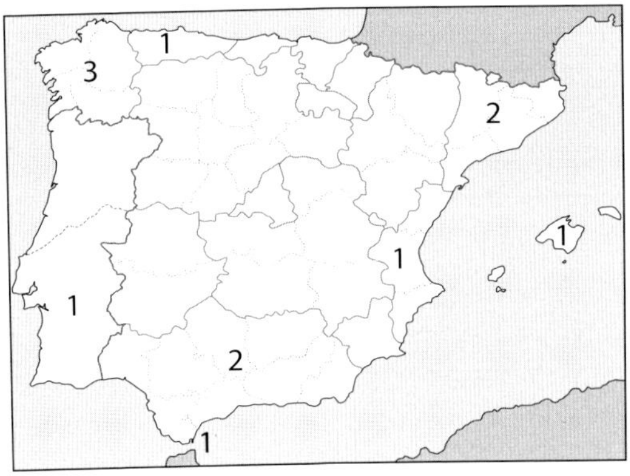

from Portugal and one from Gibraltar. All were single birds, largely at coastal sites, both Atlantic and Mediterranean (Fig. 184). The sole Balearic record was one ringed on Formentera on 29 December 2003.

There were a further five Spanish records in 2010–12: a male at Cape Touriñán, A Coruña, from at least 6 December 2010 to 6 January 2011; a male at Villaviciosa, Asturias, in November 2011; a first-winter male at the Aiguamolls de l'Empordà, Girona, in November 2011; a male at Cabo de Gata, Almería, from 13 December 2011 to 18 February 2012, and a male at Sástago, Zaragoza, from 26 December 2011 to 2 March 2012 (*Ardeola* 60: 484).

The few records give some indication of a concentration

Figure 184. Regional distribution of records, 1971–2010 (n =12)

Figure 185. Monthly distribution of records, 1971–2010 (n = 12)

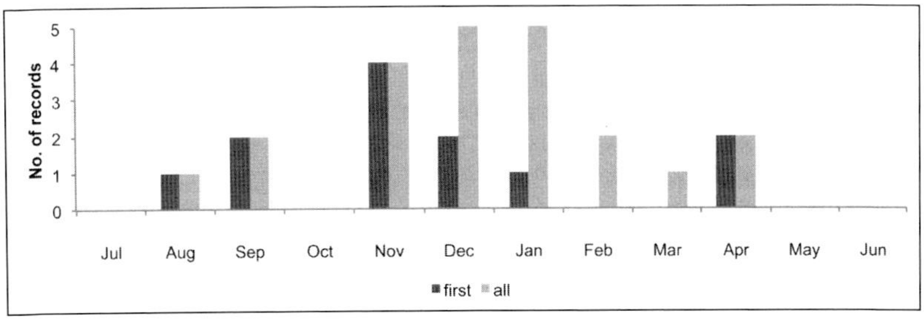

in November–January (Fig. 185). Four involved birds that wintered in the north-west: at Castrillón, Asturias, from November 1997 to March 1998; at Ponteceso, A Coruña, from November 2003 to February 2004, and again at Ponteceso, from November 2008 until January 2009. Records in Britain peak in November and wintering has also been recorded in Britain and France on rare occasions (OF, Slack 2009).

WHITE-CROWNED WHEATEAR *Oenanthe leucopyga*

(Collalba Yebélica, Chasco-de-barrete-branco)

Vagrant to Portugal. Recorded in Spain.

The only Iberian observations involve an adult male and an immature bird seen at Doñana on 28 May 1977 (Soriguer 1978), and an accepted record of a possible first-summer individual at Portimão, Faro, on 25 March 2001. A few individuals of this largely sedentary species have appeared as vagrants in Malta, Croatia, Greece, Cyprus, Turkey and even Britain (Slack 2009). One has also wandered to the Canary Islands (*Ardeola* 54: 437).

BLACK WHEATEAR *Oenanthe leucura*

(Collalba Negra, Chasco-preto)

***O. l. leucura*. A scarce resident in eastern and southeastern Spain. More local elsewhere and occurs only marginally in areas of northeastern Portugal. Vagrant to the Balearics.**

Breeding population estimates. Spain: SA1 4,000–15,000 pairs. **Portugal**: 50–500 pairs.

The Black Wheatear is characteristic of barren rocky landscapes in the warmest parts of Iberia, in the thermo- and mesomediterranean regions. It is most abundant in the Levant coastal mountains, the eastern Sierra Morena and the Betic Range. In these areas it ranges from sea level to 2,000m in the Sierra Nevada (Garzón & Henares 2012). It is also present in the arid sectors of the Ebro depression and Iberian Range, as far north as La Rioja and southern Navarra (SA2).

The breeding range is highly discontinuous elsewhere but it occurs at scattered locations in the southern meseta, Extremadura

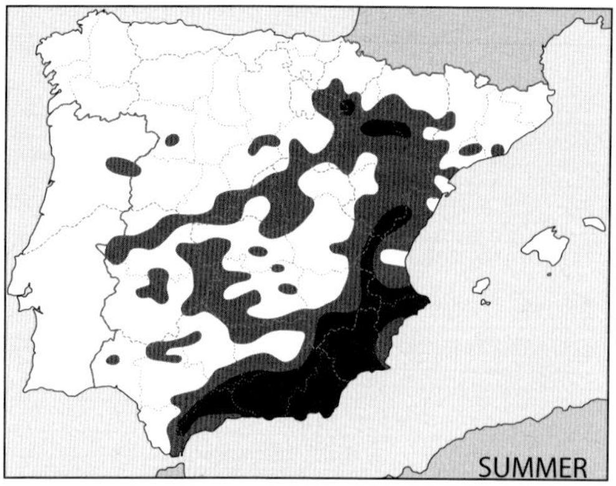

and throughout the southern slopes of the Central Range, where it reaches altitudes of 1,400m in Gredos (San Segundo 1990). It is rare in western Andalucía and may no longer breed in the vicinity of the Strait of Gibraltar. It became extinct in Gibraltar itself in the mid 20th century (BG) and has never been seen there since. It is also rare in the northern meseta but nests locally, or has nested, as far north as southern Burgos and Zamora, although it may have disappeared from these northern areas (Moreno 2006). In the past it has even been reported breeding on the southern versant of the Cantabrian Range, at Cármenes, León (Noval 1975). The range extends marginally into Portugal along the Tagus and Duero gorges on the Spanish/Portuguese border (PA2).

The Black Wheatear is strongly associated with rocky areas with only sparse vegetation. It inhabits sea cliffs, river gorges, the ramblas of the arid south-east and scree slopes. It will also nest on large buildings such as castles, ruins and dams, or within abandoned huts and caves. In the core areas, linear densities of 3.8 pairs/km have been reported from Guadix, Granada, and 4.1 pairs/km from the neighbourhood of Zaragoza (Serrano 1999).

The Spanish population has shown a long-term decline. The species has been lost from traditional sites and has shown significant decreases elsewhere, notably in western Andalucía, Navarra and Catalonia (SA2). A large decline has also been seen in Castilla y León, where up to 350 pairs were present in the 1990s (Sanz-Zuasti & Velasco 1999) but fewer than 95 pairs more recently (Moreno 2006). In Catalonia it is especially the northernmost areas of the range: Alt Empordà, Cadí and Alta Garrotxa that have been abandoned and the regional population may now be of the order of 400–600 pairs (CA2). The massif of Montserrat, Barcelona, was traditionally occupied but the Black Wheatear disappeared there during the mid 20th century, although it recolonised the site in 1995 when forest fires removed encroaching vegetation (Real 2000). This highlights the likelihood that habitat loss due to plant growth, whether through afforestation or natural regeneration following reduction in grazing, is implicated in the local displacement of some Black Wheatears.

The Portuguese population is also declining. The Black Wheatear was much more widespread there during the 19th and early 20th centuries, when it bred along the Douro basin as far west as Melres, near Porto, and perhaps also in the Serra do Gêres and around Coimbra (Tait 1887, Reis Júnior 1931, Themido 1933). It was even found nesting on the Berlengas Islands in 1939 (Lockley 1952). In the 1980s it was still present in Estremadura and Alentejo, where it no longer occurs (AP).

Most Black Wheatears are resident but some territories in the north of the range are abandoned in winter (Elósegui 1985, Carnero & Peris 1988, SWA). Stenhouse (1921) reported 'migrants' near Gibraltar in March and September–October but there are no comparable recent observations there and it may be that he was seeing family groups from the now extinct local breeding population. There are records in April, May, October and December in the Balearic Islands (Ferrer *et al.* 1986); rare observations at Doñana (Llandres & Urdiales 1990), four records from Asturias: most recently at the Salime reservoir on 18 January 2004 (*Ardeola* 52: 220); and two records from Gipuzkoa, which confirm some dispersive tendency. In general, however, observations away from the breeding sites are exceptional.

RUFOUS-TAILED ROCK THRUSH — *Monticola saxatilis*

(Roquero Rojo, Melro-das-rochas)

A scarce summer visitor to montane regions of the Peninsula and Mallorca.

Breeding population estimates. Spain: SA1 3,500–4,800 pairs. **Portugal**: 50–500 pairs. **Andorra**: 50–100 pairs.

The Rufous-tailed Rock Thrush inhabits middle and upper levels of montane areas in Spain and Portugal, as well as Andorra and the Serra de Tramuntana in Mallorca. It is generally replaced at lower levels by the Blue Rock Thrush where the ranges of both overlap. In mainland Spain it is present in the Pyrenees; the Cantabrian, Iberian, Central and Betic Ranges, and the Levant coastal and pre-coastal ranges (SA2). It is thinly distributed in the southern mountains but is more numerous in the north. In Portugal it ranges south to the Serra da Estrela (PA2, AP). However, the Portuguese breeding range has contracted since the 19th century and the species is considered endangered within the country (Tait 1887, Reis Júnior 1931, AP).

It occupies barren rocky slopes and summits interspersed with areas of herbaceous vegetation and low open scrub, with a preference for sites with a sunny south-facing aspect near rock outcrops. It nests in rock crevices but also uses holes in buildings

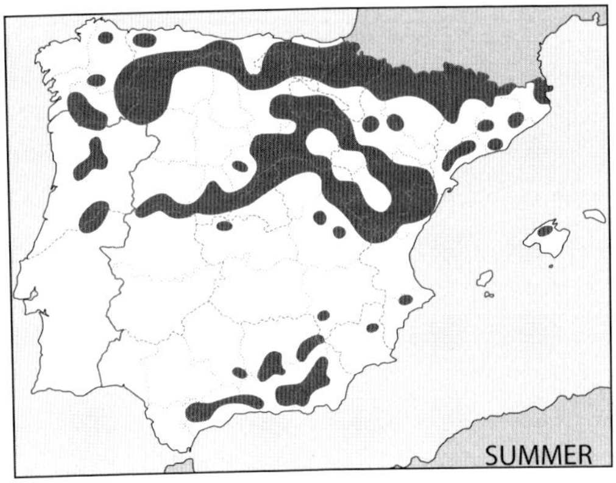

locally: even inhabited ones (Valverde 1956). It occurs down to sea level locally, for example on the Catalan coast at Cape Creus and l'Empordá (AI2, SA2), but is generally found above 600–700m in the north and progressively higher further south. Those in the eastern Pyrenees range up to 2,200m (CA1), over 2,500m in Andorra (ANA). It occurs from 1,000m to 2,100m in the Cantabrian Range and northern Iberian Range, from 800m to 2,000m in the Sierra de Gredos and from 1,500m to over 2,500m in the Sierra Nevada (de Juana 1980, San Segundo 1990, Garzón & Henares 2012). This progressive reduction in the altitude of the breeding range from south to north results in a much more extensive distribution in the northern half of the Peninsula, although the species is everywhere mainly associated with the oromediterranean bioclimatic zone (de Juana 1980).

The species is strictly migratory and the birds winter in sub-Saharan Africa. The breeding areas are occupied from mid April to September (Ferrer *et al.* 1986, AP). Some arrive as early as mid March in the south and the spring passage there continues through May, with rare stragglers observed at Gibraltar as late as mid June (BG). The return passage at Gibraltar extends from mid August to mid October. The movements anywhere are highly inconspicuous, however, and only occasional individuals are seen in any one year on passage at sites away from the breeding mountains.

BLUE ROCK THRUSH — *Monticola solitarius*

(Roquero Solitario, Melro-azul)

M. s. solitarius. **A common resident of rocky areas, mainly in Mediterranean climatic zones, in the Peninsula and Balearic Islands.**

Breeding population estimates. Spain: SA1 12,500–16,800 pairs. **Portugal**: 1,000–10,000 pairs. **Andorra**: 10 pairs. **Gibraltar**: 10–15 pairs.

The Blue Rock Thrush is predominantly associated with the warmer sectors of the Peninsula, in the thermo- and mesomediterranean climatic regions, but it may also be found in supramediterranean mountains. It is scarce in the eurosiberian zone but occurs there on south-facing slopes in the Pyrenees and, more rarely, in the Cantabrian Range, and also on sea cliffs in Galicia and the Biscay coastlands (SA2). In general it is much commoner in eastern and especially southeastern Spain than elsewhere. It nests commonly on all the main Balearic Islands and on the larger islets of the archipelago. Within Portugal it is largely absent from Ribatejo, all but easternmost Alentejo and rocky coastal areas of the centre and north but it is widespread elsewhere, including Algarve and the mountains of Estremadura (PA2). It was first observed in Andorra in 1976 and first found breeding there in 1991: it occurs in south-facing, rocky areas of Mediterranean character at 900–1,650m (ANA).

It is commonly found on sea cliffs, on mountain cliffs, in gorges and cuttings and on scree slopes, always in sunny dry areas with sparse vegetation. Many large buildings; notably castles, churches, bridges, dams and ruins, also accommodate one or more pairs and some also frequent urban rooftops, especially where there are tiled roofs and older edifices generally. It is typical of lower altitudes and seldom found much above 1,000m although some occur to 1,200m in the northern Iberian Range and to 1,500–1,600m in the Pyrenees, Sierra de Gredos and Sierra Nevada (de Juana 1980, San Segundo 1990, Garzón & Henares 2012). The Rufous-tailed Rock Thrush replaces it at higher levels in many areas. It is common in the Sierra Morena and on cliffs and gorges in Extremadura, western Andalucía, the southern meseta and along the Duero and Tagus gorges on the Portuguese/Spanish border (SA2).

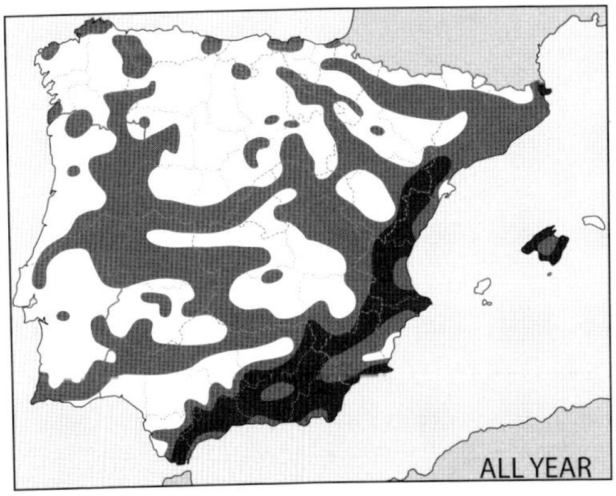

The Catalan population has been put at 4,000–5,000 pairs (CA2) but that of Castilla y León at only 700–800 pairs (Sanz-

Zuasti & Velasco 1999). SACRE (1998–2012) results suggest a moderate decline (Escandell 2013). However, a 23% increase in occupied 10km-squares was seen in Catalonia between the first and second breeding bird atlas surveys (CA1: 1983, CA2: 2004) and the Blue Rock Thrush may have benefited there from quarrying and the terrain-clearing effects of forest fires. The Portuguese population appears to be stable, although with some indications of range expansion in the north-west in the region of Peneda-Gerês National Park (PA2).

The breeding and wintering ranges in Spain largely coincide but some are seen in winter in areas where the species is not known to breed (SWA). Similarly, in Portugal, individuals that are probably migrants appear during passage periods in regions where the species does not breed, for example, in central Alentejo in October (AP). Those that nest at higher elevations only occur there seasonally: the small population of Andorra is absent in winter (ANA). There are some indications of passage at Gibraltar, where small groups are occasionally recorded in autumn, as also noted at Tarifa from mid September to early November (García Rua 1975, Tellería 1981), but very few birds seem to be involved.

WHITE'S THRUSH *Zoothera dauma*

(Zorzal Dorado del Himalaya, Tordo-dourado)

Vagrant to Spain.

There are only two old winter records from Iberia, both of individuals taken in the Balearic Islands. One on Menorca in January 1912 is still in the collection of the Ateneo de Mahón (Munn 1926; *Ardeola*. 53: 183). The other was at Cas Concos, Mallorca, in December 1965 (Nadal 1966). Only *Z. d. aurea* reaches Europe as a rare vagrant although it is very rare in the south. There are records of at least 74 in Britain, five in Ireland and 15 in France. The British records are latterly mainly in October but formerly were usually during December–February (Slack 2009).

RING OUZEL *Turdus torquatus*

(Mirlo Capiblanco, Melro-de-colar)

***T. t. alpestris*. A scarce breeding species in northern Spain, chiefly in the Pyrenees and in Andorra. More widespread on passage and locally in winter.**
***T. t. torquatus*. Occurs on passage and locally in winter.**

Breeding population estimates. Spain: SA1 6,000–7,000 pairs. **Andorra**: 150–300 pairs.

The Iberian breeding population is almost entirely confined to the central and eastern Pyrenees, from easternmost Navarra, through Aragón and Andorra to western Catalonia (SA2). Here it nests above 1,700m in open Black Pine woodlands with abundant undergrowth of rhododendron, broom and Bearberry, generally in the vicinity of montane pastures (Purroy 1974, Ferrer *et al.* 1986). Those at the western end of the Pyrenean range also occur at the upper limits of mixed beech-spruce woodland. The highest densities are reported from Pyrenean Black Pine woodlands, with a mean maximum of 4.1 birds/10ha (SA2). There are small populations in similar habitat in the central Cantabrian Range, chiefly in southern Asturias, including the Picos de Europa

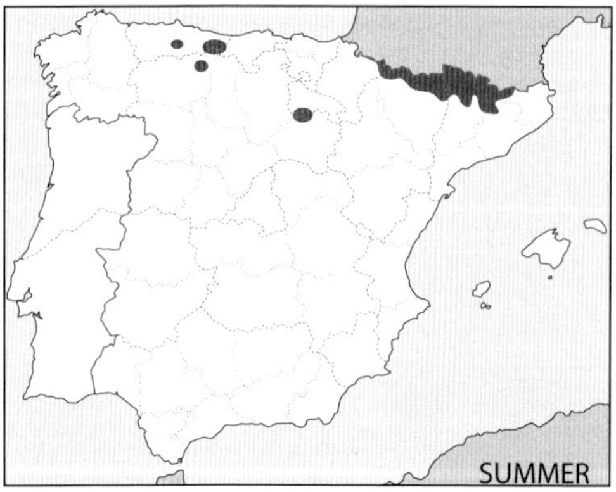

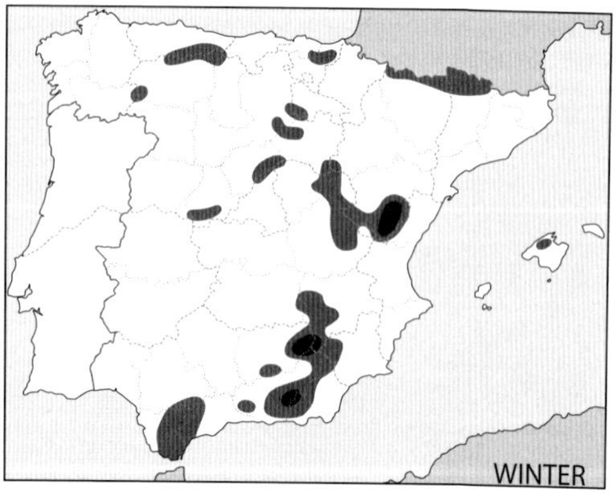

(Noval 1975). An isolated small population is known in the northern Iberian Range (de Juana 1980). A pair nested successfully at the Poljé de Abeles in the Sierra de Guara, Huesca, in 2007, a pre-Pyrenean site well to the south of the usual range in Aragón (*Ardeola* 54: 402). Saunders (1871) considered that it 'undoubtedly' bred in the Sierra Nevada, which is improbable.

Some Spanish breeding birds may winter in the Pyrenees and Cantabrian Range but most probably move further south. However, there is a probably mainly immigrant wintering population in both ranges, whose numbers and distribution are very largely determined by the year-to-year variation in the juniper berry crop, the principal species of which are the Dwarf Juniper: which forms extensive mats at high altitudes, and the Spanish Juniper: a tree of the páramos. Wintering birds in the Sierra Nevada have been found to consume quantities of Spanish Juniper berries as well as those of the Barberry (Zamora 1990). The winter distribution in the Catalan Pyrenees is principally within the Dwarf Juniper zone at 1,800–2,000m but the numbers found there range from 1,600–1,800 birds in good berry years to a few dozen in poor seasons (CA2). The winter presence of Ring Ouzels at altitude within the breeding range has only recently been revealed, probably because the high tops receive few human visitors in winter except at the ski resorts. However, the principal and most extensive Iberian juniper woodlands are in the Iberian Range and further south in Betic and Pre-Betic Ranges, rather than in the Pyrenees or the Cantabrian Range (Tellería *et al.* 2011). These latter regions, in particular the Serranía de Cuenca, Sierras de Gúdar and Javalambre, Montes Universales and the Maestrazgo in the Iberian Range; the Sierras de Cazorla and Segura in the Pre-Betic Range, and the Sierra Nevada and Serrranía de Ronda in the Betic Range, together appear to attract the largest wintering concentrations (SWA, Tellería *et al.* 2011). In fact, more than half of 38 ringing recoveries analysed by Santos (1982) were in the southern Iberian Range, most of them in Teruel. Small numbers of wintering birds have also been reported from the Basque mountains and in the Central Range: in the Sierras de Gredos and Guadarrama (Sánchez 1991, Roviralta 2004).

Both sight records and ringing recoveries show that both subspecies occur in Iberia on passage and in winter. All 47 Iberian recoveries of birds ringed elsewhere in Europe have been in Spain and most were ringed in Britain (18), Switzerland (12) or Germany (7), with the others in France (4), the Low Countries (3) or Scandinavia (3) (SMA). At least seven of the birds from Britain and two from Scandinavia (*torquatus*), and three from Switzerland (*alpestris*) were ringed there as pulli (Santos 1982). Migrants begin to arrive in the north-east from late August and in September, with peak passage in October (Santos 1982, 1985; Aymí 1989). Passage peaks in November in the Sierra de Guadarrama (Roviralta 2004). They may be observed briefly anywhere in eastern Spain, and also in the Balearics, at this time, generally attracted to fruiting shrubs. There is limited evidence that *alpestris* tends to occur somewhat further east in Iberia than the nominate race. For example, 13 of 48 birds seen by Aymí (1989) in the Montsant range, Catalonia were *alpestris* but all those found by Roviralta (2004) in Madrid were *torquatus*. The wintering birds themselves show some nomadic tendency as they seek and consume fruits, perhaps moving progressively further south as supplies become exhausted (SWA). Migrants are most often recorded in small groups, exceptionally in flocks of 40 or more. They are progressively more uncommon further west and generally rare in western Spain and Portugal at any season, although small flocks appear there in some years. A count of *c.* 180 birds in the Serra de Estrela from 19–23 October 1993 (AP) was exceptional for Portugal. Some southward passage is apparent at the Strait of Gibraltar from late September into November but only small numbers are ever recorded there.

Clearly a large proportion of the Ring Ouzels that enter the Iberian Peninsula continue further south to winter in the Atlas mountains in Morocco. They return in March–April when the Iberian wintering population also departs. There are only two foreign recoveries of birds ringed in Iberia: one ringed on the Columbretes Islands in October was in Switzerland in August, and one ringed in Barcelona in October was found in Morocco in January (*Ecología* 10: 383).

COMMON BLACKBIRD *Turdus merula*

(Mirlo Común, Melro-preto)

***T. m. merula*. An abundant resident throughout the Peninsula and in the Balearics. Large numbers of central European birds winter in the region. Birds in southern Iberia, especially the females, are darker, recalling the North-west African race *mauretanicus* (Aparicio 2011).**

Breeding population estimates. Spain: SA1 2.3–5.9 million pairs; ACRE 19 million birds. **Portugal**: 200,000–2 million pairs. **Andorra**: 400–800 pairs. **Gibraltar:** 100 pairs.

The Blackbird is ubiquitous throughout Iberia, occurring anywhere that there are trees or shrubs. As elsewhere it is often closely associated with human habitation and occurs in parks and gardens in every village, town and city. Although it is most abundant in the lowlands, and at lower and medium altitudes in mountainous regions, it frequently breeds up to the treeline, reaching 2,200m in the Central Range (Carnero & Peris 1988) and 2,300m in the Sierra Nevada (Garzón & Henares 2012). It

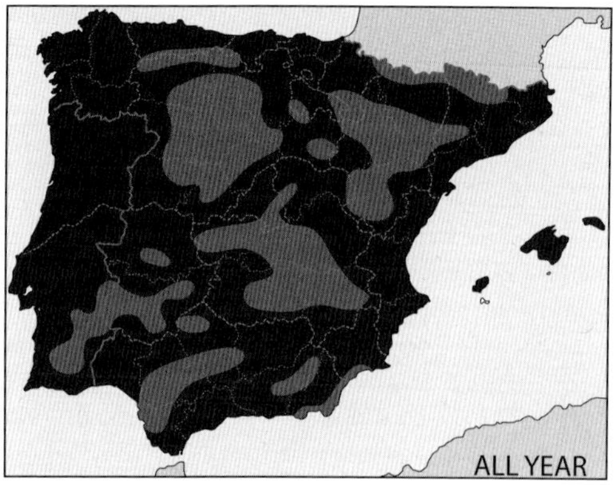

especially favours deciduous woodlands but also occurs in coniferous forests and in scrub. In otherwise dry and treeless areas it may be found in riverine woodlands and orchards.

The population showed a significant overall increase on SACRE sites during 1998–2012 (Escandell 2013). The highest reported densities in Spain (100 birds/km^2) are from thermomediterranean orchards, with up to 15–30 birds/10ha in orange groves in Valencia (Gil-Delgado 1981, Aparicio 2011), but counts of 75–95 birds/km^2 have been reported from habitats as diverse as mosaic farmland, riverine woods and pastures with trees in the Biscay coastlands, Pyrenean oakwoods, supramediterranean juniper woods, and towns and cities. The Portuguese population showed a moderate increase during 2004–09 (ACP).

The Iberian population is largely resident, although those at higher levels descend to the valleys in winter (AI2). Its winter distribution is virtually identical to that during the breeding period, showing density reductions only in extensive agricultural areas and in high mountains (SWA). Fewer than 4% of 1,651 recoveries of birds ringed in Spain as pulli or during the breeding season involved displacements further than 10km (Aparicio 2011). Large numbers of Blackbirds enter the Peninsula to spend the winter. Ringing recoveries (SMA) show that the immigrants are chiefly of central European origin: from the Low Countries, Germany, France and Switzerland, with some from as far as Scandinavia, Poland and Russia. Most arrive quite late in the year: although the autumn passage begins in September it is only really apparent from late October, peaking in November (AI2). The majority also depart early, in February.

Recoveries suggest that the wintering population is most abundant on northern and northeastern coastlands, as well as in the upper Ebro and Guadalquivir valleys. Many winter in the Balearic Islands. In contrast, relatively few of the immigrants winter in central and southern Spain or in Portugal. There is a clear tendency for Blackbirds of most northerly origin, from Scandinavia or ringed in Britain, to winter in the northern coastlands, those from central and eastern Europe wintering in the east of the Peninsula and in the Balearics (Santos 1985). Some cross-Peninsular migration to and from North Africa is assumed but has never been proved (BM).

EYEBROWED THRUSH *Turdus obscurus*

(Zorzal Rojigrís, Tordo-de-cabeça-cinzenta)

Vagrant to Portugal.

A first-winter bird was killed by hunters at Arrimal, Porto de Mós, Leiria, on 28 October 1991 (*Pardela* 5: 14). It was in a flock of Redwings. This is a rare vagrant to the western Palearctic where the records up to 2009 include 19 in Britain, 15 in France and one in Morocco.

NAUMANN'S THRUSH *Turdus naumanni*

(Zorzal de Naumann, Tordo-de-coleira-preta)

Vagrant to Spain.

A first-winter female was ringed at the Ebro Delta, where it was present on 23–24 October 2005. This is a very rare vagrant to Europe, with records from Poland, Norway, Finland, Britain, Germany, Austria, Hungary and the Czech Republic (Slack 2009).

FIELDFARE

Turdus pilaris

(Zorzal Real, Tordo-zornal)

A regular winter visitor to northern Iberia but progressively rarer in the south and scarce in the Balearic Islands.

The Fieldfare is the scarcest of the northern thrushes that enter Iberia to winter. The numbers arriving vary considerably between seasons, according to the severity and frequency of cold spells further north. Thus, for example, no more than 4,000 individuals were estimated to have wintered in Catalonia in winter 2006–07 but up to 273,000 did so the following winter (CWA). An exceptional influx of thousands into northern Portugal was noted in December 1906 (AP). Its presence in most areas can thus be considered irruptive, associated with failures or depletion of winter food supplies in regular winter quarters in central Europe (Santos 1982). In fact, 91% of the 74 ringing recoveries analysed by Santos (1982) occurred in just four of the 21 winters considered. Those recoveries (Table 54), and 35 subsequent ones (SMA), clearly indicate that most of the Iberian wintering population is of Fennoscandian origin. Recent recoveries include a few individuals ringed as far east as the Baltic States (2) and Belarus (1). Recoveries of birds ringed in Britain, where the Fieldfare rarely nests, show that birds may spend different winters, or parts of the same winter, in widely separate regions.

Table 54. Recoveries of foreign-ringed Fieldfares (n = 74). After Santos (1982).

Country of ringing	Total recoveries in Iberia to 1980 (%)	National origin confirmed
Fennoscandia	49 (66.2)	46
British Isles	11 (14.9)	0
Germany	4 (5.4)	2
Belgium and The Netherlands	4 (5.4)	0
France, Switzerland, former Czechoslovakia	6 (8.1)	1

Wintering Fieldfares are most often encountered at some elevation on the southern slopes of the Cantabrian Range, the central Pyrenees and, most of all, throughout the Iberian Range. In the last of these they are strongly attracted to the fruiting Spanish Juniper woods of the uplands (*parameras*) where mean densities of 20–60 birds/km^2 have been recorded (SWA). Favoured fruits consumed in Catalonia and probably elsewhere include those of the Alpine Juniper, Rowan, Dog Rose, Hawthorn and Blackthorn (CWA). Some also frequent meadows in mixed farmland and beech or oak woods in northern Spain (AI2). They are usually uncommon but fairly regular in northern Portugal, in the Peneda-Gerês National Park (AP). Fieldfares were regarded as not uncommon in Mallorca in winter by Munn (1921). They were described as not rare in the Balearic Islands in winter by Bernis (1958) and more recently are considered scarce visitors to Mallorca, Menorca and Ibiza (*A. O. Balears 2010*: 267). Everywhere they arrive relatively late in the season, some in October but most not until late November: they also depart early, by the end of February (Santos 1982).

Observations from southern Iberia are irregular. The ringing recoveries analysed by Santos (1982) included just three in Andalucía and one each in Extremadura, Alentejo and Algarve. A record of a 'substantial flock' in the Guadiana valley in central Extremadura on 19 February 1967 (*Ardeola* 12: 241) is noteworthy for the region. A count of 500 at Odelouca, Faro, in December 1992 (*Pardela* 6: 23) was exceptional so far south. An influx was noted in Moura district, Alentejo, in winter 2003–04, where a flock of 260 was recorded, and there is a record of a flock of 110 in the Serra da Estrela in December 2001 (*Anu. Ornitol.* 1: 46). The late dates of most of the southern records support the suggestion that Fieldfares may exhibit rather nomadic behaviour (SWA). Some descend as far as southern Andalucía in some years, where they are occasionally attracted to fruiting shrubs in the higher levels of the Cazorla, Sierra Nevada and Ronda mountains. There are a handful of records from the Strait of Gibraltar: five were recorded at Tarifa during 11–15 October (García Rua 1975) and there are two records from Gibraltar itself: on 24 February (2 birds) and 5 March 1987 (GONHS). However, the Fieldfare seldom crosses the Strait: it is considered an accidental visitor to Morocco, from where there are only 21 records (27+ birds) up to 2011.

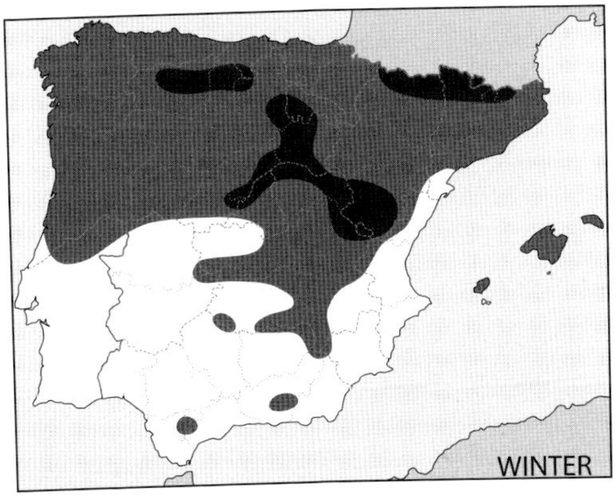

SONG THRUSH

Turdus philomelos

(Zorzal Común, Tordo-pinto)

***T. p. philomelos*. A common resident or partial migrant, chiefly in the northern third of Iberia. It does not nest in the Balearic Islands. Abundant on passage and in winter throughout the region.**
***T. p. clarkei*. A scarce migrant and winter visitor.**

Breeding population estimates. Spain: SA1 200,000–400,000 pairs; ACRE 940,000 birds. **Portugal**: 100–1,000 pairs. **Andorra**: 200–300 pairs.

The Song Thrush breeds mainly within the northern third of the Peninsula: in the eurosiberian zone, as well as in adjacent forested areas of the supramediterranean. It is common from Galicia and northern Portugal across the Biscay coastlands and Cantabrian Range and throughout the Pyrenees. It is also present in most of León, northwestern Zamora, northern Palencia and northern and eastern Burgos. In the last of these, and in La Rioja and Soria, it inhabits the Iberian Range, extending south to the Central Range in Madrid and Segovia. In the north-east its range takes in the whole of Girona and Barcelona, south to the north of Tarragona, occurring down to sea level (SA2). There are small scattered populations elsewhere in the northern half of Spain in more humid woodland patches, for example in Salamanca, the southern Iberian Range and in the Levant mountains. It is a rare, local and perhaps sporadic nester in similar humid woodlands in the southern mountains, with records from northern Extremadura, the Sierra Morena and Alicante.

The southern breeding records may represent the earlier stages of colonisation of those regions since the Song Thrush has shown a progressive southward range expansion during the 20th century that seems to be continuing. The historical literature only mentions breeding Song Thrushes in the Biscay coastlands (e.g. Irby 1883, Witherby 1922) and in the Pyrenees (Saunders 1884, Wallis 1895, Ticehurst & Whistler 1925), with the exception of Castellarnau (1877) who stated that it bred at San Ildefonso, on the northern slopes of the Central Range. Witherby (1928) could not find the species there and thought that Castellarnau may have confused it with the Mistle Thrush, although this is far from clear. Nevertheless, Gil Lletget (1945) stated that breeding in central Spain was unconfirmed and the first evidence of breeding Song Thrushes in the Iberian Range was in the 1970s (de Juana 1980).

The species has certainly spread in Catalonia, where a comparison between the first and the second breeding bird atlases shows a 22% increase in the number of occupied 10km-squares (CA1: 1983, CA2: 2004). Two pairs nested in citrus groves near the Albufera de Valencia in 2009, the first recorded instance from the coastlands of the Comunidad Valenciana (*A. O. C. Valenciana 2009*: 120–121). Southward range expansion has also been apparent in southern France: in the 1930s the Song Thrush was restricted there to the Pyrenees and the Alps, but it spread to the mountains of Provence in the 1970s, Gard in the 1980s and the Rhône valley forests in the 1990s (OF).

The breeding population in Portugal also appears to be of recent origin and it has certainly increased and expanded southwards since June 1990, when the first proved instance of nesting was reported in the Serra do Gerês (Santarém 1991, PA2). The core range is in the north of Minho and Trás-os-Montes but breeding has been reported as far south as Coimbra and Mira in Beira Litoral (AP).

Breeding birds inhabit farmland mosaics with tall hedgerows, orchards, large parks, riparian growth and, especially, woodlands: chiefly broadleaf woods of beech, oak, birch or chestnut but also of Scots Pine and Fir. In areas with a Mediterranean climate they are often restricted to riparian woodlands, including poplar plantations. Most occur at altitudes below 1,500m but some range to over 2,000m in the eastern Pyrenees, for example in Andorra (ANA).

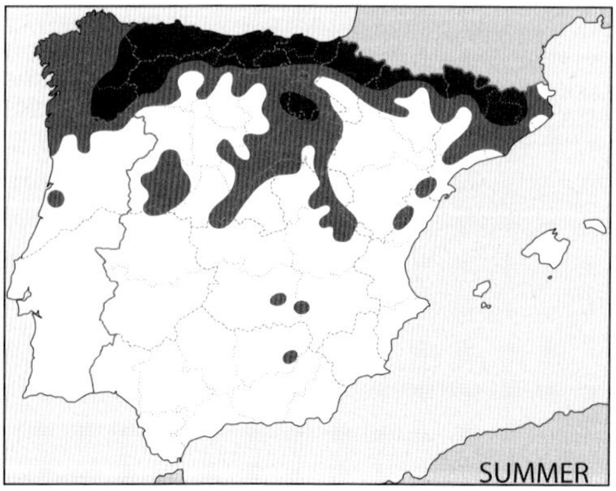

The population on SACRE sites showed a significant increase during 1998–2012 (Escandell 2013). The highest densities are in the eurosiberian zone, in pastures with trees in the Biscayan sector (13–23 birds/km^2) and in deciduous woodlands of the Cantabrian Range and Pyrenees (10–21 birds/km^2). The Catalan population has been estimated at 39,000–66,000 pairs (CA2). The Portuguese population is increasing. The Song Thrush is thinly distributed in Andorra but there too its range has expanded since the early 1990s (ANA).

The Song Thrush is far more widespread outside the breeding season (SWA, Elias *et al.* 1998, AP). The Iberian population is thought to be chiefly resident or dispersive, abandoning higher altitude locations for the winter. It is joined by a very large influx of birds from further north. These begin to appear in September but the peak arrival is in October–November (Santos 1982). Very large numbers winter in the northern coastlands as well as in the lowlands of western Portugal, the Levant coast and the southern third of the Peninsula, but relatively few do so on the Meseta and in the Ebro valley, where winters may be harsh. In the south and east great numbers are attracted to olive groves, which fruit in winter, where they may do considerable damage locally. Olive groves are indeed their principal wintering habitat in Iberia and densities of 400–500 birds/km^2 are often reported there (Muñoz-Cobo & Purroy 1980, Rey 2011). Other fruiting trees or shrubs, such as wild olive, lentisc and juniper, and holly in the north, are also attractive to them (Jordano 1985, Guitián 1989, Tellería *et al.* 2005). They are unfortunately a favourite quarry of hunters and millions are still shot or trapped every year (J. E. Gutiérrez 1991). Such hunting activity has traditionally been concentrated in the Guadalquivir valley in Andalucía, the Comunidad Valenciana and the Balearic Islands, and these three regions together account for 59% of all the recoveries analysed by Santos (1982).

A proportion of the immigrants continue south to winter in North Africa. The Song Thrush is a conspicuous migrant at the Strait of Gibraltar where the southward passage is noted from late September to mid November, peaking in late October (BG, Tellería 1981). A return passage is noted from late February to early April, chiefly in March, but is less conspicuous. The wintering population of Iberia and the Balearics departs in March–May (Santos 1982).

The many ringing recoveries of foreign-ringed birds within the Peninsula show that individuals from right across Europe head south-west to winter in Iberia. Some 77% have been found in Andalucía, Portugal and the north-east: in Catalonia, the Comunidad Valenciana and the Balearic Islands. However the breeding distribution of these immigrants is reflected spatially in their occurrence within Iberia, both during passage periods and in winter. Thus birds from the British Isles (*clarkei*), Norway and Denmark make up a greater proportion of those found in the western half of Iberia; those from Sweden, France and the Low Countries tend to winter in more central regions, and those from Germany, Poland, Finland, Russia, Austria, Italy and the former Czechoslovakia are chiefly recovered in eastern Iberia and the Balearics (Santos 1982). The 24 recoveries from coastal Algeria and one from Morocco of birds ringed in Spain, and also 23 recoveries of Spanish-ringed birds from Italy, are largely of thrushes ringed on autumn passage in the Balearics or on the Levant coast (SMA).

REDWING *Turdus iliacus*

(Zorzal Alirrojo, Tordo-ruivo)

***T. i. iliacus.* A common and locally abundant on passage and in winter throughout the Peninsula and Balearic Islands.**
***T. i. coburni.* Regular on passage and in winter throughout the region.**

The Redwing is a regular and numerous winter visitor to Iberia, where it is far more widespread and common than the Fieldfare (SWA, AP). However, as with that species, the numbers that reach the Peninsula vary considerably from year to year. The first arrivals are in mid October and continue through November (Santos 1982). The Redwing is widespread in woodlands throughout the Peninsula and in the Balearics in December–February but the largest concentrations are seen in the northern mountains, where the birds are heavily dependent on holly berries and hawthorn fruits (Guitián 1989, Guitián *et al.* 2000, SWA). It also frequents the juniper woods of the Iberian Range and the southern Spanish mountains, and large flocks are attracted to fruiting olives, both cultivated and wild, in the Levant coastlands, Andalucía and southern Portugal in some years (Soler *et al.* 1988, SWA, AP). The wintering population demonstrates a degree of nomadism as the birds seek out concentrations of fruiting trees or shrubs, but many also seek terrestrial invertebrates in fields

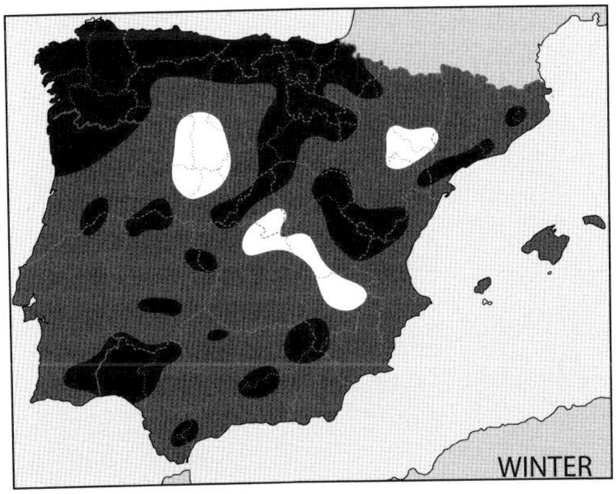

and pastures, or even in large parks or gardens (SWA). Over 900 were seen entering a roost in Viana do Castelo in December 2001 (*Anu. Ornitol.* 1: 46). The wintering population departs in March: one at Tavira, Algarve, on 19 June 2004 (*Anu. Ornitol.* 4: 42) was highly unseasonal. Some 89% of ringing recoveries are during November–February, 52% in December–January (Santos 1985).

Significant numbers cross the Peninsula to winter in North Africa, at least in some years (BM), and passage is evident at Gibraltar in autumn especially, although the Redwing is never as common there as the Song Thrush. The movements at the Strait have been noted, often heard at night, in late October–early November, with a return in February–early March (BG, Tellería 1981, Pineau & Giraud-Audine 1976). Most wintering birds originate from Scandinavia but the Icelandic race *coburni* is also well represented (SMA). The latter are especially encountered in northwestern Iberia (Santos 1982). Seven ringed in Spain have been recovered abroad: in France (5), Italy and Finland.

MISTLE THRUSH *Turdus viscivorus*

(Zorzal Charlo, Tordoveia)

***T. v. viscivorus*. A widespread resident in mainland Spain and Portugal. Some enter Iberia, including the Balearics, to winter.**

Breeding population estimates. Spain: SA1 330,000–790,000 pairs; ACRE 690,000 birds. **Portugal**: 10,000–100,000 pairs. **Andorra**: 300–600 pairs.

The Mistle Thrush is present, at various densities, throughout the Peninsula, with the exception of the most arid and treeless areas. It is thus largely absent in the breeding season from the Guadalquivir and Ebro depressions, the southeastern and southern Mediterranean coastlands, and steppe areas and treeless croplands in Extremadura, the meseta and parts of southern Portugal such as central Alentejo (SA2, PA2). It breeds down to sea level in the north and west but the bulk of the population, especially in the southern half of the Peninsula, is associated with mountainous areas above 600–700m. The highest breeding densities in Portugal are north of the Mondego river and especially in northeastern Trás-os-Montes (PA2). It occurs to 1,700m in the Navarran Pyrenees (Elósegui 1985), to 2,000m in the northern Iberian Range (de Juana 1980) and to 2,400m in the Sierra Nevada (Pleguezuelos 1992).

It nests in open coniferous and broadleaved woodlands as well as in dehesas and, locally, in areas of tall scrub (AI2). It is often found on forest edges adjoining pastureland, especially in the mountains, where it occurs on the margins of mixed beech and spruce forests in the Pyrenees and in similar habitats in the Central Range, for example. Elsewhere it may be found in riparian forests and a remarkable development has been its colonisation of fruit orchards at the Plana de Lleida, in the eastern Ebro valley, where it may now attain high densities (CA1, CA2). The Catalan population has been assessed to number some 64,000–103,000 pairs (CA2). Densities of up to 18 birds/km^2 have been reported from mesomediterranean orchards in northeastern Spain but counts of 6–10 birds/km^2 are typical of its core habitats in supramediterranean and Pyrenean coniferous woodlands (ACRE). A significant increase was seen on SACRE sites in 1998–2012 (Escandell 2013).

The Iberian birds are resident or perform altitudinal movements, descending to the valleys in winter (Santos 1982). Some remain at high levels in the Pyrenees all year round, however (ANA, CWA). Some birds from further north in Europe winter in Iberia and in the Balearics but the numbers involved are very modest, especially in comparison with the Song Thrush and

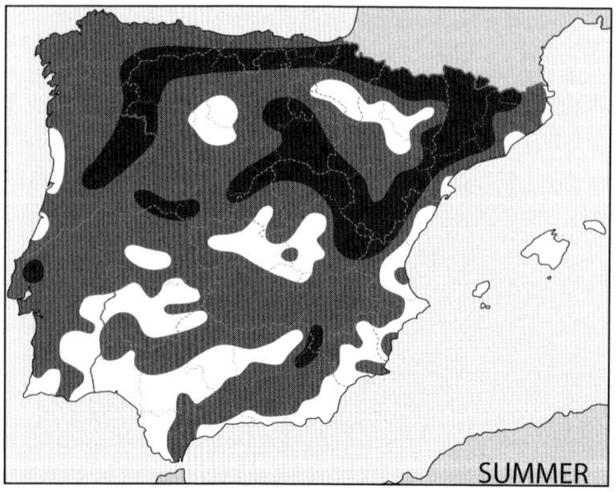

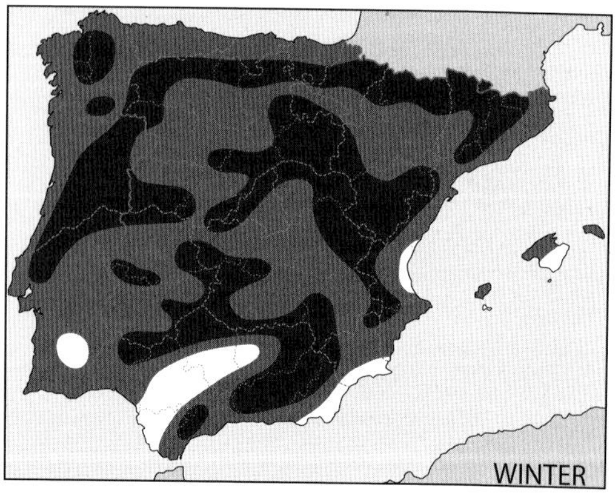

Redwing. Ringing recoveries further suggest that most immigrants winter in northeastern Spain: from the Basque Country to Catalonia, the Comunidad Valenciana and the southern Iberian Range, and few reach the west of the Peninsula. The winter range is therefore quite similar to that during the breeding season (SWA). The 48 Iberian ringing recoveries of foreign-ringed birds, all of them from Spain, are chiefly of birds ringed in the Low Countries, Switzerland Sweden and central Europe, with single birds from Austria, the Czech Republic and Scotland (SMA). The last of these was a migrant ringed on Fair Isle in September 1973 that was found the following February at Villamanrique, Sevilla. Mistle Thrushes share a predilection for fruiting junipers in winter with Fieldfares and Ring Ouzels and they too flock to the juniper woods of the Iberian Range, where they may attain average winter densities of 30–40 birds/km^2 (Tellería *et al.* 2011, SWA). There are some indications of small-scale passage to and from North Africa at the Strait of Gibraltar, southwards in October–November and northwards in February–March (BG) but very few individuals are seen in any one year.

AMERICAN ROBIN *Turdus migratorius*

(Zorzal Robín, Tordo-americano)

Vagrant to Spain.

A female or first-winter bird was at Laukiz, Vizcaya, during 10–15 December 1999. Individuals of the northern nominate race are migratory and account for rare observations in Europe, where 25 have been recorded in Britain but none in France.

CETTI'S WARBLER *Cettia cetti*

(Cetia Ruiseñor, Rouxinol-bravo)

***C. c. cetti.* A common resident in the Peninsula and the Balearics.**

Breeding population estimates. Spain: SA1 18,000–19,000 pairs; SA2 97,000+ pairs; ACRE 3.1 million birds. **Portugal**: 10,000–100,000.

Cetti's Warbler is characteristic of dense thickets, mainly along watercourses, throughout the region. The largest populations are along the Duero, Ebro and Guadalquivir basins, and, in Portugal, that of the lower Tagus. It is absent only from the driest parts of the southern meseta and the arid southeast as well as, for less obvious reasons, from much of Galicia and the adjacent coastlands of northwestern Portugal (SA2, PA2). It occurs locally up to 1,300m (SA2), and exceptionally up to 1,400m in the Sierra Nevada (Garzón & Henares 2012), but is generally absent from montane areas. In the northern Iberian Range it reaches 1,100m on the southern versant but just 800m on the northern side (de Juana 1980). It is widespread in the Balearic Islands, although local and uncommon in Ibiza.

It frequents dense riparian scrub, bramble patches and cane brakes along all manner of streams, ditches and irrigation channels. It may also be found around pools and lakes at the margins of reedbeds, where these give way to dense surrounding vegetation. The association with water is invariable in the meso- and thermomediterranean bioclimatic regions but less marked north of the Cantabrian Range. The males sing practically all year round and so its presence is readily established.

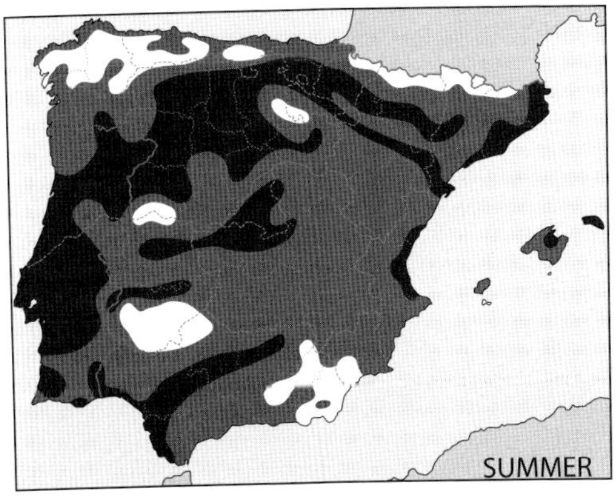

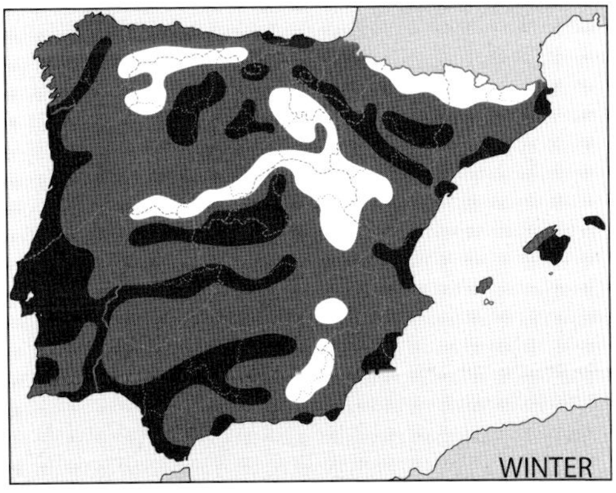

Cetti's Warbler is known to have increased and expanded its Iberian range in recent times but both Spanish atlas assessments were clearly significant underestimates. The population for Palencia alone was put at 10,000 pairs by Jubete (1997) and Román *et al.* (1996) estimated over 10,000 pairs in Burgos, whereas the Catalan population has been put at some 19,000–27,000 pairs (CA2). Densities of 28–44 birds/km² have been recorded along wooded watercourses. All populations may be subject to short-term fluctuations resulting from cold weather mortality, to which Cetti's Warbler is quite sensitive.

Many are entirely sedentary and most ringing recoveries are on-site retraps (Cantos 1992). However, some of those that breed inland winter in warmer coastal regions, those from the Ebro valley moving to Catalonia and the Comunidad Valenciana and those from the central mesetas mainly to the Comunidad Valenciana, but also to Catalonia, Andalucía, Portugal and Galicia. There are several known movements of over 400km but the longest known displacement was 605km from Valencia to Segovia. Females appear to outnumber males among those that perform such migratory movements (Bibby & Thomas 1984, AI2, AP). When compared with the breeding range, the winter distribution shows a reduced presence at higher elevations and greater concentration both in coastal regions and in the middle and lower reaches of major rivers, including the Tagus and the Guadiana in Extremadura and the Guadalquivir in Andalucía (SWA). In winter Cetti's Warblers may occur at high densities in reedbeds, where they do not breed except where trees and bushes are present (AP). They also become commoner in winter in irrigated croplands, often occupying quite small ditches (SWA). Ringing and recovery dates in Spain (Cantos 1992, AI2) as well as records from coastal sites where the species does not breed, for example in Asturias (Fig. 186), suggest that most movements occur in September–November and again in March–April.

Figure 186. Monthly distribution of Cetti's Warbler records at sites in Asturias where the species does not breed, 1993–2000 (n = 46). Source: *A. O. Asturias*

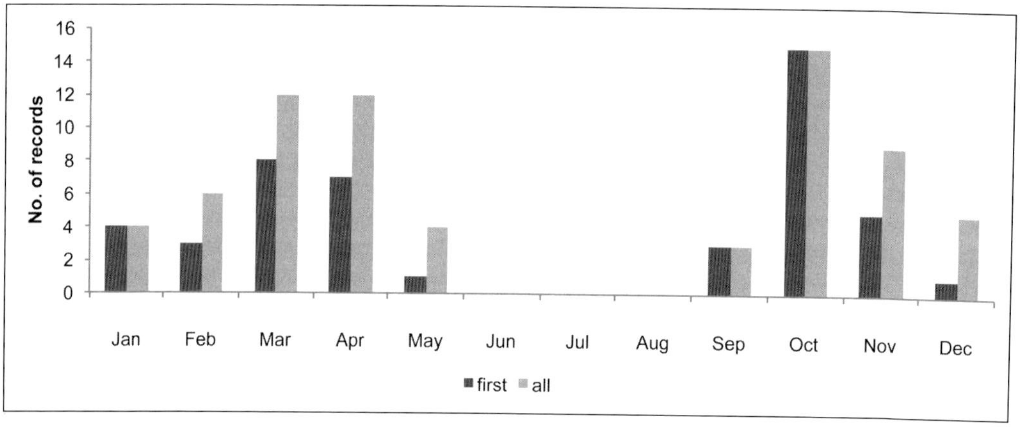

ZITTING CISTICOLA *Cisticola juncidis*

(Cistícola Buitrón, Fuinha-dos-juncos)

C. j. cisticola. **A common resident in the Peninsula and the Balearics.**

Breeding population estimates. Spain: SA1 14,300–57,000 pairs; SA2 80,000+ pairs; ACRE 1.7 million birds. **Portugal**: 50,000–500,000 pairs. **Gibraltar**: 0–3 pairs.

The Zitting Cisticola (Fan-tailed Warbler) is especially common along the Spanish east coast; in the Guadalquivir and Ebro valleys; in the farmlands of Andalucía, Extremadura and the Meseta, and latterly also along the Biscay coastline (SA2). It is widespread in Portugal except in much of Beira Alta and Trás-os-Montes, where it is scarce or absent. The highest densities in Portugal are in Estremadura, Ribatejo and Alentejo (PA2).

It occurs wherever there is extensive dense low cover in level open country. It is thus often found in the marginal scrub around lakes, estuaries, saltpans and other large water bodies but it is not otherwise particularly associated with water. Large numbers frequent farmland, especially where cereal crops shorter than about one metre are grown, as well as irrigated land and ricefields. It is everywhere associated with warmer areas, in the meso- and thermomediterranean and in the coastal areas of the eurosiberian zone, and thus occurs above 750m only locally (AI2, SA2), for example in the foothills of the Sierra Nevada,

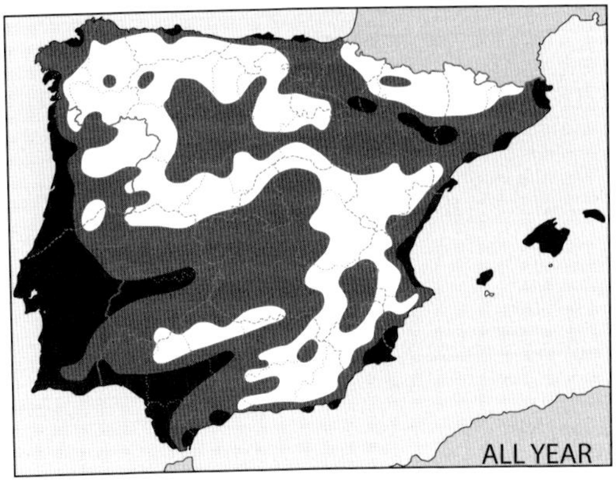

where it reaches 1,000m (Garzón & Henares 2012).

It was largely confined to southern and eastern Spain and southern Portugal at the start of the 20th century but has since spread throughout the Peninsula, with an accompanying increase in population. The more northerly historical references include those by Boxberger (1921), who saw it in the Ebro Valley, at Zaragoza and Casetas, and both Witherby (1928) and Bernis (1952), who give localities in the southern Meseta, including Aranjuez, Daimiel and Quero. Earlier, one was seen in the northern coastlands at Santander in May 1876 (Irby 1883). Colonisation (or perhaps recolonisation in view of Irby's record) of the coastal fringes of Cantabria began in the 1950s and the Cisticola was abundant there by the 1980s (SA2). However, it is highly vulnerable to prolonged episodes of very cold weather, which lead to local extinctions (Ferrer *et al.* 1986, AI2). For example, a cold snap in January 1985 reduced the Cantabrian population drastically and cold weather in southern Spain in 2005 caused severe losses, including the eradication of the small Gibraltar population (Bensusan 2008). Recovery after such setbacks is usually rapid, especially in optimum habitats, but some areas may take up to ten years to regain their lost populations (Galarza 1993).

The Spanish atlas population estimates are likely have been underestimates and an increase of 20–50% during 1970–90 was indicated by Birdlife International/European Bird Census Council (2000), including a marked range expansion in the interior of the Peninsula. The Catalan population alone has been put more recently at some 27,000–70,000 pairs (CA2). More than half of the population in the ACRE survey was in Andalucía (52%), the next largest subpopulations being in Castilla-La Mancha (12%) and Extremadura (9%). The densest populations, up to 35 birds/km^2, are in the thermomediterranean: in pastures, cereal crops and rice fields. There are also high densities on the margins of mesomediterranean wetlands (21 birds/km^2). Neither the Spanish nor the Portuguese populations have shown any clear recent trends (Escandell 2013, ACP).

The birds are largely sedentary but local movements occur and some populations are more widely dispersive, perhaps only in some years. The longest known displacement of a ringed bird is 49km, from Murcia to Alicante (*Ecología* 6: 262). Nesting habitats are often ephemeral, due to crop rotation for example, so particular areas may be abandoned and others colonised in some years. For example, in Extremadura its abundance has been linked to rainfall, the species becoming much scarcer and disappearing locally in very dry years (de Juana & García 2005). Small-scale southward migration is noted at the Strait of Gibraltar immediately after the breeding season, in July–August, with a poorly-marked return in March (BG). The Cisticola is unique among Iberian warblers in that it performs these movements by day. Migration has also been reported in the Basque Country in September–November and April (SA1) and Noval (1986) thought that part of the Asturian population was absent in winter. Tait (1924) considered it to be just a summer visitor to coastal northern Portugal.

WESTERN GRASSHOPPER WARBLER *Locustella naevia*

(Buscarla Pintoja, Cigarrinha-malhada)

L. n. naevia. **A locally common summer visitor to the Cantabrian coastal fringe. A regular migrant elsewhere.**

Breeding population estimates. Spain: SA2 5,000–10,000 pairs.

It has a marginal presence as an Iberian breeding species, being confined to the eurosiberian zone in the Biscayan coastal fringe and its hinterland. Here it is most abundant from the French border across the Basque Country and Cantabria into eastern Asturias. It becomes progressively scarcer further west but is present from western Asturias to Galicia north of A Coruña. A few nest in the extreme north of Navarra and northeasternmost Castilla y León, in Burgos (SA2). Occasional reports of breeding elsewhere in the Peninsula (AI2) have never been confirmed.

Breeding birds favour dense low cover in open situations; such as bramble patches, heathland with scattered gorse clumps and young pine or eucalyptus plantations with a dense understorey. Similar vegetation along the margins of wetlands, low scrub on sand dunes and the earliest stages of scrub formation on neglected farmland are also used (SA1, SA2). Most occupy low altitude sites, below 600m (SA2). It is vulnerable to changes in land use, particularly the development of coastal land and the afforestation of marginal farmland. The breeding population arrives from mid-March through April and departs in

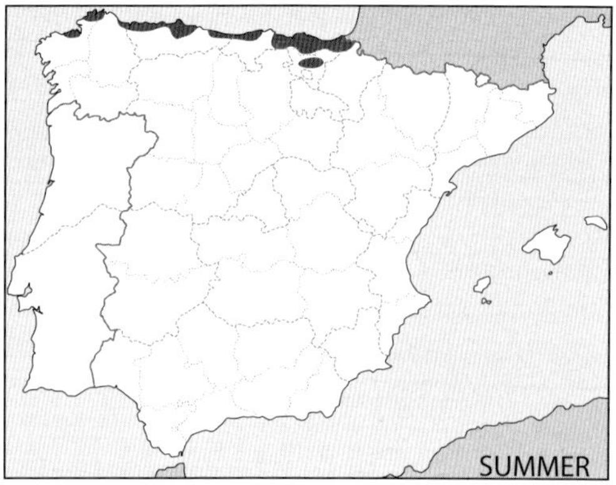

September: the mean first arrival date for a 20-year period in Asturias was 26 March (Noval 1986).

Migrants may appear almost anywhere. There are Spanish recoveries of birds ringed in Germany (2), Sweden and Belgium, and two recoveries of birds ringed in Iberia: from Belgium and Senegal. Observations and captures at Gibraltar (BG) and in northern Morocco (BM) indicate that passage occurs principally from late March to mid May in spring and in late August–October in autumn. In northeastern Spain, spring passage is in April–May, peaking from mid April to mid May, and autumn passage from mid August to late October, peaking in the first ten days of September (Ferrer *et al.* 1986). Irby (1895) reported a wintering bird in Málaga and a few may winter in Morocco (BM), so further winter records from Iberia may be expected.

RIVER WARBLER *Locustella fluviatilis*

(Buscarla Fluvial, Cigarrinha-ribeirinha)

Vagrant to Spain.

One was ringed at Castelló d'Empúries, Girona, on 22 August 1980 (Sargatal & del Hoyo 1989), and there is an accepted observation from the Miño estuary, Pontevedra, on 4 September 2003. The River Warbler extended its range in the mid 20th century into southern Sweden, Denmark and western Germany (Kennerley & Pearson 2010). It has appeared in western European countries with some frequency since then, with 40 recorded in Britain and 13 in France. There are several spring records from Algeria and one from Morocco.

SAVI'S WARBLER *Locustella luscinioides*

(Buscarla Unicolor, Cigarrinha-ruiva)

***L. l. lusciniodes*. A scarce and highly local summer visitor throughout mainland Iberia. Has bred in the Balearics.**

Breeding population estimates. Spain: SA1, SA2 2,000 pairs. **Portugal**: 500 pairs.

Savi's Warbler breeds in scattered localities that offer more or less extensive reedbeds over fresh or brackish water, these often admixed with reedmace or sedges and bordering on willow scrub. It is far more selective in its habitat choice than the widely distributed Reed and Great Reed Warblers.

In Spain the bulk of the population is at and around Doñana. Some 20% occur at the wetlands of Castilla-La Mancha: where 261 pairs were censused at Damiel in 1996. A later survey of Castilla-La Mancha (Velasco *et al.* 2004) put the regional population at 350–550 pairs distributed between 51 localities and including 150–200 pairs at Daimiel, 40–60 pairs at El Taray, 30–40 pairs at the Dehesa de Monreal, 25–35 pairs at the Castrejón reservoir and 20–25 pairs at the Villafranca de los Caballeros lagoons. A further 20% are at the Ebro delta and other east coast wetlands, including the Aiguamolls de l'Empordà; the Albufera, the Marjal del Moro and the Marjal de Pego-Oliva in Valencia, and the Alicante wetlands (SA2). There were 150 pairs at the Ebro delta in 1995 and perhaps 100 pairs at Pego-Oliva in 1993. Smaller numbers breed locally elsewhere, including at the Arrocampo reservoir, Extremadura; the Laguna de La Nava; Gallocanta; and, in Andalucía, at the lakes and reservoirs of the Upper Guadalquivir in Jaén and at various southern

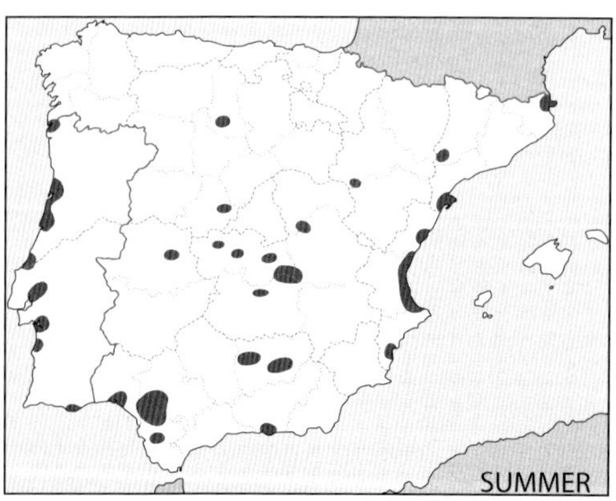

lagoons in Huelva, Sevilla, Cádiz and Granada. Birds have disappeared from some sites, such as El Hondo, in recent years and the population overall is thought to be declining slowly (LR).

Savi's Warblers breed at a number of coastal lagoons and estuaries in central and northern Portugal, as well as very locally in coastal Alentejo and Algarve and at a few inland sites (PA2). The principal sites include the Ria de Aveiro: where there were an estimated 300–400 pairs at the start of the 21st century, the Lagoa de Santo André, Setúbal, and the marshes (*pauis*) of the Baixo Mondego (AP). The Paul do Boquilobo, Santarém, held up to 30 pairs in the mid 1970s but Savi's Warbler has not bred there since 2001 (AP).

In the Balearics Munn (1934) observed it at the Albufera de Mallorca, where it was then not uncommon, and found several nests there in May 1933. It may have nested in Ibiza, in 1963 (Mester 1966) and there are later reports of possible breeding in Mallorca and Menorca (Muntaner & Congost 1984, Ferrer *et al.* 1986). Breeding birds may have returned to the Albufera in 2010 and 2011, with up to five singing males reported during April–early May 2011 (*Ardeola* 59: 448).

Many of the principal wetlands that it inhabits are at risk from the lowering of the water table due to excessive abstraction for crop irrigation, a particular problem at Daimiel, as well as from drying up in drought years. Some sites, perhaps especially the smaller ones that supported just 1–2 pairs, have been lost altogether. On the other hand, new sites such as reservoirs may be colonised eventually where suitable vegetation develops: the Arrocampo reservoir in Cáceres, was constructed in the 1980s but the species was first proved to nest there only in 2002 (*Ardeola* 51: 253).

The breeding sites are occupied from late March to September (Ferrer *et al.* 1986, AI2). Migrants, presumably including some birds from other western European populations, may occur widely at wetlands on passage, including on the Balearic islands, but generally pass unnoticed. However, there are 15 recoveries of birds ringed in France, one from Belgium and one from Senegal. Seven birds ringed on the Levant coast have been recovered abroad: in France (4), Germany and Senegal (2).

BOOTED WARBLER — *Iduna caligata*

(Zarcero Escita, Felosa-calçada)

Vagrant to Spain.

There are three records: one ringed on the islet of Dragonera, Balearics, on 24 September 1998; one seen at Colunga, Asturias, on 8 September 2004, and one seen at Ferrol, A Coruña, on 8 October 2009. This is a relatively frequent vagrant to western Europe, with 128 individuals recorded in Britain and nine in France.

WESTERN OLIVACEOUS WARBLER — *Iduna opaca*

(Zarcero Bereber, Felosa-pálida)

A scarce and local summer visitor to southern and eastern Spain, rare elsewhere. A rare migrant in Portugal and the Balearic Islands.

Breeding population estimates. Spain: SA2 5,200+ pairs. **Portugal**: <5 pairs.

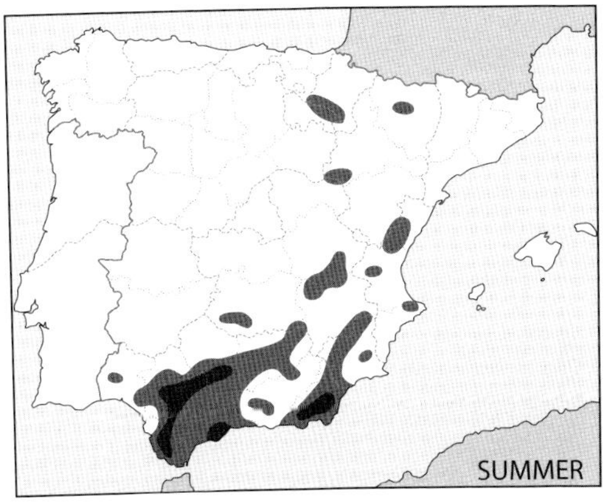

The breeding distribution is almost exclusively thermomediterranean and concentrated in the lowlands and seasonally arid watercourses of Andalucía. The Olivaceous Warbler is characteristic of the Guadalquivir valley: from where it ascends the neighbouring sierras along the lines of watercourses; the Guadalhorce and other river valleys in Málaga, and the ramblas of Almería. However, it is practically absent from Huelva and has only a limited presence in Granada. It has a more fragmented distribution in Murcia, the Comunidad Valenciana and southern Castilla-La Mancha (SA2). There is also a thinly-spread population in the upper Ebro valley in La Rioja and Navarra, and at Los Monegros and a few other sites in Aragón (Sampietro & Latorre 1998, Woutersen & Platteew

1998): many records from these regions are of singing males in breeding habitat but few nests are ever found. Singing birds were found in the Ebro Valley by Boxberger (1921) in Zaragoza and by Ticehurst & Whistler (1925) at Castejón. Breeding has been recorded sporadically elsewhere, for example in Madrid, Castilla y León, Catalonia and in Extremadura in northeastern Badajoz and on the Guadiana in Badajoz city and Mérida (SA1, SA2, *Ardeola* 58: 509, 59: 449, 60: 538). A number of sightings in central and northern Spain are questionable however since the possibility of confusion with pale Melodious Warblers has not always been considered (*Ardeola* 35: 313, Sanz-Zuasti & Velasco 1999). Birds have been reported breeding up to 1,500m in Almería (Manrique 1997) but nesting is generally below 1,000m.

The typical nesting habitat is a seasonally dry or partly dry watercourse with stretches of tamarisks and also other riparian vegetation, such as willows and cane brakes. It is also found in similar vegetation around lakes and reservoirs as well as in abandoned olive groves, palm groves, gardens and open woodlands. It is nowhere abundant and is thinly spread even in its core distribution. The highly fragmented range suggests that it may have been more widely distributed in the past and it may still be declining. Formerly it bred north to Catalonia, where it was found in the Garraf area in 1955 and 1963 and in the Empordà region during the 1960s (Wallace & Sage 1969), but no breeding birds were found by the first Catalan atlas survey in 1975–83 (CA1). Some range contraction may also have occurred in the Comunidad Valenciana (Urios *et al.* 1991), where several pairs were found breeding in Valencia city parks in 1954 (Deetjen 1955).

A Portuguese population of 5–50 pairs was suggested by Birdlife International (2004) but in reality it is next to non-existent. It may nest occasionally in the east but confirmed breeding is practically unknown, although a pair was seen carrying food at a site in the Guadiana basin in June 2004 (*Anu. Ornitol.* 4: 43; AP). Migrants are rarely recorded in Portugal, chiefly in autumn, and the reliability of many such records has been questioned (AP). It is also a rarity in the Balearic Islands, where 34 probably overshooting or displaced migrants were detected during 1990–2011, 75% being trapped by ringers, most of them on Cabrera (*A. O. Balears*). There are also four ringing records from the Columbretes islands (*A. O. C. Valenciana*).

This is one of the last summer visitors to return to Iberia from its winter quarters in West Africa. Although some appear from mid April, exceptionally as early as mid March, most arrivals are in May and into June. Departure may begin in late July and continues during August–September, with a few stragglers in the first half of October. The extreme dates for 51 birds recorded outside the breeding range: in Catalonia, the Comunidad Valenciana and the Balearic Islands, were 7 April and 16 June in spring, and 5 August and 12 October in autumn (regional bird reports).

Individuals are occasionally reported well outside the breeding range. Single birds at Somió, Asturias, on 16 August 1976 and at Mañón, A Coruña, on 20 August 2011 (Álvarez-Balbuena *et al.* 2000, *Ardeola*. 58: 509) were exceptionally far north. There are also two records from Gipuzkoa (*A. O. Gipuzkoa 1995*: 52,*1998*: 84).

MOUSTACHED WARBLER — *Acrocephalus melanopogon*

(Carricerín Real,Felosa-real)

***A. m. melanopogon*. A scarce and declining local resident and short-distance migrant in mainland Spain and in the Balearics. Vagrant to Portugal (4+).**

Breeding population estimates. Spain: SA1 2,400–3,200 pairs; 2005 census 1,017 pairs.

The Moustached Warbler has always been very locally distributed in the Iberian peninsula, where it is confined to a series of east coast wetlands and a few sites inland (SA2). It also breeds at a few sites on the Balearic Islands, where the Albufera de Mallorca now accommodates about half of the Iberian population. It first seems to have been reported at the Albufera by Munn (1921), who described it as a plentiful resident. On the mainland Arévalo (1887) recorded that it bred at the Albufera de Valencia where Witherby (1930) also found it nesting. Witherby (1930) and Ticehurst & Whistler (1930) found it at the Ebro delta. Possible nesting on the southern Meseta was first reported from Daimiel in 1951 (Bernis 1952). The principal mainland site is now the Marjal de Pego-Oliva, Alicante, with some 14% of the population in 2006. The remainder are chiefly distributed between El Hondo, Alicante; the Marjal de Xeresa, Valencia; the Prat de Cabanes-Torreblanca, Castellón; the Albufera de

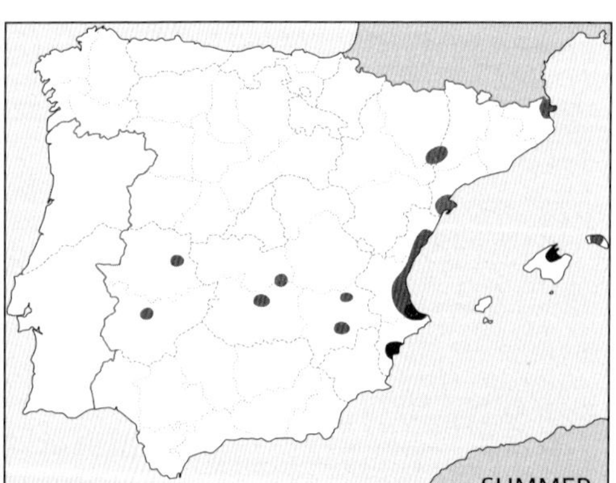

Region	2005 census	mid 1990s
Catalonia	7	141
Comunidad Valenciana	437	815
Balearic Islands	561	1,500
Castilla-La Mancha	10	166
Extremadura	2	2
Total	1,017	2,624

Table 55. Moustached Warbler breeding population (pairs). Sources: Castany & López (2006), LR.

Valencia, the Santa Pola saltpans, Alicante, and the Ebro delta on the east coast, and the Tablas de Daimiel in Castilla-La Mancha. Occasional breeding has been reported very rarely elsewhere: for example, a pair nested at a gravel pit near Mérida, Badajoz, in 1998, 1999 and 2000, and a pair bred successfully at Sariñena, Huesca, in 2006 (*Ardeola* 49: 193, *A. O. Aragón 2004–2007*: 321). The Balearic population is dominated by the important presence at the Albufera de Mallorca but local nesting also occurs in Menorca, where it was first reported in 1982 (Mejías & Amengual 2000), and possibly also in Ibiza (Martínez & Palerm 1998).

The Moustached Warbler is confined to mature reedbeds that offer an admixture of other aquatic plants, notably reedmace, fen-sedge and club-rush. Fragmented tracts offering expanses of open water also seem to be important, perhaps because it is thought to obtain much of its insect food near the water surface (Bibby 1982). It is certainly much more discriminating in habitat choice than other Iberian *Acrocephalus* species and it appears to be particularly sensitive to any decline in water quality through eutrophication (LR). It has been reported in the breeding season from some 70 localities in Spain, but was only found at 60 sites during a national census in 2005, with a total population of 1,017 pairs (95% confidence limits 484–1,777), over 60% fewer than in the mid-1990s (Table 55).

The principal finding of the 2005 census was that a severe decline had occurred over just ten years. The decline was particularly marked at the Ebro delta, where none at all were found in 2005 but where there had been 100–122 pairs in 1995 (Martínez-Vilalta *et al.* 1998). The well-studied population at the Prat de Cabanes–Torreblanca (Valencia) declined from 400–500 pairs in the mid 1990s to 150–170 pairs in 2005. Away from the east coast, the formerly significant population of Daimiel and the other lakes of La Mancha was reduced from 138–194 pairs in the mid 1990s to fewer than ten (LR), even though Velasco *et al.* (2004) had recently estimated the regional population at 125–250 pairs at 21 localities, including 25–50 pairs at Daimiel, 25–35 pairs at the Dehesa de Monreal, 20–30 pairs at El Taray, and 10–15 pairs each at the Buendía and Almoguera reservoirs. The Mallorcan population had been put at 1,000–2,000 pairs in 1995 (Chris 1997).

The decline seems to be due to a number of factors, which are exacerbated since the birds are concentrated at a limited number of sites. Drying-out of wetlands during the breeding season during drought years, or as a result of excessive water abstraction for irrigation of crops or golf courses, may lead to site abandonment in some years. Destruction of the preferred areas of mature reed, by burning or trampling by cattle, may have severe effects locally. This being so, habitat management for waterbird species at the Albufera de Mallorca that employs controlled grazing by ungulates may be proving prejudicial to the species. Declines in water quality may also lead to site abandonment. The species is listed as 'Vulnerable' in Spain (LR) but may now be better regarded as 'Endangered' (Castany & López 2006).

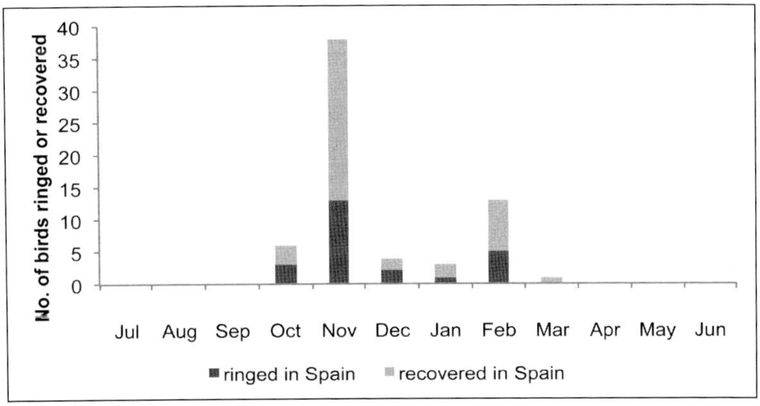

Figure 187. Monthly distribution of Moustached Warblers recorded in eastern Spain. The figure is based on recovery dates of those that were ringed in France and recovered Spain and ringing dates of those that were ringed in Spain and later recovered in France (n = 65). Source: Castany & López 2006.

The birds in Mallorca seem to be largely or entirely resident, since only one of the many individuals ringed on the islands has been reported on the mainland. In contrast, some of the mainland sites, especially those inland, appear to be entirely or partly abandoned in winter but most of the birds probably do not travel far: 75 of 82 within-Spain recoveries of birds ringed at the Prat de Cabanes-Torreblanca were within 100km of the ringing site (Castany & López 2006). Nevertheless, at least some birds are more widely dispersed outside the breeding season since there are scattered winter or passage records from localities where the species does not breed, including Doñana and the Guadalhorce and Vélez estuaries, Málaga. One landed on a fishing boat off Cape Billano, Bizkaia, on 12 May 2003 (*A. O. Bizkaia 2002–2003*: 143). There were also at least nine November–February records from the Laguna de Sariñena, Huesca, during 1993–2000 and one ringed there in November 1993 was recaptured at the same site in February 1995, December 1995 and October 1997 (*A. O. Aragón 1995–1996*: 54, *1997–1998*: 59).

Very few have reached Portugal from where there are 12 records of single birds: nine in October and one each in January, February and March (AP, RBP). These include seven that were captured by a ringing expedition of the British Trust of Ornithology at the Lagoa de Santo André, Setúbal, during 13–25 October 1977. These seven records have never been critically examined, which may call their reliability into question (AP): however, it seems unlikely that the BTO team would have misidentified Sedge Warblers, the only likely confusion species.

Overwintering birds include migrants from France, from the population of the Camargue and Mediterranean coastal areas, which has been estimated at 3,000–6,000 breeding pairs and is also partly resident there (Balança & Schaub 2005, OF). There were over 100 recoveries in total by 2010, including both French-ringed birds recovered in Spain and vice-versa, all during October–February except for one in early March (Fig. 187). These show that most immigrants of French origin move to Catalonia: especially the Ebro delta, and the Comunidad Valenciana, but one ringed at Salobreña, Granada, in December 1998 was recovered in Hérault (982km) in August 2005 (*Ecología* 21: 281). The Catalan wintering population has been put at 179–243 individuals (CWA) and that of the Comunidad Valenciana at 500–2,000 (SWA). Ringing recoveries peak in November and February, which may indicate some through passage. There are also three Spanish recoveries of birds ringed in Slovenia (SMA), including one that was recovered at the Ebro delta (1,255km) both in October 2004 and October–November 2005 (*Ecología* 20: 328). One ringed at the Ebro delta in November 1997 was found in Italy in April 1998 (*Ecología* 14: 321). Moustached Warblers are also occasionally recorded during passage periods at sites in Morocco where they do not breed (BM), which implies some migration to and from North Africa.

AQUATIC WARBLER *Acrocephalus paludicola*

(Carricerín Cejudo, Felosa-aquática)

A scarce migrant in mainland Iberia and the Balearic Islands.

Most Aquatic Warblers nest in Belarus, Ukraine and Poland. They winter in sub-Saharan West Africa and are widespread on passage across much of western Europe, including Spain (Schäffer *et al.* 2006, Walther *et al.* 2007). However, only three have been found in the Balearics and they have only been recorded in Portugal as scarce autumn migrants. Passage birds occur in a broad range of wetlands, including reedbeds and riparian growth, and much of what is known of their occurrence is due to the activity of ringers.

They are scarce but regular: 148 had been ringed in Spain by 1998, with an additional 88 sight records there. Since then a further 1,043 were ringed up to 2009, as ringing programmes proliferated. Ringers in Portugal captured 67 Aquatic Warblers during 1977–2008 (Neto *et al.* 2010) and a further 21 in total in 2011–12 (RBP). Most spring records in Spain are in April although the earliest arrivals are in early March and the latest in early May (Atienza *et al.* 2001). The spring passage tends to be routed further east than the autumn movements (Atienza *et al.* 2001). Thus 67% of 49 records from Catalonia during 1996–2009 were in spring, nearly all of them in April (*Anuari d'Ornitologia de Catalunya*). Autumn passage is more widely apparent and extends from late July to mid October but is chiefly from mid August to September. One at the Ebro delta on 8 December 2009 (*A. O. Catalunya 2009*: 213) was exceptional.

Very few locations report the species, very probably reflecting the siting of ringing activity. Up to 1999 they comprised the Galachos del Ebro, Zaragoza (33 records), the Marjal del Moro, Valencia (29), Txingudi, San Sebastian (29), the Aiguamolls de l'Empordà (27) and the Ebro delta (15) (Atienza *et al.* 2001). Subsequent work at the Laguna de La Nava has also revealed the signal importance of that site as a staging post: 371 birds were captured during 1999–2001, albeit with the aid of tape-lures (Jubete 2001). More recently, 15 were ringed at Gallocanta, during 12–29 August 2010 (*Ardeola* 57: 543): two had been ringed in France in mid-August that year, one in the north and the other on the west coast. There are three earlier August recoveries of French-ringed birds in Spain, all of which had been ringed earlier in the same month, and one recovered in León in September 2006 had been ringed in Belgium the previous month. Clearly some birds move very quickly through western Europe but individual timing varies since three ringed in August in Palencia were recovered the following year in France, also in August. An adult male ringed in Belarus in June 2006 was recovered at Castromocho, Palencia, on 11 August 2009 (CMA). Recent Spanish recoveries include two more from the Low Countries, another from Belarus and two from Poland (SMA). A few individuals ringed in Spain have been recovered in France (4) and Austria (1) and one ringed at La Nava in August 2003 was recovered at Tiguet, Senegal in January 2007 (SMA, *Ecología* 22: 209).

Practically all recent Portuguese records have involved birds caught in autumn by ringers at coastal wetlands north to the Ria de Aveiro (Neto *et al.* 2010). Up to 58 birds (87%) were caught at the Lagoa de Santo André, Setúbal, where the ringing effort has been most intensive. All captures have been in August– September, peaking late in August, adults occurring nine days ahead of juveniles on average. Adults also occur earlier in autumn than juveniles in Spain (Zumalacárregui *et al.* 2011). The species may formerly have been more numerous on (autumn) passage in Portugal since it was well known to Tait (1887, 1924) who described it as common at saltpans near Porto between mid August and mid October.

The Iberian wetlands are clearly important to migrants of this endangered species and so have an important part to play in its conservation. To this end, the key site of the Laguna de La Nava has attracted funding from EU sources aimed at expanding the reserve and managing the halophytic vegetation attractive to the Aquatic Warbler, as well as sustaining research and environmental education at the Reserve (LR).

SEDGE WARBLER *Acrocephalus schoenobaenus*

(Carricerín Común, Felosa-dos-juncos)

Scarce but regular on passage throughout the Peninsula and in the Balearics. May have nested locally.

Breeding in Iberia is sporadic at best and hardly any records of possible breeding seem to amount to more than observations of late spring migrants. However, definite breeding was said to have occurred at Muskiz, Bizkaia, in July 1966, and at the Ría del Bidasoa, Gipuzkoa, at least in the mid 1960s and possibly subsequently (*Ardeola* 44: 259, Noval 1967, SA2). Recent searches have failed to detect any breeding birds anywhere, although singing migrants have been seen at Bidasoa as late as June. The breeding range in coastal southwestern France extends relatively close to the Basque Country (Yeatman-Berthelot & Jarry 1994, OF).

The species is widespread on passage across Iberia but most published sight records refer only to a few birds. However, ringers have revealed the transitory presence of many more. Grandío (1998a, 1998b, 1999) captured 1,668 individuals in two coastal reedbeds at Bidasoa, a stopover site in the Basque Country where autumn migrants replenish fat reserves over several days, largely by consuming Plum Aphids.

As many as 95% of 300 extra-regional ringed birds recovered in Spain were ringed in the British Isles, France or the Low Countries, with a few from Germany and Scandinavia and one from Italy (SMA). Birds ringed in Iberia have been recovered in these countries with a few recoveries also in Italy (3), Switzerland, Croatia, Ceuta (North Africa), Mauritania and Senegal (2) (SMA). Spring migrants occur from March to mid-May, exceptionally earlier and later. For example, one was ringed at El Hondo on 10 February 1993 (García Peiró *et al.* 2005), there were four singing males at the Brazo del Este on 25 February 2001 (Chiclana *et al.* 2002) and the first of the season at the Prat de Cabanes-Torreblanca, Castellón, in 2009 was on 18 February (*A. O. C. Valenciana 2009*: 124). The return passage is noted from mid July to October, adults passing earlier than juveniles (Grandío 1999). In central Spain the spring passage peaks during the second half of April and the autumn passage is mainly between mid July and mid August (Bermejo & de la Puente 2002). Spring ringing recoveries range from January to April, peaking in March, and autumn recoveries are July to November, peaking in September (Cantos 1992, García Peiró *et al.* 2005). Spring recoveries have a more easterly distribution than the autumn ones, which are more evenly spread (Cantos 1992).

PADDYFIELD WARBLER *Acrocephalus agricola*

(Carricero Agrícola, Felosa-agrícola)

Vagrant to Spain and Portugal.

The first Iberian record was a bird ringed at Portimão, Faro, on 12 November 1993, since when there have been a total of 13 accepted records in Spain and two further records in Portugal, nearly all of birds captured by ringers. One ringed at the Ivars lagoon, Lleida, on 6 March 2007 was found at the Marjal de Pego-Oliva, Alicante, on 10 September same year (*Ardeola* 56: 332). Six of the 16 individuals involved were identified as adults and six as first-year birds.

The records have been between August and April. In contrast, those from Britain and France have mainly been in autumn (OF, Slack 2009). Almost two-thirds have been between November and March and some individuals were apparently overwintering. For example, one ringed at the Ebro delta on 20 December 2009 was recaptured there on 6 February 2010

(*Ardeola* 59: 395). Iberian observations have tended to be coastal and concentrated in Catalonia (9), with others in Andalucía (3) and southern Portugal (2).

The European range may be spreading westwards (Kennerley & Pearson 2010), with recent breeding records in Latvia, Finland and the Netherlands, which may partly explain a recent increase in observations in western Europe. A total of 83 individuals have been found in Britain and 41 in France, where the first record was in 1990. We are unaware of any records from North Africa, but there is one from the Azores.

BLYTH'S REED WARBLER *Acrocephalus dumetorum*

(Carricero de Blyth, Felosa-das-moitas)

Vagrant to Spain and Gibraltar.

One ringed at Cubas, Cantabria, on 24 September 1972 (*Ardeola* 19: 33) is only listed as a 'possible'. One was ringed in Gibraltar in 1973, also on 24 September (BG). Two others have been identified by ringers in Spain: one at the Brazo de la Torre, Doñana, on 2 September 2001 was recaptured there on 28 August 2002, and a first-winter was on Cabrera, Balearic Islands, on 6 October 2007 (*Ardeola* 51: 534, 56: 333). The breeding range of this species expanded westwards during the 20th century and now reaches Finland, southern Sweden and the Baltic States. Birds are not infrequently recorded nearby in Europe and 122 had been found as far west as Britain and Ireland by 2009, most in autumn. However, there are only 11 records from France and apparently none from North Africa.

MARSH WARBLER *Acrocephalus palustris*

(Carricero Políglota, Felosa-palustre)

Vagrant to Spain.

The breeding range extends from northern France and southern Scandinavia east to the West Siberian plain. All populations converge on the Middle East on passage, travelling well to the east of Iberia (Kennerley & Pearson 2010). Silent birds are only reliably distinguishable from Reed Warblers in the hand, although the complex mimetic song of the Marsh Warbler is highly distinctive.

The reliability of some early records in Spain has been questioned (Bernis 1954a, de Juana 2006, Aymí 2011) and there are no accepted records for Portugal (AP). The Marsh Warbler is claimed to have bred at the Ebro delta (Maluquer 1971) and 66 individuals were ringed in Spain up to 1994 (Cantos & Gómez-Manzaneque 1995). However, the Spanish rarities committee has considered the species since the 1990s, during which period four records have been accepted and nine rejected. The accepted records, of single birds, comprise one that was singing in a small reedbed in Alicante from 30 April to 9 May 1993, and individuals ringed at Cabrera, Balearic Islands, on 27 May 1992; at Flix, Tarragona, on 5 September 1992, and at the Ebro delta on 25 August 1996.

EURASIAN REED WARBLER *Acrocephalus scirpaceus*

(Carricero Común, Rouxinol-dos-caniços)

***A. s. scirpaceus*. A widespread and common summer visitor and migrant.**

Breeding population estimates. Spain: SA1 7,100–13,600 pairs; SA2 22,343+ pairs; ACRE 977,000 birds. **Portugal**: 1,000–10,000 pairs.

The Reed Warbler is common in waterside habitats throughout mainland Iberia and also on the Balearic Islands. It is most frequent in Spain along all the major river valleys and at the wetlands of the meseta and the Levant coast (SA2). It is widespread within the Mediterranean bioclimatic zone, including the Balearics. In contrast its distribution within the eurosiberian zone is more discontinuous and almost entirely coastal: here it nests at the major estuaries. The wider Portuguese distribution is also chiefly at coastal wetlands although some breed inland in Ribatejo and Alto Alentejo along the Tagus and Sado valleys (PA2).

The nesting habitats are chiefly at low altitudes. It frequents both large reedbeds and smaller ones along rivers, streams,

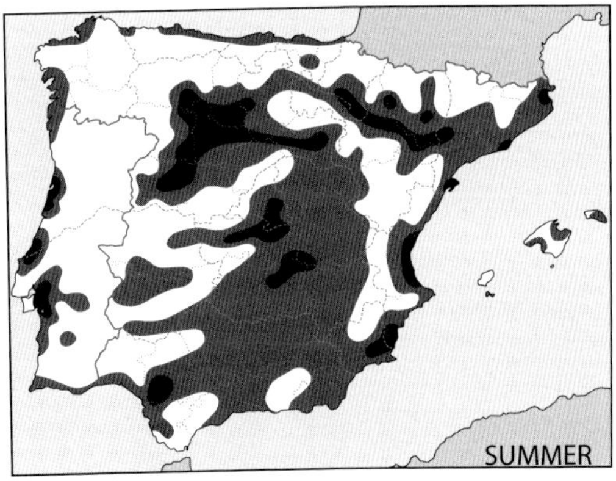

lake margins and irrigation channels. Areas where reeds are mixed with other aquatic plants, such as reedmace and sedges, are often used. Males hold small territories within the reedbeds but much activity and feeding occurs in willows and other scrub adjacent to waterbodies.

Successive estimates of the Spanish population have found progressively more birds. The highest densities are recorded in riparian growth in the mesomediterranean (39 birds/km^2). The SACRE results for 1998–2012 indicate a significant increase in the Spanish population, especially evident in the south (Escandell 2011, 2013). The Doñana population certainly appears to have enjoyed a long-term increase since Noble (1902) found it to be 'not very common' there and Valverde (1960) said it was 'scarce', whereas it is now considered abundant (Chiclana *et al.* 2000, García *et al.* 2000). Population trends in Portugal are unclear since there is a lack of historical data, but the drainage and modification of wetlands during the 20th century seems likely to have reduced its abundance. At the Paul do Boquilobo some 500 pairs in 1973 had diminished to a few dozen by 1993 (AP).

Migrants occur widely during both passage periods, when birds from elsewhere in western Europe pass through Iberia (BWP, Zwarts *et al.* 2009). Spring passage begins in March but is principally in April–May. The return passage is from mid August to mid November, peaking in September (AI2). A ringed individual is known to have overwintered at Doñana (Chiclana *et al.* 2000). However, some other winter records, especially in December, may involve late migrants. There are over 1,900 Iberian recoveries (SMA), of birds ringed mainly in the British Isles, the Low Countries, Fennoscandia, Denmark, France, Germany, Poland, the Czech Republic, as well as some from Slovakia, the Baltic States, Italy, Croatia, Austria, Hungary, Ukraine and Russia (6). There are also Spanish recoveries of Reed Warblers ringed in Morocco (4), Mauritania (4), Mali (2) and Senegal (35). Birds ringed in Iberia have been recovered in the same countries and east to Ukraine (3) and Russia (3). Spanish-ringed birds have provided 24 recoveries in Morocco, one in Algeria and 18 in sub-Saharan Africa: in Senegal (8), Mauritania (6), Mali (3) and the Ivory Coast (SMA). There are also 18 recoveries from Senegal of birds ringed in Portugal. Migrants may occur outside aquatic habitats when on passage but they are hard to detect there. However, they are regularly caught in dense olive scrub by ringers in Gibraltar: their occurrence there was unsuspected until ringing operations began in the 1970s.

GREAT REED WARBLER — *Acrocephalus arundinaceus*

(Carricero Tordal, Rouxinol-grande-dos-caniços)

***A. a. arundinaceus*. A widespread and common summer visitor and migrant.**

Breeding population estimates. Spain: SA1 9,500–12,600 pairs; SA2 10,277+ pairs; ACRE 659,000 birds. **Portugal**: 1,000–10,000 pairs.

The Great Reed Warbler is widely distributed throughout the thermo- and mesomediterranean climatic zones in mainland Iberia and also nests in Mallorca and Menorca in the Balearics. It is far more local in the eurosiberian zone, where it is much less widespread and scarcer than the Eurasian Reed Warbler and largely confined to a few coastal sites from Galicia to the Basque Country.

It is most abundant along the principal river basins, as well as in the wetlands of the east and south-west coasts. It is also widespread at lagoons and gravel pits, and along watercourses, throughout the Meseta (SA1, SA2). The highest concentrations in Portugal are coastal: at the Tagus and Sado estuaries, the Ria de Aveiro and the marshlands of the lower Mondego basin, but it also occurs widely if at low density inland in the centre and south, notably in Alentejo (PA2, AP). It selects extensive tall reedbeds, wet cane brakes and reedmace, generally with a presence of tamarisks, willows and other tall waterside shrubs. It may be found around all manner of lakes and reservoirs as well as along streams, rivers and irrigation channels, provided that these have significant well-developed marginal vegetation. It is thus a bird of the lowlands , although it occurs to 1,000m both in the northern Iberian Range (de Juana 1980) and in the Catalan Pyrenees (CA2).

Both Spanish atlases probably understimated the breeding population since Castilla y León alone was estimated to hold over 10,000 pairs by Sanz-Suasti & Velasco (1999). Catalonia supports some 1,900–3,100 pairs (CA2) and even a relatively compact

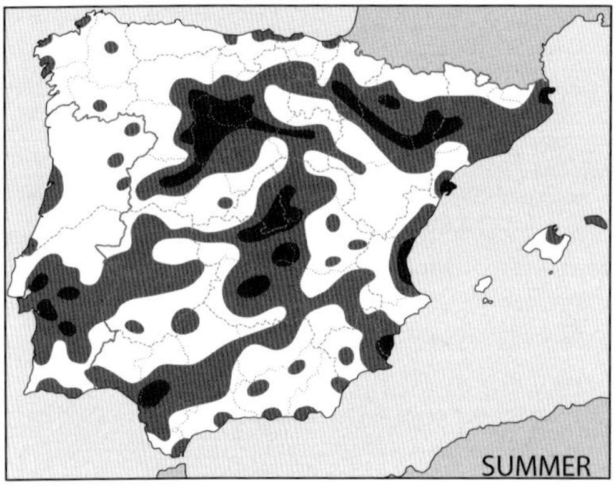

site, such as the marismas del Odiel was thought to hold over 500 pairs (Garrido 1996). However, a significant decline was seen on SACRE sites during 1998–2012 (Escandell 2013). Locally the Great Reed Warbler attains very high densities in the most extensive reedbeds: for example up to 33 pairs/10ha on the river Águeda in Salamanca (Carnero & Peris 1988) and the impression is given that it may displace the Eurasian Reed Warbler within such sites (de Juana 1980).

The breeding population is chiefly present during April–September. It winters in sub-Saharan West Africa. Migrants from northwestern Europe cross the Peninsula on passage but migration anywhere is poorly evident. There are 38 Iberian recoveries of birds ringed in France, The Netherlands, Germany, the Czech Republic, Slovakia, Hungary, Lithuania, Sweden, Italy and Morocco (2). Thirty ringed in Spain have been recovered in these countries as well as in Croatia, Switzerland and Malta, and in Morocco (4) and Mauritania (1) (SMA). Passage periods are mid-March–May, peaking during the second half of April, and July–October (Cantos 1992). Winter records are very rare but include five January observations: near Málaga airport, at Alcochete, Setúbal; and individuals captured by ringers in Madrid, Valencia and the Albufera de Mallorca (*Ardeola* 28: 163, 43: 256, 49: 193, 57: 544; *Anu. Ornitol.* 8: 81).

ICTERINE WARBLER *Hippolais icterina*

(Zarcero Icterino, Felosa-amarela)

A scarce migrant in eastern Iberia and the Balearic Islands. Vagrant to Portugal.

The range of the Icterine Warbler largely complements that of the Melodious Warbler. The nearest breeding birds to Iberia are in central France but they are only widespread from eastern France eastwards. It is a scarce but regular migrant in the Balearics and easternmost Iberia but is very rare further west. There are, for example, no known records for Madrid notwithstanding the large local concentration of birders and ringers. A singing male at Carballo, A Coruña, in May 1997 (*A. O. Galicia 1997*: 117) was most exceptional. It is also a vagrant to Portugal with only three certain records: one on the Berlengas Islands in May 1998 and single birds ringed at the Alvor estuary, Algarve, in September and October 1997 (*Pardela* 11: 17., *Anu. Ornitol.* 1: 30).

Figure 188. Temporal distribution of Icterine Warblers ringed in Eastern Spain: in Catalonia, Comunidad Valenciana, Murcia and the Balearic Islands, 1987–2010 (n = 114).

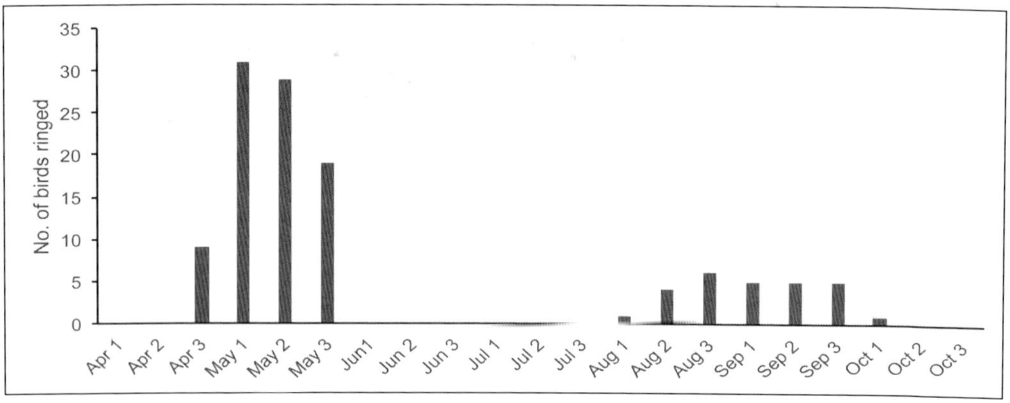

Most occurrences are in spring, from late April but chiefly in May, and many records result from the activities of ringers, so that variation in ringing effort may affect how many birds are detected each season. However, a constant ringing effort on Formentera during the springs of 1986–90 gave successive annual catches of 26, 7, 2, 4 and 8, showing considerable interannual variation in occurrence (*Ardeola* 37: 348). On the mainland it is normally only likely to be encountered on the east coast: the regional bird reports show that of 193 birds ringed in Catalonia during 1997–2009, chiefly at the Llobregat and Ebro deltas and at the Aiguamolls de l'Empordà, 171 were in spring and 22 in autumn. Comparable figures for the Comunidad Valenciana during 1988–2008 are 42 ringed in spring and only two in autumn. In Murcia six were recorded in spring and one in autumn during 2005–10. The great majority of records, again mostly of birds captured by ringers, come from the Balearic Islands. In total, 498 were ringed in spring in the Balearics during 1987–2010: chiefly on Cabrera and Illa del Aire, Menorca, but just eight were found in autumn during the same period. Nearly all spring records from the Levant coastal sites or on the Balearic Islands fall between 20 April and 1 June (Figure 188). An early exception was one found in Murcia on 13 April. The comparatively few autumn records fall between 27 July and 9 October but the majority are between mid August and the end of September. It is clearly never numerous but some probably pass unnoticed among the much larger numbers of migrant Melodious Warblers. Migrants are likely to appear in coastal and riparian scrub and in east coast gardens.

Prevailing easterlies in spring have sometimes led to occurrences further west than usual. For example, there was a locally significant influx of at least five birds at Gibraltar on 14 May 2006 and others were recorded in western Morocco at this time (*GBR 2006*). Ringing evidence shows that individuals may follow widely different routes in different years: one ringed in Cabrera in May 1992 was recovered 911km to the east on Ventotene island, off Naples, Italy, in May 1994 (*Ecología* 10: 393) and another ringed on Ventotene in May 2008 was recovered in Cabrera in May 2009 (CMA). One ringed in Menorca in May 2008 was found in Schleswig-Holstein, Germany, in June 2009 (CMA).

MELODIOUS WARBLER *Hippolais polyglotta*

(Zarcero Común, Felosa-poliglota)

A common and widespread summer visitor to mainland Iberia and a passage migrant throughout the region.

Breeding population estimates. Spain: SA1 700,000–1.5 million pairs; ACRE 4.1 million birds. **Portugal**: 50,000–500,000 pairs.

The Melodious Warbler is common and widespread as a breeding species throughout the Peninsula. It does not breed in the Balearic Islands but occurs there on passage. It frequents habitats that offer dense low cover with some trees or tall bushes, often near water. It is thus especially abundant in riparian vegetation as well as along woodland edges and in hedges. Areas of extensive cultivation, continuous forest cover and dehesas are avoided. Hence it is scarce or absent in parts of Extremadura and the southern Meseta, as well as in the most open sectors of Alentejo (SA2, PA2). Most occur below 1,000m so it is also largely absent from the Pyrenees and Cantabrian mountains (AI2, SA2) but some nest exceptionally to 1,500m and locally even higher: up to 2,000m in the Sierra Nevada (Garzón & Henares 2012). It has bred sporadically in Andorra (ANA).

In Spain the population on SACRE sites showed a significant increase during 1998–2012 (Escandell 2013) although the Portuguese population showed no clear trend during 2004–09 (ACP). At a landscape level, the highest reported densities are from mesomediterranean vineyards in the north-east (60 birds/km^2). In general, the ACRE survey found this warbler to be most abundant in the supramediterranean zone and parts of the eurosiberian. Iberia may accommodate over three-quarters of the European population (BE), France holding 300,000–500,000 pairs (OF).

This is one of the later summer visitors to arrive from its winter quarters in West Africa. First arrivals may be in late March in the south but most do not appear until April–May. The earliest observations in the north are usually in late April or early May (de Juana 1980, Ferrer *et al.* 1986, Cantos 1992, AI2). The return passage begins in mid-July and continues through August–September. In the Balearics, passage occurs between mid April and early June, peaking in May, and between late August and mid October (Garcias 1992). A study in central Spain found that adults depart ahead of juveniles in autumn (Bermejo *et al.* 2002). There are a few Iberian recoveries of birds ringed in France (5) and The Netherlands and birds ringed in Iberia have been found in France (2), Morocco (2), Italy and the Canary Islands.

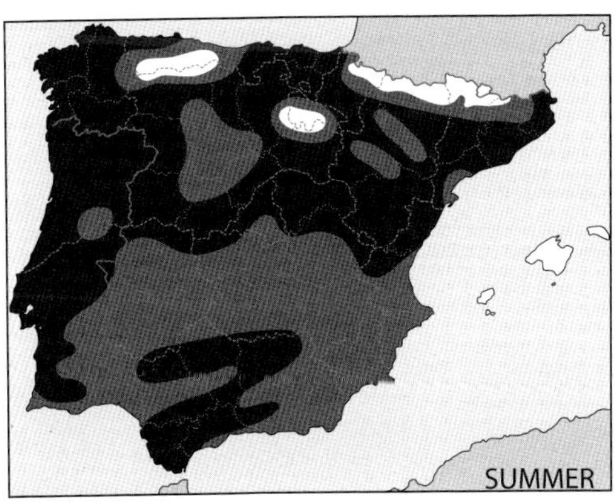

Few are recorded even in October so single birds at Villanueva de la Serena, Badajoz on 4 January 1973; Barcelona on 18 December 1994; Requena, Valencia on 25 January 1998; at the Albufera de Valencia on 17 December 1998; and at Denia, Alicante, on 20 February 2000, were most unseasonal (*Ardeola* 19: 32, 42: 229; *A. O. C. Valenciana 1998*; *A. O. Alicante 2000* 101).

MARMORA'S WARBLER *Sylvia sarda*

(Curruca Sarda, Toutinegra da Sardenha)

Vagrant to Spain and Gibraltar.

There are only four accepted records from Spain: a singing male on El Garraf massif, Barcelona, on 25 May 1997: during a spring which also saw ten individuals in southern France (OF); a first-winter male at the Llobregat delta, Barcelona, from 14–20 March 2004, and a first-winter and a second year male ringed at Illa de l'Aire, Menorca, on 8 April 2005 and 17 April 2011 respectively. An individual *S. sarda/balearica* at Sant Adrià del Besòs, Barcelona, on 2 April 2007 also seems likely to have been a Marmora's Warbler. A first-year male was trapped on Tabarca Island, Alicante, in May 2012 (RBS). Four records from Gibraltar (BG, GONHS) that antedate the rarities committee are perhaps best regarded as provisional; some of the identifications have been queried by Finlayson & Cortes (1987). Two later Gibraltar records: two birds in August–September 1984 and one on 4 September 1996 are accepted (C. Perez, pers. comm) .

Marmora's Warbler only breeds in Corsica and Sardinia. Some winter in North Africa: principally from northeastern Algeria east to northwestern Libya, although there are 12 records from Morocco. There are five records of single males found in Britain and 24 in France: these concentrated on the Mediterranean coast.

BALEARIC WARBLER *Sylvia balearica*

(Curruca Balear, Toutinegra das Baleares)

A common resident in most of the Balearic Islands.

The Balearic Warbler has been treated historically as a subspecies of Marmora's Warbler but recent morphological, behavioural and molecular biological studies have justified their separation (Shirihai *et al.* 2001). It is resident on Mallorca, Ibiza, Formentera, Cabrera and Dragonera, and their larger offshore islets (Sunyer 2010). Munn (1924) found it 'plentiful and absurdly tame' in Menorca but it no longer breeds there, having apparently been displaced by the Dartford Warbler in the 1980s (SA1).

It is a common species in its preferred habitats. It is associated with Mediterranean scrub of cistus, Rosemary and tree heath, as well as open pine, juniper and cypress woodlands with a well-developed understorey (SA2). Here it is a pioneer recolonist of recently burnt areas. It avoids oak forests and crops. In Formentera it favours more open vegetation than the Sardinian Warbler, typically nesting in junipers (Berthold & Berthold 1973). It is present all year round up to 1,200m in the Serra de Tramuntana in Mallorca, where it typically inhabits ridges covered with the grass *Ampelodesmos mauritanicus* (SA1, SWA).

The population in the 1990s was estimated at 14,000–25,000 pairs (SA1) and is currently stable. Breeding densities in the most favourable scrub habitats in Mallorca may reach six pairs/10ha, and the winter atlas found densities of up to 15 birds/km^2 in scrub and open pinewoods (SWA). Densities of 15–25 pairs/km^2 were found by Berthold & Berthold (1973) in Formentera, the island population being estimated at 375 pairs. Up to 435 pairs were estimated to inhabit Cabrera, implying a mean density of some 38 pairs/km^2 (SA2).

In the past it was suggested that this species, then regarded as conspecific with Marmora's Warbler, might have been breeding along the Spanish Mediterranean coast (Williamson 1964), for example at Jávea (Corley Smith 1959) and at several sites in the Costa Brava (Wallace & Sage 1969). A number of observations in the Ebro valley: from Aragón, Navarra and Burgos, were also attributed to it. In all these cases the likelihood of misidentification, probably involving confusion with juvenile Dartford Warblers, was then high, as noted by van Impe (1971). An unlikely record of a pair said to have been photographed at the nest at El Acebuche, Doñana, in May 1986 (Llandres & Urdiales 1990) was not corroborated. In short, the occurrence of the Balearic Warbler on the Iberian mainland has never been proved (SA1, de Juana 2006).

DARTFORD WARBLER *Sylvia undata*

(Curruca Rabilarga, Toutinegra-do-mato)

***S. u. dartfordiensis*. A common resident or partial migrant in coastal northern and northwestern Iberia. Darker than the nominate.**

***S. u. undata*. A common resident or partial migrant in most of mainland Iberia. Also in Menorca and northeastern Mallorca.**

***S. u. toni*. Birds approaching this race are common residents or partial migrants in southern Iberia. Duller above with a slight sandy brown cast.**

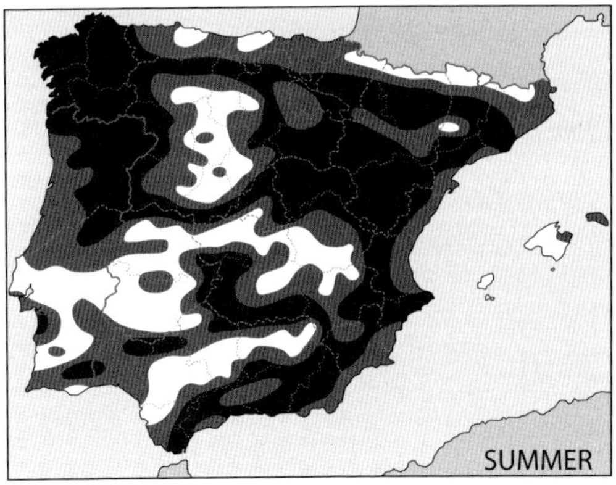

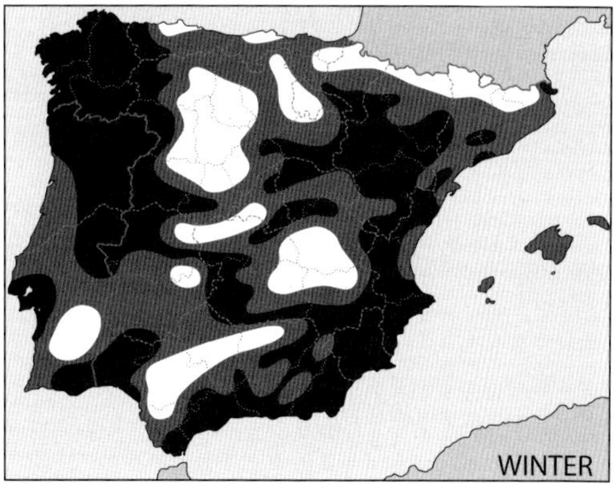

Breeding population estimates. Spain: SA1 1.7–3 million pairs; ACRE 1.3 million birds. **Portugal**: 10,000–100,000 pairs. **Andorra**: 20–30 pairs.

Irrespective of subspecies, this is a rather common bird throughout mainland Iberia. It is most abundant in areas of Mediterranean climate and also, in the north-west and west, where there is the maritime influence of the Atlantic (SA2). It is widespread in Portugal, where is most abundant in scrub in the uplands, being relatively uncommon in regions of extensive cultivation, as in Alentejo (PA2). In Spain it is also least frequent in the level and predominantly cultivated areas of the Mesetas, Extremadura and the Guadalquivir valley. It is also scarce or absent at the upper levels of the northern mountains, including the northern Iberian Range (de Juana 1980). In the Balearic Islands it colonised Menorca in the mid 1970s (Muntaner 1980) where it seems to have completely displaced the Balearic Warbler, which no longer occurs there. It subsequently colonised Mallorca, in about 1997 (Sunyer 1998), and it is thinly distributed there in the north-east.

It occurs wherever there is dense low scrub up to 1.5m tall. It particularly favours *Cistus* and *Halimium* scrub as well as gorse, *Genista*, *Erica* heath and Kermes Oak. At higher levels it also occurs in Juniper scrub, as in the Central Range, reaching altitudes up to 2,000m (AI2). Some also nest above 2,000m in the Sierra Nevada, principally on the southern versant (Garzón & Henares 2012). It readily adopts regenerating burnt areas, where it may attain high densities, of up to 17 pairs/km^2 (CA2). The highest densities (18–20 birds/km^2) in the ACRE survey were in supramediterranean matorrals. The population decreased significantly on SACRE plots during 1998–2012 (Escandell 2013). Some decline has also been reported since the 1980s in the Portuguese breeding population, especially in the coastal region between Peniche and Cape da Roca, where it was formerly widespread but is now absent (PA2).

The Spanish winter atlas found densities of over 30 birds/km^2 in the preferred scrub habitats (SWA). Many Dartford Warblers are entirely resident but there is juvenile dispersal as well as altitudinal movements in which colder mountain areas are abandoned in winter, as occurs at the upper levels of the Sierra Nevada and in Andorra (Garzón & Henares 2012, ANA). The few Spanish ringing recoveries have been local but some move to winter in the warmer parts of the Peninsula, such as the Monegros desert, the arid southeastern coastlands and the lowlands of Granada, where winter densities may be rather high (SWA) and where the birds are recorded from October to May (Paracuellos 1994, Hódar 1996). Influxes into coastal areas of Portugal, such as the Douro and Tagus estuaries, occur in winter (Tait 1887, AP). One ringed in September in Mediterranean France was recovered near Lisbon in November (Cantos 1992). Some also appear in winter in the Balearic Islands, for example on Ibiza, where the Dartford Warbler does not breed (*A. O. Balears 2010*: 267). There are clear indications of passage at Gibraltar, where it does not breed but where there are small influxes in some years in October–November, with a less evident return in March (GONHS). It thus seems that at least small-scale movements occur to and from North Africa.

TRISTRAM'S WARBLER *Sylvia deserticola*

(Curruca del Atlas, Toutinegra do Atlas)

Vagrant to **Gibraltar.**

A male was seen on Windmill Hill, Gibraltar on 10 April 1988 (J. E. Cortés pers. comm.; *British Birds* 82: 22). This endemic species of the Maghreb has appeared in the Canary Islands in October 1995 (*Ardeola* 44: 137) and there is one record from Malta (Sultana & Gauci 1982).

SPECTACLED WARBLER *Sylvia conspicillata*

(Curruca Tomillera, Toutinegra-tomilheira)

***S. c. conspicillata*. A locally common summer visitor or partial migrant.**

Breeding population estimates. Spain: SA1 140,000–300,000 pairs; ACRE 393,000 birds. **Portugal**: 250–2,500 pairs.

The Spectacled Warbler is characteristic of the Mediterranean climatic region. It is thus absent from the north and north-west. In Portugal it reaches the coast only in the south of the country. It is most common in Andalucía, the Ebro valley and the páramos of the Iberian Range. It is also widespread in scattered locations throughout the Meseta and Extremadura (SA2). The Portuguese range is highly fragmented but extends throughout the eastern half of the country with a southwesterly outpost around Sagres, Algarve (PA2). It is rare in the Balearics, with just a few breeding pairs in Menorca and Mallorca, although it may have also nested in Ibiza in the past (Ferrer *et al.* 1986, SA2). One was seen on Cabrera in October 2009 (*Ardeola* 57: 544).

It occurs in lower, sparser vegetation than any of the other scrub *Sylvia* species and, conversely, it avoids all taller formations and any but the most open woodlands. It may be found in coastal *Salicornia* and other halophytic vegetation around saltpans and lagoons, for example in Doñana. It typically inhabits sparse xerophytic scrub in arid habitats as well as degraded maquis, where it is an early colonist of burnt areas. It is usually found where scattered low shrubs are separated by bare stony or sandy terrain, where it often feeds on the ground. Hence it is typical of broken, eroded terrain. It is most abundant in the thermomediterranean, in the arid badlands of southeastern Spain, mainly in Almería province and Murcia (Pleguezuelos & Manrique 1987, Manrique 1993). It may be encountered from sea-level to mountaintops, reaching 2,000m in the Betic Range in the Sierra Nevada, but only up to about 1,000m in the northern Meseta (AI2, Garzón & Henares 2012).

The great majority are summer visitors, wintering in North Africa, including within the Sahara. The Spanish winter atlas fieldwork did not encounter any wintering birds at all during 2007–10. However, some have been found in winter in the past, particularly in southeastern Spain where the climate is then the mildest in Iberia. Thus a study population in the Guadalentín valley, Murcia was present all year round although numbers declined in winter: from 7.6 birds/ha on average during March–October to 2.2 birds/ha in November–February (Hernandez-Gil 1990). However, the Murcia Bird Reports for 2005–11 give only one midwinter record, at Cieza, on 21 January 2007 (*A. O. Murcia 2007*: 44), so any wintering population, if present, is small. Midwinter records elsewhere include single birds at Ballobar, Huesca, in January 2001; at Villanueva de Siguena, Huesca, in December 2003; at Burriana, Castellón, in January 1993, and on the Columbretes Islands in December 1996 (*A. O. Aragón 1999–2003*: 198; *A. O. C. Valenciana 1993*: 110, *1995–1997*: 207). There were two at Las Amoladeras, Almería, on 28 December 2011 (*Ardeola* 59: 191). Passage at Gibraltar is detected from March–May, with a peak in April, and then again in August–October (BG). The earliest arrivals in southern Portugal are in late February and most appear there during the second half of March (AP).

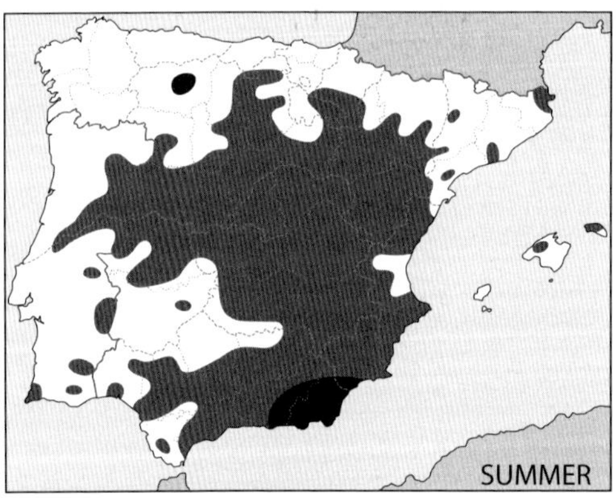

SUBALPINE WARBLER *Sylvia cantillans*

(Curruca Carrasqueña, Toutinegra-de-bigodes)

***S. c. cantillans*. A common summer visitor to the Mediterranean climatic region. Also a passage migrant. Winters in sub-Saharan Africa.**
***S. c. moltonii*. A summer visitor to the Balearic Islands.**
***S. c. inornata*. Birds in southernmost Iberia have been assigned to this race by some authorities.**
***S. c. albistriata*. Vagrant to the Balearic Islands.**

Breeding population estimates. Spain: SA1 1.1–2.3 million pairs; ACRE 5.8 million birds. **Portugal**: 10,000–100,000 pairs. **Andorra**: *c.* 100 pairs.

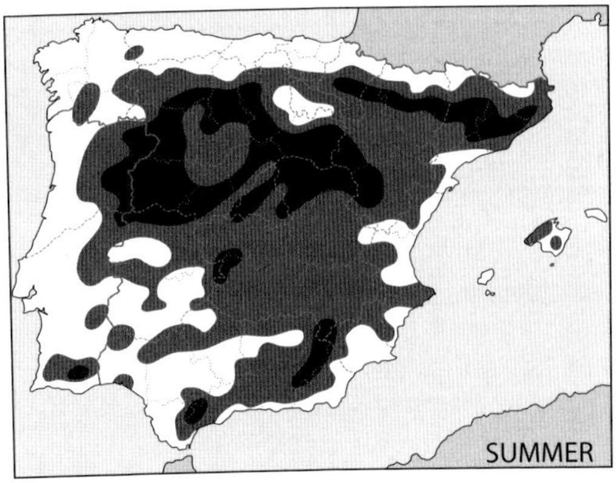

Variation largely involves underpart colour and is most evident in males. Those of *moltonii* are often conspicuously and extensively salmon-pink below, without the orange tones of *cantillans*. Males of *inornata* are deeper orange below than *cantillans* but the former is less poorly differentiated from the latter than *moltonii* and the variation is clinal. However, Svensson (2013) has proposed splitting the taxon into three species: the Western Subalpine Warbler *S. inornata* (with those of mainland Iberia becoming *S. i. iberiae*), Moltoni's Warbler *S. subalpina* and Eastern Subalpine Warbler *S. cantillans*.

The mainland population is widespread in Mediterranean Iberia but almost entirely absent from the euro-Siberian zone. However, there are small isolated breeding populations in Galicia and Asturias (SA2). A record from the Cíes islands, off Galicia, was of an adult male on 12 July 1986 (*Ardeola* 35: 314) and not a breeding record as stated by AI2. The population is quite evenly spread in the supramediterranean, but the range is much more fragmented further south, where the Subalpine Warbler is only really common in montane areas and completely avoids the thermomediterranean zone (SA2). It is widespread in the eastern half of north and central Portugal, relatively scarce in Alentejo but common in the hills of Algarve (PA2).

In the Balearics, *moltonii* nests only on Mallorca and Cabrera (*A. O. Balears 2010*: 267), where the first confirmed breeding records were in the early 1990s (*Ardeola* 40: 102, Sunyer 1997). Apparent breeding on Menorca was first reported in 2004 (*Ardeola* 52: 434). A pair carrying food in Ibiza in 1999 may possibly have been nesting there (*A. O. Balears 1999*: 109).

The Subalpine Warbler selects taller and denser habitats than other scrub warblers; trees, typically Iberian Holm or Lusitanian Oaks, are often an important element. Tall cistus and heath scrub with scattered sclerophyllous trees is a typical choice, as well as young pine plantations with a dense understorey and overgrown Holm Oak dehesas (SA1, AI2). It is particularly associated with scrubby ravines in hilly and mountainous areas, being most abundant at mid-levels but reaching 2,000m in the Sierra Nevada and 1,700m in the Catalan Pyrenees (Garzón & Henares 2012, CA2).

The Spanish population increased on SACRE plots during 1998–2012 (Escandell 2013). The highest densities, 30–60 birds/km^2, have been reported from the supramediterranean region, in Holm Oak woodlands, matorrals and juniper scrub. The breeding range in Catalonia expanded by some 31% between the early 1980s and 1999–2002, mainly within the Pyrenean foothills (CA2). Ongoing northward range expansion has also been reported from neighbouring France (OF). The Portuguese population is thought to be stable. The Andorran population inhabits south-facing slopes in the Valira valley, mainly at 1,100–1,300m (ANA): Eagle Clarke (1889) did not find the species there in May 1889.

Passage migrants occur widely and commonly appear in coastal scrub and other habitats in which they do not breed. The spring passage is much more evident at Gibraltar than the post-breeding passage. The first birds are among the earliest trans-Saharan migrants to arrive on the south coast, from late February, and passage there continues throughout March–April, with stragglers in May. The southward passage is during August–October (BG). The winter quarters are in the Sahel zone (Shirihai *et al.* 2001). The Spanish breeding areas are principally occupied during April–September (AI2). A male near Traspinedo, Valladolid, on 18 November 1990 (*Ardeola* 37: 349) was exceptionally late. There are also two January 2007 records from Portugal: near Mourão, Alentejo, and at Castro Marim, Algarve (AP).

Five males of the eastern subspecies *S. c. albistriata* have been captured in spring by ringers on the Balearic Islands: one on 2 April 1997 on Illa de l'Aire, Menorca; three in spring 2007, on Cabrera on 27 March and 10 April and in Formentera on 25 April (*Ardeola* 47: 154–155, 56: 334;). A male was seen at the Ebro delta in April 2013 (RBS).

SARDINIAN WARBLER *Sylvia melanocephala*

(Curruca Cabecinegra, Toutinegra-dos-valados)

***S. m. melanocephala*. Common and mainly resident in Mediterranean Iberia, local in the north. A common resident in the Balearic Islands.**

Breeding population estimates. Spain: SA1 0.99–1.90 million pairs; ACRE 10.1 million birds. **Portugal**: 100,000–1 million pairs. **Gibraltar**: 100 pairs.

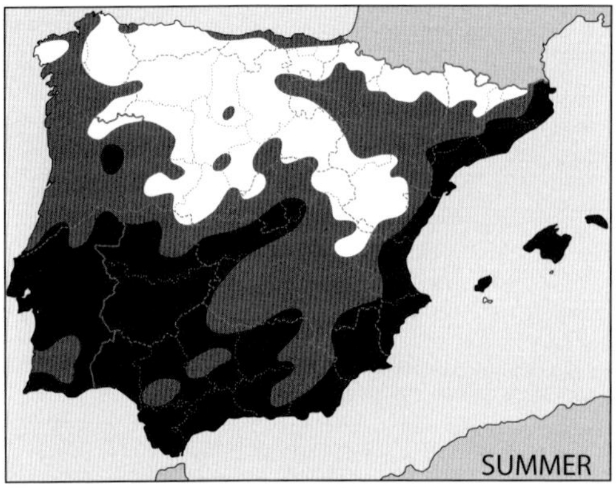

The Sardinian Warbler is ubiquitous throughout the southern half of Iberia, including the mountains of the Betic Range, and along the entire east coast. It also inhabits the Ebro valley north to La Rioja and from there it ascends the foothills of the Pyrenees and the Iberian Range. It is absent from higher levels of these ranges and of the Central and Cantabrian Ranges and is scarcely present on the northern Meseta, where winters tend to be cold (SA2). However, there are local populations where favourable microclimates permit, for example in the Sil basin, León (AI2). Sporadic breeding has also been reported in Andorra (ANA) but there is no established population there.

The range does extend north along the Duero valley and there are outpost populations in Galicia: chiefly on the coast and along the Miño valley, and along the Biscay coast in the Basque Country, Cantabria and, locally, in Asturias (SA2). Most of these northern populations are of quite recent origin. The range expanded north along the Ebro valley during the 1970s and the Sardinian Warbler was first reported in Galicia, Cantabria and the Basque Country during the 1980s (Pérez de Ana 1993). The first record from Gipuzkoa was not until 1995, at Zumaia (*A. O. Gipuzkoa 1995*: 52). It has continued to spread slowly on the north coast, although it only reached Asturias in 2001 (SA2). It has also recently become established on the southwestern French Atlantic coast in the Landes (Dubois *et al.* 2012). The historical literature shows that the Sardinian Warbler was common or abundant throughout southern and eastern Spain during the late 19th and up to the middle 20th century but it was then absent from the north and centre, although Lilford (1866) reported it as occurring rarely in Aranjuez, Madrid, and Gil Lletget (1945) included Cuenca province within its range. Bernis (1952) saw several in the Tagus valley in Toledo in July 1951. The Sardinian Warbler is now common almost throughout Portugal, where it is absent only at the higher elevations of the extreme north in Trás-os-Montes (PA2). However, it was formerly much less widespread north of the Duero and has colonised large areas there since the mid 20th century (Tait 1887, Reis Júnior 1931, AP).

It is more of a generalist than the other small *Sylvia* species and so occupies a broader range of habitats, the fundamental requirements being some tall, dense cover and the availability of fruiting shrubs: notably wild olive and lentisc, for sustenance in winter (Tejero *et al.* 1983, Cuadrado 1988). It is poorly tolerant of severe cold and so is largely restricted to areas of warm or mild climate. Here it can be found in all manner of tall scrub, riparian growth, hedgerows, olive groves, orchards, gardens and, in particular, in open pine – especially Maritime Pine – or oak woodlands with a well developed understorey. It reaches 1,000–1,200m generally in the mountains, highest on south-facing slopes, although it is recorded up to 2,100m in the Sierra Nevada (Garzón & Henares 2012).

The breeding population on SACRE sites showed a significant 15% increase during 1998–2012 (Escandell 2013). The Catalan winter population has been put at 1.4–1.9 million individuals (CWA). Maximum densities, up to 145 birds/km^2, are in thermomediterranean pinewoods and matorrals. The Sardinian Warbler is also abundant in the Balearics, where densities of 200 pairs/km^2 recorded in central Mallorca are regarded as representative of the whole island (Sunyer 2009). The Portuguese population is thought to be stable (ACP).

Cold winters produce set-backs, especially in peripheral populations that may suffer considerable mortality. These events may be common in inland areas such as Madrid, where abundance changes greatly between years (pers. obs.), and are not unknown even on the Mediterranean coast: a highly exceptional cold snap of only four days' duration at Gibraltar in January 2005, when air temperatures fell as low as 1°C, led to an 80% loss in a study population (Bensusan 2008). The numbers ringed at constant effort plots in Catalonia fell by 50% after the harsh winter of 2001–02 (CA2). Munn (1926) commented that many were found dead in Mallorca following heavy snowfall in late February 1924.

Most birds are sedentary but altitudinal movements occur in high mountain areas and there are some long-range movements, probably related to juvenile dispersal (Cantos 1992). Nevertheless, there are only four extraregional recoveries despite large-scale ringing (SMA). Influxes at Gibraltar, especially in September–October, suggest that some migrate across the Strait (BG). Migratory movements have been detected in the Balearic Islands in October–November and February–March (Shirihai *et al.* 2001) and there are two winter recoveries from Algeria of juveniles ringed in Ibiza and Teià, Barcelona (AI2). One ringed in Ceuta on 3 September 2001 was recovered on the opposite side of the Strait in Gibraltar on 1 October, a northward movement in autumn (*Ecología* 17: 274). There is a similar record of one that was ringed at Viladecans, Barcelona, on 10 October 2003 and recovered 183km to the north-east in the Pyrenées-Orientales, France, on 15 November 2003 (*Ecología* 20: 331).

RÜPPELL'S WARBLER *Sylvia rueppellii*

(Curruca de Rüppell, Toutinegra do Levante)

Vagrant to Spain.

A male was captured by ringers on Cabrera, Balearic Islands, on 18 April 2010 (*Ardeola* 59: 395). This is a very rare vagrant to western Europe, where there are records from Britain (five), Finland (two), France (two), the Faroe Islands and Denmark (Slack 2009). However, there are 40+ records from Italy, mainly in the south during March–April. The two French records were in the south-east, in March 1970 and May 1996.

AFRICAN DESERT WARBLER *Sylvia deserti*

(Curruca Sahariana, Toutinegra do Sara)

Vagrant to Spain.

The sole accepted record was one at the Clot de Galvany, Elche, Alicante, on 28 October 2011. Vagrants have appeared in Italy, Malta and the Canary islands.

ASIAN DESERT WARBLER *Sylvia nana*

(Curruca del Turquestán, Toutinegra do Turquestão)

Vagrant to Spain.

A record of one photographed at the Ría de Ajo, Cantabria, in November 2013 (RBS) was pending assessment in 2014. Vagrants have appeared in Britain (11) and France (1).

WESTERN ORPHEAN WARBLER *Sylvia hortensis*

(Curruca Mirlona, Toutinegra-real)

A widespread but scarce summer visitor to mainland Iberia. A scarce migrant in the Balearics.

Breeding population estimates. Spain: SA1 170,000–440,000 pairs; ACRE 317,000 birds. **Portugal**: 250–2,500 pairs.

This is a species of open dry woodland, with or without a scrubby understorey (AI2). Well-spaced small trees are a constant feature of its chosen habitats and it selects dehesas of Cork and Iberian Holm Oaks, orchards, olive groves, open pinewoods and Spanish Juniper woodlands, where it feeds within the canopy. It is tied to the Mediterranean climatic region and almost entirely absent from the eurosiberian as well as from coastal regions of Portugal. It is widespread elsewhere throughout the Iberian mainland but largely absent from the large tracts of the Meseta dominated by agricultural land. It is also absent at higher altitudes, above 1,000–1,300m, although exceptionally it is found to 1,900m in Almería (Manrique 1997). Those in the valley woodlands of the Sierra Nevada are generally found at 800–1,200m but sometimes up to 1,500m (Garzón & Henares 2012). The Portuguese distribution is largely in the eastern half of the country where it is generally scarce, although locally more numerous in parts of Trás-os-Montes, Beira Alta and central Algarve (PA2). Eagle Clarke (1889) saw it in Andorra in late May and singing males have since been reported there, at 900–1,500m (ANA). It does not breed in the Balearic Islands, although may have nested there early in the 20th century (Munn 1921, 1931).

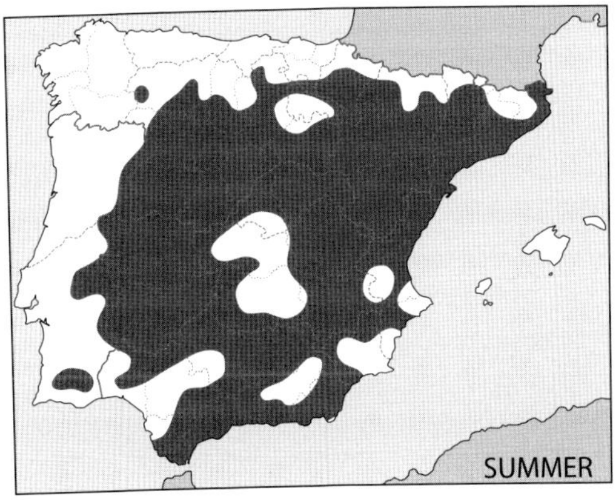

Some early authors (e.g. Saunders 1871, Jourdain 1936–1937, Witherby 1928, Gil Lletget 1945) give the impression that it may formerly have been much commoner in several regions in Spain. Nevertheless, the population showed a marked significant increase of 131% on SACRE plots during 1998–2012, which was most evident in central Spain (Escandell 2011, 2013). A similar trend had already been detected in Catalonia, where a comparison between the two breeding bird atlases shows a 34% increase in the number of occupied 10km-squares, from 104 squares in 1975–1983 to 138 in 1999–2002 138 (CA1: 1983, CA2: 2004). The Orphean Warbler occurs everywhere at low density, usually below 1 bird/10ha but up to 3.46 birds/10ha in optimum habitat (SA2). On a larger scale, the highest densities have been reported from juniper woodlands in the supra-/mesomediterranean interface, with 5.5 birds/km^2. The small Portuguese population is thought to be stable.

Arrivals in southern Spain from sub-Saharan Africa are from late March but chiefly in April, passage continuing into May (BG). It departs relatively early, from late July, the chief movements being in August–early September. Stragglers may remain into October or, rarely, early November (SA1, AP). There is an exceptional winter record of one at Alanís, Sevilla, on 13 December 1970 (*Ardeola* 19: 32).

It is a very scarce spring migrant in the Balearic Islands but it has become annual in recent years, with several captured by ringers in some springs, including an exceptional nine in 2004. Extreme spring dates are 13 April and 3 June but nearly all are found from mid-April to mid-May (*A. O. Balears*). Autumn Balearic records are very rare and there have been only three recently, all ringed in Cabrera in September (*A. O. Balears 1994*: 113, *2010*: 88; *Ardeola* 59: 449).

BARRED WARBLER *Sylvia nisoria*

(Curruca Gavilana, Toutinegra-barrada)

Vagrant to Spain.

There are only four accepted Iberian records. Two were of birds ringed on the Balearics: one in Menorca on 27 September 1983, and the other, a juvenile, in Formentera on 7 October 1989. In October 2011, a first-winter bird was ringed at Santa Pola, Alicante, on 24th, and an individual was seen at Castropol, Asturias, on 28th. A few records for peninsular Spain that antedate the rarities committee are not considered entirely reliable (Bernis 1954a, de Juana 2006). They include a possible recovery of a Finnish-ringed bird near Barcelona in October 1973 (*Ardeola* 20: 374). There was a further Spanish record in 2012: at Roses, Girona, in October (RBS).

The Barred Warbler winters in East Africa and accordingly European birds migrate south-eastwards. It is rare in southern Italy, only occasionally recorded in France and unknown in the Maghreb. However, some juveniles on post-fledging dispersal tend to appear in Britain in autumn: 199 were reported annually on average during 2000–2003.

LESSER WHITETHROAT *Sylvia curruca*

(Curruca Zarcerilla, Papa-amoras-cinzento)

Rare migrant, mainly reported in the Balearic Islands. Vagrant to Portugal and Gibraltar.

Western European populations winter in sub-Saharan Africa, from Chad eastwards. Their movements occur largely east of Iberia and they are also scarce and irregular in Morocco. There is one accepted record from the Canary Islands, in October 2010.

Irby (1895) collected an individual in his garden in Gibraltar on 19 April 1872. However, many older reports have been considered doubtful given the possibility of misidentifications involving females and juveniles of other *Sylvia* warblers (Bernis 1954a, Ferrer *et al.* 1986, AP). Reports have become increasingly frequent in Spain since the mid 20th century. One was seen in the Guadalquivir marshes on 2–3 May 1956, and another one was ringed at the Ebro delta on 30 April 1969 (*Ardeola* 3: 309–310, 15: 89). The latter was the first Lesser Whitethroat to be ringed in Spain but there have been many more since, including the four in Catalonia cited by Carrera & Martínez (1985). The only accepted record from Portugal is of a juvenile ringed at Vale Santo, Vila do Bispo, Faro, on 16 October 1994 but another was found at the same locality in September 2012 (RBP). The species has been considered by the Spanish Rarities Committee only since 2006, and 15 records were accepted by 2010, eight of them individuals captured for ringing. There are three accepted records from Gibraltar: single birds on 24 October 1995, 27 September 2001 and 20 November 2003, two of them captured by ringers (GONHS).

Figure 189. Monthly distribution of records (n = 64). Sources: Portuguese rarity reports; Spanish rarity reports 2006–10; Spanish regional bird reports; *Ardeola*; GBR.

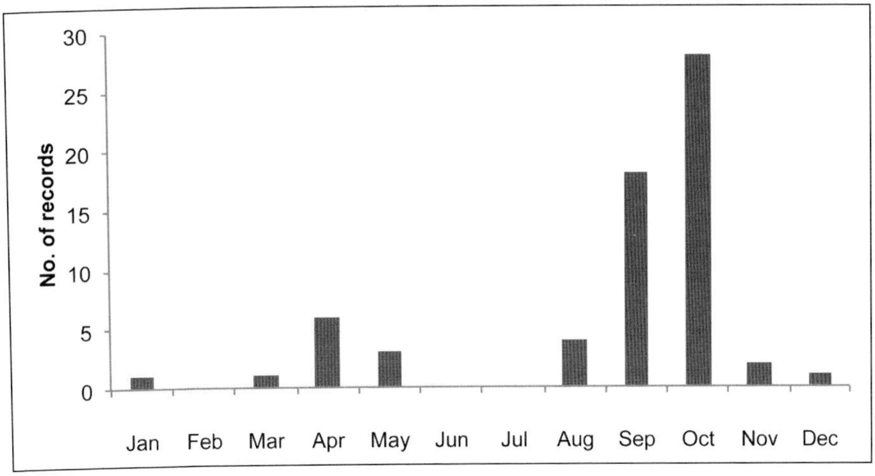

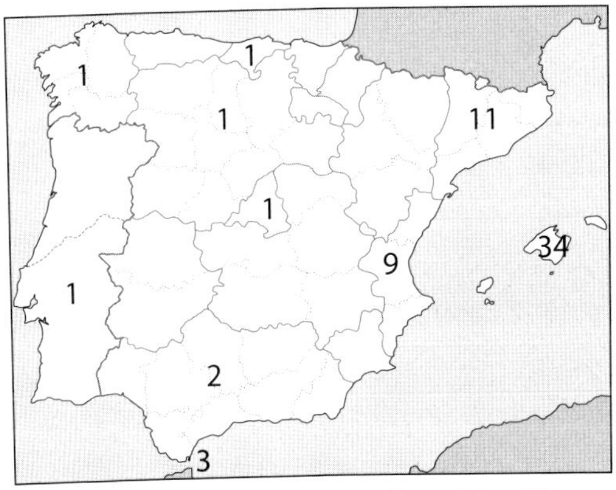

Figure 190. Geographical distribution of records (n = 64)

Given the uncertainty attached to some records, we have only considered those that have been formally accepted and/or involve birds examined in the hand, usually by ringers. This gives a sample of 64 records, all of single birds. Some 81% were in the autumn, mostly from late August to late October, with extreme dates of 4 August and 23 November (Fig. 189): among these, 22 of 44 individuals (41%) were in juvenile or first-winter plumage. Only 16% were in spring, mainly between mid April to early May, extreme dates being 27 March and 14 May. There are also two midwinter records from northern Spain: one ringed at Maliaño, Cantabria, on 30 December 1987, and another seen at O Grove, Pontevedra, from 2 January to 2 February 2006 (*Ardeola* 35: 314, 55: 279). The great majority (84%) have been found in the north-east littoral regions (Fig. 190).

COMMON WHITETHROAT *Sylvia communis*

(Curruca Zarcera, Papa-amoras-comum)

***S. c. communis*. A widespread summer visitor to mainland Iberia and a common migrant throughout.**

Breeding population estimates. Spain: SA1 450,000–600,000 pairs; ACRE 1.3 million birds. **Portugal**: 2,500–25,000 pairs.

The Common Whitethroat is indeed common in much of the northern half of mainland Spain, especially in the north coastal provinces and much of Castilla y León, although it is largely lacking from the Ebro valley. It is much less widespread in the south, where it is associated with hilly and mountainous areas, notably the Montes de Toledo, the Betic Range and the Sierra Morena, and is largely absent from the major river valleys and the coasts (SA2). The breeding range in Portugal is almost entirely north of the Tagus, although there is an outpost population in Algarve, in the Serra de Monchique (PA2). It generally avoids the thermo- and mesomediterranean zones except where the local conditions are relatively humid (AI2) but, in contrast, it also tends to avoid the most humid sectors of the eurosiberian, for example in the northern Iberian Range (de Juana 1980), reaching its highest densities usually within the supramediterranean and oromediterranean zones (AI2).

It favours scrub of moderate height, generally lower than 6m, as well as hedgerows in farmland, woodland edge habits and young tree plantations. It especially selects areas with scattered hawthorns, brambles and Dog Roses. It occurs in montane scrub to 2,000m in the major ranges and up to 2,500m in the Sierra Nevada (Zamora 1988, Garzón & Henares 2012). In the

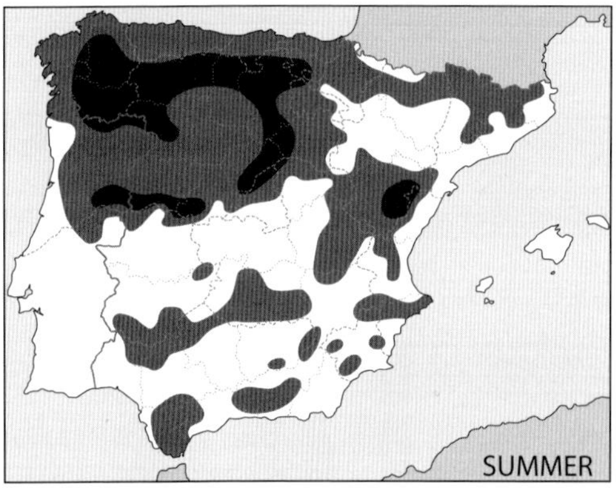

south it generally favours north-facing elevations but tends to occupy the southern slopes in the northern mountains.

The Spanish population showed a significant decline during 1998–2012, particularly in the centre and south (Escandell 2013). Maximum densities have been reported from the Cantabrian region, in matorrals (34 birds/km^2) and pastures with trees (22 birds/km^2), and in the supramediterranean, in matorrals (20 birds/km^2) and mixed farmland (13 birds/km^2). The Common Whitethroat has suffered very large declines and numerical fluctuations throughout western Europe since the mid 20th century and the Spanish historical literature gives the impression that it was formerly very much more abundant, at least locally. For example, Irby (1883) found it 'very abundant everywhere' in Cantabria in May and June 1876, Ticehurst and Whistler (1925) said it was 'by far the commonest bird about Pamplona, whether in the cultivation or on the small sierras', and Valverde (1956) considered it the commonest of the *Sylvia* warblers in the northern meseta.

The Portuguese population may now be stable although there are also indications that the Whitethroat was much more numerous there during the early 20th century, when it was described as common or abundant in the north (Tait 1924, Reis Júnior 1931) and where it remains widespread but scarce (AP). It nested in Andorra until the 1980s but not recently, despite the existence of extensive suitable habitat (ANA). Eagle Clarke (1889) said it was not uncommon about the town of Andorra.

The western populations winter in the sub-Saharan Sahel. Migrants are widespread on passage, including on the Balearic Islands, where none breed. Birds ringed in the British Isles, France, Germany, the Low Countries, Scandinavia, Switzerland and the Czech Republic, Slovakia and Poland have been recovered in Iberia, with a clear predominance of birds ringed in Britain: 51% of 217 recoveries of extra-regional origin (SMA, AP). Iberian-ringed birds have been recovered from the same European countries as well as Lithuania (1) and there are also single recoveries from Morocco and Congo. One ringed in Tunisia in spring was recovered in Portugal in autumn. The autumn passage takes place mainly west of a line between the Basque Country and the Strait of Gibraltar, whereas the spring passage is mainly through the eastern half (Cantos 1992, AI2). Spring passage extends from March–May with the postnuptial movements in August–October (AI2). Autumn passage at Doñana is seen from early August to mid October, peaking around 5–9 September (Herrera 1974b). The breeding areas are occupied from April to September. A study in the western Pyrenees recorded peak autumn passage during the second half of September (Grandío 1997). Stragglers may occur in November and there are a number of records of apparently wintering individuals (Shirihai *et al.* 2001, SWA), including three ringing recoveries from the Portuguese Atlantic coastlands and three in southern Spain (AI2).

GARDEN WARBLER *Sylvia borin*

(Curruca Mosquitera, Toutinegra-das-figueiras)

***S. b. borin*. A summer visitor to northern mainland Iberia. Common and widespread on passage.**

Breeding population estimates. Spain: SA1 400,000–700,000 pairs; ACRE 562,000 birds. **Portugal**: 1,000–10,000 pairs. **Andorra**: 80–150 pairs.

The Garden Warbler is typical of the eurosiberian and supramediterranean climatic zones and so is widespread in eastern Galicia, along the Cantabrian coast and throughout the Cantabrian Range, as well as in the Pyrenees, the Central and Iberian Ranges and the Duero basin. It also occurs in cooler and relatively humid locations throughout the northern half of the Peninsula (SA2). It is absent from drier regions of Mediterranean climate and does not breed in the Balearics: although Munn (1925) claimed to have found a nest with young in Mallorca in July 1923. Breeding records from scattered sites in the western Sierra Morena, Huelva (SA1), have not been confirmed subsequently (SA2).

In Portugal it only nests in the extreme north, bordering Galicia, and some recent range contraction may have occurred in Minho (PA2). It may have been more widespread in the historical past since around 1945 Coverley reported breeding at Braga, near Porto, and also south of the Douro by the Caima river (AP). The breeding range in Andorra has shown some recent expansion (ANA).

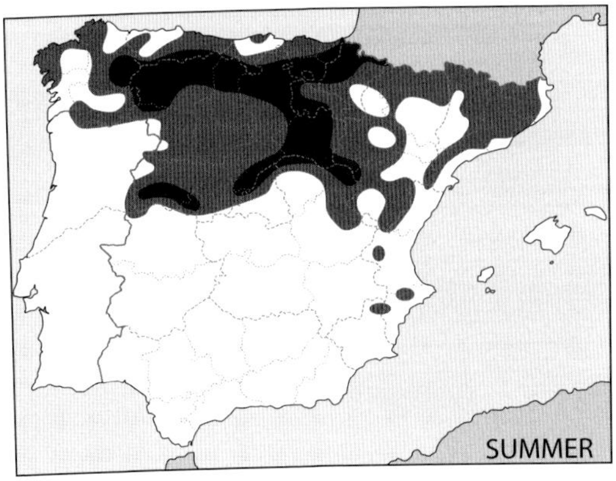

The Garden Warbler mainly selects mature deciduous woodland but it also occurs in young pine plantations and open pinewoods of Scots Pine, with a well-developed shrub layer. In Spain it favours woodlands of Pyrenean, Sessile or Lusitanian oaks, ashwoods and mixed oak/beechwoods. It is also characteristic of tall riparian woodland: especially dense willow clumps, and forest edge habitats with fringing scrub, as well as mature hedgerows with some trees. It ranges from sea level to 2,000m, occurring in broom scrub, and in Mountain Pine forests in the eastern Pyrenees, at the highest elevations (Ferrer *et al.* 1986, AI2, SA2). Some three-quarters of the Spanish population in the ACRE survey were in Castilla y León. Densities are typically low: fewer than 1 bird/10ha (SA2), but in the Cantabrian region 13–18 birds/km^2 occur in beechwoods, pastures with trees above 750m and matorrals above 500m.

Arrivals in spring from sub-Saharan West Africa are chiefly in April–May, but some appear in early March in the south and stragglers may occur on passage in June. Spring passage at the Aiguamolls de l'Empordà peaked in early May, and in mid May in Cabrera (Barriocanal & Robson 2007). The return passage is from mid August to October, peaking in September, with stragglers in early November (Herrera 1974b, Grandío 1997, Bonnin 2007, BG). The autumn passage is relatively leisurely and birds linger at sites with abundant fruiting bushes such as brambles, elders, lentiscs and figs, to accumulate fat prior to the Saharan crossing (Aparicio *et al.* 1991). On the basis of ringing recoveries, the autumn passage is much more evident in central and western Iberia, whereas the spring movement seems to have a more easterly orientation (Cantos 1992, AI2). Passage birds are nonetheless widespread, including on the Balearics. A juvenile was still present at Doñana on 6 December 1967 (Herrera 1974b). There are several winter month ringing recoveries from the southern half of the Peninsula (Cantos 1992) but such records are open to misreporting of dates and there is no other available evidence of a winter presence (SWA).

Birds ringed in the British Isles, France, Germany and the Low Countries predominate but some from Scandinavia, Finland, Lithuania, Switzerland, the Czech republic and Russia (2) have been recovered in Iberia. Birds ringed in Iberia have been recovered in the same European countries, especially in Germany, as well as in Italy (3) and Slovenia (1). There are 15 recoveries of Spanish ringed birds in North Africa: 14 in Morocco and one in Algeria, and a further six from sub-Saharan Africa: in Mali (2), Ghana, Equatorial Guinea, Mauritania and Sierra Leone.

BLACKCAP *Sylvia atricapilla*

(Curruca Capirotada, Toutinegra-de-barrete)

***S. a. atricapilla*. Common in northern mainland Iberia, where mainly resident. Abundant on passage and in winter throughout the Peninsula and Balearics.**
***S. a. heineken*. A common resident throughout most of mainland Iberia. Darker than the nominate, with a shorter, rounder wing (Finlayson 1981, Shirihai *et al.* 2001).**
***S. a. pauluccii*. A common resident on the Balearic islands. Darker below and shorter-winged than the nominate but paler above than *heineken*: with which it may integrade on the eastern mainland (Shirihai *et al.* 2001).**

Breeding population estimates. Spain: SA1 0.85–1.5 million pairs; SA2; ACRE 5.2 million birds. **Portugal**: 50,000–500,000 pairs. **Andorra**: 200–300 pairs. **Gibraltar**: 150 pairs.

This is a common breeding bird over most of the Peninsula, although it is particularly associated with the eurosiberian and supramediterranean climatic zones in the north and with montane or particularly humid areas in the south. In the Balearics it also breeds commonly on Mallorca and Menorca, but only irregularly in Ibiza. The Blackcap is a summer visitor to the valley woodlands of Andorra and is a common breeding resident in the olive scrub and gardens of Gibraltar (BG).

Birds showing the characters of the nominate form are quite uniformly distributed throughout the northern half of Iberia, although AP assigns the entire Portuguese breeding population to *heineken*. The range of *heineken*, in the southern half of the Peninsula, where the unsuitable arid areas and expanses of treeless farmland are avoided, is more discontinuous. It is absent from large areas of the southern Meseta, the central Ebro valley, the arid south-east in Almería and Murcia and the

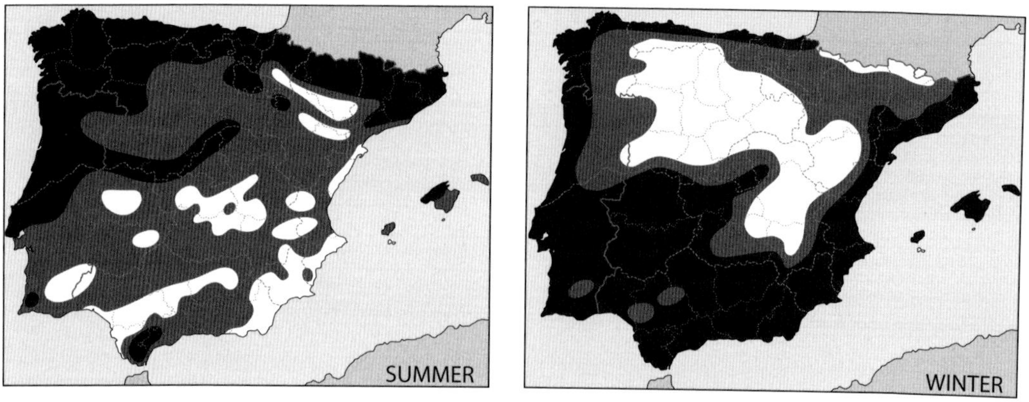

Guadalquivir valley but breeding birds may be found even in these areas in riparian vegetation, parks and gardens (SA2). The Blackcap is widespread in Portugal, except in the dry farmlands of Alentejo, but is most abundant north of the Tagus and in the more humid western half of the country (PA2).

The Blackcap selects woodlands with a good shrub layer. Mature woodlands with clearings and woodland edge habitats provide ideal conditions, as do tall hedgerows, riparian woodland and young plantations. Beech, oak and mixed woodlands are favoured but some occur in open coniferous woodland. Many also occur in the extensive eucalyptus plantations in the northern coastlands, which have a well-developed understorey of native shrubs. Suitable habitats are widespread in the eurosiberian zone but further south the species is allied to humid, shady environments often near water (Carbonell & Tellería 1998) or found in the deciduous montane woodlands of the supramediterranean zone. It ranges from sea level to over 2,000m in the Pyrenees and the Sierra Nevada, where it reaches the treeline (CA2, Garzón & Henares 2012).

The highest concentrations in the Peninsula are in southeastern Galicia, the Cantabrian mountains and the Basque mountains: all in the north and presumably race *atricapilla*, and in Los Alcornocales and Serranía de Ronda in the western Betic Range: presumably race *heineken*. In these areas densities of over 12birds/ha have been noted. On a broader scale, high densities have been reported from Pyrenean oakwoods (78 birds/km^2), and from a wide range of habitats in the Cantabrian region (37–52 birds/km^2), as well as from riparian woodlands (43 birds/km^2) (ACRE). The Spanish population increased significantly on SACRE plots during 1998–2012 (Escandell 2013) but the Portuguese population is thought to be stable (ACP).

A recent expansion of the breeding range has been documented in Catalonia, where a comparison between the first and the second atlases, dating respectively from the early 1980s and 1999–2002, showed a 6% increase in the number of occupied 10-km squares (CA2) and where an even greater spread had previously been reported (Mestre 1983, CA1). Range expansion is also apparent from a comparison of the first and second Spanish atlases, and the historical literature gives a clear impression that very significant expansion in the Blackcap breeding range took place during the latter half of the 20th century. For example, Saunders (1897) did not encounter it on the Spanish side of the Pyrenees, nor did Eagle Clarke (1889) in Andorra, nor Ticehurst & Whistler (1925) in Navarra. In Central Spain, Witherby (1928) found just a few birds in the Guadarrama, Gredos and Guadalupe mountains, as well as in Aranjuez, and Valverde (1956) stated that he had never found breeding Blackcaps on the northern Meseta. The inferred range expansion does not certainly apply to the *heineken* populations of southern Andalucía, where breeding birds were reported as plentiful in the corkwoods of Cádiz by Irby (1875).

Many Iberian birds, especially *heineken* and *pauluccii*, are resident but those of the nominate form, especially from other European populations, are migratory. Very large numbers of Blackcaps from northwestern Europe cross Iberia on passage and an unknown but large proportion remain to winter (Cantos 1995, Leal *et al.* 2004). A wintering population of 1.2–2.2 million birds has been estimated for Catalonia alone (CWA) and much higher wintering numbers can be expected to occur in some other regions, especially Andalucía and the Comunidad Valenciana. A proportion of those that continue to Africa cross the Sahara to winter in the humid northern tropics. The winter distribution in Iberia is concentrated within the thermomediterranean and, to a lesser degree, the mesomediterranean and eurosiberian climatic zones, reflecting the locations of its favoured fruiting plants (SWA). The highest winter densities are recorded from the Mediterranean coastlands, the Balearic Islands and the Guadalquivir valley. The northern mountains and most of the Castilian mesetas are largely abandoned in winter.

The wild olive, with its small, easily-swallowed fruits, is especially attractive to the Blackcap but birds also attend cultivated olive groves (Rodríguez de los Santos *et al.* 1986, Rey 1995). The olive may not be native to the Mediterranean (Blondel & Aronson 1999), although archaeological material shows that the wild olive was already present in the Iberian thermomediterranean over 10,000 years ago (Oliva & Montes 2005). Olive cultivation is more recent and has expanded greatly in modern times, so large-scale wintering in Iberia by Blackcaps may not always have been as noteworthy as it is now. The olive tree fruits in autumn and early winter so the crop is depleted by January–February. Some Blackcaps may move on south to Africa when this occurs (Shirihai *et al.* 2001) but many find other food sources locally for the remainder of the winter. Several other fruits are important in the winter diet, the Lentisc being especially significant (Jordano & Herrera 1981, Jordano 1985). Autumn migrants exploit the fruits of brambles, honeysuckle and fig trees (Jordano 1981b). An interesting if probably minor aspect of the regular winter diet of Blackcaps is nectarivory, often from exotic plant species. Birds take nectar from eucalyptus flowers in northern Spain (Tellería & Galarza 1991) and from the Tree Aloe in Gibraltar in December–January.

Winter site fidelity is well documented in this species (Cuadrado *et al.* 1995) but there are also instances of long displacements during the winter period. The influx of migrants into the Peninsula is noted from September in the north but not until mid-October in the extreme south and arrivals continue into November. Most passage is in October (Murillo & Sancho 1969, Jordano 1985, Grandío 1997, Bonnin 2007). Wintering birds depart in March–April and returning migrants from Africa cross the Peninsula during March–May, peaking in March (Leal *et al.* 2004). Males predominate among birds that remain to winter in Spain, suggesting that females migrate further south on average (Cantos 1995, Leal *et al.* 2004). Most immigrant Blackcaps recovered in Iberia: 92% of 2,026 recoveries, have been ringed in France, the Low Countries, the British Isles and Germany (Cantos 1992, SMA, AP) and most foreign recoveries of Iberian-ringed birds are also from these countries. Birds ringed in Denmark, Poland, Switzerland, Italy, Slovenia, the Czech Republic, Slovakia, Hungary, Norway, Sweden and Morocco have also been recovered in Iberia. Birds ringed in Spain have also been recovered in Italy (5), Sardinia (5), Denmark (2), Sweden, the Czech Republic, Austria, Hungary, Croatia and Cyprus, as well as in Algeria (90), Morocco (35), Ceuta, Mauritania, Mali and Sierra Leone (SMA).

GREENISH WARBLER *Phylloscopus trochiloides*

(Mosquitero Troquiloide, Felosa-verde)

Vagrant to Spain.

There are accepted records of two ringed in Spain: on the Columbretes islands, Castellón, on 6 September 1988 and at the Llobregat delta, Barcelona, on 27 September 2004. Another was ringed on Illa de l'Aire, Menorca on 25 September 2013 (RBS). The extreme scarcity of Iberian records relative to those from northwestern Europe may indicate that the arrivals there are due to post-breeding dispersion that is then followed by migration proper to the Indian sub-continent, as seems to occur with some other Siberian leaf-warblers (de Juana 2008). *P. t. viridianus* has expanded its breeding range westwards recently, some nesting as far west as northern Germany. It winters in the Indian sub-continent but it is nonetheless regular in autumn in the British Isles, where 452 had been recorded by 2005. There are only 27 records from France, however.

ARCTIC WARBLER *Phylloscopus borealis*

(Mosquitero Boreal, Felosa-boreal)

Vagrant to Spain, Portugal and Gibraltar.

Single birds were seen in Gibraltar on 30 October 1984 and at the Laguna de Medina, Cádiz, on 28 and 30 September 2008, and a juvenile was ringed at the Lagoa de Santo André, Setúbal, on 28 September 2009. This species often appears in western Europe in autumn, especially in Britain, where 319 individuals had been recorded up to 2011. There were only 15 records from France up to 2005, but none apparently from North-west Africa.

PALLAS'S WARBLER *Phylloscopus proregulus*

(Mosquitero de Pallas, Felosa-de-Pallas)

Vagrant to Spain, Portugal and Gibraltar.

The earliest Iberian observation was at Cazorla, Jaén, on 17 April 1987 (Klinkhamer & van der Maat 1990). Since then there have been another 13 records from Spain and the Balearics (*Ardeola* 60: 487), two from Portugal (AP) and one in Gibraltar (*GBR* 1: 48). These 17 records, of single birds, are concentrated in the north-east: eight in the area of Catalonia, the Comunidad

Valenciana and the Balearics, and south: five distributed between Andalucía, Gibraltar and southern Portugal, a similar pattern to that shown by the Yellow-browed Warbler. Ten were in autumn, between 18 October and 31 December, and seven in spring, between 13 February and 24 April, a very different pattern from that seen elsewhere in Europe, where most are between mid October and November and spring records are exceptional (Vinicombe & Cottridge 1996, de Juana 2008).

Pallas's Warbler is frequently observed in northern Europe in autumn, particularly in the British Isles and Fennoscandia. In Britain, at least 1,783 individuals were recorded during 1958–2003. Records further south in Europe are much scarcer, with only 107 found in France, for example. There are none from the Canary Islands but one was found in Morocco in December 1985 and other was on the Chafarinas islands in April 2001.

YELLOW-BROWED WARBLER *Phylloscopus inornatus*

(Mosquitero Bilistado, Felosa-listada)

Vagrant to Spain, Portugal and Gibraltar.

This is by far the commonest of the eastern Asian warblers that occurs regularly in Europe in autumn, when it appears in the north and north-west, from Finland, Sweden and Norway east to the British Isles. Over 9,000 were recorded in Britain alone during 1968–2003. Some numbers are also recorded in Denmark, Germany, the Low Countries and France, but relatively few appear in southern Europe. At least 1,006 were recorded in France up to 2005, although over 90% were on Ushant island, Brittany. There are 22 accepted records for the Canary Islands, five for Morocco and three for Algeria, and individuals have been found in Egypt, Libya and Senegal.

The earliest Iberian record was one seen in Gibraltar on 20 February 1932 but this record cannot now be properly assessed and is only regarded as a 'possible' (BG). The first confirmed observations involved birds ringed at Doñana on 21 and 31 October 1967 (Valverde 1968) and one seen at Tazones, Asturias, on 30 December (Álvarez-Balbuena *et al.* 2000). The Spanish rarities committee accepted 95 records for mainland Spain and the Balearic Islands during 1985–2011. There were also 20 in Portugal during 1983–2010, but only 14 of them were submitted to the Portuguese rarities committee. There are three accepted records for Gibraltar: one seen on 30 October 1984, one ringed on 26 October 1996 and an individual that was first detected in early February 2012 and remained at the same garden site until at least mid April 2012 (Finlayson & Cortes 1987, GONHS). All records have involved single birds, many of them trapped by ringers, with the exception of two seen together in Mallorca in October 2005 and up to four at Fisterra, A Coruña, in November 2011 (*Ardeola* 54: 438, 60: 487).

Figure 191. Regional distribution of records, 1967–2010 (n = 99)

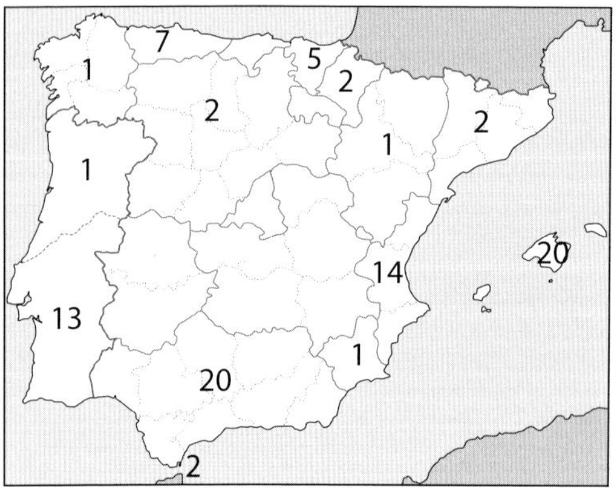

The accepted records up to 2010 are concentrated in the north-east: in Catalonia, the Comunidad Valenciana and Balearics, and in the south: in Andalucía and southern Portugal; these two regions together accounting for 78% of the records (Fig. 191). The great majority (96%) were in autumn, with extreme dates of 29 September and 8 December (Fig. 192). These records are rather later than those in Scandinavia, Britain or France (Cambi & Cambi 1986, de Juana 2008). Birds seen in the south of Iberia also tend to appear later than in the north, the north-east observations being intermediate (Fig. 193). Further south still, Morocco and the Canaries together have seen only three Yellow-browed Warblers in October but there have been eight there in November and 18 during December–March. Winter records appear to be exceptional: they comprise single birds at Los Barrios, Cádiz, on 27 January 1993; at Barcelona on 3 January 2004, and at Leioa, Bizkaia, during 9 January–17 February 2008. The Gibraltar bird mentioned above may have overwintered there and there is another accepted spring record: one at Albufeira, Faro, on 1 April 2000.

The scarcity of Iberian records relative to those in northern and northwestern Europe is not thought to be entirely due to any shortage of observers, not least because the ringing effort is now on a considerable scale. The possibility has been raised that both the Yellow-browed and Pallas's Warblers, and perhaps also some other Siberian passerines, may not strictly-speaking be vagrants to western Europe (Gilroy & Lees 2003). Their appearances may be due mainly to juvenile dispersion, the birds later returning east directly to winter in south-east Asia (de Juana 2008). This being so, only a small fraction of the hundreds, and sometimes thousands, of Yellow-browed Warblers that appear each autumn in Scandinavia and Britain ever continue south to Iberia, other parts of southern Europe or North Africa. Wintering in Europe seems to be exceptional although a few birds may remain in the Canary Islands and North Africa, which would account for the sporadic European records in spring. Unprecedented numbers, 30+ individuals, overwintered in the Canary Islands in 2013–2014, following an influx into the Peninsula in October–November 2013 in which similar numbers were involved (RBS).

Figure 192. Monthly distribution of records, 1967–2010 (n = 99)

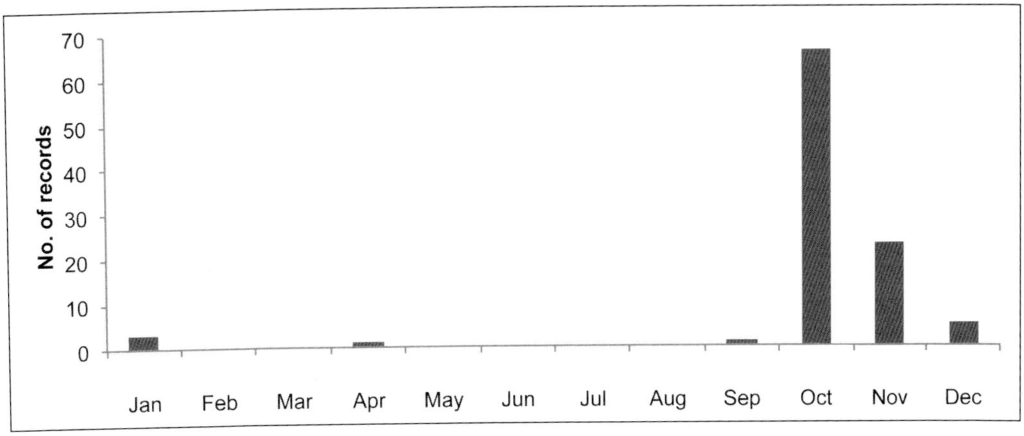

Figure 193. Changes in regional abundance of Yellow-browed Warblers in Iberia during the course of the autumn, 1967–2010 (n = 90). North = Biscay coastlands, northern Portugal, Castilla y León, Navarra & Aragón (n = 15); South = southern Portugal, Andalucía, Gibraltar & Murcia (n = 33); North-east Mediterranean (NE Medit.) = Catalonia, Comunidad Valenciana & Balearic Islands (n = 42)

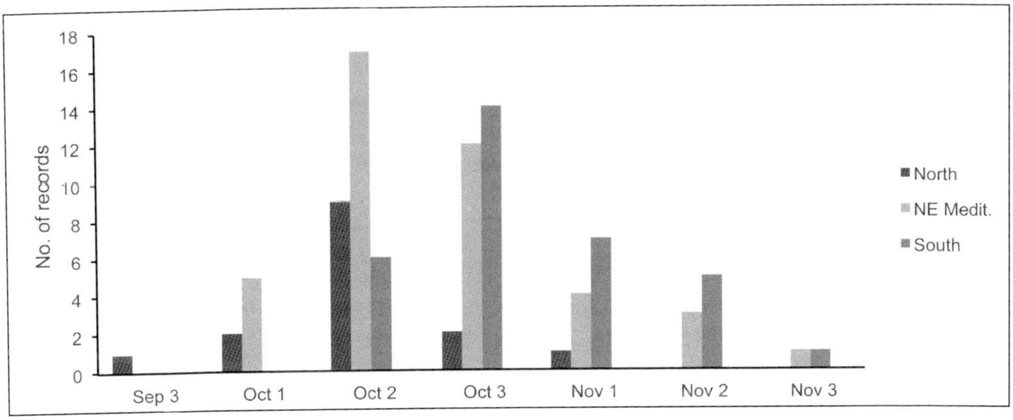

HUME'S LEAF WARBLER *Phylloscopus humei*

(Mosquitero de Hume, Felosa-listada de Hume)

Vagrant to Spain.

The sole accepted record was of one observed at Cabañas del Castillo, Cáceres, on 4 February 2006. Another was reported from Ribadeo, Lugo, in January–February 2012 (RBS). This Asian *Phylloscopus* was treated as a subspecies of the Yellow-browed Warbler until relatively recently, which probably means that earlier occurrences were often overlooked. It occurs with some frequency in northwestern Europe, chiefly in October–November although quite a few have been found in winter and some have remained until March or April (Slack 2009). Most records have been in Fennoscandia, The Netherlands and the British Isles. At least 121 have been identified in Britain and 19 in France.

RADDE'S WARBLER *Phylloscopus schwarzi*

(Mosquitero de Schwarz, Felosa de Radde)

Vagrant to Spain.

One was collected at Doñana on 7 November 1966 (Valverde 1968) and there have since been a further five accepted records for Spain, all of them birds trapped by ringers: on Cabrera island, Balearics, on 20 October 1996; on Dragonera islet, Balearics, on 7 November 2003; at Gozón, Asturias, on 15 October 2005; at Villajoyosa, Alicante, on 10 October 2008, and again on Cabrera island on 30 October 2009. Subsequently one was seen at Cape Peñas, Asturias, in October 2012 (RBS). This species is annual in northwestern Europe in autumn, especially in Sweden and the British Isles: where 274 had been recorded by 2005, but is much rarer further south, with only 12 records from France, most of them in Brittany, and none from Morocco.

DUSKY WARBLER *Phylloscopus fuscatus*

(Mosquitero sombrío, Felosa-sombría)

Vagrant to Spain, Portugal and Gibraltar.

The earliest report was of one in Gibraltar on 3 January 1989 (*Alectoris* 9: 36). Another was seen at Ponta da Erva, Lisboa, on 18 November 1990 (*Ardeola* 39: 81). Since then a further eight records have been accepted: at Portimão, Faro, on 1 November 1994; at the Ria de Alvor, Faro, on 23 October 1995; in Menorca on 4 May 1997; at Vilanova de la Barca, Lleida, on 15 November 2001; at Cabana de Bergantiños, A Coruña, on 9 November 2003; at the Ebro Delta on 19 November; at Quinta do Lago, Faro, on 3 and 4 November 2008, and at Fisterra, A Coruña, from 30 October–6 November 2011. The winter record in Gibraltar and the spring one in Menorca are unusual, and the other autumn records: seven of them in November, are rather later on average than elsewhere in Europe. All have been at coastal locations, seven in the Atlantic and three in the Mediterranean. Six were caught by ringers. There are additional records of single birds at Pals, Girona, in October–November 2012, and in Leiria in November 2012 (RBS, RBP). This species appears annually in autumn in northwestern Europe, particularly in Britain, where 310 had been recorded by 2005. There are records of 53 birds in France and one was recorded in Morocco in February 1974.

WESTERN BONELLI'S WARBLER *Phylloscopus bonelli*

(Mosquitero Papialbo Occidental, Felosa-de-papo-branco)

A locally common summer visitor to mainland Iberia and a scarce passage migrant in the Balearic Islands.

Breeding population estimates. Spain: SA1 1.1–2.7 million pairs; ACRE 4.5 million birds. **Portugal**: 10,000–100,000 pairs. **Andorra:** 250–400 pairs.

Bonelli's Warbler is widely distributed in the northern half of the Peninsula, although it occurs only locally in Galicia and in the Cantabrian coastlands and is absent from much of the central Ebro valley. The range is even patchier further south and few breed in Extremadura, western Castilla-La Mancha or western Andalucía (SA2). The Andalucian birds are concentrated along the Betic Range, being especially numerous in the Grazalema and Ronda mountains. They are also common in Cazorla and the eastern Betic mountains (SA1). Everywhere the species seems to select dry but relatively cool areas (Tellería & Santos 1994). The core distribution corresponds quite closely with the supramediterranean climatic zone, avoiding the lower ground of the southern river valleys where, in any case, much of the terrain is unsuitable. In the east and north especially, breeding birds may be found from sea level to 2,000m, in the Pyrenees (CA2). They also reach 2,200m in the south, in the Sierra Nevada (Garzón & Henares 2012). The Portuguese population is largely in the north and north-east, in an area ranging from Beira Alta to Trás-os-Montes and the Alto Douro, although there is a largely isolated population in the centre-west, in Ribatejo and western Alentejo (PA2). It does not breed in the Balearic islands.

Bonelli's Warbler is most abundant in oak woods, mostly of Pyrenean, Lusitanian and Holm Oaks in the north, but it

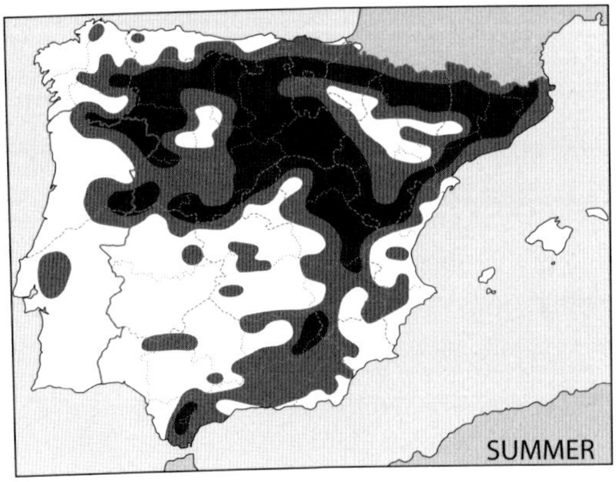

also frequents both young and mature pinewoods, especially where there is an oak understorey, as well as pine plantations and juniper woods (de Juana 1980, AI2). Densities of over 15 birds/10ha have been reported from Holm Oak woods in León, Pyrenean Oak woods in the Central Range, young pine plantations in Aragón and mixed oak-pine woods elsewhere (SA2). The Spanish population showed a significant increase on SACRE sites during 1998–2012 (Escandell 2013). Censuses in supramediterranean habitats found 25–50 birds/km^2. The Iberian population represents a very significant proportion of the overall European population (BE), that of France comprising only 100,000–300,000 pairs (OF).

Migrants cross the Peninsula on passage and are also then found in the Balearics, where they are scarce (*A. O. Balears 2010*: 268). The winter quarters are in sub-Saharan West Africa. Three Spanish-ringed birds have been recovered in France and one has been found in Morocco. Spring arrivals in the south begin in mid March but most pass in April and the last birds in May (AI2, Cantó & Pastor 2006). The return passage at Gibraltar is from mid July to September, peaking in August (BG). A few stragglers have been reported in Iberia in October.

WOOD WARBLER *Phylloscopus sibilatrix*

(Mosquitero Silbador, Felosa-assobiadeira)

A scarce passage migrant, chiefly in eastern Iberia, and a rare or sporadic nester in northernmost Spain. A rare migrant in Portugal.

The western European range barely reaches northern Iberia, where the Wood Warbler is a very rare or sporadic nester. Breeding records come from montane beechwoods, chiefly of the Cantabrian Range, but also in the Pyrenees. Breeding in the former was first proved in montane beechwoods in Asturias, at Brañagallones, Caso in 1986 and at Cangas de Narcea in 1997 (*Ardeola* 36: 261, *A. O. Asturias 1997*: 201). The Asturian Bird reports give a further 20 records of singing males heard during 1986–2000, chiefly in beechwoods at 700–1,400m in the central and western Cantabrian Range, from Somiedo and Muniellos to the Picos de Europa. Breeding is likely to have occurred on at least some of these occasions. Nests were found in the Pyrenees at Pueyo de Jaca in 1971 and at Ansó in 1992 (*Ardeola* 20: 374–375, *A. O. Aragón 1991–1992*: 54). Some birds are undoubtedly overlooked since large tracts of apparently suitable habitat are relatively inaccessible but the scarcity of the species is not in question. The Wood Warbler is known only as a passage migrant in Andorra (ANA), where Eagle Clarke (1889) found it common in the woods and pine forests in late May, although he made no mention of breeding.

Elsewhere birds are most often encountered during spring passage and the species is generally very scarce or rare in autumn, records then being much more frequent in central Mediterranean countries (BWP). The Spanish records come chiefly from the eastern seaboard. Ferrer *et al.* (1986) and Clavell (2002) describe it as a scarce but regular spring migrant in the eastern coastlands and Balearics, much rarer in autumn. Most published records are of 1–2 birds but ringers have found that some numbers pass through the Levant coastlands and, especially, the Balearics in spring. Ringers in Catalonia captured over 260 in spring between 1996 and 2009, mainly at the Ebro and Llobregat deltas and the Aiguamolls de l'Empordà. Over 171 were ringed in spring in the Comunidad Valenciana during 1988–2010, many of them on the Columbretes islands, where there was an exceptional record of 58 seen on 26 April 1996 (*A. O. C. Valenciana 1995–1997*: 211). Well over 600 were ringed in the Balearics during 1989–2010, with records throughout the archipelago but mainly from Cabrera and Illa del Aire, Menorca (*A. O. Balears*). Thunderstorms grounded some numbers on the Tarragona coast on 21 April 2011, when over 50 were seen in a 300m stretch of shrubbery (*Ardeola* 58: 510). All spring records from the Levant and the Balearic Islands fall between 1 April and 30 May, with the exception of a late bird at Valencia on 6 June 1992, the majority passing from mid April to early May.

The many fewer autumn passage records from eastern Iberia and the Balearics include 13 birds in Catalonia and 17 in the Comunidad Valenciana during 1996–2009 and 1988–2010 respectively. Most were found in September but some were reported from mid August and until mid October, with a very late bird at Calpe saltpans, Alicante, on 21 November 2000 (*A. O. Alicante 2000*: 105).

Very few Wood Warblers are reported elsewhere in Iberia. For example, there were only five records of single birds from

Madrid during 1988–2008 and only four records (five birds) from Burgos during 1981–99 (*A. O. Madrid, A. O. Burgos*). Occasional birds appear with other migrants at coastal sites in Asturias: at least 15 in total were recorded during 1988–2005, with records both in spring and autumn. There were also two autumn records from A Coruña, Galicia: single birds at the Cecebre reservoir on 18 September 1987 and at Valdoviño on 20 September 1988 (Martínez Mariño 2009).

Further south, the Wood Warbler has been reported nearly annually at Gibraltar in spring in recent years but has never been recorded there in autumn: very few are seen and just 38 were ringed during 1991–2010 (*GBR 2010*). Gibraltar records are in April–May with exceptionally early birds on 31 March 1973 (BG) and on 25 March 2011 (*GBR* 11: 36). It is recorded regularly as a scarce migrant at Doñana in April and May, and there are autumn records there from late July to late September, including a September recovery of a German-ringed bird (Llandres & Urdiales 1990).

There are only 12 records from Portugal: comprised by single birds in July and October, two in April, two in May and six in September, as well as two old specimen records (AP). Only three have been ratified by the Portuguese rarities committee: individuals seen at Vila do Bispo, Faro, on 9 and 14 May 2007 and ringed at Portimão, Faro, on 29 April 2010. The few Iberian recoveries of foreign-ringed Wood Warblers; from Germany, Belgium, Holland and Sweden, apparently include three recovered in Portugal (Cantos 1992): which seems remarkable given the apparent rarity of the species in that country. One ringed at Granada in September 1985 was found in southern Sweden the following May (*Ecología* 1: 335) and another bird from southern Andalucía was recovered in the Channel Islands (SMA).

COMMON CHIFFCHAFF *Phylloscopus collybita*

(Mosquitero Común, Felosinha-comum)

***P. c. collybita*. Locally common as a breeding species, especially in Catalonia and Andorra. Common on passage and abundant in winter throughout the region.**
***P. c. abietinus*. A scarce migrant.**
***P. c. tristis*. Vagrant to Spain and Gibraltar.**

Breeding population estimates. Spain: SA2 40,000 pairs; ACRE 1.7 million birds. **Portugal**: unknown. **Andorra**: 300–500 pairs.

The breeding distribution of the Common Chiffchaff is being gradually distinguished from that of its close relative the Iberian Chiffchaff (q.v.) but doubts remain regarding the specific identity of the breeding birds in certain areas. It seems to be the only breeding chiffchaff in the eastern Pyrenees, including Andorra: usually nesting in oakwoods, and in inland Catalonia and Valencia: always occurring in humid areas and usually in the mountains. In Catalonia up to 20 pairs/km^2 have been reported in montane broadleaved forests, with much lower densities in Holm Oak woods and pinewoods (CA2). The Common Chiffchaff is also found in scattered locations across Castilla y León, notably in the Duero valley and along both flanks of the Central Range, these generally associated with tall, well-developed riparian woodland (Balmori *et al.* 2002). There are breeding birds spread more thinly throughout the rest of Iberia, although avoiding the southern lowlands and generally favouring riverine woods. They breed in humid woodlands of the Sierra Nevada, at 1,000–2,100m, where the Iberian Chiffchaff is known only as a migrant (Garzón & Henares 2012).

The Common Chiffchaff is also almost unknown as a breeding species in Portugal, where nesting has only been confirmed in the Serra da Nogueira in the far north-east in 1999 (Patacho 2001, PA2). However, a few singing males have recently been

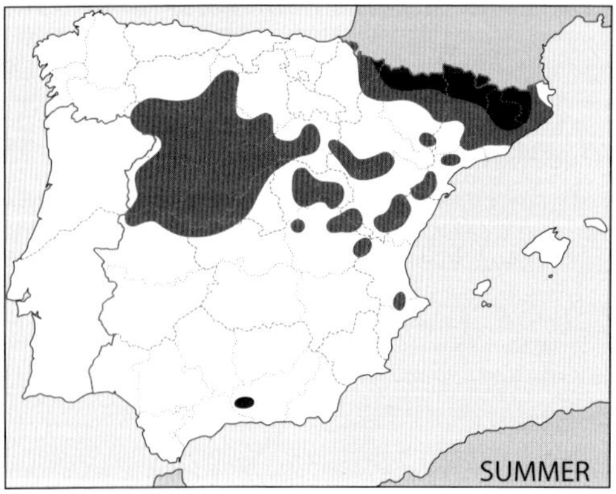

found more widely and it seems likely that a small population will be found to be present (AP).

The Common Chiffchaff was thought to be outnumbered by the Iberian Chiffchaff by ten to one (SA2). The ACRE assessment, of 1.7 million birds, implies that the Common Chiffchaff outnumbers the Iberian Chiffchaff in the breeding season by about 50% overall, although the relative frequencies of the two species clearly vary conspicuously between regions. The Catalan breeding population alone has been put at 40,000–101,000 pairs (CA2). The large discrepancy between recent and earlier estimates (Bernis 1963b, SA1, SA2) may partly reflect recent range expansion. Such a spread in the breeding range has been reported from Catalonia where a comparison between the first and second atlases, respectively from the early 1980s and 1999–2002, shows a 60% increase in the number of occupied 10-km squares, the expansion having been mainly southwards and eastwards (CA2). Nesting on the Balearic Islands was first reported in Mallorca in 2008 at Banyalbufar (Garcias 2009b) and a pair fledged at least one nestling there at Cúber in 2009 (*Ardeola* 57: 545). It is also possible that recent assessments may have taken some late passage migrants to be breeding birds, since migrant males in April commonly sing while on passage (de Juana 1980, Jubete 1997).

Individuals of the nominate race are ubiquitous in winter, occurring throughout the Peninsula and the Balearic Islands, except at higher levels, with the highest numbers in the milder Mediterranean regions (Cantos 1992, AI2, SWA). The highest densities are therefore seen in southern Portugal, Extremadura, Andalucía and along the Mediterranean coastlands, including the Balearics. Individuals ringed in the British Isles, France, the Low Countries, Germany and Switzerland comprise over 90% of 775 Common Chiffchaffs recovered in Iberia, together with a few from Scandinavia (15), Italy (3), the Czech Republic and Slovenia. Foreign recoveries of birds ringed in Iberia are from most of these countries as well as Poland (2).

The size of the wintering population is unknown but is certainly very large indeed. The numbers wintering in Catalonia alone have been put at 0.75–1.2 million individuals (CWA). The Common Chiffchaff may be found in almost any type of vegetation in winter but it shows a preference for open areas and avoids unbroken forests (SWA). Although it is a woodland bird when nesting it becomes very much more tied to lower growth in winter. Locally may it become semi-terrestrial and is often seen feeding on the ground. Some occur in winter in woodland edge and riparian habitats but many may be found far from any trees both in winter and on passage: along field margins, in fallow areas and in very low scrub. They are commonly found in such places as saltpans, marismas and estuaries, foraging in low bushes and along ditches. Parks and gardens all seem to have their own wintering Chiffchaff populations, some of the birds taking advantage of nectar offered by exotic plants. Drinking from the flowers of Tree Aloes is well known in Gibraltar in December–January (BG) and is also reported on the Costa del Sol.

Migrants may be hard to distinguish from wintering individuals but peaks in abundance at Gibraltar show that the passage there is from February to early April: with occasional stragglers as late as early May, and in October–November. The main winter arrival in southern Andalucía is from mid-October, by when the local Iberian Chiffchaffs have departed. Migrants that do not winter in Iberia do so in North Africa, south to northern sub-Saharan Africa. Autumn passage in the Balearics extends from early October to mid November, peaking in late October (Bonnin 2007). In Madrid, where few occur in mid-winter, passage is mainly from early September to early December, peaking in October–November, and again between mid February and mid April (Pinilla & de la Puente 2001). At a ringing site near Lisbon most captures were between early October and early March, and whereas there was no sex-ratio difference among the transient individuals, females comprised some two-thirds of the wintering population (Catry *et al.* 2007). In contrast, adult males predominate in northern Spain in winter (Cantos & Tellería 1994).

Clavell (2002) gives 17 east coast records (21 birds) of *P. c. abietinus*, mostly handled for ringing, especially at the Marjal del Moro but also from the Albufera de Valencia, Llobregat delta and Columbretes islands, and from Cabrera island in the Balearics. These all fall between 26 October and 18 March but most were in November and there were none in December or January, suggesting that *abietinus* only occurs on passage. One was ringed at the Marjal de Pego-Oliva (Alicante) on 16 February 2005 (*Ardeola* 53: 392). This subspecies is known from Gibraltar but there are no confirmed records from Portugal (AP), although occasional occurrences there seem likely. At least six Iberian recoveries of birds ringed in Norway or Sweden, are likely to have been of this subspecies.

The Siberian Chiffchaff *P. c. tristis* reaches western Europe, in particular the British Isles, Finland and Sweden, quite frequently (BWP, Dean *et al.* 2010) but very few have been identified in Iberia since the first record, in Gibraltar on 28 October 1978 (*Ardeola* 28: 164, *Alectoris* 3: 13). Since then there have been a further observation in Gibraltar, in winter 1988/89 (*Ardeola* 44: 137) and seven accepted records in Spain, all of single birds and most recently in 1995 (de Juana 2006). Reports of this subspecies quite often fail to satisfy the rarities committees and a series of Spanish observations have been rejected: at least 13 were pending assessment from 2011–12 alone. There are at least four reports from Portugal (AP) but none have been accepted so far. Their occurrence pattern resembles that in Britain, the extreme dates being 28 October and 26 March. The geographical distribution of *tristis* records recalls that of the Yellow-browed and Pallas's Warblers, with concentrations in the south and north-east.

IBERIAN CHIFFCHAFF

Phylloscopus ibericus

(Mosquitero Ibérico, Felosinha-ibérica)

A widespread summer visitor, chiefly to northern and southernmost Iberia. Occurs on passage in the Balearics.

Breeding population estimates. Spain: SA2 400,000 pairs; ACRE 1.1 million birds. **Portugal**: 10,000–100,000 pairs.

The breeding chiffchaffs of Iberia have long been regarded as subspecifically distinct from those nesting elsewhere in western Europe, chiefly on account of differences in their songs and other vocalisations (Ticehurst & Whistler 1928, Bernis 1945b, Salomon & Hemim 1992). A review by Helbig *et al.* (1996) recommended the elevation of the Iberian Chiffchaff from a subspecies of the Common Chiffchaff to a separate monotypic species, and this is now widely accepted. The scientific name changed from *P. brehmii* to *P. ibericus* when the holotype that was used to describe *brehmii* was found to be a Common Chiffchaff (Salomon 1997, Salomon *et al.* 2003). *P. (collybita) ibericus* was described by Ticehurst (1937) who designated as the holotype a bird collected by H. F. Witherby at Coimbra in May 1920. Where both species overlap geographically, the Iberian Chiffchaff tends to be larger and more brightly coloured, with a more pointed wing and a comparatively shorter bill and tarsi (Érard & Salomon 1989, Onrubia *et al.* 2003, Rodríguez *et al.* 2013). The early breeding bird atlases did not fully separate the two species. SA1 considered all the nesting chiffchaffs of the Peninsula to be Iberian. However, the presence of breeding Common Chiffchaffs was known by the time of SA2 but the two were not adequately separated during the fieldwork. As a result the distribution of the two species within Iberia was not adequately mapped in SA2 and is still not fully clear.

The principal populations of the Iberian Chiffchaff in Spain are split between the north and south, with few in the intervening meseta and Extremadura, on the Levant coasts or in the eastern half of the Pyrenees. The wide separation of the two population blocks and clear-cut, if minor, structural differences between them prompted Salomon *et al.* (2003) to propose the separation of the northern birds as a new subspecies, *P. i. biscayensis*, but this suggestion has yet to draw general acceptance. The northern population is widespread from Galicia through all the regions bordering the Biscay coast east to Navarra in the Pyrenees. Further east there are few definite records from Aragón, mostly involving birds trapped by ringers (*A. O. Aragón 2004–2007*: 328–329). There are a few records of singing males from Catalonia but breeding has never been proved there (CA2). The western Pyrenean breeding range extends slightly into the Pyrénées-Atlantiques, southwestern France, where a narrow secondary contact zone with the Common Chiffchaff has been described, and where some of the males, perhaps hybrids, show an intermediate type of song (Thielcke & Lisenmair 1963, Helbig *et al.* 1996 & 2001). A few 'mixed singers' have also been detected in Spain, as far west as Asturias (García-Rovés & Álvarez Laó 2003). The Iberian Chiffchaff is also found in the Iberian Range and western and central Castilla y León south to the Central Range (Bernis 1963b, de Juana 1980, Balmori *et al.* 2002, SA2). The other centre of population is in the Betic Range in southern Andalucía, in particular the western section from the hinterland of the Strait of Gibraltar and the Alcornocales to the Sierra de Grazalema, but apparentlyonly the Common Chiffchaff nests in the Sierra Nevada (Garzón & Henares 2012). There is also a thinly distributed breeding population across the Sierra Morena. It does not breed in the Balearic Islands. Breeding chiffchaffs in northwesternmost Morocco, west of the Rif mountains, are thought to be of this species (BM).

The Portuguese range is also divided between two principal centres of distribution, in Trás-os-Montes and Alto Douro in the far north and further south in an area extending from Estremadura and Ribatejo through western Alentejo and across most of Algarve (Elias 2004, PA2). The Iberian Chiffchaff is scarce or absent in most of the remainder of northern and eastern Portugal, especially the dry farmlands of eastern Alentejo.

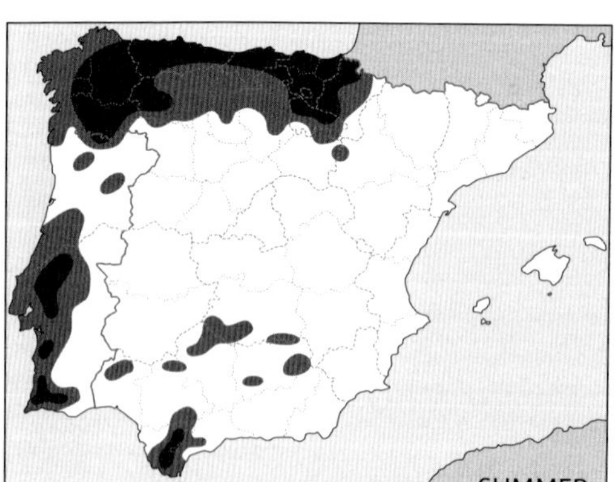

In general, the Iberian Chiffchaff is tied to the eurosiberian zone, from sea level to 1,700m in the Cantabrian Range and western Pyrenees, and to the supramediterranean region elsewhere. An exception is the southernmost population in the Betic Range and Los Alcornocales, Cádiz, which is in the meso- and thermomediterranean, but there the Iberian Chiffchaff is always found in humid microclimates, for example in the laurel growth along watercourses and in riparian growth generally. It is less associated with water in the more humid north (Balmori *et al.* 2002) but is still found in riparian woodlands there as well as in a wide range of deciduous woodlands offering well-developed undergrowth. It is less likely to be found in pine, Lusitanian Oak and Pyrenean Oak woods, the preferred habitat of Bonelli's Warbler.

Figure 194. Comparative temporal distributions of records in Catalonia (1999–2009, n = 64), the Balearics (2007–10, n = 32) and Gibraltar (1998–2010, n = 358), plotted by half-monthly periods.
Sources: Regional bird reports, Perez & Cortes 2002, GONHS records.

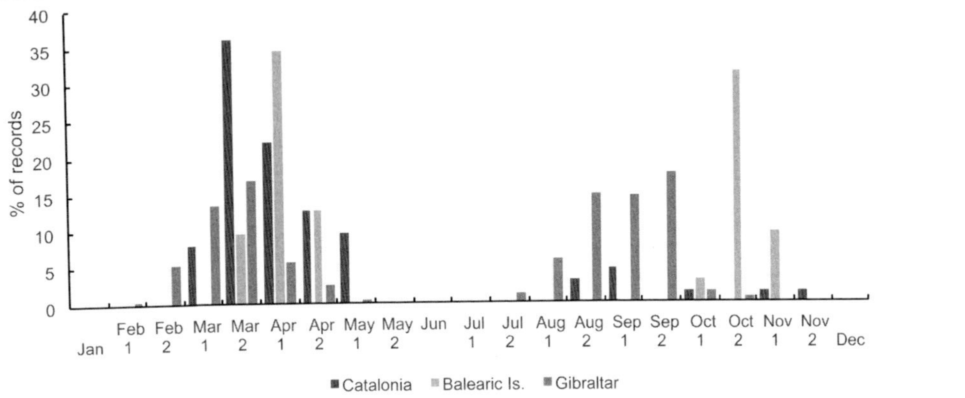

The largest regional populations are in the north: in Galicia (38%), followed by Castilla y León (29%), the Basque Country (15%) and Navarra (12%). Andalucía may account for less than 1% of the overall Spanish population. The highest reported densities are also in the northernmost woodlands of the Peninsula, most notably along wooded watercourses in the Cantabrian region (30 birds/km^2) and in oak and pine woodlands of the eurosiberian zone (12–20 birds/km^2).

The Iberian Chiffchaff has been regarded as outnumbering the Common Chiffchaff (*q.v.*) as a breeding species in Spain by ten to one (SA2) but recent studies suggest that it is the scarcer of the two species in Spain (ACRE). However, on present evidence it is by far the commoner of the two in Portugal, where there are few breeding records of the Common Chiffchaff (AP), as well as in at least Galicia, the Basque Country and Navarra (ACRE). It does not nest in Andorra or Gibraltar.

The temporal distribution of observations (Fig. 194) indicate that the Iberian Chiffchaff is a summer visitor and its morphology makes it better suited to long-distance migration that the Common Chiffchaff (Pérez-Tris *et al.* 2003). The Portuguese population is known to be migrant, and those captured in August–September show the extensive fat accumulation that is typical of a trans-Saharan migrant (Catry *et al.* 2005). The winter quarters are probably in sub-Saharan West Africa. There are records of birds on passage in Morocco (Bergier *et al.* 2013). Migrants have also been reported in northern and western Mauritania, and there are winter records from the Senegal delta (Isenmann *et al.* 2010), but the winter quarters are incompletely known (HBW).

The first spring arrivals at Gibraltar are in the second half of February, the passage peaking in early March and continuing through April. The return there is from early August to mid October, with a clear peak in the second half of September (Fig. 194). Ringing captures at a site in Álava, within the northern core population area, show a similar spring pattern but peak in July–August (Onrubia & Tellería 2013). Smaller numbers have been ringed on the Catalan coast: at the Ebro and Llobregat deltas especially and in the Balearic Islands. Gibraltar is within the main flyway for Iberian Chiffchaffs entering or leaving the Peninsula, unlike Catalonia and the Balearics. The timing of passage in these latter regions appears to differ to some extent from that in the south (Fig. 194) but the available sample sizes for the two regions are comparatively small and any differences need to be seen in this context. The records include single birds handled as late as 7 December 1997 in Valencia and 20 November 2006 in the Ebro delta (*A. O. C. Valenciana 2006–2007*: 212, *A. O. Catalunya 2006*: 171).

MOUNTAIN CHIFFCHAFF

Phylloscopus sindianus

(Mosquitero Montano, Felosa-da-montanha)

Vagrant to Gibraltar.

There is a record from Gibraltar of one that was ringed on 27 November 2001, whose biometrics suggest an individual of the nominate race (*GBR* 1: 24, 3: 48; Perez 2005). This Asian species is known as a vagrant to Iraq and Kuwait, and one was found in Israel in March 1983.

WILLOW WARBLER *Phylloscopus trochilus*

(Mosquitero Musical, Felosa-musical)

P. t. trochilus. **An extremely scarce and local breeding species in Spain. Abundant on passage throughout mainland Iberia and the Balearic Islands.**
P. t. acredula. **A scarce passage migrant.**

Breeding population estimates. Spain: SA1 20–100 pairs; SA2 60 pairs.

The Willow Warbler has a tiny relict breeding population in the eurosiberian zone, chiefly in Asturias but with breeding records also known from Galicia, Cantabria and the Basque Country. Records of possible breeding elsewhere are generally no more than observations of late singing males in spring. Irby (1875) considered that it bred in the corkwoods near Gibraltar and claimed to have seen fledged young there: it is highly likely that he mistook Iberian Chiffchaffs for this species. The tiny breeding population is known from mid-levels of the northern side of the Cantabrian Range, occupying open oak woodlands and young pine plantations: always those with a dense shrub layer, as well as woodland edge habitats.

The winter quarters are in sub-Saharan Africa and very large numbers cross the Peninsula and the Balearic archipelago during both passages. Most are of the western nominate race but the Scandinavian race *acredula* is also regular, although hard to identify in the field. Willow Warblers ringed in the British Isles, Norway and Sweden comprise over 60% of 516 extra-regional birds recovered in Iberia, with others from France, the Low Countries, Germany, the Czech Republic, Slovakia, Austria, Denmark, Finland and Switzerland and Poland as well as a few ringed as migrants in Italy, Slovenia and Malta (SMA, AP). Birds ringed in Iberia have been recovered in these countries, especially in Sweden and Britain, as well as in Morocco (7), Algeria (2), Mali (1) and Nigeria (1).

The spring passage is from mid March in the south but chiefly in April and throughout May. A few appear as early as late February but one at Alcochete, Setúbal, on 17 January 2010 was extraordinarily early (*Anu. Ornitol.* 8: 81). Returning birds appear from late July and occur until early October, with a peak in September (Asensio & Cantos 1989a, BG). One at Vila do Bispo, Algarve, on 10 November 2008 (*Anu. Ornitol.* 7: 87) was exceptionally tardy. Juveniles make up some 70% of migrants in August–September, increasing to 90% in October (Asensio & Cantos 1989a). There are a number of ringing recoveries of apparently wintering birds: AI2 records seven in the thermomediterranean and five in the mesomediterranean regions, but such instances are exceptional.

GOLDCREST *Regulus regulus*

(Reyezuelo Sencillo, Estrelinha-de-poupa)

R. r. regulus. **A common breeder in northern montane forests. More widespread in winter when some migrants from the north are also present.**

Breeding population estimates. Spain: SA1 170,000–580,000 pairs; **Andorra**: 500–1,000 pairs.

The Goldcrest breeds only in montane forests in the northern half of the Peninsula: in the Pyrenees and in the Cantabrian, Iberian and Central Ranges. Its presence in some of these ranges went undetected until relatively recently, although it is quite common locally. Thus Bernis (1954c), who found it in the southern Iberian Range in the Serranía de Cuenca in July 1951, wrote that previously it was only known as a breeding species from the Pyrenees. The first indications of breeding in the Cantabrian Range were in 1955 at Riaño (*Ardeola* 3: 190) and in the northern Iberian Range in 1974 at Abejar (*Ardeola* 20: 375–378). There are some isolated populations in such areas as Montseny in Catalonia and Moncayo in the Iberian Range (de Juana 1980, CA1). It has not been known to breed in Portugal but a singing male has been located in the Serra da Nogueira, close to the known breeding range in Galicia (AP) and so occasional nesting there in the north would not be surprising.

Breeding birds occur at altitudes of 900–2,200m, up to the treeline. The chosen habitats are predominantly subalpine coniferous forests but there are noticeable differences in habitat selection in different mountain ranges (AI2). Scots Pine is favoured in the Central and Iberian Ranges, and Mountain Pine and Silver Fir in the Pyrenees. Densities of up to 15birds/10ha have been recorded in Scots Pine and Silver Fir forests (SA2). The Cantabrian population is associated with birch, beech and oak forests, in the absence of natural pinewoods there (Álvarez & Purroy 1993). Numbers are likely to fluctuate in response to winter conditions. The Andorran population has benefited from reforestation with conifers there (ANA), as has probably occurred elsewhere.

Birds are present in the northern regions all year round but may descend to lower altitudes to avoid harsh weather in winter. Goldcrests are more widespread in winter (November–March) but are still much commoner then in the northern half of Spain, especially in the north-east. The wintering population in Catalonia averaged some 700,000 individuals in 2007–08 and

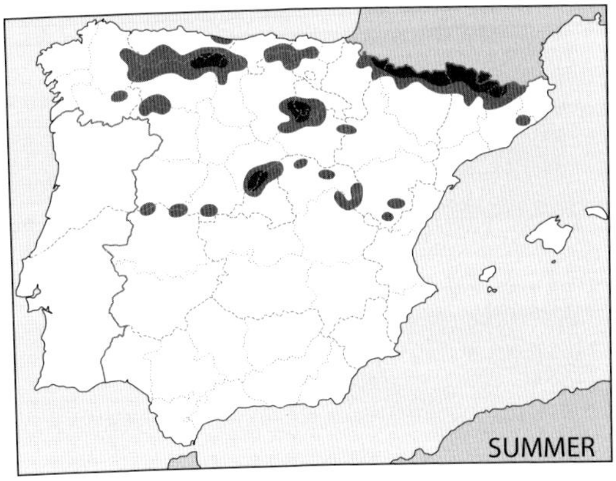

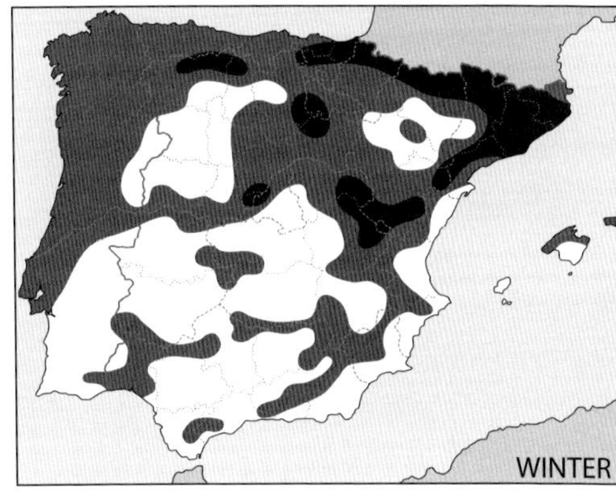

2008–09, when they were widely dispersed throughout the region, but there were only some 90,000 birds in 2006–07, when they mostly remained in the Pyrenees. Large year-to-year variation in wintering numbers has also been reported elsewhere in Spain (SWA). It thus seems that influxes from other European countries only occur in some winters (CWA).

Migrants from northern Europe make up an unknown proportion of the Iberian wintering population. There are a very few recoveries of birds ringed in Germany, the Czech Republic, Scandinavia, Finland, Poland, Russia and the Low Countries. One ringed at Bagà, Barcelona, in January 2002 was recovered in the Franche-Comté, France (652km NE) the following March (*Ecología* 17: 277). Few travel as far as Andalucía: out of 12 foreign recoveries in Iberia, seven were in the northern half of Spain, two in the Balearics, one in Extremadura and two in Andalucía: in Jaén and Cádiz (Cantos 1992). There are only eight records (nine birds) from Gibraltar, on dates from 5 November to 17 March, with one exceptionally late bird on 19 April 1973 (GONHS), and there are only four records from Morocco, where it is considered a vagrant (BM). Goldcrests occur sparsely on passage in the Balearics and some remain to winter in Mallorca (Serra 1978), where Munn (1921) considered them common in winter. The Goldcrest is a scarce winter visitor to Portugal in variable numbers: it has been recorded throughout the country but is most often seen in the north, chiefly during November–March (AP).

FIRECREST *Regulus ignicapillus*

(Reyezuelo Listado, Estrelinha-real)

***R. i. ignicapillus*. A widespread resident in northern Iberia and in the southern Iberian mountains. Also a scarce passage migrant and winter visitor.**
***R. i. balearicus*. A widespread resident in the Balearic Islands. Paler and greyer below than the nominate.**

Breeding population estimates. Spain: SA1 910,000–2,000,000 pairs; ACRE 3.0 million birds. **Portugal**: 10,000–100,000 pairs. **Andorra**: 500–1,000 pairs.

The Firecrest is widely distributed throughout the eurosiberian zone across the entire Cantabrian range and Pyrenees. Further south it is common in the supramediterranean zone, including the Iberian and Central Ranges and the Montes de Toledo, as well as in pine plantations at various levels across Castilla y León. It also inhabits some northern zones in the mesomediterranean, such as coastal Catalonia and the north of the Comunidad Valenciana (SA2). It is absent from much of Castilla-La Mancha and Extremadura. In southern Spain it is largely confined to the Sierra Morena and the Betic Range, especially in the more humid areas, descending to near sea level at the Strait and in Los Alcornocales, Cádiz. Birds occur up to 2,200m in pine plantations in the Sierra Nevada, where the Goldcrest is only a rare winterer (Garzón & Henares 2012), but elsewhere are generally most abundant at middle levels, for example at 800–1,400m in Catalonia (CA2), being largely replaced by the Goldcrest higher up.

In Portugal it is widespread in the north, northwest and centre, principally north of the Tagus (PA2). It occurs very locally in the south, notably in the Serra de São Mamede, Algarve; in southwestern Alentejo, and on the Setúbal peninsula (AP). The Firecrest is widespread at lower elevations in Andorra, being replaced by the Goldcrest higher up (ANA).

The principal habitat everywhere is conifers, both natural forests and plantations of exotic pines. Where the Firecrest and the Goldcrest coincide in the Pyrenees the Firecrest is strongly predominant numerically in Scots Pine forests whereas the Goldcrest predominates in Mountain Pines and Silver Firs (Purroy 1974). In the northern Iberian Range, however, the Goldcrest is more abundant than the Firecrest in Scots Pine forests (de Juana 1980). The Firecrest also frequents a wide range of broad-leaved woodlands,

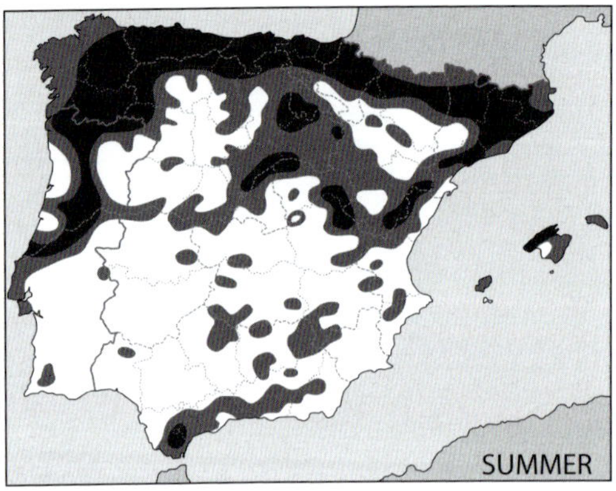

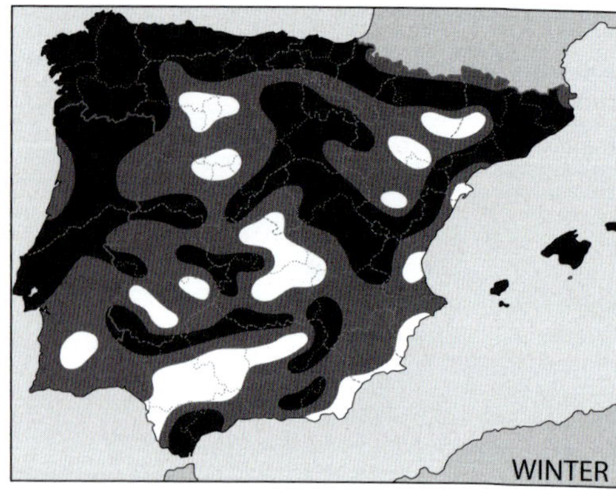

including riparian formations and hedgerows with tall trees in the northern countryside. Those in the western Betic mountains select mixed Cork Oak and Lusitanian Oak woodlands. In general, in the absence of conifers the presence of such evergreen trees or shrubs as Holly, Box, Yew or Ivy seems to favour its presence (de Juana 1980). It is much less of a habitat specialist than the Goldcrest and less associated with mountains, at least in the north, but it avoids areas that cannot provide a humid and relatively cool microclimate.

Firecrests of the race *balearica* nest commonly on all the main islands in the Balearics, in a diversity of habitats that includes parks and gardens. They also favour holm oak woods and pinewoods. They are more abundant in Mallorca and Ibiza than in Menorca (*A. O. Balears* 2010: 268). Breeding was proved on the islet of Cabrera in 2005 (*A. O. Balears 2005*: 231).

In Spain the population on SACRE sites showed a significant increase during 1988–2012 (Escandell 2013). A marked increase in the breeding population, of some 11% per year, was reported from Catalonia during 2002–08 (CA2). The upsurge in Iberian populations may be partly explained by the conifer afforestation, which are known to greatly favour the Firecrest. Some early authors certainly give an impression of former scarcity. For example, Arévalo (1887) was unaware of its breeding in Spain, Tait (1924) recorded that no nest had yet been found in Portugal and Witherby (1928) only found it breeding in the Sierra de Guadarrama but not elsewhere in central Spain. However, a breeding population in the humid ranges of southernmost Andalucía has long been known (Irby 1898). Local densities may be relatively high. The greatest numbers occur in mixed oak-pine woods, fir woods and Monterey Pine plantations, with mean maximum densities here of 18.1 birds/10ha (SA2). The Firecrest seems especially well adapted to Monterey pine plantations, where it occurs at high densities (Tellería 1992). On a broader scale, the highest reported densities are from Pyrenean mixed pine/fir woods (56 birds/km^2), with 25–39 birds/km^2 in other eurosiberian forest habitats (ACRE). The wooded watercourses of the northeastern mesomediterranean can hold 46 birds/km^2 and 29 birds/km^2 have been recorded from supramediterranean pinewoods above 1,250m. Residents descend from higher levels to the lowlands in winter and there is then some dispersal throughout the Peninsula. There is also some immigration from further north for the winter, bringing birds both to mainland Iberia and the Balearics (Cantos 1992). The Firecrest can be found in winter in most of Iberia, being scarce or absent only in deforested areas of the Mesetas; the Ebro and Guadalquivir valleys, and the arid south-east (SWA). It remains commonest in conifer forests in winter but it may then be found in a wide range of habitats, including oak woodlands and dehesas, olive groves, riparian groves and urban parks (SWA). It only appears in Gibraltar from September to March, with some indication of a passage peak in November (BG) but any cross-Strait movements are probably small. Similarly it only appears in Doñana outside the breeding season, being present from late July to late May, again chiefly in pinewoods (Llandres & Urdiales 1990). There are a few Iberian recoveries of birds ringed in the British Isles, France and Belgium. One ringed in Arlanzón, Burgos in March 1990 was found at Versailles, France in July 1991 (*Ecología* 6: 268) and another bird travelled from northeastern Spain to the Polish Baltic coast (SMA).

SPOTTED FLYCATCHER *Muscicapa striata*

(Papamoscas Gris, Taralhão-cinzento)

***M. s. striata*. A widespread summer visitor to mainland Iberia and a common migrant throughout the region.**
***M. s. balearica*. A common summer visitor to the Balearic Islands. Smaller, paler, sandier and less streaked than the nominate (Gargallo 1994).**

Breeding population estimates. Spain: SA1 640,000–690,000 pairs; ACRE 915,000 birds. **Portugal**: 500–5,000 pairs. **Andorra**: 50 pairs. **Gibraltar**: 1–2 pairs.

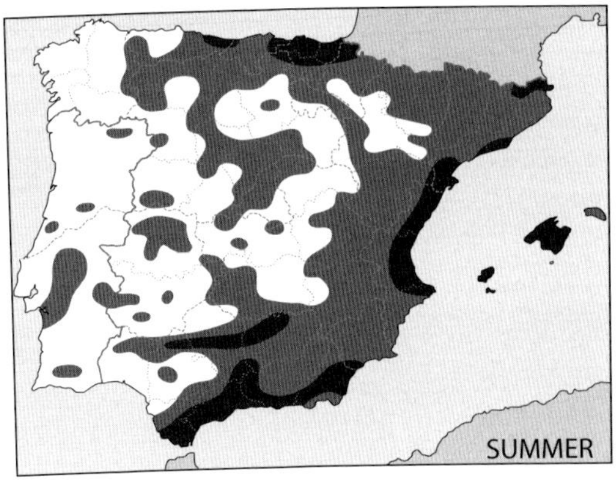

The Spotted Flycatcher is very unevenly distributed across the Peninsula. It is usually common in the Mediterranean coastlands but becomes progressively scarcer further inland. It is particularly local across the Meseta and in Extremadura and it is also very uncommon or absent in Galicia and western Asturias. However, it is common in eastern Cantabria and the Basque Country and, at low density, throughout the Pyrenees. In the south, it is common in the Sierra Morena and the Betic Range but scarce in the Guadalquivir valley and much of Huelva province (SA2). In Portugal it is widespread in parts of Ribatejo and Alto Alentejo but is very sparsely distributed elsewhere, especially north of the Tagus (PA2). The Balearic race is common throughout Mallorca, Ibiza and Formentera, and rather less so in Menorca (Mayol 1978, Muntaner & Congost 1984, *A. O. Balears 2010*: 268).

This flycatcher favours open woodlands, notably of oaks and pines, especially in the vicinity of water. Many nest in close proximity to people; in gardens, parks and golf courses. Woodland edge habitats, orchards and riparian woodlands are also selected, as well as carob, olive, orange and almond groves, but treeless and densely-forested regions are avoided (SA2). Most breed below 1,000m but some range to 1,300m in the Pyrenees (Ferrer *et al.* 1986) and to 2,000m in the Sierra Nevada (Garzón & Henares 2012). Its highest densities, up to 17 birds/km^2, are in thermomediterranean habitats, including matorrals and pinewoods, orchards and the margins of human habitation and olive groves (ACRE).

The Spotted Flycatcher has suffered recent severe declines in parts of Europe: over 70% in England for example (Brown & Grice 2005), but no numerical trends are known from Spain (Escandell 2013). It may have been commoner in Iberia in the past since late 19th and early 20th centuries authors considered it to be very common or abundant: for example at Santander (Irby 1883), in Andorra (Eagle Clarke 1889), near Gibraltar (Irby 1898, Stenhouse 1921) and in Andalucía in general (Saunders 1871, Arévalo 1887). The overall distribution then, however, may have been similar to the present one: Ticehurst & Whistler (1928) did not find it in Galicia, and Witherby (1928) only found it common in central Spain at Aranjuez: with 'a few more' seen at San Ildefonso, Segovia, and at Arenas de San Pedro and Candeleda, Ávila.

Migrants are common throughout Iberia. Birds ringed in the British Isles, the Low Countries, France and Germany comprise 91% of 137 recoveries of extra-regional origin, with others from Switzerland, Sweden, Denmark, Finland and Slovakia (SMA, AP). There are 12 recoveries in total from these countries of birds ringed in Spain. Four Spanish-ringed birds have been found in Morocco. It is relatively late in arriving from its sub-Saharan winter quarters and few appear in the south before mid-April (AI2). Northward migration continues throughout May, with stragglers in June. The return passage is in August–early October, with stragglers in November (BG, Tellería 1981, Ferrer *et al.* 1986, AI2).

There are a few records of very late birds. For example, single birds were seen at different locations in Doñana on 4, 7 and 8 December 1970 ; two were at El Rocío, Doñana, on 31 December 1984, and one was in Soria on 4 January 2007 (*Ardeola* 19: 33, 32: 423, 54: 182). There are also several ringing recoveries in January and February (AI2). All these records are highly surprising, given that the species arrives late in spring and departs promptly in autumn.

RED-BREASTED FLYCATCHER *Ficedula parva*

(Papamoscas Papirrojo, Papa-moscas-real)

Vagrant to Spain, Portugal and Gibraltar.

This flycatcher is rarely seen in Iberia and the increased frequency of recent records may reflect increased observer effort. Several old records from Spain include some of doubtful reliability (Bernis 1954a). However, a Swedish-ringed bird was captured in Castellón in October 1949 (*Ardeola* 1: 124). The *Ardeola* 'Noticiarios' give six observations during 1960–73 and one was ringed in Catalonia in 1983 (Carrera & Martínez 1985). During 1984–2010 the Spanish rarities committee accepted 30 records from mainland Spain and the Balearics and there are also 11 of single birds for Portugal and three for Gibraltar. There are thus 52 available records, involving 53 individuals.

Seventeen of 22 known-age individuals in autumn were first-year birds and five were older. The sexed individuals comprise four males and four females in autumn and two males and one female in spring. Observations show a notable concentration in the coastal north-east and very few have been found inland or in the north (Fig. 195). Some 92% were in autumn, between

31 July and 27 November, and there are only four spring observations, between 23 March and 15 May (Fig. 196).

This species frequently appears in northwestern Europe in autumn, with at least 3,367 individuals recorded in Britain during 1968–2003 and 264 in Ireland up to 2011. At least 383 were recorded in France by 2005, mostly in the north. There are ten records from Morocco, at least six from Algeria and five for the Canary Islands.

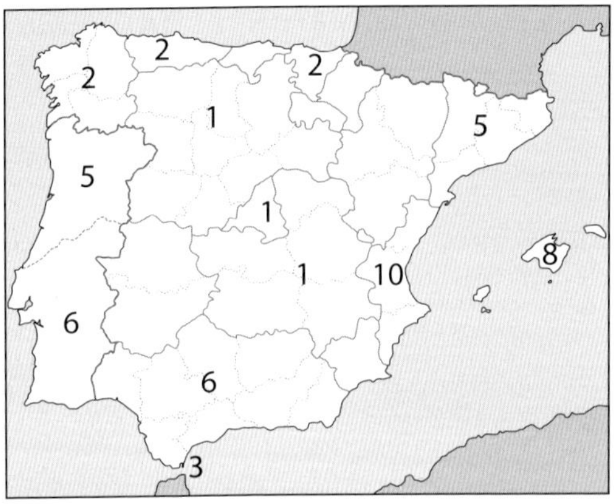

Figure 195. Regional distribution of records, 1949–2010 (n = 52)

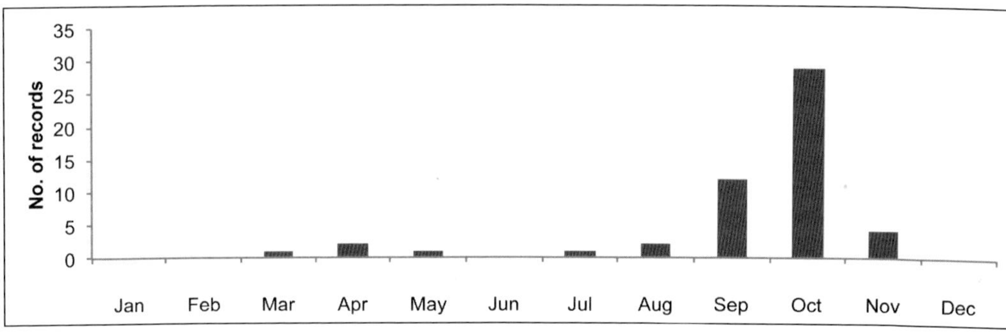

Figure 196. Monthly distribution of records, 1949–2010 (n = 52)

SEMI-COLLARED FLYCATCHER *Ficedula semitorquata*

(Papamoscas Semiacollarado, Papa-moscas-de-meio-colar)

Vagrant to Spain.

A male was ringed on Cabrera, Balearic Islands, on 17 April 2011 (*Ardeola* 60: 489). There are two records from France: on the Mediterranean island of Porquerolles in April 2004 and May 2007 (Duquet 2008), and two from Italy: on Linosa island in May 1968 and on Capri island in April 1988 (Iapichino & Massa 1989, *Riv. Ital. Orn.* 59: 271).

COLLARED FLYCATCHER *Ficedula albicollis*

(Papamoscas Acollarado, Papa-moscas-de-colar)

A rare migrant on eastern coasts, including the Balearic Islands. Vagrant elsewhere.

A record of an adult male in Sevilla in March 1868 (Saunders 1871) may be the oldest Iberian observation. Jourdain reportedly saw one in Mallorca in May 1930 (Munn 1932). More recent records are of one at Doñana on 22 May 1957 (Mountfort & Ferguson-Lees 1961) and the possible recovery of one with a Hungarian ring at Alcácer do Sal, Setúbal, on 12 March 1960 (Tait 1963). The latter is the only known occurrence in Portugal but the record is considered 'indeterminate' there since there were no supporting details (AP). There are other Iberian records that precede the estabishment of the rarities committees but some may actually have involved the Iberian race of the Pied Fycatcher, *F. h. iberiae*, whose males sometimes display a 'collar' (Potti & Merino 1995). We thus only consider accepted records, which total 45 up to 2010, excluding a believed hybrid with a Pied Flycatcher (*Ardeola* 40: 190). All but two records were of single birds. Spring records included 34 males and only nine of the less conspicuous females. There were at least four more records from eastern Spain in spring 2011 (*Ardeola* 60: 490) and there was an unprecedented influx in the same region in spring 2013, when at least 18 individuals were reported, most of them in Catalonia during the last few days of April (RBS).

Some 96% of the accepted observations up to 2010 have been in spring, between 5 April and 14 May (Fig. 197), considerably earlier than records from Britain which are nearly all between early May and mid June (Slack 2009). Male Collared Flycatchers reach their breeding areas about a week earlier than the females (Pärt & Gustafsson 1989), a pattern reflected in the Iberian records since those in April comprised 26 males and three females and those in May eight males and six females. Collared Flycatchers are difficult to distinguish from Pied Flycatchers in the field in autumn and the only Iberian records then are of individuals ringed at the Ebro delta on 27 September 1994 and 19 September 1998 (*Ardeola* 46: 145, 51: 535). There is an earlier record also of a female ringed at Doñana in October 1964 (Bernis & Fernández-Cruz 1966). Records are predominantly from the east of the region, including some 60% in the Balearic Islands, as might be expected (Fig. 198). The sole inland record is of a male at Piedrahita de Muñó, Burgos, on 1 May 2002 (*Ardeola* 51: 535).

Figure 197. Monthly distribution of records, 1957–2010 (n = 45).

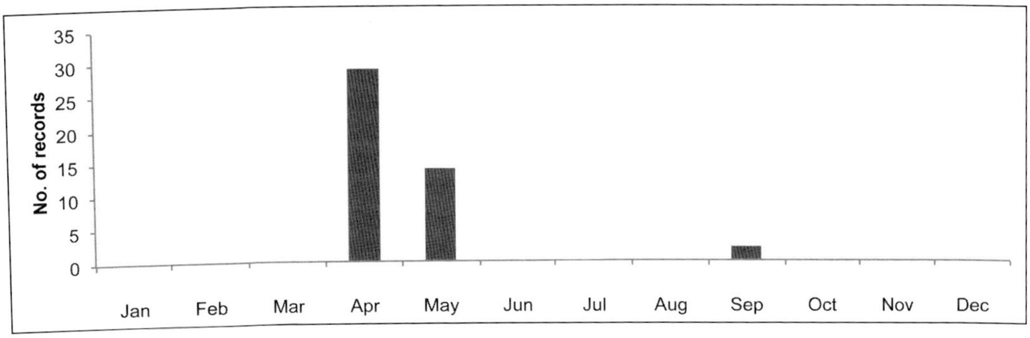

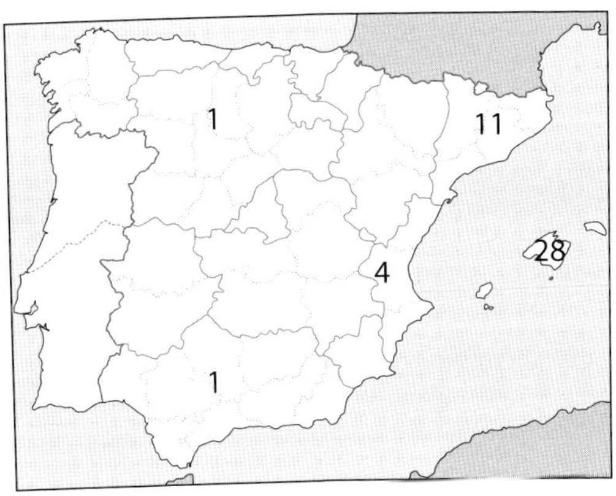

Unlike the Red-breasted Flycatcher, the Collared Flycatcher comparatively seldom reaches the European Atlantic seaboard where, for example, there are just 35 British records up to 2011. There are very few observations in France outside the breeding range and these are largely in spring in the eastern third of the country. There are eight records for Morocco, again chiefly in the east in spring, but only one from the Canary Islands.

Figure 198. Regional distribution of records, 1957–2010 (n = 45)

PIED FLYCATCHER *Ficedula hypoleuca*

(Papamoscas Cerrojillo, Papa-moscas-preto)

***F. h. iberiae*. A very local summer visitor. Widespread on passage. The frontal patch and white wing panel of the male are larger than in the nominate.**

***F. h. hypoleuca*. A numerous and widespread passage migrant.**

***F. h. speculigera*. (Atlas Flycatcher). A record of a male at Cape Creus, Girona, on 30 April 2012 was under review in 2014 (RBS).**

Breeding population estimates. Spain: SA1 130,000–350,000 pairs.

The Pied Flycatcher has a highly local distribution in Spain in the eastern sectors of the eurosiberian zone and a relict presence in mountains further south (AI2, SA1, SA2). In the former it is very thinly distributed in the Cantabrian Range, chiefly in the southern sectors in Castilla y León, but it is absent from Galicia. There are good populations in the Iberian Range. It is also well represented in the Central Range in Guadarrama and also locally further east in Gredos. It is present, but scarce and declining, in the east coast mountains in the Comunidad Valenciana, where the published atlases indicate breeding in a number of

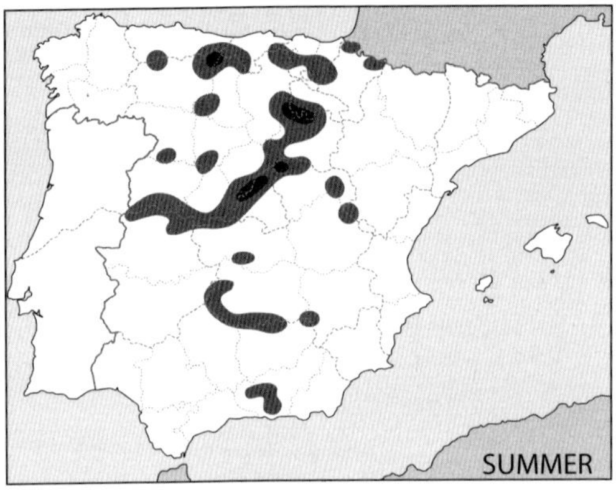

squares, mainly in Castellón and Alicante (Urios *et al.* 1991, SA1, SA2) but where the post-2000 regional bird reports do not give any conclusive breeding data. Its breeding status in northern Murcia also requires confirmation: there are no breeding records in the regional bird reports. Breeding in Catalonia seems to be irregular or sporadic (Ferrer *et al.* 1986): the 1999–2002 atlas just found singing males at two sites (CA2). It is very uncommon in the Pyrenees except in northern Navarra. There are also small populations in the Montes de Toledo, in eastern Sierra Morena and in the Betic Range: where it is almost entirely confined to high altitudes in the Sierra Nevada. It has never been found nesting in Portugal (AP). Breeding has similarly not been proved in Andorra, although it has nested nearby in Catalonia (ANA). A pair nested exceptionally in Mallorca in 1993 (*A. O. Balears 2010*: 268).

Witherby (1928) found several pairs apparently breeding in the royal gardens of San Ildefonso, Segovia. Irby (1883) had previously found it to be common in the beech forests near Potes, Cantabria, in May–June 1876. Breeding in southern Iberia was first reported by Armitage (in Jourdain 1936–37), who saw a pair at a nesthole in a tall conifer in the Generalife gardens in Granada. Bernis (1952) found the species frequent in oak woods at Fuencaliente, Ciudad Real, in the Sierra Morena, in June 1951.

The Pied Flycatcher requires mature deciduous woodlands where there are sufficient old trees to provide a supply of nest holes. It readily accepts nestboxes and these allow it to occur in younger woodlands where natural holes are lacking (de Juana 1980, Potti & Montalvo 1990). It has even been reported using nestboxes in orange groves (Urios *et al.* 1991). Its readiness to use nestboxes has allowed a considerable amount of research into its breeding and evolutionary biology in Spain (see review in Morales 2012). The more northerly populations occur in sessile oak, beech and pine woodlands. Further south it occurs in oakwoods, and also in chestnut woods in northern Extremadura. Most occur at altitudes of 1,000–1,500m but some nest up to 1,900m. Mean maximum densities in optimum habitats in Spain are 5.37 birds/10ha (SA2).

It is widespread on passage throughout the Peninsula and the Balearic Islands, and may then be found in all manner of open scrub and woodland habitats, as well as in parks and gardens. Most migrants are of the nominate race and over 75% of 1,293 recoveries of birds of extra-regional origin found in Iberia had been ringed in Britain, Germany or Fennoscandia, with a further 7.5% from as far east as Russia (SMA, AP). Others had been ringed in France, The Low Countries, Rumania, Poland, the Czech Republic, Austria, Switzerland, Italy, Belarus and Ukraine. Two of the more far-flung recoveries involve one ringed at the Evros delta, Greece, in April 2003 that was found in La Rioja the following September (*Ecología* 20: 334) and one ringed in Malta in April 2005 that was found in Lleida in August 2006 (*Ecología* 21: 290), both illustrating the tendency for the spring movements to be orientated further east than the autumn ones. There are 55 recoveries of Iberian-ringed birds from across Europe including individuals found in Ukraine, Russia (3) and Malta. Thirteen ringed in Spain have been recovered in North Africa: in Morocco (9), Algeria (3) and Tunisia, and one ringed in Portugal was found in Ghana.

The autumn passage is concentrated on the western side of the Peninsula, whereas the spring passage is more evident on the Mediterranean side and in the Balearics (Bernis 1963c, Lundberg & Alatalo 1992, Ferrer *et al.* 1986). Western Iberia is an important staging area where many birds gain significant weight prior to continuing south (Bibby & Green 1980, Veiga 1986). Concentrations of migrants are frequent, especially when birds are grounded by inclement weather, but a count of 327 at Cape Peñas, Asturias, on 21 September 2006 was exceptional (*Ardeola* 54: 182): it followed the transit of the aftermath of Hurricane Gordon.

Passage periods in the south are somewhat earlier in spring: when a few may appear in late March (Finlayson & Cortés 1987), and later in autumn, than in the Spotted Flycatcher. The first arrivals at Gibraltar are usually in early April, peaking later in that month, and passage continues throughout May. The return passage in Madrid extends from mid August to late October, peaking in the third weak of September (Villarán *et al.* 2005).There are some Gibraltar records in late August but the autumn passage there chiefly extends from mid-September to the end of October, with stragglers to mid November (BG, Tellería 1981). Passage on the Levant coast and in Portugal has similar timing to Gibraltar and November observations there are rare.

The winter quarters are in sub-Saharan Africa. BWP refers to winter ringing recoveries in Iberia that it discounts as probably resulting from inaccurate reporting arising from illegal hunting (Ojanen 1992). There are a few winter month records of single birds but there is no indication of regular wintering (SWA): they include at least four December records in southern locations and one at Jódar, Jaén on 7 January 1989 (*Ardeola* 36: 262).

BEARDED PARROTBILL *Panurus biarmicus*

(Bigotudo, Chapim-de-bigodes)

***P. b. biarmicus*. A scarce and very local resident in Spain. Absent from Portugal and the Balearic Islands.**

Breeding population estimates. Spain: LR 650–1,100 pairs.

The Bearded Parrotbill (Bearded Tit) is restricted to a very small number of locations in Spain, all of them wetlands with well developed peripheral reedbeds. There are three main subpopulations: at many of the scattered lakes of La Mancha, a stronghold of the species; on the Levant coast and in the Ebro valley/northern Meseta (SA2). The population is put at only 650–1,100 pairs (LR), more than half of them at the La Mancha wetlands. An earlier figure of 2,300–4,500 pairs (SA1) is considered to have been a great overestimate. Nevertheless, its fortunes have fluctuated, with declines to extinction at some sites, increases at others, sporadic breeding at certain locations and colonisation of new sites all recently reported.

The large population of La Mancha is distributed between the major lakes and wetlands of the region. It was estimated to comprise 580–890 pairs, at some 50 locations, by Velasco *et al.* (2004), the most important population nuclei being at Daimiel (150–200 pairs), El Taray (80–100), the Dehesa de Monreal (40–60), the Lagunas de Villafranca (35–60), the Castrejón reservoir (40–50), the Laguna de Vadoancho (40–50) and the Almoguera reservoir (30–40). There is also occasional breeding in Madrid region, as in 1974 at Arganda, 1988–89 at Rivas-Vaciamadrid and 1995 at San Martín de la Vega, and also at Velilla de San Antonio in 2000 (*Ardeola* 43: 257, *A. O. Madrid 2001*: 204). Further west, at least one pair bred in Extremadura at the Arrocampo reservoir in 2011 (*Ardeola* 58: 512).

The northern population is concentrated in the upper Ebro valley, where birds are present at the Lagunas de Pitillas and Las Cañas, Navarra, and in neighbouring areas of La Rioja: at La Grajera reservoir and at a marsh near Alfaro where 6–10 pairs bred in 2011 (*Ardeola* 59: 192). This population was increasing during the 1990s and may have given rise to the small population that has since bred in Palencia, at Ventas de Valdemudo (*Ardeola* 45: 252).

The Levant population seems to be the most vulnerable and is probably only a remnant of a larger population that inhabited all the major east coast wetlands. The Albufera de Valencia was formerly a traditional site for the species: Saunders (1871), Arévalo (1887) and Witherby (1930) all saw it there. Arévalo found it to be abundant generally at Levant wetlands. The population in Catalonia has declined dramatically. Here the Ebro delta population increased during the 1970s until the mid 1980s, when it may have approached 50 pairs (SA1), but it has since declined to fewer than ten pairs (CA2). The current Ebro delta population seems to have been of relatively recent origin since the species was first noted there in September 1961 and there was only one record in the subsequent decade (Maluquer 1971). The small population (five pairs in 1978) of the Aigüamolls de l'Empordà has disappeared. Elsewhere in Catalonia, some birds have bred recently at the Utxesa reservoir where at least two pairs were present in April 2009 (*A. O. Catalunya 2009*: 231). The population in the Comunidad Valenciana was estimated at perhaps 100 pairs at the Albufera, 70 pairs at El Hondo and 25 pairs at Santa Pola by Urios *et al.* (1991) but a more recent assessment (G. López *et al.* 2006 in Tirado 2011) put the population at 81 pairs at El Hondo, 33 at Santa Pola but only five pairs at the Albufera, where the species seems recently to have declined close to extinction. Ringing at constant effort sites at El Hondo detected an annual decline of some 10% during 1992–99 (García Peiró & López Macià 2002).

Birds from northern and central populations especially are vulnerable to icy spells in winter, which may result in high mortality. More generally, the species has suffered from the loss of wetlands, especially in La Mancha and on the east coast, and from loss of habitat quality (LR). It breeds in loose colonies, which makes it especially vulnerable to predation by rats and other predators, although the potential reproductive output is very high: up to four broods per year.

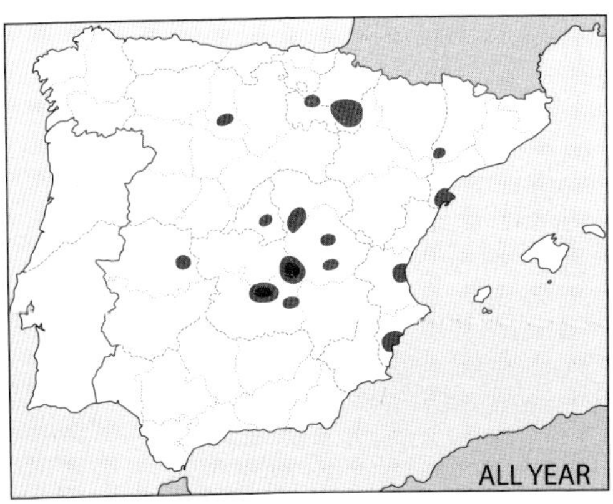

Ringing has established that the Spanish birds are sedentary or at best perform only short-distance dispersal in winter. It has not been recorded in western Andalucía, not even at the wetlands of the Guadalquivir valley, and there are no confirmed records from Portugal (AP). The longest known displacement is 260km, from Zaragoza in September 1964 to Barcelona in April 1965 (*Ardeola* 13: 81). A recovery of an adult bird ringed at Pétrola, Albacete in April and recovered 153km away at Miguel Esteban, Toledo in June of the following year was also quite exceptional (LR). Nevertheless, birds appear occasionally in winter at locations where they do not breed. In Extremadura three were seen near Cáceres on 29 December 1966 ; a flock of 15 was in the Guadiana valley near Villanueva de la Serena on 19 February 1967; three were on the Guadiana itself at Mérida on 6 March 1996; at least one was at the Arrocampo reservoir on 16 October 2002 and one was seen there during the breeding season in 2012, following breeding at that site in 2011 (*Ardeola* 12: 240–241, 59: 450; *A. O. Extremadura 2001–2003*: 284). Up to five individuals, including two juveniles, at the Laguna de Atapuerca, Burgos, in July 2012 may have represented a first breeding record for Burgos (*Ardeola* 59: 450). On the north coast one was ringed at Txingudi, Gipuzkoa, on 21 October 1996 and two were there in February 1997, and a pair were at the Ría de Villaviciosa, Asturias, on 18 March 1979 (*Ardeola* 28: 162, 44: 259).

Some birds from nearby countries are thought to join Spanish birds in winter but only one short-distance ringing recovery suggests this: one in October 1997 at the Aiguamolls de l'Empordá (Girona) was recovered in the breeding season at a coastal colony 62km away in southern France the following spring (*Ecología* 15: 408). There are only three records (five birds) from Morocco (BM) so clearly birds very rarely cross the entire Península.

RED-BILLED LEIOTHRIX *Leiothrix lutea*

(Leiotrix Piquirrojo, Rouxinol do Japão)

Native to mainly montane Asian forests, from the Himalayas east to S China. There are naturalised populations in northeastern Spain, France and Germany descended from escaped cagebirds. Small numbers have also been found recently in Portugal. Subspecies unknown. Category C in Spain, Category E1 in Portugal.

Two small populations of captive origin in Catalonia near Barcelona were first discovered in 1990 (*Ardeola* 46: 161). They favour forests, chiefly of Aleppo Pine, with dense undergrowth near watercourses. In the late 1990s the population was restricted to the Serra de Collserola and was thought to number 50–150 pairs (CA2). However, since 2002 the range has expanded into adjacent areas and the population is reported to be growing exponentially (Llimona *et al.* 2007, Herrando *et al.* 2010). Some 800–1,800 individuals have been recently estimated to be present in winter (CWA). There is an established population near Pau in the Pyrenées Atlantiques, France (Dubois 2007) a short distance from the Basque Country, from where dispersal into Spain may yet occur. The Spanish population is sedentary. Recent observations of small groups from two widely separated areas in Portugal: up to eight birds near Miranda do Corvo, Coimbra, in 2010 and up to seven at Sintra, Lisboa, in 2006–07, suggest breeding in the country (*Anu. Ornitol.* 8: 101).

LONG-TAILED TIT *Aegithalos caudatus*

(Mito Común, Chapim-rabilongo-comum)

***A. c. taiti*. A common resident in northern and northwestern Spain, in Andorra and in Portugal north of the Tagus.**
***A. c. irbii*. A common resident in southern Spain and in Portugal south of the Tagus.**

Breeding population estimates. Spain: SA1 700,000–1.65 million pairs; ACRE 7.2 million birds. **Portugal**: 50,000–500,000 pairs. **Andorra**: 100–200 pairs.

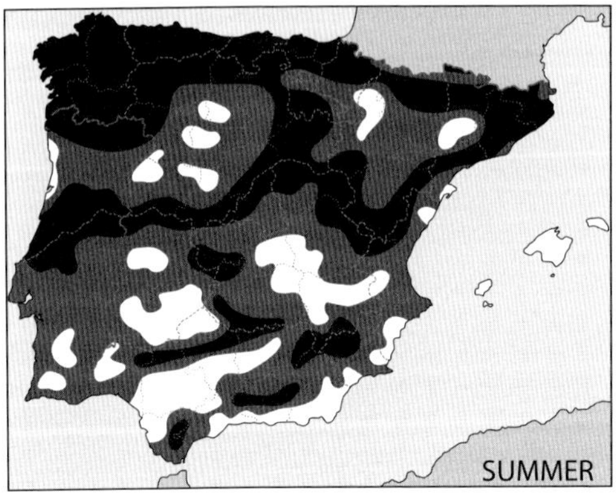

The Long-tailed Tit is widespread throughout Iberia but commonest in the northern third of the Peninsula (SA2). Both Iberian subspecies are smaller and darker than those in central Europe (*europaeus*) and the British Isles (*rosaceus*) and show dark-streaked cheeks and very little pink above. In particular, *irbii* shows almost no pink above and has a narrow dark band across the upper mantle. The two intergrade in central Iberia.

It is absent from arid and agricultural regions although it occurs there locally in riparian woodlands. It inhabits all woodland types but it is commonest in broadleaved formations, especially oak woods that have a well-developed understorey. It is scarce in beechwoods, where the undergrowth tends to be sparse. It also occurs in young pinewoods and pine plantations that have deciduous undergrowth. It is mainly found below 1,000m but it reaches 1,500–1,600m in the Central Range and eastern Pyrenees, and 2,400m in the Sierra Nevada (AI2, Garzón & Henares 2012). The ACRE estimate of 7.2 million birds in Spain is very high, particularly since the French population is put at 1.0–1.5 million pairs (OF) and the entire European population has been estimated at 5–12 million pairs (BE).

The Long-tailed Tit has only recently been recorded in the Balearics, where two were seen on Ibiza in January 1999 and two on Valldemossa, Mallorca in January 2000 (*Ardeola* 46: 161, *A. O. Balears 2000*: 44). Breeding was first confirmed in western Mallorca at Calvià in 2003 and at Banyalbufar and at Puigpunyent in 2006 (*A. O. Balears 2003*: 102, *2006*: 201). It was reported annually from western Mallorca during 2007–12 with breeding confirmed at least in 2009 and 2010. Most observations have been of flocks of up to five birds, but in 2006 there were up to 30 at Valldurgent, Calvià, and 22 at Puigpunyent. The Long-tailed Tit seems to have established itself in western Mallorca and may in time occur more widely in the archipelago – but equally it may soon disappear. The subspecies involved is not known.

Both subspecies are sedentary but family parties may combine to form small flocks that often wander in winter and may then occur outside the breeding range. A flock of ten that appeared in Gibraltar in November 2004 was only the second record there since the first in 1934 (*GBR 2004*: 37), although the species is common in the hinterland. The winter range in Spain is broadly similar to that during the breeding season, although some occupation of the eastern and southeastern littorals, interior Extremadura and the Guadalquivir valley has been detected (SWA). There is no evidence of migration and only one record from Morocco (BM). The farthest known displacement within the Peninsula was 176km, from Salamanca to León (*Ecología* 17: 278).

MARSH TIT — *Poecile palustris*

(Carbonero Palustre, Chapim-palustre)

P. p. palustris. **A moderately common resident in northern Spain.**

Breeding population estimates. Spain: SA1 82,000–96,000 pairs.

The Marsh Tit is confined to the cool montane regions of northern Spain, occurring across most of the Cantabrian Range and the Pyrenees east to Catalonia. An outpost population a little further south in the northern Sistema Ibérico, in La Demanda and Cebollera mountains, was first detected in 1969 (*Ardeola* 16: 272–273). It occurs down to sea level on the northern side of the Cantabrian Range but only down to 1,000m on the southern side (SA1, SA2). In the northern Iberian Range it is found above 1,300m on the southern slopes but occurs down to 800m on the northern ones (de Juana 1980). Nonetheless, it generally does not range above 1,400m. The Marsh Tit is restricted to cool, humid, deciduous woodlands and woodland edge habitats offering a dense shrub layer. It is particularly associated with beechwoods and sessile oak woods but also occurs in mixed woodland, open Scots Pine woodland and riparian growth (AI2, SA2).

Within Spain, the population of the Basque Country has been put at some 12,000 pairs (A. Galarza in SA2) and that of Catalonia at 3,000–18,000 pairs (CA2). Local densities range from 0.19–4 birds/10ha (SA2). The Marsh Tit seems to have withdrawn from the western extreme of the Cantabrian Range in Galicia, perhaps in response to habitat loss in which pine and eucalyptus plantations have replaced native broadleaves (SA1). There are no accepted records for Portugal

The birds are sedentary and largely confine themselves to their home ranges but short-distance winter dispersal has been recorded, due to birds descending from higher ground to winter in woodlands in the foothills. They then appear in a wider range of forest types, including Pyrenean Oak woodlands (Álvarez & Purroy 1993).

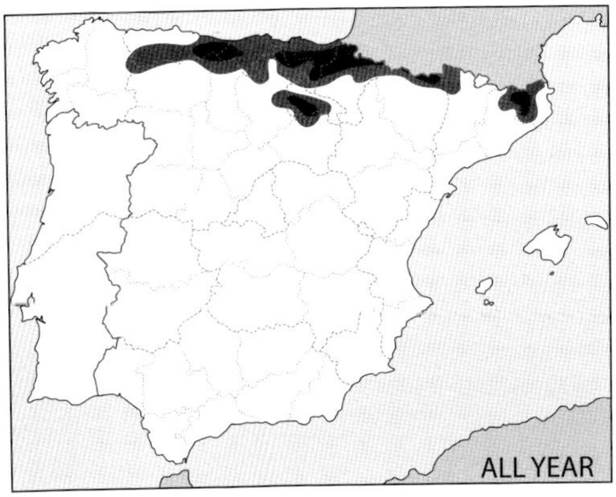

CRESTED TIT *Lophophanes cristatus*

(Herrerillo Capuchino, Chapim-de-poupa)

L. c. mitratus. **A common resident throughout northern and central mainland Iberia south to the Sierra de Guadarrama and the Comunidad Valenciana. Intergrades with *weigoldi* occur in southern Galicia, the Sierra de Gredos in the Central Range and Murcia in the south-east.**
L. c. weigoldi. **A locally common resident in Portugal and southwestern Spain. Smaller than *mitratus*, greyer above and paler below.**

Breeding population estimates. Spain: SA1 860,000–1.5 million pairs; ACRE 2.1 million birds. **Portugal**: 50,000–500,000 pairs. **Andorra**: 1,000–3,000 pairs.

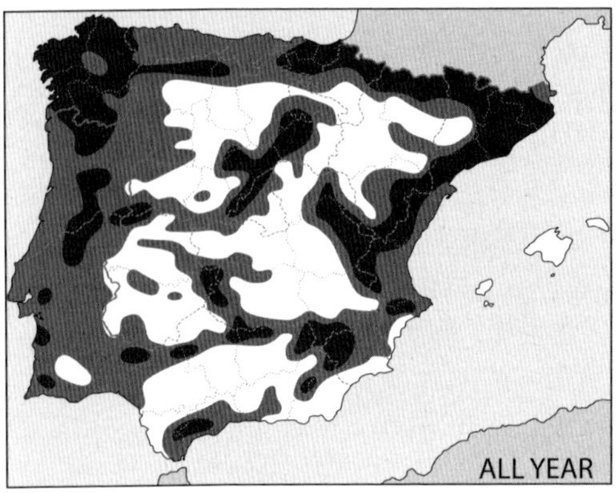

The Crested Tit is a familiar species across mainland Iberia in forests and woodlands of a diversity of native conifers, from Aleppo and Stone Pines on the coasts, to Scots Pines and firs in the Pyrenees. It also uses Monterey Pine plantations locally, as in the Basque Country. It may also be found in a diversity of other forest types, including Pyrenean Oak woods on the southern flanks of the Sierra de Gredos and Spanish Junipers in the Iberian Range: where it is quite common (AI2). The population of *weigoldi* in the Cádiz sierras, notably within Los Alcornocales, inhabits Cork Oak forest. Parks and gardens are also occupied on the outskirts of Madrid (pers. obs.). Birds may be found from sea level to over 1,500m in the mountains (SA1), although as high as 1,700m in the Sierra Nevada (Garzón & Henares 2012). Those in Catalonia are most abundant at altitudes of 1,200–2,000m (CA2). There are no confirmed records from the Balearic Islands (*A. O. Balears 2010*: 259–270). The Crested Tit was reportedly rare in Portugal in the mid 19th century and is presumed to have greatly benefited from the spread of pine plantations in the country since then (AP), as may also have been the case in Spain.

Large gaps in the range correspond with the major agricultural areas of the Peninsula and the arid southeast coastlands (SA2, PA2). It is most abundant in extensive pine forests in Galicia, the Basque Country, Catalonia, parts of the northern Meseta, the southern flanks of the Central Range, the uplands of the Levant region and Murcia and the eastern Sierra Morena, as well as in the corkwoods of the Betic Range and of the Sierra de Aracena in the western Sierra Morena (SA2).

The highest reported breeding densities are from Holm Oak woods in the Pyrenees (26 birds/km^2) and elsewhere from pinewoods: ranging from 25 birds/km^2 in the supra-mediterranean to 12 birds/km^2 in the thermomediterranean. Much higher densities have been reported locally in winter, however, up to 100–130 birds/km^2 (SWA). Some range expansion has been apparent in Catalonia, where a comparison between the first and second breeding atlases (CA1: 1983, CA2: 2004) shows a significant 8% increase in the number of occupied 10km-squares. The population is strongly resident. Single birds have been known to wander to Gibraltar on only three occasions: in December 1979, October 1990 and June 2004 (GONHS), although Crested Tits breed commonly in the Cádiz sierras nearby. There are only two old records for Morocco (BM).

COAL TIT *Periparus ater*

(Carbonero Garrapinos, Chapim-carvoeiro)

P. a. vieirae. **A locally common resident throughout mainland Iberia. Olive-brown above, less grey above than the nominate, with buffish brown flanks.**
P. a. ater. **Intergrades with *vieirae* in the Pyrenees. Some, probably few, reach Iberia in winter or during irruptive movements.**

Breeding population estimates. Spain: SA1 1.66–5.20 million pairs; ACRE 6.3 million birds. **Portugal**: 50,000–500,000 pairs. **Andorra**: 10,000–20,000 pairs.

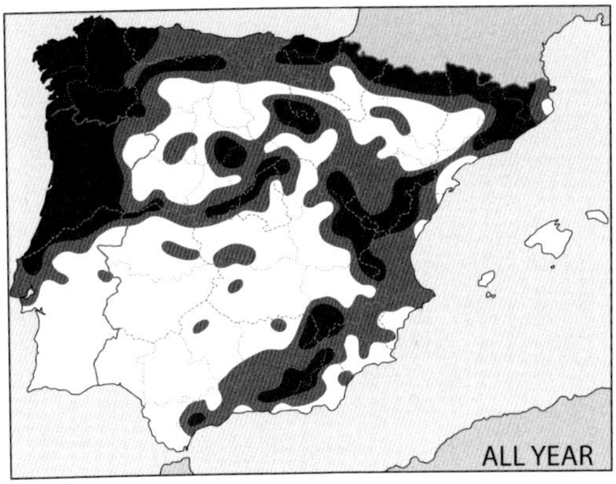

The Coal Tit is even more a bird of coniferous forests, woodlands and plantations than the Crested Tit and the two have very similar Iberian distributions. The Coal Tit is present in conifers in all the major mountain ranges. It spread elsewhere by taking advantage of conifer plantations, especially in the northern Meseta, and it is a major beneficiary of nestbox provision in such woodlands (Potti 1985, Polo 2010). It is usually reliant on pine seeds in winter, typically storing them in caches. However, it also frequents oak, mixed oak/conifer and oak/beech forests in the Cantabrian Range, as well as Pyrenean beech forests (AI2, SA2). It even occurs in eucalyptus plantations in the north-west. It may be encountered in parks and gardens in many areas. Within its range it occurs from sea level to the treeline, reaching 2,500m in the Sierra Nevada (Garzón & Henares 2012). It is most abundant at altitudes of 1,500–1,800m both in Madrid and Catalonia (CA1, Díaz *et al.* 1994).

It is abundant from Galicia across the Cantabrian range and throughout the Pyrenees and the Iberian and Central Ranges. It is also common in the Levant mountains and the Betic Range. In the latter it extends west to about Ronda, Málaga, and it is virtually absent from Cádiz province. It is also scarce in the Sierra Morena but there are scattered outpost populations there in Córdoba and Jaén. There are also isolated populations in eastern Extremadura in Las Villuercas, and in the Montes de Toledo. The Coal Tit is widespread in Portugal north of the Tagus but is almost entirely absent south of the river except at a few sites nearby (PA2).

The Catalan population alone has been put at some 225,000–445,000 pairs (CA2). Overall the population showed a significant 31% increase on SACRE plots during 1998–2012 (Escandell 2013). The highest densities, up to 19 birds/10ha, are in mature Scots Pine and Black Pine forests in the Pyrenees and Sierra de Gredos. High densities have also been reported from other Pyrenean and Cantabrian conifer woods (34–120 birds/km^2) and in Cantabrian beechwoods (38 birds/km^2).

Range expansion has been reported in several regions, including Castilla y León (Sanz-Zuasti & Velasco 1999) and Catalonia (Ferrer *et al.* 1986, CA2). The early literature shows that in Spain the Coal Tit was strictly limited to a few montane areas during the late 19th century and first half of the 20th century. In central Spain, for example, Lilford (1866) only saw it once in the Sierra de Guadarrama, at San Ildefonso, and Witherby (1928) reported it just from the Guadarrama and Gredos ranges, whereas it is now very common even in the parks and gardens of Madrid. Jourdain (1936–1937) said that it was a local resident in Andalucía, confined to the highest sierras, although it was reported to nest in the Gibraltar Botanic Gardens, where there are numerous pines, in the 19th and early 20th centuries (Watkins 1857, Jones 1900, Rait-Kerr 1934a): there are only five recent Gibraltar records, all post-1981. In the north, Irby (1883) reported only one observation in Cantabria, by Lord Lilford near Santander in November 1876. Much later Bernis (1946) found it still very scarce in the mountains of León.

The Portuguese population showed no clear trend during 2004–09 (ACP). It was formerly considered to be rare in Portugal but has benefited from pine plantations established since the mid 19th century (AP). It is one of the commonest birds in Andorra: Eagle Clarke (1889) found it abundant there, in the mountain pine forests. The Coal Tit is only a rare vagrant to the Balearic Islands, from where there are five records (birds), all since 1982 and between late September and mid January.

There are a few recoveries from north-eastern Spain of Coal Tits ringed in Germany (6), Belgium (1) and Switzerland (1) (Zink 1981, SMA). However, the longest known displacement within the Peninsula is 195km, from Teruel to Soria (AI2). The winter and breeding distributions are essentially identical, although the Spanish winter atlas fieldwork found it in some areas where it does not appear to breed, mainly in the south-west: in Huelva, Badajoz and Ciudad Real (SWA).

BLUE TIT *Cyanistes caeruleus*

(Herrerillo Común, Chapim-azul)

C. c. caeruleus. **A common resident in northern Spain. Influxes from northern and central Europe occur in some years.**

C. c. ogliastrae. **A common resident in Portugal and the southern two-thirds of Spain. Smaller than the nominate, brighter blue above and brighter yellow below and shows reduced sexual dimorphism (Harrap & Quinn 1996).**

C. c. balearicus. **A local resident on Mallorca. Similar in size to the nominate. Slightly paler and greyer on the mantle and paler yellow below, with more extensive white on the breast and belly (Harrap & Quinn 1996).**

Breeding population estimates. Spain: SA1 930,000–3.6 million pairs; ACRE 13.0 million birds. **Portugal**: 500,000–2.5 million pairs. **Andorra**: 200–300 pairs. **Gibraltar**: 30 pairs.

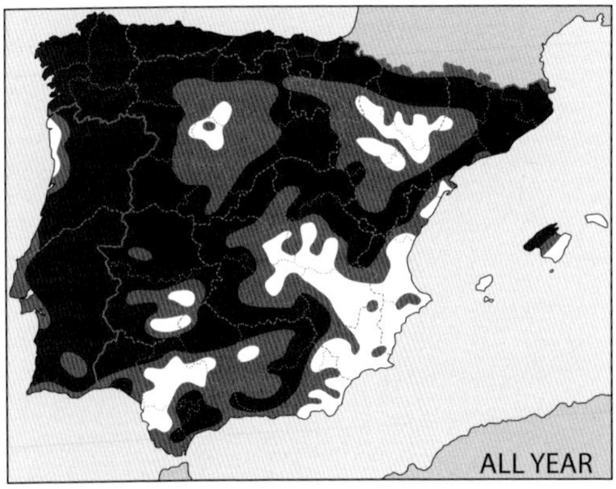

The Blue Tit may be encountered almost anywhere in Iberia where there are trees. It occurs in mixed woodlands and deciduous woods and forests of all kinds, as well as in riparian woodlands and tall hedgerows, but tends to avoid coniferous forests on the Peninsula. It largely depends on tree holes for nesting but nestboxes allow it to occupy relatively young woodlands and plantations: studies of such populations have allowed the breeding biology and feeding ecology of Iberian Blue Tits to be well known (Salvador 2005). Most occur below 1,500m but they range to 1,650m in the Pyrenees: the upper limit of beech/fir woodland, and up to 2,000m on the north side of the Sierra Nevada: the upper limit of Pyrenean Oak (Purroy 1977, Pleguezuelos 1992, AI2). Pine plantations in the Sierra Nevada have more recently allowed Blue Tits to occur as high as 2,400m there (Garzón & Henares 2012). This is a familiar species in parks and gardens, where it readily accepts birdfeeders and nestboxes (SA2, PA2).

Those in Mallorca occur largely in the north-west and may be encountered in coniferous, broadleaved and mixed woodland habitats. However they are commonest in pure conifer stands (Tato Cumming 1971), in complete contrast to the situation on the mainland (Tellería & Santos 1994). Occurrences on the other Balearic islands are very rare. The very few published observations include a first record from Ibiza in March 1974 and one ringed there in April 2001 (*Ardeola* 50: 166).

Within Spain the population of Catalonia alone has been estimated at 466,000–657,000 pairs (CA2). The population showed a significant 50% overall increase on SACRE sites during 1998–2012 (Escandell 2013). The highest densities in Spain, 75–115 birds/km^2, are in montane oak forests in the supramediterranean and Pyrenees and also in the holm oak and cork woods in the thermo- and mesomediterranean. The Portuguese population showed a moderate increase during 2004–09 (ACP). The Blue Tit is common in the deciduous woodlands at lower elevations in Andorra, and in Gibraltar.

Iberian Blue Tits are resident but altitudinal movements occur in montane areas. Although their general distribution barely alters in winter, their numbers may then decrease greatly in some mountain forests and show a corresponding increase in, for example, cultivated areas and thermomediterranean woodlands (AI2, Gainzarain 2006, SWA). Some individuals from northern European populations cross the Pyrenees in irruption years (BWP) and there are 26 Iberian recoveries of birds ringed in France (10), Germany (7), Belgium, Switzerland and as far away as Lithuania (2; 2,082km) and Russia (2,200km) (SMA). The most notable displacements of ringed birds within Iberia include from La Puebla de Montalbán, Toledo, to La Puebla del Río, Sevilla (325km; *Ecología* 2: 402) and from Valsaín, Segovia, to Gaibiel, Castellón (316km; *Ecología* 21: 292). There are no accepted Iberian records of the African Blue Tit *C. teneriffae*: the origin of one trapped in a Barcelona park on 19 February 2013 (RBS) may come to be seen as suspect but an occasional genuine occurrence would not be surprising since it breeds as near as the southern shore of the Strait.

GREAT TIT *Parus major*

(Carbonero Común, Chapim-real)

P. m. major. **A common resident in northernmost Portugal and the northern half of Spain. Intergrades with *corsus* in central Spain.**

P. m. corsus. **A common resident in most of Portugal and the southern half of Spain. Smaller, more greenish-olive above and paler yellow below than the nominate form, with grey-washed flanks.**

P. m. mallorcae. **A common resident in the Balearic islands. Smaller than the nominate form and with reduced yellow below. More grey-blue above than *corsus*.**

Breeding population estimates. Spain: SA1 1.6–4.3 million pairs; ACRE 13.2 million birds. **Portugal**: 0.5–2.5 million pairs. **Andorra**: 500–1,000 pairs. **Gibraltar** <5 pairs.

The Great Tit is almost universally distributed in nearly all types of woodland, including orchards of all types and riparian formations, throughout the Peninsula and the Balearic islands, being absent only from totally treeless regions. It avoids pure conifer stands, however, except in the Balearics where it positively selects pine forests, although it also occurs there in a wide range of habitats (Adrover *et al.* 2010). In common with the Blue Tit, it is a familiar bird in parks and gardens of any size,

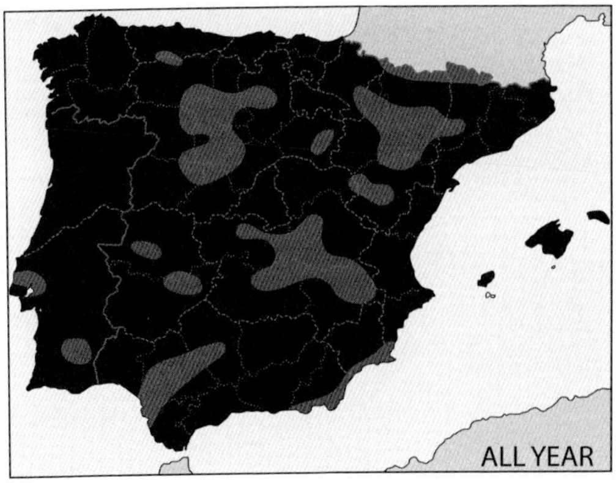

where it readily accepts food from bird feeders and also nests in nestboxes. It ranges from sea level to 1,600m in the Pyrenees and up to 2,500m in the Sierra Nevada (AI2, SA2, SWA, Garzón & Henares 2012). The Balearic race is present on all the main islands and islets. It is much commoner and more widespread than the Blue Tit in Mallorca.

Within Spain, the Catalan population has been estimated at 0.75–1 million pairs (CA2), with some 1.3–1.7 million individuals present in winter (CWA). The Spanish population showed a significant increase on SACRE plots during 1998–2012 (Escandell 2013). The highest recorded densities are in pinewoods (47–72 birds/km^2) and oakwoods (53–56 birds/km^2) across the full range of bioclimatic regions. The Portuguese population showed no clear trend during 2004–09 (ACP). It is commoner than the Blue Tit in Andorra (ANA, BE) but much scarcer than that species in Gibraltar.

Most Iberian Great Tits are highly sedentary: only five of 109 individuals ringed in Iberia and recovered elsewhere in the region had travelled further than 10km (AI2). In Portugal, a bird ringed in Mindelo, Porto, and recovered two years later 430km away in Évora (AP) was probably highly exceptional. However, there are also a few Iberian recoveries of birds ringed in Belgium (5), Switzerland (4), France (2), Poland (2), Germany, Lithuania and Russia, confirming the at least occasional occurrence in Iberia of birds of northern European origin (SMA, AP).

EURASIAN NUTHATCH *Sitta europaea*

(Trepador Azul, Trepadeira-azul)

***S. e. caesia*. A locally common resident in the Cantabrian Range and the Pyrenees.**

***S. e. hispaniensis*. A locally common resident in central and southern Spain and in Portugal. Smaller and more slender-billed than *caesia*; underparts paler buff.**

Breeding population estimates. Spain: SA1 550,000–1.2 million pairs; ACRE 1.1 million birds. **Portugal**: 10,000–100,000 pairs. **Andorra**: 100–300 pairs.

The Nuthatch is associated with mature forests in humid areas, particularly in mountainous areas and those with some Atlantic influence. It is quite common in northern Spain, except for the littoral fringe, but becomes progressively more local further south. However, it is present and often abundant in the mountains of the Iberian and Central Ranges and the Montes de Toledo, as well as in western Huelva and eastwards from there in the Sierra Morena. It is also common in the western part of the Betic Range, in the sierras of Cádiz and Málaga. It occurs from near sea level where the habitat is suitable (SA2) up to the treeline, reaching 2,200m in the eastern Pyrenees (CA2). However, in the northern Iberian Range it is seldom found below 1,000m on southern slopes or below 800m on northern ones (de Juana 1980).

It may have become more widespread in Andalucía during the 20th century since Saunders (1871) and Chapman & Buck (1893) knew it only from the Sierra Nevada (although Garzón & Henares (2012) regard it as recently arrived there) and Irby (1895) never recorded it in Andalucía. It may also have been much rarer in the past in the north: in Cantabria, Irby (1883) only recorded it in some oak and beech forests in Liébana; in Navarra, Ticehurst & Whistler (1925) detected it only at Lecumberri and Burguete in the Pyrenees, and in Galicia, Ticehurst & Whistler (1928) only met with a pair near Santiago de Compostela. In central Spain, Lilford (1865) only found it in the pinewoods at San Ildefonso, in the Guadarrama range, and Witherby (1928) also encountered it in the Guadarrama, Gredos and Guadalupe mountains. Recent breeding range expansion has been detected in Catalonia, where a comparison between the two published

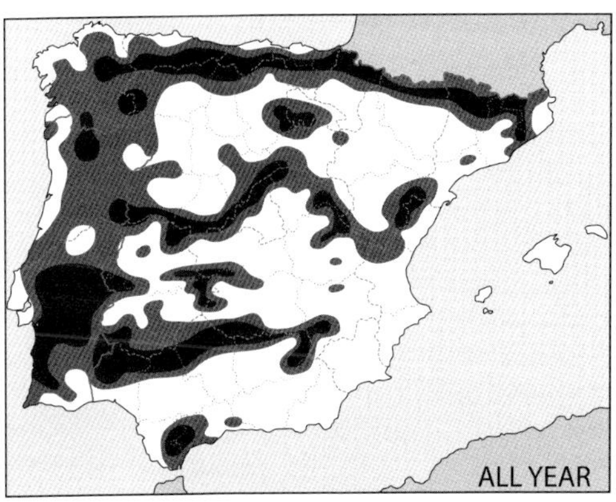

atlases, from the early 1980s and 1999–2002 respectively, shows a 13% increase in the number of occupied 10km-squares (CA2). The population on SACRE plots showed a significant increase in 1998–2012 (Escandell 2013).

The Portuguese range is also discontinuous. There the Nuthatch is most widespread and abundant south of the Tagus: in Ribatejo, Alto Alentejo and the Serra de Grândola, Setúbal. It is scarce or absent from coastal regions north of Lisbon, the Alto Douro and eastern Algarve (PA2; AP). Tait (1924) found it common in some places around Porto, but very local.

The Nuthatch requires mature trees that offer plenty of dead wood in which it can excavate its nest holes. It prefers oakwoods and beechwoods but occurs commonly in mature pine forests. It also occurs in dehesas of Cork Oaks and Iberian Holm Oaks, in Andalucía and Extremadura. Pine plantations are often unsuitable, since the trees are felled before they are old enough to offer nest sites, and eucalyptus plantations are avoided (AI2, SA2, CA2).

High densities of 4–7 birds/10ha have been reported in the Cantabrian Range, with mean maximum densities of 9.46 birds/10ha, but 2 birds/10ha would be more typical further south (SA2). Habitat-specific censuses show highest densities in supramediterranean pinewoods (22 birds/km^2) and Pyrenean beechwoods (20 birds/km^2). The Catalan breeding population has been put at just 6,400–15,000 pairs (CA2) but the winter population there has been estimated at 105,000–208,000 individuals (CWA): suggesting that many move to lower levels in the north in winter.

Both subspecies appear to be highly sedentary and there are no indications of other than local, short-distance movements, 38km being the farthest known (*Ecología* 22: 221). The Nuthatch is absent from the Balearic Islands and has never been recorded in Gibraltar.

WALLCREEPER *Tichodroma muraria*

(Treparriscos, Trepa-fragas)

***T. m. muraria*. A scarce resident in the central Cantabrian Range and the Pyrenees, more widespread in winter. Vagrant to the Balearic Islands and Portugal.**

Breeding population estimates. Spain: SA1 9,000–12,000 pairs. **Andorra**: <10 pairs.

The Iberian breeding population is confined to the middle sector of the Cantabrian Range and to the Pyrenees. Limestone areas are favoured when breeding. Nesting has been reported at a mean altitude of 1,680m in the Cantabrian mountains and 1,930m in the Pyrenees (SA1) although a pair nested at 700m at Teverga, Asturias, in 2011 (*Ardeola* 58: 512). Breeding birds have been reported in the Catalan Pyrenees up to 2,900m (CA2). The Cantabrian population is centred on the Picos de Europa, especially the central massif, and on adjacent summits in Asturias and in northeastern León and northwestern Palencia (SA1, SA2).

The Pyrenean population is most abundant in the central sector, in Huesca, and to a lesser extent further east, in Lleida and Girona (Hernández *et al.* 1992, Sampietro *et al.* 1998). It barely extends into Navarra in the west, in the Valle de Roncal and perhaps the Foz de Lumbier. The core populations in Huesca are around the Ordesa y Monte Perdido National Park, including the Añisclo valley, and, further east, in the Aigües Tortes National Park, Arán and Benasque valleys. The principal location in Girona is Cadí-Ribes (SA1, SA2). The Wallcreeper is present in Andorra but is uncommon there (ANA).

The frequency of actual observations suggests that the SA1 estimate may have been very optimistic. The Catalan population has been put at only 40–100 pairs (CA2) and there are thought to be fewer than 600 pairs in Castilla y León (Sanz-Zuasti & Velasco 1999).

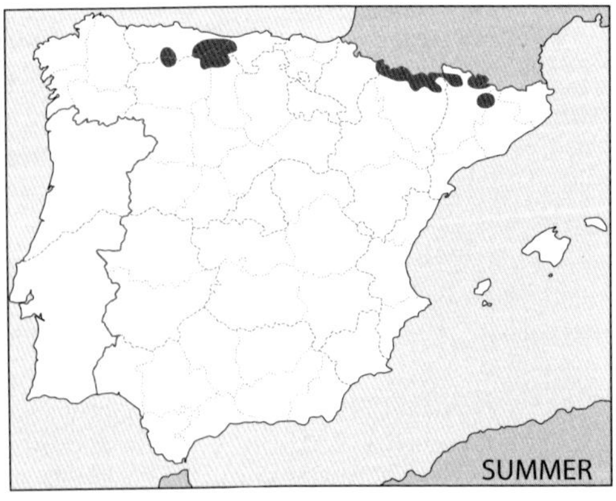

Wallcreepers may ascend to the highest levels of their breeding ranges after the nesting season, reaching up to 3,375m in the Pyrenees, apparently to exploit mountain-top habitats that are then snow-free (Hernández *et al.* 1993). The higher levels seem to be abandoned in winter when birds descend lower. There were no records above 1,800m in the Pyrenees during the fieldwork for the recent Catalan winter atlas. Most winter records are from the Pyrenees and Cantabrian Range, or from neighbouring ranges (Bernis & Maluquer 1955, Hernández *et al.* 1992, SWA), but some disperse more widely throughout Iberia. Records from eastern Spain suggest that the main movement away from the breeding sites is during late October, with birds returning from winter quarters in early April (Fig. 199). The extreme dates for the records in the Figure were 7 September and 1 May.

Figure 199. Temporal distribution of observations outside the breeding range in Catalonia (1996–2008), the Comunidad Valenciana (1986–2009), Murcia (2005–10) and the Balearic Islands (1985–2003). Sources: regional bird reports (n = 131 records: Catalonia 94, Valencia 27, Murcia 6, Balearics 4).

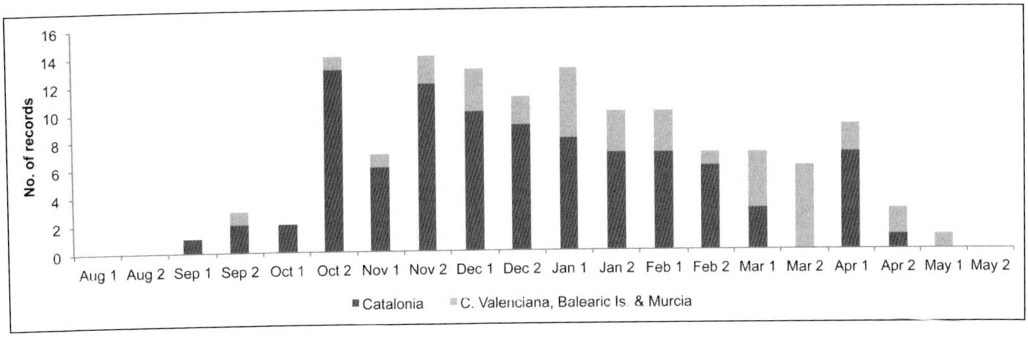

Winter records are concentrated in the north and, especially, in the northeastern quadrant of the Peninsula. Within the Mediterranean littoral ranges the Wallcreeper is relatively frequent in winter in Catalonia, but rarer in the Comunidad Valenciana, Murcia and Andalucía (SWA). One was found dead on the Columbretes islands in autumn 1973 (*A. O. C. Valenciana 1999*). There are three records of single birds from Gibraltar: on 22 November 1966, 15 November 1979 and 1 December 1988, and Irby (1875) alludes to sightings there in the 19th century. Nevertheless, there are only two records for Morocco (BM) so any movement across the Strait must be on a very small scale indeed. Wallcreepers are still attracted to limestone areas in winter but they also occur then on non-calcareous sierras, often in river gorges (Hernández *et al.* 1993). Some are occasionally seen on buildings, usually large ones such as castles, churches and dams. It is unsurprising that only 12 have ever been ringed in Iberia but of these an adult male ringed at Cordiñanes de Valdeón, in the Picos de Europa, in July 2004 was found dead at Cañamero, Cáceres (424km) in January 2005 (*Ecología* 21: 292).

The Wallcreeper is regarded as an accidental visitor to Portugal, with just 11 records to 2010, chiefly of single birds reported north of the Tagus: at least three more were reported during 2011–12 (RBP). It is also a rare winter visitor to the Balearic Islands (Ferrer *et al.* 1986). The regional bird reports give just four records: in Mallorca in December 1985; in Menorca in March 1990 and October 1994; and in Cabrera in October 2003 (*A. O. Balears 1987, 1991,1994*: 101, *2003*: 102).

EURASIAN TREECREEPER *Certhia familiaris*

(Agateador Euroasiático, Trepadeira do Norte)

***C. f. macrodactyla*. Resident in the mountains of northern Spain. Darker than the nominate race and more buffish or rufous below.**

Breeding population estimates. Spain: SA1 32,000–38,000 pairs. **Andorra**: 50–100 pairs.

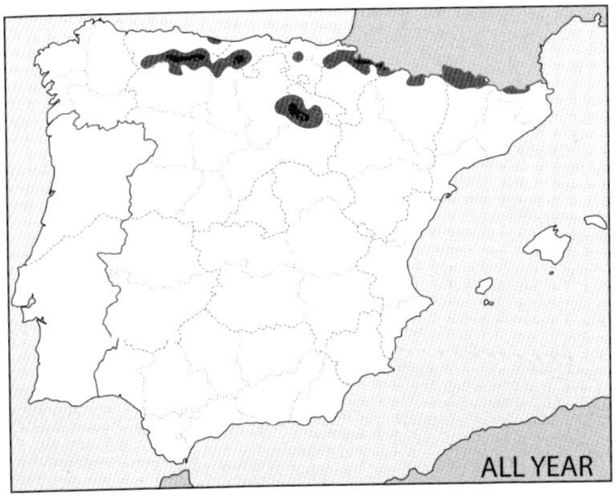

This is a montane species in the region, where it is restricted to the central Cantabrian Range: in Asturias, Cantabria and Castilla y León, and the Pyrenees, with an outpost population in the north of the Iberian Range: in the Sierras de La Demanda and Cebollera, and a presence in the Basque mountains: around the headwaters of the Nervión river. It may be encountered at 730–2,000m in the Pyrenees, 1,000–1,700m in the Iberian Range and 1,000–1,800m in the Cantabrian mountains, generally preferring shady aspects (Purroy 1973b, de Juana 1980, SA1, SA2, CA2).

It inhabits cool, humid, mature closed-canopy forests of beech, fir, birch and sessile oak: montane pine forests are mostly occupied by the Short-toed Treecreper. It is quite sedentary but some may descend to winter at lower altitudes (SA1, Gainzarain 2006, SWA). Maximum densities, 3 birds/10ha, occur in mixed beech/fir forests in the Pyrenees,. There are no clear population trends although some populations may have been lost where mature forests have been felled (Román *et al.* 1999). No long-distance movements of Iberian birds have been reported and the species is unknown in Portugal.

SHORT-TOED TREECREEPER *Certhia brachydactyla*

(Agateador Europeo, Trepadeira do Sul)

***C. b. megarhynchos*. A widespread resident in northern Portugal and northern Spain, including the western Pyrenees. Slightly paler and more rufous than the nominate race.**
***C. b. brachydactyla*. A widespread resident elsewhere in mainland Iberia, intergrading with *megarhynchos* in central Spain and southern Portugal. The two are poorly differentiated and inseparable in the field (BWP).**

Breeding population estimates. Spain: SA1 1.0–3.3 million pairs; ACRE 4.9 million birds. **Portugal:** 50,000–500,000 pairs. **Andorra:** 300–400 pairs.

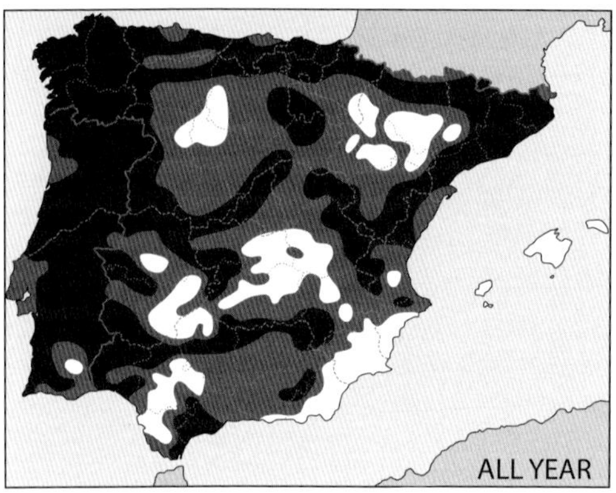

This is the common treecreeper of Iberia, a generalist that may be found anywhere in mainland Spain or Portugal where there are trees of at least 10cm trunk diameter, although it may inhabit younger plantations if nestboxes are available (SA1). It is always most abundant where there are plenty of mature trees and may thus be found in all types of oak and coniferous woodlands and in parks and gardens with large trees. It does, however, prefer cooler, wetter areas to very dry ones and is much more abundant in dense woodlands than in very open ones, such as many of the Iberian Holm Oak dehesas. It is only absent from the arid south-east and from most of the treeless croplands, although some occur even there in riparian woodlands (SA1, SA2). It may be a vagrant to Mallorca (*A. O. Balears 2010*: 268) although there are no recent accepted records.

It ranges from sea level to 1,600m in the Pyrenees, with some found higher, to 2,200m (CA2). It also occurs at higher altitudes than the Eurasian Treecreeper in Mountain Pine and Scots Pine woodlands in the Pyrenees and Iberian Range. It may also be encountered up to 1,900m in Scots Pines in the Sierra de Guadarrama and to over 2,000m in the Sierra Nevada (SA2). The two species appear to coexist widely without apparent competition in Andorra, where although the Short-toed Treecreeper favours oakwoods, Scots Pines and Mountain Pines, the two coincide in mixed pine and Silver Fir woods, where the Short-toed Treecreeper ranges regularly to 2,000–2,100m and sometimes to 2,300m (ANA). However, across Iberia generally, the Short-toed Treecreeper is generally commonest at altitudes of 500–1,500m (Carrascal 2009).

The population on SACRE plots showed a significant increase during 1998–2012 (Escandell 2013). The highest densities are in supramediterranean and eurosiberian pinewoods (30–65 birds/km^2). Densities also tend to be higher in the cooler, wetter north than in the warmer, drier south of Iberia (Carrascal 2009), which may reflect the greater abundance of large trees in well-watered regions. It is the commoner of the two treecreeper species in Andorra.

This species is largely sedentary (AI2, SWA) but some disperse in winter and may then appear in areas where they do not nest, as at Gibraltar. The great majority of over 1,300 recoveries of Spanish-ringed birds have been local and the furthest recorded displacement was 23km (AI2).

EURASIAN PENDULINE-TIT *Remiz pendulinus*

(Pajaro-moscón Europeo, Chapim-de-mascarilha)

***R. p. pendulinus*. A local resident in Spain. Has bred occasionally in Portugal. Frequent on passage and in winter throughout mainland Iberia. A scarce winter visitor to the Balearic Islands.**

Breeding population estimates. Spain: SA1 12,400–14,600 pairs. **Portugal**: rare.

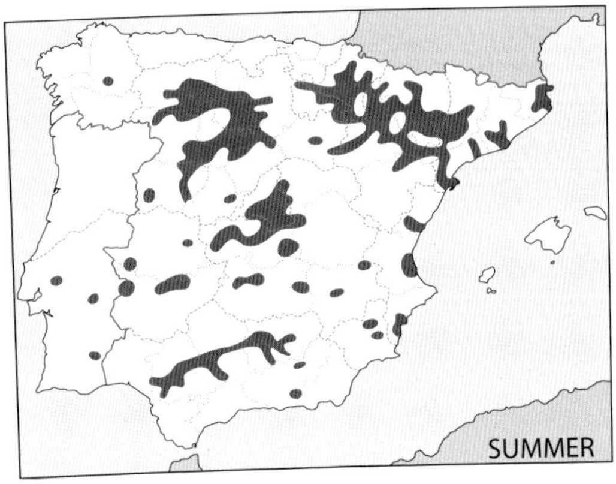

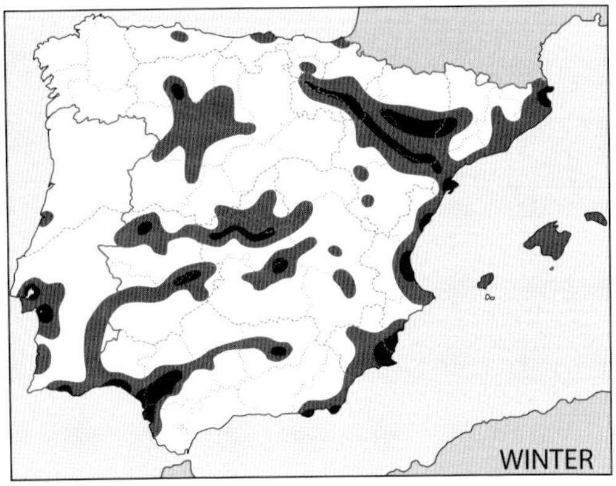

The Penduline Tit may be encountered in riparian vegetation along the major river systems of Iberia and also at scattered wetland sites, especially on the east coast. Virtually all occur within the Mediterranean climatic zone and the species is absent from the Cantabrian coastlands during the breeding season. The chief populations are in the Ebro, Tagus and Duero basins, and there are also scattered populations along the east coast rivers: including the Fluvià, Ter, Llobregat, Gaià, Palancia, Júcar, Albaida and Segura (SA2). Breeding has also occurred at El Hondo and the Albufera de Valencia in some years (*A. O. C. Valenciana 2009*: 137). There are also populations of relatively recent origin in the Guadiana and Guadalquivir basins. It does not breed in the Balearics. Only occasional nesting is known from Portugal, where the Penduline Tit was first recorded as recently as October 1976 (AP). The Portugese breeding records include single pairs reported in the south: from Setil, Cartaxo, in 1982; on the Rio Degebe, Évora, in 1985; near Lagos in 1993; in Elvas in 1994; near Cristãos Novos, Évora, in 1994, and near Mértola in 2003 (Elias & Ferreira 1994, PA2).

Breeding Penduline Tits favour riparian trees, where they chiefly nest in poplars: especially the White Poplar, and willows, and occasionally in elms and ash. They may also be encountered in lakes and marshes where poplar or willow scrub borders or is mingled with reedbeds and reedmace. Nesting sites are within the lower and middle reaches of the river systems and always below 800m (SA1).

The Spanish breeding population has been increasing steadily since the late 20th century. Saunders (1871) and Arévalo (1887) knew it only from the Levant coastal wetlands but it was found breeding along the Ebro river from Zaragoza to Castejón, Navarra, during the early 20th century (Navas 1918–20, Boxberger 1921, Ticehurst & Whistler 1925). In the 1970s the Penduline Tit was still almost entirely restricted to the east coast and the Ebro valley: from at least Lodosa, Navarra, to Flix, Tarragona (Aragüés 1964), although there were sporadic instances of breeding in Castilla y León near Valladolid and León previous to 1950 and near Santa María del Campo, Burgos, around 1960, as well as nesting in the Guadalquivir estuary near Coria del Río during the 1960s (Delibes *et al.* 1980). It subsequently spread throughout the Ebro basin and then those of the Tagus and Duero. It established itself in the Guadiana basin in the 1970s (*Ardeola* 20: 382) and more widely in the Guadalquivir basin during the 1980s, and first nested in Cádiz province in 1990 (Valera *et al.* 1990, Harrap & Quinn 1996, SA1). Further range expansion is evident when the first and the second Spanish atlases, respectively from 1975–95 and 1998–2002, are compared, especially in the Guadiana and Guadalquivir valleys. The latter atlas also recorded breeding in Galicia, in southern Lugo. The Spanish population is estimated to have increased by 50% during 1970–90 (SA2). Local increases have also been noteworthy: for example, the Madrid population was put at 500 pairs by Díaz *et al.* (1994) but the first breeding record there was only in 1971 (*Ardeola* 16: 273–274). The overall increase in breeding numbers has not been uniform and recent indications are that more recently it has halted, or even reversed, in eastern Spain. Thus the population of Catalonia showed a 13% annual decrease during 2002–09 (*A. O. Catalunya 2009*: 236), and the last known instance of nesting at the Albufera de Valencia was in 1993 (*A. O. C. Valenciana 2009*: 137).

The Penduline Tit is much more widespread on passage and, particularly, in winter (SWA). Iberia is now one of the principal wintering areas for western European populations: the numbers wintering in Catalonia alone have been estimated at 25,000–66,000 individuals (CWA). It may then be found at most of the principal wetlands throughout southern and eastern Iberia, as well as in its breeding areas. Most immigrants settle in the Tagus and Ebro valleys and in the east coast wetlands, but some occur south to the Guadalquivir. Regular wintering in Portugal has been noted since 1975 (Harrap & Quinn 1996). It also occurs as a scarce wintering species in the Balearic Islands (*A. O. Balears 2010*: 268). A significant wintering population in Cantabria has recently been discovered at the Marismas Blancas, Santander, where 100 were ringed between November 2009 and March 2010 (*Ardeola* 57: 240). Very few are known to travel beyond the Peninsula and the species was almost unknown in Morocco until the 1980s, since when it has become a local winter visitor to the north-west (BM, Amezian *et al.* 2011). The Moroccan records include the recovery of a German-ringed individual and two of birds ringed in Spain.

The high-pitched whistled call is a characteristic winter sound from reedbeds as small flocks forage for food, especially the seedheads of reedmace, which they demolish. Ringing recoveries show that migrants originate chiefly from the German, French and southern Scandinavian populations, which have also increased in recent decades as the species has expanded its range westwards. An analysis of 243 recoveries during 1971–2002 (Villarán 2003) found that most originated from central Europe, chiefly Germany (69%). Males accounted for 62% of 101 individuals whose sex was established, most of them in central Spain. At the Txingudi marshes, in the Basque Country, the proportion of males was even greater, 72%, and 94% of all birds were juveniles (Arizaga & Alonso 2005). The capture dates of foreign-ringed birds establish the wintering period as October–March (Villarán 2003). Passage was detected at Txingudi in October–November and again in March–April (Arizaga & Alonso 2005). Adults showed a higher tendency to philopatry in winter quarters than did juveniles. There are at least 147 recoveries from central Europe of birds ringed in Iberia: 55% in Germany and 27% in France. Some Spanish-ringed birds, probably juveniles, originating from the Duero and Ebro basins in the north, were recovered further south in winter, in the Tagus, Guadiana and Guadalquivir basins. The farthest of such displacements on record are from Valladolid to Faro (539km) and from León to Sevilla (571km) (*Ecología* 11: 406, 22: 221).

EURASIAN GOLDEN ORIOLE *Oriolus oriolus*

(Oropéndola Europea, Papa-figos-comum)

A widespread summer visitor to most of mainland Iberia. Frequent on passage throughout the Peninsula and a scarce migrant on the Balearic Islands.

Breeding population estimates. Spain: SA1 150,000–200,000 pairs; ACRE 2.1 million birds. **Portugal**: 10,000–100,000 pairs.

The Golden Oriole is widespread across most of the Mediterranean climatic region of Iberia but much more local in the Eurosiberian. It is absent from treeless areas but it can occupy quite small stands and so occurs in many agricultural areas in poplar plantations, shelter belts, riparian woodland and treelines along roads. In Portugal it is most frequent in the east and largely absent from Estremadura and the western coastlands between the Minho and Mondego rivers (PA2).

The Oriole is especially associated with mature trees along watercourses, notably poplars, willows and alders. Poplar plantations are generally in valleys and these too are selected. It nests widely in oakwoods but generally avoids conifers. Tall trees in parks and gardens, and orchards with large trees, are also used. It is commonest below 1,000m (de Juana 1980, Ferrer *et al.* 1986, AI2) but occurs to 1,500m in the Catalan Pyrenees (CA2) and up to 1,750m in oakwoods on the Sierra Nevada (Garzón & Henares 2012).

The SA1 figure may have been a major underestimate. However, the population of Catalonia has been estimated at just 20,000–28,000 pairs (CA2) so the ACRE estimate may yet prove to have been excessively generous. Nevertheless, a significant increase was seen on SACRE sites during1998–2012 (Escandell 2013). Mean maximum densities in preferred habitats are 3.09 birds/10ha but much higher densities occur locally in poplar plantations and riverine woodland: up to 13.3 birds/10ha have been recorded along the Duero river in Valladolid (SA2). In general, it is most abundant in the meso- and supramediterranean regions (ACRE), particularly in riparian woodlands, orchards and oakwoods, where there may be over 20 birds/km^2. There are no breeding records from Andorra (ANA).

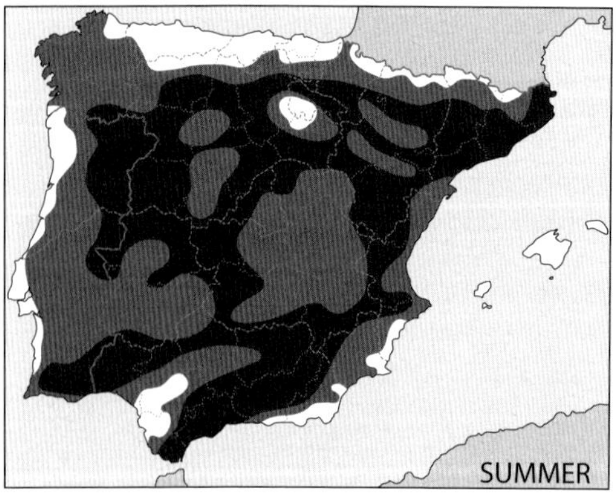

The winter quarters are in sub-Saharan West Africa. There are two December recoveries of Spanish-ringed birds from the Democratic Republic of Congo: of one ringed as a nestling in Badajoz (5,500km) and of an adult female ringed in Barcelona (6,126km). The Oriole returns to southern Iberia from mid April onwards but a few may appear there as early as mid-March. For example, three were seen arriving from the sea at Chipiona, Cádiz, on 19 March 1970 (Fernández Cruz & Sáez-Royuela 1971). In 2009 there were three extremely early records of males: one at Sertã, Castelo Branco, on 17 January, one at Odemira, Beja, on 2 February and two at Silves, Faro, on 26 February (*Anu. Ornitol.* 8: 83). Arrivals and passage throughout Iberia and the Balearics occur during April–May and into early June. The return passage is in August, peaking at the end of that month and extending into September (BG, Casaux & Aparicio 2002). Very few are recorded later: single birds at Monfragüe, Cáceres, on 22 December 2005 and at Doñana, on 6 January 2006, and three at El Rocío, Doñana, on 6 January 2009 (*Ardeola* 54: 182, 57: 240), are remarkably unseasonal records.

BLACK-CROWNED TCHAGRA *Tchagra senegalus*

(Chagra del Senegal, Picanço-assobiador)

Vagrant to Spain.

One was seen near Tarifa, Cádiz, on 15 July 1995. This is another African species whose breeding range takes it to the southern shores of the Strait of Gibraltar but which is all but unknown in Europe.

ISABELLINE SHRIKE *Lanius isabellinus*

(Alcaudón Isabel, Picanço-isabel)

Vagrant to Spain.

An adult male *phoenicuroides* was seen at the Salobrar de Campos, Mallorca, on 9 September 1994 (Willis 2002, *Ardeola* 47: 156). The other four accepted records have been at well-scattered coastal locations in peninsular Spain: at Cape Estaca de Bares, A Coruña, from 12–16 September 1997; at Roquetas de Mar, Almería, on 10 and 12 November 1999; at El Hondo, Alicante, on 6 December 2001, and at the Guadalhorce estuary, Málaga, from 1–10 February 2008. Those at Cape Bares and the Guadalhorce were first-winters. *L. i. phoenicuroides* breeds from Iran and the Caspian region to Pakistan and extreme northwestern China, *L. i. isabellinus* breeds from the Altai to Mongolia and northern China and two further races breed elsewhere in China. The first two races mainly winter in Arabia and sub-Saharan East Africa, with some *phoenicuroides* regularly reaching as far west as Senegal and southern Mauritania , and both reach western Europe with some frequency (Slack 2009). There are records of 89 individuals in Britain alone, as well as 16 in France and one in Morocco.

RED-BACKED SHRIKE *Lanius collurio*

(Alcaudón Dorsirrojo, Picanço-de-dorso-ruivo)

***L. c. collurio*. A locally common summer visitor to northern Iberia. Very scarce on passage further south and in the Balearic Islands.**

Breeding population estimates. Spain: SA1 240,000–500,000 pairs; ACRE 360,000 birds. **Portugal**: 1,100 pairs. **Andorra**: 20–50 pairs.

The Red-backed Shrike is locally common as a breeding species in the eurosiberian region: in the Biscay coastlands, the Cantabrian Range and the Pyrenees. It inhabits the more humid sectors of the southern slopes of the Cantabrian range in Castilla y León and the Pyrenean foothills in Aragón and Catalonia (SA2, AI2). The breeding distribution also extends southwards along the Iberian Range above 800m (de Juana 1980). An outpost breeding population in the Central Range is of relatively recent origin. It breeds down to sea level in the north and north-west in the Biscay coastlands and Galicia, and up to 2,000m in the eastern Pyrenees in Andorra (ANA) and Catalonia (CA2).

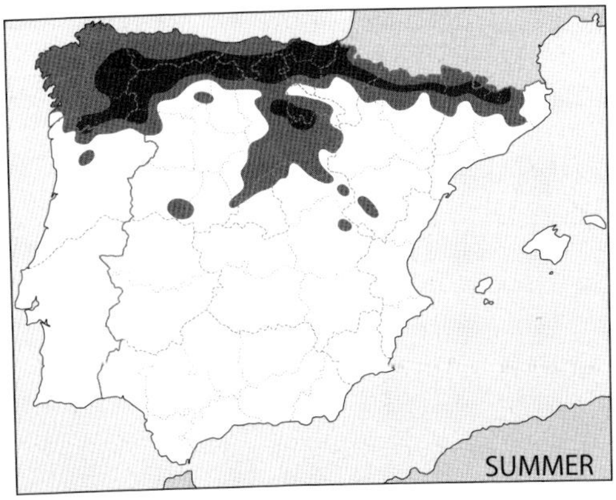

In Portugal it is almost entirely restricted to the extreme north, where the Peneda-Gerês National Park accommodates 600–800 pairs, the bulk of the Portuguese population (Reino 2001, AP). Nesting south of the Douro was confirmed in the

Serra de Montemuro, Beira Alta, in 2003, and has occasionally been suspected elsewhere in that region. A female was seen in the Serra da Estrela on 6 June 2004 (*Anu. Ornitol.* 4: 43).

The Red-backed Shrike selects habitats offering a mosaic of bushes and open land (AMI). It may be found in farmland hedgerows, in thorny scrub on the margins of forests and woodlands and in mountain pastures with scattered tall shrubs. Thorny shrubs, including hawthorn, brambles and wild roses, are favoured as nest sites and for impaling prey items in a 'larder'. In general it favours bushier, denser habitats than other Iberian shrikes (Hernández 1994).

The densest populations are those of the eurosiberian and supramediterranean regions, in matorrals (4–6 birds/km^2) and traditional mixed farmland (3–6 birds/km^2). The population on SACRE plots showed a significant decline during 1998–2012, (Escandell 2013) but the small population in north-eastern Portugal has shown no clear numerical trends, with local declines and increases both detected (AP). The breeding birds in Andorra are chiefly in the warmer areas of the centre, south and south-west (ANA, BE).

There have been quite clear changes in distribution in Spain during recent decades, with notable southward and westward range expansion. However, it has become scarcer in coastal Galicia and the Biscay coastlands (Noval 2001), perhaps as a result of habitat change and an increased use of agricultural chemicals, although it is still common locally. It has spread southwards in the Iberian Range, becoming more widespread in Burgos during the 1980s and 1990s (Valverde 1956, de Juana 1980, Román *et al.* 1996). First breeding in Segovia was reported in 1994, at Montejo de la Vega (*Ardeola* 41: 201). It was first proved to breed in the Central Range in 1990 in the Sierra de Malagón (*Ardeola* 37: 350) and was relatively widespread there by the late 1990s in the Guadarrama mountains. The first observation from Madrid province was in 1987 and the first breeding attempt was recorded there in 1994, with 50–100 pairs estimated there by 2003 (López Septiem *et al.* 2004). First breeding in neighbouring Guadalajara was first detected at Cardoso de la Sierra in 1997 (*Ardeola* 46: 161). At around this time, the Red-backed Shrike also seems to have become established in the southern Iberian Range in Teruel, where it was first found to breed in 1994: at Caudé (*Ardeola* 41: 201), and where breeding birds were located in 2004–07 at several localities, including nine pairs at Bronchales (*A. O. Aragón 2004–2007*: 335–336). Westward range expansion along the Central Range has reached at least as far as the western tip of Ávila, where several pairs were found breeding in 2009 at the Puerto de Tornavacas, bordering Extremadura (*Ardeola* 56: 365, 57: 546). Breeding in Salamanca was proved for the first time at Candelario in 2012 and three pairs nested at La Garganta de San Martín, Cáceres in 2012 (*Ardeola* 59: 450–451, 60: 541).

The winter quarters are in southern and eastern Africa. Migrants travel via the southeastern Mediterranean and thus do not cross central or southern Iberia (AMI). A recent study (Fundación Migres 2013) of two males fitted with geolocators at Castrillino, León, in spring 2011 and recovered there the following year, has shown that both birds first flew eastwards to northern Italy before heading southeastwards across the Libyan Sahara to winter eventually in savannas between Zimbawe and Mozambique. They left Spain in August but did not reach their final destinations until late October, after travelling 8,750km. Both returned by a much longer and more easterly route, reaching the Arabian Gulf and Iraq before heading westwards to Spain, a journey of 10,950km from March to late May. There are nonetheless several Iberian recoveries of birds ringed in Germany (7), Switzerland (3), France (1), Finland (1) and Denmark (1) (AI2, AP, SMA). One ringed at Grado, Asturias, in August 1968 was found at Padova, northeastern Italy, the following month (*Ardeola* 14: 56) and another bird from coastal Asturias has been recovered in Greece (SMA). Red-backed Shrikes are commonly seen in Catalonia, where they breed in the Pyrenees, but they are rare further south on the east coast in Valencia. The regional bird reports give only 19 records (20 birds) for the Comunidad Valenciana, including the Columbretes islands, during 1988–2009. Seven were in spring, between 31 March and 26 June, and ten in autumn: between 30 July and 10 October, with a very late juvenile on 23 November 2009 at Alcossebre (*A. O. C. Valenciana 2009*: 138). There is also a highly exceptional winter record from this area: a male at Santa Pola on 21 January 1990 (*A. O. C. Valenciana 1993*: 117).

It is uncommon on the Balearic Islands, although several are seen or ringed there in most years. The Balearic bird reports give 58 records (59 birds) during 1989–2010, 34 in spring and 24 in autumn. Spring records there fall between 30 April and 11 June, but 30 were in May, and autumn records are chiefly in September but fall between 28 August and 18 October. As many as 23 birds were ringed on Cabrera.

Red-backed Shrikes are only occasionally reported in the southern half of the Peninsula. For example, there are only two published records for Extremadura proper away from the Sierra de Gredos: single birds in the Sierra de San Pedro in May 1998 and in La Vera in September 2004 (*Aves de Extremadura 1998*: 242, *Ardeola* 53: 209). Further south, Chapman (1884) collected a specimen at Doñana in April and reported that a locally obtained bird was in the Jerez museum. A male was at Ponta dos Corvos, Setúbal, on 23 October 2009 (*Anu. Ornitol.* 8: 83). There are three records (four birds) from Gibraltar: three were in mid October and one in late July. In view of the easterly migration route it is unsurprising that the species is a vagrant in Morocco, from where there are just six 20th century records (BM).

LESSER GREY SHRIKE

Lanius minor

(Alcaudón Chico, Picanço-de-mascarilha)

***L. m. minor*. A very rare summer visitor with a tiny and declining population in northeastern Spain. Critically endangered.**

As far is known, this shrike has always had no more than a marginal presence in Iberia, with breeding confirmed in just three provinces in northeastern Spain: in Huesca (Aragón) and in Lleida and Girona (Catalonia) (LR). The Girona population was in the Empordà district, close to the French border, whereas those of Lleida and Huesca were almost contiguous and centred on the lower Cinca and Segre valleys. There it occupies – or rather occupied – open farmland offering a mix of herbaceous crops and areas of fallow or pasture. It also requires scattered large trees, notably poplars in which it nests.

It was already said to breed in Girona in the late 19th century (Vayreda 1883). The first specific record from its Spanish breeding area came much later, however, when an adult was taken at Termens, Lleida, on 5 June 1947 (Maluquer Sostres 1952). A nesting presence in Huesca and Girona was first reported in the late 1960s (*Ardeola* 23: 240, Wallace & Sage 1968). The total population was estimated in the 1975–95 atlas (SA1) at 45–90 pairs. It oscillated between 19 and 26 pairs during 1999–2005 (LR, Giralt *et al.* 2010). However, a precipitous decline thereafter meant that there were only 13 pairs in 2006 and nine in 2007. The last breeding record from Girona was in 2001 and it last nested in Huesca in 2009. An *ex-situ* conservation programme was initiated in 2006 and several captive-born birds have been released since 2009 (Giralt *et al.* 2010), but it will be remarkable if such measures succeed in preserving a Spanish-breeding population. Further decline has taken it to the brink of extinction: a census in 2010 found only a single breeding pair, in Lleida, and 6–8 unpaired individuals divided between Huesca, Girona, Barcelona and Lleida (Giralt *et al.* 2010). One pair, including a released bird, nested successfully in 2013 (*Ardeola* 60: 541).

The decline of the Lesser Grey Shrike in Spain is part of a wider decrease and range contraction that has affected its other western European populations. For example, it was relatively numerous in southern and central France in the 19th century but only 36 pairs remained in the extreme south in 2007 (OF) and there were just 16 pairs in 2011 (*Ornithos* 19: 318–319). It became extinct during the second half of the 20th century in several other European countries, including Germany, Switzerland and the former Czechoslovakia (BWP). The Spanish birds are on the extreme western margin of the species' range and the population is unlikely to survive unless there is an upturn in its fortunes more widely. The causes of its regional demise are uncertain. Habitat loss has been proposed as one cause (LR, Giralt *et al.* 2008, Albero *et al.* 2010) and agricultural intensification in its small breeding range has resulted in a reduction of the habitat mosaic and a loss of riparian trees. A reduction in prey densities due to insecticides has also been proposed as a threat to this and other shrike species (Lefranc 1993) and climate change may be an additional negative factor (Giralt & Valera 2007). The tiny size of the population now makes it especially vulnerable to predation. Its marginal position also means that recruitment from other populations is unlikely. Indeed conservation measures in recent years, in particular magpie control, have meant that the breeding success of the last few breeding pairs was relatively high, with 81% of nesting attempts proving successful and four young fledging on average per successful nest, but returns from winter quarters have been poor (Giralt *et al.* 2010).

The Lesser Grey Shrike is a summer visitor whose winter quarters are in southern Africa. It migrates there via the eastern Mediterranean and east Africa and is thus very rare in Iberia outside its limited breeding range (AMI). One at Paradela, Ourense, in August 1993 was especially surprising (*A. O. Galicia 1992–1993*: 38). It has been recorded on passage in the Balearics but is very rarely seen there (Ferrer *et al.* 1981). Clavell (2002) gives seven records (eight birds) for Valencia, to the south of the breeding range, during 1985–97: no fewer than five of these were in November–January, prompting doubts over identification. There is also a winter record from the Albufera de Valencia: one on 15 December 1994 that remained for a further two weeks (*Ardeola* 42: 230). In general the breeding population arrived in May and departed in August (LR), the first clutches usually being detected in late May or early June (Giralt *et al.* 2010). There are no confirmed records from Portugal (AP).

GREAT GREY SHRIKE

Lanius excubitor

(Alcaudón Norteño, Picanço-real-nortenho)

Apparently a very scarce winter visitor to northernmost Spain.

Wintering Great Grey Shrikes occur as far west in Europe as Britain and south to the Pyrenees in France and it has been assumed that at least some then also occur in the Iberian Peninsula, in particular in northern and northeastern Spain (Noval 1967, AMI, Ferrer *et al.* 1986). However, the Southern Grey Shrike was treated as a subspecies of the Great Grey Shrike until quite recently and records that make specific reference to the latter have only recently begun to emerge.

The species has been considered by the Spanish rarities committee since 2006 but the first accepted record was in 2010, at Auritz-Burguete, Navarra, on 9 December. None were observed during the Spanish Winter Atlas fieldwork and there are no Iberian recoveries of birds ringed elsewhere. There are no Portuguese records. The Iberian status of the Great Grey Shrike thus remains unclear but the regional bird reports list an increasing number of observations from the Biscay coastlands, in Galicia and, especially, Asturias: where the Southern Grey Shrike is absent, as well as a few from the north-east and one in the Balearic Islands. These fall between August and April but the overwhelming majority (92%) of first observations are between October and February (Figure 200). There were annual records in Asturias during the period covered and some individuals made extended stays there in winter, notably at the Ría de Villaviciosa, where up to three were present in some years. Records from Catalonia were much rarer and there were none at all there during 2002–09.

Figure 200. Monthly distribution of Great Grey Shrikes reported from Spain: in Galicia 1998–2003 (5), Asturias 1993–2000 (38), Catalonia 2000–01 (3), the Comunidad Valenciana 2000–02 (3) and the Balearic Islands (1: on Cabrera in 1993). Sources: regional bird reports.

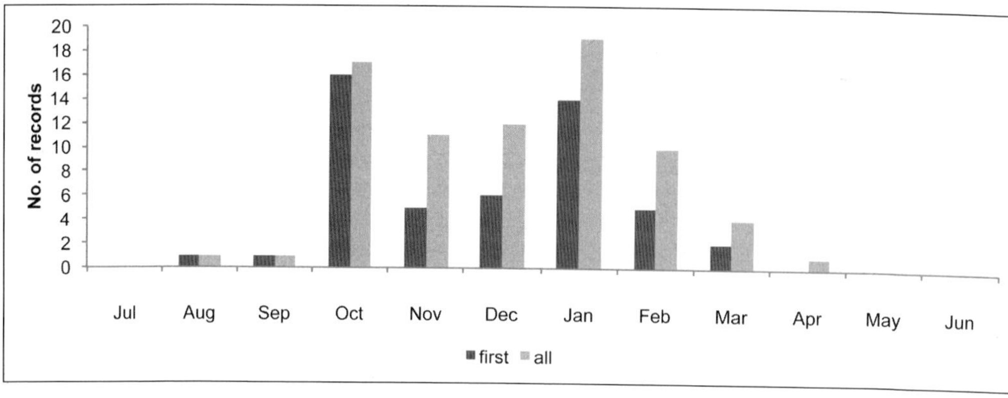

SOUTHERN GREY SHRIKE

Lanius meridionalis

(Alcaudón Real, Picanço-real)

***L. m. meridionalis*. A widespread resident in mainland Iberia. Rare in the Balearic Islands.**

Breeding population estimates. Spain: SA1 200,000–250,000 pairs; ACRE 938,000 birds. **Portugal**: 10,000–50,000 pairs.

The Southern Grey Shrike is widespread and locally common throughout the Mediterranean climatic region but much scarcer and more locally distributed in the Eurosiberian. Thus it is largely absent from Asturias, Cantabria and the coastal regions of the Basque Country and Galicia. It is also very scarce in coastal Catalonia and Valencia, in Cádiz and parts of the Guadalquivir valley in Andalucía, and on the eastern Meseta (SA2). It is widespread in the southern half of Portugal but largely absent from the west and centre north of Nazaré, Estremadura, and Ribatejo (PA2, AP). It generally breeds below 1,500m, although it reaches 1,600m in the Sierra de Gredos (San Segundo 1990) and 1,900m in Sierra Nevada (Garzón & Henares 2012). It is thus absent from much of the Pyrenees and the Cantabrian Range, where it nonetheless occupies the valley bottoms and southern foothills. This is a shrike of open country where it favours pastureland with scattered shrubs and trees, as well as open oak dehesa woodlands, areas of Mediterranean scrub, olive and almond groves and forest edges. It is also characteristic of thorny scrub in the arid south-east. It is a familiar sight perched on roadside posts and wires.

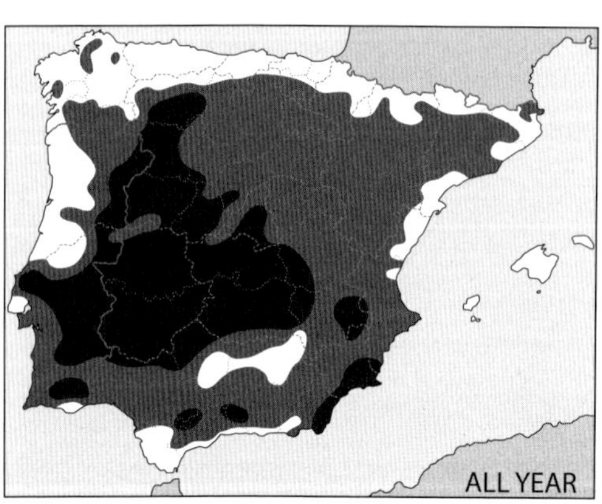

Territory sizes may be large, up to 10ha. Accordingly, population densities tend to be low even in favoured areas. Mean maximum densities in preferred habitats are 0.74birds/10ha (SA2). The highest densities, 5–10 birds/km^2, are in meso- and supramediterranean pastures and matorrals, and in mesomediterranean vineyards. The largest regional populations are those of Extremadura (26%) and Andalucía (21%).

A progressive decline was first noted in Spain in the 1970s. Habitat loss through agricultural intensification but also land abandonment, as well as pesticide use, may be responsible for declines in Catalonia, Aragón, Navarra, La Rioja, the Basque Country, the Ebro valley, parts of the northern Meseta and the Mediterranean coastlands (LR). In Catalonia, for example, a comparison between the two breeding atlases shows a 26% decrease in the number of occupied 10km-squares (CA2) whereas in Navarra a range contraction of some 46% has been seen since the early 1980s (LR). The Spanish population showed a significant overall decrease on SACRE plots during 1998–2012 (Escandell 2013). The Portuguese population is considered largely stable, with local increases reported in southern Estremadura (AP): it showed no clear trend during 2004–09 (ACP).

This is the only one of the four Iberian breeding shrikes that is not a summer visitor. It is the largest of the four and takes more vertebrate prey, in the form of small mammals, birds and, locally, lizards (Hódar 2006), which presumably facilitates winter survival when arthropods are scarce. Some pairs are clearly resident but in others the males remain on their territories all year round (Hernández 1999) and females and juveniles disperse more widely in winter, when there appears to be a higher degree of occupancy of very open areas. The breeding and wintering ranges in Spain are essentially similar (SWA) but some littoral areas in Portugal where the species does not breed are occupied during the autumn and winter (AP). One ringed as a chick at Cornalvo, Badajoz, in May was found in Portugal at Pombal, Leiria (220km) the following November (*Ardeola* 29: 140) and Bernis (AMI) refers to one ringed at Trás-os-Montes in September that was recovered a year later 165km to the west at Minho in November, but all other ringing recoveries have been local or short-distance. Nonetheless, there is evidence of very small-scale passage at the Strait in August–October (BG, Tellería 1981), when presumably some cross to Morocco, but the return movement in April is barely apparent. Very few individuals are involved and the movement is only detectable because of the absence of local nesters near the Strait.

The Southern Grey Shrike is a rare migrant and winter visitor to the Balearic Islands (*A. O. Balears 2010*: 268). The older records especially do not distinguish between it and the Great Grey Shrike, although there are no confirmed records of the latter for the islands. The Balearic Bird reports list 35 records (36 birds) of the Southern Grey Shrike during 2000–10, and six records of Southern Grey/Great Grey Shrikes during 1990–99. Together all these records fall between 29 August and 1 May, and 29 (71%) were during November–March.

WOODCHAT SHRIKE *Lanius senator*

(Alcaudón Común, Picanço-barreteiro)

***L. s. senator*. A common summer visitor to most of mainland Iberia. Also a common passage migrant.**
***L. s. badius*. A common summer visitor to the Balearic Islands.**
***L. s. niloticus*. Vagrant to Spain.**

Breeding population estimates. Spain: SA1 390,000–860,000 pairs; ACRE 4.5 million birds. **Portugal**: 10,000–100,000 pairs.

The Woodchat Shrike is widely distributed throughout the Mediterranean climatic zone. It ranges from sea level to about 1,500m, and up to 1,600m in the Sierra Nevada (Garzón & Henares 2012), but only to 1,000m in the northern Iberian Range. It is a familiar sight along roadsides, usually sitting on fence posts, wires and other prominent perches. It tends to avoid areas of Atlantic maritime influence and so is much more local in the Eurosiberian region where there are just a very few pairs in warmer spots in Galicia and in the Biscay coastlands (SA2). It is widespread south of the Tagus in Portugal but absent from coastal regions further north as well as from the densely wooded areas of the centre and north and from elevations above 1,000m (PA2). It occupies a wide range of habitats but all offer open areas and scattered shrubs or trees. It may therefore be encountered in open broadleaved and coniferous woodlands, including dehesas, as well as in hedges along field margins, riparian woodlands, patchy scrub, olive groves and other orchard crops (AMI, Hernández 1994, AI2).

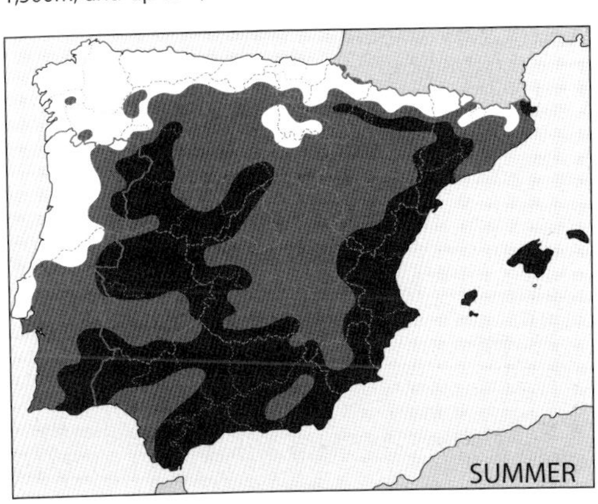

Irrespective of the high ACRE estimate, a widespread decline of at least 20% was noted in Spain during the 1970s and 1980s (SA2) and a significant 14% decline occurred on SACRE sites in 1998–2012 (Escandell 2013). In general, widespread decreases have affected the populations in the northern half of Iberia. For example, in Catalonia, where the population has been put at 19,000–45,000 pairs, there was a 15% decrease in the number of occupied 10km-squares between the first and the second atlas surveys, in the early 1980s and the early 2000s respectively (CA2). Early authors considered it very numerous in Portugal. Smith (1868) described it as 'extremely common' and Tait (1924) said that it was 'very abundant in summer all over the country in or near woods'. However, the Portuguese population too has shown at least a moderate decline (AP, ACP).

The decline of the Woodchat Shrike follows a wider decrease that has affected other European countries since the 1960s (Lefranc & Worfolk 1997, BE). It may be due to a combination of agricultural intensification and the impact of pesticides as well as to a loss of grazing land as pastures are abandoned, leading to scrub encroachment. Other factors, as yet unclear, may affect it on passage and in its African winter quarters. Despite the losses it is still common in much of Mediterranean Iberia. Its highest densities are seen in the dehesas of Andalucía and Extremadura and in the Central Range, with maxima of 16 birds/10ha and mean maximum densities of 7.45 birds/10ha in oak dehesas (SA2). In the past it was reported as very numerous or abundant in all these regions (e.g. Saunders 1871, Arévalo 1887, Witherby 1928, Gil Lletget 1945). Particularly favoured habitats are thermomediterranean pastures and woodlands of Holm or Cork Oaks in the thermo- and mesomediterranean zones, where densities may approach 40 birds/km^2 (ACRE).

The winter quarters of Iberian birds are in sub-Saharan West and Central Africa. The first birds tend to arrive in the south in mid-March followed by the main arrivals during April (AMI), although there are several recent records in February, including one at Santa Amalia, Badajoz, on 7 February 2010 (*Ardeola* 57: 240). Some passage continues throughout May and into June. Migrants may be very conspicuous on southern and eastern coasts during the spring passage especially. The post-breeding movement commences in July but the main passage is in August–September, with stragglers in October (AMI, BG). Midwinter records are very rare. They include single birds in Algarve on 27 December 1997 and 22 December 2002 (*Pardela* 7: 20, *Anu. Ornitol.* 2:45) and at Alaior, Menorca, on 12 January 2011 (*Ardeola* 59: 451). There are a few recoveries of birds ringed in Germany (6), Switzerland (2), France (2) and Slovakia (1), and single recoveries in Germany and France of birds ringed in Iberia. There are also 23 recoveries from North Africa of birds ringed in Spain: in Morocco (19), Algeria (3) and Tunisia.

L. s. badius nests on all the main Balearic Islands, where it is abundant, especially in Mallorca and Ibiza (Mayol 1978, *A. O. Balears 2010*: 268). Nominate race birds also occur there on passage. There are very few records of *badius* from the Iberian mainland, which is unsurprising since the numbers involved are small and the birds are easily overlooked. Many probably fly to and from Africa directly but some may be expected in eastern Iberia especially. Single birds were seen at Punta Sabinal, Almería, on 27 and 28 April 1960 (Glanville & Walker 1963) and one was at the Albufera de Valencia on 2 May 2006 (*Ardeola* 54: 182). There is one record from Gibraltar: two birds on 27 August 2001 (GONHS).

Four individuals of *L. s. niloticus* have been trapped and ringed on the eastern archipelagos: a female in Menorca on 4 April 1992, a male also in Menorca on 18 April 2002, a male on the Columbretes islands on 21 April 1995, and a male on Cabrera on 8 May 2011.

MASKED SHRIKE *Lanius nubicus*

(Alcaudón Núbico, Picanço da Núbia)

Vagrant to Spain.

One was recorded near Gibraltar in May 1863 (Saunders 1871, BG), two were seen in Doñana on 3 May 1956 (Mountfort & Ferguson-Lees 1961) and one was at Cazorla, Jaén, on 1 May 1962 (*Ardeola* 9: 151–152). Since then there have been two accepted records from the Balearics: an adult male at Pollença, Mallorca, from 22–26 April 1991, and a second calendar year female ringed in Cabrera on 5 September 2003. The very few western European records include two in Britain and two in France. There are also single records from Morocco and Algeria.

EURASIAN JAY *Garrulus glandarius*

(Arrendajo Euroasiático, Gaio-comum)

***G. g. glandarius*. A common resident in northernmost Spain and the Pyrenees.**
***G. g. fasciatus*. A common resident throughout most of mainland Iberia.**

Breeding population estimates. Spain: SA1 540,000–1.1 million pairs; ACRE 2.7 million birds. **Portugal:** 10,000–100,000 pairs. **Andorra:** 300–400 pairs.

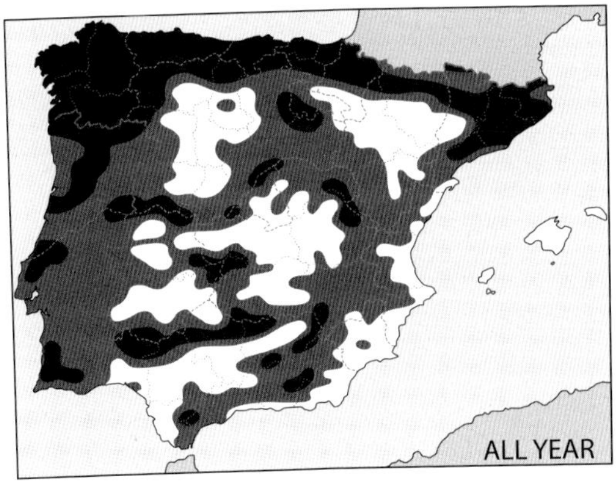

The Jay is common anywhere in the Peninsula that oakwoods occur, although it shows a marked preference for the acorns of the Iberian Holm Oak (Pons & Pausas 2007). It may also be encountered in other forest formations, including beechwoods, conifers and even eucalyptus plantations. Coniferous forests are a major habitat in the central Spanish sierras, including the Gredos, Guadarrama and Montes de Toledo mountains, as well as in Andorra (Sánchez 1991, Álvarez & Santos 1992, Díaz *et al.* 1994). The Jay ranges from sea level to the treeline, reaching 2,600m in the Catalan Pyrenees and 2,400m in the Sierra Nevada (CA2, Garzón & Henares 2012).

The highest densities occur in the Cantabrian Range and Pyrenees and also in the Central and eastern Betic Ranges (SA2). There may be over 20 birds/km^2 in pastures with trees in the Cantabrian sector of the eurosiberian zone; in eurosiberian beechwoods, oakwoods, matorrals and riparian woods, and in supramediterranean pinewoods above 1,250m (ACRE). The Catalan population has been put at 111,000–156,000 pairs in the breeding season and some 221,000 individuals during the winter (CA2, CWA). SACRE (1998–2012) recorded a significant increase that was most evident in northern Spain (Escandell 2013), perhaps the result of colonisation of the new habitats resulting from afforestation. The Jay is widespread throughout Portugal, although more numerous north of the Tagus (PA2). The Portuguese population showed no clear trend during 2004–09 (ACP).

Iberian Jays are largely resident and eruptive birds from elsewhere in Europe may account for the exceptional records in the Balearics: two were captured in Mallorca in December 1965 and another was caught there in April 1966 (*Ardeola* 11: 139, 15: 94.).

IBERIAN AZURE-WINGED MAGPIE *Cyanopica cooki*

(Rabilargo Ibérico, Charneco)

An endemic resident of Portugal and west-central Spain.

Breeding population estimates. Spain: SA1 240,000–260,000 pairs; ACRE 6.4 million birds. **Portugal**: 40,000 pairs.

The Azure-winged Magpie has long been renowned on account of its curiously disjunct distribution. The core distribution of *Cyanopica cyanus* is in southern Siberia, Mongolia, China, Korea and central Japan, where there are a total of seven subspecies (HBW). However, a distinct form, first described by Charles Lucien Bonaparte in 1850 from specimens from Madrid, is well established in the Iberian Peninsula. There are no known intermediate populations so its presence in Iberia led to conjecture that the western population is descended from introduced birds brought back to Europe by 16th century Portuguese sailors (Santos Júnior 1968). The alternative hypothesis is that the current range is a relict of a continuous Eurasian distribution that was sundered by the Pleistocene glaciations (Sacarrão 1974).

Recent discoveries strongly support the second hypothesis. Fossil remains of the Azure-winged Magpie were identified in Pleistocene deposits from two separate caves in Gibraltar, these being carbon-dated to over 44,000 years old (Cooper 1998, 2000; Cooper & Voous 1999). Moreover, the findings of molecular genetic analyses (Fok *et al.* 2002, Kryukov *et al.* 2004) show that the Iberian birds diverged from the Far Eastern ones at least one million years ago. It therefore seems apparent not only that the occurrence of the Azure-winged Magpie in Iberia is natural but also that the Iberian birds may be regarded as a separate species. We follow HBW in recognising the Iberian endemic form as *Cyanopica cooki*. It differs morphologically from the Far Eastern species, the Asian Azure-winged Magpie *C. cyanus*, in being smaller, with a smaller bill, pinkish body plumage and bluer wings and tail, and in lacking a white tail-tip.

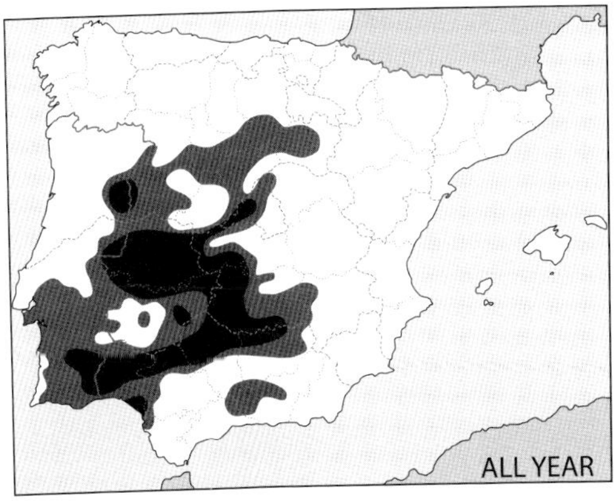

The entire Spanish population is divided between Extremadura, Andalucía, Castilla-La Mancha, Castilla y León and Madrid. The core Iberian distribution is probably the northern half of Extremadura: Cáceres province. From here it ranges southwestwards throughout much of the southern third of Portugal; southwards throughout Huelva; eastwards across the Sierra de Cazorla, Jaén, and the west of Albacete; northwards to Zamora and Valladolid and northeastwards as far as Burgos and Soria (SA2). An outpost population in northwestern Granada, Andalucía, extends into the north-east of Málaga province. The Portuguese distribution is largely south of the Tagus but the bird is also present along the eastern fringe of the country (PA2). The distribution generally extends across warm or hot regions and it avoids areas where the cooling influences of the Atlantic prevail, as on the Portuguese west coast. However, it occurs at sea level in Algarve and on the Huelva coast, including at Doñana, where it is very common.

This distribution seems essentially similar to that reported by Lilford (1866), although he said it was by then extremely rare in Valencia and Murcia and, conversely, that it was extremely abundant in the immediate neighbourhood of Madrid, where he obtained many eggs at the Casa de Campo park. It is now absent from all these areas. The extreme northern boundary of the range was reported to lie at Silos, Burgos, by Valverde (1956) and in the foothills of the northern Iberian Range by de Juana (1980). In Portugal, the northern limit was thought by Tait (1924) to be at Penamacor, Castelo Branco, but subsequently both Coverley (*c.*1945, in AP) and Santos Júnior (1968) found it common at Barca d'Alva, by the river Douro, and there has apparently been recent northward spread from there to Miranda do Douro, along the Spanish border (Sacarrão 1972, AP).

The principal habitat is open woodlands, particularly Iberian Holm Oak dehesas, but it also occurs in Stone Pine woods and locally in olive groves, orchards and riparian woodlands. It is generally found at low elevations but ranges to 1,100m in Gredos (San Segundo 1990). It does not seem to favour pure Cork Oak woodlands and so is almost entirely absent from Cádiz province, occurring there only in the Stone Pines of La Algaida, on the east bank of the River Guadalquivir. It has never been reported in Gibraltar in historical times.

Iberian Azure-winged Magpies form monogamous pairs but these are assisted by nest-helpers, presumed to be the parents' non-breeding young from previous seasons (HBW). Nests are often grouped in loose colonies. The Azure-winged Magpie has been claimed to host the Great Spotted Cuckoo on rare occasions (Valverde 1971) but these records are certainly exceptional and may not be entirely reliable (M. Soler, pers. comm.) and no recent instances have been reported.

The Spanish population showed a significant increase on SACRE sites during 1998–2012 (Escandell 2013). The Portuguese population showed no clear trend during 2004–09 (ACP). Densities may vary widely according to season, from 0.1 to 25.4 birds/10ha. Exceptionally high densities have been reported in Spain from supra- and mesomediterranean juniper scrub (121 birds/km^2) but more typical of the higher densities are those in dehesas, pinewoods and olive groves in the thermo-, meso- and supramediterranean (31–80 birds/km^2). Flocks of up to 300 birds occur outside the breeding season although most recent observations have involved smaller groups, of up to 100 individuals. However, a roost at Valdesalor, Cáceres, held a remarkable 1,500+ birds on 20 August 2007 (*Ardeola* 57; 240).

The Azure-winged Magpie is largely sedentary. Recoveries of birds ringed as adults have all been local and the farthest recovery of a bird ringed as chick was 45km (de la Cruz & Valencia 2004). Altitudinal movements have been reported in the Sierra de Gredos, where birds descend to the foothills in winter (AI2). Individuals have wandered north: to Polientes, Cantabria, in November 1967 (*Ardeola* 13: 265) and to the Pajares pass, on the Asturias/Castilla y León boundary, in autumn 1975 (Álvarez-Balbuena *et al.* 2000). One was in the far south-east at the Sierra de Espuña, Murcia, on 4 February 2006 (*A. O. Murcia 2006*: 38). There are some 30 records north-east of the breeding range, notably in Catalonia, Valencia and the Ebro valley (Ferrer 1987, Clavell 2002): recent reports there a flock of eight at Ezcaray, La Rioja, in October 2002. and one at Sant Llorenç Savall, Barcelona, in September 2005 (*A. O. La Rioja 2001–2003*: 120, *A. O. Catalunya 2002–2005*: 473). In addition, there are six records (seven birds) for France (OF), the most recent being one in Camargue in December 2005. In general, although some individuals, or even small flocks, may wander quite far, such occurrences are rare.

EURASIAN MAGPIE *Pica pica*

(Urraca Común, Pega-rabilonga)

P. p. melanotos. **A common resident in most of mainland Iberia. Absent from the Balearics.**

Breeding population estimates. Spain: SA1 220,000–1.2 million pairs; ACRE 6.1 million birds. **Portugal**: 5,000–50,000 pairs. **Andorra**: 20–35 pairs.

The Magpie is a familiar sight over almost all of the Iberian Peninsula although it is most abundant in central areas (SWA). It is however very local in parts of the southeastern coastlands, from Valencia to Almería, although there has been some recent range

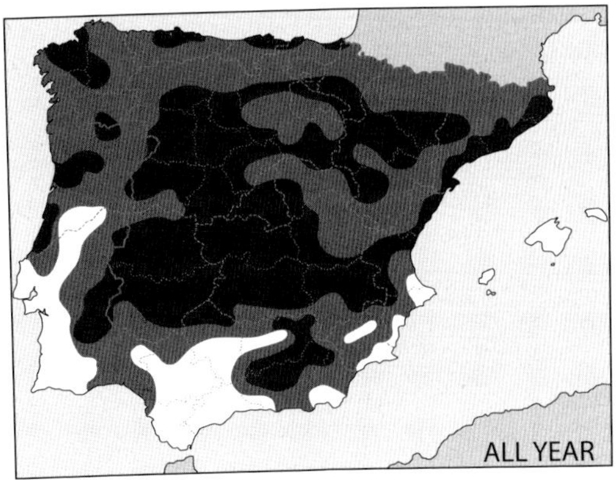

expansion there, and it has long been absent from most of the Guadalquivir valley, Cádiz and Málaga in Andalucía (SA2). Irby (1895) never saw it south of Sevilla except on the banks of the Guadalquivir as far as Coria, Boxberger (1931) said it was absent around Málaga and Jourdain (1936–37) noted its absence from Cádiz province. However, here too there has been some evident range expansion very recently. The Portuguese range is highly fragmented and the Magpie is most abundant there in Alto Alentejo, being widespread but generally scarce in the north of the country, where it also avoids higher elevations. It is entirely absent from southern Estremadura, much of Ribatejo and most of Baixo Alentejo and Algarve: although there has been some recent range expansion into southeasternmost Algarve (PA2).

Some of these distributional gaps are hard to explain since the habitats there do not seem to be unsuitable, but they may indicate regions where the Magpie has not yet become as closely dependent on human activity as elsewhere. In addition, small and sometimes temporary gaps in its range elsewhere are possibly due to local abandonment following persecution by the Great Spotted Cuckoo: the Magpie is its primary host (Soler *et al.* 1996, 1998a & b). The present distribution is certainly wider than in the historical past, when the species was persecuted more intensively by gamekeepers and at least some recent expansion may represent a recovery of lost ground. Irby (1883) described it as 'not common' at Santander in 1876 but subsequently it seems to have been absent from the Biscay coastlands, certainly in Cantabria, until the mid-1930s when it began to re-establish itself there, spreading through the region progressively during the 1940s and 1950s (de La Lama 1959b). The Magpie has also recolonised the coastlands of the Comunidad Valenciana and southern Catalonia (Rico & Gil-Delgado 1986, Urios *et al.* 1991, CA2). For example, Arévalo (1887) saw it commonly in Valencia but it clearly declined thereafter. It was little more than a vagrant at the Albufera until very recently but has become well established there and up to 17 together were observed in 2009 (*A. O. C. Valenciana 2009*: 140). Also in 2009, a winter roost at the Mijares estuary, Castellón, held 162 birds on 20 February (*A. O. C. Valenciana 2009*: 140).

The first observations in Almería date from 1979 and the first proved breeding there was not until 1983, when two nests were found near Tabernas (*Ardeola* 30: 122–123) but it is now quite widespread in that province (Manrique 1993, SA2). In recent years there have been several occurrences in Málaga as well as occasional sightings near the Strait, where it has long been been absent. For example, single birds appeared in Gibraltar in April 1976, April 2004 and April 2010 (GONHS records) and near San Roque there were two in January 1982 and 11 in September 1982 (Finlayson & Cortes 1987). One was on the coast at Málaga city in March 2007 and another at the Guadalhorce estuary in December 2010 (*Ardeola* 54: 402, 58: 513). Three or four pairs are believed to have bred at La Janda, Cádiz, in 2010 (K. Bensusan, pers. comm.) and at least two pairs were present there in April 2011.

The Magpie may be encountered in a very broad range of habitats, a principal requirement being the availability of some dense shrubs or, more usually, tall trees for nesting and roosting. Highest densities are in the mesomediterranean, where the margins of wetlands (94 birds/km^2) and riparian woodlands (70 birds/km^2) hold very high numbers locally (ACRE). Other favoured habitats are orchards, vineyards and traditional mixed farmland (39–65 birds/km^2) and the vicinity of human habitation (42–47 birds/km^2). It is especially common in farming areas, where it occupies riparian woodland, tall hedges, scrub, roadside treelines and copses, and where its large nests are often a conspicuous feature. It may also be found in open woodlands, such as dehesas, orchards and pinewoods, but it avoids forests. It is often encountered near human habitation, on the margins of towns and within parks and gardens, as occurs elsewhere in much of Europe, although this association may perhaps be less marked in southern Iberia. During the first half of the 20th century it was only present in some of the larger Madrid parks but by the late 1980s it could be found throughout the city, nesting even in street trees far from open areas (Bernis 1988b). It ranges from sea level to 1,200–1,400m but extends much higher locally, for example to 2,400m in pine plantations on the Sierra Nevada (AI2, Pleguezuelos 1992). This versatile and omnivorous species can make use of the widest range of food items (Martínez 2011). In Iberia it frequently attends carcases of large mammals, in the company of vultures and other scavengers. Those in the Madrid parks frequently kill other birds, including House Sparrows, Domestic Pigeons and Wood Pigeons (Fernández-Juricic *et al.* 2004).

Regional populations in Spain include 60,000 breeding birds in Madrid and 103,000–143,000 pairs in Catalonia (Díaz *et al.* 1994, CA2). The Spanish population showed no clear trends on SACRE sites during 1998–2012 (Escandell 2013) but the Portuguese population showed a moderate increase during 2004–09 (ACP). The small population in Andorra appears to be of recent origin since the first confirmed sighting there was in 1983 (BE, ANA).

The Magpie is apparently highly sedentary and most ringing recoveries are local. Nevertheless, rates of gene flow between populations suggest that there may be a relatively high amount of dispersal (Martínez *et al.* 1999). However, we suspect that a recovery from Morocco of one ringed as a nestling at Ferreira, Granada, (*Ecología* 11: 408) may actually have involved a young Great Spotted Cuckoo.

A specimen from Mallorca in February 1961 is of uncertain provenance (Nadal 1965). More recently, single birds have been seen there on at least eight occasions in 2004–10: one of which, at Alcúdia on 25 February 2007, has been accepted by the Balearic rarities committe (*A. O. Balears 2007*: 127), but once again there are reservations regarding whether they reached the island naturally.

NUTCRACKER *Nucifraga caryocatactes*

(Cascanueces, Quebra-nozes)

Vagrant to Spain and Portugal.

This species is prone to periodical irruptions linked to fluctuations in pine seed crops. These have often extended into western Europe and they generally involve *N. c. macrorhynchos*. Very few birds have ever travelled as far south-west as Iberia, however.

There are a couple of 19th century Portuguese records of birds taken at Estarreja, Aveiro, in 1872 and near Estremoz, Évora, in December 1888 (Tait 1924), and there is a much later record of a juvenile ringed at Alter do Chão, Portalegre, on 14 August 1962 (Isidoro 1965). There are early 20th century observations from Ávila in central Spain, at Candeleda and Hontanares, the latter in 1917 (Gil Lletget 1945). From the Basque Country there is an observation at Zumaia, Gipuzkoa, on 20 December 1966 (Noval 1967). Most Iberian records, however, come as might be expected from Catalonia. Barcelona university has two specimens taken at unspecified locations, one in 1890 and the other at some time prior to 1913 (Ferrer *et al.* 1986). Much more recently, one was taken at Puigcerdà, Girona, in early February 1954 (Pons 1956) and another at Vallbona de les Monges, Lleida, on 30 October 1977 (*Ardeola* 3: 185, 24: 264), and there are observations of single birds from near Vic, Barcelona, on 6 January 1973 (Aymerich *et al.* 1991), at Navarcles, Barcelona, on 26 January 1981 (Ferrer *et al.* 1986), at Serradell, Lleida, on 9 April 1982 (Ferrer *et al.* 1986) and in the Ribes valley, Girona, on 10 February 1985 (*Ardeola* 38: 163). The last record is the only one that has been considered (and accepted) by the Spanish rarities committee and it coincided with a large irruption elsewhere in Europe. Subsequently, a recently dead individual was found at Villarroya de los Pinares, Teruel, in October 2012 (RBS).

YELLOW-BILLED CHOUGH *Pyrrhocorax graculus*

(Chova Piquigualda, Gralha-de-bico-amarelo)

P. g. graculus. **A locally common resident in the Pyrenees and Cantabrian Range. Rare elsewhere in mainland Spain and a vagrant to the Balearics. There are no records from Portugal (AP).**

Breeding population estimates. Spain: SA1 10,000–11,000 pairs. **Andorra**: 200–300 pairs.

The Yellow-billed Chough ranges from western Catalonia to Navarra in the Pyrenees, and from the Basque mountains to central Asturias and León in the Cordillera Cántabrica. It is generally found nesting above 1,000m but does so down to 700m in Navarra (Elósegui 1985), the Cantabrian Range (García Dory 1983) and Catalonia (CA1). In 2000 a nest was found at only 610m at Peñamellera Baja, Asturias (*A. O. Asturias 2000*: 226) and a breeding pair nested at an altitude of only 250m on Peña Santullán, at Castro Urdiales, Cantabria, in both 2007 and 2008 (*A. O. Cantabria*). It nests up to 2,800m in Aragón and Andorra (Sampietro *et al.* 1998, ANA).

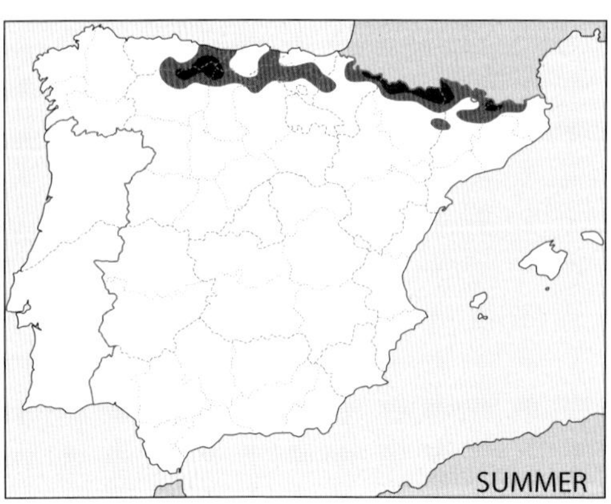

It occurs in open, rocky mountain terrain and requires steep cliffs or gorges in which to nest and roost. It may prefer limestone rocks, where breeding sites are easily found in cliffs and sinkholes. Flocks feed on the alpine moorland and pastures, often in the company of Red-billed Choughs. Here they seek invertebrates in the turf and especially within cow pats and horse dung piles, which are often previously dissected by the latter species. Fruiting trees are visited in autumn. The birds become quite confiding around ski facilities, where they scavenges food, and they also approach picnickers for scraps.

The Spanish population is considered to be stable. The largest numbers are in the Pyrenees but several thousand inhabit the Cantabrian Range, where they are a familiar sight in the Picos de Europa. The Catalan winter population has been estimated at some 3,200–4,400 individuals (CWA).

It is generally resident although altitudinal movements may bring it lower in winter (García Dory 1983). For example, in coastal Cantabria at least 30 were at Liendo in February 2009 and one was at Escalante in November 2011 (*A. O. Cantabria*). There are occasional records across the Meseta far from the nesting areas. For example, several were with Jackdaws at ruins near Pedraza, Segovia, in March 1981; two were in the Río Dulce gorge, Guadalajara, in May 1988 ; there was a flock of 75 near Pancorbo, Burgos, in January 1994 ; two were near Villanueva de Jiloca, Teruel, on in February 1994, and one was at Montejo de la Vega, Segovia, in November 1994 (*Ardeola* 28: 161, 41: 102, 201). There is a single record from Valencia: five individuals at Sueca in January 1984 (Ferrer *et al.* 1986) and four from the Balearic Islands, most recently one bird on Ibiza on 25 April 1993 (*A. O. Balears 1993*: 99). Very few birds were involved in all these sightings except at Pancorbo, which is relatively near to the breeding range.

Some observations in eastern Andalucía, at Gibraltar and in northernmost Morocco (Pineau & Giraud-Audine 1979, BM) suggest that small numbers cross the Strait, at least in some years. There are eight records from Gibraltar, including two flocks of seven birds and one of 18: all but one settled bird were seen flying north and all were between 17 March and 17 May, apart for one September sighting (GONHS). Nisbet (1960) saw four in the Sierra Nevada in April 1960 and there was a series of observations in the Málaga sierras in the 1970s (Blasco *et al.* 1980). There is also a record of a flock of six near Jaén city in February 1998 (*Ardeola* 45: 126) and a single bird was on the Sierra Blanca, Cádiz, in April 1988 (Cortes & Finlayson 1989). Clearly this species can turn up at relatively low altitudes in the south of Spain but the circumstances in which it occurs are unclear. Possibly some migrate south from their usual haunts during hard winters but some or all may originate from the substantial population of the Moroccan High Atlas.

RED-BILLED CHOUGH *Pyrrhocorax pyrrhocorax*

(Chova Piquirroja, Gralha-de-bico-vermelho)

***P. p. erythroramphos*. Widespread throughout Spain and locally common. Very local in Portugal. A rare visitor to the Balearic Islands.**

Breeding population estimates. Spain: SA1 7,000–9,800 pairs; SA2 16,943+ pairs; **Portugal**: PA2 95–198 pairs. **Andorra**: 200–300 pairs.

This sociable cliff-nesting species traditionally occurs where there are suitable mountain or sea cliffs, or river gorges, in the vicinity of pastureland, fallow areas and croplands in which it can probe for its invertebrate food. Colonies also breed locally on ruins, castle walls, bridges, abandoned mine workings and other similar human constructions. The breeding range is most continuous in the eastern half of Spain, where limestone rocks predominate, including the Pyrenees and their foothills, the Ebro valley, the Iberian Range, the eastern Central Range and the northeastern Meseta. It extends in the north across the Cantabrian Range and the Biscayan lowlands west to Galicia. Further south there are scattered populations throughout the western and southern Meseta and eastern Extremadura and it is widespread in the Betic Range, the lowlands of eastern Andalucía, Murcia and the eastern Sierra Morena (SA2, LR).

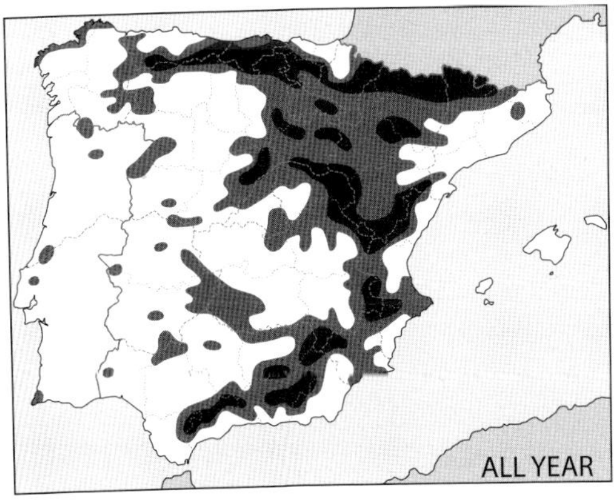

It is scarce and very locally distributed in Portugal. Most of the population there is in the east of Trás-os-Montes, in the Douro gorges. The remainder are in central Trás-os-Montes in Peneda-Gerês national park; in the limestone massif of central Estremadura and in the extreme south-west of Algarve, around Cape St Vincent (PA2).

There are colonies from sea level to high elevations, for example to 1,900m in the Pyrenees in Navarra (Elósegui 1985) and to 3,000m in the Sierra Nevada (Garzón & Henares 2012). Sea-cliff populations are characteristic of Galicia, Asturias and Cantabria and southwestern Portugal in particular but the Red-billed Chough is scarce or absent on eastern and southern Spanish coasts. The breeding range overlaps with that of the Yellow-billed Chough in the Pyrenees and the Cantabrian Range and mixed flocks are often seen there. Some Red-billed Choughs also ascend to higher levels in the mountains after the breeding season, to forage in the alpine pastures. They are occasionally parasitised by the Great Spotted Cuckoo in Andalucía (Zúñiga 1989).

Both atlas estimates of the Spanish population may be conservative. Regional estimates include 8,000 individuals in Granada, 4,000 individuals in Murcia, 3,200 individuals in Guadalajara, 2,600–5,800 pairs in Castilla y León and 3,500–5,700 pairs in Catalonia (CA2, Cuevas & Blanco 2011). Monitored populations have fluctuated in numbers. Populations in Madrid, Murcia, eastern Andalucía, parts of Extremadura and southwestern Castilla y León have seen obvious declines and the species seems to have practically disappeared from Huelva in western Andalucía. There has also been a substantial decline in the coastal populations of Galicia, of 38–48%, leading to its disappearance from some areas (LR). Increases have been reported from some areas where the birds nest in buildings, notably in Segovia where Red-billed Choughs make use of historical edifices in the old city and elsewhere. Similarly, although it seems formerly to have been rather scarce in central Aragon (Boxberger 1921) a flourishing population now nests there in the Monegros area, in abandoned farm houses (Blanco *et al.* 1996).

The Portuguese population was only some 95–198 pairs during 2002–05, 60% of them in the Alto Douro (PA2). This is a marked decline from the 700–800 pairs found in the late 1980s (Farinha & Teixeira 1989), although the population had already declined to 400–500 pairs by 1998, with the loss of some subpopulations (AP).

The species has been affected locally by agricultural intensification, including the loss of marginal land, the deleterious impact of agricultural chemicals and the displacement of traditional crops by irrigated ones (Blanco *et al.* 1998, LR). There has also been some loss of montane pastures to afforestation. Some populations have been persecuted through being deliberately excluded from nesting in historical buildings, as has been noted in Segovia.

The Red-billed Chough is largely sedentary and there are no indications of regular migration. Altitudinal movements occur, however, and some individuals, especially juveniles, may wander at times. Only 22% of 139 ringing recoveries in Spain during 1973–2009 were beyond the ringing locality, the farthest being of a bird ringed as a chick in Navarra, and recovered 227km away in Huesca (*Ecología* 15: 409). There are also two recoveries in the French Pyrenees of birds ringed as chicks 148km and 181km away in Aragón (*Ecología* 9: 329,19: 271). Such movements may account for the series of records from the Balearic Islands, where the species is a rare but perhaps increasingly frequent winter visitor and potential colonist. Munn (1925) said that it was an occasional visitor to Mallorca. Ferrer *et al.* (1986) gave ten records for the archipelago and Clavell (2002) records a further 11 records (23 birds) during 1987–98. Subsequently there were at least 11 records during 2002–09 (*Ardeola* 52: 233, 52: 435, 55: 307, 57: 546), involving a minimum of 76 birds: they included a flock of 22 in November 2002 and one of 20 in December 2003. All the post-1975 records fall between 24 October and 26 April.

There is no suggestion of passage at the Strait, unlike in the Yellow-billed Chough, and there are only two accepted records for Gibraltar: one bird in October 1997 and two in February 2004 (GONHS). Birds of the longer-billed North African race *P. p. barbarus* nest right up to the southern coast of the Strait at Jbel Mousa (Pineau & Giraud-Audine 1979) so there must be a possibility that individuals of this race might stray into Iberia occasionally, but they would be next to impossible to detect in the field.

EURASIAN JACKDAW *Corvus monedula*

(Grajilla Occidental, Gralha-de-nuca-cinzenta)

***C. m. spermologus*. A widespread resident throughout most of mainland Spain, more local in Portugal. Rare in the Balearic Islands.**
***C. m. monedula*. Vagrant to Spain.**

Breeding population estimates. Spain: SA1 423,600–533,000 pairs; ACRE 4.1 million birds. **Portugal**: 1,000–10,000 pairs.

The Jackdaw is common across most of Spain and widespread, though mainly scarce, in the eastern half of Portugal. It seldom nests much above 1,500m and so is absent from high elevations. It does not nest in Andorra and seldom visits the country, although it occurs nearby in Catalonia (ANA). There are large gaps too in its distribution in the western Guadalquivir valley and in parts of the eastern Meseta. It is very uncommon in the Eurosiberian zone, being largely absent from most of the Biscay coastlands, parts of Atlantic coastal Galicia and the western half of Portugal as far south as Sines (SA2, PA2). A coastal population inhabits the rocky coastline and onshore stacks of southwestern Portugal, between Sines and central Algarve. It is similarly local in coastal areas in the Mediterranean.

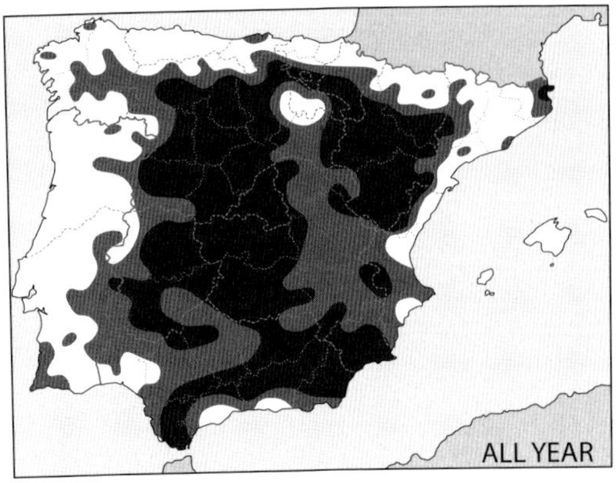

The Jackdaw is tied to agricultural land of all types, especially extensive cereal crops (Soler *et al.* 1990). It requires cavities for nesting, however, and makes use especially of cliffs and river gorges, as well as man-made equivalents in the form of holes in walls, bridges and buildings, often in castles and churches. Historic towns, such as Cáceres and Trujillo in Extremadura and Salamanca and León in Castilla y León, often have their population of roof- and wall-nesting Jackdaws but urban nesting is perhaps less widespread than in central and northern Europe (Soler 2006). In fact, the Madrid city population has declined and has tended to abandon the inner districts (Bernis 1988b). Some nest in tree cavities, especially in central and southern Spain (AI2) and even in rabbit burrows (Soler & Soler 1987). Nesting birds are most often found in small colonies of up to a few tens of pairs, occasionally more. Some 2,500 breeding pairs were estimated to occupy a 6.6km stretch of the cliffs bordering the Jarama river in southeastern Madrid (Arroyo 1977). This species is occasionally parasitised by the Great Spotted Cuckoo (Zúñiga *et al.* 1983, Soler 1990).

Notwithstanding the high ACRE assessment, which has been strongly criticised by Blanco *et al.* (2014), Jackdaws seem to be decreasing in Spain since the population on SACRE plots showed a significant overall decline of during 1998–2012, which was most marked in the north and centre (Escandell 2011, 2013). Recent range contraction has been seen in Catalonia, where a 49% decline in the number of occupied 10km-squares occurred between the first and second breeding atlas surveys, in the early 1980s and early 2000s respectively (CA2). The present Catalan population has been put at 2,700–4,600 breeding pairs and some 4,900–6,900 individuals in winter (CWA). Numbers have also dropped sharply in Granada, where the present population may now be half of that during the 1980s (Soler 2006). Highest densities are in the mesomediterranean, near wetlands (132 birds/km^2), in irrigated farmland (52 birds/km^2) and, perhaps more typically, in matorrals, urbanised areas and vineyards (24–40 birds/km^2) (ACRE).

Despite the recent negative population trends, the present range in Spain is apparently much more extensive than was the case in the 1920s and may still be expanding (AI2, SA1). Early authors, including Saunders (1871, 1897), Irby (1895), Noble (1902), Ticehurst & Whistler (1925), Witherby (1928) and Jourdain (1936–37), all described it as very locally distributed. It was not reported at all from Valencia by Arévalo (1887), nor from Aragón by Boxberger (1921) or from Galicia by Ticehurst & Whistler (1928). The enquiry on corvids in Spain by Marina & Bezares (1933) acknowledges Jackdaw range expansion in many areas of the northern and southern Mesetas around 1920–1925. It was already present at Doñana in the late 19th century (Irby 1895) but the population there later underwent a huge increase (Valverde 1960). However, Bernis (1945a) considered it to be well distributed across the two mesetas, while not denying that it might recently have been increasing there.

The Portuguese population is also thought to be declining. The breeding range in the country has contracted since the 1990s, when the Jackdaw used to nest at sites in Peneda-Gerês national park, Estremadura, Ribatejo and eastern Algarve, for example (PA2).

Iberian Jackdaws are resident but the nesting colonies are abandoned after breeding and only reoccupied in the autumn (Soler 2006). Wing-tagged juveniles marked in Madrid mainly travelled 10–20km, the maximum dispersal being 42km (Soler 2006). Large gatherings may form on farmland to forage after the harvest and large communal roosts are used outside the breeding season. Local populations may be high. For example, some 30,000 have been censused at roosts within an area of 1,400km^2 in the Hoya de Guadix, Granada (Zúñiga *et al.* 1982).

Some birds from further north in Europe join Iberian residents in winter. An influx is noted in Catalonia from late August and especially in September, with departures from late February but particularly in the second half of April–early May (Ferrer *et al.* 1986). There is a November recovery of a Swiss bird in Catalonia (BWP), another ringed in Switzerland in May was found in Navarra in November (Glutz von Blotzheim & Bauer 1993) and a bird from southern France was found in eastern Spain (SMA). There is a recent record of the Scandinavian race *C. m. monedula* in Spain: a single bird at El Franco, Asturias, during 4–6 February 2006 (*Ardeola* 56: 335). There are references to small-scale movement across the Strait of Gibraltar (Tellería 1981, BSG) but such movements may not be regular. There are only occasional records from Gibraltar itself, although a small colony bred on the north face of the Rock in the mid 1960s (BG).

The Jackdaw is a rare vagrant in the Balearics. One was collected in Mallorca in October 1965 (*Ardeola* 11: 139) and there have been at least eleven subsequent records of single birds, all within the period 29 August–25 April, most recently on Dragonera in April 2009 and at Calvià in September 2011 (*A. O. Balears 2009*: 96, *Ardeola* 59: 451).

ROOK

Corvus frugilegus

(Graja, Gralha-calva)

C. f. frugilegus. **A very local resident in León, Spain. Formerly widespread in mainland Iberia in winter but now very rare. Vagrant to Portugal and the Balearics.**

Breeding population estimates. Spain: 2011 census 1,399 pairs.

The Rook has always been an extremely local breeder in Iberia, at least in historical times. The sole population is in the north-west, in a small area of León province in the vicinity of the city of León and the town of La Bañeza, along the valleys of the river Esla and its tributaries. This population was first decribed as recently as 1953 (Valverde 1953b) but the first census was not conducted until 1976, when some 1,100 nests were found (Ena 1979). It may well be a relict of a former more extensive presence, since Martínez del Espinar, a 17th century hunter in the service of King Felipe IV, wrote that Rook ('*cuerva*') colonies in trees, such as at Aranjuez, Madrid, used to be destroyed by people (de Juana & de Juana 1984). The rookeries are sited in isolated copses of tall poplars, although in recent years there have been nests in conifers, in park trees in León city and on pylons (Olea 2000, LR). The birds forage in agricultural land, and also attend municipal rubbish tips (Olea & Baglione 2008). There is a single and somewhat surprising record of nesting in eastern Portugal: by an isolated pair in the Serra do Marão and Alvão, Douro Litoral/Alto Douro, on several occasions in the late 1970s (Fontoura 1985).

The Spanish population has fluctuated since it was first censused (García Fernández 2012). It declined to no more than 887 pairs in 1978, as a result of direct persecution, but rallied to number almost 2,200 nests in 2006 (Olea *et al.* 1997, García Fernández *et al.* 2008). However, only 1,399 nests were found in 2011 (García Fernández 2012). A mean annual increase in the number of pairs of 1.3% during 1996–2006 was followed by a mean annual decline of 6% during 2006–11. The area occupied by the breeding birds has also contracted and in 2011 it covered 824km^2, some 58% of the range in 1996. The Rook is less subject now to direct persecution but it can be harmed by agricultural intensification, notably the use of agrochemicals, and by losses of nesting sites. Many colonies disappear when the poplar plantations that they especially favour are felled for timber. Tree felling accounted for the loss of 44 of 51 former colonies (García Fernández 2012).

The Rook used to be well known in Iberia as a winter visitor, often in very large numbers, but it declined almost to the point of disappearance during the late 20th century (Román & Gutiérrez 2008). Four ringing recoveries in Spain comprise three from western Russia and one ringed as a pullus in Poland (SMA), showing that at least some of the immigrants were of eastern European origin. Until the 1930s the Rook was regular in winter across most of Spain except for the southeastern quadrant (Marina & Bezares 1933). It was then considered locally abundant in Galicia and Asturias, around Sevilla and Doñana in Andalucía, in Madrid, around Segovia in Castilla y León and in Girona in northeastern Catalonia, as well as being widely present elsewhere. Castellarnau (1877) describes how it arrived on the northern meseta in very large numbers in November and remained until March, ranging widely over farmland in flocks of many thousands of individuals. Irby (1895) saw large flocks on the Coto del Rey, Doñana, in January; Witherby (1928) saw large numbers near Aranjuez in November 1922; and Gil Lletget (1943) reported wintering flocks of a thousand or more at Candeleda, Ávila. It was also widespread in winter in Portugal during the 19th century and until at least the 1930s (AP): Tait (1879) noted that large flocks arrived in autumn and remained until February–March.

The Iberian winter range became progressively more restricted between 1933 and 1979 and it became essentially confined to the northern Meseta and the Ebro valley, although it was still locally numerous there, with observations of a flock of 800 in La Rioja and over 1,000 in Valladolid during this period (Román & Gutiérrez 2008). Large-scale passage across the western Pyrenees was reported by Iribarren (1969), who sometimes saw enormous flocks crossing the passes in autumn. Since 1979 reports have come from only nine provinces but by 2006–07 they were confined to only one, La Rioja, and during this period the wintering population fell from nearly 8,000 birds in 18 flocks to just a single flock of 50. There has only been one record in Portugal since the early 1980s: a single bird at Odelouca, Algarve, on 9 December 1987 (*Pardela* 8: 16). This dramatic decline in the wintering population has been paralleled elsewhere in southern Europe and is principally due to birds no longer travelling as far south as they used to (BWP, Román & Gutiérrez 2008). The

outcome is that the Rook has become a rarity in most of Iberia and any future observations are of interest.

The breeding population is essentially resident, with winter observations usually being no more than 10km from the breeding colonies (Ena 1979). A roost of 23 at Castronuño, Valladolid, in January 2005 (*Ardeola* 52: 435) could have involved wintering birds since Castronuño is some 100km from the nearest colony. Occasional individuals still appear outside the breeding range and the recent wintering range. For example, there was a flock of five at Mocejón, Toledo, in January 2001 (*Ardeola* 49: 194). Similarly, there was one at the Ría de Villaviciosa in January 2006 and again in December 2006 (*Ardeola* 56: 172). One was seen in flight at Sant Pere de Ribes, Barcelona on 28 May 2010 and two appeared at Motril, Granada, in December 2011 (*Ardeola* 57: 547, 59: 193). One at Cartagena in January 2013 was the first known occurrence in Murcia (*Ardeola* 60: 181).

The Rook has always been a rare visitor to the Balearic islands: Ferrer *et al.* (1986) records just three occurrences in Mallorca: pre-1866, in 1914 and in April 1933, two in Menorca: pre-1911 and during winter 1975/76, and one in Ibiza: in 1914.

CARRION CROW *Corvus corone*

(Corneja Negra, Gralha-preta)

C. c. corone. **A common resident over much of Iberia, especially the north. Very rare in the Balearics.**

Breeding population estimates. Spain: SA1 310,000–530,000 pairs; ACRE 2.1 million birds. **Portugal**: 5,000–50,000 pairs. **Andorra**: 100–200 pairs.

The Carrion Crow is widespread and common across large tracts of Iberia but there are significant gaps in its breeding distribution. It is rare or absent along the Mediterranean coast except in the extreme north of Catalonia. It is similarly absent from most of Extremadura: although there are a few pairs in the north in the foothills of the Central Range, and it is rare in southwestern Castilla-La Mancha, in Ciudad Real and Toledo. It is also lacking from western Andalucía. Its presence in the southern half of Spain is largely restricted to eastern Castilla-La Mancha, the eastern Sierra Morena, the mountains of Murcia and Almería and the east of Jaén and Granada (SA2). Although it occurs throughout much of Portugal it is absent from most of Algarve, southern Alentejo and the Guadiana valley (PA2).

In general the Carrion Crow is widespread throughout the eurosiberian zone, and also in the supramediterranean, but it avoids the meso- and thermomediterranean zones (AI2). It tends occupy the intermediate levels in mountainous regions, ranging to 1,600m in the western Pyrenees, to 1,800m in the Iberian Range and the Central Range and to 1,300m in the Sierra Nevada. This avoidance of the upper levels may be more due to its attachment to cultivation than to any aversion for high altitudes (de Juana 1980). It is very much associated with all types of agricultural land but is perhaps especially attracted by cereal crops. However, and unlike elsewhere in much of western Europe, it is not yet widespread in urbanised habitats, although a progressive recent spread into built-up areas has been reported in Portugal (AP) and may be expected in Spain. Bernis (1988b) did not include it among the 'urban' species of the Castilian mesetas and Carrascal *et al.* (2005) indicate that it avoids urban environments in Spain. It avoids forests but occurs on the margins, and in other wooded habitats, such as dehesas, open pinewoods and riparian growth, that offer suitable trees for nesting. Some forage regularly on the seashore on northern and western coasts. Nearly all nest in tall trees but pylons and cliffs may be used exceptionally. Large flocks form in open country outside the breeding season.

Recent trends on SACRE plots indicate a small decline in Spain (Escandell 2013). It is most abundant on the northern Meseta and also in Galicia. Maximum densities, up to 16 birds/km^2, have been found in human-managed habitats in the supramediterranean and eurosiberian zones: in matorrals and pastures with trees and agricultural land (ACRE). The Portuguese population has been considered stable (PA2), although a moderate increase was detected during 2004–09 (ACP). There is also some evidence of a significant recent increase in Andorra (BE, ANA).

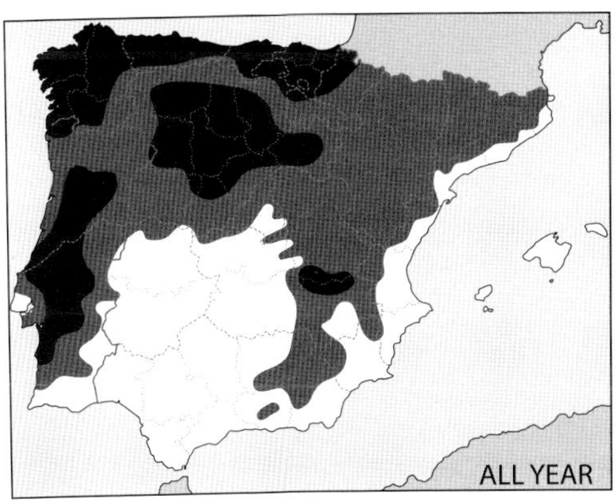

The Carrion Crow was formerly heavily persecuted in Iberia and some recovery of lost ground, as in Catalonia (Ferrer *et al.* 1986) or in coastal Portugal (AP), may reflect an easing of such pressure. The historical literature certainly gives an impression of former scarcity, especially in comparison to the Raven. Eagle Clarke (1889) only saw one pair in Andorra;

Saunders (1897) saw only two in Catalonia, near Figueres; Irby (1883) only noticed a few between Torrelavega and Unquera during his visit to Cantabria with Lord Lilford in 1876, and Castellarnau (1877) said it was uncommon in Segovia, on the northern versant of the Central Range. Somewhat later, Bernis (1946) found that it was still uncommon in the La Maragatería region in León. The only exception among the earlier accounts appears to be the observation by Ticehurst & Whistler (1928) that the Carrion Crow was exceedingly common throughout the whole of Galicia, as is still the case.

Iberian crows appear to be sedentary but some wander more widely outside the breeding season, giving rise to the few records from outside the nesting range. An observation of five at Barão de S. João in November 2010 (*Anu. Ornitol.* 8: 83) is noteworthy for western Algarve. A flock of 42 at Riomalo de Abajo, Cáceres, on 21 March 2012 was a regional record (*Ardeola* 59: 193). The only published record for the Balearics involved a flock of 20 on Ibiza in February 2000 (*A. O. Balears 2000*: 146). Although there is no evidence to prove it, the arrival of a sizable flock suggests that they might have been long-distance migrants originating from further north in Europe and such birds may also give rise to rare records of small flocks arriving from Spain on the southern shore of the Strait or occurring in winter in northern Morocco (BM). They include 36 seen arriving at Cap Ceres, south of Tarifa, on 23 October 1974 (Pineau & Giraud-Audine 1976). For all that, there are very few observations from the north side of the Strait. Llandres & Urdiales (1990) give only one record for Doñana: in late September 1986. Mountfort (1958b) saw four flying towards the Strait on the Málaga coast in October 1957. Two were seen near Algeciras in October 1977 and two near Tarifa in November 1977 (AI2). Three were near Vejer de La Frontera, Cádiz, in September 2012 (*Ardeola* 59: 451). There are only three March records of 1–2 birds from Gibraltar: in 1967, 1974 and 1986 (GONHS).

HOODED CROW *Corvus cornix*

(Corneja Cenicienta, Gralha-cinzenta)

Vagrant to Spain. Recorded once in Portugal (Category D).

Some Hooded Crow populations are partial migrants, travelling mainly short distances to winter south of the breeding range. Scandinavian birds were formerly common in France in winter but are now only rarely seen there, in the extreme north. Some are also reported in winter in Brittany and in southern France and it is thought that the former may originate from Ireland (*C. c. cornix*) and the latter from Italy (*C. c. sharpii*). There are no accepted records for Morocco but an individual of doubtful origin was present in Ceuta during May–June 2007.

There are a few records for Spain. One was seen at the Albufera de Mallorca in June 1926 (Munn 1928). Another was captured at Villaviciosa, Asturias, in September 1976 (Lobón-Cerviá & González 1982) and there are observations from Ripoll, Girona, in July 1981 (Chacón *et al.* 1985), and Osona, Barcelona, in September 1984 (Aymerich *et al.* 1991). The Spanish rarities committee has accepted a further seven records of single birds. Three were in the northern coastlands: at Arzúa, A Coruña, on 27–28 April 1990; at Tapia de Casariego, Asturias, during 18–22 November 2006, and at Meaño, Pontevedra, on 10 November 2010. Subsequently, there was one at A Frouxeira lagoon, A Coruña, on 10 May 2012 (RBS). The other observations have been in Catalonia: at El Prat de Llobregat on 19 April 1989; at the Aiguamolls del Empordà on 27 March 1993; at Pals on 4 May 2003, and on the lower Río Ter from August 2006 until mid 2008. There was another observation in this region in 2012: one at Castell-Platja d'Aro in March–April 2012 and it or another individual was at the Pals ricefields in May–June 2012 (RBS). There were two more Catalan records in March 2013: single birds at the Llobregat delta and at the Ter ricefields (RBS). The Catalan records might be associated with the frequent observations of birds of Italian origin in southern France (OF). On the other hand, the records from Galicia and Asturias could perhaps have involved birds from Ireland. The only observation in Portugal, a bird at Vila do Conde, Porto, in November–December 2010 (*Anu. Ornitol.* 8: 17) has been assigned to Category D (Matias *et al.* 2011).

COMMON RAVEN *Corvus corax*

(Cuervo Grande, Corvo-comum)

***C. c. hispanus*. A common resident throughout most of mainland Iberia and the Balearics. Smaller and shorter-winged than the nominate race, with a more arched culmen.**

Breeding population estimates. Spain: SA1 60,000–90,000 pairs; ACRE 240,000 birds. **Portugal**: 300–1,000 pairs. **Andorra**: 50–150 pairs. **Gibraltar**: 1 pair.

The Raven is resident throughout almost all of the Peninsula and on the main Balearic Islands. It is scarce or absent only

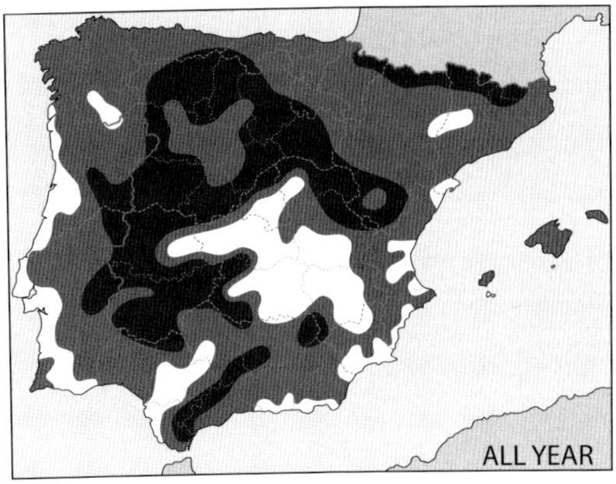

in parts of the southern Meseta and the Guadalquivir valley and in some coastal regions of Portugal (SA2, PA2). It is most characteristic of areas that offer steep rocky escarpments for nesting: including mountains, river gorges and sea cliffs, although it will nest on pylons, ruins and large trees. It may be found in an extremely wide range of habitats, from the seashore to the highest alpine pastures and including all except perhaps the most intensively managed types of farmland. It ranges to the summits of the Sierra Nevada, although it only nests there up to 2,200m (Garzón & Henares 2012). It regularly attends rubbish dumps, often in some numbers, and frequently joins vultures at carcases.

The highest densities in Spain, 1.1–1.8 birds/km^2, are in supramediterranean pastures, cereal croplands and pinewoods above 1,250m and in the eurosiberian, in pine/fir and beech woods and in pastures with trees above 750m (ACRE). The Catalan regional population has been put at 1,400–1,600 breeding pairs (CA2).

The Raven seems to have been abundant in many Iberian regions during the late 19th century and early 20th century, as testified for example by Lilford (1866), Saunders (1871), Arévalo (1887), Munn (1921), Tait (1924) and Jourdain (1936–37). More recently the population was severely affected by the indiscriminate use of poison baits for foxes and other vermin until the 1970s, when the use of strychnine in such baits was forbidden in Spain. The Raven has also suffered from direct persecution, the abandonment of traditional farming practices and disturbance, all of which still affect it to some extent. Some are killed by collisions with powerlines or electrocution on pylons. Local declines have also followed the closure of rubbish-tips and the reduction of extensive livestock grazing. A significant decline was reported on SACRE sites during 1998–2012 (Escandell 2013). Persecution is thought to have been responsible for the 50-year disappearance of the Raven from Estremadura during the 20th century: the region has since been recolonised (AP). The Portuguese population may still be decreasing, with apparent range contraction seen in some coastal regions and in central Alentejo (PA2).

Pairs are territorial and have large home ranges so population densities tend to be low. Adults are resident on their territories but young birds may wander, although their movements are poorly known. Few Ravens have been found to wander very far: one ringed at Rasquera, Tarragona, in April 2008 and recovered in Aude, France (202km) in April 2009 (ICO) is the only extra-regional recovery. Other long-distance recoveries include a bird ringed in Toledo and recovered 136km away in the same province and another ringed in Tarragona and recovered 144km away in Barcelona (*Ecología* 3: 228; ICO, CMA). Flocks of Ravens, perhaps principally comprising immature birds, may form after the breeding season and persist through the winter, when they may range through agricultural areas where few Ravens breed, notably across the Southern Meseta and in the Guadalquivir Valley (SWA). These flocks sometimes include 100 or more individuals and several flocks may combine in large communal roosts. A roosting flight of 515 over Holm Oak dehesas near Navalvillar de Pela, Badajoz, in December 1989 (*Ardeola* 37: 350) is noteworthy.

Irby (1895) claimed to have identified the North African race *C. c. tingitanus* on the Spanish side of the Strait in spring. There have been no subsequent records to confirm this but it is fairly common on the south side of the Strait (Pineau & Giraud-Audine 1979) and occasional occurrences in Spain cannot be ruled out particularly since Ravens – race unknown - are occasionally seen crossing the Strait in both directions (pers. obs). *C. c. tingitanus* is even smaller than *hispanus* with a shorter, deeper bill, longer wings, a shorter tail and very short throat hackles: the plumage may bleach brownish on the head and body (Madge & Burn 1994).

CRESTED MYNA — *Acridotheres cristatellus*

(Miná Crestado, Mainá-de-crista)

Introduced to Portugal (Category C) and recorded in Spain (Category E2).

Escaped birds have recently established a small breeding population in the Greater Lisbon area in Portugal, where they were first detected in 1997. They occur west of the city on both sides of the mouth of the Tagus and also in the south of the Setúbal peninsula around Sesimbra (PA2, AP). They are most numerous around Corroios, where they were first found, having increased there from some ten birds in 1997 to over 100 by 2002 (AP). The birds occur in urban and suburban areas offering abundant leafy tree cover, typically of palms and other exotic species. Occasional escaped individuals are recorded in Spain.

SPOTLESS STARLING *Sturnus unicolor*

(Estornino Negro, Estorninho-preto)

An abundant resident throughout most of mainland Iberia. Has bred in the Balearics.

Breeding population estimates. Spain: SA1 2.0–2.5 million pairs; ACRE 52.7 million birds. **Portugal**: 50,000–500,000 pairs. **Gibraltar**: 100 pairs.

The Spotless Starling is endemic to the western Mediterranean and has its core population in Iberia. It is now common throughout Portugal and all of Spain, following a sustained period of range expansion, although it is not fully established in the Basque coastlands. It is prone to hybridise with the Common Starling where their ranges overlap in northern and northeastern Spain but the significance of this is yet to become apparent. The two species are genetically similar (Celis *et al.* 2007, Lovette *et al.* 2008) but it remains unknown whether their hybrids are fertile (Veiga & Polo 2011).

It is mainly a vagrant to the Balearic Islands (Ferrer *et al.* 1986, *A. O. Balears 2010*: 269) and the regional bird reports give only ten records during 1986–2010, most of them of single birds. One was paired with a Common Starling at the Albufereta de Pollença, Mallorca, in spring 1994 and fledged hybrid young (*A. O. Balears 1994*: 114). There is also a record of a colony of 6–7 pairs reported from Ciutadella, Menorca, during May–June 1998 and a report of ten individuals, again at Ciutadella, in December 2010 (*A. O. Balears 1998*: 118, *2010*: 215). Most recently, two small breeding colonies were found in Mallorca in 2011, in Pollença and Palma, and four pairs nested in Ibiza in 2012 (*Ardeola* 59: 451, 60: 542).

This is the familiar starling of most of Iberia and it is a common sight and sound in cities, towns and villages, where it nests in holes in buildings and, especially, under roofs (Bernis 1989a). It is also common in agricultural land of all kinds, feeding in flocks in the fields and pastures and roosting in trees. Nesting and roosting in holes in sea cliffs also occurs locally. Large numbers occur in the dehesas of central and southern Iberia, especially those in which ungulates are grazed, where they nest in tree holes: they benefit from the short sward that results from grazing and also from the grain feed supplied to the animals. The Spotless Starling is essentially a commensal of humans and avoids densely forested areas as well as higher elevations, above 1,300–1,500m, although flocks ascend to feed alongside grazing cattle at up to 2,200m in the Central Range in summer (SA1, AI2, Veiga & Polo 2011). Mixed feeding flocks with Common Starlings are widespread in open country in winter. Both species are largely insectivorous in the breeding season and consume mainly plant material in winter. The Spotless Starling in the Salamanca dehesas has been found to forage in grassland for cranefly larvae and other insects, notably beetles, from April to June, later switching mainly to grasshoppers and crickets; in winter it takes grain, other seeds and fruits (Peris 1980).

The ACRE assessment of the mean Spanish population in spring during 2004–06, over 50 million birds, is remarkable. Such a huge figure, equivalent to the entire western Palearctic population of the Common Starling, may reflect significant methodological biases in the assessment procedure (Murgui 2011a) and thus needs to be regarded with circumspection. The population on SACRE plots increased significantly by 27% during 1998–2012 (Escandell 2013). High densities, 250–550 birds/km^2, are found in and around urbanised areas in the meso- and supramediterranean and in the eurosiberian region, as well as in mesomediterranean riparian woodlands, pastureland and irrigated croplands (ACRE).

The Portuguese population has shown both local increases and declines in numbers, for example becoming more numerous in Lisbon city but scarcer in parts of coastal Algarve (AP). The Spotless Starling occurs throughout the country but the highest concentrations are south of the Tagus and in the north-east (PA2). It is only a non-breeding, although increasingly frequent, visitor to Andorra (ANA)

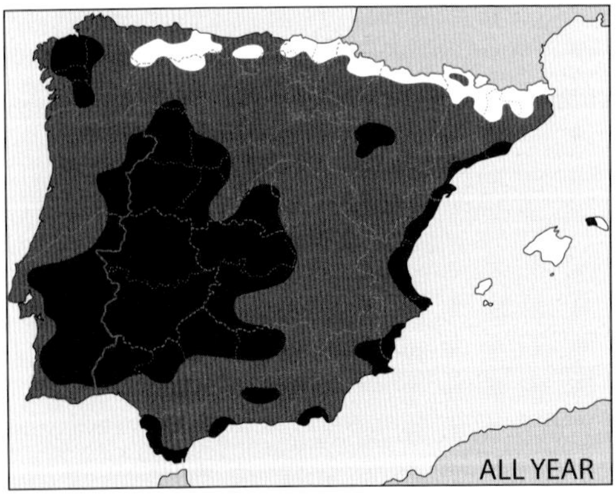

The Spotless Starling was unknown in northeastern Iberia until quite late in the 20th century and it was apparently uncommon near Gibraltar in the late 19th century and early 20th century (Irby 1875, Stenhouse 1921), although Chapman (1884) found it to be abundant in the west of Cádiz province in June 1872. It was also apparently unknown in Algarve during the 19th century (Paulino d'Oliveira 1896) although Ticehurst & Whistler (1933) recorded it there. It was chiefly found in central and western Iberia, including most of Portugal and also in Galicia: where Ticehurst and Whistler (1928) noted breeding colonies around Santiago de Compostela in May 1927. However, an ongoing process of range expansion, that began in the 1940s (Bernis 1945), took it to all corners of the Peninsula. Breeding birds became established south to the Strait of Gibraltar perhaps in the 1930s (Tellería 1981) but they only reached Gibraltar itself during the 1960s (BG). The Spotless Starling also spread into Valencia in the 1960s and moved north into southern Catalonia during the 1970s (Mestre 1978), thereafter expanding into northern Catalonia in the 1980s and 1990s at a mean rate of 4.7km/year (Ferrer *et al.* 1991, CA2). It also moved up the Ebro valley from Catalonia in the 1970s, to the southern reaches of Cantabria and the Basque Country (de Juana 1980, Elósegui 1985), continuing to the Basque coastlands in the 1980s (Álvarez *et al.* 1985). Birds also spread east from Galicia along the Biscay coastlands, across Asturias, during this period (Noval 1986). The present breeding range extends into southern France as far as Narbonne and breeding has also been reported further east at Var and Bouches-du-Rhône (OF).

Ringing recoveries show that it is largely resident, although juveniles disperse short distances (Peris 1991, Peris & Galeano 2005). Most ringing recoveries have been less than 20km from the ringing sites, although one ringed as a nestling flew 701km from Álava to Sevilla and a juvenile ringed at Villajimena, Palencia, travelled 383km to Ribeira, Pontevedra (Palacios & Asensio 1986, CMA). Nevertheless, some Spotless Starlings may perform daily foraging movements of up to several tens of kilometres year-round (Bernis 1989a). A curious example of this was recorded by Tellería (1981) who observed a coastal roost on the shoreline of the Strait of Gibraltar at Tarifa in October 1977, from which up to 400 Spotless Starlings were seen to commute daily to and from Morocco, 16km to the south.

COMMON STARLING — *Sturnus vulgaris*

(Estornino Pinto, Estorninho-malhado)

***S. v. vulgaris.* A recent colonist and increasing resident in northeastern and coastal northern Spain. Common and locally abundant in winter throughout the Peninsula and the Balearic Islands.**

Breeding population estimates. Spain: SA1 220,000–1 million pairs; SA2 400,000–1.2 million pairs.

A pair of Common Starlings is said to have nested on a house in the Llano de Vic, Barcelona, in spring 1881 (Vayreda 1883) but that record is impossible to verify (Mestre 1975). Saunders (1884) described it as common in the Pyrenees but his later writings (Saunders 1897) make it clear that this was only in winter. The first confirmed reports of Common Starlings nesting in Iberia came from northern Navarra in the late 1950s (SA2) and from 1960 in northern Girona (Balcells & Masoliver 1960). Mestre (1975) summarises the process of the early colonisation of Catalonia, where they reached the Ebro by the early 1970s. On the north coast birds spread west to Cantabria during the 1970s (Lavín 1977) and subsequently extended to central Asturias in the 1980s (AI2). In general, the species has continued to expand its breeding range slowly along the north coast. In the north-east the breeding range includes most of Catalonia and the easternmost part of Aragón. There are also scattered nesting pairs along the upper Ebro valley in La Rioja and Navarra, and occasional nesting has been reported across the northern Meseta (SA2). Further south two pairs first nested in the Comunidad Valenciana on the outskirts of Valencia city in 1995 and there have been isolated nesting attempts elsewhere in the region thereafter, including at Barracas, Castellón in 2007, but the Common Starling is not firmly established there (*A. O. C. Valenciana 2009*: 143–144). It has not yet been found nesting in Andorra although it occurs there on passage and in winter (ANA).

The birds in Catalonia are strongly associated with farmland as well as with urban areas, nesting in cavities in buildings. They occur commonly from sea level to the lower levels of the Pyrenees and coastal mountains (CA2). A comparison between the first

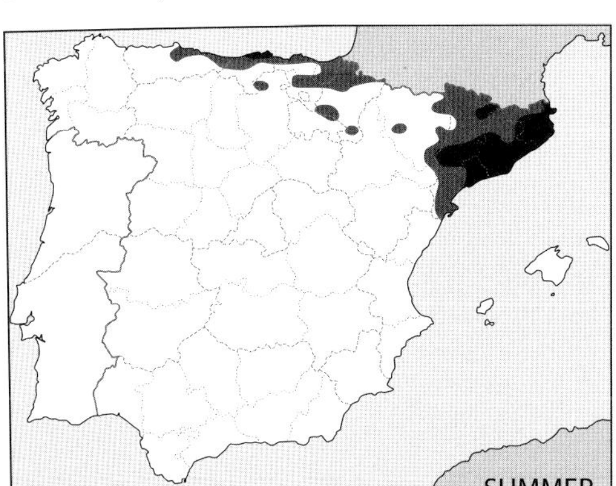

and second Catalan breeding atlases, from the early 1980s and early 2000s respectively, shows a 21% increase in occupied 10-km squares during this period. The population in Navarra favours open deciduous woodlands, where many nest in holes in trees. Further west they use riparian trees in the upper Ebro valley. They are common in the humid pastures of the Biscay coast, nesting in buildings (SA1). The ongoing increase in the Spanish breeding population is also confirmed by the significant increases, of 62%, seen on SACRE sites during 1998–2012 (Escandell 2013). The core population, of Catalonia, has been put at 345,000–598,000 pairs (CA2).

Common Starlings bred sporadically in the Balearic Islands in Mallorca during the 1980s (Ferrer *et al.* 1986). Only a few pairs were involved and the population failed to become established. However, a pair bred successfully at the Albufera de Mallorca in 2005 and 2006 (*A. O. Balears 2005*: 237, *2006*: 204).

The southward spread of this species may be countered by the northward spread of the Spotless Starling. Indeed their respective rates of spread appear to have slowed down since their ranges came into contact, perhaps due to interspecific competition (Ferrer *et al.* 1991). Moreover, the Common Starling population that became established on the Huesca plains, northern Aragón, during the 1970s and 1980s has declined since the Spotless Starling arrived there (SA1, SA2). Both species often occur together at all seasons and form mixed flocks. The two may even be found nesting in the same roof. They are very closely related and show only minimal genetic differences (Celis *et al.* 2007, Lovette *et al.* 2008) and interbreeding is fairly frequent. Hybrids were estimated to comprise 0.49–1.02% of the population in a farming area in Lleida (Motis 1992). It remains to be seen whether the hybrids have reduced fitness or whether introgression occurs between the two species.

Very large numbers of Common Starlings from central and northern Europe winter in Iberia and a minority continue south to northwestern Africa. The wintering population was estimated at ten million individuals by Peris *et al.* (1991) but it appears to have declined since the 1960s, at least in the southern half of Iberia (Peris *et al.* 1991), and certainly the numbers observed there show marked interannual variation. The Common Starling was once renowned as a pest by olive growers, since it descended in hordes on the fruiting trees in winter (Bernis 1960c), but this problem seems to have ameliorated. It used to be trapped in very large numbers in parts of western Andalucía and Extremadura, as well as in the Ebro Valley from southern Navarra to western Catalonia, and quantities of the birds were sold for food in the markets of Sevilla and elsewhere (Parsons 1960, Bernis 1960c and 1989a). Specialist hunters employed large traps to take advantage of the huge roosts that then existed at such locations as Gallur and Zaidín in Aragón, Don Benito in Extremadura, and Cantillana, Villaverde del Río and El Cuervo in Andalucía.

The former wintering range largely coincided with the distribution of olive groves so that the largest concentrations were in the Ebro Valley, the Levant and Balearic Islands, Extremadura and Western Andalucía, with only small and very scattered flocks appearing in northern Spain (Bernis 1960c). Wintering in the north seems to be an increasing trend (Peris *et al.* 1991, SWA). However, the recent Spanish wintering distribution (SWA) appears to be in some way complementary to that of the Spotless Starling, with high densities of Common Starlings frequenting both the Atlantic and Mediterranean coastlands, and the Ebro Valley, and only low densities occurring inland, especially in both Castiles and Extremadura, where the Spotless Starling is the more abundant of the two species.

The first arrivals in the north may be as early as August (Artázkoz 1995) but the main influx is in October–November, when the birds spread throughout Iberia (Bernis 1960c). The principal departures are in February–March (Bernis 1960c) but some remain until April and occasional individuals linger later. Ringing recoveries show that the winter immigrants originate from north-central Europe: especially from Germany, Switzerland, the Czech Republic and Poland, with some from as far east as the Baltic States (34), Belarus (16) and Russia (68) but hardly any from Scandinavia (5) or Britain (6) (Bernis 1960c, AI2, Artázkoz 1995, SMA, AP). One ringed at Alfaro, La Rioja, in January 1973 was found at Zolotnosha, Ukraine (2,700km) in March 1974 (ICONA, *Ardeola* 29: 147). There are also 13 North African recoveries of birds ringed in Iberia: from Algeria (10) and Morocco (3). Different populations seem also to differ to some extent in their choice of wintering area: birds from Italy and Switzerland occur in the Balearic Islands and the Levant coast; those from central and eastern Europe winter in central Iberia, and the Cantabrian coast receives French and Belgian birds (Bernis 1960c, Peris *et al.* 1991).

Wintering birds form large, occasionally enormous, roosts, sometimes mixed with Spotless Starlings and House Sparrows but more often monospecific. The roosts, which may hold hundreds of thousands of birds, are in reedbeds, riparian woodlands and, often, in parks and gardens in urban areas. A spectacular roost is sometimes reported at the Albufera de Mallorca, with an estimated two million birds in January 2002 (*Ardeola* 52: 223).

Significant numbers of Common Starlings continue south from Iberia to winter in Morocco but the numbers involved vary considerably from year to year and mid-winter movements may occur, probably in response to hard weather in Europe (BM). In general the passage is unspectacular but there are records of very large numbers on occasion. Pineau & Giraud-Audine (1976) saw enormous flocks arriving near Tangier from Spain on 29–30 November 1970 and they counted 10,000 leaving Morocco for Spain on 11 January 1973. Winter roost counts of up to one million Common Starlings in the Tangier Peninsula (Pineau & Giraud-Audine 1979) make it clear that very large numbers indeed reach Morocco in some seasons. Migrants have been seen crossing the Strait from early October but most passage is from mid October through November, with a return in February–early March. There are indications that migration into Morocco may have declined in recent decades (SWA).

ROSE-COLOURED STARLING *Pastor roseus*

(Estornino Rosado, Estorninho-rosado)

Vagrant to Spain and Portugal.

The Rose-coloured Starling often appears in western Europe, notably in the British Isles, and there are significant westward eruptions in some years, as in 2002. There are records of at least 723 in Britain during 1958–2003, 135 in Ireland up to 2011, and 390 in France up to 2011. There is also one accepted record for Morocco, in February 2007, and one for the Canary Islands in January–February 1981.

Figure 201. Regional distribution of records, 1957–2010 (n = 52)

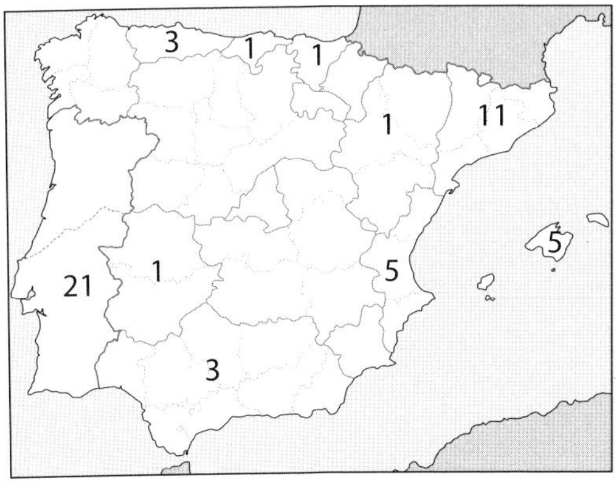

There are undated references to its presence in Spain in the 19th century (Arévalo 1887, Gil Lletget 1945, Ferrer *et al.* 1986) and specific observations of single birds at Toses pass, Girona, on 4 November 1957, near Cáceres city, on 17–18 May 1971, and at Beniure, Lleida, on 12 April 1984 (*Ardeola* 4, 193, 16: 277, 29: 190). An adult was killed in Portugal at Entroncamento, Santarém, on 3 November 1965 (Isidoro 1968). Subsequent to the establishment of the rarities committees there have been 28 accepted records for Spain and the Balearics up to 2010 and 20 for Portugal. In all there are 52 records (55 individuals). These were chiefly at coastal locations (Fig. 201), concentrated in two widely separated areas: 40% in southern Portugal and 40% in total in northeastern Spain: in Catalonia, the Balearics and the Comunidad Valenciana. A further eight individuals were reported in Spain and eight in Portugal during 2011–12 (*Ardeola* 60: 491, RBS, RBP).

Figure 202. Monthly distribution of records, 1957–2010 (n = 52)

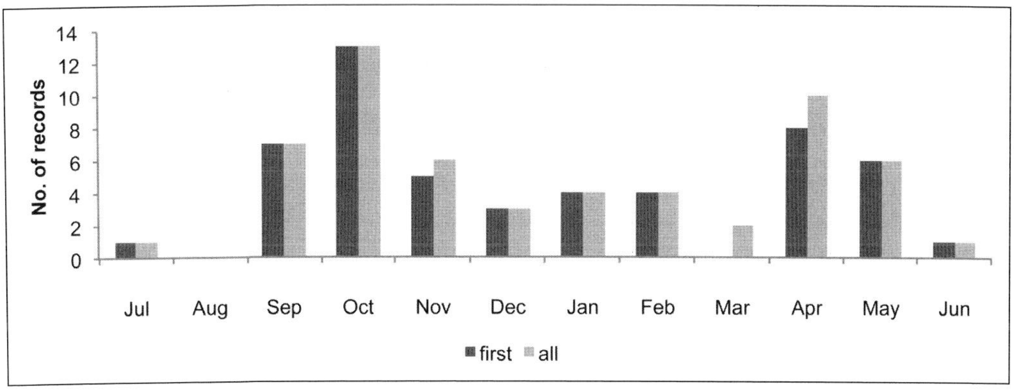

There have been records year-round but with a major peak in autumn during September–November and a lesser spring peak around April–May, and very few have been in summer (Fig 202). Most new arrivals are first-year birds, which comprised 31 of 33 individuals recorded during September–February. Autumn arrivals are most evident in the west, and winter and spring records predominate in the east (Fig. 203): September–November records make up 70% of the total in western Iberia and just 18% in eastern Iberia. Observations in France also peak in autumn and spring, although the autumn peak is earlier and the spring peak later than in Iberia, and whereas most autumn records are on the Atlantic seaboard, nearly all the spring ones are on the Mediterranean coast (OF). The French and Iberian records combined thus suggest that there is a small southwesterly movement in autumn along the Atlantic coast as far as southern Portugal, with a spring return along the Mediterranean coast. Perhaps a few winter in North-west Africa?

Figure 203. First dates of records compared between western and eastern Iberia. West = Biscay coastlands, Portugal, Extremadura & Andalucía (n = 30). East = Catalonia, Comunidad Valenciana, Balearic Islands and Aragón (n = 22).

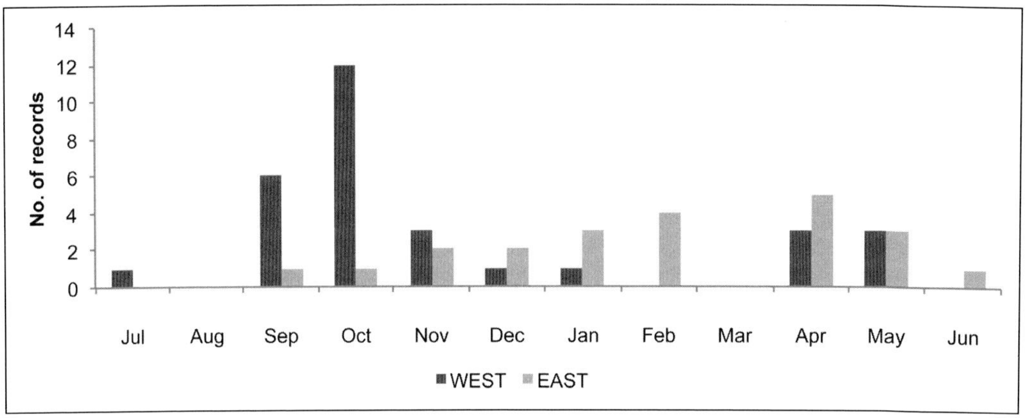

HOUSE SPARROW *Passer domesticus*

(Gorrión Común, Pardal-de-telhado)

P. d. domesticus. **An abundant resident throughout mainland Iberia.**

P. d. balearoibericus. **A common resident in the Balearic Islands.**

Breeding population estimates. Spain: SA1 9.3–10 million pairs; ACRE 164 million birds. **Portugal**: 1–10 million pairs. **Andorra**: 500–1,000 pairs. **Gibraltar** 300+ pairs.

House Sparrows are abundant throughout Iberia. They range from sea level to at least 1,600m, the altitude of the highest villages in Spain, but nest higher locally: up to 2,400m in Andorra (ANA). They are especially closely associated with human artefacts, nesting in or near cities, towns, villages or isolated settlements (Bernis 1989b). Their nests are generally placed in holes in buildings, including bridges and dams, but locally also in natural holes in trees and rocks. House Sparrows are ubiquitous in town squares, parks and gardens where they forage for scraps. The presence of parks has been positively correlated with House Sparrow density in Valencia city (Murgui 2009) but the highest densities in Madrid have been found in the most built-up districts (Bernis 1988b, 1998b). Not all House Sparrows are urbanised however and many flock in their thousands to feed in cultivated land, where they are especially abundant in irrigated croplands. Some of the birds seen in farmland may commute from towns and villages nearby, especially at harvest time, but others nest there colonially in eucalyptus or poplar plantations and in riparian trees, sometimes alongside Spanish Sparrows (pers. obs.). These build large, spherical nests of grasses and other plant material, in the outer branches. Such nests are most commonly encountered in Extremadura, Andalucía, the southern Meseta and the Ebro delta and they allow large numbers to exploit regions of steppe or farmland where few buildings are available. Some nest within the large stick nests of storks and raptors, often while these are in occupation.

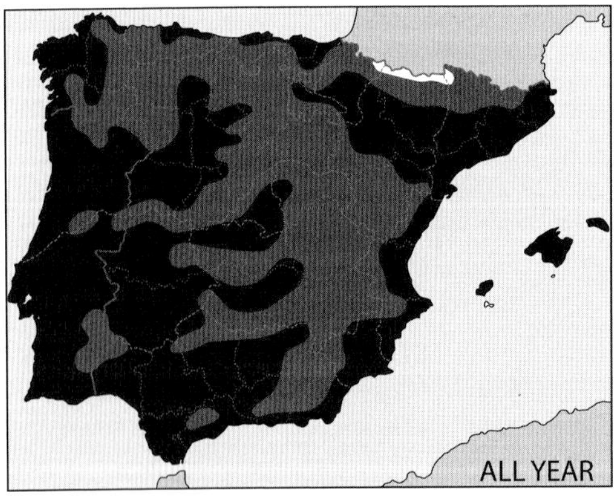

The 2004–06 ACRE estimate of *c.* 164 million birds in Spain is most remarkable since the entire European population has previously been put at just 63–130 million pairs (BE); the methodology employed is open to criticism (Murgui 2011a). More specifically, the Catalan population, some 8.6% of the Spanish birds according to the ACRE survey, has been independently estimated at 2.6–3.8 million pairs (CA2), about a quarter of the ACRE estimate. Hence the ACRE figures must be regarded as provisional.

SACRE results show that the House Sparrow is the most

abundant and most widespread bird species in Spain but they also indicate some local declines. The population on SACRE plots decreased by 5% during 1998–2012 (Escandell 2013). The alarming decreases in numbers seen in some European countries, notably in Britain (Robinson *et al.* 2005), have not yet been apparent in Iberia except on a relatively local scale (Murgui 2011b). For example, the population in orange groves in Valencia declined by 90% from 1975–2001, probably in response to more intensive management including increased use of herbicides for weed control (Gil-Delgado *et al.* 2002). The population in and around Valencia city also declined by about 70% during 1998–2007, a change attributed to reduced food availability resulting from intensified urban development (Murgui & Macías 2010). Roost counts during 1984–2003 have revealed a 60% decline in the Maresme region of Barcelona (CA2). Some decline has also been noted in parts of northern Spain, attributed to human rural depopulation and to afforestation of large areas with exotic trees. An annual decline of some 5% in Valladolid during 2002–06 is apparently highly correlated with increased electromagnetic radiation from mobile phone antennae there (Balmori & Hallberg 2007), which is interesting. This species is vulnerable to agricultural intensification and its future population trends in Iberia require monitoring, as a pointer to the overall health both of the rural and urban environments.

The highest House Sparrow concentrations in Spain are in towns and villages, and their outskirts (1,300–2,900 birds/km^2). The mean densities in agricultural areas are much lower, sometimes less than 1 bird/10ha but the results obtained may differ widely according to whether the birds are censused during the breeding season or when they are flocking in winter. High densities are reported from farmland mosaics in the thermomediterranean (941 birds/km^2).

The Portuguese population has been considered stable but there was a moderate increase during 2004–09 (ACP). It was described as abundant on the Berlengas Islands in the 19th century, when some attempts at agriculture were made there, but it no longer breeds on the archipelago and seldom visits it (AP). The population of the Andorran valleys has probably increased considerably since the late 19th century, when Eagle Clarke (1889) was only able to find a few in the capital.

Most Iberian birds are highly sedentary. Over 190,000 had been ringed in Spain by 2009, giving over 2,400 recoveries, but very few had moved any considerable distance, the three longest displacements within the Peninsula being from Málaga to Cádiz (138km), from Valencia to Cádiz (580km) and from Zaragoza to Málaga (650km). Tellería (1981) decribed a southward movement of over 20,000 birds at the Strait of Gibraltar in 1977 between late September and early November but with a clear peak in mid October: most of the individuals involved were juveniles. Such a movement has not been reported from the south side of the Strait. However, four recoveries from Morocco confirm that some birds do reach there from Spain. All were ringed during June–August and recovered during January–April. The displacements were from El Colmenar, Málaga, to Skirat-Temane (322km); Los Barrios, Cádiz, to Khenifra (213km); Antequera, Málaga, to Tangier (177km), and Castellar de la Frontera, Cádiz, to Oulad Slimane (407km). Three of the birds were ringed as fledged juveniles and the other was an adult.

SPANISH SPARROW — *Passer hispaniolensis*

(Gorrión Moruno, Pardal-espanhol)

***P. h. hispaniolensis*. A common breeder in southern and western Spain and adjacent eastern Portugal; a local or sporadic nester elsewhere. Partial migrant. A rare vagrant in the Balearics.**

Breeding population estimates. Spain: SA1 20,000–50,000 pairs; ACRE 3.2 million birds. **Portugal**: 10,000–100,000 pairs.

The traditional core nesting range is the valleys of the Guadiana and Tagus in Extremadura, southwestern Castilla-La Mancha and eastern Portugal. It now also nests commonly in the lower Guadalquivir valley and more locally elsewhere in western Andalucía: in Huelva, Cádiz, Málaga, Cordoba and western Jaén; in Madrid and Guadalajara on the southern Meseta, and in Salamanca and Zamora in Castilla y León. Sporadic nesting has been reported widely in the north-central and southern Meseta (SA2, PA2).

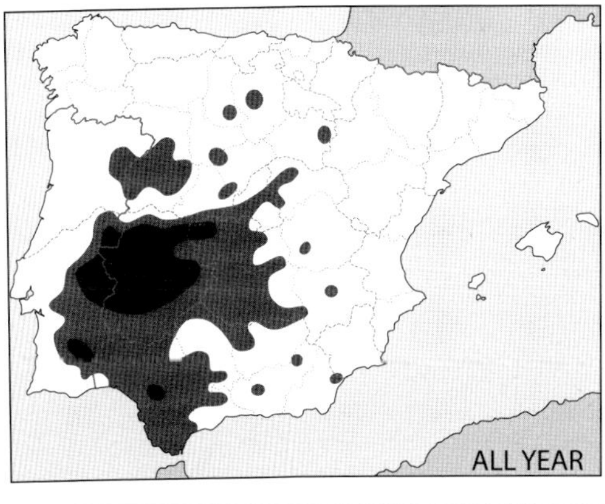

The range of the Spanish Sparrow appears to have contracted considerably during the early 20th century after which some lost ground has been recovered and new areas colonised. Range contraction was especially apparent in Andalucía, where both Saunders (1879) and Arévalo (1887) considered it abundant. However, Jourdain (1936–1937) found it to be 'not uncommon, but extremely local' in

Andalucía and during the early 20th century many ornithologists failed to find any at Doñana, so much so that Valverde (1960) thought that it was no more than a summer visitor to Spain from Morocco that performed irregular, sporadic invasions. Range contraction may also have affected Central Spain, where Lilford (1866) mentioned great numbers breeding at Aranjuez but Witherby (1928) could not find them there and only encountered a few at Candeleda, close to the border with Extremadura.

Its numbers were perhaps at their lowest point in the 1960s, when breeding birds were practically restricted to the Zújar valley in Extremadura (Alonso 1986). The range broadened by the 1980s, but was still largely confined to the Guadiana and Tagus basins in Extremadura, mainly in the irrigated farmlands of the Guadiana and Alagón rivers. Breeding birds were also widespread in westernmost Andalucía and at various points of the Levant coast by the early 1990s: mainly in the Cartagena region in Murcia and the Plana Baixa in Alicante (Urios *et al.* 1991, SA1). Breeding in the southern Levant had not previously been reported (e.g. by Arévalo 1887 or Gil Lletget 1945), but the Spanish Sparrow had formerly bred in Catalonia in at least the Penedés region, where colonies existed until 1938 at Sant Pere de Ribes and until 1954 at Canyelles (CA1). However, the expansion into the Levant seems to have been short-lived since the Spanish Sparrow has now ceased to breed in eastern Spain except in Almería.

In contrast, since the late 1990s there has been marked and continuing range expansion into the Guadalquivir valley, Castilla-La Mancha, Madrid and the northern Meseta. In Madrid the first colony was discovered in 1987 and the first nest in neighbouring Guadalajara was found in 1993 (*Ardeola* 34: 291, 41: 102). By 1999 the Spanish Sparrow had spread throughout the southern foothills of the Central Range, with an overall population of at least 1,364 pairs in Madrid province (Roviralta 2000a, b). In Castilla y León just a few small colonies were known by the 1980s in the extreme south-west, close to Extremadura (Carnero & Peris 1988), but in 1995–96 scattered colonies were found in several provinces, as far north as Burgos and León (Román *et al.* 1997). A comparison between the first and the second Spanish breeding atlases, published respectively in 1997 and 2003, shows further eastward spread, up to eastern Soria in the north-east and to Albacete and Almería in the south-east. The Portuguese range has also expanded westwards in Alentejo as far as Setúbal since the 1990s (PA2). The present range represents a relatively recent colonisation of the country, where the Spanish Sparrow was first detected early in the 20th century, in Alentejo and Alto Douro (Tait 1924) and where it was still found only in a few regions bordering Spain by the early 1950s (Themido 1952). Most of the current Portuguese range has been occupied since the 1980s (AP).

Spanish Sparrows are very rare anywhere else in Iberia. Two records of lone males from Galicia: at Moaña, in October 1996 and at Xinzo de Limia, in April 1998 (*A. O. Galicia 1996*: 120, *1998*: 70) are exceptional. A singing male was at Torre-ribera, Lleida, on 10 July 1999 (*A. O. Catalunya 1999*: 262). There are several more recent records from the Ebro valley in southern Aragón, mainly at the Juslibol oxbow lake, Zaragoza, where 1–2 males were identified during the breeding seasons in 2002–2004. Also in Aragón, one was at Laspuña, Huesca, in June 2004 and up to three pairs may possibly have nested at Andorra, Teruel (*A. O. Aragón 2004–2007*: 340).

Regional bird reports give only six records for the Balearic Islands: single birds in May 1993 in Mallorca and April 1995 in Menorca, and four records of birds ringed in Cabrera, also in April–May: in 1993 (2), 1994 and 1996.

The recent Spanish population is clearly much larger than in the early 1990s. The great majority (69%) are in Extremadura and most of the remainder (24%) in Andalucía. The highest densities in spring have been found in dry croplands of the thermomediterranean, with 55 birds/km^2. Densities of 30 birds/km^2 have also been noted at this season in riparian growth in the supramediterranean and in Holm Oak dehesas and grasslands in the mesomediterranean. The Portuguese population is also considered to be increasing.

The reasons for these fluctuations in numbers and distribution are unclear but the Spanish Sparrow has always been prone to abandoning colonies and founding others, perhaps in response to predation pressure or to changes in food availability. Nest productivity in Spain is much lower than in the House Sparrow, apparently because the relatively accessible nests of Spanish Sparrows suffer high predation, especially by Azure-winged Magpies (Alonso 1984, Marques 2003). The Spanish Sparrow is locally regarded as a serious pest in irrigated farmlands in Extremadura, notably in the Guadiana basin where large flocks gather to feed, especially at ricefields, after nesting and in winter (Roviralta 2003, SWA). Some Extremaduran colonies have been displaced by deliberate persecution but attempts to control sparrow numbers and to scare them off crops have had only limited success.

This species is much more closely linked with agricultural land than the House Sparrow and it does not nest in buildings. However, some winter roosts have been found in Badajoz and other towns in Extremadura (SWA). Most Spanish Sparrows nest in noisy colonies in riparian trees or shrubs, eucalyptus plantations and poplar stands, often in the company of House Sparrows. They build large untidy spherical nests of grasses and other herbaceous material, placed in the outer branches. Some colonies are very large: up to 30,000 nests in Extremadura (Prieta 2003), but those of a few thousand or a few hundred pairs are more usual. Colonies also occur in oak dehesas and in olive groves, pines and palms in otherwise open country. Many nest within the stick nests of storks and raptors, a habit that was first noted by Lord Lilford in 1866. Spanish Sparrows are chiefly found below 500m and rarely higher than 1,000m.

Hybridisation between Spanish and House Sparrows is not unusual but does not seem to be resulting in progressive introgression. Fewer than one in 50 male birds in sparrow samples from various Iberian sources were found to show hybrid features (Alonso 1985). Female hybrids are not safely distinguishable. Moreover, morphological differences between the two species in beak and tarsal characters were found to be greater in regions where they coexist, suggesting that character displacement strengthens their segregation (Alonso 1985).

The movements of Iberian Spanish Sparrows are not fully known but many birds are clearly nomadic outside the breeding season and an unknown proportion may be migratory and winter in North Africa. Post-breeding fat accumulation in central Spanish birds provides evidence of migratory tendencies (Asensio & Cantos 1989b). Some southward passage across the Strait of Gibraltar has been noted in late September–October, with a return in March–April (BG, AI2). There is a March recovery from Málaga of a Spanish Sparrow ringed in Moroco (BM) and one crossed the Alborán basin from coastal Spain to coastal Algeria (SMA). Nevertheless, the great majority of ringing recoveries have been local, the farthest recorded displacement being of a bird ringed in Badajoz and recovered in Madrid.

EURASIAN TREE SPARROW *Passer montanus*

(Gorrión Molinero, Pardal-montês)

***P. m. montanus*. Abundant throughout the meseta and common but more local elsewhere throughout mainland Iberia. Largely resident. A local resident in the Balearic Islands, where some also occur on passage.**

Breeding population estimates. Spain: SA1 2.5–4.1 million pairs; ACRE 4.4 million birds. **Portugal**: 10,000–100,000 pairs. **Andorra**: 20–50 pairs.

The Tree Sparrow is strongly associated with open farmland and is locally abundant across the meseta. It occurs locally throughout most of the rest of Spain below 1,000m (SA2), but it ranges to 1,300m in the Sierra Nevada (Garzón & Henares 2012), to 1,400m in the Sierra de Gredos (San Segundo 1990) and to 1,500m in the northern Iberian Range (de Juana 1980). It is widespread throughout most of Portugal, but there are notable gaps in the breeding distribution in Algarve and in the treeless farmlands of Alentejo and Beira Baixa (PA2).

The present extensive Iberian range follows marked range expansion during the second half of the 20th century. It was first found in Portugal in April 1913, at the Tagus estuary (Weigold 1922), and was still a relative rarity there until at least the 1950s, but had occupied its recent range by the 1990s (Tait 1924, Reis Junior 1931, AP, PA2). In Spain, Arévalo (1887) thought that it only bred in the north, where Saunders (1884) had reported it as common in the western Pyrenees, and that it only appeared in the south and east in winter. Saunders (1871) said that his only observation in southern Spain had been on the banks of the river Segura, Murcia, and he never saw a Tree Sparrow among the many House Sparrows brought into the southern markets. Irby (1895) never saw it close to Gibraltar. Somewhat later, Jourdain (1936–1937) noted that probably it was just an irregular winter visitor to Andalucía 'except in Almeria'. Witherby (1928) was the first to mention its presence in Central Spain, at Aranjuez. Ticehurst & Whistler (1930) found it breeding in Catalonia at Tarragona and the Ebro delta, but had earlier failed to find it during their visits to Navarra (1925) and Galicia (1928). In the 1940s, Bernis does not mention having found it at Ledesma, Salamanca, and he considered it to be scarce in the La Maragatería region of León (Bernis 1945a, 1946). Even by the late 1950s, Valverde (1960) gives only three records for Doñana, the earliest in 1956, despite the many visitors to the region since the second half of the 19th century; it is now considered abundant there (García *et al.* 2000). Castellarnau (1877), Irby (1883), Eagle Clarke (1889) and Boxberger (1921) failed to find it in Segovia, Cantabria, Andorra and Aragón respectively. Such an easily identifiable bird is very unlikely to have been widely overlooked in the past so it appears that it only occurred as a local breeding species in northern Spain until the mid 20th century, but its later range expansion has not been well documented.

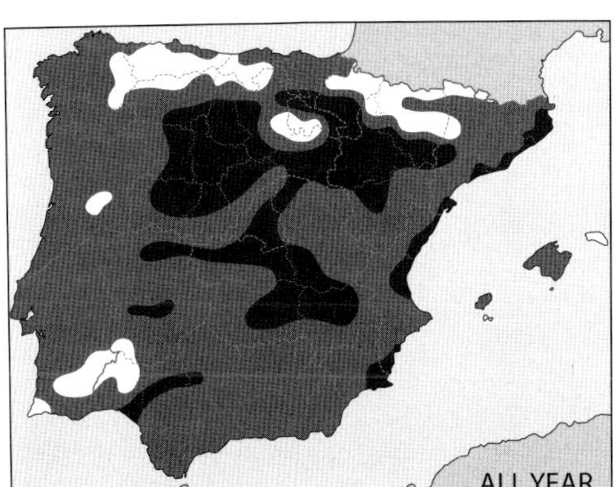

The Tree Sparrow nests in holes in mature trees as well as in walls, bridges and rural buildings, often under eaves. It readily uses nestboxes, from where it often displaces tit species (Sánchez Aguado 1984, García-Navas *et al.* 2008), and also the abandoned nests of Sand Martins. A number of pairs may nest in close proximity, forming loose colonies (de Juana 1980, Bernis 1989). Breeding birds are particularly associated with riparian woodland in farmland and other open areas offering scattered trees in the vicinity of dry or irrigated cultivation, including dehesas, olive groves and also large parks: as in Madrid (Alonso & Purroy 1979). They flock in open country in winter.

The Tree Sparrow was first reported nesting in the Balearics in 1971, when a small colony was found in Ibiza (Norman 1973). It nested in Mallorca, from 1994 onwards and was well distributed there by 2003–07 (SA1, Avellà *et al.* 1997, Adrover *et al.* 2010). It also occurs in the islands as a scarce migrant, during late October–November and mid March–early May (Ferrer *et al.* 1986).

The great majority inhabit the Meseta: 58% in Castilla y León and 20% in Castilla La Mancha. It attains particularly high densities, 85–110 birds/km^2, in orchards, irrigated farmland and in riparian trees (ACRE). The Spanish population on SACRE plots showed a significant overall decline during 1998–2012 (Escandell 2013). This decline was already apparent in some regions, notably the Basque Country (Álvarez *et al.* 1985), by the 1980s. Similarly, in Catalonia, where the breeding population has been independently estimated at 208,000–331,000 pairs, there was a 10% decline in the number of occupied 10km-squares between the first and second breeding bird atlas surveys, in 1975–83 and 1999–2002 (CA2). The declines reported in Spain are noteworthy in view of the drastic recent collapse of the Tree Sparrow population in Britain and elsewhere in northern Europe (Brown & Grice 2005), showing that the species is highly sensitive to agricultural intensification.

The Tree Sparrow is largely resident in Spain where large-scale ringing has produced almost entirely local recoveries. The farthest known displacements were by a bird ringed in March 1972 in Navarra, and recovered 330km south-southeast in Valencia, in September 1974 (*Ardeola* 29: 144), and by two ringed as chicks in León province in July 1972 that were recovered respectively in Burgos, 157km to the east, and Asturias, 85km to the north (*Ardeola* 29: 143–144). However, three Belgian-ringed birds and two from France have been recovered in Spain (SMA) and there are Portuguese recoveries of birds ringed in Belgium (5), The Netherlands and Russia, showing that some Tree Sparrows from further north in Europe reach the Peninsula in winter. There is some small scale passage at the Strait in late September–early November and March–April (BG, Tellería 1981): Tellería estimated as many 2,000 birds in mixed migrant sparrow flocks at Tarifa in autumn 1977. A record of 15 on the Columbretes islands in April 1965 (Bernis & Castroviejo 1968) is a further indication of migration, this time off the Levant coast. In Morocco it was formerly known as just an accidental visitor, mainly to the Strait area, but is now regarded as a very scarce winter visitor that has also bred occasionally in the north since 1985 (BM).

ROCK SPARROW *Petronia petronia*

(Gorrión Chillón, Pardal-francês)

***P. p. petronia*. A widespread resident in mainland Iberia and in the western Balearics.**

Breeding population estimates. Spain: SA1 825,000–1.15 million pairs; ACRE 3.2 million birds. **Portugal**: 10,000–100,000 pairs. **Andorra**: 15 pairs.

The Rock Sparrow is widespread and locally numerous in the supra- and mesomediterranean climatic regions but largely absent from the thermomediterranean and eurosiberian. The highest densities, 18–36 birds /km^2, are seen in the supramediterranean in the vicinity of human habitation, in open scrubby habitats and in farmland (ACRE). It occurs throughout the interior of the Peninsula except for parts of southern Extremadura, western Andalucía and western Castilla-La Mancha, where it is much more local. It is absent from the Guadalquivir valley and most of Cádiz province, as well as from most coastlands. There are Atlantic coastal outposts in Alentejo and, very locally, in Galicia and Asturias (SA2, Noval 2001, PA2). Breeding in coastal Galicia was noticed in May 1927 when two pairs were found nesting in coastal ruins at A Coruña (Ticehurst & Whistler 1928), where it no longer occurs. There are only a few small colonies in Andorra (ANA). The wider Portuguese range shows an absence from the entire western half of the country north of the Tagus as well as from Algarve and eastern Baixo Alentejo. A flock of 82 at Ameixial, Faro in February 2009 (*Anu. Ornitol.* 8: 83) is thus noteworthy. The Rock Sparrow is most frequent south of the Tagus in Ribatejo, Alto Alentejo and western Baixo Alentejo. It is also present further north in the east of Beira Baixa, Beira Alta, Trás-os-Montes and Alto Douro (PA2, AP).

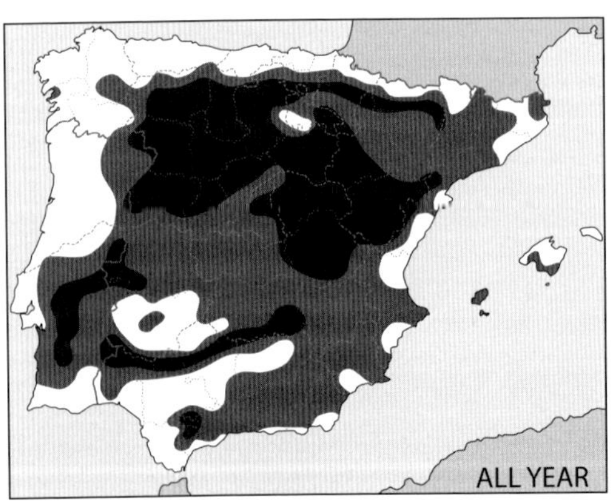

It is probably a fairly recent colonist of the Balearic Islands, where it is a common resident in Ibiza and Formentera and it occurs locally in Mallorca (Ferrer *et al.* 1986, *A. O. Balears 2010*: 269). It is absent from Menorca.

This species uses a diversity of habitats but it is most often found where rocky outcrops and scree slopes provide suitable nest sites. This may account for an apparent general predilection for limestone areas, where cavities are abundant (de Juana 1980, CA1, Bernis 1989). Holes in trees are used in open woodland areas, for example in oak dehesas in Salamanca and Cáceres (Bernis 1989). It also makes use of holes in buildings, stone walls, bridges and other artefacts. It may be found nesting within villages and towns on the mesetas, even in the town centres, although it typically forages on the outskirts or in nearby fields (Bernis 1989b). The Rock Sparrow has also been found nesting within the stick nests of large raptors (Carnero & Peris 1988) and also in Red-rumped Swallow nests and in burrows dug by Bee-eaters and Sand Martins (AI2, Baucells & Abella 2007). In general, it favours dry, sunny regions offering cultivation, steppe and other sparsely-vegetated areas, juniper and other open scrub formations or open woodlands, including Holm Oak dehesas and woods of Stone or Black Pines. It occurs as high as 1,600m in the southern Iberian Range (Bernis 1989b), 1,700m in Andorra (ANA) and 3,000m in the Sierra Nevada. Progressive colonisation of higher altitudes has been reported in the last of these where the Rock Sparrow was only recorded up to 1,900m in the early 1990s and to 2,500m a decade later but it reached 3,000m in 2005, where it nests in the infrastructure of the ski resort (Pleguezuelos 1992, Garzón & Henares 2012).

It is most often encountered in colonies of up to 20 pairs, although isolated pairs occur. Communal roosts form post-breeding and these may be of some size: for example, over 500 roosted on cranes at a building site within the city of Plasencia, Cáceres, during three consecutive winters (*Ardeola* 57: 547). Winter flocks may also be sizable but one of 1,400 birds at El Hito, Cuenca, in January 2010 and another of some 2,000 at Sepúlveda, Segovia, in December (*Ardeola* 57: 240, Bernis 1989b) were exceptional. The Rock Sparrow is very sedentary and no migratory movements have been reported. However, it becomes more widespread on agricultural land in winter, when large numbers may flock with other seed-eaters (SWA). Those that nest in mountain areas probably make some altitudinal movements since in Catalonia, for example, few are seen in winter above 900m (CWA).

WHITE-WINGED SNOWFINCH — *Montifringilla nivalis*

(Gorrión Alpino, Pardal-alpino)

M. n. nivalis. **Resident in the high Pyrenees and Cordillera Cantabrica. Rarely encountered elsewhere.**

Breeding population estimates. Spain: SA1 5,000–10,000 pairs; SA2 4,500–6,000 pairs.

The Snowfinch has two discrete populations in northern Spain, in the central and eastern Pyrenees: from eastern Navarra to western Lleida, and in the central Cantabrian Range, from Somiedo in the west to the Picos de Europa in the east. Breeding birds are generally found no lower than 1,800m but nesting at 1,680m has been observed on Pico Orhi, Navarra (SA1, SA2). Breeding has not been proved in Andorra but a few pairs may nest there (ANA). Saunders (1871) states that it was certainly found in the Sierra Nevada, and Arévalo (1887) records that several specimens were obtained there, but later visitors, from Witherby (1928) onwards, have failed to find any.

Snowfinches frequent rocky scree slopes, especially near snowfields and alpine pastures. The snowfields act as traps for flying insects and are an important source of invertebrate food. Some remain at high altitudes even in winter, when they may forage around ski centres, but often they descend to lower levels in the vicinity of the breeding areas, then occurring as low as 500m. Flocks of 100+, exceptionally over 1,000 (Noval 1988), occur in winter. The birds are more dispersed when nesting but loose colonies are often encountered.

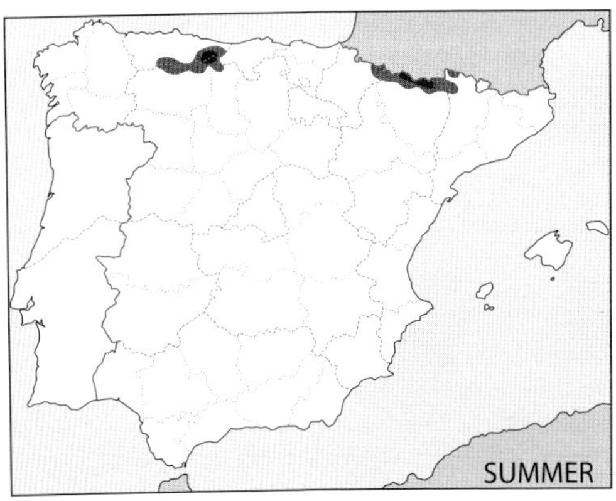

The revised estimate of the Spanish population (SA2) comprised 1,500–2,000 pairs in the Cantabrian Range and 3,000–4,000 pairs in the Pyrenees. The population is believed to be stable but it could be vulnerable to disturbance by walkers and skiers and the habitat may be damaged by ski infrastructure and also by overgrazing, the latter having a deleterious effect on the invertebrate populations in the alpine pastures that are the Snowfinches' principal food during the breeding season (SA2). However, some undoubtedly benefit from foraging around mountain resorts.

Most winter records are from or near the same mountains where the Snowfinch breeds, often at lower altitudes (SWA). In Catalonia, for example, most remain in the Pyrenees but small flocks appear in the Montseny, Montserrat, Garraf and other ranges nearby (CWA). Similarly, a wintering presence has recently been reported from the Cantabrian range in northern Burgos, with a maximum count of 110 on the Peña Amaya massif in December 2012 (*A. O. Burgos*). Otherwise, few are ever recorded at any distance from the breeding range but individuals have appeared at a diversity of locations elsewhere in the Peninsula. For example, two were at 2,469m on Pico Peñalara in the Sierra de Guadarrama, Madrid, in December 1957 and two were at 2,100m in the same area in February 1985 (*Ardeola* 4: 202, 33: 212). One was at 2,974m in the Sierra Nevada, Granada, in December 1975 (*Ardeola* 23: 243). There are nine records from the Comunidad Valenciana, all from Castellón and most recently in 2009, when up to five were present during January–February at 1,700m on Penyagolosa (1,813m), the highest peak in the province (*A. O. C. Valenciana 2009*: 146). They include a single bird seen on the Columbretes islands on 9 November 1993 (*A. O. C. Valenciana 1993*: 122). One was recorded further south on the Mediterranean coast at Cartagena in early February 1999 (Hernández 2006).

A few individuals also reach the Balearic Islands, where the species is described as rare in winter in Mallorca and accidental in Menorca, Ibiza and Formentera (*A. O. Balears 2010*: 269). The regional bird reports give 20 records (34 birds) for the archipelago during 1986–2009, four on Menorca but the remainder in Mallorca; extreme dates were 8 October and 13 May but most were during November–February. There are also four accepted records from the coastal extremes of Portugal: two birds at Cape St Vincent in January 1999, single birds at Cape da Roca in January 2005 and January 2006, and two birds also at Cape da Roca in December 2007 (AP, CPR 2008–09). Snowfinches are thus liable to occur on rare occasions probably anywhere in Iberia, even at sea level.

Observations in the eastern Pyrenees increase from November to May: the recent Catalan winter atlas estimated the wintering population there at 1,800–2,500 birds, whereas the breeding population may number fewer than 50 pairs (CWA, CA2). Some Snowfinches from the Alps possibly join the Pyrenean birds in winter (ANA, CWA). A juvenile ringed in Austria in June 2005 and recovered at Toses, Girona, in January 2006 (*Ecología* 21: 295) and one ringed at Toses in January 2009 and recovered in the Canton de Valais, Switzerland, the following March (ICO, CMA), provide some evidence of this.

BLACK-HEADED WEAVER *Ploceus melanocephalus*

(Tejedor Cabecinegro, Tecelão-de-cabeça-preta)

Introduced into Portugal (Category C).

A small naturalised population is established in southern Portugal where its range is expanding. It was first detected at Alcochete in the Tagus estuary in 1996. More recently three subpopulations have been found: around the estuary and lower course of the Tagus; at the Paul de Tornada, Caldas da Rainha, Leiria and in coastal Faro at Vilamoura, Quinta do Lago and the Lagoa dos Salgados (AP, PA2). The birds frequent cane brakes and riparian vegetation for nesting but forage in nearby farmland (AP).

YELLOW-CROWNED BISHOP *Euplectes afer*

(Obispo Coronigualdo, Arcebispo)

Introduced into Portugal (Category C) and Spain (Category E1).

The Yellow-crowned Bishop is well established at scattered locations in Portugal and even more locally in Spain, where there are breeding populations associated with rice-growing areas and freshwater marshes. The earliest known observations were in 1989 in Algarve, at the Lagoa ricefields, and a breeding population remains in that area. Breeding birds were next found at the Tagus estuary in 1990 and since then the Yellow-crowned Bishop has colonised a series of coastal locations on the Portuguese west coast and, more recently, inland. The principal population nuclei are in south-central Faro; the ricefields of the Tagus and Sado estuaries; the Paul do Boquilobo, Santarém; Quinta do Canal, Figueira da Foz, Coimbra; and the Barrinha de Esmoriz, Aveiro (AP, PA2). The Portuguese population is small but clearly increasing and flocks of up to several hundred have been reported from key sites, including over 500 at the Giganta ricefields of the Tagus estuary in December 2004 (AP).

Breeding birds are apparently well established in Spain at the Guadalquivir estuary, in the ricefields of the Brazo del Este and Isla Menor, Sevilla. One was seen there on 8 September 1998 and breeding was first reported in 2000, when 100–200 pairs were estimated to be present (*Ardeola* 46: 314, Chiclana *et al.* 2002). A flock of 400 was at the Brazo del Este on 9 December 2010 (*Ardeola* 58: 219). The species has consolidated its population there since then and has increased in numbers. Elsewhere in Spain, up to 16 were reported from Alhaurín de la Torre, Málaga, on various dates between October 1998 and March 2000 (*Ardeola* 46: 162, 47: 170, 49: 194), possible breeding has been reported in northern Huelva, and the bishop probably also nests at the Miño estuary, Pontevedra (SA2): this latter population may have been founded by birds from adjacent Portugal. A colony of about 12 birds was found in the Guadiana valley in Extremadura at Badajoz in August 2000 and up to 20 birds were in that area in September 2011 (*Ardeola* 49, 194, 58: 514). Further afield, single individuals have been reported from the Río Besòs, Barcelona, and the Albufera de Valencia in 2003, and one was found at San Martín de la Vega, Madrid, in September 2007 (*Ardeola* 51: 254, 54: 403). Breeding has also been reported from Mallorca at Pla de Saint Jordi. The Yellow-crowned Bishop remains in Category E of the Spanish List but seems likely to be admitted to Category C in the future if its principal population in the Guadalquivir ricefields continues to thrive. The Iberian populations breed relatively late in the year and males in breeding plumage may be seen from May until early November (AP).

COMMON WAXBILL *Estrilda astrild*

(Estrilda Común, Bico-de-lacre)

Locally common in Portugal (Category C) and Spain (Category C). Recorded in Mallorca. Subspecies unknown but the original introductions in Portugal are believed to have involved birds from Guinea *E. a. kempi* and the Cape Verde Islands *E. a. jagoensis* (Hagermeijer & Blair 1997).

Breeding population estimates. Spain: SA2 2,800–12,000 pairs. **Portugal**: several thousand pairs.

This is a well-established alien species in parts of mainland Iberia. It breeds no closer than southern Mali and there are no records from Morocco (BM) or even from Mauritania (Isenmann *et al.* 2010), so its natural occurrence in the Peninsula is unlikely. Individuals were first reported in the wild Portugal in 1964, when 100 were observed at the Lagóa de Óbidos, near Peniche, on the west coast. These were believed to be descended from escaped birds and so might have already bred at liberty for some years, but the first breeding record was in 1965 (Reino & Silva 1996). A further introduction may have occurred in Algarve prior to 1977.

During the 1980s Common Waxbills spread throughout the Tagus and Sado valleys and along the Portuguese coast north to Galicia, Spain, where they were present by 1989 at least and first proved to breed at the Miño estuary in 1994 (*Ardeola* 41: 202). They colonised the Guadiana valley in the mid 1980s, spreading upstream into Extremadura, where they first bred in 1986 (Guerrero *et al.* 1989). The birds consolidated their presence in northwestern Portugal during the 1990s and extended their range along the Galician west coast, the Miño valley and the Ulla estuary. The species also became established in southern Portugal, becoming common and widespread in all areas. It was more local in northern Portugal, occurring chiefly along the valleys of the Minho, Lima, Cávado, Ave and Douro. The Common Waxbill is now widespread throughout the western and southern lowlands of Portugal and, especially, in the Tagus and Sado estuaries and basins; the Ria de Aveiro, Beira Litoral, Baixo Mondego, Estremadura/Ribatejo and southern Algarve (PA2).

The range in Spain expanded during the 1980s and 1990s to include the Guadalquivir valley, chiefly around Sevilla, the Guadalhorce valley in Málaga and, in the 1990s, the wetlands of the Levant coast from Valencia to Barcelona. The east coast population may be derived from independent introductions (SA2). Northern Extremadura was colonised from 2006 onwards, when birds were reported from diverse sites in the Alagón valley and at the Arrocampo reservoir (*Ardeola* 56: 366). Sporadic breeding has been reported from Granada, Madrid, Ciudad Real, Murcia, Bizkaia, Girona and Tarragona. In all, breeding had been reported from 21 provinces in Spain by 2003 (SA2). A small population is established in the Balearic Islands, nesting at the Albufera de Mallorca since at least 2001: 51 were counted there on 24 November 2011 (*Ardeola* 59: 452) and breeding has also been reported from Ibiza. The chief Spanish population is along the Guadiana valley in Extremadura, where the birds have clearly benefited from the expansion of rice cultivation. The other major populations are those in Galicia and in the lower Guadalquivir.

The Common Waxbill is particularly associated with rank vegetation along watercourses and around water bodies. Here it inhabits reedbeds, reedmace, canebrakes and other aquatic vegetation but it also spreads into adjacent cultivation and other low, dense cover. The birds are generally seen in family parties or in larger flocks, sometimes of several hundred. The breeding season seems to be highly flexible and nesting has been reported from the Río Miño, Galicia, in every month except February (C. Vidal, pers. comm.). The distribution is clearly associated with the thermomediterranean and the milder regions of Atlantic Iberia. Ringing at the Miño suggests that the birds there are resident (SA2) but wanderers may be encountered away from the known breeding sites on occasion, generally at coastal wetlands or in river valleys. For example, they occasionally appear at the Guadiaro estuary, Cádiz, and there are three records from Gibraltar.

The Iberian populations are increasing. The rate of spread during 1969–94 has been calculated at 13km/year (Reino & Silva 1998) but has probably since slowed. This tiny tropical bird might be expected to be vulnerable to the cold snaps that affect even the mildest parts of Iberia in some winters but it has proved remarkably resilient so far. Its presence seems benign but it could become a pest to rice growers and other farmers in future.

RED AVADAVAT

Amandava amandava

(Bengalí Rojo, Bengali-vermelho)

A local resident in mainland Spain (Category C) and Portugal (Category C). Exotic. Iberian birds are believed to be of the nominate subspecies, a native of the Indian subcontinent.

The Red Avadavat is locally common in southwestern Iberia, with scattered populations elsewhere. It was first detected in the wild at Arganda del Rey, Madrid, in 1973 and then in the Guadiana Valley in 1981 (*Ardeola* 20: 385–386, 28: 165). It has since become established particularly in the Guadiana valley in Extremadura, from where it has spread to the Tiétar, Alagón and Árrago valleys, and in the Tagus-Jarama valleys in Madrid. There are other Spanish populations in Andalucía: in the lower Guadalquivir valley, the lower Guadalhorce and Vélez valleys in Málaga, and the Motril coast and Guadalfeo valley in Granada. Sporadic nesting has been reported elsewhere in Andalucía, in Huelva and Cádiz, and especially from wetlands on the Levant coast in Alicante and Catalonia, but the species does not seem to be established there.

The Avadavat has also extended its range into Portugal where it is established in the Guadiana valley at Elvas in eastern Alentejo, adjacent to Badajoz, Extremadura (Costa *et al.* 1997, AP), and in the Tagus valley at the Barroca marsh east of Lisbon (Leitão & Costa 1996). It has also been reported from Faro, Silves and Portimão in Algarve (Vowles & Vowles 1994) and from the Sado valley and Baixo Alentejo (Costa *et al.* 1997) and occasionally elsewhere. In general, the Avadavat is very local and uncommon in Portugal.

This species is associated with dense, rank, riparian growth, generally near irrigated crops including maize, rice, sugar cane and cotton. It breeds late in the year during mid July–December, a period that corresponds with the Indian monsoon. The birds occur in family parties and small flocks. The species is largely sedentary in Iberia. The Spanish population numbers a few thousand birds, the majority in the Guadiana valley in Spain, where two roosts were found to hold 1,400 individuals (de Lope *et al.* 1984). The population has fluctuated, with some declines in Extremadura associated with losses of riparian vegetation, but is generally believed to be increasing slowly. The largest numbers reported in Portugal have also been in the Guadiana valley, where at least 2,000 were present in December 1990. A few hundred occur in the Tagus valley and only small groups have been reported elsewhere (AP).

CHESTNUT MUNIA

Lonchura malacca

(Capuchino Tricolor, Capuchinho-tricolor)

Escaped birds are naturalised in Portugal (Category C) and have bred in Spain (Category E).

The Chestnut Munia has small established populations in Portugal, principally at the Ría de Aveiro, in the rice growing areas of the lower Tagus and Sado rivers, and in the Arade valley, Faro (AP, PA2). The earliest observations in the country were in June 1992 in the lower Tagus valley. It is usually encountered in small numbers but over 80 were seen at the Paul de Rilvas, Alcochete, in April 2002 (AP). The limited available information suggests that it nests in winter and early spring, from December to April (AP). Recent observations in Portugal have been very few, possibly indicating decline (Matias 2011). This species has seldom been found in Spain but has been reported to have nested in Catalonia: in Girona in 1999 and in Barcelona in 2002 (SA2).

RED-EYED VIREO

Vireo olivaceus

(Vireo Chiví, Juruviara-norte-americano)

Vagrant to Spain.

The three Iberian records are of first-year birds on Mediterranean coast: at Tarragona, on 19 October 1995; at Elche on 25 October 1995, and at Tiana, Barcelona, on 30 October 2000. Birds of the nominate form appear with some regularity in Europe, chiefly in the British Isles and nearly always in October. There are records of some 123 individuals in Britain, 56 in Ireland and 16 in France, as well as one in Morocco and 12 in the Azores.

COMMON CHAFFINCH *Fringilla coelebs*

(Pinzón Vulgar, Tentilhão-comum)

***F. c. coelebs*. An abundant resident. Large numbers from other European countries winter in Iberia. The breeding birds have shorter, more rounded wings than the wintering birds and are sometimes separated as a distinct subspecies *F. c. balearica* but they still fall within the range of variation of the nominate form (HBW).**
***F. c. africana*. Vagrant to Gibraltar.**

Breeding population estimates. Spain: SA1 2.6–6.4 million pairs; ACRE 19.3 million birds. **Portugal**: 0.5–2.5 million pairs. **Andorra**: 5,000–8,000 pairs.

The Chaffinch is ubiquitous in woodlands throughout mainland Iberia. In Spain it is only absent from some stretches of the Mediterranean coast and from major treeless expanses (SA2). Elsewhere it may be found from sea level to the treeline in the highest mountain ranges, thus reaching 2,550m in the Sierra Nevada (Garzón & Henares 2012). It is equally widespread in Portugal, where it is most abundant in the north and centre, but it is scarce or absent in greater Lisbon and the treeless farmlands of the southeast (PA2). It is a common resident in the Balearic Islands in Mallorca and Menorca but does not nest in Ibiza or Formentera (*A. O. Balears 2010*: 269). It is widespread and one of the commonest species in Andorra (ANA).

It is common and often abundant in dehesas of Cork and Iberian Holm Oaks; coniferous forests: both natural and planted, including those of exotic species such as the Monterey Pine (Carrascal & Tellería 1990); all types of oak forest; beech, chestnut and birch woods and mixed woodlands. Some occur in broom scrub at high elevations, for example in the Sierra de Gredos (Sánchez 1989). It also frequents olive groves and orchards, although it is less attracted to citrus groves. It may be found nesting in otherwise treeless agricultural land where there is riparian woodland along the watercourses.

The regional population of Catalonia has been put at 200,000–271,000 pairs (CA2). There was a significant overall increase on SACRE sites during 1998–2012 (Escandell 2013) but no clear trend in Portugal during 2004–09 (ACP). Favoured habitats have been found to hold 70–100 Chaffinches/km^2 in spring (ACRE).

Iberian breeders are largely sedentary but those nesting at high altitudes descend to lower levels later: for example, the subalpine forests of the Pyrenees are partly abandoned in winter (CWA). Large numbers enter Iberia and the Balearic Islands to spend the winter and a significant fraction continue south to North Africa. The Catalan winter atlas (CWA) put the wintering population there at some 8–14 million birds, a 23-fold increase over the breeding population. The Catalan wintering population has furthermore tended to increase, by some 5% annually between 2002–03 and 2008–09. Chaffinches may often be seen in open country in winter, generally foraging for weed seeds and crop residues in arable land. Nevertheless, they still seem to prefer areas with some tree cover, so they are less common in the widely deforested areas of both mesetas, the Ebro and Guadalquivir valleys, and certain areas of Extremadura and Alentejo (AI2, SWA). They then occur in flocks of up to several hundred individuals, sometimes mixed with Bramblings, other finches and buntings. Many of these flocks are likely to consist mainly or entirely of immigrants from further north in Europe. Females comprise a majority of the birds wintering in Spain, up to 66% in Catalonia (Asensio 1984a, CWA) and 83% at a ringing site at Villamejor, Madrid (Villarán *et al.* 2002).

Foreign Chaffinches ringed in the breeding season and recovered in Iberia originate mainly from Finland, Scandinavia, the Baltic States, Germany and Poland. There are 226 recoveries of Chaffinches ringed in Russia, where they probably also originated. There are also numerous recoveries of birds ringed outside the breeding season, and thus possibly in transit from elsewhere, especially from Switzerland (244), France (58) and the Low Countries (98), with lesser numbers from Italy and the

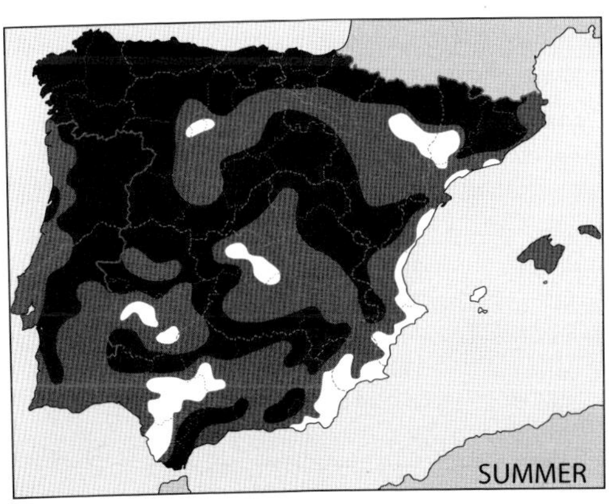

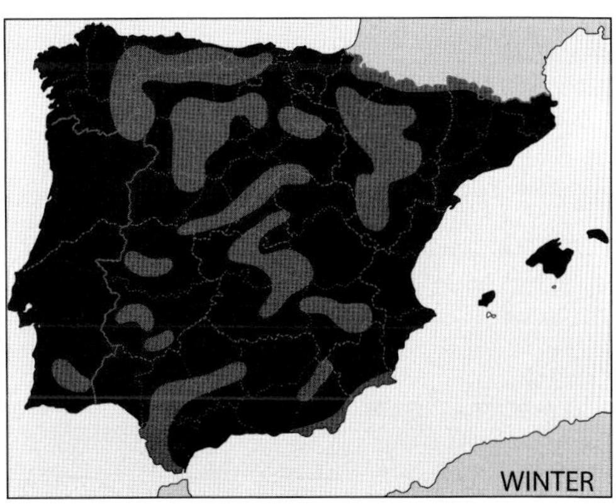

Czech republic (Asensio 1984a, SMA, AP). Recoveries in Portugal include at least 35 from the former USSR. Ringing recoveries also indicate that wintering birds of foreign origin generally avoid the Meseta, where the winters are colder, in favour of peripheral regions of Iberia (Asensio 1984a). Those from southern Europe tend to winter in the Mediterranean coastlands of northeastern Spain, those from further north settling instead in the south and south-west. Birds ringed in Iberia have been recovered in northern and central Europe, as well as in Italy (10), Morocco (3) and Ceuta (1).

Migrants enter the Peninsula during September–November and depart during February–April, chiefly in March (Asensio 1984a). Visible migration can be conspicuous at the Strait in some autumns: in October–November, with stragglers into December (Tellería 1981, BG). Intense passage is also apparent in coastal Portugal, again mainly in October and the first half of November (AP). In the Balearics the autumn passage is mainly between mid October and mid November, peaking in late October (Bonnin 2007). Most ringing recoveries in Spain have been between October and mid February (AI2).The spring passage at Gibraltar is chiefly in February–March (BG) but is much less evident than the autumn one.

The African Chaffinch *F. c. africana* nests as close to Iberia as the southern coastlands of the Strait. The males at least are conspicuously different from the Iberian birds, their green backs being the most obvious distinction. There are very few records of such birds anywhere in Europe and three records from Gibraltar, the most recent being a male ringed on 9 April 2004 (*GBR* 4: 39), seem to be the only observations from Iberia. It would be surprising if it did not occur more often in the south, given the volume of Chaffinch migration across the Strait in some years.

BRAMBLING *Fringilla montifringilla*

(Pinzón Real, Tentilhão-montês)

A widespread but generally scarce and irregular winter visitor to the Peninsula and the Balearic Islands.

Brambling winter flocks in central Europe may number millions of birds at times but usually relatively few migrate into the Iberian Peninsula. The numbers involved fluctuate from year to year but are generally low to moderate. However, there was an unprecedented influx into northeastern Spain during winter 2010–2011 when a massive roost, estimated to have peaked at around one million individuals in late November, was discovered at the Macizo del Gorbea, Bizkaia (*Ardeola* 58: 514). The roost was occupied until March 2011. The formation of this roost began in mid November and it coincided with the beginning of a very cold spell in central Europe as well as with a beechmast year in the Basque forest involved. Another very large roost in the Basque Country was discovered at Artikutza, Gipuzcoa, on 23 November 2012, when an estimated 200,000 birds were present (*Ardeola* 60: 187). Some 60 Iberian recoveries include 13 birds ringed within the breeding range: in Fennoscandia, the Baltic States or Russia, as well as others ringed Germany, France, the Low Countries, the British Isles, Swizerland, Italy and Hungary (SMA, AP). Four ringed in Spain have been recovered in Germany, Austria and Sweden (2).

Bramblings may occur from September to April but the main arrivals are in October–November departing in February–March. Nearly half of the birds recovered in Spain were in January (Asensio 1984a), which may indicate that some arrivals may occur during that month, perhaps in response to poor weather or depleted food sources further north. The numbers that reach Portugal are highly variable and very few occur in some years. However, flocks of up to a few hundred have been reported in other seasons, notably from Ribatejo in the environs of the Paul do Boquilobo (AP).

Bramblings may be found throughout the Peninsula in winter, often in mixed flocks with Chaffinches and other seed-eaters. However, they tend to be most frequent in the northern half of the region: in Galicia; the Duero basin in León, Palencia, Burgos, Segovia and Soria especially; the Upper Ebro in La Rioja, Álava, the north of Burgos and southern Cantabria; as well as in the southern Iberian Range and the Pyrenean and Pre-Pyrenean margins (SWA). The wintering population of Catalonia was estimated to range from 44,000 to 145,000 birds during the winter atlas survey there (CWA). In the Balearics it is considered scarce during winter and in migration in Mallorca and Menorca, and rare in Ibiza (*A. O. Balears 2010*: 269). Most observations involve flocks of up to ten individuals, less often 30–40 birds, and larger flocks are exceptional. They appear in a wide range of habitats ranging from alpine pastures, montane Beech and Scots Pine forests in the north, Stone Pine woods and dehesas in the south, stubble fields – particularly of maize, and fallow fields. An

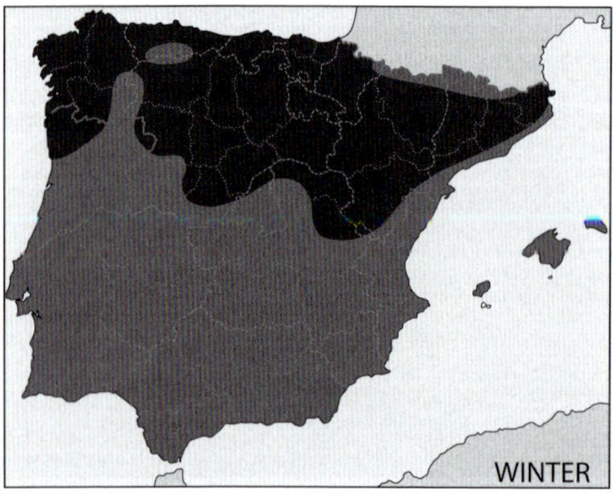

exceptional farmland gathering was seen in Badajoz, Extremadura, during winter 1977–78, when at least 4,000 were present with other finches on an unharvested sunflower field (de Lope *et al.* 1983).

Probably very few migrate to North Africa in most years. Migrants are only recorded at Gibraltar in some years and then only in very small numbers. They are scarce in Morocco, where most records involve single birds or groups of 2–4 (BM).

EUROPEAN SERIN *Serinus serinus*

(Serín Verdecillo, Milheirinha-europeia)

A common resident or partial migrant, and common winter visitor, throughout the Peninsula and in the Balearic Islands.

Breeding population estimates. Spain: SA1 4.1–6.6 million pairs; ACRE 35.7 million birds. **Portugal**: 1–5 million pairs. **Andorra**: 200–700 pairs.

The Serin is one of the commonest and most widespread birds in Iberia (SA2, PA2), where it is abundant throughout the Mediterranean climatic region. It is also common during the breeding season in the Eurosiberian. It inhabits all types of habitats offering a mixture of open areas and trees or large shrubs. It is commonest in Mediterranean woodlands and also characteristic of orchards, parks and gardens everywhere. It ranges to the treeline in the principal mountain ranges, nesting at up to 1,800m in the western Pyrenees and up to 2,300m both in Andorra and the Sierra Nevada (ANA, Garzón & Henares 2012). It is common in the Balearic Islands with the exception of Menorca, where it has only bred locally in recent years (SA2, *A. O. Balears 2010*: 269) but where Munn (1924) had found it to be as common as in Mallorca.

The population of Catalonia was estimated at 768,000–1,037,000 pairs in the 1999–2002 regional atlas (CA2). The highest densities of breeding birds in Spain, 250–360 birds/km^2 are in the thermo- and mesomediterranean, in olive groves, vineyards, open pinewoods and orchards (ACRE). The population on SACRE sites has shown a significant decline during 1998–2012 (Escandell 2013), due perhaps in part to the effects of pesticide and herbicide applications in olive groves and other favoured habitats. The Serin is also one of the commonest species in Portugal, where it is especially abundant in the centre and north (PA2, AP). The population is considered to be stable (ACP). It is also common in Andorra and sometimes nests in Gibraltar (BG).

Northern and central Iberian populations are partly migratory, wintering in the south of the Peninsula, chiefly in the coastal south-west, and in North Africa (Asensio 1985a). Thus, in winter very few remain in the northern half of the region, except on the west and east coasts and in the Ebro basin, generally abandoning areas above 600m (SWA). In addition, very large numbers of Serins from elsewhere in western Europe spend the winter in Iberia, chiefly in the northeastern Mediterranean coastal regions: Catalonia and the Comunidad Valenciana, where they are present from October to March, and where some 64% of ringing recoveries are during November–January (Asensio 1985a). In Catalonia, where only some 10% of the breeding population is thought to remain in winter, the wintering population has been put at some 1.9 million individuals (range: 1.5–2.6 million), with marked interannual oscillations (CWA). Most recoveries of birds ringed extraregionally, all of them found in Spain, are from Germany or France, with a few ringed in Switzerland, Belgium and Italy, and Spanish-ringed birds have been recovered in the same countries. Wintering birds frequent similar habitats to those used for nesting but, in addition, they may be found to a greater extent in agricultural land, feeding on weed seeds alongside other finches, sometimes in flocks of hundreds (CWA, SWA).

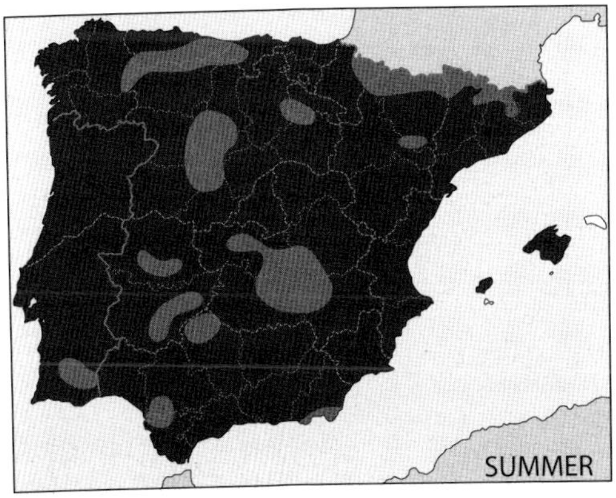

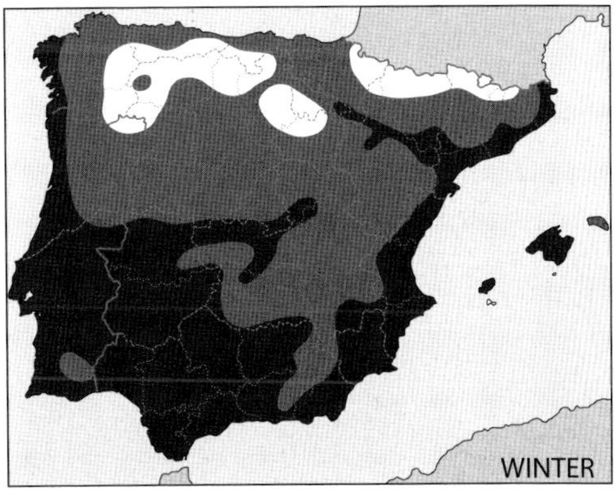

A proportion of the continental immigrants also continue further south to winter in North Africa but they may be a minority, since there are comparatively few recoveries of foreign-ringed birds in southernmost Spain (Asensio 1984b). Serins, apparently chiefly of Iberian origin, form an important and conspicuous component of the diurnal finch migration southward across the Strait of Gibraltar. Tellería (1981) counted over 15,000 flying south there at Tarifa during autumn 1977. The passage is chiefly in October and the first half of November but ranges from September to early December. The return passage is noted from late February until early April, peaking in March, with some stragglers into May (BG). There are six recoveries in Ceuta and two in Morocco of birds ringed in Spain during the breeding season, two of which had travelled a considerable distance: from Cieza, Murcia, to Sidi Slimane, Morocco (597km) and from Lles, Lleida, to Ceuta (940km) (*Ecología* 9: 331, 17: 281).

EUROPEAN GREENFINCH *Carduelis chloris*

(Verderón Común, Verdilhão-ocidental)

***C. c. aurantiiventris*. An abundant resident and partial migrant throughout most of Spain and in the Balearic Islands. Slightly smaller than the nominate race but much brighter-coloured. Adult males especially are bright yellow-green above and bright yellow on the flanks and belly.**

***C. c. vanmarli*. An abundant resident in northwestern Spain and throughout Portugal. Similar to *C. c. aurantiiventris* but duller above and with more olive flanks.**

***C. c. chloris*. Common throughout, and regionally abundant, on passage and in winter.**

Breeding population estimates. Spain: SA1 1.06–3.60 million pairs; ACRE 16.9 million birds. **Portugal**: 200,000–2 million pairs. **Andorra**: 50–100 pairs. **Gibraltar** <5 pairs.

The Greenfinch is abundant throughout Iberia, nesting in almost any habitat that offers a mixture of scattered trees and open areas. It is closely associated with human activity and is common in orchards, parks and gardens throughout the region as well as in farmland that provides some trees for nesting. Riparian woodland in agricultural areas is especially favoured. It ranges up to the treeline at 2,000m in the eastern Pyrenees and 2,200m in the Sierra Nevada but is most numerous everywhere at lower altitudes since, in general, it favours warmer regions (SA2, PA2, Tellería & Santos 1994, Garzón & Henares 2012). It is scarce in Andorra, where it is chiefly found in riparian growth in the valleys (ANA).

High breeding densities of 80–175 birds/km^2 have been recorded in orchards and olive groves and urban areas are also much favoured in the breeding season (ACRE). Up to 195 pairs/km^2 have been recorded in orange groves in Valencia (Gil-Delgado & Catalá 1989). The largest regional population is that of Andalucía (ACRE). That of Catalonia was estimated at 184,000–257,000 pairs in the 1999–2002 regional atlas (CA2). The population on SACRE sites has shown a significant increase during 1998–2012 (Escandell 2013).

Iberian Greenfinches are partial migrants, especially those in the northern half of the Peninsula, many of which move to southern Iberia and into North Africa to winter. Winter Greenfinch densities are very low in most of the eurosiberian and supramediterranean zones and, conversely, are high or very high in the thermomediterranean zone, including the Balearic Islands (SWA). For example, the wintering population of Catalonia has been put at some 562,000–1,236,000 individuals, whereas just 4,000–24,000 birds winter in Álava province, in the Basque Country (CWA). There are 18 recoveries of Spanish-ringed birds in North Africa: in Morocco (13), Algeria (4) and Ceuta (3), at least 11 of which were ringed during the breeding

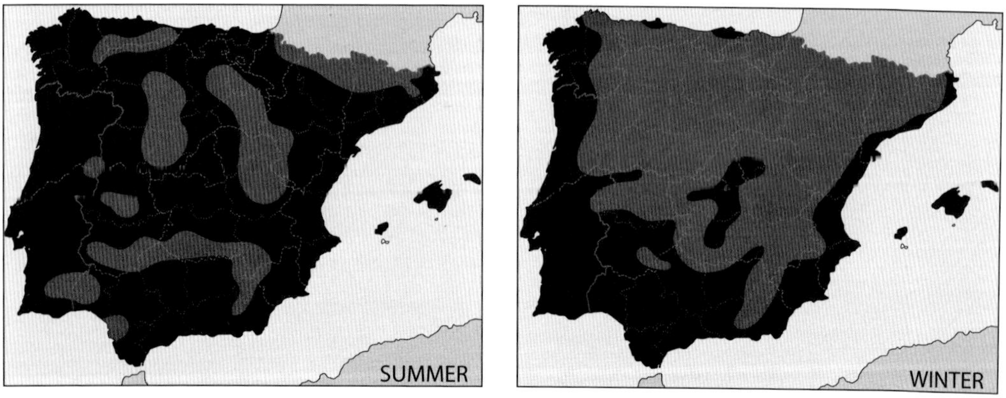

season. Iberian birds are joined by Greenfinches of the nominate race from populations in northern Europe, with recoveries especially of birds ringed in France, Germany and Belgium, with some also from The Netherlands, the British Isles, Switzerland and Scandinavia (Asensio 1984a, SMA, AP). Flocks of Greenfinches range widely in winter, when they may join other finches on stubble and fallows. However, they still favour habitats with trees in winter and many then frequent olive groves, dehesas and pinewoods, particularly in the meso- and thermomediterranean regions (AI2, SWA).

Nominate race migrants arrive from October onwards and mainly depart in March–April. Many, however, pass through the Peninsula and continue south to winter in North Africa. Large numbers cross the Strait of Gibraltar in autumn, where the peak movement extends throughout October and the first half of November (BG). There Tellería (1981) counted over 22,000 flying south at Tarifa in October–November 1977. Biometrics and ringing recoveries show that this passage includes both Iberian birds and those from further north (Asensio 1984b). The return passage is less obvious and is more protracted, extending from mid-February to mid-May, peaking in April (BG).

CITRIL FINCH *Carduelis citrinella*

(Verderón Serrano, Verdilhão Serrano)

Locally common in montane conifers of the major mountain ranges in Spain. Vagrant to Portugal and the Balearic Islands.

Breeding population estimates. Spain: SA1 225,000–230,000 pairs. **Andorra**: 1,000 pairs.

The Citril Finch inhabits subalpine pinewoods in the northern half of Spain, in the Pyrenees and Cantabrian Range, and also in the Iberian Range and Central Range: there especially in the Sierra de Guadarrama. Outpost populations in Andalucía: in Cazorla and the eastern Sierra Nevada mountains (SA2) may be of relatively recent origin. A comparison between the first and second Spanish atlases suggests that there has been some westward range expansion along the Cantabrian Range, from central to western Asturias. However, the Citril Finch remains absent from Galicia, from where there are only three confirmed records: in Abadín, northern Lugo, in June 1999; at Boiro, A Coruña, in November 1999 (*A. O. Galicia 1999*: 63) and a 2008 ringing recovery of one ringed in Teruel (CMA). There is evidence that the westernmost sector of Navarra, in the Basque mountains, was only colonised during the second half of the 20th century, following afforestation with conifers (Elósegui 1985) and some population expansion in this region and in adjacent mountains in northern Burgos has been noted since the 1980s (Román *et al.* 1996, SA2).

The preferred habitat is woods of Black and Scots Pines, generally in the vicinity of alpine pastures. However, the Citril Finch also breeds in mixed beech-fir woods and in Monterey Pine plantations in the western Pyrenees, and it occupies beech forests there and in the Cantabrian and northern Iberian ranges (Noval 1975, de Juana 1980, Elósegui 1985). Nesting birds will settle when and where pine seed crops are available. For example, they have been known to nest during February–April at 400–1,600m in the eastern Pyrenees, to take advantage of good seed crops, afterwards moving to higher altitudes to raise further broods where the pines set seed later in the season (Borràs & Senar 1991). Breeding occurs in the western Pyrenees and the Montes Vascos at altitudes as low as 700m, near the coast (Álvarez *et al.* 1985), and at 900m in the pinewoods of the northern Meseta, in Soria, (de Juana 1980). In general, though, the Citril Finch inhabits higher altitudes, above 1,500m in the eastern Pyrenees and above 1,200m in the western Pyrenees, for example, but flocks forage in pastures to 2,500m and even higher (SA1). Nesting occurs as high as 2,400m in Andorra, where the Citril Finch is widespread (ANA).

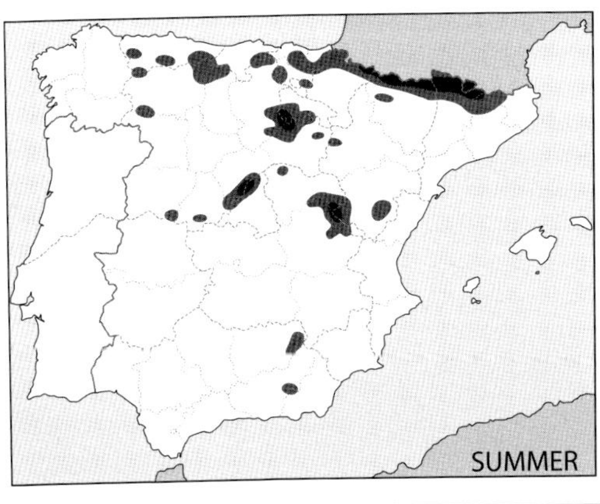

The two breeding populations in Andalucía may have been overlooked in the past but more probably are of recent origin. The capture of an adult and three juveniles in Cazorla, Jaén on 13 August 1997 (*Ardeola* 46: 162) was an early indication of breeding there. The Citril Finch is most probably a recent colonist of the Sierra Nevada (Pérez-Contreras *et al.* 2005) and there is no convincing evidence that it ever occurred there in the past, although López-Seoane (1861) said that it was present and Saunders (1869) claimed to have found a nest in the unlikely setting of the Alhambra gardens in Granada. The Sierra Nevada population has been especially noted around Puerto de la Ragua (2,030m), to the east of the main massif: seven birds were seen there in June 2003 and breeding was proved in 2004 when an adult was seen feeding a fledgling (*Ardeola* 50: 355, 51: 556). Subsequent observations of up to 60 birds in September 2007 (*Ardeola* 54: 403) and 87 birds in November 2010 (Garzón & Henares 2012), suggested that this population was well-established. Several other small breeding nuclei exist in both the Granada and Almería sectors of the Sierra Nevada and the total population of the range has recently been estimated at some 50–80 pairs (Garzón & Henares 2012). The settlement of the Puerto de la Ragua has undoubtedly been enabled by fairly recent afforestation there with Scots Pines.

The Spanish population is certainly of great conservation value since it represents some 80% of the global population (BE). The Catalan subpopulation was put at some 15,000–27,000 pairs in the 2004 regional atlas (CA2). Densities of 12–15 pairs/10ha have been recorded locally in favoured locations.

Birds descend to moderate altitudes in winter but mainly remain within the mountains where they breed. However, birds from the entire Pyrenean range, as well as from the Alps and Jura mountains, have been found wintering on the southern slopes of the Catalan Pyrenees, at altitudes of 300–1,200m (Borràs *et al.* 2010, 2011, 2013). There are recoveries within Spain of birds ringed in France, Germany, Switzerland and Italy, and one ringed in Lleida in November 1993 was found at Sarthe, France (440km) in May 1995 (*Ecología* 11: 412). At the same time, ringing has shown that some Pyrenean birds winter in the Central and Iberian ranges (Borràs *et al.* 2013). A Catalan study has highlighted the importance of mixed open forests, including Black and Scots Pines and a diversity of weed species, as a wintering habitat, warm midwinter temperatures also being necessary to allow the pine cones to open and expose their seeds (Borràs *et al.* 2010).

Citril Finches are seldom noted far outside the breeding range so a record of 40 at Cabañeros, Ciudad Real, on 14 December 2000 is noteworthy (*Ardeola* 50: 167). There are occasional records from coastal areas of Catalonia and Valencia as well as from the Balearics (Clavell 2002). A record of 20 in the Sierra Mariola in December 2009 was only the second confirmed occurrence in Alicante (*A. O. C. Valenciana 2009*: 148). Further south, two were seen at Águilas, Murcia, in September 1975 (*Ardeola* 23: 241) and there are winter records of single birds from Doñana, Cáceres, Jaén and Málaga (*Ardeola* 51: 556, Borràs *et al.* 2013).

Clavell (2002) cites eight records for the Balearic archipelago, chiefly in October, which suggests through passage, but they include single birds at Pollença, Mallorca, in August–September 1994 and March–June 1995. One was in Mallorca on 3 October 2000 (*A. O. Balears 2010*: 46). Claims of occasional nesting by single pairs in Mallorca (Henrici, in Munn 1928) seem unlikely to have been correct.

The Citril Finch is a rare vagrant in Portugal, with four accepted records: single birds at three different locations in the Serra da Estrela in March 1987 and two birds at Bragança on 20 October 1990 (*Airo* 2: 56). Two records of single birds from Ceuta, on the African side of the Strait of Gibraltar in March 1991 and April 1991 (BM), are the only ones for the African continent and a further indication that some individuals stray some distance from their natal mountains.

EUROPEAN GOLDFINCH — *Carduelis carduelis*

(Jilguero Europeo, Pintassilgo-comum

***C. c. parva*. An abundant resident and partial migrant throughout Iberia and in the Balearics. Noticeably smaller and shorter-winged than the nominate race.**
***C. c. carduelis*. A common resident and partial migrant in western continental Europe, intergrading with *parva* in the Pyrenees. Abundant on passage and in winter in Iberia and the Balearic Islands.**
***C. c. britannica*. A common winter visitor. Darker brown on the mantle, back and breast-sides than the other races and slightly smaller than the nominate form.**

Breeding population estimates. Spain: SA1 800,000–2.9 million pairs; ACRE 34.4 million birds. **Portugal**: 0.5–2.5 million pairs. **Andorra**: 400–1,000 pairs.

This is a common breeding bird throughout the region but it is especially characteristic of warmer regions, in the thermo- and mesomediterranean. It is also abundant in the Biscay coastlands, although it is somewhat scarcer in Galicia and in northwestern Portugal (SA2, PA2). Its scarcity in the Iberian north-west was already apparent early in the 20th century (Ticehurst & Whistler 1928, H. W. Coverley 1945 in AP). It frequents low-lying areas and river valleys predominantly and is scarce or absent from higher elevations, although it extends locally to 2,000m in the Pyrenees (CA1) and to 2,200m in the Sierra Nevada (Garzón & Henares 2012). In Portugal it is especially abundant in the south (PA2). The Andorran population is restricted to the valleys

It abounds where there is a ready availability of herbaceous weed seeds together with some taller cover for nesting. It is thus particularly common in the breeding season in orchards and olive groves, riparian woodland through agricultural land, the outskirts of villages, woodland edge habitats and open woodlands such as dehesas. It often nests in parks and gardens. Its propensity for nesting in loose colonies and for siting the nest in such small trees as almonds, junipers and tamarisks,

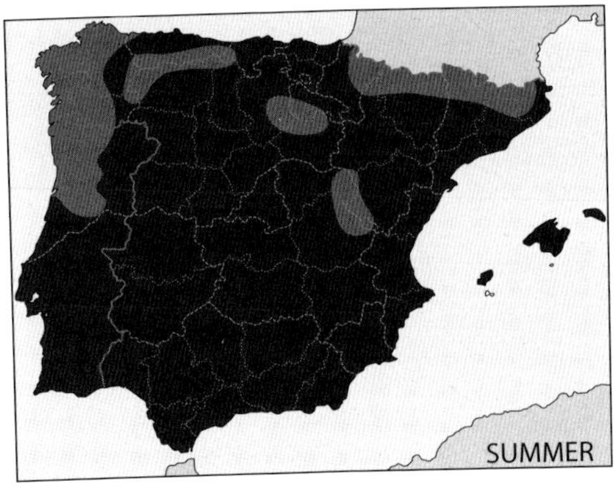

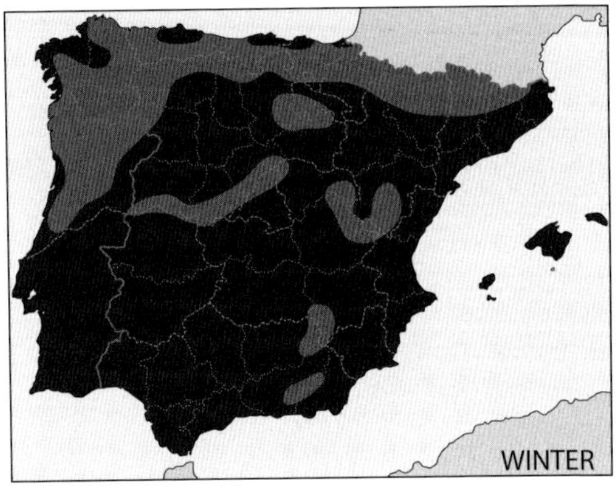

may explain its high densities even in rather treeless regions. The highest recorded densities are all in the thermo- and mesomediterranean climatic regions, with up to 205 birds/km^2 in the open pinewoods of the southern Spain and 150–200 birds/km^2 in dry croplands, orchards, olive groves and vineyards (ACRE). The largest regional population is in Andalucía (38%).

Flocks of hundreds or even several thousand form outside the breeding season and these roam over farmland, especially where there are fallows, stubble fields and rough pastures. The thistle-studded pastures of southern Spain attract great flocks in winter, comprising both Iberian birds and migrants from elsewhere. The highest winter densities are to be found in the warmest regions, especially along the Mediterranean coast and in Andalucía, Extremadura and southern Portugal (Elias *et al.* 1998, SWA). The wintering population of Catalonia alone has been put at 1.8–2.9 million birds (CWA). In contrast, wintering Goldfinches in the eurosiberian and supramediterranean zones occur at much lower densities there than during the breeding season (AI2).

Goldfinches of the two non-resident races winter in very large numbers in Iberia and in the Balearic Islands. Most enter the Peninsula from October to mid December and they depart between March and early May (Asensio 1984a, Arizaga *et al.* 2009a). Over 900 ringing recoveries show that the great majority of the immigrants originate from the British Isles (*C. c. britannica:* 25%) and from France, Germany or Switzerland (*C. c. carduelis:* 63%) but there are also recoveries of birds ringed in the Low Countries as well as a few from the Czech Republic, Austria, Slovenia, northern Italy, Estonia and Russia (Asensio 1986, SMA, AP). Most enter the Peninsula at both ends of the Pyrenees but mainly at the western end, where the Basque Country and Navarra account for over half of the Spanish recoveries of British-ringed birds. Juveniles tend to pass earlier than adults during the autumn passage there (Arizaga *et al.* 2009a). Many Iberian breeders leave the Peninsula and winter in North Africa and the biometric measurements of autumn migrants at the Strait of Gibraltar suggest that it is the small Iberia-bred birds that comprise the majority of the passage there (Asensio 1984b). There are 60 recoveries from North Africa of birds ringed in Iberia: in Morocco (47), Algeria (10) and Ceuta (3), at least 16 of which were of birds ringed in Spain during the breeding season.

The autumn Goldfinch passage across the Strait of Gibraltar is on a very large scale. The movements are during September–December, with a clear peak in the second half of October (BG). Tellería (1981) counted 60,000 birds flying south at Tarifa in October–November 1977. The return passage is less concentrated and extends from mid February until early May, with a peak in April (BG). Goldfinches were the prime quarry of finch trappers in the hinterland of the Strait, and elsewhere, and were trapped and netted in autumn on a massive scale. Most were destined for the pot and some ended up as popular cagebirds. Such trapping is now illegal, as is serving *'pajaritos'* in country ventas, but it still occurs on a significant scale.

EURASIAN SISKIN *Carduelis spinus*

(Jilguero Lúgano, Lugre-comum)

A scarce breeding species in Spain, chiefly in the northern mountains. Also breeds in Andorra. Widespread on passage and in winter throughout the Peninsula and Balearic Islands in highly variable numbers.

Breeding population estimates. Spain: SA1 500–1,000 pairs. **Andorra**: up to 20 pairs.

The Siskin is a scarce but regular breeding species in the high Pyrenees, although the breeding population is thought to fluctuate widely from year to year. Nesting elsewhere is irregular and tends to follow winters in which large numbers arrived from northern Europe. Such sporadic nesting has been reported from Galicia; the Montes de León; scattered locations within the Cantabrian Range including the Basque Mountains; the Iberian Range: particularly in the pine forests of Soria and in the

Serranía de Cuenca; the Central Range: particularly in the Sierra de Guadarrama; the Sierra de Cazorla and the Serranía de Ronda in the Betic Range; and even Mallorca in the Balearics (SA1, SA2). There are no confirmed records of nesting in Portugal (AP, PA2). Breeding is opportunist and takes advantage of good conifer seed crops. The birds nest in mixed Scots Pine and Fir woods in the Pyrenees and in mixed Scots Pine and Birch woods in the Cantabrian Range. Elsewhere nesting is usually in Scots or Black Pines and in larches, and has also been noted in Aleppo Pines and in Monterey Pine plantations.

Siskins enter the Peninsula in highly variable numbers, dependent on winter food availability further north. There are recoveries of birds ringed in most western and central European countries but the few ringed in the breeding season mainly originate from Norway, Sweden, Finland, the Baltic States and eastern Germany (Asensio 1985b). However, birds ringed in Fennoscandia and Russia comprise 34% of 741 recoveries and a high proportion of these are likely to have bred in those regions (SMA, AP). Birds ringed in Iberia have been recovered in the same regions and one from Spain was found in Algeria (1). The Levant coastlands from Alicante to Catalonia, and the eastern Pyrenees, are the main wintering area, Siskins there originating mainly from the Baltic region and central Europe (SWA). A second major wintering area is in the north-west, in Galicia, attracting birds from more westerly populations, in Scandinavia and Atlantic Europe, which arrive via the western Pyrenees and travel west along the northern coastlands.

Winter ringing recoveries in Spain (Asensio 1985b) contrast the relative importance of the Levant coastlands, where Catalonia and the Comunidad Valenciana together account for 32% of the total, with that of Galicia: with just 3.2%. Other regions that have contributed relatively high percentages of winter recoveries are Andalucía (13.8%), the Basque Country (13.0%) and Extremadura (8.1%). It seems likely there may be high interannual variation in the extent and location of the wintering range across the Peninsula, as has been noted in Catalonia (CWA). A minority, perhaps 10% of the population, behave as winter residents and settle in favourable areas for all or most of the winter. The remainder are nomadic and may range over very large areas (Senar *et al.* 1992).

Irruptions occur in some years, generally when a good breeding season has been followed by a shortage of winter food, and Siskins may then be widespread throughout the Peninsula and in the Balearic Islands, sometimes in flocks of 100 or more. Massive irruptions were noted in Spain in the winters of 1949–50, 1959–60, 1961–62, 1965–66, 1966–67 and 1972–73 and there were also major influxes in 1968–69 and 1975–76 (Asensio 1985c). Since then, and up to 2002, there were irruptions in 1981–82, 1985–86, 1988–89, 1989–90, 1990–91, 1996–97, 1997–98 and 2000–01 (SA2). Irruptions thus occur irregularly but are relatively frequent.

The first arrivals tend to be in October and continue through November. Departures are in March–April. The birds are attracted to pine and birch woods, and to alders and elms along water courses. They also associate with other finches to take seeds from thistles and other herbaceous plants in farmland. Opportunistic feeding from exotic plant species has also been noted. For example, up to 100 at the Gibraltar Botanic Gardens in January–February 2006 were attracted to seeds of the Chinese Banyan. They have also been noted in Gibraltar feeding on seeds of the False Yellowhead and eucalyptus (GONHS). Tellería & Galarza (1991) also noted flocks feeding on eucalyptus seeds in a plantation in the Basque Country.

A small proportion of the birds that enter the Peninsula cross the Strait of Gibraltar to winter in North Africa. The first arrivals at Gibraltar are in early to mid October but passage peaks late in the month and through November, with some passing in December. The numbers involved vary conspicuously from year to year but some hundreds may appear at Gibraltar in an exceptional season (BG). A return movement is detectable during February–April, peaking in March. Individuals have been recorded exceptionally at Gibraltar in May, early June and mid September (GONHS).

COMMON LINNET *Carduelis cannabina*

(Pardillo Común, Pintarroxo-de-bico-escuro)

***C. c. mediterranea*. An abundant resident or partial migrant throughout most of the Peninsula and in the Balearics (HBW). Paler and shorter-winged than the nominate, more rufous above and darker rufous on the flanks.**
***C. c. cannabina*. Breeds in northern Spain. Numerous throughout the region in winter.**

Breeding population estimates. Spain: SA1 1.7–3.3 million pairs; ACRE 21.4 million birds. **Portugal**: 100,000–1 million pairs. **Andorra**: up to 1,000 pairs.

The Linnet is abundant as a breeding species throughout the Peninsula and also on the Balearic Islands. However, it tends to be less common in the warmest lowlands: in Extremadura, western Andalucía and southern Portugal, as well as in the coastlands of northern Portugal and in the Mediterranean coastlands from Catalonia to Murcia (SA2, PA2). It nests from sea level to over 2,000m in all the major mountain ranges and up to 3,080m in the Sierra Nevada (Garzón & Henares 2012).

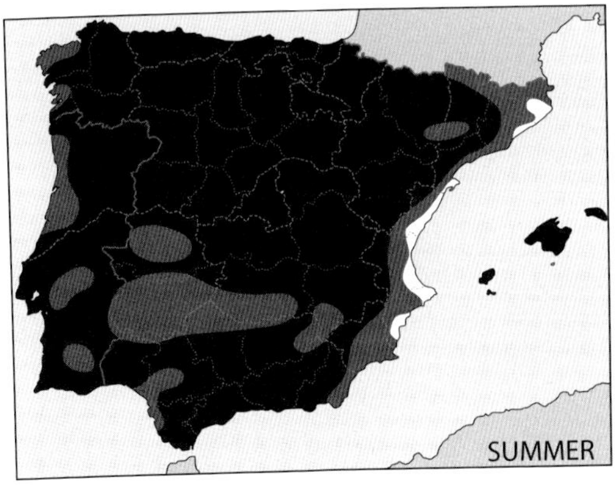

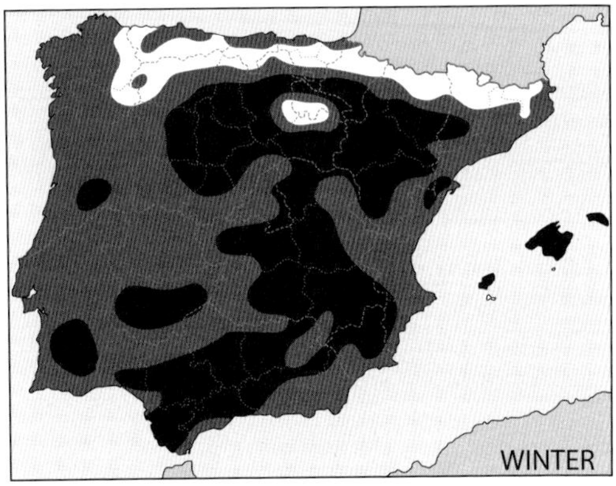

It selects open scrubby habitats and it may be found nesting in Gorse scrub in the north, *Cistus* and *Genista* scrub further south and in montane heathland and juniper scrub. It may also be encountered on sparsely vegetated rocky slopes on hillsides and mountains. It is common in agricultural areas where it frequents fallow land, field margins and vineyards, usually nesting in hedgerows or low shrubs. However, it tends to avoid olive and citrus groves, or is uncommon there. Breeding birds are often encountered in loose colonies of several pairs in close proximity. The Linnet avoids woodlands, irrigated farmland and urban areas (AI2, SA2, PA2). The highest densities are generally in Mediterranean scrub in the southern third of the Peninsula, with 96–205 birds/km^2, but higher concentrations have been reported locally in mesomediterranean vineyards (327 birds/km^2). The Linnet has suffered very large declines in northern Europe due to agricultural intensification. In Iberia it may be vulnerable to habitat loss due to afforestation and to the spread of crop irrigation.

Iberian Linnets are sedentary in many areas but some leave the north to winter further south in Iberia or in North Africa (Asensio 1984b). Those inhabiting high mountain areas descend to lower levels in winter. Linnets are thus largely absent then from the Pyrenees, the Cantabrian Range and the Basque mountains, and they are scarce in the Galician uplands and the Biscay coastlands (SWA). Flocks, often of many hundreds, roam in winter over farmland and pastures in search of weed seeds. Densities of up to 250–450 birds/km^2 are then reported from dry croplands and agricultural mosaics (SWA). The winter flocks include large numbers of continental Linnets. These appear from September onwards but the chief arrivals are in October–November. The immigrants remain to winter or cross the Peninsula to North Africa. Those that stay, tend to remain in the eastern half of Spain, since winter ringing recoveries are concentrated within the Mediterranean coastlands, including the Balearic Islands, and in eastern and central Andalucia (Asensio 1984a, AI2). The same regions also produce large numbers of recoveries during passage periods together with the Basque Country and intermediate areas south to the Strait of Gibraltar. There are some 920 Iberian recoveries of Linnets of extra-regional origin but only six of these were found in Portugal. Most had been ringed in England, northern France, the Low Countries and Germany and birds from these countries accounted for 93% of the sample. The remainder had been ringed in southern Scandinavia and Switzerland with a very few from northern Italy, southern Finland, Russia, Poland and the Czech Republic (SMA, AP). Birds ringed in Spain have been recovered in the same regions but there are also four recoveries from North Africa: in Algeria (2), Morocco (1) and Melilla (1). The Moroccan recovery was of a bird ringed at Ledanca, Guadalajara, in July 1995 and so was probably of Iberian origin: it was found at Hattane (753km) the following November (*Ecología* 11: 415).

This is one of the commonest species that participates in the autumn finch passage across the Strait of Gibraltar. The Linnet movements here are during September–November, with a peak in October. Tellería (1981) counted over 20,000 birds flying south at Tarifa in October–November 1977. The return passage is from mid March to May, with a peak in April.

TWITE *Carduelis flavirostris*

(Pardillo piquigualdo, Pintarroxo-de-bico-amarelo)

Vagrant to Spain and Portugal.

The earliest of the very few Iberian records was one captured in Murcia in November in the late 19th century (Saunders 1871). Since then one was taken at Vila do Conde, Porto, on 4 February 1964 (Santos Júnior 1980), one was seen near Baldaio, A Coruña, on 15 January 1977, another was at Cape Peñas, Asturias, on 16 December 1978, and two were taken near Huesca on 16 November 1980 (*Mustela* 4: 50–51; *Ardeola* 28: 165, 29: 190). Only one subsequent observation has been accepted but it involved an unprecedented flock of 19 seen on a beach at Gijón, Asturias, on 15 October 1993.

The Twite only breeds in Europe in the north of Britain and Ireland (*C. f. pipilans*) and from northern Norway east to the Kola Peninsula (*C. f. flavirostris*). The British populations are largely sedentary. Scandinavian birds, however, winter around the Baltic and North Seas and some reach central Europe. Only 20–130 birds winter in France and they only occur in the extreme north, although they were more widespread and abundant in the past (OF).

COMMON REDPOLL *Carduelis flammea*

(Pardillo Sizerín, Pintarrôxo-de-queixo-preto)

Vagrant to Spain and Portugal.

Alpine Redpolls are largely sedentary or perform short-distance altitudinal movements but those of Western Europe, including the British Isles, may travel further, sometimes irruptively, and these occasionally reach northern Italy, southern France and the Iberian Peninsula. There is one recent record from North Africa, one caught at Ceuta in April 1987, as well as an old report from Tangier, Morocco, and two accepted records from the Azores.

The reliability of several 19th century records from Spain has been questioned (F. Bernis, *Ardeola* 3: 9–11). More recently, one was captured near Santander, Cantabria, in December 1955 (*Ardeola* 3: 1–9) and a small influx occurred during winter 1959–60 when several were captured in Mallorca and there were also records of a female or juvenile at the Prat de Llobregat, Barcelona, in October 1959, a first-year male at La Granja, Segovia, in January 1960, a female at San Sebastian, Gipuzkoa, on 6 March 1960, and another female at Hondarribia, Gipuzkoa, on 17 April 1960 (*Ardeola* 6: 394–395, Ameztoy 1966). More recent Spanish records of single birds have come from Sopelana, Bizkaia, in October 1969; Jaca, Huesca, in January or February 1972; Igries, Huesca, on 7 February 1981, and Ponga, Asturias, during winter 1982–83 or winter 1983–84 (*Ardeola* 15: 156, 21: 447, 29: 190; Álvarez-Balbuena *et al.* 2000). Several records from Mallorca in spring 1969 may not be reliable (Magee, 1977; Ferrer *et al.*, 1986). Two were captured in Portugal at Vila do Conde, Porto, on 12 November 1977, one of which had been ringed at Brigstock, England, on 11 September the same year (Ferreira 1979); another was captured at Póvoa do Varzim, Porto, on 3 November 1985 (Ferreira 1986).

Records of the Common Redpoll are currently subject to assessment by the Iberian rarities committees. Five Spanish records (11 birds) have been accepted since 1984: a female ringed in Menorca, Balearic Islands, on 26 December 1988; one seen at Cariño, A Coruña, on 11 January 1994; a flock of five at Pasajes, Gipuzkoa, on 7 December 1997; a bird at Málaga on 11 November 2005, and at least three at Castropol, Asturias, from 6 November to 7 December 2007 (*Ardeola* 38: 163, 49: 165, 57: 509, 53: 185, 54: 439). There is one accepted Portuguese record: a bird at Cape Espichel, Setúbal, on 26 November 1994 (*Ardeola* 43: 117). In addition, there were three records (8 birds) in November–December 2012 (RBS): one at Puerto de los Alazores, Málaga; one ringed and five seen at the Algar estuary, Alicante, and one in a Pamplona park, Navarra. All Iberian records thus fall between October and April, with a peak presence in November–December. The subspecies involved has not always been identified but most if not all are Lesser Redpolls *C. f. cabaret*.

COMMON CROSSBILL *Loxia curvirostra*

(Piquituerto Común, Cruza-bico-comum)

***L. c. curvirostra*. A locally common resident in Spain and Andorra. An occasional rare breeder in Portugal. Some continental migrants enter Iberia during irruption years.**
***L. c. balearica*. A locally common resident in Ibiza and Mallorca in the Balearic Islands. Smaller, duller and shorter-billed than the nominate form.**

Breeding population estimates. Spain: SA1 140,000–190,000 pairs. **Portugal**: 0–100 pairs. **Andorra**: 300–1,000 pairs.

The nominate race is widespread in conifer forests in Spain but has only a sporadic presence in Portugal, chiefly in the north bordering Galicia but occasionally elsewhere (AP). Crossbills from southern Iberia (Alicante and Málaga) show intermediate characteristics between the nominate form and the Balearic subspecies although the southern birds are more similar to the northern Iberian birds than to the island form (Alonso *et al.* 2006).

Most of the breeding population is in the eastern half of the Peninsula, where most of the suitable forests are located. Breeding crossbills may be encountered wherever there are pinewoods, both natural and planted. They are thus present in all the major mountain systems, occurring as high as 2,400m in the Sierra Nevada (Garzón & Henares 2012), and they also occur locally at low

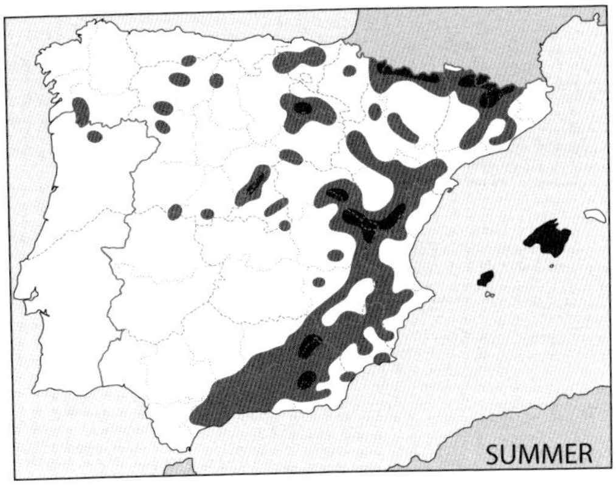

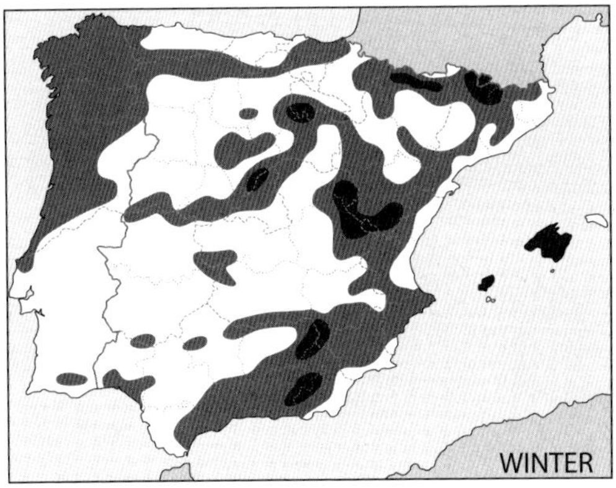

levels on the Mediterranean coasts. The precise breeding and wintering ranges are, however, fluid since Crossbills may start breeding in winter if the pine seed crop is abundant and they may be nomadic when seed is scarce (Elósegui 1985, Ferrer *et al.* 1986, SA2).

The most favoured forests are those of Black and Scots Pines, together with Mountain Pines in the Pyrenees (SA2). However, the characteristics of the pine species involved have been found to influence whether or not the birds are resident or present opportunistically (Senar *et al.* 1993). Scots Pines, and the Mountain Pines of the Pyrenees, are reliable seed sources and they retain the seeds for a year before they are shed. Crossbill populations there tend to be largely resident. The Black Pine, which is widespread in the Central, Iberian and Betic Ranges and in the Levant coastal mountains, is much more erratic in seed production and this varies considerably from year to year. As a result, crossbills are present in those forests when and where food supplies permit and they wander elsewhere in other years. One ringed in Málaga in August 1992 was found at San Martín de la Vega, Madrid (392km) in October 2000 and one ringed at Gúdar, Teruel, in July 2008 was also recovered in Madrid (254km) in January 2009 (CMA). The Aleppo Pines of the Mediterranean coastlands harbour crossbills at times, generally after irruptions, but their robust cones are hard to open and, moreover, they shed their seeds soon after production: hence they comprise a suboptimal habitat and support low densities of crossbills at best (SA2).

The Balearic race, in contrast, occupies woods of Aleppo Pines in Mallorca and Ibiza, where it is abundant. However, these pines have less sturdy cones than those on the mainland, perhaps because there are no squirrels on the islands, and so they offer a more accessible food supply to crossbills (Alonso *et al.* 2006). The species is however only a vagrant to Menorca and Formentera (*A. O. Balears 2010*: 269).

The Crossbill populations of northern Europe are prone to eruptions when periodic depletion of the seed crops occurs Asensio (1984a) noted 30 such irruptions into Spain during the 20th century. They tend to begin in late spring, in May or June, and continue through the summer and autumn. Crossbills may then occur in pinewoods almost anywhere in the Peninsula, and may breed during the following winter and early spring before returning north. There are Iberian recoveries of Crossbills ringed in Italy, Germany, Belgium, Switzerland, Sweden, Poland and Russia, and a bird ringed at Cercedilla, Madrid in November 1970 was recovered the following September at Thierrens, Switzerland (1,050km) (*Ardeola* 20: 121).

TRUMPETER FINCH — *Bucanetes githagineus*

(Camachuelo Trompetero, Trombeteiro-mourisco)

***B. g. zedlitzi*. A scarce resident in southeastern Spain but some continuing recruitment from North Africa is suggested. Rare elsewhere in mainland Spain and in the Balearic Islands. Vagrant to Portugal.**

Breeding population estimates. Spain: 1,000 pairs.

The Trumpeter Finch inhabits the arid south-east. Most of the population is in Almería, from where it has gradually extended its range to localities in Granada, Murcia and Alicante. It inhabits undulating sparsely-vegetated semi-desert terrain broken by the ramblas – usually dry watercourses. In Almería the core population inhabits the Tabernas desert, with others in the Sierra de Cabo de Gata, Sierra Alhamilla and the Huércal-Overa district. Small numbers inhabit the coastal sierras and scattered inland localities of Murcia. There is also a small population in central Alicante and in the Guadix-Baza depression in Granada (SA2). The birds are largely resident but tend to frequent sandy coastal areas in winter, where they may be found in flocks of up to 300 individuals (Manrique & Miralles 1988). The Spanish population in the mid 1990s was estimated at 200–300 pairs (SA1)

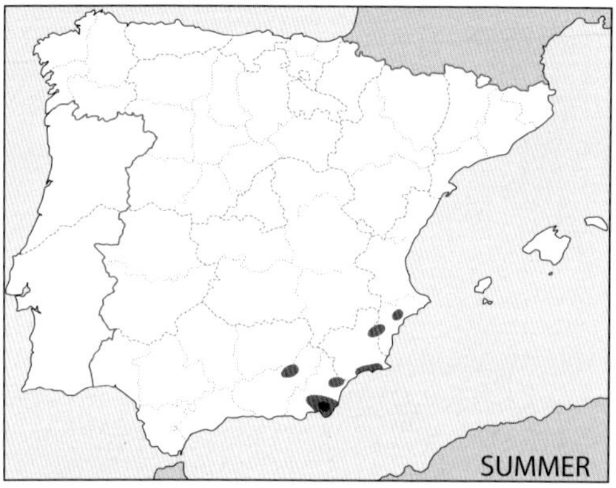

but it increased thereafter and may have numbered 500 pairs by the first years of the 21st century (LR). The Almería subpopulation has more recently been put at 800 pairs (Yanes & Delgado 2006).

Historical references to observations of the Trumpeter Finch in the Iberian Peninsula date back to the 19th century (Irby 1879, Arévalo 1887, Jourdain 1936). However, breeding was only first proved to occur in 1971, in Almería, following a series of sightings there that began in 1962 (*Ardeola* 15: 158–159, König & Cano 1971, García Rodríguez 1972). In 1964, there were three captures of single birds in Cádiz: at Sanlúcar de Barrameda in April and November and at Chipiona in April, and one was seen in Gibraltar in April 1966 (*Ardeola* 7: 268, 10: 71; Lathbury 1970). The first sighting in Murcia was at Cartagena in February 1986 (*Ardeola* 34: 292). In Granada the population in the Gorafe area was only discovered in 1997 (Fernández-Ordóñez & Pérez-Contreras 2006). The first report from Alicante, and the first for the Comunidad Valenciana, was the capture of a female in October 1988 (*A. O. C. Valenciana 1988*: 57) but breeding there was first proved only in 2000, in the Montnegre watercourse, Jijona, and the Sierra de Albatera (SA2, López Iborra *et al.* 2006). At Montnegre there were some 20 pairs in 2001 and 40 pairs in more recent years (López Iborra *et al.* 2006). An additional population in Almería, at Alboloduy-Santa Cruz de Marchena at 500m on the eastern slopes of Sierra Nevada, was first noted around 2000: the birds only seem to be present there during the breeding season and presumably winter at lower levels (López Iborra *et al.* 2006). It may thus be that the species is a relatively recent addition to the Iberian avifauna, having previously been a vagrant from North Africa. It has become progressively better established in the arid south-east and is gradually spreading northwards in the Levant region (Carrillo *et al.* 2007). One at Montserrat, Barcelona, on 31 May 2010 and up to three, including an apparent juvenile, at Belchite, Zaragoza, on 2 and 3 July 2010 (*Ardeola* 57: 548), may indicate such range expansion.

Figure 204. Monthly distribution of Iberian records away from the breeding areas (n = 48)

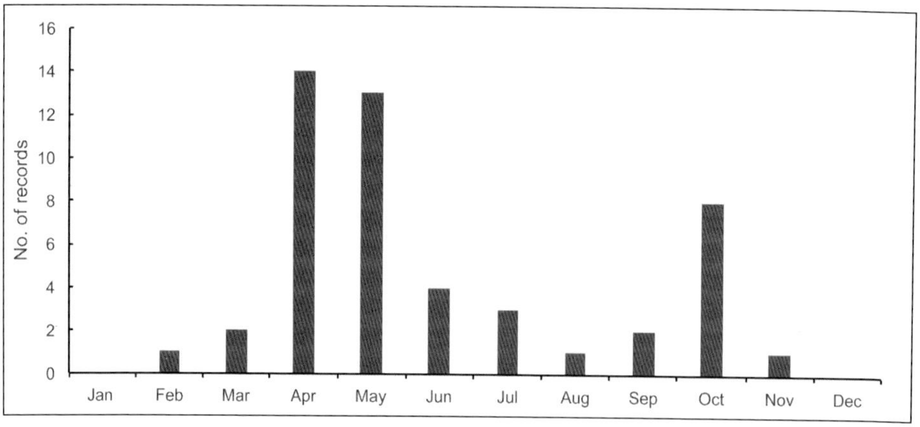

Observations away from the breeding areas are relatively frequent. There are at least 20 records of single birds from the Balearic Islands, during 1993–2011 (*A. O. Balears*; *Ardeola* 59: 453): four from Mallorca, four from Menorca and eleven on Cabrera. Eight were caught by ringers. Most (14) occurred between 26 March and 13 June, and eight of them were in May. Two others were found in July and two in October. One caught at Andratx, Mallorca, on 4 July 2010 was a female with a shrinking brood patch (Llabrés 2011). In addition, one was on the Columbretes Islands in May 2005 (*Ardeola* 53: 211).

There are ten records (nine single birds and a flock of four) from Gibraltar, the most recent in March 2011 (GONHS). One was captured at the Guadalhorce estuary, Málaga in February 1984 and another was there in October 1996 (*Ardeola* 32: 424; *A. O. Malaga 1995–1996–1997*: 128). One ringed as a chick at Jijona in May 2001 was recovered the same year at the Ebro delta (249km) in October (López Iborra *et al.* 2006). Single birds in northwestern Galicia on the Islas Cíes in April 2005; at Toro, Zamora, in summer 2006; on the Basque coast at Amezi, in June 2010, and at Cariño, La Coruña, in May 2011 (*Ardeola* 54: 183, 57: 241, 548, 58: 515) were very far from other record locations.

There are only two recent records of single birds from Portugal: at Cape Espichel, Setúbal, in August 1996 and June 1999 (*Pardela* 8: 16, *Anu. Ornitol.* 3: 11). In addition, one was collected in Portugal in 1884 by the King, Dom Carlos (AP). There are 15 records for France (OF), including Corsica: four in April, six in May, three in June, one in September and one in November; thus showing a similar pattern to the Spanish extralimital records.

Taken together (Fig. 204) all these records suggest that there is some movement to and from North Africa and the clear preponderance of April–May records over autumn ones offers some circumstantial evidence of recruitment from the North African population. In Morocco, the irruptions away from the normal breeding range are particularly noted in spring (BM). Continuing immigration from Africa may account for the relatively high genetic diversity found in the Spanish population (Barrientos *et al.* 2009). Carrillo *et al.* (2007) suggest that the increase and spread of the Trumpeter Finch in the most arid part of southwestern Europe may be due to climatic change.

SCARLET ROSEFINCH *Carpodacus erythrinus*

(Camachuelo Carminoso, Peito-carmim-comum)

Vagrant to Spain, Portugal and Gibraltar.

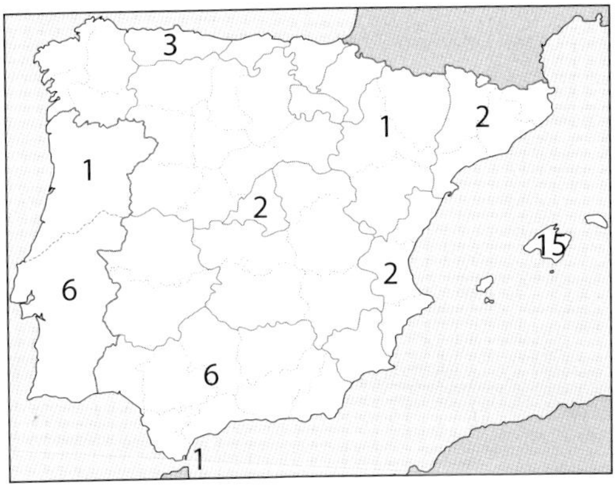

Figure 205. Regional distribution of records, 1983–2010 (n = 39)

The breeding range has spread westwards and occasional nesting is now recorded in France, the Netherlands and Britain. Individuals are often observed west of the breeding range. For example, over 3,000 were recorded in Britain in 1958–2003, including 248 in 1992, and 197 in Ireland up to 2011. Some singing males appear in northern France every spring: although they seem to have been more frequent during the 1990s, and there also regular autumn records in France, mainly of immature birds found in the northern coastlands. French records total 242 individuals up to 2011. South of Iberia, there are two records from Morocco and one in the Canary Islands.

Arévalo (1887) refers to young birds having been captured in Málaga in October in three separate years. However, the earliest known 20th century Iberian record was not until 1983, when a juvenile was ringed at Sanlúcar de Barrameda, Cádiz (*Ardeola* 44: 139). Including this observation, up to 2010 there are 31 accepted records for Spain, seven for Portugal and one for Gibraltar: on 10 October 1999. Eleven of these 39 records were during 1994–96, coinciding with the peak abundance years of singing males in France, but there have only been 22 records since then. One record involved two birds but all the others were of single individuals, often trapped for ringing. There were a further three records from eastern Spain during 2011–12 (*Ardeola* 60: 491, RBS).

Figure 206. Monthly distribution of records, 1983–2010 (n = 39)

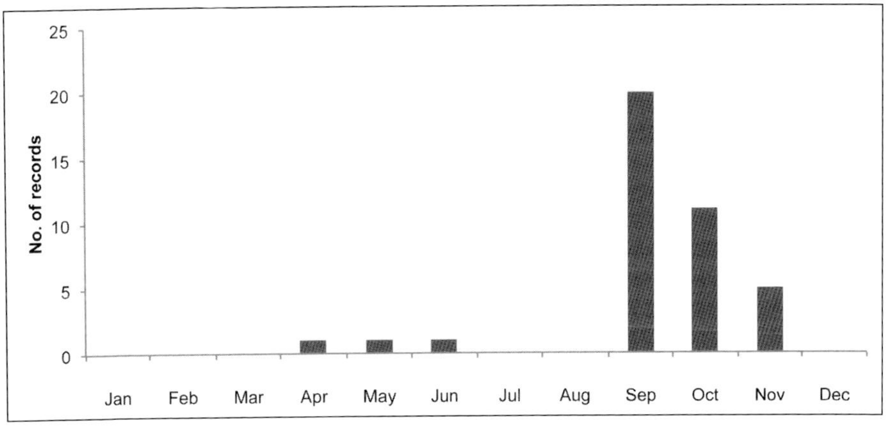

Iberian records are concentrated in northeastern Spain: in the Balearics, Catalonia and Comunidad Valenciana (49%), and in the south between Andalucía and Portugal, with very few found inland or in the northern coastlands (Fig. 205). The great majority have been in autumn (Fig 206) between 8 September and 16 November. Of these, 31 have been aged as juveniles or first-winters and only five as adults: one male and four females. There are only three spring records: a female/first-winter in Menorca on 28 April 1996; a singing male in the Valle de Arán, Lleida, on 31 May 1996, and another singing male at Somiedo pass, Asturias, on 2 June 2005 (*Ardeola* 45: 113, 47: 156, 54: 439).

COMMON BULLFINCH *Pyrrhula pyrrhula*

(Camachuelo Común, Dom-fafe-comum)

***P. p. iberiae*. A common resident in northern Iberia. Smaller, paler grey above and brighter rosy-red below than *europoea*.**
***P. p. europoea*. A generally scarce winter visitor from northwestern Europe to mainland Iberia. A rare vagrant to the Balearic Islands.**
***P. p. pyrrhula*. Recorded once in Spain: three birds by the river Segre in Lleida, on 15 January 2006 (*A. O. Catalunya 2008*: 19–20).**

Breeding population estimates. Spain: SA1 118,000–170,000 pairs; ACRE 339,000 birds. **Portugal**: 200–2,000 pairs. **Andorra**: 200–400 pairs.

Breeding birds are restricted to the eurosiberian climatic region and the more humid sectors of the northernmost supramediterranean (Bernis 1957). They are thus found in the Pyrenees, the Cantabrian Range, the Biscay coastlands and the whole of Galicia and the adjacent sector of northernmost Portugal: in Alto Minho and the north of Trás-os-Montes (SA2, PA2). Their southern limit extends from northern Portugal across the foothills of the Cantabrian Range to those of the Pyrenees, although an almost isolated population inhabits the northernmost sector of the Iberian Range: in the Sierras de Urbión, La Demanda and Cebollera (de Juana 1980). They are associated with a diversity of woodland types: including birch, beech, sessile oak, fir and pine woods, as well as hazel stands and riparian alders, but they frequent the clearings and margins rather than continuous expanses of forest (AI2, Díaz 2010). Farmland mosaics and pastures offering tall hedgerows and bramble patches are favoured in the lowlands. They are also often encountered in orchards and gardens. Nesting is believed to occur at altitudes as high as 2,400m in Andorra (ANA).

The largest fraction of the population is in Galicia (28%), closely followed by the Basque Country (24%) and then by Castilla y León (20%) and Asturias (15%) (ACRE). The Bullfinch was present in some 20% of SACRE sites and the population there showed a 21% increase during 1998–2012 (Escandell 2013). The Portuguese population also appears to be increasing, perhaps with some accompanying southward and eastward range extension (PA2, AP).

The Iberian birds are largely sedentary, although those inhabiting higher altitudes descend lower in winter (Noval 1971). Some appear to disperse further south in the Peninsula in winter and they are joined by small numbers of migrants of the race *europoea*. Three ringed on the Biscay coast have been recovered in western coastal France (Asensio 1984) and one ringed as a nestling in North Rhein-Westphalia, Germany, in June 1971 was recovered in Portugal at Vila Franca de Xira, Ribatejo, in December 1972 (Ferreira 1980). Within Iberia, one ringed in Extremadura in winter 1989 at La Haba, Badajoz, was recovered in Galicia the following spring, at Marín, Pontevedra (455km) (*Ecología* 10: 417).

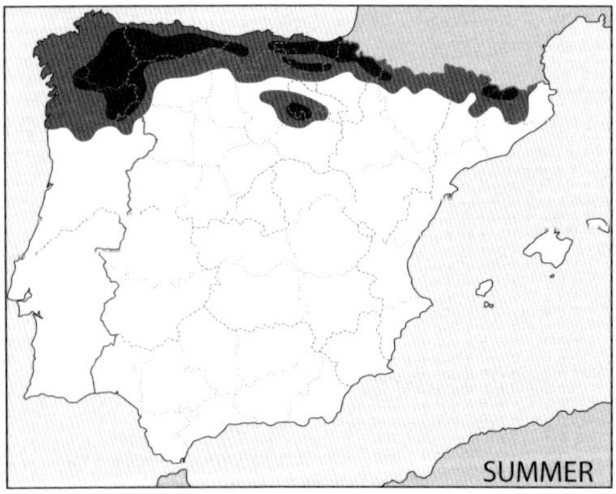

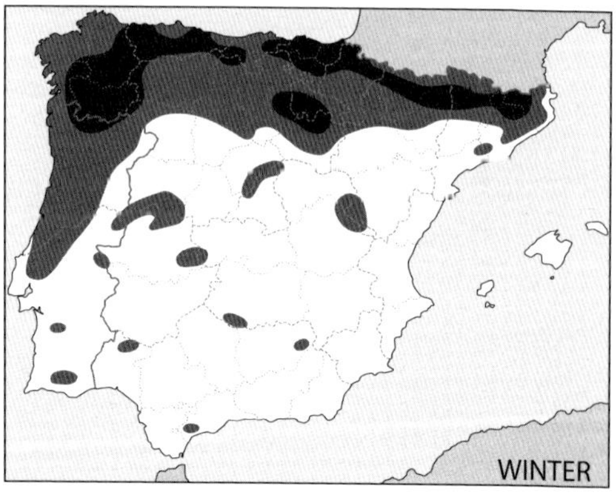

There are increasingly regular winter records from almost the whole of Iberia except for the arid south-east, where nonetheless one was found at Moratalla, Murcia in November 2010 (*A. O. Murcia 2010/1*: 57). Tellería *et al.* (1999) gave only 40 October–April records for the whole of Iberia outside the breeding range and this sparsity does not reflect the current situation. For example, wintering birds have been reported almost annually in Madrid region since 1996, usually in very small numbers, but there were ten at Guadarrama village in February 2001 and 23 at Zarzalejo in February 2003 (*A. O. Madrid 2001*: 208, *2003*: 181). Further south there is a similar wintering presence in Toledo and Ciudad Real and some are reported annually in winter in Extremadura: for example, there are six records, of at least 35 birds, for winter months in 1997 and 1998 from Extremadura (Prieta 2003) and small numbers have been reported annually in winter there since the mid 1990s, especially on the southern versant of the Sierra de Gredos (Langlois 2007). The Bullfinch also now winters in small groups in Sevilla, chiefly in the Sierra Norte and in the vicinity of Doñana (Chiclana *et al.* 2002). Most of these southern records are during November–March (Figure 207). Occasional stray Bullfinches are seen well south of the breeding range during the breeding season. Recent examples include single birds at Monfragüe on 7 July 1997 and in the Sierra de Mijas, Málaga, on 6 June 1998 (*Ardeola* 45: 128, 46: 314). The subspecies involved in all these records has not been determined.

Figure 207. Monthly distribution of records outside the Iberian breeding range (n = 179). Sources: regional bird reports; GONHS.

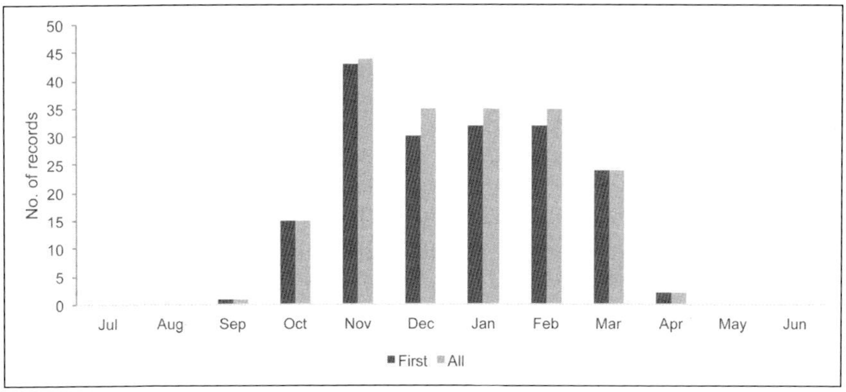

Bullfinches were reported in winter in central and southern Spain by Lilford (1866), Castellarnau (1877), Boxberger (1931), Jourdain (1936–37) and Gil Lletget (1945) so their occurrence there is not new. Nevertheless, the recent impression is that the species has become commoner in winter in southern Spain than was previously the case. Gibraltar is a well-watched location where such a conspicuous species would be easily detected if it occurred at all often. However, the first Gibraltar record was in March 1983 and there have been a further 15 records since: in all there are six records of single birds in March–April and ten records (20 birds) all falling between 30 October and 14 November, possibly suggesting passage to and from Morocco (GONHS records). Nevertheless, there are only five published records for Morocco, totalling nine birds (BM) and two from Ceuta: one in December 1994 and two birds in March 2001 (*Ardeola* 45: 128, 50: 167).

The Bullfinch is regarded as a vagrant in the Balearic Islands (*A. O. Balears 2010*: 269). Single birds in Mallorca in April 1938, July 1967 and October 1968 (Munn 1939, Nadal 1971), were regarded as escaped cagebirds. However, five birds in Menorca in January 1977 and one on Formentera in November 1988 (Muntaner & Congost 1984, *A. O. Balears 1988*: 42) seem to have been wild.

HAWFINCH *Coccothraustes coccothraustes*

(Picogordo Común, Bico-grossudo-comum)

***C. c. coccothraustes*. A scarce resident in central mainland Iberia. Also a scarce winter visitor to the Peninsula and Balearic Islands.**

Breeding population estimates. Spain: SA1 4,000–5,000 pairs; ACRE 340,000 birds. **Portugal**: 1,000–10,000 pairs.

The principal breeding range is in the supramediterranean region, avoiding oceanic climatic influences and also the warmer Mediterranean areas. It includes Extremadura, most of Castilla y León: including the southern foothills of the Cantabrian Range, eastern Castilla-La Mancha and Madrid. The Hawfinch is widespread and locally common in Andalucía across the Sierra

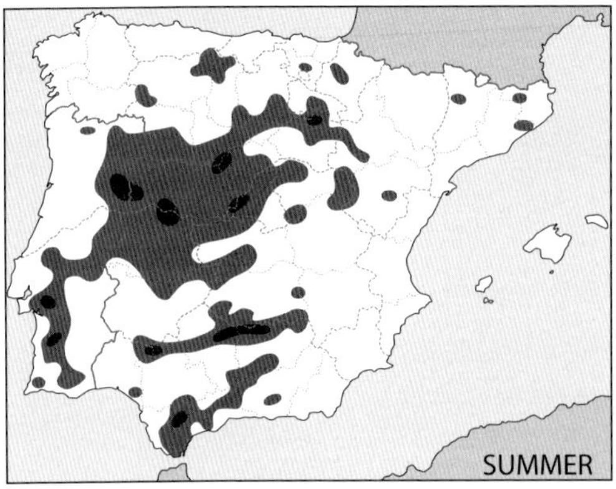

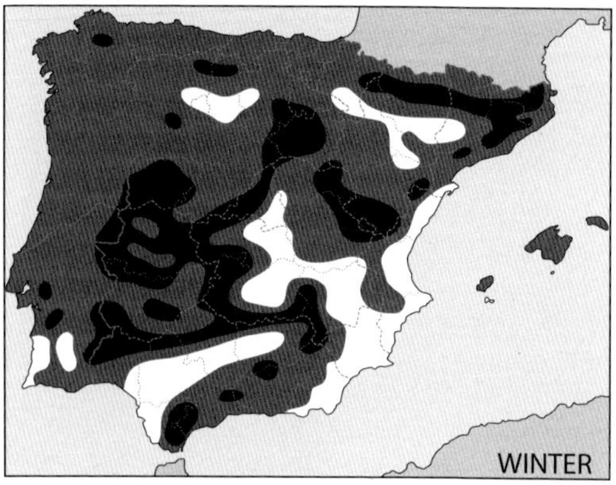

Morena, from Huelva in the west to Jaén in the east, and in the Betic Range: particularly in Los Alcornocales in Cádiz and in the Serranía de Ronda, Málaga (SA2). In southern Portugal it occurs mainly in the Tagus and Sado valleys, the Serra de Grándola in Baixo Alentejo, and Algarve; in the north it occurs almost entirely in the eastern third from Beira Baixa to Trás-os-Montes and Alto Douro (PA2).

It is almost entirely absent from the Eurosiberian region, including the northern side of the Cordillera Cantabrica, the whole of Galicia and the coastal regions of the northern two-thirds of Portugal. It is uncommon in coastal regions of southwestern Spain although there is a regular breeding presence in Doñana (Llandres & Urdiales 1990, García *et al.* 2000). Otherwise it does not breed in the lower Guadalquivir valley nor in the lower Ebro valley. It is also largely absent from a broad swathe of eastern Spain, comprising eastern Castilla-La Mancha, Almería, Murcia, the Comunidad Valenciana, most of Catalonia and the Pyrenees (SA2).

Breeding birds occur in mature deciduous, coniferous and mixed woodlands with a well-developed understorey, generally in the vicinity of water. Oakwoods, notably Cork Oak and Iberian Holm Oak dehesas are particularly favoured. They also frequent beech, ash, elm and maple woodlands, as well as riparian trees and large parks and gardens. Nesting birds are frequently encountered in loose colonies.

The ACRE estimate of the Spanish population in spring during 2004–06 is far in excess of the SA1 estimate but the earlier figures seem to have been very low. The long-term population trend in Spain is unclear, but the Portuguese population has shown some decline and range contraction since the 1990s, in Estremadura, the south-west and the north-east, attributed to intensification of orchard management in some areas (PA2).

The Iberian birds appear to be largely sedentary but some Hawfinches from elsewhere in western Europe enter the Peninsula in winter, occurring from October to April, but with higher counts occurring in December–February (Asensio & Antón 1990, Aymí & Romero 1996, Polo 2009). Hawfinch movements into Iberia have an irruptive nature (Polo 2009). Immigrants are reported mainly from Catalonia, where the wintering population has varied widely between seasons: for example, some 2,900 birds were estimated in both 2006/07 and 2007/08, but as many as 117,000 were present in winter 2008/09 (CWA). Occasional migrants and wintering birds are reported from Asturias (Noval 1986) and Hawfinches were said to be 'abundant' at numerous sites in the Pyrenees in Huesca, Aragón, in winter 1988–89 (*Ardeola* 36: 264). Ringing recoveries, and sex ratios of birds ringed in Iberia, suggest that females outnumber males among the Hawfinches that enter the Peninsula to winter (Asensio & Antón 1990). Asensio & Antón (1990) give a total of 31 Spanish recoveries of birds ringed in Germany (13), the Low Countries, Switzerland, the Czech Republic and France. There are also single Spanish recoveries of birds ringed in Austria, Hungary and Italy (SMA). Spanish recoveries are mainly in the north-east but there are also a few Portuguese recoveries: of birds ringed in Germany (4) and France (AP). There are recoveries from Italy (2), Germany and France of birds ringed in north-east Spain in winter.

Some also occur in winter in the Balearic Islands: a large influx occurred in Mallorca during winter 1947–48 and there was another in winter 2005/06, especially to Menorca, where 300 were at the Albufera des Grau on 5 December (Munn 1948a, *Ardeola* 53: 293). Winter 2005/06 also saw much higher numbers than usual both in Catalonia and in the Comunidad Valenciana (CWA, Polo 2009).

Hawfinches frequently form flocks outside the breeding season, notably to exploit local food supplies. Fruiting Nettle Trees at the Alhambra, Granada attracted up to 70 in December 2007 (*Ardeola* 55: 150). An attraction to Nettle Trees has been also noticed in Catalonia and the Comunidad Valenciana (Aymí & Romero 1996, Polo 2009). The northeasterly distribution of immigrant Hawfinches may be partly explained by the natural distribution of the Nettle Tree, which tends to avoid the cold high areas of inland Spain. In Madrid, wintering Hawfinches often appear in the parks of Madrid, especially in the Botanic Gardens there, where the large Nettle Trees were known to attract them as long ago as 1940 (Bernis 1988b). Nettle Tree groves also attract hundreds to the Arribes del Duero region in winter: for example, there were 200 at Hinojosa de Duero, Salamanca, on 30 January 2004 and 400 at the same site on 25 March 2004 (*A. O. Salamanca 2004–2006*: 174). Hawfinches in Iberia also take the fruits of various oaks, beech, hazel and wild olive.

There is evidence of very small-scale migration across the Strait of Gibraltar but the extent to which either Iberian or other European birds participate is unknown. Tellería (1981) recorded only 26 apparent migrants at the Strait during October 1977. There are only nine records from Gibraltar itself, these corresponding largely with migration periods: two records (four birds) were in March and there are four records (20 birds) in October–November.

YELLOW-RUMPED WARBLER *Setophaga coronata*

(Reinita Coronada, Mariquita-coroada)

Vagrant to Spain.

One was at the Laguna de Traba, Laxe, A Coruña, on 18 September 1982 (*Ardeola* 59: 399). This is a rare vagrant to Europe with 28 records in Great Britain and Ireland, 14 in Iceland, eight in the Azores and one each in Norway, the Netherlands and the Canary Islands.

COMMON YELLOWTHROAT *Geothlypis trichas*

(Mascarita Común, Mariquita-de-mascarilha)

Vagrant to Portugal.

A juvenile male was caught by ringers at the Parque Ambiental de Vilamoura, Loulé, Faro, on 18 October 2010 (Tipper & Hodson 2011; *Anu. Ornitol.* 8: 18–19). This American warbler is a very rare vagrant in Europe, from where there are ten records from Britain and Ireland, six from the Azores and one from Iceland.

LAPLAND BUNTING *Calcarius lapponicus*

(Escribano lapón, Escrevedeira-da-lapónia)

Vagrant to Spain and Portugal.

The nominate form breeds from Scandinavia east to northeastern Siberia and *C. l. subcalcaratus* breeds in Greenland and Canada. European birds largely winter in Russia and Ukraine but some do so around the North Sea, where several hundred occur in Britain and an average of 160 appeared in northern France in 2000–06. Records south of Vendée or in Mediterranean France are exceptional. Some of those that reach Ireland, Scotland, southern Norway and perhaps Brittany may originate from Greenland. There are no records from Morocco but it has been found in the Azores.

The earliest Iberian record seems to have been of a male that was found dead in a ringer's net on the Columbretes islands, Castellón, on 7 October 1972 (Pechuán 1973). Three were then seen at Cape Estaca de Bares, A Coruña, on 10 October 1974 (*Ardeola* 23: 242), and single males were noted in Asturias at Cape Peñas on 25 October 1980 and at Cudillero on 30 October 1984 (García-Rovés & García-Rovés 1989, Álvarez-Balbuena *et al.* 2000). The earliest Portuguese observation was of three juveniles at Cape Espichel, Lisboa, on 10 October 1990 (Moore 1992). Up to 2010 there have been 33 accepted records from Spain and eight from Portugal.

We here consider 45 Iberian records, involving at least 92 individuals. Single birds were involved in three-quarters of cases but there are instances of flocks, including up to nine at Carballo, Galicia, in December 1998, nine at the same place in November–February 2010 and up to 11 at Cape Peñas in September 2010 (*Ardeola* 41: 115, 56: 337, 59: 399). Nearly all records have been from Biscayan and Atlantic coasts between Asturias and Portugal (Fig. 208). The exceptions have been three Mediterranean sightings: two from the Columbretes islands, the record cited above and another there on 9 October 1989, and one at the Aiguamolls de l'Empordà during 24 October–9 November 2010 (*Ardeola* 38: 163, 59: 399). Most reports have been in autumn, 89% during September–November, but there are a few in winter, although none in spring (Fig. 209). Extreme dates are 12 September and 26 February. Stays are usually short but a flock was at Carballo from 2 November 2010 to 26 February 2011 (*Ardeola* 59: 399). Records in France and in western Europe generally are also much more frequent in autumn than in winter (BWP, OF).

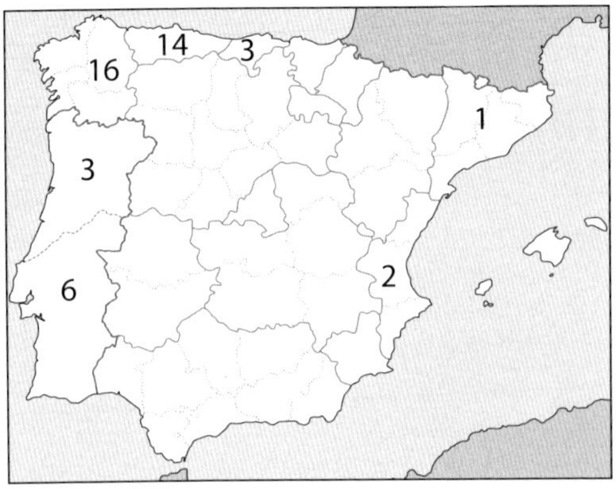

Figure 208. Regional distribution of records, 1972–2010 (n = 45)

An unprecedent irruption into Spain occurred in 2010–11 (*Ardeola* 59: 399–400, 60: 491; RBS). Almost 90 individuals were reported between 12 September 2010 and 25 March 2011. As usual, most were in the northern coastlands: 34 in Galicia, 30 in Cantabria and 19 in Asturias, but there were also two in Catalonia, one in the Comunidad Valenciana, one in the Balearics: at Alcúdia, Mallorca, in October, and one in Andalucía. The main influx was relatively early, in September, and may perhaps have involved birds from North America or Greenland. The largest reported concentration was 24 at Ribamontán al Mar, Cantabria, in January. There was one in Portugal in October, at Peniche, Leiria (RBP). Unprecedented numbers of Lapland Buntings were also noticed in 2010–11 in Britain, France and central Europe (Pennington *et al.* 2012, Lawicki 2013, Yésou 2013). At least 22 more birds were reported in Spain in winter 2011–12 and at least nine during winter 2012–13, largely from the northern coastlands (RBS).

Figure 209. Monthly distribution of records, 1972–2010 (n = 45)

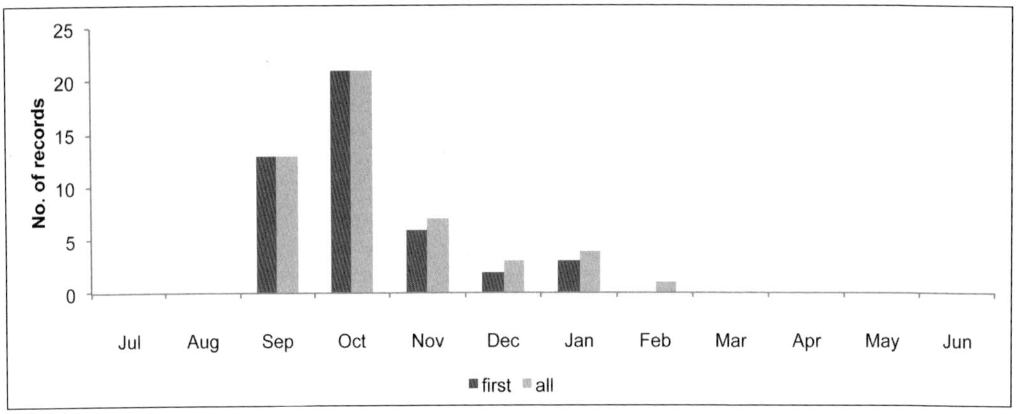

SNOW BUNTING *Plectrophenax nivalis*

(Escribano Nival, Escrevedeira-das-neves)

A scarce and local winter visitor, chiefly to northwestern coasts. Vagrant to the Balearic Islands. Presumably *P. n. nivalis* / *P. n. insulae*.

The Iberian peninsula is at the southern extreme of the European wintering range and very few individuals reach it (SWA), probably fewer than 300 in a good year. Nevertheless, de Souza (1991) found 322 records, involving 2,860 individuals, for Iberia during 1974–90. Our own survey of the Spanish regional bird reports compiles 208 records (Fig. 210). Some 69% were in Asturias and Galicia, with 63 and 48 records respectively in 1995–2000. The remainder were in the coastal north east but in this same period they comprised only eight records In Catalonia, four in the Comunidad Valenciana and one in the Balearics. Small parties appear very locally on northwestern coasts from October to March, occasionally lingering into April. The extreme dates for the records analysed by de Souza (1991) are 23 September and 1 May. The November peak in Spanish records suggests that some early arrivals later move further south, most probably into coastal Portugal. Snow Buntings frequent dune edges on sandy shorelines, field margins and coastal scrub, where they search for seeds.

Most of the birds seen in Galicia appear in the coastal north-west, where significant flocks may appear in some localities, most notably at the Baldaio and A Frouxiera beaches. Both are regular wintering sites but the numbers present vary considerably from year to year. De Souza (1991) found that the mean winter flock size at Baldaio was 12.1 birds but flocks of 31–40 occurred on 11 occasions and flocks of 41–50 on three occasions during 1975–86, with an exceptional flock of 100 in November 1975 (de Souza 1991). Adult females and first-winter birds outnumber adult males: at Baldaio full adult males comprised only 6% of birds in autumn and 14% in winter. Fewer birds have been reported in later years, especially post-2000. The Galicia bird reports for 1995–2004 give 73 records, involving at least 342 individuals, a mean flock size of 4.7 birds. The largest reported gatherings during this latter period were in A Coruña: up to 53 at Cariño in October–November 1996, 30+ at Muros on 6 February 1997, and up to 29 at Valdoviño in November 1999. The heathland at Cape Peñas, Asturias, and Liencres beach in Cantabria, are also regular sites (Noval 1986, de Souza 1991) but only very small numbers occur there. A wintering flock of up to 26 birds at Aboño, on the coast just west of Gijón, during 1988-89 was exceptional for Asturias, where flocks often comprise just a few individuals.

Most Snow Buntings are found on coasts but there are a number of records from mountaintops in Galicia and Asturias especially: the more noteworthy of these include eight on the Serra de Capelada, A Coruña, in November 2003 and one at an altitude of 1,300m at El Cordal, Quirós, Asturias, in February 2000 (*A. O. Galicia 2003*: 168, *A. O. Asturias 2000*: 241). In Catalonia one was found at an altitude of 2,100m at Vilallonga de Ter, Girona, in November 1999 (*A. O. Catalunya 2001*: 358). There is also a record of five at 2,380m in the Sierra de Béjar, Salamanca, in December 1998 (*A. O. Salamanca 1924–2003*: 134). It is unclear whether such birds are on overland passage or whether they actually winter inland: scarce winterers at high elevations are likely to go unnoticed. Other inland records comprise at least six from Castilla y León; four from Andalucía; three each Extremadura, Burgos and Navarra; two from Aragón and single observations from Castilla-La Mancha, La Rioja and Madrid. They include a quite exceptional flock of 20 at Doñana in January 1998 (*Ardeola* 52: 224).

Figure 210. Monthly distribution of records in Spain, 1995–2011 (n = 208). Sources: regional bird reports.

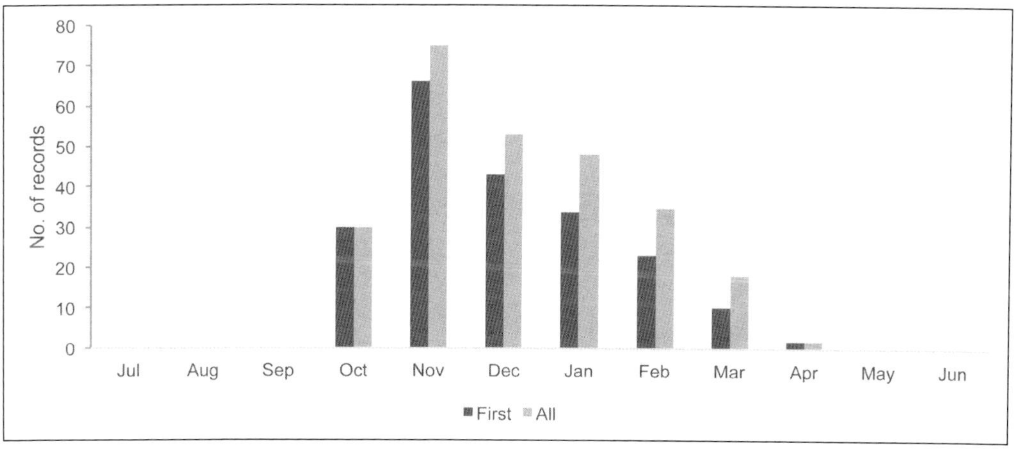

The Snow Bunting occurs more or less annually in Portugal. There were some 30 Portuguese records during 1979–2007, most of them of single birds or pairs but some of groups of up to six (AP), and a further seven records (11 birds) from 2009–11 (RBP). A flock of 14 at São Jacinto, Aveiro, Beira Litoral, in October 1984 is noteworthy but there are historical references to influxes in cold winters, as in 1928 when small flocks were common on the Beira coast. Most observations in Portugal are also coastal, with birds appearing with some regularity at São Jacinto; the Lagoa de Óbidos and Cape da Roca, Estremadura; Cape Espichel, Ribatejo; and the Algarve coast. However, here too there are some inland records and the Serra da Estrela has seen birds in at least four different winters, including a flock of seven in November 1993: sightings have generally been near the Torre, the monument marking the highest point in Portugal (1,993m) (AP).

Snow Buntings occur annually in Catalonia but singly or in twos and threes. Ferrer *et al.* (1986) give ten records for the region and there were at least 32 further records (36 birds) during 1985–2009 (Clavell 2002, *A. O. Catalunya*, CWA). There are also some 12 records (13 birds) from the Comunidad Valenciana (*A. O. C. Valenciana*). The extreme dates for these Levant records are 22 October and 'early' March but the great majority were during November–January.

Occasional individuals also appear in the Balearics in winter. Ferrer *et al.* (1986) give seven records and there were at least five records of single birds during 1998–2011, including a very late individual at Felanitx, Mallorca, on 23 May 2011 (*Ardeola* 59: 453).

PINE BUNTING *Emberiza leucocephalos*

(Escribano Cabeciblanco, Escrevedeira-de-barrete-branco)

Vagrant to Spain and Gibraltar.

One was taken at Llavaneras, Barcelona, around October 1986; a male and a female were seen in Gibraltar on 2 May 1987, and single males were recorded on the Columbretes Islands on 14 November 1989, and at Masegosa, Cuenca, on 30 March 1995. Most recently, single males were found wintering in flocks of Yellowhammers at Monóvar, Alicante: from 17 December 2005 to 17 March 2006, and at Aspa, Lleida, from 2–11 January 2009. Most Pine Buntings winter in southern Asia but small wintering populations have recently been discovered far to the west, in Israel and Italy (Shirihai 1996, Occhiato 2003). It has been recorded with some frequency elsewhere in western Europe, with 50 individuals in Britain and 43 in France. Many European records have been in October–November, but there are quite a few between February and April and some in mid winter, for example in Camargue, France.

YELLOWHAMMER *Emberiza citrinella*

(Escribano Cerillo, Escrevedeira-amarela)

***E. c. citrinella*. A common resident in the northern mountains, more widespread in winter when small numbers of continental birds are also present. Vagrant to the Balearic Islands.**

Breeding population estimates. Spain: SA1 140,000–170,000 pairs; ACRE 634,000 birds. **Portugal**: <400 pairs. **Andorra**: 50–150 pairs.

The Yellowhammer is a characteristic of the eurosiberian zone, where it is largely confined when breeding to the valleys of the Pyrenees, the Cantabrian mountains and the northern Iberian Range south to Soria (AI2). The highest densities are probably on the southern versant of the Cantabrian range, from León to the Basque Country, and in some sectors of the northern Iberian Range (de Juana 1980, SA2). It is chiefly montane in distribution, generally scarce below 500m but reaching 2,000m locally and up to 2,300m in the eastern Pyrenees (CA2). The breeding range extends into northernmost Portugal, principally on the Mourela plateau in Peneda-Gerês National Park (PA2, AP). It is also present, but rare and declining, in the coastal lowlands of Galicia and the Bay of Biscay.

Breeding birds select open terrain with plenty of scattered shrubs and are thus common in pastures and farmland with hedgerows, open scrub and woodland edges bordering farmland and grazing land. Flocks are attracted to farmland in winter when considerable numbers occur on stubble and fallow fields, often roosting in reedbeds (Calvet *et al.* 2004, Arizaga *et al.* 2007, CWA, SWA).

The two Spanish breeding bird atlases document some contraction in the breeding range in Galicia, Asturias, León and the eastern Pyrenees. In Catalonia, a comparison between the 1983 and 2004 breeding bird atlases shows a 17% reduction in the number of occupied 10km-squares (CA2). The population on SACRE sites showed a significant decline during 1998–2012 (Escandell 2013). The Portuguese population is small: the principal nucleus on the Mourela plateau may number 50–350 pairs (Pimenta & Santarém 1996). The Yellowhammer is fairly common in Andorra (BE, ANA).

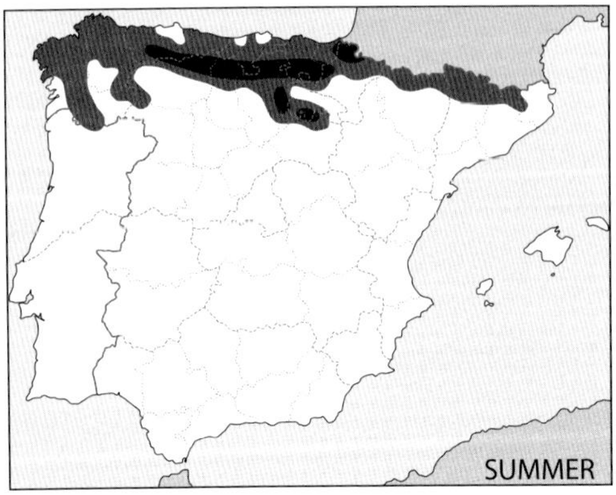

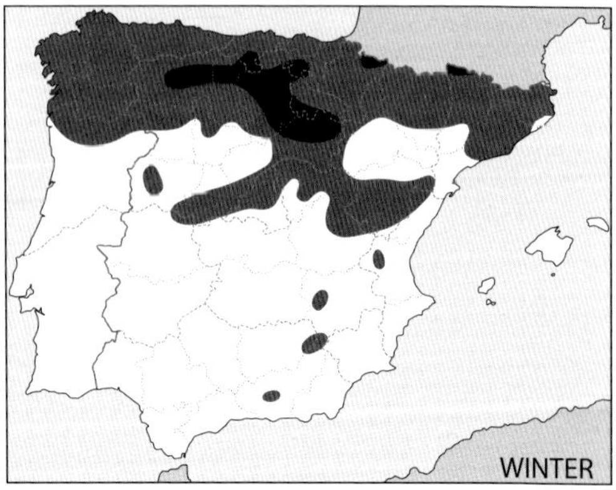

Flocks descend into the northern farmlands of the Meseta and the Ebro valley in winter, where they are mainly present from mid October to early April (Fig. 211), as reported for example from the Plana de Lleida (Calvet *et al.* 2004). Variable numbers winter as far south as Madrid and northern Castilla-La Mancha, but few range further south, where they are more likely to be encountered in the uplands. Individuals ringed in Germany, Belgium, France and Poland have been found in Spain in winter (SMA). In Catalonia, where the breeding population is rather small: 2,900–14,200 pairs, the estimated wintering numbers are much larger and they varied from some 98,000 individuals in 2006–07 to 237,000 in 2007–08, suggesting a variable seasonal influx of immigrants (CWA).

Figure 211. Monthly distribution of records outside the breeding range, 1984–2010 (n = 111).

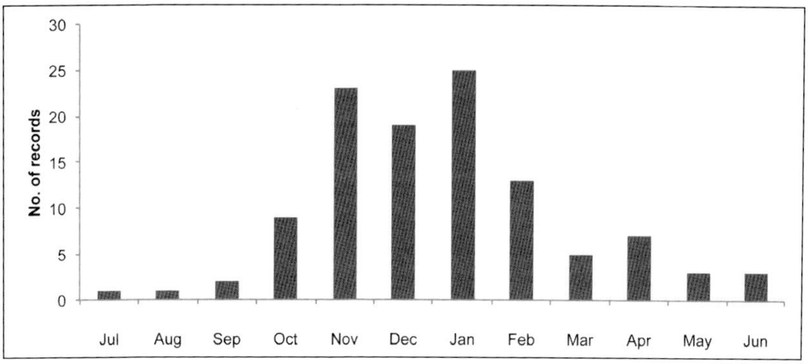

Winter records elsewhere in Iberia are much sparser. There were at least 38 records (94 birds) from the Comunidad Valenciana during 1988–2009, most of them of single birds but including up to 25 at El Toro, Castellón, in January 2008 (*A. O. Castellón 2008*: 150). Most of the records from this region are during November–February and 12 records (15 birds) involved apparent migrants found on the Columbretes islands. Further south on the east coast, one at Moratalla, Murcia, in November 2010 was exceptional (*A. O. Murcia 2010-I*: 58). There are only seven records (13 birds) for Extremadura, also in winter (J. Prieta, pers. comm.). However, it occurs locally as far south as the Sierra Nevada, where small numbers have been reported in recent winters, at altitudes above 1,000m on the northern slopes (Garzón & Henares 2012). It is virtually unknown elsewhere in the far south, from where the few available records include single birds at Gibraltar in March 1984 (GONHS), and, in Málaga, at Antequera in December 1994, at Riogordo in October 1995 and in the Sierra Tejeda in January 1997 (*A. O. Málaga 1994*: 53, *1995–1996–1997*: 129).

There are very few winter reports from Portugal, the most southerly being an individual at the Sado estuary in January 1995 (*Pardela* 2: 23). The largest flock recorded was of 36 individuals in Trás-os-Montes in February 2005 (AP). It is very rarely found in the Balearics: there were 11 records of single birds during 1985–2009, nine of them during October–March but there were unseasonal birds on 17 May 1985 and 1 June 1988 (*A. O. Balears 1988*: 43, *2010*: 269). The Yellowhammer is a vagrant in Morocco, with scattered records chiefly in the north, but an influx in 1992–93 included a flock of 18 (BM).

CIRL BUNTING *Emberiza cirlus*

(Escribano Soteño, Escrevedeira-de-garganta-preta)

***E. c. cirlus*. A common resident in mainland Iberia and on some of the Balearic Islands.**

Breeding population estimates. Spain: SA1 500,000–800,000 pairs; ACRE 2.9 million birds. **Portugal**: 10,000–100,000 pairs. **Andorra**: 20–40 pairs.

The Cirl Bunting is a common and locally numerous resident throughout much of Iberia, although most plentiful in the supramediterranean zone (AI2). It is most abundant in open woodlands and in farmlands with some tree or bush cover, and is scarce or absent from largely treeless regions and at higher altitudes, above *c.* 1,300m. Exceptionally, it occurs to 2,100m in the Sierra Nevada (Garzón & Henares 2012).

The highest densities in Spain, 20–50 birds/km^2, occur in farming areas with scattered trees, notably in vineyards, mixed farmland and orchards, as well as in riparian growth and supramediterranean scrub (ACRE). It inhabits open coniferous,

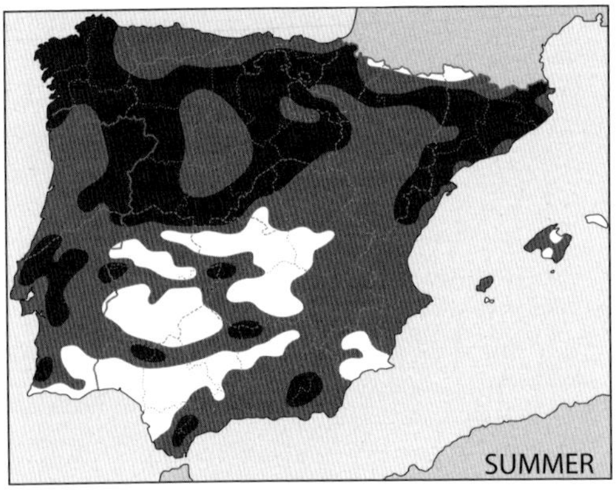

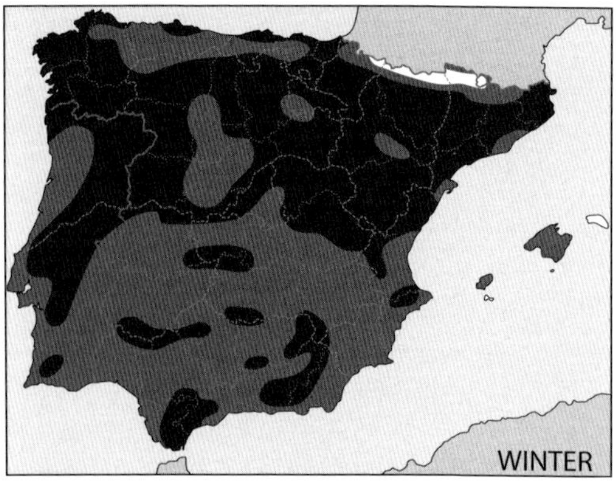

deciduous and sclerophyllous woodlands and may also be found in citrus and olive groves, city parks and hedgerows. It is reputedly scarcer in the eurosiberian region, where it may be in competition with the Yellowhammer (de Juana 1980), but it is nonetheless common in Galicia and in the farmlands along the Biscay coast. In Spain it is scarce in the central Duero and Ebro valleys, and absent from large areas of southern Extremadura, southern Castilla-La Mancha, Murcia and the lower Guadalquivir valley (SA2). Indeed, in the southern third of the Peninsula it is largely associated with the wooded hillsides of the Sierra Morena and the Betic Range, where it is common (SA2), and with the western coastal lowlands in Portugal (PA2). The wider Portuguese breeding range covers most of the country, with concentrations in the north-east: in eastern Trás-os-Montes and Beira Alta, and in the centre-south and south-west: in Estremadura, Ribatejo and western Baixo Alentejo. Tait (1887) described it as by far the commonest bunting in northern Portugal. It is largely absent from the dry south-east of Portugal, including most of eastern Baixo Alentejo and eastern Algarve (PA2). It is resident in Mallorca and Ibiza in the Balearics but is only reported as a vagrant in Menorca (*A. O. Balears 2010*: 269).

The Iberian populations are largely sedentary although some birds descend to valley bottoms and foothills in winter, when small flocks feed on weed seeds on farmland, although never far from cover. They are also common in winter in open woodlands, such as dehesas and olive groves. Short-distance cold weather movements, generally of less than 100km, have also been reported, but nearly all ringing recoveries are local. The Catalan winter population, 650,000–1,100,000 individuals (CWA), is broadly similar to the estimated post-breeding population, suggesting that any winter immigration there is on a small scale. There are three just Spanish recoveries, from Catalonia and the Basque Country (2), of Cirl Buntings ringed in France (SMA). There is also some evidence of very small-scale migration across the Strait of Gibraltar in September–November with a return passage in March–April (BG, Tellería 1981).

ROCK BUNTING *Emberiza cia*

(Escribano Montesino, Cia)

***E. c. cia.* A common resident throughout most of the Peninsula. Absent from the Balearics.**

Breeding population estimates. Spain: SA1 820,000–2 million pairs; ACRE 3.6 million birds (ACRE). **Portugal**: 50,000–500,000 pairs. **Andorra**: 200–1,000 pairs.

This is a characteristic species of broken terrain everywhere, ranging from sea level locally to the highest mountain peaks, although it is most numerous above 500m. It reaches 2,000m in the northern Iberian Range, 2,300m in the eastern Pyrenees and 2,500m in the Sierra Nevada (de Juana 1980, CA2, Garzón & Henares 2012). It is thus scarce or absent in Spain from the farmlands of the Meseta, the Guadalquivir and central Ebro valleys, and parts of the Biscay and Levant coasts (SA2). It is widespread north of the Tagus in Portugal, except in low-lying coastal areas, but in the south it is found mainly in the sierras of southern and eastern Alentejo and Algarve (PA2).

It inhabits barren rocky screes with scant vegetation, open rocky hillsides and mountains with scattered grassy areas and patches of juniper, cistus or heather scrub or open coniferous or deciduous woodland. A rocky element to the habitat is characteristic but not invariable. Patches of bare ground are always a habitat feature and the Rock Bunting often takes advantage of recently burnt areas (García-Villanueva *et al.* 1998, Herrando *et al.* 2002).

SACRE results for 1998–2012 suggest overall stability in Spain (Escandell 2013). The highest breeding densities in Spain have been found in matorral in the supramediterranean and Cantabrian regions (31–55 birds/km^2). On a local scale, up to

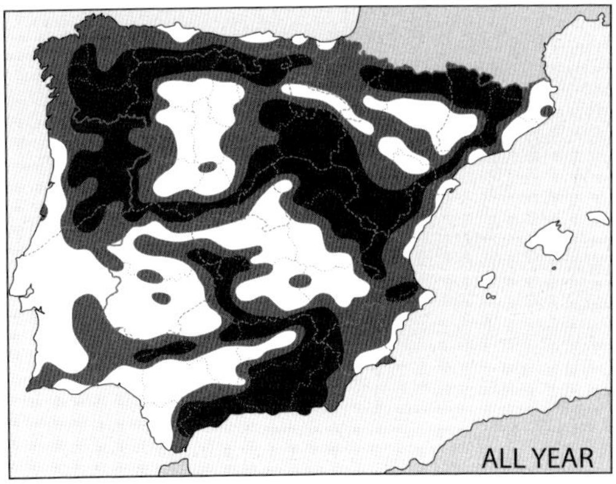

137 birds/km² have been found in broom scrub in the Sierra Nevada (Zamora 1988). A recent assessment put the regional population of Catalonia at 145,000–300,000 pairs (CA2). The Portuguese population is also considered stable (PA2) and this is a common species in Andorra (BE, ANA). Sporadic nesting has occurred in Gibraltar (BG).

The Rock Bunting is largely sedentary but altitudinal movements occur in winter, when small flocks appear at low elevations including rocky sea coasts and some agricultural areas (AI2, SWA). Some however remain at high altitudes, in the Sierra de Gredos for example (Sánchez 1991). There is little evidence of real migration but small numbers arrive at Gibraltar in autumn in some years (pers. obs.), chiefly in October–November, and winter there occasionally. There is also some evidence of small-scale migration to Morocco in the same period (Tellería 1981, BM). The two longest displacement of a ringed bird in Spain was 102km from Nerpio, Albacete, to Águilas, Murcia (*Ecología* 4: 290). There are no extra-regional ringing recoveries.

The Rock Bunting is only a rare winter visitor to the Balearic Islands (*A. O. Balears 2010*: 269). The regional bird reports give 13 records (15 birds) for the archipelago during 1985–2007. One was on 14 August 1986 in Formentera but the remainder were between 21 September and 13 March.

HOUSE BUNTING *Emberiza sahari*

(Escribano Sahariano, Escrevedeira-doméstica)

Vagrant to Spain.

This familiar bird of human settlements in northwestern Africa has recently extended its range northwards and there is now an established population at Tangier, on the Strait of Gibraltar (Amezian *et al.* 2006). There are two accepted records from southernmost Spain. A male was caught at Algeciras harbour, Cádiz, in June 1987, and another was seen at Nerja, Málaga on 13 October 2009. The rarities committee has not ruled out assisted transport in either case since the House Bunting is a popular cagebird. Conversely, the range expansion in the Maghreb may be beginning to include Iberia.

ORTOLAN BUNTING *Emberiza hortulana*

(Escribano Hortelano, Escrevedeira-amarela)

A summer visitor, chiefly to northern Iberia. A scarce migrant in the Balearic Islands.

Breeding population estimates. Spain: SA1 200,000–225,000 pairs; ACRE 521,000 birds. **Portugal**: 500–2,500 pairs.

The Ortolan is the only one of the nesting buntings that is a summer visitor, wintering in sub-Saharan Africa. It is a relatively late migrant in spring, arriving in April–May, and it abandons its breeding sites in August–September. The breeding range falls almost entirely within the northern half of the Iberian Peninsula but there is a high altitude outpost population in the Sierra Nevada in Andalucía. Reports of breeding elsewhere in Andalucía and in Murcia (SA1) are unconfirmed and seem unlikely. The core breeding distribution comprises the southern flanks of the Cantabrian, Central and Iberian ranges and the Pyrenean foothills. It is most numerous at elevations from about 1,000–2,000m; Bernis (1945) only found it at 945–1,500m in the Montes de León, but it occurs locally at sea level in Catalonia and more widely at up to 2,400m in the Pyrenees (CA2). It favours south-facing slopes everywhere except in the Sierra Nevada, where the north side is chosen, at altitudes of 2,000–2,700m (Garzón & Henares 2012). In general, it is principally a montane species within the Mediterranean climatic region and is almost entirely absent from Eurosiberian areas (SA2). It is scarce and local in Portugal, where it is confined to the sierras and plateaus of the centre-north and north-east, with concentrations in the Serra da Estrela and also in the Montemuro and Leomil mountains (PA2, AP).

Iberian Ortolans inhabit open bushy areas on sunny hillsides. Breeding sites offer plenty of areas of bare ground or scant vegetation with scattered bushes and small trees. Some favour vineyards and olive groves. Mountain heathlands with scattered

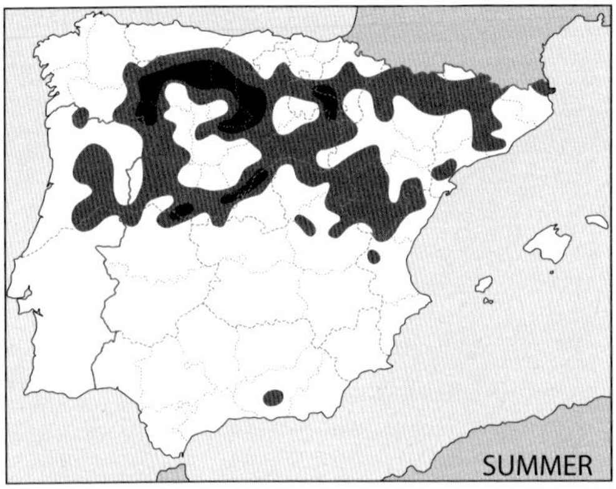

junipers or Pyrenean oaks are used at higher altitudes. Tall vegetation provides song posts but the birds feed and nest on the ground.

Large local fluctuations in abundance occur that are probably related to changes in vegetation structure. For example, Mestre (1983) recorded a strong decline at Penedès, Catalonia, from the 1950s to the 1970s whereas Muntaner *et al.* (CA1) noted its colonisation of new sites to the south of Lleida and in Tarragona. A comparison between the 1975–83 and 1999–2002 Catalan breeding bird atlases shows an 88% increase in the number of occupied 10km-squares, which is mainly attributed to the colonisation of recently burnt forest areas (CA2, Brotons *et al.* 2008). On a landscape scale, highest densities occur in the cereal farming areas (25 birds/km^2) and matorrals (7–14 birds/km^2) of the supramediterranean. More locally, the highest densities have been found on broom-covered mountain sides: up to 53 birds/km^2 in the Sierra Nevada and 61 birds/km^2 in Gredos (AI2). The SACRE results show a significant decline of 2.9% during 1998–2012 (Escandell 2012). The Portuguese population seems stable (BE, PA2). The Ortolan is rare and very local in Andorra and may not breed there regularly (ANA).

Migrants may appear almost anywhere, often in small flocks. Small numbers regularly occur at the Strait of Gibraltar and on the Mediterranean coasts, as well as in the Balearic Islands. The passage periods are relatively brief. Passage at Gibraltar in spring is in April and, especially, in May; the return is noted from late August until early October BG). Relatively few occur in the Balearics; about 100 in total were ringed during 1989–2010, mainly on Cabrera and Illa del Aire (*A. O. Balears*), with extreme spring dates of 12 April and 25 May and autumn dates of 6 September and 5 October; some two-thirds of the spring birds were in April.

Some Ortolans from elsewhere in western Europe occur in Iberia on passage since there are a few recoveries of birds ringed in Sweden, Finland, Belgium, Switzerland, and northern Italy (Zink 1985, SMA). Winter records are unknown and one at Siete Aguas, Valencia, on 16 November 1991 (*A. O. C. Valenciana 1991*: 107) was very late.

CRETZSCHMAR'S BUNTING *Emberiza caesia*

(Escribano Ceniciento, Sombria-de-capuz)

Vagrant to Spain.

A male was on Cabrera, Balearic Islands, on 1–2 May 2011. This species has been reported in many European countries, although some may have been escaped cage birds. Slack (2009) lists four records in Britain up to 2008, seven in France, about ten in Germany, four in Finland, two in the Netherlands and one each in Sweden and Austria. Many were in the 19th century and most in spring, mainly in May. It is also a vagrant to Italy (Brichetti & Massa 1998). The most recent French record was in Nice in March 1917.

RUSTIC BUNTING *Emberiza rustica*

(Escribano Rústico, Escrevedeira-rústica)

Vagrant to Spain and Portugal.

At least nine have been found in Iberia: a male at Los Palacios, Sevilla, on 21 December 1987; a first-year male at Portimão, Faro, on 12 November 1990; a male at La Albufera de Valencia on 7 October 1992; an adult male near Algeciras, Cádiz, on 6 November 2004; an adult at El Salobral, Albacete, on 12 December 2005; a first-winter male near Tarifa, Cádiz, on 17 February 2005; a female second calendar-year at Delta del Ebro, Tarragona, on 18 March 2008; a first-winter individual at Lagos, Faro, on 11 December 2008, and one at Santa Coloma de Gramenet, Barcelona, on 30 November 2011. Most records are of birds trapped for ringing. Most are from southern or eastern locations and they fall in October–December (7) or February–March (2). The Rustic Bunting appears with some frequency to the west of its breeding range, both in spring and in autumn. There were 451 records from Britain by 2005. There are 35 records from France, generally in October, but none from Morocco.

LITTLE BUNTING *Emberiza pusilla*

(Escribano Pigmeo, Escrevedeira-pequena)

Vagrant to Spain, Portugal and Gibraltar.

Little Buntings commonly appear in western Europe, particularly in autumn. As many as 838 were recorded in Britain during 1958–2003, and 41 were reported in Ireland up to 2011 and 139 in France. There are no confirmed records from the Maghreb.

The earliest Iberian record was a bird captured near Málaga on 28 December 1874 (Irby 1879, Arévalo 1887). Only one further record antedates the convening of the Spanish rarities committee: one seen at Cudillero, Asturias, on 21 December 1978 (Álvarez-Balbuena *et al.* 2000). Nevertheless, there have since been 47 Spanish records during 1987–2010, five in Portugal during 1991–2009 and two in Gibraltar: on 28 September–1 October 2011 and 15 November 2012, most of single birds captured by ringers. Three birds together were seen at the Miño estuary in February 2009 (*Ardeola* 58: 474). Five further records occurred in Spain during 2011–12 and four in Portugal in 2012 (*Ardeola* 60: 493, RBS, RBP).

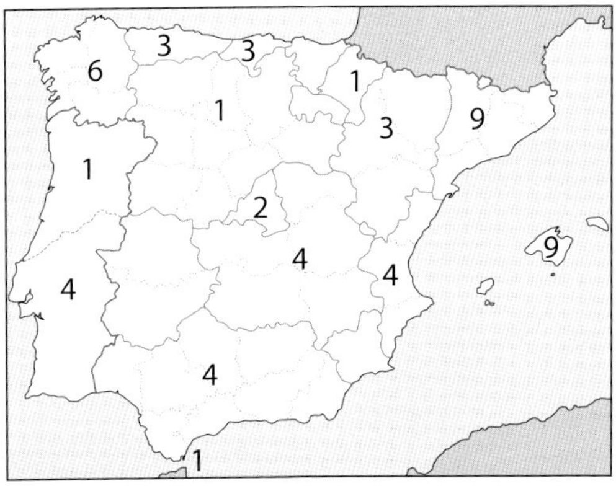

Figure 212. Regional distribution of records, 1874–2010 (n = 55)

First-year individuals appear to predominate: 24 individuals of known age caught between October and January comprised 20 first-years and four adults. Records have been widespread but there is some concentration in the north-east, with Catalonia, the Comunidad Valenciana and the Balearic Islands together accounting for 40% of observations (Fig. 212). The records fall between 26 September and 30 April, with 67% during October–December (Fig. 213). The autumn peak is rather later than in the British Isles, where it occurs in September–October (Dymond *et al.* 1989), and later too than in France, where records peak during the final third of October (OF). Winter records are comparatively frequent in Iberia, however.

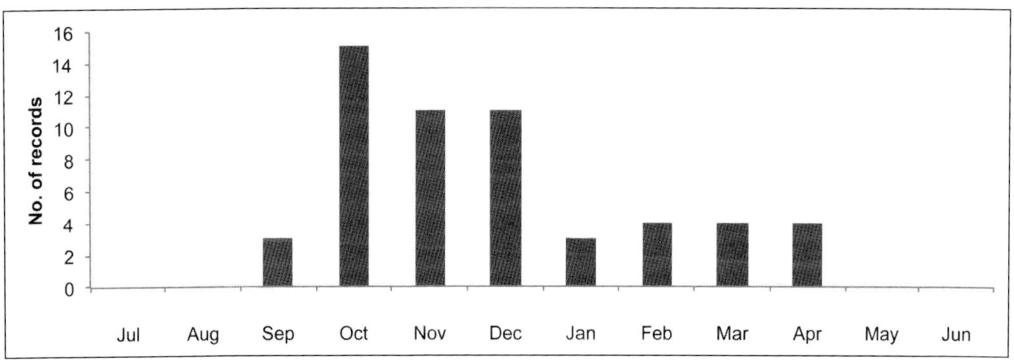

Figure 213. Monthly distribution of records, 1874–2010 (n = 55)

YELLOW-BREASTED BUNTING *Emberiza aureola*

(Escribano Aureolado, Escrevedeira-de-barriga-amarela)

Vagrant to Spain and Portugal.

Four have been recorded in Iberia: one was caught at Chipiona, Cádiz, on 3 October 1969; a male was at Belchite, Zaragoza, on 7 November 1987; a first-year bird was on Cabrera, Balearic Islands, from 4–11 October 1995, and a juvenile/first-winter was at Cape Espichel, Setúbal, on 14 September 1996. Yellow-breasted Buntings occur quite frequently in northwestern Europe; at least 236 have been recorded in Britain, but only five in Ireland and seven in France. There are no records for Morocco.

COMMON REED BUNTING *Emberiza schoeniclus*

(Escribano Palustre, Escrevedeira-dos-caniços)

***E. s. lusitanica*. Endemic to Atlantic Iberia. A critically endangered resident. The darkest race: males in particular show extensive dark streaking on the rump and underparts.**
***E. s. witherbyi*. Endemic to Mediterranean Iberia, the Balearic Islands, southern France and Atlantic Morocco. An endangered resident. Darker and larger-billed than the nominate form.**
***E. s. schoeniclus*. Common on passage and in winter.**

Breeding population estimates. Spain: <1,000 pairs. **Portugal**: <500 pairs.

The Reed Bunting is a scarce and declining breeding species in Iberia, where two distinct subspecies are resident. Both are darker than the nominate. *E. s. lusitanica* is the darkest race of this most polymorphic of buntings (Clements 2007). *E. s. witherbyi* is also distinguished by its significantly heavier bill, >5.9mm deep in *witherbyi*, <5.9mm deep in *schoeniclus* (Belda *et al.* 2009). Nominate subspecies individuals wintering in Catalonia are longer-winged than the local *witherbyi*: 76.1mm (±2.8) *v.* 74.7mm (±2.6) (Copete *et al.* 1999).

E. s. lusitanica is confined to the coastal fringes of Cantabria, as well as western Galicia and the Portuguese Atlantic coast south to Lisbon. Its population declined markedly during the 20th century and a census in Spain in 2005 found only 65–71 pairs in 14 localities, all but one of them in Galicia but with a remnant population of just three pairs in Cantabria, at the Alday marshes, near Santander (Atienza 2006). The Portuguese population is very small and local, and largely confined to the Ria de Aveiro, with some also at the Minho estuary and a few pairs nesting occasionally in the wetlands of the lower Mondego valley (PA2). The Aveiro population may number 350–400 pairs (Kvist *et al.* 2011). It is now extinct as a breeding species in the Basque Country, Cantabria and Asturias (Atienza 2009, Íñigo 2009). In Portugal it no longer seems to breed at the Tagus estuary and it has disappeared from the Sado estuary and the lower Mondego and Boquilobo marshes, where it formerly nested (PA2).

E. s. witherbyi is more numerous but it too has also suffered severe declines in both range and population. The breeding range includes the Mediterranean coast from the Ebro delta south to the north of Alicante, some wetlands of Castilla-La Mancha and a few sites along the Ebro valley. The 2005 census found only 254–360 pairs here in 21 localities, with concentrations in the Ebro delta and central Castilla-La Mancha. The Ebro delta estimate: 50–100 pairs, was actually made in 2001 (CA2) but the 2005 population was thought to be similar. In Castilla-La Mancha, 160–205 pairs were found in 16 localities in 2005, whereas Velasco *et al.* (2004) found 100–200 pairs in 36 breeding localities: 2005 was an exceptionally dry year that may have concentrated the population at fewer sites, notably at the (wet) Tablas de Daimiel, where there were 90–120 pairs that year. This subspecies now appears to be extinct in Andalucía, Castilla y León and Aragón (Atienza 2009), where it nested locally until at least the 1990s (SA1). In Catalonia it formerly nested at the Aiguamolls de l'Empordà and the Llobregat delta, and the Ebro delta population was once much larger: 321–485 pairs were estimated there in 1995 (CA2). Inland, some 20–30 pairs may breed at the Utxesa reservoir, Lleida (Calvet *et al.* 2004).

Small numbers of *witherbyi* breed in Mallorca, at the Albufera, where the population was censused at only 19 pairs in spring 2007 and 30 pairs in 2010 (*Ardeola* 55: 307). It was common there in the 1920s (Munn 1926) and estimates ranging from 250 to 1,000 pairs were made during 1990–2000 (Atienza 2006).

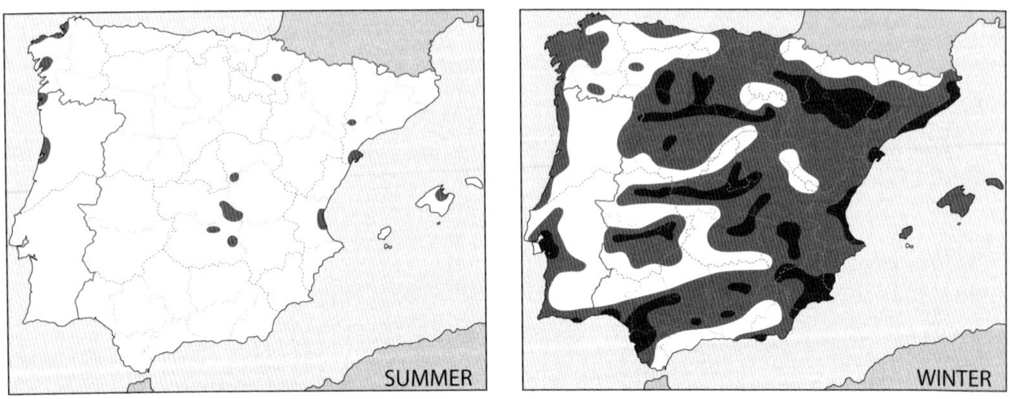

Both breeding races are closely associated with wetlands, in particular those fringed by reedbeds. However, the densest reedbeds are avoided. The preferred habitat seems to be more open reed stands, mixed low scrub, reedmace and rushes. The birds feed largely on the ground, often in cultivated land near the wetlands. The diet during the breeding season, and especially the food brought to the young, is predominantly insects. It has been suggested that the large bill of *E. s. witherbyi* is an adaptation to taking larger insects, such as caterpillars, gleaned within the reedbeds themselves (LR). Seeds are the principal diet in winter.

The Iberian birds remain restricted to wetlands when nesting, something which may be exacerbating their decline. In contrast, some other European populations have diversified their habitat selection in response to loss of wetlands and now also nest in crops. The Iberian birds are also especially vulnerable to the loss of wetlands through drought and the over-abstraction of ground water (LR). The latter has especially affected the wetlands of Castilla-La Mancha, notably those at Daimiel. A recent study has found considerably reduced genetic variation in *lusitanica* (Kvist *et al.* 2011). The same study also recognises three genetically distinct subpopulations of *witherbyi*: Ebro delta, Castilla-La Mancha and Mallorca. Reed management appears to be essential in order to sustain and reinforce populations within protected areas.

The nominate race is relatively numerous and widespread in Iberia in winter, although wintering numbers have declined by as much as 30% since the 1990s (LR). Nevertheless, Spain in particular is an important wintering area for this subspecies. Wintering birds occur in small groups that feed on seeds in farmland near wetlands. They gather together in flocks of hundreds to roost in reedbeds. They are especially associated with the courses of the major rivers and with the margins of their associated reservoirs. In Spain the largest winter concentrations have been reported at the coastal wetlands of the Mediterranean, notably at the Aiguamolls de l'Empordà, the Ebro delta, the Albufera de Valencia and the southern Alicante wetlands, as well as at the Guadalquivir estuary (SWA). The wintering Catalan population has been estimated at 238,000–598,000 individuals (CWA).

Reed Buntings of Nordic origin winter principally in Cantabria and central Iberia, notably in the Tagus valley, whereas those from central Europe principally winter on the Mediterranean coasts and in the Ebro valley (Villarán (1999b). Only one of some 1,100 Iberian recoveries of extraregional Reed Buntings was in Portugal (SMA, AP). Most birds ringed during the breeding season originate from Germany, Sweden, the Low Countries and France, and birds from these countries and Switzerland account for some 90% of all recoveries. The remainder were ringed in Finland, the Baltic States, Poland, Russia, Britain, the Czech Republic, Slovakia, Austria, Slovenia or northern Italy (SMA). Most recoveries of either group fall north of 40°N and east of 4°W, which includes the northeast quadrant of the Peninsula, bounded very approximately by a line joining Santander, Madrid and Valencia. Nevertheless, the wintering range extends to the extreme south of the Peninsula, notably the Guadalquivir valley (Chiclana *et. al.* 2002). Only small numbers continue further south to winter in Morocco (BM) and hence few are noted on passage at the Strait. Reed Buntings are also widespread in winter in coastal regions of Portugal, especially at the Tagus and Sado estuaries and in western Baixo Alentejo, but are uncommon inland due to lack of suitable habitat (AP). Over 500 European recoveries of birds ringed in Spain have largely been in central and northern Europe: two-thirds of them in France, Germany, Sweden or Switzerland (SMA). One ringed in Spain has been recovered in Algeria.

The first wintering birds arrive in early September but peak immigration is in October. The majority have departed by the end of March but some remain throughout April. Males migrate earlier than females during both passages. Both adult and immature males winter further north and east than females. Females outnumber males by two to one in the wintering population, the opposite of what occurs in more northern countries (Villarán 1999b). This was noted by Witherby (1928) who remarked on the 'tremendous preponderance of females' among the Reed Buntings that he saw in November at Aranjuez and Daimiel. Ringers in Catalonia have captured individuals showing the characteristics of *E. s. passerina* (NW Siberia) and *E. s. intermedia* (central Mediterranean) (Aymí & Martínez 1991).

PALLAS'S REED BUNTING *Emberiza pallasi*

(Escribano de Pallas, Escrevedeira-de-Pallas)

Vagrant to Portugal.

A first-winter female was seen at the Lagoa de Santo André, Setúbal, on 12 and 19 January and 4 February 1997 (Moore & Armelin 1999). There are only six accepted occurrences in Europe; four in Britain, one in Italy and one in Portugal (Slack 2009, Hudson et al. 2011).

RED-HEADED BUNTING *Emberiza bruniceps*

(Escribano Carirrojo, Escrevedeira-de-cabeça-ruiva)

Vagrant to Spain.

The not infrequent European observations, particularly in the north-west, have largely been attributed to escaped cagebirds. Nevertheless, their temporal pattern, for example in France and Italy (Yésou et al. 2003) fits the migratory calendar of the species quite well. There are five records of single males in Spain: taken at Masnou, Barcelona, on 12 November 1962 (*Ardeola* 9: 156–157); captured on board a ship off Tinamayor, Asturias, on 20 June 1967 (Noval 1986); taken at Sanlúcar de Barrameda, Cádiz, between 20 and 29 October 1967 (*Ardeola* 50: 146); seen at Castelló d'Empúries, Girona, on 10 and 16 October 1987 (*Ardeola* 36: 122); and seen at Sant Pere de Ribes, Barcelona, on 20 October 1996 (*Ardeola* 46: 146). There are thus one June observation and four in October–November.

BLACK-HEADED BUNTING *Emberiza melanocephala*

(Escribano cabecinegro, Escrevedeira-de-cabeça-preta)

Vagrant to Spain.

There are 19th century references to birds found in Spain but the earliest specific record was of a male at Torredembarra, Tarragona, on 21 May 1957 followed by another male at Zahara de los Atunes, Cádiz, on 1 September 1969 (*Ardeola* 7: 267, 15: 156). The Spanish rarities committee accepted a further ten records during 1992–2008, all of single males. The 12 records follow the frequent distribution pattern of passerines of eastern origin, being concentrated in the north-east (six in Catalonia and three in the Balearics) and to a lesser extent in Andalucía (three). Nine records have been in spring, between 30 April and 27 June, and three in autumn, between 1 September and 16 October. The Black-headed Bunting frequently wanders to western Europe. There are 201 records from Britain and 126 from France: where there have been several breeding records in the south-east since 2000. Observations in northwestern Africa are very rare: there is an early 20th century sighting near Tangier and a record from the Algerian Sahara in 1960.

CORN BUNTING *Emberiza calandra*

(Escribano Triguero, Trigueirão)

E. c. calandra. **An abundant resident and partial migrant in mainland Iberia and the Balearics. Some migrants from central Europe are present in winter.**

Breeding population estimates. Spain: SA1 1.44–4.30 million pairs; ACRE 21.8 million birds. **Portugal**: 100,000 to 1 million pairs.

The Corn Bunting is one of the most abundant, widespread and characteristic species of Iberia. Saunders (1871) considered it to be the most abundant bird in Andalucía, partly because it outnumbered all the other thousands of passerines sold in

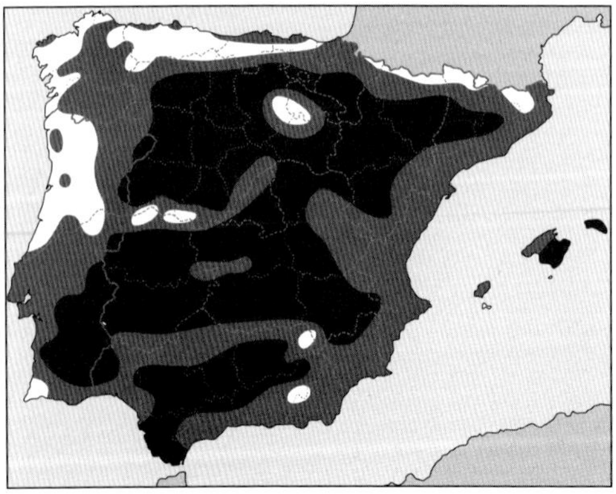

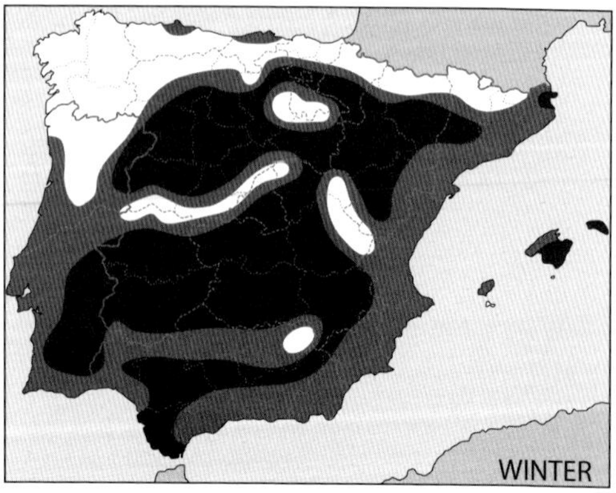

the local markets. Smith (1868) also remarked that it outnumbered all other species in Lisbon market, and Arévalo (1887) commented on the heavy persecution of the Corn Bunting, on account of its tasty flesh. Chapman (1884) described its numbers throughout Spain as 'incalculable'.

In spring its jangling song is omnipresent wherever there is open country. It is particularly associated with farmland, especially the cereal steppes of Castilla y León, Castilla-La Mancha, Andalucía, Extremadura and Alentejo, and also irrigated land, cultivated dehesas, vineyards and the more open orchards. It is scarce or absent in very arid regions, but is also relatively uncommon on the cooler, wetter, northern side of the Cantabrian mountains and in western Galicia (SA2, PA2). It ranges commonly from sea level to about 1,000m but occurs higher locally, for example to 1,600m in the Sierra de Gredos and locally to 1,800m in the Sierra Nevada (SA2, Garzón & Henares 2012).

The Spanish population on SACRE sites showed some stability overall during 1998–2012 but significant declines of about 17% were reported in both the north and in Andalucía (Escandell 2011, 2013). Comparison with the breeding range at the time of the first national breeding atlas (SA1) shows that there has been some decline in Catalonia and perhaps elsewhere in eastern Spain. The highest densities are in pasturelands (110–137 birds/km^2) and cereal crops (82–107 birds/km^2), as well as on the margins of Mediterranean wetlands (112 birds/km^2). Cultivated dehesas, where cereal crops provide food and nesting cover and the widely-spaced trees offer song posts, are also favoured, with up to 100 birds/km^2 (ACRE).

The Portuguese population is thought to be stable and showed no clear trend during 2004–09 (ACP). Breeding has occurred only on the fringes of Andorra (ANA) and is exceptional in Gibraltar (BG). Local increases in abundance have been positively linked to rainfall in steppic areas in Extremadura in the breeding season (de Juana & García 2005), perhaps because the increase in plant cover on the steppes in wetter seasons favours nest-siting. Those that nest in crops are less likely to be subject to such fluctations. In Alentejo the Corn Bunting nests more abundantly in crops than in pastures or long-term fallows, tending to avoid the latter where they are grazed (Moreira & Leitão 1996, Moreira 1999b).

Steep declines and local extinctions in northern Europe (BWP) point to this species being particularly vulnerable to agricultural intensification and it must be regarded as potentially sensitive to any changes in land use and farming practices in Iberia. Some effects of this kind have already occurred: Corn Buntings formerly also frequented olive groves but are now largely absent from them following an increased use of insecticides and herbicides (SA1).

Large flocks range widely on farmland outside the breeding season and gather nightly to roost in denser cover, such as reedbeds (Villarán *et al.* 2008). They are widespread throughout the peninsular interior in winter, although then absent from most of the northern coastlands and from higher elevations: for example from Galicia and northern Portugal where they do not return until April or May (AI2, AP, CWA, SWA). The largest winter concentrations in Spain are reported from the south-west, notably from the provinces of Toledo, Ciudad Real, Badajoz, Córdoba, Sevilla, Cádiz and Granada, and some numbers also winter in the Balearic archipelago (SWA).

Most recoveries of Iberian ringed birds are local: the exceptions include five that travelled over 300km within the Peninsula and one ringed in Palencia in July 1999 that was found in Faro (644km) in January 2002 (*Ecología* 17: 284). Nevertheless there is clear evidence of some migration across the Strait of Gibraltar, chiefly in October–November, when hundreds occur on passage at Tarifa (Tellería 1981). The spring passage at the Strait is less evident perhaps because it may be relatively prolonged, apparent migrants occurring at Gibraltar between February and early May (BG). Some Corn Buntings from central Europe occur in Iberia in winter and there are a total of eight Spanish recoveries of birds ringed in France, Belgium, Germany, Switzerland or the Czech Republic. One ringed in southern France was recovered in Mallorca in December. Corn Buntings ringed in Spain have been recovered in France (4), Morocco (2) and Algeria (2), and one ringed at Cape Bon in coastal Tunisia in May 1967 was recovered at Écija, Sevilla in July 1969 (SMA).

BOBOLINK *Dolichonyx oryzivorus*

(Charlatán, Triste-pia)

Vagrant to Gibraltar.

A male was ringed in Gibraltar, where it was present from 11–16 May 1984 (*Alectoris* 7: 56). This is the only known spring record in the Western Palearctic (Slack 2009). The Bobolink has seldom been found in Europe. Most have appeared in the British Isles, from where there are 31 records. There are three records from France, 11 accepted records from the Azores and one from the Canary Islands.

9. CATEGORY D SPECIES

Species that would otherwise appear in Category A except that there is reasonable doubt that they have ever occurred in a natural state.

The species listed below are in Category D in either Spain or Gibraltar. Portuguese Category D species all occur under Category A in Spain and so are included in the main systematic list, where their Portuguese list status is indicated. The Portuguese Category D species are the Mute Swan, Bar-headed Goose, Greater Canada Goose, Ruddy Shelduck, Falcated Duck, Great White Pelican, Marabou Stork, Lesser Flamingo, Gyr Falcon, Demoiselle Crane and Laughing Dove (*Matias et al.* 2007).

FULVOUS WHISTLING DUCK *Dendrocygna bicolor*

(Suirirí Bicolor, Marreca-caneleira)

The Fulvous Whistling Duck breeds no closer to Iberia than the Sahel in Mauritania, where its numbers fluctuate according to annual rainfall (Isenmann *et al.* 2010). It is commonly kept in European wildfowl collections, however, so observations in the wild in Europe, including two instances of nesting in France, have been attributed to escapes (OF).

There are 15 Iberian observations, all from Spain. The first was an adult female shot at Doñana in September 1971 (Castroviejo 1972). One was at the Laguna de Fuentedepiedra, Málaga, from May to July 1998 (*Ardeola* 47: 144). The remaining 13 have been recorded in the GAE reports 2004–07 (8), in regional bird reports (2) and on the Rare Birds in Spain website (3). In total, there are five records from Cataluña, four from Andalucía, two from Aragón and one each from Asturias, Galicia, Castilla y León and the Balearics. Although the sample is small, the records show peak occurrences in spring and autumn. This also applies to observations in France, where the records include a flock of six in April 2007. The first dates of observations in Morocco; two at the Massa estuary and one near Rabat, were in April, May and September (BM): they included a flock of 11. Taken together, the Western Palearctic records known to us, including those from Spain and Morocco plus 11 dated ones from France, four from Germany and one each from Italy and the Azores, show spring and autumn peaks in occurrence (Fig. 214). This pattern is very different from what would be expected for escaped individuals, which tend to appear all year round. At least some of the European occurrences may instead be natural and linked to rainfall patterns in the Sahel, where dry years may trigger long-distance dispersal in this and other aquatic species.

Figure 214. Monthly distribution of occurrences of the Fulvous Whistling Duck in Spain (n =15) and in other Western Palearctic countries (n = 19)

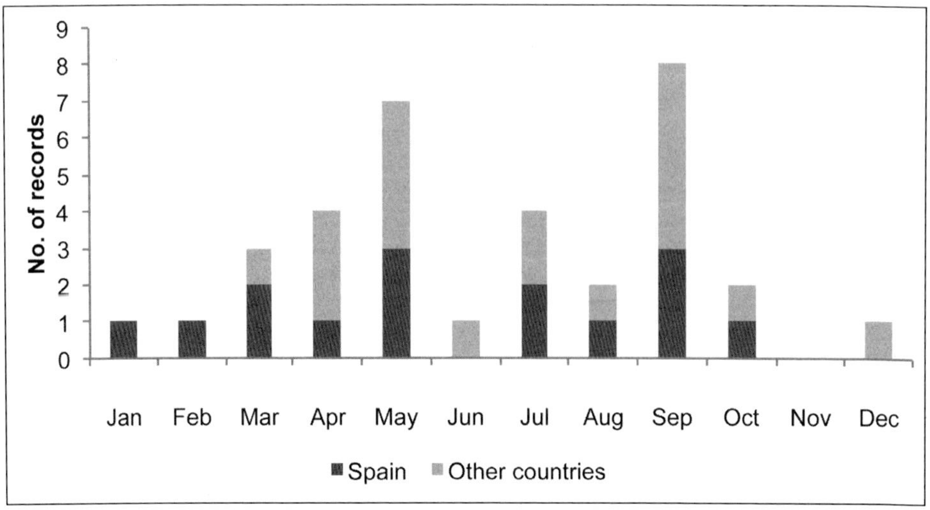

WHITE-FACED WHISTLING DUCK *Dendrocygna viduata*

(Suirirí cariblanco, Ireré)

The White-faced Whistling Duck breeds no closer to Iberia than the Mauritanian Sahel, where numbers fluctuate with rainfall but where it is more abundant than the Fulvous Whistling Duck (Isenmann *et al.* 2010). Unlike the latter, however, it has not been recorded in Morocco. It is common in European collections and occurrences in the wild there are attributed to escapes.

There are three Spanish records of birds shot by hunters: at the Prat de Llobregat in summer 1914 (Sagarra 1915), at the Albufera de Mallorca in December 1973 (*Ardeola* 22: 111) and at the Ebro delta in April 1984 (Martínez-Vilalta & Motis 1989). We are also aware of five records from France (Dubois *et al.* 2008) and four from Germany (reports of the Deutsche Seltenheitenkommission in *Limicola*, up to 2008). Considering all European records for which the exact dates are known, nine first dates fell between April and mid June and there were one each in September and December. This pattern, and the fact that all the Spanish and most of the French records were on the Mediterranean coasts, suggests possible natural arrivals from Africa.

SNOW GOOSE *Anser caerulescens*

(Ánsar Nival, Ganso-das-neves)

Some may reach Europe as rare vagrants but most European records are generally attributed to feral or escaped birds. The feral population in Britain was estimated to number at least 250 birds in 1992 and 86 in 2000 (Rowell *et al.* 2004). There are also very small breeding populations in The Netherlands and Germany, numbering just two and five pairs respectively in 2005, and some winter regularly in northern Italy (Banks *et al.* 2008).

There are a few reports from Iberia, all of them in Spain and mostly on the northern meseta in winter, especially at Villafáfila. Prior to 1984 there was one seen at Doñana on 10 January 1982 (*Ardeola* 29: 180) and there were three recoveries of ringed birds of captive origin: two from Norway and one from England (Sáez-Royuela & Santos 1985, de Juana 2006). The Spanish rarities committee accepted 12 Category D records during 1984–2005 but there have been none since then.

CACKLING CANADA GOOSE *Branta hutchinsii*

(Barnacla Canadiense Chica, Ganso-do-canadá pequeño)

The Cackling Canada Goose comprises five arctic-breeding subspecies that have formerly been treated as miniature forms of the Greater Canada Goose (Banks *et al.*, 2004; Sangster *et al.*, 2005; *AERC Western Palearctic list, March 2011*). Some subspecies are rare vagrants to Europe, with observations from the British Isles, The Netherlands and France as well as two recoveries of ringed birds. There are two Spanish records. A record of a group of three at the Lagunas de Villafáfila on 31 October 1999 was accepted at the time as 'Canada Goose showing features of subspecies *minima*'. One was with Greylags in the Vegas Altas del Guadiana, Badajoz, on 2 January 2010 and what was presumably the same bird was again found nearby on 24 and 28 January 2011.

WOOD DUCK *Aix sponsa*

(Pato Joyuyo, Pato-carolino)

Recorded in Spain (Category D) and Portugal (Category E). European observations of this Nearctic duck tend to be attributed to escaped or feral birds (Dubois 2007, Banks *et al.* 2008). Nevertheless, apparently wild birds have been reported in Iceland and perhaps in Britain (BWP, Mitchell & Young 1997). One ringed in North Carolina was found in the Azores (Dennis 1990) and four others have been recorded there (Jara *et al.* 2011). One of unknown origin was found in the Canary Islands in December 1995 (*Ardeola* 45: 100, Martín & Lorenzo 2001).

There is an accepted record of a female at the Ensenada da Ínsua, A Coruña, on 29 October 2006, a location with a long history of attracting Nearctic migrants. The GAE Annual Reports give ten further Spanish records during 2003–07. Excluding two that were in city parks, the remaining eight were all during October–December and in the northern half of the country. There are seven recent Portuguese observations: in October 2002 (2), in April and September 2004, in summer 2006, April 2007 and September 2009, two of which involved groups of three (*Anu. Ornitol.* 7: 98, 8: 96–97). Some records may involve feral birds from elsewhere in Europe but the possibility of rare natural occurrences cannot be entirely excluded.

CINNAMON TEAL *Anas cyanoptera*

(Cerceta Colorada, Marreca-carijó)

European occurrences of the Cinnamon Teal are generally attributed to escapes from captivity, although the occurrence of genuine Nearctic vagrants cannot be discounted entirely (Madge & Burn 1988, Scott 2004). There are five Spanish records but the species has not been included in the Spanish rarities list since 2006. A female or juvenile was at the Embalse de La Granda, Asturias, from 18 October to 3 November 1994. Since then the GAE Bulletins have noted two males at the Aiguamolls de l'Empordà, Girona, from 20 December 2003: one of which remained there until 13 April 2004; a male was at the Lagunas de Villafáfila, Zamora, on 12 April 2004; one was at Alfaro, La Rioja, on 18 December 2007; and a male was at the Aiguamolls de l'Empordà, from 10 May 2008.

REED CORMORANT *Phalacrocorax africanus*

(Cormorán Africano, Corvo-marinho-africano)

A small cormorant, a juvenile, was shot on the Catalan coast, probably between Barcelona and Tarragona, in or around 1855. The skin is preserved in the University of Barcelona collection and was accepted as the sole record of a Pygmy Cormorant *Phalacrocorax pygmeus* for Spain. However, a recent and meticulous examination of the specimen, including isotopic analysis of its plumage, has revealed unambiguously that it is a Reed Cormorant (Copete *et al.* 2011) and, as such, the only European record.

The Reed Cormorant breeds in Mauritania north to the Banc d'Arguin. It is described as an accidental visitor to the Cap Blanc peninsula and there are three records from further north on the Moroccan Atlantic coast (BM). Hence, although the Spanish record is currently listed under Category D (*Ardeola* 59: 406) a natural occurrence would not seem inconceivable.

YELLOW-BILLED STORK *Mycteria ibis*

(Tántalo Africano, Tântalo-africano)

The Yellow-billed Stork nests no closer to Iberia than the Senegal delta but some move further north with the rains and occur during the wet season as far as the Banc d'Arguin, Mauritania (Gowthorpe & Lamarche 1996, Isenmann *et al.* 2010). Presumed vagrants have reached Israel, Jordan, Turkey, Tunisia and Morocco. The sole known occurrence in Morocco was an immature bird on 17 May 1959 at the Moulouya estuary, on the Mediterranean coast near the Algerian border (Castan & Olier 1959).

The Spanish rarities committee has accepted three records of single adults of unknown origin: at the Albufera de Mallorca from 5–13 December 1996; at Los Canchales reservoir, Badajoz, from 24–28 August 2001, and at La Puebla del Río, Sevilla, from October 2002 probably until March 2003. Two more at the Albufera de Valencia in May 2001 were attributed to escapes from the 'Safari park' at El Vergel, Alicante (*Ardeola* 50: 153). Seven further records from the Levant coastlands that have been published by the GAE may also involve birds that originated at the Park. The provenance of an adult found on a Torremolinos beach on 20 March 2007 is unknown (GAE 2007). Many of the occasional Yellow-billed Storks found in the wild in Spain are thus clearly of captive origin but the rare occurrence of wild birds cannot be ruled out entirely.

AFRICAN SPOONBILL *Platalea alba*

(Espátula Africana, Colhereiro-africano)

The African Spoonbill is a resident breeding species at the Senegal delta, on the border with Mauritania, and several thousand occur there in January in some years (Isenmann *et al.* 2010). However, it has not (yet) been found in Morocco. Vagrants have strayed to Oman and Yemen but the recent spate of European records has been attributed to birds of captive origin. These have chiefly appeared in Spain and France but there are also records from Holland, Belgium, Denmark, the Czech Republic and Italy.

There are Iberian records from mainland Spain and the Balearic Islands but not from Portugal. The earliest was an adult at the Albufera de Mallorca, from 1 May to 24 September 1988 and a recently dead individual was found at the Ebro delta some months later, on 20 February 1989 (*Ardeola* 38: 152). The Spanish rarities committee has accepted 16 records, the most recent being in 2005 (Dies *et al.* 2007). All were of single birds, 13 of them adults, and generally accompanying Eurasian Spoonbills. The records are well distributed across the year but show something of an autumn peak, particularly in September. This pattern is more pronounced when the French and Italian records are also included, when both autumn and spring peaks appear (Fig.

215). At least three instances of over-summering have been reported in France but there are none in Spain. However, there is a winter record in Spain at Odiel, where an adult that was seen on 1 January 1999 was thought to have been present from November 1998 (*Ardeola* 48: 122).

Figure 215. Months of initial observations of African Spoonbills recorded in Spain (n = 16), France (n = 11) and Italy (n = 5).

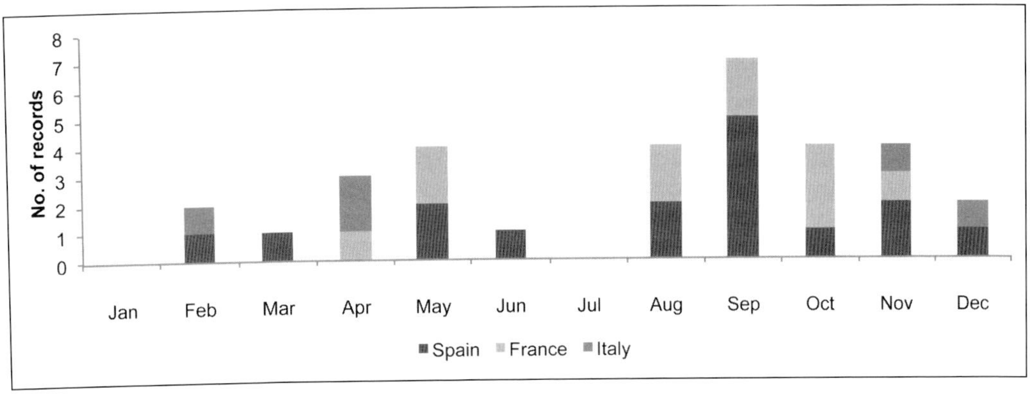

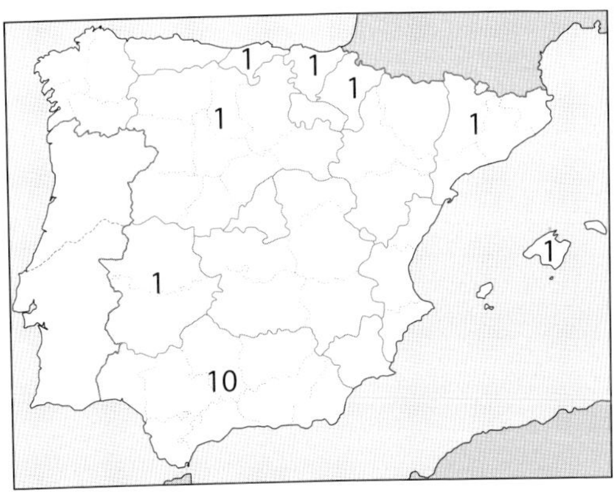

The Iberian records are concentrated in Andalucía, the remainder lying mainly along the migration route of the Eurasian Spoonbill (Fig. 216). A sizable proportion of the French records have also been along the migration routes of Dutch Eurasian Spoonbills. The temporal pattern of the French and Spanish records, coupled with the tendency of African Spoonbills to accompany migrating Eurasian Spoonbills, supports the possibility that at least some of the former are wild birds originating from West Africa (de Juana 2006).

Figure 216. Regional distribution of African Spoonbill records in Spain (n = 16)

HOODED VULTURE *Necrosyrtes monachus*

(Alimoche Sombrío, Jagudi)

This Afrotropical species ranges further north in the rainy season in West Africa, reaching the Senegal delta and valley in June–September. Individuals have strayed as far north as the Banc d'Arguin in Mauritania and also to the western Sahara and possibly Boumalne du Dadès in Morocco (Isenmann *et al.* 2010, BM).

One was identified at La Janda, Cádiz, on 16 February 2003 (*Ardeola* 53: 192). Despite the proximity of the site to the Strait a captive origin remains possible. Another at Gibraltar in spring 2004 had a chain dangling from one leg and had apparently escaped from a collection near Málaga. There was also a report of one from the Alcudia valley, Ciudad Real, in August 2004 (*Noticiario GAE 2003–2005*: 17).

COMMON BULBUL *Pycnonotus barbatus*

(Bulbul Naranjero, Engole-malagueta)

This species is common in Morocco where it occurs north to the shores of the Strait of Gibraltar. It is resident with some post-breeding dispersal occurring within the nesting range (BM). The very few Iberian records, all in southernmost Spain, include a recent instance of breeding. Some wild birds may reach the Peninsula on rare occasions (AMI) but the occurrence of escaped cagebirds is also possible.

There is a doubtful record of one near Granada in June 1858 (López-Seoane 1861). The accepted records comprise single birds in trees by the Torremolinos/Málaga road on 13, 14 and 16 November 1956; on the banks of the Guadalquivir in Sevilla city on 28 August 2005; at the Miraflores estate, Mijas, Málaga, on 5 April 2006 and, very probably the same bird, at the same estate from January to September 2007. More recently, one was present at and near Tarifa, Cádiz, from at least 25 April to 11 May 2013 and two were seen together there on 1 May. These bred successfully since two adults were seen in the same area feeding a chick on 8 July 2013 and a family party of two adults and two juveniles was then present throughout August and into September (RBS).

HOUSE CROW *Corvus splendens*

(Cuervo Indio, Gralha-indiana)

The House Crow has seldom been found in Europe although a small naturalised population has become established at the Hook of Holland since 1994 (Lever 2005). One was in Gibraltar from 28 March until 4 April 1991, following the arrival of a flotilla of warships from the Arabian Gulf, where the House Crow is abundant (*British Birds* 85: 14). Another was seen at Parada, A Coruña, on 6 March 2004 (*Ardeola* 53: 185).

DESERT FINCH *Rhodospiza obsoleta*

(Camachuelo Desertícola, Asa-rosada)

This species has very rarely been found in Europe and it is unclear whether its occurrences are due to genuine vagrancy or escapes from captivity. A female was caught by ringers on the Illa de l'Aire, Menorca, on 30 April 1997 (*Ardeola* 47: 170).

SONG SPARROW *Melospiza melodia*

(Chingolo Cantor, Tico-tico-musical)

The Song Sparrow is a Nearctic vagrant to western Europe from where there are seven records from Britain and one each from Norway, Belgium and the Netherlands (Slack 2009). One was seen and photographed at Algeciras, Cádiz, on 29 March 2013 (RBS). Since Algeciras is a major port, the possibility of ship-assisted arrival clearly exists. The record is comparable to those of the White-throated Sparrow, Dark-eyed Junco, Indigo Bunting and Bobolink from nearby Gibraltar, all of which were also in spring.

WHITE-THROATED SPARROW *Zonotrichia albicollis*

(Chingolo Coroniblanco, Chingolito)

This is a rare Nearctic vagrant in Europe, whose arrival may often be ship-assisted since some European records have been near seaports. The sole Iberian record was one that was at Gibraltar harbour during 18–25 May 1986, where it was trapped and ringed (*Alectoris* 7: 57; *Ardeola* 39: 82–83). It was in the company of a Dark-eyed Junco and so seems highly likely to have been ship-borne.

MEADOW BUNTING *Emberiza cioides*

(Escribano de Brandt, Escrevedeira de Brandt)

A very few Meadow Buntings of doubtful or captive origin have been found in Europe. They include a male trapped near Santander on 3 December 1994 (*Ardeola* 43: 117–118).

DARK-EYED JUNCO

Junco hyemalis

(Junco Pizarroso, Junco-ardoseiro)

The northern nominate race of this Nearctic species is migratory, reaching the Gulf of Mexico in winter, and is the only one identified in Europe, where most observations (36) have been in Britain (Slack 2009, Hudson *et al.* 2011). There is a recently accepted record for the Azores. The sole Iberian record involved an adult male that was at Gibraltar harbour during 18–25 May 1986, where it was trapped and ringed (*Alectoris* 7: 57; *Ardeola* 39: 83). It was in the company of a White-throated Sparrow and seems highly likely to have been ship-borne.

ROSE-BREASTED GROSBEAK

Pheucticus ludovicianus

(Picogrueso Pechirrosado, Realejo)

This is a very rare Nearctic vagrant to Europe, where the records include 23 in Britain and two in France (Hudson *et al.* 2008, Reeber *et al.* 2008, Slack 2009). There are also 11 accepted records from the Azores. Most European records have involved juveniles found in October. The sole Iberian record is no exception: a juvenile was captured at Benidorm, Alicante, on 17 October 1982 (*Ardeola* 29: 190).

INDIGO BUNTING

Passerina cyanea

(Azulillo Índigo, Mariposa-azul)

The Nearctic Indigo Bunting has very rarely been recorded in Europe and observations there have often been attributed to escapes (Mitchell & Young 1997, Slack 2009). There are records from Iceland, Ireland, England and the Netherlands, as well as 11 accepted from the Azores: including one of a flock of seven (Jara *et al.* 2008, Muchaxo *et al.* 2011). A first-summer male was caught by ringers in Gibraltar on 30 April 2004 (*GBR* 4: 44).

APPENDIX 1

CATEGORY E1 SPECIES

These are species recorded in Iberia as introductions, human-assisted transportees or escapees from captivity that have established breeding populations that may be close to becoming self-sustaining, whereupon the species concerned will be included in Category C. The Spanish version of the full list of Category E species may be consulted on the GAE (Grupo de Aves Exóticas) section of the SEO/Birdlife website. The Portuguese species are listed by Matias *et al.* (2007). Species that are in Category C in either country are included in the principal Systematic List.

California Quail *Callipepla californica*
(Colín de California, Codorniz da Califórnia)

(N America)
E1 in Spain (E3 in Portugal).

Northern Bobwhite *Colinus virginianus*
(Colín de Virginia, Codorniz da Virgínia)

(N America)
E1 in Spain.

Blue-crowned Conure *Aratinga acuticaudata*
(Aratinga Cabeciazul, Periquitão-de-cabeça-azul)

(S America)
E1 in Spain and Gibraltar (E3 in Portugal).

Mitred Conure *Aratinga mitrata*
(Aratinga Mitrada, Periquitão-mitrado)

(S America)
E1 in Spain.

Red-masked Conure *Aratinga erythrogenys*
(Aratinga de Guayaquil, Periquitão-de-cabeça-vermelha)

(S America).
E1 in Spain.

Red-vented Bulbul *Pycnonotus cafer*
(Bulbul cafre, Tuta-de-ventre-vermelho)

(Indian sub-continent)
E1 in Spain.

Red-billed Quelea *Quelea quelea*
(Quelea Común, Bico-carmim)

(Sub-Saharan Africa)
E1 in Spain (E3 in Portugal).

Orange-cheeked Waxbill *Estrilda melpoda*
(Estrilda de Carita Naranja, Faces-laranja)

(Sub-Saharan Africa)
E1 in Spain (E2 in Portugal).

Black-rumped Waxbill *Estrilda troglodytes*
(Estrilda Colinegro, Bico-de-lacre-de-uropígio-preto)

(Sub-Saharan Africa)
E1 in Spain (E2 in Portugal).

Zebra Finch *Taeniopyga guttata*
(Diamante Cebra, Mandarim)

(Australasia)
E1 in Portugal.

Scaly-breasted Munia *Lonchura punctulata*
(Capuchino Punteado, Bico-do-chumbo-malhado)

(Indian sub-continent and SE Asia)
E1 in Portugal.

APPENDIX 2

BREEDING SPECIES OF CONSERVATION CONCERN

a) Breeding species of conservation concern in Spain grouped by categories in the Spanish Red Book

Category	Species
CR Critically endangered	Balearic Shearwater, Great Bittern, Marbled Duck, Ferruginous Duck, Osprey, Small Buttonquail, Red-knobbed Coot, Common Guillemot, Lesser Grey Shrike
EN Endangered	Shag (ssp. *aristotelis*), White-headed Duck, Red Kite, Lammergeier, Egyptian Vulture, Spanish Imperial Eagle, Bonelli's Eagle, Western Capercaillie (ssp, *cantabricus* & *aquitanicus*), Dotterel, Common Snipe, Eurasian Curlew, Black Tern, Dupont's Lark, Rufous-tailed Bush Robin, Common Reed Bunting (ssp. *lusitanica* & *witherbyi*)
VU Vulnerable	Eurasian Teal, Northern Pintail, Garganey, Red-crested Pochard European Storm-petrel, Shag (ssp *desmarestii*), Black Stork, Glossy Ibis, Eurasian Spoonbill, Black Vulture, Montagu's Harrier, Lesser Kestrel, Rock Ptarmigan, Grey Partridge, Little Bustard, Great Bustard, Collared Pratincole, Kentish Plover, Black-tailed Godwit, Common Redshank, Slender-billed Gull, Audouin's Gull, Black-legged Kittiwake, Gull-billed Tern, Whiskered Tern, Black-bellied Sandgrouse, Pin-tailed Sandgrouse, European Turtle Dove, White-rumped Swift, Roller, White-backed Woodpecker, Greater Short-toed Lark, Common Redstart, Moustached Warbler, Rook
NT Near-threatened	Common Shelduck, Northern Shoveler, Black-necked Grebe, Squacco Heron, Greater Flamingo, Black-shouldered Kite, Black Kite, Golden Eagle, Hobby, Eleonora's Falcon, Oystercatcher, Stone-curlew (ssp. *oedicnemus* & *saharae*), Sandwich Tern, Common Tern, Little Tern, Short-eared Owl, Boreal Owl, Common Kingfisher, Middle Spotted Woodpecker, Lesser Short-toed Lark, Black-eared Wheatear, Savi's Warbler, Western Olivaceous Warbler, Willow Warbler, Bearded Parrotbill, Southern Grey Shrike, Red-billed Chough, Trumpeter Finch
DD Data deficient	Red-legged Partridge, Common Quail, Spotted Crake, Little Crake, Baillon's Crake, Stock Dove, Eurasian Wryneck

b) Breeding species of conservation concern in Portugal grouped by categories in the Portuguese Red Book.

* = Species whose classification differs from that in the Spanish red book

** = Species that are not included in the Spanish red book

Category	Species
CR Critically endangered	Balearic Shearwater (but does not breed in Portugal), Squacco Heron*, Red Kite*, Black Vulture*, Hen Harrier**, Spanish Imperial Eagle*, Osprey, Red-knobbed Coot, Common Snipe*, Common Redshank*, Whiskered Tern*, Common Guillemot, Pin-tailed Sandgrouse*, Roller*, Lesser Short-toed Lark*, Black Wheatear**
EN Endangered	Northern Shoveler*, Common Pochard**, Red-crested Pochard*, Purple Heron**, Egyptian Vulture, Marsh Harrier**, Montagu's Harrier*, Golden Eagle*, Bonelli's Eagle, Great Bustard*, Gull-billed Tern*, Common Tern*, Black-bellied Sandgrouse*, Short-eared Owl*, Water Pipit**, Rufous-tailed Rock Thrush**, Red-billed Chough*
VU Vulnerable	Gadwall**, Tufted Duck**, Madeiran Storm-petrel*, Shag (ssp *aristotelis*), Little Bittern**, Black Stork, Eurasian Spoonbill, European Honey-buzzard**, Short-toed Eagle**, Marsh Harrier**, Goshawk*, Lesser Kestrel, Hobby*, Peregrine Falcon**, Purple Swamphen**, Little Bustard, Stone-curlew*, Collared Pratincole, Common Sandpiper**, Audouin's Gull, Lesser Black-backed Gull**, Little Tern*, European Nightjar**, Red-necked Nightjar**, Great Spotted Cuckoo**, Black-eared Wheatear*, Whinchat**, Savi's Warbler*, Garden Warbler**, Common Crossbill**, Yellowhammer**, Reed Bunting (ssp. *lusitanica*)
NT Near-threatened	Black-necked Grebe, Black-shouldered Kite, Griffon Vulture**, Booted Eagle**, Pied Avocet**, Eagle Owl**, Alpine Swift**, Calandra Lark**, Tree Pipit**, Rufous-tailed Scrub-robin*, Song Thrush**, Reed Warbler**, Spectacled Warbler**, Western Orphean Warbler**, Spotted Flycatcher**, Woodchat Shrike**, Red-backed Shrike**, Northern Raven**
DD Data deficient	Great Bittern, Spotted Crake, Baillon's Crake, Woodcock**, Rock Dove**, Stock Dove, Scops Owl**, Long-eared Owl**, Eurasian Wryneck, Western Olivaceous Warbler*, Ortolan Bunting**
LC Least concern	Red-legged Partridge*, Common Quail*, Black Kite*, Kentish Plover*, European Turtle Dove*, Common Kingfisher*, Greater Short-toed Lark*,Common Redstart*, Southern Grey Shrike*

APPENDIX 3

SCIENTIFIC NAMES OF OTHER SPECIES MENTIONED IN THE TEXT

a) Animals other than birds

American Mink *Mustela vison*
Argentine Ant *Linepithema humile*
Atlantic Salmon *Salmo salar*
Badger *Meles meles*
Black Bullhead *Ameiurus melas*
Brine Shrimp *Artemia* sp.
Common Carp *Cyprinus carpio*
Common Vole *Microtus arvalis*
European Anchovy *Engraulis encrasicolus*
Henslow's Swimming Crab *Polybius henslowii*
Iberian Lynx *Lynx pardinus*
Mediterranean Pine Vole *Microtus duodecimcostatus*
Mole Cricket *Gryllotalpa gryllotalpa*
Mummichog *Fundulus heteroclitus*
Ocellated Lizard *Timon lepidus*
Pine Processionary Moth *Thaumetopoea pityocampa*
Plum Aphid *Hyalopterus pruni*
Rabbit *Oryctolagus cuniculus*
Red Fox *Vulpes vulpes*
Red Swamp Crayfish *Procambarus clarkii*
Sardine *Sardina pilchardus*
Signal Crayfish *Pacifastacus leniusculus*

b) Plants

Aleppo Pine *Pinus halepensis*
Alpine Juniper *Juniperus communis alpina*
Ash *Fraxinus excelsior*
Barbery *Berberis communis*
Bearberry *Arctostaphylos uva-ursi*
Beech *Fagus sylvatica*
Black Pine *Pinus nigra*
Blackthorn *Prunus spinosa*
Blaeberry *Vaccinium* spp.
Box *Buxus sempervirens*
Bramble *Rubus fruticosus*
Carob *Ceratonia siliqua*
Chinese Banyan *Ficus microcarpa*
Club-rush *Scirpus* spp.
Common Reed *Phragmites australis*
Cork Oak *Quercus suber*
Dog Rose *Rosa spp.*
Downy Oak *Quercus pubescens*
Dwarf Palm *Chamaerops humilis*
Esparto Grass *Stipa tenacissima*
False Yellowhead *Dittrichia viscosa*
Fen-sedge *Cladium* sp.
Field Maple *Acer campestre*
Fig *Ficus carica*
Friar's Cowl *Arisarum vulgare*
Giant Reed *Arundo donax*
Hawthorn *Crataegus monogyna*
Holly *Ilex aquifolium*
Honeysuckle *Lonicera* spp.
Iberian Holm Oak *Quercus ilex* subsp. *ballota (=rotundifolia)*
Ivy *Hedera helix*
Kermes Oak *Quercus coccifera*
Lentisc *Pistachia lentiscus*
Lusitanian Oak *Quercus faginea*
Maritime Pine *Pinus pinaster*
Mediterranean Mezereon *Daphne gnidium*
Monterey Pine *Pinus radiata*
Mountain Pine *Pinus uncinata*
Myrtle *Myrtus communis*
Nettle Tree *Celtis australis*
Olive *Olea europaea*
Pedunculate Oak *Quercus robur*
Prickly Pear *Opuntia ficus-indica*
Provence Broom *Cytisus purgans*
Pyrenean Oak *Quercus pyrenaica*
Reedmace *Typha* spp.
Rhododendron *Rhododendron ponticum*
Rosemary *Rosmarinus officinalis*
Rowan *Sorbus aucuparia*
Sand Crocus *Romulea bulbocodium*
Scots Pine *Pinus sylvestris*
Sessile Oak *Quercus petraea*
Silver Fir *Abies alba*
Spanish Fir *Abies pinsapo*
Spanish Juniper *Juniperus thurifera*
Stone Pine *Pinus pinea*
Sweet Chestnut *Castanea sativa*
Tree Aloe *Aloe arborescens*
Tree Heath *Erica arborea*
White Poplar *Populus alba*
Yew *Taxus baccata*

Appendix 4

SPANISH REGIONAL BIRD REPORTS CONSULTED

Region	Area	Title	Abbreviation	Published by	Period
Andalucía	Cádiz	Anuario Ornitológico de la provincia de Cádiz y Estrecho de Gibraltar	*A. O. Cádiz*	Colectivo Ornitológico Cigüeña Negra	2004–08
	Doñana	Anuario Ornitológico de Doñana	*A. O. Doñana*	Estación Biológica de Doñana	1999–2001
	Granada, coast	Aves de la Costa Granadina*	*A. O. Granada*	Asociación Buxus	2002–09
	Málaga	Anuario Ornitológico de Málaga	*A. O. Málaga*	SEO–Málaga	1994–97
	Sevilla	Anuario Ornitológico de Sevilla	*A. O. Sevilla*	Grupo Ornitológico del Sur	2000–05
Aragón	Aragón	Anuario Ornitológico de Aragón (included in *Rocín* 1–6)	*A. O. Aragón*	SEO–Aragón	1991–2007
Asturias	Asturias	Anuario Ornitológico Asturiano (included in *El Draque* 1–7)	*A. O. Asturias*	Coordinadora Ornitolóxica d'Asturies	1993–2000
Balearic Islands	Balearic Islands	Anuari Ornitològic de les Balears*	*A. O. Balears*	GOB Mallorca	1986–2010
Basque Country	Gipuzkoa	Antxeta. Anuario Ornitológico de Gipuzkoa	*A. O. Gipuzkoa*	Itsas Enara Ornitologi Elkarteta	1990–2000
	Bizkaia	Artadi. Anuario Ornitológico de Bizkaia	*A. O. Bizkaia*	Sociedad Ornitológica Lanius	2000–03
Cantabria	Cantabria	Anuario Ornitológico de Cantabria*	*A. O. Cantabria*	Máximo Sánchez Cobo	2008–12
Castilla y León	Burgos	Anuario Ornitológico de Burgos*	*A. O. Burgos*	Aula de Medio Ambiente, Caja de Burgos	2003–12
	Palencia	Anuario Ornitológico de Palencia	*A. O. Palencia*	Fernando Jubete	1998–2001
	Salamanca	Anuario Ornitológico de la provincia de Salamanca	*A. O. Salamanca*	SEO–Salamanca	1924–2006
Castilla-La Mancha	Albacete	Anuario Ornitológico de Albacete	*A. O. Albacete*	Sociedad Albacetense de Ornitología	1997–98
	Ciudad Real	Anuario Ornitológico de Ciudad Real	*A. O. Ciudad Real*	SEO–Ciudad Real	1995–2005
	Toledo	Anuario Ornitológico de Toledo	*A. O. Toledo*	Agrupación Naturalista ESPARVEL	2001–07
Catalonia	Catalonia	Anuari d'Ornitologia de Catalunya*	*A. O. Catalunya*	Institut Català d'Ornitologia	1996–2009

continued

Region	Area	Title	Abbreviation	Published by	Period
Comunidad Valenciana	Comunidad Valenciana	Anuario Ornitológico de la Comunidad Valenciana	*A. O. C. Valenciana*	Estació Ornitològica de l'Albufera (SEO)	1988–97
	Comunidad Valenciana	Aves de la Comunidad Valenciana*	*A. O. C. Valenciana*	Societat Valenciana d'Ornitologia	1998–2002
	Alicante	Las Aves en Alicante. Anuario Ornitológico	*A. O. Alicante*	SEO–Alicante	1999–2003
	Castellón	Anuario Ornitológico de Castellón*	*A. O. Castellón*	Ecologistas en Acción & Grup Au	2003–08
	Comunidad Valenciana	Anuario Ornitológico de la Comunidad Valenciana*	*A. O. C. Valenciana*	M. Tirado *et al.*	2009
Extremadura	Extremadura	Aves de Extremadura. Anuario ADENEX*	*A. O. Extremadura*	ADENEX	1998–2003
Galicia	Galicia	Anuario das Aves de Galicia*	*A. O. Galicia*	Sociedade Galega de Ornitoloxía	1992–2004
La Rioja	La Rioja	Anuario Ornitológico de La Rioja (included in *Zubía* 20–22)	*A. O. La Rioja*	Ecologistas en Acción de La Rioja	1993–2008
Madrid	Madrid	Anuario Ornitológico de Madrid	*A. O. Madrid*	SEO–Monticola	1996–2008
Murcia	Murcia	Anuario Ornitológico de la Región de Murcia*	*A. O. Murcia*	Ángel Guardiola	2002–10
Navarra	Navarra	Anuario Ornitológico de Navarra*	*A. O. Navarra*	Sociedad de Ciencias Naturales Gorosti	1993–2001

* available online (1992–2000 only for Galicia)

APPENDIX 5

LIST OF SPANISH AND PORTUGUESE RARITY REPORTS

The tables provide references to all the reports published by the Spanish rarities committee (CR-SEO) and the Portuguese rarities committee (CPR) by the end of 2013. The year(s) given are the period covered by each report, not the year of publication. Reports for 1987–94 were published jointly for both countries by the Iberian Peninsula rarities committee (CIR).

Spain		
1984	*Ardeola* 32: 137–143.	CR-SEO
1985	*Ardeola* 34: 123–133.	CR-SEO
1986	*Ardeola* 35: 167–174.	CR-SEO
1987	*Ardeola* 36: 111–123.	CIR
1988	*Ardeola* 37: 107–125.	CIR
1989	*Ardeola* 38: 149–166.	CIR
1990	*Ardeola* 39: 73–83.	CIR
1991	*Ardeola* 40: 177–192.	CIR
1992	*Ardeola* 41: 103–117.	CIR
1993	*Ardeola* 42: 97–113.	CIR
1994	*Ardeola* 43: 103–118.	CIR
1995	*Ardeola* 44: 119–141.	CR-SEO
1996	*Ardeola* 45: 97–116.	CR-SEO
1997	*Ardeola* 46: 129–148.	CR-SEO
1998	*Ardeola* 47: 141–159.	CR-SEO
1999	*Ardeola* 48: 117–136.	CR-SEO
2000	*Ardeola* 49: 141–171.	CR-SEO
2001	*Ardeola* 50: 123–149.	CR-SEO
2002	*Ardeola* 51: 515-541.	CR-SEO
2003	*Ardeola* 53: 185–206.	CR-SEO
2004	*Ardeola* 54: 163–190.	CR-SEO
2005	*Ardeola* 54: 405–446.	CR-SEO
2006	*Ardeola* 55: 259–287.	CR-SEO
2007	*Ardeola* 56: 309–334.	CR-SEO
2008	*Ardeola* 57: 481–516.	CR-SEO
2009	*Ardeola* 58: 441–480.	CR-SEO
2010	*Ardeola* 59: 353–411.	CR-SEO
2011	*Ardeola* 60: 437–506.	CR-SEO

Portugal		
1987	*Ardeola* 36: 111–123.	CIR
1988	*Ardeola* 37: 107–125.	CIR
1989	*Ardeola* 38: 149–166.	CIR
1990	*Ardeola* 39: 73–83.	CIR
1991	*Ardeola* 40: 177–192.	CIR
1992	*Ardeola* 41: 103–117.	CIR
1993	*Ardeola* 42: 97–113.	CIR
1994	*Ardeola* 43: 103–118.	CIR
1995	*Pardela* 5: 4–19.	CPR
1996	*Pardela* 8: 3–23.	CPR
1997–1998	*Pardela* 11: 3-27.	CPR
1999–2001	*Anuário Ornitológico* 1: 3–35.	CPR
2002	*Anuário Ornitológico* 2: 1–20.	CPR
2003	*Anuário Ornitológico* 3: 1–21.	CPR
2004	*Anuário Ornitológico* 4: 1–15.	CPR
2005	*Anuário Ornitológico* 5: 1–34.	CPR
2006–2007	*Anuário Ornitológico* 6: 1–45.	CPR
2008–2009	*Anuário Ornitológico* 7: 3–71.	CPR
2010	*Anuário Ornitológico* 8: 3–52.	CPR

REPORTING OBSERVATIONS OF RARITIES

Reports of rare birds found in Iberia may be sent to the relevant rarities committees at the addresses below, which also give further details on record submission.

SPAIN

Comité de Rarezas de la Sociedad Española de Ornitología (CR-SEO)

https://www.seo.org/conocenos/grupos-de-trabajo/comite-de-rarezas/

Current secretary: José Luis Copete (rarezas@seo.org)

PORTUGAL

Comité Português de Raridades (CPR)

http://www.spea.pt/pt/observar-aves/comite-portugues-raridades/

Current secretary: Joaquim Muchaxo (raridades@spea.pt)

GIBRALTAR

Gibraltar Ornithological and Natural History Society rarities panel.

http://www.gonhs.org/rarities.htm

Current secretary: Keith Bensusan (sgbo@gonhs.org)

REFERENCES

Abs, M. 1958. Contribución a la avifauna de la provincia de Salamanca. *Ardeola* 4: 7–24.

Abs, M. 1963. Vergleichende Untersuchungen au Haubenlerche (*Galerida cristata*) und Theklalerche (*Galerida theklae*). *Bonner Zoologische Beiträge* 14: 1–128.

Acha, A. & Velasco, T. 1992. La invernada de limícolas en charcas y embalses de la provincia de Cáceres. *Butll. Parc Natural del delta del Ebre* 7: 4–9.

ADN. 2002. *Atles dels ocells nidificants d'Andorra/Atlas of the breeding birds of Andorra*. Associació per a la Defensa de la Natura. Andorra.

Adrover, J. 2011. Estat de conservació de la població d'abellerol *Merops apiaster* a Mallorca. *Anuari Ornitològic de les Balears 2010*: 15–27.

Adrover, J., Muñoz, A. & Riera, J. 2003. Situació de la milana *Milvus milvus* a Mallorca. *Anuari Ornitològic de les Balears 2002*: 41–45.

Adrover, J. *et al.* 2010. *Atles dels aucells nidificants de Mallorca i Cabrera (2003–2007)*. GOB. Palma de Mallorca.

Afonso, I. & Canut, J. 2006. *El lagópodo alpino (*Lagopus muta pyrenaica*) en el Parque Nacional de Aigüestortes i Estany de Sant Maurici*. Organismo Autónomo de Parques Nacionales. Madrid.

Agencia Estatal de Meteorología. 2011. *Iberian Climate Atlas/ Atlas Climático Ibérico*. Ministerio de Medio Ambiente y Medio Rural y Marino. Madrid.

Agostini, N., Baghino, L., Coleiro, C., Corbi, F. & Premuda, G. 2002. Circuitous autumn migration in the Short-toed Eagle (*Circaetus gallicus*). *J. Raptor Research* 36: 111–114.

Agostini, N., Baghino, L., Panuccio, M. & Premuda, G. 2002. A conservative strategy in migrating Short-toed Eagles *Circaetus gallicus*. *Ardeola* 49: 297–291.

Agostini, N. & Mellone, U. 2008. Does migration flyway of Short-toed Snake-Eagles breeding in central Italy reflect the colonization history? *J. Raptor Research* 42: 158–159.

Aguilar, J. S. 1994. *Atlas de las aves marinas de Baleares*. Documents tècnics de conservació 23. Conselleria de Agricultura y Pesca del Gobierno Balear. Palma de Mallorca.

Ahmad, M. 2005. Franklin's Gulls and Laughing Gulls in Britain & Ireland in November 2005. *Birding World* 18: 461–464.

Aja, R. 2010. El Quebrantahuesos vuelve a casa. *La Nueva España*, 10 July.

Alamany, O. 1989. Situación de la Lechuza de Tengmalm en la Península Ibérica. *Quercus* 44: 8–15.

Albero, J. C., Rivas, J. L., Antor, R. J., Barrachina, P. & Alcántara, M. 2010. Selección de microhábitat y comportamiento de alimentación del alcaudón chico (*Lanius minor*) en Aragón: implicaciones para su conservación. *Rocín, Anuario Ornitológico de Aragón, 2005–2007*: 101–110.

Alberto, L. J. & Pena, J. C. 1981. Niveles de contaminantes organochlorados y metales pesados en huevos de aves de las Marismas del Guadalquivir, 1975. *Doñana Acta Vertebrata* 8: 195–206.

Alberto, L. J. & Purroy, F. J. 1981. Censos de limícolas invernantes en España (1978, 79 y 80) realizados por la Sociedad Española de Ornitología. *Ardeola* 28: 3–34.

Alberto, L. J. & Velasco, T. 1988. Limícolas invernantes en España. *In:* Tellería, J. L. (Ed.), pp. 71–78.

Alcántara, M. 2004. Desarrollo y ejecución del Plan de Conservación del Hábitat del Cernícalo Primilla en Aragón. *In:* Alcántara, M. (Ed.) *Actas del VI Congreso Nacional sobre el Cernícalo Primilla*, pp. 53–61. Departamento de Medio Ambiente del Gobierno de Aragón. Zaragoza.

Alcántara, M. 2005. Intento de cría de una pareja de águila pescadora en Aragón. *Quercus* 123: 41.

Alcobendas, G. 2009. Tamaño de población, densidad, número de nidos, nidotópica y características del hábitat de nidificación del azor común (*Accipiter gentilis*) en un área del sudoeste de Madrid. *Anuario Ornitológico de Madrid 1999*: 122–131.

Alegre, J., Hernández, A. & Purroy, F. J. 1989. Distribución altitudinal y patrones de afinidad trófica geográfica de la Lechuza Común (*Tyto alba*) en León. *Ardeola* 36: 41–54.

Allen, F. G. H. & Brudenell-Bruce, P. G. C. 1967. The White-rumped Swift *Apus affinis* in southern Spain. *Ibis* 109: 113–115.

Almeida, J. 1996. Situation and conservation perspectives of the Common Crane in Portugal. *Alytes* 8: 99–104.

Alonso, D., Arizaga, J., Miranda, R. & Hernández, M. A. 2006. Morphological diversification of Common Crossbill *Loxia curvirostra* populations within Iberia and the Balearics. *Ardea* 94: 99–107.

Alonso, J. A. 1983. Propuesta de protección para la pajarera de garcillas bueyeras del tajo de Barbate (Cádiz). *Alytes* 1: 139–151.

Alonso, J. A. & Purroy, F. J. 1979. Avifauna de los parques de Madrid. *Naturalía Hispanica* 18. OCONA. Madrid.

Alonso, J. A. & Alonso, J. C. (Eds) 1990a. *Distribución y demografía de la grulla común* (Grus grus) *en España*. Colección Técnica. ICONA. Madrid.

Alonso, J. A., Alonso, J. C. & Nowald, G. 2008. Migration and wintering patterns of a central European population of Common Cranes. *Bird Study* 55: 1–7.

Alonso, J. C. 1984. Estudio comparado de los principales parametros reproductivos de *Passer hispaniolensis* y *Passer domesticus* en España centro-occidental. *Ardeola* 30: 3–21.

Alonso, J. C. 1985. Description of intermediate phenotypes between *Passer hispaniolensis* and *Passer domesticus*.

Ardeola 32: 31–38.

Alonso, J. C. 1986. On the status and distribution of the Spanish Sparrow (*Passer hispaniolensis* Temm.) in Iberia. *International Studies on Sparrows* 13: 35–43.

Alonso, J. C. & Alonso, J. A. 1990b. Sobre el tamaño de la población de avutardas de la península Ibérica. *In:* Alonso, J. C. & Alonso, J. A. (Eds) *Parámetros demográficos, selección de hábitat y distribución de la Avutarda* (Otis tarda) *en tres regiones españolas*, pp. 81–86. ICONA. Madrid.

Alonso, J. C. & Alonso, J. A. 1996. The Great Bustard *Otis tarda* in Spain: present status, recent trends and evaluation of earlier censuses. *Biological Conservation* 77: 79–86.

Alonso, J. C., Alonso, J. A., Martínez, J. H., Avignon, S. & Petit, P. 2003. European Cranes shift their wintering area northwards: new evidences from radiotagged birds. *In:* Salvi, A. (Ed.) *Proceedings of 4th European Crane Workshop*, pp. 201–207. Verdun, France.

Alonso, J. C. & Palacín, C. 2009. Avutarda. *In:* Salvador, A., & Bautista, L. M. (Eds).

Alonso, J. C., Palacín, C. & Martín, C. A. (Eds) 2005. *La Avutarda Común en la Península Ibérica: población actual y método de censo.* SEO/Birdlife. Madrid.

Alonso, J. C., Palacín, C., Alonso, J. A. & Martín, C. A. 2009. Post-breeding migration in male great bustards: low tolerance of the heaviest Palaearctic bird to summer heat. *Behavioral Ecology and Sociobiology* 63: 1705–1715.

Alström, P., Colston, P. & Lewington, I. 1991. *A field guide to the rare birds of Britain and Europe*. HarperCollins Publishers. London.

Álvarez Laó, C. M. 2001. Procedencia de la gaviota sombría *Larus graellsii* en Asturies. *Actas de los I Encuentros Ornitológicos Asturianos*, pp. 38–43. Gijón.

Álvarez Laó, C. 2011. A review of the status of the Herring Gull *Larus argentatus* in Spain. *Seabird* 24: 72–82.

Álvarez, A. & Purroy, F. J. 1993. Comparación de las comunidades de aves nidificantes e invernantes en los medios forestales de la Cordillera Cantábrica leonesa. *Ecología* 7: 403–418.

Álvarez, C. & Méndez, M. 1995. Alimentación de la Gaviota Patiamarilla *Larus cachinnans* en dos localidades costeras asturianas. *Chioglossa* 1: 23–30.

Álvarez, D. 2009. Primera cita de un Cormorán Moñudo (*Phalacrocorax aristotelis*) ibérico en el norte de Europa. *Revista de Anillamiento* 24: 6–8.

Álvarez, D. & Pajuelo, M. A. F. 2004. Differential migration in a wintering population of Common Guillemots *Uria aalge* affected by the Prestige oil spill. *Ardeola* 51: 451–454.

Álvarez, D. & Velando, A. 2007. *El cormorán moñudo en España. Población en 2006–2007 y método de censo.* SEO/BirdLife. Madrid

Álvarez, F. 1994. A gens of Cuckoo *Cuculus canurus* parasitizing Rufous Bush Chat *Cercotrichas galactotes*. *J. Avian Biology* 25: 239–243.

Álvarez, F. & Arias de Reyna, L. 1974. Mecanismos de parasitación por *Clamator glandarius* y defensa por *Pica pica*. *Doñana, Acta Vertebrata* 1: 43–66.

Álvarez, G. & Santos, T. 1992. Efectos de la gestión del monte sobre la avifauna de una localidad mediterránea (Quintos de Mora, Montes de Toledo). *Ecología* 6: 187–198.

Álvarez, J., Bea, A., Faus, J. M. & Castién, E. 1985. *Atlas de los vertebrados continentales de Álava, Vizcaya y Guipúzcoa.* Gobierno Vasco. Vitoria-Gasteiz.

Álvarez-Balbuena, F., Vigil, A., Álvarez, C., Carballal, M. E., García, E. & García, J. A. (Eds) 2000. *Aves raras y escasas en Asturias.* Coordinadora Ornitolóxica d'Asturies. Avilés.

Alves, J. A., Lourenço, P. M., Piersma, T., Sutherland, W. J. & Gill, J. A. 2010. Population overlap and habitat segregation in wintering Black-tailed Godwits *Limosa limosa*. *Bird Study* 57: 381–391

Amat, J. A. 1979a. Captura de Cerceta Aliazul *Anas discors* en las Marismas del Guadalquivir. *Doñana, Acta Vertebrata* 6: 131.

Amat, J. A. 1979b. Captura de *Porphyrula alleni* en las Marismas del Guadalquivir. *Doñana, Acta Vertebrata* 6: 131.

Amat, J. A. 1984. Las poblaciones de aves acuaticas en las lagunas andaluzas: composición y diversidad durante el ciclo anual. *Ardeola* 31: 61–79.

Amat, J. A. 1986. Information on the diet of the Stone Curlew *Burhinus oedicnemus* in Doñana, southern Spain. *Bird Study* 33: 71–73.

Amat, J. A. 2012. Chorlitejo patinegro. *In:* Salvador, A. & Morales, M. B. (Eds).

Amat, J. A., García-Criado, B. & García-Ciudad, A. 1991. Food, feeding behaviour and nutritional ecology of wintering Greylag Geese *Anser anser*. *Ardea* 79: 271–282.

Amat, J. A., Lucientes, J. & Ferrer, X. 1987. La migración de muda del Pato Colorado (*Netta rufina*) en España. *Ardeola* 34: 79–88.

Amat, J.A. & Soriguer, R. C. 1982. Datos sobre selección de hábitat y ecología alimenticia del Porrón Pardo (*Aythya nyroca*). *Doñana, Acta Vertebrata* 9: 388–394.

Amat, J. A. & Soriguer, R. 1984. Alimentación invernal de la Polla de Agua (*Gallinula chloropus*) en las Marismas del Guadalquivir. *Ardeola* 31: 136–140.

Amezian, M., Bensusan, K., Perez, C. & Thompson, I. 2006. Is House Bunting about to colonise Europe? *Birding World* 19: 263.

Amezian, M., Thompson, I., Bensusan, K., Cortes, J., Louah, A. & Qninba, A. 2011. On regular wintering of Eurasian Penduline Tits *Remiz pendulinus* in northern Morocco. *Ostrich* 82: 39–42.

Ameztoy, I. 1966. Pardillos sizerines (*Carduelis flammea*) capturados en Guipúzcoa. *Munibe* 18: 221–222.

Andrada, J. & Franco, A. 1975. Sobre el area de invernada de *Falco naumanni* en España. *Ardeola* 21 (esp): 321–324.

Antor, R. & Pedrocchi, C. 1989. Composición, abundancia y diversidad de las comunidades de paseriformes del piso alpino pirenaico. *Pirineos* 133: 99–110.

AOB. 2011. Estatus de l'avifauna balear. *Anuari ornitològic de les Balears 2010*. Annex II: 259–270.

Aparicio, R. J. 2011. Mirlo común. *In:* Salvador, A. & Morales, M. (Eds).

Aparicio, R. J., Casaux, E. & Pérez, A. 1991. Migración posnupcial de la Curruca Mosquitera (*Sylvia borin*) en una localidad del Valle del Tajo (España Central). *Ardeola* 38: 3–10.

Aragonés, J. 1996. Incidencia de las actividades humanas en la mortalidad del Chotacabras Pardo. *Quercus* 123: 24–26.

Aragüés, A. 1964. Información sobre *Remiz pendulinus* en el valle del Ebro. *Ardeola* 9: 153–156.

Aragüés, A. & Herranz, A. 1983. Dupont's Lark in the Spanish steppes. *British Birds* 76: 57–62.

Aragüés, A. & Lucientes, J. 1980. *Fauna de Aragón: las aves.* Guara Editorial, Zaragoza.

Aragüés, A., Pérez Bujarrabal, E., Lucientes, J. & Bielsa, M. A. 1974. Observaciones estivales en Gallocanta (Zaragoza). *Ardeola* 20: 229–244.

Arambarri, R. & Rodríguez, A. F. 1996. Distribución y estima poblacional de Pico Mediano en Álava. *Ardeola* 43: 221–223.

Araújo, J. & García Rua, A. E. 1974. El censo Español de limícolos de enero de 1973. *Ardeola* 20: 151–159.

Araújo, J., Muñoz-Cobo, J. & Purroy, F. J. 1977. *Las rapaces y aves marinas del archipiélago de Cabrera.* Naturalia Hispanica No. 12. ICONA. Madrid.

ARCEA. 2005. *Estudio de la población de aves invernantes y nidificantes en el Parque Nacional marítimo-terrestre Islas Atlánticas de Galicia. Enero 2004–enero 2005.* Unpublished report to Atlantic Islands National Park.

Arcos, F., Mouriño, J., Martínez, J. M. & Sierra, F. 1995. Notas sobre ecología, mortalidad y evolución de las poblaciones de Arao común (*Uria aalge*) en el suroeste de Galicia. *Chioglossa* 1: 53–59.

Arcos, J. M. 2000. Host selection by Arctic Skuas *Stercorarius parasiticus* in the Western Mediterranean during spring migration. *Ornis Fennica* 77: 131–135.

Arcos, J. M. 2011. *International species action plan for the Balearic Shearwater,* Puffinus mauretanicus. BirdLife International. Cambridge.

Arcos, J. M. & Albiol, D. 1997. Observaciones de aves marinas en aguas de Barcelona (abril 1996–marzo 1997). *Bol. GIAM* 23: 13–16.

Arcos, J. M., Bécares, J., Rodríguez, B. & Ruiz, A. 2009. *Áreas importantes para la conservación de las aves marinas en España.* SEO/BirdLife. Madrid.

Arévalo y Baca, D. J. 1887. *Aves de España*. Memorias de la Real Academia de Ciencias Exactas, Físicas y Naturales, XI. Madrid.

Arias de Reyna, L. 1998. Coevolution of great spotted cuckoo and its hosts. *In:* Rothstein, S. I. & Robinson, S. K. (Eds) *Parasitic birds and their hosts, studies in coevolution*, pp. 129–142. Oxford University Press. Oxford.

Arizaga, J., Aldalur, A., Herrero, A., Cuadrado, J. F., Mendiburu, A. & Sanpera, C. 2010. High importance of fish prey in the diet of Yellow-legged Gull *Larus michahellis* chicks from the southeast Bay of Biscay. *Seabird* 23: 1–6.

Arizaga, J. & Alonso, D. 2005. Migración e invernada del pájaro moscón (*Remiz pendulinus*) en la marisma de Txingudi (N de España). *Munibe* 56: 145–154.

Arizaga, J., Alonso, D., Campos, F., Unamuno, J. M., Monteagudo, A., Fernández, G., Carregal, X. M. & Barba, E. 2006. ¿Muestra el pechiazul *Luscinia svecica* en España una segregación geográfica en el paso posnupcial a nivel de subespecie? *Ardeola* 53: 285–291.

Arizaga, J., Alonso, D., Mendiburu, A. & Vilches, A. 2007. Winter roosts of Yellowhammers *Emberiza citrinella* in Northern Iberia: population dynamics and biometry. *Revista Catalana d'Ornitologia* 23: 18–26.

Arizaga, J., Cuadrado, J. F. & Romero, L. 2009a. Seasonal individual and population-associated patterns of migration of goldfinches *Carduelis carduelis* through the western edge of Pyrenees. *Ardeola* 56: 57–69.

Arizaga, J., Galarza, A., Herrero, A., Hidalgo, J. & Aldalur, A. 2009b. Distribución y tamaño de la población de la Gaviota Patiamarilla *Larus michahellis lusitanius* en el País Vasco: tres décadas de estudio. *Revista Catalana d'Ornitologia* 25: 32–42.

Arizaga, J., García, J. & Suárez-Seoane, S. 2011. Ruiseñor pechiazul. *In:* Salvador, A. & Bautista, L. M. (Eds)

Arizaga, J., Herrero, A. & Aldalur, A. 2009c. Primera cita en el Mediterráneo de un ejemplar de gaviota patiamarilla (*Larus michahellis*) nacido en el Cantábrico. *Revista de Anillamiento* 23: 6–9.

Arizaga, J., Herrero, A., Galarza, A., Hidalgo, J., Aldalur, A., Cuadrado, J. F. & Ocio, G. 2010. First-year movements of Yellow-legged Gull (*Larus michahellis lusitanius*) from the Southeastern Bay of Biscay. *Waterbirds* 33: 444–450.

Armada, R., 2011. The Red-footed Booby in Spain. *Birding World* 23: 524–529.

Arratíbel, P. 2007. El Archibebe Claro *Tringa nebularia* en Navarra. *Anuario Ornitológico de Navarra 2000–2001*: 92–96.

Arribas, O. 2004. *Fauna y paisaje de los Pirineos en la Era Glaciar.* Lynx. Barcelona.

Arrigoni, E. 1902. Remarks on Audouin's Gull (*Larus audouini*). *Ibis* 44: 491–499.

Arroyo, B. 1977. La avifauna de un cantil estepárico. *Ardeola* 23: 41–47.

Arroyo, B., Ferreiro, E. & Garza, V. (Coords.) 1990a. *II Censo*

Nacional de Buitre Leonado (Gyps fulvus). *Población, distribución, demografía y conservación.* ICONA. Madrid.

Arroyo, B., Ferreiro, E. & Garza, V. 1990. *El Águila Real (*Aquila chrysaetos*) en España. Censo, distribución, reproducción y conservación.* ICONA. Madrid.

Arroyo, B., Ferreiro, E. & Garza, V. 1995. *El Águila Perdicera* (Hieraaetus fasciatus) *en España. Censo, reproducción y conservación.* ICONA. Madrid.

Arroyo, B. E. 1994. Reproductive success of Montagu's Harrier (*Circus pygargus*) and Hen Harrier (*Circus cyaneus*) in agricultural habitats. *In:* Muntaner, J. & Mayol, J. (Eds) *Biología y conservación de las rapaces mediterráneas*, pp. 459–463. SEO/BirdLife. Madrid.

Arroyo, B. E. & García, J. T. 2007. *El aguilucho cenizo y el aguilucho pálido en España. Población en 2006 y método de censo.* SEO/BirdLife. Madrid.

Arroyo, B. E. & García, J. T. 1999. Los aguiluchos Cenizo (*Circus pygargus*) y Pálido (*C. cyaneus*) en las áreas cerealistas del Jarama: resumen de 8 años de estudio. *Anuario Ornitológico de Madrid 1998*: 14–25.

Arroyo, B. E., García, J. T. & Bretagnolle, V. 2002. Conservation of Montagu's Harriers *Circus pygargus* in agricultural areas. *Animal Conservation* 5: 283–290.

Artázkoz, A. 1995. La migración e invernada del Estornino Pinto (*Sturnus vulgaris*, L.) en Navarra según los resultados del anillamiento. *Anuario Ornitológico de Navarra 1993–94*, 27–36.

Asensio, B. 1984a. *Migración de aves fringílidas (Fringillidae) a base de resultados del anillamiento.* Ediciones de la Universidad Complutense. Madrid.

Asensio, B. 1984b. Sobre los origenes de los fringilidos migrantes en el extremo sur de España. *Ardeola* 31: 128–134.

Asensio, B. 1985a. Migración en España del Verdecillo (*Serinus serinus*), según los resultados del anillamiento. *Ardeola* 32: 173–178.

Asensio, B. 1985b. Migración e invernada en España de Lúganos (*Carduelis spinus* L.) de origen Europeo. *Ardeola* 32: 179–186.

Asensio, B. 1986. La migración en España del Jilguero (*Carduelis carduelis* L.) según los resultados de anillamiento. *Ardeola* 33: 176–183.

Asensio, B. 1987. *Informe sobre la campaña de anillamiento de aves en España: año 1985.* Ministerio de Agricultura, Pesca y Alimentación. Madrid.

Asensio, B. 1992. La migración y la invernada de la Avefría (*Vanellus vanellus*) en Iberia. *Doñana, Acta Vertebrata* 19: 71–84.

Asensio, B. & Antón, C. 1990. Situación del Picogordo (*Coccothraustes coccothraustes*) en España. *Ardeola* 37: 29–35.

Asensio, B. & Cantos, F. 1989a. La migración postnupcial de *Phylloscopus trochilus* en el Mediterráneo occidental. *Ardeola* 36: 61–71.

Asensio, B. & Cantos, F. J. 1989b. Migración parcial de las poblaciones centroibéricas de *Passer hispaniolensis*. *Miscelánea Zoológica* 13: 215–217.

Asensio, B. & Cantos, F. J. 1996. La gaviota reidora o el éxito del oportunismo. *Quercus* 124: 31–37.

Asensio, B., Cantos, F. J., Fernández, A. & Vega, I. 1991. La migración del Avión Zapador (*Riparia riparia*) en España. *Ardeola* 38: 37–49.

Asensio, B. & Carrascal, L. M. 1987. Migratología de las Agachadizas Comunes (*Gallinago gallinago*, L.) invernantes en la Península Ibérica. *Ardeola* 34: 225–242.

Asensio, B. & Carrascal, L. M. 1992. Migration of Common Teal (*Anas crecca*) wintering in the Iberian Peninsula. *Folia Zoologica* 41: 45–53.

Asensio, B., Cantos, F., Fernández, A. & Fajardo, Í. 1992. La Lechuza Campestre en España. *Quercus* 76: 18–24.

Astrain, C. & Etxeberria, A. 1998. Distribución y población de la avifauna esteparia amenazada en Navarra. *Anuario Ornitológico de Navarra* 3: 45–51.

Atienza, J. C. 2006. *El escribano palustre en España. I Censo Nacional (2005).* SEO/BirdLife. Madrid.

Atienza, J. C. 2009. Ave del año 2009: Escribano Palustre. *La Garcilla* 138: 6–7.

Atienza, J. C., Pinilla, J. & Justribó, J. H. 2001. Migration and conservation of the Aquatic Warbler in Spain. *Ardeola* 48: 197–208.

Austin, J. J., Bretagnolle, V. & Pasquet, E. 2004. A global molecular phylogeny of the small *Puffinus* shearwaters and implications for systematics of the Little-Audubon's Shearwater complex. *Auk* 121: 847–864.

Avellà, F. J., Garcias, P. J., Jurado, J. R. & Muñoz, A. 1997. *Atles dels aucells nidificants de Mallorca i Cabrera (1983–1994).* GOB. Palma de Mallorca.

Avilés, J. M. 1999. Distribución de la Carraca (*Coracias garrulus*) nidificante en España. *Ardeola* 46: 223–226.

Avilés, J. M. & Costillo, E. 1998. Selection of breeding habitats by the Roller (*Coracias garrulus*) in farming areas of the southwest of the Iberian Peninsula. *Die Vogelwarte* 39: 242–247.

Avilés, J. M. & Sánchez, A. 1998. Evolución del número de parejas reproductoras de Carraca *Coracias garrulus* en cinco hábitats de Extremadura. *Butll. Grup Català d'Anellament* 14: 25–29.

Avilés, J. M., Sánchez, J. M. & Parejo, D. 2000. The Roller *Coracias garrulus* in Extremadura (southwestern Spain) does not show a preference for breeding in clean nestboxes. *Bird Study* 47: 252– 254.

Avilés, J. M., Sánchez, J. M. & Parejo, D. 2002. Food selection of wintering common cranes (*Grus grus*) in holm oak (*Quercus ilex*) dehesas in south-west Spain in a rainy season. *J. Zoology* 256: 74–79.

Avilés, J. M., Sánchez, J. M., Sánchez, Á. & Parejo, D. 1999. Breeding biology of the Roller *Coracias garrulus* in farming areas of the southwest Iberian Peninsula. *Bird*

Study 46: 217–223.

Aymerich, J., Baucells, J., Bigas, D., Camprodón, J., Estrada, J., Molist, M., Ordeix, M., Ramoneda, J. & Vigué, J. 1991. *Els Ocells d'Osona*. Lynx. Barcelona.

Aymí, R. 1989. La migración postnupcial del Mirlo Capiblanco (*Turdus torquatus*) en una área de influencia de la Sierra del Montsant (Tarragona). *Miscélanea Zoológica* 13: 211–214.

Aymí, R. 2011. Identificación del Carricero Políglota (*Acrocephalus palustris*) y estatus actual en España. *Revista de Anillamiento* 28: 13–23.

Aymí, R. & Martínez, I. 1991. Presència i situació d'algunes subespècies d'ocells a Catalunya. *Butll. Grup Català d'Anellament* 7: 45–52.

Aymí, R. & Romero, J. L. 1996. Migració i distribució estacional del Durbec *Coccothraustes coccothraustes* a Catalunya. *Butll. Grup Català d'Anellament* 13: 21–35.

Aymí, R., Rodríguez, M. & Cama, A. 2011. Influx of Kittiwakes *Rissa tridactyla* in Catalonia, NE Spain, in January 2009 and review of previous records. *Revista Catalana d'Ornitologia* 27: 17–24.

Azafzaf, H., Dlensi, H. & Feltrup-Azafzaf, C. 2010. Première observation du Flamant nain *Phoenicopterus minor* en Tunisie. *Alauda* 78: 330.

Baeta, R. 2004. Le Vanneau sociable *Chettusia gregaria*. Généralités sur l'espèce et pattern d'apparition en France, particulièrement en Région Centre: hypothèses sur l'explication du phénomène. *Le p'tit grav'* 2: 67–71.

Baillon, F. & Dubois, P. J. 1992. Nearctic gull species in Senegal and The Gambia. *Dutch Birding* 14: 49–50.

Balança G. & Schaub M. 2005. Post-breeding migration ecology of Reed *Acrocephalus scirpaceus*, Moustached *A. melanopogon* and Cetti's Warblers *Cettia cetti* at a Mediterranean stopover site. *Ardea* 93: 245–257.

Balbontín, J. & Ferrer, M. 2009. Movements of juvenile Bonelli's eagles *Aquila fasciata* during dispersal. *Bird Study* 56: 86–95.

Balbontín, J., Negro, J. J., Sarasola, J. H., Ferrero, J. J. & Rivera, D. 2008. Land-use changes may explain the recent range expansion of the Black-shouldered Kite *Elanus caeruleus* in southern Europe. *Ibis* 150: 707–716.

Balcells, E. 1961. Sobre épocas de migración y trashumancia de aves en el NE español. *Ardeola* 7: 5–58.

Balcells, E. & Masoliver, M. 1960. Sobre la nidificación de dos aves en la llamada Cataluña húmeda (*Sitta europea* y *Sturnus vulgaris*). *Ardeola* 6: 5–10.

Ballesteros, F. 1998. *Las especies de caza en España. Biología, ecología y conservación*. Estudio y Gestión del Medio. Oviedo.

Ballesteros, G., Cabrera, M., Echevarrías, J. L., Lorenzo, C. J., Raya, C., Torres-Esquivias, J. A. & Viedma, C. (Eds) 2008. *Tarro canelo, cerceta pardilla, porrón pardo, malvasía cabeciblanca y focha moruna en España. Población en 2007 y método de censo*. SEO/BirdLife. Madrid.

Balmori, A. 2003. Migración otoñal diferencial del Andarríos Chico, *Actitis hypoleucos*, en la cuenca del Duero. *Ardeola* 50: 59–66.

Balmori, A., Cuesta, M. A. & Caballero, J. M. 2002. Distribución de los mosquiteros Ibérico (*Phylloscopus brehmii*) y Europeo (*Phylloscopus collybita*) en los bosques de ribera de Castilla y León (España). *Ardeola* 49: 19–27.

Balmori, A. & Hallberg, Ö. 2007. The urban decline of the House Sparrow (*Passer domesticus*): a possible link with electromagnetic radiation. *Electromagnetic Biology and Medicine* 26: 141–151.

Banks, R. C., Cicero, C., Dunn, J. L., Kratter, A. W., Rasmussen, P. C., Remsen, J. V., Rising, J. D.& Stotz. D. F. 2004. Forty-fifth supplement to the American Ornithologists' Union Check-list of North American Birds. *Auk* 121: 985–995.

Bannerman, D. A. & Bannerman, W. M. 1966. *Birds of the Atlantic Islands. Vol. 3. A History of the Birds of the Azores*. Oliver & Boyd. Edinburgh & London.

Barbosa, A. (coord.). 1997. *Las aves limícolas en España*. Organismo Autónomo Parques Nacionales. Madrid.

Bárcena, F. 1985. Localización e inventario de las colonias de Arao Común, *Uria aalge* Pontop., en las costas de Galicia. Determinación de las posibles causas de su desaparición. *Bol. Estación Central de Ecología* 28: 19–28.

Bárcena, F. & Domínguez, J. 1986. La Tourterelle Turque (*Streptopelia decaocto*) dans la Peninsule Ibérique. *Alauda* 54: 107–120.

Bárcena, F. & Domínguez, J. 1995. *Limosa limosa*. *In:* S.G.H.N. (Ed.) *Atlas de vertebrados de Galicia*. Vol. 2. Aves, p. 417. Consello da Cultura Galega. Santiago de Compostela.

Bárcena, F., de Souza, J. A., Fernández de la Cigoña, E. & Domínguez, J. 1987. Las colonias de aves marinas de la costa occidental de Galicia. Características, censo y evolución de sus poblaciones. *Ecología* 1: 187–209.

Bárcena, F., Teixeira, A. M. & Bermejo, A. 1984. Breeding seabird populations in the Atlantic sector of the Iberian península. *In:* Croxall, J. P., Evans, P. G. H. & Schreiber, R. W. (Eds) *Status and conservation of the world's seabirds*, pp. 335–345. ICBP Technical Publication No. 2. ICBP. Cambridge.

Barrientos, R., Kvist, L., Barbosa, A., Valera, F., López-Iborra, G. & Moreno, E. 2009. Colonization patterns and genetic structure of peripheral populations of the trumpeter finch (*Bucanetes githagineus*) from Northwest Africa, the Canary Islands and the Iberian Peninsula. *J. Biogeography* 36: 210–219.

Barriocanal, C. & Robson, D. 2007. Phenology and body mass dynamics of the Garden Warbler *Sylvia borin* on the island of Cabrera (Balearic Islands) and in the Aiguamolls de l'Empordà (NE Iberian Peninsula) during spring migration. *Revista Catalana d'Ornitologia* 23: 10–17.

Barrios, L. & Rodríguez, A. 2004. Behavioural and

environmental correlates of soaring-bird mortality at on-shore wind turbines. *J. Applied Ecology* 41: 72–81.

Bartolomé, J., Fernández-Cruz, M. & Campos, F. 1996. Band recoveries of Spanish Little Egrets *Egretta garzetta*. *Colonial Waterbirds* 19: 220–225.

Bartrolí, X. 1991. An intermediate population of Italian and Spanish Yellow Wagtails *Motacilla flava iberiae* and *M. f. cinereocapilla* in northeastern Spain. *Ardeola* 38: 199–205.

Batllori, X. & Nos, R. 1985. Presencia de la Cotorrita Gris (*Myiopsitta monachus*) y de la Cotorrita de Collar (*Psittacula krameri*) en el área metropolitana de Barcelona. *Miscelánea Zoológica* 9: 407–411.

Baucells, J. & Abella, J. C. 2007. Les colònies de Pardal Roquer *Petronia petronia* a la comarca d'Osona: estat de la població i requeriments ecològics. *Revista Catalana d'Ornitologia* 23: 1–9.

Bauer, K. M. & Glutz von Blotzheim, U. N. 1966. *Handbuch der Vögel Mitteleuropas*, vol. 1. Akademische Verlagsgesellschaft. Frankfurt am Main.

Bautista, L. M. 2009. Grulla común. *In:* Salvador, A. (Ed.) *Enciclopedia Virtual de los Vertebrados Españoles*. Museo Nacional de Ciencias Naturales. Madrid. http://www.vertebradosibericos.org/

Bea, A., Beitia, R. & Fernández, J. M. 2003. The census and distribution of wintering woodpigeons *Columba palumbus* in the Iberian peninsula. *Ornis Hungarica* 12–13: 157–167.

Beaubrun, P. C. 1985a. Recensement hivernal d'oiseaux marins au Maroc, janvier 1984. *Documents de l'Institut Scientifique de Rabat* 9:119.

Beaubrun, P. C. 1985b. Arrivée massive et mortalité de Mouettes tridactyles (*Rissa tridactyla*) en janvier 1984 le long des côtes du Maroc. *Documents de l'Institut Scientifique de Rabat* 9: 21–26.

Béchet, A. 2009. *European Union Management Plan 2009–2011, Golden Plover* Pluvialis apricaria. Office for Official Publications of the European Communities. Luxembourg.

Béchet, A., Arnaud, A., Germain, C. & Johnson, A. 2009. Greater Flamingos stop breeding in the Camargue (southern France) in 2007, for the first time in 38 years; the beginning of a new era? *In:* Béchet, A., Rendón-Martos, M., Amat, J. A., Baccetti, N. & Childress, B. (Eds) *Proceedings of the IVth International Workshop on the Greater Flamingo in the Mediterranean region and northwest Africa, Antequera, Spain, 5–6 November 2007*, pp: 26–29. Wildfowl & Wetlands Trust. Slimbridge.

Belda, E. J. & Sánchez, A. 2001. Seabird mortality on longline fisheries in the Western Mediterranean • Factors affecting bycatch and proposed mitigation measures. *Biological Conservation* 98: 357–363.

Belda, E. J., Kvist, L., Monrós, J. S., Ponnikas, S. & Torralvo, C. 2009. Uso de técnicas moleculares y análisis discriminantes para differenciar mediante biometría dos subespecies de Escribano Palustre *Emberiza schoeniclus*. *Ardeola* 56: 85–94.

Belinchón, C., Grande, C., Guil, N., Olmedo, S., Ruiz, M. P. & de Juana, E. 1999. Situación de la cotorra argentina (*Myiopsitta monachus*) en la ciudad de Madrid, 1998. *Anuario Ornitológico de Madrid 1998*: 64–71.

Benson, C. W., Brooke, R. K., Irwin, M. P. S. & Steyn, P. 1968. White-rumped Swifts in Spain. *Ibis* 110: 106.

Bensusan, K. J. 2008. The impact of a cold spell on populations of Sardinian Warblers *Sylvia melanocephala* and Zitting Cisticolas *Cisticola juncidis* at Windmill Hill Flats, Gibraltar. *Gibraltar Bird Report 2007*: 47–50.

Bensusan, K. J., Garcia, E. F. J. & Cortes, J. E. 2007. Trends in abundance of migrating raptors at Gibraltar in spring. *Ardea* 95: 83–90.

Bensusan, K. & Perez, C. 2007. Rüppell's Vulture, Moroccan White Wagtail and Icterine Warbler in Gibraltar in 2006. *Gibraltar Bird Report* 2006: 47–50.

Bensusan, K. & Perez, C. 2013. Migration phenology of three nocturnal bird species at Gibraltar. *Gibraltar Bird Report* 2012: 51–53.

Bergier, P. 2004. Quelques observations d'automne dans le Sahara Atlantique et la région de Bani. *Go-South Bulletin* 1: 1–6.

Bergier, P. & Thévenot, M. 1991. Statut et écologie du Hibou du Cap nord-africain *Asio capensis tingitanus*. *Alauda* 59: 206–224.

Berián, E. 2008. *Estudio sobre el autillo en Pamplona*. Ayuntamiento de Pamplona. Pamplona.

Bermejo, A. & Carrera, E. 1983. *Premières données sur les réprises de Goélands leucophées bagués dans la Peninsule Ibérique, aux Baleares et aux Chafarinas.* In Proceedings of III Reunion du groupe de travail d'oiseaux de mer. La Rochelle.

Bermejo, A. & de la Puente, J. 2002. Stopover characteristics of Sedge Warbler (*Acrocephalus schoenobaenus*) in central Iberia. *Die Vogelwarte* 41: 181–189.

Bermejo, A. & de la Puente, J. 2004. Wintering and migration of Bluethroat *Luscinia svecica* in central Spain. *Ardeola* 51: 285–296.

Bermejo, A., de la Puente, J. & Pinilla, J. 2002. Fenología, biometría y parámetros demográficos del Zarcero Común (*Hippolais polyglotta*) en España Central. *Ardeola* 49: 75–86.

Bermejo, A., Carrera, E., de Juana, E. & Teixeira, A. M. 1986. Primer censo general de gaviotas y charranes (Laridae) invernantes en la península Ibérica (enero de 1984). *Ardeola* 33: 47–68.

Bernaldo de Quirós, J. L. 1956. El *Dryocopus martius* (L.) en el centro de España. *Ardeola* 3: 13–17.

Bernis, F. 1944. Distribución geográfica y presentaciones del *Bubulcus ibis ibis* (L.) en la Europa occidental y en especial en España. *Bol. Real Sociedad Española de*

Historia Natural 42: 451–458.

Bernis, F. 1945a. Aves de Ledesma. Con algunas consideraciones sobre avifauna ibérica. *Bol. Real Sociedad Española de Historia Natural* 43: 93–145.

Bernis, F. 1945b. Datos sobre el *Phylloscopus collybita brehmii*. *Bol. Real Sociedad Española de Historia Natural* 43: 339–348.

Bernis, F. 1946. Un mes de julio en Maragatería y montes de León. *Boletín de la Real Sociedad Española de Historia Natural* 44: 169–188 & 305–335.

Bernis, F. 1952. Sobre nidificación y presencia estival de aves en España Central (año 1951). *Bol. Real Sociedad Española de Historia Natural* 50: 5–16.

Bernis, F. 1954a. Prontuario de la avifauna española. *Ardeola* 1: 11–85.

Bernis, F. 1954b. Nuevos datos de *Hirundo daurica rufula* en el interior de España. *Ardeola* 1: 117–118.

Bernis, F. 1954c. Nota sobre los reyezuelos (*Regulus*) en España. *Ardeola* 1: 118–119.

Bernis, F. 1956a. Algunas capturas de Anatidae en Valencia y Delta del Ebro. *Ardeola* 3: 19–29.

Bernis, F. 1956b. Sobre la garceta (*Egretta*) de color negruzco aparecida en Doñana. *Ardeola* 3: 93–114.

Bernis, F. 1957. Acerca del Camachuelo Común, *Pyrrhula p.* en España. *Ardeola* 4: 243–251.

Bernis, F. 1959. La migración de las cigüeñas españolas y de las otras cigüeñas "occidentales". *Ardeola* 5: 9–80.

Bernis, F. 1960a. About wintering and migration of the Common Crane (*Grus grus*) in Spain. *In: Proceedings XII International Congress Helsinki 1958*, pp. 110–117.

Bernis, F. 1960b. Nuevas localidades con *Hirundo daurica* en Castilla y León. *Ardeola* 6: 388–390.

Bernis, F. 1960c. Migración, problema agrícola y captura del estornino pinto (*Sturnus vulgaris*). *Ardeola* 6: 11–109.

Bernis, F. 1961a. Cuatro notas sobre garzas españolas. *Ardeola* 7: 204–217.

Bernis, F. 1961b. Dos capturas de *Elanus caeruleus* en el sur de España. *Ardeola* 7: 255–256.

Bernis, F. 1961c. Sobre el comportamiento invernal del Avión roquero en España. *Ardeola* 7: 264–265.

Bernis, F. 1963a. Para el 'status' de *Tadorna tadorna* en España. *Ardeola* 8: 272–273.

Bernis, F. 1963b. Más sobre el *Phylloscopus collybita ibericus*. *Ardeola* 8: 286–287.

Bernis, F. 1963c Sobre migración de nuestros passeriformes transaharianos. *Ardeola* 8: 41–119.

Bernis, F. 1964a. La invernada y migración de nuestros ánsares (*Anser anser* y *Anser fabalis*). *Ardeola* 9: 67–109.

Bernis, F. 1964b. *Información española sobre anátidas y fochas (época invernal)*. Sociedad Española de Ornitología. Madrid.

Bernis, F. (AMI) 1966–67, 1971, 1972. *Aves Migradoras Ibéricas.*. Sociedad Española de Ornitología. Madrid.

Bernis, F. 1966a. El Buitre Negro (*Aegypius monachus*) en Iberia. *Ardeola* 12: 45–99.

Bernis, F. 1966b. *Migración en aves. Tratado teórico y práctico.* Sociedad Española de Ornitología. Madrid.

Bernis, F. 1969a. Sobre la presencia de *Phalacrocorax carbo* en aguas interiores. *Ardeola* 13: 236–237.

Bernis, F. 1969b. Sobre recientes datos de *Plegadis falcinellus* en Iberia. *Ardeola* 13: 242–243.

Bernis, F. 1971. Reconsideración del ejemplar melánico de *Egretta* spp. obtenido en Doñana en 1956. *Ardeola* 15: 107–110.

Bernis, F. 1972. La avifauna ibérica. Presente, pasado y futuro de las aves españolas. *In:* Fitter, R. & Fernández-Cruz, M. *El libro de las aves de España*, pp. 6–16. Selecciones del Readers' Digest. Madrid.

Bernis, F. 1973. Migración de Falconiformes y *Ciconia* spp. por Gibraltar, verano-otoño 1972–1973. Primera parte. *Ardeola* 19: 151–224.

Bernis, F. 1974. De la fuerte declinación sufrida por el *Falco naumanni* en Madrid y su provincia. *Ardeola* 20: 351–354.

Bernis, F. 1980. *La migración de las aves en el Estrecho de Gibraltar. Vol. 1. Aves planeadoras.* Universidad Complutense. Madrid.

Bernis, F. 1981. *La población de cigüeñas españolas. Estudios y tablas de censos, periodo 1948–1974.* Publicaciones de la Cátedra de Zoología de Vertebrados. Universidad Complutense de Madrid. Madrid.

Bernis, F. 1988a. *Los vencejos. Su biología, su presencia en las mesetas españolas como aves urbanas*. Universidad Complutense. Madrid.

Bernis, F. 1988b. Aves del medio urbano en las mesetas castellanas. *In:* Bernis, F. (Ed.) *Aves de los medios urbano y agrícola en las mesetas españolas*, pp. 2–172. Sociedad Española de Ornitología. Madrid.

Bernis, F. 1989a. *Los estorninos en las mesetas españolas.* Comunicaciones I.N.I.A. No. 54. Instituto Nacional de Investigaciones Agrarias. Madrid.

Bernis, F. 1989b. *Los gorriones. Con especial referencia a su distribución y eto-ecología en las mesetas españolas.* Comunicaciones I.N.I.A. No. 53. Instituto Nacional de Investigaciones Agrarias. Madrid.

Bernis, F. 1995. *Diccionario de nombres vernáculos de aves.* Gredos. Madrid.

Bernis, F., Asensio, B. & Benzal, J. 1985. Sobre la expansión y ecología de la Tórtola Turca (*Streptopelia decaocto*), con nuevos datos del interior de España. *Ardeola* 32: 279–294.

Bernis, F. & Castroviejo, 1968. Aves de las Islas Columbretes en primavera. *Ardeola* 12: 143–163.

Bernis, F., Díez, P. M. & Márquez, J. 1958. Someras notas sobre aves de Extremadura. *Ardeola* 4: 99–108.

Bernis, F., Díez, P. M. & Tato, J. 1958. Guión de la avifauna balear. *Ardeola* 4: 25–97.

Bernis, F. & Fernández-Cruz, M. 1966. Actividad del Centro

de Migración de Aves de la Sociedad Española de Ornitología, bienio 1963–1964. *Ardeola* 11: 5–20.

Bernis, F. & García Rua, A. 1974. Observación de *Buteo rufinus* en la provincia de Cádiz. *Ardeola* 20: 341–343.

Bernis, F. & Iribarren, J. J. 1968. Observación de Pico Dorsiblanco *Dendrocopos leucotos* en el Pirineo navarro. *Ardeola* 12: 239–240.

Bernis, F. & Maluquer, J. M. 1955. Datos sobre *Tichodroma muraria* (L.) en la Península Ibérica. *Ardeola* 2: 1–11.

Bernis, F. & Valverde, J. A. 1952. La gran colonia de garzas del Coto Doñana (año 1952). *Munibe* 4: 201–213.

Bernis, F. & Valverde, J. A. 1954a. Sobre la Garza Real (*Ardea cinerea*) en España. *Bol. Real Sociedad Española de Historia Natural* 50: 201–213.

Bernis, F. & Valverde, J. A. 1954b. Sur le Flamant rose dans la Péninsule Ibérique. *Alauda* 22: 32–39.

Bernis, F. & Valverde, J. A. 1972. El censo español de aves acuáticas del invierno 1967–68. *Ardeola* 17/18: 105–126.

Berthold, P. & Berthold, H. 1973. Zur Biologie von *Sylvia sarda balearica* und *S. melanocephala*. *J. für Ornithologie* 114: 79–95.

Bertolero, A., Genovart, M., Martínez-Abraín, A., Molina, B., Mouriño, J., Oro, D. & Tavecchia, G. (Eds) *Gaviotas cabecinegra, picofina, de Audouin y tridáctila, y gavión atlántico en España. Población en 2007 y método de censo.* SEO/BirdLife. Madrid.

Bertolero, A., Oro, D., Martínez-Vilalta, A. & López, M. À. 2005. Selection of foraging habitats by Little Terns *Sterna albifrons* at the Ebro Delta (NE Spain). *Revista Catalana d'Ornitologia* 21: 37–42.

Bibby, C. J. 1982. Studies of West Palearctic birds. 184. Moustached Warbler. *British Birds* 75: 346–359.

Bibby, C. J. & Green, R. E. 1980. Foraging behaviour of migrant Pied Flycatchers *Ficedula hypoleuca* on temporary territories. *J. Animal Ecology*. 49: 507–521.

Bibby, C. J. & Green, R. E. 1981. Autumn migration strategies of Reed and Sedge Warblers. *Ornis Scandinavica* 12: 1–12.

Bibby, C. J. & Thomas, D. K. 1984. Sexual dimorphism in size, moult and movements of Cetti's Warbler *Cettia cetti*. *Bird Study* 31: 28–34.

Biber, J. P. 1996. Action plan for the conservation of Lesser Kestrel. *In:* Heredia, B., Rose, L. & Painter, M. (Eds) *Globally threatened birds in Europe. Action Plans.* Council of Europe. Strasbourg.

Biber, O., Enggist, P., Marti, C. & Salathé, T. (Eds) 1995. *Proceedings of the International Symposium on the White Stork (Western Population) Basel 1994*. Sempach (Switzerland).

Bigas, D. & Gutiérrez, R. 2000. The Siberian Blue Robin in Spain – the second Western Palearctic record. *Birding World* 13: 414–417.

Bijleveld, M. 1974. *Birds of prey in Europe*. Macmillan. London.

Bijlsma, S., Hagemeijer, E. J. M., Verkley, G. J. M. & Zollinger, R. 1990. Reproducción del Cernícalo Primilla en Extremadura, 1987. *In:* González, J. L. & Merino, M. (Eds) *El Cernícalo Primilla (*Falco naumanni*) en la Península Ibérica*, pp. 91–99. ICONA. Madrid.

Bildstein, K. L., Bechard, M. J., Farmer, C. & Newcomb, L. 2009. Narrow sea crossings present major obstacles to migrating Griffon Vultures *Gyps fulvus*. *Ibis* 151: 382–391

Biondi, M., Corso, A. & Grussu, M. 2004. Nuovi dati sullo svernamento e la fenologia del Calandro maggiore *Anthus richardi* in Italia (1995–2004). *Gli Uccelli d'Italia* 29: 39–47.

BirdLife International. 2004. *Birds in Europe: population estimates, trends and conservation status*. Birdlife Conservation Series No. 12. Cambridge.

BirdLife International. 2012. *IUCN Red List for birds*. http://www.birdlife.org.

BirdLife International/European Bird Census Council. 2000. *European bird populations: estimates and trends*. Birdlife International Conservation Series No. 10. Cambridge.

Blanco, E. 2007. Descripción de un dormidero de Búho Chico *Asio otus* en Corella (Navarra). *Anuario Ornitológico de Navarra 2000–2001*: 71–73.

Blanco, G., Fargallo, J. A., Tella, J. L. & Cuevas, J. A. 1996. Role of buildings as nest-sites in the range expansion and conservation of choughs *Pyrrhocorax pyrrhocorax* in Spain. *Biological Conservation* 79: 117–122.

Blanco, G., Tella, J. L. & Torre, I. 1998. Traditional farming and key foraging habitats for chough *Pyrrhocorax pyrrhocorax* conservation in a Spanish pseudosteppe landscape. *J. Applied Ecology* 35: 232–239.

Blanco, G., Sergio, F., Sánchez-Zapata, J., Pérez-García, J., Botella, F., Martínez, F., Zuberogoitia, I., Frías, O., Roviralta, F., Martínez, J. & Hiraldo, F. 2012. Safety in numbers? Supplanting data quality with fanciful models in wildlife monitoring and conservation. *Biodiversity and Conservation* 21: 3269–3276.

Blanco, G., Friás, O., Cuevas, J. A., González, J. L. & Martínez, F. 2014. Commonness of the not-so-common birds: the need for baseline knowledge of actual population size for the validation of population size predictions. *Bird Study* 61: 351–360.

Blanco, J. C. & González, J. L. (Eds) 1992. *Libro Rojo de los Vertebrados de España*. ICONA. Madrid.

Blanco, J. M., González, P. P., Juárez, F. C., Höfle, U. & Suárez, A. 2011. El águila imperial cría en cautividad por primera vez: ¿cómo se hizo? *Quercus* 308: 16–24.

Blanco-Fontao, B., Quevedo, M., Obeso, J. 2011. Abandonment of traditional uses in mountain areas: typological thinking versus hard data in the Cantabrian Mountains (NW Spain). *Biodiversity and Conservation* 20: 1133–1140.

Blasco, M., Escudero, J. C. & Vargas, J. M. 1980. Presence of the Alpine Chough in Málaga (Spain). *Bird Study* 27: 56–57.

Blondel, J. & Aronson, J. 1999. *Biology and wildlife of the Mediterranean region*. Oxford University Press. Oxford.

Blondel, J. 1966. Sobre la migración primaveral en el Delta del Ebro. Ardeola 11: 79–84.

Blondel, J. & Blondel, C. 1964. Remarques sur l'hivernage des Limicoles et autres oiseaux aquatiques au Maroc (janvier 1964). *Alauda* 32: 250–279.

Blondel, J. & Isenmann, P. 1981. *Guide des oiseaux de Camargue*. Delachaux & Niestlé. Neuchâtel–Paris.

Blondel, J. & Viellard, J. 1966. Sobre migración y avifauna en Mallorca, primavera 1963. *Ardeola* 11: 85–94.

Bolt, S. & Appleton, T. 2001. Satellite tracking of Scottish and translocated Ospreys on migration. *Wings over Africa. Proceedings of the international seminar on bird migration: reserach, conservation, education and flight safety*, pp. 86–89. Israel, April 29 – May 11, 2001.

Bolton, M., Smith, A. L., Gómez-Díaz, E., Friesen, V. L., Medeiros, R., Bried, J., Roscales, J. L. & Furness, R. W. 2008. Monteiro's Storm Petrel *Oceanodroma monteiroi*: a new species from the Azores. *Ibis* 150: 717–727.

Bonaccorsi, G. 2002. Les Laridés (*Larus*, *Rhodostethia*, *Pagophila*, *Xema*, *Rissa*) remarquables en Méditerranée: une synthèse. *Alauda* 70: 327–339.

Bonaccorsi, G. 2003. Les Procellariiformes (*Diomedeidae*, *Procellariidae* et *Hydrobatidae*) non nicheurs en Méditerranée occidentale: une synthèse. *Alauda* 71: 1–7.

Bonal, R. & Aparicio, J. M. 2008. Evidence of prey depletion around lesser kestrel *Falco naumanni* colonies and its short term negative consequences. *J. Avian Biology* 39: 189–197.

Bonnin, J. 2007. La migració postnupcial al Parc Natural de sa Dragonera 1997–2006. *Anuari Ornitològic de les Balears 2006*: 15–42.

Borralho, R., Carvalho, S., Rego, F. & Pinto, P. V. 1999. Habitat correlates of Red-legged Partridge (*Alectoris rufa*) breeding density on Mediterranean farmland. *Revue de Ecologie (La Terre et La Vie)* 54: 59–69.

Borràs A. & Senar, J. C. 1991. Opportunistic breeding of the Citril Finch *Serinus citrinella*. *J. für Ornithologie* 132: 285–289.

Borràs, A., Cabrera, J., Colome, X. & Senar, J. C. 2013. Una revisión sobre las metapobaciones de Verderón Serrano: patrones y movimientos. *Revista de Anillamiento* 29–30: 8–24.

Borràs, A., Cabrera, J., Colome, X., Cabrera, T. & Senar, J. C. 2011. Patterns of connectivity in Citril Finches *Serinus citrinella*: sympatric wintering of allopatric breeding birds? *Bird Study* 58: 257–263.

Borràs, A., Senar, J. C., Alba-Sánchez, F., López-Sáez, J. A., Cabrera, J., Colomé, X. & Cabrera, T. 2010. Citril finches during the winter: patterns of distribution, the role of pines and implications for the conservation of the species. *Animal Biodiversity and Conservation* 33: 89–115.

Borrow, N. & Demey, R. 2001. *Birds of Western Africa*. Helm. London.

Boscá, A. 1916. *Fauna Valenciana*. Casa Editorial de Alberto Martín. Barcelona.

Bosch, J. & Meyburg, B.-U. 2012. The Lesser Spotted Eagle *Aquila pomarina* in Catalonia (Spain) – Breeding attempt and migration. *Vogelwelt* 133: 89–97.

Bosch, M., Oro, D. & Ruiz, X. 1994. Dependence of Yellow-legged Gull (*Larus cachinnans*) on food from human activity in two Western Mediterranean colonies. *Avocetta* 18: 135–139.

Bosch, R., Real, J., Tinto, A. & Zozaya, E. L. 2007. An adult male Bonelli's eagle (*Hieraaetus fasciatus*) eaten by a subadult golden eagle (*Aquila chrysaetos*). *J. Raptor Research* 41: 338.

Bourne, W. R. P. 1965. A trip to Gibraltar in September. *Seabird Report* 1: 34–37.

Bourne, W. R. P. 1967. Long-distance vagrancy in the petrels. *Ibis* 109: 141–167.

Bourne, W. R. P. 1984. Wrecked seabirds died in Iberian nets. *Marine Pollution Bull.* 15: 277.

Bourne, W. R. P. 1995. The movements of Bulwer's Petrel and the larger shearwaters in the Atlantic Ocean. *Sea Swallow* 44: 49–52.

Bourne, W. R. P., Mackrill, E. J., Paterson, A. M. & Yésou, P. 1988. The Yelkouan Shearwater *Puffinus (puffinus?) yelkouan*. *British Birds* 81: 306–319.

Bourne, W. R. P. & Norris, A. Y. 1966. Observaciones durante una travesía de ida y vuelta entre Gran Bretaña y Gibraltar, septiembre 1964. *Ardeola* 11: 57–63.

Boutin, J.M. 2001. Elements for a Turtle Dove (*Streptopelia turtur*) management plan. *Game and Wildlife Science* 18: 87–112.

Boxberger, L. von. 1921. Ornithologische Gelegenheitsbeobachtungen aus den östlichen Spanien. *Ornithologische Monatsschrift* 46: 65–77 & 81–91.

Boxberger, L. von. 1931. Winterbeobachtungen aus Malaga. *Ornithologische Monatsberichte* 39: 73–77.

Boxberger, L. von. 1934. Beiträge zur Brutvogelfauna der Provinz Malaga. *J. für Ornithologie* 82: 185-209.

Bradley, P. 1986. The breeding biology of Audouin's Gull at the Chafarinas Islands. *In:* MEDMARAVIS & Monbailliu, X. (Eds) *Mediterranean Marine Avifauna. Population Studies and Conservation*, NATO ASI Series, Series G: Ecological Sciences, 12: 221–230. Springer-Verlag. Berlin.

Brehm, A. E. 1857. Vorläufige Zusammenstellung der Vögel Spaniens. *Allgemeine Deutsche Naturhistorische Zeitung* III: 431–489.

Bretagnolle, V. 1995. Systematics of the Soft-plumaged Petrel *Pterodroma mollis* (Procellaridae): new insight from the study of vocalizations. *Ibis* 137: 207–218.

Bretagnolle, V., Carruthers, M. P., Cubitt, M. G., Bioret, F. & Cuillandre, J. P. 1991. Six captures of a dark-rumped, fork-tailed storm-petrel in the northeastern Atlantic. *Ibis* 133: 351–356.

Brichetti, P., Arcamone, E. & C. O. I. 1997. Comitato di Omologazione Italiano (C. O. I.). 11. *Rivista Italiana di Ornitologia* 66: 171–174.

Brichetti, P., Arcamone, E., Occhiato, D. & C. O. I. 2002. Commissione ornitologica Italiana (COI) già Comitato di Omologazione Italiano, report n. 15. *Avocetta* 26: 117–121.

Brichetti, P. & Massa, B. 1998. Chek-list degli ucelli italiani aggiornata a tutto il 1997. *Rivista Italiana di Ornitologia* 68: 129–152.

Bridges, H. C. 1938. Nesting of the Red-rumped Swallow (*Hirundo daurica rufula*) in Andalucia. *Oologists' Record* 18: 63–64.

British Ornithologists' Union. 2006. The British List: A Checklist of Birds of Britain (7th edition). *Ibis* 148: 526–563.

Brooke, M. de L. 2004. *Albatrosses and petrels across the world.* Oxford University Press. Oxford.

Brosse, J. & Jacquemard-Brosse, S. 1962. Notes sur l'avifaune de Gibraltar au moment des migrations d'automne. *L'Oiseau et R.F.O.* 32: 228–239.

Brosset, A. 1959. Les oiseaux de l'embouchure de la Moulouya (Maroc oriental): les migrateurs. *Alauda* 27: 36–60.

Brosset, A. & Olier, A. 1966. Les îsles Chaffarines, lieu de reproduction d'une importante colonie de Goelands d'Audouin *Larus audouinii*. *Alauda* 34: 187–190.

Brotons, L., Herrando, S. & Pons, P. 2008. Wildfires and the expansion of threatened farmland birds: the ortolan bunting *Emberiza hortulana* in Mediterranean landscapes. *J. Applied Ecology* 45: 1059–1066.

Brown, A. & Grice, P. 2005. *Birds in England*. T. & A. D. Poyser. London.

Brown, L., Urban, E. K. & Newman, K. B. 1982. *The birds of Africa*. Vol. I. Academic Press. London.

Brú, F. 1913. *Notas de Caza*. Imprenta de José Guix. Valencia.

Bruhn, J. F. W. & Jeffrey, B. 1958. Notas sobre aves de Murcia y Cádiz. Observaciones de fin de verano. *Ardeola* 4: 109–117.

Buckley, P. A. & Wurster, C. F. 1970. White-faced storm petrels *Pelagodroma marina* in the North Atlantic. *Bull. British Ornithologists' Club* 90: 35–38.

Bueno, A. 1995. Las aves limícolas de Aragón. *Lucas Mallada* 7: 77–104.

Bueno, J. M. 1990. Migración e invernada de pequeños turdinos en la Península Ibérica. I. Pechiazul (*Luscinia svecica*) y Ruiseñor Común (*Luscinia megarhynchos*). *Ardeola* 37: 67–73.

Bueno, J. M. 1991. Migración e invernada de pequeños turdinos en la Península Ibérica. II. Collalba gris (*Oenanthe oenanthe*), tarabilla norteña (*Saxicola rubetra*) y tarabilla común (*Saxicola torquata*). *Ardeola* 38: 117–129.

Bueno, J. M. 1992a. Migración e invernada de pequeños turdinos en la Península Ibérica. III. Colirrojo Real (*Phoenicurus phoenicurus*). *Ardeola* 39: 3–7.

Bueno, J. M. 1992b. Migración e invernada de pequeños turdinos en la Península Ibérica. IV. Colirrojo Tizón (*Phoenicurus ochruros*). *Ardeola* 39: 49–54.

Bueno, J. M. 1998. Migración e invernada de pequeños turdinos en la Península Ibérica. V. Petirrojo (*Erithacus rubecula*). *Ardeola* 45: 193–200.

Busack, S. D. & Hedges, S. B. 1984. Is the peninsular effect a red herring? *American Naturalist* 123: 266–275.

Bustamante, J. & Negro, J. J. 1994. The postfledging dependence period of the lesser kestrel (*Falco naumanni*) in Southwestern Spain. *Journal of Raptor Research* 28: 158–163.

Cabot, J. & Serrano, P. 1984. Variation in weight and gonad size in dunlins (*Calidris alpina*) in the Guadiana estuary, south-west Spain. *Doñana, Acta Vertebrata* 11: 15–24.

Cabral, M. J., Almeida, J., Almeida, P. R., Dellinger, T., Almeida, N. F., Oliveira, M. E., Palmeirim, J. M., Queiroz, A. I., Rogado, L. & Santos-Reis, M. (Eds) 2005. *Livro Vermelho dos Vertebrados de Portugal.* Instituto da Conservação da Natureza, Lisboa.

Cadahía, L., López-López, P., Urios, V., Soutullo, Á. & Negro, J. J. 2009. Natal dispersal and recruitment of two Bonelli's Eagles *Aquila fasciata*: a four-year satellite tracking study. *Acta Ornithologica*. 44: 193–198.

Cadahía, L., Urios, V. & Negro, J. J. 2005. Survival and movements of satellite tracked Bonelli's eagle *Hieraetus fasciatus* during their first winter. *Ibis* 147: 415–419.

Calado, M. 1996. Little Tern (*Sterna albifrons*) status and conservation at Ria Formosa Natural Park, Algarve, Portugal. *Colonial Waterbirds* 19 (Special Publication 1): 78–80.

Calado, M. 1998. *Sterna albifrons* no Parque Natural de Ria Formosa: alguns aspectos da ecología alimentar. *Airo* 9: 53.

Calderón, J. 1976. Segunda cita en España de *Porphyrula alleni* (Thomson, 1842). *Doñana, Acta Vertebrata* 3: 99–101.

Calderón, J., Castroviejo, J., García, L. & Ferrer, M. 1988. El Águila imperial (*Aquila adalberti*) en Doñana: dispersión de los jóvenes, estructura de edades y mortalidad. *Doñana, Acta Vertebrata* 15: 79–98.

Callaghan, D. A. (compiler) 1999. Ferruginous Duck (*Aythya nyroca*). *In*: Schäffer, N & Gallo-Orsi, U. (Eds) *European Union action plans for eight priority bird species.* Office for Official Publications of the European Communities. Luxembourg.

Calleja, D. & Gorospe, G. 2000. Aves nuevas: Focha Americana *Fulica americana*. *La Garcilla* 109: 22–23.

Calvet, J., Estrada, J., Mañosa, S., Moncasí, F. & Solans, J. (Eds) 2004. *Els ocells de la Plana de Lleida*. Pagès Editors. Lleida.

Calvo, B. 1993. Nota breve sobre la población nidificante de

Canastera *Glareola pratincola* en la provincia de Sevilla. *Ecología* 7: 455–456.

Calvo, B. & Furness, R. W. 1995. Colony and nest-site selection by Collared Pratincole *Glareola pratincola* in southwest Spain. *Colonial Waterbirds* 18: 1–10.

Cama, A., Abellana, R., Christel, I., Ferrer, X. & Vieites, D. R. 2012. Living on predictability: modelling the density distribution of efficient foraging seabirds. *Ecography* 35: 1–10.

Cama, A. Josa, P., Ferrer-Obiol, J. & Arcos, J. M. 2011. Mediterranean Gulls *Larus melanocephalus* wintering along the Mediterranean Iberian coast: numbers and activity rhythms in the species' main winter quarters. *J. Ornithology* 152: 897–907.

Cambelo, A. J. 2007. Espectacular paso de buitres en Ceuta. Mayo–Junio de 2006. *Revista Alcudón* 4: 13–18.

Cambi, D. & Cambi, L. 1986. Migrazione e fenologia del Luì forestiero, *Phylloscopus inornatus*, in Italia e nell'Europa meridionale. *Rivista Italiana di Ornitologia* 56: 79–94.

Camiña, A. 2004a. Consequences of the Bovine Spongiform Encephalopathy (BSE) on breeding success and food availability of Spanish vulture populations. *In:* Meyburg, B. U. & Chancellor, R. D. (Eds), pp. 27–44.

Camiña, A. 2004b. Griffon Vulture *Gyps fulvus* monitoring in Spain: current research and conservation projects. *In:* Meyburg, B. U. & Chancellor, R. D. (Eds), pp. 45–66.

Campora, M. & Cattaneo, G. 2005. Ageing and sexing Short-toed Eagles. *British Birds* 98: 370–376.

Campos, F. 1981. Las Ardeidas en la cuenca del Duero. *Doñana, Acta Vertebrata* 8: 177–194.

Campos, F. & Fernández-Cruz, M. 2006. La garza real en España. I. Población reproductora (1950–2000). *Publicaciones de Biología de la Universidad de Navarra* 30: 1–118.

Campos, F., Fernández-Cruz, M. & Prosper, J. 2001. Movements of Grey Herons *Ardea cinerea* to and within the Iberian Peninsula and the Balearic Islands. *Ardeola* 48: 209–215.

Campos, F. & Lekuona, J. M. 1994. La población invernante de Cormorán Grande (*Phalacrocorax carbo*) en el norte de España y suroeste de Francia. *Ardeola* 41: 13–18.

Campos, F., Santamaría, T., Corrales, L. & Santonia, F. J. 2011. Morphological and biometric features of male bluethroats *Luscinia svecica* in central Iberia. *Ardeola* 58: 267–276.

Camprodón, J., Campión, D., Martínez-Vidal, R., Onrubia, A., Robles, H., Romero, J. L. & Senosiain, A. 2007. Estatus, selección del hábitat y conservación de los pícidos ibéricos. *In:* Camprodón, J. & Plana, E. (Eds) *Conservación de la biodiversidad y gestión forestal. Su aplicación en la fauna vertebrada*, pp. 391–434. Universitat de Barcelona. Barcelona.

Cañadas, S., Castro, H., Manrique, J. & Miralles, J. M. 1982. Aportaciones al conocimiento de la ornitocenosis de la estepa litoral almeriense. *Paralelo 37, Revista de Estudios Geográficos*, año 1982: 93-104.

Cano, A. 1960. Sobre nidificación de *Erythropygia galactotes* en Pechina, Almería. *Ardeola* 6: 320–323.

Cano, A. & König, K. 1971. Der Wüstengimpel (*Rhodopechys githaginea*) Brutvogel in Europa. *J. für Ornithologie* 112: 461–462.

Cano, A. & Valverde, J. A. 1959. Notas sobre el quebrantahuesos *Gypaetus barbatus* en la sierra de Cazorla. *Ardeola* 5: 121–126.

Cano, C., Ayerza, P. & Fernández de la Hoz, J. 2008. *El veneno en España (1990–2005)*. WWF/ADENA. Madrid.

Cano Alonso, L. S. 2006. An approach to wintering of Black Stork *Ciconia nigra* in the Iberian Peninsula. *Biota* 7: 7–13.

Cano Alonso, L. S. 2012. Cigüeña negra. *In:* Salvador, A. & Morales, M. B. (Eds).

Cano Alonso, L. S. & Fernández, M. 2003. The black stork (*Ciconia nigra*) in Madrid region: status, population changes and reproduction. *Aves* 40: 38–44.

Cano Alonso, L. S., Franco, C., Doval, G., Torés, A., Carbonell, I. & Tellería, J. L. 2013. Conservation of Iberian Black Storks *Ciconia nigra* outside breeding areas: distribution, movements and mortality. *Bird Conservation International* (doi:10.1017/S0959270912000482).

Cano Alonso, L. S., Franco, C., Pacheco, C., Reis, S., Gonçalo, R. & Fernández-García, M. 2006. The breeding population of Black Stork *Ciconia nigra* in the Iberian Peninsula. *Biota* 7: 15–23.

Cano Alonso, L. S. & Tellería, J. L. 2013. Migration and winter distribution of Iberian and central European black storks *Ciconia nigra* moving to Africa across the Strait of Gibraltar: a comparative study. *J. Avian Biology* 44: 189–197.

Cantó, J. L. & Pastor, E. 2006. Fenología comparada entre poblaciones de mosquitero papialbo (*Phylloscopus bonelli*) y musical (*P. trochilus*) en el Parque Natural del Carrascal de la Font Roja (Alicante). *Revista de Anillamiento* 16–17: 67–70.

Cantos, F. J. 1992. *Migración e invernada de la familia* Sylviidae *(order* Passeriformes*, clase* Aves*) en la península Ibérica*. Ph. D. Thesis. Universidad Complutense. Madrid.

Cantos, F. J. 1995. Migración e invernada de la curruca capirotada (*Sylvia atricapilla*) en la Península Ibérica. *Ecología* 9: 425–433.

Cantos, F. J. 2009. *Gaviota reidora. In:* Molina (Ed.), pp. 17–32.

Cantos, F. J. & Asensio, B. 1989. La migración postnupcial de la Lavandera Boyera (*Motacilla flava*) a lo largo de las costas mediterráneas españolas. *Ardeola* 36: 139–147.

Cantos, F. J. & Asensio, B. 1990. Evolución de la invernada de Gaviota Reidora (*Larus ridibundus*) en Madrid. *Ardeola* 37: 305–308.

Cantos, F. J., Fernández, A. & Hernández, F. 1993. Evolución y algunos parámetros de la población invernante de

Gaviota reidora *Larus ridibundus* en Madrid (España Central). *Butll. Grup Catalá d'Anellament* 10: 25–31.

Cantos, F. J. & Serrano, M. 2009. Evolución de la invernada y fenología de las gaviotas reidora (*Larus ridibundus*) y sombría (*Larus fuscus*) en la Comunidad de Madrid. *Anuario Ornitológico de la Comunidad de Madrid 2007–2008*: 148–157.

Cantos, F. J. & Tellería, J. L. 1994. Stopover site fidelity of four migrant warblers in the Iberian Peninsula. *J. Avian Biology* 25: 131–134.

Canut, J., Martínez, R., Parellada, X. García, D. & Piqué, J. 2004. Gall fer, *Tetrao urogallus*. *In:* Estrada *et al.* (Eds), pp. 106–107.

Capó, J. & Pons, J. M. 2002. Aves nuevas: Águila Pomerana *Aquila pomarina*. *La Garcilla* 112: 24–25.

Carbajo, F. & Ferrero, J. J. 1985. The ecology and status of the Black-shouldered Kite in Extremadura, Western Spain. *In:* Newton, I. & Chancellor, R. D. (Eds) *Conservation studies on raptors*, pp. 137–141. ICBP. Cambridge.

Carbonell, M. & Muñoz-Cobo, J. 1980. Censo español de aves acuáticas. *Ardeola* 25: 3–46.

Carbonell, R. & Tellería, J. L. 1998. Selección y uso del habitat por cinco poblaciones ibéricas de Curruca Capirotada (*Sylvia atricapilla*). *Ardeola* 45: 1–10.

Carboneras, C. 1987. *Alas en la mar.* Ajuntament de Badalona. Badalona.

Carboneras, C. 1988. The auks in the western Mediterranean. *Ringing & Migration* 9: 18–26.

Carboneras, C. 2012. *Acuerdo sobre la Conservación de Albatros y Petreles. Evaluación de Especies: pardela balear* Puffinus mauretanicus. http://www.acap.aq

Carboneras, C., Aymí, R., Cama, A., Duponcheel, C., Ferrer, J., Flamant, R., García, S., Garzón, J., Gutiérrez, A., Olivé, M. & Poot, M. 2010. Mediterranean Gulls *Larus melanocephalus* wintering in Spain and Portugal: one population or several? *Airo* 20: 3–11.

Cardiel, I. E. 2006. *El milano real en España. II Censo Nacional (2004).* SEO/BirdLife. Madrid.

Carnero, I. & Peris, S. 1988. *Atlas ornitológico de la provincia de Salamanca.* Diputación de Salamanca. Salamanca.

Carrascal, L. M. 2009. Agateador común. *In:* Salvador, A. & Bautista, L. M. (Eds).

Carrascal, L. M. 2011. Data, preconceived notions and methods: the case of population sizes of common breeding birds in Spain. *Ardeola* 58: 371–385.

Carrascal, L. M. & Salvador, A. (Eds). Enciclopedia Virtual de los Vertebrados Españoles. Museo Nacional de Ciencias Naturales. Madrid. http://www.vertebradosibericos.org/

Carrascal, L. M., Bautista, L. M. & Lázaro, E. 1993. Geographical variation in the density of the White Stork (*Ciconia ciconia*) in Spain: influence of habitat structure and climate. *Biological Conservation* 65: 83–87.

Carrascal, L. M. & Palomino, D. 2008. *Las aves comunes reproductoras en España. Población en 2004–2006.* SEO/Birdlife. Madrid.

Carrascal, L. M. & Seoane, J. 2009. Linking density, productivity and trends of an endangered species: the Bonelli's eagle in Spain. *Acta Oecologica* 35: 341–348.

Carrascal, L. M. & Tellería, J. L. 1990. Impacto de las repoblaciones de *Pinus radiata* sobre la avifauna forestal del norte de España. *Ardeola* 37: 247–266.

Carrascal, L. M., Weykman, S., Palomino, D., Lobo, J. M. & Díaz, L. 2005. Corneja negra (*Corvus corone*). *In:* Salvador, A. & Bautista, L. M. (Eds).

Carrera, E. 1983 Hivernada de làrids (Aves: Laridae) al litoral del Principat i del País Valenciá (1980 i 1981). *Butll. Institució Catalana de Història Natural (Sec. Zool.)* 49: 143–150.

Carrera, E. 1987. *Gavines*. Cyan. Barcelona.

Carrera, E. & Vilagrasa, X. 1984. On the reproduction, biology, food and movements of the Yellow-legged Herring Gull (*Larus argentatus michahellis*) colony in the Medes islands. In Olivella, J. R. I. & Gilli, J. M. (Eds) Els sistemas naturals de les illes Medes. *Arxius Sec. Ciencies* 73: 291–306.

Carrera, E., Ferrer, X., Martínez-Vilalta, A. & Muntaner, J. 1981a. Invernada de láridos en el litoral Mediterráneo catalán y levantino. *Ardeola* 28: 35–50.

Carrera, E. & Martínez, A. 1985. Anellament d'ocells poc comuns à Catalunya (1975–1983). *Butll. Grup Català d'Anellament* 4: 2–7.

Carrera, E., Martínez-Vilalta, A., Motis, A. & Pardo, R. 1981b. Nota sobre la invernada de *Chlidonias hybrida* (Pallas) (Aves, Sternidae) en el Mediterráneo ibérico. *Miscelánea Zoológica* 7: 222–224.

Carrera, E., Monbailliu, X. & Torre, A. 1993. Ringing recoveries of Yellow-legged Gulls in northern Europe. *In:* Aguilar, J. S., Monbailliu X. & Paterson, A. M. (Eds) *Status and conservation of seabirds*, pp. 181–193. SEO/BirdLife. Madrid.

Carrera, E., Nebot, R. & Vilagrasa, X. 1981c. Comentaris sobre els desplaçaments erràtics de la població catalana de Gavià argentat (*Larus argentatus michahellis*). *Butll. Institució Catalana de Història Natural* 47 (Sec. Zool. 4): 143–153.

Carrera, E., Trias, J. Bermejo, A., de Juana, E. & Varela, J. 1987. Étude biométrique des populations ibériques et nord-africaine du Goéland leucophée *Larus cachinnans*. *L'Oiseau et R.F.O.* 57: 32–38.

Carrera Buergo, L. 2003. Datos preliminares sobre fenología y abundancia del Bisbita Alpino (*Anthus spinoletta*) en la campiña costera del concejo de Llanes (Asturies). *El Draque* 4: 16–18.

Carrete, M., Donázar, J .A. & Margalida, A. 2006. Density-dependent productivity depression in Pyrenean bearded vultures: implications for conservation. *Ecological Applications* 16: 1674–1682.

Carrete, M., Sánchez-Zapata, J., Calvo, J. F. & Lande, R. 2005.

Demography and habitat availability in territorial occupancy of two competing species. *Oikos* 108: 125–136.

Carrete, M., Sánchez-Zapata, J. A., Martínez, J. E., Sánchez, M. A. & Calvo, J. F. 2002. Factors influencing the decline of a Bonelli's eagle *Hieraaetus fasciatus* population in southeastern Spain: demography, habitat or competition? *Biological Conservation* 11: 975–985.

Carrillo, C., Barbosa, A., Valera, F., Barrientos, R. & Moreno, E. 2007. Northward expansion of a desert bird: effects of climate change? *Ibis* 149: 166–169.

Casado, M. A., Levassor, C. & Parra, F. 1983. Régime alimentaire estival du ganga cata *Pterocles alchata* (L.) dans le centre de l'Espagne. *Alauda* 51: 203–209.

Casaux, E. & Aparicio, R. J. 2002. Algunos datos sobre fenología, actividad diaria y biometría de la oropéndola (*Oriolus oriolus*) en época postnupcial en el centro de España. *Revista de Anillamiento* 9–10: 56–61.

Castan, R. & Olier, A. 1959. Le Tantale ibis, *Ibis ibis* (L.), dans le Sud tunisien et au Maroc oriental. *Alauda* 27: 148–150.

Castaño, J. P. 1997. Fenología de puesta y parámetros reproductivos en una población de Aguilucho Cenizo (*Circus pygargus*) en el Campo de Montiel. *Ardeola* 44: 51–60.

Castaño, J. P., Martínez, F., Pinilla, J. & Traverso, J. M. 1995. Distribución del área de cría del Aguilucho Pálido *Circus cyaneus* en España central. *Alytes* 7: 129–134.

Castany, J. & López, G. 2006. *El carricerín real en España. I Censo Nacional (2005)*. SEO/BirdLife. Madrid.

Castellanos, J. J. 1976. La Agachadiza Común (*Gallinago gallinago*) criando en España Central (provincia de Ávila). *Ardeola* 22: 133–134.

Castellarnau, J. M. 1877. Estudio ornitológico del Real Sitio de San Ildefonso. *Anales de la Real Sociedad Española de Historia Natural* 4: 151–209.

Castro, H. 1993. Las Salinas de Cabo de Gata. *Ecología y dinámica anual de las poblaciones de aves en las salinas de Cabo de Gata (Almería)*. Instituto de Estudios Almerienses. Almería.

Castroviejo, J. 1971. Primeras capturas del Ansar índico *(Anser indicus)* y de la Barnacla cuellirroja *(Branta ruficollis)* en España. *Alauda* 39: 74–77.

Castroviejo, J. 1972. Première capture du Dendrocygne fauve *Dendocygna bicolor* dans la Péninsule Ibérique. *Alauda* 40: 287.

Castroviejo, J. 1975. *El urogallo en España*. Monografías de la Estación Biológica de Doñana. CSIC.

Castroviejo, J. 1993. *Memoria del mapa del Parque Nacional de Doñana*. Consejo Superior de Investigaciones Científicas y Agencia de Medio Ambiente de la Junta de Andalucía. Sevilla.

Castroviejo, J., Delibes, M., García Dory, M. A., Garzón, J. & Junco, E. 1974. Censo de urogallos cantábricos (*Tetrao urogallus cantabricus*). *Asturnatura* 2: 53–74.

Catchot, S. 1998. Primer registre de nidificació de mussol reial *Asio otus* a l'illa de Menorca. *Anuari Ornitològic de les Balears 1997*: 109–110.

Catley, G. P. 1983. Stilt Sandpiper on Mallorca in May 1983. *Dutch Birding* 5: 89–91.

Catley, G. P. & Myers, E. 1986. Aves nuevas para la avifauna española: Correlimos Zancolín *Micropalama himantopus*. *La Garcilla* 67: 32–33.

Catry, I., Alcazar, R., Franco, A. M. A. & Sutherland, W. J. 2009. Identifying the effectiveness and constraints of conservation interventions: A case study of the endangered lesser kestrel. *Biological Conservation* 142: 2782–2791.

Catry, I., Dias, M. P., Catry, T., Afanasyev, V., Fox, J., Franco, A. M. A. & Sutherland, W. J. 2010. Individual variation in migratory movements and winter behaviour of Iberian Lesser Kestrels *Falco naumanni* revealed by geolocators *Ibis* 153: 154–164.

Catry, P. 1999. Aves nidificantes possivelmente extintas em Portugal Continental. Revisão e síntese da informação disponível. *Airo* 10: 1–13.

Catry, P., Bearhop, S. & Lecoq, M. 2007. Sex differences in settlement behavior and condition of chiffchaffs *Phylloscopus collybita* at a wintering site in Portugal. Are females doing better? *J. Ornithology* 148: 241–249.

Catry, P., Costa, H., Elias, G. & Matias, R. 2010a. *Aves de Portugal. Ornitologia do território continental*. Assírio & Alvim. Lisboa.

Catry, P., Lecoq, M., Araújo, A., Conway, G., Felgueiras, M., King, J. M. B., Rumsey, S., Salima, H. & Tenreiro, P. 2005. Differential migration of Chiffchaffs *Phylloscopus collybita* and *P. ibericus* in Europe and Africa. *J. Avian Biology* 36: 184–190.

Catry, P., Lecoq, M., Granadeiro, J. P. & Ramírez, I. 2010b. Projecto Arenaria. *Pardela* 39: 9–11.

Catry, P. & Pacheco, C. 2008. Alterações na distribuição da avifauna portuguesa. Bosquejo de algunas das grandes tendencias do século XIX ao XXI. *In:* Equipa Atlas (Eds), pp. 67–83.

Catry, T., Ramos, J. A., Catry, I., Allen-Revez, M. & Grade, N. 2004. Are salinas a suitable alternative breeding habitat for little terns *Sterna albifrons*? *Ibis* 146: 247–257.

Catry, P., Tomé, R. & Cardoso, A. C. 1997. Biologia da reproducção e estatuto da gaivina-dos-pauis *Chlidonias hybridus* no Paul do Boquilobo. *Airo* 8: 7–15.

Catusse, M., Mothe, T. & Menoni, E. 1992. La Gélinotte des bois *Bonasa bonasia* existe dans les Pyrénées. *Alauda* 60: 129–133.

Cavallin, P., Ventroux, J., Chevallier, D. & Baillon, F. 2007. Suivi par balise Argos: de la Vienne au Mali. *La Plume du Circaète* 6: 4–5.

Cayuela, E. & Mayol, J. 2002. Segona nidificació de l'àguila d'albufera, *Circus pygargus*, a l'illa de Mallorca. *Anuari Ornitològic de les Balears 2001*: 65–68.

Ceballos, A, & Ramos, J. J. 2007. Reseña sobre el paso de espátula común *(Platalea leucorodia)* en el azud de Riolobos (Salamanca) durante el año 2006. *Anuario Ornitológico de la provincia de Salamanca 2004–2006*: 30–32.

Ceballos, O. & Donázar, J. A. 1988. Selección del lugar de nidificación por el Alimoche *(Neophron percnopterus)* en Navarra. *Munibe* 40: 3–8.

Celis, P., Gil, D. & Graves, J. A. 2007. Isolation and characterization of polymorphic microsatellites isolated from the spotless starling (*Sturnus unicolor*) and cross-species amplification in the European starling (*Sturnus vulgaris*). *Molecular Ecology Notes* 7: 251–253.

Ceña, A., Ceña, J. C. & Moya, I. 1996. *Fauna de la Rioja: vertebrados.* Fundación de la Caja de Ahorros de la Rioja. Logroño.

Chacón, G., Fernández, J. & Martínez, F. 1985a. Descubrimiento de una población nidificante de Guión de Codornices (*Crex crex*) en España. *Alytes* 3: 180–182.

Chacón, G., Rusiñol, A. & Bonilla, F. 1985. Observación de *Corvus corone cornix* en Ripoll (Girona). *Alytes* 3: 186.

Chancellor, R. D., Meyburg, B.-U. & Ferrero, J. J. (Eds) *Holarctic birds of prey. Proceedings of an international conference.* ADENEX – WWGBP. Badajoz.

Chapman, A. 1884. Rough notes on Spanish ornithology. *Ibis* 26: 66–99.

Chapman, A. 1888. Winter notes in Spain. *Ibis* 30:444–461.

Chapman, A. & Buck, W. J. 1893. *Wild Spain.* Gurney & Jackson. London.

Chapman, A. & Buck, W. J. 1910. *Unexplored Spain.* Edward Arnold. London.

Chapman, K. A. 1969. White-rumped Swifts in Morocco. *British Birds* 62: 337–339.

Chevalier, D., Le Maho, Y., Brossault, P., Baillon, F. & Massemin, S. 2011. The use of stopover sites by Black Storks (*Ciconia nigra*) migrating between West Europe and West Africa as revealed by satellite telemetry. *J. Ornithology* 152: 1–13.

Cheylan, G. 1981. Sur le role determinant de l'abondance de ressources dans le succés de reproduction de l'Aigle de Bonelli, *Hieraaetus fasciatus*, en Provence. *In: Rapaces Meditérranéens*, pp. 95–99. Parc natural Régional de Corse & Centre Ornithologique de Provence.

Chiclana, F., Lama, J. A. & Salcedo, J. 2002. *Aves de la provincia de Sevilla.* Diputación Provincial de Sevilla & SEO–Sevilla. Sevilla.

Childress, B., Hughes, B., Harper, D. & van den Bossche, W. 2007. East African flyway and key site network of the Lesser Flamingo (*Phoenicopterus minor*) documented through satellite tracking. *Ostrich* 78: 463–468.

Chozas, P. 1985. Mortalidad de la población ibérica de Cigüeña Blanca (*Ciconia ciconia*). *Ardeola* 32: 119–123.

Chris, T. 1997. Estudis de boscarles a s'Albufera. *Butll. Parc Natural s'Albufera de Mallorca* 3: 45–46.

Clavell, J. 2002. *Catàleg dels ocells dels Països Catalans.* Lynx. Barcelona.

Clavell, J., López, F., Clarabuch, O., Burgas, A. Cama, A., Ferrer, J., Gil, M., Ollé, À. & Rodríguez, M. 2010. *Llista Patró dels ocells de Catalunya.* Edició 3.0, novembre de 2010. Comitè Avifaunístic de Catalunya, ICO. Barcelona.

Clements, J. F. 2007. *The Clements Checklist of the Birds of the World.* 6th edition. Christopher Helm. London.

Clergeau, P. & Yésou, P. 2006. Behavioural flexibility and numerous potential sources of introduction for the Sacred Ibis: causes of concern in Western Europe? *Biological Invasions* 8: 1381–1388.

Clergeau, P., Yésou, P. & Chadenas, C. 2005. Ibis sacré *Threskiornis aethiopicus*: état actuel et impacts potentiels des populations introduites en France métropolitaine. Rennes. www.rennes.inra.fr/scribe/document/ibis_v2.pdf.

Club Alcyon. 1961. La colonia de Ardeidas de "El Taray" (Toledo). *Ardeola* 7: 218–219.

Collar, N. J. 1978. Nesting of Black-shouldered Kites in Portugal. *British Birds* 71: 398–412.

Congreve, W. M. 1943. Two Andalucian breeding records. *Ibis* 85: 517–518.

Contel, F., Pradas, C. & Ruiz, C. L. 2010. Nuevos datos de nidificación de vencejo pálido (*Apus pallidus*) en la ciudad de Zaragoza. *Anuario Ornitológico de Aragón 205–2007*: 145–148.

Cooper, J. H. 1998. *Cyanopica* – a mystery solved. *Bull. British Ornithologists' Club* 118: 135.

Cooper, J. H. 2000. First fossil record of Azure-winged Magpie *Cyanopica cyanus* in Europe. *Ibis* 142: 150–151.

Cooper, J. H. & Voous, K. H. 1999. Iberian Azure-winged Magpies come in from the cold. *British Birds* 92: 659–665.

Copete, J. L., Mariné, R., Bigas, D. & Martínez-Vilalta, A. 1999. Differences in wing shape between sedentary and migratory Reed Buntings *Emberiza schoeniclus*. *Bird Study* 46: 100–103.

Corbacho, C., Sánchez Guzmán, J. M. & Villegas, M. A. 2009. *Pagazas, charranes y fumareles en España. Población reproductora en 2007 y método de censo.* SEO/BirdLife. Madrid.

Cordero-Tapia, P. J. 1983. *Las aves del Maresme. Catálogo, status y fenología.* Edicions Universitat de Barcelona. Barcelona.

Cordero-Tapia, P. & López de Vilar, P. 1985. Fenología de limícolas en un pequeño río litoral mediterráneo -el Tordera- (NE de España). *Ardeola* 32: 131–136.

Corley Smith, G. T. 1960. La Curruca Sarda en Jávea (Alicante). *Ardeola* 6: 390–391.

Corley Smith, G. T. 1958. Notas ornitológicas sobre cuatro especies. *Ardeola* 4: 199–201.

Corley Smith, G. T. 1959. Más sobre el Pechiazul en España. *Ardeola* 5: 127–131.

Corley Smith, G. T. & Bernis, B. 1956. Sobre el Pechiazul

(*Luscinia svecica*) y especialmente su población ibérica. *Ardeola* 3: 115–125.

Cornwallis, R. K. 1961. Four invasions of Waxwings during 1956–60. *British Birds* 54: 1–30.

Cornwallis, R. K. & Townsend, A. D. 1968. Waxwings in Britain and Europe during 1965/66. *British Birds* 61: 97–118.

Corso, A. 2001. Raptor migration across the Strait of Messina, southern Italy. *British Birds* 94: 196–202.

Corso, A. & Cardelli, C. 2004. The migration of Pallid Harrier across the central Mediterranean with particular reference to the Strait of Messina. *British Birds* 97: 238–246.

Cortés, J. A., Cobos, V. & Vidoy, I. 2002. El plumaje de los pechiazules (*Luscinia svecica*) invernantes en la provincia de Málaga. *Revista de Anillamiento* 9–10: 41–48.

Cortes, J. E., Finlayson, J. C., Garcia, E. F. J. & Mosquera, M. A. 1980. *The birds of Gibraltar*. Gibraltar Bookshop. Gibraltar.

Cortes, J. E., Shaw, E., Blair, M. & Candelin, G. 2005. The control of the Yellow-legged Gull *Larus cachinnans* in Gibraltar. *Almoraima* 31: 199–216.

Cortés-Avizanda, A., Ceballos, O. & Donázar, J. A. 2009. Long-term trends in population size and breeding success in the Egyptian Vulture (*Neophron percnopterus*) in Northern Spain. *J. Raptor Research* 43: 43–49.

Costa, H. 1994. On the overwintering of booted eagles *Hieraaetus pennatus* in Portugal. *Airo* 5: 24–27.

Costa, H., Araújo, A., Farinha, J. C., Poças, M. & Machado, M. 2000. *Nomes portugueses das aves do Paleárctico Ocidental*. Assirio & Alvim. Lisboa.

Costa, H., Lobo Elias, G. & Farinha, J. C. 1997. Exotic birds in Portugal. *British Birds* 90: 562–568.

Costa, L. 1984. Alimentación de la Pagaza Piconegra (*Gelochelidon nilotica*) en las marismas del Guadalquivir. *Doñana, Acta Vertebrata* 11: 185–195.

Costa, L. T., Nunes, M., Gerlades, P. & Costa, H. (Eds). 2003. Zonas importantes para as aves em Portugal. SPEA. Lisboa.

Costa, S. 2002. Nidificació del tudó *Columba palumbus* a l'illa de Formentera. *Anuari Ornitològic de les Balears 2001*: 57–58.

Costas, R. 1990. Aves nuevas para España: Chorlitejo Culirrojo *Charadrius vociferus*. *La Garcilla* 77: 16–17.

Costillo, E., Corbacho, C., Morán, R. & Villegas, A. 2007. The diet of the Black Vulture *Aegypius monachus* in response to environmental changes in Extremadura (1970–2000). *Ardeola* 54: 197–204.

Coverley, H. W. 1932. Breeding of snipe in Portugal. *Ibis* 74: 684–685.

Coverley, H. W. 1939. Nesting notes from Portugal, 1937–38. *Ibis* 81: 149–152.

Cramp, S. *et al.* (Eds). 1977–1994. *The birds of the Western Palearctic*. Oxford University Press.

Crawford, R. J. M., Shelton, P. A., Cooper, J. & Brooke, R. K. 1983. Distribution, population size and conservation of the Cape Gannet *Morus capensis*. *South African J. Marine Science* 1: 153–174.

Crivelli, A. J., Catsadorakis, G. & Naziridis, T. 1997. *Pelecanus onocrotalus* White Pelican. *BWP Update* Vol. 1 No. 3: 154–156.

Crochet P.-A. & Joynt G. 2012. *AERC list of Western Palearctic birds*. December 2012 version. http://www.aerc.eu/tac.html

Crozier, J. & Argelich, J. 1993. Presence du Pluvier guignard (*Eudromias morinellus*) en Principauté d'Andorre (Pyrénées) en période de nidification. *Alauda* 61: 214.

Cuadrado, M. 1988. Winter foraging behaviour of Blackcap and Sardinian Warblers in a Mediterranean scrubland. *Ardea* 76: 107–110.

Cuadrado, M., Arjona, S. & Rodríguez, M. 1985. Hivernage du Martinet pâle *Apus pallidus* dans le sud de l'Espagne. *Alauda* 53: 306–307.

Cuadrado, M., Senar, J. C. & Copete, J. L. 1995. Do all Blackcaps *Sylvia atricapilla* show winter site fidelity? *Ibis* 137: 70–75.

Cubitt, M., Carruthers, M., & Zino, F. 1992. Unravelling the mystery of the Tyne petrels. *Birding World* 5: 438–442.

Cuervo, J. J. 2003. Cigüeñuela común. *In:* Carrascal, L. M. & Salvador, A. (Eds).

Cuevas, J. A. & Blanco, G. 2009. Chova piquirroja. *In:* Salvador, A. & Bautista, L. M. (Eds).

Dalmau, J., Mariné, R., Martínez-Vidal, R., Canut, J. & García-Ferré, D. 2001. El Mussol de Tengmalm *Aegolius funereus* a la Cerdanya, el Pallars Sobirà i el Principat d'Andorra: noves localitats de cant i reproducció. *Anuari d'Ornitologia de Catalunya 1998*: 306–314.

Davies, N. B. 2000. *Cuckoos, cowbirds and other cheats*. T. & A. D. Poyser. London.

De Borbón, M. N., Barros, C. & de Juana, E. 1999. El gregarismo en las gangas ibérica y ortega. *In:* Herranz & Suárez (Eds), pp. 195–214.

De Frutos, A. & Olea, P. P. 2008. Importance of the premigratory areas for the conservation of lesser kestrel: space use and habitat selection during the post-fledging period. *Animal Conservation* 11: 224–233.

De Frutos, A., Olea, P. P., Mateo-Tomás, P. & Purroy, F. J. 2010. The role of fallow in habitat use by the lesser kestrel during the post-fledging period: inferring potential conservation implications from the abolition of obligatory set-aside. *European J. Wildlife Research* 56: 503–511.

De Juana, E. 1974. Datos invernales sobre aves de Marruecos (diciembre 1973). *Ardeola* 20: 267–286.

De Juana, E. 1980. *Atlas ornitológico de La Rioja*. Instituto de Estudios Riojanos. Logroño.

De Juana, E. 1984. The status and conservation of seabirds in the Spanish Mediterranean. *In:* Croxall, J. P., Evans, P. G. H. & Schreiber, R. W. (Eds) *Status and conservation of the world's seabirds*, pp. 347–361. ICBP. Cambridge.

De Juana, E. 1988. La Serena, una comarca extremeña de singular importancia. *La Garcilla* 71/72: 26–27.

De Juana, E. 1990. *Áreas importantes para las aves en España*. Sociedad Española de Ornitología. Madrid.

De Juana, E. 1991. El Chorlito Social (*Chettusia gregaria*) en España. *Ardeola* 38: 55–59.

De Juana, E. (Ed.) 1994. *Where to watch birds in Spain*. Lynx. Barcelona.

De Juana, E. 1999. Atención al Buitre Moteado. *La Garcilla* 106: 29–31.

De Juana, E. 2002. Aves. *In:* Reyero, J. M. (Ed.) *La naturaleza de España*, pp. 236–253. Organismo Autónomo Parques Nacionales. Madrid.

De Juana, E. 2003. Revisión del estatus de la Polluela Bastarda *Porzana parva* en España. *Ardeola* 50: 5–13.

De Juana, E. 2004. Cambios en el estado de la conservación de las aves en España, años 1954 a 2004. *Ardeola* 51: 19–50.

De Juana, E. 2006. *Aves raras de España*. Lynx Edicions. Barcelona.

De Juana, E. 2008. Where do Pallas's and yellow-browed warblers (*Phylloscopus proregulus, Ph. inornatus*) go after visiting northwest Europe in autumn? An Iberian perspective. *Ardeola* 55: 179–192.

De Juana, E. 2009. The dramatic decline of the Little Bustard *Tetrax tetrax* in Extremadura (Spain). *Ardeola* 56: 119–125.

De Juana, E. 2011. The Sociable Lapwing in Europe. *British Birds* 104: 84–90.

De Juana, E., Bradley, P. M., Varela, J. M. & Witt, H.-H. 1987. Sobre los movimientos de la Gaviota de Audouin (*Larus audouinii*). *Ardeola* 34: 15–24.

De Juana, E., Bueno, J. M., Carbonell, M., Mellado, V. P. & Varela, J. 1979. Aspectos de la alimentación y biología de reproducción de *Larus audouinii* Payr. en su gran colonia de cría de las Islas Chafarinas (año 1976). *Bol. Estación Central de Ecología* 8: 53–65.

De Juana, E. & de Juana, F. 1984. Noticia histórica sobre la reproducción de *Corvus frugilegus* en España Central. *La Garcilla* 63: 21–22.

De Juana, E., de Juana, F. & Calvo, S. 1988. La invernada de las aves de presa (O. Falconiformes) en la Península Ibérica. *In:* Tellería (Ed.), pp 97–122.

De Juana, E. & Ferrer, X. 1996. El estatus de la Garceta Grande *Egretta alba* en la Península Ibérica y las Islas Baleares. *Ardeola* 43: 225–229.

De Juana, E. & García, A. M. 2005. Fluctuaciones relacionadas con la precipitación en la riqueza y abundancia de aves de medios esteparios Mediterraneos. *Ardeola* 52: 53–66.

De Juana, E. & Garcia, E. 2010. Vagrancy or migration: why do American teals cross the Atlantic? *Ardeola* 57: 417–430.

De Juana, E. & Martínez, C. 1996. Distribution and conservation status of the Little Bustard *Tetrax tetrax* in the Iberian Peninsula. *Ardeola* 43: 157–167.

De Juana, E. & Martínez, C. 2001. *European Union Species Action Plan for the Little Bustard* (Tetrax tetrax). BirdLife International - European Commission. Brussels.

De Juana, E & Paterson, A. M. 1986. The status of the seabirds of the extreme western Mediterranean. *In:* MEDMARAVIS & Monbailliu, X. (Eds) *Mediterranean Marine Avifauna. Population Studies and Conservation*, pp. 39–106. NATO ASI Series, Series G: Ecological Sciences. Springer-Verlag. Berlin.

De Juana, E. & Varela, J. M. 1993. La población mundial reproductora de la Gaviota de Audouin. *In:* Aguilar, J. S., Monbailliu, X. & Paterson, A. M. (Eds) *Status and conservation of seabirds*, pp. 71–85. SEO/BirdLife. Madrid.

De Juana, F. 1989. Situación actual de las rapaces (Orden Falconiformes) en España. *Ecología* 3: 237–292.

De la Cruz, J. & Montoya, F. 2004. Censo coordinado de grullas invernantes, 24–25 de enero de 2004. Colectivo Ornitológico Cigüeña Negra (COCN). http://cocn.tarifainfo.com/documentos/iniciodocumentos.html

De la Cruz, C. & Valencia, J. 2004. Rabilargo. *In:* Carrascal, L. M. & Salvador, A. (Eds).

De La Lama, A. 1955. *Estudio ornitológico de la bahía de Santander*. Centro de Estudios Montañeses. Santander.

De La Lama, A. 1959a. Algunas capturas interesantes efectuadas en Santander. *Ardeola* 5: 133–141.

De La Lama, A. 1959b. Sobre la presencia y expansion de *Pica pica* (Linn.) y *Clamator glandarius* (Linn.) en la provincia de Santander. *Ardeola* 5: 201–204.

De La Lama, Arturo. 1976. Ornitología de la provincia de Santander. *Anales del Instituto de Estudios Industriales, Económicos y de Ciencias "Torres Quevedo"* 1: 331–441.

De la Puente, J., Moreno-Opo, R. & del Moral, J. C. 2007. *El buitre negro en España. Censo Nacional (2006)*. SEO/BirdLife. Madrid.

De la Puente, J., Seoane, J. & Bermejo, A. 1997. Nueva cita de invernada de carricero tordal, *Acrocephalus arundinaceus*, en Europa. *Apus* 10: 4–6.

De le Court, C. & Aguilera, E. 1997. Dispersal and migration in Eurasian Spoonbills (*Platalea leucorodia*). *Ardea* 85: 193–202.

De le Court, C. & Feria, E. 2009. *La Espátula en Andalucía. Bases para su conservación.* Manuales de conservación de la Naturaleza. No. 5. Consejería de Medio Ambiente de la Junta de Andalucía. Sevilla.

De le Court, C. & Rodríguez, R. 2005. An action Plan for Eurasian Spoonbill *Platalea leucorodia* in Spain. *Eurosite Spoonbill Network Newsletter* 3: 6–7.

De Lope, F. 1980. Biologie de la reproduction de l'Hirondelle rousselin *Hirundo daurica* en Espagne. *Alauda* 48: 99–112.

De Lope, F. 1981. La invasión de *Hirundo daurica rufula* Temm. en la península Ibérica. *Doñana, Acta Vertebrata* 8: 313–315.

De Lope, F. 1983. La reproduction d'*Hirundo rustica rustica* en Extrémadoure (Espagne). *Alauda* 51: 81–91.

De Lope, F. 1984. La invernada de la golondrina común ibérica (*Hirundo rustica rustica*) en África. *Actas de la II Reunión Iberoamericana de Conservación y Zoología de Vertebrados*, pp. 236–243. Cáceres.

De Lope, F., Fernández, A. & de la Cruz, C. 1984. Un nouvelle espèce à classer parmis les oiseaux de la Péninsule Ibérique: *Estrilda (Amandava) amandava* L. (Ploceidae, Passeriformes). *Alauda* 52: 312.

De Lope, F., Fernández, A. & de la Cruz, C. 1987. La situación del Avión Zapador (*Riparia riparia*) en Extremadura. *Alytes* 5: 121–123.

De Lope, F., Guerrero, J., García, M. E., de la Cruz, C., Carretero, J. J., Navarro, J. A., Silva, E. & Otano, J. 1983. Masiva afluencia de pinzones reales (*Fringilla montifringilla*) en la Baja Extremadura. *Alytes* 1: 393–400.

De Pablo, F. 2000. Estatus del alimoche *Neophron percnopterus* en Menorca, Islas Baleares. *Anuari Ornitològic de les Balears 2000*: 3–9.

De Pablo, F. & Capó, J. 2004. Primeros datos de cría del alcotán, *Falco subbuteo*, en Menorca. *Anuari Ornitològic de les Balears 2003*: 71-72.

De Pablo, F. & Pons, J. M. 2003. El milano real *Milvus milvus* en Menorca (Islas Baleares): situación y conservación. *Anuari Ornitològic de les Balears 2002*: 47–52.

De Seynes, A. et les coordinateurs-espèce. 2011. Les oiseaux nicheurs rares et menacés en France en 2010. *Ornithos* 18: 145–181.

De Souza, J. A. 1991. El escribano nival (*Plectrophenax nivalis*) en la Península Ibérica. *Ardeola* 38: 179–198.

De Souza, J. A., Barros, A., Sandoval, A. & Bao, R. 2010. A review of the status of Great Northern Diver *Gavia immer* in Galicia, northwest Spain. *Seabird* 23: 76–90.

De Souza, J. A. & Lorenzo, M. 2003. *Aves Acuáticas Invernantes en Galicia. Análisis de los censos de enero 1987–1999*. Dirección Xeral de Conservación da Natureza, Xunta de Galicia. Santiago de Compostela.

Dean, A., Bradshaw, C., Martin, J., Stoddart, A. & Walbridge, G. 2010. The status in Britain of 'Siberian Chiffchaff'. *British Birds* 103: 320–337.

Deán, J. I. 1996. Censos invernal y primaveral de cernícalo vulgar (*Falco tinnunculus*) en Navarra. *Ecología* 10: 457–469.

Deán, J. I. 2001. El Archibebe Común *Tringa totanus*: su presencia en Navarra. *Anuario Ornitológico de Navarra 1999*: 131–135.

Deetjen, H. 1955. Notizen über die Vogelwelt der Provinz Valencia (Spanien). *Ornithologische Mitteilungen* 7: 84–87.

Dejonghe, J.-F. 1980. Analyse de la migration prénuptiale des rapaces et des cigognes au Cap Bon (Tunisie). *L'Oiseau et R.F.O.* 50: 125–147.

Del Campo, J. C. & García-Gaona, J. F. 1983. Censo de Urogallos (*Tetrao urogallus*) en la Cordillera Cantábrica. *Naturalia Hispanica* 25. ICONA. Madrid.

Del Hoyo, J. *et al.* (Eds). 1992–2011. *Handbook of the Birds of the World.* 16 volumes. Lynx Edicions. Barcelona.

Del Junco, O. 1966. Nota sobre migración de Avefrías. *Ardeola* 11: 148–149.

Del Junco, O. 1968. *Tringa flavipes* en Andalucía. *Ardeola* 12: 237–238.

Del Junco, O. & González, B. 1966. Una nueva especie para la avifauna europea: *Apus affinis*. *Ardeola* 12: 5–9.

Del Junco, O. & González, B. 1969. La nueva especie de vencejo en el Paleártico: *Apus caffer*. *Ardeola* 13: 115–127.

Del Moral, J. C. (Ed.) 2006. *El águila perdicera en España. Población en 2005 y método de censo.*SEO/BirdLife. Madrid.

Del Moral, J. C. (Ed.) 2008. *El halcón de Eleonora en España. Población en 2004–2007 y método de censo.* SEO/BirdLife. Madrid.

Del Moral, J. C. (Ed.) 2009a. *El alimoche común en España. Población reproductora en 2008 y método de censo.* SEO/BirdLife. Madrid.

Del Moral, J. C. (Ed.) 2009b. *El buitre leonado en España. Población reproductora en 2008 y método de censo.* SEO/BirdLife. Madrid.

Del Moral, J. C. (Ed.) 2009c. *El águila real en España. Población reproductora en 2008 y método de censo.* SEO/BirdLife. Madrid.

Del Moral, J. C. 2001. Censo y seguimiento de la población de halcón peregrino (*Falco peregrinus*) en la Comunidad de Madrid. *Anuario Ornitológico de Madrid 2001*: 126–131.

Del Moral, J. C. & de la Puente, J. 2010. Buitre negro. *In:* Salvador, A. & Bautista, L. M. (Eds).

Del Moral, J. C. & de Souza, J. A. 2004. *Cormorán Grande Invernante en España. II Censo Nacional.* SEO/BirdLife. Madrid.

Del Moral, J. C. & Martí, R. (Eds) 2001. *El buitre leonado en la Península Ibérica. III Censo Nacional y I Censo Ibérico Coordinado, 1999.* SEO/BirdLife. Madrid.

Del Moral, J. C. & Martí, R. (Eds) 2002. *El alimoche común en España y Portugal (I censo coordinado). Año 2000.* SEO/BirdLife. Madrid.

Del Moral, J. C. & Molina, B. (Ed.) 2009. *El halcón peregrino en España. Población reproductora en 2008 y método de censo.* SEO/BirdLife. Madrid.

Del Moral, J. C., Molina, B., Bermejo, A. & Palomino, D. (Eds) 2012a. *Atlas de las aves en invierno en España 2007–2010.* Ministerio de Agricultura, Alimentación y Medio Ambiente & SEO/BirdLife, Madrid.

Del Moral, J. M., Molina, B. & Garrido, E. 2012b. Las garzas recuperan antiguos territorios. *Aves y naturaleza* 9: 30–33.

Delany, S., Scott, D., Dodman, T. & Stroud, D. (Eds) 2009. *An atlas of wader populations in Africa and western Eurasia.* Wetlands International. Wageningen.

Delgado, A. & Moreira, F. 2000. Bird assemblages of an Iberian cereal steppe. *Agriculture, Ecosystems and Environment* 78: 65–76.

Delgado, A. & Moreira, F. 2010. Between-year variations in Little Bustard *Tetrax tetrax* population densities are influenced by agricultural intensification and rainfall. *Ibis* 152: 633–642.

Delgado, M. P., Morales, M. B., Traba, J. & García de la Morena, E. L. 2009. Determining the effects of habitat management and climate on the population trends of a declining steppe bird. *Ibis* 151: 440–451.

Delibes, M. 1978. Ecología alimenticia del águila imperial ibérica (*Aquila adalberti*) en el Coto Doñana durante la crianza de los pollos. *Doñana, Acta Vertebrata* 5: 35–60.

Delibes, M., Calderón, J. & Hiraldo, F. 1975. Selección de presa y alimentación en España del Águila Real (*Aquila chrysaetos*). *Ardeola* 21: 285–302.

Delibes, M., Costa, L., Gisbert, J., Llamas, O. & Tirados, I. 1980. Sobre la expansion reciente del Pajaro Moscón (*Remiz pendulinus*) en la Península Ibérica. *Ardeola* 25: 193–206.

Dennis, J. V. 1981. A summary of banded North American birds encountered in Europe. *North American Bird Bander* 6: 88–96.

Dennis, J. V. 1987. Additional recoveries of banded North American birds in Europe. *North American Bird Bander* 12: 11–12.

Dennis, J. V. 1990. Banded North American birds encountered in Europe: an update. *North American Bird Bander* 15: 130–133.

Dennis, J. V. 1994. Transatlantic migration by ringed birds from North America. *Dutch Birding* 16: 235–237.

Dias, P. C. 1991. Les ardéidés nicheurs au Portugal. Distribution, biologie, conservation. *Alauda* 59: 23–26.

Díaz, J. 2006. *El Águila Calzada y su conservación en la Comunidad de Madrid.* Fondo para la Investigación y Conservación de los Animales Salvajes y su Hábitat. Madrid.

Díaz, L. 2010. *Camachuelo común. In* Salvador & Bautista (Eds).

Díaz, M., Asensio, B. & Tellería, J. L. 1996. *Aves Ibéricas I. No Paseriformes.* J. M. Reyero. Madrid.

Díaz, M., Martí, R., Gómez-Manzaneque, Á. & Sánchez, A. (Eds) 1994. *Atlas de las aves nidificantes en Madrid.* Sociedad Española de Ornitología y Agencia de Medio Ambiente de la Comunidad de Madrid. Madrid.

Díaz, M., Naveso, M. A. & Rebollo, E. 1993. Respuesta de las comunidades nidificantes de aves a la intensificación agrícola en cultivos cerealistas de la Meseta Norte (Valladolid-Palencia, España). *Aegypius* 11: 1–6.

Dierschke, V. 1997. Die Vorkommen des Sumpfläufers *Limicola falcinellus* in Deustchland. *Limicola* 11: 217–230.

Dies, B., Dies, J. I., Oltra, C., García i Gans, F. J. & Català, F. J. 1999. *Las aves de l'Albufera de Valencia.* Vaersa. Valencia.

Dies, J. I. 2000. Habitat management for gulls and terns at l'Albufera de València (West Mediterranean). Preliminary results. *In:* Yésou, P. & Sultana, J. (Eds) *Monitoring and conservation of birds, mammals and sea turtles of the Mediterranean and Black Sea*, pp. 250–261. Environment Protection Department. Malta.

Dies, J. I. & Dies, B. 1991. Presencia de *Stercorariidae* en aguas litorales del golfo de Valencia (Mediterráneo ibérico). *Medi Natural* 2: 85–96.

Dies, J. I. & Dies, B. 1998. Hybridisation between Lesser Crested and Sandwich Tern in Valencia, Spain, and plumage of offspring. *British Birds* 91: 165–170.

Dies, J. I., Marín, J. & Pérez, C. 2005. Diet of nesting Gull-billed Terns in Eastern Spain. *Waterbirds* 28: 106–109.

Dies, J. I., Prosper, J. & Dies, B. 2001. Occasional breeding by Western Reef Egret in eastern Spain. *British Birds* 94: 382–386.

Díez, J. F., González, P., Telletxea, I. & Pérez, J. A. 2010. Es posible conocer con exactitud los movimientos migratorios de la becada? *Locustella* 7: 107–114.

Díez Ponce de León, P. M. & Sáez-Royuela, R. 1955. Aves invernantes en los estanques de la Casa de Campo. *Ardeola* 2: 23–30.

Docampo, F. & Velando, A. 1995. Comentarios sobre el origen y cambios de la población nidificante de Gaviota Tridáctila (*Rissa tridactyla*) en la Península Ibérica. *In:* Munilla, I. & Mouriño, J. (Eds) *Actas do II Congreso Galego de Ornitoloxía*, pp. 139–151. Universidad de Santiago de Compostela. Santiago de Compostela.

Docavo, I. 1979. *La Albufera de Valencia: sus peces y sus aves.* Edicions Alfons el Magnànim, Diputació Provincial de València. Valencia.

Domènech, J., Carrillo, J. & Senar, J. C. 2003. Population size of the Monk Parakeet *Myiopsitta monachus* in Catalonia. *Revista Catalana d'Ornitologia* 20: 1–9.

Domínguez, F. & Cuadrado, M. 1994. Recurrencia y tasa de supervivencia en el Alzacola *Cercotrichas galactotes. Butll. Grup Catalá d'Anellament* 11: 45–49.

Domínguez, J., Bárcena, F., de Souza, J. A. & Villarino, A. 1987. Breeding waders in Galicia, north-west Spain. *Wader Study Group Bull.* 50: 28–29.

Domínguez, J. & Lorenzo, M. 1992. Waders wintering on the open shores of Galicia (NW Spain). *Wader Study Group Bull.* 64: 73–77.

Domínguez, J. & Rabuñal, J. L. 1989. Migrating waders on the Atlantic coast of Galicia (NW Spain). *Miscelánea Zoológica* 13: 141–151.

Domínguez, J. & Vidal, M. 2003. Influencia del investigador en el éxito reproductor del chorlitejo patinegro *Charadrius alexandrinus*. *Ardeola* 50: 15–19.

Domínguez, S. 1986. Descripción general de las características biológicas de la Laguna de Gallocanta y su cuenca. *In:* Da Cruz, H. (Ed.), *Guía de las zonas húmedas de la Península Ibérica y Baleares*, pp. 79–92.

Federación de Amigos de la Tierra. Madrid.

Donázar, J. A. 1990. Unidades reproductoras inusuales: trios poliándricos. *In:* Heredia & Heredia (Eds), pp. 39–46.

Donázar, J. A. 1992. Muladares y basureros en la biología y conservación de las aves en España. *Ardeola* 39: 29–40.

Donázar, J. A. 1993. *Los buitres ibéricos. Biología y conservación*. J. M. Reyero. Madrid.

Donázar, J. A., Blanco, G., Hiraldo, F., Soto-Largo, E. & Oria, J. 2002. Effects of forestry and other land-use practices on the conservation of Cinereous Vultures. *Ecological Applications* 12: 1445–1456.

Donázar, J. A., Ceballos, O. & Tella, J. L. 1994. Communal roosts of Egyptian Vulture (*Neophron percnopterus*): dynamics and implications for the species conservation. *In:* Muntaner, J. & Mayol, J. (Eds) *Biología y conservación de las rapaces mediterráneas*, pp. 189–201. SEO/BirdLife, Madrid.

Donázar, J. A., Negro, J. J. & Hiraldo, F. 1993. Foraging habitat selection, land-use changes and population decline in the lesser kestrel *Falco naumanni*. *J. Applied Ecology* 30: 515–522.

Dorka, V. 1960. Observaciones ornitológicas en Cataluña en agosto de 1958. *Ardeola* 6: 111–123.

Dubois, P. J. 2007. Les oiseaux allochtones en France: statut et interactions avec les espèces indigènes. *Ornithos* 14: 329–364.

Dubois, P. J. & CHN 1994. La Sterne élégante en France. *Ornithos* 1–2: 74–79.

Dubois, P. J. & Duhautois, L. 1977. Notes sur l'ornithologie Marocaine. *Alauda* 45: 285–291.

Dubois, P. J. & Duquet, M. 2009. Joris, Klaus et la mouette blanche. Les tempêtes de janvier 2009 en France. *Ornithos* 16: 81–89.

Dubois, P. J., Duquet, M., Fossé, A., Le Maréchal, P., Olioso, G. & Yésou, P. 2012. Notes d'ornithologie française. Première mise à jour du nouvel inventaires des oiseaux de France. *Ornithos* 19: 2–41.

Dubois, P. J., Le Maréchal, P., Oliso, G. & Yésou, P. 2008. *Nouvel inventaire des oiseaux de France*. Delachaux et Niestlé. Paris.

Dubois, P. J. & Yésou, P. 1995. Identification of Western Reef Egrets and dark Little Egrets. *British Birds* 88: 307–319.

Duchateau, S. 2007. Le statut de l'Aigle ibérique *Aquila adalberti* en France. *Alauda* 75: 33–42.

Duquet, M. 1997. Important afflux de grèbes esclavons *Podiceps auritus* en France en février 1996. *Ornithos* 4: 41–43.

Duquet, M. 2008. Premières mentions du Gobemouche à demi-collier *Ficedula semitorquata* en France. *Ornithos* 15: 64–67.

Durany, E., García, S. & Santaeufemia, D. 2003. Los cernícalos urbanos de Barcelona. *Quercus* 204: 24–27.

Dymond, J. N., Fraser, P. A. & Gantlett, S. J. M. 1989. *Rare birds in Britain and Ireland.* T. & A. D. Poyser. Calton.

Eagle Clarke, W. 1889. On the Ornithology of the Valleys of Andorra and the Upper Ariège, and other Contributions to the Avifauna of the Eastern Pyrenees. *Ibis* 31: 520–552

Eastham, A. 1968. The avifauna of Gorham's Cave, Gibraltar. *Bull. Institute of Archaeology* 7: 37–42.

Ebels, E. & van der Laan, J. 1994. Occurrence of Blue-cheeked Bee-eater in Europe. *Dutch Birding* 16: 95–101.

Editors Ardeola. 1976. Sobre movimientos invernales en Iberia de *Prunella collaris*. *Ardeola* 23: 231–235.

Elias, G. 2004. Aspects of Iberian chiffchaff *Phylloscopus ibericus* distribution in Spain and Portugal. *Ibis* 146: 685–686.

Elias, G. & Ferreira, L. F. 1994. Nidificação do chapim-de-faças-pretas *Remiz pendulinus* em Portugal. *Airo* 5: 27–29.

Elias, G. L., Reino, L. M., Silva, T., Tomé, R. & Geraldes, P. 1998. *Atlas das aves invernantes do Baixo Alentejo*. Sociedade Portuguesa para o Estudo das Aves. Lisboa.

Elkins, N. & Etheridge, B. 1974. The Crag Martin in winter quarters at Gibraltar. *British Birds* 67: 376–389.

Elkins, N. & Etheridge, B. 1977. Further studies of wintering Crag Martins. *Ringing & Migration* 1: 158–165.

Elorriaga, J. & Muñoz, A.-R. 2010. First breeding record of North African Long-legged Buzzard *Buteo rufinus cirtensis* in continental Europe. *British Birds* 103: 399–401.

Elósegui, J. 1985. *Navarra. Atlas de aves nidificantes*. Caja de Ahorros de Navarra. Pamplona.

Ena, V. & Purroy, F. J. 1982. *Censos invernales de aves acuáticas en España (enero 1978, 1979,1980)*. ICONA. Madrid.

Ena, V. & Purroy, F. J. 1984. Resultados del censo de aves acuáticas (enero 1983). *Ardeola* 30: 100–105.

Encarnação, V. 1992. Recenseamentos de limícolas no Parque Natural da Ría Formosa. *Airo* 3: 67–68.

England, M. D. 1963. Observations on the Black-winged Kite in Portugal with preliminary notes on its status. *British Birds* 56: 444–452.

Equipa Atlas. 2008. *Atlas das aves nidificantes em Portugal (1999–2005)*. Instituto da Conservação da Natureza e da Biodiversidade, Sociedade Portuguesa para o Estudo das Aves, Parque Natural da Madeira e Secretaria Regional do Ambiente e do Mar. Assirio & Alvim. Lisboa.

Érard, C. 1967. L'irruption de jaseurs *Bombycilla garrulus* (L.) en France en 1965–66. *Alauda* 35: 203–233, 270–287.

Érard, C. & Salomon, M. 1989. Essai de caractérisation morphologique du Pouillot véloce *Phylloscopus collybita brehmii* (Homeyer). *L'Oiseau et R.F.O* 59: 26–44.

Érard, C. & Vielliard, J. 1966. Comentarios sobre avifauna invernal en el oriente español. *Ardeola* 11: 95–100.

Escala, C., Alonso, D., Mazuelas, D., Mendiburu, A., Vilches, A. & Arizaga, J. 2009. Winter diet of Long-eared Owls *Asio otus* in the Ebro Valley (NE Iberia). *Revista Catalana d'Ornitologia* 25: 49–53.

Escandell, V. 2011. Programa Sacre primavera. *In:* Escandell,

V., Palomino, D., Molina, B., Leal, A., Remacha, C., Bermejo, A., de la Puente, J. & del Moral, J. C. (Eds) *Programas de seguimiento de SEO/BirdLife en 2009–2010*, pp. 4–13. SEO/Birdlife. Madrid.

Escandell, V. 2012. Programa Sacre. *In:* del Moral, J. C., Escandell, V., Bermejo, A., Molina, B., Palomino, D., de la Puente, J. & Leal, A. (Eds) *Programas de seguimiento de SEO/BirdLife en 2011*, pp. 4–9. SEO/Birdlife. Madrid

España Cantos, J. 1948. *Album con los datos de aves anilladas recuperadas en España hasta 31 de diciembre de 1948.* Federación Nacional de Caza. Madrid.

Esteban, L. & Campos, F. 1995. Reproducción del Mirlo Acuático (*Cinclus cinclus*) en Navarra. *Anuario Ornitológico de Navarra 1993–94*: 8–16.

Esteban, L. & Campos, F. 1998. Distribución del Mirlo Acuático (*Cinclus cinclus*) en Navarra. *Anuario Ornitológico de Navarra 1997*: 29–34.

Estrada, J., Pedrocchi, V., Broton, L. & Herrando, S. (Eds) 2004. *Atles dels Ocells Nidificants de Catalunya 1999–2002.* Institut Català d'Ornitologia & Lynx. Barcelona.

Estrella, S. M. & Masero, J. A. 2010. Prey and prey size selection by the Near-threatened Black-tailed Godwit foraging in non-tidal areas during migration. *Waterbirds* 33: 293–299.

Fábián, G. & Sterbetz, I. 1965. Black Little Egrets (*Egretta g. garzetta* L.) in Europe. *Aquila* 69–70: 99–112.

Fajardo, Í. 1990. Mortalidad de la lechuza común (*Tyto alba*) en España central. *Ardeola* 37: 101–106.

Fajardo, Í. 1995. Situación de las rapaces nocturnas ibéricas. *In:* Mikkola, H. *Rapaces nocturnas de Europa*, pp. 357–368. Perfils. Lleida.

Fajardo, Í. & Pividal, V. 1994. Alarmante declive de la lechuza común en la Comunidad de Madrid. *Quercus* 97: 34.

Fajardo, Í., Pividal, V., Trigo, M. & Jiménez, M. 1998. Habitat selection, activity peaks and strategies to avoid road mortality by the Little Owl *Athene noctua*. *Alauda* 66: 49–60.

Fargallo, J. A., Blanco, G., Potti, J. & Viñuela, J. 2001. Nestbox provisioning in a rural population of Eurasian Kestrels: breeding performance, nest predation and parasitism. *Bird Study* 48: 236–244.

Fargallo, J. A., Martínez-Padilla, J., Viñuela, J., Blanco, G., Torre, I., Vergara, P. & de Neve, L. 2009. Kestrel-prey dynamic in a Mediterranean region: the effect of generalist predation and climatic factors. *PLoS ONE* 4: e4311.

Farinha, J. C. & Costa, H. 1999. *Aves aquáticas de Portugal.* Instituto da Conservação da Natureza. Lisboa.

Farinha, J. C. & Encarnação, V. 1996. Estado actual das colónias de colhereiro *Platalea leucorodia* em Portugal. *In:* Farinha, J. C., Almeida, J. & Costa, H. (Eds) *Actas do I Congresso de Ornitologia*, pp. 70–71. SPEA. Lisboa.

Farinha, J. C. & Teixeira, A. 1989. The Chough in Portugal – status and distribution. *In:* Bignal, E. & Curtis, D. J. (Eds) *Choughs and Land-use in Europe*: 25–28. Scottish Chough Study Group. Argyll.

Farrobo, A. & Leitão, D. 1997. A perdiz-do-mar em Portugal. *Pardela* 6: 8–11.

Fasola, M., Canova, L. & Saino, N. 1996. Rice fields support a large portion of herons breeding in the Mediterranean Region. *Colonial Waterbirds* 19: 129–134.

Fasola, M., Rubolini, D., Merli, E., Boncompagni, E. & Bressan, U. 2010. Long-term trends of heron and egret populations in Italy, and the effects of climate, human-induced mortality, and habitat on population dynamics. *Population Ecology* 52: 59–72.

Federación Nacional de Caza. 1947. *Album con los datos de aves anilladas recuperadas en España hasta 31 de diciembre de 1947.* Federación Nacional de Caza. Madrid.

Fenech, N. & Sammut, M. 2013. Increased numbers of wintering Moussier's Redstarts in Malta. *British Birds* 106: 42–43.

Ferguson-Lees, I. J. 1967. Studies of less familiar birds 145. White-rumped or Little Swift. *British Birds* 60: 286–290.

Fernández, A., Román, J., de la Torre, J. A., Ansola, L. M., Santa María, J., Ventosa, R., Román, F. & Palma, C. 1998. Demografía y conservación de una población de Águila Perdicera *Hieraaetus fasciatus* en declive. *In:* Chancellor *et al.* (Eds), pp. 305–322.

Fernández, C. & Azkona, P. 1996. Influence of forest structure on the density and distribution of the White-backed Woodpecker *Dendrocopos leucotos* and the Black Woodpecker *Dryocopus martius* in Quinto Real (Spanish Western Pyrenees). *Bird Study* 43: 305–313.

Fernández, C., Azkona, P. & Donázar, J. A. 1996. Density-dependent effects on productivity in the Griffon Vulture *Gyps fulvus*: the role of interference and habitat heterogeneity. *Ibis* 140: 64–69.

Fernández, C., Azkona, P. & Lorente, L. 1994. Corología y caracterización del hábitat del Pico Dorsiblanco (*Dendrocopos leucotos lilfordi*) en el Pirineo occidental español. *Ardeola* 41: 135–140.

Fernández, C. & Insausti, J. A. 1990. Golden eagles take up territories abandoned by Bonelli's Eagles in Northern Spain. *J. Raptor Research* 24: 124–125.

Fernández, C. & Purroy, F. J. 1990. Tendencias geográficas en la alimentación del Águila Real (*Aquila chrysaetos* L.) en Navarra. *Ardeola* 37: 197–206.

Fernández, J. 2004. *50 años en defensa de las aves. Sociedad Española de Ornitología (SEO) 1954–2004.* SEO/BirdLife. Madrid.

Fernández Cordeiro, A. & Costas, R. 1995. Aspectos de la fenología y selección de hábitat de *Sterna sandvicensis* en las Rías Bajas (Galicia. NW Península Ibérica). *Chioglossa* vol. especial 1: 47–52.

Fernández Cordeiro, A. & Domínguez, J. 1991. Catálogo provisional de aves nidificantes amenazadas de Galicia.

Actas I Congreso Galego de Ornitoloxía: 9–17. Universidad de Santiago de Compostela. Santiago de Compostela.

Fernández de la Cigoña, E. 1983. Barnaclas carinegras en el estuario del Miño en época estival. *Cyanopica* III (1): 117–118.

Fernández de la Cigoña, E. 1986. Presença do cisne de Bewick (*Cygnus bewickii*) na Galiza e no norte de Portugal. *Cyanopica* III (4): 759–760.

Fernández Palacios, J. & Raya, C. 1991. Biología de la Focha cornuda (*Fulica cristata*) en Cádiz y otros humedales del Bajo Guadalquivir. *In:* Fernández Palacios, J. M. & Martos, M. J. (Eds) *Plan Rector de Uso y Gestión de las Reservas Naturales de las Lagunas de Cádiz*, pp. 97–117. Agencia de Medio Ambiente de la Junta de Andalucía. Sevilla.

Fernández-Alcázar, G. & Fernández-Cruz, M. 1991. Situación actual de las garzas coloniales en España. *Quercus* 60: 8–16.

Fernández-Calvo, I. C. & González-Sánchez, F. 2008. El charrán común (*Sterna hirundo*) en la bahía de Santander: evolución de la colonia y propuestas de conservación. *Actas del XIX Congreso Español de Ornitología*. SEO/BirdLife. Santander.

Fernández-Cruz, M. 1969. Abundancia de *Phalacrocorax carbo* en un embalse del interior. *Ardeola* 13: 235.

Fernández-Cruz, M. 1975. Revisión de las actuales colonias de Ardeidas de España. *Ardeola* vol. especial: 65–126.

Fernández-Cruz, M. (Coord.) 1981. La migración e invernada de la grulla común (*Grus grus*) en España. Resultados del Proyecto Grus. *Ardeola* 26–27: 11–64.

Fernández-Cruz, M. 2004. La migración otoñal de la Cigüeña Blanca (*Ciconia ciconia*) por el estrecho de Gibraltar. *In:* Molina, B. & del Moral, J. C. (Eds) 2005. *La Cigüeña Blanca en España. VI Censo Internacional (2004)*, pp. 162–201 SEO/Birdlife. Madrid.

Fernández-Cruz, M. & Farinha, J. C. 1992. Primer censo de Ardeidas invernantes en la Península Ibérica y Baleares (1991–92). *Airo* 3: 41–54.

Fernández-Cruz, M., Fernández-Alcázar, G. & Campos, F. 1993. Principales problemas de conservación de las Ardeidas coloniales españolas. *Alytes* 6: 453–460.

Fernández-Cruz, M., Fernández-Alcázar, G., Campos, E. & Días, P. C. 1992. Colonies of ardeids in Spain and Portugal. *In:* Finlayson, C. M., Hollis, G. E. & Davis, T. J. (Eds) *Managing Mediterranean wetlands and their birds*, pp. 76–78. IWRB. Slimbridge.

Fernández-Cruz, M., Martín-Novellas, C., Paris, M., Fernández-Alcázar, G., Sánchez, E. G., Nevado, J. C., Rendón, M. & Rubio, J. C. 1991. Dinámica de la población del Flamenco (*Phoenicopterus ruber roseus*) en España. *In:* Agencia de Medio Ambiente de la Junta de Andalucía (Ed.) *Reunión técnica sobre la situación y problemática del flamenco rosa* (Phoenicopterus ruber roseus) *en el Mediterráneo Occidental y Africa Noroccidental*, pp. 11–46. Junta de Andalucía. Sevilla.

Fernández-Cruz, M. & Sáez-Royuela, R. 1971. Comisión de Fenología. Encuesta sobre primeras llegadas y paso primaveral (Año 1970). *Ardeola* 15: 51–78.

Fernández-Delgado, C., Arenas, R., Persson, G. & Midtlyng, P. J. 2008. The eradication of introduced fish species for restoration of the waterfowl fauna of Laguna de Zoñar, Andalusia (Southern Spain). *In:* Gherardi, F. & Aquiloni, L. (Eds) *Managing alien species for sustainable development of aquaculture and fisheries*, p. 40. IMPASSE Project Final Conference. Book of Abstracts. University of Florence. Florence.

Fernández-Juricic, E., Jokimaki, J., McDonald, J. C., Melado, F., Toledano, A., Mayo, C., Martín, B., Fresneda, I. & Martin, V. 2004. Effects of opportunistic predation on anti-predator behavioural responses in a guild of ground foragers. *Oecologia* 140: 183–190.

Fernández-Ordóñez, J. C. & Pérez-Contreras, J. 2006. Situación del camachuelo trompetero (*Bucanetes githagineus*) en la provincia de Granada (SE España). *Revista de Anillamiento* 16–17: 71–72.

Ferns, P. N. & Hinsley, S. A. 1999. El comportamiento de las gangas (*Pterocles alchata* y *Pterocles orientalis*), con especial atención a los bebederos y al efecto de los predadores. *In:* Herranz & Suárez (Eds), pp. 177–194.

Ferreira, A. C. 1979. O *Acanthis flammea* (L.), ave acidental no nosso país. *Cyanopica* II (1): 94–101.

Ferreira, A. C. 1980. Recapturas em Portugal de aves anilhadas no estrangeiro (1972–1974). *Cyanopica* II (2): 57–94.

Ferreira, A. C. 1986. O *Acanthis flammea* (L.) em Portugal (novos elementos). *Cyanopica* III (4): 774–778.

Ferrer, M. & Calderón, J. 1990. The Spanish imperial eagle (*Aquila adalberti*) in Doñana National Park: a study on population dynamics. *Biological Conservation* 51: 151–161.

Ferrer, M., de la Riva, M. & Castroviejo, J. 1991. Electrocution of raptors on power lines in southwestern Spain. *J. Field Ornithology* 62: 181–190.

Ferrer, M. & Negro, J. J. 2004. The near extinction of two large European predators: super specialists pay a price. *Conservation Biology* 18: 344–349.

Ferrer, M. & Oliveros, F. 2005. El águila pescadora nidifica en el Parque Natural de Los Alcornocales. *Quercus* 234: 66–67.

Ferrer, X. 1977. Introducció ornitològica al delta de l'Ebre. *Treballs de la Institució Catalana d'Història Natural* 8: 227–302.

Ferrer, X. 1987. Presencia de Rabilargo (*Cyanopica cyana*) en el cuadrante NE de la Península Ibérica. *Ardeola* 34: 110–113.

Ferrer, X. & Martínez-Vilalta, A. 1986. Fluctuations of the gull and tern populations in the Ebro delta, north-east spain (1960–85). *In:* MEDMARAVIS & Monbailliu, X. (Eds) *Mediterranean Marine Avifauna. Population Studies and*

Conservation, pp. 273–284. NATO ASI Series, Series G: Ecological Sciences. Springer-Verlag. Berlin.

Ferrer, X., Martínez-Vilalta, A. & Muntaner, J. (Eds) 1986. *Història Natural dels Països Catalans. Vol. 12. Ocells*. Enciclopèdia Catalana. Barcelona.

Ferrer, X., Motis, A. & Peris, S. J. 1991. Changes in the breeding range of starlings in the Iberian Peninsula during the last 30 years: competition as a limiting factor. *J. Biogeography* 18: 631–636.

Ferrero, J. J. 1995. La población ibérica de aguilucho cenizo *Circus pygargus*. *Alytes* 7: 539–560.

Ferrero, J. J. & Onrubia, A. 1998. Expansión del área de cría y distribución actual del Elanio Azul *Elanus caeruleus* en España. *In:* Chancellor *et al.* (Eds), pp. 159–171.

Figuerola, J. & Bertolero, A. 1996a. Aves nuevas: Chorlitejo Mongol Grande *Charadrius leschenaultii*. *La Garcilla*, 96: 30–31.

Figuerola, J. & Bertolero, A. 1996b. Differential autumn migration of Curlew Sandpiper (*Calidris ferruginea*) through the Ebro delta, northeast Spain. *Ardeola* 43: 169–175.

Filippi-Codaccioni, O., Moussus, J-P., Urcun, J-P. & Jiguet F. (2011) Advanced autumn migration of the Common Crane *Grus grus* over Western Pyrenean passes. *Acta Ornithologica* 46: 37–45.

Finlayson, J. C. 1979. *The ecology and behaviour of closely related species at Gibraltar (with special reference to swifts and warblers)*. D. Phil. Thesis. University of Oxford. Oxford.

Finlayson, J. C. 1980. The recurrence in winter quarters at Gibraltar of some scrub passerines. *Ringing & Migration* 3: 32–34.

Finlayson, J. C. 1981. The morphology of Sardinian Warblers (*Sylvia melanocephala*) and Blackcaps (*Sylvia atricapilla*) resident on Gibraltar. *Bull. British Ornithologists' Club* 101: 299–304

Finlayson, J. C. 1992. *The birds of the Strait of Gibraltar*. T. & A. D. Poyser. London.

Finlayson, J. C. & Cortes, J. E. 1984. The migration of Gannets (*Sula bassana*) past Gibraltar in spring. *Seabird* 7: 19–22.

Finlayson, J. C. & Cortes, J. E. 1987. *The birds of the Strait of Gibraltar: its waters and the northern shore. Alectoris* 6. Special edition. GONHS. Gibraltar.

Fjeldsa, J. 1975. A downy chick of the Little Crake *Porzana parva* taken in Spain, with notes on its diagnostic features. *Bull. British Ornithologists' Club* 95: 81–82.

Flood, R. L. & Fisher, E. A. 2010. Wilson's Storm-petrels off the Isles of Scilly: a ten-year analysis, 2000–09. *British Birds* 103: 396–399.

Flood, R. L. 2009. 'All-dark' *Oceanodroma* storm-petrels in the Atlantic and neighbouring seas. *British Birds* 102: 365–385.

Fok, K. W., Wade, C. M. & Parkin, D. T. 2002. Inferring the phylogeny of disjunct populations of the Azure-winged Magpie *Cyanopica cyanus* from mitochondrial control region sequences. *Proceedings Royal Society of London* Series B 269: 1671–1679.

Fombellida, I., Herrera, A. G. & Rollán, M. 2009. Cantabria, bastión del pico mediano. *Quercus* 275: 14–20.

Font Tullot, I. 1983. *El clima de España y Portugal*. Instituto Nacional de Meteorología. Madrid.

Fontoura, A. P. 1985. Observação de aves das serras do Marão, Alvão e regiões limítrofes. *Cyanopica* III (3): 444–448.

Förschler, M. I., Borràs, A., Kalko, E. K. V., Cabrera, J., Cabrera, T. & Senar, J. C. 2006. Inter-locality variation in breeding phenology and nesting habitat of the citril finch *Carduelis citrinella* in the Catalonian Pre-Pyrenees. *Ardeola* 53: 115–126.

Forsman, D. 1999. *The raptors of Europe and the Middle East. A handbook of field identification*. T. & A. D. Poyser. London.

Forsman, D. 2005. Rüppell's Vultures in Spain. *Birding World* 18: 435–438.

Forsman, D. & Erterius, D. 2012. Pallid Harriers in northwest Europe and the identification of presumed Pallid Harrier x Hen Harrier hybrids. *Birding World* 25: 68–75.

Fortuna, M. A. 2002. Selección de hábitat de la Perdiz Roja *Alectoris rufa* en período reproductor en relación con las características del paisaje de un agrosistema de La Mancha (España). *Ardeola* 49: 59–66.

Fouarge, J. P. 1992. Observation de deux aigles imperiaux iberiques (*Aquila heliaca adalberti*) dans la region de Chechaouen. *Porphyrio* 4: 25–28.

Fouces, V. & Estrada, J. 1992. Evaluación de las posibilidades de censo de la población reproductora de avetorillo (*I. minutus*) del delta del Ebro. *Butll. Parc Natural Delta de l'Ebre* 2: 7–12.

Fournier, O. & Fournier, S. 1972. Anátidas, fochas y limícolos en las costas cantábricas y gallegas, enero 1969. *Ardeola* 17–18: 79–98.

Fox, T. A. D., Christensen, T. K., Bearhop, S. & Newton, J. 2007. Using stable isotope analysis of multiple feather tracts to identify moulting provenance of vagrant birds: a case study of Baikal Teal *Anas formosa* in Denmark. *Ibis* 149: 622–625.

Fraga, R. M. & Amat, J. A. 1996. Breeding biology of a Kentish Plover (*Charadrius alexandrinus*) population in an inland saline lake. *Ardeola* 43: 69–85.

Franco, A. & Andrada, J. 1977. Alimentación y selección de presa en *Falco naumanni*. *Ardeola* 23: 137–187.

Franco, A. & Rodríguez de los Santos, M. (Eds): *Libro rojo de los vertebrados amenazados de Andalucía*, pp. 115–116. Consejería de Medio Ambiente de la Junta de Andalucía. Sevilla.

Franco, A. M. A. & Sutherland, W. J. 2004. Modelling the foraging habitat selection of lesser kestrels: conservation implications of European Agricultural

Policies. *Biological Conservation* 120: 63–74.

Fray, R., Pennington, M., Riddington, R., Meek, E., Higson, P., Forsyth, A., Leitch, A., Scott, M., Marr, T., Rheinallt, T. ap & Olofson, S. 2012. An unprecedented influx of Iceland Gulls in the northeastern Atlantic in January/February 2012. *British Birds* 105: 263–272

Frederiksen, M. *et al.* 2011. Multicolony tracking reveals the winter distribution of a pelagic seabird on an ocean basin scale. *Diversity and Distributions* 18: 1–13.

Fry, C. H. 1984. *The bee-eaters*. T. & A. D. Poyser. Calton.

Fry, C. H., Fry, K. & Harris, A. 1992. *Kingfishers, bee-eaters and rollers*. Christopher Helm. London.

Fundación Migres. 2013. La enigmática migración de los alcaudones dorsirrojos: de León a Mozambique y vuelta a casa. *Quercus* 328: 58–59.

Furness, R. W. & Monteiro, L. R. 1995. Red-billed tropicbird *Phaethon aethereus* in the Azores: first breeding record for Europe. *Bull. British Ornithologists' Club* 115: 6–8.

Fuset Tubia, J. 1913. Aves de Cataluña. *Memorias de la Real Sociedad Española de Historia Natural* VII: 455–609.

Gacio, H. & Sayago, J. M. 2011. *Guía de las aves del Paraje Natural Marismas del Odiel (Huelva)*. Algakon. Sevilla.

Gainzarain, J. A. 2006. *Atlas de las Aves Invernantes en Álava (2002–2005)*. Diputación Foral de Álava. Vitoria-Gasteiz.

Gainzarain, J. A., Arambarri, R. & Rodríguez, A. F. 2002. Population size and factors affecting the density of the peregrine falcon (*Falco peregrinus*) in Spain. *Ardeola* 49: 67–74.

Galarza, A. 1984. Fenología de las aves acuáticas en el estuario de Gernika (Golfo de Vizcaya). *Ardeola* 31: 17–25.

Galarza, A. 1986. Migración de la Espátula (*Platalea leucorodia* Linn.) por la Península Ibérica. *Ardeola* 33: 195–201.

Galarza, A. 1993. Selección de hábitat en una población de Buitrón (*Cisticola juncidis*) tras su desaparición por efecto de una ola de frío. *Ardeola* 40: 169–171.

Galván I., Marchamalo J., Bakken V. & Traverso, J. M. 2003. The origin of Lesser Black-backed Gulls *Larus fuscus* in central Iberia. *Ringing & Migration* 21: 209–214.

Gámez, I. & Gutiérrez, C. 1995. Reproducción del Aguilucho Lagunero *Circus aeruginosus* en La Rioja. *Alytes* 7: 79–80.

Gantlett, S. & Millington, S. 2009. The Glossy Ibis invasion in autumn 2009. *Birding World* 22: 424–434.

Ganusevich, S. A., Maechtle, T. L., Seegar, W. S., Yates, M. A., Mcgrady, M. J., Fuller, M., Schueck, L., Dayton, J. & Henny, C. J. 2004. Autumn migration and wintering areas of Peregrine Falcons *Falco peregrinus* nesting on the Kola Peninsula, northern Russia. *Ibis* 146: 291–297.

García, A. M. & Cervera, F. 2001. Notas sobre la variación estacional y geográfica de la dieta del Búho Chico *Asio otus*. *Ardeola* 48: 75–80.

García, A. M., Cervera, F. & Rodríguez, A. 2011. Trophic and nesting ecology of the long-eared owl in Mediterranean environments. *In:* Zuberogoitia, Í. & Martínez, J. E. (Eds) *Ecología y Conservación de las Rapaces Forestales Europeas*, pp. 261–267. Diputación Foral de Bizkaia. Bilbao.

García, D., Quevedo, M., Obeso, J. R. & Abajo, A. 2005. Fragmentation patterns and protection of montane forest in the Cantabrian range (NW Spain). *Forest Ecology and Management* 208: 29–43.

Garcia, E. F. J. 1973. Seabird activity in the Strait of Gibraltar: a progress report. *Seabird Report* 1971: 30–36.

Garcia, E. F. J. 2004. Field identification of Northwest African Long-legged Buzzards *Buteo rufinus cirtensis*. *Gibraltar Bird Report 2002*: 40–46.

Garcia, E. F. J. 2006. The Eagle Owl *Bubo bubo* in Gibraltar. *Gibraltar Bird Report* 5: 46–52.

Garcia, E. F. J. & Bensusan, K. J. 2006. Northbound migrant raptors in June and July at the Strait of Gibraltar. *British Birds* 99: 569–575.

Garcia, E. F. J. & Paterson, A. 2008. *Where to watch birds in southern and western Spain*. 3rd edition. Christopher Helm. London.

García, F. J. & Calvo, J. F. 1987. El Zampullín Cuellinegro en la Laguna de La Mata (Alicante). *Ardeola* 34: 102–105.

García, J. 2000. Dispersión premigratoria del Cernícalo Primilla *Falco naumanni* en España. *Ardeola* 47: 197–202.

García, J. & Álvarez, E. 2000. Expansión altitudinal del Avión Zapador *Riparia riparia* en la Cordillera Cantábrica. *Ardeola* 47: 247–250.

García, J. T. & Arroyo, B. E. 1998. Migratory movements of Montagu's Harriers *Circus pygargus*: a review. *Bird Study* 45: 188–194.

García, J. T., Suárez, F., Garza, V., Hernández, J., Oñate, J. J., Hervás, I., Calero, M. & García de la Morena, E. 2008. Assessing the distribution, hábitat, and population size of the threatened Dupont's Lark *Chersophilus duponti* in Morocco: lessons for conservation. *Oryx* 42: 592–599.

García, J., Ramos, L. A. & Vázquez, X. 2008. *Atlas de las aves nidificantes de la provincia de León*. Edilesa. León.

García, L., Calderón, J. & Castroviejo, J. 1989. *Las aves de Doñana y su entorno*. Estación Biológica de Doñana/ Cooperativa Marismas del Rocío.

García Rodríguez, L. 1972. Primera nidificación de *Rhodopechys githaginea* en el sur-oeste de Europa. *Ardeola* 16: 215–222.

García, L., Viada, C., Moreno-Opo, R., Carboneras, C., Alcalde, A. & González, F. 2003. *Impacto de la marea negra del "Prestige" sobre las aves marinas*. SEO/BirdLife, Madrid.

García, L., Ibáñez, F., Garrido, H., Arroyo, J. L., Máñez, M. & Calderón, J. 2000. Prontuario de las aves de Doñana. *Anuario Ornitológico de Doñana Diciembre 2000*. Estación Biológica de Doñana y Ayuntamiento de Almonte. Almonte.

García, L., Máñez, M., Rodríguez, R., Garrido, H., Arroyo, J.

L., del Valle, J. L., Ibáñez, F., Martínez, A., Chico, A. & San Martín, I. 2009. *Espátula Común (Platalea leucorodia). In:* Máñez & Rendón-Martos (Eds), pp. 33–56.

García Sánchez, E. & García Cañal, J. A. 1992. Migración otoñal de aves marinas frente a la costa asturiana en 1991. *Bol. Ciencias Naturales R.I.D.E.A.* 42: 115–151.

García de la Morena, E. L., Bota, G., Ponjoan, A. & Morales, M. B. 2006. *El sisón común en España. I Censo Nacional (2005).* SEO/BirdLife. Madrid.

García Dory, M. A. 1983. Datos sobre la ecología del género *Pyrrhocorax* (*P. pyrrhocorax* y *P. graculus*) en el Parque Nacional de la Montaña de Covadonga. *Alytes* 1: 411–448.

García Peiró, I. & López Macià, M. 2002. Evolución de la abundancia del Bigotudo *Panurus biarmicus* en carrizales del Parque Natural de El Hondo (SE de España). *Revista Catalana d'Ornitologia* 19: 11–16.

García Peiró, I., Robledano, F. & Esteve, M. A. 2005. Fenología del Carricerín común *Acrocephalus schoenobaenus* (Linnaeus, 1758) y la Buscarla unicolor *Locustella luscinioides* (Savi, 1824) en el parque natural de El Hondo (Alicante, SE España). *Anales de Biología* 27: 15–22.

García Rua, A. E. 1975. Migrantes y migración visible en la zona del Estrecho de Gibraltar (años 1972–1974). *Ardeola* 21: 627–655.

García Sánchez, E. 1993. Puesta al día de las observaciones de la Gaviota de Delaware *Larus delawarensis* en Asturias (hasta diciembre 1992). *In:* Gorospe Rombouts, G. (Ed.) *Aves marinas ibéricas. Actas del IV Congreso del Grupo Ibérico de Aves Marinas*, pp. 55–63. Itsas Enara Ornitologi Elkartea. San Sebastián.

García-Barcelona, S. 2002. Observaciones de aves a bordo de pesqueros en el Mediterráneo occidental. *Bol. GIAM* 25: 12–32.

García-Barcelona, S. 2011. Gaviota cabecinegra. *In:* Salvador, A. & Morales, M. B. (Eds).

García-Barcelona, S., Macías, D., Ortiz de Urbina, J. M., Estrada, A., Real, R. & Báez, J. C. 2010. Modelling abundance and distribution of seabird by-catch in the Spanish Mediterranean longline fishery. *Ardeola* 57: 65–78.

García-Dios, I. S. 2009. Aguililla calzada. In: Salvador, A. & Bautista, L. M. (Eds).

García Dory, M. A. 1982. La Perdiz Nival debe ser declarada especie protegida. *Quercus* 4: 28–29.

García Fernández, J. 2000. La expansión de la Garcilla Bueyera *Bubulcus ibis* en el noroeste de la Península Ibérica. *Ecología* 14: 269–273.

García Fernández, J. 2012. *La graja en España. Población reproductora en 2011 y método de censo.* SEO/BirdLife. Madrid.

García Fernández, J., Álvarez, E. & Falagán, J. 2002. El pico mediano *Dendrocopos medius* en la provincia de León: cambios en la distribución y tamaño poblacional. *Ecología* 16: 335–342.

García Fernández, J., Ramos, L. A. & Vázquez, X. 2008. *Atlas de las aves reproductoras de León*. Diputación de León. León.

García-Navas, V. 2008. Gorrión molinero. In: Carrascal, L. M. & Salvador, A. (Eds).

García-Navas, V., Arroyo, L., Sanz, J. J. & Díaz, M. 2008. Effect of nestbox type on occupancy and breeding biology of Tree Sparrows *Passer montanus* in central Spain. *Ibis* 150: 356–364.

García-Petit, J., Martí, M. E. & Tomás, X. 1986. Colonització de la ciutat de Barcelona pel Gavià Argentat (*Larus cachinnans*). *Miscel.lània Zoològica* 10: 401–403.

García-Ripollés, C., López-López, P. & Urios, V. 2010. First description of migration and wintering of adult Egyptian Vultures *Neophron percnopterus* tracked by GPS satellite telemetry. *Bird Study* 57: 261–265.

García-Rovés, J. F. & Álvarez Laó, C. 2003. Mosquiteros (*Phylloscopus ibericus/ collybita*) con canto mixto en Asturies. *El Draque* 4: 20–22.

García-Rovés, J. F. & García-Rovés, J. G. 1989. *Las aves en el concejo de Cudillero*. Servicio de Publicaciones del Principado de Asturias. Oviedo.

García-Villanueva, J. A., Tárrega, R., Ena, V. & Serrano, M. 1998. Bird community from a fire degraded pinewood. *Proceedings 3rd International Conference on Forest Fire Research* 2: 1717–1725.

Garcias, P. 1992. Contribució al coneixement del statut de les bosquetes (Gènere *Hippolais*) a les Illes Balears. *Anuari Ornitològic de les Balears 1991*: 43–47.

Garcias, P. 1996. Situació del tiruril.lo camanegra (*Charadrius alexandrinus*) a Mallorca. *Anuari Ornitològic de les Balears 1995*: 25–33.

Garcias, P. 1997. Primera observació per a les Balears d'un Camagroga gros (*Tringa melanoleuca*). *Anuari Ornitològic de les Balears 1997*: 41–44.

Garcias, P. 2009a. Primera cita de cria de la subespecie *cinereocapilla* del xàtxero groc *Motacilla flava* a Mallorca. *Anuari Ornitològic de les Balears 2008*: 27–35.

Garcias, P. 2009b. Primer registre de nidificació d'ull de bou *Phylloscopus collybita* a les Balears. *Anuari Ornitològic de les Balears 2008*: 43–46.

Garcias, P. 2011. Revisió de l'estatus del formiguer *Jynx torquilla* a Mallorca. *Anuari Ornitològic de les Balears 2010:* 29–41.

Gargallo, G. 1994. Identificació de la subspècie balear de menjamosques *Muscicapa striata balearica*. *Anuari Ornitològic de les Balears 1993*: 21–30.

Garmendia, A., Cárcamo, S. & Schwendtner, Ó. 2006. Forest management considerations for conservation of Black Woodpecker *Dryocopus martius* and White-backed Woodpecker *Dendrocopos leucotos* populations in Quinto Real (Spanish Western Pyrenees). *Biodiversity and Conservation* 15: 1399–1415.

Garrido, H. 1996. *Aves de las Marismas del Odiel y su entorno.*

Rueda. Alcorcón.

Garrido, H. & Gutiérrez, C. 2010. El torillo andaluz y la conservación de papel. *Aves y Naturaleza* 2: 32–35.

Garrido, J. R. 2008. Censo de la población reproductora de cernícalo primilla *Falco naumanni* en la provincia de Sevilla. Años 2004 y 2005. *In: Anuario Ornitológico de la provincia de Sevilla, años 2000–2005*, pp. 17–22.

Garrido, J. R & del Moral, J. C. (Eds) 2012. *Las garzas en España, población reproductora e invernante en 2010–2011 y método de censo.* SEO/BirdLife. Madrid.

Garrido, M., Alba, E. & González, J. M. 1985. *Las aves de las zonas húmedas malagueñas*. Diputación Provincial de Málaga. Málaga.

Garrod, D. A. E., Buxton, L. H. D., Elliot Smith, G. & Bate, D. M. A. 1928. Excavation of a Mousterian rock-shelter at Devil's Tower, Gibraltar. *J. Royal Anthropological Institute* 58: 33–115.

Garza, V. & Suárez, F. 1988. *La alondra de dupont (*Chersophilus duponti*) en la Península Ibérica: población, distribución y problemática de conservación.* Report by SEO to the E.E.C. Unpublished.

Garza, V. & Suárez, F. 1990. Distribución, población y selección de hábitat de la Alondra de Dupont (*Chersophilus duponti*) en la Península Ibérica. *Ardeola* 37: 3–12.

Garza, V., Traba, J. & Suárez, F. 2003. Is the European population of Dupont's Lark *Chersophilus duponti* adequately estimated? *Bird Study* 50: 309–311.

Garzón, J. 1974. Contribucción al estudio del status, alimentación y protección de las Falconiformes en España central. *Ardeola* 19: 279–330.

Garzón, J. 1977. Birds of prey in Spain: their present situation. *In:* Chancellor, R. D. (Ed.) *Proceedings I World Conference on Birds of Prey 1975*, pp. 159–170. ICBP. Vienna.

Garzón, J. & Henares, I. 2012. *Las Aves de Sierra Nevada.* Consejería de Medio Ambiente de la Junta de Andalucía. Granada.

Genovart, M., Bertolero, A. & Martínez-Abraín, A. 2008. *La Gaviota de Audouin. In:* Bertolero *et al.* (Eds), pp. 44–68.

Genovart, M., Juste, J., Contreras-Díaz, H. & Oro, D. 2012. Genetic and phenotypic differentiation between the critically endangered Balearic Shearwater and neighboring colonies of its sibling species. *J. Heredity* 103: 330-334.

Genovart, M., Juste, J. & Oro, D. 2005. Two sibling species sympatrically breeding: a new conservation concern for the critically endangered Balearic shearwater. *Conservation Genetics* 6: 601–606.

Genovart, M., Oro, D., Juste, J., Louzao, M. & Bertorelle, G. 2011. Pardelas de Menorca, *¿yelkouan o mauretanicus? In:* Valeiras, X., Muñoz, G., Bermejo, A., Arcos, J. M. & Paterson, A. M. (Eds) 2011. *Actas del 6° Congreso del GIAM y el Taller internacional sobre la Ecología de Paiños y Pardelas en el sur de Europa*, pp. 159–162.

Géroudet, P. 1965. Notes sur les oiseaux du Maroc. *Alauda* 33: 294–308.

Gil, J. A., Lorente, L., Bonada, A., López, A., Potrony, D., Saló, R. & Badosa, E. 2008. Primera recuperación de larga distancia de una hembra reproductora de mochuelo boreal (*Aegolius funereus*) en el Pirineo. *Revista de Anillamiento 21–22*: 101–103.

Gil Lletget, A. 1943. Observaciones ornitológicas en El Rincón (Candeleda), verificadas durante las vacaciones de Navidad de 1941–1942. *Bol. Real Sociedad Española de Historia Natural* 41: 269–280.

Gil Lletget, A. 1945. *Sinopsis de las aves de España y Portugal.* Trabajos del Instituto de Ciencias Naturales 'José de Acosta', Vol. 1. Madrid.

Gil-Delgado, J. A. & Catalá, M. C. 1989. El Verderón Común (*Carduelis chloris*): población nidificante, tamaño de la puesta y éxito reproductivo. Un ejemplo de estudios a largo plazo. *Mediterránea, Ser. Biol.* 5: 121–132.

Gil-Delgado, J. A. 1981. Breeding bird communities in orange groves. *In:* Purroy, F. (Ed.) *Bird census and Mediterranean landscape*, pp. 100–106. Universidad de León. León.

Gil-Delgado, J. A., Vives-Ferrándiz, C. & Tapiero, A. 2002. Tendencia decreciente de una población de Gorrión Común *Passer domesticus* en los naranjales del este de España. *Ardeola* 49: 195–209.

Gil-Sánchez, J. M., Moleón, M., Otero, M. & Bautista, J. 2004. A nine-year study of successful breeding in a Bonelli's eagle population in southeast Spain: a basis for conservation. *Biological Conservation* 118: 685–694.

Gilroy, J. J. & Lees, A. C. 2003. Vagrancy theories: are autumn vagrants really reverse migrants? *British Birds* 96: 427–438.

Giménez, M. & Benavente, J. M. 2005. La Albufera de Valencia, un humedal en busca de su sostenibilidad hídrica. *La Garcilla* 123: 6–7.

Giralt, D., Brotons, L., Valera, F. & Kristín, A. 2008. The role of natural habitats in agricultural systems for bird conservation: the case of the threatened Lesser Grey Shrike. *Biodiversity and Conservation* 17: 1997–2012.

Giralt, D., Rivas, J. L. & Albero, J. C. (Eds) 2010. *El alcaudón chico en España. Población reproductora en 2010 y método de censo.* SEO/Birdlife. Madrid.

Giralt, D. & Valera, F. 2007. Population trends and spatial synchrony in peripheral populations of the endangered Lesser Grey Shrike in response to environmental change. *Biodiversity and Conservation* 16: 841–856.

Giraud-Audine, M. & Pineau, J. 1973. *Emberiza striolata* et *Vanellus gregarius* dans le Tangérois. *Alauda* 41: 317.

Glanville, D. & Walker, C. 1963. Migración primaveral en Almería, año 1960. *Ardeola* 8: 131–141.

Glutz von Blotzheim, U. N. & Bauer, K. M. 1993. *Handbuch der Vögel Mitteleuropas.* Band 13/III. Passeriformes (4 Teil).

Aula-Verlag. Wiesbaden.

Glutz von Blotzheim, U. N., Bauer, K. M. & Bezzel, E. 1973. *Handbuch der Vögel Mitteleuropas*. Vol. 5. Aula-Verlag. Wiesbaden.

Glutz von Blotzheim, U. N., Bauer, K. M. & Bezzel, E. 1977. *Handbuch der Vögel Mitteleuropas*. Vol. 7. Aula-Verlag. Wiesbaden.

Gómez, L. & de Juana, E. 1984. Aspectos de la invernada de *Larus ridibundus* en Madrid. *Ardeola* 31: 123–128.

Gómez de Segura, A., Martínez, J. M. & Alcántara, M. 2012. Population size of the endangered Bearded Vulture *Gypaetus barbatus* in Aragón (Spain): an approximation to the Pyrenean population. *Ardeola* 59: 43–55.

Gómez-Díaz, E., González-Solís, J., Peinado, M. A. & Page, R. D. M. 2006. Phylogeography of the *Calonectris* shearwaters using molecular and morphometric data. *Molecular Phylogenetics & Evolution* 41: 322–332.

Gómez-Manzaneque, Á. 1992. *Situation actuelle de la population de la Cigogne blanche en Espagne. In:* Mériaux, J. L., Schierer, A., Tombal, C. & Tombal, J. C. (Eds) *Les cigognes d'Europe*, pp. 183–188. Metz.

Gómez-Manzaneque, A. 1997. El pechiazul (*Luscinia svecica*) en la Comunidad de Madrid. *Anuario Ornitológico de Madrid 1996*: 60–69.

Gonçalves, A. 2009–10. Nidificação de Gaivotão-real *Larus marinus* em casal misto com Gaivota-de-patas-amarelas *L. michahellis* – primeiros registos por Portugal. *Anuário Ornitológico* 7: 126.

González, G., Santiago, J. M., Fernández, L. 1992. *El Águila Pescadora (*Pandion haliaetus*) en España*. ICONA. Madrid.

González, J. L. 1987. *Inventario de la población española de Cigüeña Negra*. ICONA. Unpublished report.

González, J. L. 1991. *El aguilucho lagunero* Circus aeruginosus *(L. 1748) en España. Situación, biología de la reproducción, alimentación y conservación*. ICONA. Madrid.

González, J. L. & Merino, M. 1990. *El Cernícalo Primilla (*Falco naumanni*) en la Península Ibérica. Situación, problemática y aspectos biológicos*. ICONA. Madrid.

González, J. V., Gómez, J. R. & Muñoz, B. 2002. *El Torcecuello. Determinación de la edad y el sexo, reproducción y fenología en el Noroeste peninsular ibérico*. Sociedad Asturiana de Historia Natural y Ecología. Gijón.

González, L. M. 1990. Situación de las poblaciones de Águila Imperial y Buitre Negro en España. *Quercus* 58: 16–22.

González, L. M. 1991. *Historia natural del águila imperial ibérica (*Aquila adalberti *Brehm, 1861). Taxonomía, población, análisis de la distribución geográfica, alimentación, reproducción y conservación*. ICONA. Madrid.

González, L. M. 2008. Origin and formation of the Spanish imperial eagle (*Aquila adalberti*). *J. Ornithology* 149: 151–159.

González, L. M. 2012a. Águila imperial ibérica. *In:* Salvador, A. & Bautista, L. M. (Eds).

González, L. M. 2012b. *El Águila Imperial Ibérica. El resurgir de una especie amenazada*. Fundación BBVA. Madrid.

González, L. M., Bustamante, J. & Hiraldo, F. 1992. Nesting habitat selection by the Spanish Imperial Eagle *Aquila adalberti*. *Biological Conservation* 59: 45–50.

González, L. M., Hiraldo, F., Delibes, M. & Calderón, J. 1989. Reduction in the range of Spanish Imperial Eagle (*Aquila adalberti*) since AD 1850. *J. Biogeography* 16: 305–315.

González, L. M. & Margalida, A. (Eds) 2008. *Biología de la conservación del Águila imperial ibérica* (*Aquila adalberti*). Organismo Autónomo Parques Nacionales. Madrid.

González, L. M., Margalida, A., Mañosa, S., Sánchez, R., Oria, J., Molina, J. I., Caldera, J., Aranda, A. & Prada, L. 2007. Causes and spatio-temporal variations of non-natural mortality in the endangered Spanish Imperial Eagle (*Aquila adalberti*) during a recovery period. *Oryx* 41: 495–502.

González, L. M., Oria, J., Margalida, A., Sánchez, R., Prada, L., Caldera, J., Aranda, A. & Molina, J. I. 2006. Effective natal dispersal and age of maturity in the threatened Spanish Imperial Eagle *Aquila adalberti*: conservation implications. *Bird Study* 53: 285–293.

González, M. A., Olea, P. P., Robles, L. & Ena. V. 2010. The Mediterranean *Quercus pyrenaica* oak forest: a new habitat for the capercaillie? *J. Ornithology* 151: 901–906.

González, R. & Pérez-Aranda, D. 2011. *Las aves acuáticas en España, 1980–2009*. SEO/Birdlife. Madrid.

González, S. & Villarino, A. 1997. Nidotópica y situación actual del Avión Zapador *Riparia riparia* (L. 1758) en la provincia de Ourense (NO de España). *Ardeola* 44: 41–49.

González Díez, M. 1956. Cuatro interesantes capturas en Andalucía. *Ardeola* 3: 182.

González Díez, M. 1958. Captura de *Alectoris barbara* en Cádiz.

González Villalba, C., Zumalacárregui, C. & Jubete, F. 2011. Migración postnupcial del Ruiseñor Pechiazul (*Luscinia svecica*) en el noroeste de España (Laguna de La Nava, Cuenca del Duero). *Revista de Anillamiento* 27: 32–39.

González-Solís, J., Croxall, J.-P, Oro, D. & Ruiz, X. 2007. Trans-equatorial migration and mixing in the wintering areas of a pelagic seabird. *Frontiers in Ecology and the Environment* 5: 297–301.

Gordo, O., Ferrer, X. & Reig-Ferrer, A. 2006. Unlikely occurrence of the Andalusian Hemipode (*Turnix sylvatica*) in Catalonia (NE Spain) one century ago. *Revista Catalana d'Ornitologia* 22: 30–34.

Gorospe, G. 1996. Aves nuevas: Gaviota de Ross *Rhodostethia rossea*. *La Garcilla* 97: 22–23.

Gowthorpe, P. & Lamarche, B. 1996. *Guide des oiseaux du Parc National du Banc d'Arguin (Mauritanie)*. Parc National du Banc d'Arguin. Nouakchott.

Granadeiro, J. P., Morais, L., Santos, C., Lecoq, M., Silva, M. & Grande, N. 1998. Breeding Madeiran storm-petrels

Oceanodroma castro on the Farilões Islands, Portugal. *In:* Walmsley, J. G., Goutner, V., El Hili, A. & Sultana, J. (Eds) *Ecologie des oiseaux marins et gestion intégrée du littoral en Méditerranée. Proceedings du Quatrième Symposium Méditerranéen des Oiseaux Marins*, pp. 184–195. Arcs Éditions. Tunis.

Granados, M. 1987. *Transformaciones históricas de los ecosistemas del Parque Nacional de Doñana*. Ph. D. Thesis. Universidad de Sevilla.

Grande, J. M., Carrete, M. & Donázar, J. A. 2010. El alimoche (*Neophron percnopterus*) en el valle del Ebro: principales resultados del seguimiento a largo plazo. *Anuario Ornitológico de Aragón 2004–2007*: 33–56.

Grandío, J. M. 1997. Sedimentación y fenología otoñal de tres especies de currucas (*Sylvia* spp.) en el extremo occidental del Pirineo. *Ardeola* 44: 163–171.

Grandío, J. M. 1998a. Comparación del peso y su incremento, del tiempo de estancia y de la abundancia del Carricerín Común (*Acrocephalus schoenobaenus*) entre dos zones de la marisma de Txingudi (N de España). *Ardeola* 45: 137–142.

Grandío, J. M. 1998b. Consideraciones sobre la estrategia migratoria postnupical del Carricerín Común *Acrocephalus schoenobaenus* en la península Ibérica. *Ardeola* 45: 213–216.

Grandío, J. M. 1999. Migración postnupcial diferencial del Carricerín Común (*Acrocephalus schoenobaenus*) en la marisma de Txingudi (N de España). *Ardeola* 46: 171–178.

Green, A. J. 1998. Clutch size, brood size and brood emergence in the Marbled Teal *Marmaronetta angostirostris* in the Marismas del Guadalquivir, southwest Spain. *Ibis* 140: 670–675.

Green, A. J. 2010. Cerceta pardilla. *In:* Salvador, A. & Bautista, L. M. (Eds).

Green, A. J. & Navarro, J. D. 1997. National censuses of the Marbled Teal *Marmaronetta angustirostris* in Spain. *Bird Study* 44: 80–87.

Gretton, A. 1991. *The ecology and conservation of the Slender-billed Curlew* (*Numenius tenuirostris*). ICBP Monograph N° 6. ICBP. Cambridge.

Griesinger, J. 1998. Juvenile dispersion and migration among Griffon Vultures (*Gyps fulvus*) in Spain. *In:* Chancellor *et al.* (Eds), pp. 613–621.

Grupo de Ornitoloxía da Delegación da Coruña da S. G. H. N. 1980. Láridos non comuns nas costas galegas. *Braña* ano IV 1–2: 151–160.

Grussu, M 1993. Il problema delle Garzette scure in Europa e in Italia. *Rivista Italiana di Birdwatching* 1: 11–34.

Grussu, M. & Biondi, M. 2004. Record numbers of wintering Richard's Pipits in the Western Palearctic. *British Birds* 97: 231–237.

Guerrero, J., de Lope, F. & de la Cruz, C. 1989. Un nouvel Estrildidae nicheur dans le sud-ouest de l'Espagne: *Estrilda astrild*. *Alauda* 57: 234.

Guillem, R. 2009. *Yellow-legged Gull (Larus michahellis) Gibraltar breeding census 2009*. Unpublished report to GONHS.

Guillem, R. 2010. *Yellow-legged Gull (Larus michahellis) Gibraltar breeding census 2010*. Unpublished report to GONHS.

Guillou, J. J. 1969. Sur la signification possible de la première observation du Flamant nain (*Phoeniconaias minor*) dans le domaine paléarctique. *Alauda* 37: 160–161.

Guirao, Á. 1859. Catálogo metódico de las aves observadas en una gran parte de la provincia de Murcia. *Memorias de la Real Academia de Ciencias Exactas, Físicas y Naturales*, tomo IV, parte 3.

Guitián, J. 1989. Consumo de frutos de Acebo (*Ilex aquifolium* L.) y movilización de semillas por passeriformes en las montañas Cantábricas occidentales, noroeste de España. *Ardeola* 36: 73–82.

Guitián, J., Guitián, P., Munilla, I., Guitián, J., Bermejo, T., Larrinaga, A. R., Navarro, L. & López, B. 2000. *Zorzales, espinos y serbales. Un estudio sobre el consumo de frutos silvestres de las aves migradoras en la costa occidental europea*. Universidade de Santiago de Compostela. Santiago de Compostela.

Guitián, J., Sánchez, J. L., de Castro, A. & Bas, S. 1980. Sobre *Hirundo rupestris* (Scop.) en Galicia. *Ardeola* 25: 181–192.

Gunnarsson, T. G., Gill, J. A., Atkinson, P. M., Philip, W., Croger, R. E., Guillaume, G., Gardarsson, A. & Sutherland, W. J. 2005. Estimating population size in Black-tailed Godwits *Limosa limosa islandica* by colour-marking. *Bird Study* 52: 153–158.

Gutiérrez Expósito, C., Copete, J. L., Crochet, P.-A., Qninba, A. & Garrido. H. 2011. History, status and distribution of Andalusian Buttonquail in the WP. *Dutch Birding* 33: 75–93.

Gutiérrez Expósito, C. & Qninba, A 2010. Identificación de rastros e indicios del torillo andaluz. *Quercus* 289: 14–19.

Gutiérrez, J. E. 1991. La caza de paseriformes en España durante la temporada 89–90. *La Garcilla* 80–81: 12–23.

Gutiérrez, R. 1991. Aves nuevas para España: Abejaruco Papirrojo *Merops superciliosus*. *La Garcilla* 82: 19–20.

Gutiérrez, R. 1998. Elegant Tern in Llobregat delta, Spain, in April 1993. *Dutch Birding* 20: 1–5.

Gutiérrez, R. 2001. The first breeding record of Cream-coloured Courser in Europe. *Birding World* 14: 323–325.

Gutiérrez, R. 2003. Occurrence of Rüppell's Griffon Vulture in Europe. *Dutch Birding* 25: 289–303.

Gutiérrez, R. 2004. Identification of Yelkouan, Balearic and Manx Shearwaters. *Birding World* 17: 111–122.

Gutiérrez, R., de Juana, E. & Lorenzo, J. A. 2012. *Lista de las aves de España*. SEO/Birdlife.

Gutiérrez, R., Esteban, P. & Santaeufemia, F. X. 1995. *Els ocells del delta del Llobregat*. Lynx. Barcelona.

Gutiérrez, R. & Figuerola, J. 1995. Wintering distribution of the Balearic Shearwater (*Puffinus yelkouan mauretanicus*, Lowe 1921) off the northeastern coast of Spain. *Ardeola* 42: 161–166.

Gutiérrez, R. & Figuerola, J. 1997. Estimating the size of Little Grebe (*Tachybaptus ruficollis*) breeding populations. *Ardeola* 44: 157–161.

Gutiérrez, R. & Guinart, E. 2008. The Ebro Delta Audouin's colony and vagrancy potential to northwest Europe. *British Birds* 101: 443–447.

Gutiérrez, R., López, F., Ramal, A. & Guinnart, E. 2007. Coastal Storm-petrel populations: isolated small breeding sites or outlying subcolonies of larger breeding colonies? *Atlantic Seabirds* 8: 31–40.

Gutiérrez, R. & Lorenzo, J. A. 2010. Swinhoe's Storm-petrels in Spanish waters. *British Birds* 103: 66.

Gutiérrez, R. & Muntané, O. 1997. Aves nuevas: Charrán Elegante *Sterna elegans*. *La Garcilla* 99: 26–27.

Gutiérrez, R. & Sales, S. 2001. Aves nuevas: Coliazul Cejiblanco *Tarsiger cyanurus*. *La Garcilla* 111: 20–21.

Guyomarc'h, J. C. 1992. Structure, fonctionnement et microévolution des populations de caille des blés (*Coturnix c. coturnix*) dans le paléarctique occidental. *Gibier Faune Sauvage* 9: 387–401.

Guyomarc'h, J. C., Combreau, O., Puigcerver, M., Fontoura, P., Aebischer, N. & Wallace, D. I. M. 1998. Quail, *Coturnix coturnix*. *BWP Update* 2: 27–46.

Guyomarc'h, C. & Guyomarc'h, J. C. 1985. La communication inter-spécifique chez les cailles: importance de la cadence interne des chants sur la sexualité des femelles japonaises. *Gibier Faune Sauvage* 4: 5–20.

Guyomarc'h, J.-C. & Perennou, C. 2009. *European Union Management Plan 2009–2011, Common Quail* Coturnix coturnix. Office for Official Publications of the European Communities. Luxembourg.

Guyot, I. 1993. Breeding distribution and numbers of Shag (*Phalacrocorax aristotelis desmarestii*) in the Mediterranean. *In:* Aguilar, J. S., Monbailliu, X & A. M. Paterson (Eds) *Estatus y conservación de aves marinas*, pp. 37–45. SEO/BirdLife. Madrid.

Guzmán, J. L., Ferrand, Y. & Arroyo, B. 2011. Origin and migration of Woodcock *Scolopax rusticola* wintering in Spain. *European J. Wildlife Research* 57: 647–655.

Haas, M. 2012. *Extremely rare birds in the Western Palearctic*. Lynx. Barcelona.

Haas, M. & Crochet, P.-A. 2009. Western Palearctic list updates: Cape Petrel. *Dutch Birding* 31: 24–28.

Hagemeijer, E. J. M. & Blair, M. J. 1997. *The EBBC Atlas of European breeding birds. Their distribution and abundance*. T. & A. D. Poyser. London.

Hake, M., Kjellén, N. & Alerstam, T. 2001. Satellite tracking of Swedish Ospreys *Pandion haliaetus*: autumn migration routes and orientation. *J. Avian Biology* 32: 47–56.

Handrinos, G. & Akriotis, T. 1997. *The Birds of Greece*. Christopher Helm. London.

Harrap, S. & Quinn, D. 1996. *Tits, Nuthatches and Treecreepers*. Christopher Helm. London.

Harris, A., Shirihai, H. & Christie, D. 1996. *The Macmillan birder's guide to European and Middle Eastern birds*. Macmillan. London & Basingstoke.

Harrop, H. 2003. A Masked Booby in the Bay of Biscay. *Birding World* 16: 388–390.

Hashmi, D. K. 1988. Forster's Tern *Sterna forsteri* in the Straits of Gibraltar. *Verhandlungen der Ornithologische Gesellschaft in Bayern* 24: 761–762.

Hashmi, D. K. 1993. Importance of the Mediterranean for wintering Northern Gannets (*Sula bassana*). *In:* Aguilar, J. S., Montbailliu, X. & Paterson, A. M. (Eds) *Estatus y Conservación de Aves Marinas*, pp. 383–386. SEO/BirdLife. Madrid.

Hashmi, D. K. & Fliege, G. 1994. Herbstzug der sturmschwalbe (*Hydrobates pelagicus*) in der Meerenge von Gibraltar. *J. für Ornithologie* 135: 203–207.

Haverschmidt, F. 1978. *Die Trauerseeschwalbe* – Chlidonias niger. A. Ziemsen Verlag. Wittenberg, Lutherstadt.

Hazevoet, C. J. & Haafkens, L. B. 1988. Aves nuevas para España: Lavandera Cetrina *Motacilla citreola*. *La Garcilla* 73: 27.

Heim de Balsac, H. & Mayaud, N. 1962. *Les oiseaux du nord-ouest de l'Afrique*. Lechevalier. Paris.

Helbig, A. J., Martens, J., Seibold, I., Henning, F., Schottler, B. & Wink, M. 1996. Polygeny and species limits in the Palearctic Chiffchaff *Phylloscopus collybita* complex: mitochondrial differentiation and bioacoustic evidence. *Ibis* 138: 650–666.

Helbig, A. J., Salomon, M., Bensch, S. & Seibold, I. 2001. Male-biased gene flow across an avian hybrid zone: evidence from mitochondrial and microsatellite DNA. *J. Evolutionary Biology* 14: 277–287.

Helbig, J. 1985. Occurrence of White-tailed Plover in Europe. *Dutch Birding* 7: 79–84.

Henderson, I. 2009. Progress of the UK Ruddy Duck eradication programme. *British Birds* 102: 680–690.

Hepburn, I. 1983. Hunting bags and population of Woodcock in Europe. *In:* H. Kalchreuter (Ed.), *Proceeding of the Second European Woodcock and Snipe Workshop*, pp. 138–145. IWRB. Slimbridge.

Heredia, B. 1992. Reintroducción de especies y reforzamiento de poblaciones en la conservación de aves en España. *Ardeola* 39: 41–47.

Heredia, B., Hiraldo, F., González, L. M. & González, J. L. 1988. Status, ecology and conservation of the Peregrine Falcon in Spain. *In:* Cade, T. J., Enderson, J. H., Thelander, C. G. & White, C. M. (Eds) *Peregrine Falcon Populations:Their Management and Recovery*, pp. 219–226. The Peregrine Fund Inc. Boise, Idaho.

Heredia, B. & Máñez, M. 1985. Primera cita del Chorlito social (*Vanellus gregarius*) en las Marismas del Guadalquivir.

Doñana, Acta Vertebrata 12: 182–183.

Heredia, B. & Soto-Largo, E. 1996. *Estatus de la población y conservación de la Cigüeña Negra en España.* II Conferencia Internacional sobe la Cigüeña Negra, marzo 1996. ADENEX. Trujillo.

Heredia, R. & Heredia, B. (Eds) 1991. *El quebrantahuesos* (Gypaetus barbatus) *en los Pirineos. Características ecológicas y biología de la conservación.* ICONA. Madrid.

Heredia, R. & Margalida, A. 2003. Luces y sombras en la recuperación del Quebrantahuesos. *La Garcilla* 116: 10–14.

Heredia, R. & Razin, M. 1999. Ecology and conservation of the Bearded Vulture: the case of the Spanish and French Pyrenees. *In:* Sakoulis, M., Probonas, M. & Xirouchakis, S. (Eds) *Proceedings of the 4th Bearded Vulture workshop 1998*, pp. 29–45. Heraklion, Crete.

Hernández, Á. 1994. Selección de hábitat en tres especies simpátricas de alcaudones (real, *Lanius excubitor* L., dorsirrojo, *Lanius collurio* L., y común, *Lanius senator* L.): segregación intraespecífica. *Ecología* 8: 395–413.

Hernández, Á. 1999. La migración de los alcaudones *Lanius* spp. en la Península Ibérica. *Ecología* 13: 289–302.

Hernández, Á., Alegre, J., Velasco, T. & Casas, V. M. 1992. El treparriscos *Tichodroma muraria* en la península Ibérica. *Quercus* 71: 16–22.

Hernández, Á., Alegre, J., Velasco, T. & Casas, V. M. 1993. Hábitat del Treparriscos *Tichodroma muraria* en la península Ibérica e islas Baleares. *Butll. Grup Català d'Anellament* 10: 39–45.

Hernández, A. J. 2006. Listado actualizado de las aves de la comarca de Cartagena. *In:* ANSE, *Aves de la Comarca de Cartagena.* Asociación de Naturalistas del Sureste. http://archivo.asociacionanse.org/naturalista-digital/indexb.html/

Hernández, F. & Pelayo, E. 1987. Sobre comunidades de aves esteparias en planices del valle medio del Ebro. *In: Actas del I Congreso Internacional sobre Aves Esteparias*, pp. 369–393. Junta de Castilla y León. León.

Hernández, J. 2010. www.murciaenclaveambiental.es/junio-2010.html

Hernández, M. 1988. Road mortality of the Little Owl (*Athene noctua*) in Spain. *J. Raptor Research* 22: 81–84.

Hernández, M., Campos, F., Arizaga, J. & Alonso, D. 2003. Migración del Pechiazul *Luscinia svecica* en la Península Ibérica. *Ardeola* 50: 259–263.

Hernández, M. & Margalida, A. 2009. Poison-related mortality effects in the endangered Egyptian vulture (*Neophron percnopterus*) population in Spain. *European J. Wildlife Research* 55: 415–423.

Hernández-Carrasquilla, F. 1993. Catálogo provisional de los yacimientos con aves del Cuaternario de la Península Ibérica. *Archaeofauna* 2: 231–275.

Hernandez-Gil, V. 1990. Sedentarismo en la Curruca Tomillera (*Sylvia conspicillata*) en Murcia (SE España). *Ardeola* 37: 93–97.

Herrando, S., Brotons, L., del Amo R. & Llacuna, S. 2002. Bird community succession after fire in a dry Mediterranean shrubland. *Ardea* 90: 303–310.

Herrando, S., Brotons, L., Estrada, J., Cuallar, S. & Anton, M. (Eds) 2011. *Atles dels ocells de Catalunya a l'hivern 2006–2009.* Institut Català d'Ornitologia & Lynx. Barcelona.

Herrando, S., Llimona, F., Brotons, L. & Quesada, J. 2010. A new exotic bird in Europe: recent spread and potential range of Red-billed Leiothrix *Leiothrix lutea* in Catalonia (northeast Iberian Peninsula). *Bird Study* 57: 226–235.

Herranz, J., Manrique, J., Yanes, M. & Suárez, F. 1994. The breeding biology of Dupont´s Lark, *Chersophilus duponti*, in Europe. *Avocetta* 18: 141–146.

Herranz, J. & Suárez, F. (Eds) *La Ganga Ibérica (*Pterocles alchata*) y la Ganga Ortega (*Pterocles orientalis*) en España: Distribución, abundancia, biología y conservación.* Organismo Autónomo Parques Nacionales. Madrid.

Herrera, C. M. 1974a. La captura de carnívoros por las Strigiformes. *Ardeola* 19: 439–444.

Herrera, C. M. 1974b. El paso otoñal de *Sylvia borin* y *Sylvia communis* en la Reserva de Doñana. *Doñana, Acta Vertebrata* 1: 83–119

Herrera, C. M. 1981. Fruit food of Robins wintering in southern Spanish Mediterranean scrubland. *Bird Study* 28: 115–122.

Herrera, C. M. & Alcaide, J. L. 1977. Posible nidificación de *Jynx torquilla* en Sierra Morena occidental. *Ardeola* 22: 148–149.

Herrera, C. M. & Hidalgo, J. 1974. Sobre la presencia invernal de *Clamator glandarius* en Andalucía. *Ardeola* 20: 307–311.

Herrera, C. M. & Hiraldo, J. 1971. Sobre un antiguo ejemplar de *Aquila pomarina* en Andalucía. *Ardeola* 15: 133–134.

Herrera, C. M. & Rodríguez, M. 1979. Year to year site constancy among three passerine species wintering at a southern Spanish locality. *Ringing & Migration* 2: 160.

Hidalgo, J. 1989. Sobre el Tarro Canelo en las Marismas del Guadalquivir. *Ardeola* 36: 237.

Hidalgo, J. 1991. The Marbled Teal in the Marismas del Guadalquivir, Spain. *IWRB Threatened Waterfowl Research Group Newsletter* 1: 6.

Hidalgo, J. M. 1998. Investigación. *La Garcilla* 103: 34–35.

Hidalgo, S. & Rocha, G. 2001. Distribución y fenología de la becada durante la invernada en Extremadura. *Zoologica baetica* 12: 37–48.

Hillcoat, B., Keijl, G. O. & Wallace, D. I. M. 1998. Species new to the Western Palearctic. Accidentals. *Morus capensis* Cape Gannet. *BWP Update* 2: 67–68.

Hiraldo, F. 1969. Observaciones de *Tringa stagnatilis* en las Marismas del Guadalquivir. *Ardeola* 15. 136–139.

Hiraldo, F. 1971a. Observaciones y capturas en las Marismas del Guadalquivir de *Anatidae* raras o excepcionales en la

Península Ibérica. *Ardeola*, 15: 121–122.

Hiraldo, F. 1971b. Primera captura segura de *Egretta gularis gularis* en España. *Ardeola* 15: 103–107.

Hiraldo, F. 1974. *Colonias de cría y censo de los buitres negros (*Aegypius monachus*) en España*. ICONA. Madrid.

Hiraldo, F., Delibes, M. & Calderón, J. 1979. *El Quebrantahuesos: sistemática, taxonomía, biología, distribución y protección*. Monografía No. 22. ICONA. Madrid.

Hódar, J. A. 1996. Temporal variations in two shrub-steppe bird assemblages in southeastern Spain: the importance of wintering for non-steppe birds. *In* Fernández Gutiérrez & Sanz-Zuasti (Eds), Pp 137–151.

Hódar, J. A. 2006. Diet composition and prey choice of the Southern grey shrike *Lanius meridionalis* (L.) in south-eastern Spain: the importance of vertebrates in the diet. *Ardeola* 53: 237–249.

Holt-White, R. 1901. *The life and letters of Gilbert White of Selbourne*. John Murray. London.

Homeyer, A. von. 1862–1863. Die Balearen. *J. für Ornithologie* 10: 1–23, 10: 241–285, 10: 417–434 & 11: 81–99.

Hoogendoorn, W., Adriaens, P., Cederroth, C., de Smet, G. & Lindholm, A. 2003. Three American Herring Gulls at Porto, Portugal, in March-April 2001. *Dutch Birding* 24: 235–246.

Hoogendoorn, W. & Mackrill, E. J. 1987. Audouin's Gull in the southwestern Palearctic. *Dutch Birding* 9: 99–107.

Hoogendoorn, W. & Steinhaus, G. H. 1990. Nearctic gulls in the Western Palearctic. *Dutch Birding* 12: 109–164.

Hortas, F. 1990. Phenology of waders in Salinas La Tapa, Cádiz Bay, Southwest Spain. *Wader Study Group Bull.* 59: 10.

Hortas, F. 2001. Chorlitejo Patinegro *Charadrius alexandrinus*. In: *Libro Rojo de los vertebrados amenazados de Andalucía*, pp. 157–158. Consejería de Medio Ambiente de la Junta de Andalucía. Sevilla.

Hortas, F., Arroyo, G. M. & Pérez-Hurtado, A. (Coords). 2000. *Breeding waders in Spain*. Unpublished report for Wader Study Group project: breeding waders in Europe 2000.

Howell, S. N. G. 2002. A Black-capped Petrel off the Bay of Biscay: the fourth for the Western Palearctic. *Birding World* 15: 219–220.

Hudson, N. & the Rarities Committee. 2012. Report on rare birds in Great Britain in 2011. *British Birds* 105: 556–625.

Hudson, R. 1974. Allen's Gallinule in Britain and the Palearctic. *British Birds* 67: 405–413.

Hughes, B., Criado, J., Delany, S., Gallo-Orsi, U., Green, A. J., Grussu, M., Perennou, C. & Torres, J. A. 1999. *The Status of the North American Ruddy Duck* Oxyura jamaicensis *in the Western Palearctic: towards an action plan for eradication 1999–2002*. Council of Europe. Strasbourg.

Hughes, B., Robinson, J. A., Green, A. J., Li, Z. W. D. & Mundkur, T. (compilers). 2006. *International Single Species Action Plan for the Conservation of the White-headed Duck* Oxyura leucocephala. CMS Technical Series No. 13 & AEWA Technical Series No. 8. Bonn.

Huyskens, G. & Maas, P. 1966. Migración de aves marinas al NW de España, septiembre de 1965. *Ardeola* 11: 65–68.

Huyskens, G. & Maes, P. 1971. Migración de aves marinas en el NW de España. *Ardeola* vol. especial: 155–180.

Iapichino, C. & Massa, B. 1989. *The Birds of Sicily*. British Ornithologists' Union. Tring.

Ibáñez, J. L., López Jurado, L. F., McIvor, J. & Talavera, P. A. 1974. Primer dato de reproducción de Flamencos (*Phoenicopterus ruber*) en Alicante. *Ardeola* 20: 328–330.

Ibarra, W. & Martín, J. A. 1996. Origin and distribution of the Great Cormorant *Phalacrocorax carbo* recovered in the Iberian Peninsula. *WI–CRG Bull.* 2: 14–20.

Iglesias, L. 1927. *As Aves de Galicia*. Arquivos do Seminario de Estudos Galegos, tomo I. Santiago de Compostela.

Iglesias, L. 1952. *Fauna de Galicia III. Aves de Galicia existentes en el Museo Nacional de Historia Natural de la Facultad de Ciencias de la Universidad de Santiago de Compostela*. Universidad de Santiago. Santiago de Compostela.

Igual, J. M., Forero, M. G., Gómez, T., Orueta, J. F. & Oro, D. 2006. Rat control and breeding performance in Cory's shearwater (*Calonectris diomedea*): effects of poisoning effort and habitat features. *Animal Conservation* 9: 59–65.

IKT. 2008. *Censo y estado de conservación de las poblaciones nidificantes de gaviota sombría* Larus fuscus *y gaviota patiamarilla* L. michahellis *en la Comunidad Autónoma del País Vasco. Temporada 2007*. Departamento de Medio Ambiente y Ordenación del Territorio. Vitoria.

Íñigo, A. 2009. El escribano palustre, a la espera de nuevas ZEPA y planes de gestión. *La Garcilla* 140: 38.

Íñigo, A. & Barov, B. 2010. *Action plan for the lesser kestrel* Falco naumanni *in the European Union*. BirdLife International. Cambridge.

Íñigo, A., Garza, V., Tella, J. L., Laiolo, P., Suárez, F. & Barrow, B. 2008. *Action Plan for the Dupont´s lark* Chersophilus duponti *in the European Union*. European Commission. Brussels.

Irby, L. H. 1872. (Letter on addenda to Saunders' list.) *Ibis* 14: 199–201.

Irby, L. H. 1873. (Second letter on addenda to Saunders' list.) *Ibis* 15: 96–98.

Irby, L. H. 1875. *The ornithology of the Straits of Gibraltar*. Taylor & Francis. London.

Irby, L. H. 1879. Notes on the birds of the Straits of Gibraltar. *Ibis* 21: 342–346.

Irby, L. H. 1883. Notes on the birds of Santander, northern Spain. *Ibis* 25: 173–190.

Irby, L. H. 1895. *The ornithology of the Straits of Gibraltar*. 2nd edition. Taylor & Francis. London.

Iribarren, J. J. 1969. *Nuestras aves. Navarra*. Diputación Foral de Navarra. Pamplona.

Isenmann, P. 1976. Note sur l'stationement hivernal de Larides sur la côte méditerranéenne d'Espagne. *L' Oiseaux et R.F.O.* 46: 135–142.

Isenmann, P., Benmergui, M., Browne, P., Ba, A. D., Diagana, C. H., Diawara, Y. & El Abidine ould Sidaty, Z. 2010. *Oiseaux de Mauritanie/Birds of Mauritania*. Société d'Études Ornithologiques de France. Paris.

Isenmann, P., Gaultier, T., El Hili, A., Azafzaf, H., Dlensi, H. & Smart, M. 2005. *Oiseaux de Tunisie. Birds of Tunisia.* Société d'Études Ornithologiques de France. Paris.

Isenmann, P. & Moali, A. 2000. *Oiseaux d'Algerie – Birds of Algeria.* Société d'Études Ornithologiques de France. Paris.

Isidoro, A. F. 1965. Contribuição para o estudo da Ornitologia do Alto Alentejo (1962, 1963 e 1964). *Anais da Faculdade de Ciências do Porto* 47: 273–289.

Isidoro, A. E. 1968. Aparecimento acidental de aves na fauna ornitológica portuguesa. *Cyanopica* I (1): 64–70.

Izhaki, I., Shmueli, M., Arad, Z., Steinberg, Y. & Crivelli, A. 2002. Satellite tracking of migratory and ranging behavior of immature Great White Pelicans. *Waterbirds* 25: 295–304.

Janni, O. & Fracasso, G. 2009. Commissione Ornitologica Italiana (COI) – Report 22. *Avocetta* 33: 117–146.

Janni, O. & Fracasso, G. 2011. Commissione Ornitologica Italiana (COI) – Report 24. *Avocetta* 36: 81-88.

Jehl, J. R. 1999. Population studies of Wilson's Phalaropes at fall staging areas, 1980–1997: A challenge for monitoring. *Waterbirds* 22: 37–46.

Jensen, F. P. & Lutz, M. (compilers) 2007. *Management Plan for Velvet Scoter* (Melanitta fusca) *2007–2009*. Technical Report/008/2007. Office for Official Publications of the European Communities. Luxembourg.

Jesus, J., Menezes, D., Gomes, S., Oliveira, P., Nogales, M. & Brehm, A. 2009. Phylogenetic relationships of gadfly petrels *Pterodroma* spp. from the Northeastern Atlantic Ocean: molecular evidence for specific status of Bugio and Cape Verde petrels and implications for conservation. *Bird Conservation International* 19: 199–214.

Jiguet, F., Doxa, A. & Robert, A. 2008. The origin of out-of-range pelicans in Europe: wild bird dispersal or zoo escapes? *Ibis* 150: 606–618.

Jiménez, J. & Carda, J. 1997. Invernada de la Gaviota de Audouin *Larus audouinii* en las Islas Columbretes. *Ardeola* 44: 183–189.

Jiménez, R., Hódar, J. A. & Camacho, I. 1992. La alimentación estival de la Tórtola Común (*Streptopelia turtur*) en el sur de España. *Gibier Faune Sauvage* 9: 311–319.

Jiménez García-Herrera, J., del Moral, A., Murillo, C. & Sánchez, M. J. 1992. *Las aves del parque nacional de las Tablas de Daimiel y otros humedales manchegos.* Lynx. Barcelona.

Johnsen, A., Andersson, S., García Fernández, J., Kempenaers, B., Pavel, V., Questiau, S., Raess, M., Rindal, E. & Lifjeld, J. T. 2006. Molecular and phenotypic divergence in the bluethroat (*Luscinia svecica*) subspecies complex. *Molecular Ecology* 15: 4033–4047.

Johnson, A. R. 1991. An overview of the distribution, numbers, and movements of flamingos in the western Mediterranean and north-west Africa. Agencia de Medio Ambiente de la Junta de Andalucía (Ed): *Reunión técnica sobre la situación y problemática del flamenco rosa* (Phoenicopterus ruber roseus) *en el Mediterráneo Occidental y Africa Noroccidental,* pp. 63–81. Junta de Andalucía. Sevilla.

Johnson, A. R. & Cézilly, F. 2007. *The Greater Flamingo*. T. & A. D. Poyser. London.

Jones, K. H. 1900. Ornithological notes from south-western Europe. *Zoologist* 4: 448–457.

Jones, C. W. 1857. Birds of Andalusia. *Zoologist* 15: 5312–5316.

Jönsson, O. 2011. Great Black-backed Gulls breeding at Khniffis Lagoon, Morocco and the status of Cape Gull in the Western Palearctic. *Birding World* 24: 68–76.

Jordan, R., Bonaccorsi, G. & Jordan, J.-P. 2000. Le statut des labbes *Stercorariidae* en Corse et en Méditerranée occidentale: une mise au point. *Alauda* 68: 238–243.

Jordano, P. 1981a. Relaciones interespecíficas y coexistencia entre el Águila real (*Aquila chrysaetos*) y el Águila perdicera (*Hieraaetus fasciatus*) en Sierra Morena Central. *Ardeola*, 28: 67–88.

Jordano, P. 1981b. Alimentación y relaciones tróficas entre los paseriformes en paso otoñal por una localidad de Andalucía central. *Doñana Acta Vertebrata* 8: 103–124.

Jordano, P. 1985. El ciclo anual de los paseriformes frugívoros en el matorral mediterráneo del sur de España: importancia de su invernada y variaciones interanuales. *Ardeola* 32: 69–94.

Jordano, P. 1989. Variación de la dieta frugívora otoño-invernal del Petirrojo (*Erithacus rubecula*): efectos sobre la condición corporal. *Ardeola* 36: 161–183.

Jordano, P. & Herrera, C. M. 1981. The frugivorous diet of the Blackcap *Sylvia atricapilla* populations wintering in southern Spain. *Ibis* 123: 502–507.

Jourdain, F. C. R. 1936. The birds of southern Spain. Passeres. *Ibis* 78: 725–763.

Jourdain, F. C. R. 1937. The birds of southern Spain. Passeres (concluded). *Ibis* 79: 110–152.

Juan, M. 2002. El Porrón Pardo (*Aythya nyroca*) en la Comunidad de Madrid. *Anuario Ornitológico de Madrid 2001*: 22–33.

Juan, M., González, D., Marchamalo, J. & Ruiz, J. M. 2007. Fenología migratoria de la gaviota cabecinegra (*Larus melanocephalus*) en Madrid (centro de España). *Anuario Ornitológico de Madrid 2006*: 24–39.

Juan, M., Ruiz, J. M., González, D. & Marchamalo, J. 2009.

Fenología migratoria y estatus regional de las gaviotas escasas y raras en Madrid. *Anuario Ornitológico de Madrid 2007–2008*: 26–35.

Jubete, F. 1997. *Atlas de las aves nidificantes de la provincia de Palencia*. Asociación de Naturalistas Palentinos. Palencia.

Jubete, F. 2001. La migración del carricerín cejudo en España y en la laguna palentina de La Nava. *Quercus* 184: 18–23.

Jubete, F. & Martín, C. M. 2009. *Aves de las lagunas de La Nava, Boada y Pedraza y su entorno*. Obra Social de Caja España. Palencia.

Jubete, F., Martínez, F. & Ortega, A. 1995. Censo invernal del Aguilucho Lagunero *Circus aeruginosus* en la península Ibérica. *Alytes*, 7: 21–36.

Jubete, F., Onrubia, A. & Román, J. 1996. La Lechuza Campestre en España: de invernante a reproductor. *Quercus* 119: 19–22.3

Jutglar, F. & Masó, A. 1999. *Aves de la Península Ibérica*. Planeta. Barcelona.

Kennerley, P. & Pearson, D. 2010. *Reed & bush warblers*. T. & A. D. Poyser. London.

Kiss, J. B., Szabó, L. & Duquet, M. 2001. Le Vanneau à queue blanche *Vanellus leucurus*, une nouvelle espèce nicheuse pour l'Europe. *Ornithos* 8: 100–107.

Klaassen, R. H. G., Strandberg, R., Hake, M., Olofsson, P., Tøttrup, A. P. & Alerstam, T. 2010. Loop migration in adult marsh harriers *Circus aeruginosus*, as revealed by satellite telemetry. *J. Avian Biology* 41: 200–207.

Klaassen, R. H. G., Ens, B. J., Shamoun-Baranes, J., Exo, K.-M. & Bairlein, F. 2012. Migration strategy of a flight generalist, the Lesser Black-backed Gull *Larus fuscus*. *Behavioral Ecology* 23: 58–68.

Klinkhamer, T. A. & van der Maat, S. 1990. Aves nuevas para España: Mosquitero de Pallas *Phylloscopus proregulus*. *La Garcilla* 78: 15.

Knox, A. G., Collinson, M., Parkin, D. T., Sangster, G. & Svensson, L. 2008. Taxonomic recommendations for British birds: Fifth report. *Ibis* 150: 833–835.

König, C. & Weick, F. 2008. *Owls of the world*. 2nd edition. Christopher Helm. London.

König, I. & König, C. 1973. Nueva contribución para el conocimiento de la avifauna de la Costa Brava septentrional. *Ardeola* 19: 49–55.

König, K. & Cano, A. 1971. Der Wüstengimpel *Rhodopechys githaginea* Brutvogel in Europa. *J. für Ornithologie* 112: 461.

Koppejan, T. 2001. Madeiran Storm-petrel off Denia, south-eastern Spain, in June 1997. *Dutch Birding* 23: 82.

Kryukov, A., Iwasa, M. A., Kakizawa, R., Suzuki, H., Pinsker, W. & Haring, E. 2004. Synchronic east-west divergence in azure-winged magpies (*Cyanopica cyanus*) and magpies (*Pica pica*). *J. Zoological Systematics and Evolutionary Research* 42: 342–351.

Kumerloeve, H. 1963. Le Lanier *Falco biarmicus* Temminck nichait-il en France et en Espagne il y a quelques siècles? *Alauda* 21: 1–4.

Kvist, L., Ponnikas, S., Belda, E. J., Encabo, I., Martínez, E., Onrubia, A., Hernández, J. M., Vera, P., Neto, J. & Monrós, J. S. 2011. Endangered subspecies of the reed bunting (*Emberiza schoeniclus witherbyi* and *E. s. lusitanica*) in Iberian Peninsula have different genetic structures. *J. Ornithology* 152: 681–693.

Lack, D. & Lack, E. 1953. Visible migration through the Pyrenees: an autumn reconnaissance. *Ibis* 95: 271–309.

Lack, P. (Ed.) 1986. *The atlas of wintering birds in Britain and Ireland*. T. & A. D. Poyser. Calton.

Laiolo, P. & Tella J. L. 2005. Habitat fragmentation affects culture tranmission: patterns of song matching in Dupont´s Lark. *J. Applied Ecology* 42: 1183–1193.

Langlois, D. 2007. Invernada de camachuelo (*Pyrrhula pyrrhula*) en el este de La Vera, Cáceres, 2002–03 y 2003–04. *Aves de Extremadura. Anuario ADENEX 2001–2003*: 65–69.

Lathbury, G. 1970. A review of the birds of Gibraltar and its surrounding waters. *Ibis* 112: 25–43.

Lautensach, H. 1967. *Geografía de España y Portugal*. Vicens-Vives. Barcelona.

Lavín, J. 1977. Sobre distribución y cría del *Sturnus vulgaris* en la provincia de Santander. *Ardeola* 24: 245–248.

Lawicki, L. 2013. The Lapland Bunting influx in central Europe in 2010/11. *British Birds* 106: 45–47.

Lázaro, E., Chozas, P. & Fernández-Cruz, M. 1986. Demografía de la Cigüeña Blanca (*Ciconia ciconia*) en España. Censo Nacional de 1984. *Ardeola* 33: 131–169.

Le Grand, G. 1983. Bilan des observations sur les oiseaux d'origine néarctique effectuées aux Açores (jusqu›en Janvier 1983). *Arquipélago, Série Ciências da Natureza* 4: 73–83.

Leal, A., Monrós, J. S. & Barba, E. 2004. Migration and wintering of blackcaps *Sylvia atricapilla* in eastern Spain. *Ardeola* 51: 345–355.

Lees, A. C. & Gilroy, J. J. 2004. Pectoral Sandpipers in Europe: vagrancy patterns and the influx of 2003. *British Birds* 97: 638–646.

Lefranc, N. 1993. *Les Pie-grièches d'Europe, d'Afrique du Nord et du Moyen-Orient*. Delachaux et Niestlé. Lausanne & Paris.

Lefranc, N. & Worfolk, T. 1997. *Shrikes: A Guide to the Shrikes of the world*. Pica Press. Sussex.

Leitão, D. 2005. *Eurasian Golden Plover (*Pluvialis apricaria*) wintering in Portugal: recent trend and estimates*. SPEA. Lisboa. Unpublished report, available online.

Leitão, D. & Costa, H. 1996. Populações das aves exóticas no Paul da Barroca d'Alva. *In:* Farinha, J. C., Almeida, J. & Costa, H. (Eds) *I Congresso de Ornitologia da SPEA*: 78–80. SPEA. Lisboa.

Leitão, D. & Costa, L. T. 2001. First approach to the study of the non-breeding abundance and habitat use by the little bustard *Tetrax tetrax* in the lower Tejo grasslands (South Portugal). *Airo* 11: 37–43.

Leitão, D. & Peris, S. 2003. Distribução e abundância do abibe *Vanellus vanellus* e da tarabola-dourada *Pluvialis apricaria* em Portugal. *Airo* 13: 3–16.

Leitão, D. & Peris, S. 2004. The origin of Lapwings *Vanellus vanellus* and Golden Plovers *Pluvialis apricaria* wintering in Portugal. *Ornis Fennica* 81: 49–64.

Lekuona, J. M. 2001. *Uso del espacio por la avifauna y control de la mortalidad de aves en los parques eólicos de Navarra*. Gobierno de Navarra. Pamplona.

Lescouret, F. & Gériard, M. 1982. Première nidification prouvée de Pluvier guignard *Eudromias morinellus* dans les Pyrénées Francaises. *L'Oiseau et R.F.O.* 52: 367.

Lever, C. 2005. *Naturalised birds of the world.* T. & A. D. Poyser. London.

Lilford, Lord. 1865. Notes on the ornithology of Spain. *Ibis* 7: 166–177.

Lilford, Lord. 1866. Notes on the ornithology of Spain. *Ibis* 8: 173–187, 377–392.

Lilford, Lord. 1873. (Letter.) *Ibis* 15: 98.

Limiñana, R., Romero, M., Mellone, U. & Urios, V. 2012. Mapping the migratory routes and wintering areas of Lesser Kestrels *Falco naumanni*: new insights from satellite telemetry. *Ibis* 154: 389–399.

Limiñana R., Soutullo A., López-López P. & Urios V. 2008. Pre-migratory movements of adult Montagu's Harriers *Circus pygargus*. *Ardea* 96: 81–90.

Limiñana, R., Soutullo, A. & Urios, V. 2007. Autumn migration of Montagu's harriers *Circus pygargus* tracked by satellite telemetry. *J. Ornithology* 148: 517–523.

Limiñana, R., Surroca, M., Miralles, S., Urios, V. & Jiménez, J. 2006. Population trend and breeding biology of Montagu's Harrier *Circus pygargus* in a natural vegetation site in northeast Spain. *Bird Study* 53: 126–131.

Llabrés, X. 2011. Cria posible de pinsà trompeter *Bucanetes githagineus* a Mallorca. *Anuari Ornitològic de les Balears 2010*: 51–55.

Llandres, C. & Urdiales, C. 1990. *Las aves de Doñana*. Lynx Edicions. Barcelona.

Llimona, F. & Martínez-Vilalta, A. 1987. Cens d'ardèids nidificants al Delta de l'Ebre 1986. *Butll. Parc. Natural Delta de l'Ebre* 2: 14–22.

Llimona, F. & Matheu, E. 1986. Aves nuevas para la avifauna española: *Larus delawarensis*. *La Garcilla* 66: 43–44.

Llimona, F., Sol, D., Sales, S., Tenés, A. & Matheu, E. 2007. El ruiseñor del Japón invade el parque de Collserola. Proceso de colonización de una especie exótica. *Quercus* 258: 16–20.

Lo Valvo, M. & Massa, B. 1988. Considerations of a specimen of Cory's Shearwater ringed at Selvagem Grande and recovered in the central Mediterranean. *Bocagiana* 124: 1–5.

Lobón-Cerviá, J. & González, L. M. 1982. Cita de corneja cenicienta (*Corvus corone cornix*) en Asturias. *Miscel.lània Zoològica* 6: 161.

Lockley, R. M. 1952. Notes of the birds of the Berlengas (Portugal), the Desertas and Baixo (Madeira) and the Salvages. *Ibis* 94: 144–157.

Lockwood, M. W. 1999. Possible anywhere. Fork-tailed Flycatcher. *Birding* 31: 126–139.

López, C. B., Potrony, D., López, A., Badosa, E., Bonada, A. & Saló, R. 2010. Nest-box use by boreal owls (*Aegolius funereus*) in the Pyrenees mountains in Spain. *J. Raptor Research* 44: 40–49.

López, Z. & Guitián, J. (Eds) 1983. *Atlas provisional de los vertebrados terrestres de Galicia. Parte II. Aves nidificantes*. Universidad de Santiago de Compostela.

López de Carrión, M. 2009. Aproximación a la situación del pico menor *Dendrocopos minor* en la provincia de Toledo. *Anuario Ornitológico de Toledo 2002–2007*: 106–115.

López Gordo, J. L. 1975. Sobre la migración posnupcial del Abejaruco (*Merops apiaster*) en el Estrecho de Gibraltar. *Ardeola* 21 (especial): 615–625.

López Iborra, G. 1989. Variación de la colocación y orientación del nido del Alzacola (*Cercotrichas galactotes*) en dos especies de árboles. *Doñana, Acta Vertebrata* 16: 57–67.

López Iborra, G. & Gil-Delgado, J. A. 1988. Aspects of the breeding ecology of Rufous Bush Robins *Cercotrichas galactotes* in southeast Spain. *Bird Study* 35: 85–90.

López Iborra, G., Rodríguez Jerez, M. A., Gómez Picazo, J. A., Gómez Ríos, J. A. & Zaragozí, T. 2006. El Camachuelo Trompetero en la provincia de Alicante. *Las Aves en Alicante. Anuario Ornitológico 2001–2002–2003*: 244–249.

López Martos, J. M. 1991. Antonio Cano en el paraje natural del Sabinar (Punta Entinas, Punta del Sabinar) (Almería). *Bol. Instituto de Estudios Almerienses* 9–10: 173–199.

López Septiem, J. A., Corraleche, L., Domínguez, C., Moreno, F., Roviralta, F., Soto, T., Tajuelo, J. & Tello, J. 2004. Expansión y situación actual del alcaudón dorsirrojo (*Lanius collurio*) en la Comunidad de Madrid. *Anuario Ornitológico de Madrid 2003*: 48–51.

López Velasco, D. 2010. Identification of the first Pacific Diver for Spain. *Birding World* 23: 14–19.

López-Jurado, L. F. & Ibáñez, J. M. 1981. Captura de Barnacla Carinegra, *Branta bernicla*, en la costa mediterránea española. *Doñana, Acta Vertebrata* 8: 306.

López-López, P., Ferrer, M., Madero, A., Casado, E. & McGrady, M. 2011. Solving man-induced large-scale conservation problems: the Spanish Imperial Eagle and power lines. *PLoS ONE* 6: e17196.

López-López, P., Limiñana, R., Mellone, U. & Urios, V. 2010. From the Mediterranean Sea to Madagascar. Are there ecological barriers for the long-distance migrant Eleonora's falcon? *Landscape Ecology* 25: 803–813.

López-Seoane, V. 1861. Catálogo de las aves observadas en Andalucía. *Revista del Progreso de las Ciencias Exactas, Físicas y Naturales*: 1–62.

Lorente, L., Ascaso, J. C. & Chéliz, G. 2007. Estudio y conservación del mochuelo boreal en el Pirineo aragonés. Historia de un descubrimiento reciente. *Quercus* Especial Aragón: 14–15.

Lorenzo, M. & Planelles, P. 2010. *La agachadiza común en España. Población reproductora en 2009 y método de censo.* SEO/BirdLife. Madrid.

Lourenço, P. M., Kentie, R., Schroeder, J., Alves, J. A., Groen, N. M., Hooijmeijer, J. C. E. W. & Piersma, T. 2010. Phenology, stopover dynamics and population size of migrating Black-tailed Godwits *Limosa limosa limosa* in Portuguese rice plantations. *Ardea* 98: 35–42.

Lourenço, P. M. & Piersma, T. 2008. Stopover ecology of Black-tailed Godwits *Limosa limosa limosa* in Portuguese rice fields: a guide on where to feed in winter. *Bird Study* 55: 194–202.

Lovette, I. J., McCleery, B. V., Talaba, A. L. & Rubenstein, D. R. 2008. A complete species-level molecular phylogeny for the "Eurasian" starlings (Sturnidae: *Sturnus, Acridotheres*, and allies): recent diversification in a highly social and dispersive avian group. *Molecular Phylogenetics and Evolution* 47: 251–260.

Lozano, S. 1993. Aportación a la fenología de limícolas en los Llanos de Cáceres. *Airo* 3: 57–65.

LPO 2007. Un gigantesque dortoir de Crécerellettes découvert au Sénégal. *Ornithos* 14: 134.

Lucientes, J. 1971. Observación de *Podiceps auritus* en Zaragoza. *Ardeola* 15: 100–101.

Lucientes, J. 1977. Breves notas sobre Anatidae en Aragón. *Ardeola* 23: 223–226.

Lucio, A. J. & Purroy, F. J. 1985. Migración del Chorlito Carambolo (*Charadrius morinellus*) por España: revisión bibliográfica y observaciones en Castilla y León. *Ardeola* 32: 400–404.

Lucio, A. J. & Purroy, F. J. 1992. Red-legged partridge (*Alectoris rufa*) habitat selection in northwest Spain. *Gibier et Faune Sauvage* 9: 417–429.

Lucio, A. J., Purroy, F. J. & Sáenz de Buruaga, M. 1992. *La Perdiz Pardilla (*Perdix perdix hispaniensis*) en España: distribución, tendencias demográficas, estado de conservación y pautas de gestión.* ICONA. Madrid.

Lucio, A. J. & Sáenz de Buruaga, M. 2000. *La becada en España.* Federación Española de Caza & Fedenca. Madrid.

Lundberg, A. & Alatalo, R. A. 1992. *The Pied Flycatcher.* T. & A. D. Poyser. London.

Lutz, M. 2007. Management plan for Turtle Dove (*Streptopelia turtur*) 2007–2009. Office for official publications of the European Communities.

Lynes, H. 1912. Bird-notes in two Andalucian sierras. *Ibis* 54: 454–489.

Macías, M., Green, A. J. & Sánchez, M. I. 2004. The diet of the Glossy Ibis during the breeding season in Doñana, southwest Spain. *Waterbirds* 27: 234–239.

Madero, A. & Ferrer, M. 2002. *Proyecto de reintroducción del águila imperial ibérica en la provincia de Cádiz.* Consejería de Medio Ambiente de la Junta de Andalucía. Sevilla.

Madge, S. & Burn, H. 1994. *Crows and Jays.* Christopher Helm. London.

Madge, S. & Burn, H. 1988. *Wildfowl: an identifcation guide to the ducks, geese and swans of the world.* Christopher Helm. London.

Madroño, A., González, C. & Atienza, J. C. (Eds) 2004. *Libro rojo de las aves de España.* Dirección General para la Biodiversidad – SEO/Birdlife. Madrid.

Madroño, A., Palacios, C. J. & de Juana, E. 1992. La migración de la Cigüeña Negra *Ciconia nigra* a través de España peninsular. *Ardeola* 39: 9–13.

Magee, J. D. 1977. Algunas observaciones de otoño (1968) y primavera (1969) en Mallorca. *Ardeola* 22: 97–100.

Magnusdottir, E., Leat, E. H. K., Bourgeon, S., Strøm, H., Petersen, A., Phillips, R. A., Hanssen, S. A., Bustnes, J. O., Hersteinsson, P. & Furness, R. W. 2012. Wintering areas of Great Skuas *Stercorarius skua* breeding in Scotland, Iceland and Norway. *Bird Study* 59: 1–9.

Makatsch, W. 1969. Studies of less familiar birds. Audouin's Gull. *British Birds* 62: 230–232.

Malling Olsen, K. & Larsson, H. 1997. *Skuas and jaegers: a guide to the skuas and jaegers of the world.* Pica Press. Mountfield, Sussex.

Malling Olsen, K. & Larsson, H. 2003. *Gulls of Europe, Asia and North America.* Christopher Helm. London.

Maluquer Maluquer, S. 1960. La avifauna primaveral de la Isla de Buda. *Ardeola* 6: 151–210.

Maluquer Maluquer, S. 1964a. Cisnes en Cataluña durante el invierno 1962–1963. *Ardeola* 9: 111–119.

Maluquer Maluquer, S. 1964b. Primera captura de *Botaurus lentiginosus* en la Península Ibérica. *Ardeola* 9: 136–137.

Maluquer Maluquer, S. 1971. La avifauna del Delta del Ebro en primavera–verano. *Ardeola* vol. especial: 191–334.

Maluquer Maluquer, S. & Pons, J. R. 1961. La avifauna de la isla de Buda en primavera-verano de 1961. *Ardeola* 7: 79–111.

Maluquer Sostres, J. 1952. Notas para el estudio de las aves de Lérida. *Pirineos* 24: 421–470.

Maluquer Sostres, J. 1971. Las aves de la Albufera de Valencia y del Delta del Ebro en la bibliografía antigua. *Ardeola* volumen especial: 335–380.

Maluquer Sostres, J. 1973. *Els ocells de les terres catalanes.* 2nd edition. Barcino. Barcelona.

Maluquer Sostres, J. 1981. *Els ocells de les terres catalanes.* 3rd edition. Barcino. Barcelona.

Martorell, M. 1966. Nota sobre la avifauna de la laguna del Hondo, en Elche. *Ardeola* 11: 129–136.

Máñez, M. 1983. Espectro alimenticio del Mochuelo Común (*Athene noctua*) en España. *Alytes* 1: 275–290.

Máñez, M. 2001. Martinete común *Nycticorax nycticorax*. *In:* Franco, A. & Rodríguez de los Santos, M. (Eds): *Libro Rojo de los Vertebrados Amenazados de Andalucía*. Consejería de Medio Ambiente de la Junta de Andalucía. Sevilla.

Máñez, M. 2001. Quebrantahuesos *Gypaetus barbatus*. *In:* Franco, A. & Rodríguez de los Santos, M. (Eds): *Libro rojo de los vertebrados amenazados de Andalucía*, pp. 115–116. Consejería de Medio Ambiente de la Junta de Andalucía. Sevilla.

Máñez, M. & Rendón-Martos, M. (Eds) 2009. *El morito, la espátula y el flamenco en España. Población en 2007 y método de censo*. SEO/BirdLife. Madrid.

Máñez, M., Tortosa, F. S., Barcell, M. & Garrido, H. 1994. La invernada de la Cigüeña Blanca en el suroeste de España. *Quercus* 105: 10–12.

Mañosa, S. 1994. Goshawk feeding habits in a Mediterranean area of northeastern Spain. *J. Raptor Research* 28: 84–92.

Mañosa, S. 2001. Strategies to identify dangerous electricity pylons for birds. *Biodiversity and Conservation* 10: 1997–2012.

Mañosa, S., Real, J. & Codina, J. 1998. Selection of settlement areas by juvenile Bonelli's Eagle in Catalonia. *J. Raptor Research* 32: 208–214.

Manrique, J. 1993. *Las aves de Almería*. Instituto de Estudios Almerienses. Almería.

Manrique, J. 1997. *Corología y ecografía de las aves nidificantes en la provincia de Almería (sureste ibérico)*. Doctoral Thesis, University of Granada.

Manrique, J. & Miralles, J. M. 1988. El camachuelo trompetero. *Quercus* 32: 34–36.

Marchamalo, J. 1995. La invernada de la Cigüeña Blanca en España. *In:* Biber, O., Enggist, P., Marti, C. & Salathé, T. (Eds) *Proceedings of the International Symposium on the White Stork (Western Population) Basel 1994*, pp. 77–78. Sempach (Switzerland).

Marchamalo, J. 2002. Cigüeña Blanca (*Ciconia ciconia*). *In:* del Moral, J. C., Molina, B., de la Puente, J. & Pérez-Tris, J. (Eds) *Atlas de las aves invernantes de Madrid 1999–2001*, pp. 84–85. SEO-Monticola y Comunidad de Madrid. Madrid.

Margalida, A. 2011. Quebrantahuesos. *In:* Salvador, A. & Bautista, L. M. (Eds).

Margalida, A., Bertran, J. & Heredia, R. 2009. Diet and food preferences of the endangered bearded vulture *Gypaetus barbatus*: a basis for their conservation. *Ibis* 151: 235–243.

Margalida, A. & Heredia, R. (Eds) 2005. *Biología de la conservación del Quebrantahuesos (*Gypaetus barbatus*) en España*. Organismo Autónomo Parques Nacionales. Madrid.

Margalida, A., Heredia, R., Razin, M. & Hernández, M. 2008. Sources of variability in mortality of the Bearded Vulture *Gypaetus barbatus* in Europe. *Bird Conservation International* 18: 1–10.

Margalida, A., García, D., Bertran, J. & Heredia, R. 2003. Breeding biology and success of the Bearded Vulture *Gypaetus barbatus* in the eastern Pyrenees. *Ibis* 145: 244–252.

Margalida, A., González, L. M., Sánchez, R., Oria, J., Prada, L., Caldera, J., Aranda, A. & Molina, J. I. 2007. A long-term scale study of the breeding biology of Spanish Imperial Eagles. *J. Ornithology* 148: 309–322.

Marina, G. & Bezares, E. 1933. Información sobre los cuervos en España. *Instituto Forestal de Investigaciones y Experiencias* 12: 1–47.

Mariné, R. & Dalmau, J. 2000. Uso del hábitat por el mochuelo boreal (*Aegolius funereus*) en Andorra (Pirineo oriental) durante el período reproductor. *Ardeola* 47: 29–36.

MARM. 2008. *Anuario Estadístico Forestal 2008*. Madrid.

Maroñas, J. A. 2010. El Vencejo Pálido, nueva especie reproductora para Cantabria. *Locustella* 7: 94–98

Marques, A. T., Henriques, I., Catry, I. & Moreira, M. I. 2005. Distribution of the Roller *Coracias garrulus* in Portugal, an historical approach. *Ardeola* 52: 173–176.

Marques, P. M. 2003. Spanish Sparrow *Passer hispaniolensis* breeding success in 2002 in southern Portugal. *Ardeola* 50: 251–253.

Martí, R. 1999. Results of the 5th International White Stork Census (1994) in Spain. *In:* Schulz, H. (Ed.): White storks on the up? *Proceedings of the International Symposium on the White Stork (Hamburg, 1996)*, pp. 61–68. NABU. Bonn.

Martí, R. & del Moral, J. C. (Eds) 2002. *La invernada de aves acuáticas en España*. Organismo Autónomo Parques Nacionales. Madrid.

Martí, R. & del Moral, J. C. (Eds) 2003. *Atlas de las Aves Reproductoras de España*. Dirección General de Conservación de la Naturaleza-Sociedad Española de Ornitología. Madrid.

Martí, R., Gómez-Manzaneque, A. & Perales, J. A. 1988. Invernada del Acentor Alpino (*Prunella collaris* Scop.) en los altos niveles de la Sierra de Gredos (Sistema Central). *Ardeola* 35: 143–146.

Martí, R., Gómez-Manzaneque, A. & Perales, J. A. 1989. Diferencias según edad y sexo en los movimientos dispersivos de una población de Acentor Alpino (*Prunella collaris*) en España central. *Ardeola* 36: 224–226.

Martí, R., Perales, A. J. & Gómez-Manzaneque, A. 1986. Notas sobre la dieta del pollo del Acentor Alpino (*Prunella collaris* Scop.) en la sierra de Gredos. *Ardeola* 33: 189–195.

Martí, R. & Ruiz, A. 2001. La Pardela Balear. *La Garcilla* 110: 14–17.

Martín, A. & Lorenzo, J. A. 2001. *Aves del Archipiélago Canario*. Francisco Lemus. La Laguna (Tenerife).

Martín, C. A. 2011. Ganga ibérica. *In:* Salvador, A. & Morales, B. M. (Eds).

Martín, C. A., Alonso, J. C., Alonso, J. A., Palacín, C., Magaña, M. & Martín, B. 2007. Sex-biased juvenile survival in a bird with extreme size dimorphism, the great bustard (*Otis tarda*). *J. Avian Biology* 38: 335–346.

Martín, C. A., Alonso, J. C., Alonso, J. A., Palacín, C., Magaña, M. & Martín, B. 2008. Natal dispersal in great bustards: the effect of sex, local population size and spatial isolation. *J. Animal Ecology* 77: 326–334.

Martín, C. A., Casas, F., Mougeot, F., García, J. T. & Viñuela, J. 2010a. Seasonal variations in habitat preferences of the pin-tailed sandgrouse in agrarian pseudo-steppes. *Ardeola* 57: 191–198.

Martín, C. A., Casas, F., Mougeot, F., García, J. T. & Viñuela, J. 2010b. Positive interactions between vulnerable species in agrarian pseudo-steppes: habitat use by pin-tailed sandgrouse depends on its association with the little bustard. *Animal Conservation* 13: 383–389.

Martín, C. A., Martínez, C., Bautista, L. M. & Martín, B. 2012. Population increase of the great bustard *Otis tarda* in its main distribution area in relation to changes in farming practices. *Ardeola* 59: 31–42.

Martín, G. 2002. Distribución temporal y preferencias de hábitat de los limícolas escasos en la Comunidad de Madrid. I. Géneros *Recurvirostra*, *Charadrius*, *Calidris* y *limícolas accidentales*. *Anuario Ornitológico de Madrid 2001*: 34–51.

Martín, G. 2004. Distribución temporal y preferencias de hábitat de los limícolas escasos en la Comunidad de Madrid. II. Géneros *Philomachus*, *Limosa*, *Numenius* y *Tringa*. *Anuario Ornitológico de Madrid 2003*: 20–39.

Martín, J. A. & Pérez, A. 1990. Movimientos del Martín Pescador (*Alcedo atthis*, L.) en España. *Ardeola* 37: 13–18.

Martín, M. 2001. *Paloma zurita*. *In:* Franco & Rodríguez de los Santos (Eds).

Martín Fernández, J. A., Ibarra, W. & Martín García, P. 1997. Alimentación invernal del Cormorán Grande (*Phalacrocorax carbo sinensis*) en zonas del interior de la Península Ibérica, *In:* Manrique, J., Sánchez, A., Suárez, F. & Yanes, M. (Eds) *Actas de las XII Jornadas Ornitológicas Españolas*, pp. 157–163. Instituto de Estudios Almerienses. Almería.

Martín Pajares, M. 2006. La Cotorra Argentina (*Myiopsitta monachus*) en la ciudad de Madrid: expansión y hábitos de nidificación. *Anuario Ornitológico de Madrid 2005*: 76–95.

Martínez, C. 1988. Size and sex composition of Great Bustard (*Otis tarda*) flocks in Villafáfila, northwest Spain. *Ardeola* 35: 125–133.

Martínez, C. & de Juana, E. 1996. Breeding bird communities of cereal crops in Spain: hábitat requirements. *In:* Fernández Gutiérrez & Sanz-Zuasti (Eds).

Martínez, C. & Tapia, G. G. 2002. Density of the Little Bustard *Tetrax tetrax* in relation to agricultural intensification in Central Spain. *Ardeola* 49: 301–304.

Martínez, F., Ortega, A. & Jubete, F. 1994. Situación actual del Aguilucho Lagunero (*Circus aeruginosus*) en España. Reproducción e invernada. *In:* Muntaner, J. & Mayol, J. (Eds) *Biología y Conservación de las Rapaces Mediterráneas, 1994*, pp. 451–458. SEO/BirdLife. Madrid.

Martínez, I. & Carrera E. 1983. Nova colònia de gavina corsa *Larus audouinii* Payr. a l'Estat Espanyol. *Butll. Institució Catalana de Història Natural* 49: 159–161.

Martínez, J., Viñuela, J. & Villafuerte, R. 2002. *Socioeconomic aspects of gamebird hunting, hunting bags, and assessment of the status of gamebird populations in REGHAB countries*. Unpublished report to REGHAB project. European Commission.

Martínez, J. A. & López, G. 1995. Dispersal and causes of mortality of the Barn Owl (*Tyto alba*) in Spain. *Ardeola* 42: 29–37.

Martínez, J. A. & Zuberogoitia, Í. 2001. The response of the Eagle Owl (*Bubo bubo*) to an outbreak of the rabbit haemorrhagic disease. *J. Ornitholology* 142: 204–211.

Martínez, J. A. & Zuberogoitia, Í. 2004a. Habitat preferences and causes of population decline for Barn Owls *Tyto alba*: a multi-scale approach. *Ardeola* 51: 303–317.

Martínez, J. A. & Zuberogoitia, Í. 2004b. Effects of habitat loss on perceived and actual abundance of the Little Owl *Athene noctua* in Eastern Spain. *Ardeola* 51: 215–219.

Martínez, J. A. & Zuberogoitia, Í. 2004c. Habitat preferences for Long-eared Owls (*Asio otus*) and Little Owls (*Athene noctua*) in semi-arid environments at three spatial scales. *Bird Study* 51: 163–169.

Martínez, J. A., Zuberogoitia, Í. & Alonso, R. 2002. *Rapaces nocturnas. Guía para la determinación del sexo y la edad de las rapaces nocturnas Ibéricas*. Monticola Ediciones. Madrid.

Martínez, J. A., Martínez, J. E., Pérez, E., Zuberogoitia, I. & Izquierdo, A. 2003b. Possible first record of multiple brooding of the Eagle Owl *Bubo bubo*. *Ardeola* 50: 77–79.

Martínez, J. A., Serrano, D. & Zuberogoitia, Í. 2003a. Predictive models of habitat preferences for the Eurasian eagle owl *Bubo bubo*: a multi-scale approach. *Ecography* 26: 21–28.

Martínez, J. E. & Calvo, J. F. 2001. Diet and breeding success of the Eagle Owl in southwestern Spain: effect of rabbit haemorrhagic disease. *J. Raptor Research* 35: 259–262.

Martínez, J. E. & Sánchez-Zapata, J. A. 1999. Invernada de Aguililla Calzada (*Hieraaetus pennatus*) y Culebrera Europea (*Circaetus gallicus*) en España. *Ardeola* 46: 93–96.

Martínez, J. E., González, C. & Calvo, J. F. 2005. Cooperative nesting by a trio of Booted Eagles (*Hieraaetus pennatus*). *J. Raptor Research* 39: 92–94.

Martínez, J. E., Martínez, J. A., Zuberogoitia, I., Zabala, J., Redpath, S. M. & Calvo, J. F. 2008. The effect of intra- and interspecific interactions on the large-scale distribution of cliff-nesting raptors. *Ornis Fennica* 85: 13–21.

Martínez, J. G. 2011. Urraca. *In:* Salvador, A. & Morales, M. B. (Eds).

Martínez, J. G., Soler, J. J., Soler, M., Møller A. P. & Burke, T. 1999. Comparative population structure and gene flow of a brood parasite, the great spotted cuckoo (*Clamator glandarius*) and its primary host, the magpie (*Pica pica*). *Evolution* 53: 269–278.

Martínez, J. L. & Suárez, M. 2008. Fenología migratoria y movimientos primaverales de la codorniz común *Coturnix coturnix* en las Islas Baleares. *Anuari Ornitològic de les Balears* 2007: 49–58.

Martínez, O. & Palerm, J. C. 1998. Avifauna se Ses Freixes d'Eivissa. *Anuari Ornithològic de Balears* 13: 21–30.

Martínez Gámez, V. 1906. *Apuntes para la ornitología andaluza y de España en general.* Imprenta Ricardo Rojas. Madrid.

Martínez Mariño, J. M. (coord.) 2009. *Anuario Histórico das Aves de Galicia.* Sociedade Galega de Ornitoloxía. Santiago de Compostela.

Martínez-Abraín, A., Díes, J. I., Gómez, J. A., Villuendas, E., Sarzo, B., Ramón, N. & Carda, J. 2003. Una nueva colonia de Gaviota de Audouin se instala en la Albufera. *Quercus* 212: 12–13.

Martínez-Abraín, A., Sánchez, A. & Oro, D. 2002a. Atlantic Cory's Shearwaters breeding in a colony of Mediterranean Cory's Shearwaters. *Waterbirds* 25: 221–224.

Martínez-Abraín, A., Oro, D., Carda, J. & del Señor, X. 2002b. Movements of Yellow-legged Gulls (*Larus (cachinnans) michahellis*) from two small western Mediterranean colonies. *Atlantic Seabirds* 4: 101–108.

Martínez-Abraín, A., Velando, A., Oro, D., Genovart, M., Gerique, C., Bartolomé, M.A., Villuendas, E. & Sarzo, B. 2006. Sex-specific mortality of European shags during an oil spill: demographic implications for the recovery of colonies. *Marine Ecology Progress Series* 318: 271–276.

Martínez-Abraín, A., Viedma, C., Bartolomé, M. A., Gómez, J. A. & Oro, D. 2007. Hunting sites as ecological traps for coots in southern Europe: implications for the conservation of a threatened species. *Endangered Species Research* 3: 69–76.

Martínez-Cabello, A., Soler, M. & Soler, J. J. 1991. Alimentación del Acentor Común (*Prunella modularis*) durante su invernada en el sureste de la Península Ibérica. *Ardeola* 38: 305–315.

Martínez-Cruz, B. 2011. Conservation genetics of Iberian raptors. *Animal Biodiversity and Conservation* 34: 341–353.

Martínez-Cruz, B. & Godoy, J. A. 2007. Genetic evidence for a recent divergence and subsequent gene flow between Spanish and Eastern imperial eagles. *BMC Evolutionary Biology* 7: 170.

Martínez-Padilla, J. 2012. Cernícalo Vulgar. *In:* Salvador, A. & Morales, M. B. (Eds).

Martínez-Vilalta, A. 1988. Cens de gavines i xatracs nidificants al delta de l'Ebre. *Butll. Parc Natural del delta del Ebre* 3: 36–39.

Martínez-Vilalta, A. 1991. Primer censo nacional de limícolas coloniales y pagaza piconegra, 1989. *Ecología* 5: 321–327.

Martínez-Vilalta, A., Bertolero, A., Bigas, D. & Paquet, J-Y. 1998. Notes faunístiques. *Butll. Parc Natural del delta del Ebre* 10: 45.

Martínez-Vilalta, A. & Motis, A. 1985. Aves nuevas para la avifauna española: *Calidris tenuirostris*. *La Garcilla* 65: 41–42.

Martínez-Vilalta, A. & Motis, A. 1989. *Els ocells del Delta de l'Ebre.* Lynx. Barcelona.

Martínez-Vilalta, A., Ferrer, X. & Carboneras, C. 1984. Situación de los págalos (*Stercorarius* sp.) en el litoral catalán. *Miscel-lània Zoològica* 8: 217–223.

Martín-Vivaldi, M., Marín, J. M. & Villar, M. 1995. Selección de habitat, tamaño de bando y movimientos locales del Acentor Alpino (*Prunella collaris*) en Sierra Nevada (SE de España). *Ardeola* 42: 11–20.

Martorell, M. 1966. Nota sobre la avifauna de la laguna del Hondo, Elche. *Ardeola* 11: 129–136.

Mas, R. E. 2002. Primeres dades de nidificació de llambritja *Sterna hirundo* a les Illes Balears. *Anuari Ornitològic de les Balears 2001*: 61–63.

Masero, J. A. 2008. Correlimos común *Calidris alpina*. *In:* Palacios, M. J. *et al.* (Eds) *Catálogo regional de especies amenazadas de Extremadura. Fauna II. Clase Aves*, pp. 192–193. Junta de Extremadura. Mérida.

Masero, J. A., Pérez-Hurtado, A., Castro, M. & Arroyo. G. M. 2000. Complementary use of intertidal mudflats and adjacent salinas by foraging waders. *Ardea* 88: 177–191.

Masero, J. A. & Santiago-Quesada, F. 2008. Chorlitejo patinegro *Charadrius alexandrinus*. *In:* Palacios, M. J. et al. (Eds) *Catálogo regional de especies amenazadas de Extremadura. Fauna II. Clase Aves*, pp. 186–187. Junta de Extremadura. Mérida.

Masero, J. A., Santiago-Quesada, F., Sánchez-Guzmán, J. M., Abad-Gómez, J. M., Villegas, A. & Albano, N. 2009. Origen geográfico, tasas de retorno y movimientos de la casi amenazada aguja colinegra *Limosa limosa* en una importante zona de parada en Iberia. *Ardeola* 56: 253–258.

Masero, J. A., Santiago-Quesada, F., Sánchez-Guzmán, J. M., Villegas, A., Abad-Gómez, J. M., Lopes, R. J., Encarnação, V., Corbacho, C. & Morán, R. 2011. Long lengths of stay, large numbers, and trends of the Black-tailed Godwit

Limosa limosa in rice fields during spring migration. *Bird Conservation International* 21: 12–24.

Mateo, R., Estrada, J., Paquet, J.-Y., Riera, X., Domínguez, L., Guitart, R. & Martínez-Vilalta, A. 1999. Lead shot ingestion by marsh harriers *Circus aeruginosus* from the Ebro delta, Spain. *Environmental Pollution* 104: 435–440.

Mather, J. H. 2010. Pacific Diver in Yorkshire: new to Britain and the Western Palearctic. *British Birds* 103: 539–545.

Matias, R. 2009. Removal of Black Crake *Amaurornis flavirostris* from the Western Palearctic list, and the first record of Lesser Moorhen *Gallinula angulata* for Madeira. *Bull. British Ornithologists' Club* 129: 116–119.

Matias, R. 2010. Aves exóticas em Portugal: anos de 2005–2008. *Anuário Ornitológico* 7: 95–108.

Matias, R. 2011. Aves exóticas em Portugal: anos de 2009–2010. *Anuário Ornitológico* 8: 94–104.

Matias, R., Alfrey, P., Costa, H., Jara, J., Moore, C. C., Lima Santos, J. & Tipper, R. 2011. Adições e alterações à lista das aves de Portugal Continental: primeira actualização. *Anuário Ornitológico* 8: 105–117.

Matias, R., Catry, P., Costa, H., Elias, G., Jara, J., Moore, C. C. & Tomé, R. 2007. Lista sistemática das aves de Portugal Continental. *Anuário Ornitológico* 5: 74–132.

Mayaud, N. 1936. *Inventaire des oiseaux de France*. Société d'Études Ornithologiques. Paris.

Mayaud, N. 1941. Commentaires sur l'ornithologie française (suite). *L'Oiseau et R.F.O.* No. spécial: LIX-CXXXVI.

Mayaud, N. 1970. Additions et contributions à l'avifaune du Nord-Ouest de l'Afrique. *Alauda* 38: 27–43.

Mayol, J. 1977a. Contribución al conocimiento del Buitre Negro (*Aegypius monachus*) en Mallorca. *Bol. Sociedad de Historia Natural de las Baleares* 22: 150–178.

Mayol, J. 1977b. La ornitofagia tradicional en la Isla de Ibiza. *Vida Silvestre* 24: 242–247.

Mayol, J. 1977c. Estudios sobre el Halcón de Eleonor *Falco eleonorae* en las Islas Baleares. *Ardeola* 23: 103–136.

Mayol, J. 1978. *Els aucells de les Balears*. Moll. Mallorca.

Mayol, J., Jurado, J., Montaner, J. C. & Muntaner, J. 2003. Tendències demogràfiques de l'avifauna de les Balears del 1975 al 2000. *Anuari Ornitològic de les Balears* 17: 21–29.

Mayol, J. & Viada, C. 2011. *El projecte per a la reintroducció de l'àguila coabarrada a Mallorca*. Conselleria d'Agricultura, Medi Ambient i Territori, Palma de Mallorca.

Mazzoleni, S., di Pascuale, G., Mulligan, M., di Martino, P. & Rego, F. (Eds) 2004. *Recent dynamics of the Mediterranean vegetation and landscape*. John Wiley. Chichester.

McGowan, R. Y. & Massa, B. 1990. Evidence for breeding of the Lanner Falcon *Falco biarmicus erlangeri* in Spain in the 19th century. *Bull. British Ornithologists' Club* 110: 65–65.

Mead, C. J. 1974. The results of ringing auks in Britain and Ireland. *Bird Study* 21: 45–86.

Mead, C. J. & Harrison, J. D. 1979. Overseas movements of British and Irish Sand Martins. *Bird Study* 26: 87–98.

Meirinho, A., Martins, R., Hilton, G., Ceia, R., Fagundes, A. I., Leal, A., Marques, A. T., Santos, C. Tavares, J., Costa, L. & Leitão, D. 2011. *O estado das aves comuns em Portugal 2010: Relatório do projecto Censo de Aves Comuns*. SPEA. Lisboa.

Mejías, R. & Amengual, J. 2000. *Libro Rojo de los Vertebrados de las Baleares (2ª edición)*. Conselleria de Medi Ambient, Govern de les Illes Balears. Palma de Mallorca.

Melendro, J. & Rodríguez, A. 1977. Censo de laridos y pelecánidos en la costa de Pontevedra (Diciembre 1974). *Ardeola* 22: 75–83.

Mellone, U., López-López, P., Limiñana, R. & Urios, V. 2012. Wintering habitats of Eleonora's Falcons *Falco eleonorae* in Madagascar. *Bird Study* 59: 29–36.

Mertens, R. & Stehle, W. 1960. De un viaje ornitológico por España en mayo–junio de 1959. *Ardeola* 6: 358–361.

Merzoug, A., Rouibi, A. & Houhamdi, M. 2010. Première observation du Flamant nain *Phoenicopterus minor* en Algérie. *Alauda* 78: 26.

Mester, H. 1966. Nota sobre aves de Ibiza y Formentera. *Ardeola* 11: 136–137.

Mestre, P. 1975. Las zonas de cría del Estornino Pinto (*Sturnus vulgaris*) en Cataluña. *Ardeola* 21: 457–467.

Mestre, P. 1978. Nidificación de *Sturnus unicolor* en Cataluña. *Ardeola* 24: 242–244.

Mestre, P. 1980. *Ocells del Penedès. Ocells nidificants*. Museu de Vilafranca. Vilafranca del Penedès.

Mestre, P. 1983. Variaciones sobre el estatus de algunas especies de aves durante los veinticinco últimos años en la comarca del Penedès (Cataluña). *Alytes* 1: 191–224.

Mestre, P., Peris, S., Santos, T., Suárez, F. & Soler, B. 1987. The decrease of the Black-eared Wheatear *Oenanthe hispanica* on the Iberian Peninsula. *Bird Study* 34: 239–243.

Meyburg, B. U. & Chancellor, R. D. (Eds) *Raptors Worldwide*. World Working Group on Birds of Prey/MME. Hungary.

Meyburg, B.-U., Gallardo, M., Meyburg, C. & Dimitrova, E. 2004. Migrations and sojourn in Africa of Egyptian vultures (*Neophron percnopterus*) tracked by satellite. *J. Ornithology* 145: 273–280.

Meyburg, B.-U., Howey, P. W., Meyburg, C. & Fiuczynski, K. D. 2011. Two complete migration cycles of an adult Hobby tracked by satellite. *British Birds* 104: 2–15.

Meyburg, B.-U., Meyburg, C. & Barbraud, J. C. 1998. Migration strategies of an adult short-toed eagle *Circaetus gallicus* tracked by satellite. *Alauda* 66: 39–48.

Meyburg, B.-U., Meyburg, C. & Pacteau, C. 1996. Migration autumnale d'un Circaete Jean-le-Blanc *Circaetus gallicus* suivi par satellite. *Alauda* 64: 339–344.

Mínguez, E. 1994. Censo, cronología de puesta y éxito reproductor del Paíño Común (*Hydrobates pelagicus*) en la Isla de Benidorm (Alicante, E de España). *Ardeola* 41: 3–11.

Mitchell, D. & Young, S. 1997. *Photographic handbook of the rare birds of Britain and Europe*. New Holland. London.

Moleón, M., Sánchez-Zapata, J., Real, J., García-Charton, J., Gil-Sánchez, J. M., Palma, L., Bautista, J. & Bayle, P. 2009. Large-scale spatio-temporal shifts in the diet of a predator mediated by an emerging infectious disease of its main prey. *J. Biogeography* 36: 1502–1515.

Molina, B. 2008. *Gaviota Cabecinegra. In:* Bertolero *et al.* (Eds), pp. 10–20.

Molina, B. (Ed.) 2009. *Gaviota reidora, sombría y patiamarilla en España. Población en 2007–2009 y método de censo.* SEO/BirdLife. Madrid.

Molina, B. & Bermejo, A. 2009. *La gaviota patiamarilla. In:* Molina (Ed.), pp 50–117.

Molina, B. & del Moral, J. C. 2005. *La Cigüeña Blanca en España. VI Censo Internacional (2004).* SEO/Birdlife. Madrid.

Molina, B. & Martínez, F. 2008. *El aguilucho lagunero en España. Población en 2006 y método de censo.* SEO/BirdLife. Madrid.

Moll, J. 1957. *Las aves de Menorca.* Estudio General Luliano. Palma de Mallorca.

Møller, A. P. 1975. Ynglebestanden af Sandterne *Gelochelidon nilotica Gmelin* 1789 i Europa, Afrika ogVestasien, med et tilbageblik over bestandsaendringer i dettearhundrede. *Dansk Ornithologisk Forenings Tidsskrift* 69: 1–8.

Montoya, F. 2000. Nuevos avances en el estudio del Torillo. *Quercus* 167: 64–65.

Moore, C. C. 1982. Notes on a first-year Iceland Gull (*Larus glaucoides*) at Peniche (Portugal), 1981. *Cyanopica* II (4): 42–43.

Moore, C. C. 1983. Première observation du Goéland a bec cerclé (*Larus delawarensis*) au Portugal. *Alauda* 51: 233.

Moore, C. C. 1990. Red-billed Tropicbird off Portugal in August 1988. *Dutch Birding* 12: 12–14.

Moore, C. C. 1992. Lapland Bunting *Calcarius lapponicus* and Dusky Warbler *Phylloscopus fuscatus*: two new species for Portugal. *Airo* 3: 21–24.

Moore, C. C. 1996. Nidificação de andorinhão-cafre em Portugal. *Pardela* 2: 20–21.

Moore, C. C. 2000a. Movimentações invulgares de aves marinhas junto ao cabo Carvoeiro. *Pardela* 13: 7–10.

Moore, C. C. 2000b. Alvéola-amarela *Motacilla (flava) cinereocapilla*. Notas sobre um novo *taxon* para Portugal, com comentários sobre as variações observadas relativamente à forma *M. f. iberiae*. *Pardela* 12: 18–20.

Moore, C. C. & Armelin, M. 1999. Pallas's Reed Bunting *Emberiza pallasi*: the first Portuguese and Iberian record. *Airo* 10: 34–36.

Moore, C. C. & Elias, G. 1997. Sobre a ocorrência invernal de Petinha-dos-campos *Anthus campestris* e de Petinha de Richard *Anthus novaeseelandiae* em Portugal. *Airo* 8: 44–47.

Moore, C. C., Noivo, C. & Faria, P. 2007. Pacific Golden Plover *Pluvialis fulva*: a new addition to the Portuguese avifauna list. *Anuário Ornitológico* 5: 137–139.

Morales, J. 2012. Papamoscas cerrojillo. In: Salvador, A. & Morales, M. B. (Eds).

Morales, M. B., Suárez, F., García, E. L. & de Juana, E. 2002. Movimientos estacionales e invernada del Sisón. *Quercus* 193: 34–39.

Moreira, F. 1994. Diet, prey-size selection and intake rates of black-tailed godwits. *Ibis* 136: 349–355.

Moreira, F. 1999a. On the use by birds of intertidal areas of the Tagus estuary: implications for management. *Aquatic Ecology* 33: 301–309.

Moreira, F. 1999b. Relationships between vegetation structure and breeding bird densities in fallow cereal steppes in Castro Verde, Portugal. *Bird Study* 46: 309–318.

Moreira, F. & Leitão, D. 1996. A preliminary study of the breeding bird community of fallows of cereal steppes in southern Portugal. *Bird Conservation International* 6: 255–259.

Moreno, J. 2006. Collalba negra. *In:* Carrascal, L. M. & Salvador, A. (Eds).

Moreno-Opo, R., Arredondo, A., Soria, C., Guil, F., Higuero, R. & Guzmán, J. 2009. La cigüeña negra *Ciconia nigra* en concentraciones postnupciales y migratorias ibéricas: fenología, actividad y estructura de edades. *Ecología* 22: 127–134.

Moreno-Opo, R. & Guil, F. (Eds) 2007. *Manual de gestión del hábitat y de las poblaciones de buitre negro en España.* Dirección General para la Biodiversidad, Ministerio de Medio Ambiente. Madrid.

Moreno-Opo, R. & Seoane, J. 2004. Descripción y cuantificación de la comunidad de aves reproductoras en los espacios gestionados por Patrimonio Nacional en el centro de España. *Anuario Ornitológico de Madrid 2003*: 52–73.

Moreno-Rueda, G. 2011. Mirlo acuático. *In:* Salvador, A. & Bautista, L. M. (Eds).

Mörzer Bruyns, M. F. 1975. The Marismas of the Guadalquivir as wintering area of the Grey Lag Geese (*Anser anser* L.). *Ardeola* 21: 127–131.

Mosquera, M. 2011. An exceptionally heavy passage of Black Kites *Milvus migrans* over Gibraltar on August 2nd 2011. *Gibraltar Bird Report* 10: 51–53.

Motis, A. 1992. Mixed breeding pairs of European Starling *Sturnus vulgaris* and Spotless Starling *Sturnus unicolor* in the north-east of Spain. *Butll. Grup Català d'Anellament* 9: 19–23.

Motjé, L. 1988. Nota sobre la presència de *Pelecanus onocrotalus* L. a Catalunya. *Butll. Institució Catalana de Història Natural* 55 (Sec. Zool.) 7: 119–120.

Mountfort, Guy. 1954. The larks of Andalucía. *Ibis* 96: 111–115.

Mountfort, G. 1958a. *Portrait of a Wilderness. The story of the Coto Doñana expeditions.* Hutchinson. London.

Mountfort, G. 1958b. Migración visible en las costas de Málaga. *Ardeola* 4: 139–143.

Mountfort, G. & Ferguson-Lees, I. J. 1961. The birds of the Coto Doñana. *Ibis* 103: 86–109.

Mullarney, K. 1999. The Elegant Tern in County Wexford. *Birding World* 12: 275–280.

Munilla, I. 1997a. Henslow's swimming crab (*Polybius henslowii*) as an important food for yellow-legged gulls (*Larus cachinnans*) in NW Spain. *ICES Journal of Marine Science* 54: 631–634.

Munilla, I. 1997b. Desplazamientos de la Gaviota Patiamarilla *Larus cachinnans* en poblaciones del norte de la Península Ibérica. *Ardeola* 44: 19–26.

Munilla, I., Díez, C. & Velando, A. 2007. Are edge bird populations doomed to extinction? A retrospective analysis of the common guillemot collapse in Iberia. *Biological Conservation* 137: 359–371.

Munn, P. W. 1921. Notes on the birds of Alcudia, Majorca. *Ibis* 63: 672–719.

Munn, P. W. 1924. Notes on the birds of Minorca. *Ibis* 66: 446–467.

Munn, P. W. 1925. Additional notes on the birds of Alcudia, Majorca. *Ibis* 67: 39–47.

Munn, P. W. 1926. Additional notes on the birds of the Balearic Islands. *Ibis* 68: 467–477.

Munn, P. W. 1928. Further notes on the birds of the Balearic Islands. *Ibis* 70: 17–22.

Munn, P. W. 1930. Further notes on the birds of the Balearic Islands. *Ibis* 72: 123–128.

Munn, P. W. 1931. The birds of the Balearic Islands. *Nov. Zool.* 37: 53–132.

Munn, P. W. 1932. Further notes on the birds of the Balearic Islands. *Ibis* 74: 262–266.

Munn, P. W. 1934. Notes on the birds of the Balearic Islands. *Ibis* 76: 717–723.

Munn, P. W. 1936. Further notes on the birds of the Balearic Islands. *Ibis* 78: 591–594.

Munn, P. W. 1939. Further notes on the birds of the Balearic Islands. *Ibis* 81: 159–161.

Munn, P. W. 1943. Notes on the birds of the Balearic Islands. *Ibis* 85: 341–343.

Munn, P. W. 1948a. Notes on the birds of the Balearic Islands. *Ibis* 90: 593–595.

Munn, P. W. 1948b. The nesting of Kentish and Little Ringed Plovers in Mallorca. *Ibis* 90: 595.

Muntaner, J. 1980. Sur la colonisation récente de l'ile de Minorque (Baléares) par la fauvette pitchou. *Alauda* 48: 185–192.

Muntaner, J. 1998. La gaviota de Audouin *Larus audouinii* en las islas Baleares y sus movimientos. *Anuari Ornitològic de les Balears 1997*: 35–65.

Muntaner, J. 2012. Comprobada la reproducción del buitre leonado en Mallorca. *Quercus* 315: 58–59.

Muntaner, J. & Congost, J. 1984. *Avifauna de Menorca.* Second edition. Treballs del Museu de Zoologia, Ajuntamente de Barcelona. Barcelona.

Muntaner, J. & Ferrer, X., 1983. Invernada del Porrón bastardo *Aythya marila* en España. *Miscelánea Zoológica* 7: 139–144.

Muntaner, J., Ferrer, X. & Martínez-Villalta, A. (Eds) 1983. *Atlas dels ocells nidificants de Catalunya i Andorra.* Ketres. Barcelona.

Muntaner, J., Ferrer, X. & Rodá, F. 1974. Censo de aves acuáticas en el Delta del Ebro. Invierno 1973–1974. *Ardeola* 20: 179–185.

Muñoz, A. R. & Palomo, L. J. 1996. Distribución actual y expansión de la Cotorra Argentina (*Myiopsitta monachus*) en la provincia de Málaga (sur de España). *Anuario Ornitológico de Málaga 1994*: 70–75.

Muñoz, A. R. & Real, R. 2006. Assessing the potential range expansion of the exotic monk parakeet in Spain. *Diversity and Distributions* 12: 656–665.

Muñoz Fernández, J. D. 2008. La familia Rallidae en las marismas de Sevilla: una contribución a su conocimiento. *In:* Barragán, A., Fernández-Mejías, J., Plata A. & Pinilla, J. (Eds) *Anuario Ornitológico de la provincia de Sevilla, años 2000–2005*: 27–32. GOSUR, Sevilla.

Muñoz Jiménez, J. & Sanz Herráiz, C. 1995. *Guía física de España, 5. Las montañas.* Alianza Editorial. Madrid.

Muñoz-Cobo, J. & Purroy, F. J. 1980. Wintering bird communities in the olive tree plantations of Spain. *In:* Oelke, H. (Ed.) *Bird census work and nature conservation*, pp. 185–189. Dachverband Deutscher Avifaunisten. Göttingen.

Muñoz-Cobo, J. 1990. Evolución de la avifauna nidificante en olivares viejos de Jaén. *Testudo* 1: 99–117.

Muñoz-Fuentes, V., Green, A. J., Sorensen, M. D., Negro, J. J. & Vilà, C. 2006. The Ruddy Duck *Oxyura jamaicensis* in Europe: natural colonization or human introduction? *Molecular Ecology* 15: 1441–1453.

Muñoz-Pulido, R., Sierra, J. & Robles, J. L. 1990. First successful breeding record of the Tufted Duck (*Aythya fuligula*) in Spain. *Doñana, Acta Vertebrata* 17: 103–104.

Murgui, E. 2009. Seasonal patterns of habitat selection of the House Sparrow *Passer domesticus* in the urban landscape of Valencia (Spain). *J. Ornithology* 150: 85–94.

Murgui, E. 2011a. How many common breeding birds are there in Spain? A comparison of census methods and national population size estimates. *Ardeola* 58: 343–364.

Murgui, E. 2011b. Gorrión común. *In:* Salvador, A. & Morales, M. B. (Eds).

Murgui, E. & Macias, A. 2010. Changes in the House Sparrow *Passer domesticus* population in Valencia (Spain) from 1998 to 2008. *Bird Study* 57: 281–288.

Muriel, R., Ferrer, M., Casado, E., Madero, A. & Pérez Calabuig,

C. 2011. Settlement and successful breeding of reintroduced Spanish imperial eagles *Aquila adalberti* in the province of Cádiz (Spain). *Ardeola* 58: 323–334.

Muriel, R., Ferrer, M., Casado, E. & Pérez, C. 2010. First successful breeding of reintroduced Ospreys *Pandion haliaetus* in mainland Spain. *Ardeola* 57: 175–180.

Muriel, R., Ferrer, M., Casado, E. & Schmidt, D. 2006. First breeding success of Osprey (*Pandion haliaetus*) in mainland Spain since 1981 using cross-fostering. *J. Raptor Research* 40: 303–304.

Murillo, F. & Sancho, F. 1969. Migración de *Sylvia atricapilla* y *Erithacus rubecula* en Doñana según datos de capturas. *Ardeola* 13: 129–137.

Musgrove, A. J., Austin, G. E., Hearn, R. D., Holt, C. A., Stroud, D. A. & Wotton, Simon. R. 2011. Overwinter population estimates of British waterbirds. *British Birds* 104: 364–397.

Nadal, B. 1961. Lista sistemática de aves recientemente capturadas en la Isla de Mallorca. *Ardeola* 7: 234.

Nadal, B. 1965. Aves poco comunes capturadas en Mallorca de 1961 a 1964. *Ardeola* 10: 55.

Nadal, B. 1966. Capturas interesantes para la ornitología balear, año 1965. *Ardeola* 11: 139.

Nadal, B. 1971. Lista de capturas de aves raras o poco comunes en Mallorca, periodo 1966–1969. *Ardeola* 15: 97.

Navarrete, J. 2008. Migración postnupcial de la Pardela Cenicienta *Calonectris diomedea* en Ceuta. *Bol. GIAM* 31: 2–6.

Navarro Medina, J. D. & Navarro García, J. 1984. La avifauna de los embalses de 'El Hondo' (Alicante). *Mediterranea* Ser. Biol. 7: 117–137.

Navarro, J. D. & Robledano, F. 1995. *La Cerceta Pardilla* Marmaronetta angustirostris *en España*. ICONA. Madrid.

Navarro, M., San Segundo, C., Caballero, J. & Rodríguez, J. L. 1990. Segunda puesta de lechuza común (*Tyto alba*) asociada al aumento de consumo del topillo campesino (*Microtus arvalis*). *Doñana, Acta Vertebrata* 17: 106–108.

Navas, L. 1918–1920. Pájaros de Aragón. *Revista de la Academia de Ciencias Exactas, Físicas y Naturales de Zaragoza* 3: 7–69 & 5: 65–130.

Navedo, J. G., Masero, J. A. & Juanes, J. A. 2007. Updating waterbird population estimates within the East Atlantic flyway: status and trends of migratory waterbirds in Santoña marshes. *Ardeola* 54: 237–249.

Negro, J. J., de la Riva, M. & Bustamante, J. 1991. Patterns of winter distribution and abundance of Lesser Kestrel (*Falco naumanni*) in Spain. *J. Raptor Research* 25: 30–35.

Negro, J. J., Donázar, J. A., Hiraldo, F., Hernández, L. M. & Fernández, M. A. 1993. Organochlorine and heavy-metal contamination in nonviable eggs and its relation to breeding success in a Spanish population of lesser kestrels (*Falco naumanni*). *Environmental Pollution* 82: 201–205.

Negro, J. J., Torres, M. J. & Godoy, J. A. 2001. RAPD analysis for detection and eradication of hybrid partridges (*Alectoris rufa* x *A. graeca*) in Spain. *Biological Conservation* 98: 19–24.

Neto, J. M., Encarnação, V. & Fearon, P. 2010. Distribution, phenology and condition of aquatic warblers *Acrocephalus paludicola* migrating through Portugal. *Ardeola* 57: 181–190.

Nisbet, I. C. T. 1960. Sobre algunas aves observadas en Andalucía en abril de 1960. *Ardeola* 6: 211–219.

Nisbet, I. C. T., Evans, P. R. & Feeny, P. P. 1961. Migration from Morocco into southwest Spain in relation to weather. *Ibis* 103: 349–372.

Noble, H. 1902. Forty-four days' nesting in Andalucia. *Ibis* 44: 67–89.

Norman, R. 1971. Resultados de una visita al delta del Ebro, primavera 1969. *Ardeola* 15: 85–90.

Norman, R. K. 1973. Una visita a Ibiza en la primavera de 1971. *Ardeola* 19: 60–63.

Noval, A. 1967. Estudio de la avifauna de Guipúzcoa. *Munibe* 1–2: 5–78.

Noval, A. 1971. Movimientos estacionales y distribución del Camachuelo Común (*Pyrrhula pyrrhula iberiae*) en el norte de España. *Ardeola* vol. especial: 491–507.

Noval, A. 1975. *El libro de la fauna Ibérica. Aves.* Naranco. Oviedo.

Noval, A. 1981. *Enciclopedia Temática de Asturias*. Tomo 2: Vertebrados. Silverio Cañada editor. Gijón.

Noval, A. 1986. *Guía de las Aves de Asturias.* Alfredo Noval editor. Gijón.

Noval, A. 1988. *Introducción al estudio de la comunidad de vertebrados*. In: Echevarri, J. (Ed.) *Naturaleza y vida en los Picos de Europa*, pp. 122-155. INCAFO. Madrid.

Noval, A. 2001. *Guía de las Aves de Asturias.* 3rd edition. Alfredo Noval editor. Gijón.

Novoa, C. 2008. La perdix grise de montagne *Perdix perdix hispaniensis*. *In:* ONFC & FNC (Ed.) *Atlas de la biodiversité de la faune sauvage - tout le gibier de France*, pp. 99–104. Hachette. Paris.

Nuñez, J. C. 2000. *Aves de Extremadura. Anuario Adenex 1998*: 242.

Obeso, J. R. & Bañuelos, M. J. 2003. *El urogallo (Tetrao urogallus cantabricus) en la Cordillera Cantábrica*. Ministerio de Medio Ambiente. Madrid.

Occhiato, D. 2003. Pine Buntings in Italy: status and distribution. *Dutch Birding* 25: 32–39.

Ocio, G. & Astigarraga, J. G. 2008. Distribución espacio-temporal de las aves marinas en el Cantábrico oriental. *Artadi, Anuario Ornitológico de Bizkaia 2002–2003*: 31–39.

Ocio, G. & Sánchez, C. 2007. Presencia de la Gaviota de Audouin *Larus audouinii* fuera de su area biogeográfica tradicional de reproducción, migración e invernada en la Península Ibérica y Europa Occidental. *Bol. GIAM* 28: 2–6.

Ojanen, M. 1982. Var övervintrar den svartvita flugsnapparen *Ficedula hypoleuca*? Kommentar till Sten Österlöfs artikel i Vår Fågelvärld 1979. *Vår Fågelvärld* 41: 29–30.

Olea, P. P. 2000. *Ecología de la población reproductora de la Graja* Corvus frugilegus *en la provincia de León*. Universidad de León. León.

Olea, P. P. 2001. Postfledging dispersal in the endangered lesser kestrel *Falco naumanni*. *Bird Study* 48: 110–115.

Olea, P. P. & Baglione, V. 2008. Population trends of Rooks *Corvus frugilegus* in Spain and the importance of refuse tips. *Ibis* 150: 98–109.

Olea, P. P., García, J., Fernández, J., Román, J., González-Vélez, M. & Gutiérrez, C. 1997. Situación actual de la población reproductora de la Graja (*Corvus frugilegus* Linnaeus, 1758) en España. *Ardeola* 44: 37.

Oliva, M. & Montes, E. 2005. On the origin and domestication of *Olea europaea* L. (olive) in Andalucía, Spain, based on the biogeographical distribution of its finds. *Vegetation History and Archaeobotany 14*: 551–561.

Onrubia, A., Arroyo, G. M., Barrios, L., Muñoz. A. R., de la Cruz, A., Ramírez, J., González, M. & Cuenca, D. 2009. Migración diurna visible de pequeñas aves en el Estrecho de Gibraltar. Año 2008. *Migres, Revista de Ecología* 1: 65–72.

Onrubia, A., Arroyo, J. L., Andrés, T., Gómez, J., Unamuno, J. M. & Zufiaur, F. 2003. El Mosquitero Ibérico (*Phylloscopus ibericus*): identificación, biometría y apuntes sobre su migratología. *Revista de Anillamiento* 12: 18–29.

Onrubia, A. & Jubete, F. 1998. Status reproductor de la Lechuza Campestre *Asio flammeus* en España (1993–1994). *In:* Chancellor *et al.* (Eds), pp. 459–466.

Onrubia, A., Muñoz, A.-R., Arroyo, G.M., Ramírez, J., de la Cruz, A., Barrios, L., Meyburg, B.-U., Meyburg, C. & Langgemach, T. 2011. Autumn migration of Lesser Spotted Eagle *Aquila pomarina* in the Strait of Gibraltar: accidental or regular? *Ardea* 99: 113–116.

Onrubia, A. & Tellería, J. L. 2013. Fenología migratoria del mosquitero Ibérico *Phylloscopus ibericus* en la Península Ibérica: una comparación con los mosquiteros común *P.collybita* y musical *P. trochilus*. Pp. 87–92 in Rodríguez, N., García, J. & Copete, J. L. (Coords). 2013. *El Mosquitero Ibérico*. Grupo Ibérico de Anillamiento, León.

Ontiveros, D. 2010. Águila perdicera. *In:* Salvador, A. & Bautista, L. M. (Eds).

Ontiveros, D. 2011. Culebrera europea. *In:* A. Salvador & M. B. Morales (Eds).

Ontiveros, D. & Pleguezuelos, J. M. 2003. Influence of climate on Bonelli's eagle (*Hieraaetus fasciatus* V.) breeding success through the Western Mediterranean. *J. Biogeography* 30: 755–760.

Ontiveros, D., Pleguezuelos, J. M. & Caro, J., 2005. Prey density, prey detectability and food habits: the case of Bonelli's eagle and the conservation measures. *Biological Conservation* 123: 19–25.

Ontiveros, D., Real, J., Balbontín, J., Carrete, M., Ferreiro, E., Ferrer, M., Mañosa, S., Pleguezuelos, J. M. & Sánchez-Zapata, J. A. 2004. Biología de la conservación del Águila Perdicera *Hieraaetus fasciatus* en España: investigación científica y gestión. *Ardeola* 51: 461– 470.

Oro, D., Aguilar, J. S., Igual, J. M. & Louzao, M. 2004. Modelling demography and extinction risk in the endangered Balearic shearwater. *Biological Conservation* 116: 93–102.

Oro, D., Bosch, M. & Ruiz, X. 1995. Effects of a trawling moratorium on the breeding success of the Yellow-legged Gull *Larus cachinnans*. *Ibis* 137: 547–549.

Oro, D., Louzao, M. & Genovart, M. 2009. Pardela balear. In: Salvador, A. & Bautista, L. M. (Eds).

Oro, D. & Martínez-Abraín, A. 2007. Deconstructing myths on large gulls and their impact on threatened sympatric waterbirds. *Animal Conservation* 10: 117–126.

Oro, D., Martínez-Abraín, A., Villuendas, E., Sarzo, B., Mínguez, E., Carda, J. & Genovart, M. 2011. Lessons from a failed translocation program with a seabird species: determinants of success and conservation value. *Biological Conservation* 144: 851–858.

Oro, D. & Martínez-Vilalta, A. 1994a. Migration and dispersal of Audouin's Gull *Larus audouinii* from the Ebro Delta colony. *Ostrich* 65: 225–230.

Oro, D. & Martínez-Vilalta, A. 1994b. Factors affecting kleptoparasitism and predation rates upon a colony of Audouin's Gull (*Larus audouinii*) by Yellow-legged Gulls (*Larus cachinnans*) in Spain. *Colonial Waterbirds* 17: 35–41.

Oro, D. & Muntaner, J. 2000. La gaviota de Audouin en Cabrera. *In:* Pons, G. X. (Ed.) *Las aves del Parque Nacional marítimo-terrestre del archipiélago de Cabrera* (*Islas Baleares, España*), pp. 95–112. Ministerio de Medio Ambiente. Madrid.

Oro, D., Pradel, R. & Lebreton, J.-D. 1999. Food availability and nest predation influence life history traits in Audouin's Gull, *Larus audouinii*. *Oecologia* 118: 438–445.

Oro, D. & Ruxton, G. D. 2001. The formation and growth of seabird colonies: Audouin's Gull as a case study. *J. Animal Ecology* 70: 527–535.

Ortega, E., Mañosa, S., Margalida, A., Sánchez, R., Oria, J. & González, L. M. 2009. A demographic description of the recovery of the Vulnerable Spanish imperial eagle *Aquila adalberti*. *Oryx* 43: 113–121.

Ortega, F., Raya, C., Paracuellos, M. & Guerrero, F. 2007. Reintroducción de la Focha Moruna (*Fulica cristata*): valoración de diferentes técnicas en la formación de nuevos núcleos de reproducción en el Mediterráneo Occidental. *In:* Paracuellos, M. (Ed.) *Ambientes mediterráneos. Funcionamiento, biodiversidad y conservación de los ecosistemas mediterráneos*, pp. 173–183. Instituto de Estudios Almerienses. Almería.

Ortega, F. J., Molina, B. & Martínez, F. 2006. El alimoche común (*Neophron percnopterus*) en la Comunidad de Madrid: distribución histórica, estatus y fenología. *Anuario Ornitológico de Madrid 2005*: 44–67.

Ortego, J. 2004. A possible case of double brooding of eagle-owls (*Bubo bubo*) in Spain. *J. Raptor Research* 38: 378–379.

Ortego, J. 2010. Cernícalo primilla. *In:* Salvador, A. & Bautista, L. M. (Eds).

Ortego, J., Aparicio, J. M., Calabuig, G. & Cordero, P. J. 2007. Increase of heterozygosity in a growing population of lesser kestrels. *Biology Letters* 3: 585–588.

Ortego, J. & Calvo, J A. 2004. Distribución y estatus poblacional y de conservación del Búho Real (*Bubo bubo hispanus*) en el sector central de la provincia de Toledo: resultados preliminares. *Anuario Ornitológico de Toledo. Revisión Histórica / 2001*: 148–159.

Ortego, J., Cordero, P. J., Aparicio, J. M. & Calabuig, G. 2010. Parental genetic characteristics and hatching success in a recovering population of Lesser Kestrels. *J. Ornithology* 151: 155–162.

Osborne, P. E. & Suárez-Seoane, S. 2007. Identifying core areas in a species' range using temporal suitability analysis: an example using little bustards *Tetrax tetrax* L. in Spain. *Biodiversity and Conservation* 16: 3505–3518.

Österlof, S. 1977. Migration, wintering areas, and site tenacity of the European Osprey *Pandion haliaetus* (L.). *Ornis Scandinavica* 8: 61–78.

Otero, C. 1985. Techniques for the capture of Little Bustards during the autumn/winter season. *Bustard Studies* 2: 171–172.

Pacheco, C. & McGregor, P. K. 2004. Conservation of the Purple Gallinule (*Porphyrio porphyrio* L.) in Portugal: causes of decline, recovery and expansion. *Biological Conservation* 119: 115–120.

Paepegaey, B. 2012. Vague de froid de décembre 2010: les anatidés abandonnés aux tirs des chasseurs du nord de la France. *Ornithos* 19: 42–56.

Paevskii, V. A. 1971. (*Atlas of bird migration according to data on birds ringed in Courland Bay, Kaliningrad*) (In Russian). Science Editorial. Leningrad.

Pagezy, H. & Trotignon, J. 1972. Breve reseña de un censo de aves acuáticas en las costas de Portugal, invierno 1968–1969. *Ardeola* 17–18: 99–104.

Pajuelo, L., de Lope, F. & da Silva, E. 1992. Biología de la reproducción del Avión Común (*Delichon urbica*) en Badajoz, España. *Ardeola* 39: 15–23.

Palacín, C. 1990. Biología y situación del Alcotán en España. *Quercus* 53: 6–10.

Palacín, C. 2007. *Comportamiento migratorio de la Avutarda Común* Otis tarda *en la Peninsula Ibérica*. Doctoral thesis. Universidad Complutense, Madrid.

Palacín, C. 2010. Alcotán europeo. In: Salvador, A. & Bautista, L. M. (Eds).

Palacín, C. & Alonso, J. C. 2008. An updated estimate of the world status and population trends of the Great Bustard *Otis tarda*. *Ardeola* 55: 13–25.

Palacín, C., Alonso, J. C., Alonso, J. A., Martín, C. A., Magaña, M. & Martín, B. 2009. Differential migration by sex in the Great Bustard: possible consequences of an extreme sexual size dimorphism. *Ethology* 115: 1–10.

Palacios, B. & Asensio, B. 1986. *Informe sobre la campaña de anillamiento de aves en España: años 1983 y 1984.* Ministerio de Agricultura, Pesca y Alimentación. Madrid.

Palacios, F., Garzón, J. & Castroviejo, J. 1975. La alimentación de la Avutarda (*Otis tarda*) en España, especialmente en primavera. *Ardeola* 21: 347–406.

Palacios, J. & Rodríguez, M. 1993. *Guía de la Reserva de las Lagunas de Villafáfila (Fauna. Aves)*. Junta de Castilla y León. Valladolid.

Palaus, F. J. 1960. Notas ornitológicas del Nor-Este de España (con algunos datos de Levante). *Ardeola* 6: 221–234.

Palaus, F. J. 1962. Algunas capturas interesantes en Cataluña, Levante y Baleares. *Ardeola* 8: 241–242.

Palma, C. 1985. The present situation of birds of prey in Portugal. *In:* Newton, I. & Chancellor, R. E. (Eds.) *Conservation studies on raptors*, pp. 3–14. ICBP. Cambridge.

Palma, L. 2001. The Osprey *Pandion haliaetus* on the Portuguese coast: past, present and recovery potential. *Vogelwelt* 122: 179–180.

Palma, L., Beja, P., Pais, M., & da Fonseca, L. C. 2006. Why do raptors take domestic prey? The case of Bonelli's eagles and pigeons. *J. Applied Ecology* 43: 1075–1086.

Palma, L., Onofre, N. & Pombal, E. 1999. Revised distribution and status of diurnal birds of prey in Portugal. *Avocetta* 23: 3–18.

Palomares, V. 2009. Un Pélican blanc *Pelecanus onocrotalus* cohabitant et migrant avec des Cigognes blanches *Ciconia ciconia*. *Ornithos* 16: 346–348.

Palomino, D. 2006. *El milano negro en España. I censo nacional (2005)*. Seguiminto de aves: Monografía no.11. SEO/Birdlife. Madrid.

Palomino, D. 2011. Factors determining the large-scale seasonal abundance of the common kestrel in Central Spain. *Ardeola* 58: 87–101.

Palomino, D. 2012. Lavandera boyera. *In:* Salvador, A. & Morales, M. B. (Eds).

Palomino, D. & Molina, B. (Eds) 2009. *Aves acuáticas reproductoras en España. Población en 2007 y método de censo.* SEO/BirdLife. Madrid.

Palomino, D. & Valls, J. 2011. *Las rapaces forestales en España. Población reproductora en 2009–2010 y método de censo.* SEO/Birdlife. Madrid.

Papaconstantinou, C. 2007. *Eleonora´s Falcon, ruling the Aegean skies*. Helenic Ornithological Society. Athens.

Paracuellos, M. 1994. Dinánima anual de la comunidad de paseriformes en un saladar litoral del sudoeste ibérico. *Doñana Acta Vertebrata* 21: 119–130.

Paracuellos, M. & Nevado, J. C. 1995. Nidificación de láridos en la provincia de Almería (SE Ibérico). *Doñana Acta Vertebrata* 22: 102–106.

Paracuellos, M. & Nevado, J. C. 2010. Culling Yellow-legged Gulls *Larus michahellis* benefits Audouin's Gulls *Larus audouinii* at a small and remote colony. *Bird Study* 57: 26–30.

Pardo, L. 1922. La *Havelda glacialis* (Stph.) en Valencia. *Bol. Real Sociedad Española de Historia Natural* 22: 251.

Parsons, J. J. 1960. Sobre la caza a gran escala del estornino pinto (*Sturnus vulgaris*) en España. *Ardeola* 6: 235–241.

Pärt, T. & Gustafsson, L. 1989. Causes and consequences of breeding dispersal in the Collared Flycatcher *Ficedula albicollis*. *J. Animal Ecology* 58: 305–320.

Pascual, P. & Aparicio, R. J. 1998. Aves marinas en el litoral oriental de Almería, sureste de España. *Bol. GIAM* 24: 13–20.

Patacho, D. 2001. Felosa-comum *Phylloscopus collybita* – uma nova espécie nidificante em Portugal. *Pardela* 15: 14–16.

Paterson, A. M. 1987. A study of seabirds in Málaga Bay, Spain. *Ardeola* 34: 167–192.

Paterson, A. M. 1990. Seasonal evolution of the gull populations in Málaga, Spain. *Ardeola* 37: 19–27.

Paterson, A. M. 1997. *Las aves marinas de España y Portugal.* Lynx. Barcelona.

Paterson, A. M., Martínez-Vilalta, A. & Dies, J. I. 1992. Partial breeding failure of Audouin's Gull in two Spanish breeding colonies in 1991. *British Birds* 85: 97–100.

Paterson, A. M. & Riddiford, N. J. 1990. Does the Cape Gannet enter European eaters? *British Birds* 83: 519–526.

Paul, J. P. & Olioso, G. 2006. Afflux mémorable de Jaseurs boréaux en France dans l'hiver 2004– 2005. *Ornithos* 13: 2–11.

Paulino d'Oliveira, M. 1896. *Aves da Peninsula Ibérica e especialmente de Portugal.* Coimbra University Press.

Pavón, D., Limiñana, R., Urios, V., Izquierdo, A., Yáñez, B., Ferrer, M. & de la Vega, A. 2010. Autumn migration of juvenile Short-toed Eagles *Circaetus gallicus* from southeastern Spain. *Ardea* 98: 113–117.

Pechuán, L. 1965. Sobre Charranes, Fumareles y Ardeidas que nidifican en la Albufera de Valencia. *Ardeola* 10: 39–41.

Pechuán, L. 1969. Captura de un *Alopochen aegyptiacus* en la Albufera de Valencia. *Ardeola* 13: 244.

Pechuán, L. 1973. Algunas observaciones y capturas en las Islas Columbretes. *Ardeola* 19: 56.

Pechuán, L. 1974. La colonia de *Larus audouinii* en las Islas Columbretes. *Ardeola* 20: 358–359.

Peinado Lorca, M. & Rivas Martínez, S. (Eds) 1987. *La vegetación de España.* Universidad de Alcalá. Alcalá de Henares.

Peiró, I. G. 1997. A study of migrant and wintering bluethroats *Luscinia svecica* in southeastern Spain. *Ringing and Migration* 18: 18–24.

Peiró, V. 1990. Aspectos de la reproducción de la Tórtola Común (*Streptopelia turtur* L.) en Madrid. *Mediterránea* Ser. Biol. 12: 89–96.

Pelayo, E., Sampietro, F. J., Rivas, J. L., Sanz, J., Albero, J. C. & Alcántara, M. 2009. Seguimiento de los pasos migratorios e invernada de la grulla común (*Grus grus*) en la Laguna de Gallocanta (Aragón, España). *IV Congreso Ibérico de Ornitología*. Elvas.

Penas, X. M. Pedreira, C. & Silvar, C. 1991. *Guía das aves de Galicia*. Bahía Ediciones. La Coruña.

Pennington, M. G., Riddington, R. & Miles, W. T. S. 2012. The Lapland Bunting influx in Britain & Ireland in 2010/11. *British Birds* 105: 654–673.

Penteriani, V. & Delgado, M. M. 2010. Búho real. *In:* Salvador, A. & Bautista, L. M. (Eds).

Penteriani, V., Delgado, M. M., Maggio, C., Aradis, A. & Sergio, F. 2005. Development of chicks and predispersal behaviour of young in the Eagle Owl *Bubo bub*. *Ibis* 147: 155–168.

Penteriani, V., Lourenço, R. & Delgado, M. 2012. Eagle Owls in Doñana: a conservation dilemma or not? *British Birds* 105: 88–105.

Perea, D. F., Ruiz-Tapiador, F. J. & Ruiz-Tapiador, J. 2004. La expansión de la Gaviota Patiamarilla (*Larus michahellis*) en aguas interiores y su nidificación en Toledo. *Anuario Ornitológico de Toledo. Revisión Histórica/2001*: 126–139.

Perea, J. L., Morales, M. y Velasco J. 1990. *El alimoche (*Neophron percnopterus*) en España. Población, distribución, problemática y conservación.* ICONA. Madrid.

Pereira, C. & Campos, A. C. 2000. A narceja-galega, uma espécie poco conhecida em Portugal. *Pardela* 12: 5–7.

Perennou, C. 2003. La protection de l'Érismature à tête blanche *Oxyura leucocephala* en France. *Ornithos* 10: 281–283.

Perez, C. & Cortes, J. 2002. Passage of the Iberian Chiffchaff *Phylloscopus ibericus* at Gibraltar. *Gibraltar Bird Report* 2001: 29–31.

Perez, C. 2005. Mountain Chiffchaff *Phylloscopus sindianus* Brooks, 1880, in Gibraltar. *Iberis* 1: 3–10.

Pérez, F. F. & Boscolo, R. (Eds) 2010. *Climate in Spain: past, present and future.* CLIVAR Spain & Ministerio de Medio Ambiente y Medio Rural y Marino. Madrid.

Pérez, J. M., Sánchez-Zapata, J. A., Botella, F., Donázar, J. A., Hiraldo, F. & Sayago, J. M. 2010. Broods of five fledglings in the Eurasian Eagle-Owl (*Bubo bubo*). *J. Raptor Research* 44: 161–163

Pérez Chiscano, J. L. 1965. Nidificación de canasteras y charrancitos en el río Guadiana. *Ardeola* 10: 37–38.

Pérez Chiscano, J. L. 1971. Captura de *Psittacula k. krameri* (Scopoli) en el SE de la provincia de Cáceres. *Ardeola* 15:

144–145.

Pérez Chiscano, J. L. 1975. Avifauna de los cultivos de regadíos del Guadiana. *Ardeola* 21: 753–794.

Pérez Chiscano, J. L., Ferrero, J. J. & Román, J. A. 1984. Los Vencejos Reales *Apus melba* en la cuenca extremeña del Guadiana. *Alytes* 2: 107–133.

Pérez de Ana, J. M. 1993. Distribución y expansión de la Curruca Cabecinegra (*Sylvia melanocephala*) en la Península Ibérica. *Ardeola* 40: 81–85.

Pérez-Barbería, F. J. 1991. Influencia de la variación latitudinal en la contribución de los murciélagos (Chiroptera) a la dieta de la Lechuza Común (*Tyto alba*). *Ardeola* 38: 61–68.

Pérez-Contreras, J., Garzón, J., García, I. & Tamayo, A. 2005. Situación del verderón serrano *Serinus citrinella* (Pallas, 1746) en Sierra Nevada (España). *Zoologia baetica* 16: 171–176.

Pérez-García, J. M., Sánchez-Zapata, J. A. & Botella, F. 2012. Distribution and breeding performance of a high density Eagle Owl population in SE Spain. *Bird Study* 59: 22–28.

Pérez-Tris, J. 2000. Evolución de la población de Cormorán Grande (*Phalacrocorax carbo*) invernante en España según los resultados del anillamiento. *In:* Carbonell, R. & Juliá, M. (Eds) *Actas de las XIII Jornadas Ornitológicas Españolas*, pp. 215–217. SEO/BirdLife. Madrid.

Pérez-Tris, J. & Asensio, B. 1997a. Migración e invernada de la Lavandera Boyera (*Motacilla flava*) en la Península Ibérica. *Ardeola* 44: 71–78.

Pérez-Tris, J. & Asensio, B. 1997b. Migración e invernada de las lavanderas Cascadeña *Motacilla cinerea* y Blanca *M. alba* en la Península Ibérica e Islas Baleares. *Doñana, Acta Vertebrata* 24: 79–89.

Pérez-Tris, J., Ramírez, Á. & Tellería, J. L. 2003. Are Iberian Chiffchaffs *Phylloscopus (collybita) brehmii* long-distance migrants? An analysis of flight-related morphology. *Bird Study* 50: 146–152.

Peris, S. J. 1980. Biología del estornino negro (*Sturnus unicolor* Temm.). Alimentación y variación de la dieta. *Ardeola* 25: 207–240.

Peris, S. J. 1991. Ringing recoveries of the Spotless Starling *Sturnus unicolor* in Spain. *Ringing & Migration* 12: 124–125.

Peris, S. J. & Galeano, R. E. 2005. Recuperaciones de anillamientos en el estornino negro (*Sturnus unicolor*): años 1989–1998. *Revista de Anillamiento* 15: 10–15.

Peris, S. J., Motis, A., Martínez-Vilalta, A. & Ferrer, X. 1991. Winterquartiere und Winterbestand des Stars (*Sturnus vulgaris*) in Spanien während der lezten 30 Jahre. *J. für Ornithologie* 132: 445–449.

Perktas, U., Barrowclough, G. F. & Groth, J. G. 2011. Phylogeography and species limits in the green woodpecker complex (Aves: Picidae): multiple Pleistocene refugia and range expansion across Europe and the Near East. *Biological J. Linnean Society* 104: 710–723.

Persson, H. 2004. Lesser White-fronted Geese shot in Spain in the winters 1985/86–2000/01. *Fennoscandian Lesser White-fronted Goose conservation project. Report 2001–2003*, p. 50. WWF Finland & Norsk Ornitologisk Forening. Oslo.

Petronilho, J. M. S. & Vingada, J. V. 2002. First data on feeding ecology of Goshawk *Accipiter gentilis* during the breeding season in the Natura 2000 site Dunas de Mira, Gândara e Gafanhas (Beira Litoral, Portugal). *Airo* 11: 12–16.

Pimenta, M. & Santarém, M. 1996. *Atlas das aves do Parque Nacional da Peneda-Gerês*. Parque Nacional da Peneda-Gerês. ICN. Lisboa.

Pineau, J. 1975. Première donnée sur l'Aigle criard *Aquila clanga* au Maroc. *Alauda* 43: 196.

Pineau, J. & Giraud-Audine, M. 1974. Notes sur les migrateurs traversant l'extrême nord-ouest du Maroc. *Alauda* 42: 159–188.

Pineau, J. & Giraud-Audine, M. 1976. Notes sur les oiseaux hivernant dans l'extreme nord-ouest du Maroc et sur leurs mouvements. *Alauda* 44: 47–75.

Pineau, J. & Giraud-Audine, M. 1979. *Les oiseaux de la Péninsule Tingitane.* Travaux de l'Institut Scientifique. Rabat.

Pinilla, J., Arambarri, R. & Rodríguez, A. F. 1994. Distribución actual y estima poblacional del aguilucho pálido (*Circus cyaneus*) en España. *Ardeola* 41: 177–181.

Pinilla, J. & de la Puente, J. 2001. Migración e invernada del Mosquitero Común (*Phylloscopus collybita*) en el sur de Madrid durante el 2000: resultados del anillamiento en dos estaciones de esfuerzo constante. *Anuario Ornitológico de Madrid 2000*: 102–109.

Pinilla, J. & de la Puente, J. 2003. Fenología del Petirrojo (*Erithacus rubecula*) en el tramo bajo del río Jarama. *Anuario Ornitológico de Madrid 2002*: 86–91.

Pizarro, J. M. & Ferrero, J. J. 1999. Cigüeña Negra, símbolo de lo agreste. *Biológica* 33: 36–44.

Pleguezuelos, J. M. 1992. *Avifauna nidificante de las sierras béticas orientales y depresiones de Guadix, Baza y Granada.* Universidad de Granada. Granada.

Pleguezuelos, J. M. & Manrique, J. 1987. Distribución y estatus de las aves esteparias nidificantes en el SE de la Península Ibérica. *Actas del I Congreso Internacional sobre Aves Esteparias, León*: 349–358.

Pocino, N., Giralt, N. & Ferrer, X. 2005. Colonització i expansió de la Tórtora Turca *Streptopelia decaocto* a Catalunya. *Revista Catalana d'Ornitologia* 21: 1–10.

Polo, T. 2009. Fenología, irrupciones y distribución del Picogordo (*Coccothraustes coccothraustes*) en la Comunidad Valenciana. *El Serenet* 7: 1–8.

Polo, V. 2010. Carbonero garrapinos. *In:* Salvador, A. & Bautista, L. M. (Eds).

Ponce, C., Calleja, J. A. & Magaña, Ó. 2008. Migración del chotacabras cuellirrojo (*Caprimulgus ruficollis*) en la Comunidad de Madrid. *Revista de Anillamiento* 21–22: 71–76.

Pons, J. & Pausas, J. G. 2007. Not only size matters: acorn selection by the European jay (*Garrulus glandarius*). *Acta Oecologica* 31: 353–360.

Pons, J.-M., Crochet, P. A., Thery, M. & Bermejo, A. 2004. Geographical variation in the yellow-legged gull: introgression or convergence from the herring gull? *J. Zoological Systematics and Evolutionary Research* 42: 245–256.

Pons, J.-M., Olioso, G., Cruaud, C. & Fuchs, J. 2011. Phylogeography of the Eurasian green woodpecker (*Picus viridis*). *J. Biogeography* 38: 311–325.

Poot, M. & Flamant, R. 2006. Numbers, behaviour and origin of Mediterranean Gulls *Larus melanocephalus* wintering along the west coast of southern Portugal. *Airo* 15: 43–50.

Potti, J. 1985. La sucesión de las comunidades de aves en los pinares repoblados de *Pinus sylvestris* del Macizo de Ayllón (Central Range). *Ardeola* 32: 253–277.

Potti, J. & Merino, S. 1995. Some male Pied Flycatchers *Ficedula hypoleuca* in Iberia become collared with age. *Ibis* 137: 405–409.

Potti, J. & Montalvo, S. 1990. Ocupación de areas con nidales por el papamoscas cerrojillo (*Ficedula hypoleuca*). *Ardeola* 37: 75–84.

Preiswerk, G. 1981a. Protocol of the observation of a Laughing Gull (*Larus atricilla*) at Porto (Portugal). *Cyanopica* II (3): 40–42.

Preiswerk, G. 1981b. Protocol of the observation of an Iceland Gull (*Larus glaucoides*) at Porto (Portugal). *Cyanopica* II (3): 37–39.

Premuda, G. & Baghino, L. 2004. La migrazione autunnale dell' Aquila minore, *Hieraaetus pennatus*, attraverso la Peninsola Italiana. *Rivista Italiana di Ornitologia* 74: 103–115.

Premuda, G., Baghino, L., Guillosson, T., Jardin, M., Tirado, M. & Esteller, V. 2007. A remarkable case of circuitous autumn migration of the Booted Eagle *Hieraaetus pennatus* through the western and central Mediterranean. *Ardeola* 54: 349–357.

Prieta, J. (Ed.) 2003. *Aves de Extremadura. Anuario Adenex* 1999–2000. ADENEX. Mérida.

Prieta, J. & del Moral, J. C. 2008. *La grulla común invernante en España. Población en 2007 y método de censo.* SEO/BirdLife. Madrid.

Prieta, J. & Molina, B. 2003. Selección de hábitat por el pico menor (*Dendrocopos minor*) en el Valle del Jerte, Cáceres. *Aves de Extremadura. Anuario ADENEX 1999–2000*: 77–87.

Proyecto Eremita. 2012. Poster at International Workshop on *Proper design of avian reintroduction projects: trying to establish guidelines*. Baeza, Spain. 15–17 October 2012. Fundación Migres website.

Puigcerver, M., Rodríguez-Teijeiro, J. D. & Gallego, S. 1995. Migration patterns of the quail (*Coturnix c. coturnix*): new data. *In: Proceedings of the XXII International Union of Game Biologists Congress*, Sofia, September 1995.

Puigcerver, M., Rodríguez-Teijeiro, J. D. & Gallego, S. 1999. The effects of rainfall on wild populations of Common Quail (*Coturnix coturnix*). *J. für Ornithologie* 140: 335–340.

Purroy, F. J. 1972. El Pico Dorsiblanco *Dendrocopos leucotos* del Pirineo. *Ardeola* 20: 145–158.

Purroy, F. J. 1973a. El Vencejo Real (*Apus melba*) en los Pirineos. *Ardeola* 19: 89–95.

Purroy, F. J. 1973b. La répartition des deux grimperaux dans les Pyrénées. *L'Oiseau et R.F.O.* 43: 205–211.

Purroy, F. J. 1974. Contribución al conocimiento ornitológico de los pinares pirenaicos. *Ardeola* 20: 245–261.

Purroy, F. J. 1976. *Fauna navarra en peligro de extinción*. Diario de Navarra. Pamplona.

Purroy, F. J. 1988. Sobre la invernada de la Paloma Torcaz (*Columba palumbus*) en Iberia. *In:* Tellería, J. L. (Ed.), pp. 137–151.

Purroy, F. J. 1995. La tórtola común. *La Garcilla* 94: 22–23.

Purroy, F. J. (Ed.) 1997. *Atlas de las Aves de España (1975–1995)*. SEO-Birdlife. Lynx. Barcelona.

Purroy, F. 1999. El Urogallo desaparece de las montañas españolas. *La Garcilla* 104: 10–14.

Purroy, F. J., Álvarez, A. & Pettersson, B. 1984. La población de Pico Mediano *Dendrocopos medius* (L.) de la Cordillera Cantábrica. *Ardeola* 31: 81–90.

Purroy, F. J., Onrubia, A. & Robles, J. L. 1997. Estima numérica de las aves nidificantes en la España peninsular e Islas Baleares. *In:* Manrique, J. *et al.* (Eds) *Actas de las XII Jornadas Ornitológicas Españolas*, pp. 207–217. Instituto de Estudios Almerienses. Almería.

Purroy, F. J. & Rodero, M. 1984. *Migración e invernada de la paloma torcaz* (Columba palumbus) *en Europa y España*. Fundación José María Blanc. Madrid.

Purroy, F. J., Rodero, M. & L. Tomiałojc. 1984. The ecology of woodpigeons *Columba palumbus* wintering on the Iberian Peninsula. *Acta Ornithologica* 20: 111–146.

Pütz, K., Rahbek, C., Saurola, P., Pedersen, K. T., Juvaste, R. & Helbig, A. J. 2007. Satellite tracking of the migratory pathways of first-year Lesser Black-backed Gulls *Larus fuscus* departing from the breeding grounds of different subspecies. *Vogelwelt* 128: 141–148.

Quevedo, M., Bañuelos, M. J., Sáez, O. & Obeso, J. R. 2006. Habitat selection by Cantabrian capercaillie *Tetrao urogallus cantabricus* at the edge of the species' distribution. *Wildlife Biology* 12: 267–276.

Quevedo, M. A., López, J. M. & Aguilera, E. 2007. Update of Proyecto eremita. *In:* Boehm, C., Bowden, C. G. R., Jordan, M. & King, C. (Eds) *Northern Bald Ibis Conservation and*

Reintroduction workshop. 2nd IAGNBI Meeting, Vejer, Spain, September 2006, pp. 54–63. RSPB. Sandy.

Quintana, M. 2009. La subespecie *namnetum* del ruiseñor pechiazul (*Luscinia svecica*) en el litoral cantábrico. *Revista de Anillamiento* 23: 35–38.

Rabuñal, J. L. 1986. Observación de *Botaurus lentiginosus* en la laguna de Traba en el mes de octubre de 1982. *Mustela* 2: 40–43.

Rabuñal, J. L. 1990. Aves nuevas para España: Chorlitejo Semipalmeado *Charadrius semipalmatus*. *La Garcilla* 79: 27–28.

Rabuñal. J. 2000. Aves nuevas: Porrón Islándico *Bucephala islandica*. *La Garcilla* 108: 22–23.

Rait-Kerr, H. 1934 a-d. The birds of Gibraltar. What to look for on the Rock. *The Gibraltar Chronicle*, 2, 23 & 30 July 1934, 8 October 1934.

Ramírez, A. & Tellería, J. L. 2003. Efectos geográficos y ambientales sobre la distribución de las aves forestales ibéricas. *Graellsia* 59: 219–231.

Ramírez, J. 2009. Noticiario ornitológico del Estrecho de Gibraltar 2008. *Migres* 1: 175–190.

Ramos, A., Dopico, M. & Romay, C. D. 2011. Edad, sexo y condición física de *Uria aalge* orillados en el noroeste de Galicia. *Chioglossa* 3: 75–81.

Ramos, R., Ramírez, F., Sanpera, C., Jover, L. & Ruiz, X. 2009. Diet of Yellow-legged Gull (*Larus michahellis*) chicks along the Spanish Western Mediterranean coast: the relevance of refuse dumps. *J. Ornithology* 150: 265–272.

Real, J. 2000. Los incendios pueden favorecer la recolonización de la Collalba Negra *Oenanthe leucura*. *Ardeola* 47: 93–96.

Real, J., Grande, J. M., Mañosa, S. & Sánchez-Zapata, J. A. 2001. Causes of death in different areas for Bonelli's eagle *Hieraaetus fasciatus* in Spain. *Bird Study* 48: 221–228.

Real, J. & Mañosa, S. 1997. Demography and conservation of Western European Bonelli´s Eagle (*Hieraaetus fasciatus*) populations. *Biological Conservation* 79: 59–66.

Real, J. & Mañosa, S. 2001. Dispersal of juvenile and immature Bonelli's Eagle in northeastern Spain. *J. Raptor Research* 35: 9–14.

Rebane, M. & Garcia, E. 2008. *Where to watch birds in Northern and Eastern Spain*. Christopher Helm. London.

Ree, V. 1973. Dagens avifaunistiske situasjon i Las Marismas i sør-Spania. (Birds of the Guadalquivir delta, South Spain.) *Sterna* 12: 225–268.

Reeber, S. & CHN. 2009. Les oiseaux rares en France en 2008. *Ornithos* 16: 273-315.

Reeber, S. & CHN. 2010. Les oiseaux rares en France en 2009. *Ornithos* 17: 361–405.

Reeber, S. & CHN. 2011. Les oiseaux rares en France en 2010. *Ornithos* 18: 325–367.

Reeber, S. & CHN. 2012. Les oiseaux rares en France en 2011. *Ornithos* 19: 353–395.

Reeber, S., Frémont, J.-Y., Flitti, A. & CHN. 2008. Les oiseaux rares en France en 2006–2007. *Ornithos* 16: 273–315.

Reid, S. G. 1871–1890. *Stray notes on ornithology. A record of ornithological observations in Gibraltar, Bermuda, Great Britain and Ireland, New York, South Africa, Natal, the Canary Islands & Spain*. Manuscript notebooks in BMNH Library, Tring, UK.

Reig, A. 2004a. Los viajes ornitológicos de Alfred Brehm a España. *Quercus* 219: 22–26.

Reig, A. 2004b. El descubrimiento del Pico Mediano en León y algunas notas sobre su presencia histórica en la Península Ibérica. *Argutorio* 12: 37–39.

Reino, L. M. 2001. O picanço-de-dorso-ruivo *Lanius collurio*: uma espécie rara e pouco conhecida em Portugal. *Pardela* 15: 10–13.

Reino, L. M. & Silva, T. 1996. Distribuçao e expançao do Bico-de-lacre em Portugal. *In:* Farinha, J. C., Almeida, J. & Costa, H. (Eds) *I Congresso de Ornitología da SPEA*, pp. 77–80. SPEA. Lisboa.

Reino, L. M. & Silva, T. 1998. The distribution and expansion of the Common Waxbill (*Estrilda astrild*) in the Iberian Peninsula. *Biologia e Conservacione de la Fauna* 102: 16–22.

Reis Júnior, J. A. 1931. *Catálogo Sistemático e Analítico das Aves de Portugal*. Araújo & Sobrinho e Sucessores. Porto.

Rendón, M. A., Garrido, A., Ramírez, J. M., Rendón-Martos, M. & Amat, J. A. 2011. Despotic establishment of breeding colonies of greater flamingos, *Phoenicopterus ruber*, in southern Spain. *Behavioral Ecology & Sociobiology* 50: 55–60.

Rendón, M. A., Green, A. J., Aguilera, E. & Almaraz, P. 2008. Status, distribution and long term changes in the waterbird community wintering in Doñana, south-west Spain. *Biological Conservation* 141: 1371–1388.

Rendón-Martos, M., Garrido, A., Chávez, J., Méndez, J. M. & Sayago, J. M. 2008. Odiel Marshes: New breeding site for Greater Flamingo (*Phoenicopterus roseus*) in Spain. *In:* Childress, B., Arengo, F. & Béchet, A. (Eds) *Flamingo*, Bulletin of the IUCN-SSC/ Wetlands International Flamingo Specialist Group, No. 16, December 2008, pp. 23–24. Wildfowl & Wetlands Trust. Slimbridge.

Rendón-Martos, M., Garrido, A., Rendón, M. A. & Amat, J. A. 2009b. Flamenco Común (*Phoenicopterus roseus*). *In:* Máñez & Rendón-Martos (Eds), pp. 57–93.

Rendón-Martos, M., Garrido, A., Rendón, M. A. & Ramírez, J. M. 2009a. Greater Flamingo monitoring and conservation at Fuente de Piedra Lake. *In:* Béchet, A., Rendón-Martos, M., Amat, J. A., Baccetti, N. & Childress, B. (Eds): *Proceedings of the IVth International Workshop on the Greater Flamingo in the Mediterranean region and northwest Africa, Antequera, Spain, 5–6 November 2007*, pp: 1–11. Wildfowl & Wetlands Trust. Slimbridge.

Rendón-Martos, M., Vargas, J. M., Rendón, M. A., Garrido, A.

& Ramírez, J. M. 2000. Nocturnal movements of breeding greater flamingos in southern Spain. *Waterbirds* 23 (Special Publication 1): 9–19.

Rey Benayas, J. M., Guzmán, J., Barbancho, S. & Gullick, T. 2012. Reintroducción de especies: ¿por qué unas sí y otras no? *Quercus* 322: 24–28.

Rey, P. J. 1995. Spatio-temporal variation in fruit and frugivorous bird abundance in olive orchards. *Ecology* 76: 1625–1635.

Rey, P. J. 2011. Preserving frugivorous birds in agro-ecosystems: lessons from Spanish olive orchards. *J. Applied Ecology* 48: 228–237.

Reyes-González, J. M. & González-Solís, J. 2012. Parcela cenicienta. In: Salvador, A. & Morales, M. (Eds).

Ribas, J. 2000. *Els ocells del Vallès Oriental*. Lynx. Barcelona.

Rico, L. & Gil-Delgado, J. L. 1986. *Aves nidificantes en la provincia de Alicante*. Instituto Juan Gil-Albert, Diputación de Alicante. Alicante.

Riddell, W. H. 1944. The Buff-backed Heron *Ardeola ibis ibis* (Linnaeus). *Ibis* 86: 503–511.

Ríos Naceyro, F. 1850. Catálogo de las aves observadas en las cercanías de Santiago y otros puntos de Galicia. *Memorias de la Real Academia de Ciencias Exactas, Físicas y Naturales* 1: 98–116.

Rivas Martínez, S. 1987. *Memoria y mapa de las series de vegetación de España. 1:400.000.* ICONA. Madrid.

Robb, M., Mullarney, K. & the Sound Approach. 2008. *Petrels night and day.* The Sound Approach. Poole, Dorset.

Robinson, R., Siriwardena, G. & Crick, H. 2005. Size and trends of the House Sparrow *Passer domesticus* population in Great Britain. *Ibis* 147: 552–562.

Robledano, F. & Calvo, J. F. 1989. La expansión del Tarro Blanco, *Tadorna tadorna* (L.) como reproductor en España. *Ardeola* 36: 91–95.

Robles, H., Ciudad, C., Vera, R., Olea, P. P., Purroy, F. J. & Matthysen, E. 2007. Sylvopastoral management and conservation of the middle spotted woodpecker at the south-western edge of its distribution range. *Forest Ecology and Management* 242: 343–352.

Robles, H. & Olea, P. P. 2003. Distribution and abundance of the Middle Spotted Woodpecker *Dendrocopos medius* in a southern population of the Cantabrian Mountains. *Ardeola* 50: 275–280.

Robles, L., Ballesteros, F. & Canut, J. (Eds) 2006. *El urogallo en España, Andorra y Pirineos franceses. Situación actual (2005).* SEO/BirdLife. Madrid.

Rocha, G. & Hidalgo, S. J. 2002a. The spread of the Collared Dove *Streptopelia decaocto* in Europe: colonization patterns in the west of the Iberian Peninsula. *Bird Study* 49: 11–16.

Rocha, G. & Hidalgo, S. J. 2002b. *La tórtola común (Streptopelia turtur). Análisis de los factores que afectan a su status.* Universidad de Extremadura. Cáceres.

Rocha, P. A., Catry, I. & Reis, S. 2002. Censo nacional do francelho-das-torres *Falco naumanni*. *Airo* 12: 3–9.

Rodrigues, T. M. & Gonçalves, D. 2013. The occurrence of two allopatric snipe *Gallinago* spp. in the Azores islands. *Ardeola* 60: 113–121.

Rodríguez, A. & Muntaner, J. 2004. Primeros resultados del marcado de gaviota patiamarilla *Larus michahellis* con anillas de lectura en las islas Baleares. *Anuari Ornitològic de les Balears 2004*: 69–77.

Rodríguez, A., García. A. M., Cervera, F. & Palacios, V. 2006. Landscape and anti-predation determinants of nest-site selection, nest distribution and productivity in a Mediterranean population of long-eared owls *Asio otus*. *Ibis* 148: 133–145.

Rodríguez, A., Negro, J. J., Bustamante, J., Fox, J. W. & Afanasyev, V. 2009. Geolocators map the wintering grounds of threatened Lesser Kestrels in Africa. *Diversity and Distributions* 15: 1010–1016.

Rodríguez, C. & Bustamante, J. 2008. Patterns of Orthoptera abundance and lesser kestrel conservation in arable landscapes. *Biodiversity and Conservation* 17: 1753–1764.

Rodríguez, C., Tapia, L., Kieny, F. & Bustamante, J. 2010. Temporal changes in lesser kestrel (*Falco naumanni*) diet during the breeding season in southern Spain. *J. Raptor Research* 44: 120–128.

Rodríguez, G., Elorriaga, J. & Ramírez, J. 2013. Identification of Atlas Long-legged Buzzard and its status in Europe. *Birding World* 26: 147–173.

Rodríguez, J. 2000. Evolución de la población de tórtola turca (*Streptopelia decaocto*) en la Comunidad de Madrid. *Anuario Ornitológico de Madrid 1999*: 42–47.

Rodríguez, J., Avilés, J. M. & Parejo, D. 2011.The value of nestboxes in the conservation of Eurasian Rollers *Coracias garrulus* in southern Spain. *Ibis* 153: 735–745.

Rodríguez, M. & Palacios, J. 1991. *El Ansar Campestre y el Ansar Común en Castilla y León.* Junta de Castilla y León. Valladolid.

Rodríguez, M. & Palacios, J. 2009. *Aguja Colinegra. In:* Palomino & Molina (Eds), pp. 126–130.

Rodríguez, N., García, J. & Copete, J. L. (Coords). 2013. *El Mosquitero Ibérico.* Grupo Ibérico de Anillamiento, León.

Rodríguez de los Santos, M. & Rubio, J. C. 1986. Biology and biometry of the Pallid Swift (*Apus pallidus*) in southern Spain. *Le Gerfaut* 76: 19– 30.

Rodríguez de los Santos, M., Cuadrado, M. & Arjona, S. 1986. Variation in the abundance of Blackcaps wintering in an olive (*Olea europea*) orchard in southern Spain. *Bird Study* 33: 81–86.

Rodríguez Silvar, J. & Bermejo, A. 1975. Primera nidificación de la Gaviota Tridáctila (*Rissa tridactyla*) en el SW de Europa. *Ardeola* 21: 409–414.

Rodríguez-Molina, A. & McMinn-Grivé, M. 2005. Population and distribution of the breeding colonies of the Balearic Shearwater *Puffinus mauretanicus*. *2nd International Manx Shearwater Workshop Abstracts*. Belfast.

Rodríguez-Muñoz, R. 2011. Urogallo común. In: Salvador, A. & Morales, M. B. (Eds).

Rodríguez-Pastor, R., Senar, J. C., Ortega, A., Faus, J., Uribe, F. & Montalvo, T., 2012. Distribution patterns of invasive Monk parakeets (*Myiopsitta monachus*) in an urban habitat. *Animal Biodiversity & Conservation* 35: 107–117.

Rodríguez-Teijeiro, J. D., Barroso, A., Gallego, S., Puigcerver, M. & Vinyoles, D. 2006. Orientation-cage experiments with the European quail during breeding season and autumn migration. *Canadian J. Zoology* 84: 887–894.

Rodríguez-Teijeiro, J. D., Sardà-Palomera, F., Nadal, J. Ferrer, X., Ponz, C. & Puigcerver, M. 2009. The effects of mowing and agricultural landscape management on population movements of the common quail. *J. Biogeography* 36: 1891–1898.

Román, J. & Gutiérrez, C. 2008. La Graja *Corvus frugilegus* deja de invernar en España: ¿Un nuevo caso de acortamiento en las migraciones? *Ardeola* 55: 229–235.

Román, J., Onrubia, A., Roviralta, F., Balmori, A. Fernández, J., Sanz-Zuasti, J., Gutiérrez, C., Jubete, F., Román, F., García, J. & Olea, P. P. 1997. Sobre el status del Gorrión Moruno, *Passer hispaniolensis* (Temmick, 1820), en la submeseta norte. *Ecología* 11: 453–456.

Román, J., Román, F., Ansola, L. M., Palma, C. & Ventosa, R. 1996. *Atlas de las aves nidificantes de la provincia de Burgos*. Caja de Ahorros del Círculo Católico. Burgos.

Romero, J. L. & Gálvez, M. 2001. Distribución del Pico menor (*Dendrocopos minor*) en Cataluña (NE de España). *Butll. Grup Catalá d'Anellament* 18: 17–28.

Rosa, G., Pacheco, C., Monteiro, A., Carvalho, A. & Araújo, A. 2001. Situação da Cegonha preta *Ciconia nigra* em Portugal: Receseamento da população nidificante (1995–1997). *Airo* 11: 15–22.

Rouco, M. 2007. *Las aves del azud de Riolobos y de su entorno*. SEO-Salamanca. Salamanca.

Roviralta, F. 2000a. El gorrión moruno (*Passer hispaniolensis*) en la Comunidad de Madrid. *Anuario Ornitológico de Madrid 1999*: 42–47.

Roviralta, F. 2000b. La expansión del Gorrión Moruno en la Península Ibérica. *Quercus* 175: 10–14.

Roviralta F. 2003. Gorrión moruno. *In:* Carrascal, L. M. & Salvador, A. (Eds).

Roviralta, F. 2004. Migración e invernada del mirlo capiblanco (*Turdus torquatus*) en la Comunidad de Madrid. *Anuario Ornitológico de Madrid 2003*: 40–47.

Rowell, H., Ward, R., Hall, C. & Cranswick, P. 2004. *The naturalised goose survey.* Wildfowl and Wetlands Trust. Slimbridge.

Rufino, R. (Ed.) 1989a. *Atlas das aves que nidificam em Portugal Continental*. CEMPA. Lisboa.

Rufino, R. (Ed.) 1989b. *Contagens de aves aquáticas – Jan–Fev 1989*. CEMPA. Lisboa.

Rufino, R. & Araújo, A. 1987. Seasonal variation in wader numbers and distribution at the Ria de Faro. *Wader Study Group Bull.* 51: 48–53.

Rufino, R., Araújo, A. & Abreu, M. 1985. Breeding raptors in Portugal: Distribution and Population estimates. *ICBP Technical Publication No. 5*: 15–28.

Rufino, R. & Neves, R. 1986. *Contagens de aves aquáticas invernantes. Janeiro de 1985*. CEMPA. Lisboa.

Rufino, R. & Neves, R. 1987. *Contagens de aves aquáticas em Portugal – Janeiro de 1986*. CEMPA. Lisboa.

Rufino, R. & Neves, R. 1990. Invernada de pato-negro *Melanitta nigra* na costa de Aveiro: Janeiro de 1990. *Airo* 2: 1–2.

Rufino, R. & Neves, R. 1991a. Recenseamento da população de Perna-longa *Himantopus himantopus* em Portugal. *Airo* 2: 10–11.

Rufino, R. & Neves, R. 1991b. Snipe on wet grasslands in Portugal. *Wader Study Group Bull.* 61 Supplement: 31–32.

Rufino, R. & Neves, R. 1995. Black-winged Stilt (*Himantopus himantopus*) wintering population: recent changes in range and numbers. *Wader Study Group Bull. 76*: 40–42.

Ruiz, A. & Martí, R. 2004. *La Pardela Balear*. SEO/BirdLife & Conselleria de Medi Ambient del Govern de les Illes Balears. Madrid.

Ruiz, X., Oro, D., Martínez-Vilalta, A. & Jover, L. 1996. Feeding ecology of Audouin's Gull *Larus audouinii* in the Ebro Delta. *Colonial Waterbirds* 19: 68–74.

Rydzewski, W. 1958. Aves anilladas recuperadas en las islas del Atlántico oriental. *Ardeola* 4: 149–155.

Sacarrão, G. F. 1972. Contribution à l'étude de la distribution de *Cyanopica cyanus cooki* BP (Aves–Corvidae) dans la Péninsule Ibérique. *Arquivos do Museo Bocage* (2ª série) III (11): 347–354.

Sacarrão, G. F. 1974. Acerca de alguns aspectos problemáticos da ecologia geográfica de *Cyanopica cyanus* (Pallas) (Aves: Corvidae). *Estudos da Fauna Portuguesa* 1: 1–28.

Sacarrão, G. F. & Soares, A. A. 1985. Sobre o estatus de *Elanus caeruleus* em Portugal (1975–1983). *Cyanopica* III (3): 339–350.

Sáez-Royuela, R. 1952. Captura en España de *Eremophila alpestris flava* (Gmelin, 1788) anillada en Suecia. *Bol. Real Sociedad Española de Historia Natural* 50: 153–154.

Sáez-Royuela, R. 1958. *Saxicola rubetra* criando en las afueras de Burgos. *Ardeola* 4: 204–205.

Sáez-Royuela, R. & Santos, T. 1985. *Estudio sobre la biología migratoria del orden Anseriformes (Aves) en España*. ICONA. Madrid.

Sagarra, I. 1915. Noves ornitològiques *Dendrocygna viduata* (L.). *Treballs de la Institució Catalana d'Història Natural* 1: 165–168.

Sainz Ollero, H., Sánchez de Dios, R. & García-Cervigón, A. 2010. La cartografía sintética de los paisajes vegetales españoles: una asignatura pendiente en geobotánica. *Ecología* 23: 249–272.

Sala, I. 1958. Tres recientes capturas en la Albufera de Valencia. *Ardeola* 4: 46–48.

Salas, M. A., Ferrer, X., Rovira, J. V. 2005. Distribución del Pico Mediano en la Península Ibérica. Revisión histórica crítica (siglos XIX y XX). *Observatorio Medioambiental* 8: 175–195.

Salomon, M. 1997. Quel status taxonomique donner au Pouillot véloce ibérique? *Alauda* 65: 63–81.

Salomon, M. & Hemim, Y. 1992. Song variation in the Chiffchaffs (*Phylloscopus collybita*) of the western Pyrenees – The contact zone between *collybita* and *brehmii* forms. *Ethology* 92: 265–282.

Salomon, M., Voisin, J.-F. & Bried, J. 2003. On the taxonomic status and denomination of the Iberian Chiffchaffs. *Ibis* 145: 87–97.

Salvador, A. 2005. Herrerillo común. *In:* Carrascal, L. M. & Salvador, A. (Eds).

Salvador, A. & Bautista, L. M. (Eds) *Enciclopedia Virtual de los Vertebrados Españoles.* Museo Nacional de Ciencias Naturales. Madrid. http://www.vertebradosibericos.org/

Salvador, A. & Morales, M. B. (Eds) *Enciclopedia Virtual de los Vertebrados Españoles.* Museo Nacional de Ciencias Naturales. Madrid. http://www.vertebradosibericos.org/

Salvadores, R., Vidal, C. & Mosquera, I. 1994. Aves nuevas: Arao Aliblanco *Cepphus grylle*. *La Garcilla* 91: 26–27.

Sampietro, F. J., Pelayo, E., Hernández, F., Cabrera, M. & Guiral, J. 1998. *Aves de Aragón. Atlas de especies nidificantes.* Diputación General de Aragón e Ibercaja. Zaragoza.

Samraoui, F., Alfarhan, A. H., Al-Rasheid, K. A. S. & Samraoui, B. 2011. An appraisal of the status and distribution of waterbirds of Algeria: indicators of global changes? *Ardeola* 58: 137–163.

San Segundo, C. 1990. *Atlas de las aves nidificantes de la provincia de Ávila y Sierra de Gredos.* Ph. D. Thesis. Universidad Complutense. Madrid.

San Segundo, C. 1992. *Areas de concentración e invernada de la Cigüeña Negra en España.* ICONA. Unpublished report.

San Segundo, C., Fernández, J. M. & Traverso, J. M. 1994. Recuento de cigüeñas negras en migración otoñal por Gibraltar. *Quercus* 102: 13–16.

Sánchez, A. 1989. Distribución altitudinal de la avifauna de la Sierra de Gredos. *Bol. Universitario UNED Ávila* 7: 119–127.

Sánchez, A. 1991. Estructura y estacionalidad de las comunidades de aves de la sierra de Gredos. *Ardeola* 38: 207–232.

Sánchez, Á. & Sánchez, J. M. 1991. La reproducción de la Pagaza Piconegra *Gelochelidon nilotica* en las Vegas Altas del Guadiana en la primavera de 1989. *Ardeola* 38: 131–135.

Sánchez, B., González, L. M. & Barov, B. 2008. *Action plan for the Spanish Imperial Eagle* Aquila adalberti *in the European Union.* Birdlife International. Cambridge.

Sánchez, I. 2007. Evidence of the historic presence of the Northern Bald Ibis (*Geronticus eremita*) in Spain. *In:* Boehm, C., Bowden, C. G. R., Jordan, M. & King, C. (Eds) *Northern Bald Ibis Conservation and Reintroduction workshop.* 2nd IAGNBI Meeting, Vejer, Spain, September 2006, pp. 105–110. RSPB. Sandy.

Sánchez, R., Margalida, A., González, L. M. & Oria, J. 2009. Temporal and spatial differences in the feeding ecology of the Spanish imperial eagle *Aquila adalberti* during the non-breeding season: effects of the rabbit population crash. *Acta Ornithologica* 44: 53–58.

Sánchez Aguado, F. J. 1984. Fenología de la reproducción y tamaño de la puesta en el Gorrión Molinero (*Passer montanus* L.). *Ardeola* 31: 33–45.

Sánchez Corominas, T., González-Quirós, P. & Vázquez, V. M. 2009. El pico mediano (*Dendrocopos medius*), el pico menor (*Dendrocopos minor*) y el picamaderos negro (*Dryocopus martius*) (Picidae, Aves) en el Principado de Asturias (España). *Bol. Ciencias Naturales del Real Instituto de Estudios Asturianos* 50: 281–302.

Sánchez Guzmán, J. M., Avilés, J. M., Medina, F. J. & Sánchez, Á. 1998. Status and trends of the Common Crane *Grus grus* on the western route. *Bird Conservation International* 8: 269–279.

Sánchez Guzmán, J. M., Sánchez, Á., Fernández, A. & Muñoz del Viejo, A. 1993. *La Grulla común (*Grus grus*) en Extremadura. Status y relación con el uso del suelo.* Universidad de Extremadura. Badajoz.

Sánchez Guzmán, J. M. & Fasola, M. 2002. *Gelochelidon nilotica* Gull-billed Tern. *BWP Update* 4: 1–13.

Sánchez Guzmán, J. M. & Sánchez, Á. 1991. La reproducción de la pagaza piconegra *Gelochelidon nilotica* en las Vegas Altas del Guadiana en la primavera de 1989. *Ardeola* 38: 131–135.

Sánchez Guzmán, J. M., Muñoz del Viejo, A., Corbacho, C., Costillo, E. & Fuentes, C. 2004. Status and trends of Gull-billed Tern *Gelochelidon nilotica* in Europe and Africa. *Bird Conservation International* 14: 335–351.

Sánchez Marco, A. 1996. The presence of the Waldrapp *Geronticus eremita* (Plataleidae) in the Plio-Pleistocene boundary in Spain. *Ibis* 138: 560–561.

Sánchez Sanz, J. M. 1997. Censo de Cigüeñuelas (*Himantopus himantopus*) nidificantes en cultivos de arroz en Aragón. *Rocín, Anuario Ornitológico de Aragón 1993–1994*: 107–111.

Sánchez-Lafuente, A. M. 2010. Calamón común. *In:* Salvador, A. & Bautista, L. M. (Eds).

Sánchez-Lafuente, M., Rey, P., Valera, F. & Muñoz-Cobo, J. 1992. Past and current distribution of the Purple Swamphen *Porphyrio porphyrio* L. in the Iberian Peninsula. *Biological Conservation* 61: 23–30.

Sandoval, A., Hevia, R. & Fernández, D. 2009. *Bol. Estación Ornitológica de Estaca de Bares.* No. 1., año 2008. C.M.A. Galicia/Terranova Interpretacion y Gestión Ambiental S.

L. La Coruña.

Sandoval, A., Hevia, R. & Fernández, D. 2010. *Bol. Estación Ornitológica de Estaca de Bares*. No. 2., año 2009. C.M.A. Galicia/Terranova Interpretacion y Gestión Ambiental S. L. La Coruña.

Sangster, G., Collinson, J. M., Crochet, P.-A., Knox, A. G., Parkin, D. T. & Votier, S. C. 2012. Taxonomic recommendations for British birds: eighth report. *Ibis* 154: 874–883.

Sangster, G., Collinson, J. M., Helbig, A. J., Knox, A. G. & Parkin, D. T. 2002a. The specific status of Balearic and Yelkouan Shearwaters. *British Birds* 95: 636–639.

Sangster, G., Knox, A. G., Helbig, A. J. & Parkin, D. T. 2002b. Taxonomic recommendations for European birds. *Ibis* 144: 153–159.

Sangster, G., Collinson, J. M., Helbig, A. J., Knox, A. G. & Parkin, D. T. 2005. Taxonomic recommendations for British birds: third report. *Ibis* 147: 821–826.

Sanpera, C., Ruiz, X., Moreno, R., Jover, L. & Waldron, S. 2007. Mercury and stable isotopes in feathers of Audouin's Gulls as indicators of feeding habits and migratory connectivity. *Condor* 109: 268–275.

Santarém, M. 1991. Primeiros dados sobre a nidificação do tordo-comum *Turdus philomelos* em Portugal. *Airo* 2: 52–53.

Santoro, S., Máñez, M., Green, A. J. & Figuerola, J. 2010. Formation and growth of a heronry in a managed wetland in Doñana, southwest Spain. *Bird Study* 57: 515–524.

Santos Júnior, J. R. 1959. New data about the *Hirundo daurica* in Portugal. *Publicações do Instituto de Zoologia 'Dr. Augusto Nobre'* 69: 126–129.

Santos Júnior, J. R. 1968. The colony of azure-winged magpies in the Barca d'Alva región. *Cyanopica* I (1): 9–28.

Santos Júnior, J. R. 1979a. Colónia de rolas-turcas em plena cidade do Porto. *Cyanopica* II (1): 57–68.

Santos Júnior, J. R. 1979b. Cinco campanhas de anilhagem de rolas na Reserva Ornitológica de Mindelo, 1974 a 1978. *Cyanopica* II (1): 29–50.

Santos Júnior, J. R. dos 1980. Um *Acanthis flavirostris* em Mindelo. *Cyanopica* II (2): 95–98.

Santos, T. 1982. *Migración e invernada de zorzales y mirlos (Genus Turdus) en la península Ibérica*. Universidad Complutense. Madrid.

Santos, T. 1985. *Estudio sobre la biología migratoria de la tribu Turdini (Aves) en España*. ICONA. Madrid.

Santos, T. & Suárez, F. 2005. Biogeography and population trends of Iberian steppe birds. *In:* Bota, G., Morales, M. B., Mañosa, S. & Camprodon, J. (Eds) *Ecology and conservation of steppe-land birds*, pp. 69–102. Lynx. Barcelona.

Santos, T., Suárez, F. & Tellería, J. L. 1983. The bird communities of Iberian Juniper woodlands (*Juniperus thurifera* L.). *In:* Purroy (Ed.) *Bird census and Mediterranean landscape*, pp. 79–88. Universidad de León. León.

Sanz, T. 1997. Migración e invernada del águila pescadora en España. *Quercus* 139: 14–15.

Sanz-Zuasti, J. & Velasco, T. 1999. *Guía de las aves de Castilla y León*. Carlos Sánchez Alonso editor. Medina del Campo.

Sarasa, C. G., Bartolomé, J., Fernández-Cruz, M. & Farinha, J. C. 1993. Segundo censo de Ardeidas invernantes en la Península Ibérica y Baleares (1992–1993). *Airo* 4: 41–50.

Sarasa, C. G., Garrido, J. R., Bartolomé, J., Igual, J. M. & Fernández-Cruz, M. 1997. Movimientos poblacionales de la Garcilla Bueyera (*Bubulcus ibis ibis*, L. 1758) en la Península Ibérica. *In:* Manrique, J., Sánchez, A., Suárez, F. & Yanes, M. (Eds) *Actas de las XII Jornadas Ornitológicas Españolas*, pp. 239–244. Instituto de Estudios Almerienses. Almería.

Sargatal, J. & del Hoyo, J. 1989. *Els ocells del Parc Natural dels Aiguamolls de l'Empordà*. Lynx. Barcelona.

Sargatal, J. & Llinàs, R. 1978. *Els ocells de l'Empordà*. Centre Excursionista Empordanès. Figueres.

Saunders, H. 1869. Notes on the ornithology of Italy and Spain. *Ibis* 11: 391–403.

Saunders, H. 1871. A list of the birds of southern Spain. *Ibis* 13: 54–68, 205–225, 384–402.

Saunders, H. 1884. Notes on the birds of the Pyrenees. *Ibis* 26: 365–392.

Saunders, H. 1897. Further notes on the birds of the Pyrenees. *Ibis* 39: 64–89.

Sayago, J. M. 2002a. Seguimiento de la población invernante de Águila Pescadora (*Pandion haliaetus*) en el Paraje Natural Marismas del Odiel y el litoral onubense. *Revista de Anillamiento* 9–10: 23–24.

Sayago, J. M. 2002b. Seguimiento de la población reproductora de Búho Real (*Bubo bubo*) entre 1997–2001 en la provincia de Huelva. *Revista de Anillamiento* 9–10: 25–26.

Sayago, J. M. 2008. La invernada del águila pescadora en la provincia de Huelva. *Quercus* 272: 22–26.

Schäffer, N., Walther, B. A., Gutteridge, K. & Rahbek, C. 2006. The African migration and wintering grounds of the Aquatic Warbler *Acrocephalus paludicola*. *Bird Conservation International* 16: 33–56.

Schmaljohann, H. & Thoma, M. 2005. First record of American Golden Plover *Pluvialis dominica* for Mauritania, and its status in western Africa. *Bull. African Bird Club* 12: 158–161.

Schulz, H. (Ed.) 1999. *White storks on the up?* Proceedings of the International Symposium on the White Stork (Hamburg, 1996). NABU. Bonn.

Scott, D. A. & Rose, P. M. 1996. *Atlas of Anatidae populations in Africa and Western Eurasia*. Wetlands International. Wageningen.

Scott, M. 2004. The Cinnamon Teal on the Outer Hebrides – a new Western Palearctic bird. *Birding World* 17: 200–201.

Seibold, I., Helbig, A. J., Meyburg, B. U., Negro, J. J. & Wink, M. 1996. Genetic differentiation and molecular phylogeny of European *Aquila* eagles according to cytochrome b nucleotide sequences. *In:* Meyburg, B. U. & Chancellor, R. D. (Eds) *Eagle Studies*, pp. 1–15. World Working Group on Birds of Prey. London.

Senar, J. C., Borrás, A., Cabrera, T. & Cabrera, J. 1993. Testing for the relationship between coniferous crop stability and Common Crossbill residence. *J. Field Ornithology* 64: 464–469.

Senar, J. C., Burton, P. J. K. & Metcalfe, N. B. 1992. Variation in the nomadic tendency of a wintering finch *Carduelis spinus* and its relationship with body condition. *Ornis Scandinavica* 23: 63–72.

Senar, J. C. & Sol, D. 1991. Censo de palomas *Columba livia* var. de la ciudad de Barcelona: aplicación del muestreo estratificado con factor de corrección. *Butll. Grup Catalá d'Anellament* 8: 19–24.

Senosiain, A. 1978. Observaciones de Pico Dorsiblanco *Dendrocopos leucotos* en el Pirineo navarro y primera nidificación comprobada en la Península Ibérica. *Ardeola* 24: 236–242.

SEO 1981. Primer censo de buitreras (1979). *Ardeola* 26/27: 256–259.

SEO/Birdlife. 2001. *Recuperación de* Puffinus mauretanicus *en las ZEPAS de las Islas Baleares.* Memoria Final Proyecto Life b-4/3200/97/246. Unpublished report to Conselleria de Medi Ambient, Govern de les Illes Baleares.

SEO/Birdlife. 2003. *La catástrofe del petrolero Prestige y su impacto sobre las aves marinas.* Tercer informe (unpublished). Enero–Febrero 2003. SEO/Birdlife.

SEO/Birdlife. 2010. *Estado de conservación de las aves en España.* SEO/Birdlife. Madrid.

SEO/BirdLife 2012. *Análisis preliminar del banco de datos de anillamiento de aves del Ministerio de Agricultura, Alimentación y Medio Ambiente, para la realización de un atlas de migración de aves de España.* SEO/BirdLife-Fundación Biodiversidad. Madrid. Available online at www.migraciondeaves.org only.

SEO/BirdLife-EOA. 2000. *Censos de aves acuáticas nidificantes del la Comunidad Valenciana, 2000.* Unpublished report to Conselleria de Medi Ambient, Generalitat Valenciana.

Seoane, J. 2005. *El Alzacola en España. I Censo nacional (2004).* SEO/Birdlife. Madrid.

Seoane, J., Justribó, J. H., García, F., Retamar, J., Rabadán, C. & Atienza, J. C. 2005. Habitat-suitability modelling to assess the effects of land-use changes on Dupont´s lark *Chersophilus duponti*: A case study in the Layna Important Bird Area. *Biological Conservation* 128: 241–252.

Serra, J. M. 1978. *Els aucells de les Balears.* Palma de Mallorca.

Serradilla, J. & Calvo, J. M. 2001. Distribución del Pico menor (*Dendrocopos minor*) en Salamanca. *Butll. Grup Catalá d'Anellament* 18: 29–41.

Serrano, D. 1999. Contribución al conocimiento de la densidad reproductora de la Collalba Negra (*Oenanthe leucura*) en el valle medio del Ebro. *Rocín, Anuario Ornitológico de Aragón 1995–1996*: 107–111.

Serrano, D. 2000. Relationship between raptors and rabbits in the diet of Eagle Owls in southwestern Europe: competition removal or food stress? *J. Raptor Research* 34: 305–310.

Serrano, D. 2001. Parámetros reproductivos y fecha de puesta del Búho Real (*Bubo bubo*) en el valle del Ebro. *Rocín, Anuario Ornitológico de Aragón 1997–1998*: 65–73.

Serrano, D., Tella, J. L. & Ursúa, E. 2005. Proximate causes and fitness consequences of hatching failure in lesser kestrels *Falco naumanni*. *J. Avian Biology* 36: 242–250.

Serrano, M. & Cantos, F. J. 2013. Gaviota sombría, la nueva reina del invierno en Madrid. *Quercus* 331: 16–23.

Shirihai, H. 1996. *The birds of Israel.* Academic Press. London.

Shirihai, H., Bretagnolle, V. & Zino, F. 2010. Identification of Fea's, Desertas and Zino's Petrels at sea. *Birding World* 23: 239–275.

Shirihai, H., Gargallo, G. & Helbig, A. J. 2001. *Sylvia warblers.* Christopher Helm. London.

Silva, J. P., Pinto, M. & Palmeirim, J. M. 2004. Managing landscapes for the little bustard *Tetrax tetrax*: lessons from the study of winter habitat selection. *Biological Conservation* 117: 521–528.

Silva, P. & Beja, P. 2001. Breeding habitat of the Black-shouldered Kite (*Elanus caeruleus*) in a rural area of Southern Portugal. *In: Abstracts 4th Eurasian Congress on Raptors*, p. 175. Estación Biológica de Doñana-CSIC & Raptor Research Foundation. Sevilla.

Simón, M. A. 2003. Centro de Cría Guadalentín: una experiencia exitosa en Cazorla. *La Garcilla* 116: 15–17.

Simpson, G. G. 1964. Species density of North American recent mammals. *Systematic Zoology* 13: 57–73.

Skinner, K. L. 1925. Details of some nests and eggs of the Black Vulture *Aegypius monachus* (L.) from the province of Caceres, Spain. *The Oologist's Record* 5: 41–43.

Slack, R. 2009. *Rare birds: where and when. An analysis of status and distribution in Britain & Ireland.* Vol. 1. Russell Slack Rare Bird Books. York.

Smith, A. C. 1868. A sketch of the birds of Portugal. *Ibis* 10: 428–460.

Smith, K. D. 1965. On the birds of Morocco. *Ibis* 107: 493–526.

Snow, D. W., Owen, D. F. & Moreau, R. E. 1955. Land and sea-bird migration in north-west Spain, autumn 1954. *Ibis* 97: 557–571.

Soares, A. A. 1970. Rapináceos de Portugal. I – Falconiformes. *Arquivos do Museo Bocage* 2ª série, vol II (15): 203–318.

Sol, D., Arcos, J. M. & Senar, J. C. 1995. The influence of rubbish tips on the winter distribution of the Yellow-legged Gull *Larus cachinnans*. *Bird Study* 42: 216–221.

Soler, J. J. 2008. Críalo europeo. In: Carrascal, L. M. & Salvador, A. (Eds).

Soler, M. & Soler, J. J. 1987. Selección de lugar de nidificación en Grajilla (*Corvus monedula*). *Alytes* 5: 8798.

Soler, M. 1990. Relationships between the Great Spotted Cuckoo *Clamator glandarius* and its corvid hosts in a recently colonized area. *Ornis Scandinavica* 21: 212–223.

Soler, M. 2006. Grajilla. In: Carrascal, L. M. & Salvador, A. (Eds).

Soler, M., Alcalá, N. & Soler, J. J. 1990. Alimentación de la Grajilla *Corvus monedula* en tres zonas del sur de España. *Doñana, Acta Vertebrata* 17: 17–48.

Soler, M., Martínez, J. G. & Soler, J. J. 1996. Effects of brood parasitism by the Great Spotted Cuckoo on the breeding success of the Magpie host. *Ardeola* 43: 87–96.

Soler, M., Palomino, J. J., Martínez, J. G. & Soler, J. J. 1994. Activity, survival, independence and migration of fledgling great spotted cuckoos. *Condor* 96: 802–805.

Soler, M., Pérez-González, J. A., Tejero, E. & Camacho, I. 1988. Alimentación del Zorzal Alirrojo (*Turdus iliacus*) durante su invernada en olivares de Jaén (sur de España). *Ardeola* 35: 183–196.

Soler, M., Soler, J. J. & Martínez, J. G. 1998a. Duration of sympatry and coevolution between the great spotted cuckoo (*Clamator glandarius*) and its primary host, the magpie (*Pica pica*). *In:* Rothstein, S. I. & Robinson, S. K. (Eds) *Parasitic birds and their hosts, studies in coevolution*, pp. 113–128. Oxford University Press. Oxford.

Soler, M., Soler, J. J., Martínez, J. G., Pérez-Contreras, T. & Møller, A. P. 1998b. Micro-evolutionary change and population dynamics of a brood parasite and its primary host: the intermittent arms race hypothesis. *Oecologia* 117: 381–390.

Soriguer, R. C. 1978. Primera cita de la Collalba yebélica (*Oenanthe leucopyga*) en la Península Ibérica. *Doñana, Acta Vertebrata* 5: 109–110.

Southern, J. B. 1958. Notas sobre aves de Puerto Pollensa y su comarca (Mallorca). *Ardeola* 4: 176–178.

Soutullo, A., Urios, V., Ferrer, M. & Peñarrubia, S. G. 2006a. Dispersal of Golden Eagles *Aquila chrysaetos* during their first year of life. *Bird Study* 53: 258–264.

Soutullo, A., Urios, V., Ferrer, M. & Peñarrubia, S. G. 2006b. Post-fledging behaviour in Golden Eagles: onset of the juvenile dispersal and progressive distancing from the nest. *Ibis* 148: 307–312.

SOVON 2002. *Atlas van der Nederlandse Broedvogels 1998–2000*. Leiden.

Stead, P. J. 1959. Notas ornitológicas de Andalucía. *Ardeola* 5: 179–184.

Stenhouse, J. H. 1921. Bird notes from southern Spain. *Ibis* 63: 573–594.

Strandberg, R., Klaassen, R. H. G., Hake, M., Olofsson, P., Thorup, K. & Alerstam, T. 2008. Complex timing of Marsh Harrier *Circus aeruginosus* migration due to pre- and post-migratory movements. *Ardea* 96: 159–171.

Strandberg, R, Klaassen, R. H. G., Olofsson, P. & Alerstam, T. 2009. Daily travel schedules of adult Eurasian Hobbies *Falco subbuteo*—variability in flight hours and migration speed along the route. *Ardea* 97: 287–295.

Suárez, F. 1977. Nidificación y puesta en la Collalba Rubia (*Oenanthe hispanica*). *Ardeola* 23: 63–79.

Suárez, F. 1987. *Historia Natural de la Collalba Rubia (*Oenanthe hispanica *L.) durante la época de reproducción.* Ph. D. Thesis. Universidad Complutense. Madrid.

Suárez, F. 1990. El dimorfismo de la Collalba Rubia (*Oenanthe hispanica* L.): variaciones geográficas y entre hábitats. *Ardeola* 37: 291–298.

Suárez, F. (Ed.) 2010. *La alondra ricotí (*Chersophilus duponti*).* Organismo Autónomo Parques Nacionales. Madrid.

Suárez, F., del Pozo, R., García, J. T. & Justribó, J. H. 2008. Differential use of home range by Dupont's Lark *Chersophilus duponti* in relation to cereal fields during the post-moulting period. *Ardeola* 55: 97–102.

Suárez, F., García, J. T., Sampietro, F. J. & Garza, V. 2006b. The non-breeding distribution of Dupont's Lark *Chersophilus duponti* in Spain. *Bird Conservation International* 16: 1–7.

Suárez, F. & Garza, V. 1989. La invernada de la Alondra de Dupont, *Chersophilus duponti*, en la Península Ibérica. *Ardeola* 36: 107–110.

Suárez, F. Garza, V. & Morales, M. B. 2002. Habitat use of two sibling species, the Short-toed *Calandrella brachydactyla* and the Lesser Short-toed *C. rufescens* larks, in mainland Spain. *Ardeola* 49: 259–272.

Suárez, F., Garza, V. & Morales, M. B. 2003. The role of extensive cereal crops, dry pasture and shrub-steppe in determining skylark *Alauda arvensis* densities in the Iberian peninsula. *Agriculture Ecosystems & Environment* 95: 551–557.

Suárez, F., Herranz, J., Martínez, C., Manrique, J., Astrain, C., Echeverria, A., Curcó, A., Estrada, J. & Yanes, M. 1999a. Utilización y selección de hábitat de las gangas ibérica y ortega en la Península Ibérica. *In:* Herranz & Suárez (Eds), pp. 127–156.

Suárez, F., Hervás, I. & Herranz, J. 2009. *Las alondras de España peninsular.* Organismo Autónomo Parques Nacionales. Madrid.

Suárez, F., Hervás, I., Herranz, J. & del Moral, J. C. 2006a. *La ganga ibérica y la ganga ortega en España: población en 2005 y método de censo.* SEO/BirdLife. Madrid.

Suárez, F., Hervás, I., Levassor, C. & Casado, M. A. 1999b. La alimentación de la Ganga Ibérica y la Ganga Ortega. *In:* Herranz & Suárez (Eds), pp. 215–229.

Suárez, F., Martínez, C., Herranz, J. &, Yanes, M. 1997. Conservation status and farmland requirements of Pin-tailed Sandgrouse *Pterocles alchata* and Black-bellied Sandgrouse *Pterocles orientalis* in Spain. *Biological Conservation* 82: 73–80.

Suárez, F., Oñate, J. & Herranz, J. 1999c. Estado y problemática de conservación de las gangas en España. *In:* Herranz & Suárez (Eds), pp. 273–302

Suárez, F. & Sáez-Royuela, C. 1987. Variación estacional de la estructura y demanda energética de dos comunidades de paseriformes de zonas semiáridas. *Studia Oecologica* 6: 181–203.

Suárez, F., Santos T. & Tellería, J. L. 1982. The status of Dupont's Lark, *Chersophilus duponti*, in the Iberian Peninsula. *Le Gerfaut* 72: 231–235.

Suárez, M. 2001. Recuperació errònia d'un raspinell pirenaic *Certhia familiaris. Anuari Ornitològic de les Balears 2000*: 33–34.

Suárez-Seoane, S. & García-Rovés, P. 2004. Do disturbances in surrounding areas affect a core population of Cantabrian Capercaillie (*Tetrao urogallus cantabricus*)? The case of the natural reserve of Muniellos (Asturias, NW Spain). *Ardeola* 51: 395–409.

Sultana, J. & Gauci, C. A. 1982. *A new guide to the birds of Malta*. The Ornithological Society. Valletta, Malta.

Sunyer, C. & Viñuela, J. 1990. Migración e invernada del Esmerejón en España. *Ardeola* 37: 279–290.

Sunyer, C. & Vinuela, J. 1994. Invernada de rapaces (O. Falconiformes) en España peninsular y Baleares. In. Muntaner, J. & Mayol, J. (Eds). *Biología y Conservación de las Rapaces Mediterráneas*. SEO/BirdLife. Madrid.

Sunyer, J. R. 1997. Noves dades de cria de busqueret garriguer (*Sylvia cantillans*) a Mallorca. *Anuari Ornitològic de les Balears 1996*: 35–39.

Sunyer, J. R. 1998. Nidificació del bosqueret roig collarga *Sylvia undata* a Mallorca. *Anuari Ornithològic de les Balears* 12: 117–123.

Sunyer, J. R. 2009. Sobre la nidificació del busqueret capnegre *Sylvia melanocephala* al centre de Mallorca. *Anuari Ornitològic de les Balears 2008*: 57–61.

Sunyer, J. R. 2010. Busqueret coallarg, *Sylvia balearica. In:* Grup Balear d'Ornitologia i Defensa de la Naturalesa, *Atles dels aucells nidificants de Mallorca i Cabrera, 2003–2007*, pp. 228–229. GOB. Palma, Mallorca.

Svensson, L. 2013. A taxonomic revision of the Subalpine Warbler *Sylvia cantillans. Bull. B.O.C.* 133: 240–248.

Szijj, J. 1975. Probleme des Anatidenzuges, dargestellt an den verlagerunden des europäischen Kolbenentenbestandes. *Ardeola* 21: 153–171.

Tait, G. M. 1960. Aves anilladas recuperadas en Portugal (abril 1957–diciembre 1958). *Ardeola* 6: 259–278.

Tait, G. M. 1963. Aves anilladas recuperadas en Portugal. Años 1960 y 1961. *Ardeola* 8: 189–222.

Tait, W. C. 1887. A list of the birds of Portugal. *Ibis* 29: 77–96,182–201, 302–314, 372–400.

Tait, W. C. 1924. *The birds of Portugal*. Witherby. London.

Talamelli, A. 2007. First record of Eurasian Black Vulture *Aegypius monachus* for Senegal. *Bull. African Bird Club* 14: 203.

Tamas, E. A. 2011. Longevity and survival of the black stork *Ciconia nigra* based on ring recoveries. *Biologia* 66: 912–915.

Tapia, L. 2010. Busardo ratonero. In: Salvador, A. & Bautista, L. M. (Eds).

Tato Cumming, J. J. 1971. Datos para una ecología estival de los bosques de la isla de Mallorca y sus aves. *Ardeola* vol. especial: 405–456.

Tébar, J. 1998. *Las aves de la Bahía de Cádiz*. GONHS. Gibraltar.

Teixeira, A. M. 1983. Seabirds breeding at the Berlengas, forty-two years after Lockley's visit. *Ibis* 125: 417–420.

Teixeira, A. M. 1986. Razorbill *Alca torda* losses in Portuguese nets. *Seabird* 9: 11–14.

Teixeira, A. M. & Moore, C. C. 1983. The breeding of Madeira Storm-petrel *Oceanodroma castro* on Farilhão Grande, Portugal. *Ibis* 125: 382–384.

Tejero, E., Soler, M. & Camacho, I. 1983. La alimentación de la curruca cabecinegra (*Sylvia melanocephala*, Gmelin 1788) en olivares de la provincial de Jaén (otoño–invierno). *Doñana, Acta Vertebrata* 10: 133–153.

Tella, J. L. 2004. Situación de las poblaciones de Cernícalo Primilla: pasado, presente y necesidades de futuro. *In:* M. Alcántara (Ed.), *Actas del VI Congreso Nacional sobre el Cernícalo Primilla*, pp. 26–34. Departamento de Medio Ambiente del Gobierno de Aragón. Zaragoza.

Tella, J. L., Forero, M. G., Hiraldo, F. & Donázar, J. A. 1998. Conflicts between lesser kestrel conservation and European agricultural policies as identified by habitat use analyses. *Conservation Biology* 12: 593–604.

Tella, J. L. & Mañosa, S. 1993. Eagle owl predation on Egyptian vulture and northern goshawk: a possible effect of a decrease in European rabbit availability. *J. Raptor Research* 27: 111–112.

Tella, J. L., Vögeli, M., Serrano, D. & Carrete, M. 2005. Current status of the threatened Dupont's Lark in Spain: overestimation, decline and extinction of local populations. *Oryx* 39: 90–94.

Tellería, J. L. 1979. La migration posnupciale du Guêpier d'Europe *Merops apiaster* L. au détroit de Gibraltar en 1977. *Alauda* 47: 139–150.

Tellería, J. L. 1980. Autumn migration of Cory's Shearwater through the Straits of Gibraltar. *Bird Study* 27: 21–26.

Tellería, J. L. 1981. *La migración de las Aves en el Estrecho de Gibraltar (Epoca posnupcial). Vol. 2. Aves no planeadoras.* Universidad Complutense. Madrid.

Tellería, J. L. (Ed.) 1988. *Invernada de aves en la Península Ibérica*. Sociedad Española de Ornitología. Madrid.

Tellería, J. L. 1992. Gestión forestal y conservación de las aves en España peninsular. *Ardeola* 39: 99–114.

Tellería, J. L. 2009. Overlap between wind power plants and Griffon Vultures *Gyps fulvus* in Spain. *Bird Study* 56: 268–271.

Tellería, J. L., Asensio, B. & Díaz, M. 1999. *Aves Ibéricas II. Paseriformes*. J. M. Reyero Ed. Madrid.

Tellería, J. L., de La Hera, I., Ramírez, A. & Santos, T. 2011. Conservation opportunities in Spanish Juniper *Juniperus thurifera* woodlands: the case of migratory thrushes *Turdus* spp. *Ardeola* 58: 57–70.

Tellería, J. L. & Galarza, A. 1991. Avifauna invernal en un eucaliptal del norte de España. *Ardeola* 38: 239–247.

Tellería, J. L., Pérez-Tris, J., Ramírez, A., Fernández-Juricic, E. & Carbonell, R. 2001. Distribution of Robins (*Erithacus rubecula*) in wintering grounds: effects of conspecific density, migratory status and age. *Ardea* 89: 363–373.

Tellería, J. L., Ramírez, A. & Pérez-Tris, J. 2005. Conservation of seed dispersing migrant birds in Mediterranean habitats: Shedding light on patterns to preserve processes. *Biological Conservation* 124: 493–502.

Tellería, J. L. & Santos, T. 1994. Factors involved in the distribution of forest birds in the Iberian Peninsula. *Bird Study* 41: 161–169.

Tellería, J. L., Santos, T., Álvarez, G. & Sáez-Royuela, C. 1988. Avifauna de los campos de cereales del interior de España. *In:* Bernis, F. (Ed.) *Aves de los medios urbano y agrícola en las mesetas españolas*, pp. 173–319. Sociedad Española de Ornitología. Madrid.

Terrasse, M. 2006. Évolution des déplacements du Vautour Fauve (*Gyps fulvus*) en France et en Europe. *Ornithos* 13: 273–299.

Tewes, E. 2004. Situación del buitre negro *Aegypius monachus* en Mallorca. *Anuari Ornitològic de les Balears* 19: 99–108.

Teyssèdre, A. 1983. Étude compare de quatre populations de Goélands argentés à pàttes jaunes d'Europe occidentale. *L'Oiseau et R.F.O.* 53: 43–52.

Themido, A. A. 1933. Aves de Portugal (Catálogo do Museu Zoológico de Coimbra). *Memórias e Estudos do Museu Zoológico da Universidade de Coimbra* Series 1. 65: 1–281.

Themido, A. A. 1952. Aves de Portugal (chaves para a sua determinação). *Memórias e Estudos do Museu Zoológico da Universidade de Coimbra* 213: 1–241.

Thévenot, M., Vernon, R. & Bergier, P. 2003. *The birds of Morocco*. BOU. Tring.

Thibault, J.-C. 1995. Effect of predation by the Black Rat *Rattus rattus* on the breeding success of Cory's shearwater *Calonectris diomedea* in Corsica. *Marine Ornithology* 23: 1–10.

Thibault, J.-C. & Bretagnolle, V. 1988. A Mediterranean breeding colony of Cory's Shearwater *Calonectris diomedea* in which individuals show behavioural and biometric characters of the Atlantic subspecies. *Ibis* 140: 523–528.

Thibault, J.-C. & Delattre, D. 1969. Más noticias sobre aves de Mallorca (años 1967–1968. *Ardeola* 13: 201–204.

Thibault, J.-C., Triay, R., Beaubrun, P., Boukhalfa, D., Dominici, J. M. & Torre, A. 1996. Osprey (*Pandion haliaetus*) in Mediterranean: characteristics of a resident population with a patchy distribution. *In:* Muntaner, J. & Mayol, J. (Eds) *Biología y Conservación de las Rapaces Mediterráneas, 1994*, pp. 135–144. SEO/BirdLife. Madrid.

Thielcke, G. & Linsenmair, K. E. 1963. Zur geographischen Variation des Gesanges des Zilpzalps, *Phylloscopus collybita*, in Mittel- und Südwesteuropa mit einem Vergleich des Gesanges des Fitis, *Phylloscopus trochilus*. *J. für Ornithologie* 104: 372–402.

Thiollay, J.-M. 1977. Distribution saisonnière des rapaces diurnes en Afrique occidentale. *L'Oiseau et R.F.O.* 47: 253–294.

Thiollay, J.-M. 2006. The decline of raptors in West Africa: long-term assessment and the role of protected areas. *Ibis* 148: 240–254.

Thiollay, J.-M. & Bretagnolle, V. (Coords.) 2004. *Rapaces nicheurs de France. Distribution, effectifs et conservation.* Delachaux & Niestlé. Paris.

Thiollay, J.-M. & Perthuis, A. 1975. La migration d'automne a Gibraltar (1er au 20 Octobre 1974): Analyse et interpretation. *Ardeola* 21 (especial): 595–614.

Thorup, O. 2006. *Breeding waders in Europe 2000*. International Wader Studies 14. Wader Study Group. Thetford.

Ticehurst, C. B. 1937. *Phylloscopus collybita ibericus, subsp. nov. Bull. British Ornithologists' Club* 57: 63–64.

Ticehurst, C. B. & Whistler, H. 1925. A contribution to the ornithology of Navarre, northern Spain. *Ibis* 67: 443–460.

Ticehurst, C. B. & Whistler, H. 1928. On the avifauna of Galicia. *Ibis* 70: 663–683.

Ticehurst, C. B. & Whistler, H. 1930. A spring tour in eastern Spain and the Pityusae islands. *Ibis* 72: 663–683.

Ticehurst, C. B. & Whistler, H. 1935. Some further notes on the birds of Portugal. *Ibis* 77: 554–563.

Tirado, M. (Ed.) 2011. *Anuario Ornitológico de la Comunidad Valenciana 2009*. Internatura. Castellón.

Tirado, M. 2011. Algunos datos sobre la invernada del avión roquero (*Ptyonoprogne rupestris*) en Castellón. *In:* Tirado, M. (Ed.), pp. 193–197.

Toral, G. M., Stillman, R. A., Santoro, S. & Figuerola, J. 2012. The importance of rice fields for glossy ibis (*Plegadis falcinellus*): Management recommendations derived from an individual-based model. *Biological Conservation* 148: 19–27.

Torres-Esquivias, J. A. 1982. Informe sobre la reproducción de la población española de malvasías (*Oxyura leucocephala*). *Boletín Circular SEO* 62: 9.

Torres-Esquivias, J. A. 2002. *Las Malvasias Cordobesas 25 años despues*. Servicio de publicaciones de la Exma. Diputación Provincial de Córdoba.

Torres-Esquivias, J. A. 2008. *La malvasía cabeciblanca. In:* Ballesteros *et al.* (Eds), pp. 54–69.

Torres-Esquivias, J. A. & Moreno-Arroyo, B. 2000. La recuperación de la Malvasía Cabeciblanca (*Oxyura leucocephala*) en España durante el último decenio del siglo XX. *Oxyura* 10: 5–52.

Traverso, J. M. 2000. Invernada de la Gaviota Sombría (*Larus fuscus*) en Extremadura. Diciembre, 1998. *Aves de Extremadura. Anuario ADENEX 1998*: 97–101.

Tréca, B. 1994. Diet of Ruffs and Black-tailed Godwits in Senegal. *Ostrich* 65: 256–263.

Triay, R. 1995. Reproducción del águila pescadora (*Pandion haliaetus*) en la Isla de Menorca (Mediterráneo Occidental). *Ardeola* 42: 21–28.

Triay, R. 2002. Seguimiento por satellite de tres juveniles de Águila Pescadora nacidos en la Isla de Menorca. *Ardeola* 49: 249–257.

Triay, R. & Siverio, M. (Eds) 2008. *El águila pescadora en España. Población en 2008 y método de censo.* SEO/BirdLife. Madrid.

Trierweiler, C, Koks, B. J., Drent, R. H., Exo, K.-M., Komdeur, J., Dijkstra, C. & Bairlein, F. 2007. Satellite tracking of two Montagu's Harriers (*Circus pygargus*): dual pathways during autumn migration. *J. Ornithology* 148: 513–516.

Trigo de Yarto, E. 1971. *La Avutarda en España.* XVIII Triennial General Meeting of the International Council for Hunting. Unpublished report. Federación Española de Caza. Madrid.

Triplet, P., Overdijk, O., Smart, M., Nagy, S., Schneider-Jacoby, M., Karauz, E. S., Pigniczki, C., Baha El Din, S., Kralj, J., Sandor, A. & Navedo, J. G. (Eds) 2008. *International Single Species Action Plan for the Conservation of the Eurasian Spoonbill* Platalea leucorodia. AEWA Technical Series No. 35. UNEP/AEWA Secretariat. Bonn.

Trotignon, E., Trotignon, J., Baillon, M., Dejonghe, J. F., Duhautois, L. & Lecomte, M. 1980. Recensement hivernal des Limicoles et autres oiseaux aquatiques sur le Banc d'Arguin (Mauritanie) (Hiver 1978/1979). *L'Oiseau et R.F.O.* 50: 323–343.

Troya, A. & Bernués, M. 1990. *Censo de acuáticas invernantes en España (Enero 1989).* ICONA. Madrid.

Trujillo, D., Barone, R. & Siverio, F. 1994. Aves nuevas: Zampullín Picogrueso *Podilymbus podiceps*. *La Garcilla* 94: 30–31.

Tucker, G. M. & Heath, M. F. 1994. *Birds in Europe. Their conservation status.* BirdLife Conservation Series No. 3. BirdLife International. Cambridge.

Tuke, A. J. S. 1953. *An introduction to the birds of southern Spain and Gibraltar.* Gibraltar Chronicle. Gibraltar.

Urcun, J.-P. 2003. The autumn migration of the Black Stork (*Ciconia nigra*) through the Pyrenees. *Aves* 40: 140–154.

Urcun, J.-P. & Bried, J. 1998a. *The autumn migration of the Red Kite (Milvus milvus) through the Pyrenees. In:* Chancellor *et al.* (Eds), pp. 641–654.

Urcun, J.-P. & Bried, J. 1998b. *The autumn migration of raptors through the Pyrenees. In:* Chancellor *et al.* (Eds), pp. 655–680.

Urios, V., Escobar, J. V., Pardo, R. & Gómez, J. A. 1991. *Atlas de las aves nidificantes de la Comunidad Valenciana.* Conselleria d'Agricultura i Pesca. Generalitat Valenciana. Valencia.

Valencia, J., de la Cruz, C., Carranza, J. & Solís, E. 2005. No brood parasitism by the Great Spotted Cuckoo *Clamator glandarius* on the Azure-winged Magpie *Cyanopica cyanus*. *Bird Study* 52: 204–209.

Valera, F. 2009. Abejaruco europeo. *In:* Salvador, A. & Bautista, L. M. (Eds).

Valera, F., Rey, P., Sánchez-Lafuente, M. S. & Muñoz-Cobo, J. 1990. The situation of Penduline Tit (*Remiz pendulinus*) in southern Europe: a new stage of its expansion. *Journal für Ornithologie* 131: 413–420.

Väli, Ü, & Lõhmus, A. 2004. Nestling characteristics and identification of the lesser spotted eagle *Aquila pomarina*, the greater spotted eagle *A. clanga* and their hybrids. *J. Ornithology* 145: 256–263.

Valverde, J. A. 1953a. Le Guêpier d'Europe dans le bassin du Duero (Espagne). *Nos Oiseaux* 22: 7.

Valverde, J. A. 1953b. Le Corbeau freux nicheur en Espagne. *Nos Oiseaux* 22: 78–82.

Valverde, J. A. 1956. Notas ornitológicas sobre Santo Domingo de Silos (Burgos). *Munibe* vol. especial: 3–31.

Valverde, J. A. 1957. *Aves del Sahara Español. Estudio ecológico del desierto.* Instituto de Estudios Africanos, CSIC. Madrid.

Valverde, J. A. 1958. Las aves esteparias de la Península Ibérica. *Publicaciones del Instituto de Biología Aplicada* 27: 41–48.

Valverde, J. A. 1960. Vertebrados de las marismas del Guadalquivir: introducción a su estudio ecológico. *Archivos del Instituto de Aclimatación de Almería* 9: 1–168.

Valverde, J. A. 1964a. Datos sobre Cerceta Pardilla (*Anas angustirostris*) en Las Marismas. *Ardeola* 9: 121–132.

Valverde, J. A. 1964b. La reproducción de flamencos en Andalucía en el año 1963. *Ardeola* 9: 55–65.

Valverde, J. A. 1968. Dos especies de *Phylloscopus* nuevas para la avifauna ibérica. *Ardeola* 12: 117–120.

Valverde, J. A. 1971. Notas sobre la biología de reproducción del Críalo *Clamator glandarius* (L.). *Ardeola* vol. especial: 591–647.

Valverde, J. A. 1992. Rugidera. *In: Gran Enciclopedia Extremeña*, vol. IX, p. 68. Ediciones Extremeñas. Mérida.

Valverde, J. A. 2005. *Reyes, osos, lobos, espátulas y otros bichos. Memorias de un biólogo heterodoxo.* Vol. VI. CSIC. Madrid

Valverde, J. A. & Bernis, F. 1960. Sur l'écologie de *Gyps fulvus* en Espagne. *Proceedings XII International Ornithological Congress, Helsinki 1958* 1: 110–117.

Valverde, J. A. & Weickert, P. 1956. Sobre la migración de varias garzas españolas. *Munibe* 8: 1–23.

Van den Berg, A. B. & Bosman C. A. W. 1999. *Rare birds of the Netherlands, with a complete list of all the species.* Pica Press. Mountfield.

Van den Berg, A. B., Bosman, C. A. W., Haase, B. J. M., Van der Niet, A. M. & Smeenk, C. 1991. Wilson's Petrel off south-western Europe and northern Africa in June 1984 and July 1985. *Dutch Birding* 13: 9–11.

Van der Vliet, R. E. 1998. Little Swift possibly breeding in Spain in 1996–98. *Dutch Birding* 20: 287–288.

Van Impe, J. 1971. Notes ornithologiques de la région de l'Aragón, juin 1967 et juillet–août 1970. *Ardeola* 15: 82–85.

Vaquero, J. M., García-Herrera, R., Wheeler, D., Chenoweth M. & Mock, C. J. 2008. A historical analog of 2005 Hurricane Vince. *Bull. American Meteorological Society* February 2008: 191–201.

Varela, J. M. 2007. *Aves amenazadas de España.* Lynx. Barcelona.

Varo, N. & Amat, J. A. 2008. Differences in foraging behaviour of sympatric coots with different conservation status. *Wildlife Research* 35: 612–616.

Varo, N. 2007. *Ecología de las Fochas Moruna,* Fulica cristata, *y Común,* Fulica atra, *en un área de simpatría.* Ph. D. Thesis. Universidad de Sevilla. Sevilla.

Varo, N., Green, A. J., Sánchez, M. I., Ramo, C., Gómez, J. & Amat, J. A. 2011. Behavioural and population responses to changing availability of *Artemia* prey by moulting black-necked grebes, *Podiceps nigricollis. Hydrobiologia* 664: 163–171.

Vaughan, R. 1971. Breve nota sobre aves acuáticas en el Delta del Ebro, 10–17 agosto 1969. *Ardeola* 15: 90.

Vaurie, C. 1959. *The Birds of the Palearctic Fauna: Order Passeriformes.* H. F. & G. Witherby. London.

Vayreda i Vila, E. 1883. *Fauna ornitológica de la provincia de Gerona.* Imprenta y Librería de Paciano Torres. Gerona.

Veiga, J. P. 1986. Settlement and fat accumulation by migrant pied flycatchers in Spain. *Ringing & Migration* 7: 85–98.

Veiga, J. P. & Polo, V. 2011. Estornino negro. In: Salvador, A. & Morales, M. B. (Eds).

Velando, A. & Freire, J. 2002. Population modelling of European shags (*Phalacrocorax aristotelis*) at their southern limit: conservation implications. *Biological Conservation* 107: 59–69.

Velasco, T. 1992a. Migración e invernada del Correlimos de Temminck (*Calidris temminckii*) en España. *Ecología* 6: 207–214.

Velasco, T. 1992b. El Archibebe Fino (*Tringa stagnatilis*) en España. *Ardeola* 39: 55–61.

Velasco, T. 1999. *La Mancha húmeda en 1997. Situación de su avifauna acuática. In:* Otero, C. (Ed) *Patrimonio natural y propiedad rural en España*, pp. 183–192. Exlibris. Madrid.

Velasco, T. 2002. Nota sobre la abundancia otoñal del Falaropo Picogrueso (*Phalaropus fulicarius*) en los humedales de La Mancha (España Central). Año 2000. *Anuario Ornitológico de Ciudad Real 1995–2001*: pp. 153–154.

Velasco, T. 2009. Algunos datos sobre la presencia de alzacola *Cercotrichas galactotes* en la provincia de Toledo. *Anuario Ornitológico de Toledo 2002–2007*: 116–115.

Velasco, T., Acha, A. & Marchamalo, J. 1999. Nidificación de la gaviota patiamarilla (*Larus michahellis*) en España Central. *Butll. Grup Català d'Anellament* 16: 41–46.

Velasco, T. & Alberto, L. J. 1993. Numbers, main locations and distribution maps of waders wintering in Spain. *Wader Study Group Bull.* 70: 33–41.

Velasco, T. & Alberto, L. J. 1994. Wintering of Kentish Plover *Charadrius alexandrinus* in Spain: numbers, distribution and habitat selection. *Wader Study Group Bull.* 73: 54–55.

Velasco, T., Bustamante, P., Galindo, A. J. & Torralvo, C. 2002. Situación de la Terrera Marismeña (*Calandrella rufescens*) en La Mancha (Ciudad Real-Cuenca-Toledo). Primeros datos de nidificación y localidades de invernada. *Anuario Ornitológico de Ciudad Real 1995–2001*: 107–116.

Velasco, T., Torralvo, C., Carrasco, M., Pérez, J. L., Abarca, L. J. & Ruiz, J. M. 2007. Primeros datos sobre reproducción de garceta grande *Egretta alba*, espátula común *Platalea leucorodia* y ¿focha moruna *Fulica cristata*? en Castilla-La Mancha, con nueva información sobre la nidificación de garcilla cangrejera *Ardeola ralloides. Anuario Ornitológico de Ciudad Real 2004–2005*: 211–218.

Velasco, T., Torralvo, C., Picazo, J., López, M., Cañizares, D., Cañizares, J. A. & Lara, A. J. 2004. Nuevos datos sobre distribución y estimación poblacional de los efectivos reproductores de cuatro especies de paseriformes palustres escasas en Castilla-La Mancha. *Anuario Ornitológico de Ciudad Real 2002–2003*: 161–170.

Venâncio, L., Silva, J. P. & Catry, T. 2002. Acções de conservação dirigidas à perdiz-do-mar e à andorinha-do-mar-anã nã albufeira do Caia. *Pardela* 16:10–11.

Ventoso, L. & Muntaner, J. 2011. Important hivernada i migració prenupcial d'aligot *Buteo buteo* a Mallorca 2010–2011. *Anuari Ornitològic de Balears 2010*: 57–61.

Verducci, D., Biondi, M., Sighele, M. & Norante, N. 2012. Revisione degli avvistamenti e delle catture di Corrione biondo *Cursorius cursor* in Italia con cenni sul suo status in Europa. *Gli Uccelli d'Italia*, 37: 16–32.

Verner, W. C. 1909. *My life among the wild birds in Spain.* John Bale, sons & Danielsson Ltd. London.

Viada, C. 1994. Re-catalogación y estatus del búho chico (*Asio otus*) en Mallorca. *Ardeola*, 41: 59–62.

Viada, C. & de Pablo, F. 2009. Cens d'àguila calçada *Hieraaetus pennatus* a Balears al 2009 i estat de conservació. *Anuari Ornitològic de les Balears* 2009: 1–15.

Vicens, P. 2004. Evolució de la població dels Ardeids al Parc Natural de s'Albufera de Mallorca. *Anuari Ornitològic de les Balears 2003*: 27–36.

Vicente, R. O. 1974. Nidificação da Garça-cinzenta (*Ardea cinerea*) em Portugal. *Cyanopica* I (4): 99–100.

Vidal, E., Medail, F. & Tatoni, T. 1998. Is the yellow-legged gull a superabundant bird species in the Mediterranean? Impact on fauna and flora, conservation measures and research priorities. *Biodiversity and Conservation* 7: 1013–1026.

Vilá Valentí, J. 1968. *La Península Ibérica*. Ariel. Barcelona.

Vilagrasa, X., Carrera, E. & Pardo, R. 1982. Hivernage de la Sterne caugek (*Sterna sandvicensis*) sur les côtes de la Catalogne et du Levant (Espagne). *Alauda* 50: 108–113.

Villacampa, A. G. 1993. Cita de Grévol (*Tetrastes bonasia*) en el Parque Nacional de Ordesa y Monte Perdido. *Lucas Mallada* 5: 199.

Villafuerte, R., Viñuela, J. & Blanco, J. C. 1997. Extensive predator persecution caused by population crash in a game species: the case of red kites and rabbits in Spain. *Biological Conservation* 84: 181–188.

Villarán, A. 1999b. Migración e invernada del Escribano Palustre (*Emberiza palustris*) en España. *Ardeola* 46: 71–80.

Villarán, A. 2003. La invernada del Pajaro Moscón *Remiz pendulinus* en España. *Ardeola* 50: 245–250.

Villarán, A., Medina, C., Pascual-Parra, J. & Mezquida, E. T. 2005. Paso postnupcial de la curruca mosquitera (*Sylvia borin*) y el papamoscas cerrojillo (*Ficedula hypoleuca*) en el monte de Valdelatas (2003–2004): fenología y selección de hábitat. *Anuario Ornitológico de Madrid 2004*: 120–133.

Villarán, A., Mezquida, T. & Pascual-Parra, J. 2008. Análisis de un dormidero comunal de triguero (*Emberiza calandra*) en un carrizal del valle del Tajo (España central). *Revista de Anillamiento* 21–22: 77–83.

Villarán, A., Pascual-Parra, J., Mezquida, E. T., Alonso, B., Domínguez-Méndez, J. y Medina, C. 2002. Invernada de fringílidos en un carrizal del centro ibérico. *Revista de Anillamiento* 9–10: 62–69.

Villarino, A., González, S. & Bárcena, F. 2002. *Vertebrados de A Limia: dende a lagoa de Antela ós nosos días. I. Aves: Gaviiformes a Piciformes*. Ed. Limaia Produccións. A Coruña.

Villers, A., Millon, A., Jiguet, F., Lett, J. M., Attie, C., Morales, M. B. & Bretagnolle, V. 2010. Migration of wild and captive-bred Little Bustards *Tetrax tetrax*: releasing birds from Spain threatens attempts to conserve declining French populations. *Ibis* 152: 254–261.

Viney, C. 2002. New breeding site for white-rumped swift in Portugal. *Dutch Birding* 24: 94–95.

Vinicombe, K. E. 1985. Ring-billed Gulls in Britain and Ireland. *British Birds* 78: 327–337.

Vinicombe, K. E. & Cottridge, D. M. 1996. *Rare birds in Britain and Ireland. A photographic record*. HarperCollins. London

Vinicombe, K. E. & Harrop, A. H. J. 1999. Ruddy Shelducks in Britain and Ireland 1986-1994. *British Birds* 92: 225–255.

Viñuela, J., Martí, R. & Ruiz, A. (Eds) 1999. *El milano real en España*. SEO/BirdLife. Madrid.

Violani, C. G. & Massa, B. 1993. Extinction of the Andalusian Hemipode *Turnix s. sylvatica* (Desf.) in the Mediterranean region. *Bull. British Ornithologists' Club* 113: 225–229.

Von Jordans, A. 1928. Die Ergebnisse meiner dritten Reise nach den Balearen. Nebst einer Uebersicht aller bisher von der Inselgruppe der Balearen und Pityusen bekannten Vogelarten. *Novitates Zoologicae* 34: 262–336.

Von Jordans, A. 1933. Ueber einige Vogelrassen der Nord-Pyrenaen und Nordostspanien. *Anzeiger der Ornithologischen Gesellschaft in Bayern* 2: 250–266.

Voous, K. H. 1960. *Atlas of European birds*. Nelson. London.

Voous, K. H. & Bijleveld, M. F. I. J. 1968. Algunos datos de paso o primeras llegadas en Doñana y comarcas vecinas (2–6 marzo de 1967). *Ardeola* 12: 208–210.

Votier, S. C., Bowen, G. J., & Newton, J. 2009. Stable hydrogen isotope analyses suggest natural vagrancy of Baikal Teal to Britain. *British Birds* 102: 697–699

Votier, S. C., Harrop, A. H. J. & Denny, M. 2003. A review of the status and identification of American Wigeon in Britain and Ireland. *British Birds* 96: 2–22.

Vowles, G. A. & Vowles, R. S. 1994. *Birds of the Algarve*. Gloucestershire. England.

Walker, F. J. 1994. Seabird movements off Cape St. Vincent, Portugal (November–December 1992, January–June and August–November 1993, January–March 1994). *Bol. GIAM* 18: 2–9.

Wallace, D. I. M. 1969. Observations on Audouin's Gulls in Majorca. *British Birds* 62: 223–229.

Wallace, D. I. M. & Sage, B. L. 1969. Las aves de la Costa Brava (Cataluña). *Ardeola* 13: 151–170, 14: 143–157.

Wallis, H. M. 1895. Notes on the birds of the Central Pyrenees. *Ibis* 37: 64–85.

Walter H. 1979. *Eleonora's Falcon. Adaptations to prey and habitat in a social raptor*. University of Chicago Press. Chicago.

Walther, B. A., Schäffer, N., van Niekerk, A., Thuiller, W., Rahbek, C. & Chown, S. L. 2007. Modelling the winter distribution of a rare and endangered migrant, the Aquatic Warbler *Acrocephalus paludicola*. *Ibis* 149: 701–714.

Weickert, P. 1963. Espátulas (*Platalea leucorodia*) y otras aves en la 'Pajarera' de Doñana en 1962. *Ardeola* 8: 258–261.

Weigold, H. 1922. *Frühjahr 1913 in Portugal, Spanien und Tanger*. Lämmle und Müllerschön. Winnenden.

Wernham, C. V., Toms, M. P., Marchant, J. H., Clark, J. A., Siriwardena, G. M. & Baillie, S. R. (Eds) 2002. *The Migration Atlas: movements of the birds of Britain and Ireland*. T. & A. D. Poyser. London.

Westernhagen, W. von. 1956. Ornithologische

Beobachtungen aus dem Ebrodelta. *Ornithologische Mitteilungen* 8: 181–183.

Westernhagen, W. von. 1958. Sobre algunas aves de Mallorca durante los últimos cien años. *Ardeola* 4: 157.

Westernhagen, W. von. 1962. Greifvögel in Andalusien. *Ornithologische Mitteilungen* 14: 41–48.

Westernhagen, W. von & Pons, J. R. 1966. Ornithologische notizen aus dem Ebrodelta. *J. für Ornithologie* 107: 154–166.

Williamson, K. 1964. *Identification for ringers 3. The genus* Sylvia. BTO. Tring.

Willis, N. F. 2002. Aves nuevas: Alcaudón Isabel *Lanius isabellinus*. *La Garcilla* 113: 22–23.

Witherby, H. F. 1922. Note on birds from Northern Spain and description of a new form of *Dryobates*. *Bull. British Ornithologists' Club* 42: 47–49.

Witherby, H. F. 1928. On the birds of central Spain, with some notes on those of south-east Spain. *Ibis* 70: 385–436, 587–663.

Witherby, H. F. 1930. (Notes on a journey in eastern Spain in the company of A. Gil Lletget, in May and June 1929.) *Bull. British Ornithologists' Club*. 1: 75–78.

Witt, H.-H., Crespo, J., de Juana, E. & Varela, J. 1981. Comparative feeding ecology of Audouin's Gull *Larus audouinii* and the Herring Gull *L. argentatus* in the Mediterranean. *Ibis* 123: 519–526.

Woutersen, K. 1993. Seevogels bij Cabo Finisterre, NW Spanje. *Sula* 7: 121–132.

Woutersen, K. & Platteew, M. 1998. *Atlas de las aves de Huesca*. Kees Wouterson Publiaciones. Huesca.

Wüst, W. 1983. Küstenreiher (*Egretta gularis schistacea*) aus Pakistan nach Mittelfranken importiert. *Ornithologische Mitteilungen* 35: 132–133.

WWF. 2006. *Stormy Europe: the power sector and extreme weather.* WWF International, Gland, Switzerland (www.panda.org/powerswitch/stormyeurope).

Yanes, M. 2000. *La depredación en nido de aláudidos almerienses*. Instituto de Estudios Almerienses. Almería.

Yanes, M. & Delgado, J. M. 2006. *Aves esteparias en Andalucía. Bases para su conservación*. Consejería de Medio Ambiente de la Junta de Andalucía. Sevilla.

Ybarra, L. 1966. Observación de Zarapito Fino (*Numenius tenuirostris*) en Doñana..

Yeatman, L. 1976. *Atlas des oiseaux nicheurs de France*. SOF, Paris.

Yeatman-Berthelot, D. & Jarry, G. 1994. *Nouvel atlas des oiseaux nicheurs de France, 1985–1989*. Société Ornithologique de France. Paris.

Yésou, P. 1985. Le cycle de présence du Goéland leucophée *Larus cachinnans michahellis* sur le litoral atlantique: l'example des marais d'Olonne. *L'Oiseau et R.F.O.* 55: 93–105.

Yésou, P. 2013. The Lapland Bunting influx in France in 2010/11. *British Birds* 106: 45.

Yésou, P. & Clergeau, P. 2005. Sacred Ibis: a new invasive species in Europe. *Birding World* 18: 517–526.

Yésou, P., Duquet, M. & Corso, A. 2003. Le Bruant à tête rousse *Emberiza bruniceps* en France et en Italie: statut et origine. *Ornithos* 10: 249–251.

Zabala, J., Zuberogoitia, I., Martínez, J. A., Martínez, J. E., Azkona, A., Hidalgo, S. & Iraeta, A. 2006. Occupancy and abundance of Little Owl (*Athene noctua*) in an intensively managed forest area in Biscay. *Ornis Fennica* 83: 97–107.

Zamora, R. 1987. Variaciones altitudinales en la composición de las comunidades nidificantes de aves de Sierra Nevada (sur de España). *Doñana, Acta Vertebrata* 14: 83–106.

Zamora, R. 1988. Composición y estructura de las comunidades de paseriformes de alta montaña de Sierra Nevada. *Ardeola* 35: 197–220.

Zamora, R. 1990. Le regime frugivore des Merles à plastron hivernants en Sierra Nevada. *Alauda* 58: 67–70.

Zink, G. 1975. *Der Zug europaïscher Singvögel* Vol. 2. Vogelzug-Verlag. Möggingen.

Zink, G. 1981. *Der Zug europäischer Singvögel* Vol. 3. Vogelzug-Verlag, Möggingen.

Zink, R. M. 1985. *Der Zug europäischer Singvögel*, Vol. 4. Vogelwarte Radolzfell. Radolzfell.

Zino, F., Brown, R. & Biscoito, M. 2008. The separation of *Pterodroma madeira* (Zino's Petrel) from *Pterodroma feae* (Fea's Petrel). *Ibis* 150: 326–333.

Zino, F., Phillips, R. & Biscoito, M. 2011. Zino's Petrel movements at sea – a preliminary analysis of datalogger results. *Birding World* 24: 216–219.

Zino, F., Phillips, R. & Biscoito, M. 2012. Bulwer's Petrel movements at sea – a preliminary analysis of datalogger results from Selvagem Grande. *Birding World* 26: 79–81.

Zuberogoitia, Í. 2011. Weather influence on breeding success of the tawny owl on the southwest limit of Eurosiberian region. *In:* Zuberogoitia, I. & Martínez, J. E. (Eds) *Ecology and conservation of European forest-dwelling raptors*, pp. 184–189. Diputación Foral de Bizkaia. Bilbao.

Zuberogoitia, Í & Campos, L. F. 1997. Intensive census of nocturnal raptors in Biscay. *Munibe* 49: 117–127.

Zuberogoitia, Í. & Campos, L. F. 1999. Hibridación de lechuzas, *Tyto alba alba* y *T. a. guttata*, en el norte de la Península Ibérica. *Estudios del Museo de Ciencias Naturales de Álava* 14: 187–192.

Zuberogoitia, Í. & Martínez, J. A. 2000. Methods for surveying Tawny Owl *Strix aluco* populations in large areas. *Biota* 1: 137–146.

Zuberogoitia, Í., Martínez, J. A., Azkona, A., Martínez, J. E., Castillo, I. & Zabala, J. 2009. Using recruitment age, territorial fidelity and dispersal as decisive tools in the conservation and management of peregrine falcon (*Falco peregrinus*) populations: the case of a healthy population in Northern Spain. *J. of Ornithology* 150: 95–101.

Zuberogoitia, Í. & Torres, J. J. 1998. *Aves acuaticas de Bizkaia*. BBK. Bilbao.

Zuberogoitia, Í., Zabala, J., Martínez, J. A., Hidalgo, S., Martínez, J. E., Azkona, A. & Castillo, I., 2007. Seasonal dynamics in social behaviour and spacing patterns of the Little Owl *Athene noctua*. *Ornis Fennica*, 84: 173–180.

Zucca, M. & CMR. 2010. Les observations d'oiseaux migrateurs rares en France. 6e. rapport du CMR (année 2007). *Ornithos* 17: 2–43.

Zucca, M. & Jiguet, F. 2002. La Gorgebleue à miroir (*Luscina svecica*) en France: nidification, migration et hivernage. *Ornithos* 9: 242–252.

Zumalacárregui, C., González, C., Herrero, A., Jubete, F. & Bello, A. 2011. Características del paso posnupcial del Carricerín Cejudo (*Acrocephalus paludicola*) en un complejo de zonas húmedas del noroeste de España. *Revista de Anillamiento* 28: 24–29.

Zúñiga, J. M. 1989. An integral study of the chough in Andalucia (S. Spain) from 1978 to 1985. *In:* Bigna, E. & Curtis, D. J. (Eds) *Choughs and Land-use in Europe*: 65–69. Scottish Chough Study Group. Argyll.

Zúñiga, J. M., Soler, M. & Camacho, I. 1982. Status de la avifauna terrestre de la hoya de Guadix. Aspectos ecológicos. *Trabajos y Monografías del Departamento de Zoología de la Universidad de Granada* 5: 17–51.

Zúñiga, J. M., Soler, M. & Camacho, I. 1983. Nota sobre nuevas especies parasitadas por el críalo (*Clamator glandarius*) en España. *Doñana, Acta Vertebrata* 10: 207–209.

Zwaaneveld, J. & Ebels, E. B. 1994. Roze Pelikanen in Nederland in november 1987–februari 1988 en mei 1990. *Dutch Birding* 16: 106–110.

Zwarts, L., Bijlsma, R.J., van der Kamp, J. & Wymenga, E. 2009. *Living on the edge. Wetlands and birds in a changing Sahel*. KNNV Publishers. Utrecht.

INDEX OF ENGLISH NAMES

INDEX OF SCIENTIFIC NAMES